INTRODUCTION TO MANUFACTURING PROCESSES

INTRODUCTION TO MANUFACTURING PROCESSES

Mikell P. Groover

Professor Emeritus of Industrial and Systems Engineering
Lehigh University

JOHN WILEY & SONS, INC.

VP & EXECUTIVE PUBLISHER	Don Fowley
EXECUTIVE EDITOR	Linda Ratts
EDITORIAL ASSISTANT	Christopher Teja
MARKETING MANAGER	Clay Stone
PRODUCTION MANAGER	Janis Soo
SENIOR PRODUCTION EDITOR	Joyce Poh
DESIGNER	Seng Ping Ngieng

Cover image: Photo courtesy of Sandvik Coromant

This book was set in 9.5/11.5 Times Roman by Thomson Digital and printed and bound by Quad/Graphics. The cover was printed by Quad/Graphics.

This book is printed on acid free paper.

Founded in 1807, John Wiley & Sons, Inc. has been a valued source of knowledge and understanding for more than 200 years, helping people around the world meet their needs and fulfill their aspirations. Our company is built on a foundation of principles that include responsibility to the communities we serve and where we live and work. In 2008, we launched a Corporate Citizenship Initiative, a global effort to address the environmental, social, economic, and ethical challenges we face in our business. Among the issues we are addressing are carbon impact, paper specifications and procurement, ethical conduct within our business and among our vendors, and community and charitable support. For more information, please visit our website: www.wiley.com/go/citizenship.

Library of Congress Cataloging-in-Publication Data

Groover, Mikell P., 1939-

Introduction to manufacturing processes / Mikell P. Groover.

p. cm.

Includes index.

ISBN 978-0-470-63228-4 (pbk.)

1. Manufacturing processes. 2. Production engineering. I. Title.

TS183.G785 2012

670—dc23

2011025938

Printed in the United States of America

V10006662_120518

PREFACE

Introduction to Manufacturing Processes is designed for a first course in manufacturing at the junior level in mechanical, industrial, and manufacturing engineering curricula. It may also be appropriate for technology programs related to these engineering disciplines. The book is based largely on my other manufacturing book, ***Fundamentals of Modern Manufacturing: Materials, Processes, and Systems***. That book is approximately 1000 pages long, and it competes with other manufacturing textbooks that are also very long. Complaints are sometimes leveled that these books include more content than can possibly be covered in a one-semester course. The counterargument to that complaint is that these very comprehensive volumes will serve as valuable references for students in their future professions, assuming those professions are related to design and/or manufacturing.

With this new book, we are attempting to provide an offering that is significantly shorter than the other texts (about 700 pages versus a thousand or more pages). To decide on the coverage of this text, John Wiley & Sons conducted a survey of faculty who have adopted the ***Fundamentals*** book or its competitors to determine which topics they considered most important in their respective courses. Based on the feedback from that survey, we developed the topical content of the current book, which focuses on manufacturing processes. Coverage of engineering materials has been reduced from eight chapters to two, and coverage of production systems has been reduced from five chapters to three. The two chapters dealing with electronics manufacturing have been eliminated because our survey showed that many mechanical engineering instructors do not feel the need to include this area in their courses. Finally, there are several instances in which I combined chapters. All of these changes have resulted in a new book that contains a total of 30 chapters, compared with 42 chapters for the fourth edition of the ***Fundamentals*** book.

The chapters on manufacturing processes are taken almost verbatim from ***Fundamentals***. In some cases, I have shortened the coverage by omitting certain processes or details about processes that seemed appropriate for the more comprehensive text but not for this introductory version. The emphasis on manufacturing science and mathematical modeling of processes remains an important attribute of the new book. Readers will notice that the Historical Notes and end-of-chapter Multiple-Choice Questions have been eliminated from the new book. End-of-chapter Review Questions and Problems have been retained, but the number of problems has been reduced. All of these changes have been made to reduce page count, which results in a textbook that includes most of the topics that are covered by most instructors who teach courses in manufacturing. For instructors who require a more comprehensive treatment of the subject than is contained in this new book, we hope they will continue to adopt the ***Fundamentals*** book.

SUPPORT MATERIALS FOR INSTRUCTORS

For instructors who adopt this book for their courses, the following support materials are available:

- A ***Solutions Manual*** (in digital format) covering all problems and review questions.
- A complete set of Powerpoint slides for all chapters.

These support materials and others may be found at the website www.wiley.com/college/groover. Evidence that the book has been adopted as the main textbook for the course must be verified. Individual questions or comments may be directed to the author personally at Mikell.Groover@Lehigh.edu.

ABOUT THE AUTHOR

Mikell P. Groover is Professor Emeritus of Industrial and Systems Engineering at Lehigh University. He received his B.A. in Arts and Science (1961), B.S. in Mechanical Engineering (1962), M.S. in Industrial Engineering (1966), and Ph.D. (1969), all from Lehigh. He is a Registered Professional Engineer in Pennsylvania. His industrial experience includes several years as a manufacturing engineer with Eastman Kodak Company. Since joining Lehigh, he has done consulting, research, and project work for a number of industrial companies.

His teaching and research areas include manufacturing processes, production systems, automation, material handling, facilities planning, and work systems. He has received a number of teaching awards at Lehigh University, as well as the ***Albert G. Holzman Outstanding Educator Award*** from the Institute of Industrial Engineers (1995) and the ***SME Education Award*** from the Society of Manufacturing Engineers (2001). He is a Fellow of IIE (1987) and SME (1996). His publications include over 75 technical articles and eleven books (listed below). His books are used throughout the world and have been translated into French, German, Spanish, Portuguese, Russian, Japanese, Korean, and Chinese. The first edition of ***Fundamentals of Modern Manufacturing*** received the ***IIE Joint Publishers Award*** (1996) and the ***M. Eugene Merchant Manufacturing Textbook Award*** from the Society of Manufacturing Engineers (1996).

PREVIOUS BOOKS BY THE AUTHOR

Automation, Production Systems, and Computer-Aided Manufacturing, Prentice Hall, 1980.

CAD/CAM: Computer-Aided Design and Manufacturing, Prentice Hall, 1984 (co-authored with E. W. Zimmers, Jr.).

Industrial Robotics: Technology, Programming, and Applications, McGraw-Hill Book Company, 1986 (co-authored with M. Weiss, R. Nagel, and N. Odrey).

Automation, Production Systems, and Computer Integrated Manufacturing, Prentice Hall, 1987.

Fundamentals of Modern Manufacturing: Materials, Processes, and Systems, originally published by Prentice Hall in 1996, and subsequently published by John Wiley & Sons, Inc., 1999.

Automation, Production Systems, and Computer Integrated Manufacturing, Second Edition, Prentice Hall, 2001.

Fundamentals of Modern Manufacturing: Materials, Processes, and Systems, Second Edition, John Wiley & Sons, Inc., 2002.

Work Systems and the Methods, Measurement, and Management of Work, Pearson Prentice Hall, 2007.

Fundamentals of Modern Manufacturing: Materials, Processes, and Systems, Third Edition, John Wiley & Sons, Inc., 2007.

Automation, Production Systems, and Computer Integrated Manufacturing, Third Edition, Pearson Prentice Hall, 2008.

Fundamentals of Modern Manufacturing: Materials, Processes, and Systems, Fourth Edition, John Wiley & Sons, Inc., 2010.

ACKNOWLEDGMENTS

I would like to express my appreciation to the following people who served as participants in our survey that resulted in the decisions on content for this book: Yuan-Shin Lee, North Carolina State University; Ko Moe Hun, University of Hawaii; Ronald Huston, University of Cincinnati; Ioan Marinescu, University of Toledo; Val Marinov, North Dakota State University; Victor Okhuysen, California Polytechnic University, Pomona; John M. Usher, Mississippi State University; Daniel Waldorf, California Polytechnic State University; Allen Yi, Ohio State University; Jack Zhou, Drexel University; and Brian Thompson, Michigan State University.

In addition, it seems appropriate to acknowledge my colleagues at Wiley in Hoboken, New Jersey: Executive Editor Linda Ratts, Editorial Assistants Renata Marcionne and Christopher Teja, and Production Editor Micheline Frederick. Last but certainly not least, I appreciate the thorough efforts of editor Joyce Poh at Wiley in Singapore.

CONTENTS

1 INTRODUCTION AND OVERVIEW OF MANUFACTURING 1

1.1 What Is Manufacturing? 2
 1.1.1 Manufacturing Defined 2
 1.1.2 Manufacturing Industries and Products 3
 1.1.3 Manufacturing Capability 5
 1.1.4 Materials in Manufacturing 6
1.2 Manufacturing Processes 8
 1.2.1 Processing Operations 8
 1.2.2 Assembly Operations 12
 1.2.3 Production Machines and Tooling 12
1.3 Organization of the Book 13

Part I Engineering Materials and Product Attributes 15

2 ENGINEERING MATERIALS 15

2.1 Metals and Their Alloys 16
 2.1.1 Steels 17
 2.1.2 Cast Irons 22
 2.1.3 Nonferrous Metals 23
 2.1.4 Superalloys 28
2.2 Ceramics 29
 2.2.1 Traditional Ceramics 29
 2.2.2 New Ceramics 31
 2.2.3 Glass 32
2.3 Polymers 35
 2.3.1 Thermoplastic Polymers 37
 2.3.2 Thermosetting Polymers 39
 2.3.3 Elastomers 40
2.4 Composites 42
 2.4.1 Technology and Classification of Composite Materials 43
 2.4.2 Composite Materials 44

3 PROPERTIES OF ENGINEERING MATERIALS 49

3.1 Stress–Strain Relationships 50
 3.1.1 Tensile Properties 50
 3.1.2 Compression Properties 57
 3.1.3 Bending and Testing of Brittle Materials 59
 3.1.4 Shear Properties 60
3.2 Hardness 62
 3.2.1 Hardness Tests 62
 3.2.2 Hardness of Various Materials 64
3.3 Effect of Temperature on Mechanical Properties 66
3.4 Fluid Properties 67
3.5 Viscoelastic Behavior of Polymers 69
3.6 Volumetric and Melting Properties 72
 3.6.1 Density and Thermal Expansion 72
 3.6.2 Melting Characteristics 73
3.7 Thermal Properties 74
 3.7.1 Specific Heat and Thermal Conductivity 74
 3.7.2 Thermal Properties in Manufacturing 76

4 DIMENSIONS, TOLERANCES, AND SURFACES 80

4.1 Dimensions and Tolerances 80
 4.1.1 Dimensions and Tolerances 81
 4.1.2 Other Geometric Attributes 81
4.2 Surfaces 81
 4.2.1 Characteristics of Surfaces 82
 4.2.2 Surface Texture 83
 4.2.3 Surface Integrity 85
4.3 Effect of Manufacturing Processes 86

Appendix A4: Measurement of Dimensions and Surfaces 89

A4.1 Conventional Measuring Instruments and Gages 89
 A4.1.1 Precision Gage Blocks 89
 A4.1.2 Measuring Instruments for Linear Dimensions 90
 A4.1.3 Comparative Instruments 93
 A4.1.4 Angular Measurements 94
A4.2 Measurement of Surfaces 94
 A4.2.1 Measurement of Surface Roughness 95
 A4.2.2 Evaluation of Surface Integrity 96

Part II Solidification Processes 97

5 FUNDAMENTALS OF METAL CASTING 97

5.1 Overview of Casting Technology 98
 5.1.1 Casting Processes 98
 5.1.2 Sand-Casting Molds 100
5.2 Heating and Pouring 101
 5.2.1 Heating the Metal 101
 5.2.2 Pouring the Molten Metal 101
 5.2.3 Engineering Analysis of Pouring 102

5.3 Solidification and Cooling 103
5.3.1 Solidification of Metals 104
5.3.2 Solidification Time 106
5.3.3 Shrinkage 107
5.3.4 Directional Solidification 108
5.3.5 Riser Design 109

6 METAL CASTING PROCESSES 113

6.1 Sand Casting 113
6.1.1 Patterns and Cores 114
6.1.2 Molds and Mold Making 115
6.1.3 The Casting Operation 117
6.2 Other Expendable-Mold Casting Processes 117
6.2.1 Shell Molding 117
6.2.2 Expanded Polystyrene Process 118
6.2.3 Investment Casting 119
6.2.4 Plaster-Mold and Ceramic-Mold Casting 121
6.3 Permanent-Mold Casting Processes 122
6.3.1 The Basic Permanent-Mold Process 122
6.3.2 Variations of Permanent-Mold Casting 124
6.3.3 Die Casting 125
6.3.4 Squeeze Casting and Semisolid Metal Casting 127
6.3.5 Centrifugal Casting 128
6.4 Foundry Practice 129
6.4.1 Furnaces 129
6.4.2 Pouring, Cleaning, and Heat Treatment 132
6.5 Casting Quality 133
6.6 Metals for Casting 135
6.7 Product Design Considerations 137

7 GLASSWORKING 140

7.1 Raw Materials Preparation and Melting 140
7.2 Shaping Processes in Glassworking 141
7.2.1 Shaping of Piece Ware 142
7.2.2 Shaping of Flat and Tubular Glass 144
7.2.3 Forming of Glass Fibers 145
7.3 Heat Treatment and Finishing 146
7.3.1 Heat Treatment 146
7.3.2 Finishing 147
7.4 Product Design Considerations 147

8 SHAPING PROCESSES FOR PLASTICS 149

8.1 Properties of Polymer Melts 150
8.2 Extrusion 152
8.2.1 Process and Equipment 152
8.2.2 Analysis of Extrusion 154
8.2.3 Die Configurations and Extruded Products 158
8.2.4 Defects in Extrusion 160
8.3 Production of Sheet and Film 162
8.4 Fiber and Filament Production (Spinning) 165
8.5 Coating Processes 166
8.6 Injection Molding 167
8.6.1 Process and Equipment 167
8.6.2 The Mold 168
8.6.3 Shrinkage and Defects in Injection Molding 171
8.6.4 Other Injection Molding Processes 173
8.7 Compression and Transfer Molding 174
8.7.1 Compression Molding 174
8.7.2 Transfer Molding 176
8.8 Blow Molding and Rotational Molding 177
8.8.1 Blow Molding 177
8.8.2 Rotational Molding 180
8.9 Thermoforming 181
8.10 Casting 185
8.11 Polymer Foam Processing and Forming 185
8.12 Product Design Considerations 186

9 SHAPING PROCESSES FOR RUBBER AND POLYMER-MATRIX COMPOSITES 191

9.1 Rubber Processing and Shaping 192
9.1.1 Production of Rubber 192
9.1.2 Compounding 193
9.1.3 Mixing 193
9.1.4 Shaping and Related Processes 194
9.1.5 Vulcanization 196
9.2 Manufacture of Tires and Other Rubber Products 197
9.2.1 Tires 197
9.2.2 Other Rubber Products 200
9.2.3 Processing of Thermoplastic Elastomers 200
9.3 PMC Shaping Processes and Materials 200
9.3.1 Starting Materials for PMCs 201
9.3.2 Combining Matrix and Reinforcement 202
9.4 Open Mold Processes 203
9.4.1 Hand Lay-Up 204
9.4.2 Spray-Up 205
9.4.3 Automated Tape-Laying Machines 206
9.4.4 Curing 206
9.5 Closed Mold Processes 207
9.5.1 Compression Molding PMC Processes 207
9.5.2 Transfer Molding PMC Processes 208
9.5.3 Injection Molding PMC Processes 208
9.6 Filament Winding 209
9.7 Pultrusion Processes 210
9.7.1 Pultrusion 210
9.7.2 Pulforming 211
9.8 Other PMC Shaping Processes 212

Part III Particulate Processing of Metals and Ceramics 215

10 POWDER METALLURGY 215

10.1 Production of Metallic Powders 217
10.1.1 Atomization 217
10.1.2 Other Production Methods 217
10.2 Conventional Pressing and Sintering 219
10.2.1 Blending and Mixing of the Powders 219
10.2.2 Compaction 220
10.2.3 Sintering 222
10.2.4 Secondary Operations 224
10.3 Alternative Pressing and Sintering Techniques 225
10.3.1 Isostatic Pressing 225
10.3.2 Powder Injection Molding 226
10.3.3 Powder Rolling, Extrusion, and Forging 226
10.3.4 Combined Pressing and Sintering 227
10.3.5 Liquid Phase Sintering 228
10.4 Materials and Products for PM 228
10.5 Design Considerations in Powder Metallurgy 229

Appendix A10 Characterization of Engineering Powders 234

A10.1 Geometric Features 234
A10.2 Other Features 236

11 PROCESSING OF CERAMICS AND CERMETS 238

11.1 Processing of Traditional Ceramics 238
11.1.1 Preparation of the Raw Material 239
11.1.2 Shaping Processes 241
11.1.3 Drying 244
11.1.4 Firing (Sintering) 245
11.2 Processing of New Ceramics 245
11.2.1 Preparation of Starting Materials 245
11.2.2 Shaping 246
11.2.3 Sintering 247
11.2.4 Finishing 248
11.3 Processing of Cermets 248
11.3.1 Cemented Carbides 248
11.3.2 Other Cermets and Ceramic-Matrix Composites 250
11.4 Product Design Considerations 250

Part IV Metal Forming and Sheet Metalworking 252

12 FUNDAMENTALS OF METAL FORMING 252

12.1 Overview of Metal Forming 252
12.2 Material Behavior in Metal Forming 255
12.3 Temperature in Metal Forming 256
12.4 Friction and Lubrication in Metal Forming 258

13 BULK DEFORMATION PROCESSES IN METALWORKING 261

13.1 Rolling 262
13.1.1 Flat Rolling and Its Analysis 263
13.1.2 Shape Rolling 267
13.1.3 Rolling Mills 267
13.1.4 Other Processes Related to Rolling 269
13.2 Forging 271
13.2.1 Open-Die Forging 272
13.2.2 Impression-Die Forging 275
13.2.3 Flashless Forging 276
13.2.4 Forging Hammers, Presses, and Dies 277
13.2.5 Other Processes Related to Forging 280
13.3 Extrusion 283
13.3.1 Types of Extrusion 283
13.3.2 Analysis of Extrusion 286
13.3.3 Extrusion Dies and Presses 289
13.3.4 Other Extrusion Processes 291
13.3.5 Defects in Extruded Products 293
13.4 Wire and Bar Drawing 293
13.4.1 Analysis of Drawing 294
13.4.2 Drawing Practice 297

14 SHEET METALWORKING 304

14.1 Cutting Operations 305
14.1.1 Shearing, Blanking, and Punching 306
14.1.2 Engineering Analysis of Sheet-Metal Cutting 306
14.1.3 Other Sheet-Metal-Cutting Operations 309
14.2 Bending Operations 310
14.2.1 V-Bending and Edge Bending 311
14.2.2 Engineering Analysis of Bending 311
14.2.3 Other Bending and Forming Operations 313
14.3 Drawing 314
14.3.1 Mechanics of Drawing 314
14.3.2 Engineering Analysis of Drawing 317
14.3.3 Other Drawing Operations 319
14.3.4 Defects in Drawing 320
14.4 Other Sheet-Metal-Forming Operations 320
14.4.1 Operations Performed with Metal Tooling 320
14.4.2 Rubber Forming Processes 322
14.5 Dies and Presses for Sheet-Metal Processes 323
14.5.1 Dies 323
14.5.2 Presses 325

14.6 Sheet-Metal Operations Not Performed on Presses 329
14.6.1 Stretch Forming 330
14.6.2 Roll Bending and Roll Forming 330
14.6.3 Spinning 331
14.6.4 High-Energy-Rate Forming 332

Part V Material Removal Processes 336

15 THEORY OF METAL MACHINING 336

15.1 Overview of Machining Technology 338
15.2 Theory of Chip Formation in Metal Machining 341
15.2.1 The Orthogonal Cutting Model 341
15.2.2 Actual Chip Formation 343
15.3 Force Relationships and the Merchant Equation 345
15.3.1 Forces in Metal Cutting 345
15.3.2 The Merchant Equation 347
15.4 Power and Energy Relationships in Machining 350
15.5 Cutting Temperature 352
15.5.1 Analytical Methods to Compute Cutting Temperatures 352
15.5.2 Measurement of Cutting Temperature 353

16 MACHINING OPERATIONS AND MACHINE TOOLS 357

16.1 Machining and Part Geometry 357
16.2 Turning and Related Operations 360
16.2.1 Cutting Conditions in Turning 360
16.2.2 Operations Related to Turning 361
16.2.3 The Engine Lathe 363
16.2.4 Other Lathes and Turning Machines 365
16.2.5 Boring Machines 366
16.3 Drilling and Related Operations 369
16.3.1 Cutting Conditions in Drilling 369
16.3.2 Operations Related to Drilling 370
16.3.3 Drill Presses 371
16.4 Milling 373
16.4.1 Types of Milling Operations 373
16.4.2 Cutting Conditions in Milling 376
16.4.3 Milling Machines 378
16.5 Machining Centers and Turning Centers 380
16.6 Other Machining Operations 383
16.6.1 Shaping and Planing 383
16.6.2 Broaching 385
16.6.3 Sawing 386
16.7 High-Speed Machining 387
16.8 Tolerances and Surface Finish 388
16.8.1 Tolerances in Machining 388
16.8.2 Surface Finish in Machining 389
16.9 Product Design Considerations in Machining 392

17 CUTTING-TOOL TECHNOLOGY AND RELATED TOPICS 398

17.1 Tool Life 398
17.1.1 Tool Wear 399
17.1.2 Tool Life and the Taylor Tool Life Equation 400
17.2 Tool Materials 404
17.2.1 High-Speed Steel and Its Predecessors 406
17.2.2 Cast Cobalt Alloys 407
17.2.3 Cemented Carbides, Cermets, and Coated Carbides 407
17.2.4 Ceramics 410
17.2.5 Synthetic Diamonds and Cubic Boron Nitride 411
17.3 Tool Geometry 411
17.3.1 Single-Point Tool Geometry 412
17.3.2 Multiple-Cutting-Edge Tools 415
17.4 Cutting Fluids 417
17.4.1 Types of Cutting Fluids 417
17.4.2 Application of Cutting Fluids 419
17.5 Machinability 420
17.6 Machining Economics 422
17.6.1 Selecting Feed and Depth of cut 422
17.6.2 Cutting Speed 423

18 GRINDING AND OTHER ABRASIVE PROCESSES 433

18.1 Grinding 433
18.1.1 The Grinding Wheel 434
18.1.2 Analysis of the Grinding Process 437
18.1.3 Application Considerations in Grinding 443
18.1.4 Grinding Operations and Grinding Machines 444
18.2 Related Abrasive Processes 450
18.2.1 Honing 450
18.2.2 Lapping 451
18.2.3 Superfinishing 452
18.2.4 Polishing and Buffing 452

19 NONTRADITIONAL MACHINING PROCESSES 456

19.1 Mechanical Energy Processes 457
19.1.1 Ultrasonic Machining 457
19.1.2 Processes Using Water Jets 458
19.1.3 Other Nontraditional Abrasive Processes 459
19.2 Electrochemical Machining Processes 460
19.2.1 Electrochemical Machining 461

19.2.2 Electrochemical Deburring and Grinding 463
19.3 Thermal Energy Processes 464
19.3.1 Electric Discharge Processes 465
19.3.2 Electron Beam Machining 468
19.3.3 Laser Beam Machining 469
19.4 Chemical Machining 470
19.4.1 Mechanics and Chemistry of Chemical Machining 470
19.4.2 CHM Processes 472
19.5 Application Considerations 476

Part VI Property Enhancing and Surface Processing Operations 480

20 HEAT TREATMENT OF METALS 480

20.1 Annealing 481
20.2 Martensite Formation in Steel 481
20.2.1 The Time-Temperature-Transformation Curve 482
20.2.2 The Heat Treatment Process 483
20.2.3 Hardenability 484
20.3 Precipitation Hardening 485
20.4 Surface Hardening 487

21 SURFACE PROCESSING OPERATIONS 489

21.1 Industrial Cleaning Processes 489
21.1.1 Chemical Cleaning 490
21.1.2 Mechanical Cleaning and Surface Treatments 491
21.2 Diffusion and Ion Implantation 493
21.2.1 Diffusion 493
21.2.2 Ion Implantation 493
21.3 Plating and Related Processes 494
21.3.1 Electroplating 494
21.3.2 Electroforming 496
21.3.3 Electroless Plating 497
21.3.4 Hot Dipping 497
21.4 Conversion Coating 498
21.4.1 Chemical Conversion Coatings 498
21.2.4 Anodizing 499
21.5 Vapor Deposition Processes 499
21.5.1 Physical Vapor Deposition 499
21.5.2 Chemical Vapor Deposition 501
21.6 Organic Coatings 504
21.6.1 Application Methods 505
21.6.2 Powder Coating 506

Part VII Joining and Assembly Processes 509

22 FUNDAMENTALS OF WELDING 509

22.1 Overview of Welding Technology 510
22.1.1 Types of Welding Processes 510
22.1.2 Welding as a Commercial Operation 511
22.2 The Weld Joint 513
22.2.1 Types of Joints 513
22.2.2 Types of Welds 513
22.3 Physics of Welding 515
22.3.1 Power Density 516
22.3.2 Heat Balance in Fusion Welding 517
22.4 Features of a Fusion-Welded Joint 519

23 WELDING PROCESSES 522

23.1 Arc Welding 522
23.1.1 General Technology of Arc Welding 523
23.1.2 AW Processes—Consumable Electrodes 525
23.1.3 AW Processes—Nonconsumable Electrodes 529
23.2 Resistance Welding 531
23.2.1 Power Source in Resistance Welding 531
23.2.2 Resistance-Welding Processes 532
23.3 Oxyfuel Gas Welding 537
23.3.1 Oxyacetylene Welding 537
23.3.2 Alternative Gases for Oxyfuel Welding 538
23.4 Other Fusion-Welding Processes 539
23.5 Solid-State Welding 541
23.5.1 General Considerations in Solid-State Welding 542
23.5.2 Solid State-Welding Processes 542
23.6 Weld Quality 547
23.7 Design Considerations in Welding 550

24 BRAZING, SOLDERING, AND ADHESIVE BONDING 554

24.1 Brazing 554
24.1.1 Brazed Joints 555
24.1.2 Filler Metals and Fluxes 557
24.1.3 Brazing Methods 558
24.2 Soldering 560
24.2.1 Joint Designs in Soldering 560
24.2.2 Solders and Fluxes 560
24.2.3 Soldering Methods 562
24.3 Adhesive Bonding 564
24.3.1 Joint Design 564
24.3.2 Adhesive Types 566
24.3.3 Adhesive Application Technology 566

25 MECHANICAL ASSEMBLY 569

25.1 Threaded Fasteners 570
25.1.1 Screws, Bolts, and Nuts 570
25.1.2 Other Threaded Fasteners and Related Hardware 572

25.1.3 Stresses and Strengths in Bolted Joints 573
25.1.4 Tools and Methods for Threaded Fasteners 575
25.2 Rivets 575
25.3 Assembly Methods Based on Interference Fits 576
25.4 Other Mechanical Fastening Methods 579
25.5 Molding Inserts and Integral Fasteners 580
25.6 Design for Assembly 581
25.6.1 General Principles of DFA 583
25.6.2 Design for Automated Assembly 583

Part VIII Special Processing and Assembly Technologies 587

26 RAPID PROTOTYPING 587

26.1 Fundamentals of Rapid Prototyping 588
26.2 Rapid Prototyping Technologies 589
26.2.1 Liquid-Based Rapid Prototyping Systems 590
26.2.2 Solid-Based Rapid Prototyping Systems 593
26.2.3 Powder-Based Rapid Prototyping Systems 594
26.3 Application Issues in Rapid Prototyping 596

27 MICROFABRICATION AND NANOFABRICATION TECHNOLOGIES 599

27.1 Microsystem Products 600
27.1.1 Types of Microsystem Devices 600
27.1.2 Microsystem Applications 601
27.2 Microfabrication Processes 603
27.2.1 Silicon Layer Processes 604
27.2.2 LIGA Process 607
27.2.3 Other Microfabrication Processes 608
27.3 Nanotechnology Products 611
27.4 Scanning Probe Microscopes 614
27.5 Nanofabrication Processes 615
27.5.1 Top-Down Processing Approaches 615
27.5.2 Bottom-Up Processing Approaches 616

Part IX Systems Topics in Manufacturing 623

28 PRODUCTION SYSTEMS AND PROCESS PLANNING 623

28.1 Overview of Production Systems 623
28.1.1 Production Facilities 624
28.1.2 Manufacturing Support Systems 626
28.2 Process Planning 627
28.2.1 Traditional Process Planning 628
28.2.2 Make or Buy Decision 632
28.2.3 Computer-Aided Process Planning 633
28.2.4 Problem Solving and Continuous Improvement 635
28.3 Concurrent Engineering and Design for Manufacturability 635
28.3.1 Design for Manufacturing and Assembly 636
28.3.2 Concurrent Engineering 637

29 SURVEY OF AUTOMATION AND MANUFACTURING SYSTEMS 640

29.1 Computer Numerical Control 641
29.1.1 The Technology of Numerical Control 641
29.1.2 Analysis of NC Positioning Systems 643
29.1.3 NC Part Programming 648
29.1.4 Applications of Numerical Control 650
29.2 Cellular Manufacturing 651
29.2.1 Part Families 651
29.2.2 Machine Cells 653
29.3 Flexible Manufacturing Systems and Cells 655
29.3.1 Integrating the FMS Components 656
29.3.2 Applications of Flexible Manufacturing Systems 658
29.4 Lean Production 659
29.4.1 Just-in-Time Production Systems 660
29.4.2 Other Approaches in Lean Production 661
29.5 Computer-Integrated Manufacturing 662

30 QUALITY CONTROL AND INSPECTION 667

30.1 Product Quality 667
30.2 Process Capability and Tolerances 668
30.3 Statistical Process Control 669
30.3.1 Control Charts for Variables 670
30.3.2 Control Charts for Attributes 672
30.3.3 Interpreting the Charts 673
30.4 Quality Programs in Manufacturing 674
30.4.1 Total Quality Management 674
30.4.2 Six Sigma 675
30.4.3 ISO 9000 677
30.5 Inspection Principles 678
30.6 Modern Inspection Technologies 680
30.6.1 Coordinate Measuring Machines 680
30.6.2 Machine Vision 681
30.6.3 Other Noncontact Inspection Techniques 684

INDEX 687

1 INTRODUCTION AND OVERVIEW OF MANUFACTURING

Chapter Contents

1.1 What Is Manufacturing?
- 1.1.1 Manufacturing Defined
- 1.1.2 Manufacturing Industries and Products
- 1.1.3 Manufacturing Capability
- 1.1.4 Materials in Manufacturing

1.2 Manufacturing Processes
- 1.2.1 Processing Operations
- 1.2.2 Assembly Operations
- 1.2.3 Production Machines and Tooling

1.3 Organization of the Book

Making things has been an essential activity of human civilizations since before recorded history. Today, the term ***manufacturing*** is used for this activity. For technological and economic reasons, manufacturing is important to the welfare of the United States and most other developed and developing nations. ***Technology*** can be defined as the application of science to provide society and its members with those things that are needed or desired. Technology affects our daily lives, directly and indirectly, in many ways. Consider the list of products in Table 1.1. They represent various technologies that help society and its members to live better. What do all these products have in common? They are all manufactured. These technological wonders would not be available to society if they could not be manufactured. Manufacturing is the critical factor that makes technology possible.

Economically, manufacturing is an important means by which a nation creates material wealth. In the United States, the manufacturing industries account for about 12% of gross domestic product (GDP). A country's natural resources, such as agricultural lands, mineral deposits, and oil reserves, also create wealth. In the United States, agriculture, mining, and similar industries account for less than 5% of GDP (agriculture alone is only about 1%). Construction and public utilities make up around 5%. The rest is service industries, which include retail, transportation, banking, communication, education, and government. The service sector accounts for more than 75% of U.S. GDP. Government alone accounts for about as much of GDP as the manufacturing sector; however, government services do not create wealth. In the modern global economy, a nation must have a strong manufacturing base (or it must have significant natural resources) if it is to provide a strong economy and a high standard of living for its people.

In this opening chapter, we consider some general topics about manufacturing. What is manufacturing? How is it organized in industry? What are the processes by which it is accomplished?

TABLE 1.1 Products representing various technologies, most of which affect nearly all of us.

Athletic shoes	Digital video disc player	One-piece molded plastic patio chair
Automatic teller machine	E-book reader	Optical scanner
Automatic dishwasher	Fax machine	Personal computer (PC)
Ballpoint pen	Flat-screen high-definition television	Photocopying machine
Bicycle	Global positioning system	Pull-tab beverage can
Cellular telephone	Hand-held electronic calculator	Quartz crystal wrist watch
Compact disc (CD)	Hybrid gas-electric automobile	Self-propelled mulching lawnmower
Compact disc player	Industrial robot	Supersonic aircraft
Compact fluorescent light bulb	Ink-jet color printer	Tennis racket of composite materials
Contact lens	Integrated circuit	Tire
Digital camera	Magnetic resonance imaging (MRI) machine for medical diagnosis	Video game player
Digital video disc (DVD)	Microwave oven	Washing machine and dryer

Compiled from generally available product data sources.

1.1 WHAT IS MANUFACTURING?

The word ***manufacture*** is derived from two Latin words, ***manus*** (hand) and ***factus*** (make); the combination means made by hand. The English word *manufacture* is several centuries old, and "made by hand" accurately described the manual methods used when the word was first coined.[1] Most modern manufacturing is accomplished by automated and computer-controlled machinery.

1.1.1 MANUFACTURING DEFINED

As a field of study in the modern context, manufacturing can be defined two ways, one technologic, and the other economic. Technologically, ***manufacturing*** is the application of physical and chemical processes to alter the geometry, properties, and/or appearance of a given starting material to make parts or products; manufacturing also includes assembly of multiple parts to make products. The processes to accomplish manufacturing involve a combination of machinery, tools, power, and labor, as depicted in Figure 1.1(a). Manufacturing is almost always carried out as a sequence of operations. Each operation brings the material closer to the desired final state.

Economically, ***manufacturing*** is the transformation of materials into items of greater value by means of one or more processing and/or assembly operations, as depicted in Figure 1.1(b). The key point is that manufacturing ***adds value*** to the material by changing its shape or properties, or by combining it with other materials that have been similarly altered. The material has been made more valuable through the manufacturing operations performed on it. When iron ore is converted into steel, value is added. When sand is transformed into glass, value is added. When petroleum is refined into plastic, value is added. And when plastic is molded into the complex geometry of a patio chair, it is made even more valuable.

The words *manufacturing* and *production* are often used interchangeably. The author's view is that production has a broader meaning than manufacturing. To illustrate, one might speak of "crude oil production," but the phrase "crude oil manufacturing" seems out of place. Yet when used in the context of products such as metal parts or automobiles, either word seems okay.

[1] As a noun, the word ***manufacture*** first appeared in English around 1567 A.D. As a verb, it first appeared around 1683 A.D.

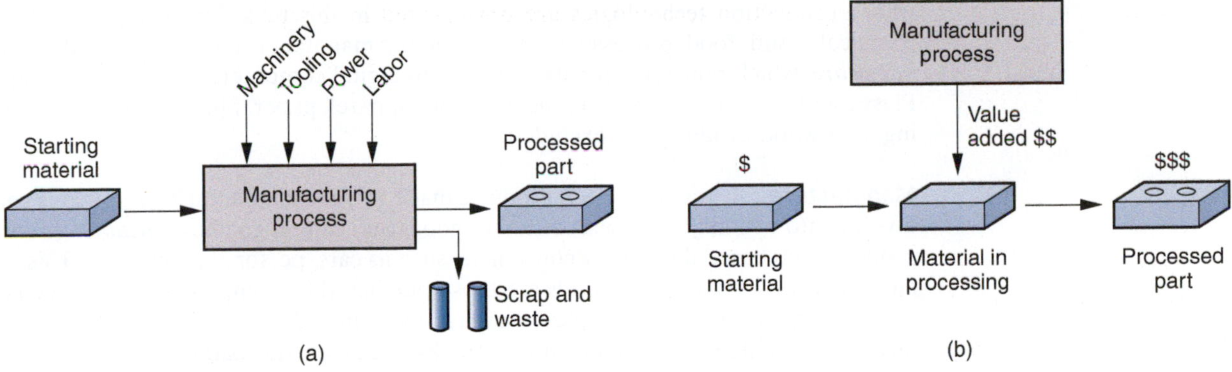

FIGURE 1.1 Two ways to define manufacturing: (a) as a technical process, and (b) as an economic process. (Credit: *Fundamentals of Modern Manufacturing*, 4th Edition by Mikell P. Groover, 2010. Reprinted with permission of John Wiley & Sons, Inc.)

1.1.2 MANUFACTURING INDUSTRIES AND PRODUCTS

Manufacturing is an important commercial activity performed by companies that sell products to customers. The type of manufacturing done by a company depends on the kind of product it makes. Let us explore this relationship by first examining the types of industries in manufacturing, and then identifying the products they make.

Manufacturing Industries Industry consists of enterprises and organizations that produce or supply goods and services. Industries can be classified as primary, secondary, or tertiary. ***Primary industries*** cultivate and exploit natural resources, such as agriculture and mining. ***Secondary industries*** take the outputs of the primary industries and convert them into consumer and capital goods. Manufacturing is the principal activity in this category, but construction and power utilities are also included. ***Tertiary industries*** constitute the service sector of the economy. A list of specific industries in these categories is presented in Table 1.2.

In this book, we are concerned with the secondary industries in Table 1.2, which include the companies engaged in manufacturing. However, the International Standard Industrial Classification (ISIC) used to compile Table 1.2 includes several industries

TABLE 1.2 Specific industries in the primary, secondary, and tertiary categories.

Primary	Secondary		Tertiary (service)	
Agriculture	Aerospace	Fabricated metals	Banking	Insurance
Forestry	Apparel	Food processing	Communications	Legal
Fishing	Automotive	Glass, ceramics	Education	Real estate
Livestock	Basic metals	Heavy machinery	Entertainment	Repair and maintenance
Quarries	Beverages	Paper	Financial services	Restaurant
Mining	Building materials	Petroleum refining	Government	Retail trade
Petroleum	Chemicals	Pharmaceuticals	Health and medical	Tourism
	Computers	Plastics (shaping)	Hotel	Transportation
	Construction	Power utilities	Information	Wholesale trade
	Consumer appliances	Publishing		
	Electronics	Textiles		
	Equipment	Tire and rubber		
		Wood and furniture		

Compiled from generally available industry data sources.

whose production technologies are not covered in this text; for example, beverages, chemicals, and food processing. In this book, manufacturing means production of ***hardware***, which ranges from nuts and bolts to digital computers and military weapons. Plastic and ceramic products are included, but apparel, paper, pharmaceuticals, publishing, and wood products are excluded.

Manufactured Products Final products made by the manufacturing industries can be divided into two major classes: consumer goods and capital goods. ***Consumer goods*** are products purchased directly by consumers, such as cars, personal computers, TVs, tires, and tennis rackets. ***Capital goods*** are those purchased by companies to produce goods and/or provide services. Examples of capital goods include aircraft, computers, communication equipment, medical apparatus, trucks and buses, railroad locomotives, machine tools, and construction equipment. Most of these capital goods are purchased by the service industries. We noted in the introduction that manufacturing accounts for about 12% of gross domestic product and services more than 75% of GDP in the United States. Yet the manufactured capital goods purchased by the service sector are the enablers of that sector. Without the capital goods, the service industries could not function.

In addition to final products, other manufactured items include the ***materials***, ***components***, and ***supplies*** used by the companies that make the final products. Examples of these items include sheet steel, bar stock, metal stampings, machined parts, plastic moldings and extrusions, cutting tools, dies, molds, and lubricants. Thus, the manufacturing industries consist of a complex infrastructure with various categories and layers of intermediate suppliers with whom the final consumer never deals.

This book is generally concerned with ***discrete items***—individual parts and assembled products rather than items produced by ***continuous processes***. A metal stamping is a discrete item, but the sheet-metal coil from which it is made is continuous (almost). Many discrete parts start out as continuous or semicontinuous products, such as extrusions and electrical wire. Long sections made in almost continuous lengths are cut to desired size. An oil refinery is a better example of a continuous process.

Production Quantity and Product Variety The quantity of products made by a factory has an important influence on the way its people, facilities, and procedures are organized. Annual production quantities can be classified into three ranges: (1) ***low*** production, quantities in the range 1 to 100 units per year; (2) ***medium*** production, from 100 to 10,000 units annually; and (3) ***high*** production, 10,000 to millions of units. The boundaries between the three ranges are somewhat arbitrary (author's judgment). Depending on the kinds of products, these boundaries may shift by an order of magnitude or so.

Production quantity refers to the number of units produced annually of a particular product type. Some plants produce a variety of different product types, each type being made in low or medium quantities. Other plants specialize in high production of only one product type. It is instructive to identify product variety as a parameter distinct from production quantity. Product variety refers to different product designs or types that are produced in the plant. Different products have different shapes and sizes; they perform different functions; they are intended for different markets; some have more components than others; and so forth. The number of different product types made each year can be counted. When the number of product types made in the factory is high, this indicates high product variety.

There is an inverse correlation between product variety and production quantity in terms of factory operations. If a factory's product variety is high, then its production quantity is likely to be low; but if production quantity is high, then product variety will be low, as depicted in Figure 1.2. Manufacturing plants tend to specialize in a combination of production quantity and product variety that lies somewhere inside the diagonal band in Figure 1.2.

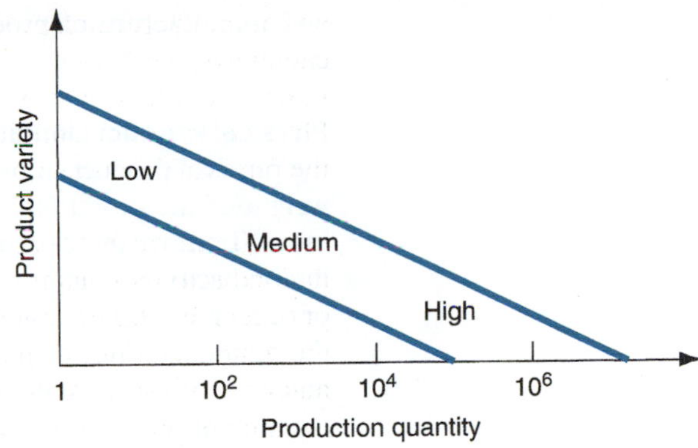

FIGURE 1.2 Relationship between product variety and production quantity in discrete product manufacturing. (Credit: *Fundamentals of Modern Manufacturing*, 4th Edition by Mikell P. Groover, 2010. Reprinted with permission of John Wiley & Sons, Inc.)

Although product variety has been identified as a quantitative parameter (the number of different product types made by the plant or company), this parameter is much less exact than production quantity because details on how much the designs differ are not captured simply by the number of different designs. Differences between an automobile and an air conditioner are far greater than between an air conditioner and a heat pump. Within each product type, there are differences among specific models.

The extent of the product differences may be small or great, as in the automotive industry. Each of the U.S. automotive companies produces cars with two or three different nameplates in the same assembly plant, although the body styles and other design features are virtually the same. In different plants, the company builds heavy trucks. The terms "soft" and "hard" might be used to describe these differences in product variety. ***Soft product variety*** is when there are only small differences among products, such as the differences among car models made on the same production line. In an assembled product, soft variety is characterized by a high proportion of common parts among the models. ***Hard product variety*** is when the products differ substantially, and there are few common parts, if any. The difference between a car and a heavy truck illustrates hard variety.

1.1.3 MANUFACTURING CAPABILITY

A company engaged in manufacturing cannot do everything. It must do only certain things, and it must do those things well in order to remain competitive in its industry. ***Manufacturing capability*** refers to the technical and physical limitations of a manufacturing firm and each of its plants. Several dimensions of this capability can be identified: (1) technological processing capability, (2) physical size and weight of product, and (3) production capacity.

Technological Processing Capability The technological processing capability of a plant (or company) is its available set of manufacturing processes. Certain plants perform machining operations, others roll steel billets into sheet stock, and others build automobiles. A machine shop cannot roll steel, and a rolling mill cannot build cars. The underlying feature that distinguishes these plants is the processes they can perform. Technological processing capability is closely related to material type. Certain manufacturing processes are suited to certain materials, whereas other processes are suited to other materials. By specializing in a certain process or group of processes, the plant is simultaneously specializing in certain material types. Technological processing capability includes not only the physical processes, but also the expertise possessed by plant personnel in these processing technologies. Companies must concentrate on the design and manufacture of products that are compatible with their technological processing capability.

Physical Product Limitations A second aspect of manufacturing capability is imposed by the physical product. A plant with a given set of processes is limited in terms of the size and weight of the products that can be accommodated. Large, heavy products are difficult to move. To move these products about, the plant must be equipped with cranes of the required load capacity. Smaller parts and products made in large quantities can be moved by conveyor or other means. The limitation on product size and weight extends to the physical capacity of the manufacturing equipment as well. Production machines come in different sizes. Larger machines must be used to process larger parts. The production and material handling equipment must be planned for products that lie within a certain size and weight range.

Production Capacity A third limitation on a plant's manufacturing capability is the production quantity that can be produced in a given time period (e.g., month or year). This quantity limitation is commonly called ***plant capacity***, or ***production capacity***, defined as the maximum rate of production that a plant can achieve under assumed operating conditions. The operating conditions refer to number of shifts per week, hours per shift, direct labor manning levels in the plant, and so on. These factors represent inputs to the manufacturing plant. Given these inputs, how much output can the factory produce?

Plant capacity is usually measured in terms of output units, such as annual tons of steel produced by a steel mill, or number of cars produced by a final assembly plant. In these cases, the outputs are homogeneous. In cases in which the output units are not homogeneous, other factors may be more appropriate measures, such as available labor hours of productive capacity in a machine shop that produces a variety of parts.

1.1.4 MATERIALS IN MANUFACTURING

Most engineering materials can be classified into one of three basic categories: (1) metals, (2) ceramics, and (3) polymers. Their chemistries are different, their mechanical and physical properties are different, and these differences affect the manufacturing processes that can be used to produce products from them. In addition to the three basic categories, there are (4) ***composites***—nonhomogeneous mixtures of the other three basic types rather than a unique category. In this section, we briefly survey these four material categories. Chapter 2 covers them in more detail.

Metals Metals used in manufacturing are usually ***alloys***, which are composed of two or more elements, with at least one being a metallic element. Metals and alloys can be divided into two basic groups: ferrous and nonferrous.

Ferrous metals are based on iron; the group includes steel and cast iron. These metals constitute the most important group commercially, more than three-fourths of the metal tonnage throughout the world. Pure iron has limited commercial use, but when alloyed with carbon, iron has more uses and greater commercial value than any other metal. Alloys of iron and carbon form steel and cast iron.

Steel can be defined as an iron-carbon alloy containing 0.02% to 2.11% carbon. It is the most important category within the ferrous metal group. Its composition often includes other alloying elements as well, such as manganese, chromium, nickel, and molybdenum, to enhance the properties of the metal. Applications of steel include construction (e.g., bridges, I-beams, and nails), transportation (trucks, rails, and rolling stock for railroads), and consumer products (automobiles and appliances).

Cast iron is an alloy of iron and carbon (2% to 4%) used in casting (primarily sand casting). Silicon is also present in the alloy (in amounts from 0.5% to 3%), and other elements are often added also, to obtain desirable properties in the cast part. Cast iron is available in several different forms, of which gray cast iron is the most common; its applications include blocks and heads for internal combustion engines.

Nonferrous metals include the other metallic elements and their alloys. In almost all cases, the alloys are more important commercially than the pure metals. The nonferrous metals include the pure metals and alloys of aluminum, copper, gold, magnesium, nickel, silver, tin, titanium, zinc, and other metals.

Ceramics A ceramic is defined as a compound containing metallic (or semimetallic) and nonmetallic elements. Typical nonmetallic elements are oxygen, nitrogen, and carbon. Ceramics include a variety of traditional and modern materials. Traditional ceramics, some of which have been used for thousands of years, include: ***clay*** (abundantly available, consisting of fine particles of hydrous aluminum silicates and other minerals used in making brick, tile, and pottery); ***silica*** (the basis for nearly all glass products); and ***alumina*** and ***silicon carbide*** (two abrasive materials used in grinding). Modern ceramics include some of the preceding materials, such as alumina, whose properties are enhanced in various ways through modern processing methods. Newer ceramics include: ***carbides***—metal carbides such as tungsten carbide and titanium carbide, which are widely used as cutting tool materials; and ***nitrides***—metal and semimetal nitrides such as titanium nitride and boron nitride, used as cutting tools and grinding abrasives.

For processing purposes, ceramics can be divided into crystalline ceramics and glasses. Different methods of manufacturing are required for the two categories. Crystalline ceramics are formed in various ways from powders and then fired (heated to a temperature below the melting point to achieve bonding between the powders). The glass ceramics (namely, glass) can be melted and then formed in processes such as traditional glass blowing.

Polymers A polymer is a compound formed of repeating structural units called ***mers***, whose atoms share electrons to form very large molecules. Polymers usually consist of carbon plus one or more other elements such as hydrogen, nitrogen, oxygen, and chlorine. Polymers are divided into three categories: (1) thermoplastic polymers, (2) thermosetting polymers, and (3) elastomers.

Thermoplastic polymers can be subjected to multiple heating and cooling cycles without substantially altering the molecular structure of the polymer. Common thermoplastics include polyethylene, polystyrene, polyvinylchloride, and nylon. ***Thermosetting polymers*** chemically transform (cure) into a rigid structure on cooling from a heated plastic condition; hence the name thermosetting. Members of this type include phenolics, amino resins, and epoxies. Although the name "thermosetting" is used, some of these polymers cure by mechanisms other than heating. ***Elastomers*** are polymers that exhibit significant elastic behavior; hence, the name elastomer. They include natural rubber, neoprene, silicone, and polyurethane.

Composites Composites do not really constitute a separate category of materials; they are mixtures of the other three types. A ***composite*** is a material consisting of two or more phases that are processed separately and then bonded together to achieve properties superior to those of its constituents. The term ***phase*** refers to a homogeneous mass of material, such as an aggregation of grains of identical unit cell structure in a solid metal. The usual structure of a composite consists of particles or fibers of one phase mixed in a second phase, called the ***matrix***.

Composites are found in nature (e.g., wood), and they can be produced synthetically. The synthesized type is of greater interest here, and it includes glass fibers in a polymer matrix, such as fiber-reinforced plastic; polymer fibers of one type in a matrix of a second polymer, such as an epoxy-Kevlar composite; and ceramic in a metal matrix, such as a tungsten carbide in a cobalt binder to form a cemented carbide cutting tool.

Properties of a composite depend on its components, the physical shapes of the components, and the way they are combined to form the final material. Some composites combine high strength with light weight and are suited to applications such as aircraft components, car bodies, boat hulls, tennis rackets, and fishing rods. Other composites are strong, hard, and capable of maintaining these properties at elevated temperatures, for example, cemented carbide cutting tools.

1.2 MANUFACTURING PROCESSES

A manufacturing process is a designed procedure that results in physical and/or chemical changes to a starting work material with the intention of increasing the value of that material. A manufacturing process is usually carried out as a ***unit operation***, which means that it is a single step in the sequence of steps required to transform the starting material into a final product. A unit operation is generally performed on a single piece of equipment that runs independently of other operations in the plant.

Manufacturing operations can be divided into two basic types: (1) processing operations and (2) assembly operations. A ***processing operation*** transforms a work material from one state of completion to a more advanced state that is closer to the final desired product. It adds value by changing the geometry, properties, or appearance of the starting material. In general, processing operations are performed on discrete workparts, but certain processing operations are also applicable to assembled items (e.g., painting a spot-welded car body). An ***assembly operation*** joins two or more components to create a new entity, called an assembly, subassembly, or some other term that refers to the joining process (e.g., a welded assembly is called a ***weldment***). A classification of manufacturing processes is presented in Figure 1.3.

1.2.1 PROCESSING OPERATIONS

A processing operation uses energy to alter a workpart's shape, physical properties, or appearance to add value to the material. The forms of energy include mechanical, thermal, electrical, and chemical. The energy is applied in a controlled way by means of machinery and tooling. Human energy may also be required, but the human workers are generally employed to control the machines, oversee the operations, and load and unload parts before and after each cycle of operation. A general model of a processing operation is illustrated in Figure 1.1(a). Material is fed into the process, energy is applied by the machinery and tooling to transform the material, and the completed workpart exits the process. Most production operations produce waste or scrap, either as a natural aspect of the process (e.g., removing material as in machining) or in the form of occasional defective pieces. It is an important objective in manufacturing to reduce waste in either of these forms.

More than one processing operation is usually required to transform the starting material into final form. The operations are performed in the particular sequence required to achieve the geometry and condition defined by the design specification.

Three general types of processing operations are distinguished: (1) shaping operations, (2) property-enhancing operations, and (3) surface processing operations. ***Shaping operations*** alter the geometry of the starting work material by various methods. Common shaping processes include casting, forging, and machining. ***Property-enhancing operations*** add value to the material by improving its physical properties without changing its shape. Heat treatment is the most common example. ***Surface processing operations*** are performed to clean, treat, coat, or deposit material onto the exterior surface of the work. Common examples of coating are plating and painting.

Shaping Processes Most shape processing operations apply heat, mechanical force, or a combination of these to effect a change in geometry of the work material. There are

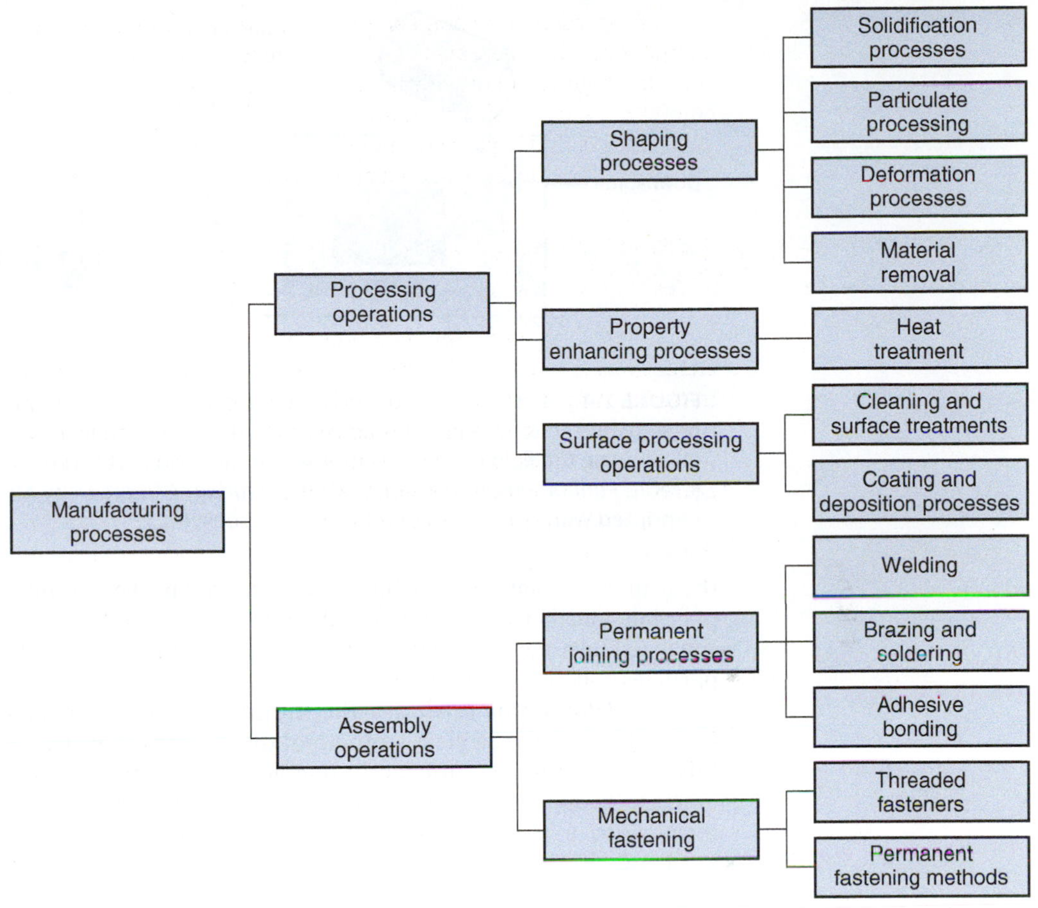

FIGURE 1.3 Classification of manufacturing processes. (Credit: *Fundamentals of Modern Manufacturing*, 4th Edition by Mikell P. Groover, 2010. Reprinted with permission of John Wiley & Sons, Inc.)

various ways to classify the shaping processes. The classification used in this book is based on the state of the starting material, by which we have four categories: (1) ***solidification processes***, in which the starting material is a heated ***liquid*** or ***semifluid*** that cools and solidifies to form the part geometry; (2) ***particulate processing***, in which the starting material is a ***powder***, and the powders are formed and heated into the desired geometry; (3) ***deformation processes***, in which the starting material is a ***ductile solid*** (commonly metal) that is deformed to shape the part; and (4) ***material removal processes***, in which the starting material is a ***solid*** (ductile or brittle), from which material is removed so that the resulting part has the desired geometry.

In the first category, the starting material is heated sufficiently to transform it into a liquid or highly plastic (semifluid) state. Nearly all materials can be processed in this way. Metals, ceramic glasses, and plastics can all be heated to sufficiently high temperatures to convert them into liquids. With the material in a liquid or semifluid form, it can be poured or otherwise forced to flow into a mold cavity and allowed to solidify, thus taking a solid shape that is the same as the cavity. Most processes that operate this way are called casting or molding. ***Casting*** is the name used for metals, and ***molding*** is the common term used for plastics. This category of shaping process is depicted in Figure 1.4.

In ***particulate processing***, the starting materials are powders of metals or ceramics. Although these two materials are quite different, the processes to shape

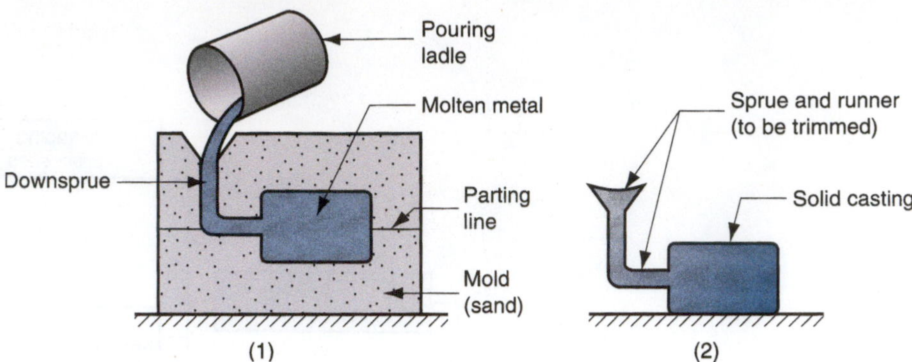

FIGURE 1.4 Casting and molding processes start with a work material heated to a fluid or semifluid state. The process consists of (1) pouring the fluid into a mold cavity and (2) allowing the fluid to solidify, after which the solid part is removed from the mold. (Credit: *Fundamentals of Modern Manufacturing*, 4th Edition by Mikell P. Groover, 2010. Reprinted with permission of John Wiley & Sons, Inc.)

them in particulate processing are quite similar. The common technique involves pressing and sintering, illustrated in Figure 1.5, in which the powders are first squeezed into a die cavity under high pressure and then heated to bond the individual particles together.

In ***deformation processes***, the starting workpart is shaped by the application of forces that exceed the yield strength of the material. For the material to be formed in this way, it must be sufficiently ductile to avoid fracture during deformation. To increase ductility (and for other reasons), the work material is often heated before forming to a temperature below the melting point. Deformation processes are associated most closely with metalworking and include operations such as ***forging*** and ***extrusion***, shown in Figure 1.6.

Material removal processes are operations that remove excess material from the starting workpiece so that the resulting shape is the desired geometry. The most

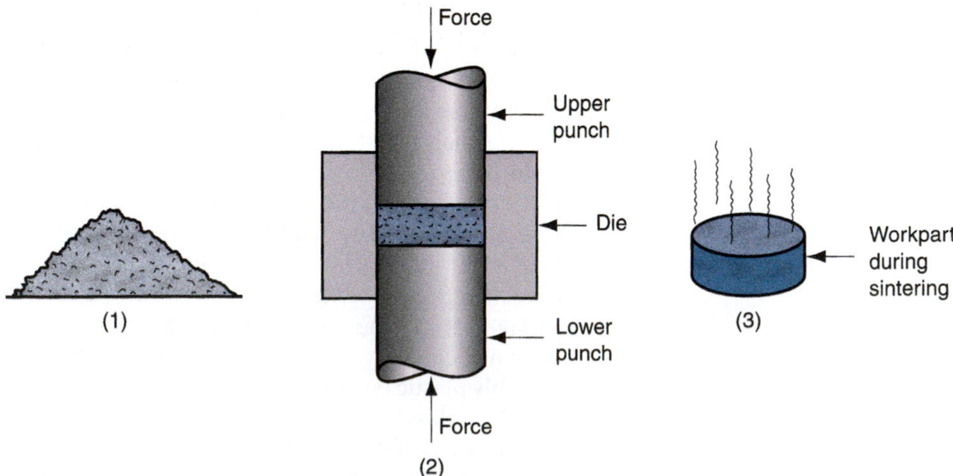

FIGURE 1.5 Particulate processing: (1) the starting material is powder; the usual process consists of (2) pressing and (3) sintering. (Credit: *Fundamentals of Modern Manufacturing*, 4th Edition by Mikell P. Groover, 2010. Reprinted with permission of John Wiley & Sons, Inc.)

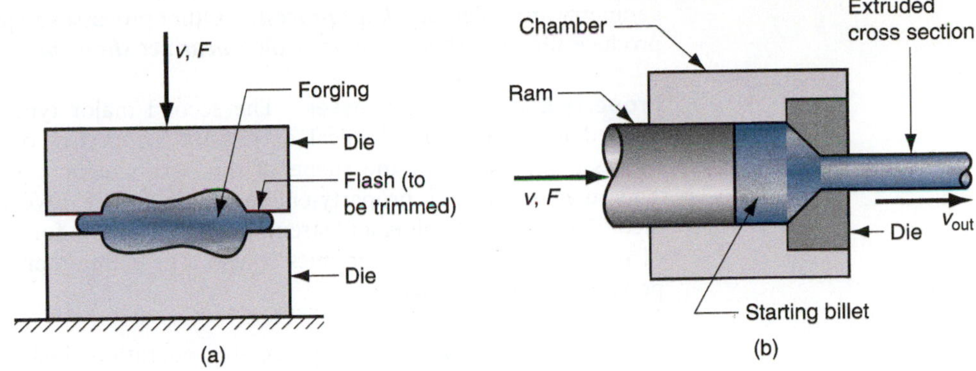

FIGURE 1.6 Some common deformation processes: (a) forging, in which two halves of a die squeeze the workpart, causing it to assume the shape of the die cavity; and (b) extrusion, in which a billet is forced to flow through a die orifice, thus taking the cross-sectional shape of the orifice. (Credit: *Fundamentals of Modern Manufacturing*, 4th Edition by Mikell P. Groover, 2010. Reprinted with permission of John Wiley & Sons, Inc.)

important processes in this category are ***machining*** operations such as ***turning***, ***drilling***, and ***milling***, shown in Figure 1.7. These cutting operations are most commonly applied to solid metals, performed using cutting tools that are harder and stronger than the work metal. ***Grinding*** is another common process in this category. Other material removal processes are known as ***nontraditional processes*** because they use lasers, electron beams, chemical erosion, electric discharges, and electrochemical energy to remove material rather than cutting or grinding tools.

It is desirable to minimize waste and scrap in converting a starting workpart into its subsequent geometry. Certain shaping processes are more efficient than others in terms of material conservation. Material removal processes (e.g., machining) tend to be wasteful of material, simply by the way they work. The material removed from the starting shape is waste, at least in terms of the unit operation. Other processes, such as certain casting and molding operations, often convert close to 100% of the starting material into final product. Manufacturing processes that transform nearly all of the starting material into product and require no subsequent machining to achieve final part

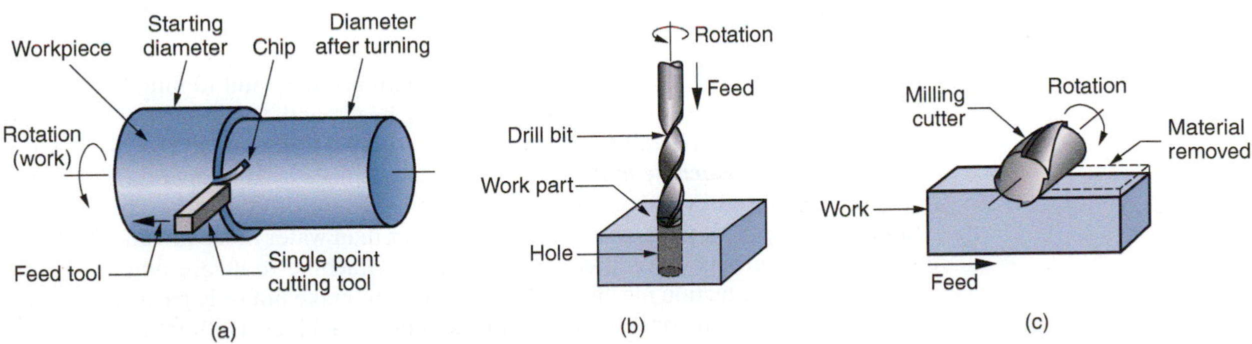

FIGURE 1.7 Common machining operations: (a) turning, in which a single-point cutting tool removes metal from a rotating workpiece to reduce its diameter; (b) drilling, in which a rotating drill bit is fed into the work to create a round hole; and (c) milling, in which a workpart is fed past a rotating cutter with multiple edges. (Credit: *Fundamentals of Modern Manufacturing*, 4th Edition by Mikell P. Groover, 2010. Reprinted with permission of John Wiley & Sons, Inc.)

geometry are called ***net shape processes***. Other processes require minimum machining to produce the final shape and are called ***near net shape processes***.

Property-Enhancing Processes The second major type of part processing is performed to improve mechanical or physical properties of the work material. These processes do not alter the shape of the part, except unintentionally in some cases. The most important property-enhancing processes involve ***heat treatments***, which include various annealing and strengthening processes for metals and glasses. ***Sintering*** of powdered metals and ceramics is also a heat treatment that strengthens a pressed powder metal workpart.

Surface Processing Surface processing operations include (1) cleaning, (2) surface treatments, and (3) coating and thin film deposition processes. ***Cleaning*** includes both chemical and mechanical processes to remove dirt, oil, and other contaminants from the surface. ***Surface treatments*** include mechanical working such as shot peening and sand blasting, and physical processes such as diffusion and ion implantation. ***Coating*** and ***thin film deposition*** processes apply a coating of material to the exterior surface of the workpart. Common coating processes include ***electroplating***, ***anodizing*** of aluminum, and organic ***coating*** (call it ***painting***). Thin film deposition processes include ***physical vapor deposition*** and ***chemical vapor deposition*** to form extremely thin coatings of various substances.

1.2.2 ASSEMBLY OPERATIONS

The second basic type of manufacturing operation is ***assembly***, in which two or more separate parts are joined to form a new entity. Components of the new entity are connected either permanently or semipermanently. Permanent joining processes include ***welding***, ***brazing***, ***soldering***, and ***adhesive bonding***. They form a joint between components that cannot be easily disconnected. Certain ***mechanical assembly*** methods are available to fasten two (or more) parts together in a joint that can be conveniently disassembled. The use of screws, bolts, and other ***threaded fasteners*** are important traditional methods in this category. Other mechanical assembly techniques form a more permanent connection; these include ***rivets***, ***press fitting***, and ***expansion fits***.

1.2.3 PRODUCTION MACHINES AND TOOLING

Manufacturing operations are accomplished using machinery and tooling (and people). The extensive use of machinery in manufacturing began with the Industrial Revolution. It was at that time that metal cutting machines started to be developed and widely used. These were called ***machine tools***—power-driven machines used to operate cutting tools previously operated by hand. Modern machine tools are described by the same basic definition, except that the power is electrical rather than water or steam, and the level of precision and automation is much greater today. Machine tools are among the most versatile of all production machines. They are used to make not only parts for consumer products, but also components for other production machines. Both in a historic and a reproductive sense, the machine tool is the mother of all machinery.

Other production machines include ***presses*** for stamping operations, ***forge hammers*** for forging, ***rolling mills*** for rolling sheet metal, ***welding machines*** for welding, and ***insertion machines*** for inserting electronic components into printed circuit boards. The name of the equipment usually follows from the name of the process.

TABLE 1.3 Production equipment and tooling used for various manufacturing processes.

Process	Equipment	Special tooling (function)
Casting	[a]	Mold (cavity for molten metal)
Molding	Molding machine	Mold (cavity for hot polymer)
Rolling	Rolling mill	Roll (reduce work thickness)
Forging	Forge hammer or press	Die (squeeze work to shape)
Extrusion	Press	Extrusion die (reduce cross section)
Stamping	Press	Die (shearing, forming sheet metal)
Machining	Machine tool	Cutting tool (material removal)
		Fixture (hold workpart)
		Jig (hold part and guide tool)
Grinding	Grinding machine	Grinding wheel (material removal)
Welding	Welding machine	Electrode (fusion of work metal)
		Fixture (hold parts during welding)

Compiled from generally available production data sources.
[a]Various types of casting setups and equipments (Chapter 11).

Production equipment can be general purpose or special purpose. ***General-purpose equipment*** is more flexible and adaptable to a variety of jobs. It is commercially available for any manufacturing company to invest in. ***Special-purpose equipment*** is usually designed to produce a specific part or product in very large quantities. The economics of mass production justify large investments in special-purpose machinery to achieve high efficiencies and short cycle times. This is not the only reason for special-purpose equipment, but it is the dominant one. Another reason may be because the process is unique and commercial equipment is not available. Some companies with unique processing requirements develop their own special-purpose equipment.

Production machinery usually requires ***tooling*** that customizes the equipment for the particular part or product. In many cases, the tooling must be designed specifically for the part or product configuration. When used with general-purpose equipment, it is designed to be exchanged. For each workpart type, the tooling is fastened to the machine and the production run is made. When the run is completed, the tooling is changed for the next workpart type. When used with special-purpose machines, the tooling is often designed as an integral part of the machine. Because the special-purpose machine is likely being used for mass production, the tooling may never need changing except for replacement of worn components or for repair of worn surfaces.

The type of tooling depends on the type of manufacturing process. Table 1.3 lists examples of special tooling used in various operations. Details are provided in the chapters that discuss these processes.

1.3 ORGANIZATION OF THE BOOK

Section 1.2 provides an introduction to the manufacturing processes covered in this book. The remaining 29 chapters are organized into 9 parts. Part I, titled Engineering Materials and Product Attributes, consists of three chapters. Chapters 2 and 3 discuss the important categories and properties of the materials that are used in the processes covered in the book. Chapter 4 provides a survey of product specifications, namely dimensions, tolerances, and surface characterization. It includes an appendix on the measurement of these attributes.

Most of the processes and operations included in our text are identified in Figure 1.3. Part II begins our coverage of the four categories of shaping processes. Part II consists of five chapters on the solidification processes that include casting of metals, glassworking, and polymer shaping. In Part III, the particulate processing of metals and ceramics is covered in two chapters. Part IV includes three chapters that deal with metal deformation processes such as rolling, forging, extrusion, and sheet metalworking. Finally, Part V examines the material removal processes. Three chapters are devoted to conventional machining, and two chapters cover grinding and nontraditional material removal technologies. The other processing operations, property enhancing (heat treatment) and surface processing (e.g., cleaning, electroplating, and painting), are covered in two chapters in Part VI.

Joining and assembly processes are considered in Part VII, which is organized into four chapters on welding, brazing, soldering, adhesive bonding, and mechanical assembly.

Several unique processes that do not neatly fit into the classification scheme of Figure 1.3 are covered in Part VIII, titled Special Processing and Assembly Technologies. Its two chapters cover rapid prototyping, microfabrication, and nanofabrication.

Part IX includes three chapters on systems topics related to manufacturing. These topics can be divided into two categories: (1) the technologies and equipment that are located in the factory and accomplish the manufacturing operations and (2) manufacturing support systems such as process planning and quality control.

REFERENCES

[1] Black, J., and Kohser, R. ***DeGarmo's Materials and Processes in Manufacturing***, 10th ed. John Wiley & Sons, Inc., Hoboken, New Jersey, 2008.

[2] Flinn, R. A., and Trojan, P. K. ***Engineering Materials and Their Applications***, 5th ed. John Wiley & Sons, Inc., New York, 1995.

[3] Groover, M. P. ***Automation, Production Systems, and Computer Integrated Manufacturing***, 3rd ed. Pearson Prentice-Hall, Upper Saddle River, New Jersey, 2008.

[4] Kalpakjian, S., and Schmid S. R. ***Manufacturing Processes for Engineering Materials***, 6th ed. Pearson Prentice Hall, Upper Saddle River, New Jersey, 2010.

REVIEW QUESTIONS

1.1. About what percentage of the U.S. gross domestic product (GDP) is accounted for by the manufacturing industries?

1.2. Define manufacturing.

1.3. The manufacturing industries are considered part of which of the following industry classifications: (a) primary, (b) secondary, or (c) tertiary?

1.4. What is the difference between a consumer good and a capital good? Give some examples in each category.

1.5. What is the difference between soft product variety and hard product variety, as these terms are defined in the text?

1.6. One of the dimensions of manufacturing capability is technological processing capability. Define technological processing capability.

1.7. What are the four categories of engineering materials used in manufacturing?

1.8. What is the definition of steel?

1.9. What are some of the typical applications of steel?

1.10. What is the difference between a thermoplastic polymer and a thermosetting polymer?

1.11. Manufacturing processes are usually accomplished as unit operations. Define unit operation.

1.12. In manufacturing processes, what is the difference between a processing operation and an assembly operation?

1.13. One of the three general types of processing operations is shaping operations, which are used to create or alter the geometry of the workpart. What are the four categories of shaping operations?

1.14. What is the difference between net shape processes and near net shape processes?

1.15. Identify the four types of permanent joining processes used in assembly.

1.16. What is a machine tool?

1.17. What is the difference between special-purpose and general-purpose production equipment?

Part I Engineering Materials and Product Attributes

2 ENGINEERING MATERIALS

Chapter Contents

2.1 Metals and Their Alloys
- 2.1.1 Steels
- 2.1.2 Cast Irons
- 2.1.3 Nonferrous Metals
- 2.1.4 Superalloys

2.2 Ceramics
- 2.2.1 Traditional Ceramics
- 2.2.2 New Ceramics
- 2.2.3 Glass

2.3 Polymers
- 2.3.1 Thermoplastic Polymers
- 2.3.2 Thermosetting Polymers
- 2.3.3 Elastomers

2.4 Composites
- 2.4.1 Technology and Classification of Composite Materials
- 2.4.2 Composite Materials

In Chapter 1, manufacturing was defined as a transformation process. It is the material that is transformed; it is the behavior of the material when subjected to the particular forces, temperatures, and other physical parameters of the process that determines the success of the operation. We find that certain materials respond well to certain manufacturing processes and poorly, or not at all, to others. What are the characteristics and properties of materials that determine their capacity to be transformed by the different processes?

Part I of this book consists of three chapters that address this question and related issues. In the current chapter, we discuss the four types of engineering materials that are used in the manufacturing processes covered in the remaining chapters of the book. The four types are (1) metals, (2) ceramics, (3) polymers, and (4) composites. Chapter 3 discusses the mechanical and physical properties of these materials that are relevant in manufacturing. Of course, these properties are also important in product design. Chapter 4 is concerned with several part and product attributes that are specified in product design and achieved in manufacturing: dimensions, tolerances, and surface finish. The appendix to Chapter 4 describes how these attributes are measured.

2.1 METALS AND THEIR ALLOYS

Metals are the most important engineering materials. A ***metal*** is a category of materials generally characterized by properties of ductility, malleability, luster, and high electrical and thermal conductivity. The category includes both metallic elements and their alloys. Metals have properties that satisfy a wide variety of design requirements. The manufacturing processes by which they are shaped into products have been developed and refined over many years.

The technological and commercial importance of metals is due to the following general properties possessed by virtually all of the common metals:

- ***High stiffness and strength***. Metals can be alloyed for high rigidity, strength, and hardness; thus, they are used to provide the structural framework for most engineered products.
- ***Toughness***. Metals have the capacity to absorb energy better than other classes of materials.
- ***Good electrical conductivity***. Metals are conductors because of their metallic bonding that permits the free movement of electrons as charge carriers.
- ***Good thermal conductivity***. Metallic bonding also explains why metals generally conduct heat better than ceramics or polymers.

In addition, certain metals have specific properties that make them attractive for specialized applications. Many common metals are available at relatively low cost per unit weight and are often the material of choice simply because of their low cost.

Although some metals are important as pure elements (e.g., gold, silver, copper), most engineering applications require the improved properties obtained by alloying. An ***alloy*** is a metal composed of two or more elements, at least one of which is metallic. Through alloying, it is possible to enhance strength, hardness, and other properties compared to pure metals.

The mechanical properties of metals can be altered by ***heat treatment***, which refers to several types of heating and cooling cycles performed on a metal to beneficially change its properties. They operate by altering the basic microstructure of the metal, which in turn determines mechanical properties. Some heat treatment operations are applicable only to certain types of metals; for example, the heat treatment of steel to form martensite is somewhat specialized since martensite is unique to steel. Heat treatments for metals are discussed in Chapter 20.

Metals are converted into parts and products using a variety of manufacturing processes. The starting form of the metal differs, depending on the process. The major categories are (1) ***cast metal***, in which the initial form is a casting; (2) ***wrought metal***, in which the metal has been worked or can be worked (e.g., rolled or otherwise formed) after casting; better mechanical properties are generally associated with wrought metals compared to cast metals; and (3) ***powdered metal***, in which the metal is purchased in the form of very small powders for conversion into parts using powder metallurgy techniques. Most metals are available in all three forms. In this chapter, our discussion will focus on categories (1) and (2), which are of greatest commercial and engineering interest. Powder metallurgy techniques are examined in Chapter 10.

Metals are generally classified into two major groups: (1) ***ferrous***—those based on iron; and (2) ***nonferrous***—all other metals. The ferrous group can be further subdivided into steels and cast irons. Our discussion in the present section is organized into four topics: (1) steels, (2) cast irons, (3) nonferrous metals, and (4) superalloys. The superalloys include high-performance metals that can be ferrous or nonferrous.

2.1.1 STEELS

Steel is one of the two categories of ferrous alloys, which are based on iron (Fe). The other is cast iron (Section 2.1.2). Together, they constitute approximately 85% of the metal tonnage in the United States [10]. Let us begin our discussion of the ferrous metals by examining the iron-carbon phase diagram, shown in Figure 2.1. Pure iron melts at 1539°C (2802°F). During the rise in temperature from ambient, it undergoes several solid phase transformations as indicated in the diagram. Starting at room temperature the phase is alpha (α), also called ***ferrite***. At 912°C (1674°F), ferrite transforms to gamma (γ), called ***austenite***. This, in turn, transforms at 1394°C (2541°F) to delta (δ), which remains until melting occurs.

Solubility limits of carbon in iron are low in the ferrite phase—only about 0.022% at 723°C (1333°F). Austenite can dissolve up to about 2.1% carbon at a temperature of 1130°C (2066°F). This difference in solubility between alpha and gamma leads to opportunities for strengthening by heat treatment, but let us leave that for Chapter 20. Even without heat treatment, the strength of iron increases dramatically as carbon content increases, and we enter the region in which the metal is called steel. Precisely, ***steel*** is defined as an iron-carbon alloy containing from 0.02% to 2.11% carbon.[1] Most steels range between 0.05% to 1.1% C.

In addition to the phases mentioned, one other phase is prominent in the iron-carbon alloy system. This is Fe_3C, also known as ***cementite***. It is a metallic compound of iron and carbon that is hard and brittle. At room temperature under equilibrium conditions, iron-carbon alloys form a two-phase system at carbon levels even slightly above zero. The carbon content in steel ranges between these very low levels and about 2.1% C. Above 2.1% C, up to about 4% or 5%, the alloy is defined as ***cast iron***.

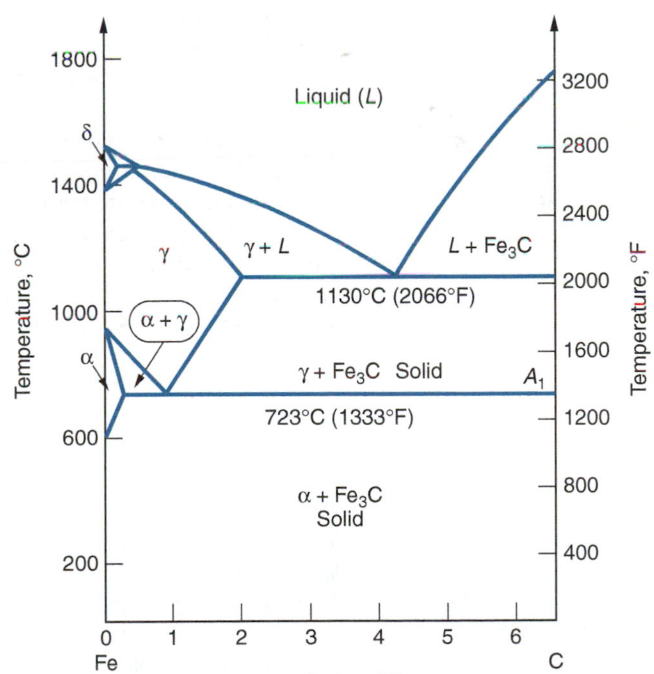

FIGURE 2.1 Phase diagram for iron-carbon system, up to about 6% carbon. (Credit: *Fundamentals of Modern Manufacturing*, 4th Edition by Mikell P. Groover, 2010. Reprinted with permission of John Wiley & Sons, Inc.)

[1] This is the conventional definition of steel, but exceptions exist. A recently developed steel for sheet-metal forming, called ***interstitial-free steel***, has a carbon content of only 0.005%.

Steel often includes other alloying ingredients, such as manganese, chromium, nickel, and/or molybdenum, but it is the carbon content that turns iron into steel. There are hundreds of compositions of steel available commercially. For purposes of organization here, the vast majority of commercially important steels can be grouped into the following categories: (1) plain carbon steels, (2) low alloy steels, (3) stainless steels, and (4) tool steels.

Plain Carbon Steels These steels contain carbon as the principal alloying element, with only small amounts of other elements (about 0.4% manganese plus lesser amounts of silicon, phosphorous, and sulphur). The strength of plain carbon steels increases with carbon content. A typical plot of the relationship is illustrated in Figure 2.2. As seen in Figure 2.1, steel at room temperature is a mixture of ferrite (α) and cementite (Fe_3C). The cementite particles distributed throughout the ferrite act as barriers to deformation; more carbon leads to more barriers, and more barriers mean stronger and harder steel.

According to a designation scheme developed by the American Iron and Steel Institute (AISI) and the Society of Automotive Engineers (SAE), plain carbon steels are specified by a four-digit number system: 10XX, where 10 indicates that the steel is plain carbon, and XX indicates the percent of carbon in hundredths of percentage points. For example, 1020 steel contains 0.20% C. The plain carbon steels are typically classified into three groups according to their carbon content:

1. ***Low carbon steels*** contain less than 0.20% C and are by far the most widely used steels. Typical applications are automobile sheet-metal parts, plate steel for fabrication, and railroad rails. These steels are relatively easy to form, which accounts for their popularity where high strength is not required. Steel castings usually fall into this carbon range, also.
2. ***Medium carbon steels*** range in carbon between 0.20% and 0.50% and are specified for applications requiring higher strength than the low-C steels. Applications include machinery components and engine parts such as crankshafts and connecting rods.
3. ***High carbon steels*** contain carbon in amounts greater than 0.50%. They are specified for still higher strength applications and where stiffness and hardness are needed. Springs, cutting tools and blades, and wear-resistant parts are examples.

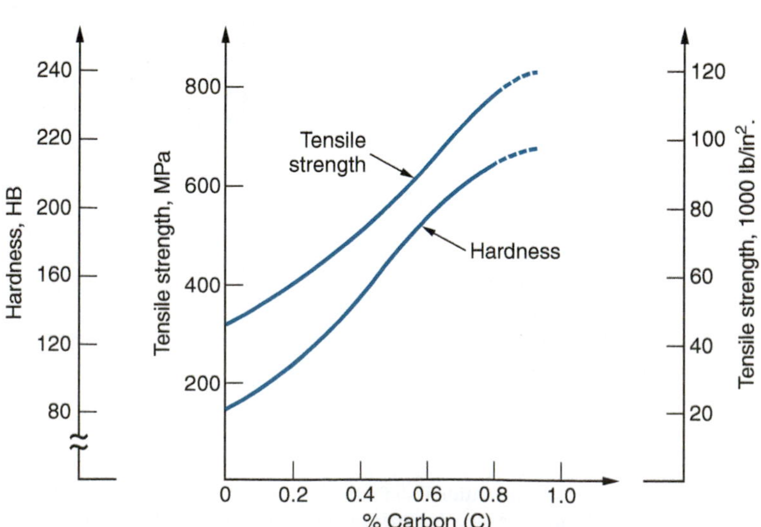

FIGURE 2.2 Tensile strength and hardness as a function of carbon content in plain carbon steel (hot-rolled, not heat-treated). (Credit: *Fundamentals of Modern Manufacturing*, 4th Edition by Mikell P. Groover, 2010. Reprinted with permission of John Wiley & Sons, Inc.)

Increasing carbon content strengthens and hardens the steel, but its ductility is reduced. Also, high carbon steels can be heat treated to form martensite, making the steel very hard and strong (Section 20.2).

Low Alloy Steels Low alloy steels are iron-carbon alloys that contain additional alloying elements in amounts totaling less than about 5% by weight. Owing to these additions, low alloy steels have mechanical properties that are superior to those of the plain carbon steels for given applications. Superior properties usually mean higher strength, hardness, hot hardness, wear resistance, toughness, and more desirable combinations of these properties. Heat treatment is often required to achieve these improved properties.

Common alloying elements added to steel are chromium, manganese, molybdenum, nickel, and vanadium, sometimes individually but usually in combinations. These elements typically form solid solutions with iron and metallic compounds with carbon (carbides), assuming sufficient carbon is present to support a reaction. We can summarize the effects of the principal alloying ingredients as follows:

- ***Chromium*** (Cr) improves strength, hardness, wear resistance, and hot hardness. It is one of the most effective alloying ingredients for increasing hardenability (Section 20.2.3). In significant proportions, Cr improves corrosion resistance.
- ***Manganese*** (Mn) improves the strength and hardness of steel. When the steel is heat treated, hardenability is improved with increased manganese. Because of these benefits, manganese is a widely used alloying ingredient in steel.
- ***Molybdenum*** (Mo) increases toughness and hot hardness. It also improves hardenability and forms carbides for wear resistance.
- ***Nickel*** (Ni) improves strength and toughness. It increases hardenability but not as much as some of the other alloying elements in steel. In significant amounts it improves corrosion resistance and is the other major ingredient (besides chromium) in certain types of stainless steel.
- ***Vanadium*** (V) inhibits grain growth during elevated temperature processing and heat treatment, which enhances strength and toughness of steel. It also forms carbides that increase wear resistance.

The AISI-SAE designations of some of the low alloy steels are presented in Table 2.1, which indicates nominal chemical analysis. As before, carbon content is specified by XX in 1/100% of carbon. For completeness, we include plain carbon steels (10XX). Properties of the various steels and other metals are defined and tabulated in Chapter 3.

Low alloy steels are not easily welded, especially at medium and high carbon levels. Since the 1960s, research has been directed at developing low carbon, low alloy steels that have better strength-to-weight ratios than plain carbon steels but are more weldable than low alloy steels. The products developed out of these efforts are called ***high-strength low-alloy*** (HSLA) steels. They generally have low carbon contents (in the range 0.10–0.30% C) plus relatively small amounts of alloying ingredients (usually only about 3% total). A typical chemistry is 0.12 C, 0.60 Mn, 1.1 Ni, 1.1 Cr, 0.35 Mo, and 0.4 Si. HSLA steels are hot-rolled under controlled conditions designed to provide improved strength compared to plain C steels, yet with no sacrifice in formability or weldability. Strengthening is achieved through alloying; heat treatment of HSLA steels is not feasible due to the low carbon content.

Stainless Steels Stainless steels are a group of highly alloyed steels designed to provide high corrosion resistance. The principal alloying element in stainless steel is chromium, usually above 15%. The chromium in the alloy forms a thin, impervious oxide film in an oxidizing atmosphere, which protects the surface from corrosion. Nickel is

TABLE 2.1 AISI-SAE designations of steels.

		Nominal Chemical Analysis, %							
Code	**Name of Steel**	**Cr**	**Mn**	**Mo**	**Ni**	**V**	**P**	**S**	**Si**
10XX	Plain carbon		0.4				0.04	0.05	
11XX	Resulfurized		0.9				0.01	0.12	0.01
12XX	Resulfurized, rephosphorized		0.9				0.10	0.22	0.01
13XX	Manganese		1.7				0.04	0.04	0.3
20XX	Nickel steels		0.5		0.6		0.04	0.04	0.2
31XX	Nickel–chrome	0.6			1.2		0.04	0.04	0.3
40XX	Molybdenum		0.8	0.25			0.04	0.04	0.2
41XX	Chrome–molybdenum	1.0	0.8	0.2			0.04	0.04	0.3
43XX	Ni–Cr–Mo	0.8	0.7	0.25	1.8		0.04	0.04	0.2
46XX	Nickel–molybdenum		0.6	0.25	1.8		0.04	0.04	0.3
47XX	Ni–Cr–Mo	0.4	0.6	0.2	1.0		0.04	0.04	0.3
48XX	Nickel–molybdenum		0.6	0.25	3.5		0.04	0.04	0.3
50XX	Chromium	0.5	0.4				0.04	0.04	0.3
52XX	Chromium	1.4	0.4				0.02	0.02	0.3
61XX	Cr–Vanadium	0.8	0.8			0.1	0.04	0.04	0.3
81XX	Ni–Cr–Mo	0.4	0.8	0.1	0.3		0.04	0.04	0.3
86XX	Ni–Cr–Mo	0.5	0.8	0.2	0.5		0.04	0.04	0.3
88XX	Ni–Cr–Mo	0.5	0.8	0.35	0.5		0.04	0.04	0.3
92XX	Silicon–Manganese		0.8				0.04	0.04	2.0
93XX	Ni–Cr–Mo	1.2	0.6	0.1	3.2		0.02	0.02	0.3
98XX	Ni–Cr–Mo	0.8	0.8	0.25	1.0		0.04	0.04	0.3

Source: [16].

another alloying ingredient used in certain stainless steels to increase corrosion protection. Carbon is used to strengthen and harden the metal; however, increasing the carbon content has the effect of reducing corrosion protection because chromium carbide forms to reduce the amount of free Cr available in the alloy.

In addition to corrosion resistance, stainless steels are noted for their combination of strength and ductility. Although these properties are desirable in many applications, they generally make these alloys difficult to work with in manufacturing. Also, stainless steels are significantly more expensive than plain C or low alloy steels.

Stainless steels are traditionally divided into three groups, named for the predominant phase present in the alloy at ambient temperature:

1. ***Austenitic stainless steels*** have a typical composition of around 18% Cr and 8% Ni and are the most corrosion resistant of the three groups. Owing to this composition, they are sometimes identified as 18-8 stainless. They are nonmagnetic and very ductile; but they show significant work hardening. The nickel has the effect of enlarging the austenite region in the iron-carbon phase diagram, making it stable at room temperature. Austenitic stainless steels are used to fabricate chemical and food-processing equipment, as well as machinery parts requiring high corrosion resistance.
2. ***Ferritic stainless steels*** have around 15% to 20% chromium, low carbon, and no nickel. This provides a ferrite phase at room temperature. Ferritic stainless steels are magnetic and are less ductile and corrosion resistant than the austenitics. Parts made of ferritic stainless range from kitchen utensils to jet engine components.
3. ***Martensitic stainless steels*** have a higher carbon content than ferritic stainlesses, thus permitting them to be strengthened by heat treatment (Section 20.2). They have

TABLE 2.2 Compositions of selected stainless steels.

Type	Chemical Analysis, %					
	Fe	Cr	Ni	C	Mn	Other[a]
Austenitic						
301	73	17	7	0.15	2	
302	71	18	8	0.15	2	
304	69	19	9	0.08	2	
309	61	23	13	0.20	2	
316	65	17	12	0.08	2	2.5 Mo
Ferritic						
405	85	13	—	0.08	1	
430	81	17	—	0.12	1	
Martensitic						
416	85	13	—	0.15	1	
440	81	17	—	0.65	1	

Compiled from [16].
[a]All of the grades in the table contain about 1% (or less) Si plus small amounts (well below 1%) of phosphorous and sulfur and other elements such as aluminum.

as much as 18% Cr but no Ni. They are strong, hard, and fatigue resistant, but not generally as corrosion resistant as the other two groups. Typical products include cutlery and surgical instruments.

Most stainless steels are designated by a three-digit AISI numbering scheme. The first digit indicates the general type, and the last two digits give the specific grade within the type. Table 2.2 lists some of the common stainless steels with typical compositions.

Tool Steels Tool steels are a class of (usually) highly alloyed steels designed for use as industrial cutting tools, dies, and molds. To perform in these applications, they must possess high strength, hardness, hot hardness, wear resistance, and toughness under impact. To obtain these properties, tool steels are heat treated. Principal reasons for the high levels of alloying elements are (1) improved hardenability, (2) reduced distortion during heat treatment, (3) hot hardness, (4) formation of hard metallic carbides for abrasion resistance, and (5) enhanced toughness.

The tool steels are classified according to application and composition. To identify the tool steel type, the AISI uses a classification scheme based on a prefix letter, defined in the following list:

T, M — ***High-speed tool steels*** are used as cutting tools in machining processes (Section 17.2.1). They are formulated for high wear resistance and hot hardness. The original high-speed steels (HSS) were developed around 1900. They permitted dramatic increases in cutting speed compared to previously used tools; hence, their name. The two AISI designations indicate the principal alloying element: T for tungsten and M for molybdenum.

H — ***Hot-working tool steels*** are intended for hot-working dies in forging, extrusion, and die-casting.

D — ***Cold-work tool steels*** are die steels used for cold working operations such as sheet-metal pressworking, cold extrusion, and certain forging operations. The designation D stands for die. Closely related AISI designations are A and O. A and O stand for air- and oil-hardening. They all provide good wear resistance.

W **_Water-hardening tool steels_** have high carbon with little or no other alloying elements. They can only be hardened by fast quenching in water. They are widely used because of low cost, but they are limited to low temperature applications. Dies to form the heads on nails and bolts are a typical application.

S **_Shock-resistant tool steels_** are intended for use in applications where high toughness is required, as in many sheet-metal shearing, punching, and bending operations.

P **_Mold steels_** are used to make molds for molding plastics and rubber.

L **_Low-alloy tool steels_** are generally reserved for special applications.

Tool steels are not the only tool materials. In our coverage of the manufacturing processes, a variety of tools will be described, as well as the materials out of which they are fabricated. The materials include plain carbon and low alloy steels, cast irons, and ceramics.

2.1.2 CAST IRONS

Cast iron is an iron alloy containing from 2.1% to about 4% carbon and from 1% to 3% silicon. Its composition makes it highly suitable as a casting metal. In fact, the tonnage of cast iron castings is several times that of all other cast metal parts combined (excluding cast ingots made during steelmaking that are subsequently rolled into bars, plates, and similar stock). The overall tonnage of cast iron is second only to steel among metals.

There are several types of cast iron, the most important being gray cast iron. Other types include ductile iron, white cast iron, and malleable iron. Ductile and malleable irons possess chemistries similar to the gray and white cast irons, respectively, but result from special treatments to be described below. Table 2.3 presents a listing of chemistries for the principal types.

Gray Cast Iron Gray cast iron accounts for the largest tonnage among the cast irons. It has a composition in the range 2.5% to 4% carbon and 1% to 3% silicon. This chemistry

TABLE 2.3 Compositions of selected cast irons.

	Typical Composition, %				
Type	Fe	C	Si	Mn	Other[a]
Gray cast irons					
ASTM Class 20	93.0	3.5	2.5	0.65	
ASTM Class 30	93.6	3.2	2.1	0.75	
ASTM Class 40	93.8	3.1	1.9	0.85	
ASTM Class 50	93.5	3.0	1.6	1.0	0.67 Mo
Ductile irons					
ASTM A395	94.4	3.0	2.5		
ASTM A476	93.8	3.0	3.0		
White cast iron					
Low-C	92.5	2.5	1.3	0.4	1.5Ni, 1Cr, 0.5Mo
Malleable irons					
Ferritic	95.3	2.6	1.4	0.4	
Pearlitic	95.1	2.4	1.4	0.8	

Compiled from [16]. Cast irons are identified by various systems. We have attempted to indicate the particular cast iron grade using the most common identification for each type.

[a]Cast irons also contain phosphorous and sulfur usually totaling less than 0.3%.

results in the formation of graphite (carbon) flakes distributed throughout the cast product upon solidification. The structure causes the surface of the metal to have a gray color when fractured; hence, the name gray cast iron. The dispersion of graphite flakes accounts for two attractive properties: (1) good vibration damping, which is desirable in engines and other machinery; and (2) internal lubricating qualities, which makes the cast metal machinable.

The American Society for Testing of Materials (ASTM) uses a classification method for gray cast iron that is intended to provide a minimum tensile strength (*TS*) specification for the various classes: Class 20 gray cast iron has a *TS* of 138 MPa (20,000 lb/in^2), Class 30 has a *TS* of 207 MPa (30,000 lb/in^2), and so forth. Compressive strength of gray cast iron is significantly greater than its tensile strength. Properties of the casting can be controlled to some extent by heat treatment. Ductility of gray cast iron is very low; it is a brittle material. Products made from gray cast iron include automotive engine blocks and heads, motor housings, and machine tool bases.

Ductile Iron This cast iron has the composition of gray iron but the molten metal is chemically treated before pouring to cause graphite spheroids to form rather than flakes. This results in a stronger and more ductile iron, and hence, its name. Applications include machinery components requiring high strength and good wear resistance.

White Cast Iron This cast iron has less carbon and silicon than gray cast iron. It is formed by more rapid cooling of the molten metal after pouring, thus causing the carbon to remain chemically combined with iron in the form of cementite (Fe_3C), rather than precipitating out of solution in the form of flakes. When fractured, the surface has a white crystalline appearance that gives the iron its name. Owing to the cementite, white cast iron is hard and brittle, and its wear resistance is excellent. These properties make white cast iron suitable for applications where wear resistance is required. Railway brake shoes are an example.

Malleable Iron When castings of white cast iron are heat treated to separate the carbon out of solution and form graphite aggregates, the resulting metal is called malleable iron. The new microstructure can possess substantial ductility compared to the metal from which it was transformed. Typical products made of malleable cast iron include pipe fittings and flanges, certain machine components, and railroad equipment parts.

2.1.3 NONFERROUS METALS

The nonferrous metals include metal elements and alloys not based on iron. The most important engineering metals in the nonferrous group are aluminum, magnesium, copper, nickel, titanium, and zinc, and their alloys. Although the nonferrous metals as a group cannot match the strength of the steels, certain nonferrous alloys have corrosion resistance and/or strength-to-weight ratios that make them competitive with steels in moderate-to-high stress applications. In addition, many of the nonferrous metals have properties other than mechanical that make them ideal for applications in which steel would be quite unsuitable. For example, copper has one of the lowest electrical resistivities among metals and is widely used for electrical wire. Aluminum is an excellent thermal conductor, and its applications include heat exchangers and cooking pans. It is also one of the most readily formed metals and is valued for that reason also. Zinc has a relatively low melting point, so zinc is widely used in die-casting operations. The common nonferrous metals have their own combination of properties that make them attractive in a variety of applications. In the following paragraphs, we discuss the nonferrous metals that are commercially and technologically the most important.

TABLE 2.4(a) Designations of wrought and cast aluminum alloys.

Alloy Group	Wrought Code	Cast Code
Aluminum, 99.0% or higher purity	1XXX	1XX.X
Aluminum alloys, by major element(s):		
Copper	2XXX	2XX.X
Manganese	3XXX	
Silicon + copper and/or magnesium		3XX.X
Silicon	4XXX	4XX.X
Magnesium	5XXX	5XX.X
Magnesium and silicon	6XXX	
Zinc	7XXX	7XX.X
Tin		8XX.X
Other	8XXX	9XX.X

Source: [17].

Aluminum and Its Alloys Aluminum and magnesium are light metals, and they are often specified in engineering applications for this feature. Both elements are abundant on Earth, aluminum on land (the principal ore is ***bauxite***) and magnesium in the sea, although neither is easily extracted from their natural states.

Aluminum has high electrical and thermal conductivity, and its resistance to corrosion is excellent due to the formation of a hard, thin oxide surface film. It is a very ductile metal and is noted for its formability. Pure aluminum is relatively low in strength, but it can be alloyed and heat-treated to compete with some steels, especially when weight is an important consideration.

The designation system for aluminum alloys is a four-digit code number. The system has two parts, one for wrought aluminums and the other for cast aluminums. The difference is that a decimal point is used following the third digit for cast aluminums. The designations are presented in Table 2.4(a).

Because properties of aluminum alloys are so influenced by work hardening and heat treatment, the temper (strengthening treatment, if any) must be designated in addition to the composition code. The principal temper designations are presented in Table 2.4(b). This designation is attached to the preceding four-digit number, separated from it by a hyphen, to indicate the treatment or absence thereof; for example, 2024-T3.

TABLE 2.4(b) Temper designations for aluminum alloys.

Temper	Description
F	As fabricated—no special treatment.
H	Strain hardened (wrought aluminums). H is followed by two digits, the first indicating a heat treatment, if any; and the second indicating the degree of work hardening remaining; for example: H1X No heat treatment after strain hardening, and X = 1 to 9, indicating degree of work hardening.
O	Annealed to relieve strain hardening and improve ductility; reduces strength to lowest level.
T	Thermal treatment to produce stable tempers other than F, H, or O. It is followed by a digit to indicate specific treatments; for example: T1 = cooled from elevated temperature, naturally aged; T2 = cooled from elevated temperature, cold worked, naturally aged; T3 = solution heat treated, cold worked, naturally aged; and so on.
W	Solution heat treatment, applied to alloys that age harden in service; it is an unstable temper.

Source: [17].

TABLE 2.5 Compositions of selected aluminum alloys.

Code	Typical Composition, %[a]					
	Al	Cu	Fe	Mg	Mn	Si
1050	99.5		0.4			0.3
1100	99.0		0.6			0.3
2024	93.5	4.4	0.5	1.5	0.6	0.5
3004	96.5	0.3	0.7	1.0	1.2	0.3
4043	93.5	0.3	0.8			5.2
5050	96.9	0.2	0.7	1.4	0.1	0.4

Compiled from [17].

[a]In addition to elements listed, alloy may contain trace amounts of other elements such as copper, magnesium, manganese, vanadium, and zinc.

Of course, temper treatments that specify strain hardening do not apply to the cast alloys. Compositions of selected aluminum alloys are presented in Table 2.5.

Magnesium and Its Alloys Magnesium (Mg) is the lightest of the structural metals; its specific gravity is 1.74. Magnesium and its alloys are available in both wrought and cast forms. It is relatively easy to machine. However, in all processing of magnesium, small particles of the metal (such as small metal-cutting chips) oxidize rapidly, and care must be taken to avoid fire hazards.

As a pure metal, magnesium is relatively soft and lacks sufficient strength for most engineering applications. However, it can be alloyed and heat-treated to achieve strengths comparable to aluminum alloys. In particular, its strength-to-weight ratio is an advantage in aircraft and missile components.

The designation scheme for magnesium alloys uses a three-to-five character alphanumeric code. The first two characters are letters that identify the principal alloying elements (up to two elements can be specified in the code, in order of decreasing percentages, or alphabetically if equal percentages). For example, A = aluminum (Al), K = zirconium (Zr), M = manganese (Mn), and Z = zinc (Zn). The letters are followed by a two-digit number that indicates, respectively, the amounts of the two alloying ingredients to the nearest percent. Finally, the last symbol is a letter that indicates some variation in composition, or simply the chronological order in which it was standardized for commercial availability. Magnesium alloys also require specification of a temper, and the same basic scheme presented in Table 2.4(b) for aluminum is used for magnesium alloys. To illustrate the designation scheme, some examples of magnesium alloys, are presented in Table 2.6.

TABLE 2.6 Compositions of selected magnesium alloys.

Code	Typical Composition, %					
	Mg	Al	Mn	Si	Zn	Other
AZ10A	98.0	1.3	0.2	0.1	0.4	
AZ80A	91.0	8.5			0.5	
ZK21A	97.1				2.3	6 Zr
AM60	92.8	6.0	0.1	0.5	0.2	0.3 Cu
AZ63A	91.0	6.0			3.0	

Compiled from [17].

TABLE 2.7 Compositions of selected copper alloys.

	Typical Composition, %				
Code	Cu	Be	Ni	Sn	Zn
C10100	99.99				
C11000	99.95				
C17000	98.0	1.7	[a]		
C24000	80.0				20.0
C26000	70.0				30.0
C52100	92.0			8.0	
C71500	70.0		30.0		

Compiled from [17].
[a]Small amounts of Ni and Fe + 0.3 Co.

Copper and Its Alloys Pure copper (Cu) has a distinctive reddish-pink color, but its most distinguishing engineering property is its low electrical resistivity—one of the lowest of all elements. Because of this property, and its relative abundance in nature, commercially pure copper is widely used as an electrical conductor (we should note here that the conductivity of copper decreases significantly as alloying elements are added). Cu is also an excellent thermal conductor. Copper is one of the noble metals (gold and silver are also noble metals), so it is corrosion resistant. All of these properties combine to make copper one of the most important metals.

On the downside, strength and hardness of copper are relatively low, especially when weight is taken into account. Accordingly, to improve strength (as well as for other reasons), copper is frequently alloyed. ***Bronze*** is an alloy of copper and tin (typically about 90% Cu and 10% Sn) that is still widely used today despite its ancient ancestry (i.e., Bronze Age in early history). Additional bronze alloys have been developed, based on other elements than tin; these include aluminum bronzes and silicon bronzes. ***Brass*** is another familiar copper alloy, composed of copper and zinc (typically around 65% Cu and 35% Zn). The highest strength alloy of copper is heat-treated beryllium-copper (only about 2% Be), which is used for springs.

The designation of copper alloys is based on the Unified Numbering System for Metals and Alloys (UNS), which uses a five-digit number preceded by the letter C (C for copper). The alloys are processed in wrought and cast forms, and the designation system includes both. Some copper alloys with compositions are presented in Table 2.7.

Nickel and Its Alloys Nickel (Ni) is similar to iron in many respects. It is magnetic, and its stiffness is virtually the same as that of iron and steel. However, it is much more corrosion resistant, and the high temperature properties of its alloys are generally superior. Because of its corrosion-resistant characteristics, it is widely used as an alloying element in steel, such as stainless steel, and as a plating metal on other metals such as plain carbon steel.

Alloys of nickel are commercially important in their own right and are noted for corrosion resistance and high-temperature performance. Compositions of some of the nickel alloys are given in Table 2.8. In addition, a number of superalloys are based on nickel (Section 2.1.4).

Titanium and Its Alloys Titanium (Ti) is fairly abundant in nature, constituting about 1% of Earth's crust (aluminum, the most abundant, is about 8%). The specific gravity of Ti is 4.7, midway between aluminum and iron. Its importance has grown in recent decades due to its aerospace applications where its light weight and good strength-to-weight ratio are exploited.

TABLE 2.8 Compositions of selected nickel alloys.

	Typical Composition, %						
Code	Ni	Cr	Cu	Fe	Mn	Si	Other
270	99.9		[a]	[a]			
200	99.0		0.2	0.3	0.2	0.2	C, S
400	66.8		30.0	2.5	0.2	0.5	C
600	74.0	16.0	0.5	8.0	1.0	0.5	
230	52.8	22.0		3.0	0.4	0.4	[b]

Compiled from [17].
[a]Trace amounts.
[b]Other alloying ingredients in grade 230 are 5% Co, 2% Mo, 14% W, 0.3% Al, and 0.1% C.

Thermal expansion of titanium is relatively low among metals. It is stiffer and stronger than aluminum, and it retains good strength at elevated temperatures. Pure titanium is reactive, which presents problems in processing, especially in the molten state. However, at room temperature it forms a thin adherent oxide coating (TiO_2) that provides excellent corrosion resistance. These properties give rise to two principal application areas for titanium: (1) in the commercially pure state, Ti is used for corrosion resistant components, such as marine components and prosthetic implants; and (2) titanium alloys are used as high-strength components in temperatures ranging from ambient to above 550°C (1000°F), especially where its excellent strength-to-weight ratio is exploited. These latter applications include aircraft and missile components. Alloying elements used with titanium include aluminum, manganese, tin, and vanadium. Some compositions for several Ti alloys are presented in Table 2.9.

Zinc and Its Alloys Its low melting point makes zinc (Zn) attractive as a casting metal. It also provides corrosion protection when coated onto steel or iron; ***galvanized steel*** is steel that has been coated with zinc.

Several alloys of zinc are listed in Table 2.10, with data on composition and applications. Zinc alloys are widely used in die casting to mass produce components for the automotive and appliance industries. Another major application of zinc is in galvanized steel, in which steel is coated with zinc for corrosion resistance. A third important use of zinc is in brass. As previously indicated in our discussion of copper, this alloy consists of copper and zinc, in the ratio of about 2/3 Cu to 1/3 Zn. Finally, readers may be interested to know that the U.S. one-cent piece is mostly zinc. The penny is coined out of zinc and then electroplated with copper, so that the final proportions are 97.5% Zn and 2.5% Cu. It costs the U.S. Mint about 1.5 cents to produce each penny.

TABLE 2.9 Compositions of selected titanium alloys.

	Typical Composition, %					
Code[a]	Ti	Al	Cu	Fe	V	Other
R50250	99.8			0.2		
R56400	89.6	6.0		0.3	4.0	[b]
R54810	90.0	8.0			1.0	1 Mo,[b]
R56620	84.3	6.0	0.8	0.8	6.0	2 Sn,[b]

Compiled from [1] and [17].
[a]United Numbering System (UNS).
[b]Traces of C, H, O.

TABLE 2.10 Compositions and applications of selected zinc alloys.

Code[a]	Typical Composition, %					Application
	Zn	Al	Cu	Mg	Fe	
Z33520	95.6	4.0	0.25	0.04	0.1	Die casting
Z35540	93.4	4.0	2.5	0.04	0.1	Die casting
Z35635	91.0	8.0	1.0	0.02	0.06	Foundry alloy
Z35840	70.9	27.0	2.0	0.02	0.07	Foundry alloy
Z45330	98.9		1.0	0.01		Rolled alloy

Compiled from [17].
[a]UNS—Unified Numbering System for metals.

2.1.4 SUPERALLOYS

Superalloys constitute a category that straddles the ferrous and nonferrous metals. Some of them are based on iron, while others are based on nickel and cobalt. In fact, many of the superalloys contain substantial amounts of three or more metals, rather than consisting of one base metal plus alloying elements. Although the tonnage of these metals is not significant compared to most of the other metals we have discussed, they are nevertheless commercially important because they are very expensive; and they are technologically important because of what they can do.

The ***superalloys*** are a group of high-performance alloys designed to meet very demanding requirements for strength and resistance to surface degradation (corrosion and oxidation) at high service temperatures. Conventional room temperature strength is usually not the important criterion for these metals, and most of them possess room temperature strength properties that are good but not outstanding. Their high-temperature performance is what distinguishes them; tensile strength, hot hardness, creep resistance, and corrosion resistance at very elevated temperatures are the mechanical properties of interest. Operating temperatures are often in the vicinity of 1100°C (2000°F). These metals are widely used in gas turbines—jet and rocket engines, steam turbines, and nuclear power plants—systems in which operating efficiency increases with higher temperatures.

The superalloys are usually divided into three groups, according to their principal constituent: iron, nickel, or cobalt:

➢ ***Iron-based alloys*** have iron as the main ingredient, although in some cases the iron is less than 50% of the total composition. Typical alloying elements include nickel, cobalt, and chromium.

➢ ***Nickel-based alloys*** generally have better high-temperature strength than alloy steels. Nickel is the base metal. The principal alloying elements are chromium and cobalt; lesser elements include aluminum, titanium, molybdenum, niobium (Nb), and iron.

➢ ***Cobalt-based alloys*** consist of cobalt (40% to 50%) and chromium (20% to 30%) as their main components. Other alloying elements include nickel, molybdenum, and tungsten.

In virtually all of the superalloys, including those based on iron, strengthening is accomplished by precipitation hardening (Section 20.3). The iron-based superalloys do not use martensite formation for strengthening.

2.2 CERAMICS

The importance of ceramics as engineering materials derives from their abundance in nature and their mechanical and physical properties, which are quite different from those of metals. A ***ceramic*** is an inorganic compound consisting of a metal (or semimetal) and one or more nonmetals. Important examples of ceramic materials are ***silica***, or silicon dioxide (SiO_2), the main ingredient in most glass products; ***alumina***, or aluminum oxide (Al_2O_3), used in applications ranging from abrasives to artificial bones; and more complex compounds such as hydrous aluminum silicate ($Al_2Si_2O_5(OH)_4$), known as ***kaolinite***, the principal ingredient in most clay products (e.g., bricks and pottery). The elements in these compounds are the most common in the Earth's crust. The ceramics group includes many additional compounds, some of which occur naturally while others are manufactured.

The general properties that make ceramics useful in engineered products are high hardness, good electrical and thermal insulating characteristics, chemical stability, and high melting temperatures. Some ceramics are translucent—window glass being the clearest example. They are also brittle and possess virtually no ductility, which can cause problems in both processing and performance of ceramic products.

For purposes of organization, we classify ceramic materials into three basic types: (1) ***traditional ceramics***—silicates used for clay products such as pottery and bricks, common abrasives, and cement; (2) ***new ceramics***—more recently developed ceramics based on nonsilicates such as oxides and carbides, and generally possessing mechanical or physical properties that are superior or unique compared to traditional ceramics; and (3) ***glasses***—based primarily on silica and distinguished from the other ceramics by their noncrystalline structure. In addition to the three basic types, we have ***glass ceramics***—glasses that have been transformed into a largely crystalline structure by heat treatment. The manufacturing processes for these materials are covered in Chapters 7 (glass-working) and 11 (particulate processing of traditional and new ceramics).

2.2.1 TRADITIONAL CERAMICS

These materials are based on mineral silicates, silica, and mineral oxides. The primary products are fired clay (pottery, tableware, brick, and tile), cement, and natural abrasives such as alumina. These products, and the processes used to make them, date back thousands of years. Glass is also a silicate ceramic material and is often included within the traditional ceramics group [12], [13]. However, glass is covered in a later section because it has an amorphous or vitreous structure (the term ***vitreous*** means glassy, or possessing the characteristics of glass), which differs from the above crystalline materials.

Raw Materials Mineral silicates, such as clays of various compositions, and silica, such as quartz, are among the most abundant substances in nature and constitute the principal raw materials for traditional ceramics. Clays are the most widely used raw materials in ceramics. They consist of fine particles of hydrous aluminum silicate that become a plastic substance that is formable and moldable when mixed with water. The most common clays are based on the mineral ***kaolinite*** ($Al_2Si_2O_5(OH)_4$). Other clay minerals vary in composition, both in terms of proportions of the basic ingredients and through additions of other elements such as magnesium, sodium, and potassium.

Besides its plasticity when mixed with water, a second characteristic of clay that makes it so useful is that it fuses into a dense, rigid material when heated to a sufficiently elevated temperature. The heat treatment is known as ***firing***. Suitable firing temperatures depend on clay composition. Thus, clay can be shaped while wet and soft, and then fired to obtain the final hard ceramic product.

Silica (SiO_2) is another major raw material for the traditional ceramics. It is the principal component in glass, and an important ingredient in other ceramic products including whiteware, refractories, and abrasives. Silica is available naturally in various forms, the most important of which is ***quartz***. The main source of quartz is ***sandstone***. The abundance of sandstone and its relative ease of processing means that silica is low in cost; it is also hard and chemically stable. These features account for its widespread use in ceramic products. It is generally mixed in various proportions with clay and other minerals to achieve the appropriate characteristics in the final product. Feldspar is one of the other minerals often used. ***Feldspar*** refers to any of several crystalline minerals that consist of aluminum silicate combined with either potassium, sodium, calcium, or barium. Mixtures of clay, silica, and feldspar are used to make stoneware, china, and other tableware.

Still another important raw material for traditional ceramics is ***alumina***. Most alumina is processed from the mineral ***bauxite***, which is an impure mixture of hydrous aluminum oxide and aluminum hydroxide plus similar compounds of iron or manganese. Bauxite is also the principal ore in the production of aluminum metal. A purer but less common form of Al_2O_3 is the mineral ***corundum***, which contains alumina in massive amounts. Slightly impure forms of corundum crystals are the colored gemstones sapphire and ruby. Alumina ceramic is used as an abrasive in grinding wheels and as a refractory brick in furnaces.

Silicon carbide, also used as an abrasive, does not occur naturally. Instead, it is produced by heating mixtures of sand (source of silicon) and coke (carbon) to a temperature of around 2200°C (3900°F), so that the resulting chemical reaction forms SiC and carbon monoxide.

Traditional Ceramic Products The minerals discussed above are the ingredients for a variety of ceramic products. We organize our coverage here by major categories of traditional ceramic products. We limit our coverage to materials commonly used in manufactured products, thus omitting certain commercially important ceramics such as cement.

- ***Pottery and Tableware*** This category is one of the oldest, dating back thousands of years; yet it is still one of the most important. It includes tableware products that we all use: earthenware, stoneware, and china. The raw materials for these products are clay usually combined with other minerals such as silica and feldspar. The wetted mixture is shaped and then fired to produce the finished piece.
- ***Brick and Tile*** Building brick, clay pipe, unglazed roof tile, and drain tile are made from various low-cost clays containing silica and gritty matter widely available in natural deposits. These products are shaped by pressing (molding) and firing at relatively low temperatures.
- ***Refractories*** Refractory ceramics, often in the form of bricks, are critical in many industrial processes that require furnaces and crucibles to heat and/or melt materials. The useful properties of refractory materials are high-temperature resistance, thermal insulation, and resistance to chemical reaction with the materials (usually molten metals) being heated. As mentioned earlier, alumina is often used as a refractory ceramic. Other refractory materials include magnesium oxide (MgO) and calcium oxide (CaO).
- ***Abrasives*** Traditional ceramics used for abrasive products, such as grinding wheels and sandpaper, are ***alumina*** and ***silicon carbide***. Although SiC is harder, most grinding wheels are based on Al_2O_3 because it gives better results when grinding

steel, the most widely used metal. The abrasive particles (grains of ceramic) are distributed throughout the wheel using a bonding material such as shellac, polymer resin, or rubber. Grinding wheel technology is presented in Chapter 18.

2.2.2 NEW CERAMICS

The term *new ceramics* refers to ceramic materials that have been developed synthetically over the last several decades and to improvements in processing techniques that have provided greater control over the structures and properties of ceramic materials. In general, new ceramics are based on compounds other than variations of aluminum silicate (which form the bulk of the traditional ceramic materials). New ceramics are usually simpler chemically than traditional ceramics; for example, oxides, carbides, nitrides, and borides. The dividing line between traditional and new ceramics is sometimes fuzzy, because aluminum oxide and silicon carbide are included among the traditional ceramics. The distinction in these cases is based more on methods of processing than chemical composition.

We organize the new ceramics into chemical compound categories: oxides, carbides, and nitrides, discussed in the following sections. More complete coverage of new ceramics is presented in references [9], [12], and [18].

Oxide Ceramics The most important oxide new ceramic is ***alumina***. Although also discussed in the context of traditional ceramics, alumina is today produced synthetically from bauxite, using an electric furnace method. Through control of particle size and impurities, refinements in processing methods, and blending with small amounts of other ceramic ingredients, strength and toughness of alumina have been improved substantially compared to its natural counterpart. Alumina also has good hot hardness, low thermal conductivity, and good corrosion resistance. This is a combination of properties that promote a wide variety of applications, including [20]: abrasives (grinding wheel grit), bioceramics (artificial bones and teeth), electrical insulators, electronic components, alloying ingredients in glass, refractory brick, cutting tool inserts (Section 17.2.4), spark plug barrels, and engineering components (see Figure 2.3).

Carbides The carbide ceramics include silicon carbide (SiC), tungsten carbide (WC), titanium carbide (TiC), tantalum carbide (TaC), and chromium carbide (Cr_3C_2). Silicon

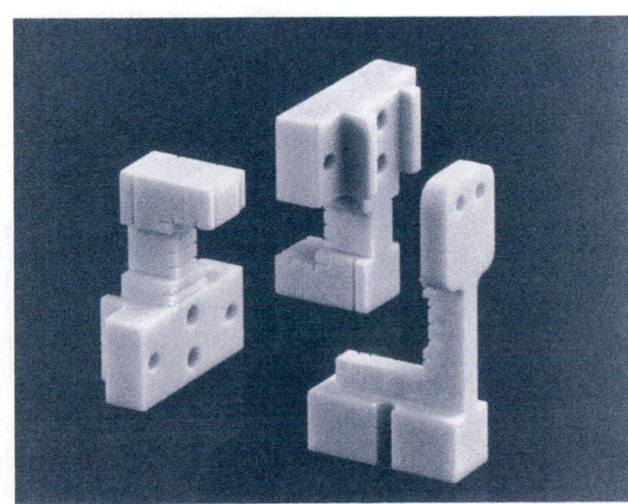

FIGURE 2.3 Alumina ceramic components. (Photo courtesy of Insaco Inc.) (Credit: *Fundamentals of Modern Manufacturing*, 4th Edition by Mikell P. Groover, 2010. Reprinted with permission of John Wiley & Sons, Inc.)

carbide was discussed previously. Although it is a man-made ceramic, the methods for its production were developed a century ago, and therefore it is generally included in the traditional ceramics group. In addition to its use as an abrasive, other SiC applications include resistance heating elements and additives in steelmaking.

WC, TiC, and TaC are valued for their hardness and wear resistance in cutting tools (Section 17.2.3) and other applications requiring these properties. Tungsten carbide was the first to be developed and is the most important and widely used material in the group. Chromium carbide is more suited to applications where chemical stability and oxidation resistance are important.

Except for SiC, the carbides must be combined with a metallic binder such as cobalt or nickel in order to fabricate a useful solid product. In effect, the carbide powders bonded in a metal framework creates what is known as a ***cemented carbide***—a composite material, specifically a ***cermet*** (reduced from ***cer***amic and ***met***al). We examine cemented carbides and other cermets in Section 2.4.2. The carbides have little engineering value except as constituents in a composite system.

Nitrides The important nitride ceramics are silicon nitride (Si_3N_4), boron nitride (BN), and titanium nitride (TiN). As a group, the nitride ceramics are hard and brittle, and they melt at high temperatures (but not generally as high as the carbides). They are usually electrically insulating, except for TiN.

Silicon nitride shows promise in high-temperature structural applications. It has low thermal expansion, good resistance to thermal shock and creep, and resists corrosion by molten nonferrous metals. These properties have motivated applications in gas turbines, rocket engines, and melting crucibles.

Boron nitride exists in several structures, similar to carbon. The important forms of BN are (1) hexagonal, similar to graphite, and (2) cubic, same as diamond; in fact, its hardness is comparable to that of diamond. This latter structure goes by the names ***cubic boron nitride*** and ***borazon***, symbolized cBN. Owing to its extreme hardness, the principal applications of cBN are in cutting tools (Section 17.2.5) and abrasive wheels (Section 18.1.1). Interestingly, it does not compete with diamond cutting tools and grinding wheels. Diamond is suited to non-steel machining and grinding, while cBN is appropriate for steel.

Titanium nitride has properties similar to those of other nitrides in this group, except for its electrical conductivity; it is a conductor. TiN has high hardness, good wear resistance, and a low coefficient of friction with the ferrous metals. This combination of properties makes TiN an ideal material as a surface coating on cutting tools. The coating is only around 0.006 mm (0.0003 in) thick, so the amounts of material used in this application are low.

2.2.3 GLASS

The term *glass* is somewhat confusing because it describes a state of matter as well as a type of ceramic. As a state of matter, the term refers to an amorphous, or noncrystalline, structure of a solid material. The glassy state occurs in a material when insufficient time is allowed during cooling from the molten condition for the crystalline structure to form. It turns out that all three categories of engineering materials (metals, ceramics, and polymers) can assume the glassy state, although the circumstances for metals to do so are quite rare.

As a type of ceramic, ***glass*** is an inorganic, nonmetallic compound (or mixture of compounds) that cools to a rigid condition without crystallizing; it is a ceramic that is in the glassy state as a solid material.

Chemistry and Properties of Glass The principal ingredient in virtually all glasses is ***silica*** (SiO_2), most commonly found as the mineral quartz in sandstone and silica sand. Quartz occurs naturally as a crystalline substance; but when melted and then cooled, it forms vitreous silica. Silica glass has a very low thermal expansion coefficient and is therefore quite resistant to thermal shock. These properties are ideal for elevated temperature applications; accordingly, chemical glassware designed for heating are made with high proportions of silica glass.

In order to reduce the melting point of glass for easier processing, and to control properties, the composition of most commercial glasses includes other oxides as well as silica. Silica remains as the main component in these glass products, usually comprising 50% to 75% of total chemistry. The reason SiO_2 is used so widely in these compositions is because it naturally transforms into a glassy state upon cooling from the liquid, whereas most ceramics crystallize upon solidification. Table 2.11 lists typical chemistries for some common glasses. The additional ingredients are contained in a solid solution with SiO_2, and each has a function, such as: (1) promoting fusion during heating; (2) increasing fluidity in the molten glass for processing; (3) retarding ***devitrification***—the tendency to crystallize from the glassy state; (4) reducing thermal expansion in the final product; (5) improving the chemical resistance against attack by acids, basic substances, or water; (6) adding color to the glass; and (7) altering the index of refraction for optical applications (e.g., lenses).

Glass Products Following is a list of the major categories of glass products. We examine the roles played by the different ingredients in Table 2.11 as we discuss these products.

- ***Window Glass*** This glass is represented by two chemistries in Table 2.11: soda-lime glass and window glass. The soda-lime formula dates back to the glass-blowing industry of the 1800s and before. It was (and is) made by mixing soda (Na_2O) and lime (CaO) with silica (SiO_2). The mixture has evolved to achieve a balance between achieving chemical durability and avoiding crystallization during cooling. Modern window glass and the techniques for making it have required slight adjustments in composition. Magnesia (MgO) has been added to reduce devitrification.
- ***Containers*** In previous times, the same basic soda-lime composition was used in manual glass blowing to make bottles and other containers. Modern processes for shaping glass containers cool the glass more rapidly than older methods, and changes

TABLE 2.11 Typical compositions of selected glass products.

	Chemical Composition (by weight to nearest %)								
Product	**SiO_2**	**Na_2O**	**CaO**	**Al_2O_3**	**MgO**	**K_2O**	**PbO**	**B_2O_3**	**Other**
Soda-lime glass	71	14	13	2					
Window glass	72	15	8	1	4				
Container glass	72	13	10	2[a]	2	1			
Light bulb glass	73	17	5	1	4				
Laboratory glass:									
Vycor	96			1				3	
Pyrex	81	4		2				13	
E-glass (fibers)	54	1	17	15	4			9	
S-glass (fibers)	64			26	10				
Optical glasses:									
Crown glass	67	8				12		12	ZnO
Flint glass	46	3				6	45		

Compiled from [10], [12], [19], and other sources.
[a]May include Fe_2O_3 with Al_2O_3.

in composition have attempted to optimize the proportions of lime (CaO) and soda (Na_2O_3). Lime promotes fluidity. It also increases devitrification, but since cooling is more rapid, this effect is not as important as in prior times when slower cooling rates were used. Soda increases chemical stability and insolubility.

- ***Light Bulb Glass*** Glass used in light bulbs and other thin glass items (e.g., drinking glasses, Christmas ornaments) is high in soda and low in lime; it also contains small amounts of magnesia and alumina. The raw materials are inexpensive and suited to the continuous melting furnaces used today in mass production of light bulbs.
- ***Laboratory Glassware*** These products include containers for chemicals (e.g., flasks, beakers, glass tubing). The glass must be resistant to chemical attack and thermal shock. Glass that is high in silica is suitable because of its low thermal expansion. The trade name "Vicor" is used for this high-silica glass. Additions of boric oxide also produce a glass with low thermal expansion, so some glass for laboratory ware contains B_2O_3. The trade name "Pyrex" is used for the borosilicate glass.
- ***Optical Glasses*** Applications for these glasses include lenses for eyeglasses and optical instruments such as cameras, microscopes, and telescopes. To achieve their function, the lenses must have different refractive indices. Optical glasses are generally divided into crowns and flints. ***Crown glass*** has a low index of refraction, while ***flint glass*** contains lead oxide (PbO) that gives it a high index of refraction.
- ***Glass Fibers*** Glass fibers are manufactured for a number of important applications, including fiberglass reinforced plastics, insulation wool, and fiber optics. The compositions vary according to function. The most commonly used glass reinforcing fibers in plastics are E-glass. Another glass fiber material is S-glass, which is stronger but not as economical as E-glass. Insulating fiberglass wool can be manufactured from regular soda-lime-silica glasses. The glass for fiber optics consists of a long, continuous core of glass with high refractive index surrounded by a sheath of lower refractive glass. The inside glass must have a very high light transmittance for long-distance communication.

Glass-Ceramics Glass-ceramics are a class of ceramic material produced by conversion of vitreous glass into a polycrystalline structure through heat treatment. The proportion of crystalline phase in the final product typically ranges between 90% and 98%, with the remainder being unconverted vitreous material. Grain size is usually between 0.1 and 1.0 μm (4 and 40 μ-in), significantly smaller than the grain size of conventional ceramics. This fine crystal microstructure makes glass-ceramics much stronger than the glasses from which they are derived. Also, due to their crystal structure, glass-ceramics are opaque (usually gray or white) rather than clear.

The processing sequence for glass-ceramics is as follows: (1) The first step involves heating and forming operations used in glassworking (Section 7.2) to create the desired product geometry. Glass shaping methods are generally more economical than pressing and sintering to shape traditional and new ceramics made from powders. (2) The product is cooled. (3) The glass is reheated to a temperature sufficient to cause a dense network of crystal nuclei to form throughout the material. It is the high density of nucleation sites that inhibits grain growth of individual crystals, thus leading ultimately to the fine grain size in the glass-ceramic material. The key to the propensity for nucleation is the presence of small amounts of nucleating agents in the glass composition. Common nucleating agents are TiO_2, P_2O_5, and ZrO_2. (4) Once nucleation is initiated, the heat treatment is continued at a higher temperature to cause growth of the crystalline phases.

The significant advantages of glass-ceramics include (1) efficiency of processing in the glassy state, (2) close dimensional control over the final product shape, and (3) good mechanical and physical properties. Properties include high strength (stronger than glass), absence of porosity, low coefficient of thermal expansion, and high resistance to thermal shock. These properties have resulted in applications in cooking ware, heat exchangers, and missile radomes. Certain formulations also possess high electrical resistance, suitable for electrical and electronics applications.

2.3 POLYMERS

Nearly all of the polymeric materials used in engineering today are synthetic (the exception is natural rubber). The materials themselves are made by chemical processing, and most of the products are made by solidification processes. A ***polymer*** is a compound consisting of long-chain molecules, each molecule made up of repeating units connected together. There may be thousands, even millions of units in a single polymer molecule. The word is derived from the Greek words ***poly***, meaning many, and *meros* (reduced to ***mer***), meaning part. Most polymers are based on carbon and are therefore considered organic chemicals.

Polymers can be separated into ***plastics*** and ***rubbers***. For our purposes in covering polymers as a technical subject, it is appropriate to divide them into the following three categories, where (1) and (2) are plastics and (3) is the rubber category:

1. ***Thermoplastic polymers***, also called ***thermoplastics*** (TP), are solid materials at room temperature, but they become viscous liquids when heated to temperatures of only a few hundred degrees. This characteristic allows them to be easily and economically shaped into products. They can be subjected to this heating and cooling cycle repeatedly without significant degradation.
2. ***Thermosetting polymers***, or ***thermosets*** (TS), cannot tolerate repeated heating cycles as thermoplastics can. When initially heated, they soften and flow for molding, but the elevated temperatures also produce a chemical reaction that hardens the material into an infusible solid. If reheated, thermosetting polymers degrade and char rather than soften.
3. ***Elastomers*** (E) are polymers that exhibit extreme elastic extensibility when subjected to relatively low mechanical stress. Although their properties are quite different from thermosets, they share a similar molecular structure that is different from the thermoplastics.

Thermoplastics are commercially the most important of the three types, constituting around 70% of the tonnage of all synthetic polymers produced. Thermosets and elastomers share the other 30% about evenly. Common TP polymers include polyethylene, polyvinylchloride, polypropylene, polystyrene, and nylon. Examples of TS polymers are phenolics, epoxies, and certain polyesters. The most common example given for elastomers is natural (vulcanized) rubber; however, synthetic rubbers exceed the tonnage of natural rubber.

Although the classification of polymers into the TP, TS, and E categories suits our purposes for organizing the topic in this section, we should note that the three types sometimes overlap. Certain polymers that are normally thermoplastic can be made into thermosets. Some polymers can be either thermosets or elastomers (we indicated that their molecular structures are similar). And some elastomers are thermoplastic. However, these are exceptions to the general classification scheme.

The growth in applications of synthetic polymers has been impressive. There are several reasons for the commercial and technological importance of polymers:

- Plastics can be formed by molding into intricate part geometries, usually with no further processing required. They are very compatible with ***net shape*** processing.

- Plastics possess an attractive list of properties for many engineering applications where strength is not a factor: (1) low density relative to metals and ceramics; (2) good strength-to-weight ratios for certain (but not all) polymers; (3) high corrosion resistance; and (4) low electrical and thermal conductivity.
- On a volumetric basis, polymers are cost competitive with metals.
- On a volumetric basis, polymers generally require less energy to produce than metals. This is generally true because the temperatures for working these materials are much lower than for metals.
- Certain plastics are translucent and/or transparent, which makes them competitive with glass in some applications.
- Polymers are widely used in composite materials (Section 2.4).

On the negative side, polymers in general have the following limitations: (1) strength is low relative to metals and ceramics; (2) modulus of elasticity or stiffness is also low—in the case of elastomers, of course, this may be a desirable characteristic; (3) service temperatures are limited to only a few hundred degrees because of the softening of thermoplastic polymers or degradation of thermosetting polymers and elastomers; and (4) some polymers degrade when subjected to sunlight and other forms of radiation.

Polymers are synthesized by joining many small molecules together to form very large molecules, called ***macromolecules***, that possess a chain-like structure. The small units, called ***monomers***, are generally simple unsaturated organic molecules such as ethylene C_2H_4. The atoms in these molecules are held together by covalent bonds; and when joined to form the polymer, the same covalent bonding holds the links of the chain together. Thus, each large molecule is characterized by strong primary bonding. Synthesis of the polyethylene molecule is depicted in Figure 2.4.

As described here, polymerization yields macromolecules of a chain-like structure, called a ***linear polymer***. This is the characteristic structure of a thermoplastic polymer. Other structures are portrayed in Figure 2.5. One possibility is for side branches to form along the chain, resulting in the ***branched polymer*** shown in Figure 2.5(b). In polyethylene, this occurs because hydrogen atoms are replaced by carbon atoms at random points along the chain, initiating the growth of a branch chain at each location. For certain polymers, primary bonding occurs between branches and other molecules at certain connection points to form ***cross-linked polymers*** as pictured in Figure 2.5(c) and (d). Cross-linking occurs because a certain proportion of the monomers used to form the polymer are capable of bonding to adjacent monomers on more than two sides, thus allowing branches from other molecules to attach. Lightly cross-linked structures are characteristic of elastomers. When the polymer is highly cross-linked we refer to it as having a ***network structure***, as in (d); in effect, the entire mass is one gigantic macromolecule. Thermosetting plastics take this structure after curing.

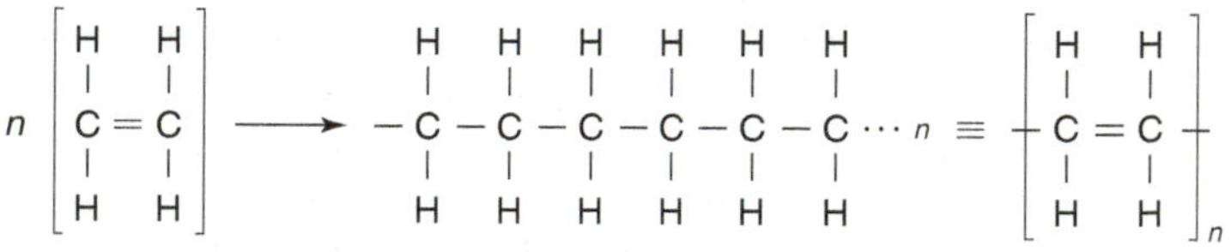

$$
n\begin{bmatrix} \mathrm{H} & \mathrm{H} \\ | & | \\ \mathrm{C} = & \mathrm{C} \\ | & | \\ \mathrm{H} & \mathrm{H} \end{bmatrix} \longrightarrow
\begin{matrix} & \mathrm{H} & & \mathrm{H} & & \mathrm{H} & & \mathrm{H} & & \mathrm{H} & & \mathrm{H} & \\ & | & & | & & | & & | & & | & & | & \\ - & \mathrm{C} & - & \mathrm{C} & - & \mathrm{C} & - & \mathrm{C} & - & \mathrm{C} & - & \mathrm{C} & \cdots n \\ & | & & | & & | & & | & & | & & | & \\ & \mathrm{H} & & \mathrm{H} & & \mathrm{H} & & \mathrm{H} & & \mathrm{H} & & \mathrm{H} & \end{matrix}
\equiv \begin{bmatrix} \mathrm{H} & \mathrm{H} \\ | & | \\ -\mathrm{C} = & \mathrm{C}- \\ | & | \\ \mathrm{H} & \mathrm{H} \end{bmatrix}_n
$$

(1) (2a) (2b)

FIGURE 2.4 Synthesis of polyethylene from ethylene monomers: (1) n ethylene monomers yields (2a) polyethylene of chain length n; (2b) concise notation for depicting the polymer structure of chain length n. (Credit: *Fundamentals of Modern Manufacturing*, 4th Edition by Mikell P. Groover, 2010. Reprinted with permission of John Wiley & Sons, Inc.)

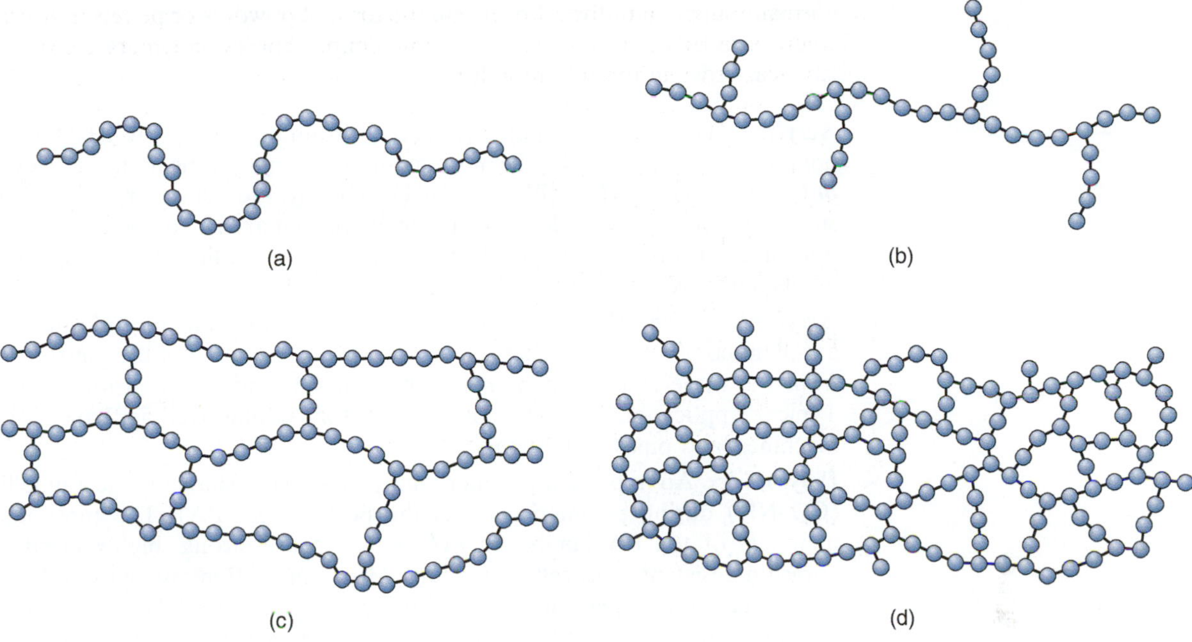

FIGURE 2.5 Various structures of polymer molecules: (a) linear, characteristic of thermoplastics; (b) branched; (c) loosely cross-linked as in an elastomer; and (d) tightly cross-linked or networked structure as in a thermoset. (Credit: *Fundamentals of Modern Manufacturing*, 4th Edition by Mikell P. Groover, 2010. Reprinted with permission of John Wiley & Sons, Inc.)

The presence of branching and cross-linking in polymers has a significant effect on properties. It is the basis of the difference between the three categories of polymers: TP, TS, and E. Thermoplastic polymers always possess linear or branched structures, or a mixture of the two. Branching increases entanglement among the molecules, usually making the polymer stronger in the solid state and more viscous at a given temperature in the plastic or liquid state.

Thermosetting plastics and elastomers are cross-linked polymers. Cross-linking causes the polymer to become chemically set; the reaction cannot be reversed. The effect is to permanently change the structure of the polymer; upon heating, it degrades or burns rather than melts. Thermosets possess a high degree of cross-linking, while elastomers possess a low degree of cross-linking. Thermosets are hard and brittle, while elastomers are elastic and resilient.

2.3.1 THERMOPLASTIC POLYMERS

The defining property of a thermoplastic polymer is that it can be heated from a solid state to a viscous liquid state and then cooled back down to solid, and that this heating and cooling cycle can be applied multiple times without degrading the polymer. The reason for this property is that TP polymers consist of linear (and/or branched) macromolecules that do not cross-link when heated. Compared to metals and ceramics, the typical thermoplastic at room temperature is characterized by the following: (1) much lower stiffness, (2) lower strength, (3) much lower hardness, and (4) greater ductility.

Thermoplastic products include molded and extruded items, fibers, films, sheets, packaging materials, paints, and varnishes. The starting raw materials for these products are normally supplied to the fabricator in the form of powders or pellets in bags, drums, or larger loads by truck or rail car. The most important TP polymers are discussed in alphabetical order in the following list:

- ***Acrylics***. The acrylics are polymers derived from acrylic acid ($C_3H_4O_2$) and compounds originating from it. The most important thermoplastic in the acrylics group is ***polymethylmethacrylate*** (PMMA) or Plexiglas (trade name for PMMA). Its outstanding property is excellent transparency, which makes it competitive with glass in optical applications. Examples include automotive tail-light lenses, optical instruments, and aircraft windows.
- ***Acrylonitrile–Butadiene–Styrene***. ABS is called an engineering plastic due to its excellent combination of mechanical properties. The name of the plastic is derived from the three starting monomers, which may be mixed in various proportions. Typical applications include components for automotive, appliances, business machines; and pipes and fittings.
- ***Polyamides***. An important polymer family that forms characteristic amide linkages (CO–NH) during polymerization is the polyamides (PA). The most important members of the PA family are ***nylons***, which are strong, highly elastic, tough, abrasion resistant, and self-lubricating. Most applications of nylon (about 90%) are in fibers for carpets, apparel, and tire cord. The rest (10%) are in engineering components, such as bearings, gears, and similar parts where strength and low friction are needed. A second group of polyamides is the ***aramids*** (aromatic polyamides) of which ***Kevlar*** (DuPont trade name) is important as a fiber in reinforced plastics. Kevlar is of interest because its strength is the same as steel at 20% of the weight.
- ***Polycarbonate***. Polycarbonate (PC) is noted for its generally excellent mechanical properties, which include high toughness and good creep resistance. In addition, it is heat resistant, transparent, and fire resistant. Applications include molded machinery parts, housings for business machines, pump impellers, safety helmets, and compact disks (e.g., audio and video). It is also widely used in window and windshield applications.
- ***Polyesters***. The polyesters form a family of polymers made up of the characteristic ester linkages (CO–O). They can be either thermoplastic or thermosetting, depending on whether cross-linking occurs. Of the thermoplastic polyesters, a representative example is ***polyethylene terephthalate*** (PET). Significant applications include blow-molded beverage containers, photographic films, and magnetic recording tape. In addition, PET fibers are widely used in apparel.
- ***Polyethylene***. Polyethylene (PE) was first synthesized in the 1930s, and today it accounts for the largest volume of all plastics. The features that make PE attractive as an engineering material are low cost, chemical inertness, and easy processing. Polyethylene is available in several grades, the most common of which are ***low-density polyethylene*** (LDPE) and ***high-density polyethylene*** (HDPE). The low-density grade is a highly branched. Applications include squeezable bottles, frozen food bags, sheets, film, and wire insulation. HDPE is a more linear structure, with higher density. These differences make HDPE stiffer and give it a higher melting temperature. HDPE is used for bottles, pipes, and house wares.
- ***Polypropylene***. Polypropylene (PP) is a major plastic, especially for injection molding. It is the lightest of the plastics, and its strength-to-weight ratio is high. PP is frequently compared with HDPE because its cost and many of its properties are similar. However, the high melting point of polypropylene allows certain applications that preclude use of polyethylene—for example, components that must be sterilized. Other applications are injection molded parts for automotive and houseware, and fiber products for carpeting.

- ***Polystyrene***. There are several polymers based on the monomer styrene (C_8H_8), of which polystyrene (PS) is used in the highest volume. It is a linear polymer that is generally noted for its brittleness. PS is transparent, easily colored, and readily molded, but degrades at elevated temperatures and dissolves in various solvents. Some PS grades contain 5% to 15% rubber to improve toughness, and the term ***high-impact polystyrene*** (HIPS) is used for these types. In addition to injection molding applications (e.g., toys), polystyrene also finds uses in packaging in the form of PS foams.
- ***Polyvinylchloride***. Polyvinylchloride (PVC) is a widely used plastic whose properties can be varied by combining additives with the polymer. Thermoplastics ranging from rigid PVC to flexible PVC can be achieved. The range of properties makes PVC a versatile polymer, with applications that include rigid pipe (used in construction, water and sewer systems, irrigation), fittings, wire and cable insulation, film, sheets, food packaging, flooring, and toys.

2.3.2 THERMOSETTING POLYMERS

Thermosetting (TS) polymers are distinguished by their highly cross-linked structure. In effect, the formed part (e.g., pot handle or electrical switch cover) becomes one large macromolecule. Owing to differences in chemistry and molecular structure, properties of thermosetting plastics are different from those of thermoplastics. In general, thermosets are (1) more rigid, (2) brittle, (3) less soluble in common solvents, (4) capable of higher service temperatures, and (5) not capable of being remelted—instead they degrade or burn.

The differences in properties of the TS plastics are attributable to cross-linking, which forms a thermally stable, three-dimensional, covalently bonded structure within the molecule. The chemical reactions associated with cross-linking are called ***curing*** or ***setting***. Curing is accomplished in three ways, depending on the starting ingredients: (1) temperature-activated systems, in which curing is caused by heating; (2) catalyst-activated systems, in which small amounts of a catalyst are added to a liquid polymer to cause curing; and (3) mixing-activated systems, in which two starting ingredients are mixed, resulting in a chemical reaction that forms the cross-linked polymer. Curing is done at the fabrication plants that shape the parts rather than the chemical plants that supply the starting materials to the fabricator.

Thermosetting plastics are not as widely used as the thermoplastics, perhaps because of the added processing complications involved in curing the TS polymers. The leading thermosets are phenolic resins, but their annual volume is less than 20% of the volume of polyethylene, the leading thermoplastic. The following list presents the most important thermosets and their typical applications:

- ***Amino Resins***. Amino plastics, characterized by the amino group (NH_2), consist of two thermosetting polymers, urea-formaldehyde and melamine-formaldehyde, which are produced by the reaction of formaldehyde (CH_2O) with either urea ($CO(NH_2)_2$) or melamine ($C_3H_6N_6$), respectively. ***Urea-formaldehyde*** is used as a plywood and particle-board adhesive. The resins are also used as a molding compound. ***Melamine-formaldehyde*** plastic is water resistant and is used for dishware and as a coating in laminated table and counter tops (trade name: Formica).
- ***Epoxies***. Epoxy resins are based on a chemical group called the ***epoxides***. Epichlorohydrin (C_3H_5OCl) is a widely used epoxide for producing epoxy resins. Cured epoxies are noted for strength, adhesion, and heat and chemical resistance. Applications include surface coatings, industrial flooring, glass fiber-reinforced composites, and adhesives. Insulating properties of epoxy thermosets make them useful as a lamination material for printed circuit boards.

- ***Phenolics***. Phenol (C_6H_5OH) is an acidic compound that can be reacted with aldehydes (dehydrogenated alcohols), formaldehyde (CH_2O) being the most reactive. ***Phenol-formaldehyde*** is the most important of the phenolic polymers. It is brittle and possesses good thermal, chemical, and dimensional stability. Applications include molded parts, printed circuit boards, counter tops, adhesives for plywood, and bonding material for brake linings and abrasive wheels.
- ***Polyesters***. Polyesters, which contain the characteristic ester linkages (CO–O), can be thermosetting as well as thermoplastic. Thermosetting polyesters are used largely in reinforced plastics (composites) to fabricate large items such as pipes, tanks, boat hulls, auto body parts, and construction panels. They can also be used in various molding processes to produce smaller parts.
- ***Polyurethanes***. This includes a large family of polymers, all characterized by the urethane group (NHCOO) in their structure. Many paints, varnishes, and similar coatings are based on urethane systems. Through variations in chemistry, cross-linking, and processing, polyurethanes can be thermoplastic, thermosetting, or elastomeric materials, the latter two being the most important commercially. The largest application of polyurethane is in foams. These can range between elastomeric and rigid, the latter being more highly cross-linked (polyurethane elastomers are covered in Section 2.3.3). Rigid foams are used as a filler material in hollow construction panels and refrigerator walls.

2.3.3 ELASTOMERS

Elastomers are polymers capable of large elastic deformation when subjected to relatively low stresses. Some elastomers can withstand extensions of 500% or more and still return to their original shape. The more popular term for elastomer is, of course, rubber. Elastomers can be divided into two categories: (1) natural rubber, derived from biological plants; and (2) synthetic rubbers, produced by polymerization processes similar to those used for TP and TS polymers.

Curing is required to effect cross-linking in most elastomers. The term for curing used in the context of natural rubber (and certain synthetic rubbers) is ***vulcanization***, which involves the formation of chemical cross-links between the polymer chains. Typical cross-linking in rubber is 1 to 10 links per hundred carbon atoms in the linear polymer chain, depending on the degree of stiffness desired in the material. This is considerably less than the cross-linking in thermosets.

Natural Rubber Natural rubber (NR) consists primarily of polyisoprene, a polymer of isoprene (C_5H_8). It is derived from latex, a milky substance produced by various plants, the most important of which is the rubber tree (*Hevea brasiliensis*) that grows in tropical climates. Latex is a water emulsion of polyisoprene (about one-third by weight), plus various other ingredients. Rubber is extracted from the latex by various methods that remove the water.

Natural crude rubber (without vulcanization) is sticky in hot weather, but stiff and brittle in cold weather. To form an elastomer with useful properties, natural rubber must be vulcanized. Traditionally, vulcanization was accomplished by mixing the crude rubber with small amounts of sulfur and heating. The effect of vulcanization is cross-linking, which increases strength and stiffness, yet maintains extensibility. The dramatic change in properties caused by vulcanization can be seen in the stress-strain curves of Figure 2.6. Sulfur alone can cause cross-linking, but the process is slow, taking hours to complete. In modern practice, other chemicals are added to sulfur during vulcanization to accelerate the process and serve other beneficial functions. Also, rubber can be vulcanized using

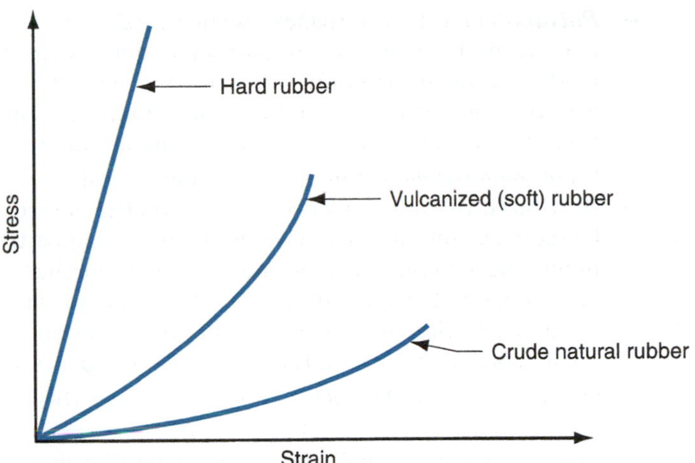

FIGURE 2.6 Increase in stiffness as a function of strain for three grades of rubber: natural rubber, vulcanized rubber, and hard rubber. (Credit: *Fundamentals of Modern Manufacturing*, 4th Edition by Mikell P. Groover, 2010. Reprinted with permission of John Wiley & Sons, Inc.)

chemicals other than sulfur. Today, curing times have been reduced significantly compared to the original sulfur curing of years ago.

As an engineering material, vulcanized rubber is noted among elastomers for its high strength, tear resistance, resilience (capacity to recover shape after deformation), and resistance to wear and fatigue. Its weaknesses are that it degrades when subjected to heat, sunlight, oxygen, ozone, and oil. Some of these limitations can be reduced through the use of additives.

The largest single market for natural rubber is automotive tires. In tires, carbon black is an important additive; it reinforces the rubber, increasing strength and resistance to tearing and abrasion. Other products made of NR include shoe soles, bushings, seals, and shock-absorbing components.

Synthetic Rubbers Today, the tonnage of synthetic rubbers is more than three times that of natural rubber. Development of these synthetic materials was motivated largely by the world wars when NR was difficult to obtain. As with most other polymers, the predominant raw material for the synthetic rubbers is petroleum. The synthetic rubbers of greatest commercial importance are discussed in the following:

- ***Butadiene Rubber***. ***Polybutadiene*** (BR) is important mainly in combination with other rubbers. It is compounded with natural rubber and with styrene (styrene-butadiene rubber is discussed later) in the production of automotive tires. Without compounding, the tear resistance, tensile strength, and ease of processing of polybutadiene are less than desirable.
- ***Butyl Rubber***. Butyl rubber consists of polyisobutylene (98% to 99%) and polyisoprene. It can be vulcanized to provide a rubber with very low air permeability, which has led to applications in inflatable products such as inner tubes, liners in tubeless tires, and sporting goods.
- ***Chloroprene Rubber***. Commonly known as ***Neoprene***, chloroprene rubber (CR) is an important special-purpose rubber. It is more resistant to oils, weather, ozone, heat, and flame than NR, but somewhat more expensive. Its applications include fuel hoses (and other automotive parts), conveyor belts, and gaskets, but not tires.
- ***Ethylene-Propylene Rubber***. Polymerization of ethylene and propylene with small proportions of a diene monomer produces ethylene-propylene-diene (EPDM), a useful synthetic rubber. Applications are for parts mostly in the automotive industry other than tires. Other uses are wire and cable insulation.

- ***Polyurethanes***. Polyurethanes (Section 2.3.2) with minimum cross-linking are elastomers, most commonly produced as flexible foams. In this form, they are widely used as cushion materials for furniture and automobile seats. Unfoamed polyurethane can be molded into products ranging from shoe soles to car bumpers, with cross-linking adjusted to achieve the desired properties for the application.
- ***Styrene-Butadiene Rubber***. SBR is the largest tonnage elastomer, totaling about 40% of all rubbers produced (natural rubber is second in tonnage). Its attractive features are low cost, resistance to abrasion, and better uniformity than NR. When reinforced with carbon black and vulcanized, its characteristics and applications are very similar to those of natural rubber. A comparison of properties reveals that most of its mechanical properties except wear resistance are inferior to NR, but its resistance to heat aging, ozone, weather, and oils is superior. Applications include automotive tires, footwear, and wire and cable insulation.

Thermoplastic Elastomers A thermoplastic elastomer (TPE) is a thermoplastic that behaves like an elastomer. It constitutes a family of polymers that is a fast-growing segment of the elastomer market. TPEs derive their elastomeric properties not from chemical cross-links, but from physical connections between soft and hard phases that make up the material. The chemistry and structure of these materials are generally complex, involving two materials that are incompatible so that they form distinct phases whose room temperature properties are different. Owing to their thermoplasticity, the TPEs cannot match conventional cross-linked elastomers in elevated temperature strength and creep resistance. Typical applications include footwear, rubber bands, extruded tubing, wire coating, and molded parts for automotive and other uses in which elastomeric properties are required. TPEs are not suitable for tires.

2.4 COMPOSITES

In addition to metals, ceramics, and polymers, a fourth material category can be distinguished: composites. A ***composite material*** is a material system composed of two or more physically distinct phases whose combination produces aggregate properties that are different from those of its constituents. The technological and commercial interest in composite materials derives from the fact that their properties are not just different from their components but are often far superior. Some of the possibilities include:

- Composites can be designed that are very strong and stiff, yet very light in weight, giving them strength-to-weight and stiffness-to-weight ratios several times greater than steel or aluminum. These properties are highly desirable in applications ranging from aircraft to sports equipment.
- Fatigue properties are generally better than for the common engineering metals. Toughness is often greater, too.
- Composites can be designed that do not corrode like steel; this is important in automotive and other applications.
- With composite materials, it is possible to achieve combinations of properties not attainable with metals, ceramics, or polymers alone.

Along with the advantages, there are disadvantages and limitations associated with composite materials. These include: (1) properties of many important composites are anisotropic, which means the properties differ depending on the direction in which they are measured; (2) many of the polymer-based composites are subject to attack by chemicals or solvents, just as the polymers themselves are susceptible to attack; (3) composite materials are generally expensive; and (4) some of the manufacturing methods for shaping composite materials are slow and costly.

2.4.1 TECHNOLOGY AND CLASSIFICATION OF COMPOSITE MATERIALS

As noted in our definition, a composite material consists of two or more distinct phases. The term ***phase*** indicates a homogeneous material, such as a metal or ceramic in which all of the grains have the same crystal structure, or a polymer with no fillers. By combining the phases, using methods yet to be described, a new material is created with aggregate performance exceeding that of its parts. The effect is synergistic.

Components in a Composite Material In the simplest manifestation of our definition, a composite material consists of two phases: a primary phase and a secondary phase. The primary phase forms the ***matrix*** within which the secondary phase is imbedded. The imbedded phase is sometimes referred to as a ***reinforcing agent*** (or similar term), because it usually serves to strengthen the composite. The reinforcing phase may be in the form of fibers or particles. The phases are generally insoluble in each other, but strong adhesion must exist at their interfaces.

The matrix phase can be any of three basic material types: polymers, metals, or ceramics. The secondary phase may also be one of the three basic materials, or it may be an element such as carbon or boron. Certain combinations are not feasible, such as a polymer in a ceramic matrix. The possibilities include two-phase structures consisting of components of the same material type, such as fibers of Kevlar (polymer) in a plastic (polymer) matrix.

The classification system for composite materials used in this book is based on the matrix phase. We list the classes here and discuss them in Section 2.4.2:

1. ***Metal-Matrix Composites*** (MMCs) include mixtures of ceramics and metals, such as cemented carbides and other cermets, as well as aluminum or magnesium reinforced by strong, high stiffness fibers.
2. ***Ceramic-Matrix Composites*** (CMCs) are the least common category. Aluminum oxide and silicon carbide are materials that can be imbedded with fibers for improved properties, especially in high temperature applications.
3. ***Polymer-Matrix Composites*** (PMCs). Thermosetting resins are the most widely used polymers in PMCs. Epoxy and polyester are commonly mixed with fiber reinforcement, and phenolic is mixed with powders. Thermoplastic molding compounds are often reinforced, usually with powders.

The matrix material serves several functions in the composite. First, it provides the bulk form of the part or product made of the composite material. Second, it holds the imbedded phase in place, usually enclosing and concealing it. Third, when a load is applied, the matrix shares the load with the secondary phase, in some cases deforming so that the stress is essentially born by the reinforcing agent.

It is important to understand that the role played by the secondary phase is to reinforce the primary phase. The imbedded phase is most commonly one of three shapes: fibers, particles, or flakes.

Fibers are filaments of reinforcing material, generally circular in cross section. Diameters range from less than 0.0025 mm (0.0001 in) to about 0.13 mm (0.005 in), depending on material. Fiber reinforcement provides the greatest opportunity to strengthen composite structures. In fiber-reinforced composites, the fiber is often considered to be the principal constituent since it bears the major share of the load.

Fibers are of interest as reinforcing agents because the filament form of most materials is significantly stronger than the bulk form. As diameter is reduced, the material becomes oriented in the direction of the fiber axis and the probability of defects in the structure decreases significantly. As a result, tensile strength increases dramatically.

Fibers used in composites can be either continuous or discontinuous. ***Continuous fibers*** are very long; in theory, they offer a continuous path by which a load can be carried by the composite part. In reality, this is difficult to achieve due to variations in the fibrous material and processing. ***Discontinuous fibers*** (chopped sections of continuous fibers) are short lengths. Various materials are used as fibers in fiber-reinforced composites. They include glass (E-glass and S-glass, Table 2.12), carbon, boron, Kevlar, aluminum oxide, and silicon carbide.

A second common shape of the imbedded phase is ***particulate***—powders ranging in size from microscopic to macroscopic. Particles are an important material form for metals and ceramics; the characterization and production of engineering powders are discussed in Chapters 10 and 11. The distribution of particles in the composite matrix is random, and therefore strength and other properties of the composite material are usually isotropic.

Flakes are basically two-dimensional particles—small flat platelets. Two examples of this shape are the minerals mica (silicate of K and Al) and talc ($Mg_3Si_4O_{10}(OH)_2$), used as reinforcing agents in plastics. They are generally lower cost materials than polymers, and they add strength and stiffness to plastic molding compounds.

Properties of a Composite Material In the selection of a composite material, an optimum combination of properties is usually being sought, rather than one particular property. For example, the fuselage and wings of an aircraft must be lightweight as well as strong, stiff, and tough. Finding a monolithic material that satisfies these requirements is difficult. Several fiber-reinforced polymers possess this combination of properties.

Another example is rubber. Natural rubber is a relatively weak material. In the early 1900s, it was discovered that by adding significant amounts of carbon black (almost pure carbon) to natural rubber, its strength is increased dramatically. The two ingredients interact to provide a composite material that is significantly stronger than either one alone. Rubber, of course, must also be vulcanized to achieve full strength.

Rubber itself is a useful additive in polystyrene. One of the distinctive and disadvantageous properties of polystyrene is its brittleness. Although most other polymers have considerable ductility, PS has virtually none. Rubber (natural or synthetic) can be added in modest amounts (5% to 15%) to produce high-impact polystyrene, which has much superior toughness and impact strength.

Fibers illustrate the importance of geometric shape. Most materials have tensile strengths several times greater in a fibrous form than in bulk. However, applications of fibers are limited by surface flaws, buckling when subjected to compression, and the inconvenience of the filament geometry when a solid component is needed. By imbedding the fibers in a polymer matrix, a composite material is obtained that avoids the problems of fibers but utilizes their strengths. The matrix provides the bulk shape to protect the fiber surfaces and resist buckling; and the fibers lend their high strength to the composite. When a load is applied, the low-strength matrix deforms and distributes the stress to the high-strength fibers, which then carry the load. If individual fibers break, the load is redistributed through the matrix to other fibers.

2.4.2 COMPOSITE MATERIALS

The three types of composite materials and their applications are discussed in this section: (1) metal-matrix composites, (2) ceramic-matrix composites, and (3) polymer-matrix composites.

Metal-Matrix Composites Metal-matrix composites (MMCs) consist of a metal matrix reinforced by a second phase. Common reinforcing phases include (1) particles of ceramic and (2) fibers of ceramics, carbon, and boron. MMCs of the first type are commonly called cermets.

A ***cermet*** is a composite material in which a ceramic is contained in a metallic matrix. The ceramic often dominates the mixture, sometimes ranging up to 96% by volume. Bonding can be enhanced by slight solubility between phases at the elevated temperatures used in processing these composites. An important category of cermet is cemented carbide.

Cemented carbides are composed of one or more carbide compounds bonded in a metallic matrix using particulate processing techniques (Section 10.2). The common cemented carbides are based on tungsten carbide (WC), titanium carbide (TiC), and chromium carbide (Cr_3C_2). Tantalum carbide (TaC) and others are less common. Typical metallic binders are cobalt and nickel. We have previously discussed the carbide ceramics (Section 2.2.2); they constitute the principal ingredient in cemented carbides, typically ranging in content from 80% to 95% of total weight.

Cutting tools are an important application of cemented carbides based on ***tungsten carbide***. Other applications of WC–Co cemented carbides include wire drawing dies, rock-drilling bits and other mining tools, dies for powder metallurgy, indenters for hardness testers, and other applications where hardness and wear resistance are critical requirements. ***Titanium carbide*** cermets are used principally for high temperature applications. Applications include gas-turbine nozzle vanes, valve seats, thermocouple protection tubes, torch tips, and hot-working spinning tools. TiC–Ni is also used as a cutting tool material in machining.

Fiber-reinforced metal-matrix composites are of interest because they combine the high tensile strength and modulus of elasticity of a fiber with metals of low density, thus achieving good strength-to-weight and stiffness-to-weight ratios in the resulting composite material. Typical metals used as the low-density matrix are aluminum, magnesium, and titanium. Some of the important fiber materials used in the composite include Al_2O_3, boron, carbon, and SiC.

Ceramic-Matrix Composites Ceramics have certain attractive properties: high stiffness, hardness, hot hardness, and compressive strength; and relatively low density. Ceramics also have several faults: low toughness and bulk tensile strength, and susceptibility to thermal cracking. Ceramic-matrix composites (CMCs) represent an attempt to retain the desirable properties of ceramics while compensating for their weaknesses. CMCs consist of a ceramic primary phase imbedded with a secondary phase. To date, most development work has focused on the use of fibers as the secondary phase. Success has been elusive. Technical difficulties include thermal and chemical compatibility of the constituents in CMCs during processing. Also, as with any ceramic material, limitations on part geometry must be considered.

Ceramic materials used as matrices include alumina (Al_2O_3), boron carbide (B_4C), boron nitride (BN), silicon carbide (SiC), silicon nitride (Si_3N_4), titanium carbide (TiC), and several types of glass. Some of these materials are still in the development stage as CMC matrices. Fiber materials in CMCs include carbon, SiC, and Al_2O_3.

Polymer-Matrix Composites A polymer-matrix composite (PMC) consists of a polymer primary phase in which a secondary phase is imbedded in the form of fibers, powders, or flakes. Commercially, PMCs are the most important of the three classes of composite materials. They include most plastic molding compounds, rubber reinforced with carbon black, and fiber-reinforced polymers (FRPs).

A ***fiber-reinforced polymer*** is a composite material consisting of a polymer matrix imbedded with high-strength fibers. The polymer matrix is usually a thermosetting plastic such as polyester or epoxy, but thermoplastic polymers, such as nylons (polyamides), polycarbonate, polystyrene, and polyvinylchloride, are also used. In addition, elastomers are also reinforced by fibers for rubber products such as tires and conveyor belts.

Fibers in PMCs come in various forms: discontinuous (chopped), continuous, or woven as a fabric. Principal fiber materials in FRPs are glass, carbon, and Kevlar 49. Less common fibers include boron, SiC, and Al_2O_3, and steel. Glass (in particular E-glass) is the most common fiber material in today's FRPs; its use to reinforce plastics dates from around 1920.

The most widely used form of the FRP itself is a laminar structure, made by stacking and bonding thin layers of fiber and polymer until the desired thickness is obtained. By varying the fiber orientation among the layers, a specified level of anisotropy in properties can be achieved in the laminate. This method is used to form parts of thin cross section, such as aircraft wing and fuselage sections, automobile and truck body panels, and boat hulls.

There are a number of attractive features that distinguish fiber-reinforced plastics as engineering materials. Most notable are (1) high strength-to-weight ratio, (2) high stiffness-to-weight ratio, and (3) low specific gravity. A typical FRP weighs only about one-fifth as much as steel; yet strength and modulus are comparable in the fiber direction.

During the last three decades there has been a steady growth in the application of fiber-reinforced polymers in products requiring high strength and low weight, often as substitutions for metals. The aerospace industry is one of the biggest users of composites. Designers are continually striving to reduce aircraft weight to increase fuel efficiency and payload capacity. Applications of composites in both military and commercial aircraft have increased steadily. Much of the structural weight of today's airplanes and helicopters consists of FRPs. The new Boeing 787 Dreamliner features 50% (by weight) composite (carbon fiber-reinforced plastic). That's about 80% of the volume of the aircraft. Composites are used for the fuselage, wings, tail, doors, and interior. By comparison, Boeing's 777 has only about 12% composites (by weight).

The automotive industry is another important user of FRPs. The most obvious applications are body panels for cars and truck cabs. FRPs have been widely adopted for sports and recreational equipment. Fiberglass reinforced plastic has been used for boat hulls since the 1940s. Fishing rods were another early application. Today, FRPs are represented in a wide assortment of sports products, including tennis rackets, golf club shafts, football helmets, bows and arrows, skis, and bicycle wheels.

In addition to FRPs, other polymer-matrix composites contain particles, flakes, and short fibers. Ingredients of the secondary phase are called ***fillers*** when used in polymer-molding compounds. Fillers divide into two categories: (1) reinforcements and (2) extenders. ***Reinforcing fillers*** serve to strengthen or otherwise improve mechanical properties of the polymer. Common examples include: wood flour and powdered mica in phenolic and amino resins to increase strength, abrasion resistance, and dimensional stability; and carbon black in rubber to improve strength, wear, and tear resistance. ***Extenders*** simply increase the bulk and reduce the cost-per-unit weight of the polymer, but have little or no effect on mechanical properties. Extenders may be formulated to improve molding characteristics of the resin.

Foamed polymers (Section 8.11) are a form of composite in which gas bubbles are imbedded in a polymer matrix. Styrofoam and polyurethane foam are the most common examples. The combination of near-zero density of the gas and relatively low density of the matrix makes these materials extremely lightweight. The gas mixture also imparts very low thermal conductivity for applications in which heat insulation is required.

REFERENCES

[1] Bauccio, M. (ed.). ***ASM Metals Reference Book***, 3rd ed. ASM International, Materials Park, Ohio, 1993.

[2] Black, J.,and Kohser, R. ***DeGarmo's Materials and Processes in Manufacturing***, 10th ed. John Wiley & Sons, Inc. Hoboken, New Jersey, 2008.

[3] Brandrup, J.,and Immergut, E. E. (eds.), ***Polymer Handbook***, 4th ed. John Wiley & Sons, Inc. New York, 2004.

[4] Carter, C. B.,and Norton, M. G., ***Ceramic Materials: Science and Engineering***. Springer, New York, 2007.

[5] Chanda, M.,and Roy, S. K., ***Plastics Technology Handbook***, 4th ed. CRC Taylor & Francis, Boca Raton, Florida, 2006.

[6] Chawla, K. K., ***Composite Materials: Science and Engineering***, 3rd ed. Springer-Verlag, New York, 2008.

[7] ***Engineering Materials Handbook***, Vol. **1**, ***Composites***. ASM International, Materials Park, Ohio, 1987.

[8] ***Engineering Materials Handbook***, Vol. **2**, ***Engineering Plastics***. ASM International, Materials Park, Ohio, 2000.

[9] ***Engineered Materials Handbook***, Vol. **4**, ***Ceramics and Glasses***. ASM International, Materials Park, Ohio, 1991.

[10] Flinn, R. A.,and Trojan, P. K. ***Engineering Materials and Their Applications***, 5th ed. John Wiley & Sons, Inc.New York, 1995.

[11] Groover, M. P., ***Fundamentals of Modern Manufacturing: Materials, Processes, and Systems***, 4th ed. John Wiley & Sons, Inc. Hoboken, New Jersey, 2010.

[12] Hlavac, J. ***The Technology of Glass and Ceramics***. Elsevier Scientific Publishing Company, New York, 1983.

[13] Kingery, W. D., Bowen, H. K.,and Uhlmann, D. R. ***Introduction to Ceramics***, 2nd ed. John Wiley & Sons, Inc.New York, 1995.

[14] Margolis, J. M., ***Engineering Plastics Handbook***. McGraw-Hill, New York, 2006.

[15] Mark, J. E.,and Erman, B. (eds.), ***Science and Technology of Rubber***, 3rd ed. Academic Press, Orlando, Florida, 2005.

[16] Metals Handbook, Vol. **1**, ***Properties and Selection: Iron, Steels, and High Performance Alloys***. ASM International, Metals Park, Ohio, 1990.

[17] Metals Handbook, Vol. **2**, ***Properties and Selection: Nonferrous Alloys and Special Purpose Materials***. ASM International, Metals Park, Ohio, 1990.

[18] Richerson, D. W. ***Modern Ceramic Engineering: Properties, Processing, and Use in Design***, 3rd ed. CRC Taylor & Francis, Boca Raton, Florida, 2006.

[19] Scholes, S. R.,and Greene, C. H., ***Modern Glass Practice***, 7th ed. CBI Publishing Company, Boston, Massachusetts, 1993.

[20] Somiya, S. (ed.). ***Advanced Technical Ceramics***. Academic Press, Inc.San Diego, California, 1989.

[21] Tadmor, Z.,and Gogos, C. G. ***Principles of Polymer Processing***. Wiley-Interscience, Hoboken, New Jersey, 2006.

[22] www.wikipedia.org/wiki/Boeing_787

[23] Young, R. J.,and Lovell, P. ***Introduction to Polymers***, 3rd ed. CRC Taylor and Francis, Boca Raton, Florida, 2008.

REVIEW QUESTIONS

2.1. What are some of the general properties that distinguish metals from ceramics and polymers?

2.2. What are the two major groups of metals? Define them.

2.3. What is an alloy?

2.4. What is the range of carbon percentages that defines an iron-carbon alloy as a steel?

2.5. What is the range of carbon percentages that defines an iron-carbon alloy as cast iron?

2.6. Identify some of the common alloying elements other than carbon in low alloy steels.

2.7. What is the predominant alloying element in all of the stainless steels?

2.8. Why is austenitic stainless steel called by that name?

2.9. Besides high carbon content, what other alloying element is characteristic of the cast irons?

2.10. Identify some of the properties for which aluminum is noted.

2.11. What are some of the noteworthy properties of magnesium?

2.12. What is the most important engineering property of copper that determines most of its applications?

2.13. What elements are traditionally alloyed with copper to form (a) bronze and (b) brass?

2.14. What are some of the important applications of nickel?
2.15. What are the noteworthy properties of titanium?
2.16. Identify some of the important applications of zinc.
2.17. The superalloys divide into three basic groups, according to the base metal used in the alloy. Name the three groups.
2.18. What is so special about the superalloys? What distinguishes them from other alloys?
2.19. What is a ceramic?
2.20. What is the difference between the traditional ceramics and the new ceramics?
2.21. What is the feature that distinguishes glass from the traditional and new ceramics?
2.22. What are the general mechanical properties of ceramic materials?
2.23. What do bauxite and corundum have in common?
2.24. What is clay, used in making ceramic products?
2.25. What are some of the principal applications of cemented carbides, such as WC–Co?
2.26. What is one of the important applications of titanium nitride, as mentioned in the text?
2.27. What is the primary mineral in glass products?
2.28. What does the term *devitrification* mean?
2.29. What is a polymer?
2.30. What are the three basic categories of polymers?
2.31. How do the properties of polymers compare with those of metals?
2.32. What is cross-linking in a polymer, and what is its significance?
2.33. The nylons are members of which polymer group?
2.34. What is the chemical formula of ethylene, the monomer for polyethylene?
2.35. How do the properties of thermosetting polymers differ from those of thermoplastics?
2.36. Cross-linking (curing) of thermosetting plastics is accomplished by one of three ways. Name the three ways.
2.37. Elastomers and thermosetting polymers are both cross-linked. Why are their properties so different?
2.38. What is the primary polymer ingredient in natural rubber?
2.39. How do thermoplastic elastomers differ from conventional rubbers?
2.40. What is a composite material?
2.41. Identify some of the characteristic properties of composite materials.
2.42. What does the term *anisotropic* mean?
2.43. Name the three basic categories of composite materials.
2.44. What are the common forms of the reinforcing phase in composite materials?
2.45. What is a cermet?
2.46. Cemented carbides are what class of composites?
2.47. What are some of the weaknesses of ceramics that might be corrected in fiber-reinforced ceramic-matrix composites?
2.48. What is the most common fiber material in fiber-reinforced plastics?
2.49. Identify some of the important properties of fiber-reinforced plastic composite materials.
2.50. Name some of the important applications of FRPs.

3 PROPERTIES OF ENGINEERING MATERIALS

Chapter Contents

3.1 Stress–Strain Relationships
- 3.1.1 Tensile Properties
- 3.1.2 Compression Properties
- 3.1.3 Bending and Testing of Brittle Materials
- 3.1.4 Shear Properties

3.2 Hardness
- 3.2.1 Hardness Tests
- 3.2.2 Hardness of Various Materials

3.3 Effect of Temperature on Mechanical Properties

3.4 Fluid Properties

3.5 Viscoelastic Behavior of Polymers

3.6 Volumetric and Melting Properties
- 3.6.1 Density and Thermal Expansion
- 3.6.2 Melting Characteristics

3.7 Thermal Properties
- 3.7.1 Specific Heat and Thermal Conductivity
- 3.7.2 Thermal Properties in Manufacturing

The properties of a given engineering material determine its response to the various forms of energy that are used in manufacturing processes. If the material responds well to the forces, temperatures, and other physical parameters generated in a particular process, the result is a successful operation that produces a high-quality part or product.

Material properties can be divided into two categories: mechanical and physical. The mechanical properties of a material determine its behavior when subjected to mechanical stresses. These properties include stiffness, ductility, hardness, and various measures of strength. Mechanical properties are important in design because the function and performance of a product depend on its capacity to resist deformation under the stresses encountered in service. In design, the usual objective is for the product and its components to withstand these stresses without significant change in geometry. This capability depends on properties such as elastic modulus and yield strength. In manufacturing, the objective is just the opposite. Here, we want to apply stresses that exceed the yield strength of the material to alter its shape. Mechanical processes such as forming and machining succeed by developing forces that exceed the material's resistance to deformation. Thus, we have the following dilemma: Mechanical properties that are desirable to the designer, such as high strength, usually make the manufacture of the product more difficult.

Physical properties define the behavior of materials in response to physical forces other than mechanical. The properties include volumetric and thermal properties as well as melting characteristics. Physical properties are important in manufacturing because they often influence the performance of the process. Melting characteristics are important in metal casting operations. Metals with higher melting temperatures require more heat input before pouring the molten metal into the mold. In machining, thermal properties of the work material determine the cutting temperature, which affects how long the tool can be used before it fails.

In this chapter, we discuss the properties of engineering materials that are most relevant to the manufacturing processes covered in this book. Mechanical properties are discussed in Sections 3.1 through 3.5, and physical properties are discussed in the remaining sections.

3.1 STRESS–STRAIN RELATIONSHIPS

There are three types of static mechanical stresses to which materials can be subjected: tensile, compressive, and shear. Tensile stresses tend to stretch the material, compressive stresses tend to squeeze it, and shear involves stresses that tend to cause adjacent portions of the material to slide against each other. The stress–strain curve is the basic relationship that describes the mechanical properties of materials for all three types of stresses.

3.1.1 TENSILE PROPERTIES

The tensile test is the most common procedure for studying the stress–strain relationship, particularly for metals. In the test, a force is applied that pulls the material, tending to elongate it and reduce its diameter, as shown in Figure 3.1(a). Standards by ASTM (American Society for Testing and Materials) specify the preparation of the test specimen and the conduct of the test itself. The typical specimen and general setup of the tensile test is illustrated in Figure 3.1(b) and (c), respectively.

The starting test specimen has an original length L_o and area A_o. The length is measured as the distance between the gage marks, and the area is measured as the (usually round) cross section of the specimen. During the testing of a metal, the specimen stretches, then necks, and finally fractures, as shown in Figure 3.2. The load and the change in length of the specimen are recorded as testing proceeds, to provide the data required to determine the stress–strain relationship. There are two different types of stress–strain curves: (1) engineering stress–strain and (2) true stress–strain. The first is more important in design, and the second is more important in manufacturing.

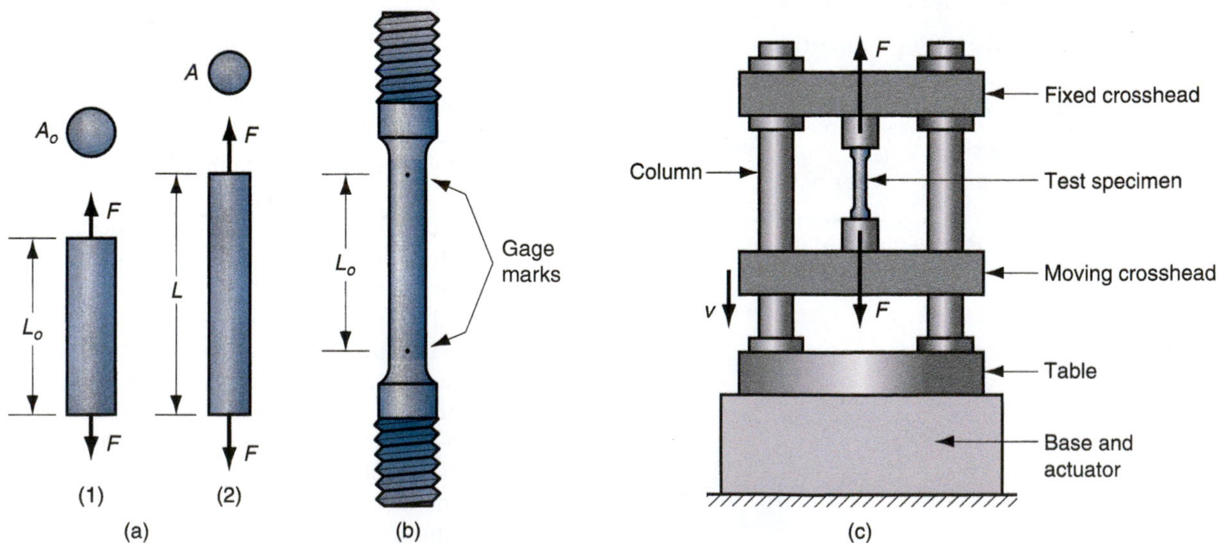

FIGURE 3.1 Tensile test: (a) tensile force applied in (1) and (2) resulting elongation of material; (b) typical test specimen; and (c) setup of the tensile test. (Credit: *Fundamentals of Modern Manufacturing*, 4th Edition by Mikell P. Groover, 2010. Reprinted with permission of John Wiley & Sons, Inc.)

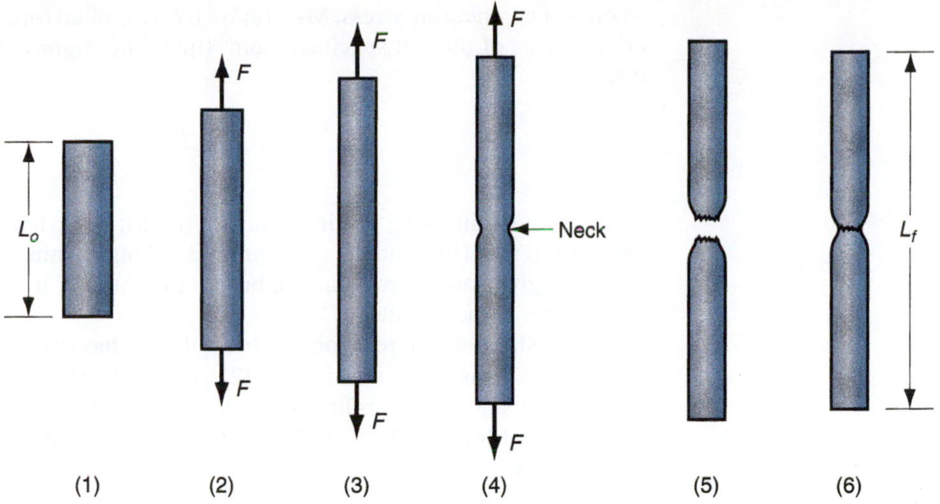

FIGURE 3.2 Typical progress of a tensile test: (1) beginning of test, no load; (2) uniform elongation and reduction of cross-sectional area; (3) continued elongation, maximum load reached; (4) necking begins, load begins to decrease; and (5) fracture. If pieces are put back together as in (6) , final length can be measured. (Credit: *Fundamentals of Modern Manufacturing,* 4th Edition by Mikell P. Groover, 2010. Reprinted with permission of John Wiley & Sons, Inc.)

Engineering Stress–Strain The engineering stress and strain in a tensile test are defined relative to the original area and length of the test specimen. These values are of interest in design because the designer expects that the strains experienced by any component of the product will not significantly change its shape. The components are designed to withstand the anticipated stresses encountered in service.

A typical engineering stress–strain curve from a tensile test of a metallic specimen is illustrated in Figure 3.3. The ***engineering stress*** at any point on the curve is defined as the force divided by the original area:

$$s = \frac{F}{A_o} \tag{3.1}$$

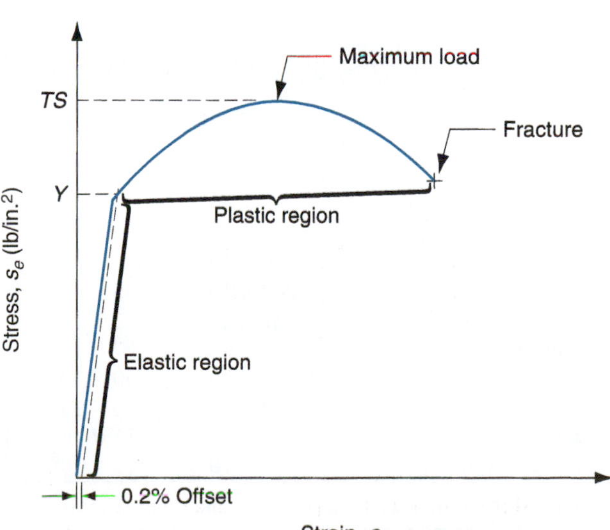

FIGURE 3.3 Typical engineering stress–strain plot in a tensile test of a metal. (Credit: *Fundamentals of Modern Manufacturing,* 4th Edition by Mikell P. Groover, 2010. Reprinted with permission of John Wiley & Sons, Inc.)

where s = engineering stress, MPa (lb/in^2), F = applied force in the test, N (lb), and A_o = original area of the test specimen, mm^2 (in^2). The ***engineering strain*** at any point in the test is given by

$$e = \frac{L - L_o}{L_o} \tag{3.2}$$

where e = engineering strain, mm/mm (in/in); L = length at any point during the elongation, mm (in); and L_o = original gage length, mm (in). The units of engineering strain are given as mm/mm (in/in), but we can think of it as representing elongation per unit length, without units.

The stress–strain relationship in Figure 3.3 has two regions, indicating two distinct forms of behavior: (1) elastic and (2) plastic. In the elastic region, the relationship between stress and strain is linear, and the material exhibits elastic behavior by returning to its original length when the load (stress) is released. The relationship is defined by ***Hooke's law*** :

$$s = Ee \tag{3.3}$$

where E = ***modulus of elasticity***, MPa (lb/in^2), a measure of the inherent stiffness of a material. It is a constant of proportionality whose value is different for different materials. Table 3.1 presents typical values for several materials, metals and nonmetals.

As stress increases, some point in the linear relationship is finally reached at which the material begins to yield. This ***yield point*** Y of the material can be identified in the figure by the change in slope at the end of the linear region. Because the start of yielding is usually difficult to see in a plot of test data (it does not usually occur as an abrupt change in slope), Y is typically defined as the stress at which a strain offset of 0.2% from the straight line has occurred. More specifically, it is the point where the stress–strain curve for the material intersects a line that is parallel to the straight portion of the curve but offset from it by a strain of 0.2%. The yield point is a strength characteristic of the material, and is therefore often referred to as the ***yield strength*** (other names include ***yield stress*** and ***elastic limit***).

The yield point marks the transition to the plastic region and the start of plastic deformation of the material. The relationship between stress and strain is no longer guided by Hooke's law. As the load is increased beyond the yield point, elongation of the

TABLE 3.1 Elastic modulus for selected materials.

	Modulus of Elasticity			Modulus of Elasticity	
Metals	**MPa**	**lb/in^2**	**Ceramics and polymers**	**MPa**	**lb/in^2**
Aluminum and alloys	69×10^3	10×10^6	Alumina	345×10^3	50×10^6
Cast iron	138×10^3	20×10^6	Diamond[a]	1035×10^3	150×10^6
Copper and alloys	110×10^3	16×10^6	Plate glass	69×10^3	10×10^6
Iron	209×10^3	30×10^6	Silicon carbide	448×10^3	65×10^6
Lead	21×10^3	3×10^6	Tungsten carbide	552×10^3	80×10^6
Magnesium	48×10^3	7×10^6	Nylon	3.0×10^3	0.40×10^6
Nickel	209×10^3	30×10^6	Phenol formaldehyde	7.0×10^3	1.00×10^6
Steel	209×10^3	30×10^6	Polyethylene (low density)	0.2×10^3	0.03×10^6
Titanium	117×10^3	17×10^6	Polyethylene (high density)	0.7×10^3	0.10×10^6
Tungsten	407×10^3	59×10^6	Polystyrene	3.0×10^3	0.40×10^6

Compiled from [8], [11], [12], [16], [17], and other sources.
[a]Although diamond is not a ceramic, it is often compared with the ceramic materials.

TABLE 3.2 Yield strength and tensile strength for selected metals.

Metal	Yield Strength		Tensile Strength		Metal	Yield Strength		Tensile Strength	
	MPa	lb/in²	MPa	lb/in²		MPa	lb/in²	MPa	lb/in²
Aluminum, annealed	28	4,000	69	10,000	Nickel, annealed	150	22,000	450	65,000
Aluminum, CW[a]	105	15,000	125	18,000	Steel, low C[a]	175	25,000	300	45,000
Aluminum alloys[a]	175	25,000	350	50,000	Steel, high C[a]	400	60,000	600	90,000
Cast iron[a]	275	40,000	275	40,000	Steel, alloy[a]	500	75,000	700	100,000
Copper, annealed	70	10,000	205	30,000	Steel, stainless[a]	275	40,000	650	95,000
Copper alloys[a]	205	30,000	410	60,000	Titanium, pure	350	50,000	515	75,000
Magnesium alloys[a]	175	25,000	275	40,000	Titanium alloy	800	120,000	900	130,000

Compiled from [8], [11], [12], [17], and other sources.
[a]Values given are typical. For alloys, there is a wide range in strength values depending on composition and treatment (e.g., heat treatment, work hardening).

specimen proceeds, but at a much faster rate than before, causing the slope of the curve to change dramatically, as shown in Figure 3.3. Elongation is accompanied by a uniform reduction in cross-sectional area, consistent with maintaining constant volume. Finally, the applied load F reaches a maximum value, and the engineering stress calculated at this point is called the ***tensile strength*** or ***ultimate tensile strength*** of the material. It is denoted as TS where $TS = F_{max}/A_o$. TS and Y are important strength properties in design calculations (they are also used in manufacturing calculations). Some typical values of yield strength and tensile strength are listed in Table 3.2 for selected metals. Conventional tensile testing of ceramics is difficult, and an alternative test is used to measure the strength of these brittle materials (Section 3.1.3). Polymers differ in their strength properties from metals and ceramics due to viscoelasticity (Section 3.5).

To the right of the tensile strength on the stress–strain curve, the load begins to decline, and the test specimen typically begins a process of localized elongation known as ***necking***. Instead of continuing to strain uniformly throughout its length, straining becomes concentrated in one small section of the specimen. The area of that section narrows down (necks) significantly until failure occurs. The stress calculated immediately before failure is known as the ***fracture stress***.

The amount of strain that the material can endure before failure is also a mechanical property of interest in many manufacturing processes. The common measure of this property is ***ductility***, the ability of a material to plastically strain without fracture. This measure can be taken as either elongation or area reduction. Elongation is defined as

$$EL = \frac{L_f - L_o}{L_o} \tag{3.4}$$

where EL = elongation, often expressed as a percent; L_f = specimen length at fracture, mm (in), measured as the distance between gage marks after the two parts of the specimen have been put back together; and L_o = original specimen length, mm (in). Area reduction is defined as

$$AR = \frac{A_o - A_f}{A_o} \tag{3.5}$$

where AR = area reduction, often expressed as a percent; A_f = area of the cross section at the point of fracture, mm^2 (in^2); and A_o = original area, mm^2 (in^2). There are problems with both of these ductility measures because of necking that occurs in metallic test

TABLE 3.3 Ductility as % elongation (typical values) for various selected materials.

Material	Elongation	Material	Elongation
Metals		***Metals, continued***	
Aluminum, annealed	40%	Steel, low C[a]	30%
Aluminum, cold worked	8%	Steel, high C[a]	10%
Aluminum alloys, annealed[a]	20%	Steel, alloy[a]	20%
Aluminum alloys, heat-treated[a]	8%	Steel, stainless, austenitic[a]	55%
Aluminum alloys, cast[a]	4%	Titanium, nearly pure	20%
Cast iron, gray[a]	0.6%	Zinc alloy	10%
Copper, annealed	45%	***Ceramics***	0[b]
Copper, cold worked	10%	***Polymers***	
Copper alloy: brass, annealed	60%	Thermoplastic polymers	100%
Magnesium alloys[a]	10%	Thermosetting polymers	1%
Nickel, annealed	45%	Elastomers (e.g., rubber)	1%[c]

Compiled from [8], [11], [12], [17], and other sources.
[a]Values given are typical. For alloys, there is a range of ductility that depends on composition and treatment (e.g., heat treatment, degree of work hardening).
[b]Ceramic materials are brittle; they withstand elastic strain but virtually no plastic strain.
[c]Elastomers endure significant elastic strain, but their plastic strain is very limited, only around 1% being typical.

specimens and the associated nonuniform effect on elongation and area reduction. Despite these difficulties, percent elongation and percent area reduction are the most commonly used measures of ductility in engineering practice. Some typical values of percent elongation for various materials (mostly metals) are listed in Table 3.3.

True Stress–Strain Thoughtful readers may be troubled by the use of the original area of the test specimen to calculate engineering stress, rather than the actual (instantaneous) area that becomes increasingly smaller as the test proceeds. If the actual area were used, the calculated stress value would be higher. The stress value obtained by dividing the instantaneous value of area into the applied load is defined as the ***true stress***:

$$\sigma = \frac{F}{A} \tag{3.6}$$

where σ = true stress, MPa (lb/in^2); F = force, N (lb); and A = actual (instantaneous) area resisting the load, mm^2 (in^2).

Similarly, ***true strain*** provides a more realistic assessment of the "instantaneous" elongation per unit length of the material. The value of true strain in a tensile test can be estimated by dividing the total elongation into small increments, calculating the engineering strain for each increment on the basis of its starting length, and then adding up the strain values. In the limit, true strain is defined as

$$\epsilon = \int_{L_o}^{L} \frac{dL}{L} = \ln \frac{L}{L_o} \tag{3.7}$$

where L = instantaneous length at any moment during elongation. At the end of the test (or other deformation), the final strain value can be calculated using $L = L_f$.

When the engineering stress–strain data in Figure 3.3 are plotted using the true stress and strain values, the resulting curve appears as in Figure 3.4. In the elastic region, the plot is virtually the same as before. Strain values are small, and true strain is nearly

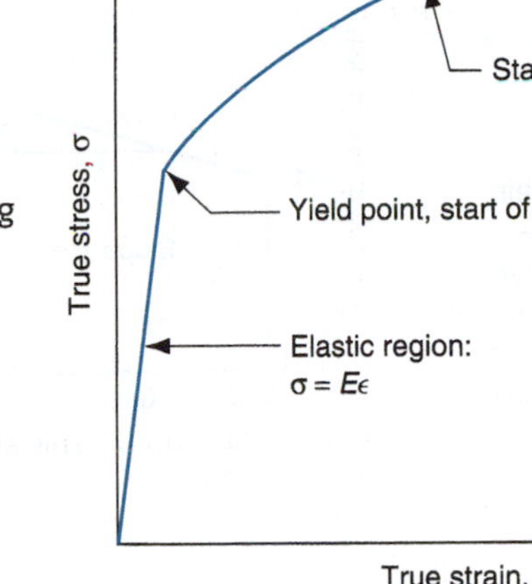

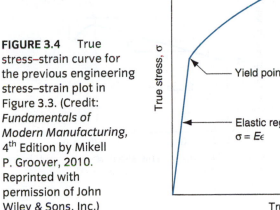

FIGURE 3.4 True stress–strain curve for the previous engineering stress–strain plot in Figure 3.3. (Credit: *Fundamentals of Modern Manufacturing*, 4th Edition by Mikell P. Groover, 2010. Reprinted with permission of John Wiley & Sons, Inc.)

equal to engineering strain for most metals of interest. The respective stress values are also very close to each other. The reason for these near equalities is that the cross-sectional area of the test specimen is not significantly reduced in the elastic region. Thus, Hooke's law can be used to relate true stress to true strain: $\sigma = E\epsilon$.

The difference between the true stress–strain curve and its engineering counterpart occurs in the plastic region. The stress values are higher in the plastic region because the instantaneous cross-sectional area of the specimen, which has been continuously reduced during elongation, is now used in the computation. As in the previous curve, a downturn finally occurs as a result of necking. A dashed line is used in the figure to indicate the projected continuation of the true stress–strain plot if necking had not occurred.

As strain becomes significant in the plastic region, the values of true strain and engineering strain diverge. True strain can be related to the corresponding engineering strain by

$$\epsilon = \ln(1 + e) \tag{3.8}$$

Similarly, true stress and engineering stress can be related by the expression

$$\sigma = s(1 + e) \tag{3.9}$$

In Figure 3.4, note that stress increases continuously in the plastic region until necking begins. When this happened in the engineering stress–strain curve, its significance was lost because an admittedly erroneous area value was used to calculate stress. Now when the true stress also increases, it cannot be dismissed so lightly. What it means is that the metal is becoming stronger as strain increases. This is the property called ***strain hardening*** that most metals exhibit to a greater or lesser degree.

Strain hardening, or ***work hardening*** as it is often called, is an important factor in certain manufacturing processes, particularly metal forming. Consider the behavior of a metal as it is affected by this property. If the portion of the true stress–strain curve representing the plastic region were plotted on a log–log scale, the result would be a linear relationship, as shown in Figure 3.5. Because it is a straight line in this transformation of the data, the relationship between true stress and true strain in the plastic region can be expressed as

$$\sigma = K\epsilon^n \tag{3.10}$$

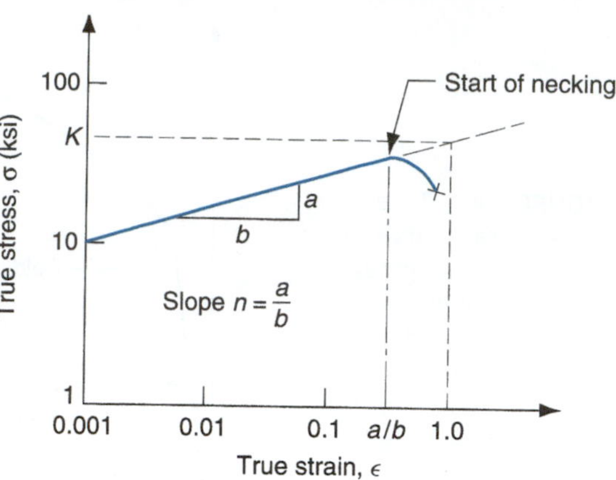

FIGURE 3.5 True stress–strain curve plotted on log–log scale. (Credit: *Fundamentals of Modern Manufacturing*, 4th Edition by Mikell P. Groover, 2010. Reprinted with permission of John Wiley & Sons, Inc.)

This equation is called the ***flow curve***, and it provides a good approximation of the behavior of metals in the plastic region, including their capacity for strain hardening. The constant K is called the ***strength coefficient***, MPa (lb/in^2), and it equals the value of true stress at a true strain value equal to one. The parameter n is called the ***strain hardening exponent***, and it is the slope of the line in Figure 3.5. Its value is directly related to a metal's tendency to work harden. Typical values of K and n for selected metals are given in Table 3.4.

Types of Stress–Strain Relationships Much information about elastic–plastic behavior is provided by the true stress–strain curve. As indicated, Hooke's law ($\sigma = E\epsilon$) governs the metal's behavior in the elastic region, and the flow curve ($\sigma = K\epsilon^n$) determines the behavior in the plastic region. Three basic forms of stress–strain relationship describe the behavior of nearly all types of solid materials, shown in Figure 3.6:

(a) ***Perfectly elastic.*** The behavior of this material is defined completely by its stiffness, indicated by the modulus of elasticity E. It fractures rather than yielding to plastic flow. Brittle materials such as ceramics, many cast irons, and thermosetting polymers possess stress–strain curves that fall into this category. These materials are not good candidates for forming operations.

TABLE 3.4 Typical values of strength coefficient *K* and strain hardening exponent *n* for selected metals.

Material	Strength Coefficient, *K*		Strain Hardening Exponent, *n*	Material	Strength Coefficient, *K*		Strain Hardening Exponent, *n*
	MPa	lb/in^2			MPa	lb/in^2	
Aluminum, pure, annealed	175	25,000	0.20	Steel, low C, annealed[a]	500	75,000	0.25
Aluminum alloy, annealed[a]	240	35,000	0.15	Steel, high C, annealed[a]	850	125,000	0.15
Aluminum alloy, heat-treated	400	60,000	0.10	Steel, alloy, annealed[a]	700	100,000	0.15
Copper, pure, annealed	300	45,000	0.50	Steel, stainless, austenitic, annealed	1200	175,000	0.40
Copper alloy: brass[a]	700	100,000	0.35				

Compiled from [10], [11], [12], and other sources.

[a]Values of K and n vary according to composition, heat treatment, and work hardening.

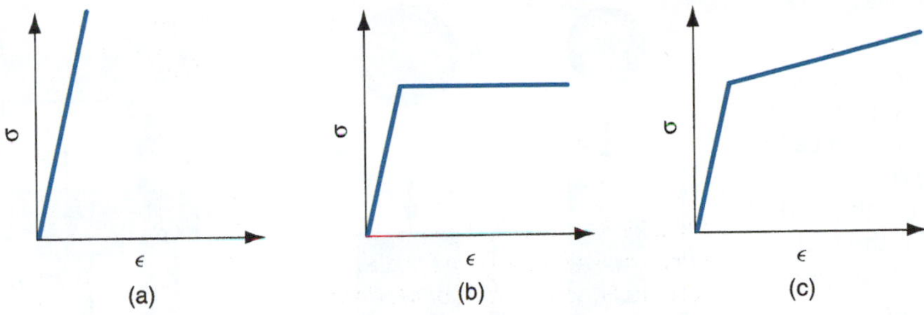

FIGURE 3.6 Three categories of stress–strain relationship: (a) perfectly elastic, (b) elastic and perfectly plastic, and (c) elastic and strain hardening. (Credit: *Fundamentals of Modern Manufacturing*, 4th Edition by Mikell P. Groover, 2010. Reprinted with permission of John Wiley & Sons, Inc.)

(b) ***Elastic and perfectly plastic***. This material has a stiffness defined by E. Once the yield strength Y is reached, the material deforms plastically at the same stress level. The flow curve is given by $K = Y$ and $n = 0$. Metals behave in this fashion when they have been heated to sufficiently high temperatures that they recrystallize rather than strain harden during deformation. Lead exhibits this behavior at room temperature because room temperature is above the recrystallization point for lead.

(c) ***Elastic and strain hardening***. This material obeys Hooke's law in the elastic region. It begins to flow at its yield strength Y. Continued deformation requires an ever-increasing stress, given by a flow curve whose strength coefficient K is greater than Y and whose strain-hardening exponent n is greater than zero. The flow curve is generally represented as a linear function on a natural logarithmic plot. Most ductile metals behave this way when cold worked.

Manufacturing processes that deform materials through the application of tensile stresses include wire and bar drawing (Section 13.4) and stretch forming (Section 14.6.1).

3.1.2 COMPRESSION PROPERTIES

A compression test applies a load that squeezes a cylindrical specimen between two platens, as illustrated in Figure 3.7. As the specimen is compressed, its height is reduced and its cross-sectional area is increased. Engineering stress is defined as

$$s = \frac{F}{A_o} \tag{3.11}$$

where A_o = original area of the specimen. This is the same definition of engineering stress used in the tensile test. The engineering strain is defined as

$$e = \frac{h - h_o}{h_o} \tag{3.12}$$

where h = height of the specimen at a particular moment into the test, mm (in); and h_o = starting height, mm (in). Because the height is decreased during compression, the value of e will be negative. The negative sign is usually ignored when expressing values of compression strain.

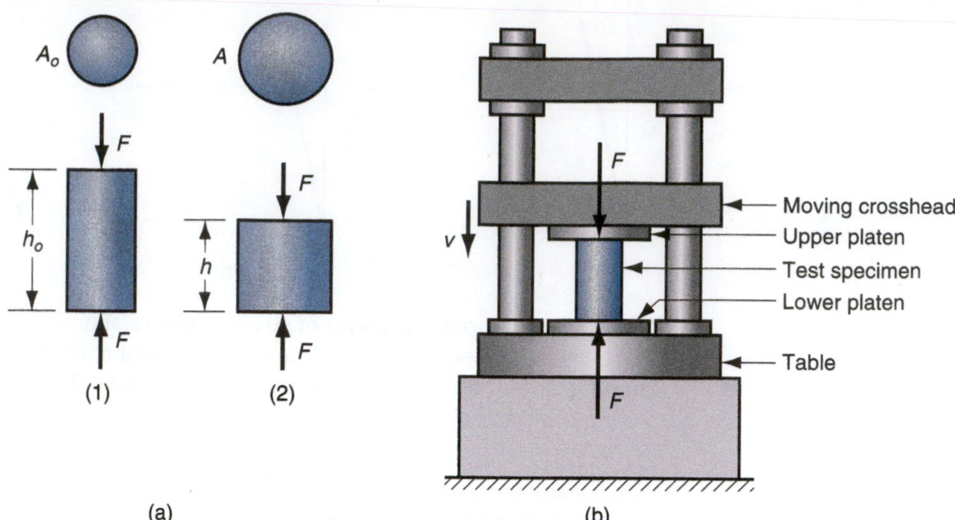

FIGURE 3.7 Compression test: (a) compression force applied to test piece in (1), and (2) resulting change in height; and (b) setup for the test, with size of test specimen exaggerated. (Credit: *Fundamentals of Modern Manufacturing*, 4th Edition by Mikell P. Groover, 2010. Reprinted with permission of John Wiley & Sons, Inc.)

When engineering stress is plotted against engineering strain in a compression test, the results appear as in Figure 3.8. The curve is divided into elastic and plastic regions, as before, but the shape of the plastic portion of the curve is different from its tensile test complement. Because compression causes the cross section to increase (rather than decrease as in the tensile test), the load increases more rapidly than previously. This results in a higher value of calculated engineering stress.

Something else happens in the compression test that contributes to the increase in stress. As the cylindrical specimen is squeezed, friction at the surfaces in contact with the platens tends to prevent the ends of the cylinder from spreading. Additional energy is consumed by this friction during the test, and this results in a higher applied force. It also shows up as an increase in the computed engineering stress. Hence, owing to the increase in cross-sectional area and friction between the specimen and the platens, we obtain the characteristic engineering stress–strain curve in a compression test as seen in the figure.

Another consequence of the friction between the surfaces is that the material near the middle of the specimen is permitted to increase in area much more than at the ends. This results in the characteristic ***barreling*** of the specimen, as seen in Figure 3.9.

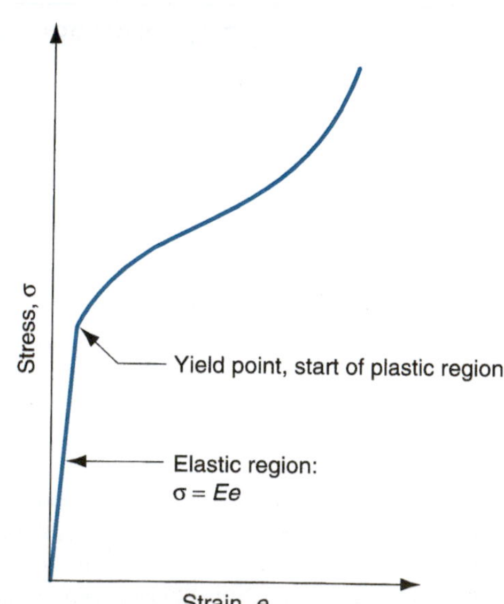

FIGURE 3.8 Typical engineering stress–strain curve for a compression test. (Credit: *Fundamentals of Modern Manufacturing*, 4th Edition by Mikell P. Groover, 2010. Reprinted with permission of John Wiley & Sons, Inc.)

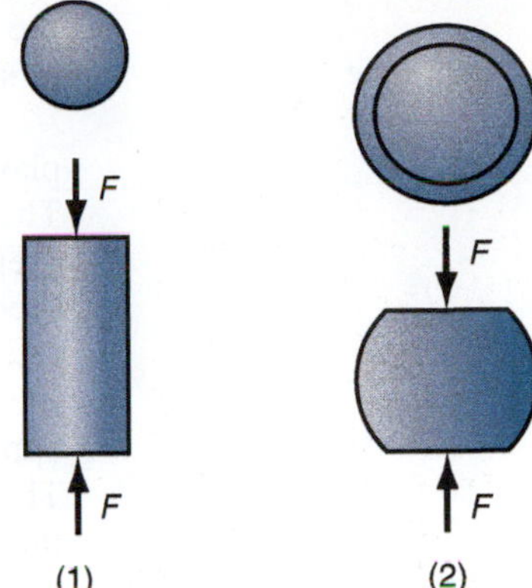

FIGURE 3.9 Barreling effect in a compression test: (1) start of test; and (2) after considerable compression has occurred. (Credit: *Fundamentals of Modern Manufacturing*, 4th Edition by Mikell P. Groover, 2010. Reprinted with permission of John Wiley & Sons, Inc.)

Although differences exist between the engineering stress–strain curves in tension and compression, when the respective data are plotted as true stress–strain, the relationships are nearly identical (for almost all materials). Because tensile test results are more abundant in the literature, values of the flow curve parameters (K and n) can be derived from tensile test data and applied with equal validity to a compression operation. What must be done in using the tensile test results for a compression operation is to ignore the effect of necking, a phenomenon that is peculiar to straining induced by tensile stresses. In compression, there is no corresponding collapse of the work. In previous plots of tensile stress–strain curves, the data were extrapolated beyond the point of necking by means of the dashed lines. The dashed lines better represent the behavior of the material in compression than the actual tensile test data.

Compression operations in metal forming are much more common than stretching operations. Important compression processes in industry include rolling, forging, and extrusion (Chapter 13).

3.1.3 BENDING AND TESTING OF BRITTLE MATERIALS

Bending operations are used to form metal plates and sheets. As shown in Figure 3.10, the process of bending a rectangular cross section subjects the material to tensile stresses (and strains) in the outer half of the bent section and compressive stresses (and strains) in

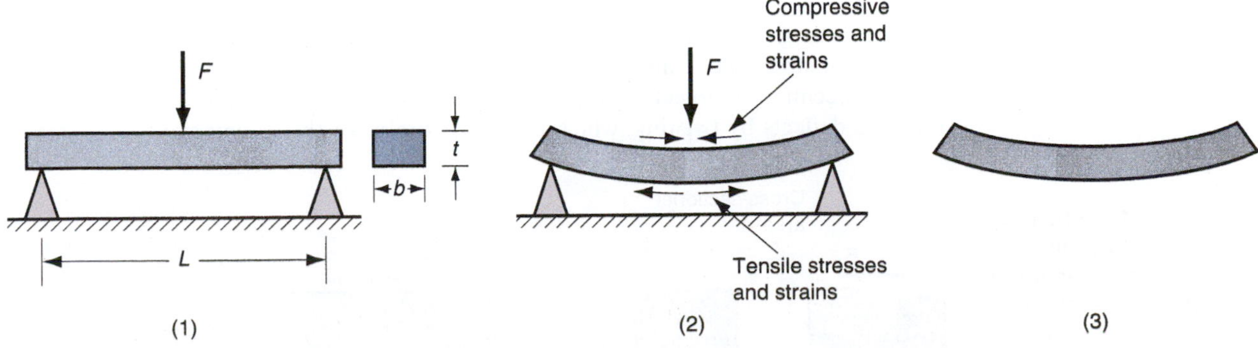

FIGURE 3.10 Bending of a rectangular cross section results in both tensile and compressive stresses in the material: (1) initial loading; (2) highly stressed and strained specimen; and (3) bent part. (Credit: *Fundamentals of Modern Manufacturing*, 4th Edition by Mikell P. Groover, 2010. Reprinted with permission of John Wiley & Sons, Inc.)

the inner half. If the material does not fracture, it becomes permanently (plastically) bent as shown in (3) of Figure 3.10.

Hard, brittle materials (e.g., ceramics), which possess elasticity but little or no plasticity, are often tested by a method that subjects the specimen to a bending load. These materials do not respond well to traditional tensile testing because of problems in preparing the test specimens and possible misalignment of the press jaws that hold the specimen. The ***bending test*** (also known as the ***flexure test***) is used to test the strength of these materials, using a setup illustrated in (1) of Figure 3.10. In this procedure, a specimen of rectangular cross section is positioned between two supports, and a load is applied at its center. In this configuration, the test is called a three-point bending test. These brittle materials do not flex to the exaggerated extent shown in Figure 3.10; instead they deform elastically until immediately before fracture. Failure usually occurs because the ultimate tensile strength of the outer fibers of the specimen has been exceeded. The strength value derived from this test is called the ***transverse rupture strength***, calculated from the formula

$$TRS = \frac{1.5FL}{bt^2} \tag{3.13}$$

where TRS = transverse rupture strength, MPa (lb/in^2); F = applied load at fracture, N (lb); L = length of the specimen between supports, mm (in); and b and t are the dimensions of the cross section of the specimen as shown in the figure, mm (in).

3.1.4 SHEAR PROPERTIES

Shear involves application of stresses in opposite directions on either side of a thin element to deflect it as shown in Figure 3.11. The shear stress is defined as

$$\tau = \frac{F}{A} \tag{3.14}$$

where τ = shear stress, MPa (lb/in^2); F = applied force, N (lb); and A = area over which the force is applied, mm^2 (in^2). Shear strain can be defined as

$$\gamma = \frac{\delta}{b} \tag{3.15}$$

where γ = shear strain, mm/mm (in/in); δ = the deflection of the element, mm (in); and b = the orthogonal distance over which deflection occurs, mm (in).

Shear stress and strain are commonly tested in a ***torsion test***, in which a thin-walled tubular specimen is subjected to a torque as shown in Figure 3.12. As torque is increased, the tube deflects by twisting, which is a shear strain for this geometry.

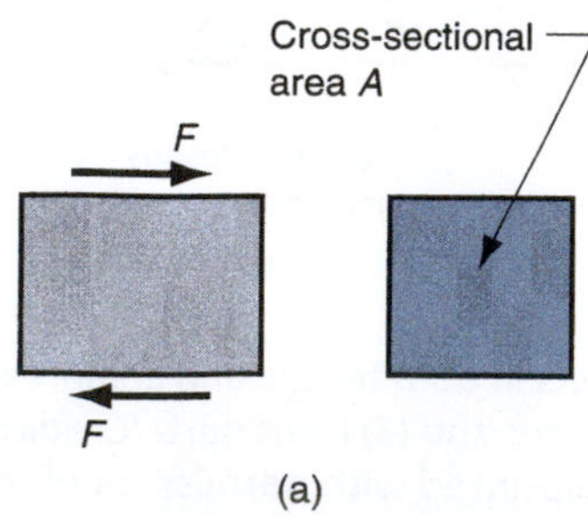

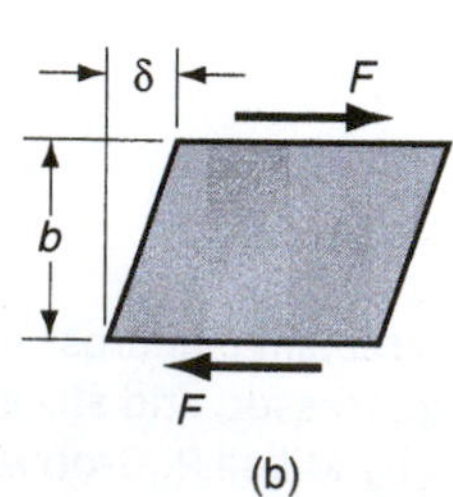

FIGURE 3.11 Shear (a) stress and (b) strain. (Credit: *Fundamentals of Modern Manufacturing*, 4th Edition by Mikell P. Groover, 2010. Reprinted with permission of John Wiley & Sons, Inc.)

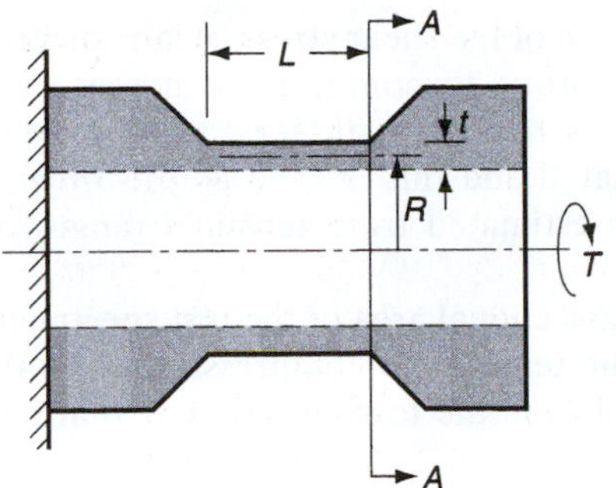

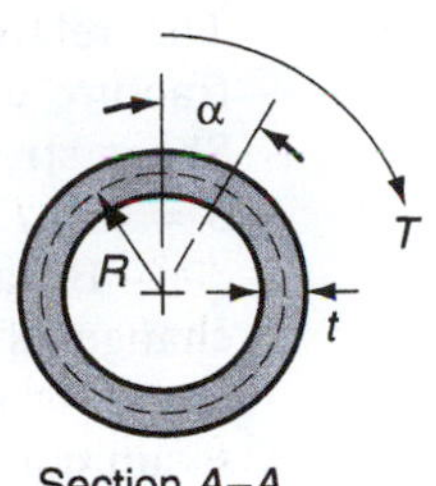

FIGURE 3.12 Torsion test setup. (Credit: *Fundamentals of Modern Manufacturing*, 4th Edition by Mikell P. Groover, 2010. Reprinted with permission of John Wiley & Sons, Inc.)

The shear stress can be determined in the test by the equation

$$\tau = \frac{T}{2\pi R^2 t} \tag{3.16}$$

where T = applied torque, N-mm (lb-in); R = radius of the tube measured to the neutral axis of the wall, mm (in); and t = wall thickness, mm (in). The shear strain can be determined by measuring the amount of angular deflection of the tube, converting this into a distance deflected, and dividing by the gauge length L. Reducing this to a simple expression,

$$\gamma = \frac{R\alpha}{L} \tag{3.17}$$

where α = the angular deflection (radians).

A typical shear stress–strain curve is shown in Figure 3.13. In the elastic region, the relationship is defined by

$$\tau = G\gamma \tag{3.18}$$

where G = the ***shear modulus***, or ***shear modulus of elasticity***, MPa (lb/in^2). For most materials, the shear modulus can be approximated by $G = 0.4E$, where E is the conventional elastic modulus.

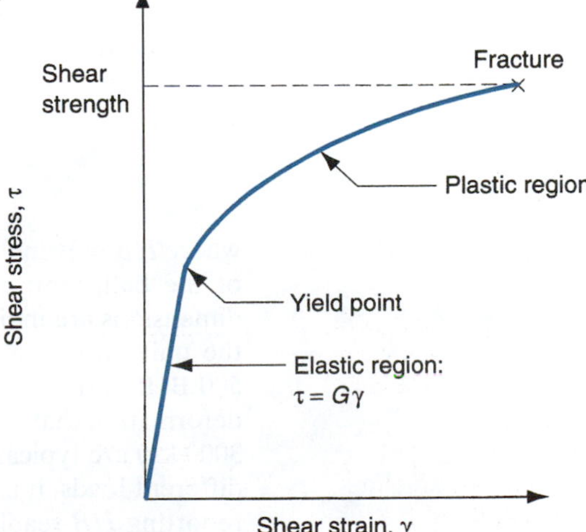

FIGURE 3.13 Typical shear stress–strain curve from a torsion test. (Credit: *Fundamentals of Modern Manufacturing*, 4th Edition by Mikell P. Groover, 2010. Reprinted with permission of John Wiley & Sons, Inc.)

In the plastic region of the shear stress–strain curve, the material strain hardens to cause the applied torque to continue to increase until fracture finally occurs. The relationship in this region is similar to the flow curve. The shear stress at fracture can be calculated and this is used as the ***shear strength*** S of the material. Shear strength can be estimated from tensile strength data by the approximation: $S = 0.7(TS)$.

Because the cross-sectional area of the test specimen in the torsion test does not change as it does in the tensile and compression tests, the engineering stress–strain curve for shear derived from the torsion test is virtually the same as the true stress–strain curve.

Shear processes are common in industry. Shearing action is used to cut sheet metal in blanking, punching, and other cutting operations (Section 14.1). In machining, the material is removed by the mechanism of shear deformation (Section 15.2).

3.2 HARDNESS

The hardness of a material is defined as its resistance to permanent indentation. Good hardness generally means that the material is resistant to scratching and wear. For many engineering applications, including most of the tooling used in manufacturing, scratch and wear resistance are important characteristics. As the reader shall see later in this section, there is a strong correlation between hardness and strength.

3.2.1 HARDNESS TESTS

Hardness tests are commonly used for assessing material properties because they are quick and convenient. However, a variety of testing methods are appropriate because of differences in hardness among different materials. The best-known hardness tests are Brinell and Rockwell.

Brinell Hardness Test The Brinell Hardness Test is widely used for testing metals and nonmetals of low to medium hardness. It is named after the Swedish engineer who developed it around 1900. In the test, a hardened steel (or cemented carbide) ball of 10-mm diameter is pressed into the surface of a specimen using a load of 500, 1500, or 3000 kg. The load is then divided into the indentation area to obtain the Brinell Hardness Number (BHN). In equation form,

$$HB = \frac{2F}{\pi D_b\left(D_b - \sqrt{D_b^2 - D_i^2}\right)} \tag{3.19}$$

where HB = Brinell Hardness Number (BHN); F = indentation load, kg; D_b = diameter of the ball, mm; and D_i = diameter of the indentation on the surface, mm. These dimensions are indicated in Figure 3.14(a). The resulting BHN has units of kg/mm^2, but the units are usually omitted in expressing the number. For harder materials (above 500 BHN), the cemented carbide ball is used because the steel ball experiences elastic deformation that compromises the accuracy of the reading. Also, higher loads (1500 and 3000 kg) are typically used for harder materials. Because of differences in results under different loads, it is considered good practice to indicate the load used in the test when reporting HB readings.

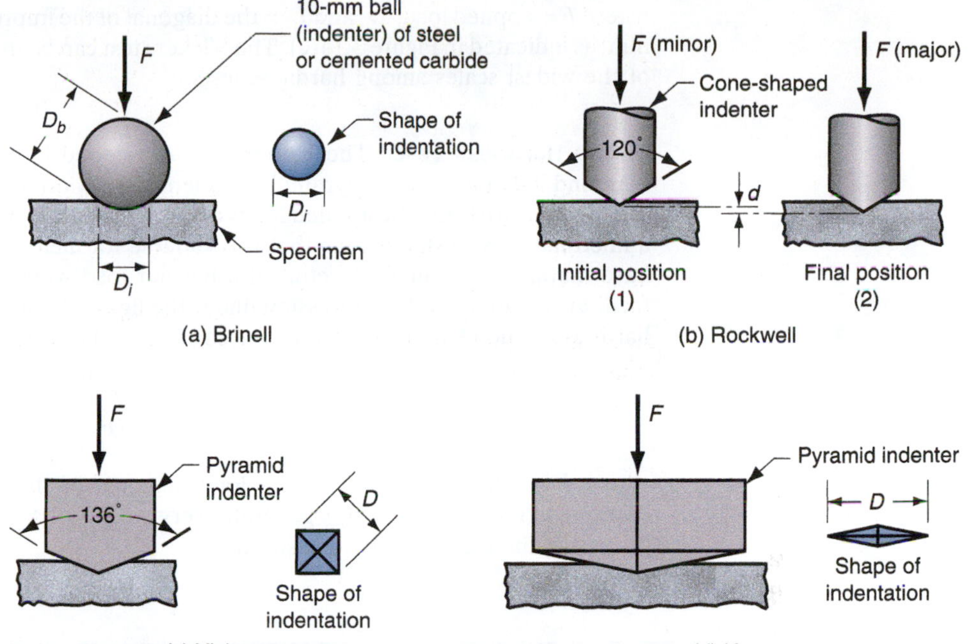

FIGURE 3.14 Hardness testing methods: (a) Brinell; (b) Rockwell: (1) initial minor load and (2) major load, (c) Vickers, and (d) Knoop. (Credit: *Fundamentals of Modern Manufacturing*, 4th Edition by Mikell P. Groover, 2010. Reprinted with permission of John Wiley & Sons, Inc.)

Rockwell Hardness Test This is another widely used test, named after the metallurgist who developed it in the early 1920s. It is convenient to use, and several enhancements over the years have made the test adaptable to a variety of materials.

In the Rockwell Hardness Test, a cone-shaped indenter or small-diameter ball, with diameter = 1.6 or 3.2 mm (1/16 or 1/8 in) is pressed into the specimen using a minor load of 10 kg, thus seating the indenter in the material. Then, a major load of 150 kg (or other value) is applied, causing the indenter to penetrate into the specimen a certain distance beyond its initial position. This additional penetration distance d is converted into a Rockwell hardness reading by the testing machine. The sequence is depicted in Figure 3.14(b). Differences in load and indenter geometry provide various Rockwell scales for different materials. The most common scales are indicated in Table 3.5.

Vickers Hardness Test This test, also developed in the early 1920s, uses a pyramid-shaped indenter made of diamond. It is based on the principle that impressions made by this indenter are geometrically similar regardless of load. Accordingly, loads of various size are applied, depending on the hardness of the material to be measured. The Vickers hardness value (HV) is then determined from the formula

$$HV = \frac{1.854\,F}{D^2} \tag{3.20}$$

TABLE 3.5 Common Rockwell hardness scales.

Rockwell Scale	Hardness Symbol	Indenter	Load (kg)	Typical Materials Tested
A	HRA	Cone	60	Carbides, ceramics
B	HRB	1.6 mm ball	100	Nonferrous metals
C	HRC	Cone	150	Ferrous metals, tool steels

Source: [8].

where F = applied load, kg, and D = the diagonal of the impression made by the indenter, mm, as indicated in Figure 3.14(c). The Vickers test can be used for all metals and has one of the widest scales among hardness tests.

Knoop Hardness Test The Knoop test, developed in 1939, uses a pyramid-shaped diamond indenter, but the pyramid has a length-to-width ratio of about 7:1, as indicated in Figure 3.14(d), and the applied loads are generally lighter than in the Vickers test. It is a microhardness test, meaning that it is suitable for measuring small, thin specimens or hard materials that might fracture if a heavier load were applied. The indenter shape facilitates reading of the impression under the lighter loads used in this test. The Knoop hardness value (HK) is determined according to the formula

$$HK = 14.2\frac{F}{D^2} \tag{3.21}$$

where F = load, kg; and D = the long diagonal of the indenter, mm. Because the impression made in this test is generally very small, considerable care must be taken in preparing the surface to be measured.

3.2.2 HARDNESS OF VARIOUS MATERIALS

This section compares the hardness values of some common materials in the three engineering material classes: metals, ceramics, and polymers.

Metals The Brinell and Rockwell Hardness Tests were developed at a time when metals were the principal engineering materials. A significant amount of data has been collected using these tests on metals. Table 3.6 lists hardness values for selected metals.

For most metals, hardness is closely related to strength. Because the method of testing for hardness is usually based on resistance to indentation, which is a form of

TABLE 3.6 Typical hardness of selected metals.

Metal	Brinell Hardness, HB	Rockwell Hardness, HR[a]	Metal	Brinell Hardness, HB	Rockwell Hardness, HR[a]
Aluminum, annealed	20		Magnesium alloys, hardened[b]	70	35B
Aluminum, cold worked	35		Nickel, annealed	75	40B
Aluminum alloys, annealed[b]	40		Steel, low C, hot rolled[b]	100	60B
Aluminum alloys, hardened[b]	90	52B	Steel, high C, hot rolled[b]	200	95B, 15C
Aluminum alloys, cast[b]	80	44B	Steel, alloy, annealed[b]	175	90B, 10C
Cast iron, gray, as cast[b]	175	10C	Steel, alloy, heat-treated[b]	300	33C
Copper, annealed	45		Steel, stainless, austenitic[b]	150	85B
Copper alloy: brass, annealed	100	60B	Titanium, nearly pure	200	95B
Lead	4		Zinc	30	

Compiled from [11], [12], [17], and other sources.

[a]HR values are given in the B or C scale as indicated by the letter designation. Missing values indicate that the hardness is too low for Rockwell scales.

[b]HB values given are typical. Hardness values will vary according to composition, heat treatment, and degree of work hardening.

TABLE 3.7 Hardness of selected ceramics and other hard materials, arranged in ascending order of hardness.

Material	Vickers Hardness, HV	Knoop Hardness, HK	Material	Vickers Hardness, HV	Knoop Hardness, HK
Hardened tool steel[a]	800	850	Titanium nitride, TiN	3000	2300
Cemented carbide (WC – Co)[a]	2000	1400	Titanium carbide, TiC	3200	2500
Alumina, Al_2O_3	2200	1500	Cubic boron nitride, BN	6000	4000
Tungsten carbide, WC	2600	1900	Diamond, sintered polycrystal	7000	5000
Silicon carbide, SiC	2600	1900	Diamond, natural	10,000	8000

Compiled from [15], [17], and other sources.
[a]Hardened tool steel and cemented carbide are the two materials commonly used in the Brinell Hardness Test.

compression, one would expect a good correlation between hardness and strength properties determined in a compression test. However, strength properties in a compression test are nearly the same as those from a tension test, after allowances for changes in cross-sectional area of the respective test specimens; so the correlation with tensile properties should also be good.

Brinell hardness value (HB) exhibits a close correlation with the ultimate tensile strength TS of steels, leading to the relationship [10], [16]:

$$TS = K_h(HB) \tag{3.22}$$

where K_h is a constant of proportionality. If TS is expressed in MPa, then $K_h = 3.45$; and if TS is in lb/in^2, then $K_h = 500$.

Ceramics The Brinell Hardness Test is not appropriate for ceramics because the materials being tested are often harder than the indenter ball. The Vickers and Knoop Hardness Tests are used to test these hard materials. Table 3.7 lists hardness values for several ceramics and hard materials. For comparison, the Rockwell C hardness for hardened tool steel is 65 HRC. The HRC scale does not extend high enough to be used for the harder materials.

Polymers Polymers have the lowest hardness among the three types of engineering materials. Table 3.8 lists several of the polymers on the Brinell hardness scale, although this testing method is not normally used for these materials. It does, however, allow comparison with the hardness of metals.

TABLE 3.8 Hardness of selected polymers.

Polymer	Brinell Hardness, HB	Polymer	Brinell Hardness, HB
Nylon	12	Polypropylene	7
Phenol formaldehyde	50	Polystyrene	20
Polyethylene, low density	2	Polyvinyl-chloride	10
Polyethylene, high density	4		

Compiled from [5], [8], and other sources.

3.3 EFFECT OF TEMPERATURE ON MECHANICAL PROPERTIES

Temperature has a significant effect on the mechanical properties of a material. It is important for the designer to know the material properties at the operating temperatures of the product when in service. It is also important to know how temperature affects mechanical properties in manufacturing. At elevated temperatures, materials are lower in strength and higher in ductility. The general relationships for metals are depicted in Figure 3.15. Thus, most metals can be formed with lower forces and power at elevated temperatures than when they are cold.

Hot Hardness A property often used to characterize strength and hardness at elevated temperatures is hot hardness. ***Hot hardness*** is simply the ability of a material to retain hardness at elevated temperatures; it is usually presented as either a listing of hardness values at different temperatures or as a plot of hardness versus temperature, as in Figure 3.16. Steels can be alloyed to achieve significant improvements in hot hardness, as shown in the figure. Ceramics exhibit superior properties at elevated temperatures. These materials are often selected for high temperature applications, such as turbine parts, cutting tools, and refractory applications. The outside skin of a shuttle spacecraft is

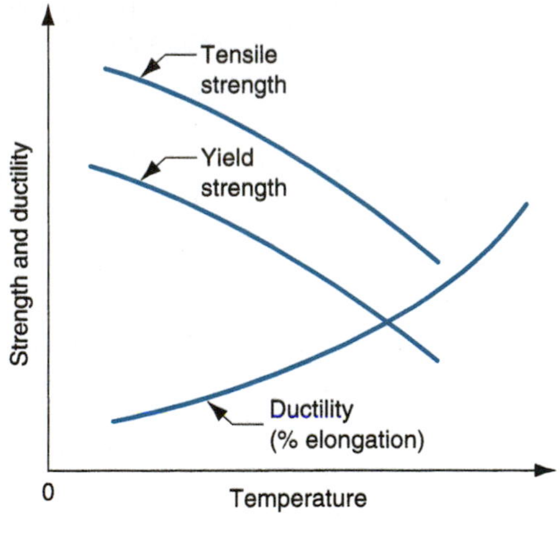

FIGURE 3.15 General effect of temperature on strength and ductility. (Credit: *Fundamentals of Modern Manufacturing*, 4th Edition by Mikell P. Groover, 2010. Reprinted with permission of John Wiley & Sons, Inc.)

Ceramic
High-alloy
steel
Hardness
High C
steel
(HT)
Low C
steel (HT)
0
250
500
Temperature, °C

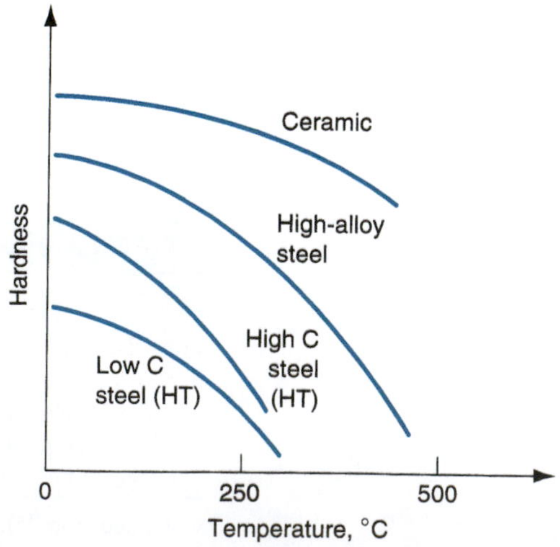

FIGURE 3.16 Hot hardness—typical hardness as a function of temperature for several materials. (Credit: *Fundamentals of Modern Manufacturing*, 4th Edition by Mikell P. Groover, 2010. Reprinted with permission of John Wiley & Sons, Inc.)

lined with ceramic tiles to withstand the friction heat of high-speed reentry into the atmosphere.

Good hot hardness is also desirable in the tooling materials used in many manufacturing operations. Significant amounts of heat energy are generated in most metalworking processes, and the tools must be capable of withstanding the high temperatures involved.

Recrystallization Most metals behave at room temperature according to the flow curve in the plastic region. As the metal is strained, it increases in strength due to strain hardening (strain-hardening exponent $n > 0$). However, if the metal is heated to a sufficiently elevated temperature and then deformed, strain hardening does not occur. Instead, new grains are formed that are free of strain, and the metal behaves as a perfectly plastic material; that is, with a strain-hardening exponent $n = 0$. The formation of new strain-free grains is a process called ***recrystallization***, and the temperature at which it occurs is about one-half the melting point ($0.5\ T_m$), as measured on an absolute scale (R or K). This is called the ***recrystallization temperature***. Recrystallization takes time. The recrystallization temperature for a particular metal is usually specified as the temperature at which complete formation of new grains requires about 1 hour.

Recrystallization is a temperature-dependent characteristic of metals that can be exploited in manufacturing. By heating the metal to the recrystallization temperature before deformation, the amount of straining that the metal can endure is substantially increased, and the forces and power required to carry out the process are significantly reduced. Forming metals at temperatures above the recrystallization temperature is called ***hot working*** (Section 12.3).

3.4 FLUID PROPERTIES

Fluids behave quite differently than solids. A fluid flows; it takes the shape of the container that holds it. A solid does not flow; it possesses a geometric form that is independent of its surroundings. Fluids include liquids and gases; the interest in this section is on the former. Many manufacturing processes are accomplished on materials that have been converted from solid to liquid state by heating. Metals are cast in the molten state; glass is formed in a heated and highly fluid state; and polymers are almost always shaped as thick fluids.

Viscosity Although flow is a defining characteristic of fluids, the tendency to flow varies for different fluids. Viscosity is the property that determines fluid flow. Roughly, ***viscosity*** can be defined as the resistance to flow that is characteristic of a fluid. It is a measure of the internal friction that arises when velocity gradients are present in the fluid—the more viscous the fluid is, the higher the internal friction and the greater the resistance to flow. The reciprocal of viscosity is ***fluidity***—the ease with which a fluid flows.

Viscosity is defined more precisely with respect to the setup in Figure 3.17, in which two parallel plates are separated by a distance d. One of the plates is stationary, while the other is moving at a velocity v, and the space between the plates is occupied by a fluid. Orienting these parameters relative to an axis system, d is in the y-axis direction and v is in the x-axis direction. The motion of the upper plate is resisted by force F that results from the shear viscous action of the fluid. This force can be reduced to a shear stress by dividing F by the plate area A:

$$\tau = \frac{F}{A} \tag{3.23}$$

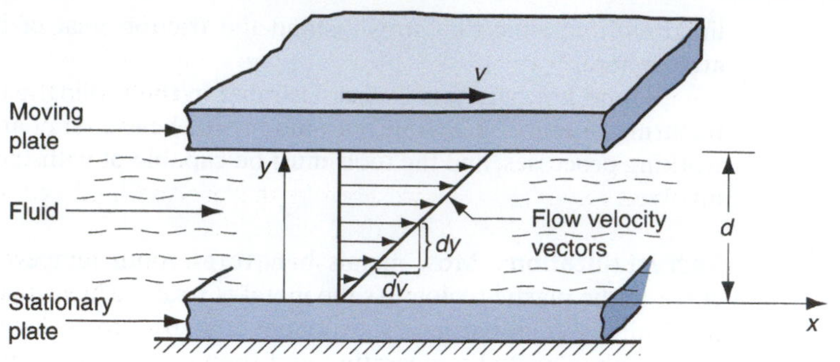

FIGURE 3.17 Fluid flow between two parallel plates, one stationary and the other moving at velocity *v*. (Credit: *Fundamentals of Modern Manufacturing*, 4th Edition by Mikell P. Groover, 2010. Reprinted with permission of John Wiley & Sons, Inc.)

where τ = shear stress, N/m^{2}or Pa (lb/in^2). This shear stress is related to the rate of shear, which is defined as the change in velocity dv relative to dy. That is,

$$\dot{\gamma} = \frac{dv}{dy} \tag{3.24}$$

where $\dot{\gamma}$ = shear rate, 1/s; dv = incremental change in velocity, m/s (in/sec); and dy = incremental change in distance y, m (in). The shear viscosity is the fluid property that defines the relationship between F/A and dv/dy; that is,

$$\frac{F}{A} = \eta \frac{dv}{dy} \quad \text{or} \quad \tau = \eta \dot{\gamma} \tag{3.25}$$

where η = a constant of proportionality called the coefficient of viscosity, Pa-s (lb-sec/in^2). Rearranging Eq. (3.25), the coefficient of viscosity can be expressed as follows:

$$\eta = \frac{\tau}{\dot{\gamma}} \tag{3.26}$$

Thus, the viscosity of a fluid can be defined as the ratio of shear stress to shear rate during flow, where shear stress is the frictional force exerted by the fluid per unit area, and shear rate is the velocity gradient perpendicular to the flow direction. The viscous characteristics of fluids defined by Eq. (3.26) were first stated by Newton. He observed that viscosity was a constant property of a given fluid, and such a fluid is referred to as a ***Newtonian fluid***. Some typical values of coefficient of viscosity for various fluids are given in Table 3.9. One can observe in several of the materials listed that viscosity varies with temperature.

TABLE 3.9 Viscosity values for selected fluids.

Material	Coefficient of Viscosity		Material	Coefficient of Viscosity	
	Pa-s	lb-sec/in^2		Pa-s	lb-sec/in^2
Glass[b], 540°C (1000°F)	10^{12}	10^8	Pancake syrup (room temp)	50	73×10^{-4}
Glass[b], 815°C (1500°F)	10^5	14	Polymer[a], 151°C (300°F)	115	167×10^{-4}
Glass[b], 1095°C (2000°F)	10^3	0.14	Polymer[a], 205°C (400°F)	55	80×10^{-4}
Glass[b], 1370°C (2500°F)	15	22×10^{-4}	Polymer[a], 260°C (500°F)	28	41×10^{-4}
Mercury, 20°C (70°F)	0.0016	0.23×10^{-6}	Water, 20°C (70°F)	0.001	0.15×10^{-6}
Machine oil (room temp.)	0.1	0.14×10^{-4}	Water, 100°C (212°F)	0.0003	0.04×10^{-6}

Compiled from various sources.
[a]Low-density polyethylene is used as the polymer example here; most other polymers have slightly higher viscosities.
[b]Glass composition is mostly SiO_2; compositions and viscosities vary; values given are representative.

Viscosity in Manufacturing Processes For many metals, the viscosity in the molten state compares with that of water at room temperature. Certain manufacturing processes, notably casting and welding, are performed on metals in their molten state, and success in these operations requires low viscosity so that the molten metal fills the mold cavity or weld seam before solidifying. In other operations, such as metal forming and machining, lubricants and coolants are used in the process, and again the success of these fluids depends to some extent on their viscosities.

Glass ceramics exhibit a gradual transition from solid to liquid states as temperature is increased; they do not suddenly melt as pure metals do. The effect is illustrated by the viscosity values for glass at different temperatures in Table 3.9. At room temperature, glass is solid and brittle, exhibiting no tendency to flow; for all practical purposes, its viscosity is infinite. As glass is heated, it gradually softens, becoming less and less viscous (more and more fluid), until it can finally be formed by blowing or molding at around 1100°C (2000°F).

Most polymer-shaping processes are performed at elevated temperatures, at which the material is in a liquid or highly plastic condition. Thermoplastic polymers represent the most straightforward case, and they are also the most common polymers. At low temperatures, thermoplastic polymers are solid; as temperature is increased, they typically transform first into a soft rubbery material, and then into a thick fluid. As temperature continues to rise, viscosity decreases gradually, as in Table 3.9 for polyethylene, the most widely used thermoplastic polymer. However, with polymers the relationship is complicated by other factors. For example, viscosity is affected by flow rate. The viscosity of a thermoplastic polymer is not a constant. A polymer melt does not behave in a Newtonian fashion. Its relationship between shear stress and shear rate can be seen in Figure 3.18. A fluid that exhibits this decreasing viscosity with increasing shear rate is called ***pseudoplastic***. This behavior complicates the analysis of polymer shaping.

3.5 VISCOELASTIC BEHAVIOR OF POLYMERS

Another property that is characteristic of polymers is viscoelasticity. ***Viscoelasticity*** is the property of a material that determines the strain it experiences when subjected to

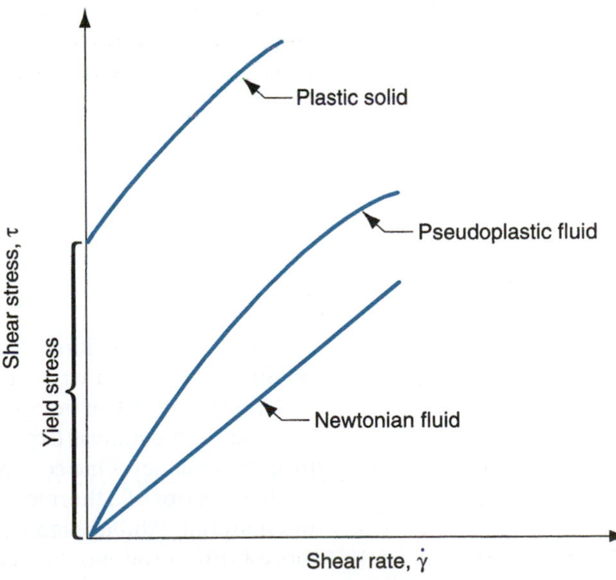

FIGURE 3.18 Viscous behaviors of Newtonian and pseudoplastic fluids. Polymer melts exhibit pseudoplastic behavior. For comparison, the behavior of a plastic solid material is shown. (Credit: *Fundamentals of Modern Manufacturing*, 4th Edition by Mikell P. Groover, 2010. Reprinted with permission of John Wiley & Sons, Inc.)

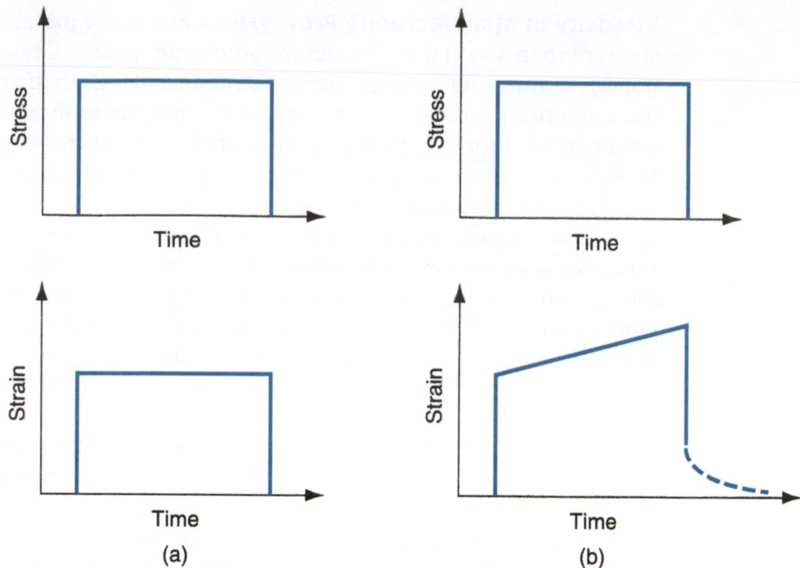

FIGURE 3.19 Comparison of elastic and viscoelastic properties: (a) perfectly elastic response of material to stress applied over time; and (b) response of a viscoelastic material under same conditions. The material in (b) takes a strain that is a function of time and temperature. (Credit: *Fundamentals of Modern Manufacturing,* 4th Edition by Mikell P. Groover, 2010. Reprinted with permission of John Wiley & Sons, Inc.)

combinations of stress and temperature over time. As the name suggests, it is a combination of viscosity and elasticity. Viscoelasticity can be explained with reference to Figure 3.19. The two parts of the figure show the typical response of two materials to an applied stress below the yield point during some time period. The material in (a) exhibits perfect elasticity; when the stress is removed, the material returns to its original shape. By contrast, the material in (b) shows viscoelastic behavior. The amount of strain gradually increases over time under the applied stress. When stress is removed, the material does not immediately return to its original shape; instead, the strain decays gradually. If the stress had been applied and then immediately removed, the material would have returned immediately to its starting shape. However, time has entered the picture and played a role in affecting the behavior of the material.

A simple model of viscoelasticity can be developed using the definition of elasticity as a starting point. Elasticity is concisely expressed by Hooke's law, $\sigma = E\epsilon$, which simply relates stress to strain through a constant of proportionality. In a viscoelastic solid, the relationship between stress and strain is time dependent; it can be expressed as

$$\sigma(t) = f(t)\epsilon \tag{3.27}$$

The time function $f(t)$ can be conceptualized as a modulus of elasticity that depends on time. It might be written $E(t)$ and referred to as a viscoelastic modulus. The form of this time function can be complex, sometimes including strain as a factor. Without getting into the mathematical expressions for it, we can nevertheless explore the effect of the time dependency. One common effect can be seen in Figure 3.20, which shows the stress–strain behavior of a thermoplastic polymer under different strain rates. At low strain rate, the material exhibits significant viscous flow. At high strain rate, it behaves in a much more brittle fashion.

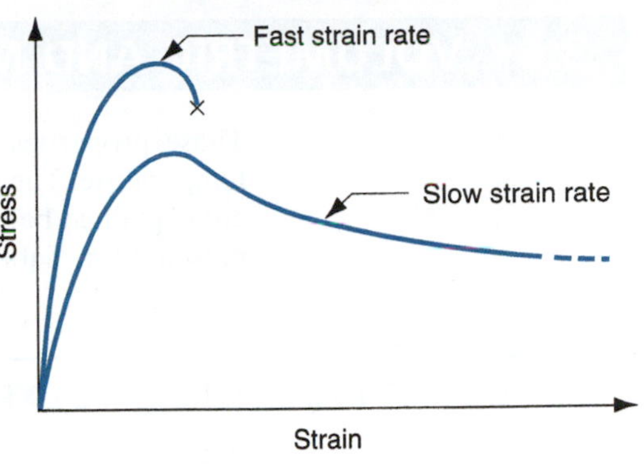

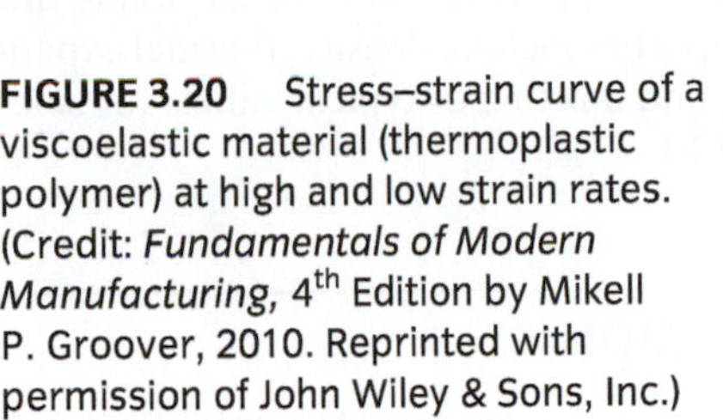

FIGURE 3.20 Stress–strain curve of a viscoelastic material (thermoplastic polymer) at high and low strain rates. (Credit: *Fundamentals of Modern Manufacturing*, 4th Edition by Mikell P. Groover, 2010. Reprinted with permission of John Wiley & Sons, Inc.)

Temperature is a factor in viscoelasticity. As temperature increases, the viscous behavior becomes more and more prominent relative to elastic behavior. The material becomes more like a fluid. Figure 3.21 illustrates this temperature dependence for a thermoplastic polymer. At low temperatures, the polymer shows elastic behavior. As T increases above the glass transition temperature T_g, the polymer becomes viscoelastic. As temperature increases further, it becomes soft and rubbery. At still higher temperatures, it exhibits viscous characteristics. The temperatures at which these modes of behavior are observed vary, depending on the plastic. Also, the shapes of the modulus versus temperature curve differ according to the proportions of crystalline and amorphous structures in the thermoplastic. Thermosetting polymers and elastomers behave differently than shown in the figure; after curing, these polymers do not soften as thermoplastics do at elevated temperatures. Instead, they degrade (char) at high temperatures.

Viscoelastic behavior manifests itself in polymer melts in the form of shape memory. As the thick polymer melt is transformed during processing from one shape to another, it "remembers" its previous shape and attempts to return to that geometry. For example, a common problem in extrusion of polymers is die swell, in which the profile of the extruded material grows in size, reflecting its tendency to return to its larger cross section in the extruder barrel immediately before being squeezed through the smaller die opening. The properties of viscosity and viscoelasticity are examined in more detail in our discussion of plastic shaping (Chapter 8).

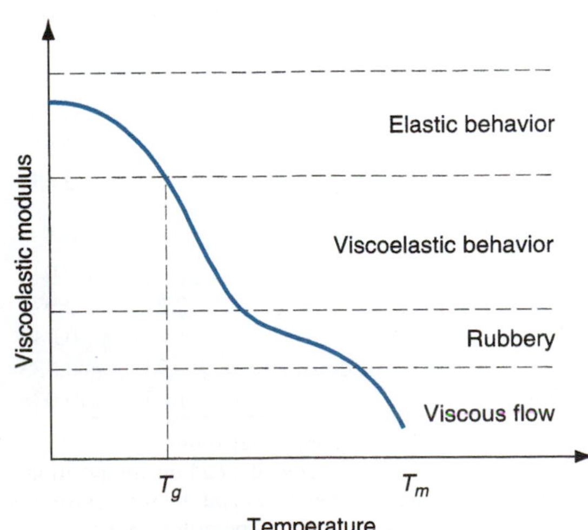

FIGURE 3.21 Viscoelastic modulus as a function of temperature for a thermoplastic polymer. (Credit: *Fundamentals of Modern Manufacturing*, 4th Edition by Mikell P. Groover, 2010. Reprinted with permission of John Wiley & Sons, Inc.)

3.6 VOLUMETRIC AND MELTING PROPERTIES

These properties are related to the volume of solids and how they are affected by temperature. The properties include density, thermal expansion, and melting point. They are explained below, and a listing of typical values for selected engineering materials is presented in Table 3.10.

3.6.1 DENSITY AND THERMAL EXPANSION

In engineering, the density of a material is its weight per unit volume. Its symbol is ρ, and typical units are g/cm^3 (lb/in^3). The density of an element is determined by its atomic number and other factors such as atomic radius and atomic packing. The term ***specific gravity*** expresses the density of a material relative to the density of water and is therefore a ratio with no units.

Density is an important consideration in the selection of a material for a given application, but it is generally not the only property of interest. Strength is also important, and the two properties are often related in a ***strength-to-weight ratio***, which is the tensile strength of the material divided by its density. The ratio is useful in comparing materials for structural applications in aircraft, automobiles, and other products in which weight and energy are a concern.

TABLE 3.10 Volumetric properties in U.S. customary units for selected engineering materials.

	Density, ρ		Coefficient of Thermal Expansion, α		Melting Point, T_m	
Material	**g/cm^3**	**lb/in^3**	**$°C^{-1} \times 10^{-6}$**	**$°F^{-1} \times 10^{-6}$**	**°C**	**°F**
Metals						
Aluminum	2.70	0.098	24	13.3	660	1220
Copper	8.97	0.324	17	9.4	1083	1981
Iron	7.87	0.284	12.1	6.7	1539	2802
Lead	11.35	0.410	29	16.1	327	621
Magnesium	1.74	0.063	26	14.4	650	1202
Nickel	8.92	0.322	13.3	7.4	1455	2651
Steel	7.87	0.284	12	6.7	[a]	[a]
Tin	7.31	0.264	23	12.7	232	449
Tungsten	19.30	0.697	4.0	2.2	3410	6170
Zinc	7.15	0.258	40	22.2	420	787
Ceramics						
Glass	2.5	0.090	1.8–9.0	1.0–5.0	[b]	[b]
Alumina	3.8	0.137	9.0	5.0	NA	NA
Silica	2.66	0.096	NA	NA	[b]	[b]
Polymers						
Phenol resins	1.3	0.047	60	33	[c]	[c]
Nylon	1.16	0.042	100	55	[b]	[b]
Teflon	2.2	0.079	100	55	[b]	[b]
Natural rubber	1.2	0.043	80	45	[b]	[b]
Polyethylene (low density)	0.92	0.033	180	100	[b]	[b]
Polystyrene	1.05	0.038	60	33	[b]	[b]

Compiled from, [8], [11], and other sources.

[a]Melting characteristics of steel depend on composition.

[b]Softens at elevated temperatures and does not have a well-defined melting point.

[c]Chemically degrades at high temperatures. NA = not available; value of property for this material could not be obtained.

The density of a material is a function of temperature. The general relationship is that density decreases with increasing temperature. Put another way, the volume per unit weight increases with temperature. Thermal expansion is the name given to this effect that temperature has on density. It is usually expressed as the ***coefficient of thermal expansion***, which measures the change in length per degree of temperature, as mm/mm/°C (in/in/°F). It is a length ratio rather than a volume ratio because this is easier to measure and apply. It is consistent with the usual design situation in which dimensional changes are of greater interest than volumetric changes. The change in length corresponding to a given temperature change is given by:

$$L_2 - L_1 = \alpha L_1(T_2 - T_1) \tag{3.28}$$

where α = coefficient of thermal expansion, $°C^{-1}(°F^{-1})$; and L_1 and L_2 are lengths, mm (in), corresponding, respectively, to temperatures T_1 and T_2, °C (°F).

Values of coefficient of thermal expansion given in Table 3.10 suggest that it has a linear relationship with temperature. This is only an approximation. Not only is length affected by temperature, but the thermal expansion coefficient itself is also affected. For some materials it increases with temperature; for other materials it decreases. These changes are usually not significant enough to be of much concern, and values like those in the table are quite useful in design calculations for the range of temperatures contemplated in service. Changes in the coefficient are more substantial when the metal undergoes a phase transformation, such as from solid to liquid, or from one crystal structure to another.

In manufacturing operations, thermal expansion is put to good use in shrink fit and expansion fit assemblies (Section 25.3.2), in which a part is heated to increase its size or cooled to decrease its size to permit insertion into some other part. When the part returns to ambient temperature, a tightly fitted assembly is obtained. Thermal expansion can be a problem in heat treatment (Chapter 20) and welding (Section 23.6) due to thermal stresses that develop in the material during these processes.

3.6.2 MELTING CHARACTERISTICS

For a pure element, the ***melting point*** T_m is the temperature at which the material transforms from solid to liquid state. The reverse transformation, from liquid to solid, occurs at the same temperature and is called the ***freezing point***. For crystalline elements, such as metals, the melting and freezing temperatures are the same. A certain amount of heat energy, called the ***heat of fusion***, is required at this temperature to accomplish the transformation from solid to liquid.

Melting of a metal element at a specific temperature, as described here, assumes equilibrium conditions. Exceptions occur in nature; for example, when a molten metal is cooled, it may remain in the liquid state below its freezing point if nucleation of crystals does not initiate immediately. When this happens, the liquid is said to be ***supercooled***.

There are other variations in the melting process—differences in the way melting occurs in different materials. For example, unlike pure metals, most metal alloys do not have a single melting point. Instead, melting begins at a certain temperature, called the ***solidus***, and continues as the temperature increases until finally converting completely to the liquid state at a temperature called the ***liquidus***. Between the two temperatures, the alloy is a mixture of solid and molten metals, the amounts of each being inversely proportional to their relative distances from the liquidus and solidus. Although most alloys behave in this way, exceptions are eutectic alloys that melt (and freeze) at a single temperature.

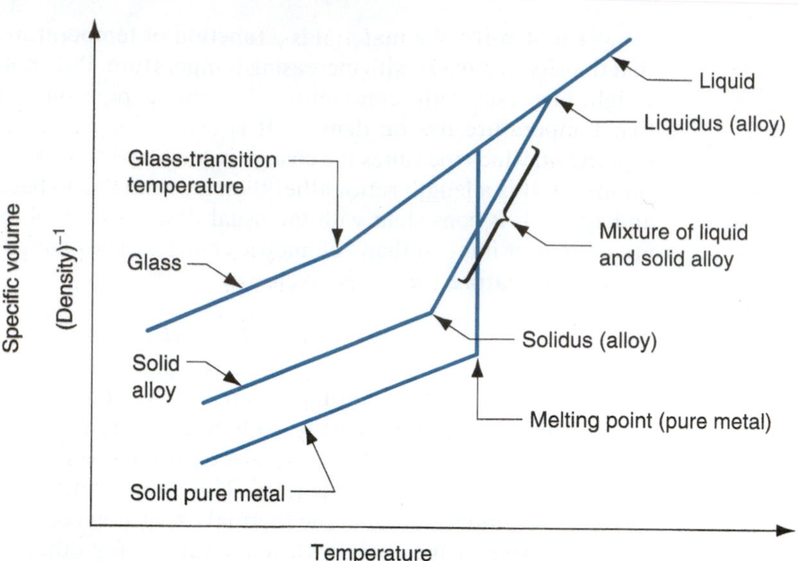

FIGURE 3.22 Changes in volume per unit weight (1/density) as a function of temperature for a hypothetical pure metal, alloy, and glass; all exhibiting similar thermal expansion and melting characteristics. (Credit: *Fundamentals of Modern Manufacturing*, 4th Edition by Mikell P. Groover, 2010. Reprinted with permission of John Wiley & Sons, Inc.)

Another difference in melting occurs with noncrystalline materials (glasses). In these materials, there is a gradual transition from solid to liquid states. The solid material gradually softens as temperature increases, finally becoming liquid at the melting point. During softening, the material has a consistency of increasing plasticity (increasingly like a fluid) as it gets closer to the melting point.

These differences in melting characteristics among pure metals, alloys, and glass are portrayed in Figure 3.22. The plots show changes in density as a function of temperature for three hypothetical materials: a pure metal, an alloy, and glass. Plotted in the figure is the volumetric change, which is the reciprocal of density.

The importance of melting in manufacturing is obvious. In metal casting (Chapters 5 and 6), the metal is melted and then poured into a mold cavity. Metals with lower melting points are generally easier to cast, but if the melting temperature is too low, the metal loses its applicability as an engineering material. Melting characteristics of polymers are important in plastic molding and other polymer shaping processes (Chapter 8). Sintering of powdered metals and ceramics requires knowledge of melting points. Sintering does not melt the materials, but the temperatures used in the process must approach the melting point to achieve bonding of the powders.

3.7 THERMAL PROPERTIES

The previous section is concerned with the effects of temperature on volumetric properties of materials. The current section examines several additional thermal properties—ones that relate to the storage and flow of heat within a substance. The usual properties of interest are specific heat and thermal conductivity, values of which are compiled for selected materials in Table 3.11.

3.7.1 SPECIFIC HEAT AND THERMAL CONDUCTIVITY

The specific heat C of a material is defined as the quantity of heat energy required to increase the temperature of a unit mass of the material by one degree. Some typical

TABLE 3.11 Values of common thermal properties for selected materials. Values are at room temperature, and these values change for different temperatures.

Material	Specific Heat	Thermal Conductivity		Material	Specific Heat	Thermal Conductivity	
	Cal/g °C[a] or Btu/lbm °F	J/s mm °C	Btu/hr in °F		Cal/g °C[a] or Btu/lbm °F	J/s mm °C	Btu/hr in °F
Metals				***Ceramics***			
Aluminum	0.21	0.22	9.75	Alumina	0.18	0.029	1.4
Cast iron	0.11	0.06	2.7	Concrete	0.2	0.012	0.6
Copper	0.092	0.40	18.7	***Polymers***			
Iron	0.11	0.072	2.98	Phenolics	0.4	0.00016	0.0077
Lead	0.031	0.033	1.68	Polyethylene	0.5	0.00034	0.016
Magnesium	0.25	0.16	7.58	Teflon	0.25	0.00020	0.0096
Nickel	0.105	0.070	2.88	Natural rubber	0.48	0.00012	0.006
Steel	0.11	0.046	2.20	***Other***			
Stainless steel[b]	0.11	0.014	0.67	Water (liquid)	1.00	0.0006	0.029
Tin	0.054	0.062	3.0	Ice	0.46	0.0023	0.11
Zinc	0.091	0.112	5.41				

Compiled from [8], [16], and other sources.
[a]Specific heat has the same numerical value in Btu/lbm-F or Cal/g-C. 1.0 Calory = 4.186 Joule.
[b]Austenitic (18-8) stainless steel.

values are listed in Table 3.11. To determine the amount of energy needed to heat a certain weight of a metal in a furnace to a given elevated temperature, the following equation can be used:

$$H = CW(T_2 - T_1) \tag{3.29}$$

where H = amount of heat energy, J (Btu); C = specific heat of the material, J/kg °C (Btu/lb °F); W = its weight, kg (lb); and $(T_2 - T_1)$ = change in temperature, °C (°F).

The volumetric heat storage capacity of a material is often of interest. This is simply density multiplied by specific heat ρC. Thus, ***volumetric specific heat*** is the heat energy required to raise the temperature of a unit volume of material by one degree, J/mm^3 °C (Btu/in^3 °F).

Conduction is a fundamental heat-transfer process. It involves transfer of thermal energy within a material from molecule to molecule by purely thermal motions; no transfer of mass occurs. The thermal conductivity of a substance is therefore its capability to transfer heat through itself by this physical mechanism. It is measured by the ***coefficient of thermal conductivity*** k, which has typical units of J/s mm °C (Btu/in hr °F). The coefficient of thermal conductivity is generally high in metals, low in ceramics and plastics.

The ratio of thermal conductivity to volumetric specific heat is frequently encountered in heat-transfer analysis. It is called the ***thermal diffusivity*** K and is determined as

$$K = \frac{k}{\rho C} \tag{3.30}$$

This property is used to calculate cutting temperatures in machining (Section 15.5.1).

3.7.2 THERMAL PROPERTIES IN MANUFACTURING

Thermal properties play an important role in manufacturing because heat generation is common in so many processes. In some operations heat is the energy that accomplishes the process; in others heat is generated as a consequence of the process.

Specific heat is of interest for several reasons. In processes that require heating of the material (e.g., casting, heat-treating, and hot metal forming), specific heat determines the amount of heat energy needed to raise the temperature to a desired level, according to Eq. (3.29).

In many processes carried out at ambient temperature, the mechanical energy to perform the operation is converted to heat, which raises the temperature of the workpart. This is common in machining and cold forming of metals. The temperature rise is a function of the metal's specific heat. Coolants are often used in machining to reduce these temperatures, and here the fluid's heat capacity is critical. Water is almost always employed as the base for these fluids because of its high heat-carrying capacity.

Thermal conductivity functions to dissipate heat in manufacturing processes, sometimes beneficially, sometimes not. In mechanical processes such as metal forming and machining, much of the power required to operate the process is converted to heat. The ability of the work material and tooling to conduct heat away from its source is highly desirable in these processes.

On the other hand, high thermal conductivity of the work metal is undesirable in fusion welding processes such as arc welding. In these operations, the heat input must be concentrated at the joint location so that the metal can be melted. For example, copper is generally difficult to weld because its high thermal conductivity allows heat to be conducted from the energy source into the work too rapidly, inhibiting heat buildup for melting at the joint.

REFERENCES

[1] Avallone, E. A., and Baumeister III, T. (eds.). ***Mark's Standard Handbook for Mechanical Engineers***, 11th ed. McGraw-Hill Book Company, New York, 2006.

[2] Beer, F. P., Russell, J. E., Eisenberg, E., and Mazurek, D. ***Vector Mechanics for Engineers: Statics***, 9th ed. McGraw-Hill Book Company, New York, 2009.

[3] Black, J. T., and Kohser, R. A. ***DeGarmo's Materials and Processes in Manufacturing***, 10th ed. John Wiley & Sons, Inc., Hoboken, New Jersey, 2008.

[4] Budynas, R. G. ***Advanced Strength and Applied Stress Analysis***, 2nd ed. McGraw-Hill Book Company, New York, 1998.

[5] Chandra, M., and Roy, S. K. ***Plastics Technology Handbook***, 4th ed. CRC Press, Inc., Boca Raton, Florida, 2006.

[6] Dieter, G. E. ***Mechanical Metallurgy***, 3rd ed. McGraw-Hill Book Company, New York, 1986.

[7] **Engineered Materials Handbook**, Vol. 2, ***Engineering Plastics***. ASM International, Materials Park, Ohio, 1987.

[8] Flinn, R. A., and Trojan, P. K. ***Engineering Materials and Their Applications***, 5th ed. John Wiley & Sons, Inc., Hoboken, New Jersey, 1995.

[9] Guy, A. G., and Hren, J. J. ***Elements of Physical Metallurgy***, 3rd ed. Addison-Wesley Publishing Company, Reading, Massachusetts, 1974.

[10] Kalpakjian, S., and SchmidS. R. ***Manufacturing Processes for Engineering Materials***, 6th ed. Pearson Prentice Hall, Upper Saddle River, New Jersey, 2010.

[11] ***Metals Handbook***, Vol. **1**, **Properties and Selection: Iron, Steels, and High Performance Alloys**. ASM International, Materials Park, Ohio, 1990.

[12] ***Metals Handbook***, Vol. **2**, **Properties and Selection: Nonferrous Alloys and Special Purpose Materials**. ASM International, Materials Park, Ohio, 1991.

[13] ***Metals Handbook***, Vol. **8**, **Mechanical Testing and Evaluation, ASM International**. Materials Park, Ohio, 2000.

[14] Morton-Jones, D. H. ***Polymer Processing***. Chapman and Hall, London, 2008.

[15] Schey, J. A. ***Introduction to Manufacturing Processes***, 3rd ed. McGraw-Hill Book Company, New York, 2000.

[16] VanVlack, L. H. ***Elements of Materials Science and Engineering***, 6th ed. Addison-Wesley Publishing Company, Reading, Massachusetts, 1991.

[17] Wick, C., and Veilleux, R. F. (eds.). ***Tool and Manufacturing Engineers Handbook***, 4th ed. Vol. 3, Materials, Finishing, and Coating. Society of Manufacturing Engineers, Dearborn, Michigan, 1985.

REVIEW QUESTIONS

3.1. What is the dilemma between design and manufacturing in terms of mechanical properties?

3.2. What are the three types of static stresses to which materials are subjected?

3.3. State Hooke's law.

3.4. What is the difference between engineering stress and true stress in a tensile test?

3.5. Define tensile strength of a material.

3.6. Define yield strength of a material.

3.7. Why can a direct conversion not be made between the ductility measures of elongation and reduction in area using the assumption of constant volume?

3.8. What is work hardening?

3.9. Under what circumstances does the strength coefficient have the same value as the yield strength?

3.10. How does the change in cross-sectional area of a test specimen in a compression test differ from its counterpart in a tensile test specimen?

3.11. Tensile testing is not appropriate for hard brittle materials such as ceramics. What is the test commonly used to determine the strength properties of such materials?

3.12. How is the shear modulus of elasticity G related to the tensile modulus of elasticity E, on average?

3.13. How is shear strength S related to tensile strength TS, on average?

3.14. What is hardness, and how is it generally tested?

3.15. Why are different hardness tests and scales required?

3.16. Define the recrystallization temperature for a metal.

3.17. Define viscosity of a fluid.

3.18. What is the defining characteristic of a Newtonian fluid?

3.19. What is viscoelasticity, as a material property?

3.20. Define density as a material property.

3.21. What is the difference in melting characteristics between a pure metal element and an alloy metal?

3.22. Define specific heat as a material property.

3.23. What is thermal conductivity as a material property?

3.24. Define thermal diffusivity.

PROBLEMS

3.1. A tensile test uses a test specimen that has a gage length of 50 mm and an area = 200 mm^2. During the test the specimen yields under a load of 98,000 N. The corresponding gage length = 50.23 mm. This is the 0.2% yield point. The maximum load of 168,000 N is reached at a gage length = 64.2 mm. Determine (a) yield strength, (b) modulus of elasticity, and (c) tensile strength. (d) If fracture occurs at a gage length of 67.3 mm, determine the percent elongation. (e) If the specimen necked to an area = 92 mm^2, determine the percent reduction in area.

3.2. A test specimen in a tensile test has a gage length of 2.0 in and an area = 0.5 in^2. During the test the specimen yields under a load of 32,000 lb. The corresponding gage length = 2.0083 in. This is the 0.2% yield point. The maximum load of 60,000 lb is reached at a gage length = 2.60 in. Determine (a) yield strength, (b) modulus of elasticity, and (c) tensile strength. (d) If fracture occurs at a gage length of 2.92 in, determine the percent elongation. (e) If the specimen necked to an area = 0.25 in^2, determine the percent reduction in area.

3.3. In a tensile test on a metal specimen, true strain = 0.08 at a stress = 265 MPa. When true stress = 325 MPa, true strain = 0.27. Determine the strength coefficient and the strain-hardening exponent in the flow curve equation.

3.4. During a tensile test, a metal has a true strain = 0.10 at a true stress = 37,000 lb/in^2. Later, at a true stress = 55,000 lb/in^2, true strain = 0.25. Determine the strength coefficient and strain-hardening exponent in the flow curve equation.

3.5. A tensile test for a certain metal provides flow curve parameters: strain-hardening exponent is 0.3 and strength coefficient is 600 MPa. Determine (a) the flow stress at a true strain = 1.0 and (b) true strain at a flow stress = 600 MPa.

3.6. The flow curve for a certain metal has a strain-hardening exponent of 0.22 and strength coefficient of 54,000 lb/in^2. Determine (a) the flow stress at a true strain = 0.45 and (b) the true strain at a flow stress = 40,000 lb/in^2.

3.7. A metal is deformed in a tension test into its plastic region. The starting specimen had a gage length = 2.0 in and an area = 0.50 in^2. At one point in the tensile test, the gage length = 2.5 in, and the corresponding engineering stress = 24,000 lb/in^2; at another point in the test prior to necking, the gage length = 3.2 in, and the corresponding engineering stress = 28,000 lb/in^2. Determine the strength coefficient and the strain-hardening exponent for this metal.

3.8. A tensile specimen is elongated to twice its original length. Determine the engineering strain and true strain for this test. If the metal had been strained in compression, determine the final compressed length of the specimen such that (a) the engineering strain is equal to the same value as in tension (it will be negative value because of compression), and (b) the true strain would be equal to the same value as in tension (again, it will be negative value because of compression). Note that the answer to part (a) is an impossible result. True strain is therefore a better measure of strain during plastic deformation.

3.9. Show that true strain = $\ln(1 + e)$, where e = engineering strain.

3.10. A copper wire of diameter 0.80 mm fails at an engineering stress = 248.2 MPa. Its ductility is measured as 75% reduction of area. Determine the true stress and true strain at failure.

3.11. A steel tensile specimen with starting gage length = 2.0 in and cross-sectional area = 0.5 in^2 reaches a maximum load of 37,000 lb. Its elongation at this point is 24%. Determine the true stress and true strain at this maximum load.

3.12. A metal alloy has been tested in a tensile test with the following results for the flow curve parameters: strength coefficient = 620.5 MPa and strain-hardening exponent = 0.26. The same metal is now tested in a compression test in which the starting height of the specimen = 62.5 mm and its diameter = 25 mm. Assuming that the cross section increases uniformly, determine the load required to compress the specimen to a height of (a) 50 mm and (b) 37.5 mm.

3.13. The flow curve parameters for a certain stainless steel are strength coefficient = 1100 MPa and strain-hardening exponent = 0.35. A cylindrical specimen of starting cross-sectional area = 1000 mm^2 and height = 75 mm is compressed to a height of 58 mm. Determine the force required to achieve this compression, assuming that the cross section increases uniformly.

3.14. A bend test is used for a certain hard material. If the transverse rupture strength of the material is known to be 1000 MPa, what is the anticipated load at which the specimen is likely to fail, given that its width = 15 mm, thickness = 10 mm, and length = 60 mm?

3.15. A special ceramic specimen is tested in a bend test. Its width = 0.50 in and thickness = 0.25 in. The length of the specimen between supports = 2.0 in. Determine the transverse rupture strength if failure occurs at a load = 1700 lb.

3.16. A torsion test specimen has a radius = 25 mm, wall thickness = 3 mm, and gage length = 50 mm. In testing, a torque of 900 N-m results in an angular deflection = 0.3°. Determine (a) the shear stress, (b) shear strain, and (c) shear modulus, assuming the specimen had not yet yielded. (d) If failure of the specimen occurs at a torque = 1200 N-m and a corresponding angular deflection = 10°, what is the shear strength of the metal?

3.17. In a torsion test, a torque of 5000 ft-lb is applied, which causes an angular deflection = 1° on a thin-walled tubular specimen whose radius = 1.5 in, wall thickness = 0.10 in, and gage length = 2.0 in. Determine (a) the shear stress, (b) shear strain, and (c) shear modulus, assuming the specimen had not yet yielded. (d) If the specimen fails at a torque = 8000 ft-lb and an angular deflection = 23°, calculate the shear strength of the metal.

3.18. In a Brinell Hardness Test, a 1500-kg load is pressed into a specimen using a 10-mm-diameter hardened steel ball. The resulting indentation has a diameter = 3.2 mm. (a) Determine the Brinell hardness number for the metal. (b) If the specimen is steel, estimate the tensile strength of the steel.

3.19. One of the inspectors in the quality control department has frequently used the Brinell and Rockwell Hardness Tests, for which equipment is available in the company. He claims that the Rockwell Test is based on the same principle as the Brinell Test, which is that hardness is always measured as the applied load divided by the area of the impressions made by an indenter. Is he correct? If not, how is the Rockwell test different?

3.20. A batch of annealed steel has just been received from the vendor. It is supposed to have a tensile strength in the range 60,000 – 70,000 lb/in^2. A Brinell Hardness Test in the receiving department yields a value of HB = 118. Does the steel meet the specification on tensile strength?

3.21. Two flat plates, separated by a space of 4 mm, move relative to each other at a velocity of 5 m/sec. The space between them is occupied by a fluid of unknown viscosity. The motion of the plates is resisted by a shear stress of 10 Pa due to the viscosity of the fluid. Assuming that the velocity gradient of the fluid is constant, determine the coefficient of viscosity of the fluid.

3.22. Two parallel surfaces, separated by a space of 0.5 in that is occupied by a fluid, move relative to each other at a velocity of 25 in/sec. The motion is resisted by a shear stress of 0.3 lb/in^2 due to the viscosity of the fluid. If the velocity gradient in the space between the surfaces is constant, determine the viscosity of the fluid.

3.23. The starting diameter of a shaft is 25.00 mm. This shaft is to be inserted into a hole in an expansion fit assembly operation. To be readily inserted, the shaft must be reduced in diameter by cooling. Determine the temperature to which the shaft must be reduced from room temperature (20°C) in order to reduce its diameter to 24.98 mm. Refer to Table 3.10.

3.24. Aluminum has a density of 2.70 g/cm^3 at room temperature (20°C). Determine its density at 650°C, using data in Table 3.10 as a reference.

3.25. With reference to Table 3.10, determine the increase in length of a steel bar whose length = 10.0 in, if the bar is heated from room temperature of 70°F to 500°F.

3.26. With reference to Table 3.11, determine the quantity of heat required to increase the temperature of an aluminum block that is 10 cm × 10 cm × 10 cm from room temperature (21°C) to 300°C.

4 DIMENSIONS, TOLERANCES, AND SURFACES

Chapter Contents

4.1 Dimensions and Tolerances
- 4.1.1 Dimensions and Tolerances
- 4.1.2 Other Geometric Attributes

4.2 Surfaces
- 4.2.1 Characteristics of Surfaces
- 4.2.2 Surface Texture
- 4.2.3 Surface Integrity

4.3 Effect of Manufacturing Processes

Appendix A4 Measurement of Dimensions and Surfaces

A4.1 Conventional Measuring Instruments and Gages
- A4.1.1 Precision Gage Blocks
- A4.1.2 Measuring Instruments for Linear Dimensions
- A4.1.3 Comparative Instruments
- A4.1.4 Angular Measurements

A4.2 Measurement of Surfaces
- A4.2.1 Measurement of Surface Roughness
- A4.2.2 Evaluation of Surface Integrity

In addition to the properties of engineering materials, other factors that determine the performance of a manufactured product include the dimensions and surfaces of its components. ***Dimensions*** are the linear or angular sizes of a component specified on the part drawing. Dimensions are important because they determine how well the components of a product fit together during assembly. When fabricating a given component, it is nearly impossible and very costly to make the part to the exact dimension given on the drawing. Instead a limited variation from the dimension is allowed, and that allowable variation is called a ***tolerance***.

The surfaces of a component are also important. They affect product performance, assembly fit, and aesthetic appeal that a potential customer might have for the product. A ***surface*** is the exterior boundary of an object with its surroundings, which may be another object, a fluid, or space, or combinations of these. The surface encloses the object's bulk mechanical and physical properties.

In this chapter we discuss dimensions, tolerances, and surfaces—three attributes specified by the product designer and determined by the manufacturing processes used to make the parts and products. In Appendix A4, we consider how these attributes are assessed using measuring and gaging devices. A closely related topic is inspection, covered in Chapter 30.

4.1 DIMENSIONS AND TOLERANCES

The basic parameters used by design engineers to specify sizes of geometric features on a part drawing are defined in this section. The parameters include dimensions and tolerances, flatness, roundness, and angularity.

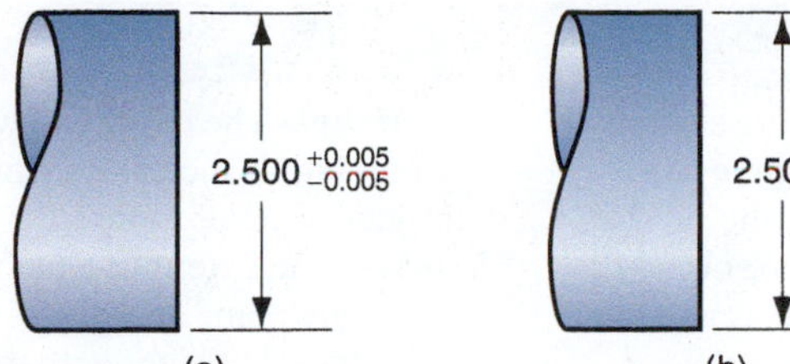

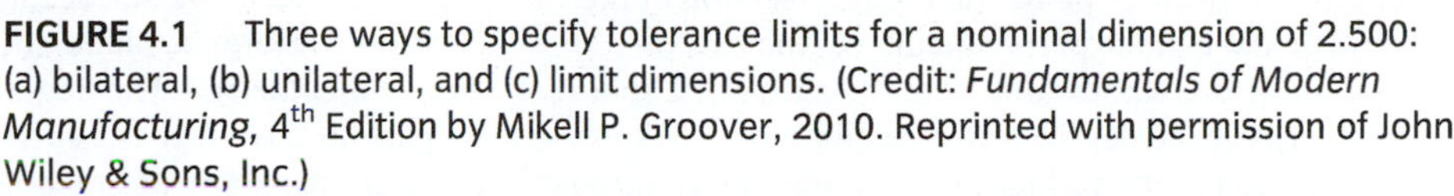
FIGURE 4.1 Three ways to specify tolerance limits for a nominal dimension of 2.500: (a) bilateral, (b) unilateral, and (c) limit dimensions. (Credit: *Fundamentals of Modern Manufacturing*, 4th Edition by Mikell P. Groover, 2010. Reprinted with permission of John Wiley & Sons, Inc.)

4.1.1 DIMENSIONS AND TOLERANCES

ANSI [3] defines a ***dimension*** as "a numerical value expressed in appropriate units of measure and indicated on a drawing and in other documents along with lines, symbols, and notes to define the size or geometric characteristic, or both, of a part or part feature." Dimensions on part drawings represent nominal or basic sizes of the part and its features. These are the values that the designer would like the part size to be, if the part could be made to an exact size with no errors or variations in the fabrication process. However, there are variations in the manufacturing process, which are manifested as variations in the part size. Tolerances are used to define the limits of the allowed variation. Quoting again from the ANSI standard [3], a ***tolerance*** is "the total amount by which a specific dimension is permitted to vary. The tolerance is the difference between the maximum and minimum limits."

Tolerances can be specified in several ways, illustrated in Figure 4.1. Probably most common is the ***bilateral tolerance***, in which the variation is permitted in both positive and negative directions from the nominal dimension. For example, in Figure 4.1(a), the nominal dimension = 2.500 linear units (e.g., mm, in), with an allowable variation of 0.005 units in either direction. Parts outside these limits are unacceptable. It is possible for a bilateral tolerance to be unbalanced; for example, 2.500 +0.010, −0.005 units. A ***unilateral tolerance*** is one in which the variation from the specified dimension is permitted in only one direction, either positive, as in Figure 4.1(b), or negative. ***Limit dimensions*** are an alternative method to specify the permissible variation in a part feature size; they consist of the maximum and minimum allowable dimensions, as in Figure 4.1(c).

4.1.2 OTHER GEOMETRIC ATTRIBUTES

Dimensions and tolerances are normally expressed as linear (length) values. There are other geometric attributes of parts that are also important, such as flatness of a surface, roundness of a shaft or hole, parallelism between two surfaces, and so on. Definitions of these terms are listed in Table 4.1.

4.2 SURFACES

A surface is what we touch when holding an object such as a manufactured part. The designer specifies the part dimensions, relating the various surfaces to each other. These ***nominal surfaces***, representing the intended surface contour of the part, are defined by lines in the engineering drawing. The nominal surfaces appear as absolutely straight lines,

TABLE 4.1 Definitions of geometric attributes of parts.

Angularity—The extent to which a part feature such as a surface or axis is at a specified angle relative to a reference surface. If the angle = 90°, then the attribute is called perpendicularity or squareness.
Circularity—For a surface of revolution such as a cylinder, circular hole, or cone, circularity is the degree to which all points on the intersection of the surface and a plane perpendicular to the axis of revolution are equidistant from the axis. For a sphere, circularity is the degree to which all points on the intersection of the surface and a plane passing through the center are equidistant from the center.
Concentricity—The degree to which any two (or more) part features such as a cylindrical surface and a circular hole have a common axis.
Cylindricity—The degree to which all points on a surface of revolution such as a cylinder are equidistant from the axis of revolution.
Flatness—The extent to which all points on a surface lie in a single plane.
Parallelism—The degree to which all points on a part feature such as a surface, line, or axis are equidistant from a reference plane or line or axis.
Perpendicularity—The degree to which all points on a part feature such as a surface, line, or axis are 90° from a reference plane or line or axis.
Roundness—Same as circularity.
Squareness—Same as perpendicularity.
Straightness—The degree to which a part feature such as a line or axis is a straight line.

Source: [16].

ideal circles, round holes, and other edges and surfaces that are geometrically perfect. The actual surfaces of a manufactured part are determined by the processes used to make it. The variety of processes available in manufacturing result in wide variations in surface characteristics, and it is important for engineers to understand the technology of surfaces.

Surfaces are commercially and technologically important for a number of reasons, different reasons for different applications: (1) Aesthetic reasons—surfaces that are smooth and free of scratches and blemishes—are more likely to give a favorable impression to the customer. (2) Surfaces affect safety. (3) Friction and wear depend on surface characteristics. (4) Surfaces affect mechanical and physical properties; for example, surface flaws can be points of stress concentration. (5) Assembly of parts is affected by their surfaces; for example, the strength of adhesively bonded joints (Section 24.3) is increased when the surfaces are slightly rough. (6) Smooth surfaces make better electrical contacts.

Surface technology is concerned with (1) defining the characteristics of a surface, (2) surface texture, (3) surface integrity, and (4) the relationship between manufacturing processes and the characteristics of the resulting surface. The first three topics are covered in this section; the final topic is presented in Section 4.3.

4.2.1 CHARACTERISTICS OF SURFACES

A microscopic view of a part's surface reveals its irregularities and imperfections. The features of a typical surface are illustrated in the highly magnified cross section of the surface of a metal part in Figure 4.2. Although the discussion here is focused on metallic

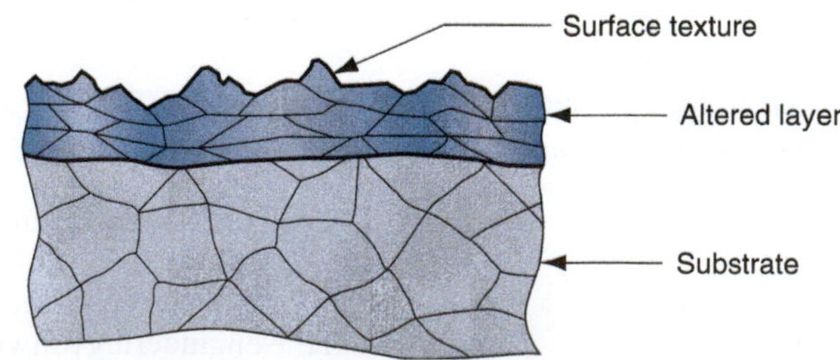

FIGURE 4.2 A magnified cross section of a typical metallic part surface. (Credit: *Fundamentals of Modern Manufacturing*, 4th Edition by Mikell P. Groover, 2010. Reprinted with permission of John Wiley & Sons, Inc.)

surfaces, the comments apply to ceramics and polymers, with modifications owing to differences in structure of these materials. The bulk of the part, referred to as the ***substrate***, has a grain structure that depends on previous processing of the metal; for example, the metal's substrate structure is affected by its chemical composition, the casting process originally used on the metal, and any deformation operations and heat treatments performed on the casting.

The exterior of the part is a surface whose topography is anything but straight and smooth. In this highly magnified cross section, the surface has roughness, waviness, and flaws. Although not shown here, it also possesses a pattern and/or direction resulting from the mechanical process that produced it. All of these geometric features are included in the term ***surface texture***.

Just below the surface is a layer of metal whose structure differs from that of the substrate. This is called the ***altered layer***, and it is a manifestation of the actions that have been visited upon the surface during its creation and afterward. Manufacturing processes involve energy, usually in large amounts, which operates on the part against its surface. The altered layer may result from work hardening (mechanical energy), heating (thermal energy), chemical treatment, or even electrical energy. The metal in this layer is affected by the application of energy, and its microstructure is altered accordingly. This altered layer falls within the scope of ***surface integrity***, which is concerned with the definition, specification, and control of the surface layers of a material (most commonly metals) in manufacturing and subsequent performance in service. The scope of surface integrity is usually interpreted to include surface texture as well as the altered layer beneath.

In addition, most metal surfaces are coated with an ***oxide film***, given sufficient time after processing for the film to form. Aluminum forms a hard, dense, thin film of Al_2O_3 on its surface (which protects the substrate from corrosion), and iron forms oxides of several chemistries on its surface (rust, which provides virtually no protection at all). There is also likely to be moisture, dirt, oil, adsorbed gases, and other contaminants on the part's surface.

4.2.2 SURFACE TEXTURE

Surface texture consists of the repetitive and/or random deviations from the nominal surface of an object; it is defined by four features: roughness, waviness, lay, and flaws, shown in Figure 4.3. ***Roughness*** refers to the small, finely spaced deviations from the nominal surface that are determined by the material characteristics and the process that formed the surface. ***Waviness*** is defined as the deviations of much larger spacing; they

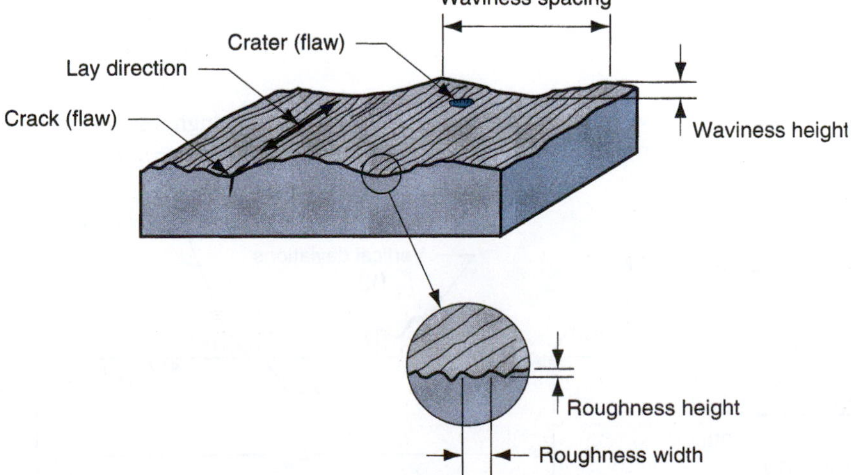

FIGURE 4.3 Surface texture features. (Credit: *Fundamentals of Modern Manufacturing*, 4th Edition by Mikell P. Groover, 2010. Reprinted with permission of John Wiley & Sons, Inc.)

Lay symbol	Surface pattern	Description
=		Lay is parallel to line representing surface to which symbol is applied.
⊥		Lay is perpendicular to line representing surface to which symbol is applied.
X		Lay is angular in both directions to line representing surface to which symbol is applied.
C		Lay is circular relative to center of surface to which symbol is applied.
R		Lay is approximately radial relative to the center of the surface to which symbol is applied.
P		Lay is particulate, nondirectional, or protuberant.

FIGURE 4.4 Possible lays of a surface. (Source: [1]). (Credit: *Fundamentals of Modern Manufacturing,* 4th Edition by Mikell P. Groover, 2010. Reprinted with permission of John Wiley & Sons, Inc.)

occur because of work deflection, vibration, heat treatment, and similar factors. Roughness is superimposed on waviness. ***Lay*** is the predominant direction or pattern of the surface texture. It is determined by the manufacturing method used to create the surface, usually from the action of a cutting tool. Figure 4.4 presents most of the possible lays a surface can take, together with the symbol used by a designer to specify them. Finally, ***flaws*** are irregularities that occur occasionally on the surface; these include cracks, scratches, inclusions, and similar defects in the surface. Although some of the flaws relate to surface texture, they also affect surface integrity (Section 4.2.3).

Surface Roughness and Surface Finish Surface roughness is a measurable characteristic based on the roughness deviations defined above. ***Surface finish*** is a more subjective term denoting smoothness and general quality of a surface. In popular usage, surface finish is often used as a synonym for surface roughness.

The most commonly used measure of surface texture is surface roughness. With respect to Figure 4.5, ***surface roughness*** can be defined as the average of the vertical deviations from the nominal surface over a specified surface length. An arithmetic average (AA) is generally used, based on the absolute values of the deviations, and this roughness value is referred to by the name ***average roughness***. In equation form,

$$R_a = \int_0^{L_m} \frac{|y|}{L_m} dx \tag{4.1}$$

where R_a = arithmetic mean value of roughness, m (in); y = the vertical deviation from nominal surface (converted to absolute value), m (in); and L_m = the specified distance

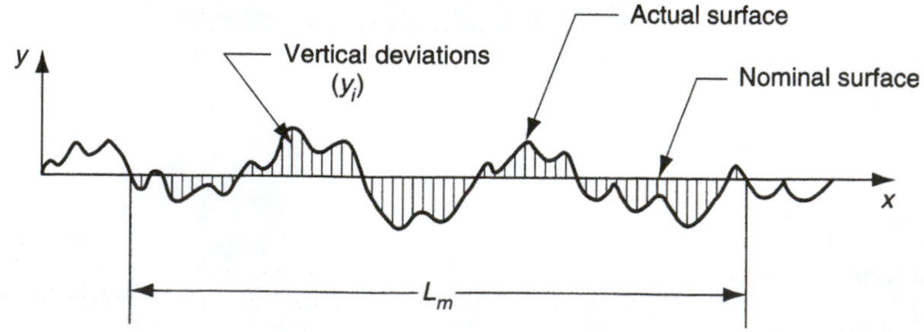

FIGURE 4.5 Deviations from nominal surface used in the definition of surface roughness. (Credit: *Fundamentals of Modern Manufacturing,* 4th Edition by Mikell P. Groover, 2010. Reprinted with permission of John Wiley & Sons, Inc.)

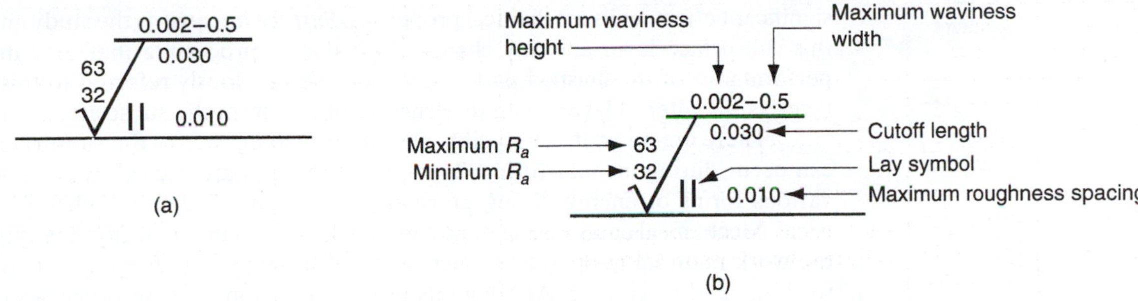

FIGURE 4.6 Surface texture symbols in engineering drawings: (a) the symbol, and (b) symbol with identification labels. Values of R_a are given in microinches; units for other measures are given in inches. Designers do not always specify all of the parameters on engineering drawings. (Credit: *Fundamentals of Modern Manufacturing*, 4th Edition by Mikell P. Groover, 2010. Reprinted with permission of John Wiley & Sons, Inc.)

over which the surface deviations are measured. An approximation of Eq. (4.1), perhaps easier to comprehend, is given by

$$R_a = \sum_{i=1}^{n} \frac{|y_i|}{n} \tag{4.2}$$

where R_a has the same meaning as above; y_i = vertical deviations converted to absolute value and identified by the subscript i, m (in); and n = the number of deviations included in L_m. The units in these equations are meters and inches. In fact, the scale of the deviations is very small, so more appropriate units are μm (μm = m × 10^{-6} = mm × 10^{-3}) or μ-in (μ-in = inch × 10^{-6}). These are the units commonly used to express surface roughness.

Surface roughness suffers the same kinds of deficiencies of any single measure used to assess a complex physical attribute. For example, it fails to account for the lay of the surface pattern; thus, surface roughness may vary significantly depending on the direction in which it is measured.

Another deficiency is that waviness can be included in the R_a computation. To deal with this problem, a parameter called the ***cutoff length*** is used as a filter that separates the waviness in a measured surface from the roughness deviations. In effect, the cutoff length is a sampling distance along the surface. A sampling distance shorter than the waviness width will eliminate the vertical deviations associated with waviness and only include those associated with roughness. The most common cutoff length used in practice is 0.8 mm (0.030 in). The measuring length L_m is normally set at about five times the cutoff length.

The limitations of surface roughness have motivated the development of additional measures that more completely describe the topography of a given surface. These measures include three-dimensional graphical renderings of the surface, as described in [17].

Symbols for Surface Texture Designers specify surface texture on an engineering drawing by means of symbols as in Figure 4.6. The symbol to designate surface texture is a check mark (looks like a square root sign), with entries as indicated for average roughness, waviness, cutoff, lay, and maximum roughness spacing. The symbols for lay are from Figure 4.4.

4.2.3 SURFACE INTEGRITY

Surface texture alone does not completely describe a surface. There may be metallurgical or other changes in the material immediately beneath the surface that can have a significant effect on its mechanical properties. ***Surface integrity*** is the study and control of this subsurface layer and any changes in it due to processing that may influence the performance of the finished part or product. We previously referred to this subsurface layer as the altered layer when its structure differs from the substrate, as in Figure 4.2.

There are a variety of possible alterations and injuries to the subsurface layer that can occur during manufacturing. The surface changes are caused by the application of various forms of energy during processing—mechanical, thermal, chemical, and electrical. Mechanical energy is the most common form in manufacturing; it is applied against the work material in operations such as metal forming (e.g., forging, extrusion), pressworking, and machining. Although its primary function in these processes is to change the workpart geometry, mechanical energy can also cause residual stresses, work hardening, and cracks in the surface layers.

4.3 EFFECT OF MANUFACTURING PROCESSES

The ability to achieve a certain tolerance or surface is a function of the manufacturing process. In this section, we describe the general capabilities of various processes in terms of tolerance and surface roughness.

Some manufacturing processes are inherently more accurate than others. Most machining processes are quite accurate, capable of tolerances of ±0.05 mm (±0.002 in) or better. By contrast, sand castings are generally inaccurate, and tolerances of 10 to 20 times those used for machined parts should be specified. In Table 4.2, we list a variety of manufacturing processes and indicate the typical tolerances for each process. The tolerances are based on the process capability for the particular manufacturing operation, as defined in Section 30.2. The tolerance should be specified as a function of part size; larger parts require more generous tolerances. Our table lists a typical tolerance value for moderately sized parts in each processing category.

The manufacturing process determines surface finish and surface integrity. Some processes are capable of producing better surfaces than others. In general, processing cost increases with improvement in surface finish. This is because additional operations

TABLE 4.2 Typical tolerance limits, based on process capability (Section 30.2), for various manufacturing processes.[a]

Process	Typical Tolerance, mm (in)	Process	Typical Tolerance, mm (in)
Sand casting		Abrasive	
Cast iron	±1.3 (±0.050)	Grinding	±0.008 (±0.0003)
Steel	±1.5 (±0.060)	Lapping	±0.005 (±0.0002)
Aluminum	±0.5 (±0.020)	Honing	±0.005 (±0.0002)
Die casting	±0.12 (±0.005)	Nontraditional and thermal	
Plastic molding:		Chemical machining	±0.08 (±0.003)
Polyethylene	±0.3 (±0.010)	Electric discharge	±0.025 (±0.001)
Polystyrene	±0.15 (±0.006)	Electrochem. grind	±0.025 (±0.001)
Machining:		Electrochem. machine	±0.05 (±0.002)
Drilling, 6 mm (0.25 in)	+0.08/−0.03 (+0.003/−0.001)	Electron beam cutting	±0.08 (±0.003)
		Laser beam cutting	±0.08 (±0.003)
Milling	±0.08 (±0.003)	Plasma arc cutting	±1.3 (±0.050)
Turning	±0.05 (±0.002)		

[a]Compiled from [4], [7], and other sources. For each process category, tolerances vary depending on process parameters. Also, tolerances increase with part size.

TABLE 4.3 Surface roughness values produced by the various manufacturing processes.[a]

Process	Typical Finish	Roughness Range[b]	Process	Typical Finish	Roughness Range[b]
Casting:			Abrasive:		
Die casting	Good	1–2 (30–65)	Grinding	Very good	0.1–2 (5–75)
Investment	Good	1.5–3 (50–100)	Honing	Very good	0.1–1 (4–30)
Sand casting	Poor	12–25 (500–1000)	Lapping	Excellent	0.05–0.5 (2–15)
Metal forming:			Polishing	Excellent	0.1–0.5 (5–15)
Cold rolling	Good	1–3 (25–125)	Superfinish	Excellent	0.02–0.3 (1–10)
Sheet metal draw	Good	1–3 (25–125)	Nontraditional:		
Cold extrusion	Good	1–4 (30–150)	Chemical milling	Medium	1.5–5 (50–200)
Hot rolling	Poor	12–25 (500–1000)	Electrochemical	Good	0.2–2 (10–100)
Machining:			Electric discharge	Medium	1.5–15 (50–500)
Boring	Good	0.5–6 (15–250)	Electron beam	Medium	1.5–15 (50–500)
Drilling	Medium	1.5–6 (60–250)	Laser beam	Medium	1.5–15 (50–500)
Milling	Good	1–6 (30–250)	Thermal:		
Reaming	Good	1–3 (30–125)	Arc welding	Poor	5–25 (250–1000)
Shaping and planing	Medium	1.5–12 (60–500)	Flame cutting	Poor	12–25 (500–1000)
Sawing	Poor	3–25 (100–1000)	Plasma arc cutting	Poor	12–25 (500–1000)
Turning	Good	0.5–6 (15–250)			

[a]Compiled from [1], [2], and other sources.

[b]Roughness range values are given, μm (μ-in). Roughness can vary significantly for a given process, depending on process parameters.

and more time are usually required to obtain increasingly better surfaces. Processes noted for providing superior finishes include honing, lapping, polishing, and superfinishing (Chapter 18). Table 4.3 indicates the usual surface roughness that can be expected from various manufacturing processes.

REFERENCES

[1] American National Standards Institute, Inc. ***Surface Texture***, ANSI B46.1-1978. American Society of Mechanical Engineers, New York, 1978.

[2] American National Standards Institute, Inc. ***Surface Integrity***, ANSI B211.1-1986. Society of Manufacturing Engineers, Dearborn, Michigan, 1986.

[3] American National Standards Institute, Inc. ***Dimensioning and Tolerancing***, ANSI Y14.5M-2009. American Society of Mechanical Engineers, New York, 2009.

[4] Bakerjian, R., and Mitchell, P. ***Tool and Manufacturing Engineers Handbook,*** 4th ed. Vol. VI, ***Design for Manufacturability***. Society of Manufacturing Engineers, Dearborn, Michigan, 1992.

[5] Brown & Sharpe. ***Handbook of Metrology***. North Kingston, Rhode Island, 1992.

[6] Curtis, M. ***Handbook of Dimensional Measurement***, 4th ed. Industrial Press Inc., New York, 2007.

[7] Drozda, T. J., and Wick, C. ***Tool and Manufacturing Engineers Handbook***, 4th ed. Vol. **I**, ***Machining***. Society of Manufacturing Engineers, Dearborn, Michigan, 1983.

[8] Farago, F. T. ***Handbook of Dimensional Measurement***, 3rd ed. Industrial Press Inc., New York, 1994.

[9] ***Machining Data Handbook***, 3rd ed., Vol. II. Machinability Data Center, Cincinnati, Ohio, 1980, Ch. 18.

[10] Mummery, L. ***Surface Texture Analysis—The Handbook***. Hommelwerke GmbH, Germany, 1990.

[11] Oberg, E., Jones, F. D., Horton, H. L., and Ryffel, H. ***Machinery's Handbook***, 26th ed. Industrial Press Inc., New York, 2000.

[12] Schaffer, G. H. "The Many Faces of Surface Texture," Special Report 801, ***American Machinist and Automated Manufacturing***, June 1988 pp. 61–68.

[13] Sheffield Measurement, a Cross & Trecker Company, ***Surface Texture and Roundness Measurement Handbook***, Dayton, Ohio, 1991.

[14] Spitler, D., Lantrip, J., Nee, J., and Smith, D. A. ***Fundamentals of Tool Design***, 5th ed. Society of Manufacturing Engineers, Dearborn, Michigan, 2003.

[15] S. Starrett Company, ***Tools and Rules***. Athol, Massachusetts, 1992.

[16] Wick, C., and Veilleux, R. F. ***Tool and Manufacturing Engineers Handbook***, 4th ed. Vol. **IV**, ***Quality Control and Assembly***. Society of Manufacturing Engineers, Dearborn, Michigan, 1987.

[17] Zecchino, M. "Why Average Roughness Is Not Enough," ***Advanced Materials & Processes***, March 2003, pp. 25–28.

REVIEW QUESTIONS

4.1. What is a tolerance?

4.2. What is the difference between a bilateral tolerance and a unilateral tolerance?

4.3. What is accuracy in measurement?

4.4. What is precision in measurement?

4.5. What are some of the reasons why surfaces are important?

4.6. Define nominal surface.

4.7. Define surface texture.

4.8. How is surface texture distinguished from surface integrity?

4.9. Within the scope of surface texture, how is roughness distinguished from waviness?

4.10. Surface roughness is a measurable aspect of surface texture; what does ***surface roughness*** mean?

4.11. Indicate some of the limitations of using surface roughness as a measure of surface texture.

4.12. What causes the various types of changes that occur in the altered layer just beneath the surface?

4.13. Name some manufacturing processes that produce very poor surface finishes.

4.14. Name some manufacturing processes that produce very good or excellent surface finishes.

APPENDIX A4: MEASUREMENT OF DIMENSIONS AND SURFACES

Measurement is a procedure in which an unknown quantity is compared with a known standard, using an accepted and consistent system of units. Two systems of units have evolved in the world: (1) the U.S. customary system (U.S.C.S.), and (2) the International System of Units (or SI, for Systeme Internationale d'Unites), more popularly known as the metric system. The metric system is widely accepted in nearly every part of the industrialized world except the United States, which has stubbornly clung to its U.S.C.S. Gradually, the United States is adopting SI.

Measurement provides a numerical value of the quantity of interest, within certain limits of accuracy and precision. ***Accuracy*** is the degree to which the measured value agrees with the true value of the quantity of interest. A measurement procedure is accurate when it is absent of systematic errors, which are positive or negative deviations from the true value that are consistent from one measurement to the next. ***Precision*** is the degree of repeatability in the measurement process. Good precision means that random errors in the measurement procedure are minimized. Random errors are usually associated with human participation in the measurement process. Examples include variations in the setup, imprecise reading of the scale, round-off approximations, and so on. Nonhuman contributors to random error include temperature changes, gradual wear and/or misalignment in the working elements of the device, and other variations.

A4.1 CONVENTIONAL MEASURING INSTRUMENTS AND GAGES

In this section of the appendix, we consider the variety of manually operated measuring devices used to evaluate dimensions such as length and diameter, as well as features such as angles, straightness, and roundness. These types of equipment are found in metrology labs, inspection departments, and tool rooms. The logical starting topic is precision gage blocks.

A4.1.1 PRECISION GAGE BLOCKS

Precision gage blocks are the standards against which other dimensional measuring instruments are compared. Gage blocks are usually square or rectangular. The measuring surfaces are finished to be dimensionally accurate and parallel to within several millionths of an inch and are polished to a mirror finish. Several grades of precision gage blocks are available, with closer tolerances for higher precision grades. The highest grade—the ***master laboratory standard***—is made to a tolerance of ±0.000,03 mm (±0.000,001 in). Depending on degree of hardness desired and price the user is willing to pay, gage blocks can be made out of any of several hard materials, including tool steel, chrome-plated steel, chromium carbide, or tungsten carbide.

Precision gage blocks are available in certain standard sizes or in sets, the latter containing a variety of different-sized blocks. The sizes in a set are systematically determined so they can be stacked to achieve virtually any dimension desired to within 0.0025 mm (0.0001 in).

For best results, gage blocks must be used on a flat reference surface, such as a surface plate. A ***surface plate*** is a large solid block whose top surface is finished to a flat plane. Most surface plates today are made of granite. Granite has the advantage of being hard, nonrusting, nonmagnetic, long wearing, thermally stable, and easy to maintain.

Gage blocks and other high-precision measuring instruments must be used under standard conditions of temperature and other factors that might adversely affect the measurement. By international agreement, 20°C (68°F) has been established as the standard temperature. Metrology labs operate at this standard. If gage blocks or other measuring instruments are used in a factory environment where the temperature differs from this standard, corrections for thermal expansion or contraction may be required. Also, working gage blocks used for inspection in the shop are subject to wear and must be calibrated periodically against more precise laboratory gage blocks.

A4.1.2 MEASURING INSTRUMENTS FOR LINEAR DIMENSIONS

Measuring instruments can be divided into two types: graduated and nongraduated. ***Graduated measuring devices*** include a set of markings (called ***graduations***) on a linear or angular scale to which the object's feature of interest can be compared for measurement. ***Nongraduated measuring devices*** possess no such scale and are used to make comparisons between dimensions or to transfer a dimension for measurement by a graduated device.

The most basic of the graduated measuring devices is the ***rule*** (made of steel, and often called a ***steel rule***), used to measure linear dimensions. Rules are available in various lengths. Metric rule lengths include 150, 300, 600, and 1000 mm, with graduations of 1 or 0.5 mm. Common U.S. sizes are 6, 12, and 24 in, with graduations of 1/32, 1/64, or 1/100 in.

Calipers are available in either nongraduated or graduated styles. A nongraduated caliper (referred to simply as a ***caliper***) consists of two legs joined by a hinge mechanism, as in Figure A4.1. The ends of the legs are made to contact the surfaces of the object being measured, and the hinge is designed to hold the legs in position during use. The contacts point either inward or outward. When they point inward, as in Figure A4.1, the instrument is an ***outside caliper*** and is used for measuring outside dimensions such as a diameter. When the contacts point outward, it is an ***inside caliper***, which is used to measure the distance between two internal surfaces. An instrument similar in configuration to the caliper is a ***divider***, except that both legs are straight and terminate in hard,

FIGURE A4.1 Two sizes of outside calipers. (Courtesy of L.S. Starrett Co.) (Credit: *Fundamentals of Modern Manufacturing*, 4th Edition by Mikell P. Groover, 2010. Reprinted with permission of John Wiley & Sons, Inc.)

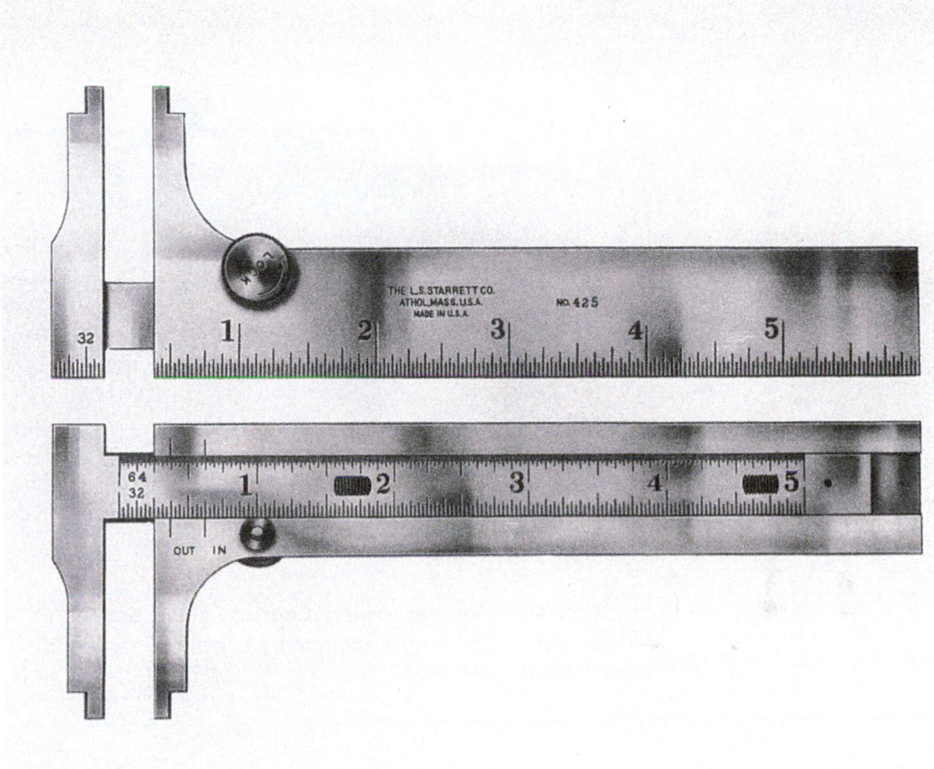

FIGURE A4.2 Slide caliper, opposite sides of instrument shown. (Courtesy of L.S. Starrett Co.) (Credit: *Fundamentals of Modern Manufacturing*, 4th Edition by Mikell P. Groover, 2010. Reprinted with permission of John Wiley & Sons, Inc.)

sharply pointed contacts. Dividers are used for scaling distances between two points or lines on a surface, and for scribing circles or arcs onto a surface.

A variety of graduated calipers are available for various measurement purposes. The simplest is the ***slide caliper***, which consists of a steel rule to which two jaws are added, one fixed at the end of the rule and the other movable, shown in Figure A4.2. Slide calipers can be used for inside or outside measurements, depending on whether the inside or outside jaw faces are used. In use, the jaws are forced into contact with the part surfaces to be measured, and the location of the movable jaw indicates the dimension of interest. Slide calipers permit more accurate and precise measurements than simple rules. A refinement of the slide caliper is the ***vernier caliper***, shown in Figure A4.3. In this device, the movable jaw includes a vernier scale, named after P. Vernier (1580–1637), a French mathematician who invented it. The vernier provides graduations of 0.01 mm in the SI (and 0.001 inch in the U.S. customary scale), much more precise than the slide caliper.

The ***micrometer*** is a widely used and very accurate measuring device, the most common form of which consists of a spindle and a C-shaped anvil, as in Figure A4.4. The spindle is moved relative to the fixed anvil by means of an accurate screw thread. On a typical U.S. micrometer, each rotation of the spindle provides 0.025 in of linear travel. Attached to the spindle is a thimble graduated with 25 marks around its circumference, each mark corresponding to 0.001 in. The micrometer sleeve is usually equipped with a vernier, allowing resolutions as close as 0.0001 in. On a micrometer with metric scale, graduations are 0.01 mm. Modern micrometers (and graduated calipers) are available

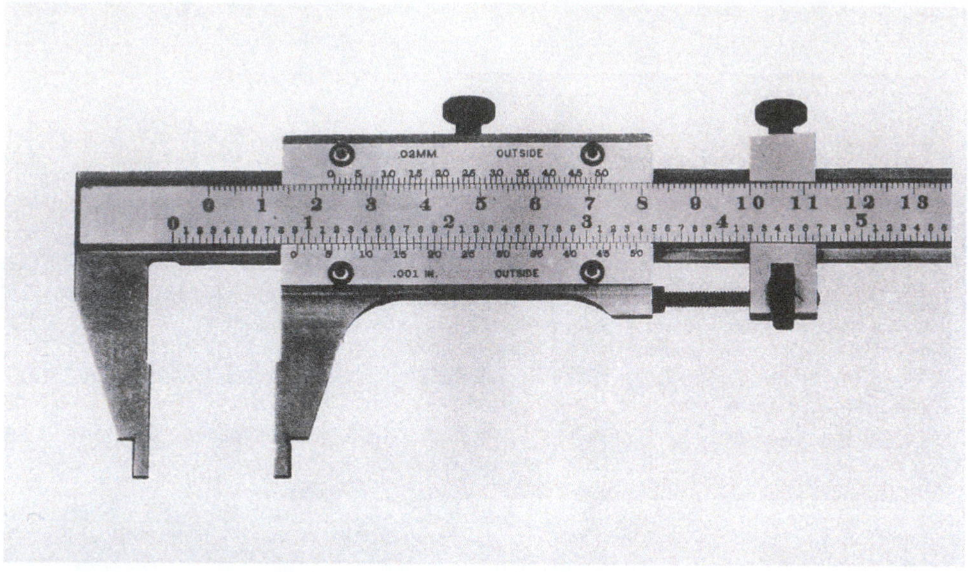

FIGURE A4.3 Vernier caliper. (Courtesy of L.S. Starrett Co.) (Credit: *Fundamentals of Modern Manufacturing,* 4th Edition by Mikell P. Groover, 2010. Reprinted with permission of John Wiley & Sons, Inc.)

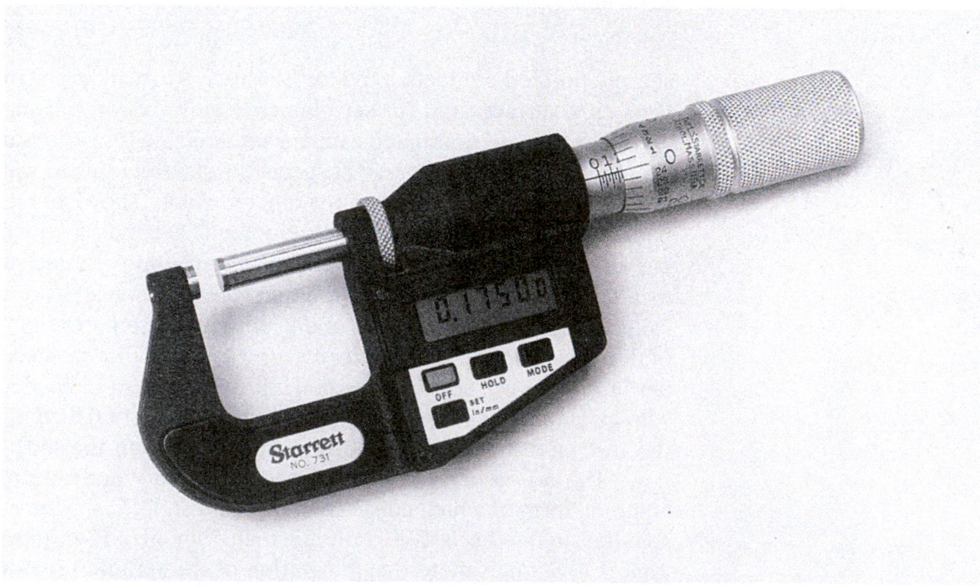

FIGURE A4.4 External micrometer, standard 1-in size with digital readout. (Courtesy of L. S. Starrett Co.) (Credit: *Fundamentals of Modern Manufacturing,* 4th Edition by Mikell P. Groover, 2010. Reprinted with permission of John Wiley & Sons, Inc.)

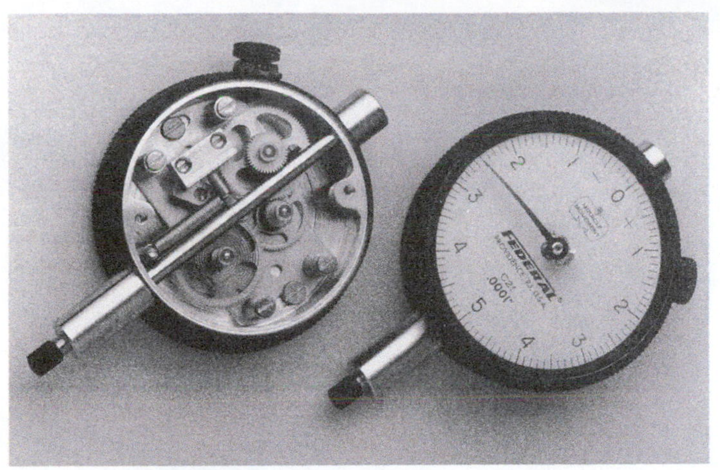

FIGURE A4.5 Dial indicator: right view shows dial and graduated face; left view shows rear of instrument with cover plate removed. (Courtesy of Federal Products Co., Providence, RI.) (Credit: *Fundamentals of Modern Manufacturing*, 4th Edition by Mikell P. Groover, 2010. Reprinted with permission of John Wiley & Sons, Inc.)

with electronic devices that display a digital readout of the measurement (as in our figure). These instruments are easier to read and eliminate much of the human error associated with reading conventional graduated devices.

The most common micrometer types are (1) ***external micrometer***, Figure A4.4, also called an ***outside micrometer***, which comes in a variety of standard anvil sizes; (2) ***internal micrometer***, or ***inside micrometer***, which consists of a head assembly and a set of rods of different lengths to measure various inside dimensions that might be encountered; and (3) ***depth micrometer***, similar to an inside micrometer but adapted to measure hole depths.

A4.1.3 COMPARATIVE INSTRUMENTS

Comparative instruments are used to make dimensional comparisons between two objects, such as a workpart and a reference surface. They are usually not capable of providing an absolute measurement of the quantity of interest; instead, they measure the magnitude and direction of the deviation between two objects. Instruments in this category include mechanical and electronic gages.

Mechanical Gages: Dial Indicators ***Mechanical gages*** are designed to mechanically magnify the deviation to permit observation. The most common instrument in this category is the ***dial indicator***, Figure A4.5, which converts and amplifies the linear movement of a contact pointer into rotation of a dial needle. The dial is graduated in small units such as 0.01 mm (or 0.001 in). Dial indicators are used in many applications to measure straightness, flatness, parallelism, squareness, roundness, and runout. A typical setup for measuring runout is illustrated in Figure A4.6.

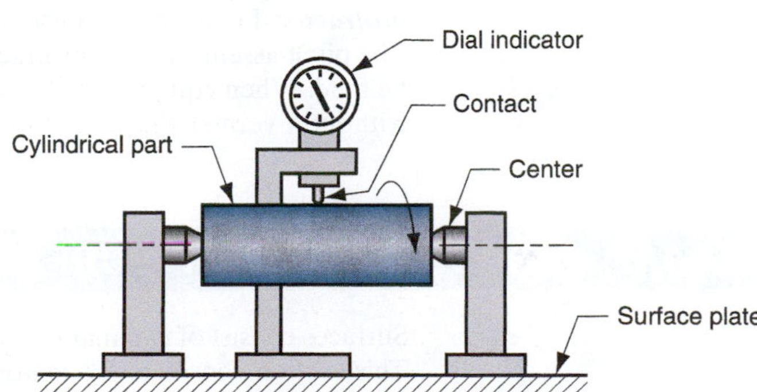

FIGURE A4.6 Dial indicator setup to measure runout; as part is rotated about its center, variations in outside surface relative to center are indicated on the dial.

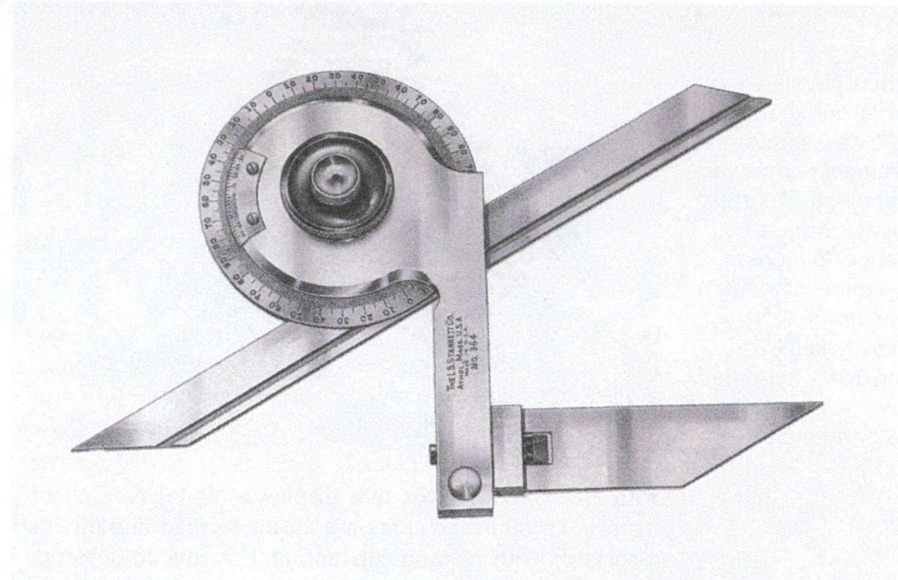

FIGURE A4.7 Bevel protractor with vernier scale. (Courtesy of L.S. Starrett Co.) (Credit: *Fundamentals of Modern Manufacturing*, 4th Edition by Mikell P. Groover, 2010. Reprinted with permission of John Wiley & Sons, Inc.)

Electronic Gages Electronic gages are a family of measuring instruments based on transducers capable of converting a linear displacement into an electrical signal. The electrical signal is then amplified and transformed into a suitable data format such as a digital readout, as in Figure A4.4. Applications of electronic gages have grown rapidly in recent years, driven by advances in microprocessor technology. They are gradually replacing many of the conventional measuring devices. Advantages of electronic gages include (1) good sensitivity, accuracy, precision, repeatability, and speed of response; (2) ability to sense very small dimensions—down to 0.025 μm (1 μ-in.); (3) ease of operation; (4) reduced human error; (5) electrical signal that can be displayed in various formats; and (6) capability to be interfaced with computer systems for data processing.

A4.1.4 ANGULAR MEASUREMENTS

Angles can be measured using any of several styles of ***protractor***. A ***simple protractor*** consists of a blade that pivots relative to a semicircular head that is graduated in angular units (e.g., degrees, radians). To use, the blade is rotated to a position corresponding to some part angle to be measured, and the angle is read off the angular scale. A ***bevel protractor***, Figure A4.7, consists of two straight blades that pivot relative to each other. The pivot assembly has a protractor scale that permits the angle formed by the blades to be read. When equipped with a vernier, the bevel protractor can be read to about 5 min; without a vernier the resolution is only about 1 degree.

A4.2 MEASUREMENT OF SURFACES

Surfaces consist of two main characteristics: (1) surface texture and (2) surface integrity. This section considers the measurement of these characteristics.

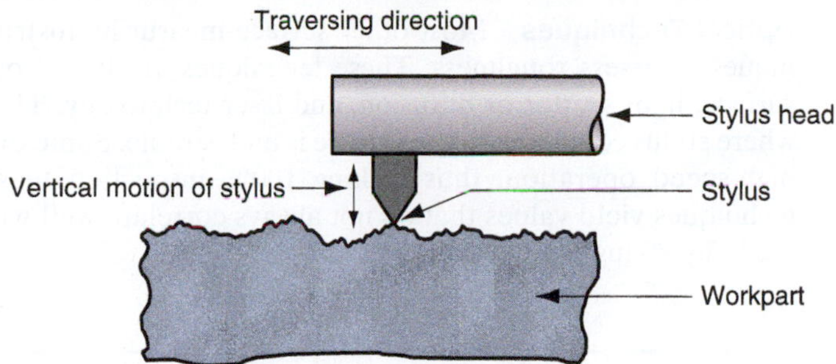

FIGURE A4.8 Sketch illustrating the operation of stylus-type instrument. Stylus head traverses horizontally across surface, while stylus moves vertically to follow surface profile. Vertical movement is converted into either (1) a profile of the surface or (2) the average roughness value. (Credit: *Fundamentals of Modern Manufacturing*, 4th Edition by Mikell P. Groover, 2010. Reprinted with permission of John Wiley & Sons, Inc.)

A4.2.1 MEASUREMENT OF SURFACE ROUGHNESS

The various methods are used to assess surface roughness can be divided into three categories: (1) subjective comparison with standard test surfaces, (2) stylus electronic instruments, and (3) optical techniques.

Standard Test Surfaces Sets of standard surface finish blocks are available, produced to specified roughness values.[1] To estimate the roughness of a given test specimen, the surface is compared with the standard visually and by the "fingernail test." In this test, the user gently scratches the surfaces of the specimen and the standards, judging which standard is closest to the specimen. Standard test surfaces are a convenient way for a machine operator to obtain an estimate of surface roughness. They are also useful for design engineers in judging what value of surface roughness to specify on a part drawing.

Stylus Instruments The fingernail test is subjective. Several stylus-type instruments are commercially available to measure surface roughness—similar to the fingernail test, but more scientific. In these electronic devices, a cone-shaped diamond stylus with point radius of about 0.005 mm (0.0002 in) and 90° tip angle is traversed across the test surface at a constant slow speed. The operation is depicted in Figure A4.8. As the stylus head is traversed horizontally, it also moves vertically to follow the surface deviations. The vertical movement is converted into an electronic signal that represents the topography of the surface. This can be displayed as either a profile of the actual surface or an average roughness value. ***Profiling devices*** use a separate flat plane as the nominal reference against which deviations are measured. The output is a plot of the surface contour along the line traversed by the stylus. This type of system can identify both roughness and waviness in the test surface. ***Averaging devices*** reduce the roughness deviations to a single value R_a. They use skids riding on the actual surface to establish the nominal reference plane. The skids act as a mechanical filter to reduce the effect of waviness in the surface; in effect, these averaging devices electronically perform the computations in Eq. (4.1).

[1] In the U.S.C.S., the typical blocks have surfaces with roughness values of 2, 4, 8, 16, 32, 64, and 128 microinches.

Optical Techniques Most other surface-measuring instruments employ optical techniques to assess roughness. These techniques are based on light reflectance from the surface, light scatter or diffusion, and laser technology. They are useful in applications where stylus contact with the surface is undesirable. Some of the techniques permit very-high-speed operation, thus making 100% inspection feasible. However, the optical techniques yield values that do not always correlate well with roughness measurements made by stylus-type instruments.

A4.2.2 EVALUATION OF SURFACE INTEGRITY

Surface integrity is more difficult to assess than surface roughness. Some of the techniques to inspect for subsurface changes are destructive to the material specimen. Evaluation techniques for surface integrity include the following:

- ***Surface texture***. Surface roughness, designation of lay, and other measures provide superficial data on surface integrity. This type of testing is relatively simple to perform and is always included in the evaluation of surface integrity.
- ***Visual examination***. Visual examination can reveal various surface flaws such as cracks, craters, laps, and seams. This type of assessment is often augmented by fluorescent and photographic techniques.
- ***Microstructural examination***. This involves standard metallographic techniques for preparing cross sections and obtaining photomicrographs to examine microstructure in the surface layers compared with the substrate.
- ***Microhardness profile***. Hardness differences near the surface can be detected using microhardness measurement techniques such as Knoop and Vickers (Section 3.2.1). The part is sectioned, and hardness is plotted against distance below the surface to obtain a hardness profile of the cross section.
- ***Residual stress profile***. X-ray diffraction techniques can be employed to measure residual stresses in the surface layers of a part.

Part II Solidification Processes

5 FUNDAMENTALS OF METAL CASTING

Chapter Contents

5.1 Overview of Casting Technology
- 5.1.1 Casting Processes
- 5.1.2 Sand-Casting Molds

5.2 Heating and Pouring
- 5.2.1 Heating the Metal
- 5.2.2 Pouring the Molten Metal
- 5.2.3 Engineering Analysis of Pouring

5.3 Solidification and Cooling
- 5.3.1 Solidification of Metals
- 5.3.2 Solidification Time
- 5.3.3 Shrinkage
- 5.3.4 Directional Solidification
- 5.3.5 Riser Design

In this part of the book, we consider those manufacturing processes in which the starting work material is either a liquid or is in a highly plastic condition, and a part is created through solidification of the material. Casting and molding processes dominate this category of shaping operations. The solidification processes can be classified according to the engineering material that is processed: (1) metals, (2) ceramics, specifically glasses,[1] and (3) polymers and polymer-matrix composites (PMCs). Casting of metals is covered in this and the following chapter. Glassworking is covered in Chapter 7, and polymer and PMC processing are treated in Chapters 8 and 9.

Casting is a process in which molten metal flows by gravity or other force into a mold where it solidifies in the shape of the mold cavity. The term ***casting*** is also applied to the part that is made by this process. It is one of the oldest shaping processes, dating back 6,000 years. The principle of casting seems simple: Melt the metal, pour it into a mold, and let it cool and solidify; yet there are many factors and variables that must be considered in order to accomplish a successful casting operation.

Casting includes both the casting of ingots and the casting of shapes. The term ***ingot*** is usually associated with the primary metals industries; it describes a large casting that is simple in shape and intended for subsequent reshaping by processes such as rolling or forging. ***Shape casting***

[1] Among the ceramics, only glass is processed by solidification; traditional and new ceramics are shaped using particulate processes (Chapter 11).

involves the production of more complex geometries that are much closer to the final desired shape of the part or product. It is with the casting of shapes rather than ingots that this chapter and the next are concerned.

A variety of shape casting methods are available, thus making it one of the most versatile of all manufacturing processes. Among its capabilities and advantages are the following:

- Casting can be used to create complex part geometries, including both external and internal shapes.
- Some casting processes are capable of producing parts to ***net shape***. No further manufacturing operations are required to achieve the required geometry and dimensions of the parts. Other casting processes are ***near net shape***, for which some additional shape processing is required (usually machining) in order to achieve accurate dimensions and details.
- Casting can be used to produce very large parts. Castings weighing more than 100 tons have been made.
- The casting process can be performed on any metal that can be heated to the liquid state.
- Some casting methods are quite suited to mass production.

There are also disadvantages associated with casting—different disadvantages for different casting methods. These include limitations on mechanical properties, porosity, poor dimensional accuracy and surface finish for some casting processes, safety hazards to humans when processing hot molten metals, and environmental problems.

Parts made by casting processes range in size from small components weighing only a few ounces up to very large products weighing tons. The list of parts includes dental crowns, jewelry, statues, wood-burning stoves, engine blocks and heads for automotive vehicles, machine frames, railway wheels, frying pans, pipes, and pump housings. All varieties of metals can be cast, ferrous and nonferrous.

Casting can also be used on other materials such as polymers and ceramics; however, the details are sufficiently different that we postpone discussion of the casting processes for these materials until later chapters. This chapter and the next deal exclusively with metal casting. Here we discuss the fundamentals that apply to virtually all casting operations. In the following chapter, the individual casting processes are described, along with some of the product design issues that must be considered when making parts out of castings.

5.1 OVERVIEW OF CASTING TECHNOLOGY

As a production process, casting is usually carried out in a foundry. A ***foundry*** is a factory equipped for making molds, melting and handling metal in molten form, performing the casting process, and cleaning the finished casting. The workers who perform the casting operations in these factories are called ***foundrymen***.

5.1.1 CASTING PROCESSES

Discussion of casting logically begins with the mold. The ***mold*** contains a cavity whose geometry determines the shape of the cast part. The actual size and shape of the cavity must be slightly oversized to allow for shrinkage that occurs in the metal during solidification and cooling. Different metals undergo different amounts of shrinkage,

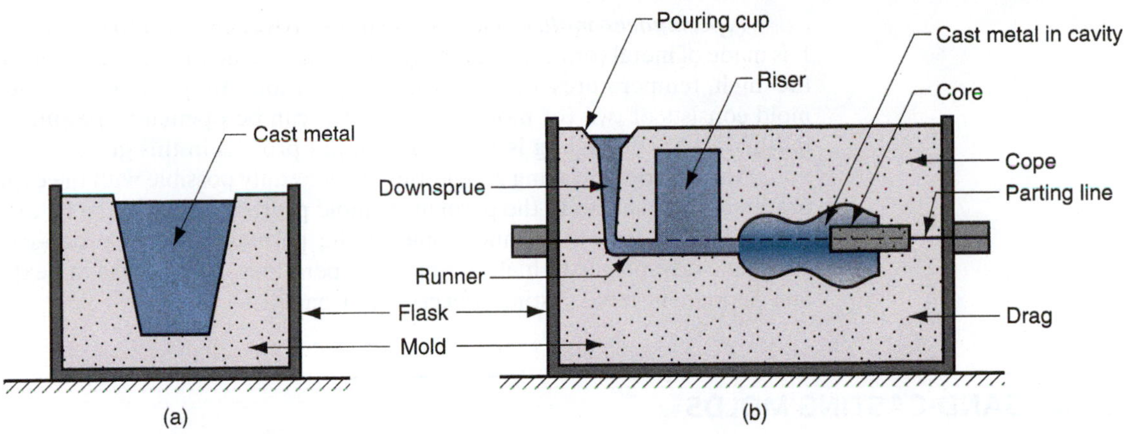

FIGURE 5.1 Two forms of mold: (a) open mold, simply a container in the shape of the desired part; and (b) closed mold, in which the mold geometry is more complex and requires a gating system (passageway) leading into the cavity. (Credit: *Fundamentals of Modern Manufacturing*, 4th Edition by Mikell P. Groover, 2010. Reprinted with permission of John Wiley & Sons, Inc.)

so the mold cavity must be designed for the particular metal to be cast if dimensional accuracy is critical. Molds are made of a variety of materials, including sand, plaster, ceramic, and metal. The various casting processes are often classified according to these different types of molds.

To accomplish a casting operation, the metal is first heated to a temperature high enough to completely transform it into a liquid state. It is then poured, or otherwise directed, into the cavity of the mold. In an ***open mold***, Figure 5.1(a), the liquid metal is simply poured until it fills the open cavity. In a ***closed mold***, Figure 5.1(b), a passageway, called the gating system, is provided to permit the molten metal to flow from outside the mold into the cavity. The closed mold is by far the more important category in production casting operations.

As soon as the molten metal is in the mold, it begins to cool. When the temperature drops sufficiently (e.g., to the freezing point for a pure metal), solidification begins. Solidification involves a change of phase of the metal. Time is required to complete the phase change, and considerable heat is given up in the process. It is during this step in the process that the metal assumes the solid shape of the mold cavity and many of the properties and characteristics of the casting are established.

Once the casting has cooled sufficiently, it is removed from the mold. Depending on the casting method and metal used, further processing may be required. This may include trimming the excess metal from the actual cast part, cleaning the surface, inspecting the product, and heat treatment to enhance properties. In addition, machining (Chapter 16) may be required to achieve closer tolerances on certain part features and to remove the cast surface.

Casting processes divide into two broad categories, according to type of mold used: expendable-mold casting and permanent-mold casting. An ***expendable-mold*** means that the mold in which the molten metal solidifies must be destroyed in order to remove the casting. These molds are made out of sand, plaster, or similar materials, whose form is maintained by using binders of various kinds. Sand casting is the most important example of expendable-mold processes. In sand casting, the liquid metal is poured into a mold made of sand. After the metal hardens, the mold must be sacrificed in order to recover the casting.

A ***permanent-mold*** is one that can be used over and over to produce many castings. It is made of metal (or, less commonly, a ceramic refractory material) that can withstand the high temperatures of the casting operation. In permanent-mold casting, the mold consists of two (or more) sections that can be opened to permit removal of the finished part. Die casting is the most familiar process in this group.

More intricate casting geometries are generally possible with the expendable-mold processes. Part shapes in the permanent-mold processes are limited by the need to open the mold. On the other hand, some of the permanent-mold processes have certain economic advantages in high production operations. We discuss the expendable-mold and permanent-mold casting processes in Chapter 6.

5.1.2 SAND-CASTING MOLDS

Sand casting is by far the most important casting process. A sand-casting mold will be used to describe the basic features of a mold. Many of these features and terms are common to the molds used in other casting processes. Figure 5.1(b) shows the cross-sectional view of a typical sand-casting mold, indicating some of the terminology. The mold consists of two halves: cope and drag. The ***cope*** is the upper half of the mold, and the ***drag*** is the bottom half. These two mold parts are contained in a box, called a ***flask***, which is also divided into two halves, one for the cope and the other for the drag. The two halves of the mold separate at the ***parting line***.

In sand casting (and other expendable-mold processes) the mold cavity is formed by means of a ***pattern***, which is made of wood, metal, plastic, or other material and has the shape of the part to be cast. The cavity is formed by packing sand around the pattern, about half each in the cope and drag, so that when the pattern is removed, the remaining void has the desired shape of the cast part. The pattern is usually made oversized to allow for shrinkage of the metal as it solidifies and cools. The sand for the mold is moist and contains a binder to maintain its shape.

The cavity in the mold provides the external surfaces of the cast part. In addition, a casting may have internal surfaces. These surfaces are determined by means of a ***core***, a form placed inside the mold cavity to define the interior geometry of the part. In sand casting, cores are generally made of sand, although other materials can be used, such as metals, plaster, and ceramics.

The ***gating system*** in a casting mold is the channel, or network of channels, by which molten metal flows into the cavity from outside the mold. As shown in the figure, the gating system typically consists of a ***downsprue*** (also called simply the ***sprue***), through which the metal enters a ***runner*** that leads into the main cavity. At the top of the downsprue, a ***pouring cup*** is often used to minimize splash and turbulence as the metal flows into the downsprue. It is shown in our diagram as a simple cone-shaped funnel. Some pouring cups are designed in the shape of a bowl, with an open channel leading to the downsprue.

In addition to the gating system, any casting in which shrinkage is significant requires a riser connected to the main cavity. The ***riser*** is a reservoir in the mold that serves as a source of liquid metal for the casting to compensate for shrinkage during solidification. The riser must be designed to freeze after the main casting in order to satisfy its function.

As the metal flows into the mold, the air that previously occupied the cavity, as well as hot gases formed by reactions of the molten metal, must be evacuated so that the metal will completely fill the empty space. In sand casting, for example, the natural porosity of the sand mold permits air and gases to escape through the walls of the cavity. In permanent metal molds, small vent holes are drilled into the mold or machined into the parting line to permit removal of air and gases.

5.2 HEATING AND POURING

To perform a casting operation, the metal must be heated to a temperature somewhat above its melting point and then poured into the mold cavity to solidify. In this section, we consider several aspects of these two steps in casting.

5.2.1 HEATING THE METAL

Heating furnaces of various kinds (Section 6.4.1) are used to heat the metal to a molten temperature sufficient for casting. The heat energy required is the sum of (1) the heat to raise the temperature to the melting point, (2) the heat of fusion to convert it from solid to liquid, and (3) the heat to raise the molten metal to the desired temperature for pouring. This can be expressed:

$$H = \rho V\{C_s(T_m - T_o) + H_f + C_l(T_p - T_m)\} \tag{5.1}$$

where H = total heat required to raise the temperature of the metal to the pouring temperature, J (Btu); ρ = density, g/cm^3 (lbm/in^3); C_s = weight specific heat for the solid metal, J/g-C (Btu/lbm-F); T_m = melting temperature of the metal, °C (°F); T_o = starting temperature—usually ambient, °C (°F); H_f = heat of fusion, J/g (Btu/lbm); C_l = weight specific heat of the liquid metal, J/g-C (Btu/lbm-F); T_p = pouring temperature, °C (°F); and V = volume of metal being heated, cm^3 (in^3).

Equation (5.1) is of conceptual value, but its computational value is limited because of the following factors: (1) Specific heat and other thermal properties of a solid metal vary with temperature, especially if the metal undergoes a change of phase during heating. (2) A metal's specific heat may be different in the solid and liquid states. (3) Most casting metals are alloys, and most alloys melt over a temperature range between a solidus and liquidus rather than at a single melting point; thus, the heat of fusion cannot be applied so simply as indicated above. (4) The property values required in the equation for a particular alloy are not readily available in most cases. (5) There are significant heat losses to the environment during heating.

5.2.2 POURING THE MOLTEN METAL

After heating, the metal is ready for pouring. Introduction of molten metal into the mold, including its flow through the gating system and into the cavity, is a critical step in the casting process. For this step to be successful, the metal must flow into all regions of the mold before solidifying. Factors affecting the pouring operation include pouring temperature, pouring rate, and turbulence.

The ***pouring temperature*** is the temperature of the molten metal as it is introduced into the mold. What is important here is the difference between the temperature at pouring and the temperature at which freezing begins (the melting point for a pure metal or the liquidus temperature for an alloy). This temperature difference is sometimes referred to as the ***superheat***. This term is also used for the amount of heat that must be removed from the molten metal between pouring and when solidification commences [7].

Pouring rate refers to the volumetric rate at which the molten metal is poured into the mold. If the rate is too slow, the metal will chill and freeze before filling the cavity. If the pouring rate is excessive, turbulence can become a serious problem. ***Turbulence*** in fluid flow is characterized by erratic variations in the magnitude and direction of the velocity throughout the fluid. The flow is agitated and irregular rather than smooth and streamlined, as in laminar flow. Turbulent flow should be avoided during pouring for several reasons. It tends to accelerate the formation of metal oxides that can become entrapped during solidification, thus degrading the quality of the casting. Turbulence also aggravates ***mold erosion***, the gradual wearing away of the mold surfaces due to impact of the flowing molten metal. The densities of most molten metals are much higher than water and other fluids we normally deal with. These molten metals are also much more chemically reactive than at room temperature. Consequently, the wear caused by the flow of these metals in the mold is significant, especially under turbulent conditions. Erosion is especially serious when it occurs in the main cavity because the geometry of the cast part is affected.

5.2.3 ENGINEERING ANALYSIS OF POURING

There are several relationships that govern the flow of liquid metal through the gating system and into the mold. An important relationship is ***Bernoulli's theorem***, which states that the sum of the energies (head, pressure, kinetic, and friction) at any two points in a flowing liquid are equal. This can be written in the following form:

$$h_1 + \frac{p_1}{\rho} + \frac{v_1^2}{2g} + F_1 = h_2 + \frac{p^2}{\rho} + \frac{v_2^2}{2g} + F_2 \tag{5.2}$$

where h = head, cm (in), p = pressure on the liquid, N/cm^2 (lb/in^2); ρ = density, g/cm^3 (lbm/in^3); v = flow velocity, cm/s (in/sec); g = gravitational acceleration constant, 981 cm/s/s (32.2 × 12 = 386 in/sec/sec); and F = head losses due to friction, cm (in). Subscripts 1 and 2 indicate any two locations in the liquid flow.

Bernoulli's equation can be simplified in several ways. If we ignore friction losses (to be sure, friction will affect the liquid flow through a sand mold), and assume that the system remains at atmospheric pressure throughout, then the equation can be reduced to

$$h_1 + \frac{v_1^2}{2g} = h_2 + \frac{v_2^2}{2g} \tag{5.3}$$

This can be used to determine the velocity of the molten metal at the base of the sprue. Let us define point 1 at the top of the sprue and point 2 at its base. If point 2 is used as the reference plane, then the head at that point is zero ($h_2 = 0$) and h_1 is the height (length) of the sprue. When the metal is poured into the pouring cup and overflows down the sprue, its initial velocity at the top is zero ($v_1 = 0$). Hence, Eq. (5.3) further simplifies to

$$h_1 = \frac{v_2^2}{2g}$$

which can be solved for the flow velocity:

$$v = \sqrt{2gh} \tag{5.4}$$

where v = the velocity of the liquid metal at the base of the sprue, cm/s (in/sec); g = 981 cm/s/s (386 in/sec/sec); and h = the height of the sprue, cm (in).

Another relationship of importance during pouring is the ***continuity law***, which states that the volume rate of flow remains constant throughout the liquid. The volume flow rate is equal to the velocity multiplied by the cross-sectional area of the flowing liquid. The continuity law can be expressed:

$$Q = v_1A_1 = v_2A_2 \tag{5.5}$$

where Q = volumetric flow rate, cm^3/s (in^3/sec); v = velocity as before; A = cross-sectional area of the liquid, cm^2 (in^2); and the subscripts refer to any two points in the flow system. Thus, an increase in area results in a decrease in velocity, and vice versa.

Equations (5.4) and (5.5) indicate that the sprue should be tapered. As the metal accelerates during its descent into the sprue opening, the cross-sectional area of the channel must be reduced; otherwise, as the velocity of the flowing metal increases toward the base of the sprue, air can be aspirated into the liquid and conducted into the mold cavity. To prevent this condition, the sprue is designed with a taper, so that the volume flow rate vA is the same at the top and bottom of the sprue.

Assuming that the runner from the sprue base to the mold cavity is horizontal (and therefore the head h is the same as at the sprue base), then the volume rate of flow through the gate and into the mold cavity remains equal to vA at the base. Accordingly, we can estimate the time required to fill a mold cavity of volume V as

$$T_{MF} = \frac{V}{Q} \tag{5.6}$$

where T_{MF} = mold filling time, s (sec); V = volume of mold cavity, cm^3 (in^3); and Q = volume flow rate, as before. The mold filling time computed by Eq. (5.6) must be considered a minimum time. This is because the analysis ignores friction losses and possible constriction of flow in the gating system; thus, the mold filling time will be longer than what is given by Eq. (5.6).

Example 5.1 Pouring Calculations

A mold sprue is 20 cm long, and the cross-sectional area at its base is 2.5 cm^2. The sprue feeds a horizontal runner leading into a mold cavity whose volume is 1560 cm^3. Determine: (a) velocity of the molten metal at the base of the sprue, (b) volume rate of flow, and (c) time to fill the mold.

Solution: (a) The velocity of the flowing metal at the base of the sprue is given by Eq. (5.4):

$$v = \sqrt{2(981)(20)} = 198.1 \text{ cm/s}$$

(b) The volumetric flow rate is

$$Q = (2.5 \text{ cm}^2)(198.1 \text{ cm/s}) = 495 \text{ cm}^3/\text{s}$$

(c) Time required to fill a mold cavity of 1560 cm^3 at this flow rate is

$$T_{MF} = 1560/495 = 3.2\text{s}$$

■

5.3 SOLIDIFICATION AND COOLING

After pouring into the mold, the molten metal cools and solidifies. In this section we examine the physical mechanism of solidification that occurs during casting. Issues associated with solidification include the time for a metal to freeze, shrinkage, directional solidification, and riser design.

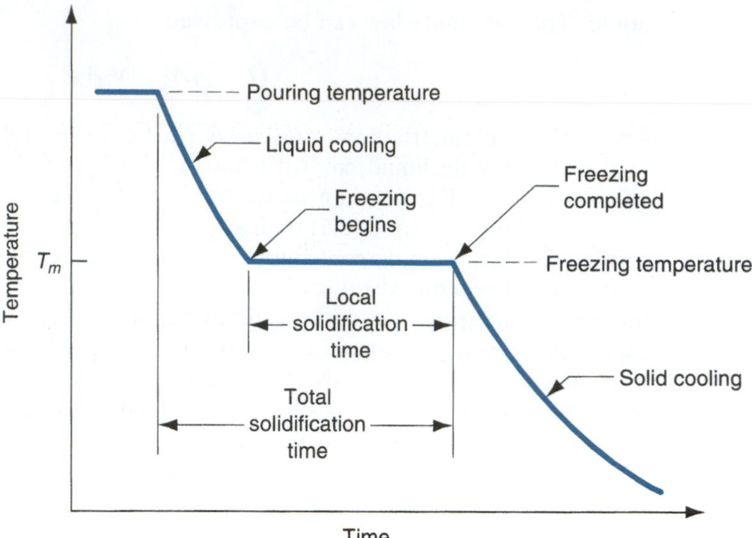

FIGURE 5.2 Cooling curve for a pure metal during casting. (Credit: *Fundamentals of Modern Manufacturing*, 4th Edition by Mikell P. Groover, 2010. Reprinted with permission of John Wiley & Sons, Inc.)

5.3.1 SOLIDIFICATION OF METALS

Solidification involves the transformation of the molten metal back into the solid state. The solidification process differs depending on whether the metal is a pure element or an alloy.

Pure Metals A pure metal solidifies at a constant temperature equal to its freezing point, which is the same as its melting point. The melting points of pure metals are well known and documented (Table 3.10). The process occurs over time as shown in the plot of Figure 5.2, called a cooling curve. The actual freezing takes time, called the ***local solidification time*** in casting, during which the metal's latent heat of fusion is released into the surrounding mold. The ***total solidification time*** is the time taken between pouring and complete solidification. After the casting has completely solidified, cooling continues at a rate indicated by the downward slope of the cooling curve.

Because of the chilling action of the mold wall, a thin skin of solid metal is initially formed at the interface immediately after pouring. Thickness of the skin increases to form a shell around the molten metal as solidification progresses inward toward the center of the cavity. The rate at which freezing proceeds depends on heat transfer into the mold, as well as the thermal properties of the metal.

It is of interest to examine the metallic grain formation and growth during this solidification process. The metal that forms the initial skin has been rapidly cooled by the extraction of heat through the mold wall. This cooling action causes the grains in the skin to be fine and randomly oriented. As cooling continues, further grain formation and growth occur in a direction away from the heat transfer. Since the heat transfer is through the skin and mold wall, the grains grow inwardly as needles or spines of solid metal. As these spines enlarge, lateral branches form, and as these branches grow, further branches form at right angles to the first branches. This type of grain growth is referred to as ***dendritic growth***, and it occurs not only in the freezing of pure metals but alloys as well. These treelike structures are gradually filled in during freezing, as additional metal is continually deposited onto the dendrites until complete solidification has occurred. The grains resulting from this dendritic growth take on a preferred orientation, tending to be coarse, columnar grains aligned toward the center of the casting. The resulting grain formation is illustrated in Figure 5.3.

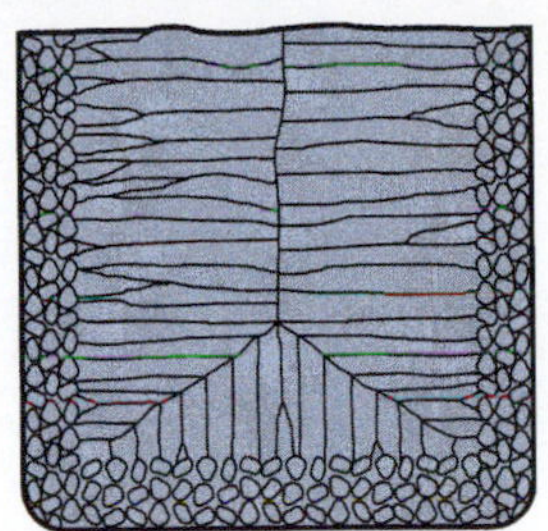

FIGURE 5.3 Characteristic grain structure in a casting of a pure metal, showing randomly oriented grains of small size near the mold wall, and large columnar grains oriented toward the center of the casting. (Credit: *Fundamentals of Modern Manufacturing*, 4th Edition by Mikell P. Groover, 2010. Reprinted with permission of John Wiley & Sons, Inc.)

Most Alloys Most alloys freeze over a temperature range rather than at a single temperature. The exact range depends on the alloy system and the particular composition. Solidification of an alloy can be explained with reference to Figure 5.4, which shows the phase diagram for a particular alloy system and the cooling curve for a given composition. As temperature drops, freezing begins at the temperature indicated by the ***liquidus*** and is completed when the ***solidus*** is reached. The start of freezing is similar to that of the pure metal. A thin skin is formed at the mold wall due to the large temperature gradient at this surface. Freezing then progresses as before through the formation of dendrites that grow away from the walls. However, owing to the temperature spread between the liquidus and solidus, the nature of the dendritic growth is such that an advancing zone is formed in which both liquid and solid metal coexist. The solid portions are the dendrite structures that have formed sufficiently to trap small islands of liquid metal in the matrix. This solid–liquid region has a soft consistency that has motivated its name as the ***mushy zone***. Depending on the conditions of freezing, the mushy zone can be relatively narrow, or it can exist throughout most of the casting. The latter condition is promoted by factors such as slow heat transfer out of the hot metal and a wide difference between liquidus and solidus temperatures. Gradually, the liquid islands in the dendrite matrix solidify as the temperature of the casting drops to the solidus for the given alloy composition.

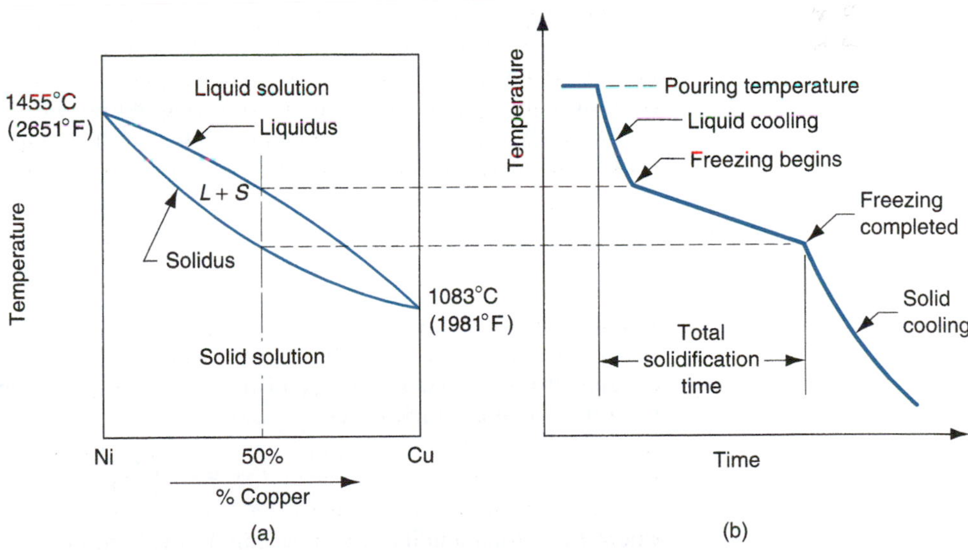

FIGURE 5.4 (a) Phase diagram for a copper–nickel alloy system and (b) associated cooling curve for a 50%Ni–50%Cu composition during casting. (Credit: *Fundamentals of Modern Manufacturing*, 4th Edition by Mikell P. Groover, 2010. Reprinted with permission of John Wiley & Sons, Inc.)

FIGURE 5.5 Characteristic grain structure in an alloy casting, showing segregation of alloying components in the center of casting. (Credit: *Fundamentals of Modern Manufacturing*, 4th Edition by Mikell P. Groover, 2010. Reprinted with permission of John Wiley & Sons, Inc.)

Another factor complicating solidification of alloys is that the composition of the dendrites as they start to form favors the metal with the higher melting point. As freezing continues and the dendrites grow, there develops an imbalance in composition between the metal that has solidified and the remaining molten metal. This composition imbalance is finally manifested in the completed casting in the form of segregation of the elements. The segregation is of two types, microscopic and macroscopic. At the microscopic level, the chemical composition varies throughout each individual grain. This is due to the fact that the beginning spine of each dendrite has a higher proportion of one of the elements in the alloy. As the dendrite grows in its local vicinity, it must expand using the remaining liquid metal that has been partially depleted of the first component. Finally, the last metal to freeze in each grain is that which has been trapped by the branches of the dendrite, and its composition is even further out of balance. Thus, we have a variation in chemical composition within single grains of the casting.

At the macroscopic level, the chemical composition varies throughout the entire casting. Since the regions of the casting that freeze first (at the outside near the mold walls) are richer in one component than the other, the remaining molten alloy is deprived of that component by the time freezing occurs at the interior. Thus, there is a general segregation through the cross section of the casting, sometimes called ***ingot segregation***, as illustrated in Figure 5.5.

Eutectic Alloys Eutectic alloys constitute an exception to the general process by which alloys solidify. A ***eutectic alloy*** is a particular composition in an alloy system for which the solidus and liquidus are at the same temperature. Hence, solidification occurs at a constant temperature (called the ***eutectic temperature***) rather than over a temperature range. Examples of eutectic alloys encountered in casting include aluminum–silicon (11.6% Si) and cast iron (4.3% C).

5.3.2 SOLIDIFICATION TIME

Whether the casting is pure metal or alloy, solidification takes time. The total solidification time is the time required for the casting to solidify after pouring. This time is dependent on the size and shape of the casting by an empirical relationship known as ***Chvorinov's rule***, which states:

$$T_{TS} = C_m \left(\frac{V}{A}\right)^n \tag{5.7}$$

where T_{TS} = total solidification time, min; V = volume of the casting, cm^3 (in^3); A = surface area of the casting, cm^2 (in^2); n is an exponent usually taken to have a value = 2; and C_m is the ***mold constant***. Given that $n = 2$, the units of C_m are min/cm^2 (min/in^2), and its value depends on the particular conditions of the casting operation, including mold material (e.g., specific heat, thermal conductivity), thermal properties of the cast metal (e.g., heat of fusion, specific heat, thermal conductivity), and pouring temperature relative to the melting point of the metal. The value of C_m for a given casting operation can be based on experimental data from previous operations carried out using the same mold material, metal, and pouring temperature, even though the shape of the part may be quite different.

Chvorinov's rule indicates that a casting with a higher volume-to-surface area ratio will cool and solidify more slowly than one with a lower ratio. This principle is put to good use in designing the riser in a mold. To perform its function of feeding molten metal to the main cavity, the metal in the riser must remain in the liquid phase longer than the casting. In other words, the T_{TS} for the riser must exceed the T_{TS} for the main casting. Since the mold conditions for both riser and casting are the same, their mold constants will be equal. By designing the riser to have a larger volume-to-area ratio, we can be fairly sure that the main casting solidifies first and that the effects of shrinkage are minimized. Before considering how the riser might be designed using Chvorinov's rule, let us consider the topic of shrinkage, which is the reason why risers are needed.

5.3.3 SHRINKAGE

Our discussion of solidification has neglected the impact of shrinkage that occurs during cooling and freezing. Shrinkage occurs in three steps: (1) liquid contraction during cooling prior to solidification; (2) contraction during the phase change from liquid to solid, called ***solidification shrinkage;*** and (3) thermal contraction of the solidified casting during cooling to room temperature. The three steps can be explained with reference to a cylindrical casting made in an open mold, as shown in Figure 5.6. The molten metal immediately after pouring is shown in part (0) of the series. Contraction of the liquid metal

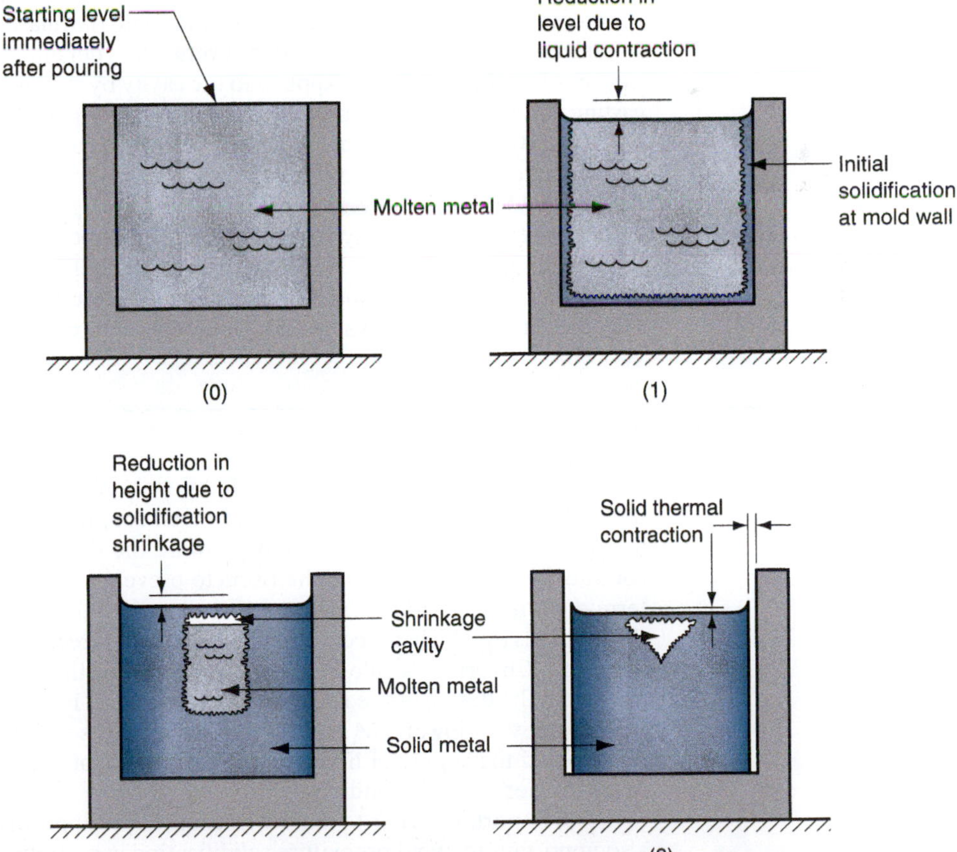

FIGURE 5.6 Shrinkage of a cylindrical casting during solidification and cooling: (0) starting level of molten metal immediately after pouring; (1) reduction in level caused by liquid contraction during cooling; (2) reduction in height and formation of shrinkage cavity caused by solidification shrinkage; and (3) further reduction in height and diameter due to thermal contraction during cooling of the solid metal. For clarity, dimensional reductions are exaggerated in our sketches. (Credit: *Fundamentals of Modern Manufacturing,* 4th Edition by Mikell P. Groover, 2010. Reprinted with permission of John Wiley & Sons, Inc.)

TABLE 5.1 Typical linear shrinkage values for different casting metals due to solid thermal contraction.

Metal	Linear shrinkage	Metal	Linear shrinkage	Metal	Linear shrinkage
Aluminum alloys	1.3%	Magnesium	2.1%	Steel, chrome	2.1%
Brass, yellow	1.3% – 1.6%	Magnesium alloy	1.6%	Tin	2.1%
Cast iron, gray	0.8% – 1.3%	Nickel	2.1%	Zinc	2.6%
Cast iron, white	2.1%	Steel, carbon	1.6% – 2.1%		

Compiled from [10].

during cooling from pouring temperature to freezing temperature causes the height of the liquid to be reduced from its starting level as in (1) of the figure. The amount of this liquid contraction is usually around 0.5%. Solidification shrinkage, seen in part (2), has two effects. First, contraction causes a further reduction in the height of the casting. Second, the amount of liquid metal available to feed the top center portion of the casting becomes restricted. This is usually the last region to freeze, and the absence of metal creates a void in the casting at this location. This shrinkage cavity is called a ***pipe*** by foundrymen. Once solidified, the casting experiences further contraction in height and diameter while cooling, as in (3). This shrinkage is determined by the solid metal's coefficient of thermal expansion, which in this case is applied in reverse to determine contraction.

Solidification shrinkage occurs in nearly all metals because the solid phase has a higher density than the liquid phase. The phase transformation that accompanies solidification causes a reduction in the volume per unit weight of metal. The exception is cast iron containing high carbon content, whose solidification during the final stages of freezing is complicated by a period of graphitization, which results in expansion that tends to counteract the volumetric decrease associated with the phase change [7]. Compensation for solidification shrinkage is achieved in several ways depending on the casting operation. In sand casting, liquid metal is supplied to the cavity by means of risers (Section 5.3.5). In die casting (Section 6.3.3), the molten metal is applied under pressure.

Pattern-makers account for thermal contraction by making the mold cavities oversized. The amount by which the mold must be made larger relative to the final casting size is called the ***pattern shrinkage allowance***. Although the shrinkage is volumetric, the dimensions of the casting are expressed linearly, so the allowances must be applied accordingly. Special "shrink rules" with slightly elongated scales are used to make the patterns and molds larger than the desired casting by the appropriate amount. Table 5.1 lists typical values of linear shrinkage for various cast metals; these values can be used to determine shrink rule scales.

5.3.4 DIRECTIONAL SOLIDIFICATION

In order to minimize the damaging effects of shrinkage, it is desirable for the regions of the casting most distant from the liquid metal supply to freeze first and for solidification to progress from these remote regions toward the riser(s). In this way, molten metal will continually be available from the risers to prevent shrinkage voids during freezing. The term ***directional solidification*** is used to describe this aspect of the freezing process and the methods by which it is controlled. The desired directional solidification is achieved by observing Chvorinov's rule in the design of the casting itself, its orientation within the mold, and the design of the riser system that feeds it. For example, by locating sections of the casting with lower V/A ratios away from the riser, freezing will occur first in these regions and the supply of liquid metal for the rest of the casting will remain open until these bulkier sections solidify.

As important as it is to initiate freezing in the appropriate regions of the cavity, it is also important to avoid premature solidification in sections of the mold nearest the riser.

Of particular concern is the passageway between the riser and the main cavity. This connection must be designed in such a way that it does not freeze before the casting, which would isolate the casting from the molten metal in the riser. Although it is generally desirable to minimize the volume in the connection (to reduce wasted metal), the cross-sectional area must be sufficient to delay the onset of freezing. This goal is usually aided by making the passageway short in length, so that it absorbs heat from the molten metal in the riser and the casting.

5.3.5 RISER DESIGN

As described earlier, a riser, Figure 5.1(b), is used in a sand-casting mold to feed liquid metal to the casting during freezing in order to compensate for solidification shrinkage. To function, the riser must remain molten until after the casting solidifies. Chvorinov's rule can be used to compute the size of a riser that will satisfy this requirement. The following example illustrates the calculation.

Example 5.2 Riser Design Using Chvorinov's Rule

A cylindrical riser must be designed for a sand-casting mold. The casting itself is a steel rectangular plate with dimensions 7.5 cm × 12.5 cm × 2.0 cm. Previous observations have indicated that the total solidification time (T_{TS}) for this casting = 1.6 min. The cylinder for the riser will have a diameter-to-height ratio = 1.0. Determine the dimensions of the riser so that its $T_{TS} = 2.0$ min.

Solution: First, determine the V/A ratio for the plate. Its volume $V = 7.5 \times 12.5 \times 2.0 = 187.5\ \text{cm}^3$, and its surface area $A = 2(7.5 \times 12.5 + 7.5 \times 2.0 + 12.5 \times 2.0) = 267.5\ \text{cm}^2$. Given that $T_{TS} = 1.6$ min, we can determine the mold constant C_m from Eq. (5.7), using a value of $n = 2$ in the equation.

$$C_m = \frac{T_{TS}}{(V/A)^2} = \frac{1.6}{(187.5/267.5)^2} = 3.26\ \text{min/cm}^2$$

Next we must design the riser so that its total solidification time is 2.0 min, using the same value of mold constant. The volume of the riser is given by

$$V = \frac{\pi D^2 h}{4}$$

and the surface area is given by $A = \pi Dh + \frac{2\pi D^2}{4}$

Since we are using a D/H ratio = 1.0, then $D = H$. Substituting D for H in the volume and area formulas, we get

$$V = \pi D^3/4$$

and

$$A = \pi D^2 + 2\pi D^2/4 = 1.5\pi D^2$$

Thus the V/A ratio = $D/6$. Using this ratio in Chvorinov's equation, we have

$$T_{TS} = 2.0 = 3.26\left(\frac{D}{6}\right)^2 = 0.09056\,D^2$$

$$D^2 = 2.0/0.09056 = 22.086\ \text{cm}^2$$

$$D = 4.7\ \text{cm}$$

Since $H = D$, then $H = 4.7$ cm also. ■

The riser represents waste metal that will be separated from the cast part and remelted to make subsequent castings. It is desirable for the volume of metal in the riser to be a minimum. Since the geometry of the riser is normally selected to maximize the V/A ratio, this tends to reduce the riser volume as much as possible. Note that the volume of the riser in our example problem is $V = \pi(4.7)^3/4 = 81.5\ \text{cm}^3$, only 44% of the volume of the plate (casting), even though its total solidification time is 25% longer.

Risers can be designed in different forms. The design shown in Figure 5.1(b) is a ***side riser***. It is attached to the side of the casting by means of a small channel. A ***top riser*** is one that is connected to the top surface of the casting. Risers can be open or blind. An ***open riser*** is exposed to the outside at the top surface of the cope. This has the disadvantage of allowing more heat to escape, promoting faster solidification. A ***blind riser*** is entirely enclosed within the mold, as in Figure 5.1(b).

REFERENCES

[1] Amstead, B. H., Ostwald, P. F., and Begeman, M. L. ***Manufacturing Processes***. John Wiley & Sons, Inc., New York, 1987.

[2] Beeley, P. R. ***Foundry Technology***. Butterworths-Heinemann, Oxford, England, 2001.

[3] Black, J., and Kohser, R. ***DeGarmo's Materials and Processes in Manufacturing***, 10th ed. John Wiley & Sons, Inc., Hoboken, New Jersey, 2008.

[4] Datsko, J. ***Material Properties and Manufacturing Processes***. John Wiley & Sons, Inc., New York, 1966.

[5] Edwards, L., and Endean, M. ***Manufacturing with Materials***. Open University, Milton Keynes, and Butterworth Scientific Ltd., London, 1990.

[6] Flinn, R. A. ***Fundamentals of Metal Casting***. American Foundrymen's Society, Inc., Des Plaines, Illinois, 1987.

[7] Heine, R. W., Loper, Jr., C. R., and Rosenthal, C. ***Principles of Metal Casting***, 2nd ed. McGraw-Hill Book Co., New York, 1967.

[8] Kotzin, E. L. (ed.). ***Metalcasting and Molding Processes***. American Foundrymen's Society, Inc., Des Plaines, Illinois, 1981.

[9] Lessiter, M. J., and Kirgin, K. "*Trends in the Casting Industry*," **Advanced Materials & Processes**, January 2002, pp. 42–43.

[10] ***Metals Handbook***, Vol. 15: ***Casting***. ASM International, Materials Park, Ohio, 2008.

[11] Mikelonis, P. J. (ed.). ***Foundry Technology***. American Society for Metals, Metals Park, Ohio, 1982.

[12] Niebel, B. W., Draper, A. B., and Wysk, R. A. ***Modern Manufacturing Process Engineering***. McGraw-Hill Book Co., New York, 1989.

[13] Taylor, H. F., Flemings, M. C., and Wulff, J. ***Foundry Engineering***, 2nd ed. American Foundrymen's Society, Inc., Des Plaines, Illinois, 1987.

[14] Wick, C., Benedict, J. T., and Veilleux, R. F. ***Tool and Manufacturing Engineers Handbook***, 4th ed. Vol. II, ***Forming***. Society of Manufacturing Engineers, Dearborn, Michigan, 1984.

REVIEW QUESTIONS

5.1. Identify some of the important advantages of shape-casting processes.

5.2. What are some of the limitations and disadvantages of casting?

5.3. What is a factory that performs casting operations usually called?

5.4. What is the difference between an open mold and a closed mold?

5.5. Name the two basic mold types that distinguish casting processes.

5.6. Which casting process is the most important commercially?

5.7. What is the difference between a pattern and a core in sand molding?

5.8. What is meant by the term *superheat*?

5.9. Why should turbulent flow of molten metal into the mold be avoided?

5.10. What is the continuity law as it applies to the flow of molten metal in casting?

5.11. What does heat of fusion mean in casting?

5.12. How does solidification of alloys differ from solidification of pure metals?

5.13. What is a eutectic alloy?

5.14. What is the relationship known as Chvorinov's rule in casting?

5.15. Identify the three sources of contraction in a metal casting after pouring.

PROBLEMS

5.1. The downsprue leading into the runner of a certain mold has a length = 175 mm. The cross-sectional area at the base of the sprue is 400 mm^2. The mold cavity has a volume = 0.001 m^3. Determine (a) the velocity of the molten metal flowing through the base of the downsprue, (b) the volume rate of flow, and (c) the time required to fill the mold cavity.

5.2. A mold has a downsprue of length = 6.0 in. The cross-sectional area at the bottom of the sprue is 0.5 in^2. The sprue leads into a horizontal runner, which feeds the mold cavity, whose volume = 75 in^3. Determine (a) the velocity of the molten metal flowing through the base of the downsprue, (b) the volume rate of flow, and (c) the time required to fill the mold cavity.

5.3. The flow rate of liquid metal into the downsprue of a mold = 1 liter/sec. The cross-sectional area at the top of the sprue = 800 mm^2, and its length = 175 mm. What area should be used at the base of the sprue to avoid aspiration of the molten metal?

5.4. The volume rate of flow of molten metal into the downsprue from the pouring cup is 50 in^3/sec. At the top where the pouring cup leads into the downsprue, the cross-sectional area = 1.0 in^2. Determine what the area should be at the bottom of the sprue if its length = 8.0 in. It is desired to maintain a constant flow rate, top and bottom, in order to avoid aspiration of the liquid metal.

5.5. Molten metal can be poured into the pouring cup of a sand mold at a steady rate of 1000 cm^3/s. The molten metal overflows the pouring cup and flows into the downsprue. The cross section of the sprue is round, with a diameter at the top = 3.4 cm. If the sprue is 25 cm long, determine the proper diameter at its base so as to maintain the same volume flow rate.

5.6. Determine the shrink rule to be used by patternmakers for white cast iron. Using the shrinkage value in Table 5.1, express your answer in terms of decimal fraction inches of elongation per foot of length compared to a standard one-foot scale.

5.7. Determine the shrink rule to be used by moldmakers for die casting of zinc. Using the shrinkage value in Table 5.1, express your answer in terms of decimal mm of elongation per 300 mm of length compared to a standard 300-mm scale.

5.8. A flat plate is to be cast in an open mold whose bottom has a square shape that is 200 mm by 200 mm. The mold is 40 mm deep. A total of 1,000,000 mm^3 of molten aluminum is poured into the mold. Solidification volumetric shrinkage is known to be 6.0%. Table 5.1 lists the linear shrinkage due to thermal contraction after solidification to be 1.3%. If the availability of molten metal in the mold allows the square shape of the cast plate to maintain its 200-mm × 200-mm dimensions until solidification is completed, determine the final dimensions of the plate.

5.9. In the casting of steel under certain mold conditions, the mold constant in Chvorinov's rule is known to be 4.0 min/cm^2, based on previous experience. The casting is a flat plate whose length = 30 cm, width = 10 cm, and thickness = 20 mm. Determine how long it will take for the casting to solidify.

5.10. Solve for total solidification time in the previous problem only using an exponent value of 1.9 in Chvorinov's rule instead of 2.0. What adjustment must be made in the units of the mold constant?

5.11. A disk-shaped part is to be cast out of aluminum. The diameter of the disk = 500 mm and its thickness = 20 mm. If the mold constant = 2.0 sec/mm^2 in Chvorinov's rule, how long will it take the casting to solidify?

5.12. In casting experiments performed using a certain alloy and type of sand mold, it took 155 sec for a cube-shaped casting to solidify. The cube was 50 mm on a side. (a) Determine the value of the mold constant in Chvorinov's rule. (b) If the same alloy and mold type were used, find the total solidification time for a cylindrical casting in which the diameter = 30 mm and length = 50 mm.

5.13. A steel casting has a cylindrical geometry with 4.0 in diameter and weighs 20 lb. This casting takes 6.0 min to completely solidify. Another cylindrical-shaped casting with the same diameter-to-length ratio weighs 12 lb. This casting is made of the same steel, and the same conditions of mold and pouring were used. Determine: (a) the mold constant in Chvorinov's rule, (b) the dimensions, and (c) the total solidification time of the lighter casting. The density of steel = 490 lb/ft^3.

5.14. The total solidification times of three casting shapes are to be compared: (1) a sphere, (2) a cylinder, in which the length-to-diameter ratio = 1.0, and (3) a cube. For all three geometries, the volume = 1000 cm^3. The same casting alloy is used in the three cases. (a) Determine the relative solidification times for each geometry. (b) Based on the results of part (a), which geometric element would make the best riser? (c) If the mold constant = 3.5 min/cm^2 in Chvorinov's rule, compute the total solidification time for each casting.

5.15. A cylindrical riser is to be used for a sand-casting mold. For a given cylinder volume, determine the diameter-to-length ratio that will maximize the time to solidify.

5.16. A cylindrical riser is to be designed for a sand-casting mold. The length of the cylinder is to be 1.25 times its diameter. The casting is a square plate, each side = 10 in and thickness = 0.75 in. If the metal is cast iron, and the mold constant = 16.0 min/in^2 in Chvorinov's rule, determine the dimensions of the riser so that it will take 30% longer for the riser to solidify.

6 METAL CASTING PROCESSES

Chapter Contents

6.1 Sand Casting
- 6.1.1 Patterns and Cores
- 6.1.2 Molds and Mold Making
- 6.1.3 The Casting Operation

6.2 Other Expendable-Mold Casting Processes
- 6.2.1 Shell Molding
- 6.2.2 Expanded Polystyrene Process
- 6.2.3 Investment Casting
- 6.2.4 Plaster-Mold and Ceramic-Mold Casting

6.3 Permanent-Mold Casting Processes
- 6.3.1 The Basic Permanent-Mold Process
- 6.3.2 Variations of Permanent-Mold Casting
- 6.3.3 Die Casting
- 6.3.4 Squeeze Casting and Semisolid Metal Casting
- 6.3.5 Centrifugal Casting

6.4 Foundry Practice
- 6.4.1 Furnaces
- 6.4.2 Pouring, Cleaning, and Heat Treatment

6.5 Casting Quality

6.6 Metals for Casting

6.7 Product Design Considerations

Metal casting processes divide into two categories, based on mold type: (1) expendable-mold and (2) permanent-mold. In expendable-mold casting operations, the mold is sacrificed in order to remove the cast part. Because a new mold is required for each new casting, production rates in expendable-mold processes are often limited by the time required to make the mold rather than the time to make the casting itself. However, for certain part geometries, sand molds can be produced and castings made at rates of 400 parts per hour and higher. In permanent-mold casting processes, the mold is fabricated out of metal (or other durable material) and can be used many times to make many castings. Accordingly, these processes possess a natural advantage in terms of higher production rates.

Our discussion of casting processes in this chapter is organized as follows: (1) sand casting, (2) other expendable-mold casting processes, and (3) permanent-mold casting processes. The chapter also includes casting equipment and practices used in foundries. Another section deals with inspection and quality issues. Product design guidelines are presented in the final section.

6.1 SAND CASTING

Sand casting is the most widely used casting process, accounting for a significant majority of the total tonnage cast. Nearly all casting alloys can be sand cast; indeed, it is one of the few processes that can be used for metals with high melting temperatures, such as steels, nickels, and titaniums. Its versatility permits the casting of parts ranging in size from small to very large and in production quantities from one to millions.

Sand casting, also known as ***sand-mold casting***, consists of pouring molten metal into a sand mold, allowing the metal to solidify, and then breaking up the mold to remove the casting. The casting must then be cleaned and

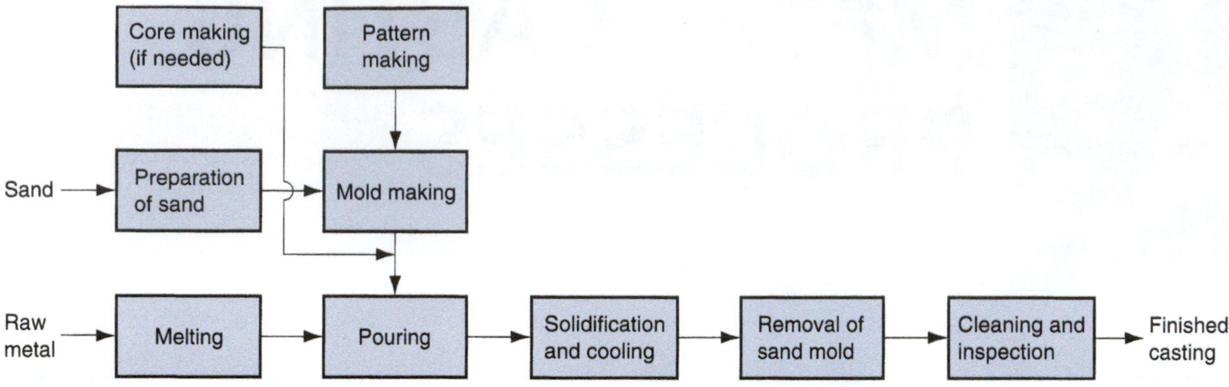

FIGURE 6.1 Steps in the production sequence in sand casting. The steps include not only the casting operation but also pattern making and mold making. (Credit: *Fundamentals of Modern Manufacturing*, 4th Edition by Mikell P. Groover, 2010. Reprinted with permission of John Wiley & Sons, Inc.)

inspected, and heat treatment is sometimes required to improve metallurgical properties. The cavity in the sand mold is formed by packing sand around a pattern (an approximate duplicate of the part to be cast), and then removing the pattern by separating the mold into two halves. The mold also contains the gating and riser system. In addition, if the casting is to have internal surfaces (e.g., hollow parts or parts with holes), a core must be included in the mold. Because the mold is sacrificed to remove the casting, a new sand mold must be made for each part that is produced. From this brief description, sand casting is seen to include not only the casting operation itself, but also the fabrication of the pattern and the making of the mold. The production sequence is outlined in Figure 6.1.

6.1.1 PATTERNS AND CORES

Sand casting requires a ***pattern***—a full-sized model of the part, enlarged to account for shrinkage and machining allowances in the final casting. Materials used to make patterns include wood, plastics, and metals. Wood is a common pattern material because it is easily shaped. Its disadvantages are that it tends to warp, and it is abraded by the sand being compacted around it, thus limiting the number of times it can be reused. Metal patterns are more expensive to make, but they last much longer. Plastics represent a compromise between wood and metal. Selection of the appropriate pattern material depends to a large extent on the total quantity of castings to be made.

There are various types of patterns, as illustrated in Figure 6.2. The simplest is made of one piece, called a ***solid pattern***—same geometry as the casting, adjusted in size for shrinkage and machining. Although it is the easiest pattern to fabricate, it is not the easiest to use in making the sand mold. Determining the location of the parting line between the two halves of the mold for a solid pattern can be a problem, and incorporating the gating system and sprue into the mold is left to the judgment and skill of the foundry worker. Consequently, solid patterns are generally limited to very low production quantities.

Split patterns consist of two pieces, dividing the part along a plane coinciding with the parting line of the mold. Split patterns are appropriate for complex part geometries and moderate production quantities. The parting line of the mold is predetermined by the two pattern halves, rather than by operator judgment.

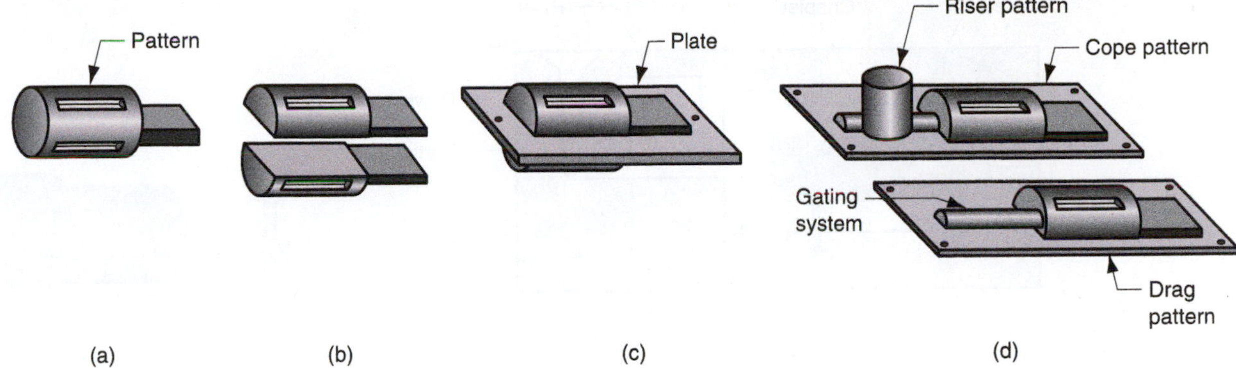

FIGURE 6.2 Types of patterns used in sand casting: (a) solid pattern, (b) split pattern, (c) match-plate pattern, and (d) cope-and-drag pattern. (Credit: *Fundamentals of Modern Manufacturing*, 4th Edition by Mikell P. Groover, 2010. Reprinted with permission of John Wiley & Sons, Inc.)

For higher production quantities, match-plate patterns or cope-and-drag patterns are used. In ***match-plate*** patterns, the two pieces of the split pattern are attached to opposite sides of a wood or metal plate. Holes in the plate allow the top and bottom (cope and drag) sections of the mold to be aligned accurately. ***Cope-and-drag patterns*** are similar to match-plate patterns except that split pattern halves are attached to separate plates, so that the cope and drag sections of the mold can be fabricated independently, instead of using the same tooling for both. Part (d) of the figure includes the gating and riser system in the cope-and-drag patterns.

Patterns define the external shape of the cast part. If the casting is to have internal surfaces, a core is required. A ***core*** is a full-scale model of the interior surfaces of the part. It is inserted into the mold cavity prior to pouring, so that the molten metal will flow and solidify between the mold cavity and the core to form the casting's external and internal surfaces. The core is usually made of sand, compacted into the desired shape. As with the pattern, the actual size of the core must include allowances for shrinkage and machining. Depending on the geometry of the part, the core may or may not require supports to hold it in position in the mold cavity during pouring. These supports, called ***chaplets***, are made of a metal with a higher melting temperature than the casting metal. For example, steel chaplets would be used for cast iron castings. On pouring and solidification, the chaplets become bonded into the casting. A possible arrangement of a core in a mold using chaplets is sketched in Figure 6.3. The portion of the chaplet protruding from the casting is subsequently cut off.

6.1.2 MOLDS AND MOLD MAKING

Foundry sands are silica (SiO_2) or silica mixed with other minerals. The sand should possess good refractory properties—capacity to stand up under high temperatures without melting or otherwise degrading. Other important features of the sand include grain size, distribution of grain size in the mixture, and shape of the individual grains. Small grain size provides a better surface finish on the cast part, but large grain size is more permeable (to allow escape of gases during pouring). Molds made from grains of irregular shape tend to be stronger than molds of round grains because of interlocking, yet interlocking tends to restrict permeability.

In making the mold, the grains of sand are held together by a mixture of water and bonding clay. A typical mixture (by volume) is 90% sand, 3% water, and 7% clay. Other

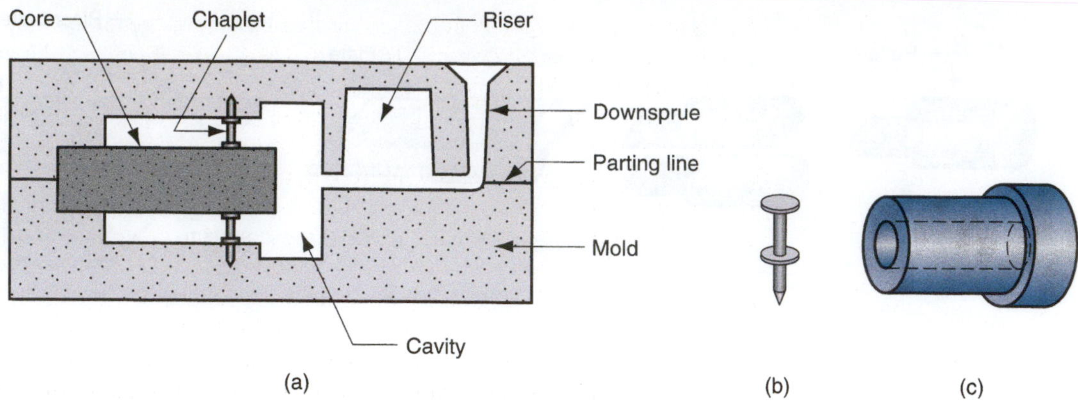

FIGURE 6.3 (a) Core held in place in the mold cavity by chaplets, (b) possible chaplet design, and (c) casting with internal cavity. (Credit: *Fundamentals of Modern Manufacturing*, 4th Edition by Mikell P. Groover, 2010. Reprinted with permission of John Wiley & Sons, Inc.)

bonding agents can be used in place of clay, including organic resins (e.g., phenolic resins) and inorganic binders (e.g., sodium silicate and phosphate). Besides sand and binder, additives are sometimes combined with the mixture to enhance properties such as strength and/or permeability of the mold.

To form the mold cavity, the traditional method is to compact the molding sand around the pattern for both cope and drag in a container called a ***flask***. The packing process is performed by various methods. The simplest is hand ramming, accomplished manually by a foundry worker. In addition, various machines have been developed to mechanize the packing procedure. These machines operate by any of several mechanisms, including (1) squeezing the sand around the pattern by pneumatic pressure; (2) a jolting action in which the sand, contained in the flask with the pattern, is dropped repeatedly to pack it into place; and (3) a slinging action, in which the sand grains are impacted against the pattern at high speed.

An alternative to traditional flasks for each sand mold is ***flaskless molding***, which refers to the use of one master flask in a mechanized system of mold production. Each sand mold is produced using the same master flask. Mold production rates up to 600 per hour are claimed for this more automated method [8].

Several indicators are used to determine the quality of the sand mold [7]: (1) ***strength***—the mold's ability to maintain its shape and resist erosion caused by the flow of molten metal; it depends on grain shape and adhesive qualities of the binder; (2) ***permeability***—capacity of the mold to allow hot air and gases from the casting operation to pass through the voids in the sand; (3) ***thermal stability***—ability of the sand at the surface of the mold cavity to resist cracking and buckling upon contact with the molten metal; (4) ***collapsibility***—ability of the mold to give way and allow the casting to shrink without cracking the casting; it also refers to the ability to remove the sand from the casting during cleaning; and (5) ***reusability***—can the sand from the broken mold be reused to make other molds? These measures are sometimes incompatible; for example, a mold with greater strength is less collapsible.

Sand molds are often classified as green-sand, dry-sand, or skin-dried molds. ***Green-sand molds*** are made of a mixture of sand, clay, and water, the word ***green*** referring to the fact that the mold contains moisture at the time of pouring. Green-sand molds possess sufficient strength for most applications, good collapsibility, good permeability, good reusability, and are the least expensive of the molds. They are the most widely used mold type, but they are not without problems. Moisture in the sand can cause defects in some castings, depending on the metal and geometry of the part. A ***dry-sand***

mold is made using organic binders rather than clay, and the mold is baked in a large oven at temperatures ranging from 200°C to 320°C (400°F to 600°F) [8]. Oven baking strengthens the mold and hardens the cavity surface. A dry-sand mold provides better dimensional control in the cast product, compared with green-sand molding. However, dry-sand molding is more expensive, and production rate is reduced because of drying time. Applications are generally limited to medium and large castings in low to medium production rates. In a ***skin-dried mold***, the advantages of a dry-sand mold are partially achieved by drying the surface of a green-sand mold to a depth of 10 to 25 mm (0.4–1 in) at the mold cavity surface, using torches, heating lamps, or other means. Special bonding materials must be added to the sand mixture to strengthen the cavity surface.

The preceding mold classifications refer to the use of conventional binders consisting of either clay-and-water or ones that require heating to cure. In addition to these classifications, chemically bonded molds have been developed that are not based on either of these traditional binder ingredients. Some of the binder materials used in these "no-bake" systems include furan resins (consisting of furfural alcohol, urea, and formaldehyde), phenolics, and alkyd oils. No-bake molds are growing in popularity due to their good dimensional control in high production applications.

6.1.3 THE CASTING OPERATION

After the core is positioned (if one is used) and the two halves of the mold are clamped together, then casting is performed. Casting consists of pouring, solidification, and cooling of the cast part (Sections 5.2 and 5.3). The gating and riser system in the mold must be designed to deliver liquid metal into the cavity and provide for a sufficient reservoir of molten metal during solidification shrinkage. Air and gases must be allowed to escape.

After solidification and cooling, the sand mold is broken away from the casting to retrieve the part. Next, the part is cleaned, which consists of separating the gating and riser system, removing sand from the surface, and inspecting the casting.

6.2 OTHER EXPENDABLE-MOLD CASTING PROCESSES

As versatile as sand casting is, other casting processes have been developed to meet special needs. The differences between these methods are in the composition of the mold material, or the manner in which the mold is made, or in the way the pattern is made.

6.2.1 SHELL MOLDING

Shell molding is a casting process in which the mold is a thin shell (typically 9 mm or 3/8 in) made of sand held together by a thermosetting resin binder. Developed in Germany during the early 1940s, the process is described and illustrated in Figure 6.4.

There are many advantages to the shell-molding process. The surface of the shell-mold cavity is smoother than a conventional green-sand mold, and this smoothness permits easier flow of molten metal during pouring and better surface finish on the final casting. Finishes of 2.5 μm (100 μ-in) can be obtained. Good dimensional accuracy is also achieved, with tolerances of ±0.25 mm (±0.010 in) possible on small-to-medium-sized parts. The good finish and accuracy often precludes the need for further machining. Collapsibility of the mold is generally sufficient to avoid tearing and cracking of the casting.

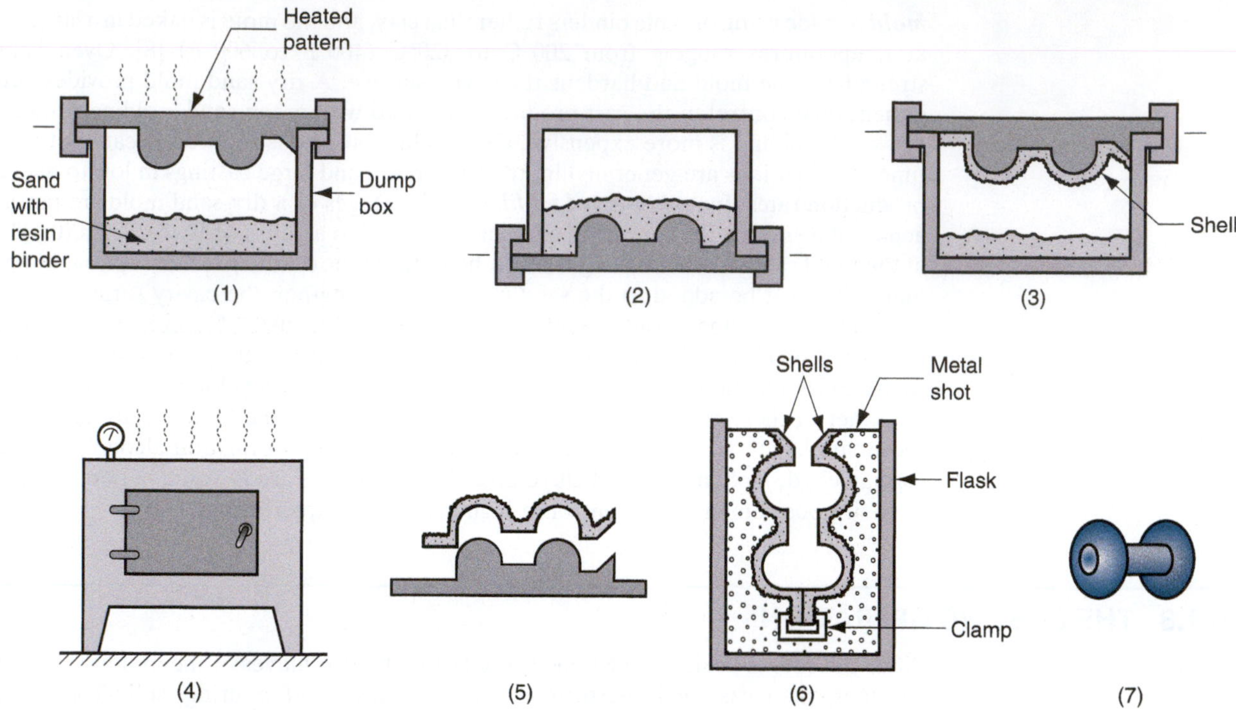

FIGURE 6.4 Steps in shell molding: (1) a match-plate or cope-and-drag metal pattern is heated and placed over a box containing sand mixed with thermosetting resin; (2) box is inverted so that sand and resin fall onto the hot pattern, causing a layer of the mixture to partially cure on the surface to form a hard shell; (3) box is repositioned so that loose, uncured particles drop away; (4) sand shell is heated in oven for several minutes to complete curing; (5) shell mold is stripped from the pattern; (6) two halves of the shell mold are assembled, supported by sand or metal shot in a box, and pouring is accomplished. The finished casting with sprue removed is shown in (7). (Credit: *Fundamentals of Modern Manufacturing*, 4th Edition by Mikell P. Groover, 2010. Reprinted with permission of John Wiley & Sons, Inc.)

Disadvantages of shell molding include a more expensive metal pattern than the corresponding pattern for green-sand molding. This makes shell molding difficult to justify for small quantities of parts. Shell molding can be mechanized for mass production and is very economical for large quantities. It seems particularly suited to steel castings of less than 20 lb. Examples of parts made using shell molding include gears, valve bodies, bushings, and camshafts.

6.2.2 EXPANDED POLYSTYRENE PROCESS

The expanded polystyrene casting process uses a mold of sand packed around a polystyrene foam pattern that vaporizes when the molten metal is poured into the mold. The process and variations of it are known by other names, including ***lost-foam process***, ***lost pattern process***, ***evaporative-foam casting***, and ***full-mold process*** (the last being a trade name). The foam pattern includes the sprue, risers, and gating system, and it may also contain internal cores (if needed), thus eliminating the need for a separate core in the mold. Also, because the foam pattern itself becomes the cavity in the mold, considerations of draft and parting lines can be ignored. The mold does not have to be opened into cope and drag sections. The sequence in this casting process is illustrated and described in Figure 6.5. Various methods for making the pattern can be used, depending

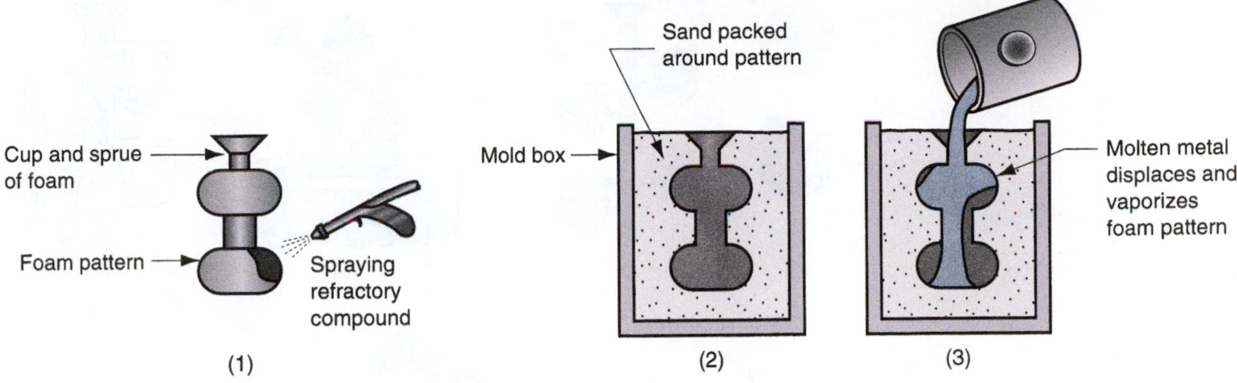

FIGURE 6.5 Expanded polystyrene casting process: (1) pattern of polystyrene is coated with refractory compound; (2) foam pattern is placed in mold box, and sand is compacted around the pattern; and (3) molten metal is poured into the portion of the pattern that forms the pouring cup and sprue. As the metal enters the mold, the polystyrene foam is vaporized ahead of the advancing liquid, thus allowing the resulting mold cavity to be filled. (Credit: *Fundamentals of Modern Manufacturing*, 4th Edition by Mikell P. Groover, 2010. Reprinted with permission of John Wiley & Sons, Inc.)

on the quantities of castings to be produced. For one-of-a-kind castings, the foam is manually cut from large strips and assembled to form the pattern. For large production runs, an automated molding operation can be set up to mold the patterns prior to making the molds for casting. The pattern is normally coated with a refractory compound to provide a smoother surface on the pattern and to improve its high-temperature resistance. Molding sands usually include bonding agents. However, dry sand is used in certain processes in this group, which aids recovery and reuse.

A significant advantage for this process is that the pattern need not be removed from the mold. This simplifies and expedites mold making. In a conventional green-sand mold, two halves are required with proper parting lines, draft allowances must be provided in the mold design, cores must be inserted, and the gating and riser system must be added. With the expanded polystyrene process, these steps are built into the pattern itself. A new pattern is needed for every casting, so the economics of the expanded polystyrene casting process depend largely on the cost of producing the patterns. The process has been applied to mass produce castings for automobile engines, in which automated systems are used to mold the polystyrene foam patterns.

6.2.3 INVESTMENT CASTING

In investment casting, a pattern made of wax is coated with a refractory material to make the mold, after which the wax is melted away prior to pouring the molten metal. The term ***investment*** comes from one of the less familiar definitions of the word ***invest***, which is "to cover completely," this referring to the coating of the refractory material around the wax pattern. It is a precision casting process, because it is capable of making castings of high accuracy and intricate detail. The process dates back to ancient Egypt and is also known as the ***lost-wax process***, because the wax pattern is lost from the mold prior to casting.

Steps in investment casting are described in Figure 6.6. Because the wax pattern is melted off after the refractory mold is made, a separate pattern must be made for every casting. Pattern production is usually accomplished by a molding operation—pouring or injecting the hot wax into a ***master die*** that has been designed with proper allowances for shrinkage of both wax and subsequent metal casting. In cases where the part geometry is complicated, several separate wax pieces must be joined to make the pattern. In

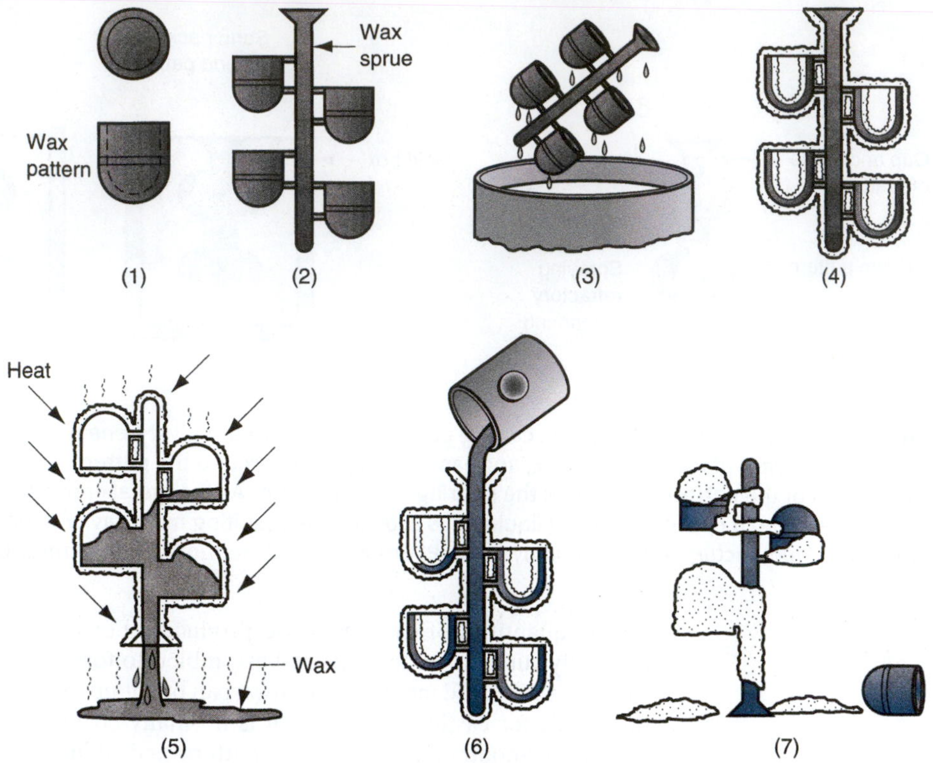

FIGURE 6.6 Steps in investment casting: (1) wax patterns are produced; (2) several patterns are attached to a sprue to form a pattern tree; (3) the pattern tree is coated with a thin layer of refractory material; (4) the full mold is formed by covering the coated tree with sufficient refractory material to make it rigid; (5) the mold is held in an inverted position and heated to melt the wax and permit it to drip out of the cavity; (6) the mold is preheated to a high temperature, which ensures that all contaminants are eliminated from the mold; it also permits the liquid metal to flow more easily into the detailed cavity; the molten metal is poured; it solidifies; and (7) the mold is broken away from the finished casting. Parts are separated from the sprue. (Credit: *Fundamentals of Modern Manufacturing*, 4th Edition by Mikell P. Groover, 2010. Reprinted with permission of John Wiley & Sons, Inc.)

high-production operations, several patterns are attached to a sprue, also made of wax, to form a ***pattern tree***; this is the geometry that will be cast out of metal.

Coating with refractory (step 3) is usually accomplished by dipping the pattern tree into a slurry of very fine-grained silica or other refractory (almost in powder form) mixed with plaster to bond the mold into shape. The small grain size of the refractory material provides a smooth surface and captures the intricate details of the wax pattern. The final mold (step 4) is accomplished by repeatedly dipping the tree into the refractory slurry or by gently packing the refractory around the tree in a container. The mold is allowed to air-dry for about 8 hours to harden the binder.

Advantages of investment casting include: (1) parts of great complexity and intricacy can be cast; (2) close dimensional control—tolerances of ±0.075 mm (±0.003 in) are possible; (3) good surface finish is possible; (4) the wax can usually be recovered for reuse; and (5) additional machining is not normally required—this is a net shape process. Because many steps are involved in this casting operation, it is a relatively expensive process. Investment castings are normally small in size, although parts with complex geometries weighing up to 34 kg (75 lb) have been successfully cast.

FIGURE 6.7 A one-piece compressor stator with 108 separate airfoils made by investment casting. Photo courtesy of Alcoa Howmet. (Credit: *Fundamentals of Modern Manufacturing*, 4th Edition by Mikell P. Groover, 2010. Reprinted with permission of John Wiley & Sons, Inc.)

All types of metals, including steels, stainless steels, and other high temperature alloys, can be investment cast. Examples of parts include complex machinery parts, blades, and other components for turbine engines, jewelry, and dental fixtures. Shown in Figure 6.7 is a part that illustrates the intricate features possible with investment casting.

6.2.4 PLASTER-MOLD AND CERAMIC-MOLD CASTING

Plaster-mold casting is similar to sand casting except that the mold is made of plaster of Paris (gypsum—$CaSO_4$–$2H_2O$) instead of sand. Additives such as talc and silica flour are mixed with the plaster to control contraction and setting time, reduce cracking, and increase strength. To make the mold, the plaster mixture combined with water is poured over a plastic or metal pattern in a flask and allowed to set. Wood patterns are generally unsatisfactory due to the extended contact with water in the plaster. The fluid consistency permits the plaster mixture to readily flow around the pattern, capturing its details and surface finish. Thus, the cast product in plaster molding is noted for these attributes.

Curing of the plaster-mold is one of the disadvantages of this process, at least in high production. The mold must set for about 20 minutes before the pattern is stripped. The mold is then baked for several hours to remove moisture. Even with the baking, not all of the moisture content is removed from the plaster. The dilemma faced by foundrymen is that mold strength is lost when the plaster becomes too dehydrated, and yet moisture content can cause casting defects in the product. A balance must be achieved between these undesirable alternatives. Another disadvantage with the plaster mold is that it is not permeable, thus limiting escape of gases from the mold cavity. This problem can be solved in a number of ways: (1) evacuating air from the mold cavity before pouring; (2) aerating the plaster slurry prior to mold making so that the resulting hard plaster contains finely dispersed voids; and (3) using a special mold composition and treatment known as the ***Antioch process***. This process involves using about 50% sand mixed with the plaster, heating the mold in an autoclave (an oven that uses superheated steam under pressure), and then drying. The resulting mold has considerably greater permeability than a conventional plaster-mold.

Plaster-molds cannot withstand the same high temperatures as sand molds. They are therefore limited to the casting of lower-melting-point alloys, such as aluminum, magnesium, and some copper-base alloys. Applications include metal molds for plastic and rubber molding, pump and turbine impellers, and other parts of relatively intricate geometry. Casting sizes range from about 20 g (less than 1 oz) to more than 100 kg (220 lb). Parts weighing less than about 10 kg (22 lb) are most common. Advantages of plaster-mold casting for these applications are good surface finish and dimensional accuracy and the capability to make thin cross sections in the casting.

Ceramic-mold casting is similar to plaster-mold casting, except that the mold is made of refractory ceramic materials that can withstand higher temperatures than plaster. Thus, ceramic molding can be used to cast steels, cast irons, and other high-temperature alloys. Its applications (relatively intricate parts) are similar to those of plaster-mold casting except for the metals cast. Its advantages (good accuracy and finish) are also similar.

6.3 PERMANENT-MOLD CASTING PROCESSES

The economic disadvantage of any of the expendable-mold processes is that a new mold is required for every casting. In permanent-mold casting, the mold is reused many times. In this section, we treat permanent-mold casting as the basic process in the group of casting processes that all use reusable metal molds. Other members of the group include die casting and centrifugal casting.

6.3.1 THE BASIC PERMANENT-MOLD PROCESS

Permanent-mold casting uses a metal mold constructed of two sections that are designed for easy, precise opening and closing. These molds are commonly made of steel or cast iron. The cavity, with gating system included, is machined into the two halves to provide accurate dimensions and good surface finish. Metals commonly cast in permanent-molds include aluminum, magnesium, copper-base alloys, and cast iron. However, cast iron requires a high pouring temperature, 1250°C to 1500°C (2300°F to 2700°F), which takes a heavy toll on mold life. The very high pouring temperatures of steel make permanent molds unsuitable for this metal, unless the mold is made of refractory material.

Cores can be used in permanent-molds to form interior surfaces in the cast product. The cores can be made of metal, but either their shape must allow for removal from the casting or they must be mechanically collapsible to permit removal. If withdrawal of a metal core would be difficult or impossible, sand cores can be used, in which case the casting process is often referred to as ***semipermanent-mold casting***.

Steps in the basic permanent-mold casting process are described in Figure 6.8. In preparation for casting, the mold is first preheated and one or more coatings are sprayed

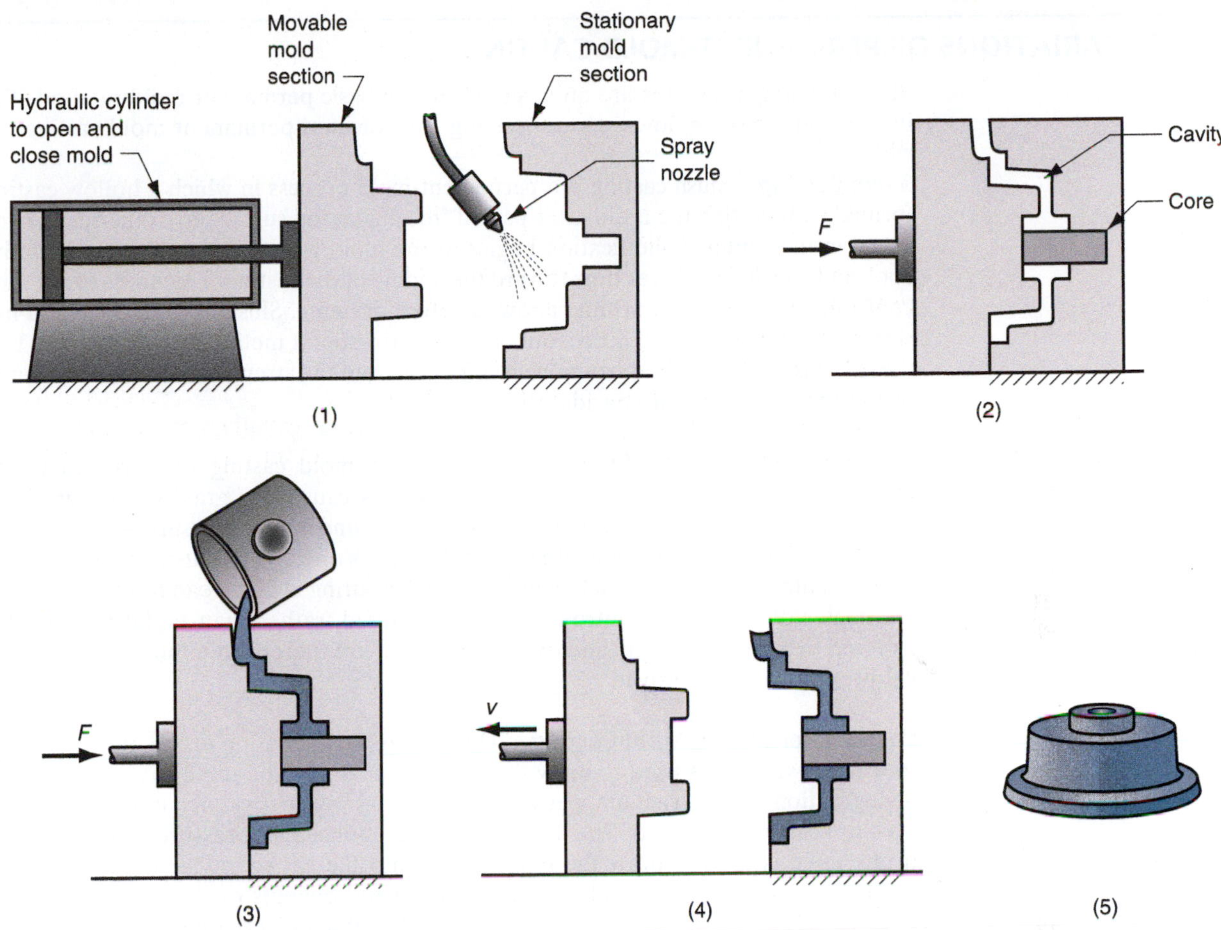

FIGURE 6.8 Steps in permanent-mold casting: (1) Mold is preheated and coated; (2) cores (if used) are inserted, and mold is closed; (3) molten metal is poured into the mold; and (4) mold is opened. Finished part is shown in (5). (Credit: *Fundamentals of Modern Manufacturing*, 4th Edition by Mikell P. Groover, 2010. Reprinted with permission of John Wiley & Sons, Inc.)

on the cavity. Preheating facilitates metal flow through the gating system and into the cavity. The coatings aid heat dissipation and lubricate the mold surfaces for easier separation of the cast product. After pouring, as soon as the metal solidifies, the mold is opened and the casting is removed. Unlike expendable-molds, permanent-molds do not collapse, so the mold must be opened before appreciable cooling contraction occurs in order to prevent cracks from developing in the casting.

Advantages of permanent-mold casting include good surface finish and close dimensional control, as previously indicated. In addition, more rapid solidification caused by the metal mold results in a finer grain structure, so stronger castings are produced. The process is generally limited to metals of lower melting points. Other limitations include simple part geometries compared to sand casting (because of the need to open the mold), and the expense of the mold. Because mold cost is substantial, the process is best suited to high-volume production and can be automated accordingly. Typical parts include automotive pistons, pump bodies, and certain castings for aircraft and missiles.

6.3.2 VARIATIONS OF PERMANENT-MOLD CASTING

Several casting processes are quite similar to the basic permanent-mold method. These include slush casting, low-pressure casting, and vacuum permanent-mold casting.

Slush Casting Slush casting is a permanent-mold process in which a hollow casting is formed by inverting the mold after partial freezing at the surface to drain out the liquid metal in the center. Solidification begins at the mold walls because they are relatively cool, and it progresses over time toward the middle of the casting. Thickness of the shell is controlled by the length of time allowed before draining. Slush casting is used to make statues, lamp pedestals, and toys out of low-melting-point metals such as zinc and tin. In these items, the exterior appearance is important, but the strength and interior geometry of the casting are minor considerations.

Low-Pressure Casting In the basic permanent-mold casting process and in slush casting, the flow of metal into the mold cavity is caused by gravity. In low-pressure casting, the liquid metal is forced into the cavity under low pressure—approximately 0.1 MPa (15 lb/in^2)—from beneath so that the flow is upward, as illustrated in Figure 6.9. The advantage of this approach over traditional pouring is that clean molten metal from the center of the ladle is introduced into the mold, rather than metal that has been exposed to air. Gas porosity and oxidation defects are thereby minimized, and mechanical properties are improved.

Vacuum Permanent-Mold Casting This process is a variation of low-pressure casting in which a vacuum is used to draw the molten metal into the mold cavity. The general configuration of the vacuum permanent-mold casting process is similar to the low-pressure casting operation. The difference is that reduced air pressure from the vacuum in the mold is used to draw the liquid metal into the cavity, rather than forcing it by positive air pressure from below. There are several benefits of the vacuum technique relative to low-pressure casting: air porosity and related defects are reduced, and greater strength is given to the cast product.

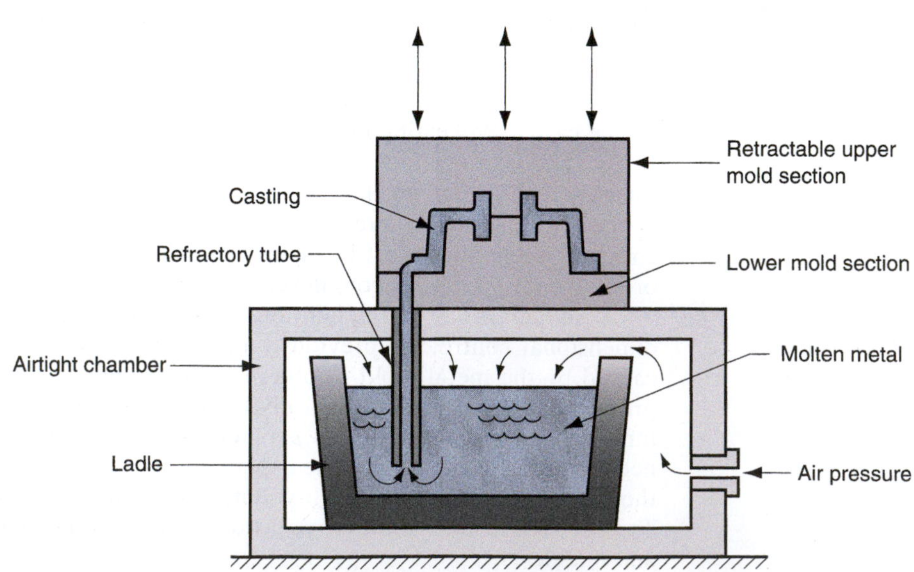

FIGURE 6.9 Low-pressure casting. The diagram shows how air pressure is used to force the molten metal in the ladle upward into the mold cavity. Pressure is maintained until the casting has solidified. (Credit: *Fundamentals of Modern Manufacturing*, 4th Edition by Mikell P. Groover, 2010. Reprinted with permission of John Wiley & Sons, Inc.)

6.3.3 DIE CASTING

Die casting is a permanent-mold casting process in which the molten metal is injected into the mold cavity under high pressure. Typical pressures are 7 to 350 MPa (1000 to 50,000 lb/in^2). The pressure is maintained during solidification, after which the mold is opened and the part is removed. Molds in this casting operation are called dies; hence, the name die casting. The use of high pressure to force the metal into the die cavity is the most notable feature that distinguishes this process from others in the permanent-mold category.

Die-casting operations are carried out in special die-casting machines that are designed to hold and accurately close the two halves of the mold and keep them closed while the liquid metal is forced into the cavity. The general configuration is shown in Figure 6.10. There are two main types of die-casting machines: (1) hot-chamber and (2) cold-chamber, differentiated by how the molten metal is injected into the cavity.

In ***hot-chamber machines***, the metal is melted in a container attached to the machine, and a piston is used to inject the liquid metal under high pressure into the die. Typical injection pressures are 7 to 35 MPa (1000 to 5000 lb/in^2). The casting cycle is summarized in Figure 6.11. Production rates up to 500 parts per hour are not uncommon. Hot-chamber die casting imposes a special hardship on the injection system because much of it is submerged in the molten metal. The process is therefore limited in its applications to low-melting-point metals that do not chemically attack the plunger and other mechanical components. The metals include zinc, tin, lead, and sometimes magnesium.

In ***cold-chamber die-casting machines***, molten metal is poured into an unheated chamber from an external melting container, and a piston is used to inject the metal under high pressure into the die cavity. Injection pressures used in these machines are typically 14 to 140 MPa (2000–20,000 lb/in^2). The production cycle is explained in Figure 6.12. Compared to hot-chamber machines, cycle rates are not usually as fast because of the need to ladle the liquid metal into the chamber from an external source. Nevertheless, this casting process is a high production operation. Cold-chamber machines are typically used for casting aluminum, brass, and magnesium alloys. Low-melting-point alloys (zinc, tin, lead) can also be cast on cold-chamber machines, but the advantages of the hot-chamber process usually favor its use on these metals.

Molds used in die-casting operations are usually made of tool steel, mold steel, or maraging steel. Tungsten and molybdenum with good refractory qualities are also being used, especially in attempts to die cast steel and cast iron. Dies can be single-cavity or

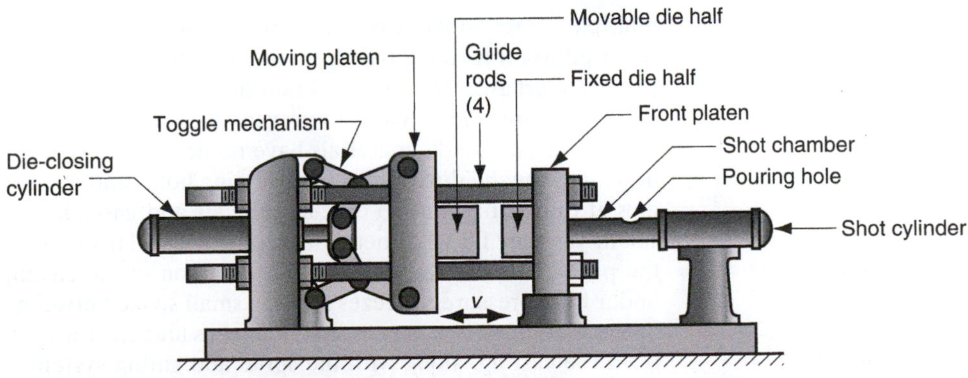

FIGURE 6.10 General configuration of a (cold-chamber) die-casting machine. (Credit: *Fundamentals of Modern Manufacturing*, 4th Edition by Mikell P. Groover, 2010. Reprinted with permission of John Wiley & Sons, Inc.)

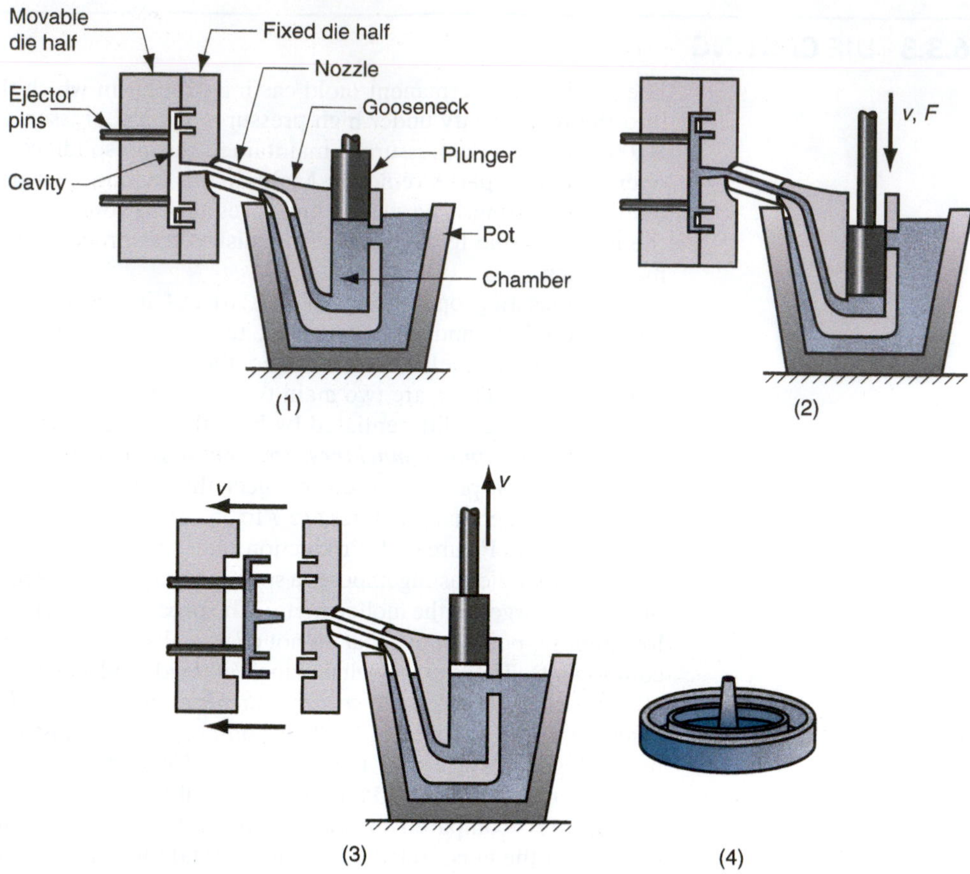

FIGURE 6.11 Cycle in hot-chamber casting: (1) With die closed and plunger withdrawn, molten metal flows into the chamber; (2) plunger forces metal in chamber to flow into die, maintaining pressure during cooling and solidification; and (3) plunger is withdrawn, die is opened, and solidified part is ejected. Finished part is shown in (4). (Credit: *Fundamentals of Modern Manufacturing*, 4th Edition by Mikell P. Groover, 2010. Reprinted with permission of John Wiley & Sons, Inc.)

multiple-cavity (single-cavity dies are shown in Figures 6.11 and 6.12). Ejector pins are required to remove the part from the die when it opens, as in our diagrams. These pins push the part away from the mold surface so that it can be removed. Lubricants must also be sprayed into the cavities to prevent sticking.

Because the die materials have no natural porosity and the molten metal rapidly flows into the die during injection, venting holes and passageways must be built into the dies at the parting line to evacuate the air and gases in the cavity. The vents are quite small; yet they fill with metal during injection. This metal must later be trimmed from the part. Also, formation of ***flash*** is common in die casting, in which the liquid metal under high pressure squeezes into the small space between the die halves at the parting line or into the clearances around the cores and ejector pins. This flash must be trimmed from the casting, along with the sprue and gating system.

Advantages of die casting include (1) high production rates possible; (2) economical for large production quantities; (3) close tolerances possible, on the order of ±0.076 mm (±0.003 in) for small parts; (4) good surface finish; (5) thin sections are possible, down to about 0.5 mm (0.020 in); and (6) rapid cooling provides small grain size

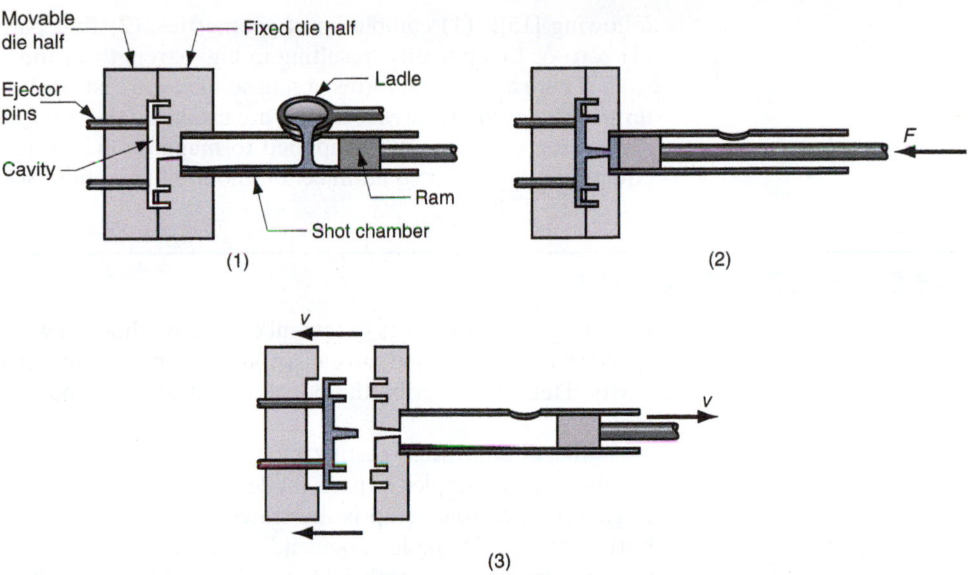

FIGURE 6.12 Cycle in cold-chamber casting: (1) With die closed and ram withdrawn, molten metal is poured into the chamber; (2) ram forces metal to flow into die, maintaining pressure during cooling and solidification; and (3) ram is withdrawn, die is opened, and part is ejected. (Gating system is simplified.) (Credit: *Fundamentals of Modern Manufacturing*, 4th Edition by Mikell P. Groover, 2010. Reprinted with permission of John Wiley & Sons, Inc.)

and good strength to the casting. The limitation of this process, in addition to the metals cast, is the shape restriction. The part geometry must allow for removal from the die cavity.

6.3.4 SQUEEZE CASTING AND SEMISOLID METAL CASTING

These are two processes that are often associated with die casting. ***Squeeze casting*** is a combination of casting and forging (Section 13.2) in which a molten metal is poured into a preheated lower die, and the upper die is closed to create the mold cavity after solidification begins. This differs from the usual permanent-mold casting process in which the die halves are closed prior to pouring or injection. Owing to the hybrid nature of the process, it is also known as ***liquid-metal forging***. The pressure applied by the upper die in squeeze casting causes the metal to completely fill the cavity, resulting in good surface finish and low shrinkage. The required pressures are significantly less than in forging of a solid metal billet and much finer surface detail can be imparted by the die than in forging. Squeeze casting can be used for both ferrous and nonferrous alloys, but aluminum and magnesium alloys are the most common due to their lower melting temperatures. Automotive parts are a common application.

Semisolid metal casting is a family of net-shape and near net-shape processes performed on metal alloys at temperatures between the liquidus and solidus (Section 5.3.1). Thus, the alloy is a mixture of solid and molten metals during casting, like a slurry; it is in the mushy state. In order to flow properly, the mixture must consist of solid metal globules in a liquid rather than the more typical dendritic solid shapes that form during freezing of a molten metal. This is achieved by forcefully stirring the slurry to prevent dendrite formation and instead encourage the spherical shapes, which in turn reduces the viscosity of the work metal. Advantages of semisolid metal casting include the following [15]: (1) complex part geometries, (2) thin part walls, (3) close tolerances, (4) zero or low porosity, resulting in high strength of the casting.

There are several forms of semisolid metal casting. When applied to aluminum, the terms ***thixocasting*** and ***rheocasting*** are used, and the production equipment is similar to a die-casting machine. When applied to magnesium, the term is ***thixomolding***, and the equipment is similar to an injection-molding machine (Section 8.6.1).

6.3.5 CENTRIFUGAL CASTING

Centrifugal casting refers to several casting methods in which the mold is rotated at high speed so that centrifugal force distributes the molten metal to the outer regions of the die cavity. Here we describe the process used to cast tubular parts, called true centrifugal casting.

In true centrifugal casting, molten metal is poured into a rotating mold to produce a tubular part. Examples of parts made by this process include pipes, tubes, bushings, and rings. One possible setup is illustrated in Figure 6.13. Molten metal is poured into a horizontal rotating mold at one end. In some operations, mold rotation commences after pouring has occurred rather than beforehand. The high-speed rotation results in centrifugal forces that cause the metal to take the shape of the mold cavity. Thus, the outside shape of the casting can be round, octagonal, hexagonal, and so on. However, the inside shape of the casting is (theoretically) perfectly round, due to the radially symmetric forces at work.

Orientation of the axis of mold rotation can be either horizontal or vertical, the former being more common. Let us consider how fast the mold must rotate in ***horizontal centrifugal casting*** for the process to work successfully. Centrifugal force is defined by this physics equation:

$$F = \frac{mv^2}{R} \tag{6.1}$$

where F = force, N (lb); m = mass, kg (lbm); v = velocity, m/s (ft/sec); and R = inside radius of the mold, m (ft). The force of gravity is its weight $W = mg$, where W is given in kg (lb), and g = acceleration of gravity, 9.8 m/s^2 (32.2 ft/sec^2). The so-called G-factor GF is the ratio of centrifugal force divided by weight:

$$GF = \frac{F}{W} = \frac{mv^2}{Rmg} = \frac{v^2}{Rg} \tag{6.2}$$

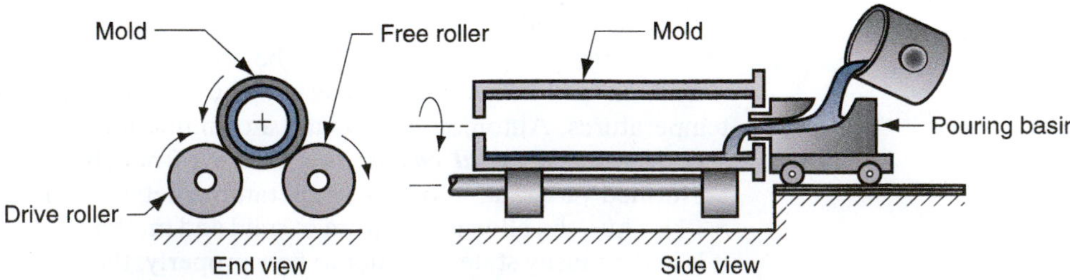

FIGURE 6.13 Setup for true centrifugal casting. (Credit: *Fundamentals of Modern Manufacturing,* 4th Edition by Mikell P. Groover, 2010. Reprinted with permission of John Wiley & Sons, Inc.)

Velocity v can be expressed as $2\pi RN/60 = \pi RN/30$, where N = rotational speed, rev/min. Substituting this expression into Eq. (6.2), we obtain

$$GF = \frac{R\left(\frac{\pi N}{30}\right)^2}{g} \quad (6.3)$$

Rearranging this to solve for rotational speed N, and using diameter D rather than radius in the resulting equation, we have

$$N = \frac{30}{\pi}\sqrt{\frac{2gGF}{D}} \quad (6.4)$$

where D = inside diameter of the mold, m (ft). If the G-factor is too low in centrifugal casting, the liquid metal will not remain forced against the mold wall during the upper half of the circular path but will "rain" inside the cavity. Slipping occurs between the molten metal and the mold wall, which means that the rotational speed of the metal is less than that of the mold. On an empirical basis, values of GF = 60 to 80 are found to be appropriate for horizontal centrifugal casting [2], although this depends to some extent on the metal being cast.

Example 6.1 Rotation Speed in True Centrifugal Casting

A true centrifugal casting operation is to be performed horizontally to make copper tube sections with OD = 25 cm and ID = 22.5 cm. What rotational speed is required if a G-factor of 65 is used to cast the tubing?

Solution: The inside diameter of the mold $D = OD$ of the casting = 25 cm = 0.25 m. We can compute the required rotational speed from Eq. (6.4) as follows:

$$N = \frac{30}{\pi}\sqrt{\frac{2(9.8)(26)}{0.25}} = \mathbf{681.7\ rev/min}.$$

■

In ***vertical centrifugal casting***, the effect of gravity acting on the liquid metal causes the casting wall to be thicker at the base than at the top. The inside profile of the casting wall takes on a parabolic shape. Consequently, part lengths made by vertical centrifugal casting are usually no more than about twice their diameters. This is quite satisfactory for bushings and other parts that have large diameters relatively to their lengths, especially if machining is used to accurately size the inside diameter.

Castings made by true centrifugal casting are characterized by high density, especially in the outer regions of the part where centrifugal force is greatest. Solidification shrinkage at the exterior of the cast tube is not a factor, because the centrifugal force continually reallocates molten metal toward the mold wall during freezing. Any impurities in the casting tend to be on the inner wall and can be removed by machining if necessary.

6.4 FOUNDRY PRACTICE

In all casting processes, the metal must be heated to the molten state to be poured or otherwise forced into the mold. Heating and melting are accomplished in a furnace. This section covers the various types of furnaces used in foundries and the pouring practices for delivering the molten metal from furnace to mold.

6.4.1 FURNACES

The types of furnaces most commonly used in foundries are (1) cupolas, (2) direct fuel-fired furnaces, (3) crucible furnaces, (4) electric-arc furnaces, and (5) induction furnaces.

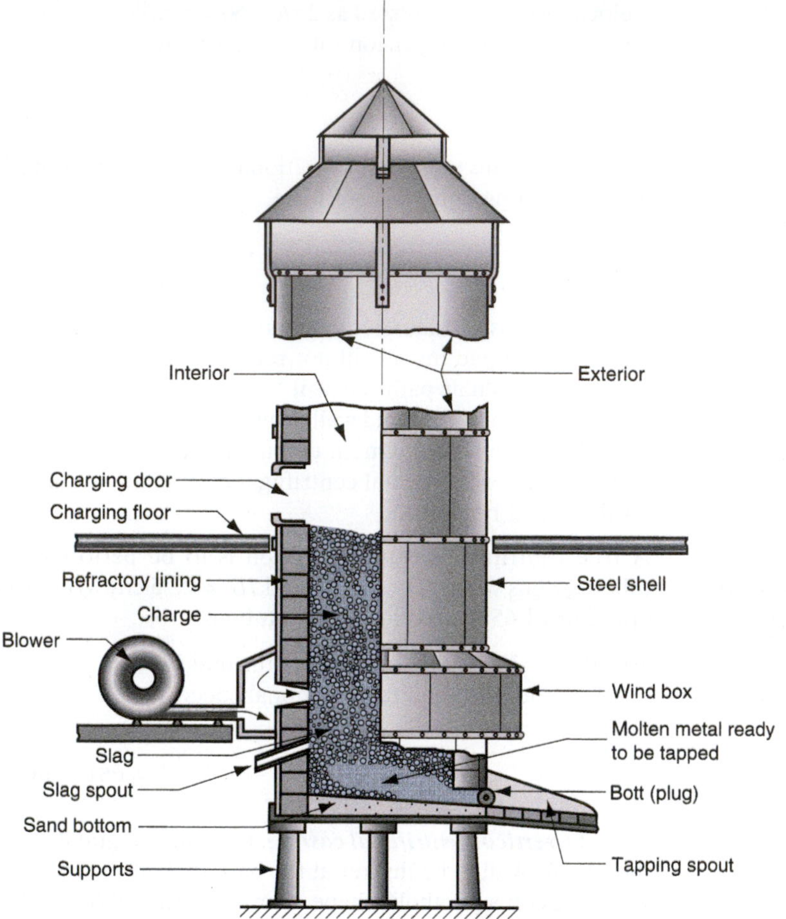

FIGURE 6.14 Cupola used for melting cast iron. Furnace shown is typical for a small foundry and omits details of emissions control system required in a modern cupola. (Credit: *Fundamentals of Modern Manufacturing*, 4th Edition by Mikell P. Groover, 2010. Reprinted with permission of John Wiley & Sons, Inc.)

Selection of the most appropriate furnace type depends on factors such as the casting alloy; its melting and pouring temperatures; capacity requirements of the furnace; costs of investment, operation, and maintenance; and environmental pollution considerations.

Cupolas A cupola is a vertical cylindrical furnace equipped with a tapping spout near its base. Cupolas are used only for melting cast irons, and although other furnaces are also used, the largest tonnage of cast iron is melted in cupolas. General construction and operating features of the cupola are illustrated in Figure 6.14. It consists of a large shell of steel plate lined with refractory. The "charge," consisting of iron, coke, flux, and possible alloying elements, is loaded through a charging door located less than halfway up the height of the cupola. The iron is usually a mixture of pig iron and scrap (including risers, runners, and sprues left over from previous castings). Coke is the fuel used to heat the furnace. Forced air is introduced through openings near the bottom of the shell for combustion of the coke. The flux is a basic compound such as limestone that reacts with coke ash and other impurities to form slag. The slag serves to cover the melt, protecting it from reaction with the environment inside the cupola and reducing heat loss. As the mixture is heated and melting of the iron occurs, the furnace is periodically tapped to provide liquid metal for the pour.

Direct Fuel-Fired Furnaces A direct fuel-fired furnace contains a small open hearth, in which the metal charge is heated by fuel burners located on the side of the furnace. The

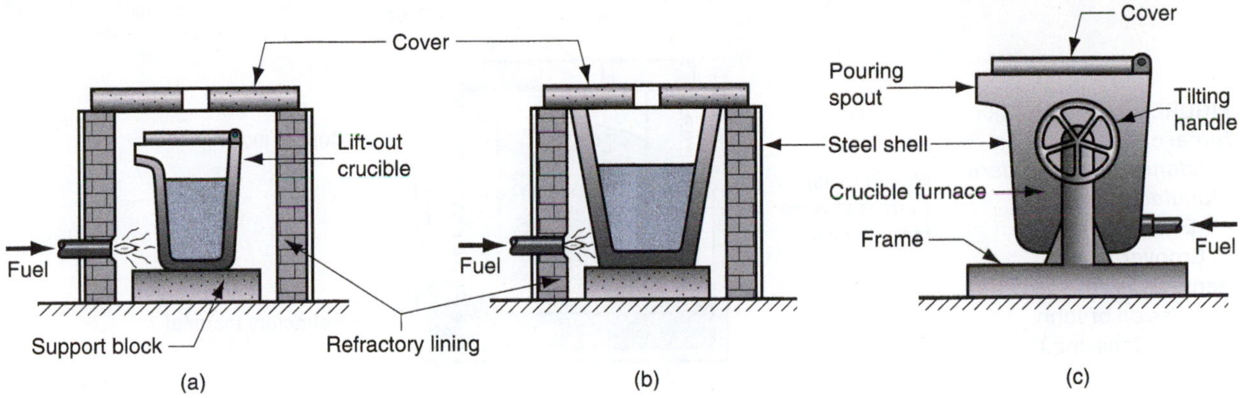

FIGURE 6.15 Three types of crucible furnaces: (a) lift-out crucible, (b) stationary pot, and (c) tilting-pot furnace. (Credit: *Fundamentals of Modern Manufacturing*, 4th Edition by Mikell P. Groover, 2010. Reprinted with permission of John Wiley & Sons, Inc.)

roof of the furnace assists the heating action by reflecting the flame down against the charge. Typical fuel is natural gas, and the combustion products exit the furnace through a stack. At the bottom of the hearth is a tap hole to release the molten metal. Direct fuel-fired furnaces are generally used in casting for melting nonferrous metals such as copper-base alloys and aluminum.

Crucible Furnaces These furnaces melt the metal without direct contact with a burning fuel mixture. For this reason, they are sometimes called ***indirect fuel-fired furnaces***. Three types of crucible furnaces are used in foundries: (a) lift-out type, (b) stationary, and (c) tilting, illustrated in Figure 6.15. They all utilize a container (the crucible) made out of a suitable refractory material (e.g., a clay–graphite mixture) or high-temperature steel alloy to hold the charge. In the ***lift-out crucible furnace***, the crucible is placed in a furnace and heated sufficiently to melt the metal charge. Oil, gas, or powdered coal are typical fuels for these furnaces. When the metal is melted, the crucible is lifted out of the furnace and used as a pouring ladle. The other two types, sometimes referred to as ***pot furnaces***, have the heating furnace and container as one integral unit. In the ***stationary pot furnace***, the furnace is stationary and the molten metal is ladled out of the container. In the ***tilting-pot furnace***, the entire assembly can be tilted for pouring. Crucible furnaces are used for nonferrous metals such as bronze, brass, and alloys of zinc and aluminum. Furnace capacities are generally limited to several hundred pounds.

Electric-Arc Furnaces In this furnace type, the charge is melted by heat generated from an electric arc that flows between two or three electrodes and the charge metal. Power consumption is high, but electric-arc furnaces can be designed for high melting capacity (23,000 to 45,000 kg/hr or 25 to 50 tons/hr), and they are used primarily for casting steel.

Induction Furnaces An induction furnace uses alternating current passing through a coil to develop a magnetic field in the metal, and the resulting induced current causes rapid heating and melting of the metal. Features of an induction furnace for foundry operations are illustrated in Figure 6.16. The electromagnetic force field causes a mixing action to occur in the liquid metal. Also, because the metal does not come in direct contact with the heating elements, the environment in which melting takes place can be closely controlled. All of this results in molten metals of high quality and purity, and induction furnaces are used for nearly any casting alloy when these requirements are

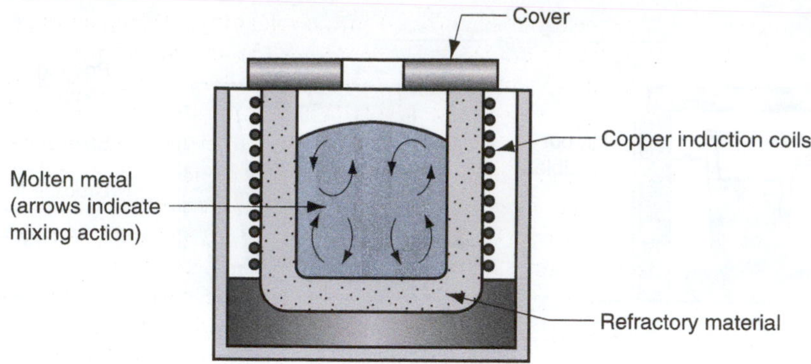

FIGURE 6.16 Induction furnace. (Credit: *Fundamentals of Modern Manufacturing*, 4th Edition by Mikell P. Groover, 2010. Reprinted with permission of John Wiley & Sons, Inc.)

important. Melting steel, cast iron, and aluminum alloys are common applications in foundry work.

6.4.2 POURING, CLEANING, AND HEAT TREATMENT

Moving the molten metal from the melting furnace to the mold is sometimes done using crucibles. More often, the transfer is accomplished by ***ladles*** of various kinds. These ladles receive the metal from the furnace and allow for convenient pouring into the molds. Two common ladles are illustrated in Figure 6.17, one for handling large volumes of molten metal using an overhead crane, and the other a "two-man ladle" for manually moving and pouring smaller amounts.

One of the problems in pouring is that oxidized molten metal can be introduced into the mold. Metal oxides reduce product quality, perhaps rendering the casting defective, so measures are taken to minimize the entry of these oxides into the mold during pouring. Filters are sometimes used to catch the oxides and other impurities as the metal is poured from the spout, and fluxes are used to cover the molten metal to retard oxidation. In addition, ladles have been devised to pour the liquid metal from the bottom, because the top surface is where the oxides accumulate.

After the casting has solidified and been removed from the mold, a number of additional steps are usually required. These operations include (1) trimming,

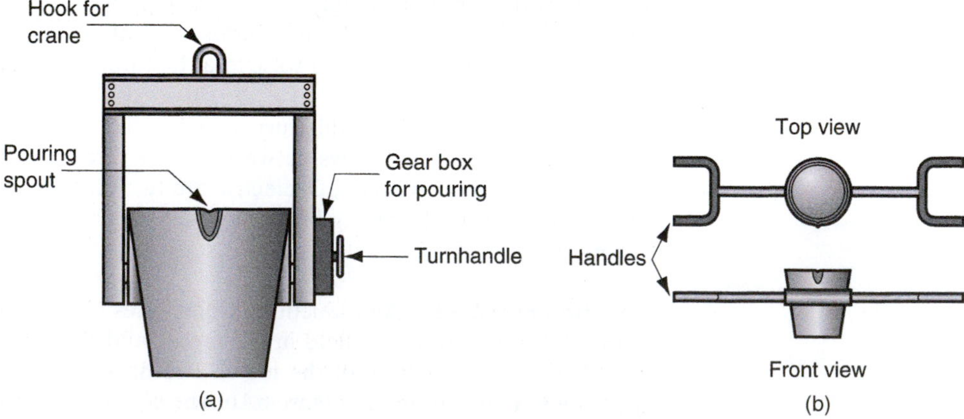

FIGURE 6.17 Two common types of ladles: (a) crane ladle and (b) two-man ladle. (Credit: *Fundamentals of Modern Manufacturing*, 4th Edition by Mikell P. Groover, 2010. Reprinted with permission of John Wiley & Sons, Inc.)

(2) removing the core, (3) surface cleaning, (4) inspection, (5) repair, if required, and (6) heat treatment. Steps (1) through (5) are collectively referred to in foundry work as "cleaning." The extent to which these additional operations are required varies with casting processes and metals. When required, they are usually labor-intensive and costly.

Trimming involves removal of sprues, runners, risers, parting-line flash, fins, chaplets, and any other excess metal from the cast part. In the case of brittle casting alloys and when the cross sections are relatively small, these appendages on the casting can be broken off. Otherwise, hammering, shearing, hack-sawing, band-sawing, abrasive wheel cutting, or various torch cutting methods are used.

If cores have been used to cast the part, they must be removed. Most cores are chemically bonded or oil-bonded sand, and they often fall out of the casting as the binder deteriorates. In some cases, they are removed by shaking the casting, either manually or mechanically. In rare instances, cores are removed by chemically dissolving the bonding agent used in the sand core. Solid cores must be hammered or pressed out.

Surface cleaning is most important in the case of sand casting. In many of the other casting methods, especially the permanent-mold processes, this step can be avoided. ***Surface cleaning*** involves removal of sand from the surface of the casting and otherwise enhancing the appearance of the surface. Methods used to clean the surface include tumbling, air-blasting with coarse sand grit or metal shot, wire brushing, and chemical pickling (Chapter 21).

Defects are possible in casting, and inspection is needed to detect their presence. We consider these quality issues in the following section.

Castings are often heat-treated (Chapter 20) to enhance their properties, either for subsequent processing operations such as machining or to bring out the desired properties for application of the part.

6.5 CASTING QUALITY

There are numerous opportunities for things to go wrong in a casting operation, resulting in quality defects in the cast product. In this section, we compile a list of the common defects that occur in casting and indicate the inspection procedures to detect them.

Casting Defects Some defects are common to any and all casting processes. These defects are illustrated in Figure 6.18 and briefly described in the following:

(a) ***Misruns*** are castings that solidify before completely filling the mold cavity. Typical causes include (1) fluidity of the molten metal is insufficient, (2) pouring temperature is too low, (3) pouring is done too slowly, and/or (4) cross section of the mold cavity is too thin.

(b) ***Cold shuts*** occur when two portions of the metal flow together but there is a lack of fusion between them due to premature freezing. Its causes are similar to those of a misrun.

(c) ***Cold shots*** result from splattering during pouring, causing the formation of solid globules of metal that become entrapped in the casting. Pouring procedures and gating system designs that avoid splattering can prevent this defect.

(d) ***Shrinkage cavity*** is a depression in the surface or an internal void in the casting, caused by solidification shrinkage that restricts the amount of molten metal available in the last region to freeze. It often occurs near the top of the casting, in which case it is referred to as a "pipe." See Figure 5.6(3). The problem can often be solved by proper riser design.

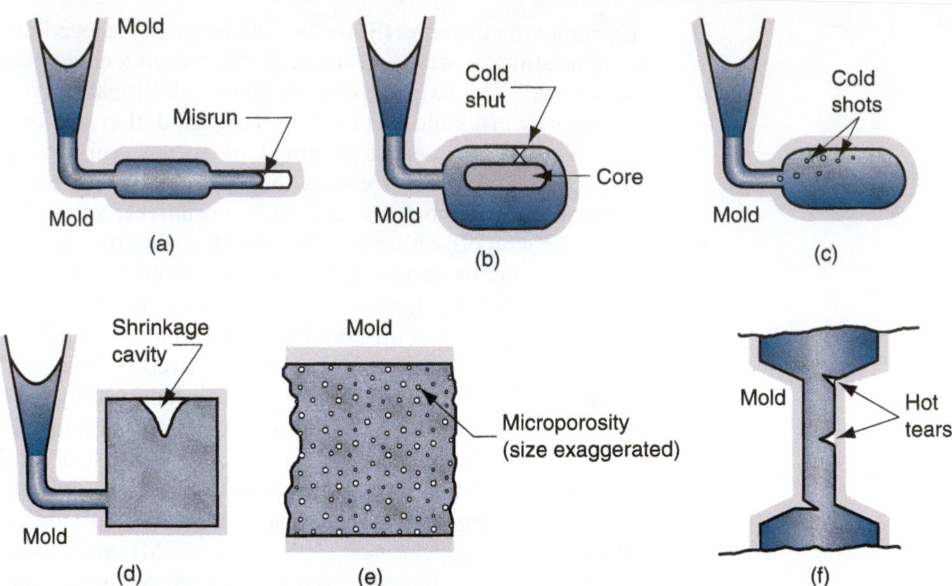

FIGURE 6.18 Some common defects in castings: (a) misrun, (b) cold shut, (c) cold shot, (d) shrinkage cavity, (e) microporosity, and (f) hot tearing. (Credit: *Fundamentals of Modern Manufacturing*, 4th Edition by Mikell P. Groover, 2010. Reprinted with permission of John Wiley & Sons, Inc.)

(e) ***Microporosity*** consists of a network of small voids distributed throughout the casting caused by localized solidification shrinkage of the final molten metal in the dendritic structure. The defect is usually associated with alloys, because of the protracted manner in which freezing occurs in these metals.

(f) ***Hot tearing***, also called ***hot cracking***, occurs when the casting is restrained from contraction by an unyielding mold during the final stages of solidification or early stages of cooling after solidification. The defect is manifested as a separation of the metal (hence, the terms ***tearing*** and ***cracking***) at a point of high tensile stress caused by the metal's inability to shrink naturally. In sand casting and other expendable-mold processes, it is prevented by compounding the mold to be collapsible. In permanent-mold processes, hot tearing is reduced by removing the part from the mold immediately after solidification.

Some defects are related to the use of sand molds, and therefore they occur only in sand castings. To a lesser degree, other expendable-mold processes are also susceptible to these problems. Defects found primarily in sand castings are shown in Figure 6.19 and described here:

(a) ***Sand blow*** is a defect consisting of a balloon-shaped gas cavity caused by release of mold gases during pouring. It occurs at or below the casting surface near the top of the casting. Low permeability, poor venting, and high moisture content of the sand mold are the usual causes.

(b) ***Pinholes***, also caused by release of gases during pouring, consist of many small gas cavities formed at or slightly below the surface of the casting.

(c) ***Sand wash*** is an irregularity in the surface of the casting that results from erosion of the sand mold during pouring, and the contour of the erosion is formed in the surface of the final cast part.

(d) ***Scabs*** are rough areas on the surface of the casting due to encrustations of sand and metal. They are caused by portions of the mold surface flaking off during solidification and becoming imbedded in the casting surface.

(e) ***Penetration*** is a surface defect that occurs when the fluidity of the liquid metal is high, and it penetrates into the sand mold or sand core. Upon freezing, the casting surface

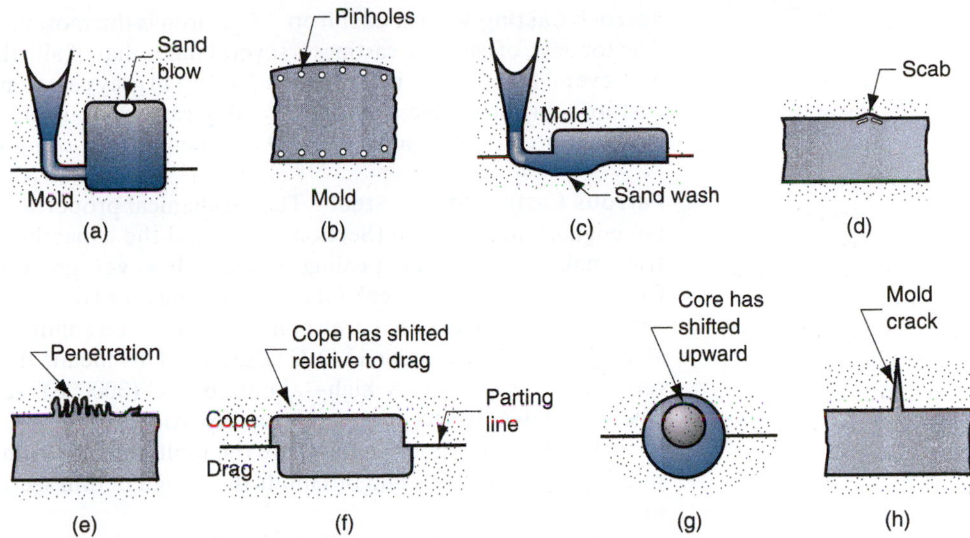

FIGURE 6.19 Common defects in sand castings: (a) sand blow, (b) pin holes, (c) sand wash, (d) scabs, (e) penetration, (f) mold shift, (g) core shift, and (h) mold crack. (Credit: *Fundamentals of Modern Manufacturing*, 4th Edition by Mikell P. Groover, 2010. Reprinted with permission of John Wiley & Sons, Inc.)

consists of a mixture of sand grains and metal. Harder packing of the sand mold helps to alleviate this condition.

(f) ***Mold shift*** refers to a defect caused by a sidewise displacement of the mold cope relative to the drag, the result of which is a step in the casting at the parting line.

(g) ***Core shift*** is similar to mold shift, but it is the core that is displaced, and the displacement is usually vertical. Core shift is caused by buoyancy of the molten metal and its tendency to lift the core, which is lighter.

(h) ***Mold crack*** occurs when mold strength is insufficient, and a crack develops, into which liquid metal can seep to form a "fin" on the final casting.

Inspection Methods Foundry inspection procedures include (1) visual inspection to detect obvious defects such as misruns, cold shuts, and severe surface flaws; (2) dimensional measurements to ensure that tolerances have been met; and (3) metallurgical, chemical, physical, and other tests concerned with the inherent quality of the cast metal [7]. Tests in category (3) include: (a) pressure testing—to locate leaks in the casting; (b) radiographic methods, magnetic particle tests, use of fluorescent penetrants, and supersonic testing—to detect either surface or internal defects in the casting; and (c) mechanical testing to determine properties such as tensile strength and hardness. If defects are discovered but are not too serious, it is often possible to save the casting by welding, grinding, or other salvage methods to which the customer has agreed.

6.6 METALS FOR CASTING

Most commercial castings are made of alloys rather than pure metals. Alloys are generally easier to cast, and properties of the resulting product are better. Casting alloys can be classified as ferrous or nonferrous. The ferrous category is subdivided into cast iron and cast steel.

Ferrous Casting Alloys: Cast Iron Cast iron is the most important of all casting alloys. The tonnage of cast iron castings is several times that of all other metals combined. There are several types of cast iron (Section 2.1.2): (1) gray cast iron, (2) nodular iron, (3) white cast iron, (4) malleable iron, and (5) alloy cast irons. Typical pouring temperatures for cast iron are around 1400°C (2500°F), depending on composition.

Ferrous Casting Alloys: Steel The mechanical properties of steel make it an attractive engineering material (Section 2.1.1), and the capability to create complex geometries makes casting an appealing process. However, great difficulties are faced by the foundry specializing in steel. First, the melting point of steel is considerably higher than for most other metals that are commonly cast. The solidification range for low-carbon steels begins at just under 1540°C (2800°F). This means that the pouring temperature required for steel is very high—about 1650°C (3000°F). At these high temperatures, steel is chemically very reactive. It readily oxidizes, so special procedures must be used during melting and pouring to isolate the molten metal from air. Also, molten steel has relatively poor fluidity, and this limits the design of thin sections in components cast out of steel.

Several characteristics of steel castings make it worth the effort to solve these problems. Tensile strength is higher than for most other casting metals, ranging upward from about 410 MPa (60,000 lb/in^2) [9]. Steel castings have better toughness than most other casting alloys. The properties of steel castings are isotropic; strength is virtually the same in all directions. By contrast, mechanically formed parts (e.g., rolling, forging) exhibit directionality in their properties. Depending on the requirements of the product, isotropic behavior of the material may be desirable. Another advantage of steel castings is ease of welding. They can be readily welded without significant loss of strength to repair the casting, or to fabricate structures with other steel components.

Nonferrous Casting Alloys Nonferrous casting metals include alloys of aluminum, magnesium, copper, tin, zinc, nickel, and titanium (Section 2.1.3). ***Aluminum alloys*** are generally considered to be very castable. The melting point of pure aluminum is 660°C (1220°F), so pouring temperatures for aluminum casting alloys are low compared to cast iron and steel. Their properties make them attractive for castings: light weight, wide range of strength properties attainable through heat treatment, and ease of machining. ***Magnesium alloys*** are the lightest of all casting metals. Other properties include corrosion resistance, as well as high strength-to-weight and stiffness-to-weight ratios.

Copper alloys include bronze, brass, and aluminum bronze. Properties that make them attractive include corrosion resistance, attractive appearance, and good bearing qualities. The high cost of copper is a limitation on the use of its alloys. Applications include pipe fittings, marine propeller blades, pump components, and ornamental jewelry.

Tin has the lowest melting point of the casting metals. ***Tin-based alloys*** are generally easy to cast. They have good corrosion resistant but poor mechanical strength, which limits their applications to pewter mugs and similar products not requiring high strength. ***Zinc alloys*** are commonly used in die casting. Zinc has a low melting point and good fluidity, making it highly castable. Its major weakness is low creep strength, so its castings cannot be subjected to prolonged high stresses.

Nickel alloys have good hot strength and corrosion resistance, which make them suited to high-temperature applications such as jet engine and rocket components, heat shields, and similar components. Nickel alloys also have a high melting point and are not easy to cast. ***Titanium alloys*** for casting are corrosion resistant and possess high strength-to-weight ratios. However, titanium has a high melting point, low fluidity, and a propensity to oxidize at high temperatures. These properties make it and its alloys difficult to cast.

6.7 PRODUCT DESIGN CONSIDERATIONS

If casting is selected by the product designer as the primary manufacturing process for a particular component, then certain guidelines should be followed to facilitate production of the part and avoid many of the defects enumerated in Section 6.5. Some of the important guidelines and considerations for casting are presented here.

- ***Geometric simplicity***. Although casting is a process that can be used to produce complex part geometries, simplifying the part design will improve its castability. Avoiding unnecessary complexities simplifies mold making, reduces the need for cores, and improves the strength of the casting.
- ***Corners***. Sharp corners and angles should be avoided, because they are sources of stress concentrations and may cause hot tearing and cracks in the casting. Generous fillets should be designed on inside corners, and sharp edges should be blended.
- ***Section thicknesses*** should be uniform to avoid shrinkage cavities. Thicker sections create ***hot spots*** in the casting, because greater volume requires more time for solidification and cooling. These are likely locations of shrinkage cavities.
- ***Draft***. Part sections that project into the mold should have a draft or taper, as defined in Figure 6.20. In expendable-mold casting, the purpose of this draft is to facilitate removal of the pattern from the mold. In permanent-mold casting, its purpose is to aid in removal of the part from the mold. Similar tapers should be allowed if solid cores are used in the casting process. The required draft need only be about 1° for sand casting and 2° to 3° for permanent-mold processes.
- ***Use of cores***. Minor changes in part design can often reduce the need for coring, as shown in Figure 6.20.
- ***Dimensional tolerances***. There are significant differences in the dimensional accuracies that can be achieved in castings, depending on which process is used. Table 6.1 provides a compilation of typical part tolerances for various casting processes and metals.
- ***Surface finish***. Typical surface roughness achieved in sand casting is around 6 μm (250 μ-in). Similarly poor finishes are obtained in shell molding, while plaster-mold and investment casting produce much better roughness values: 0.75 μm (30 μ-in). Among the permanent-mold processes, die casting is noted for good surface finishes at around 1 μm (40 μ-in).
- ***Machining allowances***. Tolerances achievable in many casting processes are insufficient to meet functional needs in many applications. Sand casting is the most

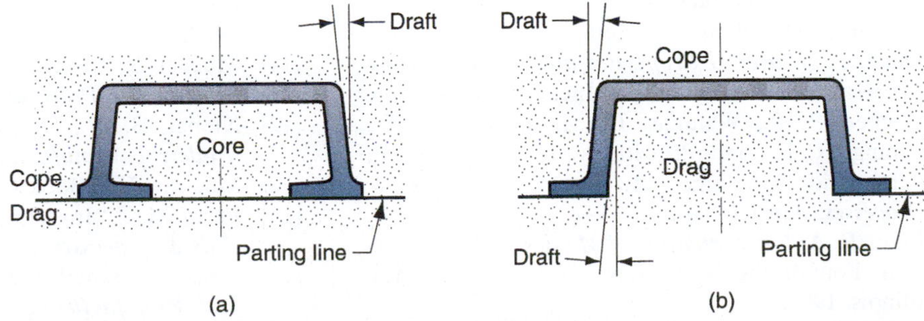

FIGURE 6.20 Draft defined. Also, design change to eliminate the need for using a core: (a) original design and (b) redesign. (Credit: *Fundamentals of Modern Manufacturing*, 4th Edition by Mikell P. Groover, 2010. Reprinted with permission of John Wiley & Sons, Inc.)

TABLE 6.1 Typical dimensional tolerances for various casting processes and metals

Casting Process	Part Size	Tolerance mm	Tolerance in	Casting Process	Part Size	Tolerance mm	Tolerance in
Sand casting				Permanent mold			
Aluminum[a]	Small	±0.5	±0.020	Aluminum[a]	Small	±0.25	±0.010
Cast iron	Small	±1.0	±0.040	Cast iron	Small	±0.8	±0.030
	Large	±1.5	±0.060	Copper alloys	Small	±0.4	±0.015
Copper alloys	Small	±0.4	±0.015	Steel	Small	±0.5	±0.020
Steel	Small	±1.3	±0.050	Die casting			
	Large	±2.0	±0.080	Aluminum[a]	Small	±0.12	±0.005
Shell molding				Copper alloys	Small	±0.12	±0.005
Aluminum[a]	Small	±0.25	±0.010	Investment			
Cast iron	Small	±0.5	±0.020	Aluminum[a]	Small	±0.12	±0.005
Copper alloys	Small	±0.4	±0.015	Cast iron	Small	±0.25	±0.010
Steel	Small	±0.8	±0.030	Copper alloys	Small	±0.12	±0.005
Plaster mold	Small	±0.12	±0.005	Steel	Small	±0.25	±0.010
	Large	±0.4	±0.015				

Compiled from [7], [14], and other sources.
[a]Values for aluminum also apply to magnesium.

prominent example of this deficiency. In these cases, portions of the casting must be machined to the required dimensions. Almost all sand castings must be machined to some extent in order for the part to be made functional. Therefore, additional material, called the ***machining allowance***, is left on the casting for machining those surfaces where necessary. Typical machining allowances for sand castings range between 1.5 mm and 3 mm (1/16 in and 1/4 in).

REFERENCES

[1] Amstead, B. H., Ostwald, P. F., and Begeman, M. L. ***Manufacturing Processes***. John Wiley & Sons, Inc., New York, 1987.

[2] Beeley, P. R. ***Foundry Technology***. Newnes-Butterworths, London, 1972.

[3] Black, J., and Kohser, R. ***DeGarmo's Materials and Processes in Manufacturing***, 10th ed. John Wiley & Sons, Inc., Hoboken, New Jersey, 2008.

[4] Datsko, J. ***Material Properties and Manufacturing Processes***. John Wiley & Sons, Inc., New York, 1966.

[5] Decker, R. F., Walukas, D. M., LeBeau, S. E., Vining, R. E., and Prewitt, N. D. "Advances in Semi-Solid Molding," ***Advanced Materials & Processes***, April 2004, pp. 41–42.

[6] Flinn, R. A. ***Fundamentals of Metal Casting***. American Foundrymen's Society, Inc., Des Plaines, Illinois, 1987.

[7] Heine, R. W., Loper, Jr., C.R., and Rosenthal, C. ***Principles of Metal Casting***, 2nd ed. McGraw-Hill Book Co., New York, 1967.

[8] Kotzin, E. L. ***Metalcasting & Molding Processes***. American Foundrymen's Society, Inc., Des Plaines, Illinois, 1981.

[9] Metals Handbook, Vol. **15**, ***Casting***. ASM International, Materials Park, Ohio, 2008.

[10] Mikelonis, P. J. (ed.). ***Foundry Technology***. American Society for Metals, Materials Park, Ohio, 1982.

[11] Mueller, B. "Investment Casting Trends," ***Advanced Materials & Processes***, March 2005, pp. 30–32.

[12] Niebel, B. W., Draper, A. B., and Wysk, R. A. ***Modern Manufacturing Process Engineering***. McGraw-Hill Book Co., New York, 1989.

[13] Perry, M. C. "Investment Casting," ***Advanced Materials & Processes***, June 2008, pp. 31–33.

[14] Wick, C., Benedict, J. T., and Veilleux, R. F. ***Tool and Manufacturing Engineers Handbook***, 4th ed. Vol. II, ***Forming***. Society of Manufacturing Engineers, Dearborn, Michigan, 1984, Chapter 16.

[15] www.wikipedia.org/wiki/semi-solid_metal_casting.

REVIEW QUESTIONS

6.1. Name the two basic categories of casting processes.

6.2. There are various types of patterns used in sand casting. What is the difference between a split pattern and a match-plate pattern?

6.3. What is a chaplet?

6.4. What properties determine the quality of a sand mold for sand casting?

6.5. What is the Antioch process?

6.6. What are the most common metals used in die casting?

6.7. Which die-casting machines usually have a higher production rate, cold-chamber or hot-chamber, and why?

6.8. What is flash in die casting?

6.9. What is a cupola?

6.10. What are some of the operations required in sand casting after the casting is removed from the mold?

6.11. What are some of the general defects encountered in casting processes? Name and briefly describe three.

PROBLEMS

6.1. A horizontal true centrifugal casting operation will be used to make copper tubing. The lengths will be 1.5 m with outside diameter = 15.0 cm, and inside diameter = 12.5 cm. If the rotational speed of the pipe = 1000 rev/min, determine the G-factor.

6.2. A true centrifugal casting operation is performed in a horizontal configuration to make cast iron pipe sections. The sections will have a length = 42.0 in, outside diameter = 8.0 in, and wall thickness = 0.50 in. If the rotational speed of the pipe = 500 rev/min, determine the G-factor. Is the operation likely to be successful?

6.3. A horizontal true centrifugal casting process is used to make brass bushings with the following dimensions: length = 10 cm, outside diameter = 15 cm, and inside diameter = 12 cm. (a) Determine the required rotational speed in order to obtain a G-factor of 70. (b) When operating at this speed, what is the centrifugal force per square meter (Pa) imposed by the molten metal on the inside wall of the mold? The density of brass is 8.62 g/cm^3.

6.4. True centrifugal casting is performed horizontally to make large diameter copper tube sections. The tubes have a length = 1.0 m, diameter = 0.25 m, and wall thickness = 15 mm. (a) If the rotational speed of the pipe = 700 rev/min, determine the G-factor on the molten metal. (b) Is the rotational speed sufficient to avoid "rain"? (c) What volume of molten metal must be poured into the mold to make the casting if solidification shrinkage and contraction after solidification are considered? Solidification shrinkage for copper = 4.5%, and solid thermal contraction = 7.5%.

6.5. If a true centrifugal casting operation were to be performed in a space station circling the Earth, how would weightlessness affect the process?

6.6. A horizontal true centrifugal casting process is used to make aluminum rings with the following dimensions: length = 5 cm, outside diameter = 65 cm, and inside diameter = 60 cm. (a) Determine the rotational speed that will provide a G-factor = 60. (b) Suppose that the ring were made out of steel instead of aluminum. If the rotational speed computed in part (a) were used in the steel casting operation, determine the G-factor and (c) centrifugal force per square meter (Pa) on the mold wall. (d) Would this rotational speed result in a successful operation? The density of steel = 7.87 g/cm^3.

6.7. For the steel ring of preceding Problem 6.6(b), determine the volume of molten metal that must be poured into the mold, given that the liquid shrinkage is 0.5%, solidification shrinkage = 3%, and solid contraction after freezing = 7.2%.

6.8. A horizontal true centrifugal casting process is used to make lead pipe for chemical plants. The pipe has length = 0.5 m, outside diameter = 70 mm, and wall thickness = 6.0 mm. Determine the rotational speed that will provide a G-factor = 60.

6.9. The housing for a certain machinery product is made of two components, both aluminum castings. The larger component has the shape of a dish sink, and the second component is a flat cover that is attached to the first component to create an enclosed space for the machinery parts. Sand casting is used to produce the two castings, both of which are plagued by defects in the form of misruns and cold shuts. The foreman complains that the parts are too thin, and that is the reason for the defects. However, it is known that the same components are cast successfully in other foundries. What other explanation can be given for the defects?

6.10. A large steel sand casting shows the characteristic signs of penetration defect: a surface consisting of a mixture of sand and metal. (a) What steps can be taken to correct the defect? (b) What other possible defects might result from taking each of these steps?

7 GLASSWORKING

Chapter Contents

7.1 Raw Materials Preparation and Melting

7.2 Shaping Processes in Glassworking
- 7.2.1 Shaping of Piece Ware
- 7.2.2 Shaping of Flat and Tubular Glass
- 7.2.3 Forming of Glass Fibers

7.3 Heat Treatment and Finishing
- 7.3.1 Heat Treatment
- 7.3.2 Finishing

7.4 Product Design Considerations

Glass products are commercially manufactured in an almost unlimited variety of shapes. Many are produced in very large quantities, such as light bulbs, beverage bottles, and window glass. Others, such as giant telescope lenses, are made individually.

Glass is one of three basic types of ceramics (Section 2.2). It is distinguished by its noncrystalline (vitreous) structure, whereas the other ceramic materials have a crystalline structure. The methods by which glass is shaped into useful products are quite different from those used for the other types. In glassworking, the principal starting material is silica (SiO_2); this is usually combined with other oxide ceramics that form glasses. The starting material is heated to transform it from a hard solid into a viscous liquid; it is then shaped into the desired geometry while in this highly plastic or fluid condition. When cooled and hard, the material remains in the glassy state rather than crystallizing.

The typical manufacturing sequence in glassworking consists of the steps pictured in Figure 7.1. Shaping is accomplished by various processes, including casting, pressing and blowing (to produce bottles and other containers), and rolling (to make plate glass). A finishing step is required for certain products.

7.1 RAW MATERIALS PREPARATION AND MELTING

The main component in nearly all glasses is silica, the primary source of which is natural quartz in sand. The sand must be washed and classified. Washing removes impurities such as clay and certain minerals that would cause undesirable coloring of the glass. ***Classifying*** the sand means grouping the grains according to size. The

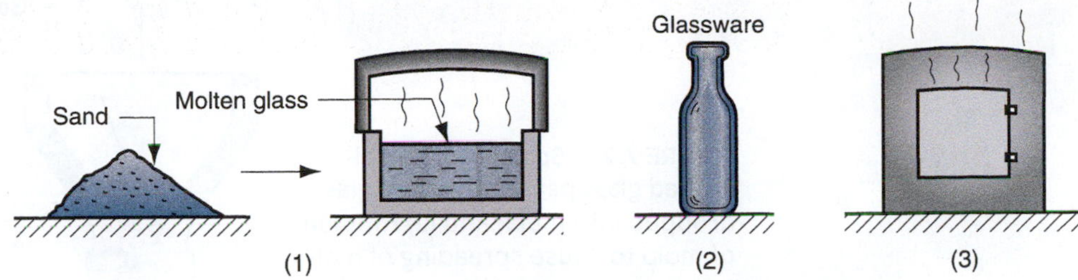

FIGURE 7.1 The typical process sequence in glassworking: (1) preparation of raw materials and melting, (2) shaping, and (3) heat treatment. (Credit: *Fundamentals of Modern Manufacturing*, 4th Edition by Mikell P. Groover, 2010. Reprinted with permission of John Wiley & Sons, Inc.)

most desirable particle size for glassmaking is in the range of 0.1 to 0.6 mm (0.004–0.025 in) [3]. The various other components, such as soda ash (source of Na_2O), limestone (source of CaO), aluminum oxide, potash (source of K_2O), and other minerals, are added in the proper proportions to achieve the desired composition. The mixing is usually done in batches, in amounts that are compatible with the capacities of available melting furnaces.

Recycled glass is usually added to the mixture in modern practice. In addition to preserving the environment, recycled glass facilitates melting. Depending on the amount of waste glass available and the specifications of the final composition, the proportion of recycled glass may be up to 100%.

The batch of starting materials to be melted is referred to as a ***charge***, and the procedure of loading it into the melting furnace is called ***charging*** the furnace. Glass-melting furnaces can be divided into the following types [3]: (1) ***pot furnaces***—ceramic pots of limited capacity in which melting occurs by heating the walls of the pot; (2) ***day tanks***—larger capacity vessels for batch production in which heating is done by burning fuels above the charge; (3) ***continuous tank furnaces***—long tank furnaces in which raw materials are fed in one end, and melted as they move to the other end where molten glass is drawn out for high production; and (4) ***electric furnaces*** of various designs for a wide range of production rates.

Glass melting is generally carried out at temperatures around 1500°C to 1600°C (2700°F to 2900°F). The melting cycle for a typical charge takes 24 to 48 hours. This is the time required for all of the sand grains to become a clear liquid and for the molten glass to be refined and cooled to the appropriate temperature for working. Molten glass is a viscous liquid, the viscosity being inversely related to temperature. Because the shaping operation immediately follows the melting cycle, the temperature at which the glass is tapped from the furnace depends on the viscosity required for the subsequent process.

7.2 SHAPING PROCESSES IN GLASSWORKING

The major categories of glass products are window glass, containers, light bulbs, laboratory glassware, glass fibers, and optical glass. Despite the variety represented by this list, the shaping processes to fabricate these products can be grouped into only three categories: (1) discrete processes for piece ware, which includes bottles, light bulbs, and other individual items; (2) continuous processes for making flat glass (sheet and plate glass for windows) and tubing (for laboratory ware and fluorescent lights); and (3) fiber-making processes to produce fibers for insulation, fiberglass composite materials, and fiber optics.

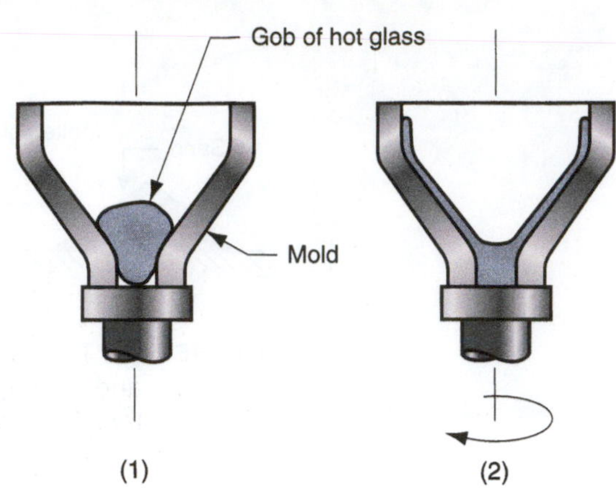

FIGURE 7.2 Spinning of funnel-shaped glass parts: (1) gob of glass dropped into mold; and (2) rotation of mold to cause spreading of molten glass on mold surface. (Credit: *Fundamentals of Modern Manufacturing*, 4th Edition by Mikell P. Groover, 2010. Reprinted with permission of John Wiley & Sons, Inc.)

7.2.1 SHAPING OF PIECE WARE

The ancient methods of hand-working glass, such as glass blowing, are still employed today to make glassware items of high value in small quantities. Most of the processes discussed in this section are highly mechanized technologies for producing discrete pieces such as jars, bottles, and light bulbs in high quantities.

Spinning Glass spinning is similar to ***centrifugal casting*** of metals, and is also known by that name in glassworking. It is used to produce funnel-shaped components. The setup is pictured in Figure 7.2. A gob of molten glass is dropped into a conical mold made of steel. The mold is rotated so that centrifugal force causes the glass to flow upward and spread itself on the mold surface.

Pressing This is a widely used process for mass-producing glass pieces such as dishes, bakeware, headlight lenses, and similar items that are relatively flat. The process is illustrated and described in Figure 7.3. The large quantities of most pressed products justify a high level of automation in this production sequence.

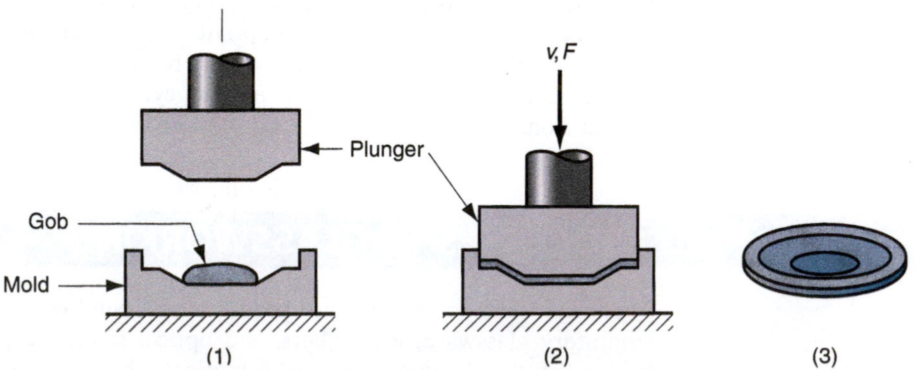

FIGURE 7.3 Pressing of a flat glass piece: (1) a gob of glass fed into mold from the furnace; (2) pressing into shape by plunger; and (3) plunger is retracted and the finished product is removed. Symbols v and F indicate motion (v = velocity) and applied force, respectively. (Credit: *Fundamentals of Modern Manufacturing*, 4th Edition by Mikell P. Groover, 2010. Reprinted with permission of John Wiley & Sons, Inc.)

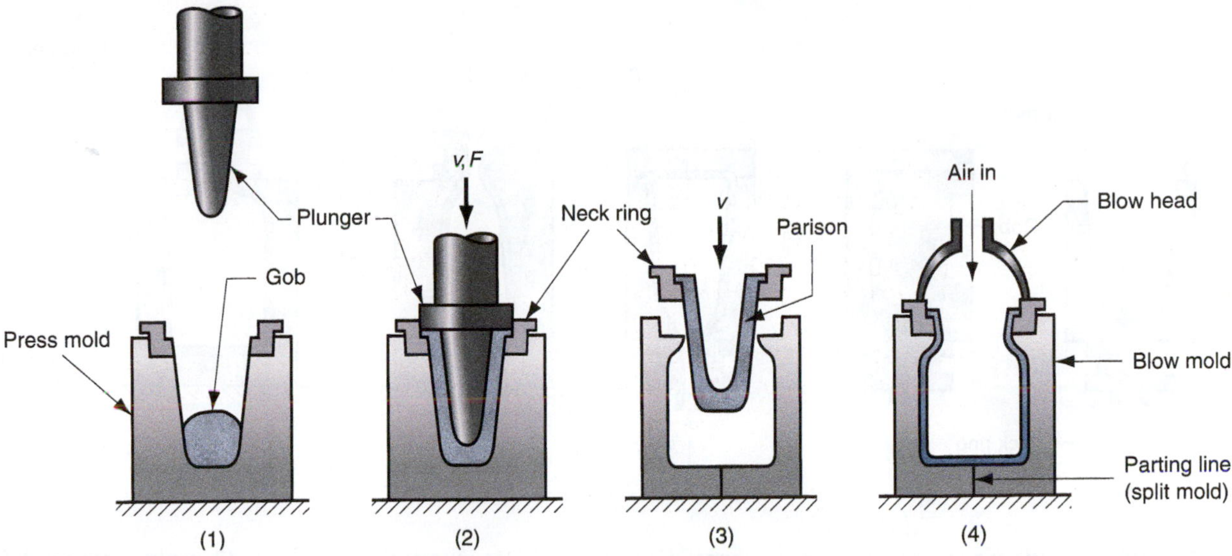

FIGURE 7.4 Press-and-blow forming sequence: (1) molten gob is fed into mold cavity; (2) pressing to form a ***parison***; (3) the partially formed parison, held in a neck ring, is transferred to the blow mold; and (4) blown into final shape. Symbols *v* and *F* indicate motion (v = velocity) and applied force, respectively. (Credit: *Fundamentals of Modern Manufacturing*, 4th Edition by Mikell P. Groover, 2010. Reprinted with permission of John Wiley & Sons, Inc.)

Blowing Several shaping sequences include blowing as one or more of the steps. Instead of a manual operation, blowing is performed on highly automated equipment. The two sequences we describe here are the press-and-blow and blow-and-blow methods.

As the name indicates, the ***press-and-blow*** method is a pressing operation followed by a blowing operation, as portrayed in Figure 7.4. The process is suited to the production of wide-mouth containers. A split mold is used in the blowing operation for part removal.

The ***blow-and-blow*** method is used to produce smaller-mouthed bottles. The sequence is similar to the preceding, except that two (or more) blowing operations are used rather than pressing and blowing. There are variations to the process, depending on the geometry of the product, with one possible sequence shown in Figure 7.5. Reheating is sometimes required between blowing steps. Duplicate and triplicate molds are sometimes used along with matching gob feeders to increase production rates. Press-and-blow and blow-and-blow methods are used to make jars, beverage bottles, incandescent light bulb enclosures, and similar geometries.

Casting If the molten glass is sufficiently fluid, it can be poured into a mold. Relatively massive objects, such as astronomical lenses and mirrors, are made by this method. These pieces must be cooled very slowly to avoid internal stresses and possible cracking due to temperature gradients that would otherwise be set up in the glass. After cooling and solidifying, the piece must be finished by lapping and polishing (Chapter 18). Casting is not much used in glassworking except for these kinds of special jobs. Not only is cooling and cracking a problem, but also molten glass is relatively viscous at normal working temperatures, and does not flow through small orifices or into small sections as well as molten metals or heated thermoplastics. Smaller lenses are usually made by pressing, discussed previously.

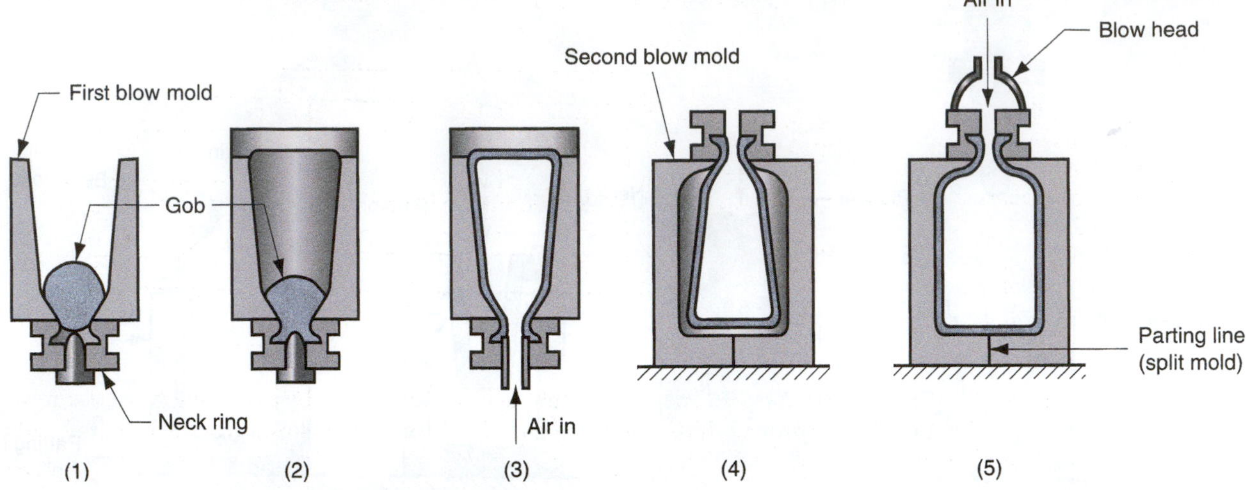

FIGURE 7.5 Blow-and-blow forming sequence: (1) Gob is fed into inverted mold cavity; (2) mold is covered; (3) first blowing step; (4) partially formed piece is reoriented and transferred to second blow mold; and (5) blown to final shape. (Credit: *Fundamentals of Modern Manufacturing,* 4th Edition by Mikell P. Groover, 2010. Reprinted with permission of John Wiley & Sons, Inc.)

7.2.2 SHAPING OF FLAT AND TUBULAR GLASS

Here we describe two methods for making plate glass and one method for producing tube stock. They are continuous processes, in which long sections of flat window glass or glass tubing are made and later cut into appropriate sizes and lengths.

Rolling of Flat Plate Flat plate glass can be produced by rolling, as illustrated in Figure 7.6. The starting glass, in a suitably plastic condition from the furnace, is squeezed through opposing rolls whose separation determines the thickness of the sheet. The rolling operation is usually set up so that the flat glass is moved directly into an annealing furnace. The rolled glass sheet must later be ground and polished for parallelism and smoothness.

Float Process This process was developed in the late 1950s. Its advantage over other methods such as rolling is that it obtains smooth surfaces that need no subsequent finishing. In the float process, illustrated in Figure 7.7, the glass flows directly from its melting furnace onto the surface of a molten tin bath. The highly fluid glass spreads evenly across the molten tin surface, achieving a uniform thickness and smoothness. After moving into a cooler region of the bath, the glass hardens and travels through an annealing furnace, after which it is cut to size.

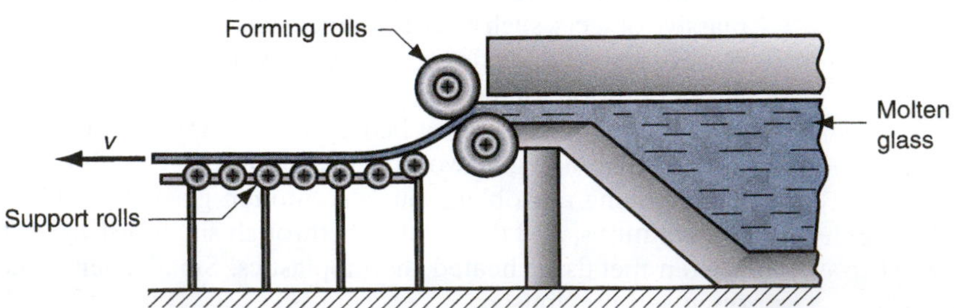

FIGURE 7.6 Rolling of flat glass. (Credit: *Fundamentals of Modern Manufacturing,* 4th Edition by Mikell P. Groover, 2010. Reprinted with permission of John Wiley & Sons, Inc.)

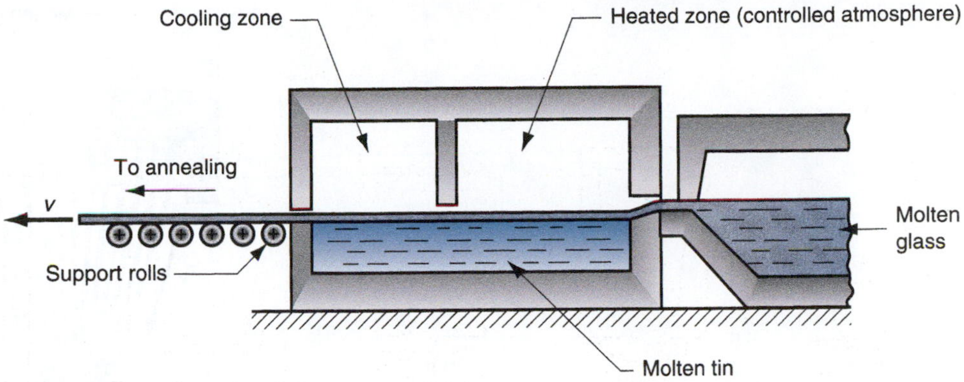

FIGURE 7.7 The float process for producing sheet glass. (Credit: *Fundamentals of Modern Manufacturing*, 4th Edition by Mikell P. Groover, 2010. Reprinted with permission of John Wiley & Sons, Inc.)

Drawing of Glass Tubes Glass tubing is manufactured by a drawing process known as the ***Danner process***, illustrated in Figure 7.8. Molten glass flows around a rotating hollow mandrel through which air is blown while the glass is being drawn. The air temperature and its volumetric flow rate, as well as the drawing velocity, determine the diameter and wall thickness of the tubular cross section. During hardening, the glass tube is supported by a series of rollers extending about 30 m (100 ft) beyond the mandrel. The continuous tubing is then cut into standard lengths. Tubular glass products include laboratory glassware, fluorescent light tubes, and thermometers.

7.2.3 FORMING OF GLASS FIBERS

Glass fibers are used in applications ranging from insulation wool to fiber optics communications lines. Glass fiber products can be divided into two categories [6]: (1) fibrous glass for thermal insulation, acoustical insulation, and air filtration, in which the fibers are in a random, wool-like condition; and (2) long, continuous filaments suitable for fiber-reinforced plastics, yarns and fabrics, and fiber optics. Different production methods are used for the two categories; we describe two methods next, representing each of the product categories, respectively.

Centrifugal Spraying In a typical process for making glass wool, molten glass flows into a rotating bowl with many small orifices around its periphery. Centrifugal force causes the glass to flow through the holes to become a fibrous mass suitable for thermal and acoustical insulation.

Drawing of Continuous Filaments In this process, illustrated in Figure 7.9, continuous glass fibers of small diameter (lower size limit is around 0.0025 mm or 0.0001 in) are

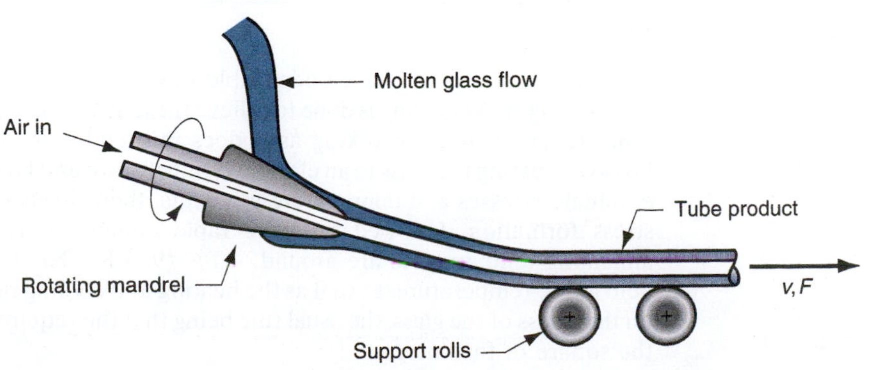

FIGURE 7.8 Drawing of glass tubes by the Danner process. Symbols v and F indicate motion (v = velocity) and applied force, respectively. (Credit: *Fundamentals of Modern Manufacturing*, 4th Edition by Mikell P. Groover, 2010. Reprinted with permission of John Wiley & Sons, Inc.)

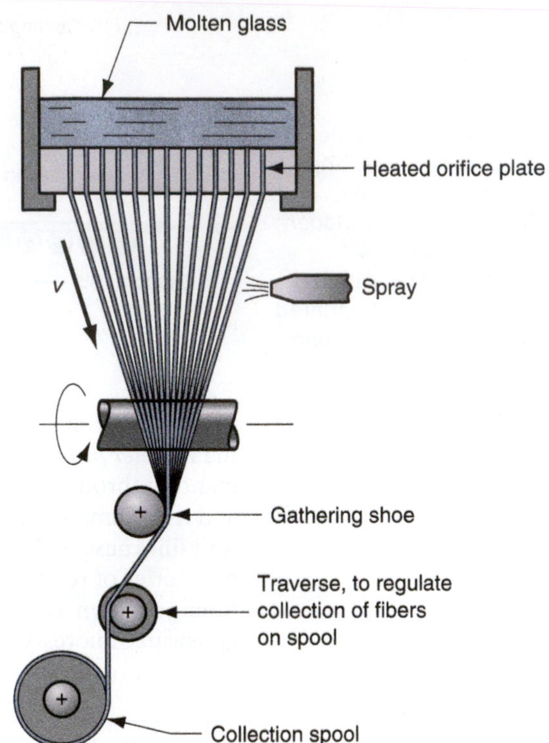

FIGURE 7.9 Drawing of continuous glass fibers. (Credit: *Fundamentals of Modern Manufacturing*, 4th Edition by Mikell P. Groover, 2010. Reprinted with permission of John Wiley & Sons, Inc.)

produced by drawing (pulling) strands of molten glass through small orifices in a heated plate made of a platinum alloy. The plate may have several hundred holes, each making one fiber. The individual fibers are collected into a strand by reeling them onto a spool. Prior to spooling, the fibers are coated with various chemicals to lubricate and protect them. Drawing speeds of around 50 m/s (10,000 ft/min) or more are not unusual.

7.3 HEAT TREATMENT AND FINISHING

Heat treatment of the glass product is the third step in the glassworking sequence. For some products, additional finishing operations are performed.

7.3.1 HEAT TREATMENT

Glass products usually have undesirable internal stresses after forming, which reduce their strength. Annealing is done to relieve these stresses; the treatment therefore has the same function in glassworking as it does in metalworking (Section 20.1). ***Annealing*** involves heating the glass to an elevated temperature and holding it for a certain period to eliminate stresses and temperature gradients, then slowly cooling the glass to suppress stress formation, followed by more rapid cooling to room temperature. Common annealing temperatures are around 500°C (900°F). The length of time the product is held at the temperature, as well as the heating and cooling rates during the cycle, depend on thickness of the glass, the usual rule being that the required annealing time varies with the square of thickness.

Annealing in modern glass factories is performed in tunnel-like furnaces, called ***lehrs***, in which the products flow slowly through the hot chamber on conveyors. Burners are located only at the front end of the chamber, so that the glass experiences the required heating and cooling cycle.

A beneficial internal stress pattern can be developed in glass products by a heat treatment known as ***tempering***, and the resulting material is called ***tempered glass***. As in the treatment of hardened steel (Section 20.2.2), tempering increases the toughness of glass. The process involves heating the glass to a temperature somewhat above its annealing temperature and into the plastic range, followed by quenching of the surfaces, usually with air jets. When the surfaces cool, they contract and harden while the interior is still plastic and compliant. As the internal glass slowly cools, it contracts, thus putting the hard surfaces in compression. Like other ceramics, glass is much stronger when subjected to compressive stresses than tensile stresses. Accordingly, tempered glass is much more resistant to scratching and breaking because of the compressive stresses on its surfaces. Applications include windows for tall buildings, all-glass doors, safety glasses, and other products requiring toughened glass.

When tempered glass fails, it does so by shattering into numerous small fragments, which are less likely to cut someone than conventional (annealed) window glass. Interestingly, automobile windshields are not made of tempered glass, due to the danger posed to the driver by this fragmentation. Instead, conventional glass is used; however, it is fabricated by sandwiching two pieces of glass on either side of a tough polymer sheet. Should this ***laminated glass*** fracture, the glass splinters are retained by the polymer sheet and the windshield remains relatively transparent.

7.3.2 FINISHING

Finishing operations are sometimes required for glassware products. These secondary operations include grinding, polishing, and cutting. When glass sheets are produced by drawing and rolling, the opposite sides are not necessarily parallel, and the surfaces contain defects and scratch marks caused by the use of hard tooling on soft glass. The glass sheets must be ground and polished for most commercial applications. In pressing and blowing operations when split dies are used, polishing is often required to remove the seam marks from the container product.

In continuous glassworking processes, such as plate and tube production, the continuous sections must be cut into smaller pieces. This is accomplished by first scoring the glass with a glass-cutting wheel or cutting diamond and then breaking the section along the score line. Cutting is generally done as the glass exits the annealing lehr.

Decorative and surface processes are performed on certain glassware products. These processes include mechanical cutting and polishing operations; sandblasting; chemical etching (with hydrofluoric acid, often in combination with other chemicals); and coating (for example, coating of plate glass with aluminum or silver to produce mirrors).

7.4 PRODUCT DESIGN CONSIDERATIONS

Glass possesses special properties that make it desirable in certain applications. The following design recommendations are compiled from Bralla [1] and other sources.

- ➢ Glass is transparent and has certain optical properties that are unusual, if not unique, among engineering materials. For applications requiring transparency, light transmittance, magnification, and similar optical properties, glass is likely to be the material of choice. Certain polymers are transparent and may be competitive, depending on design requirements.

- Glass is several times stronger in compression than in tension; components should be designed so that they are subjected to compressive stresses, not tensile stresses.
- Ceramics, including glass, are brittle. Glass parts should not be used in applications that involve impact loading or high stresses that might cause fracture.
- Certain glass compositions have very low thermal expansion coefficients and are therefore tolerant of thermal shock. These glasses can be selected for applications where this characteristic is important.
- Outside edges and corners on glass parts should have large radii or chamfers; likewise, inside corners should have large radii. Both outside and inside corners are potential points of stress concentration.
- Unlike parts made of traditional and new ceramics, threads may be included in the design of glass parts; they are technically feasible with the press-and-blow shaping processes. However, the threads should be coarse.

REFERENCES

[1] Bralla, J. G.(Editor in Chief). ***Design for Manufacturability Handbook***, 2nd ed. McGraw-Hill Book Company, New York, 1998.

[2] Flinn, R. A., and Trojan, P. K. ***Engineering Materials and Their Applications***, 5th ed. John Wiley & Sons, Inc., New York, 1995.

[3] Hlavac, J. ***The Technology of Glass and Ceramics***. Elsevier Scientific Publishing Company, New York, 1983.

[4] McLellan, G., and Shand, E. B. ***Glass Engineering Handbook***, 3rd ed. McGraw-Hill Book Company, New York, 1984.

[5] McColm, I. J. ***Ceramic Science for Materials Technologists***. Chapman and Hall, New York, 1983.

[6] Mohr, J. G., and Rowe, W. P. ***Fiber Glass***. Krieger Publishing Company, New York, 1990.

[7] Scholes, S. R., and Greene, C. H. ***Modern Glass Practice***, 7th ed. TechBooks, Marietta, Georgia, 1993.

REVIEW QUESTIONS

7.1. Glass is classified as a ceramic material; yet glass is different from the traditional and new ceramics. What is the difference?

7.2. What is the predominant chemical compound in almost all glass products?

7.3. What are the three basic steps in the glassworking sequence?

7.4. Describe the spinning process in glassworking.

7.5. What is the main difference between the press-and-blow and the blow-and-blow shaping processes in glassworking?

7.6. There are several ways of shaping plate or sheet glass. Name and briefly describe one of them.

7.7. Describe the Danner process.

7.8. Name and briefly describe the two processes for forming glass fibers are discussed in the text.

7.9. What is the purpose of annealing in glassworking?

7.10. Describe how a piece of glass is heat treated to produce tempered glass.

7.11. Describe the type of material that is commonly used to make windshields for automobiles.

8 SHAPING PROCESSES FOR PLASTICS

Chapter Contents

8.1 Properties of Polymer Melts

8.2 Extrusion
- 8.2.1 Process and Equipment
- 8.2.2 Analysis of Extrusion
- 8.2.3 Die Configurations and Extruded Products
- 8.2.4 Defects in Extrusion

8.3 Production of Sheet and Film

8.4 Fiber and Filament Production (Spinning)

8.5 Coating Processes

8.6 Injection Molding
- 8.6.1 Process and Equipment
- 8.6.2 The Mold
- 8.6.3 Shrinkage and Defects in Injection Molding
- 8.6.4 Other Injection Molding Processes

8.7 Compression and Transfer Molding
- 8.7.1 Compression Molding
- 8.7.2 Transfer Molding

8.8 Blow Molding and Rotational Molding
- 8.8.1 Blow Molding
- 8.8.2 Rotational Molding

8.9 Thermoforming

8.10 Casting

8.11 Polymer Foam Processing and Forming

8.12 Product Design Considerations

Plastics can be shaped into a wide variety of products, such as molded parts, extruded sections, films and sheets, insulation coatings on electrical wires, and fibers for textiles. In addition, plastics are often the principal ingredient in other materials, such as paints and varnishes; adhesives; and various polymer matrix composites. In this chapter, we consider the technologies by which these products are shaped, postponing paints and varnishes, adhesives, and composites until later chapters. Many plastic-shaping processes can be adapted to rubbers and polymer-matrix composites (Chapter 9).

The commercial and technological significance of these shaping processes derives from the growing importance of plastics, whose applications have increased at a much faster rate than either metals or ceramics during the last 50 years. Indeed, many parts previously made of metals are today being made of plastics and plastic composites. The same is true of glass; plastic containers have been largely substituted for glass bottles and jars in product packaging. The total volume of polymers (plastics and rubbers) now exceeds that of metals. We can identify several reasons why the plastic-shaping processes are important:

- The variety of shaping processes, and the ease with which polymers can be processed, allows an almost unlimited variety of part geometries to be formed.
- Many plastic parts are formed by molding, which is a ***net shape*** process. Further shaping is generally not needed.
- Although heating is usually required to form plastics, ***less energy*** is required than for metals because the processing temperatures are much lower.
- Because lower temperatures are used in processing, handling of the product is simplified during production. Because many plastic-processing methods are one-step

operations (e.g., molding), the amount of product handling required is substantially reduced compared with metals.

- Finishing by painting or plating is not required (except in unusual circumstances) for plastics.

As discussed in Section 2.3, the two types of plastics are ***thermoplastics*** and ***thermosets***. The difference is that thermosets undergo a curing process during heating and shaping, which causes a permanent chemical change (cross-linking) in their molecular structure. Once they have been cured, they cannot be melted through reheating. By contrast, thermoplastics do not cure, and their chemical structure remains basically unchanged upon reheating even though they transform from solid to fluid. Of the two types, thermoplastics are by far the more important type commercially, comprising more than 80% of the total plastics tonnage.

Plastic-shaping processes can be classified according to the resulting product geometry as follows: (1) continuous extruded products with constant cross section other than sheets, films, and filaments; (2) continuous sheets and films; (3) continuous filaments (fibers); (4) molded parts that are mostly solid; (5) hollow molded parts with relatively thin walls; (6) discrete parts made of formed sheets and films; (7) castings; and (8) foamed products. This chapter examines each of these categories. The most important processes commercially are those associated with thermoplastics; the two processes of greatest significance are extrusion and injection molding. We begin our coverage by examining the properties of polymer melts, because nearly all of the thermoplastic shaping processes share the common step of heating the plastic so that it flows.

8.1 PROPERTIES OF POLYMER MELTS

To shape a thermoplastic polymer, it must be heated so that it softens to the consistency of a liquid. In this form, it is called a ***polymer melt***. Polymer melts exhibit several unique properties, two of which are examined in this section: viscosity and viscoelasticity.

Viscosity Because of its high molecular weight, a polymer melt is a thick fluid with high viscosity. As we defined the term in Section 3.4, viscosity is a fluid property that relates the shear stress experienced during flow of the fluid to the rate of shear. Viscosity is important in polymer processing because most of the shaping methods involve flow of the polymer melt through small channels or die openings. The flow rates are often large, thus leading to high rates of shear; and the shear stresses increase with shear rate, so that significant pressures are required to accomplish the processes.

Figure 8.1 shows viscosity as a function of shear rate for two types of fluids. For a ***Newtonian fluid*** (which includes most simple fluids such as water and oil), viscosity is a

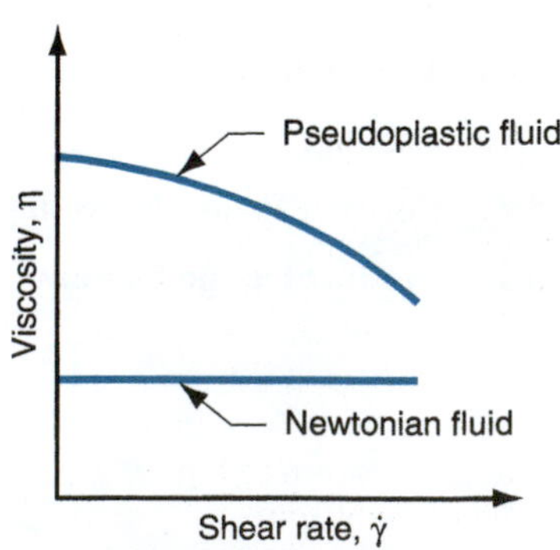

FIGURE 8.1 Viscosity relationships for Newtonian fluid and typical polymer melt. (Credit: *Fundamentals of Modern Manufacturing*, 4th Edition by Mikell P. Groover, 2010. Reprinted with permission of John Wiley & Sons, Inc.)

constant at a given temperature; it does not change with shear rate. The relationship between shear stress and shear strain is proportional, with viscosity as the constant of proportionality:

$$\tau = \eta\dot{\gamma} \quad \text{or} \quad \eta = \frac{\tau}{\dot{\gamma}} \tag{8.1}$$

where τ = shear stress, Pa (lb/in^2); η = coefficient of shear viscosity, Ns/m^2, or Pa-s (lb-sec/in^2); and $\dot{\gamma}$ = shear rate, 1/s (1/sec). However, for a polymer melt, viscosity decreases with shear rate, indicating that the fluid becomes thinner at higher rates of shear. This behavior is called ***pseudoplasticity*** and can be modeled to a reasonable approximation by the expression

$$\tau = k(\dot{\gamma})^n \tag{8.2}$$

where k = a constant corresponding to the viscosity coefficient and n = flow behavior index. For $n = 1$, the equation reduces to the previous Eq. (8.1) for a Newtonian fluid, and k becomes η. For a polymer melt, values of n are less than 1.

In addition to the effect of shear rate (fluid flow rate), viscosity of a polymer melt is also affected by temperature. Like most fluids, the value decreases with increasing temperature. This is shown in Figure 8.2 for several common polymers at a shear rate of 10^3 s^{-1}, which is approximately the same as the rates encountered in injection molding and high-speed extrusion. Thus, we see that the viscosity of a polymer melt decreases with increasing values of shear rate and temperature. Eq. (8.2) could be applied, except that k depends on temperature as shown in Figure 8.2.

Viscoelasticity The second property of interest here is viscoelasticity. We discussed this property in the context of solid polymers in Section 3.5. However, polymer melts exhibit it also. A good example is ***die swell*** in extrusion, in which the hot plastic expands when exiting the die opening. The phenomenon, illustrated in Figure 8.3, can be explained by noting that the polymer was contained in a much larger cross section before entering the narrow die channel. In effect, the extruded material "remembers" its former shape and attempts to return to it after leaving the die orifice. More technically,

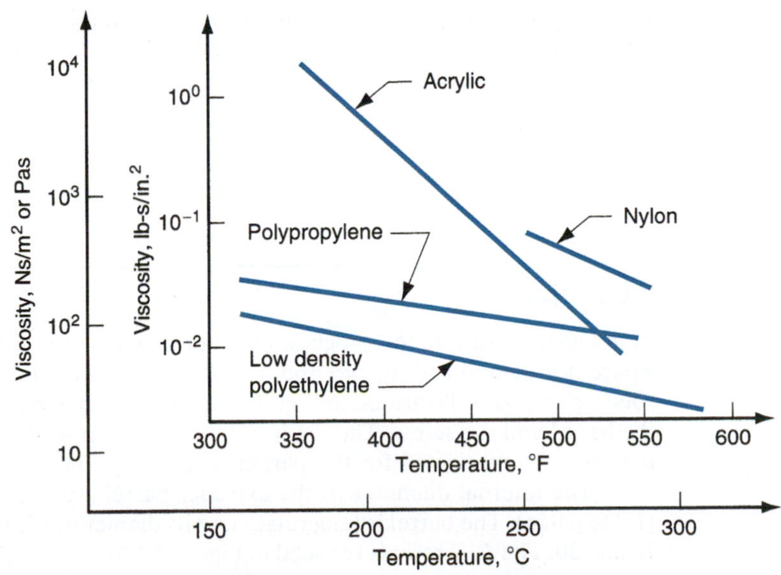

FIGURE 8.2 Viscosity as a function of temperatures for selected polymers at a shear rate of 10^3 s^{-1}. Data compiled from [12]. (Credit: *Fundamentals of Modern Manufacturing*, 4th Edition by Mikell P. Groover, 2010. Reprinted with permission of John Wiley & Sons, Inc.)

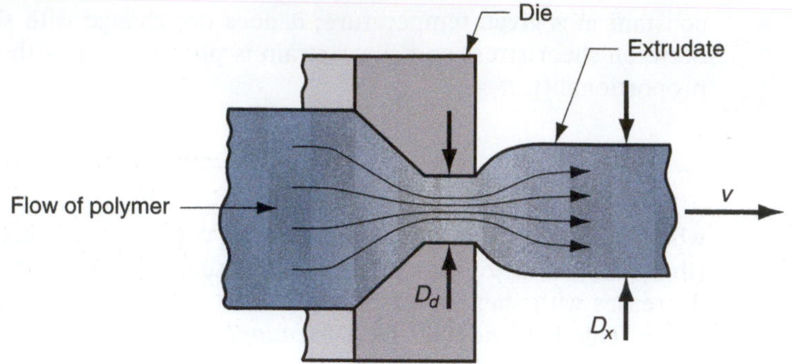

FIGURE 8.3 Die swell, a manifestation of viscoelasticity in polymer melts, as depicted here on exiting an extrusion die. (Credit: *Fundamentals of Modern Manufacturing,* 4th Edition by Mikell P. Groover, 2010. Reprinted with permission of John Wiley & Sons, Inc.)

the compressive stresses acting on the material as it enters the small die opening do not relax immediately. When the material subsequently exits the orifice and the restriction is removed, the unrelaxed stresses cause the cross section to expand.

Die swell can be most easily measured for a circular cross section by means of the ***swell ratio***, defined as

$$r_s = \frac{D_x}{D_d} \tag{8.3}$$

where r_s = swell ratio; D_x = diameter of the extruded cross section, mm (in); and D_d = diameter of the die orifice, mm (in). The amount of die swell depends on the time the polymer melt spends in the die channel. Increasing the time in the channel, by means of a longer channel, reduces die swell.

8.2 EXTRUSION

Extrusion is one of the fundamental shaping processes, for metals and ceramics as well as polymers. Extrusion is a compression process in which material is forced to flow through a die orifice to provide long continuous product whose cross-sectional shape is determined by the shape of the orifice. As a polymer-shaping process, it is widely used for thermoplastics and elastomers (but rarely for thermosets) to mass-produce items such as tubing, pipes, hose, structural shapes (such as window and door molding), sheet and film, continuous filaments, and coated electrical wire and cable. For these types of products, extrusion is carried out as a continuous process; the ***extrudate*** (extruded product) is subsequently cut into desired lengths. This section covers the basic extrusion process, and the following three sections examine processes based on extrusion.

8.2.1 PROCESS AND EQUIPMENT

In polymer extrusion, feedstock in pellet or powder form is fed into an extrusion barrel where it is heated and melted and forced to flow through a die opening by means of a rotating screw, as illustrated in Figure 8.4. The two main components of the extruder are the barrel and the screw. The die is not a component of the extruder; it is a special tool that must be fabricated for the particular profile to be produced.

The internal diameter of the extruder barrel typically ranges from 25 to 150 mm (1.0 to 6.0 in). The barrel is long relative to its diameter, with L/D ratios usually between 10 and 30. The L/D ratio is reduced in Figure 8.4 for clarity of drawing. The higher ratios

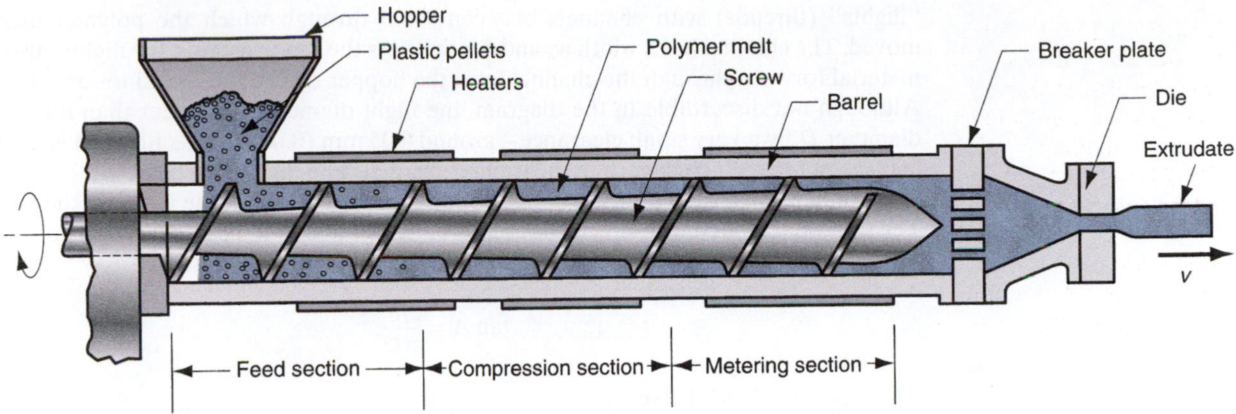

FIGURE 8.4 Components and features of a (single-screw) extruder for plastics and elastomers. (Credit: *Fundamentals of Modern Manufacturing*, 4th Edition by Mikell P. Groover, 2010. Reprinted with permission of John Wiley & Sons, Inc.)

are used for thermoplastic materials, whereas lower L/D values are for elastomers. A hopper containing the feedstock is located at the end of the barrel opposite the die. The pellets are fed by gravity onto the rotating screw whose turning moves the material along the barrel. Electric heaters are used to initially melt the solid pellets; subsequent mixing and mechanical working of the material generate additional heat, which maintains the melt. In some cases, enough heat is supplied through the mixing and shearing action that external heating is not required. Indeed, in some cases the barrel must be externally cooled to prevent overheating of the polymer.

The material is conveyed through the barrel toward the die opening by the action of the extruder screw, which rotates at about 60 rev/min. The screw serves several functions and is divided into sections that correspond to these functions. The sections and functions are (1) the ***feed section***, in which the stock is moved from the hopper port and preheated; (2) the ***compression section***, where the polymer is transformed into liquid consistency, air entrapped amongst the pellets is extracted from the melt, and the material is compressed; and (3) the ***metering section***, in which the melt is homogenized and sufficient pressure is developed to pump it through the die opening.

The operation of the screw is determined by its geometry and speed of rotation. Typical extruder screw geometry is depicted in Figure 8.5. The screw consists of spiraled

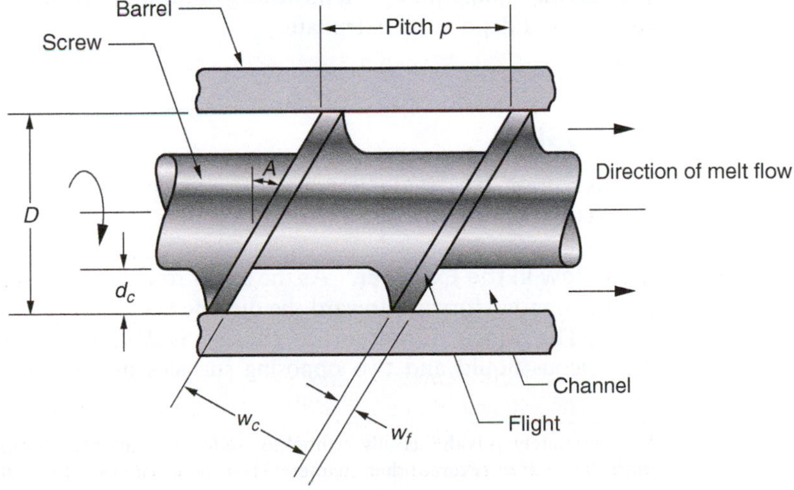

FIGURE 8.5 Details of an extruder screw inside the barrel. (Credit: *Fundamentals of Modern Manufacturing*, 4th Edition by Mikell P. Groover, 2010. Reprinted with permission of John Wiley & Sons, Inc.)

"flights" (threads) with channels between them through which the polymer melt is moved. The channel has a width w_c and depth d_c. As the screw rotates, the flights push the material forward through the channel from the hopper end of the barrel toward the die. Although not discernible in the diagram, the flight diameter is smaller than the barrel diameter D by a very small clearance—around 0.05 mm (0.002 in). Its function is to limit leakage of the melt backward to the trailing channel. The flight land has a width w_f and is made of hardened steel to resist wear as it turns and rubs against the inside of the barrel. The screw has a pitch whose value is usually close to the diameter D. The flight angle A is the helix angle of the screw and can be determined from the relation

$$\tan A = \frac{p}{\pi D} \tag{8.4}$$

where p = pitch of the screw.[1]

The increase in pressure applied to the polymer melt in the three sections of the barrel is determined largely by the channel depth d_c. In Figure 8.4, d_c is relatively large in the feed section to allow large amounts of granular polymer to be admitted into the barrel. In the compression section, d_c is gradually reduced, thus applying increased pressure on the polymer as it melts. In the metering section, d_c is small and pressure reaches a maximum as flow is restrained at the die end of the barrel. The three sections of the screw are shown as being about equal in length in Figure 8.4; this is appropriate for a polymer that melts gradually, such as low-density polyethylene (LDPE). For other polymers, the optimal section lengths are different. For crystalline polymers such as nylon, melting occurs rather abruptly at a specific melting point, and therefore a short compression section is appropriate. Amorphous polymers such as polyvinylchloride (PVC) melt more slowly than LDPE, and the compression zone for these materials must take almost the entire length of the screw. Although the optimal screw design for each material type is different, it is common practice to use general-purpose screws. These designs represent a compromise among the different materials, and they avoid the need to make frequent screw changes, which result in costly equipment downtime.

Progress of the polymer along the barrel leads ultimately to the die zone. Before reaching the die, the melt passes through a screen pack—a series of wire meshes supported by a stiff plate (called a ***breaker plate***) containing small axial holes. The screen pack assembly functions to (1) filter contaminants and hard lumps from the melt; (2) build pressure in the metering section; and (3) straighten the flow of the polymer melt and remove its "memory" of the circular motion imposed by the screw. This last function is concerned with the polymer's viscoelastic property; if the flow were left unstraightened, the polymer would play back its history of turning inside the extrusion chamber, tending to twist and distort the extrudate.

8.2.2 ANALYSIS OF EXTRUSION

In this section, we develop mathematical models to describe, in a simplified way, several aspects of polymer extrusion.

Melt Flow in the Extruder As the screw rotates inside the barrel, the polymer melt is forced to move forward toward the die; the system operates much like an Archimedian screw. The principal transport mechanism is ***drag flow***, resulting from friction between the viscous liquid and two opposing surfaces moving relative to each other: (1) the

[1] Unfortunately, p is the natural symbol to use for two variables in this chapter. It represents the screw pitch here and in several other chapters. We use the same symbol p for pressure later in the chapter.

stationary barrel and (2) the channel of the turning screw. The arrangement can be likened to the fluid flow that occurs between a stationary plate and a moving plate separated by a viscous liquid, as illustrated in Figure 3.17. Given that the moving plate has a velocity v, it can be reasoned that the average velocity of the fluid is $v/2$, resulting in a volume flow rate of

$$Q_d = 0.5vdw \tag{8.5}$$

where Q_d = volume drag flow rate, m^3/s (in^3/sec.); v = velocity of the moving plate, m/s (in/sec.); d = distance separating the two plates, m (in); and w = the width of the plates perpendicular to velocity direction, m (in). These parameters can be compared with those in the channel defined by the rotating extrusion screw and the stationary barrel surface.

$$v = \pi DN \cos A \tag{8.6}$$

$$d = d_c \tag{8.7}$$

$$\text{and } w = w_c = (\pi D \tan A - w_f) \cos A \tag{8.8}$$

where D = screw flight diameter, m (in); N = screw rotational speed, rev/s; d_c = screw channel depth, m (in); w_c = screw channel width, m (in); A = flight angle; and w_f = flight land width, m (in). If we assume that the flight land width is negligibly small, then the last of these equations reduces to

$$w_c = \pi D \tan A \cos A = \pi D \sin A \tag{8.9}$$

Substituting Eqs. (8.6), (8.7), and (8.9) into Eq. (8.5), and using several trigonometric identities, we get

$$Q_d = 0.5\,\pi^2 D^2 N\, d_c \sin A \cos A \tag{8.10}$$

If no forces were present to resist the forward motion of the fluid, this equation would provide a reasonable description of the melt flow rate inside the extruder. However, compressing the polymer melt through the downstream die creates a ***back pressure*** in the barrel that reduces the material moved by drag flow in Eq. (8.10). This flow reduction, called the ***back-pressure flow***, depends on the screw dimensions, viscosity of the polymer melt, and pressure gradient along the barrel. These dependencies can be summarized in this equation [12]:

$$Q_b = \frac{\pi D d_c^3 \sin^2 A}{12\eta}\left(\frac{dp}{dl}\right) \tag{8.11}$$

where Q_b = back pressure flow, m^3/s (in^3/sec); η = viscosity, N-s/m^2 (lb-sec/in^2); dp/dl = the pressure gradient, MPa/m (lb/in^2/in); and the other terms were previously defined. The actual pressure gradient in the barrel is a function of the shape of the screw over its length; a typical pressure profile is given in Figure 8.6. If we assume as an approximation that the profile is a straight line, indicated by the dashed line in the figure, then the pressure gradient becomes a constant p/L, and the previous equation reduces to

$$Q_b = \frac{p\pi D d_c^3 \sin^2 A}{12\eta L} \tag{8.12}$$

where p = head pressure in the barrel, MPa (lb/in^2); and L = length of the barrel, m (in). Recall that this back-pressure flow is really not an actual flow by itself; it is a reduction in

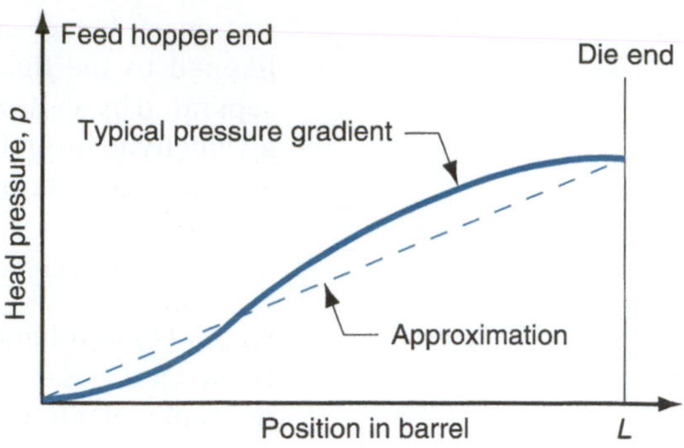

FIGURE 8.6 Typical pressure gradient in an extruder; dashed line indicates a straight line approximation to facilitate computations. (Credit: *Fundamentals of Modern Manufacturing*, 4th Edition by Mikell P. Groover, 2010. Reprinted with permission of John Wiley & Sons, Inc.)

the drag flow. Thus, we can compute the magnitude of the melt flow in an extruder as the difference between the drag flow and back-pressure flow:

$$Q_x = Q_d - Q_b$$
$$Q_x = 0.5\,\pi^2 D^2 N\, d_c \sin A \cos A - \frac{p\pi D d_c^3 \sin^2 A}{12\eta L} \tag{8.13}$$

where Q_x = the resulting flow rate of polymer melt in the extruder. Equation (8.13) assumes that there is minimal ***leak flow*** through the clearance between flights and barrel. Leak flow of melt is small compared with drag and back-pressure flow, except in badly worn extruders.

Equation (8.13) contains many parameters, which can be divided into two types: (1) design parameters, and (2) operating parameters. The design parameters are those that define the geometry of the screw and barrel: diameter D, length L, channel depth d_c, and helix angle A. For a given extruder operation, these factors cannot be changed during the process. The operating parameters are those that can be changed during the process to affect output flow; they include rotational speed N, head pressure p, and melt viscosity η. Of course, melt viscosity is controllable only to the extent to which temperature and shear rate can be manipulated to affect this property. Let us see how the parameters play out their roles in the following example.

Example 8.1 Extrusion Flow Rates

An extruder barrel has a diameter $D = 75$ mm. The screw rotates at $N = 1$ rev/s. Channel depth $d_c = 6.0$ mm and flight angle $A = 20°$. Head pressure at the end of the barrel $p = 7.0\times10^6$ Pa, length of the barrel $L = 1.9$ m, and viscosity of the polymer melt is assumed to be $\eta = 100$ Pa-s. Determine the volume flow rate of the plastic in the barrel Q_x.

Solution: Using Eq. (8.13) we can compute the drag flow and opposing back pressure flow in the barrel.

$$Q_d = 0.5\,\pi^2(75\times10^{-3})^2(1.0)(6\times10^{-3})(\sin 20)(\cos 20) = 53{,}525\,(10^{-9})\ \text{m}^3/\text{s}$$
$$Q_b = \frac{\pi(7\times10^6)(75\times10^{-3})(6\times10^{-3})^3(\sin 20)^2}{12(100)(1.9)} = 18.276(10^{-6}) = 18{,}276(10^{-9})\ \text{m}^3/\text{s}$$
$$Q_x = Q_d - Q_b = (53{,}525 - 18{,}276)(10^{-9}) = \mathbf{35{,}249(10^{-9})\ m^3/s}$$

Extruder and Die Characteristics If back pressure is zero, so that melt flow is unrestrained in the extruder, then the flow would equal drag flow Q_d given by Eq. (8.10). Given the design and operating parameters (D, A, N, etc.), this is the maximum possible flow capacity of the extruder. Let us denote it as Q_{max}:

$$Q_{max} = 0.5\pi^2 D^2 N\, d_c \sin A \cos A \tag{8.14}$$

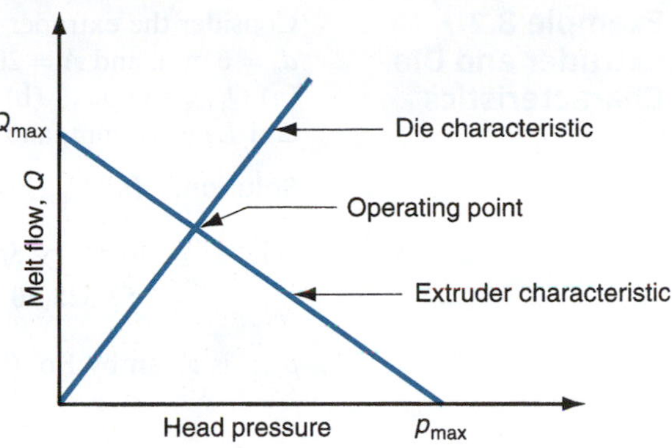

FIGURE 8.7 Extruder characteristic (also called the screw characteristic) and die characteristic. The extruder operating point is at intersection of the two lines. (Credit: *Fundamentals of Modern Manufacturing*, 4th Edition by Mikell P. Groover, 2010. Reprinted with permission of John Wiley & Sons, Inc.)

On the other hand, if back pressure were so great as to cause zero flow, then back pressure flow would equal drag flow; that is,

$$Q_x = Q_d - Q_b = 0, \text{ so } Q_d = Q_b$$

Using the expressions for Q_d and Q_b in Eq. (8.13), we can solve for p to determine what this maximum head pressure p_{max} would have to be to cause no flow in the extruder:

$$p_{\max} = \frac{6\pi DNL\eta cotA}{d_c^2} \tag{8.15}$$

The two values $Q_{\max}$ and $p_{\max}$ are points along the axes of a diagram known as the ***extruder characteristic*** (or ***screw characteristic***), as in Figure 8.7. It defines the relationship between head pressure and flow rate in an extrusion machine with given design and operating parameters.

With a die in the machine and the extrusion process underway, the actual values of Q_x and p will lie somewhere between the extreme values, the location determined by the characteristics of the die. Flow rate through the die depends on the size and shape of the opening and the pressure applied to force the melt through it. This can be expressed:

$$Q_x = K_s p \tag{8.16}$$

where Q_x = flow rate, m^3/s (in^3/sec.); p = head pressure, Pa (lb/in^2); and K_s = shape factor for the die, m^5/Ns (in^5/lb-sec). For a circular die opening of a given channel length, the shape factor can be computed [12] as

$$K_s = \frac{\pi D_d^4}{128\eta L_d} \tag{8.17}$$

where D_d = die opening diameter, m (in) η = melt viscosity, N-s/m^2 (lb-sec/in^2); and L_d = die opening length, m (in). For shapes other than round, the die shape factor is less than for a round of the same cross-sectional area, meaning that greater pressure is required to achieve the same flow rate.

The relationship between Q_x and p in Eq. (8.16) is called the ***die characteristic***. In Figure 8.7, this is drawn as a straight line that intersects with the previous extruder characteristic. The intersection point identifies the values of Q_x and p that are known as the ***operating point*** for the extrusion process.

Example 8.2 Extruder and Die Characteristics

Consider the extruder from Example 8.1, in which $D = 75$ mm, $L = 1.9$ m, $N = 1$ rev/s, $d_c = 6$ mm, and $A = 20°$. The plastic melt has a shear viscosity $\eta = 100$ Pa-s. Determine (a) Q_{max} and p_{max}, (b) shape factor K_s for a circular die opening in which $D_d = 6.5$ mm and $L_d = 20$ mm, and (c) values of Q_x and p at the operating point.

Solution: (a) Q_{max} is given by Eq. (8.14).

$$\begin{aligned} Q_{max} &= 0.5\pi^2 D^2 N d_c \sin A \cos A = 0.5\,\pi^2(75 \times 10^{-3})^2(1.0)(6 \times 10^{-3})(\sin 20)(\cos 20) \\ &= \mathbf{53{,}525(10^{-9})\ m^3/s} \end{aligned}$$

p_{max} is given by Eq. (8.15).

$$p_{max} = \frac{6\pi DNL\eta cotA}{d_c^2} = \frac{6\pi(75 \times 10^{-3})(1.9)(1.0)(100)\text{cot}20}{(6 \times 10^{-3})^2} = \mathbf{20{,}499{,}874\ Pa}$$

These two values define the intersection with the ordinate and abscissa for the extruder characteristic.

(b) The shape factor for a circular die opening with $D_d = 6.5$ mm and $L_d = 20$ mm can be determined from Eq. (8.17).

$$K_s = \frac{\pi(6.5 \times 10^{-3})^4}{128(100)(20 \times 10^{-3})} = \mathbf{21.9(10^{-12})\ m^5/Ns}$$

This shape factor defines the slope of the die characteristic.

(c) The operating point is defined by the values of Q_x and p at which the screw characteristic intersects with the die characteristic. The extruder characteristic can be expressed as the equation of the straight line between Q_{max} and p_{max}, which is

$$\begin{aligned} Q_x &= Q_{max} - (Q_{max}/p_{max})p \\ &= 53,525(10^{-9}) - (53,525(10^{-9})/20,499,874)p = 53,525(10^{-9}) - 2.611(10^{-12})p \end{aligned} \tag{8.18}$$

The die characteristic is given by Eq. (8.16) using the value of K_s computed in part (b).

$$Q_x = 21.9(10^{-12})p$$

Setting the two equations equal, we have

$$53,525(10^{-9}) - 2.611(10^{-12})p = 21.9(10^{-12})p$$

$$p = \mathbf{2.184(10^{-6})\ Pa}$$

Solving for Q_x using one of the starting equations, we obtain

$$Q_x = 53.525(10^{-6}) - 2.611(10^{-12})(2.184)(10^6) = 47.822(10^{-6})\ \text{m}^3/\text{s}$$

Checking this with the other equation for verification,

$$Q_x = 21.9(10^{-12})(2.184)(10^6) = 47.82(10^{-6})\ \text{m}^3/\text{s}$$ ■

8.2.3 DIE CONFIGURATIONS AND EXTRUDED PRODUCTS

The shape of the die orifice determines the cross-sectional shape of the extrudate. We can enumerate the common die profiles and corresponding extruded shapes as

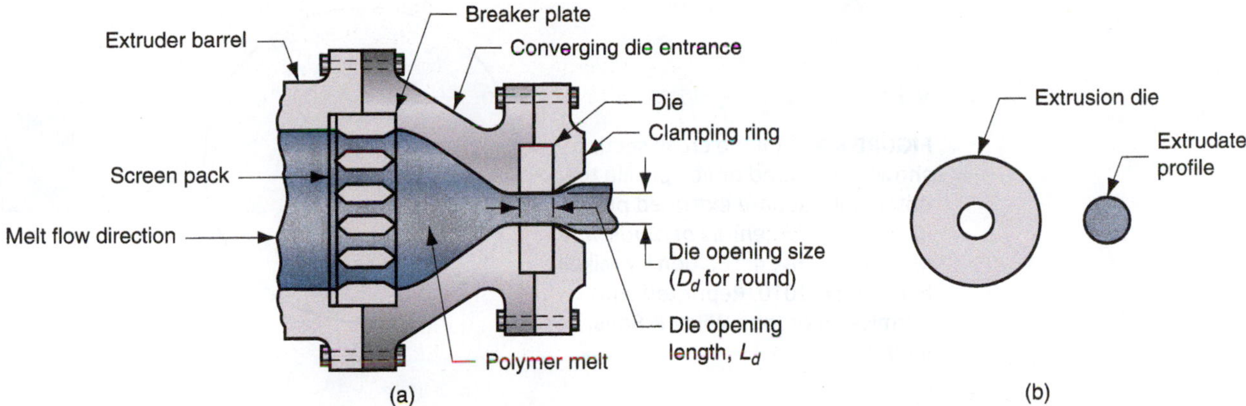

FIGURE 8.8 (a) Side view cross section of an extrusion die for solid regular shapes, such as round stock; (b) front view of die, with profile of extrudate. Die swell is evident in both views. (Some die construction details are simplified or omitted for clarity.) (Credit: *Fundamentals of Modern Manufacturing*, 4th Edition by Mikell P. Groover, 2010. Reprinted with permission of John Wiley & Sons, Inc.)

follows: (1) solid profiles; (2) hollow profiles, such as tubes; (3) wire and cable coating; (4) sheet and film; and (5) filaments. The first three categories are covered in the present section. Methods for producing sheet and film are examined in Section 8.3; and filament production is discussed in Section 8.4. These latter shapes sometimes involve forming processes other than extrusion.

Solid Profiles Solid profiles include regular shapes such as rounds and squares and irregular cross sections such as structural shapes, door and window moldings, automobile trim, and house siding. The side-view cross section of a die for these solid shapes is illustrated in Figure 8.8. Just beyond the end of the screw and before the die, the polymer melt passes through the screen pack and breaker plate to straighten the flow lines. Then it flows into a (usually) converging die entrance, the shape designed to maintain laminar flow and avoid dead spots in the corners that would otherwise be present near the orifice. The melt then flows through the die opening itself.

When the material exits the die, it is still soft. Polymers with high melt viscosities are the best candidates for extrusion, because they hold shape better during cooling. Cooling is accomplished by air blowing, water spray, or passing the extrudate through a water trough. To compensate for die swell, the die opening is made long enough to remove some of the memory in the polymer melt. In addition, the extrudate is often drawn (stretched) to offset expansion from die swell.

For shapes other than round, the die opening is designed with a cross section that is slightly different from the desired profile, so that the effect of die swell is to provide shape correction. This correction is illustrated in Figure 8.9 for a square cross section. Because different polymers exhibit varying degrees of die swell, the shape of the die profile depends on the material to be extruded. Considerable skill and judgment are required by the die designer for complex cross sections.

Hollow Profiles Extrusion of hollow profiles, such as tubes, pipes, hoses, and other cross sections containing holes, requires a mandrel to form the hollow shape. A typical die configuration is shown in Figure 8.10. The mandrel is held in place using a spider, seen in Section A-A of the figure. The polymer melt flows around the legs supporting the mandrel to reunite into a monolithic tube wall. The mandrel often includes an air channel through which air is blown to maintain the hollow form of the extrudate during hardening. Pipes and tubes are cooled using open water troughs or by pulling the

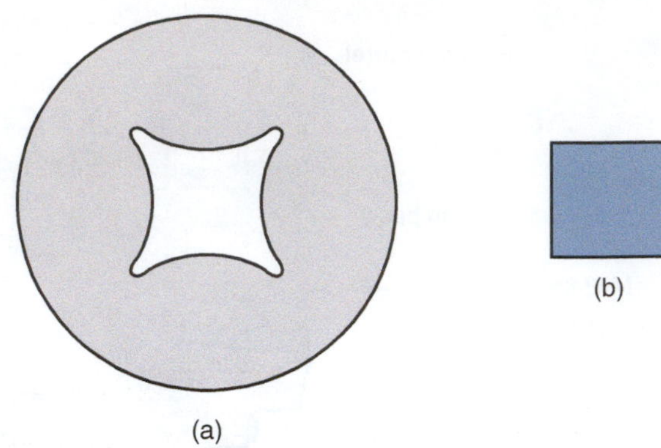

FIGURE 8.9 (a) Die cross section showing required orifice profile to obtain (b) a square extruded profile. (Credit: *Fundamentals of Modern Manufacturing,* 4th Edition by Mikell P. Groover, 2010. Reprinted with permission of John Wiley & Sons, Inc.)

soft extrudate through a water-filled tank with sizing sleeves that limit the OD of the tube while air pressure is maintained on the inside.

Wire and Cable Coating The coating of wire and cable for insulation is one of the most important polymer extrusion processes. As shown in Figure 8.11 for wire coating, the polymer melt is applied to the bare wire as it is pulled at high speed through a die. A slight vacuum is drawn between the wire and the polymer to promote adhesion of the coating. The taught wire provides rigidity during cooling, which is usually aided by passing the coated wire through a water trough. The product is wound onto large spools at speeds of up to 50 m/s (10,000 ft/min).

8.2.4 DEFECTS IN EXTRUSION

A number of defects can afflict extruded products. One of the worst is ***melt fracture***, in which the stresses acting on the melt immediately before and during its flow through the

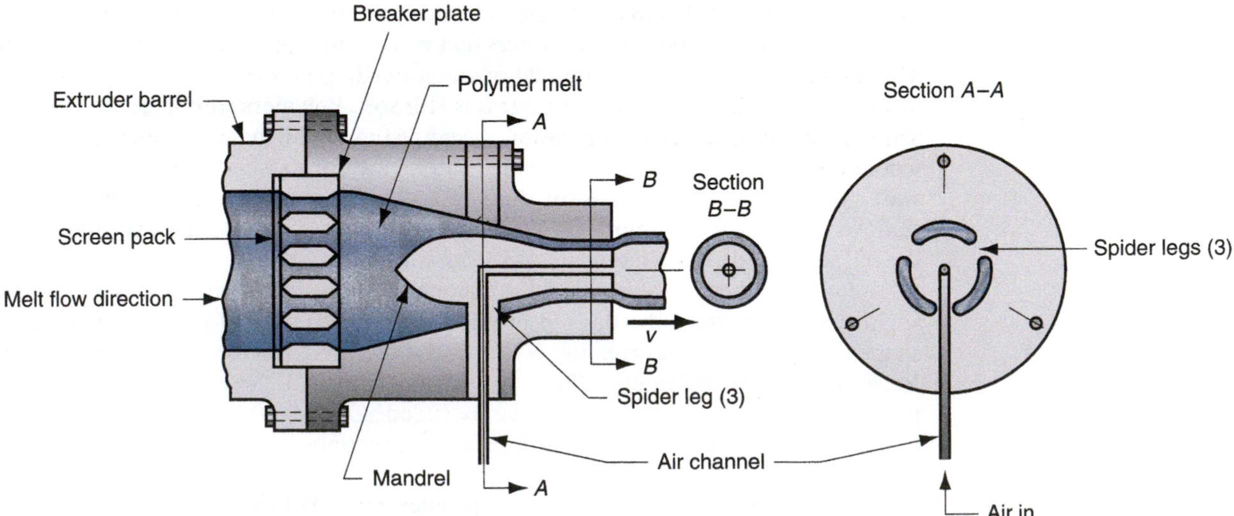

FIGURE 8.10 Side view cross section of extrusion die for shaping hollow cross sections such as tubes and pipes; Section A-A is a front view cross section showing how the mandrel is held in place; Section B-B shows the tubular cross section just prior to exiting the die; die swell causes an enlargement of the diameter. (Some die construction details are simplified.) (Credit: *Fundamentals of Modern Manufacturing,* 4th Edition by Mikell P. Groover, 2010. Reprinted with permission of John Wiley & Sons, Inc.)

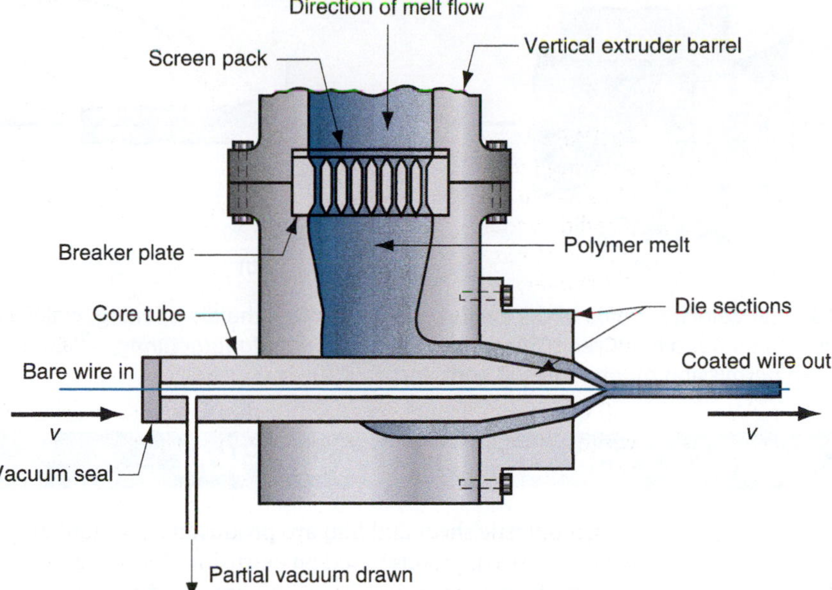

FIGURE 8.11 Side view cross section of die for coating of electrical wire by extrusion. (Some die construction details are simplified.) (Credit: *Fundamentals of Modern Manufacturing,* 4th Edition by Mikell P. Groover, 2010. Reprinted with permission of John Wiley & Sons, Inc.)

die are so high as to cause failure, manifested in the form of a highly irregular surface on the extrudate. As suggested by Figure 8.12, melt fracture can be caused by a sharp reduction at the die entrance, causing turbulent flow that breaks up the melt. This contrasts with the streamlined, laminar flow in the gradually converging die in Figure 8.8.

A more common defect in extrusion is ***sharkskin***, in which the surface of the product becomes roughened upon exiting the die. As the melt flows through the die opening, friction at the interface results in a velocity profile across the cross section, Figure 8.13. Tensile stresses develop at the surface as this material is stretched to keep up with the faster moving center core. These stresses cause minor ruptures that roughen the surface. If the velocity gradient becomes extreme, prominent marks occur on the surface, giving it the appearance of a bamboo pole; hence, the name ***bambooing*** for this more severe defect.

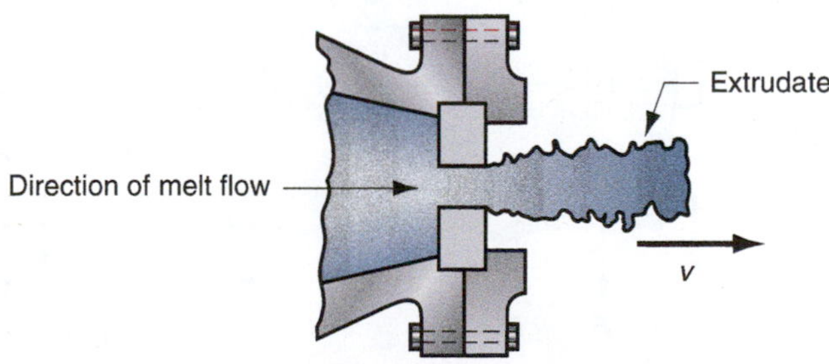

FIGURE 8.12 Melt fracture, caused by turbulent flow of the melt through a sharply reduced die entrance. (Credit: *Fundamentals of Modern Manufacturing,* 4th Edition by Mikell P. Groover, 2010. Reprinted with permission of John Wiley & Sons, Inc.)

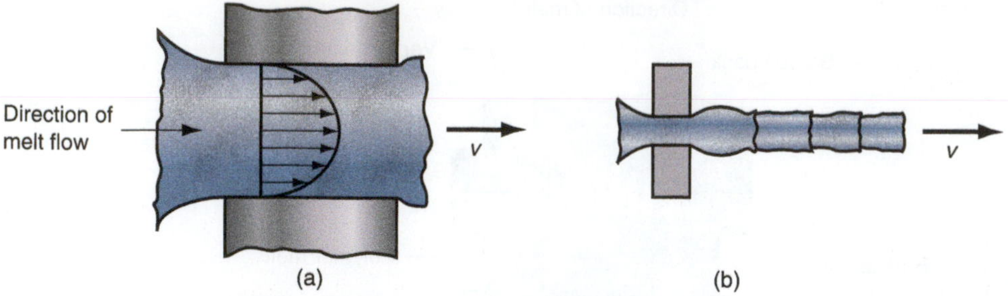

FIGURE 8.13 (a) Velocity profile of the melt as it flows through the die opening, which can lead to defects called sharkskin and (b) bambooing. (Credit: *Fundamentals of Modern Manufacturing*, 4th Edition by Mikell P. Groover, 2010. Reprinted with permission of John Wiley & Sons, Inc.)

8.3 PRODUCTION OF SHEET AND FILM

Thermoplastic sheet and film are produced by a number of processes, most important of which are two methods based on extrusion. The term ***sheet*** refers to stock with a thickness ranging from 0.5 mm (0.020 in) to about 12.5 mm (0.5 in) and used for products such as flat window glazing and stock for thermoforming (Section 8.9). ***Film*** refers to thicknesses below 0.5 mm (0.020 in). Thin films are used for packaging (product wrapping material, grocery bags, and garbage bags); thicker film applications include covers and liners (pool covers and liners for irrigation ditches).

All of the processes covered in this section are continuous, high-production operations. More than half of the films produced today are polyethylene, mostly low-density PE. The principal other materials are polypropylene, polyvinylchloride, and regenerated cellulose (cellophane). These are all thermoplastic polymers.

Slit-Die Extrusion of Sheet and Film Sheet and film of various thickness are produced by conventional extrusion, using a narrow slit as the die opening. The slit may be up to 3 m (10 ft) wide and as narrow as around 0.4 mm (0.015 in). One possible die configuration is illustrated in Figure 8.14. The die includes a manifold that spreads

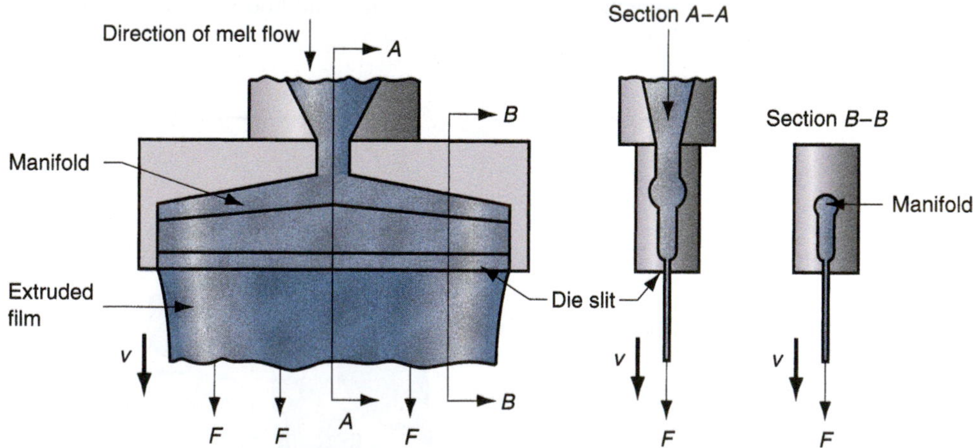

FIGURE 8.14 One of several die configurations for extruding sheet and film. (Credit: *Fundamentals of Modern Manufacturing*, 4th Edition by Mikell P. Groover, 2010. Reprinted with permission of John Wiley & Sons, Inc.)

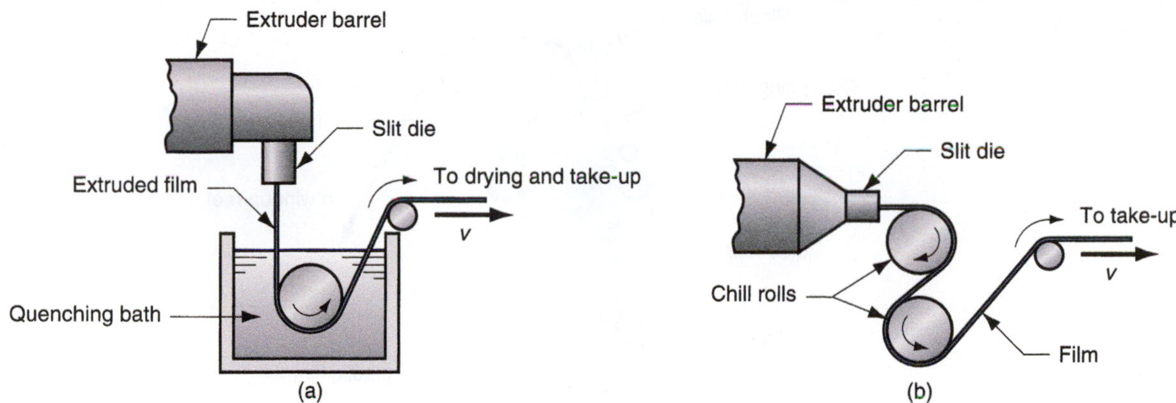

FIGURE 8.15 Use of (a) water-quenching bath or (b) chill rolls to achieve fast solidification of the molten film after extrusion. (Credit: *Fundamentals of Modern Manufacturing,* 4th Edition by Mikell P. Groover, 2010. Reprinted with permission of John Wiley & Sons, Inc.)

the polymer melt laterally before it flows through the slit (die orifice). One of the difficulties in this extrusion method is uniformity of thickness throughout the width of the stock. This results from the drastic shape change experienced by the polymer melt during its flow through the die and also to temperature and pressure variations in the die. Usually, the edges of the film must be trimmed because of thickening at the edges.

To achieve high production rates, an efficient method of cooling and collecting the film must be integrated with the extrusion process. This is usually done by immediately directing the extrudate into a quenching bath of water or onto chill rolls, as shown in Figure 8.15. The chill roll method seems to be the more important commercially. Contact with the cold rolls quickly quenches and solidifies the extrudate; in effect, the extruder serves as a feeding device for the chill rolls that actually form the film. The process is noted for very high production speeds—5 m/s (1000 ft/min). In addition, close tolerances on film thickness can be achieved. Owing to the cooling method used in this process, it is known as ***chill-roll extrusion***.

Blown-Film Extrusion Process This is the other widely used process for making thin polyethylene film for packaging. It is a complex process, combining extrusion and blowing to produce a tube of thin film; it is best explained with reference to the diagram in Figure 8.16. The process begins with the extrusion of a tube that is immediately drawn upward while still molten and simultaneously expanded in size by air inflated into it through the die mandrel. A "frost line" marks the position along the upward moving bubble where solidification of the polymer occurs. Air pressure in the bubble must be kept constant to maintain uniform film thickness and tube diameter. The air is contained in the tube by pinch rolls that squeeze the tube back together after it has cooled. Guide rolls and collapsing rolls are also used to restrain the blown tube and direct it into the pinch rolls. The flat tube is then collected onto a windup reel.

The effect of air inflation is to stretch the film in both directions as it cools from the molten state. This results in isotropic strength properties, which is an advantage over other processes in which the material is stretched primarily in one direction. Other advantages include the ease with which extrusion rate and air pressure can be changed to control stock width and gage. Comparing this process with slit-die extrusion, the blown-film method produces stronger film (so that a thinner film can be used to package a product), but thickness control and production rates are lower. The final blown film can

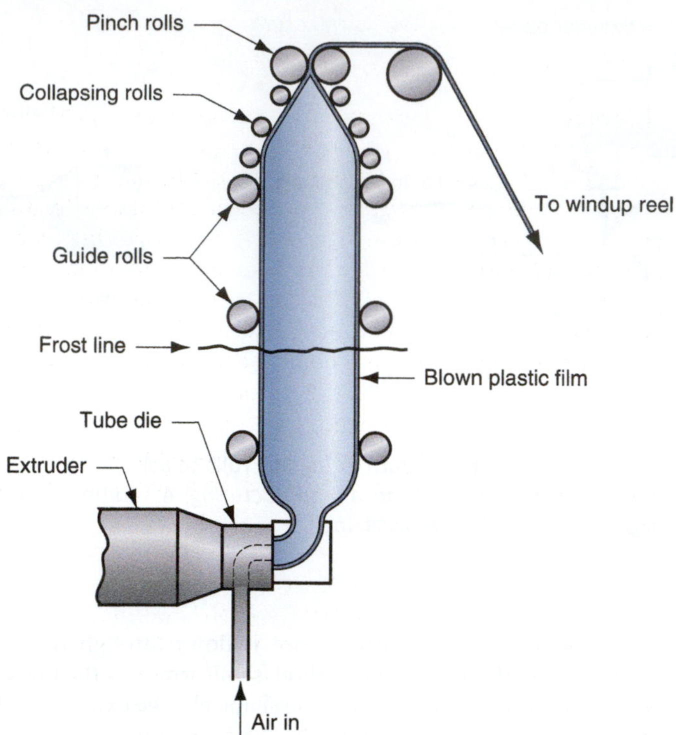

FIGURE 8.16 Blown-film process for high production of thin tubular film. (Credit: *Fundamentals of Modern Manufacturing,* 4th Edition by Mikell P. Groover, 2010. Reprinted with permission of John Wiley & Sons, Inc.)

be left in tubular form (e.g., for garbage bags), or it can be subsequently cut at the edges to provide two parallel thin films.

Calendering Calendering is a process for producing sheet and film stock out of rubber (Section 9.1.4) or rubbery thermoplastics such as plasticized PVC. In the process, the initial feedstock is passed through a series of rolls to work the material and reduce its thickness to the desired gage. A typical setup is illustrated in Figure 8.17. The equipment is expensive, but production rate is high; speeds approaching 2.5 m/s (500 ft/min) are possible. Close control is required over roll temperatures, pressures, and rotational speed. The process is noted for its good surface finish and high gage accuracy in the film. Plastic products made by the calendering process include PVC floor covering, shower curtains, vinyl table cloths, pool liners, and inflatable boats and toys.

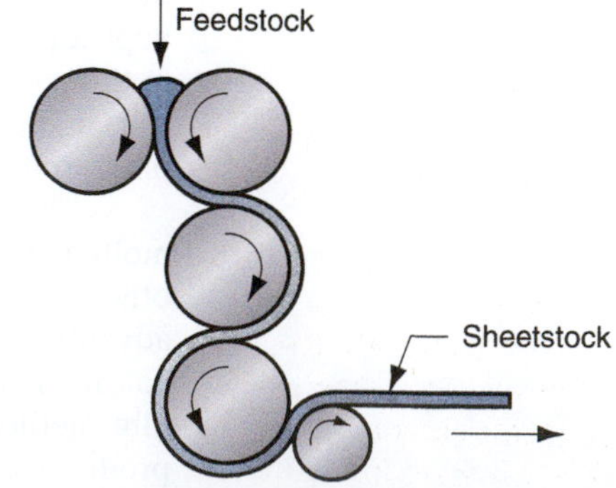

FIGURE 8.17 A typical roll configuration in calendering. (Credit: *Fundamentals of Modern Manufacturing,* 4th Edition by Mikell P. Groover, 2010. Reprinted with permission of John Wiley & Sons, Inc.)

8.4 FIBER AND FILAMENT PRODUCTION (SPINNING)

The most important application of polymer fibers and filaments is in textiles. Their use as reinforcing materials in plastics (composites) is a growing application, but still small compared with textiles. A ***fiber*** can be defined as a long, thin strand of material whose length is finite. A ***filament*** is a strand of continuous length.

Fibers can be natural or synthetic. Synthetic fibers constitute about 75% of the total fiber market today, polyester being the most important, followed by nylon, acrylics, and rayon. Natural fibers are about 25% of the total produced, with cotton by far the most important staple (wool production is significantly less than cotton).

The term ***spinning*** is a holdover from the methods used to draw and twist natural fibers into yarn or thread. In the production of synthetic fibers, the term refers to the process of extruding a polymer melt or solution through a ***spinneret*** (a die with multiple small holes) to make filaments that are then drawn and wound onto a bobbin. There are three principal variations in the spinning of synthetic fibers, depending on the polymer being processed: (1) melt spinning, (2) dry spinning, and (3) wet spinning.

Melt spinning is used when the starting polymer can best be processed by heating to the molten state and pumping through the spinneret, much in the manner of conventional extrusion. A typical spinneret is 6 mm (0.25 in) thick and contains approximately 50 holes of diameter 0.25 mm (0.010 in); the holes are countersunk, so that the resulting bore has an L/D ratio of only 5/1 or less. The filaments that emanate from the die are drawn and simultaneously air cooled before being collected together and spooled onto the bobbin, as shown in Figure 8.18. Significant extension and thinning of the filaments occur while the polymer is still molten, so that the final diameter wound onto the bobbin may be only 1/10 of the extruded size. Melt spinning is used for polyesters and nylons. Because these

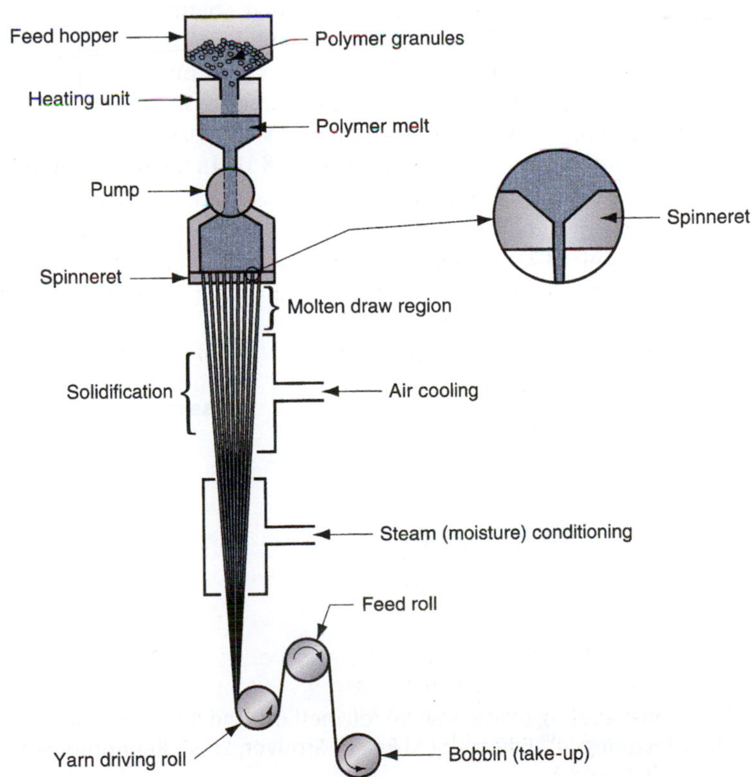

FIGURE 8.18 Melt spinning of continuous filaments. (Credit: *Fundamentals of Modern Manufacturing*, 4th Edition by Mikell P. Groover, 2010. Reprinted with permission of John Wiley & Sons, Inc.)

are the most important synthetic fibers, melt spinning is the most important of the three processes for synthetic fibers.

In ***dry spinning***, the starting polymer is in solution and the solvent can be separated by evaporation. The extrudate is pulled through a heated chamber that removes the solvent; otherwise the sequence is similar to the previous. Fibers of cellulose acetate and acrylic are produced by this process. In ***wet spinning***, the polymer is also in solution—only the solvent is nonvolatile. To separate the polymer, the extrudate must be passed through a liquid chemical that coagulates or precipitates the polymer into coherent strands that are then collected onto bobbins. This method is used to produce rayon (regenerated cellulose fibers).

Filaments produced by any of the three processes are usually subjected to further cold drawing to align the crystal structure along the direction of the filament axis. Extensions of 2 to 8 are typical [13]. This has the effect of significantly increasing the tensile strength of the fibers. Drawing is accomplished by pulling the thread between two spools, where the winding spool is driven at a faster speed than the unwinding spool.

8.5 COATING PROCESSES

Plastic (or rubber) coating involves application of a layer of the given polymer onto a substrate material. Three categories are distinguished [6]: (1) wire and cable coating; (2) planar coating, which involves the coating of a flat film; and (3) contour coating—the coating of a three-dimensional object. We have already examined wire and cable coating (Section 8.2.3); it is basically an extrusion process. The other two categories are surveyed in the following paragraphs. In addition, there is the technology of applying paints, varnishes, lacquers, and other similar coatings (Section 21.6).

Planar coating is used to coat fabrics, paper, cardboard, and metal foil; these items are major products for some plastics. The important polymers include polyethylene and polypropylene, with lesser applications for nylon, PVC, and polyester. In most cases, the coating is only 0.01 to 0.05 mm (0.0005 to 0.002 in) thick. The two major planar coating techniques are illustrated in Figure 8.19. In the ***roll method***, the polymer coating material

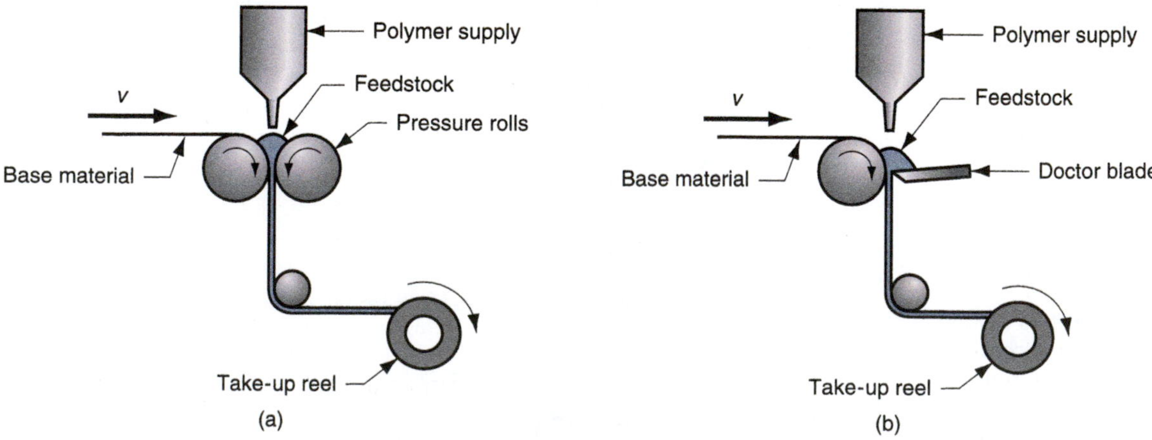

FIGURE 8.19 Planar coating processes: (a) roll method, and (b) doctor-blade method. (Credit: *Fundamentals of Modern Manufacturing*, 4th Edition by Mikell P. Groover, 2010. Reprinted with permission of John Wiley & Sons, Inc.)

is squeezed against the substrate by means of opposing rolls. In the ***doctor blade method***, a sharp knife edge controls the amount of polymer melt that is coated onto the substrate. In both cases, the coating material is supplied either by a slit-die extrusion process or by calendering.

Contour coating of three-dimensional objects can be accomplished by dipping or spraying. ***Dipping*** involves submersion of the object into a suitable bath of polymer melt or solution, followed by cooling or drying. ***Spraying*** (such as spray-painting) is an alternative method for applying a polymer coating to a solid object.

8.6 INJECTION MOLDING

Injection molding is a process in which a polymer is heated to a highly plastic state and forced to flow under high pressure into a mold cavity, where it solidifies. The molded part, called a ***molding***, is then removed from the cavity. The process produces discrete components that are almost always net shape. The production cycle time is typically in the range 10 to 30 sec, although cycles of 1 min or longer are not uncommon for large parts. Also, the mold may contain more than one cavity, so that multiple moldings are produced each cycle.

Complex and intricate shapes are possible with injection molding. The challenge in these cases is to fabricate a mold whose cavity is the same geometry as the part and also allows for part removal. Part size can range from about 50 g (2 oz) up to about 25 kg (more than 50 lb), the upper limit represented by components such as refrigerator doors and automobile bumpers. The mold determines the part shape and size and is the special tooling in injection molding. For large, complex parts, the mold can cost hundreds of thousands of dollars. For small parts, the mold can be built to contain multiple cavities, also making the mold expensive. Thus, injection molding is economical only for large production quantities.

Injection molding is the most widely used molding process for thermoplastics. Some thermosets and elastomers are injection molded, with modifications in equipment and operating parameters to allow for cross-linking of these materials. We discuss these and other variations of injection molding in Section 8.6.4.

8.6.1 PROCESS AND EQUIPMENT

Equipment for injection molding evolved from metal die casting. As illustrated in Figure 8.20, an injection molding machine consists of two principal components: (1) the plastic injection unit and (2) the mold clamping unit. The ***injection unit*** is much like an extruder. It consists of a barrel that is fed from one end by a hopper containing a supply of plastic pellets. Inside the barrel is a screw whose operation surpasses that of an extruder screw in the following respect: in addition to turning for mixing and heating the polymer, it also acts as a ram that rapidly moves forward to inject molten plastic into the mold. A nonreturn valve mounted near the tip of the screw prevents the melt from flowing backward along the screw threads. Later in the molding cycle the ram retracts to its former position. Because of its dual action, it is called a ***reciprocating screw***. To summarize, the functions of the injection unit are to melt and homogenize the polymer, and then inject it into the mold cavity.

The ***clamping unit*** is concerned with the operation of the mold. Its functions are to (1) hold the two halves of the mold in proper alignment with each other; (2) keep the mold closed during injection by applying a clamping force sufficient to resist the injection force; and (3) open and close the mold at the appropriate times in the molding cycle. The

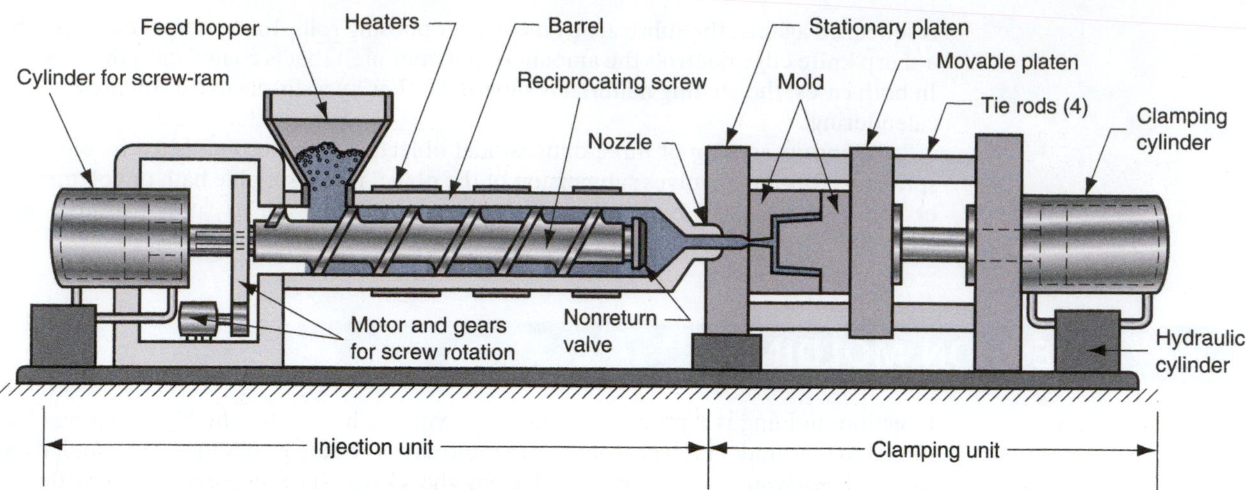

FIGURE 8.20 Diagram of an injection molding machine, reciprocating screw type (some mechanical details are simplified). (Credit: *Fundamentals of Modern Manufacturing*, 4th Edition by Mikell P. Groover, 2010. Reprinted with permission of John Wiley & Sons, Inc.)

clamping unit consists of two platens, a fixed platen and a movable platen, and a mechanism for translating the latter. The mechanism is basically a power press that is operated by hydraulic piston or mechanical toggle devices of various types. Clamping forces of several thousand tons are available on large machines.

The cycle for injection molding of a thermoplastic polymer proceeds in the following sequence, illustrated in Figure 8.21. Let us pick up the action with the mold open and the machine ready to start a new molding: (1) The mold is closed and clamped. (2) A ***shot*** of melt, which has been brought to the right temperature and viscosity by heating and the mechanical working of the screw, is injected under high pressure into the mold cavity. The plastic cools and begins to solidify when it encounters the cold surface of the mold. Ram pressure is maintained to pack additional melt into the cavity to compensate for contraction during cooling. (3) The screw is rotated and retracted with the nonreturn valve open to permit fresh polymer to flow into the forward portion of the barrel. Meanwhile, the polymer in the mold has completely solidified. (4) The mold is opened, and the part is ejected and removed.

8.6.2 THE MOLD

The mold is the special tool in injection molding; it is custom-designed and fabricated for the given part to be produced. When the production run for that part is finished, the mold is replaced with a new mold for the next part. In this section we examine several types of mold for injection molding.

Two-Plate Mold The conventional two-plate mold, illustrated in Figure 8.22, consists of two halves fastened to the two platens of the molding machine's clamping unit. When the clamping unit is opened, the two mold halves open, as shown in (b). The most obvious feature of the mold is the ***cavity***, which is usually formed by removing metal from the mating surfaces of the two halves. Molds can contain a single cavity or multiple cavities to produce more than one part in a single shot. The figure shows a mold with two cavities. The ***parting surfaces*** (or ***parting line*** in a cross-sectional view of the mold) are where the mold opens to remove the part(s).

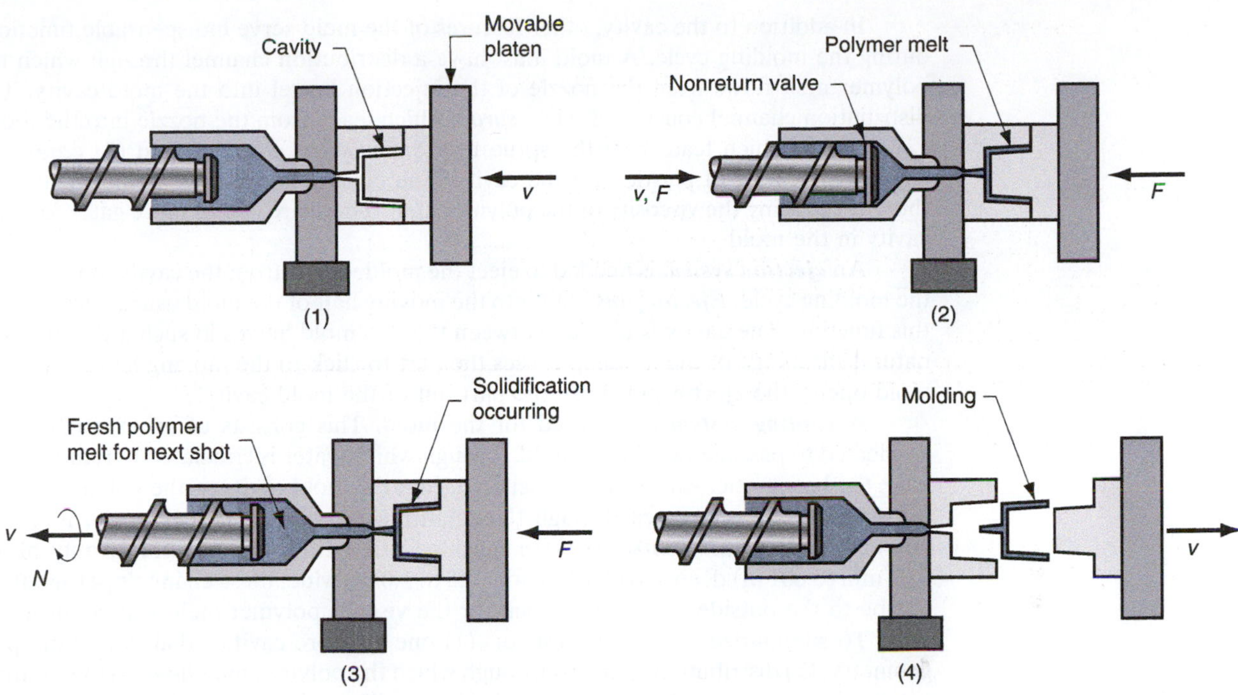

FIGURE 8.21 Typical molding cycle: (1) mold is closed, (2) melt is injected into cavity, (3) screw is retracted, and (4) mold opens, and part is ejected. (Credit: *Fundamentals of Modern Manufacturing,* 4th Edition by Mikell P. Groover, 2010. Reprinted with permission of John Wiley & Sons, Inc.)

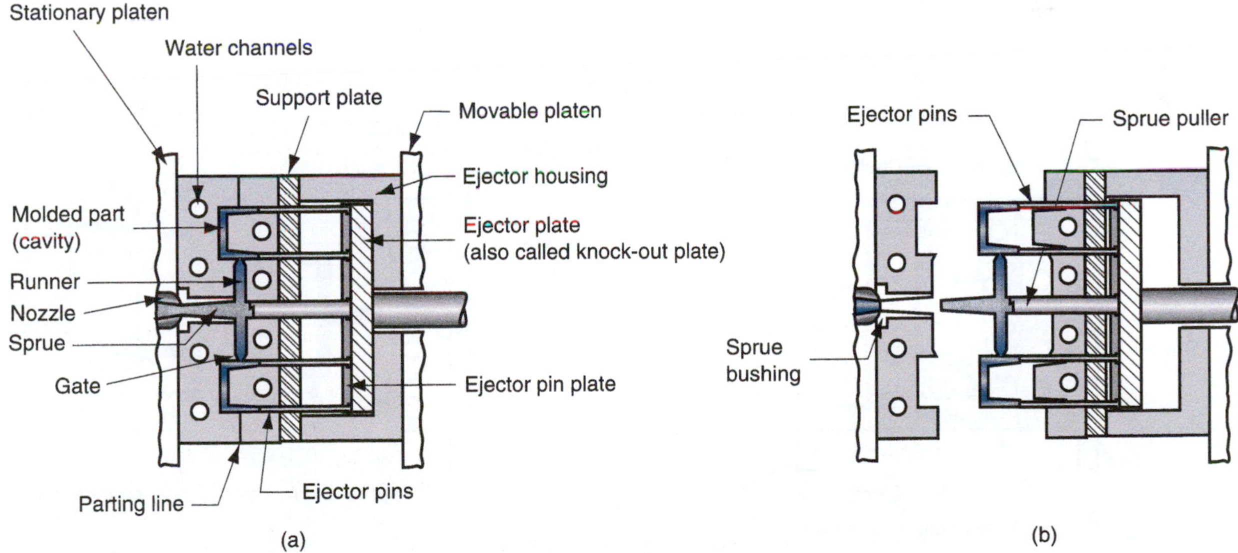

FIGURE 8.22 Details of a two-plate mold for thermoplastic injection molding: (a) closed and (b) open. Mold has two cavities to produce two cup-shaped parts (cross section shown) with each injection shot. (Credit: *Fundamentals of Modern Manufacturing,* 4th Edition by Mikell P. Groover, 2010. Reprinted with permission of John Wiley & Sons, Inc.)

In addition to the cavity, other features of the mold serve indispensable functions during the molding cycle. A mold must have a distribution channel through which the polymer melt flows from the nozzle of the injection barrel into the mold cavity. The distribution channel consists of (1) a ***sprue***, which leads from the nozzle into the mold; (2) ***runners***, which lead from the sprue to the cavity (or cavities); and (3) ***gates*** that constrict the flow of plastic into the cavity. The constriction increases the shear rate, thereby reducing the viscosity of the polymer melt. There are one or more gates for each cavity in the mold.

An ***ejection system*** is needed to eject the molded part from the cavity at the end of the molding cycle. ***Ejector pins*** built into the moving half of the mold usually accomplish this function. The cavity is divided between the two mold halves in such a way that the natural shrinkage of the molding causes the part to stick to the moving half. When the mold opens, the ejector pins push the part out of the mold cavity.

A ***cooling system*** is required for the mold. This consists of an external pump connected to passageways in the mold, through which water is circulated to remove heat from the hot plastic. Air must be evacuated from the mold cavity as the polymer rushes in. Much of the air passes through the small ejector pin clearances in the mold. In addition, narrow ***air vents*** are often machined into the parting surface; only about 0.03 mm (0.001 in) deep and 12 to 25 mm (0.5 to 1.0 in) wide, these channels permit air to escape to the outside but are too small for the viscous polymer melt to flow through.

To summarize, a mold consists of (1) one or more cavities that determine part geometry, (2) distribution channels through which the polymer melt flows to the cavities, (3) an ejection system for part removal, (4) a cooling system, and (5) vents to permit evacuation of air from the cavities.

Other Mold Types An alternative to the two-plate mold is a ***three-plate mold***, shown in Figure 8.23, for the same part geometry as before. There are several advantages to this mold design. First, the flow of molten plastic is through a gate located at the base of the cup-shaped part, rather than at the side. This allows more even distribution of melt into the sides of the cup. In the side-gate design in the two-plate mold of Figure 8.22, the plastic must flow around the core and join on the opposite side, possibly creating a

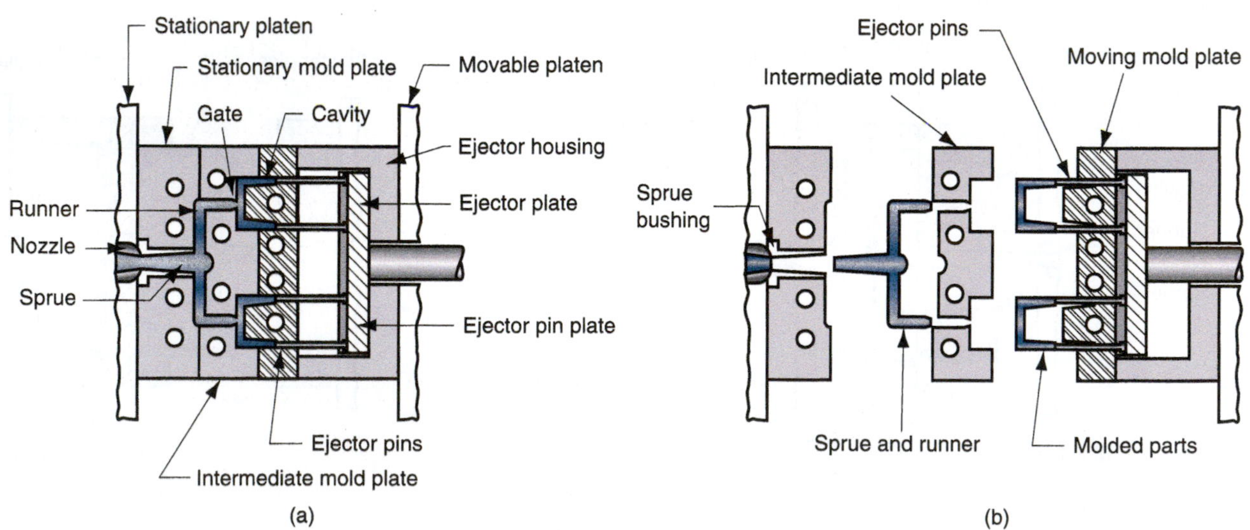

FIGURE 8.23 Three-plate mold: (a) closed, and (b) open. (Credit: *Fundamentals of Modern Manufacturing*, 4th Edition by Mikell P. Groover, 2010. Reprinted with permission of John Wiley & Sons, Inc.)

weakness at the weld line. Second, the three-plate mold allows more automatic operation of the molding machine. As the mold opens, it divides into three plates with two openings between them. This action separates the runner from the parts, which drop by gravity into containers beneath the mold.

The sprue and runner in a conventional two- or three-plate mold represent waste material. In many instances they can be ground and reused; however, in some cases the product must be made of "virgin" plastic (plastic that has not been previously molded). The ***hot-runner mold*** eliminates the solidification of the sprue and runner by locating heaters around the corresponding runner channels. Although the plastic in the mold cavity solidifies, the material in the sprue and runner channels remains molten, ready to be injected into the cavity in the next cycle.

8.6.3 SHRINKAGE AND DEFECTS IN INJECTION MOLDING

Polymers have high thermal expansion coefficients, and significant shrinkage can occur during cooling of the plastic in the mold. Contraction of crystalline plastics tends to be greater than for amorphous polymers. Shrinkage is usually expressed as the reduction in linear size that occurs during cooling to room temperature from the molding temperature for the given polymer. Appropriate units are therefore mm/mm (in/in) of the dimension under consideration. Typical values for selected polymers are given in Table 8.1.

Fillers in the plastic tend to reduce shrinkage. In commercial molding practice, shrinkage values for the specific molding compound should be obtained from the producer before making the mold. To compensate for shrinkage, the dimensions of the mold cavity must be made larger than the specified part dimensions. The following formula can be used [14]:

$$D_c = D_p + D_p S + D_p S^2 \tag{8.19}$$

where D_c = dimension of cavity, mm (in); D_p = molded part dimension, mm (in), and S = shrinkage values obtained from Table 8.1. The third term on the right-hand side corrects for shrinkage that occurs in the shrinkage.

Example 8.3 Shrinkage in Injection Molding

The nominal length of a part made of polyethylene is to be 80 mm. Determine the corresponding dimension of the mold cavity that will compensate for shrinkage.

Solution: From Table 8.1, the shrinkage for polyethylene is $S = 0.025$. Using Eq. (8.19), the mold cavity diameter should be:

$$\begin{aligned} D_c &= 80.0 + 80.0(0.025) + 80.0(0.025)^2 \\ &= 80.0 + 2.0 + 0.05 = 82.05 \text{ mm} \end{aligned}$$

■

TABLE 8.1 Typical values of shrinkage for moldings of selected thermoplastics.

Plastic	Shrinkage, mm/mm (in/in)	Plastic	Shrinkage, mm/mm (in/in)
ABS	0.006	Polyethylene	0.025
Nylon-6,6	0.020	Polystyrene	0.004
Polycarbonate	0.007	PVC	0.005

Compiled from [14].

Because of differences in shrinkage among plastics, mold dimensions must be determined for the particular polymer to be molded. The same mold will produce different part sizes for different polymer types.

Values in Table 8.1 represent a gross simplification of the shrinkage issue. In reality, shrinkage is affected by a number of factors, any of which can alter the amount of contraction for a given polymer. The most important factors are injection pressure, compaction time, molding temperature, and part thickness. As injection pressure is increased, forcing more material into the mold cavity, shrinkage is reduced. Increasing compaction time has a similar effect, assuming the polymer in the gate does not solidify and seal off the cavity; maintaining pressure forces more material into the cavity while shrinkage is taking place. Net shrinkage is thereby reduced.

Molding temperature refers to the temperature of the polymer in the cylinder immediately prior to injection. One might expect that a higher polymer temperature would increase shrinkage, on the reasoning that the difference between molding and room temperatures is greater. However, shrinkage is actually lower at higher molding temperatures. The explanation is that higher temperatures significantly lower the viscosity of the polymer melt, allowing more material to be packed into the mold; the effect is the same as higher injection pressures. Thus, the effect on viscosity more than compensates for the larger temperature difference.

Finally, thicker parts show greater shrinkage. A molding solidifies from the outside; the polymer in contact with the mold surface forms a skin that grows toward the center of the part. At some point during solidification, the gate solidifies, isolating the material in the cavity from the runner system and from compaction pressure. When this happens, the molten polymer inside the skin accounts for most of the remaining shrinkage that occurs in the part. A thicker part section experiences greater shrinkage because it contains a higher proportion of molten material.

In addition to the shrinkage issue, other things can also go wrong. Here are some of the common defects in injection molded parts:

- ***Short shots***. As in casting, a short shot is a molding that has solidified before completely filling the cavity. The defect can be corrected by increasing temperature and/or pressure. The defect may also result from use of a machine with insufficient shot capacity, in which case a larger machine is needed.
- ***Flashing***. Flashing occurs when the polymer melt is squeezed into the parting surface between mold plates; it can also occur around ejection pins. The defect is usually caused by (1) vents and clearances in the mold that are too large; (2) injection pressure too high compared to clamping force; (3) melt temperature too high; or (4) excessive shot size.
- ***Sink marks and voids***. These are defects usually related to thick molded sections. A ***sink mark*** occurs when the outer surface on the molding solidifies, but contraction of the internal material causes the skin to be depressed below its intended profile. A ***void*** is caused by the same basic phenomenon; however, the surface material retains its form and the shrinkage manifests itself as an internal void because of high tensile stresses on the still-molten polymer. These defects can be addressed by increasing the packing pressure after injection. A better solution is to design the part to have uniform section thicknesses and thinner sections.
- ***Weld lines***. Weld lines occur when polymer melt flows around a core or other convex detail in the mold cavity and meets from opposite directions; the boundary thus formed is called a weld line, and it may have mechanical properties that are inferior to those in the rest of the part. Higher melt temperatures, higher injection pressures, alternative gating locations on the part, and better venting are ways of dealing with this defect.

8.6.4 OTHER INJECTION MOLDING PROCESSES

The vast majority of injection molding applications involve thermoplastics. Several variants of the process are described in this section.

Thermoplastic Foam Injection Molding Plastic foams have a variety of applications, and we discuss these materials and their processing in Section . One of the processes, sometimes called ***structural foam molding***, is appropriate to discuss here because it involves injection molding. It involves the molding of thermoplastic parts that possess a dense outer skin surrounding a lightweight foam center. Such parts have high stiffness-to-weight ratios suitable for structural applications.

A structural foam part can be produced either by introducing a gas into the molten plastic in the injection unit or by mixing a gas-producing ingredient with the starting pellets. During injection, an insufficient amount of melt is forced into the mold cavity, where it expands (foams) to fill the mold. The foam cells in contact with the cold mold surface collapse to form a dense skin, while the material in the core retains its cellular structure. Items made of structural foam include electronic cases, business machine housings, furniture components, and washing machine tanks. Advantages cited for structural foam molding include lower injection pressures and clamping forces, and thus the capability to produce large components, as suggested by the preceding list. A disadvantage of the process is that the resulting part surfaces tend to be rough, with occasional voids. If good surface finish is needed for the application, then additional processing is required, such as sanding, painting, and adhesion of a veneer.

Injection Molding of Thermosets Injection molding can be used for thermosetting (TS) plastics, with certain modifications in equipment and operating procedure to allow for cross-linking. The machines for thermoset injection molding are similar to those used for thermoplastics. They use a reciprocating-screw injection unit, but the barrel length is shorter to avoid premature curing and solidification of the TS polymer. For the same reason, temperatures in the barrel are kept at relatively low levels, usually 50°C to 125°C (120°F to 260°F), depending on the polymer. The plastic, usually in the form of pellets or granules, is fed into the barrel through a hopper. Plasticizing occurs by the action of the rotating screw as the material is moved forward toward the nozzle. When sufficient melt has accumulated ahead of the screw, it is injected into a mold that is heated to 150°C to 230°C (300°F to 450°F), where cross-linking occurs to harden the plastic. The mold is then opened, and the part is ejected and removed. Molding cycle times typically range from 20 sec to 2 min, depending on polymer type and part size. Curing is the most time-consuming step in the cycle.

The principal thermosets for injection molding are phenolics, unsaturated polyesters, melamines, epoxies, and urea-formaldehyde. More than 50% of the phenolic moldings currently produced in the United States are made by this process [11], representing a shift away from compression and transfer molding, the traditional processes used for thermosets (Section 8.7). Most of the TS molding materials contain large proportions of fillers (up to 70% by weight), including glass fibers, clay, wood fibers, and carbon black. In effect, these are composite materials that are being injected molded.

Reaction Injection Molding Reaction injection molding (RIM) involves the mixing of two highly reactive liquid ingredients and immediately injecting the mixture into a mold cavity, where chemical reactions leading to solidification occur. Urethanes, epoxies, and urea-formaldehyde are examples of these polymer systems. RIM was developed with polyurethane to produce large automotive components such as bumpers, spoilers, and fenders. These kinds of parts still constitute the major application of the process.

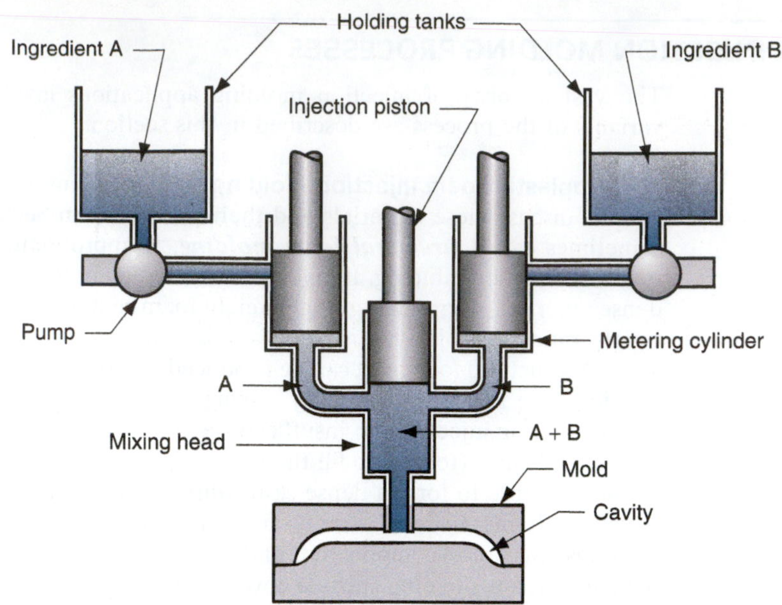

FIGURE 8.24 Reaction injection molding (RIM) system, shown immediately after ingredients A and B have been pumped into the mixing head prior to injection into the mold cavity (some details of processing equipment omitted). (Credit: *Fundamentals of Modern Manufacturing*, 4th Edition by Mikell P. Groover, 2010. Reprinted with permission of John Wiley & Sons, Inc.)

RIM-molded polyurethane parts typically possess a foam internal structure surrounded by a dense outer skin.

As shown in Figure 8.24, liquid ingredients are pumped in precisely measured amounts from separate holding tanks into a mixing head. The ingredients are rapidly mixed and then injected into the mold cavity at relatively low pressure where polymerization and curing occur. A typical cycle time is around 2 min. For relatively large cavities the molds for RIM are much less costly than corresponding molds for conventional injection molding. This results from the low clamping forces required in RIM and the opportunity to use lightweight components in the molds. Other advantages of RIM include (1) low energy is required in the process; (2) equipment costs are less than injection molding; (3) a variety of chemical systems are available that enable specific properties to be obtained in the molded product; and (4) the production equipment is reliable, and the chemical systems and machine relationships are well understood [17].

8.7 COMPRESSION AND TRANSFER MOLDING

Discussed in this section are two molding techniques widely used for thermosetting polymers and elastomers. For thermoplastics, these techniques cannot match the efficiency of injection molding, except for very special applications.

8.7.1 COMPRESSION MOLDING

Compression molding is an old and widely used molding process for thermosetting plastics. Its applications also include rubber tires and various polymer matrix composite parts. The process, illustrated in Figure 8.25 for a TS plastic, consists of (1) loading a precise amount of molding compound, called the ***charge***, into the bottom half of a heated mold; (2) bringing the mold halves together to compress the charge, forcing it to flow and conform to the shape of the cavity; (3) heating the charge by means of the hot mold to

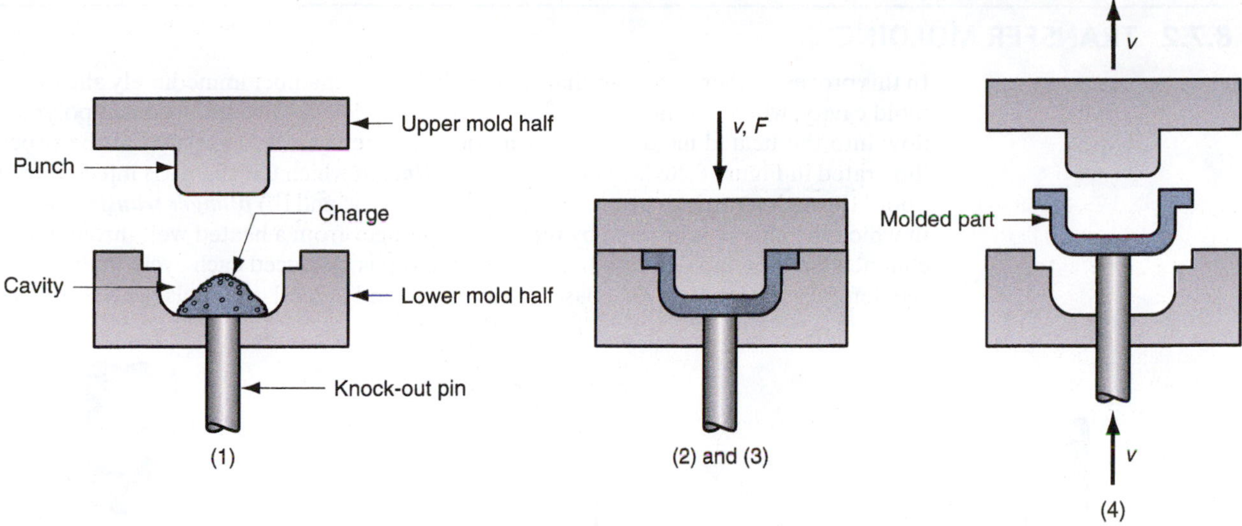

FIGURE 8.25 Compression molding for thermosetting plastics: (1) charge is loaded; (2) and (3) charge is compressed and cured; and (4) part is ejected and removed (some details omitted). (Credit: *Fundamentals of Modern Manufacturing*, 4th Edition by Mikell P. Groover, 2010. Reprinted with permission of John Wiley & Sons, Inc.)

polymerize and cure the material into a solidified part; and (4) opening the mold halves and removing the part from the cavity.

The initial charge of molding compound can be any of several forms, including powders or pellets, liquid, or preform (partially shaped blank). The amount of polymer must be precisely controlled to obtain repeatable consistency in the molded product. It has become common practice to preheat the charge before placing it into the mold; this softens the polymer and shortens the production cycle time. Preheating methods include infrared heaters, convection heating in an oven, and use of a heated rotating screw in a barrel. The latter technique (borrowed from injection molding) is also used to meter the amount of the charge.

Compression molding presses are oriented vertically and contain two platens to which the mold halves are fastened. The presses involve either of two types of actuation: (1) upstroke of the bottom platen or (2) downstroke of the top platen, the former being the more common machine configuration. They are generally powered by a hydraulic cylinder that can be designed to provide clamping capacities up to several hundred tons.

Molds for compression molding are generally simpler than their injection mold counterparts. There is no sprue and runner system in a compression mold, and the process itself is generally limited to simpler part geometries because of the lower flow capabilities of the starting TS materials. However, provision must be made for heating the mold, usually accomplished by electric resistance heating, steam, or hot oil circulation. Compression molds can be classified as ***hand molds***, used for trial runs; ***semiautomatic***, in which the press follows a programmed cycle but the operator manually loads and unloads the press; and ***automatic***, which operate under a fully automatic press cycle (including automatic loading and unloading).

Materials for compression molding include phenolics, melamine, urea-formaldehyde, epoxies, urethanes, and elastomers. Typical moldings include electric plugs and sockets, pot handles, and dinnerware plates. Advantages of compression molding in these applications include (1) molds that are simpler and less expensive, (2) less scrap, and (3) low residual stresses in the molded parts. A typical disadvantage is longer cycle times and therefore lower production rates than injection molding.

8.7.2 TRANSFER MOLDING

In this process, a thermosetting charge is loaded into a chamber immediately ahead of the mold cavity, where it is heated; pressure is then applied to force the softened polymer to flow into the heated mold where curing occurs. There are two variants of the process, illustrated in Figure 8.26: (a) ***pot transfer molding***, in which the charge is injected from a "pot" through a vertical sprue channel into the cavity; and (b) ***plunger transfer molding***, in which the charge is injected by means of a plunger from a heated well through lateral channels into the mold cavity. In both cases, scrap is produced each cycle in the form of the leftover material in the base of the well and lateral channels, called the ***cull***.

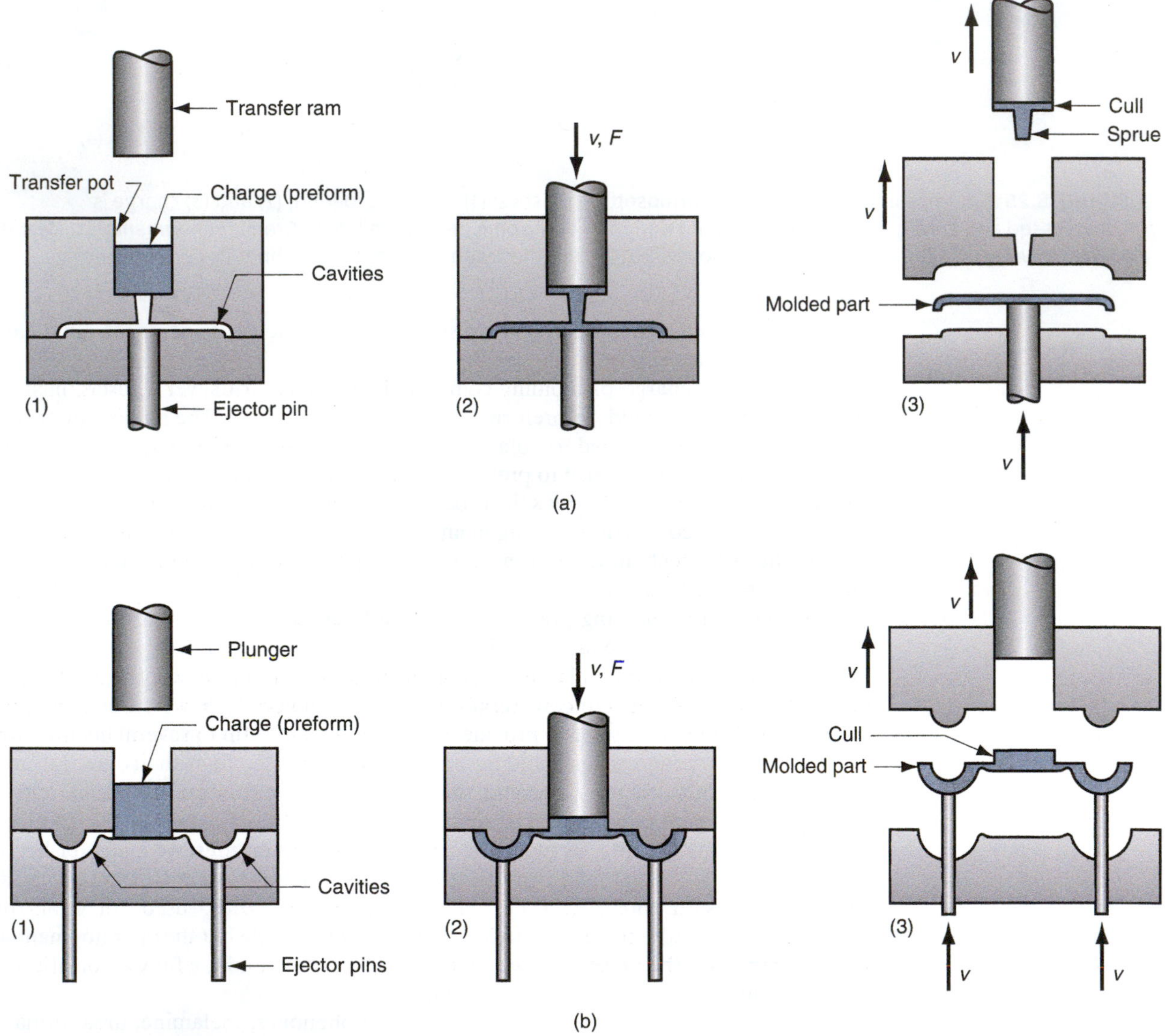

FIGURE 8.26 (a) Pot transfer molding, and (b) plunger transfer molding. Cycle in both processes is: (1) charge is loaded into pot, (2) softened polymer is pressed into mold cavity and cured, and (3) part is ejected. (Credit: *Fundamentals of Modern Manufacturing*, 4th Edition by Mikell P. Groover, 2010. Reprinted with permission of John Wiley & Sons, Inc.)

In addition, the sprue in pot transfer is scrap material. Because the polymers are thermosetting, the scrap cannot be recovered.

Transfer molding is closely related to compression molding, because it is used on the same polymer types (thermosets and elastomers). One can also see similarities to injection molding, in the way the charge is preheated in a separate chamber and then injected into the mold. Transfer molding is capable of molding part shapes that are more intricate than compression molding but not as intricate as injection molding. Transfer molding also lends itself to molding with inserts, in which a metal or ceramic insert is placed into the cavity before injection, and the heated plastic bonds to the insert during molding.

8.8 BLOW MOLDING AND ROTATIONAL MOLDING

Both of these processes are used to make hollow, seamless parts out of thermoplastic polymers. Rotational molding can also be used for thermosets. Parts range in size from small plastic bottles of only 5 mL (0.15 oz) to large storage drums of 38,000-L (10,000-gal) capacity. Although the two processes compete in certain cases, generally they have found their own niches. Blow molding is more suited to the mass production of small disposable containers, whereas rotational molding is favored for large, hollow shapes.

8.8.1 BLOW MOLDING

Blow molding is a molding process in which air pressure is used to inflate soft plastic inside a mold cavity. It is an important industrial process for making one-piece hollow plastic parts with thin walls, such as bottles and similar containers. Because many of these items are used for consumer beverages for mass markets, production is typically organized for very high quantities. The technology is borrowed from the glass industry (Section 7.2) with which plastics compete in the disposable and recyclable bottle market.

Blow molding is accomplished in two steps: (1) fabrication of a starting tube of molten plastic, called a ***parison*** (same term as in glass blowing); and (2) inflation of the tube to the desired final shape. Forming the parison is accomplished by either extrusion or injection molding.

Extrusion Blow Molding This form of blow molding consists of the cycle illustrated in Figure 8.27. In most cases, the process is organized as a very high production operation for making plastic bottles. The sequence is automated and often integrated with downstream operations such as bottle filling and labeling.

It is usually a requirement that the blown container be rigid, and rigidity depends on wall thickness among other factors. We can relate wall thickness of the blown container to the starting extruded parison [12], assuming a cylindrical shape for the final product. The effect of die swell on the parison is shown in Figure 8.28. The mean diameter of the tube as it exits the die is determined by the mean die diameter D_d. Die swell causes expansion to a mean parison diameter D_p. At the same time, wall thickness swells from t_d to t_p. The swell ratio of the parison diameter and wall thickness is given by

$$r_s = \frac{D_p}{D_d} = \frac{t_p}{t_d} \tag{8.20}$$

When the parison is inflated to the blow mold diameter D_m, there is a corresponding reduction in wall thickness to t_m. Assuming constant volume of cross section, we have

$$\pi D_p t_p = \pi D_m t_m \tag{8.21}$$

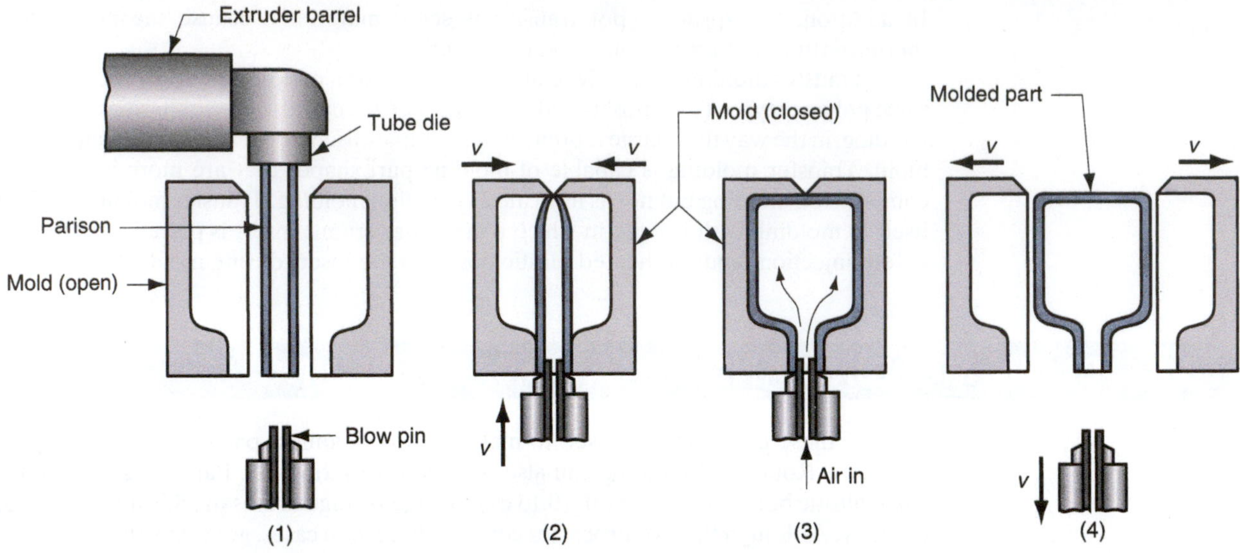

FIGURE 8.27 Extrusion blow molding: (1) extrusion of parison; (2) parison is pinched at the top and sealed at the bottom around a metal blow pin as the two halves of the mold come together; (3) the tube is inflated so that it takes the shape of the mold cavity; and (4) mold is opened to remove the solidified part. (Credit: *Fundamentals of Modern Manufacturing*, 4th Edition by Mikell P. Groover, 2010. Reprinted with permission of John Wiley & Sons, Inc.)

Solving for t_m, we obtain

$$t_m = \frac{D_p t_p}{D_m}$$

Substituting Eq. (8.20) into this equation, we get

$$t_m = \frac{r_s^2 t_d D_d}{D_m} \tag{8.22}$$

The amount of die swell in the initial extrusion process can be measured by direct observation; and the dimensions of the die are known. Thus, we can determine the wall thickness on the blow-molded container.

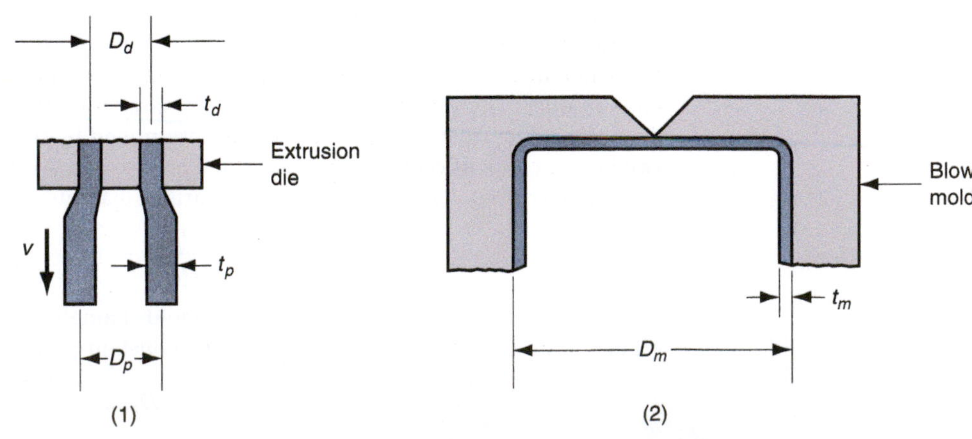

FIGURE 8.28 (1) Dimensions of extrusion die, showing parison after die swell; and (2) final blow-molded container in extrusion blow molding. (Credit: *Fundamentals of Modern Manufacturing*, 4th Edition by Mikell P. Groover, 2010. Reprinted with permission of John Wiley & Sons, Inc.)

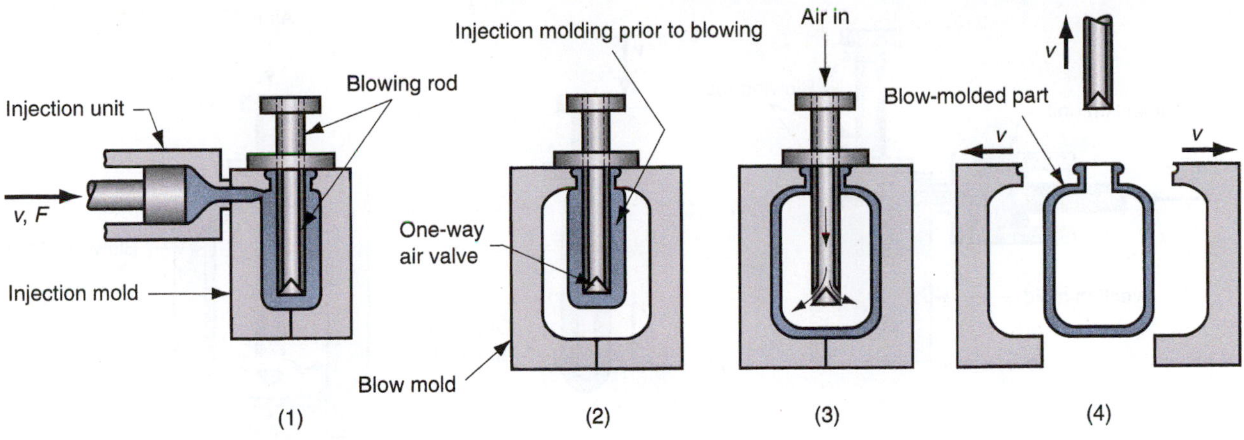

FIGURE 8.29 Injection blow molding: (1) parison is injected molded around a blowing rod; (2) injection mold is opened and parison is transferred to a blow mold; (3) soft polymer is inflated to conform to the blow mold; and (4) blow mold is opened, and blown product is removed. (Credit: *Fundamentals of Modern Manufacturing*, 4th Edition by Mikell P. Groover, 2010. Reprinted with permission of John Wiley & Sons, Inc.)

Injection Blow Molding In this process, the starting parison is injection molded rather than extruded. A simplified sequence is outlined in Figure 8.29. Compared to its extrusion-based competitor, injection blow molding usually has the following advantages: (1) higher production rate, (2) greater accuracy in the final dimensions, (3) lower scrap rates, and (4) less wasteful of material. On the other hand, larger containers can be produced with extrusion blow molding because the mold in injection molding is so expensive for large parisons. Also, extrusion blow molding is technically more feasible and economical for double-layer bottles used for storing certain medicines, personal care products, and various chemical compounds.[2]

In a variation of injection blow molding, called ***stretch blow molding*** (Figure 8.30), the blowing rod extends downward into the injection-molded parison during step 2, thus stretching the soft plastic and creating a more favorable stressing of the polymer than conventional injection blow molding or extrusion blow molding. The resulting structure is more rigid, with higher transparency and better impact resistance.

Materials and Products Blow molding is limited to thermoplastics. Polyethylene is the most widely used polymer—in particular, high-density and high-molecular-weight polyethylene (HDPE and HMWPE). In comparing their properties with those of low-density PE, given the requirement for stiffness in the final product, it is more economical to use these more expensive materials because the container walls can be made thinner. Other blow moldings are made of polypropylene and polyvinylchloride. The most widely used material for stretch blow molding is polyethylene terephthalate (PET), a polyester that has very low permeability and is strengthened by the stretch-blow-molding process. The combination of properties makes it ideal as a container for carbonated beverages (e.g., 2-L soda bottles).

Disposable containers for packaging liquid consumer goods constitute the major share of products made by blow molding; but they are not the only products. Other items include large shipping drums (55 gal) for liquids and powders, large storage tanks

[2] The author is indebted to Tom Walko, former student and at time of writing plant manager at one of Graham Packaging Company's blow molding plants, for providing the comparisons between extrusion and injection blow molding.

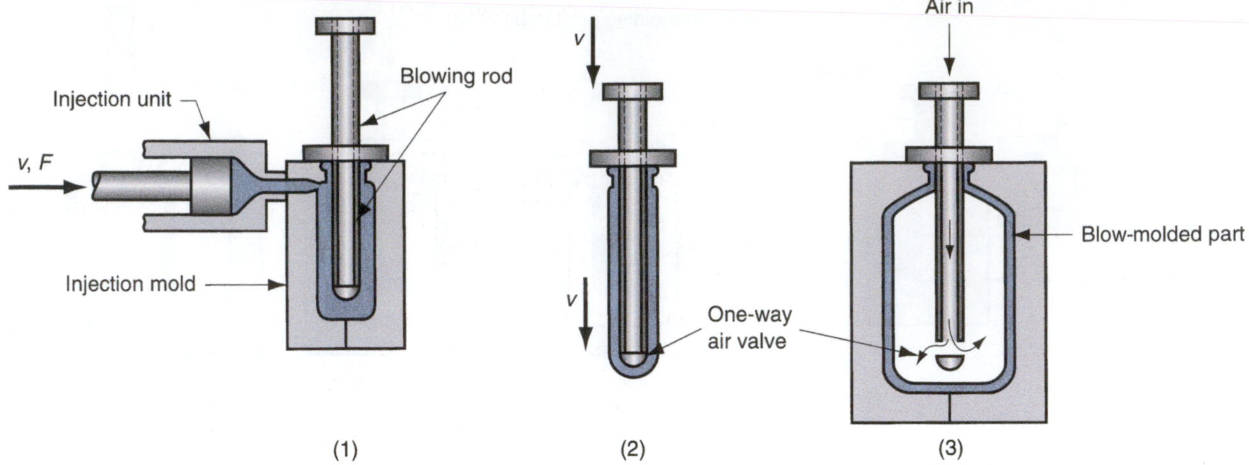

FIGURE 8.30 Stretch blow molding: (1) injection molding of parison, (2) stretching, and (3) blowing. (Credit: *Fundamentals of Modern Manufacturing,* 4th Edition by Mikell P. Groover, 2010. Reprinted with permission of John Wiley & Sons, Inc.)

(2000 gal), automotive gasoline tanks, toys, and hulls for sail boards and small boats. In the latter case, two boat hulls are made in a single blow molding and subsequently cut into two open hulls.

8.8.2 ROTATIONAL MOLDING

Rotational molding uses gravity inside a rotating mold to achieve a hollow form. Also called ***rotomolding***, it is an alternative to blow molding for making large, hollow shapes. It is used principally for thermoplastic polymers, but applications for thermosets and elastomers are becoming more common. Rotomolding tends to be more appropriate for complex external geometries, larger parts, and lower production quantities than blow molding. The process consists of the following steps: (1) A predetermined amount of polymer powder is loaded into the cavity of a split mold. (2) The mold is then heated and simultaneously rotated on two perpendicular axes, so that the powder impinges on all internal surfaces of the mold, gradually forming a fused layer of uniform thickness. (3) While still rotating, the mold is cooled so that the plastic skin solidifies. (4) The mold is opened, and the part is unloaded. Rotational speeds used in the process are relatively slow. It is gravity, not centrifugal force, that causes uniform coating of the mold surfaces.

Molds in rotational molding are simple and inexpensive compared with injection molding or blow molding, but the production cycle is much longer, lasting perhaps 10 min or more. To balance these advantages and disadvantages in production, rotational molding is often performed on a multicavity indexing machine, such as the three-station machine shown in Figure 8.31. The machine is designed so that three molds are indexed in sequence through three workstations. Thus, all three molds are working simultaneously. The first workstation is an unload–load station where the finished part is unloaded from the mold, and the powder for the next part is loaded into the cavity. The second station consists of a heating chamber where hot-air convection heats the mold while it is simultaneously rotated. Temperatures inside the chamber are around 375°C (700°F), depending on the polymer and the item being molded. The third station cools the mold, using forced cold air or water spray, to cool and solidify the plastic molding inside.

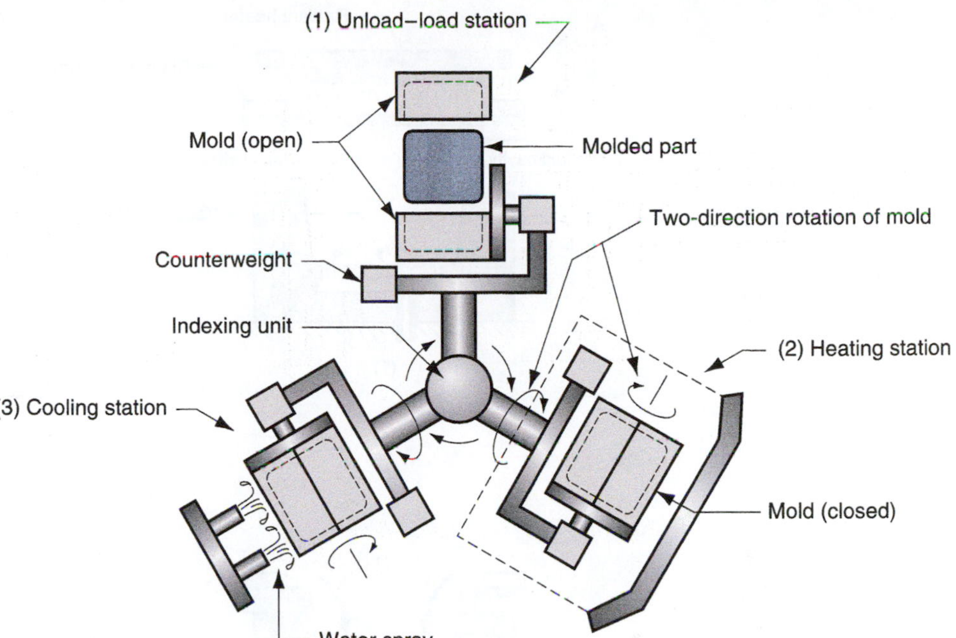

FIGURE 8.31 Rotational molding cycle performed on a three-station indexing machine: (1) unload–load station; (2) heat and rotate mold; (3) cool the mold. (Credit: *Fundamentals of Modern Manufacturing*, 4th Edition by Mikell P. Groover, 2010. Reprinted with permission of John Wiley & Sons, Inc.)

A fascinating variety of articles are made by rotational molding. The list includes hollow toys such as hobby horses and playing balls; boat and canoe hulls, sandboxes, small swimming pools; buoys and other flotation devices; truck body parts, automotive dashboards, fuel tanks; luggage pieces, furniture, garbage cans; fashion mannequins; large industrial barrels, containers, and storage tanks; portable outhouses, and septic tanks. The most popular molding material is polyethylene, especially HDPE. Other plastics include polypropylene, acrylonitrile-butadiene-styrene, and high-impact polystyrene.

8.9 THERMOFORMING

Thermoforming is a process in which a flat thermoplastic sheet is heated and deformed into the desired shape. The process is widely used for packaging of consumer products and fabricating large items such as bathtubs, contoured skylights, and internal door liners for refrigerators.

Thermoforming consists of two main steps: heating and forming. Heating is usually accomplished by radiant electric heaters, located on one or both sides of the starting plastic sheet at a distance of roughly 125 mm (5 in). Duration of the heating cycle to sufficiently soften the sheet depends on polymer thickness and color. Methods by which forming is accomplished can be classified into three basic categories: (1) vacuum thermoforming, (2) pressure thermoforming, and (3) mechanical thermoforming. In our discussion of these methods, we describe the forming of sheet stock, but in the packaging industry most thermoforming operations are performed on thin films.

Vacuum Thermoforming This was the first thermoforming process (simply called ***vacuum forming*** when it was developed in the 1950s). Negative pressure is used to draw a preheated sheet into a mold cavity. The process is explained in Figure 8.32 in its most basic form. The holes for drawing the vacuum in the mold are on the order of 0.8 mm (0.031 in) in diameter, so their effect on the plastic surface is minor.

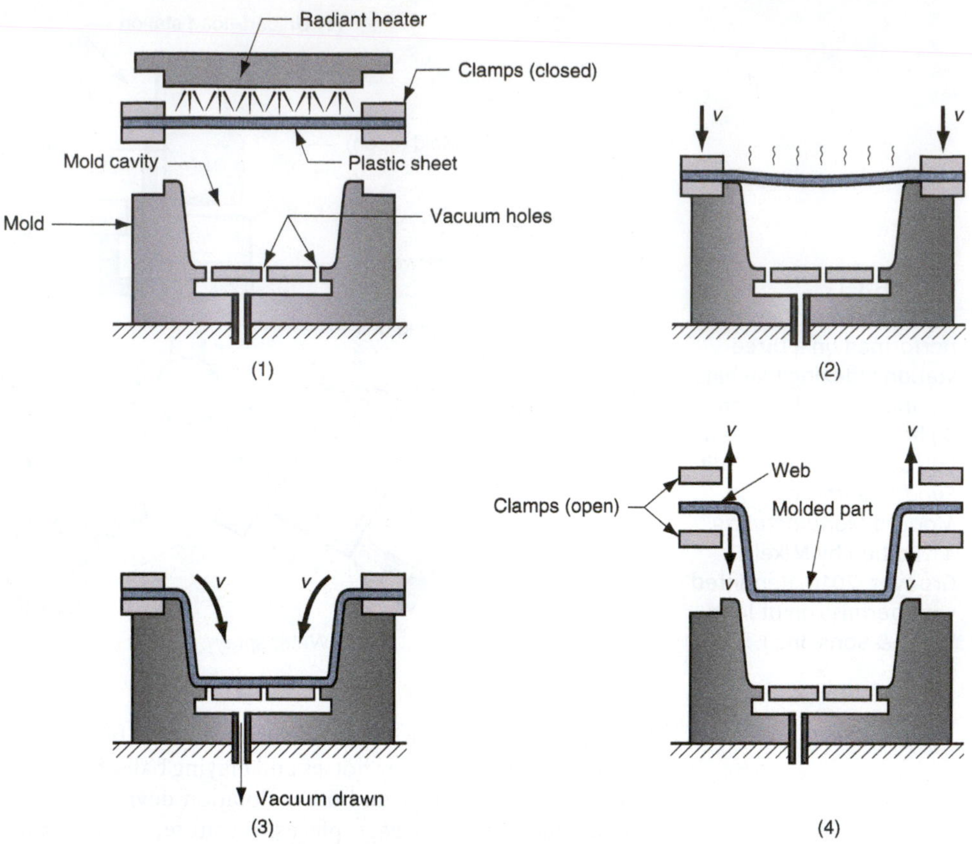

FIGURE 8.32 Vacuum thermoforming: (1) a flat plastic sheet is softened by heating; (2) the softened sheet is placed over a concave mold cavity; (3) a vacuum draws the sheet into the cavity; and (4) the plastic hardens on contact with the cold mold surface, and the part is removed and subsequently trimmed from the web. (Credit: *Fundamentals of Modern Manufacturing*, 4th Edition by Mikell P. Groover, 2010. Reprinted with permission of John Wiley & Sons, Inc.)

Pressure Thermoforming An alternative to vacuum forming involves positive pressure to force the heated plastic into the mold cavity. This is called ***pressure thermoforming*** or ***blow forming***; its advantage over vacuum forming is that higher pressures can be developed because the latter is limited to a theoretical maximum of 1 atm. Blow-forming pressures of 3 to 4 atm are common. The process sequence is similar to the previous, the difference being that the sheet is pressurized from above into the mold cavity. Vent holes are provided in the mold to exhaust the trapped air. The forming portion of the sequence (steps 2 and 3) is illustrated in Figure 8.33.

At this point it is useful to distinguish between negative and positive molds. The molds shown in Figures 8.32 and 8.33 are ***negative molds*** because they have concave cavities. A ***positive mold*** has a convex shape. Both types are used in thermoforming. In the case of the positive mold, the heated sheet is draped over the convex form and negative or positive pressure is used to force the plastic against the mold surface. A positive mold is shown in Figure 8.34 for vacuum thermoforming.

The difference between positive and negative molds may seem unimportant, because the part shapes are the same in the diagrams. However, if the part is drawn into the negative mold, then its exterior surface will have the exact surface contour of the

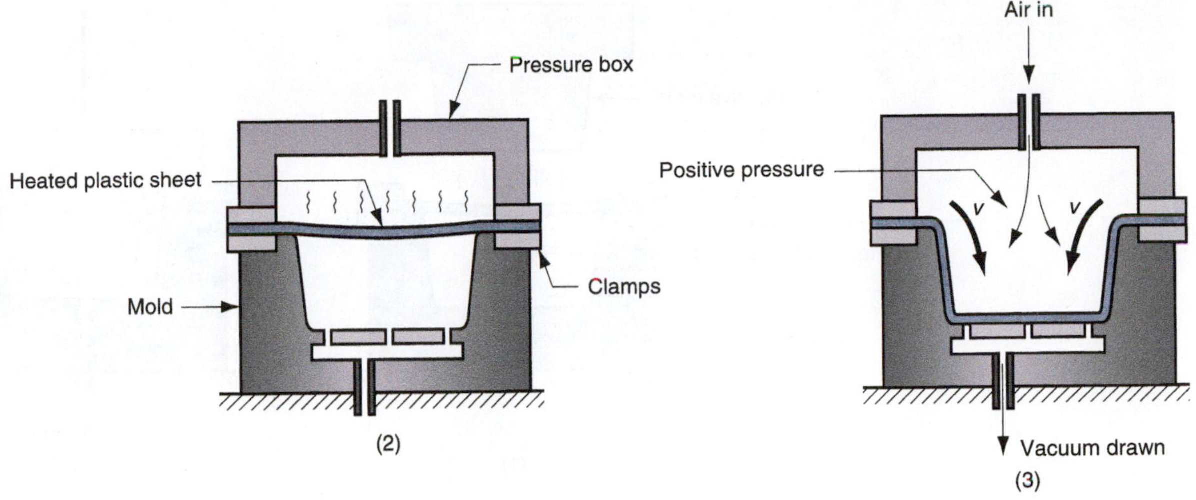

FIGURE 8.33 Pressure thermoforming. The sequence is similar to the previous figure, the difference being: (2) sheet is placed over a mold cavity; and (3) positive pressure forces the sheet into the cavity. (Credit: *Fundamentals of Modern Manufacturing,* 4th Edition by Mikell P. Groover, 2010. Reprinted with permission of John Wiley & Sons, Inc.)

mold cavity. The inside surface will be an approximation of the contour and will possess a finish corresponding to that of the starting sheet. By contrast, if the sheet is draped over a positive mold, then its interior surface will be identical to that of the convex mold; and its outside surface will follow approximately. Depending on the requirements of the product, this distinction might be important.

Another difference is in the thinning of the plastic sheet, one of the problems in thermoforming. Unless the contour of the mold is very shallow, there will be significant thinning of the sheet as it is stretched to conform to the mold contour. Positive and negative molds produce a different pattern of thinning in a given part. Consider the tub-shaped part in our figures. In the positive mold, as the sheet is draped over the convex form, the portion making contact with the top surface (corresponding to the base of

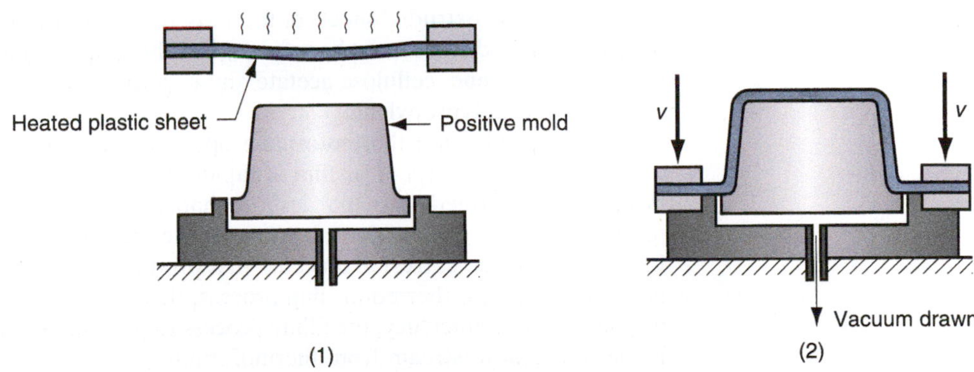

FIGURE 8.34 Use of a positive mold in vacuum thermoforming: (1) the heated plastic sheet is positioned above the convex mold and (2) the clamp is lowered into position, draping the sheet over the mold as a vacuum forces the sheet against the mold surface. (Credit: *Fundamentals of Modern Manufacturing,* 4th Edition by Mikell P. Groover, 2010. Reprinted with permission of John Wiley & Sons, Inc.)

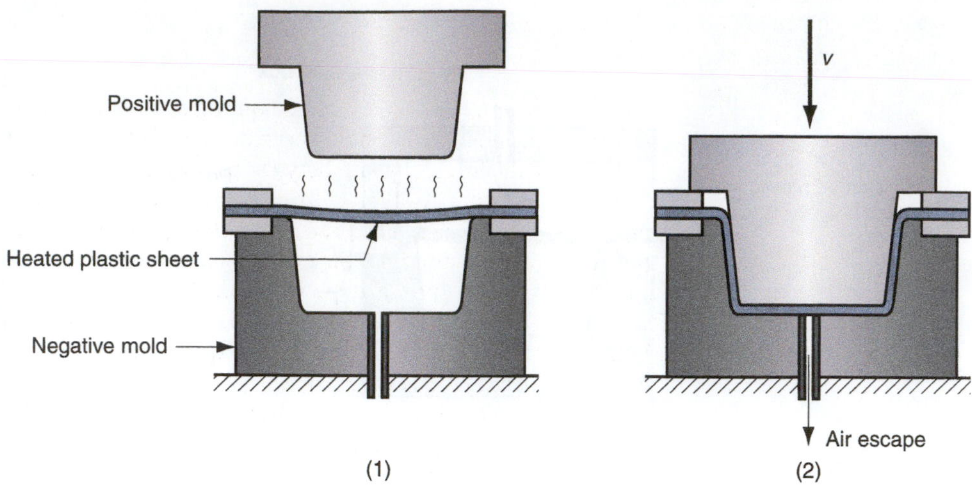

FIGURE 8.35 Mechanical thermoforming: (1) heated sheet placed above a negative mold, and (2) mold is closed to shape the sheet. (Credit: *Fundamentals of Modern Manufacturing*, 4th Edition by Mikell P. Groover, 2010. Reprinted with permission of John Wiley & Sons, Inc.)

the tub) solidifies quickly and experiences virtually no stretching. This results in a thick base but significant thinning in the walls of the tub. By contrast, a negative mold results in a more even distribution of stretching and thinning in the sheet before contact is made with the cold surface.

Mechanical Thermoforming The third method, called mechanical thermoforming, uses matching positive and negative molds that are brought together against the heated plastic sheet, forcing it to assume their shape. In pure mechanical forming, air pressure is not used at all. The process is illustrated in Figure 8.35. Its advantages are better dimensional control and the opportunity for surface detailing on both sides of the part. Its disadvantage is that two mold halves are required; therefore, the molds are more costly.

Applications Thermoforming is a secondary shaping process, the primary process being production of the sheet or film (Section 8.3). Only thermoplastics can be thermoformed, because extruded sheets of thermosetting or elastomeric polymers have already been cross-linked and cannot be softened by reheating. Common thermoforming plastics include polystyrene, cellulose acetate and cellulose acetate butyrate, ABS, PVC, polyethylene, and polypropylene.

Mass production thermoforming operations are performed in the packaging industry. The starting sheet or film is rapidly fed through a heating chamber and then mechanically formed into the desired shape. The operations are often designed to produce multiple parts with each stroke of the press using molds with multiple cavities. In some cases, the extrusion machine that produces the sheet or film is located directly upstream from the thermoforming process, thereby eliminating the need to reheat the plastic. For best efficiency, the filling process to put the item into the container is placed immediately downstream from thermoforming.

Thin film packages that are mass-produced by thermoforming include blister packs and skin packs. They offer an attractive way to display certain commodity products such as cosmetics, toiletries, small tools, and fasteners (nails, screws, etc.). Thermoforming applications include large parts that can be produced from thicker sheet stock. Examples include covers for business machines, boat hulls, shower stalls, diffusers for lights, advertising displays and signs, bathtubs, internal door liners for refrigerators, and certain toys.

8.10 CASTING

In polymer shaping, casting involves pouring of a liquid resin into a mold, using gravity to fill the cavity, and allowing the polymer to harden. Both thermoplastics and thermosets are cast. Examples of the former include acrylics, polystyrene, polyamides (nylons), and vinyls (PVC).

Thermosetting polymers for casting include polyurethane, unsaturated polyesters, phenolics, and epoxies. The process involves pouring the liquid ingredients that form the thermoset into a mold so that polymerization and cross-linking occur. Heat and/or catalysts may be required depending on the resin system. The reactions must be sufficiently slow to allow mold pouring to be completed. Fast-reacting thermosetting systems, such as certain polyurethane systems, require alternative shaping processes like reaction injection molding (Section 8.6.4).

Advantages of casting over alternative processes such as injection molding include: (1) the mold is simpler and less costly, (2) the cast item is relatively free of residual stresses and viscoelastic memory, and (3) the process is suited to low production quantities. Focusing on advantage (2), acrylic sheets (Plexiglas, Lucite) are generally cast between two pieces of highly polished plate glass. The casting process permits a high degree of flatness and desirable optical qualities to be achieved in the clear plastic sheets. Such flatness and clarity cannot be obtained by flat sheet extrusion. A disadvantage in some applications is significant shrinkage of the cast part during solidification. For example, acrylic sheets undergo a volumetric contraction of about 20% when cast. This is much more than in injection molding in which high pressures are used to pack the mold cavity in order to reduce shrinkage.

An important application of casting in electronics is ***encapsulation***, in which items such as transformers, coils, connectors, and other electrical components are encased in plastic by casting.

8.11 POLYMER FOAM PROCESSING AND FORMING

A polymer foam is a polymer-and-gas mixture, which gives the material a porous or cellular structure. The most common polymer foams are polystyrene (Styrofoam, a trademark) and polyurethane. Other polymers used to make foams include natural rubber ("foamed rubber") and polyvinylchloride (PVC).

The characteristic properties of a foamed polymer include (1) low density, (2) high strength per unit weight, (3) good thermal insulation, and (4) good energy-absorbing qualities. The elasticity of the base polymer determines the corresponding property of the foam. Polymer foams can be classified [6] as (1) ***elastomeric***, in which the matrix polymer is a rubber, capable of large elastic deformation; (2) ***flexible***, in which the matrix is a highly plasticized polymer such as soft PVC; and (3) ***rigid***, in which the polymer is a stiff thermoplastic such as polystyrene or a thermosetting plastic such as a phenolic. Depending on chemical formulation and degree of cross-linking, polyurethanes can range over all three categories.

The properties of polymer foams, and the ability to control their elastic behavior through selection of the base polymer, make these materials highly suitable for certain types of applications, including hot-beverage cups, heat-insulating structural materials and cores for structural panels, packaging materials, cushion materials for furniture and bedding, padding for automobile dashboards, and products requiring buoyancy.

Common gases used in polymer foams are air, nitrogen, and carbon dioxide. The proportion of gas can range up to 90% or more. The gas is introduced into the polymer by several methods, called foaming processes. These include (1) mixing a liquid resin with air by ***mechanical agitation***, then hardening the polymer by means of heat or chemical reaction; (2) mixing a ***physical blowing agent*** with the polymer—a gas such as nitrogen (N_2) or pentane (C_5H_{12}), which can be dissolved in the polymer melt under pressure, so that the gas comes out of solution and expands when the pressure is subsequently reduced; and (3) mixing the polymer with chemical compounds, called ***chemical blowing agents***, that decompose at elevated temperatures to liberate gases such as CO_2 or N_2 within the melt.

There are many shaping processes for polymer foam products. Because the two most important foams are polystyrene and polyurethane, our discussion is limited to shaping processes for these materials. Because polystyrene is thermoplastic and polyurethane can be either thermoset or elastomer, the processes covered here for these materials are representative of those used for other polymer foams.

Polystyrene foams are shaped by extrusion and molding. In ***extrusion***, a physical or chemical blowing agent is fed into the polymer melt near the die end of the extruder barrel; thus, the extrudate consists of the expanded polymer. Large sheets and boards are made this way and subsequently cut to size for heat insulation panels and sections.

Several molding processes are available for polystyrene foam. We previously discussed ***structural foam molding*** (Section 8.6.4). A more widely used process is ***expandable foam molding***, in which the molding material usually consists of prefoamed polystyrene beads. The prefoamed beads are produced from pellets of solid polystyrene that have been impregnated with a physical blowing agent. Prefoaming is performed in a large tank by applying steam heat to partially expand the pellets, simultaneously agitating them to prevent fusion. Then, in the molding process, the prefoamed beads are fed into a mold cavity, where they are further expanded and fused together to form the molded product. Hot-beverage cups of polystyrene foam are produced this way. In some processes, the prefoaming step is omitted, and the impregnated beads are fed directly into the mold cavity where they are heated, expanded, and fused. In other operations, the expandable foam is first formed into a flat sheet by the ***blown-film extrusion process*** (Section 8.3) and then shaped by ***thermoforming*** (Section 8.9) into packaging containers such as egg cartons.

Polyurethane foam products are made in a one-step process in which the two liquid ingredients (polyol and isocyanate) are mixed and immediately fed into a mold or other form, so that the polymer is synthesized and the part geometry is created at the same time. Shaping processes for polyurethane foam can be divided into two basic types [11]: spraying and pouring. ***Spraying*** involves use of a spray gun into which the two ingredients are continuously fed, mixed, and then sprayed onto a target surface. The reactions leading to polymerization and foaming occur after application on the surface. This method is used to apply rigid insulating foams onto construction panels, railway cars, and similar large items. ***Pouring*** involves dispensing the ingredients from a mixing head into an open or closed mold in which the reactions occur. An open mold can be a container with the required contour (e.g., for an automobile seat cushion) or a long channel that is slowly moved past the pouring spout to make long, continuous sections of foam. The closed mold is a completely enclosed cavity into which a certain amount of the mixture is dispensed. Expansion of the reactants completely fills the cavity to shape the part. For fast-reacting polyurethanes, the mixture must be rapidly injected into the mold cavity using ***reaction injection molding*** (Section 8.6.4). The degree of cross-linking, controlled by the starting ingredients, determines the relative stiffness of the resulting foam.

8.12 PRODUCT DESIGN CONSIDERATIONS

Plastics are an important design material, but the designer must be aware of their limitations. This section lists some design guidelines for plastic components, beginning with those that apply in general, and then ones applicable to extrusion and molding (injection molding, compression molding, and transfer molding).

Several guidelines apply irrespective of the shaping process. They are mostly limitations of plastic materials that must be considered by the designer.

- ***Strength and stiffness***. Plastics are not as strong or stiff as metals. They should not be used in applications in which high stresses will be encountered. Creep resistance is also a limitation. Strength properties vary significantly among plastics, and strength-to-weight ratios for some plastics are competitive with metals in certain applications.
- ***Impact resistance***. The capacity of plastics to absorb impact is generally good; plastics compare favorably with most metals.
- ***Service temperatures*** of plastics are limited relative to engineering metals and ceramics.
- ***Thermal expansion*** is greater for plastics than metals; so dimensional changes due to temperature variations are much more significant than for metals.
- Many types of plastics are subject to ***degradation*** from sunlight and certain other forms of radiation. Also, some plastics degrade in oxygen and ozone atmospheres. Finally, plastics are soluble in many common solvents. On the positive side, plastics are resistant to conventional corrosion mechanisms that afflict many metals. The weaknesses of specific plastics must be taken into account by the designer.

Extrusion is one of the most widely used plastic-shaping processes. Several design recommendations are presented here for conventional extrusion (compiled mostly from [3]).

- ***Wall thickness***. Uniform wall thickness is desirable in an extruded cross section. Variations in wall thickness result in nonuniform plastic flow and uneven cooling that tend to warp the extrudate.
- ***Hollow sections*** complicate die design and plastic flow. It is desirable to use extruded cross sections that are not hollow yet satisfy functional requirements.
- ***Corners***. Sharp corners, inside and outside, should be avoided in the cross section, because they result in uneven flow during processing and stress concentrations in the final product.

The following guidelines apply to injection molding, compression molding, and transfer molding (compiled from [3], [10], and other sources).

- ***Economic production quantities***. Each molded part requires a unique mold, and the mold for any of these processes can be costly, particularly for injection molding. Minimum production quantities for injection molding are usually around 10,000 pieces; for compression molding, minimum quantities are around 1,000 parts, because a simpler mold design is involved. Transfer molding lies between the other two.
- ***Part complexity***. Although more complex part geometries mean more costly molds, it may nevertheless be economical to design a complex molding if the alternative involves many individual components assembled together. An advantage of plastic molding is that it allows multiple functional features to be combined into one part.
- ***Wall thickness***. Thick cross sections are generally undesirable; they are wasteful of material, more likely to cause warping due to shrinkage, and take longer to harden. ***Reinforcing ribs*** can be used in molded plastic parts to achieve increased stiffness without excessive wall thickness. The ribs should be made thinner than the walls they reinforce, to minimize sink marks on the outside wall.

TABLE 8.2 Typical tolerances on molded parts for selected plastics.

Plastic	Tolerances for:[a] 50-mm dimension	Tolerances for:[a] 10-mm hole	Plastic	Tolerances for:[a] 50-mm dimension	Tolerances for:[a] 10-mm hole
Thermoplastic:			Thermosetting:		
ABS	±0.2 mm (±0.007 in)	±0.08 mm (±0.003 in)	Epoxies	±0.15 mm (±0.006 in)	±0.05 mm (±0.002 in)
Polyethylene	±0.3 mm (±0.010 in)	±0.13 mm (±0.005 in)	Phenolics	±0.2 mm (±0.008 in)	±0.08 mm (±0.003 in)
Polystyrene	±0.15 mm (±0.006 in)	±0.1 mm (±0.004 in)			

Values represent typical commercial molding practice. Compiled from [3], [7], [14], and [19].
[a]For smaller sizes, tolerances can be reduced. For larger sizes, more generous tolerances are required.

- ***Corner radii and fillets***. Sharp corners, both external and internal, are undesirable in molded parts; they interrupt smooth flow of the melt, tend to create surface defects, and cause stress concentrations in the finished part.
- ***Holes*** are quite feasible in plastic moldings, but they complicate mold design and part removal. They also cause interruptions in melt flow.
- ***Draft***. A molded part should be designed with a draft on its sides to facilitate removal from the mold. This is especially important on the inside wall of a cup-shaped part because the molded plastic contracts against the positive mold shape. The recommended draft for thermosets is around 1/2° to 1°; for thermoplastics it usually ranges between 1/8° and 1/2°. Suppliers of plastic-molding compounds provide recommended draft values for their products.
- ***Tolerances*** specify the allowable manufacturing variations for a part. Although shrinkage is predictable under closely controlled conditions, generous tolerances are desirable for injection moldings because of variations in process parameters that affect shrinkage. Table 8.2 lists typical tolerances for molded part dimensions of selected plastics.

REFERENCES

[1] Baird, D. G., and Collias, D. I. ***Polymer Processing Principles and Design***. John Wiley & Sons, Inc., New York, 1998.

[2] Billmeyer, Fred, W., Jr. ***Textbook of Polymer Science***, 3rd ed. John Wiley & Sons, New York, 1984.

[3] Bralla, J. G.(Editor in Chief). ***Design for Manufacturability Handbook***, 2nd ed. McGraw-Hill Book Company, New York, 1998.

[4] Briston, J. H. ***Plastic Films***, 3rd ed. Longman Group U.K., Ltd., Essex, England, 1989.

[5] Chanda, M., and Roy, S. K. ***Plastics Technology Handbook***. Marcel Dekker, Inc., New York, 1998.

[6] Charrier, J-M. ***Polymeric Materials and Processing***. Oxford University Press, New York, 1991.

[7] ***Engineering Materials Handbook***, Vol. 2, ***Engineering Plastics***. ASM International, Materials Park, Ohio, 1988.

[8] Hall, C. ***Polymer Materials***, 2nd ed. John Wiley & Sons, New York, 1989.

[9] Hensen, F. (ed.). ***Plastic Extrusion Technology***, Hanser Publishers, Munich, FRG, 1988. (Distributed in United States by Oxford University Press, New York.)

[10] McCrum, N. G., Buckley, C. P., and Bucknall, C. B. ***Principles of Polymer Engineering***, 2nd ed. Oxford University Press, Oxford, United Kingdom, 1997.

[11] ***Modern Plastics Encyclopedia***, Modern Plastics. McGraw-Hill, Inc., Hightstown, New Jersey, 1991.

[12] Morton-Jones, D. H. ***Polymer Processing***. Chapman and Hall, London, United Kingdom, 1989.

[13] Pearson, J. R. A. ***Mechanics of Polymer Processing***. Elsevier Applied Science Publishers, London, 1985.

[14] Rubin, I. I. ***Injection Molding: Theory and Practice***. John Wiley & Sons, New York, 1973.

[15] Rudin, A. ***The Elements of Polymer Science and Engineering***, 2nd ed. Academic Press, Inc., Orlando, Florida, 1999.

[16] Strong, A. B. ***Plastics: Materials and Processing***, 3rd ed. Pearson Educational, Upper Saddle River, New Jersey, 2006.

[17] Sweeney, F. M. ***Reaction Injection Molding Machinery and Processes.*** Marcel Dekker, Inc., New York, 1987.

[18] Tadmor, Z., and Gogos, C. G. ***Principles of Polymer Processing***. John Wiley & Sons, New York, 1979.

[19] Wick, C., Benedict, J. T., and Veilleux, R. F. ***Tool and Manufacturing Engineers Handbook***, 4th ed. Vol. II: ***Forming***. Society of Manufacturing Engineers, Dearborn, Michigan, 1984, Chapter 18.

REVIEW QUESTIONS

8.1. What are some of the reasons why plastic-shaping processes are important?

8.2. Identify the main categories of plastic-shaping processes, as classified by the resulting product geometry.

8.3. Viscosity is an important property of a polymer melt in plastic-shaping processes. Upon what parameters does viscosity depend?

8.4. How does the viscosity of a polymer melt differ from most fluids that are Newtonian.

8.5. What does viscoelasticity mean, when applied to a polymer melt?

8.6. Define die swell in extrusion.

8.7. Briefly describe the plastic extrusion process.

8.8. The barrel and screw of an extruder are generally divided into three sections; identify the sections.

8.9. What are the functions of the screen pack and breaker plate at the die end of the extruder barrel?

8.10. What are the various forms of extruded shapes and corresponding dies?

8.11. What is the distinction between plastic sheet and film?

8.12. What is the blown-film process for producing film stock?

8.13. Describe the calendering process.

8.14. Polymer fibers and filaments are used in several applications. What is the most important commercial application?

8.15. What is the technical difference between a fiber and a filament?

8.16. Briefly describe the injection-molding process.

8.17. An injection-molding machine is divided into two principal components. Name them.

8.18. What is the function of gates in injection molds?

8.19. What are the advantages of a three-plate mold over a two-plate mold in injection molding?

8.20. Discuss some of the defects that can occur in plastic injection molding.

8.21. What are the significant differences in the equipment and operating procedures between injection molding of thermoplastics and injection molding of thermosets?

8.22. What is reaction injection molding?

8.23. What kinds of products are produced by blow molding?

8.24. What is the form of the starting material in thermoforming?

8.25. What is the difference between a positive mold and a negative mold in thermoforming?

8.26. Why are the molds generally more costly in mechanical thermoforming than in pressure or vacuum thermoforming?

8.27. What are some of the general considerations that product designers must keep in mind when designing components out of plastics?

PROBLEMS

8.1. The diameter of an extruder barrel is 65 mm and its length = 1.75 m. The screw rotates at 55 rev/min. The screw channel depth = 5.0 mm, and the flight angle = 18°. The head pressure at the die end of the barrel is 5.0×10^6 Pa. The viscosity of the polymer melt is given as 100 Pa-s. Find the volume flow rate of the plastic in the barrel.

8.2. An extruder barrel has a diameter of 110 mm and a length of 3.0 m. The screw channel depth = 7.0 mm, and its pitch = 95 mm. The viscosity of the polymer melt is 105 Pa-s, and the head pressure in the barrel is 4.0 MPa. What rotational speed of the screw is required to achieve a volumetric flow rate of 90 cm^3/s?

8.3. An extruder has diameter = 80 mm and length = 2.0 m. Its screw has a channel depth = 5 mm, flight angle = 18 degrees, and it rotates at 1 rev/sec. The plastic melt has a shear viscosity = 150 Pa-s. Determine the extruder characteristic by computing Q_{max} and p_{max} and then finding the equation of the straight line between them.

8.4. Determine the helix angle A such that the screw pitch p is equal to the screw diameter D. This is called the "square" angle in plastics extrusion—the angle that provides a flight advance equal to one diameter for each rotation of the screw.

8.5. An extruder barrel has a diameter of 2.5 in. The screw rotates at 60 rev/min; its channel depth = 0.20 in, and its flight angle = 17.5°. The head pressure at the die end of the barrel is 800 lb/in^2 and the length of the barrel is 50 in. The viscosity of the polymer melt is 122 × 10^{-4} lb-sec/in^2. Determine the volume flow rate of the plastic in the barrel.

8.6. An extruder barrel has a diameter of 4.0 in and an L/D ratio of 28. The screw channel depth = 0.25 in, and its pitch = 4.8 in. It rotates at 60 rev/min. The viscosity of the polymer melt is 100 × 10^{-4} lb-sec/in^2. What head pressure is required to obtain a volume flow rate = 150 in^3/min?

8.7. An extruder has barrel diameter and length of 100 mm and 2.8 m, respectively. The screw rotational speed = 50 rev/min, channel depth = 7.5 mm, and flight angle = 17°. The plastic melt has a shear viscosity = 175 Pa-s. Determine: (a) the extruder characteristic, (b) the shape factor K_s for a circular die opening with diameter = 3.0 mm and length = 12.0 mm, and (c) the operating point (Q and p).

8.8. Consider an extruder in which the barrel diameter = 4.5 in and length = 11 ft. The extruder screw rotates at 60 rev/min; it has channel depth = 0.35 in and flight angle = 20°. The plastic melt has a shear viscosity = 125×10^{-4} lb-sec/in^2. Determine: (a) Q_{max} and p_{max}; (b) the shape factor K_s for a circular die opening in which D_d = 0.312 in and L_d = 0.75 in; and (c) the values of Q and p at the operating point.

8.9. The specified dimension = 225.00 mm for a certain injection-molded part made of ABS. Compute the corresponding dimension to which the mold cavity should be machined, using the value of shrinkage given in Table 8.1.

8.10. The part dimension for a certain injection-molded part made of polycarbonate is specified as 3.75 in. Compute the corresponding dimension to which the mold cavity should be machined, using the value of shrinkage given in Table 8.1.

8.11. The foreman in the injection-molding department says that a polyethylene part produced in one of the operations has greater shrinkage than the calculations indicate it should have. The important dimension of the part is specified as 112.5 ±0.25 mm. However, the actual molded part measures 112.02 mm. (a) As a first step, the corresponding mold cavity dimension should be checked. Compute the correct value of the mold dimension, given that the shrinkage value for polyethylene is 0.025 (from Table 8.1). (b) What adjustments in process parameters could be made to reduce the amount of shrinkage?

8.12. The extrusion die for a polyethylene parison used in blow molding has a mean diameter of 18.0 mm. The size of the ring opening in the die is 2.0 mm. The mean diameter of the parison is observed to swell to a size of 21.5 mm after exiting the die orifice. If the diameter of the blow-molded container is to be 150 mm, determine (a) the corresponding wall thickness of the container and (b) the wall thickness of the parison.

8.13. A parison is extruded from a die with outside diameter = 11.5 mm and inside diameter = 7.5 mm. The observed die swell is 1.25. The parison is used to blow mold a beverage container whose outside diameter = 112 mm (a standard size 2-liter soda bottle). (a) What is the corresponding wall thickness of the container? (b) Obtain an empty 2-liter plastic soda bottle and (carefully) cut it across the diameter. Using a micrometer, measure the wall thickness to compare with your answer in (a).

8.14. An extrusion operation is used to produce a parison whose mean diameter = 27 mm. The inside and outside diameters of the die that produced the parison are 18 mm and 22 mm, respectively. If the minimum wall thickness of the blow-molded container is to be 0.40 mm, what is the maximum possible diameter of the blow mold?

8.15. A rotational molding operation is to be used to mold a hollow playing ball out of polypropylene. The ball will be 1.25 ft in diameter and its wall thickness should be 3/32 in. What weight of PP powder should be loaded into the mold in order to meet these specifications? The specific gravity of the PP grade is 0.90, and the density of water is 62.4 lb/ft^3.

8.16. The problem in a certain thermoforming operation is that there is too much thinning in the walls of the large cup-shaped part. The operation is conventional pressure thermoforming using a positive mold, and the plastic is an ABS sheet with an initial thickness of 3.2 mm. (a) Why is thinning occurring in the walls of the cup? (b) What changes could be made in the operation to correct the problem?

9 SHAPING PROCESSES FOR RUBBER AND POLYMER-MATRIX COMPOSITES

Chapter Contents

9.1 Rubber Processing and Shaping
- 9.1.1 Production of Rubber
- 9.1.2 Compounding
- 9.1.3 Mixing
- 9.1.4 Shaping and Related Processes
- 9.1.5 Vulcanization

9.2 Manufacture of Tires and Other Rubber Products
- 9.2.1 Tires
- 9.2.2 Other Rubber Products
- 9.2.3 Processing of Thermoplastic Elastomers

9.3 PMC Shaping Processes and Materials
- 9.3.1 Starting Materials for PMCs
- 9.3.2 Combining Matrix and Reinforcement

9.4 Open Mold Processes
- 9.4.1 Hand Lay-Up
- 9.4.2 Spray-Up
- 9.4.3 Automated Tape-Laying Machines
- 9.4.4 Curing

9.5 Closed Mold Processes
- 9.5.1 Compression Molding PMC Processes
- 9.5.2 Transfer Molding PMC Processes
- 9.5.3 Injection Molding PMC Processes

9.6 Filament Winding

9.7 Pultrusion Processes
- 9.7.1 Pultrusion
- 9.7.2 Pulforming

9.8 Other PMC Shaping Processes

Many of the shaping processes used for plastics (Chapter 8) are also applicable to rubbers and polymer-matrix composites. However, the shaping processes must often be adapted because of differences in these materials. We discuss the adaptations and differences in this chapter.

The rubber industry is largely separate from the plastics industry, and goods made of rubber are dominated by one product: tires. Tires are used in large numbers for automobiles, trucks, aircraft, and bicycles. Rubber technology can be traced to the discovery by Charles Goodyear in 1839 of vulcanization, the process by which raw natural rubber is transformed into a usable material through cross-linking of the polymer molecules. During its first century, the rubber industry was concerned only with the processing of natural rubber. Around World War II, synthetic rubbers were developed; today they account for the majority of rubber production. Tires and many other rubber products are actually polymer-matrix composites because they contain carbon black as a reinforcing phase. Tires and rubber conveyors are also composite structures because they include wires of steel or other materials to limit the amount of extension that is experienced by the product. We discuss the rubber-processing technology in Section 9.1 and the manufacture of tires and other rubber products in Section 9.2.

Our coverage in this chapter also includes the manufacturing processes by which polymer-matrix composites are shaped into useful components and products. A ***polymer-matrix composite*** (PMC) is a composite material consisting of a polymer embedded with a reinforcing phase such as fibers or powders (Section 2.4). The technological and commercial importance of the PMC processes derives from the growing use of this class of material, especially ***fiber-reinforced polymers*** (FRPs). In popular usage, PMC generally refers to fiber-reinforced polymers. FRP composites can be designed with very high strength-to-weight and stiffness-to-weight ratios. These features make them attractive in aircraft, cars, trucks, boats, and sports equipment. Shaping processes for PMCs are discussed in Sections 9.3 through 9.8.

9.1 RUBBER PROCESSING AND SHAPING

Production of rubber goods can be divided into two basic steps: (1) production of the rubber itself, and (2) processing of the rubber into finished goods. Production of rubber differs, depending on whether it is natural or synthetic. Natural rubber (NR) is produced as an agricultural crop, whereas most synthetic rubbers are made from petroleum.

Production of rubber is followed by processing into final products; this consists of (1) compounding, (2) mixing, (3) shaping, and (4) vulcanizing. Processing techniques for natural and synthetic rubbers are virtually the same, differences being in the chemicals used to effect vulcanization (cross-linking). This sequence does not apply to thermoplastic elastomers, whose shaping techniques are the same as for other thermoplastic polymers.

Several distinct industries are involved in the production and processing of rubber. Production of raw natural rubber might be classified as farming because latex, the starting ingredient for natural rubber, is grown on large plantations located in tropical climates. By contrast, synthetic rubbers are produced by the petrochemical industry. Finally, the processing of these materials into tires, shoe soles, and other rubber products occurs at processor (fabricator) plants. The processors are commonly known as the rubber industry. Some of the great names in this industry include Goodyear, B. F. Goodrich, and Michelin. The importance of the tire is reflected in these names.

9.1.1 PRODUCTION OF RUBBER

In this section we briefly survey the production of rubber before it goes to the processor. Our coverage distinguishes natural rubber and synthetic rubber.

Natural Rubber Natural rubber is tapped from rubber trees (***Hevea brasiliensis***) as latex. The trees are grown on plantations in Southeast Asia and other parts of the world. Latex is a colloidal dispersion of solid particles of the polymer polyisoprene (Section 2.3.3) in water. Polyisoprene is the chemical substance that comprises rubber, and its content in the emulsion is about 30%. The latex is collected in large tanks, thus blending the yield of many trees together.

The preferred method of recovering rubber from the latex involves coagulation. The latex is first diluted with water to about half its natural concentration. An acid such as formic acid (HCOOH) or acetic acid (CH_3COOH) is added to cause the latex to coagulate after about 12 hours. The coagulum, now in the form of soft solid slabs, is then squeezed through a series of rolls that drive out most of the water and reduce the thickness to about 3 mm (1/8 in). The final rolls have grooves that impart a criss-cross pattern to the resulting sheets. The sheets are then draped over wooden frames and dried in smokehouses. The hot smoke contains creosote, which prevents mildew and oxidation of the rubber. Several days are normally required to complete the drying process. The resulting rubber, now in a form called ***ribbed smoked sheet***, is folded into large bales for shipment to the processor. This raw rubber has a characteristic dark brown color. In some cases, the sheets are dried in hot air rather than smokehouses, and the term ***air-dried sheet*** is applied; this is considered to be a better grade of rubber. A still better grade, called ***pale crepe*** rubber, involves two coagulation steps; the first removes undesirable components of the latex, then the resulting coagulum is subjected to a more involved washing and mechanical working procedure, followed by warm air drying. The color of pale crepe rubber approaches a light tan.

Synthetic Rubber The various types of synthetic rubber are identified in Section 2.3.3. Most synthetics are produced from petroleum by the same polymerization techniques used to synthesize other polymers. However, unlike thermoplastic and thermosetting polymers, which are normally supplied to the fabricator as pellets or liquid resins, synthetic rubbers are supplied to rubber processors in the form of large bales. The industry has developed a long tradition of handling natural rubber in these unit loads.

9.1.2 COMPOUNDING

Rubber is always compounded with additives. It is through compounding that the specific rubber is designed to satisfy a given application in terms of properties, cost, and processability. Compounding adds chemicals for vulcanization. Sulfur has traditionally been used for this purpose. The vulcanization process is discussed in Section 9.1.5.

Additives include fillers that act either to enhance the rubber's mechanical properties (reinforcing fillers) or to extend the rubber to reduce cost (nonreinforcing fillers). The single most important reinforcing filler in rubber is ***carbon black***, a colloidal form of carbon, black in color, obtained from the thermal decomposition of hydrocarbons (soot). Its effect is to increase tensile strength and resistance to abrasion and tearing of the final rubber product. Carbon black also provides protection from ultraviolet radiation. These enhancements are especially important in tires. Most rubber parts are black in color because of their carbon black content.

Although carbon black is the most important filler, others are also used. They include china clays—hydrous aluminum silicates ($Al_2Si_2O_5(OH)_4$), which provide less reinforcing than carbon black but are used when the black color is not acceptable; calcium carbonate ($CaCO_3$), which is a nonreinforcing filler; and silica (SiO_2), which can serve reinforcing or nonreinforcing functions, depending on particle size; and other polymers, such as styrene, PVC, and phenolics. Reclaimed (recycled) rubber is also added as a filler in some rubber products, but usually not in proportions exceeding 10%.

Other additives compounded with the rubber include antioxidants, to retard aging by oxidation; fatigue- and ozone-protective chemicals; coloring pigments; plasticizers and softening oils; blowing agents in the production of foamed rubber; and mold-release compounds.

Many products require filament reinforcement to reduce extensibility but retain the other desirable properties of rubber. Tires and conveyor belts are notable examples. Filaments used for this purpose include cellulose, nylon, and polyester. Fiberglass and steel are also used as reinforcements (e.g., steel-belted radial tires). These continuous fiber materials must be added as part of the shaping process; they are not mixed with the other additives.

9.1.3 MIXING

The additives must be thoroughly mixed with the base rubber to achieve uniform dispersion of the ingredients. Uncured rubbers possess high viscosity. Mechanical working experienced by the rubber can increase its temperature up to 150°C (300°F). If vulcanizing agents were present from the start of mixing, premature vulcanization

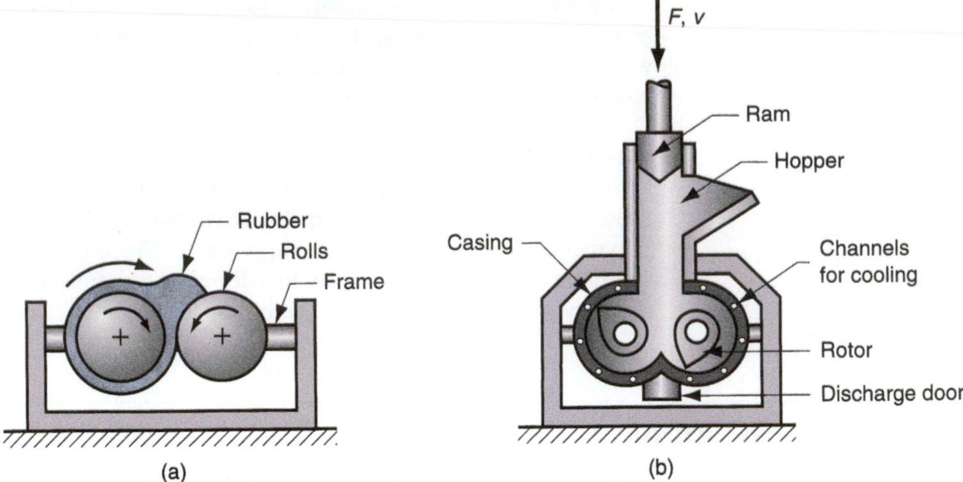

FIGURE 9.1 Mixers used in rubber processing: (a) two-roll mill and (b) Banbury-type internal mixer. These machines can also be used for mastication of natural rubber. (Credit: *Fundamentals of Modern Manufacturing*, 4th Edition by Mikell P. Groover, 2010. Reprinted with permission of John Wiley & Sons, Inc.)

would result—the rubber processor's nightmare [15]. Accordingly, a two-stage mixing process is usually employed. In the first stage, carbon black and other nonvulcanizing additives are combined with the raw rubber. The term ***masterbatch*** is used for this first-stage mixture. After thorough mixing has been accomplished, and time for cooling has been allowed, the second-stage is carried out in which the vulcanizing agents are added.

Equipment for mixing includes the two-roll mill and internal mixers such as the Banbury mixer, Figure 9.1. The ***two-roll mill*** consists of two parallel rolls, supported in a frame so they can be brought together to obtain a desired "nip" (gap size), and driven to rotate at the same or slightly different speeds. An ***internal mixer*** has two rotors encased in a jacket, as in Figure 9.1(b) for the Banbury-type internal mixer. The rotors have blades and rotate in opposite directions at different speeds, causing a complex flow pattern in the contained mixture.

9.1.4 SHAPING AND RELATED PROCESSES

Shaping processes for rubber products can be divided into four basic categories: (1) extrusion, (2) calendering, (3) coating, and (4) molding and casting. Most of these processes are discussed in the previous chapter. In this section, we examine the special issues that arise when they are applied to rubber. Some products require several basic processes plus assembly work in their manufacture, for example, tires.

Extrusion Extrusion of polymers is discussed in Section 8.2. Screw extruders are generally used for extrusion of rubber. As with extrusion of thermosetting plastics, the L/D ratio of the extruder barrels is less than for thermoplastics, typically in the range 10 to 15, to reduce the risk of premature cross-linking. Die swell occurs in rubber extrudates, because the polymer is in a highly plastic condition and exhibits the memory property. It has not yet been vulcanized.

Calendering This process involves passing rubber stock through a series of gaps of decreasing size made by a stand of rotating rolls (Section 8.3). Equipment used in the rubber industry is of heavier construction than that used for thermoplastics, because rubber is more viscous and harder to form. The output of the process is a rubber sheet of thickness determined by the final roll gap; again, swelling occurs in the sheet, causing its thickness to be slightly greater than the gap size.

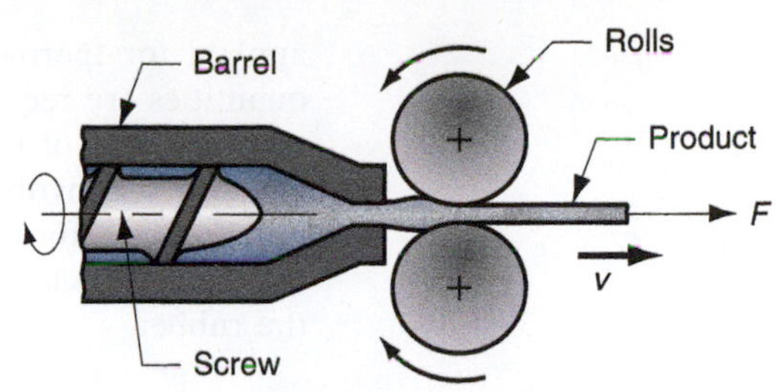

FIGURE 9.2 Roller die process: rubber extrusion followed by rolling. (Credit: *Fundamentals of Modern Manufacturing,* 4th Edition by Mikell P. Groover, 2010. Reprinted with permission of John Wiley & Sons, Inc.)

There are problems in producing thick sheet by either extrusion or calendering. Thickness control is difficult in the former process, and air entrapment occurs in the latter. These problems are largely solved when extrusion and calendering are combined in the ***roller die*** process (Figure 9.2). The extruder die is a slit that feeds the calender rolls.

Coating Coating or impregnating fabrics with rubber is an important process in the rubber industry. These composite materials are used in automobile tires, conveyor belts, inflatable rafts, and waterproof cloth for tarpaulins, tents, and rain coats. The ***coating*** of rubber onto substrate fabrics includes a variety of processes. Calendering is one of the coating methods. Figure 9.3 illustrates one possible way in which the fabric is fed into the calendering rolls to obtain a reinforced rubber sheet.

Alternatives to calendering include skimming, dipping, and spraying. In the ***skimming*** process, a thick solution of rubber compound in an organic solvent is applied to the fabric as it is unreeled from a supply spool. The coated fabric passes under a doctor blade that skims the solvent to the proper thickness, and then moves into a steam chamber where the solvent is driven off by heat. As its name suggests, ***dipping*** involves temporary immersion of the fabric into a highly fluid solution of rubber, followed by drying. Likewise, in ***spraying***, a spray gun is used to apply the rubber solution.

Molding and Casting Molded articles include shoe soles and heels, gaskets and seals, suction cups, and bottle stops. Many foamed rubber parts are produced by molding. In addition, molding is an important process in tire production. Principal molding processes for rubber are (1) compression molding, (2) transfer molding, and (3) injection molding. Compression molding is the most important technique because of its use in tire manufacture. Curing (vulcanizing) is accomplished in the mold in all three processes, this representing a departure from the shaping methods already discussed, which require a separate vulcanizing step. With injection molding of rubber, there are risks of premature curing similar to those faced in the same process when applied to thermosetting plastics. Advantages of injection molding over traditional methods for producing rubber parts include better dimensional control, less scrap, and shorter cycle times. In addition to its use in the molding of conventional rubbers, injection molding is also

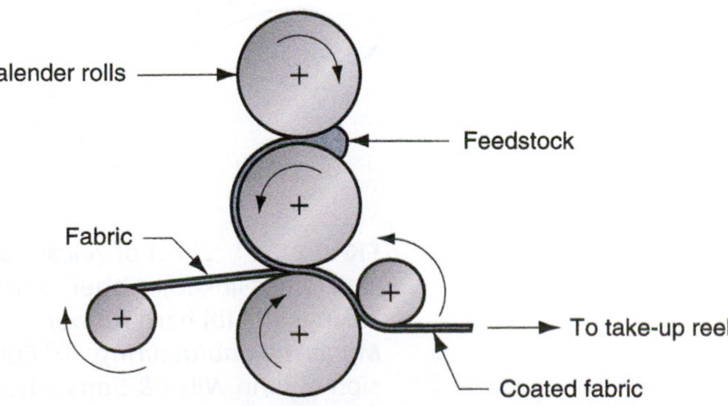

FIGURE 9.3 Coating of fabric with rubber using a calendering process. (Credit: *Fundamentals of Modern Manufacturing,* 4th Edition by Mikell P. Groover, 2010. Reprinted with permission of John Wiley & Sons, Inc.)

applied for thermoplastic elastomers. Because of high mold costs, large production quantities are required to justify injection molding.

A form of casting, called ***dip casting***, is used for producing rubber gloves and overshoes. It involves submersion of a positive mold in a liquid polymer (or a heated form into plastisol) for a certain duration (the process may involve repeated dippings) to form the desired thickness. The coating is then stripped from the form and cured to cross-link the rubber.

9.1.5 VULCANIZATION

Vulcanization is the treatment that accomplishes cross-linking of elastomer molecules, so that the rubber becomes stiffer and stronger but retains extensibility. It is a critical step in the rubber-processing sequence. On a submicroscopic scale, the process can be pictured as in Figure 9.4, in which the long-chain molecules of the rubber become joined at certain tie points, the effect of which is to reduce the ability of the elastomer to flow. A typical soft rubber has one or two cross-links per thousand units (mers). As the number of cross-links increases, the polymer becomes stiffer and behaves more like a thermosetting plastic (hard rubber).

Vulcanization, as it was first invented by Goodyear, involved the use of sulfur (about 8 parts by weight of S mixed with 100 parts of natural rubber) at a temperature of 140°C (280°F) for about 5 hours. No other chemicals were included in the process. Vulcanization with sulfur alone is no longer used as a commercial treatment today, due to the long curing times. Various other chemicals, including zinc oxide (ZnO) and stearic acid ($C_{18}H_{36}O_2$), are combined with smaller doses of sulfur to accelerate and strengthen the treatment. The resulting cure time is 15 to 20 minutes for a typical passenger car tire. In addition, various nonsulfur vulcanizing treatments have been developed.

In rubber-molding processes, vulcanization is accomplished in the mold by maintaining the mold temperature at the proper level for curing. In the other forming processes, vulcanization is performed after the part has been shaped. The treatments generally divide between batch processes and continuous processes. Batch methods include the use of an ***autoclave***, a steam-heated pressure vessel; and ***gas curing***, in which a heated inert gas such as nitrogen cures the rubber. Many of the basic shaping processes make a continuous product, and if the output is not cut into discrete pieces, continuous vulcanization is appropriate. Continuous methods include ***high-pressure steam***, suited to

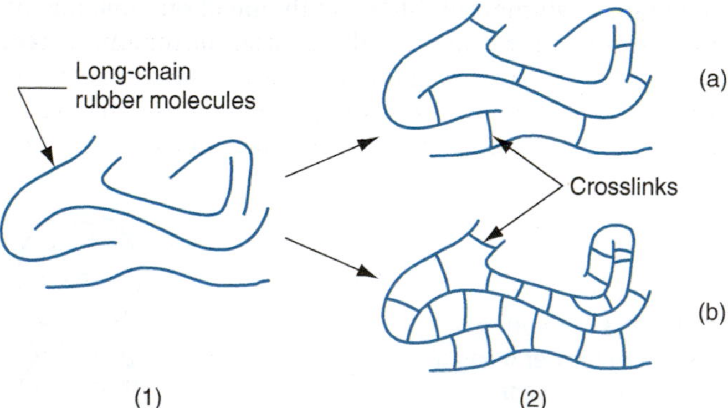

FIGURE 9.4 Effect of vulcanization on the rubber molecules: (1) raw rubber; (2) vulcanized (cross-linked) rubber. Variations of (2) include (a) soft rubber, low degree of cross-linking; and (b) hard rubber, high degree of cross-linking. (Credit: *Fundamentals of Modern Manufacturing*, 4th Edition by Mikell P. Groover, 2010. Reprinted with permission of John Wiley & Sons, Inc.)

the curing of rubber-coated wire and cable; ***hot-air tunnel***, for cellular extrusions and carpet underlays [5]; and ***continuous drum cure***, in which continuous rubber sheets (e.g., belts and flooring materials) pass through one or more heated rolls to effect vulcanization.

9.2 MANUFACTURE OF TIRES AND OTHER RUBBER PRODUCTS

Tires are the principal product of the rubber industry, accounting for about three-fourths of total tonnage. Other important products include footwear, hose, conveyor belts, seals, shock-absorbing components, foamed rubber products, and sports equipment.

9.2.1 TIRES

Pneumatic tires are critical components of the vehicles on which they are used. They are used on automobiles, trucks, buses, farm tractors, earth-moving equipment, military vehicles, bicycles, motorcycles, and aircraft. Tires support the weight of the vehicle and the passengers and cargo on board; they transmit the motor torque to propel the vehicle (except on aircraft); and they absorb vibrations and shock to provide a comfortable ride.

Tire Construction and Production Sequence A tire is an assembly of many parts, whose manufacture is unexpectedly complex. A passenger car tire consists of about 50 individual pieces; a large earthmover tire may have as many as 175. To begin with, there are three basic tire constructions: (a) diagonal ply, (b) belted bias, and (c) radial ply, pictured in Figure 9.5. In all three cases, the internal structure of the tire, known as the ***carcass***, consists of multiple layers of rubber-coated cords, called ***plies***. The cords are strands of various materials such as nylon, polyester, fiberglass, and steel, which provide inextensibility to reinforce the rubber in the carcass. The ***diagonal ply tire*** has the cords running diagonally, but in perpendicular directions in adjacent layers. A typical diagonal ply tire may have four plies. The ***belted bias tire*** is constructed of diagonal plies with opposite bias but adds several more layers around the outside periphery of the carcass. These ***belts*** increase the stiffness of the tire in the tread area and limit its diametric expansion during inflation. The cords in the belt also run diagonally, as indicated in the sketch.

A ***radial tire*** has plies running radially rather than diagonally; it also uses belts around the periphery for support. A ***steel-belted radial*** is a tire in which the circumferential belts have cords made of steel. The radial construction provides a more flexible sidewall, which tends to reduce stress on the belts and treads as they continually deform on contact with the flat road surface during rotation. This effect is accompanied by greater tread life, improved cornering and driving stability, and a better ride at high speeds.

In each construction, the carcass is covered by solid rubber that reaches a maximum thickness in the tread area. The carcass is also lined on the inside with a rubber coating. For tires with inner tubes, the inner liner is a thin coating applied to the innermost ply during its fabrication. For tubeless tires, the inner liner must have low permeability because it holds the air pressure; it is generally a laminated rubber.

Tire production can be summarized in three steps: (1) preforming of components, (2) building the carcass and adding rubber strips to form the sidewalls and treads, and (3) molding and curing the components into one integral piece. The descriptions of these steps that follow are typical; there are variations in processing depending on construction, tire size, and type of vehicle on which the tire will be used.

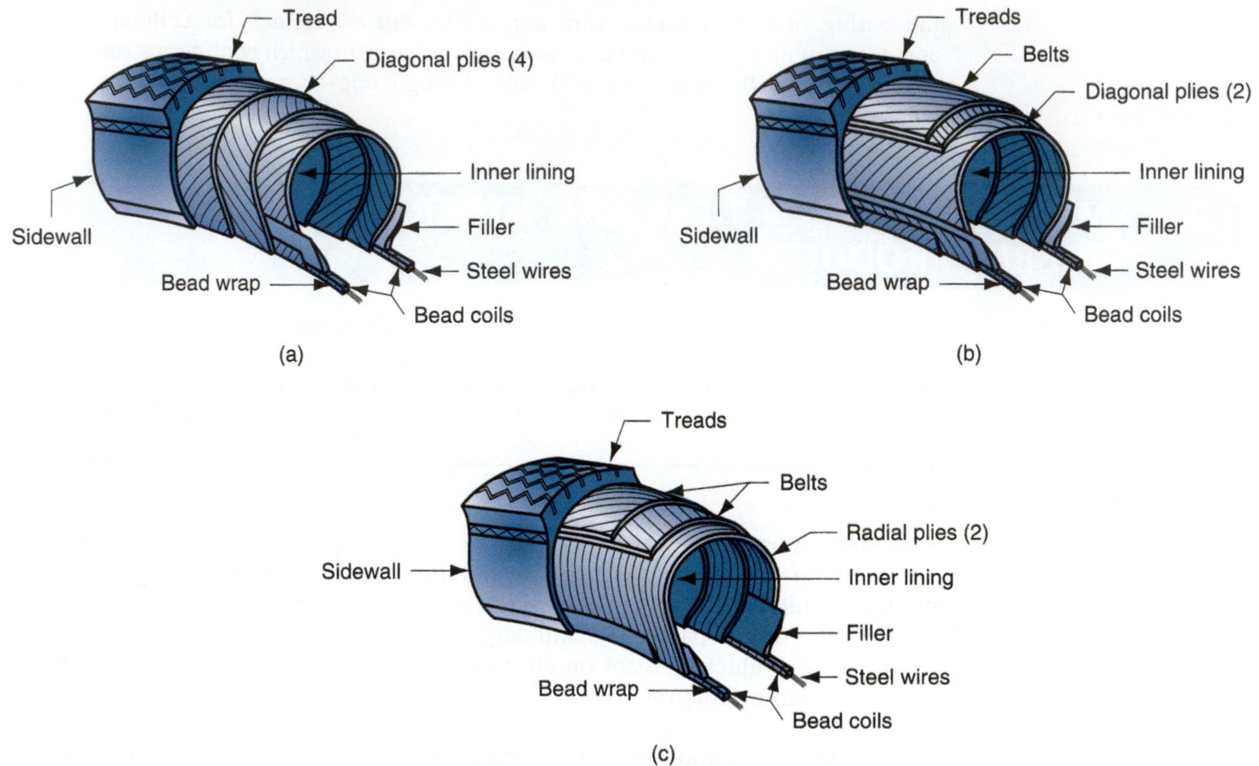

FIGURE 9.5 Three principal tire constructions: (a) diagonal ply, (b) belted bias, and (c) radial ply. (Credit: *Fundamentals of Modern Manufacturing*, 4th Edition by Mikell P. Groover, 2010. Reprinted with permission of John Wiley & Sons, Inc.)

Preforming of Components As Figure 9.5 shows, the carcass consists of a number of separate components, most of which are rubber or reinforced rubber. These, as well as the sidewall and tread rubber, are produced by continuous processes and then precut to size and shape for subsequent assembly. The components, labeled in Figure 9.5, and the preforming processes to fabricate them are:

- ***Bead coil***. Continuous steel wire is rubber coated, cut, coiled, and the ends joined.
- ***Plies***. Continuous fabric (textile, nylon, fiber glass, steel) is rubber coated in a calendering process and precut to size and shape.
- ***Inner lining***. For tube tires, the inner liner is calendered onto the innermost ply. For tubeless tires, the liner is calendered as a two-layer laminate.
- ***Belts***. Continuous fabric is rubber coated (similar to plies), but cut at different angles for better reinforcement; then made into a multi-ply belt.
- ***Tread***. Extruded as continuous strip; then cut and preassembled to belts.
- ***Sidewall***. Extruded as continuous strip; then cut to size and shape.

Building the Carcass The carcass is traditionally assembled using a machine known as a ***building drum***, whose main element is a cylindrical arbor that rotates. Precut strips that form the carcass are built up around this arbor in a step-by-step procedure. The layered plies that form the cross section of the tire are anchored on opposite sides of the rim by two bead coils. The ***bead coils*** consist of multiple strands of high-strength steel wire. Their function is to provide a rigid support when the finished tire is mounted on the wheel

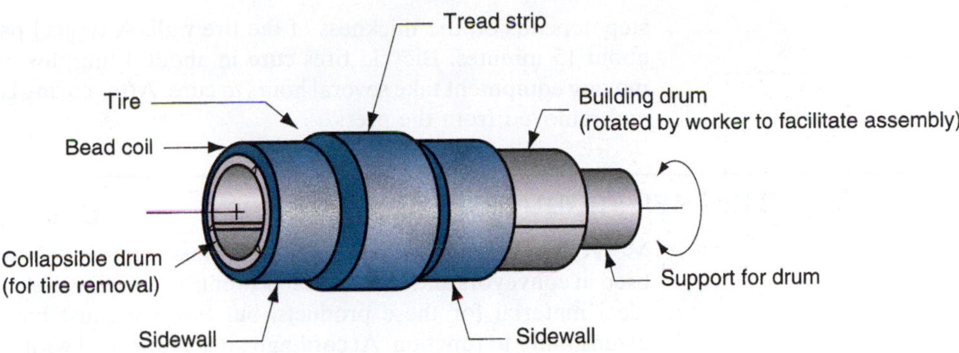

FIGURE 9.6 Tire just before removal from building drum, prior to molding and curing. (Credit: *Fundamentals of Modern Manufacturing*, 4th Edition by Mikell P. Groover, 2010. Reprinted with permission of John Wiley & Sons, Inc.)

rim. Other components are combined with the plies and bead coils. These include various wrappings and filler pieces to give the tire the proper strength, heat resistance, air retention, and fitting to the wheel rim. After these parts are placed around the arbor and the proper number of plies have been added, the belts are applied. This is followed by the outside rubber that will become the sidewall and tread.[1] At this point in the process, the treads are rubber strips of uniform cross section—the tread design is added later in molding. The building drum is collapsible, so that the unfinished tire can be removed when finished. The form of the tire at this stage is roughly tubular, as portrayed in Figure 9.6.

Molding and Curing Tire molds are usually two-piece construction (split molds) and contain the tread pattern to be impressed on the tire. The mold is bolted into a press, one half attached to the upper platen (the lid) and the bottom half fastened to the lower platen (the base). The uncured tire is placed over an expandable diaphragm and inserted between the mold halves, as in Figure 9.7. The press is then closed and the diaphragm expanded, so that the soft rubber is pressed against the cavity of the mold. This causes the tread pattern to be imparted to the rubber. At the same time, the rubber is heated, both from the outside by the mold and from the inside by the diaphragm. Circulating hot water or steam under pressure are used to heat the diaphragm. The duration of this curing

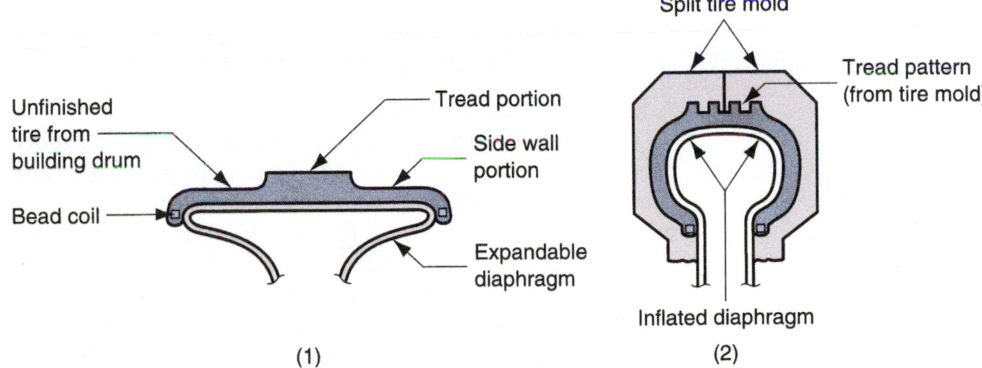

FIGURE 9.7 Tire molding (tire is shown in cross-sectional view): (1) the uncured tire is placed over expandable diaphragm; (2) the mold is closed and the diaphragm is expanded to force uncured rubber against mold cavity, impressing tread pattern into rubber; mold and diaphragm are heated to cure rubber. (Credit: *Fundamentals of Modern Manufacturing*, 4th Edition by Mikell P. Groover, 2010. Reprinted with permission of John Wiley & Sons, Inc.)

[1] Technically, the tread and sidewall are not usually considered to be components of the carcass.

step depends on the thickness of the tire wall. A typical passenger tire can be cured in about 15 minutes. Bicycle tires cure in about 4 minutes, whereas tires for large earth moving equipment take several hours to cure. After curing is completed, the tire is cooled and removed from the press.

9.2.2 OTHER RUBBER PRODUCTS

Most other rubber products are made by less complex processes. ***Rubber belts*** are widely used in conveyors and mechanical power transmission systems. As with tires, rubber is an ideal material for these products, but the belt must have flexibility and little or no extensibility to function. Accordingly, it is reinforced with fibers, commonly polyester or nylon. Fabrics of these polymers are usually coated in calendering operations, assembled together to obtain the required number of plies and thickness, and subsequently vulcanized by continuous or batch heating processes.

Rubber hose can be either plain or reinforced. Plain hose is extruded tubing. Reinforced tube consists of an inner tube, a reinforcing layer (sometimes called the carcass), and a cover. The internal tubing is extruded of a rubber that has been compounded for the particular substance that will flow through it. The reinforcement layer is applied to the tube in the form of a fabric, or by spiraling, knitting, braiding, or other application method. The outer layer is compounded to resist environmental conditions. It is applied by extrusion, using rollers, or other techniques.

Footwear components include soles, heels, rubber overshoes, and certain upper parts. Various rubbers are used to make footwear components. Molded parts are produced by injection molding, compression molding, and certain special molding techniques developed by the shoe industry. The rubbers include both solid and foamed varieties. In some cases, for low-volume production, manual methods are used to cut rubber from flat stock.

Rubber is widely used in sports equipment and supplies, including Ping-Pong paddle surfaces, golf club grips, football bladders, and sports balls of various kinds. Tennis balls, for example, are made in significant numbers. Production of these sports products relies on the various shaping processes discussed in Section 9.1.4, as well as special techniques that have been developed for particular items.

9.2.3 PROCESSING OF THERMOPLASTIC ELASTOMERS

A thermoplastic elastomer (TPE) is a thermoplastic polymer that possesses the properties of a rubber (Section 2.3); the term ***thermoplastic rubber*** is also used. TPEs can be processed like thermoplastics, but their applications are those of an elastomer. The most common shaping processes are injection molding and extrusion, which are generally more economical and faster than the traditional processes used for rubbers that must be vulcanized. Molded products include shoe soles, athletic footwear, and automotive components such as fender extensions and corner panels (but not tires—TPEs have been found to be unsatisfactory for that application). Extruded items include insulation coating for electrical wire, tubing for medical applications, conveyor belts, sheet and film stock. Other shaping techniques for TPEs include blow molding and thermoforming (Sections 8.8 and 8.9); these processes cannot be used for vulcanized rubbers.

9.3 PMC SHAPING PROCESSES AND MATERIALS

Some of the PMC shaping processes described in the following sections are slow and labor intensive. In general, techniques for shaping composites are less efficient than processes for other materials. There are two reasons for this: (1) Composite materials are more complex than other materials, consisting as they do of two or more phases and the need to orient the reinforcing phase in the case of fiber-reinforced plastics; and (2) processing technologies for composites have not been the object of improvement and refinement over as many years as processes for other materials.

The variety of shaping methods for fiber-reinforced polymers is sometimes bewildering on first reading. FRP composite shaping processes can be divided into five categories: (1) open mold processes, (2) closed mold processes, (3) filament winding, (4) pultrusion processes, and (5) other. Open mold processes include some of the original manual procedures for laying resins and fibers onto forms. Closed mold processes are much the same as those used in plastic molding; the reader will recognize the names—compression molding, transfer molding, and injection molding—although the names are sometimes changed and modifications are sometimes made for PMCs. In ***filament winding***, continuous filaments that have been dipped in liquid resin are wrapped around a rotating mandrel; when the resin cures, a rigid, hollow, generally cylindrical shape is created. ***Pultrusion*** is a shaping process for producing long, straight sections of constant cross section; it is similar to extrusion, but adapted to include continuous fiber reinforcement. The "other" category includes several operations that do not fit into the previous categories.

Some of these processes are used to shape composites with continuous fibers, whereas others are used for short-fiber PMCs. Let us begin our coverage by exploring how the individual phases in a PMC are produced and how these phases are combined into the starting materials for shaping.

9.3.1 STARTING MATERIALS FOR PMCs

In a PMC, the starting materials are a polymer and a reinforcing phase. They are processed separately before becoming phases in the composite. This section considers how these materials are produced before being combined.

Polymer Matrix All three of the basic polymer types—thermoplastics, thermosets, and elastomers—are used as matrices in PMCs. Thermosetting (TS) polymers are the most common matrix materials. The principal TS polymers are phenolics, unsaturated polyesters, and epoxies. Phenolics are associated with the use of particulate reinforcing phases, whereas polyesters and epoxies are more closely associated with FRPs. Thermoplastic (TP) polymers are also used in PMCs, and in fact, most molding compounds are composite materials that include fillers and/or reinforcing agents. As mentioned earlier, most elastomers are composite materials because nearly all rubbers are reinforced with carbon black. In this and the following sections, coverage is limited to the processing of PMCs that use TS and TP polymers as the matrix. Although many of the polymer-shaping processes discussed in Chapter 8 are applicable to polymer-matrix composites, combining polymer and reinforcing agent sometimes complicates the operations.

The Reinforcing Agent The reinforcing phase can be any of several geometries and materials. The geometries include fibers, particles, and flakes, and the materials are ceramics, metals, other polymers, or elements such as carbon or boron.

Common fiber materials in FRPs are glass, carbon, and the polymer Kevlar. Fibers of these materials are produced by various techniques, some of which we have covered in other chapters. Glass fibers are produced by drawing through small orifices (Section 7.2.3). For carbon, a series of heating treatments is performed to convert a precursor filament containing a carbon compound into a more pure carbon form. The precursor can be any of several substances, including polyacrylonitrile (PAN), pitch (a black carbon resin formed in the distillation of coal tar, wood tar, petroleum, etc.), or rayon (cellulose).

Kevlar fibers are produced by extrusion combined with drawing through small orifices in a spinneret (Section 8.4).

Starting as continuous filaments, the fibers are combined with the polymer matrix in any of several forms, depending on the properties desired in the material and the processing method to be used to shape the composite. In some fabrication processes, the filaments are continuous, whereas in others, they are chopped into short lengths. In the continuous form, individual filaments are usually available as rovings. A ***roving*** is a collection of untwisted (parallel) continuous strands; this is a convenient form for handling and processing. Rovings typically contain from 12 to 120 individual strands. By contrast, a ***yarn*** is a twisted collection of filaments. Continuous rovings are used in several PMC processes, including filament winding and pultrusion.

The most familiar form of continuous fiber is a ***cloth***—a fabric of woven yarns. Very similar to a cloth, but distinguished here, is a ***woven roving***, a fabric consisting of untwisted filaments rather than yarns. Woven rovings can be produced with unequal numbers of strands in the two directions so that they possess greater strength in one direction than the other. Such unidirectional woven rovings are often preferred in laminated FRP composites.

Fibers can also be prepared in the form of a ***mat***—a felt consisting of randomly oriented short fibers held loosely together with a binder, sometimes in a carrier fabric. Mats are commercially available as blankets of various weights, thicknesses, and widths. Mats can be cut and shaped for use as ***preforms*** in some of the closed mold processes. During molding, the resin impregnates the preform and then cures, thus yielding a fiber-reinforced molding.

Particles and Flakes Particles and flakes are really in the same class. Flakes are particles whose length and width are large relative to thickness. The characterization of engineering powders is discussed in the Chapter 10 appendix. Production methods for metal powders are discussed in Section 10.1, and techniques for producing ceramic powders are discussed in Section 11.1.

9.3.2 COMBINING MATRIX AND REINFORCEMENT

Incorporation of the reinforcing agent into the polymer matrix either occurs during the shaping process or beforehand. In the first case, the starting materials arrive at the fabricating operation as separate entities and are combined into the composite during shaping. Examples of this case are filament winding and pultrusion. The starting reinforcement in these processes consists of continuous fibers. In the second case, the two component materials are combined into some preliminary form that is convenient for use in the shaping process. Nearly all of the thermoplastics and thermosets used in plastic shaping processes are really polymers combined with fillers. The fillers are either short fibers or particulate (including flakes).

Of greatest interest in this chapter are the starting forms used in processes designed for FRP composites. We might think of the starting forms as prefabricated composites that arrive ready for use at the shaping process. These forms are molding compounds and prepregs.

Molding Compounds Molding compounds are similar to those used in plastic molding. They are designed for use in molding operations, and so they must be capable of flowing. Most molding compounds for composite processing are thermosetting polymers. Accordingly, they have not been cured prior to shape processing. Curing is done during and/or after final shaping. FRP composite molding compounds consist of the resin matrix with short, randomly dispersed fibers. They come in several forms.

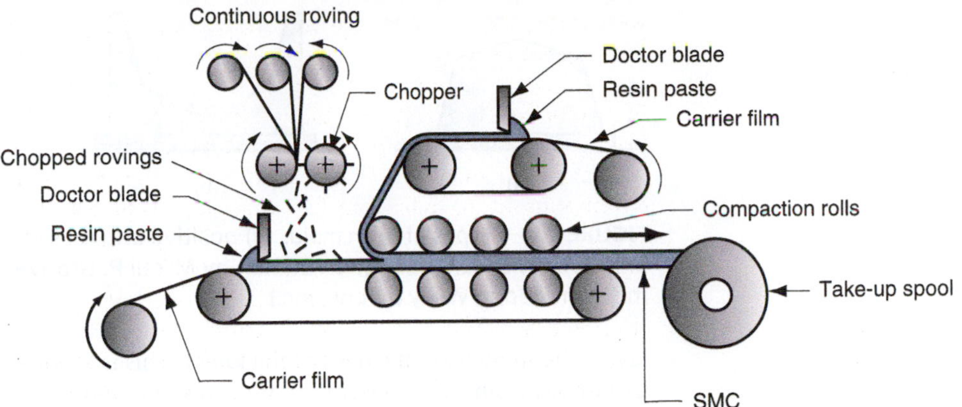

FIGURE 9.8 Process for producing sheet molding compound (SMC). (Credit: *Fundamentals of Modern Manufacturing*, 4th Edition by Mikell P. Groover, 2010. Reprinted with permission of John Wiley & Sons, Inc.)

Sheet molding compound (SMC) is a combination of TS polymer resin, fillers and other additives, and chopped glass fibers (randomly oriented) all rolled into a sheet of typical thickness = 6.5 mm (0.250 in). The most common resin is unsaturated polyester; fillers are usually mineral powders such as talc, silica, limestone; and the glass fibers are typically 12 to 75 mm (0.5 to 3.0 in) long and account for about 30% of the SMC by volume. SMCs are very convenient for handling and cutting to proper size as molding charges. Sheet molding compounds are generally produced between thin layers of polyethylene to limit evaporation of volatiles from the thermosetting resin. The protective coating also improves surface finish on subsequent molded parts. The process for fabricating continuous SMC sheets is depicted in Figure 9.8.

Bulk molding compound (BMC) consists of similar ingredients as those in SMC, but the compounded polymer is in billet form rather than sheet. The fibers in BMC are shorter, typically 2 to 12 mm (0.1 to 0.5 in), because greater fluidity is required in the molding operations for which these materials are designed. Billet diameter is usually 25 to 50 mm (1 to 2 in). The process for producing BMC is similar to that for SMC, except extrusion is used to obtain the final billet form. BMC is also known as ***dough molding compound*** (DMC), because it has a dough-like consistency. Other FRP molding compounds include ***thick molding compound*** (TMC), similar to SMC but thicker—up to 50 mm (2 in); and ***pelletized molding compounds***—basically conventional plastic molding compounds containing short fibers.

Prepregs Another prefabricated form for FRP shaping operations is ***prepreg***, which consists of fibers impregnated with partially cured thermosetting resins to facilitate shape processing. Completion of curing must be accomplished during and/or after shaping. Prepregs are available in the form of tapes or cross-plied sheets or fabrics. The advantage of prepregs is that they are fabricated with continuous filaments rather than chopped random fibers, thus increasing strength and modulus of the final product. Prepreg tapes and sheets are associated with composites that are reinforced with boron, carbon/graphite, Kevlar, or fiberglass.

9.4 OPEN MOLD PROCESSES

The distinguishing feature of this family of FRP shaping processes is its use of a single positive or negative mold surface (Figure 9.9) to produce laminated FRP structures. Other names for open mold processes include ***contact lamination*** and ***contact molding***. The starting materials (resins, fibers, mats, and woven rovings) are applied to the mold in

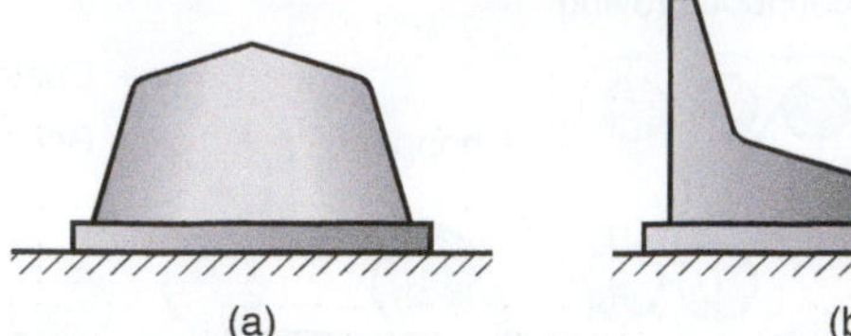

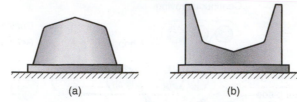

FIGURE 9.9 Types of open mold: (a) positive and (b) negative. (Credit: *Fundamentals of Modern Manufacturing*, 4th Edition by Mikell P. Groover, 2010. Reprinted with permission of John Wiley & Sons, Inc.)

layers, building up to the desired thickness. This is followed by curing and part removal. Common resins are unsaturated polyesters and epoxies, using fiberglass as the reinforcement. The moldings are usually large (e.g., boat hulls). The advantage of using an open mold is that the mold costs much less than if two matching molds were used. The disadvantage is that only the part surface in contact with the mold surface is finished; the other side is rough. For the best possible part surface on the finished side, the mold itself must be very smooth.

There are several important open mold FRP processes. The differences are in the methods of applying the laminations to the mold, alternative curing techniques, and other variations. In this section we describe three open mold processes for shaping fiber-reinforced plastics: (1) hand lay-up, (2) spray-up, and (3) automated tape-laying machines. We treat hand lay-up as the base process and the others as modifications and refinements.

9.4.1 HAND LAY-UP

Hand lay-up is the oldest open mold method for FRP laminates, dating to the 1940s when it was first used to fabricate boat hulls. It is also the most labor-intensive method. As the name suggests, hand lay-up is a shaping method in which successive layers of resin and reinforcement are manually applied to an open mold to build the laminated FRP composite structure. The basic procedure consists of five steps, illustrated in Figure 9.10. The finished molding must usually be trimmed with a power saw to size the outside edges. In general, these same five steps are required for all of the open mold processes, with the differences between methods occurring in steps 3 and 4.

In step 3, each layer of fiber reinforcement is dry when placed onto the mold. The liquid (uncured) resin is then applied by pouring, brushing, or spraying. Impregnation of resin into the fiber mat or fabric is accomplished by hand rolling. This approach is referred to as ***wet lay-up***. An alternative approach is to use ***prepregs***, in which the impregnated layers of fiber reinforcement are first prepared outside the mold and then laid onto the mold surface. Advantages cited for the prepregs include closer control over fiber-resin mixture and more efficient methods of adding the laminations [17].

Molds for open mold contact laminating can be made of plaster, metal, glass fiber–reinforced plastic, or other materials. Selection of material depends on economics, surface quality, and other technical factors. For prototype fabrication, in which only one part is produced, plaster molds are usually adequate. For medium quantities, the mold can be made of fiberglass-reinforced plastic. High production generally requires metal molds. Aluminum, steel, and nickel are used, sometimes with surface hardening on the mold face to resist wear. An advantage of metal, in addition to durability, is its high thermal conductivity that can be used to implement a heat-curing system, or simply to dissipate heat from the laminate while it cures at room temperature.

Products suited to hand lay-up are generally large in size but low in production quantity. In addition to boat hulls, other applications include swimming pools, large

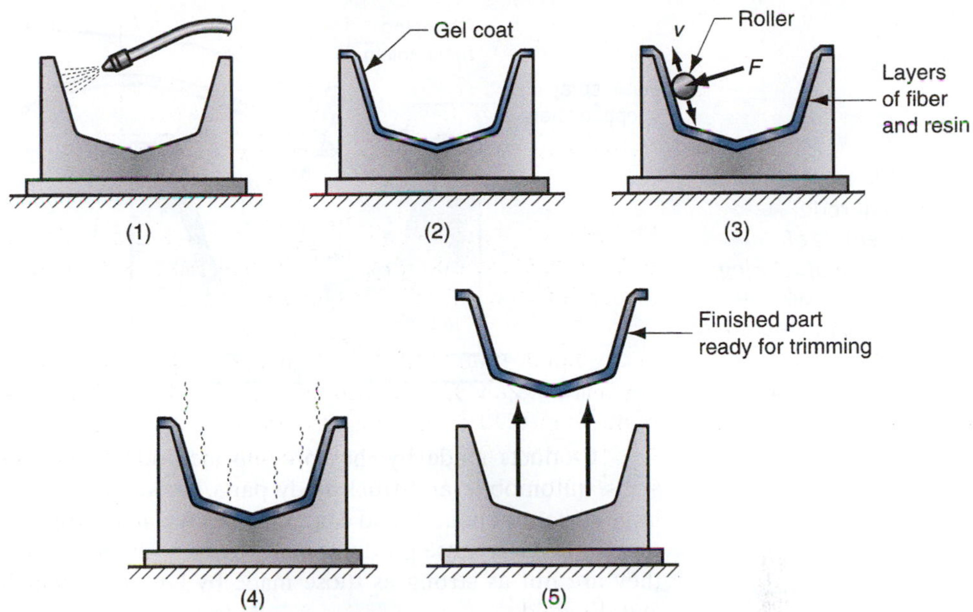

FIGURE 9.10 Hand lay-up procedure: (1) mold is cleaned and treated with a mold release agent; (2) a thin gel coat (resin, possibly pigmented to color) is applied, which will become the outside surface of the molding; (3) when the gel coat has partially set, successive layers of resin and fiber are applied, the fiber being in the form of mat or cloth; each layer is rolled to fully impregnate the fiber with resin and remove air bubbles; (4) the part is cured; and (5) the fully hardened part is removed from the mold. (Credit: *Fundamentals of Modern Manufacturing,* 4th Edition by Mikell P. Groover, 2010. Reprinted with permission of John Wiley & Sons, Inc.)

container tanks, stage props, radomes, and other formed sheets. Automotive parts have also been made, but the method is not economical for high production. The largest moldings ever made by this process were ship hulls for the British Royal Navy: 85 m (280 ft) long [3].

9.4.2 SPRAY-UP

This represents an attempt to mechanize the application of resin-fiber layers and to reduce the time for lay-up. It is an alternative for step 3 in the hand lay-up procedure. In the spray-up method, liquid resin and chopped fibers are sprayed onto an open mold to build successive FRP laminations, as in Figure 9.11. The spray gun is equipped with a chopper mechanism that feeds in continuous filament rovings and cuts them into fibers of length 25 to 75 mm (1 to 3 in) that are added to the resin stream as it exits the nozzle. The mixing action results in random orientation of the fibers in the layer—unlike hand lay-up, in which the filaments can be oriented if desired. Another difference is that the fiber content in spray-up is limited to about 35%, compared with a maximum of around 65% in hand lay-up. This is a shortcoming of the spraying and mixing process.

Spraying can be accomplished manually using a portable spray gun or by an automated machine in which the path of the spray gun is preprogrammed and computer controlled. The automated procedure is advantageous for labor efficiency and environmental protection. Some of the volatile emissions from the liquid resins are hazardous, and the path-controlled machines can operate in sealed-off areas without humans present. However, rolling is generally required for each layer, as in hand lay-up.

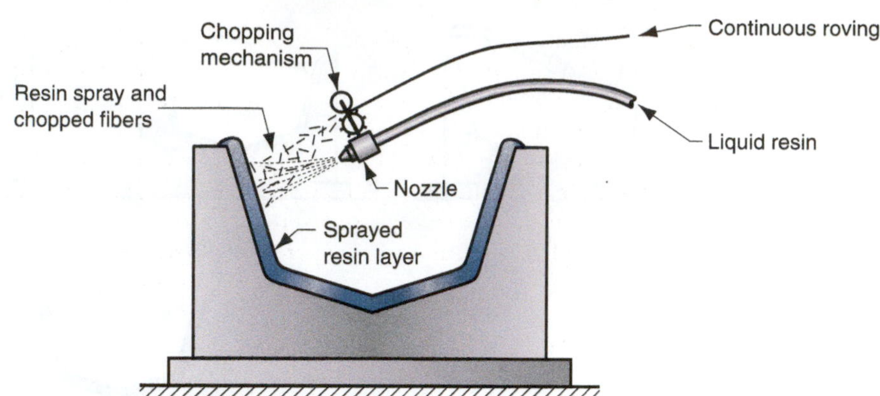

FIGURE 9.11 Spray-up method. (Credit: *Fundamentals of Modern Manufacturing*, 4th Edition by Mikell P. Groover, 2010. Reprinted with permission of John Wiley & Sons, Inc.)

Products made by the spray-up method include boat hulls, bathtubs, shower stalls, automobile and truck body parts, recreational vehicle components, furniture, large structural panels, and containers. Movie and stage props are sometimes made by this method. Because products made by spray-up have randomly oriented short fibers, they are not as strong as those made by lay-up, in which the fibers are continuous and directed.

9.4.3 AUTOMATED TAPE-LAYING MACHINES

This is another attempt to automate and accelerate step 3 in the lay-up procedure. Automated tape-laying machines operate by dispensing a prepreg tape onto an open mold following a programmed path. The typical machine consists of an overhead gantry, to which is attached the dispensing head. The gantry permits x-y-z travel of the head, for positioning and following a defined continuous path. The head itself has several rotational axes, plus a shearing device to cut the tape at the end of each path. Prepreg tape widths are commonly 75 mm (3 in), although 300 mm (12 in) widths have been reported [16]; thickness is around 0.13 mm (0.005 in). The tape is stored on the machine in rolls, which are unwound and deposited along the defined path. Each lamination is placed by following a series of back-and-forth passes across the mold surface until the parallel rows of tape complete the layer.

Much of the work to develop automated tape-laying machines has been pioneered by the aircraft industry, which is eager to save labor costs and at the same time achieve the highest possible quality and uniformity in its manufactured components. The disadvantage of this and other computer numerically controlled machines is that it must be programmed, and programming takes time.

9.4.4 CURING

Curing (step 4) is required of all thermosetting resins used in FRP-laminated composites. Curing accomplishes cross-linking of the polymer, transforming it from its liquid or highly plastic condition into a hardened product. There are three principal process parameters in curing: time, temperature, and pressure.

Curing normally occurs at room temperature for the TS resins used in hand lay-up and spray-up procedures. Moldings made by these processes are often large, and heating would be difficult for such parts. In some cases, days are required before room temperature curing is sufficiently complete to remove the part. If feasible, heat is added to speed the curing reaction.

Heating is accomplished by several means. Oven curing provides heat at closely controlled temperatures; some curing ovens are equipped to draw a partial vacuum. Infrared heating can be used in applications in which it is impractical or inconvenient to place the molding in an oven.

Curing in an autoclave provides control over both temperature and pressure. An ***autoclave*** is an enclosed chamber equipped to apply heat and/or pressure at controlled levels. In FRP composites processing, it is usually a large horizontal cylinder with doors at either end. The term ***autoclave molding*** is sometimes used to refer to the curing of a prepreg laminate in an autoclave. This procedure is used extensively in the aerospace industry to produce advanced composite components of very high quality.

9.5 CLOSED MOLD PROCESSES

These molding operations are performed in molds consisting of two sections that open and close during each molding cycle. One might think that a closed mold is about twice the cost of a comparable open mold. However, tooling cost is even greater because the equipment is more complex in these processes. Despite their higher cost, advantages of a closed mold are (1) good finish on all part surfaces, (2) higher production rates, (3) closer control over tolerances, and (4) more complex three-dimensional shapes are possible.

We divide the closed mold processes into three classes based on their counterparts in conventional plastic molding, even though the terminology is often different when polymer-matrix composites are molded: (1) compression molding, (2) transfer molding, and (3) injection molding.

9.5.1 COMPRESSION MOLDING PMC PROCESSES

In compression molding of conventional molding compounds (Section 8.7.1), a charge is placed in the lower mold section, and the sections are brought together under pressure, causing the charge to take the shape of the cavity. The mold halves are heated to cure the thermosetting polymer. When the molding is sufficiently cured, the mold is opened and the part is removed. There are several shaping processes for PMCs based on compression molding; the differences are mostly in the form of the starting materials. The flow of the resin, fibers, and other ingredients during the process is a critical factor in compression molding of FRP composites.

SMC, TMC, and BMC Molding Several of the FRP molding compounds, namely sheet molding compound (SMC), bulk molding compound (BMC), and thick molding compound (TMC), can be cut to proper size and used as the starting charge in compression molding. Refrigeration is often required to store these materials prior to shape processing. The names of the molding processes are based on the starting molding compound (i.e., ***SMC molding*** is when the starting charge is precut sheet molding compound; ***BMC molding*** uses bulk molding compound cut to size as the charge; and so on).

Preform Molding Another form of compression molding, called ***preform molding*** [17], involves placement of a precut mat into the lower mold section along with a polymer resin charge (e.g., pellets or sheet). The materials are then pressed between heated mold halves, causing the resin to flow and impregnate the fiber mat to produce a fiber-reinforced molding. Variations of the process use either thermoplastic or thermosetting polymers.

Elastic Reservoir Molding The starting charge in elastic reservoir molding (ERM) is a sandwich consisting of a center of polymer foam between two dry fiber layers. The foam core is commonly open-cell polyurethane, impregnated with liquid resin such as epoxy or polyester, and the dry fiber layers can be cloth, woven roving, or other starting fibrous form. The sandwich is placed in the lower mold section and pressed at moderate pressure—around 0.7 MPa (100 lb/in^2). As the core is compressed, it releases the resin to wet the dry surface layers. Curing produces a lightweight part consisting of a low-density core and thin FRP skins.

9.5.2 TRANSFER MOLDING PMC PROCESSES

In conventional transfer molding (Section 8.7.2), a charge of thermosetting resin is placed in a pot or chamber, heated, and squeezed by ram action into one or more mold cavities. The mold is heated to cure the resin. The name of the process derives from the fact that the fluid polymer is transferred from the pot into the mold. It can be used to mold TS resins in which the fillers include short fibers to produce a FRP composite part. Another form of transfer molding for PMCs is called ***resin transfer molding*** (RTM) [7], [18]; it refers to a closed mold process in which a preform mat is placed in the lower mold section, the mold is closed, and a thermosetting resin (e.g., polyester resin) is transferred into the cavity under moderate pressure to impregnate the preform. To confuse matters, RTM is sometimes called ***resin injection molding*** [7], [18] (the distinction between transfer molding and injection molding is blurry anyway, as the reader may have noted in Chapter 8). RTM has been used to manufacture such products as bathtubs, swimming pool shells, bench and chair seats, and hulls for small boats.

9.5.3 INJECTION MOLDING PMC PROCESSES

Injection molding is noted for low-cost production of plastic parts in large quantities. Although it is most closely associated with thermoplastics, the process can also be adapted to thermosets (Section 8.6.4).

Conventional Injection Molding In PMC shape processing, injection molding is used for both TP- and TS-type FRPs. In the TP category, virtually all thermoplastic polymers can be reinforced with fibers. Chopped fibers must be used; if continuous fibers were used, they would be reduced anyway by the action of the rotating screw in the barrel. During injection from the chamber into the mold cavity, the fibers tend to become aligned during their journey through the nozzle. Designers can sometimes exploit this feature to optimize directional properties through part design, location of gates, and cavity orientation relative to the gate [14].

Whereas TP molding compounds are heated and then injected into a cold mold, TS polymers are injected into a heated mold for curing. Control of the process with thermosets is trickier due to the risk of premature cross-linking in the injection chamber. Subject to the same risk, injection molding can be applied to fiber-reinforced TS plastics in the form of pelletized molding compound and dough molding compound.

Reinforced Reaction Injection Molding Some thermosets cure by chemical reaction rather than heat; these resins can be molded by reaction injection molding. In RIM, two reactive ingredients are mixed and immediately injected into a mold cavity where curing and solidification of the chemicals occur rapidly. A closely related process includes reinforcing fibers, typically glass, in the mixture. In this case, the process is called reinforced reaction injection molding (RRIM). Its advantages are similar to those in RIM, with the added benefit of fiber reinforcement. RRIM is used extensively in auto body and truck cab applications for bumpers, fenders, and other body parts.

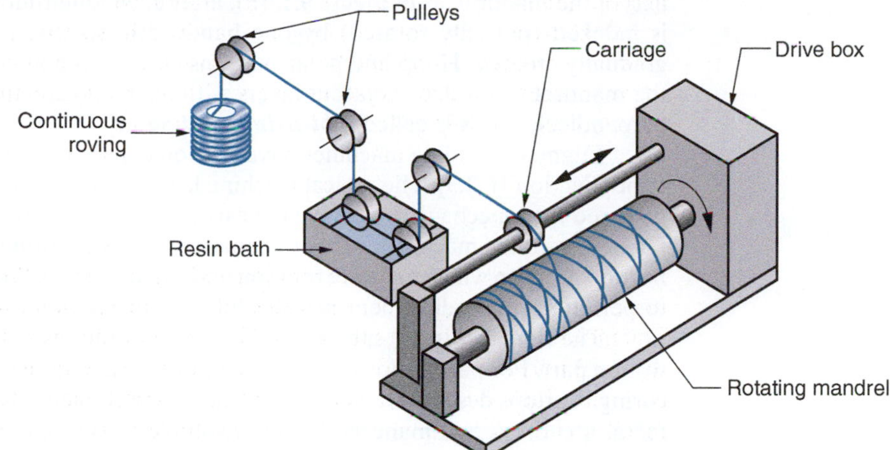

FIGURE 9.12 Filament winding. (Credit: *Fundamentals of Modern Manufacturing,* 4th Edition by Mikell P. Groover, 2010. Reprinted with permission of John Wiley & Sons, Inc.)

9.6 FILAMENT WINDING

Filament winding is a process in which resin-impregnated continuous fibers are wrapped around a rotating mandrel that has the internal shape of the desired FRP product. The resin is subsequently cured and the mandrel removed. Hollow axisymmetric components (usually circular in cross section) are produced, as well as some irregular shapes. The most common form of the process is depicted in Figure 9.12. A band of fiber rovings is pulled through a resin bath immediately before being wound in a helical pattern onto a cylindrical mandrel. Continuation of the winding pattern finally completes a surface layer of one filament thickness on the mandrel. The operation is repeated to form additional layers, each having a criss-cross pattern with the previous, until the desired part thickness has been obtained.

There are several methods by which the fibers can be impregnated with resin: (1) ***wet winding***, in which the filament is pulled through the liquid resin just before winding, as in the figure; (2) ***prepreg winding*** (also called ***dry winding***), in which filaments preimpregnated with partially cured resin are wrapped around a heated mandrel; and (3) ***postimpregnation***, in which filaments are wound onto a mandrel and then impregnated with resin by brushing or other technique.

Two basic winding patterns are used in filament winding: (a) helical and (b) polar (Figure 9.13). In ***helical winding***, the filament band is applied in a spiral pattern around the mandrel, at a helix angle θ. If the band is wrapped with a helix angle approaching 90°, so that the winding advance is one bandwidth per revolution and the filaments form nearly circular rings around the mandrel, this is referred to as a ***hoop winding***; it is a

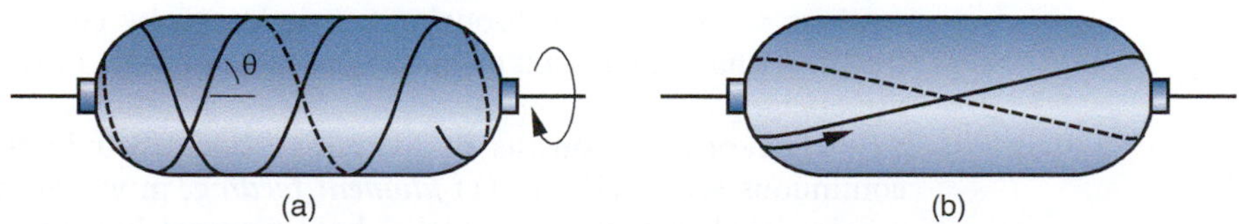

FIGURE 9.13 Two basic winding patterns in filament winding: (a) helical and (b) polar. (Credit: *Fundamentals of Modern Manufacturing,* 4th Edition by Mikell P. Groover, 2010. Reprinted with permission of John Wiley & Sons, Inc.)

special case of helical winding. In ***polar winding***, the filament is wrapped around the long axis of the mandrel, as in Figure 9.13(b); after each longitudinal revolution, the mandrel is indexed (partially rotated) by one bandwidth, so that a hollow enclosed shape is gradually created. Hoop and polar patterns can be combined in successive windings of the mandrel to produce adjacent layers with filament directions that are approximately perpendicular; this is called a ***bi-axial winding*** [3].

Filament winding machines have motion capabilities similar to those of an engine lathe (Section 16.2.3). The typical machine has a drive motor to rotate the mandrel and a powered feed mechanism to move the carriage. Various types of control are available in filament winding machines. Modern equipment uses ***computer numerical control*** (CNC, Section 29.1), in which mandrel rotation and carriage speed are independently controlled to permit greater adjustment and flexibility in the relative motions.

The ***mandrel*** is the special tooling that determines the geometry of the filament-wound part. For part removal, mandrels must be capable of collapsing after winding and curing. Various designs are possible, including inflatable/deflatable mandrels, collapsible metal mandrels, and mandrels made of soluble salts or plasters.

Applications of filament winding are often classified as aerospace or commercial [16], the engineering requirements being more demanding in the first category. Aerospace applications include rocket-motor cases, missile bodies, radomes, helicopter blades, and airplane tail sections and stabilizers. These parts are commonly made of epoxy resins reinforced with fibers of carbon, boron, Kevlar, and glass. Commercial applications include storage tanks, reinforced pipes and tubing, drive shafts, wind-turbine blades, and lightning rods; these are made of conventional FRPs. Polymers include polyester, epoxy, and phenolic resins; glass is the common reinforcing fiber.

9.7 PULTRUSION PROCESSES

The basic pultrusion process was developed around 1950 for making fishing rods of glass fiber–reinforced polymer (GFRP). The process is similar to extrusion (hence, the similarity in name), but it involves pulling of the workpiece (so the prefix "pul-" is used in place of "ex-"). Like extrusion, pultrusion produces continuous, straight sections of constant cross section. A related process, called pulforming, can be used to make parts that are curved and may have variations in cross section throughout their lengths.

9.7.1 PULTRUSION

Pultrusion is a process in which continuous fiber rovings are dipped into a resin bath and pulled through a shaping die where the impregnated resin cures. The setup is sketched in Figure 9.14, which shows the cured product being cut into long, straight sections. The sections are reinforced throughout their length by continuous fibers. Like extrusion, the pieces have a constant cross section, whose profile is determined by the shape of the die opening.

The process consists of five steps (identified in the sketch) performed in a continuous sequence [3]: (1) ***filament feeding***, in which the fibers are unreeled from a creel (shelves with skewers that hold filament bobbins); (2) ***resin impregnation***, in which the fibers are dipped in the uncured liquid resin; (3) ***pre-die forming***—the collection of filaments is gradually shaped into the approximate cross section desired; (4) ***shaping and curing***, in which the impregnated fibers are pulled through the heated

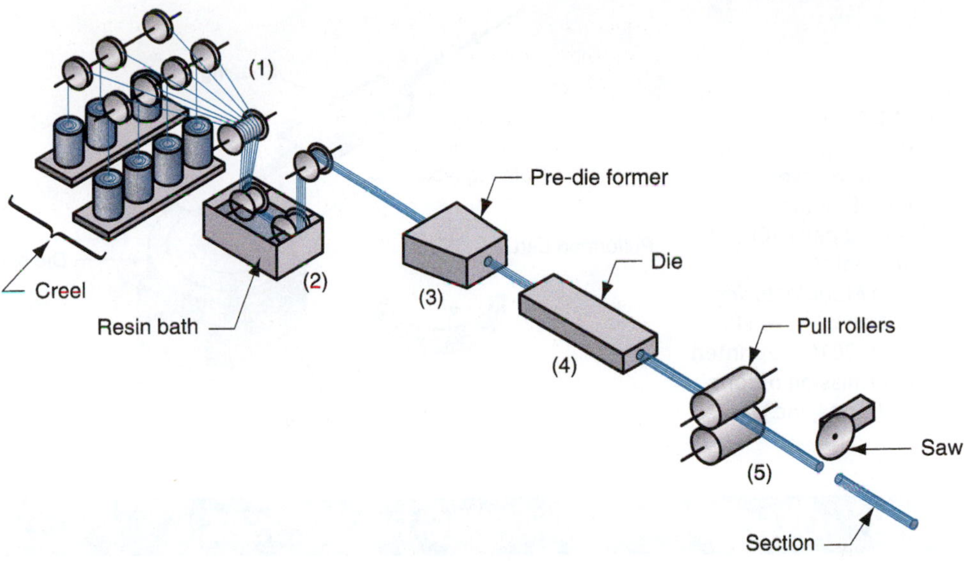

FIGURE 9.14 Pultrusion process (see text for interpretation of sequence numbers). (Credit: *Fundamentals of Modern Manufacturing*, 4th Edition by Mikell P. Groover, 2010. Reprinted with permission of John Wiley & Sons, Inc.)

die whose length is 1 to 1.5 m (3 to 5 ft) and whose inside surfaces are highly polished; and (5) ***pulling and cutting***—pullers are used to draw the cured length through the die, after which it is cut by a cut-off wheel with SiC or diamond grits.

Common resins used in pultrusion are unsaturated polyesters, epoxies, and silicones, all thermosetting polymers. There are difficulties in processing with epoxy polymers due to sticking on the die surface. Glass is by far the most widely used reinforcing material; proportions range from 30% to 70%. Modulus of elasticity and tensile strength increase with reinforcement content. Products made by pultrusion include solid rods, tubing, long and flat sheets, structural sections (such as channels, angled and flanged beams), tool handles for high-voltage work, and third-rail covers for subways.

9.7.2 PULFORMING

The pultrusion process is limited to straight sections of constant cross section. There is also a need for long parts with continuous fiber reinforcement that are curved rather than straight and whose cross sections may vary throughout the length. The pulforming process is suited to these less-regular shapes. Pulforming can be defined as pultrusion with additional steps to form the length into a semicircular contour and alter the cross section at one or more locations along the length. A sketch of the equipment is illustrated in Figure 9.15. After exiting the shaping die, the continuous workpiece is fed into a rotating table with negative molds positioned around its periphery. The work is forced into the mold cavities by a die shoe, which squeezes the cross section at various locations and forms the curvature in the length. The diameter of the table determines the radius of the part. As the work leaves the die table, it is cut to length to provide discrete parts. Resins and fibers similar to those for pultrusion are used in pulforming. An important application of the process is production of automobile leaf springs.

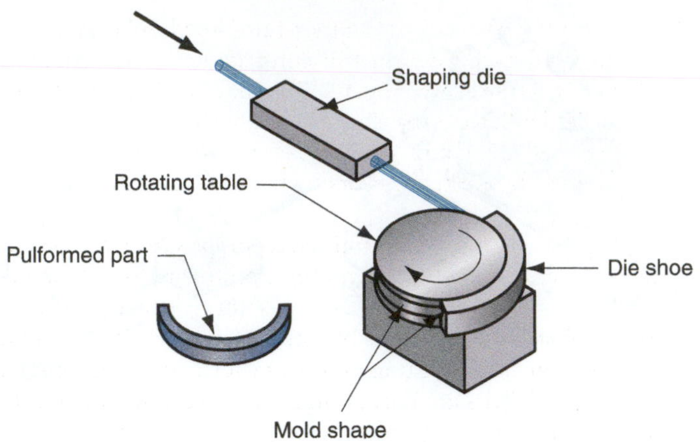

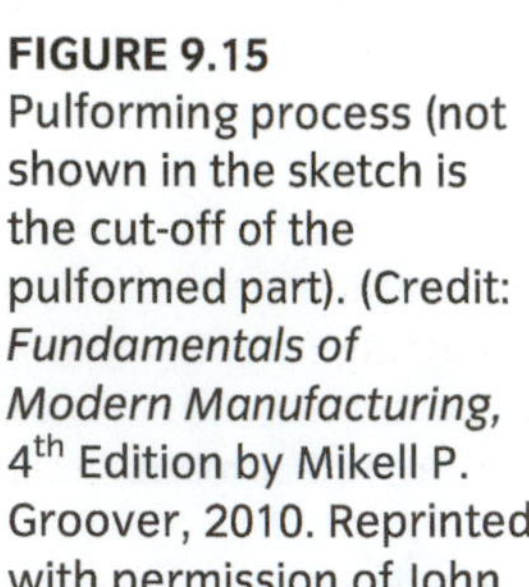

FIGURE 9.15 Pulforming process (not shown in the sketch is the cut-off of the pulformed part). (Credit: *Fundamentals of Modern Manufacturing*, 4th Edition by Mikell P. Groover, 2010. Reprinted with permission of John Wiley & Sons, Inc.)

9.8 OTHER PMC SHAPING PROCESSES

Additional PMC shaping processes worth noting include centrifugal casting, tube rolling, continuous laminating, and cutting. In addition, many of the traditional thermoplastic shaping processes are applicable to (short-fiber) FRPs based on TP polymers; these include blow molding, thermoforming, and extrusion.

Centrifugal Casting This process is ideal for cylindrical products such as pipes and tanks. The process is the same as its counterpart in metal casting (Section 6.3.5). Chopped fibers combined with liquid resin are poured into a fast-rotating cylindrical mold. Centrifugal force presses the ingredients against the mold wall, where curing takes place. The resulting inner surfaces are quite smooth. Part shrinkage or use of split molds permits part removal.

Tube Rolling FRP tubes can be fabricated from prepreg sheets by a rolling technique [12], shown in Figure 9.16. Such tubes are used in bicycle frames and space trusses. In the process, a precut prepreg sheet is wrapped around a cylindrical mandrel several times to obtain a tube wall of multiple sheet thicknesses. The rolled sheets are then encased in a heat-shrinking sleeve and oven cured. As the sleeve contracts, entrapped gases are squeezed out the ends of the tube. When curing is complete, the mandrel is removed to yield a rolled FRP tube. The operation is simple, and tooling cost is low. There are variations in the process, such as using different wrapping methods or using a steel mold to enclose the rolled prepreg tube for better dimensional control.

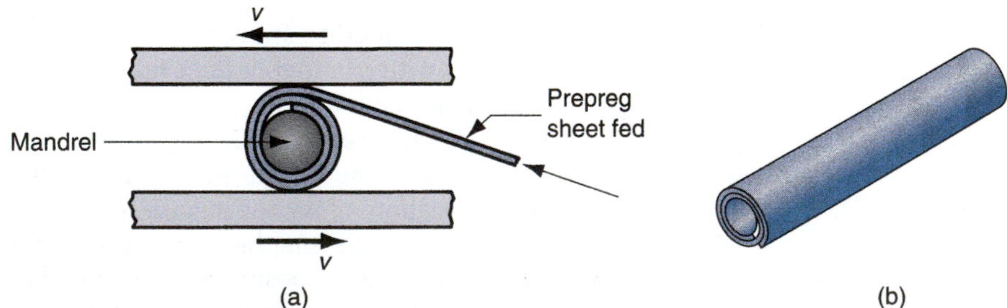

FIGURE 9.16 Tube rolling, showing (a) one possible means of wrapping FRP prepregs around a mandrel, and (b) the completed tube after curing and removal of mandrel. (Credit: *Fundamentals of Modern Manufacturing*, 4th Edition by Mikell P. Groover, 2010. Reprinted with permission of John Wiley & Sons, Inc.)

Continuous Laminating Fiber-reinforced plastic panels, sometimes translucent and/or corrugated, are used in construction. The process to produce them consists of (1) impregnating layers of glass fiber mat or woven fabric by dipping in liquid resin or by passing beneath a doctor blade; (2) gathering between cover films (cellophane, polyester, or other polymer); and (3) compacting between squeeze rolls and curing. Corrugation (4) is added by formed rollers or mold shoes.

Cutting Methods FRP-laminated composites must be cut in both uncured and cured states. Uncured materials (prepregs, preforms, SMCs, and other starting forms) must be cut to size for lay-up, molding, and so on. Typical cutting tools include knives, scissors, power shears, and steel-rule blanking dies. Also used are nontraditional cutting methods, such as laser beam cutting and water jet cutting (Chapter 19).

Cured FRPs are hard, tough, abrasive, and difficult to cut, but cutting is necessary in many FRP shaping processes to trim excess material, cut holes and outlines, and so on. For fiberglass-reinforced plastics, cemented carbide cutting tools and high-speed steel saw blades must be used. For some advanced composites (e.g., boron-epoxy), diamond cutting tools obtain best results. Water jet cutting is also used with good success on cured FRPs; this process reduces the dust and noise problems associated with conventional sawing methods.

REFERENCES

[1] Alliger, G., and Sjothun, I. J. (eds.). ***Vulcanization of Elastomers***. Krieger Publishing Company, New York, 1978.

[2] ***ASM Handbook***, Vol. **21**, ***Composites***. ASM International, Materials Park, Ohio, 2001.

[3] Bader, M. G., Smith, W., Isham, A. B., Rolston, J. A., and Metzner, A. B. ***Delaware Composites Design Encyclopedia***. Vol. 3, ***Processing and Fabrication Technology***. Technomic Publishing Co., Inc., Lancaster, Pennsylvania, 1990.

[4] Billmeyer, Fred, W., Jr. ***Textbook of Polymer Science***, 3rd ed. John Wiley & Sons, New York, 1984.

[5] Blow, C. M., and Hepburn, C. ***Rubber Technology and Manufacture***, 2nd ed. Butterworth-Heinemann, London, 1982.

[6] Bralla, J. G. (ed.). ***Design for Manufacturability Handbook***, 2nd ed. McGraw-Hill Book Company, New York, 1999.

[7] Charrier, J-M. ***Polymeric Materials and Processing***. Oxford University Press, New York, 1991.

[8] Chawla, K. K. ***Composite Materials: Science and Engineering***, 3rd ed. Springer-Verlag, New York, 2008.

[9] Coulter, J. P. "Resin Impregnation During the Manufacture of Composite Materials," ***PhD Dissertation***. University of Delaware, 1988.

[10] ***Engineering Materials Handbook***. Vol. **1**, ***Composites***. ASM International, Materials Park, Ohio, 1987.

[11] Hofmann, W. ***Rubber Technology Handbook***. Hanser-Gardner Publications, Cincinnati, Ohio, 1989.

[12] Mallick, P. K. ***Fiber-Reinforced Composites: Materials, Manufacturing, and Design***, 2nd ed. Marcel Dekker, Inc., New York, 1993.

[13] Mark, J. E., and Erman, B. (eds.). ***Science and Technology of Rubber***, 3rd ed. Academic Press, Orlando, Florida, 2005.

[14] McCrum, N. G., Buckley, C. P., and Bucknall, C. B. ***Principles of Polymer Engineering***. Oxford University Press, Inc., Oxford, United Kingdom, 1988.

[15] Morton-Jones, D. H. ***Polymer Processing***. Chapman and Hall, London, United Kingdom, 1989.

[16] Schwartz, M. M. ***Composite Materials Handbook***, 2nd ed. McGraw-Hill Book Company, New York, 1992.

[17] Strong, A. B. ***Fundamentals of Composites Manufacturing: Materials, Methods, and Applications***, 2nd ed. Society of Manufacturing Engineers, Dearborn, Michigan, 2007.

[18] Wick, C., Benedict, J. T., and Veilleux, R. F. (eds.). ***Tool and Manufacturing Engineers Handbook***, 4th ed. Vol. **II**, ***Forming***. Society of Manufacturing Engineers, Dearborn, Michigan, 1984.

[19] Wick, C.,and Veilleux, R. F. (eds.). ***Tool and Manufacturing Engineers Handbook***, 4th ed. Vol. III, ***Materials, Finishing, and Coating***. Society of Manufacturing Engineers, Dearborn, Michigan, 1985.

REVIEW QUESTIONS

9.1. How is the rubber industry organized?

9.2. What is the sequence of processing steps required to produce finished rubber goods?

9.3. What are some of the additives that are combined with rubber during compounding?

9.4. Name the four basic categories of processes used to shape rubber.

9.5. What does vulcanization do to the rubber?

9.6. Name the three basic tire constructions and briefly identify the differences in their construction.

9.7. What are the three basic steps in the manufacture of a pneumatic tire?

9.8. What is the purpose of the bead coil in a pneumatic tire?

9.9. What is a TPE?

9.10. What are the principal polymers used in fiber-reinforced polymers?

9.11. What is the difference between a roving and a yarn?

9.12. In the context of fiber reinforcement, what is a mat?

9.13. Why are flakes considered to be members of the same basic class of reinforcing material as particles?

9.14. What is sheet molding compound (SMC)?

9.15. How is a prepreg different from a molding compound?

9.16. Why are laminated FRP products made by the spray-up method not as strong as similar products made by hand lay-up?

9.17. What is an autoclave?

9.18. What are some of the advantages of the closed mold processes for PMCs relative to open mold processes?

9.19. Identify some of the different forms of polymer-matrix composite molding compounds.

9.20. What is preform molding?

9.21. Describe reinforced reaction injection molding (RRIM).

9.22. What is filament winding?

9.23. Describe the pultrusion process.

9.24. How does pulforming differ from pultrusion?

9.25. How are FRPs cut?

Part III Particulate Processing of Metals and Ceramics

10 POWDER METALLURGY

Chapter Contents

10.1 Production of Metallic Powders
- 10.1.1 Atomization
- 10.1.2 Other Production Methods

10.2 Conventional Pressing and Sintering
- 10.2.1 Blending and Mixing of the Powders
- 10.2.2 Compaction
- 10.2.3 Sintering
- 10.2.4 Secondary Operations

10.3 Alternative Pressing and Sintering Techniques
- 10.3.1 Isostatic Pressing
- 10.3.2 Powder Injection Molding
- 10.3.3 Powder Rolling, Extrusion, and Forging
- 10.3.4 Combined Pressing and Sintering
- 10.3.5 Liquid Phase Sintering

10.4 Materials and Products for PM

10.5 Design Considerations in Powder Metallurgy

Appendix 10 Characterization of Engineering Powders
- A10.1 Geometric Features
- A10.2 Other Features

This part of the book is concerned with the processing of metals and ceramics that are in the form of powders —very small particulate solids. In the case of traditional ceramics, the powders are produced by crushing and grinding common materials found in nature, such as silicate minerals (clay) and quartz. In the case of metals and the new ceramics (those based mostly on oxides and carbides), the powders are produced by a variety of industrial processes. We cover the powder-making processes as well as the methods used to shape products out of powders in two chapters: Chapter 10 on powder metallurgy and Chapter 11 on particulate processing of ceramics and cermets.

Powder metallurgy (PM) is a metal-processing technology in which parts are produced from metallic powders. In the usual PM production sequence, the powders are compressed into the desired shape and then heated to cause bonding of the particles into a hard, rigid mass. Compression, called ***pressing***, is accomplished in a press-type machine using tools designed specifically for the part to be produced. The tooling, which typically consists of a die and one or more punches, can be expensive, and PM is therefore most appropriate for medium and high production. The heating treatment, called ***sintering***, is performed at a temperature below the melting point of the metal. Considerations that make powder metallurgy an important commercial technology include:

- PM parts can be mass-produced to ***net shape*** or ***near net shape***, eliminating or reducing the need for subsequent shape processing.
- The PM process itself involves very little waste of material; about 97% of the starting powders are converted to product. This compares favorably to casting processes in which sprues, runners, and risers are wasted material in the production cycle.
- Owing to the nature of the starting material in PM, parts having a specified level of porosity can be made. This feature lends itself to the production of porous metal parts, such as filters, and oil-impregnated bearings and gears.
- Certain metals that are difficult to fabricate by other methods can be shaped by powder metallurgy. Tungsten is an example; tungsten filaments used in incandescent lamp bulbs are made using PM technology.
- Certain metal alloy combinations and cermets can be formed by PM that cannot be produced by other methods.
- PM compares favorably to most casting processes in terms of dimensional control of the product. Tolerances of ±0.13 mm (±0.005 in) are held routinely.
- PM production methods can be automated for economical production.

There are limitations and disadvantages associated with PM processing. These include the following: (1) tooling and equipment costs are high, (2) metallic powders are expensive, and (3) there are difficulties with storing and handling metal powders (such as degradation of the metal over time, and fire hazards with particular metals). Also, (4) there are limitations on part geometry because metal powders do not readily flow laterally in the die during pressing, and allowances must be provided for ejection of the part from the die after pressing. In addition, (5) variations in material density throughout the part may be a problem in PM, especially for complex part geometries.

Although parts as large as 22 kg (50 lb) can be produced, most PM components are less than 2.2 kg (5 lb). A collection of typical PM parts is shown in Figure 10.1. The largest tonnage of metals for PM are alloys of iron, steel, and aluminum. Other PM metals include copper, nickel, and refractory metals such as molybdenum and tungsten. Metallic carbides such as

FIGURE 10.1
A collection of powder metallurgy parts. Courtesy of Dorst America, Inc. (Credit: *Fundamentals of Modern Manufacturing*, 4th Edition by Mikell P. Groover, 2010. Reprinted with permission of John Wiley & Sons, Inc.)

tungsten carbide are often included within the scope of powder metallurgy; however, since these materials are ceramics, we defer their consideration until the next chapter.

Success in powder metallurgy depends to a large degree on the characteristics of the starting powders; we discuss characterization of engineering powders in the appendix to this chapter. In ceramics (except glass), the starting material is also powder, so the methods for characterizing the ceramic powders are closely related to those in PM.

10.1 PRODUCTION OF METALLIC POWDERS

To begin with, it should be noted that the producers of metallic powders are not the same companies as those that make PM parts. The powder producers are the suppliers; the plants that make components out of powder metals are the customers. The processes used by the suppliers are discussed in this section; the processes used by the PM parts producers are discussed in Sections 10.2 and 10.3.

Virtually any metal can be made into powder form. There are three principal methods by which metallic powders are commercially produced, each of which involves energy input to increase the surface area of the metal. The methods are (1) atomization, (2) chemical, and (3) electrolytic [13]. In addition, mechanical methods are occasionally used to reduce powder sizes; however, these methods are much more commonly associated with ceramic powder production and we treat them in the next chapter.

10.1.1 ATOMIZATION

This method involves the conversion of molten metal into a spray of droplets that solidify into powders. It is the most versatile and popular method for producing metal powders today, applicable to almost all metals, alloys as well as pure metals. There are multiple ways of creating the molten metal spray, several of which are illustrated in Figure 10.2. Two of the methods shown are based on ***gas atomization***, in which a high-velocity gas stream (air or inert gas) is utilized to atomize the liquid metal. In Figure 10.2(a), the gas flows through an expansion nozzle, siphoning molten metal from the melt below and spraying it into a container. The droplets solidify into powder form. In a closely related method shown in Figure 10.2(b), molten metal flows by gravity through a nozzle and is immediately atomized by air jets. The resulting metal powders, which tend to be spherical, are collected in a chamber below.

The approach shown in Figure 10.2(c) is similar to (b), except that a high-velocity water stream is used instead of air. This is known as ***water atomization*** and is the most common of the atomization methods, particularly suited to metals that melt below 1000°C (2900°F). Cooling is more rapid, and the resulting powder shape is irregular rather than spherical. The disadvantage of using water is oxidation on the particle surface. A recent innovation involves the use of synthetic oil rather than water to reduce oxidation. In both air and water atomization processes, particle size is controlled largely by the velocity of the fluid stream; particle size is inversely related to velocity.

Several methods are based on ***centrifugal atomization***. In one approach, the ***rotating disk method*** shown in Figure 10.2(d), the liquid metal stream pours onto a rapidly rotating disk that sprays the metal in all directions to produce powders.

10.1.2 OTHER PRODUCTION METHODS

Other metal powder production methods include various chemical reduction processes, precipitation methods, and electrolysis.

Chemical reduction includes a variety of chemical reactions by which metallic compounds are reduced to elemental metal powders. A common process involves liberation

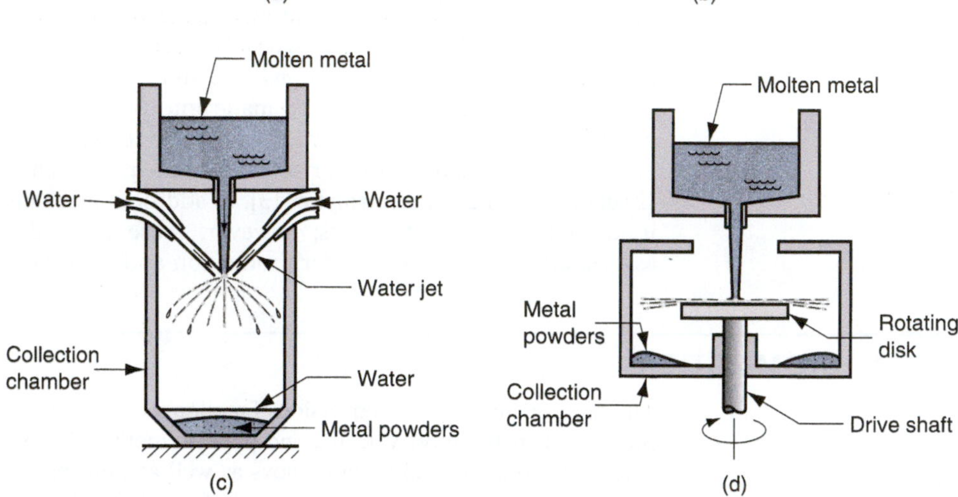

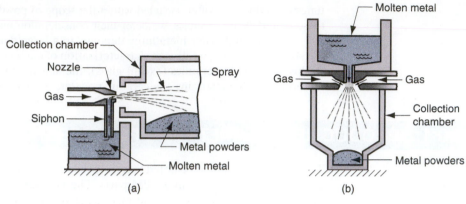

FIGURE 10.2 Several atomization methods for producing metallic powders: (a) and (b) two gas atomization methods; (c) water atomization; and (d) centrifugal atomization by the rotating disk method. (Credit: *Fundamentals of Modern Manufacturing*, 4th Edition by Mikell P. Groover, 2010. Reprinted with permission of John Wiley & Sons, Inc.)

of metals from their oxides by use of reducing agents such as hydrogen or carbon monoxide. The reducing agent is made to combine with the oxygen in the compound to free the metallic element. This approach is used to produce powders of iron, tungsten, and copper. Another chemical process for iron powders involves the decomposition of iron pentacarbonyl ($Fe(Co)_5$) to produce spherical particles of high purity. Powders produced by this method are illustrated in the photomicrograph of Figure 10.3. Other chemical processes include

FIGURE 10.3 Iron powders produced by water atomization. Photo courtesy of T. F. Murphy and Hoeganaes Corporation.

precipitation of metallic elements from salts dissolved in water. Powders of copper, nickel, and cobalt can be produced by this approach.

In ***electrolysis***, an electrolytic cell is set up in which the source of the desired metal is the anode. The anode is slowly dissolved under an applied voltage, transported through the electrolyte, and deposited on the cathode. The deposit is removed, washed, and dried to yield a metallic powder of very high purity. The technique is used for producing powders of beryllium, copper, iron, silver, tantalum, and titanium.

10.2 CONVENTIONAL PRESSING AND SINTERING

After the metallic powders have been produced, the conventional PM sequence used by parts manufacturers consists of three steps: (1) blending and mixing of the powders; (2) compaction, in which the powders are pressed into the desired part shape; and (3) sintering, which involves heating to a temperature below the melting point to cause solid-state bonding of the particles and strengthening of the part. The three steps, sometimes referred to as primary operations in PM, are portrayed in Figure 10.4. In addition, secondary operations are sometimes performed to improve dimensional accuracy, increase density, and for other reasons.

10.2.1 BLENDING AND MIXING OF THE POWDERS

To achieve successful results in compaction and sintering, the metallic powders must be thoroughly homogenized beforehand. The terms *blending* and *mixing* are both used in this context. ***Blending*** refers to when powders of the same chemical composition but different particle sizes are intermingled. Different particle sizes are often blended to reduce porosity. ***Mixing*** refers to powders of different chemistries being combined. An advantage of PM technology is the opportunity to mix various metals into alloys that would be difficult or impossible to produce by other means. The distinction between blending and mixing is not always precise in industrial practice.

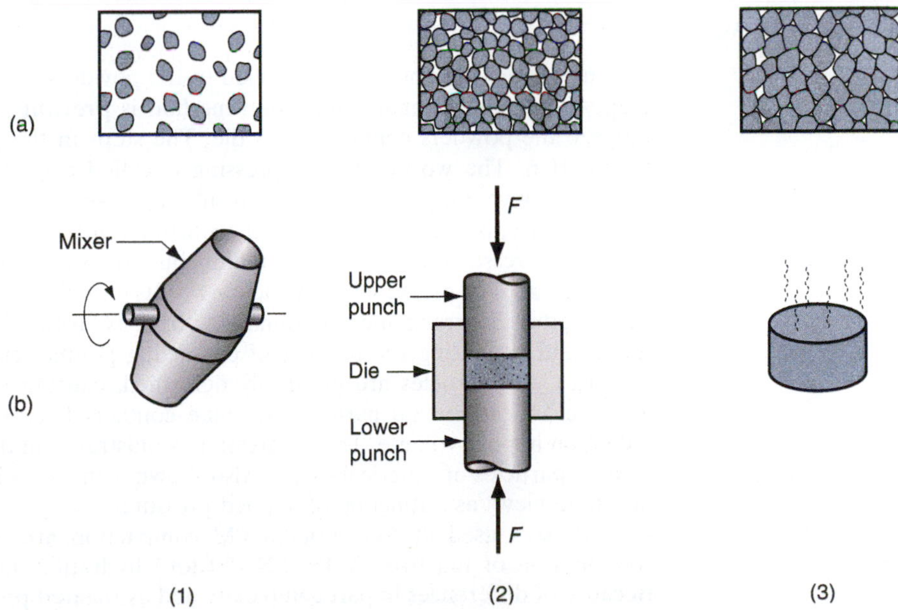

FIGURE 10.4 The conventional powder metallurgy production sequence: (1) blending, (2) compacting, and (3) sintering; (a) shows the condition of the particles, while (b) shows the operation and/or workpart during the sequence. (Credit: *Fundamentals of Modern Manufacturing*, 4th Edition by Mikell P. Groover, 2010. Reprinted with permission of John Wiley & Sons, Inc.)

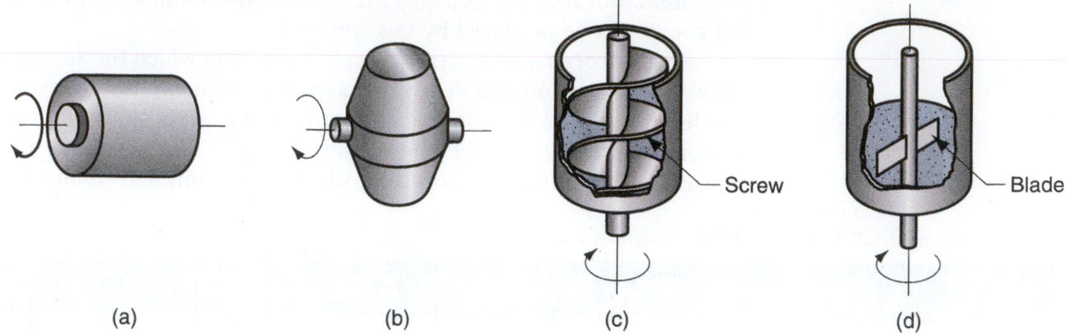

FIGURE 10.5 Several blending and mixing devices: (a) rotating drum, (b) rotating double-cone, (c) screw mixer, and (d) blade mixer. (Credit: *Fundamentals of Modern Manufacturing*, 4th Edition by Mikell P. Groover, 2010. Reprinted with permission of John Wiley & Sons, Inc.)

Blending and mixing are accomplished by mechanical means. Four alternatives are illustrated in Figure 10.5: (a) rotation in a drum; (b) rotation in a double-cone container; (c) agitation in a screw mixer; and (d) stirring in a blade mixer. There is more science to these devices than one would suspect. Best results seem to occur when the container is between 20% and 40% full. The containers are usually designed with internal baffles or other ways of preventing free-fall during blending of powders of different sizes, because variations in settling rates between sizes result in segregation—just the opposite of what is wanted in blending. Vibration of the powder is undesirable, since it also causes segregation.

Other ingredients are usually added to the metallic powders during the blending and/or mixing step. These additives include (1) ***lubricants***, such as stearates of zinc and aluminum, in small amounts to reduce friction between particles and at the die wall during compaction; (2) ***binders***, which are required in some cases to achieve adequate strength in the pressed but unsintered part; and (3) ***deflocculants***, which inhibit agglomeration of powders for better flow characteristics during subsequent processing.

10.2.2 COMPACTION

In compaction, high pressure is applied to the powders to form them into the required shape. The conventional compaction method is ***pressing***, in which opposing punches squeeze the powders contained in a die. The steps in the pressing cycle are shown in Figure 10.6. The workpart after pressing is called a ***green compact***, the word ***green*** meaning not yet fully processed. As a result of pressing, the density of the part, called the ***green density***, is much greater than the starting bulk density. The ***green strength*** of the part when pressed is adequate for handling but far less than that achieved after sintering.

The applied pressure in compaction results initially in repacking of the powders into a more efficient arrangement, eliminating "bridges" formed during filling, reducing pore space, and increasing the number of contacting points between particles. As pressure increases, the particles are plastically deformed, causing interparticle contact area to increase and additional particles to make contact. This is accompanied by a further reduction in pore volume. The progression is illustrated in three views in Figure 10.7 for starting particles of spherical shape. Also shown is the associated density represented by the three views as a function of applied pressure.

Presses used in conventional PM compaction are mechanical, hydraulic, or a combination of the two. A 450 kN (50 ton) hydraulic unit is shown in Figure 10.8. Because of differences in part complexity and associated pressing requirements, presses

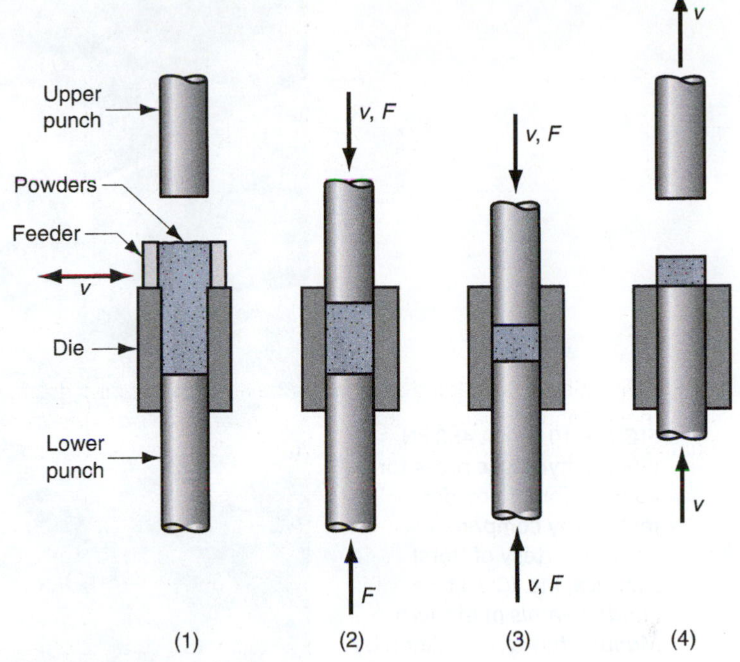

FIGURE 10.6 Pressing, the conventional method of compacting metal powders in PM: (1) filling the die cavity with powder, done by automatic feed in production, (2) initial, and (3) final positions of upper and lower punches during compaction, and (4) ejection of part. (Credit: *Fundamentals of Modern Manufacturing*, 4th Edition by Mikell P. Groover, 2010. Reprinted with permission of John Wiley & Sons, Inc.)

can be distinguished as (1) pressing from one direction, referred to as single-action presses; or (2) pressing from two directions, any of several types including opposed ram, double-action, and multiple action. Current available press technology can provide up to 10 separate action controls to produce parts of significant geometric complexity. We examine part complexity and other design issues in Section 10.5.

The capacity of a press for PM production is generally given in tons or kN or MN. The required force for pressing depends on the projected area of the PM part (area in the horizontal plane for a vertical press) multiplied by the pressure needed to compact the given metal powders. Reducing this to equation form,

$$F = A_p p_c \tag{10.1}$$

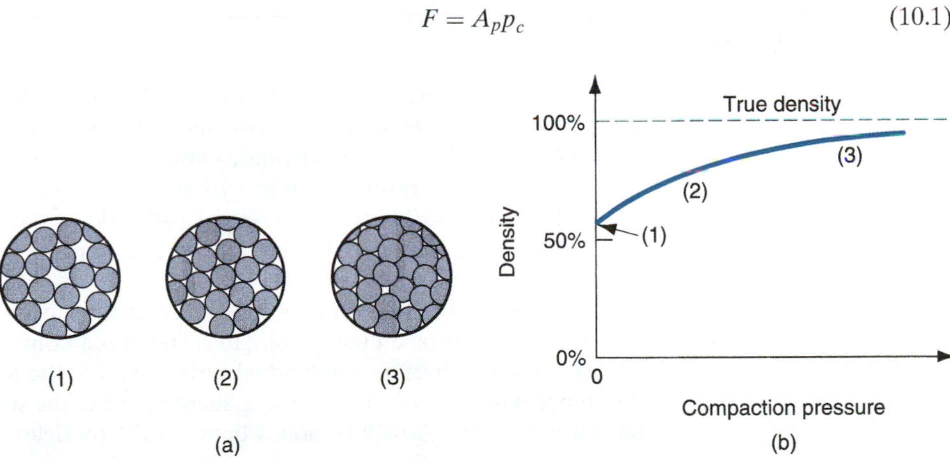

FIGURE 10.7 (a) Effect of applied pressure during compaction: (1) initial loose powders after filling, (2) repacking, and (3) deformation of particles; and (b) density of the powders as a function of pressure. The sequence here corresponds to steps 1, 2, and 3 in Figure 10.6. (Credit: *Fundamentals of Modern Manufacturing*, 4th Edition by Mikell P. Groover, 2010. Reprinted with permission of John Wiley & Sons, Inc.)

FIGURE 10.8 A 450-kN (50-ton) hydraulic press for compaction of powder metallurgy components. Photo courtesy of Dorst America, Inc. (Credit: *Fundamentals of Modern Manufacturing*, 4th Edition by Mikell P. Groover, 2010. Reprinted with permission of John Wiley & Sons, Inc.)

where F = required force, N (lb); A_p = projected area of the part, mm^2 (in^2); and p_c = compaction pressure required for the given powder material, MPa (lb/in^2). Compaction pressures typically range from 70 MPa (10,000 lb/in^2) for aluminum powders to 700 MPa (100,000 lb/in^2) for iron and steel powders.

10.2.3 SINTERING

After pressing, the green compact lacks strength and hardness; it is easily crumbled under low stresses. ***Sintering*** is a heat treatment operation performed on the compact to bond its metallic particles, thereby increasing strength and hardness. The treatment is usually carried out at temperatures between 0.7 and 0.9 of the metal's melting point (absolute scale). The terms ***solid-state sintering*** or ***solid-phase sintering*** are sometimes used for this conventional sintering because the metal remains unmelted at these treatment temperatures.

It is generally agreed among researchers that the primary driving force for sintering is reduction of surface energy [6], [16]. The green compact consists of many distinct particles, each with its own individual surface, and so the total surface area contained in the compact is very high. Under the influence of heat, the surface area is reduced through the formation and growth of bonds between the particles, with associated reduction in surface energy. The finer the initial powder size, the higher the total surface area, and the greater the driving force behind the process.

The series of sketches in Figure 10.9 shows on a microscopic scale the changes that occur during sintering of metallic powders. Sintering involves mass transport to create the necks and transform them into grain boundaries. The principal mechanism by which this occurs is diffusion; other possible mechanisms include plastic flow. Shrinkage of the

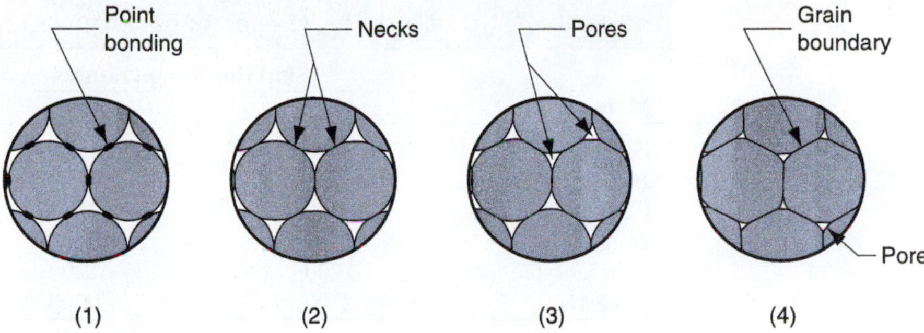

FIGURE 10.9 Sintering on a microscopic scale: (1) particle bonding is initiated at contact points; (2) contact points grow into "necks"; (3) the pores between particles are reduced in size; and (4) grain boundaries develop between particles in place of the necked regions. (Credit: *Fundamentals of Modern Manufacturing*, 4th Edition by Mikell P. Groover, 2010. Reprinted with permission of John Wiley & Sons, Inc.)

bulk part occurs during sintering as a result of pore size reduction. This depends to a large extent on the density of the green compact, which depends on the pressure during compaction. Shrinkage is generally predictable when processing conditions are closely controlled.

Since PM applications usually involve medium-to-high production, most sintering furnaces are designed with mechanized flow-through capability for the workparts. The heat treatment consists of three steps, accomplished in three chambers in these continuous furnaces: (1) preheat, in which lubricants and binders are burned off; (2) sinter; and (3) cool down. The treatment is illustrated in Figure 10.10. Typical sintering temperatures and times are given for selected metals in Table 10.1.

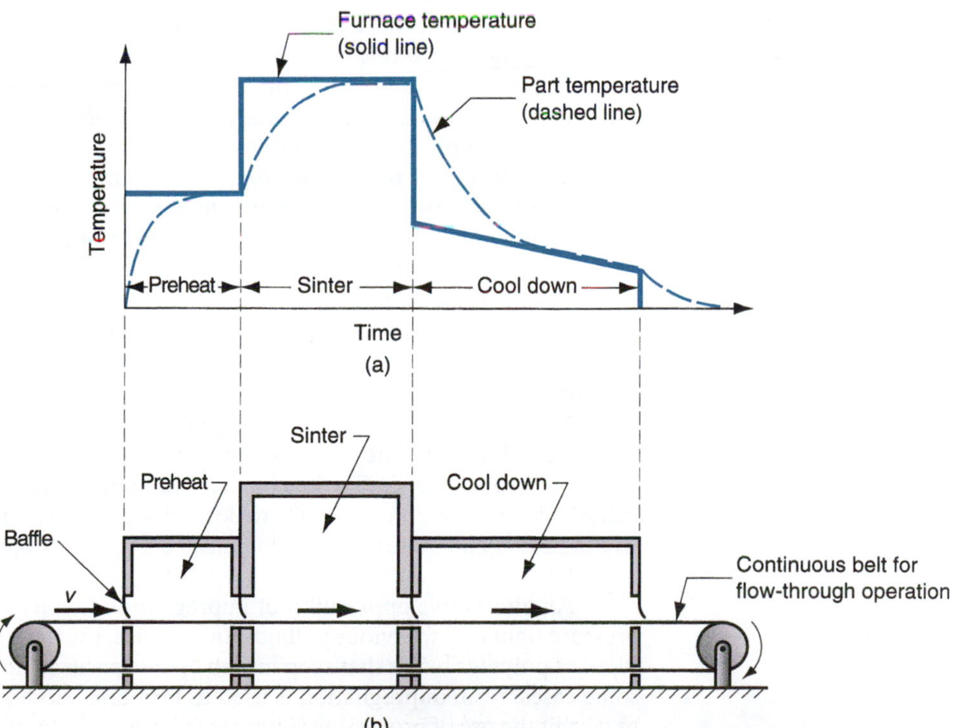

FIGURE 10.10 (a) Typical heat treatment cycle in sintering; and (b) schematic cross section of a continuous sintering furnace. (Credit: *Fundamentals of Modern Manufacturing*, 4th Edition by Mikell P. Groover, 2010. Reprinted with permission of John Wiley & Sons, Inc.)

TABLE 10.1 Typical sintering temperatures and times for selected powder metals.

Metal	Sintering Temperatures °C	Sintering Temperatures °F	Typical Time
Brass	850	1600	25 min
Bronze	820	1500	15 min
Copper	850	1600	25 min
Iron	1100	2000	30 min
Stainless steel	1200	2200	45 min
Tungsten	2300	4200	480 min

Compiled from [10] and [17].

In modern sintering practice, the atmosphere in the furnace is controlled. The purposes of a controlled atmosphere include (1) protection from oxidation, (2) providing a reducing atmosphere to remove existing oxides, (3) providing a carburizing atmosphere, and (4) assisting in removing lubricants and binders used in pressing. Common sintering furnace atmospheres are inert gas, nitrogen-based, dissociated ammonia, hydrogen, and natural gas [6]. Vacuum atmospheres are used for certain metals, such as stainless steel and tungsten.

10.2.4 SECONDARY OPERATIONS

PM secondary operations may be required to complete the part; they include densification, sizing, impregnation, infiltration, heat treatment, and finishing.

Densification and Sizing A number of secondary operations can be performed on the pressed-and-sintered part to increase density, improve accuracy, or accomplish additional shaping. ***Repressing*** is a pressing operation in which the part is squeezed in a closed die to increase density and improve physical properties. ***Sizing*** is the pressing of a sintered part to improve dimensional accuracy. ***Coining*** is a pressworking operation on a sintered part to press details into its surface.

Some PM parts require ***machining*** after sintering. Machining is rarely done to size the part, but rather to create geometric features that cannot be achieved by pressing, such as internal and external threads, side holes, and other details.

Impregnation and Infiltration Porosity is a unique and inherent characteristic of powder metallurgy technology. It can be exploited to create special products by filling the available pore space with oils, polymers, or metals that have lower melting temperatures than the base powder metal.

Impregnation is the term used when oil or other fluid is permeated into the pores of a sintered PM part. The most common products of this process are oil-impregnated bearings, gears, and similar machinery components. Self-lubricating bearings, usually made of bronze or iron with 10% to 30% oil by volume, are widely used in the automotive industry. The treatment is accomplished by immersing the sintered parts in a bath of hot oil.

An alternative application of impregnation involves PM parts that must be made pressure tight or impervious to fluids. In this case, the parts are impregnated with various types of polymer resins that seep into the pore spaces in liquid form and then solidify. In some cases, resin impregnation is used to facilitate subsequent processing, for example, to permit the use of processing solutions (such as plating chemicals) that would otherwise soak into the pores and degrade the product, or to improve machinability of the PM workpart.

Infiltration is an operation in which the pores of the PM part are filled with a molten metal. The melting point of the filler metal must be below that of the PM part. The process involves heating the filler metal in contact with the sintered component so that capillary action draws the filler into the pores. The resulting structure is relatively nonporous, and the infiltrated part has a more uniform density, as well as improved toughness and strength. An application of the process is copper infiltration of iron PM parts.

Heat Treatment and Finishing Powder metal components can be heat treated (Chapter 20) and finished (electroplated or painted, Chapter 21) by most of the same processes used on parts produced by casting and other metalworking processes. Special care must be exercised in heat treatment because of porosity; for example, salt baths are not used for heating PM parts. Plating and coating operations are applied to sintered parts for appearance purposes and corrosion resistance. Again, precautions must be taken to avoid entrapment of chemical solutions in the pores; impregnation and infiltration are frequently used for this purpose. Common platings for PM parts include copper, nickel, chromium, zinc, and cadmium.

10.3 ALTERNATIVE PRESSING AND SINTERING TECHNIQUES

The conventional press-and-sinter sequence is the most widely used shaping technology in powder metallurgy. Additional methods for processing PM parts are discussed in this section.

10.3.1 ISOSTATIC PRESSING

A feature of conventional pressing is that pressure is applied uniaxially. This imposes limitations on part geometry, since metallic powders do not readily flow in directions perpendicular to the applied pressure. Uniaxial pressing also leads to density variations in the compact after pressing. In ***isostatic pressing***, pressure is applied from all directions against the powders that are contained in a flexible mold; hydraulic pressure is used to achieve compaction. Isostatic pressing takes two alternative forms: (1) cold isostatic pressing and (2) hot isostatic pressing.

Cold isostatic pressing (CIP) involves compaction performed at room temperature. The mold, made of rubber or other elastomer material, is oversized to compensate for shrinkage. Water or oil is used to provide the hydrostatic pressure against the mold inside the chamber. Figure 10.11 illustrates the processing sequence in cold isostatic pressing. Advantages of CIP include more uniform density, less expensive tooling, and greater applicability to shorter production runs. Good dimensional accuracy is difficult to achieve in isostatic pressing, due to the flexible mold. Consequently, subsequent finish shaping operations are often required to obtain the required dimensions, either before or after sintering.

Hot isostatic pressing (HIP) is carried out at high temperatures and pressures, using a gas such as argon or helium as the compression medium. The mold in which the powders are contained is made of sheet metal to withstand the high temperatures. HIP accomplishes pressing and sintering in one step. Despite this apparent advantage, it is a relatively expensive process and its applications seem to be concentrated in the aerospace industry. PM parts made by HIP are characterized by high density (porosity near zero), thorough interparticle bonding, and good mechanical strength.

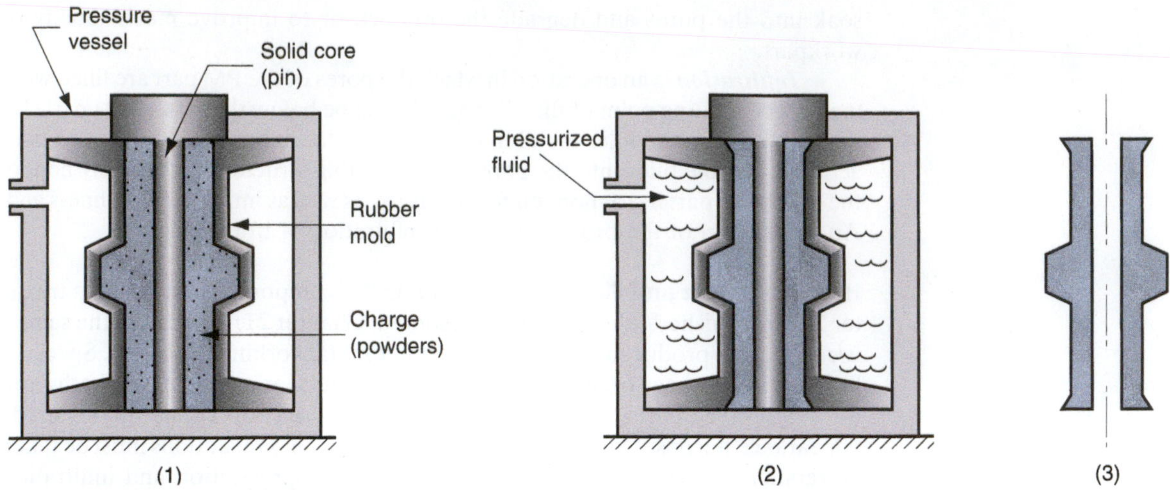

FIGURE 10.11 Cold isostatic pressing: (1) powders are placed in the flexible mold; (2) hydrostatic pressure is applied against the mold to compact the powders; and (3) pressure is reduced and the part is removed. (Credit: *Fundamentals of Modern Manufacturing,* 4th Edition by Mikell P. Groover, 2010. Reprinted with permission of John Wiley & Sons, Inc.)

10.3.2 POWDER INJECTION MOLDING

Injection molding is closely associated with the plastics industry (Section 8.6). The same basic process can be applied to form parts of metal or ceramic powders, the difference being that the starting polymer contains a high content of particulate matter, typically from 50% to 85% by volume. When used in powder metallurgy, the term ***metal injection molding*** (MIM) is used. The more general process is ***powder injection molding*** (PIM), which includes both metal and ceramic powders. The steps in MIM proceed as follows [7]: (1) Metallic powders are mixed with an appropriate binder; (2) granular pellets are formed from the mixture. (3) The pellets are heated to molding temperature, injected into a mold cavity, and the part is cooled and removed from the mold. (4) The part is processed to remove the binder using any of several thermal or solvent techniques. (5) The part is sintered. (6) Secondary operations are performed as appropriate.

The binder in powder injection molding acts as a carrier for the particles. Its functions are to provide proper flow characteristics during molding and to hold the powders in the molded shape until sintering. The five basic types of binders in PIM are (1) thermosetting polymers, such as phenolics, (2) thermoplastic polymers, such as polyethylene, (3) water, (4) gels, and (5) inorganic materials [7]. Polymers are the most frequently used.

Powder injection molding is suited to part geometries similar to those in plastic injection molding. It is not cost competitive for simple axisymmetric parts, because the conventional press-and-sinter process is quite adequate for these cases. PIM seems most economical for small, complex parts of high value. Dimensional accuracy is limited by the shrinkage that accompanies densification during sintering.

10.3.3 POWDER ROLLING, EXTRUSION, AND FORGING

Rolling, extrusion, and forging are familiar bulk metal forming processes (Chapter 13). We describe them here in the context of powder metallurgy.

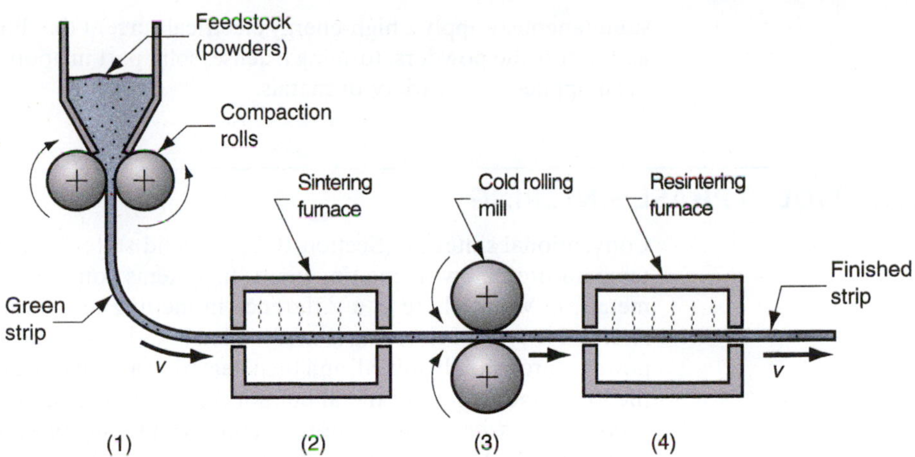

FIGURE 10.12 Powder rolling: (1) powders are fed through compaction rolls to form a green strip; (2) sintering; (3) cold rolling; and (4) resintering. (Credit: *Fundamentals of Modern Manufacturing*, 4th Edition by Mikell P. Groover, 2010. Reprinted with permission of John Wiley & Sons, Inc.)

Powder Rolling Powders can be compressed in a rolling mill operation to form metal strip stock. The process is usually set up to run continuously or semicontinuously, as shown in Figure 10.12. The metallic powders are compacted between rolls into a green strip that is fed directly into a sintering furnace. It is then cold rolled and resintered.

Powder Extrusion Extrusion is one of the basic manufacturing processes. In PM extrusion, the starting powders can be in different forms. In the most popular method, powders are placed in a vacuum-tight sheet metal can, heated, and extruded with the container. In another variation, billets are preformed by a conventional press-and-sinter process, and then the billet is hot extruded. These methods achieve a high degree of densification in the PM product.

Powder Forging Forging is an important metal forming process (Section 13.2). In powder forging, the starting workpart is a powder metallurgy part preformed to proper size by pressing and sintering. Advantages of this approach are: (1) densification of the PM part, (2) lower tooling costs and fewer forging "hits" (and therefore higher production rate) because the starting workpart is preformed, and (3) reduced material waste.

10.3.4 COMBINED PRESSING AND SINTERING

Hot isostatic pressing (Section 10.3.1) accomplishes compaction and sintering in one step. Other techniques that combine the two steps are hot pressing and spark sintering.

Hot Pressing The setup in uniaxial hot pressing is very similar to conventional PM pressing, except that heat is applied during compaction. The resulting product is generally dense, strong, hard, and dimensionally accurate. Despite these advantages, the process presents certain technical problems that limit its adoption. Principal among these are (1) selecting a suitable mold material that can withstand the high sintering temperatures; (2) longer production cycle required to accomplish sintering; and (3) heating and maintaining atmospheric control in the process [2]. Hot pressing has found some application in the production of sintered carbide products using graphite molds.

Spark Sintering An alternative approach that combines pressing and sintering but overcomes some of the problems in hot pressing is spark sintering. The process consists of two basic steps [2], [17]: (1) powder or a green compacted preform is placed in a die; and (2) upper and lower punches, which also serve as electrodes, compress the part and simultaneously apply a high-energy electrical current that burns off surface contaminants and sinters the powders, forming a dense, solid part in about 15 seconds. The process has been applied to a variety of metals.

10.3.5 LIQUID PHASE SINTERING

Conventional sintering (Section 10.2.3) is solid-state sintering; the metal is sintered at a temperature below its melting point. In systems consisting of a mixture of two powder metals, in which there is a difference in melting temperature between the metals, an alternative type of sintering is used, called liquid phase sintering. In this process, the two powders are initially mixed, and then heated to a temperature that is high enough to melt the lower-melting-point metal but not the other. The melted metal thoroughly wets the solid particles, creating a dense structure with strong bonding between the metals upon solidification. Depending on the metals involved, prolonged heating may result in alloying of the metals by gradually dissolving the solid particles into the liquid melt and/or diffusion of the liquid metal into the solid. In either case, the resulting product is fully densified (no pores) and strong. Examples of systems that use liquid phase sintering include Fe–Cu, W–Cu, and Cu–Co [6].

10.4 MATERIALS AND PRODUCTS FOR PM

The raw materials for PM processing are more expensive than for other metalworking because of the additional energy required to reduce the metal to powder form. Accordingly, PM is competitive only in a certain range of applications. In this section we identify the materials and products that seem most suited to powder metallurgy.

PM Materials From a chemistry standpoint, metal powders can be classified as either elemental or pre-alloyed. ***Elemental*** powders consist of a pure metal and are used in applications where high purity is important. For example, pure iron might be used where its magnetic properties are important. The most common elemental powders are those of iron, aluminum, and copper.

Elemental powders are also mixed with other metal powders to produce special alloys that are difficult to formulate using conventional processing methods. Tool steels are an example; PM permits blending of ingredients that is difficult or impossible by traditional alloying techniques. Using mixtures of elemental powders to form an alloy provides a processing benefit, even where special alloys are not involved. Since the powders are pure metals, they are not as strong as pre-alloyed metals. Therefore, they deform more readily during pressing, so that density and green strength are higher than with pre-alloyed compacts.

In ***pre-alloyed*** powders, each particle is an alloy composed of the desired chemical composition. Pre-alloyed powders are used for alloys that cannot be formulated by mixing elemental powders; stainless steel is an important example. The most common pre-alloyed powders are certain copper alloys, stainless steel, and high-speed steel.

The commonly used elemental and pre-alloyed powdered metals, in approximate order of tonnage usage, are: (1) iron, by far the most widely used PM metal, frequently mixed with graphite to make steel parts, (2) aluminum, (3) copper and its alloys, (4) nickel, (5) stainless steel, (6) high-speed steel, and (7) other PM materials such as tungsten, molybdenum, titanium, tin, and precious metals.

PM Products A substantial advantage offered by PM technology is that parts can be made to near net shape or net shape; little or no additional shaping is required after PM processing. Some of the components commonly manufactured by powder metallurgy are gears, bearings, sprockets, fasteners, electrical contacts, cutting tools, and various machinery parts. When produced in large quantities, metal gears and bearings are particularly well suited to PM for two reasons: (1) the geometry is defined principally in two dimensions, so the part has a top surface of a certain shape, but there are no features along the sides; and (2) there is a need for porosity in the material to serve as a reservoir for lubricant. More complex parts with true three-dimensional geometries are also feasible in powder metallurgy, by adding secondary operations such as machining to complete the shape of the pressed and sintered part, and by observing certain design guidelines such as those outlined in the following section.

10.5 DESIGN CONSIDERATIONS IN POWDER METALLURGY

Use of PM techniques is generally suited to a certain class of production situations and part designs. In this section we attempt to define the characteristics of this class of applications for which powder metallurgy is most appropriate. We first present a classification system for PM parts, and then offer some guidelines on component design.

The Metal Powder Industries Federation (MPIF) defines four classes of powder metallurgy part designs, by level of difficulty in conventional pressing. The system is useful because it indicates some of the limitations on shape that can be achieved with conventional PM processing. The four part classes are illustrated in Figure 10.13.

The MPIF classification system provides some guidance concerning part geometries that are suited to conventional PM pressing techniques. Additional advice is offered in the following design guidelines, compiled from [3], [13], and [17].

- Economics of PM processing usually require large part quantities to justify the cost of equipment and special tooling required. Minimum quantities of 10,000 units are suggested [17], although exceptions exist.
- Powder metallurgy is unique in its capability to fabricate parts with a controlled level of porosity. Porosities up to 50% are possible.
- PM can be used to make parts out of unusual metals and alloys—materials that would be difficult, if not impossible, to fabricate by other means.

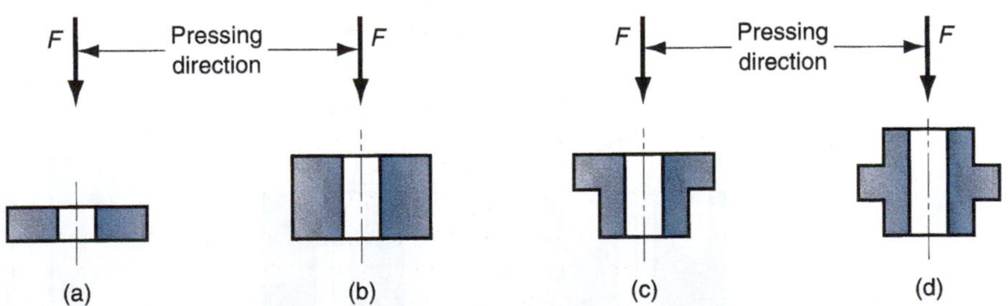

FIGURE 10.13 Four classes of PM parts, side view shown; cross section is circular (a) Class I—simple thin shapes that can be pressed from one direction; (b) Class II—simple but thicker shapes that require pressing from two directions; (c) Class III—two levels of thickness, pressed from two directions; and (d) Class IV—multiple levels of thickness, pressed from two directions, with separate controls for each level to achieve proper densification throughout the compact. (Credit: *Fundamentals of Modern Manufacturing*, 4th Edition by Mikell P. Groover, 2010. Reprinted with permission of John Wiley & Sons, Inc.)

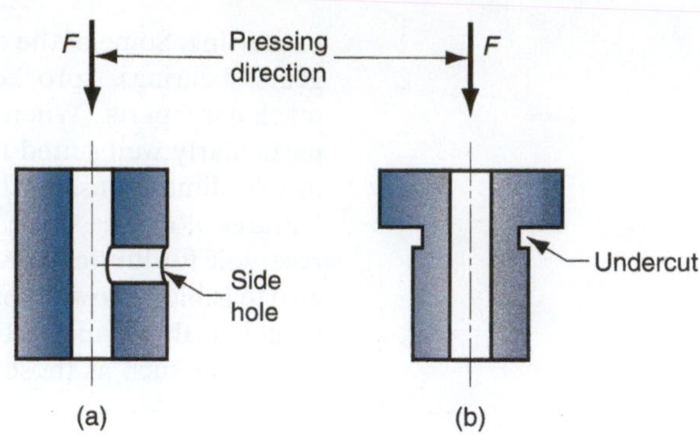

FIGURE 10.14 Part features to be avoided in PM: (a) side holes and (b) side undercuts. Part ejection is impossible. (Credit: *Fundamentals of Modern Manufacturing*, 4th Edition by Mikell P. Groover, 2010. Reprinted with permission of John Wiley & Sons, Inc.)

- The geometry of the part must permit ejection from the die after pressing; this generally means that the part must have vertical or near-vertical sides, although steps in the part are permissible as suggested by the MPIF classification system (Figure 10.13). Design features such as undercuts and holes on the part sides, as shown in Figure 10.14, must be avoided. Vertical undercuts and holes, as in Figure 10.15, are permissible because they do not interfere with ejection. Vertical holes can be of cross-sectional shapes other than round (e.g., squares, keyways) without significant increases in tooling or processing difficulty.
- Screw threads cannot be fabricated by PM pressing; if required, they must be machined into the PM component after sintering.
- Chamfers and corner radii are possible by PM pressing, as shown in Figure 10.16. Problems are encountered in punch rigidity when angles are too acute.
- Wall thickness should be a minimum of 1.5 mm (0.060 in) between holes or a hole and the outside part wall, as indicated in Figure 10.17. Minimum recommended hole diameter is 1.5 mm (0.060 in).

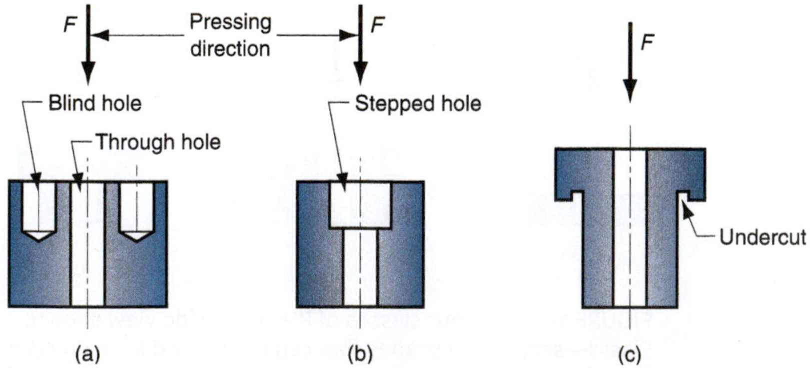

FIGURE 10.15 Permissible part features in PM: (a) vertical hole, blind and through, (b) vertical stepped hole, and (c) undercut in vertical direction. These features allow part ejection. (Credit: *Fundamentals of Modern Manufacturing*, 4th Edition by Mikell P. Groover, 2010. Reprinted with permission of John Wiley & Sons, Inc.)

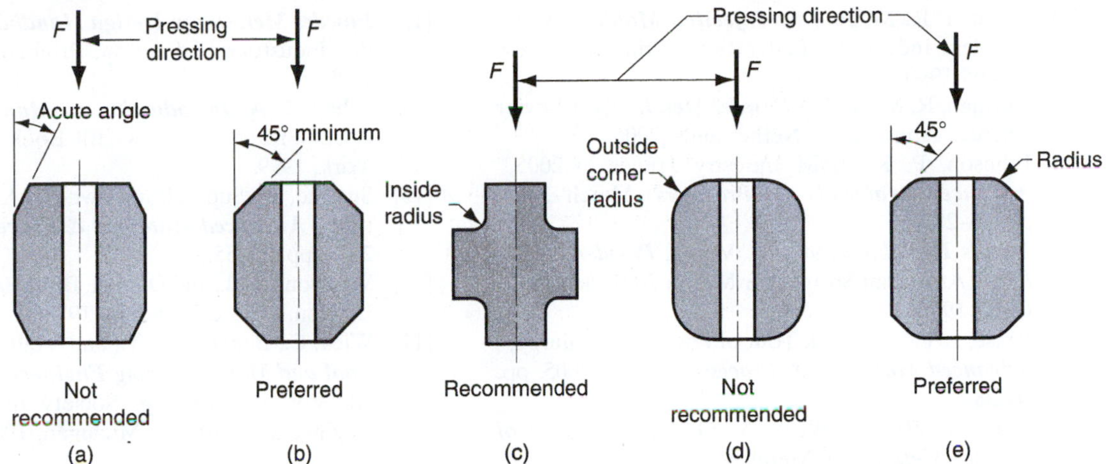

FIGURE 10.16 Chamfers and corner radii are accomplished but certain rules should be observed: (a) avoid acute chamfer angles; (b) larger angles are preferred for punch rigidity; (c) small inside radius is desirable; (d) full outside corner radius is difficult because punch is fragile at corner's edge; (e) outside corner problem can be solved by combining radius and chamfer. (Credit: *Fundamentals of Modern Manufacturing*, 4th Edition by Mikell P. Groover, 2010. Reprinted with permission of John Wiley & Sons, Inc.)

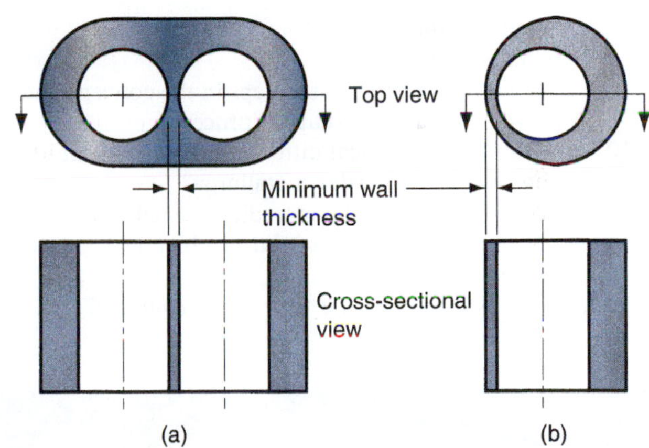

FIGURE 10.17 Minimum recommended wall thickness (a) between holes or (b) between a hole and an outside wall should be 1.5 mm (0.060 in). (Credit: *Fundamentals of Modern Manufacturing*, 4th Edition by Mikell P. Groover, 2010. Reprinted with permission of John Wiley & Sons, Inc.)

REFERENCES

[1] ***ASM Handbook***, Vol. 7, ***Powder Metal Technologies and Applications***. ASM International, Materials Park, Ohio, 1998.

[2] Amstead, B. H., Ostwald, P. F., and Begeman, M. L. ***Manufacturing Processes***, 8th ed. John Wiley & Sons, New York, 1987.

[3] Bralla, J. G. (ed.). ***Design for Manufacturability Handbook***, 2nd ed. McGraw-Hill Book Company, New York, 1998.

[4] Bulger, M. "Metal Injection Molding," ***Advanced Materials & Processes***, March 2005, pp. 39–40.

[5] Dixon, R. H. T., and Clayton, A. ***Powder Metallurgy for Engineers***. The Machinery Publishing Co. Ltd., Brighton, United Kingdom, 1971.

[6] German, R. M. ***Powder Metallurgy Science***, 2nd ed. Metal Powder Industries Federation, Princeton, New Jersey, 1994.

[7] German, R. M. ***Powder Injection Molding***. Metal Powder Industries Federation, Princeton, New Jersey, 1990.

[8] German, R. M. ***A-Z of Powder Metallurgy***. Elsevier Science, Amsterdam, Netherlands, 2006.

[9] Johnson, P. K. "P/M Industry Trends in 2005," ***Advanced Materials & Processes***, March 2005, pp. 25–28.

[10] ***Metals Handbook***, 9th ed. Vol. 7, ***Powder Metallurgy***. American Society for Metals, Materials Park, Ohio, 1984.

[11] Pease, L. F. "A Quick Tour of Powder Metallurgy," ***Advanced Materials & Processes***, March 2005, pp. 36–38.

[12] Pease, L. F., and West, W. G. ***Fundamentals of Powder Metallurgy***. Metal Powder Industries Federation, Princeton, New Jersey, 2002.

[13] ***Powder Metallurgy Design Handbook***. Metal Powder Industries Federation, Princeton, New Jersey, 1989.

[14] Schey, J. A. ***Introduction to Manufacturing Processes***, 3rd ed. McGraw-Hill Book Company, New York, 1999.

[15] Smythe, J. "Superalloy Powders: An Amazing History," ***Advanced Materials & Processes***, November 2008, pp. 52–55.

[16] Waldron, M. B., and Daniell, B. L. ***Sintering***. Heyden, London, United Kingdom, 1978.

[17] Wick, C., Benedict, J. T., and Veilleux, R. F. (eds.). ***Tool and Manufacturing Engineers Handbook***, 4th ed. Vol. II, ***Forming***. Society of Manufacturing Engineers, Dearborn, Michigan, 1984.

REVIEW QUESTIONS

10.1. Name some of the reasons for the commercial importance of powder metallurgy technology.

10.2. What are some of the disadvantages of PM methods?

10.3. What are the principal methods used to produce metallic powders?

10.4. What are the three basic steps in the conventional powder metallurgy shaping process?

10.5. What is the technical difference between blending and mixing in powder metallurgy?

10.6. What are some of the ingredients usually added to the metallic powders during blending and/or mixing?

10.7. What is meant by the term *green compact*?

10.8. Describe what happens to the individual particles during compaction.

10.9. What are the three steps in the sintering cycle in PM?

10.10. What are some of the reasons why a controlled-atmosphere furnace is desirable in sintering?

10.11. What is the difference between impregnation and infiltration in PM?

10.12. How is isostatic pressing distinguished from conventional pressing and sintering in PM?

10.13. Describe liquid phase sintering.

10.14. What are the two basic classes of metal powders as far as chemistry is concerned?

10.15. Why is PM technology so well suited to the production of gears and bearings?

PROBLEMS

10.1. In a certain pressing operation, the metallic powder fed into the open die has a packing factor of 0.5. The pressing operation reduces the powders to two-thirds of their starting volume. In the subsequent sintering operation, shrinkage amounts to 10% on a volume basis. Given that these are the only factors that affect the structure of the finished part, determine its final porosity.

10.2. A bearing of simple geometry is to be pressed out of bronze powders, using a compacting pressure of 207 MPa. The outside diameter = 44 mm, the inside diameter = 22 mm, and the length of the bearing = 25 mm. What is the required press tonnage to perform this operation?

10.3. The part shown in Figure P10.3 is to be pressed of iron powders using a compaction pressure of 75,000 lb/in^2. Dimensions are inches. Determine (a) the most appropriate pressing direction, (b) the required press tonnage to perform this operation, and (c) the final weight of the part if the porosity is 10%. Assume shrinkage during sintering can be neglected.

10.4. For each of the four part drawings in Figure P10.4, indicate which PM class the parts belong to, whether the part must be pressed from one or two directions, and how many levels of press control will be required? Dimensions are mm.

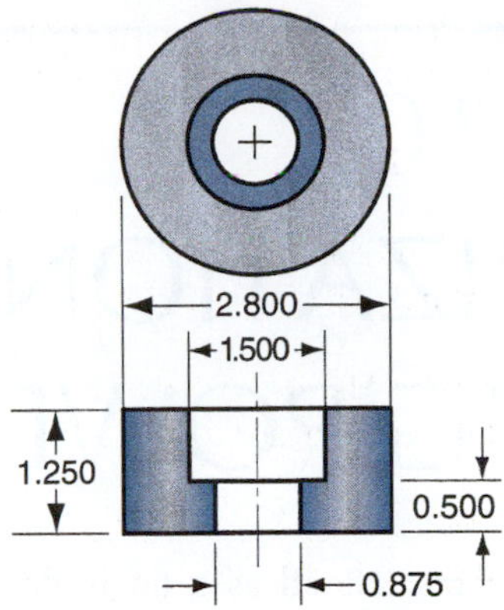

FIGURE P10.3 Part for Problem 10.3 (dimensions in inches). (Credit: *Fundamentals of Modern Manufacturing,* 4th Edition by Mikell P. Groover, 2010. Reprinted with permission of John Wiley & Sons, Inc.)

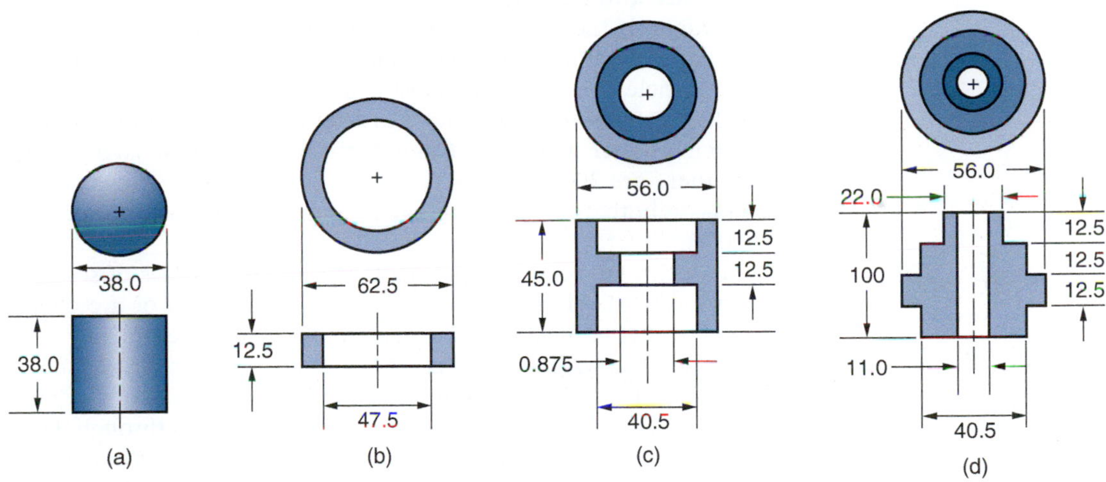

FIGURE P10.4 Parts for Problem 10.4 (dimensions in mm). (Credit: *Fundamentals of Modern Manufacturing,* 4th Edition by Mikell P. Groover, 2010. Reprinted with permission of John Wiley & Sons, Inc.)

APPENDIX A10: CHARACTERIZATION OF ENGINEERING POWDERS

A ***powder*** can be defined as a finely divided particulate solid. In this appendix we characterize metallic powders. Most of our discussion applies to ceramic powders as well.

A10.1 GEOMETRIC FEATURES

The geometry of the individual powders can be defined by the following attributes: (1) particle size and distribution, (2) particle shape and internal structure, and (3) surface area.

Particle Size and Distribution Particle size refers to the dimensions of the individual powders. If the particle shape is spherical, a single dimension is adequate. For other shapes, two or more dimensions are needed. Various methods are available to obtain particle size data. The most common method uses screens of different mesh sizes. The term ***mesh count*** is used to refer to the number of openings per linear inch of screen. Higher mesh count indicates smaller particle size. A mesh count of 200 means there are 200 openings per linear inch. Since the mesh is square, the count is the same in both directions, and the total number of openings per square inch is $200^2 = 40{,}000$.

Particles are sorted by passing them through a series of screens of progressively smaller mesh size. The powders are placed on a screen of a certain mesh count and vibrated so that particles small enough to fit through the openings pass through to the next screen below. The second screen empties into a third, and so forth, so that the particles are sorted according to size. A certain powder size might be called size 230 through 200, indicating that the powders have passed through the 200 mesh, but not 230. To make the specification easier, we simply say that the particle size is 200. The procedure of separating the powders by size is called ***classification***.

The openings in the screen are less than the reciprocal of the mesh count because of the thickness of the wire in the screen, as illustrated in Figure A10.1. Assuming that the limiting dimension of the particle is equal to the screen opening, we have

$$PS = \frac{1}{MC} - t_w \tag{A10.1}$$

where PS = particle size, in; MC = mesh count, openings per linear inch; and t_w = wire thickness of screen mesh, in. The figure shows how smaller particles would pass through the openings, while larger powders would not. Variations occur in the powder sizes sorted by screening due to differences in particle shapes, the range of sizes between mesh count steps, and variations in screen openings within a given mesh count. Also, the screening method has a practical upper limit of $MC = 400$ (approximately), due to the difficulty in making such fine screens and because of agglomeration of the small powders. Other methods to measure particle size include microscopy and X-ray techniques.

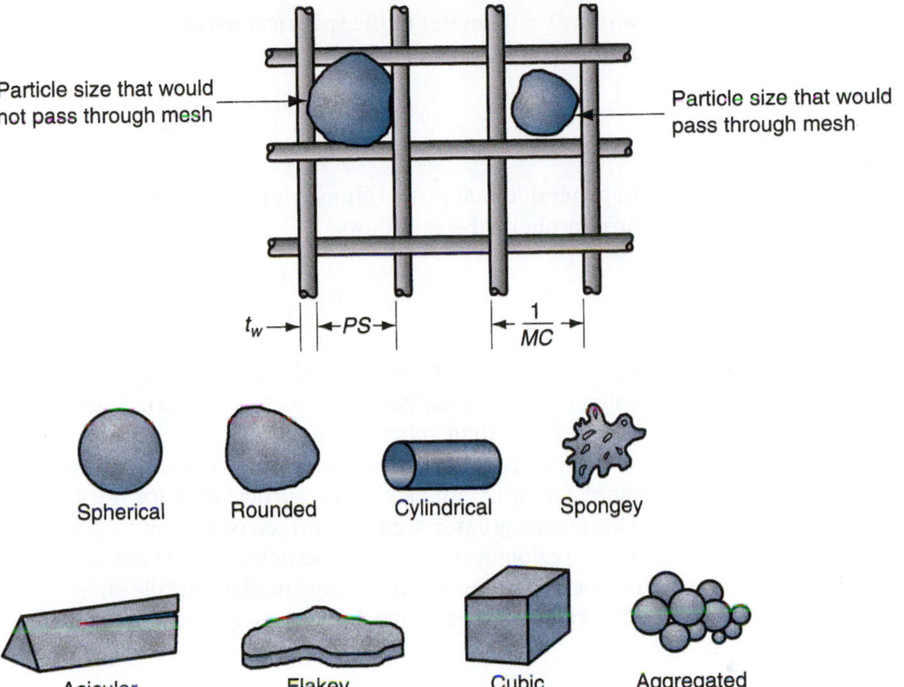

FIGURE A10.1 Screen mesh for sorting particle sizes. (Credit: *Fundamentals of Modern Manufacturing*, 4th Edition by Mikell P. Groover, 2010. Reprinted with permission of John Wiley & Sons, Inc.)

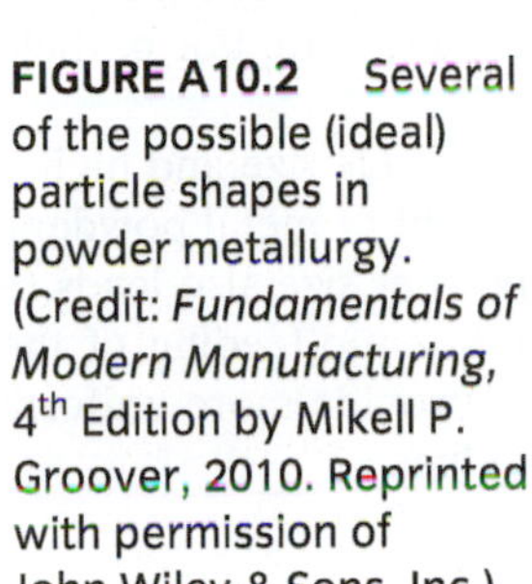

FIGURE A10.2 Several of the possible (ideal) particle shapes in powder metallurgy. (Credit: *Fundamentals of Modern Manufacturing*, 4th Edition by Mikell P. Groover, 2010. Reprinted with permission of John Wiley & Sons, Inc.)

Typical particle sizes used in conventional powder metallurgy (press and sinter) range between 25 and 300 μm (0.001 and 0.012 in).[1] The high end of this range corresponds to a mesh count of around 65. The low end of the range is too small to be measured by the mesh count method.

Particle Shape and Internal Structure Metal powder shapes can be cataloged into various types, several of which are illustrated in Figure A10.2. There will be a variation in the particle shapes in a collection of powders, just as the particle size will vary. A simple and useful measure of shape is the aspect ratio—the ratio of maximum dimension to minimum dimension for a given particle. The aspect ratio for a spherical particle is 1.0, but for an acicular grain the ratio might be 2 to 4. Microscopic techniques are required to determine shape characteristics.

Any volume of loose powders will contain pores between the particles. These are called ***open pores*** because they are external to the individual particles. Open pores are spaces into which a fluid such as water, oil, or a molten metal, can penetrate. In addition, there are ***closed pores***—internal voids in the structure of an individual particle. The existence of these internal pores is usually minimal, and their effect when they do exist is minor, but they can influence density measurements, as we shall see later.

Surface Area Assuming that the particle shape is a perfect sphere, its area A and volume V are given by

$$A = \pi D^2 \tag{A10.2}$$

$$V = \frac{\pi D^3}{6} \tag{A10.3}$$

[1] These values are provided by Prof. Wojciech Misiolek, my colleague in Lehigh's Department of Materials Science and Engineering. Powder metallurgy is one of his research areas.

where D = diameter of the spherical particle, mm (in). The area-to-volume ratio A/V for a sphere is then given by

$$\frac{A}{V} = \frac{6}{D} \tag{A10.4}$$

In general, the area-to-volume ratio can be expressed for any particle shape—spherical or nonspherical—as follows:

$$\frac{A}{V} = \frac{K_s}{D} \quad \text{or} \quad K_s = \frac{AD}{V} \tag{A10.5}$$

where K_s = shape factor; D in the general case = the diameter of a sphere of equivalent volume as the nonspherical particle, mm (in). Thus, $K_s = 6.0$ for a sphere. For particle shapes other than spherical, $K_s > 6$.

We can infer the following from these equations. Smaller particle size and higher shape factor (K_s) mean higher surface area for the same total weight of metal powders. This means greater area for surface oxidation to occur. Small powder size also leads to more agglomeration of the particles, which is a problem in automatic feeding of the powders. The reason for using smaller particle sizes is that they provide more uniform shrinkage and better mechanical properties in the final PM product.

A10.2 OTHER FEATURES

Other features of engineering powders include interparticle friction, flow characteristics, packing, density, porosity, chemistry, and surface films.

Interparticle Friction and Flow Characteristics Friction between particles affects the ability of a powder to flow readily and pack tightly. A common measure of interparticle friction is the ***angle of repose***, which is the angle formed by a pile of powders as they are poured from a narrow funnel, as in Figure A10.3. Larger angles indicate greater friction between particles. Smaller particle sizes generally show greater friction and steeper angles. Spherical shapes result in the lowest interparticle friction; as shape deviates more from spherical, friction between particles tends to increase.

Flow characteristics are important in die filling and pressing. Automatic die filling depends on easy and consistent flow of the powders. In pressing, resistance to flow increases density variations in the compacted part; these density gradients are generally undesirable. A common measure of flow is the time required for a certain amount of

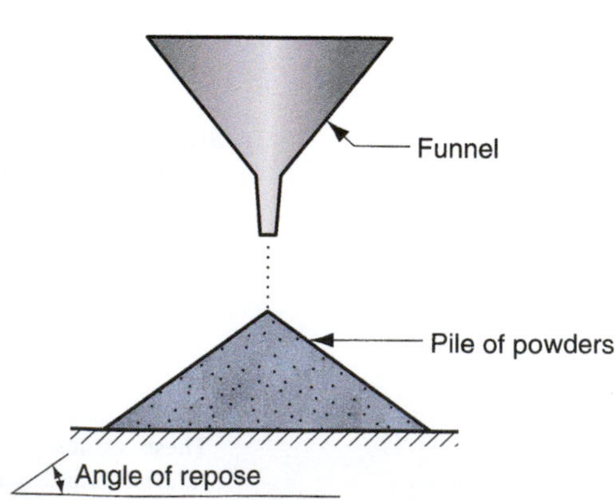

FIGURE A10.3 Interparticle friction as indicated by the angle of repose of a pile of powders poured from a narrow funnel. Larger angles indicate greater interparticle friction. (Credit: *Fundamentals of Modern Manufacturing*, 4th Edition by Mikell P. Groover, 2010. Reprinted with permission of John Wiley & Sons, Inc.)

powder (by weight) to flow through a standard-sized funnel. Smaller flow times indicate easier flow and lower interparticle friction. To reduce interparticle friction and facilitate flow during pressing, lubricants are often added to the powders in small amounts.

Packing, Density, and Porosity Packing characteristics depend on two density measures. First, ***true density*** is the density of the true volume of the material. This is the density when the powders are melted into a solid mass, values of which are given in Table 3.10. Second, ***bulk density*** is the density of the powders in the loose state after pouring, which includes the effect of pores between particles. Because of the pores, bulk density is less than true density.

The ***packing factor*** is the bulk density divided by the true density. Typical values for loose powders range between 0.5 and 0.7. The packing factor depends on particle shape and the distribution of particle sizes. If powders of various sizes are present, the smaller powders will fit into the interstices of the larger ones that would otherwise be taken up by air, thus resulting in a higher packing factor. Packing can also be increased by vibrating the powders, causing them to settle more tightly. Finally, we should note that external pressure, as applied during compaction, greatly increases packing of powders through rearrangement and deformation of the particles.

Porosity represents an alternative way of considering the packing characteristics of a powder. ***Porosity*** is defined as the ratio of the volume of the pores (empty spaces) in the powder to the bulk volume. In principle,

$$\text{Porosity} + \text{Packing factor} = 1.0 \tag{A10.6}$$

The issue is complicated by the possible existence of closed pores in some of the particles. If these internal pore volumes are included in the above porosity, then the equation is exact.

Chemistry and Surface Films Characterization of the powder would not be complete without an identification of its chemistry. Metallic powders are classified as either elemental, consisting of a pure metal, or pre-alloyed, wherein each particle is an alloy. These classes and the metals commonly used in PM are discussed in Section 10.4.

Surface films are a problem in powder metallurgy because of the large area per unit weight of metal when dealing with powders. The possible films include oxides, silica, adsorbed organic materials, and moisture [6]. Generally, these films must be removed prior to shape processing.

11 PROCESSING OF CERAMICS AND CERMETS

Chapter Contents

11.1 Processing of Traditional Ceramics
- 11.1.1 Preparation of the Raw Material
- 11.1.2 Shaping Processes
- 11.1.3 Drying
- 11.1.4 Firing (Sintering)

11.2 Processing of New Ceramics
- 11.2.1 Preparation of Starting Materials
- 11.2.2 Shaping
- 11.2.3 Sintering
- 11.2.4 Finishing

11.3 Processing of Cermets
- 11.3.1 Cemented Carbides
- 11.3.2 Other Cermets and Ceramic-Matrix Composites

11.4 Product Design Considerations

Ceramic materials divide into three categories (Section 2.2): (1) traditional ceramics, (2) new ceramics, and (3) glasses. The processing of glass involves solidification primarily and is covered in Chapter 7. In the present chapter, we consider the particulate processing methods used for traditional and new ceramics. We also cover the processing of metal-matrix composites and ceramic-matrix composites.

Traditional ceramics are made from minerals occurring in nature. The products include pottery, porcelain, bricks, and cement. New ceramics are made from synthetically produced raw materials and embrace a wide spectrum of items such as cutting tools, artificial bones, nuclear fuels, and substrates for electronic circuits. The starting material for both categories is powder. In the case of the traditional ceramics, the powders are usually mixed with water to temporarily bind the particles together and achieve the proper consistency for shaping. For new ceramics, other substances are used as binders during shaping. After shaping, the green parts are sintered. This is often called ***firing*** in ceramics, but the function is the same as in powder metallurgy: to effect a solid-state reaction that bonds the material into a hard solid mass.

The processing methods discussed in this chapter are commercially and technologically important because virtually all ceramic products are formed by these methods (except, of course, glass products). The manufacturing sequence is similar for traditional and new ceramics because the form of the starting material is the same: powder. However, the processing methods for the two categories are sufficiently different that we discuss them separately.

11.1 PROCESSING OF TRADITIONAL CERAMICS

In this section we describe the production technology used to make traditional ceramic products such as pottery, stoneware and other dinnerware, bricks, tile, and ceramic

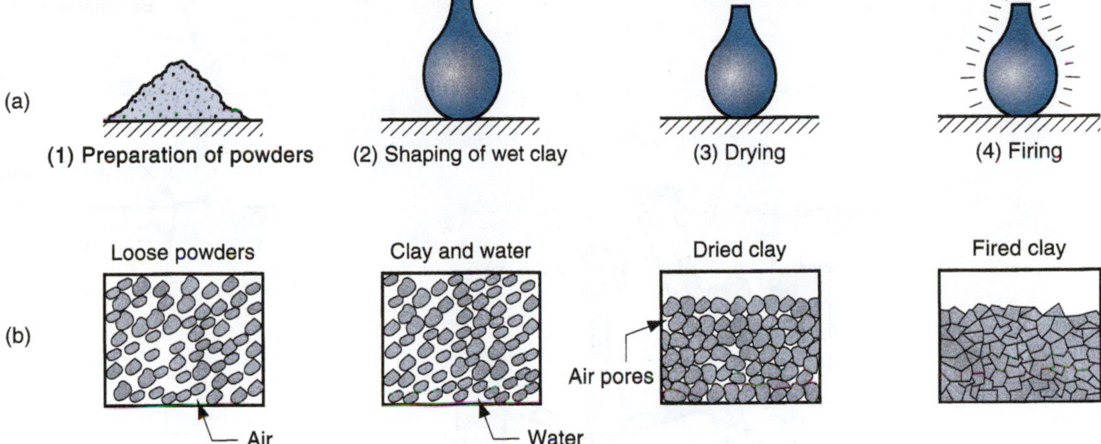

FIGURE 11.1 Usual steps in traditional ceramics processing: (1) preparation of raw materials, (2) shaping, (3) drying, and (4) firing. Part (a) shows the workpart during the sequence, while (b) shows the condition of the powders. (Credit: *Fundamentals of Modern Manufacturing*, 4th Edition by Mikell P. Groover, 2010. Reprinted with permission of John Wiley & Sons, Inc.)

refractories. Bonded grinding wheels are also produced by the same basic methods. What these products have in common is that their raw materials consist primarily of silicate ceramics—clays. The processing sequence for most of the traditional ceramics consists of the steps depicted in Figure 11.1.

11.1.1 PREPARATION OF THE RAW MATERIAL

The shaping processes for traditional ceramics require that the starting material be in the form of a plastic paste. This paste is made of fine ceramic powders mixed with water, and its consistency determines the ease of forming the material and the quality of the final product. The raw ceramic material usually occurs in nature as rocky lumps, and reduction to powder is the purpose of the preparation step in ceramics processing.

Techniques for reducing particle size in ceramics processing involve mechanical energy in various forms, such as impact, compression, and attrition. The term ***comminution*** is used for these techniques, which are most effective on brittle materials, including cement, metallic ores, and brittle metals. Two general categories of comminution operations are distinguished: crushing and grinding.

Crushing refers to the reduction of large lumps from the mine to smaller sizes for subsequent further reduction. Several stages may be required (e.g., primary crushing, secondary crushing), the reduction ratio in each stage being in the range 3 to 6. Crushing of minerals is accomplished by compression against rigid surfaces or by impact against surfaces in a rigid constrained motion [1]. Figure 11.2 shows several types of equipment used to perform crushing: (a) jaw crushers, in which a large jaw toggles back and forth to crush lumps against a hard, rigid surface; (b) gyratory crushers, which use a gyrating cone to compress lumps against a rigid surface; (c) roll crushers, in which the ceramic lumps are squeezed between rotating rolls; and (d) hammer mills, which use rotating hammers impacting the material to break up the lumps.

Grinding, in our context here, refers to the operation of reducing the small pieces produced by crushing into a fine powder. Grinding is accomplished by abrasion and impact of the crushed mineral by the free motion of unconnected hard media such as

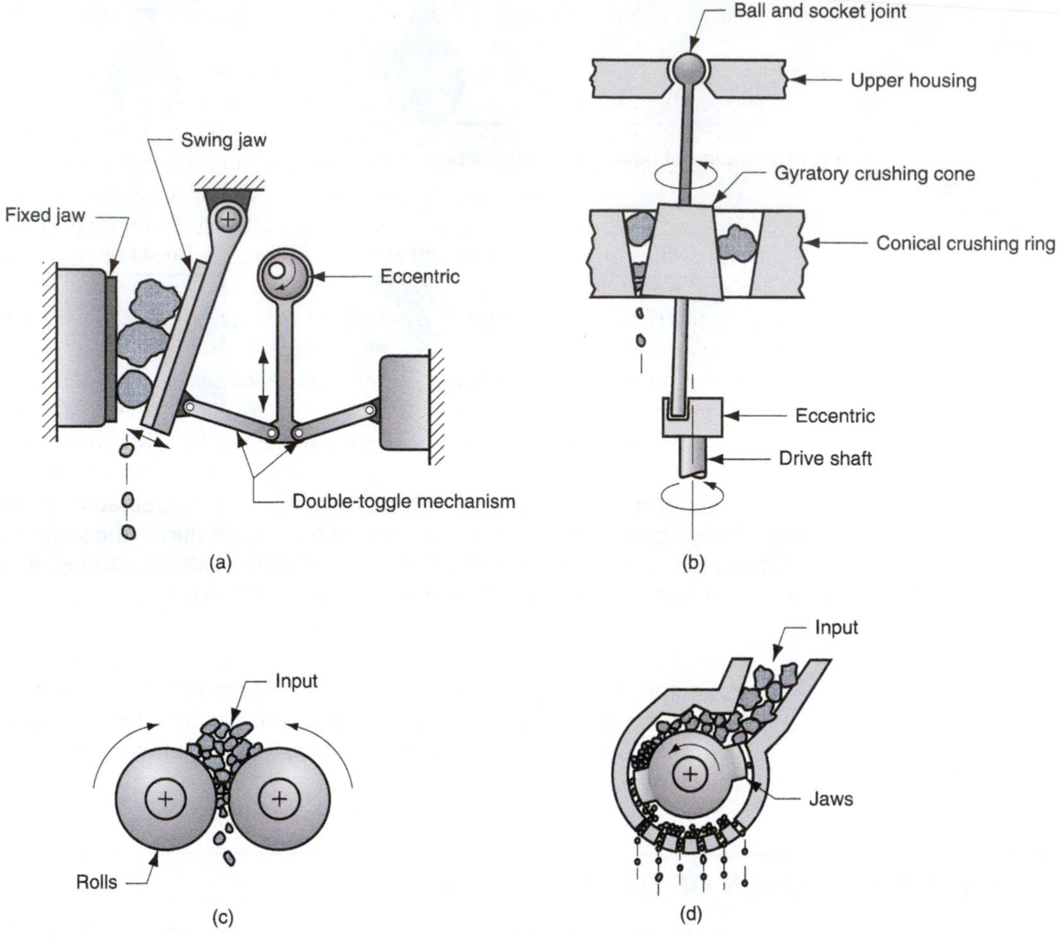

FIGURE 11.2 Crushing operations: (a) jaw crusher, (b) gyratory crusher, (c) roll crusher, and (d) hammer mill. (Credit: *Fundamentals of Modern Manufacturing,* 4th Edition by Mikell P. Groover, 2010. Reprinted with permission of John Wiley & Sons, Inc.)

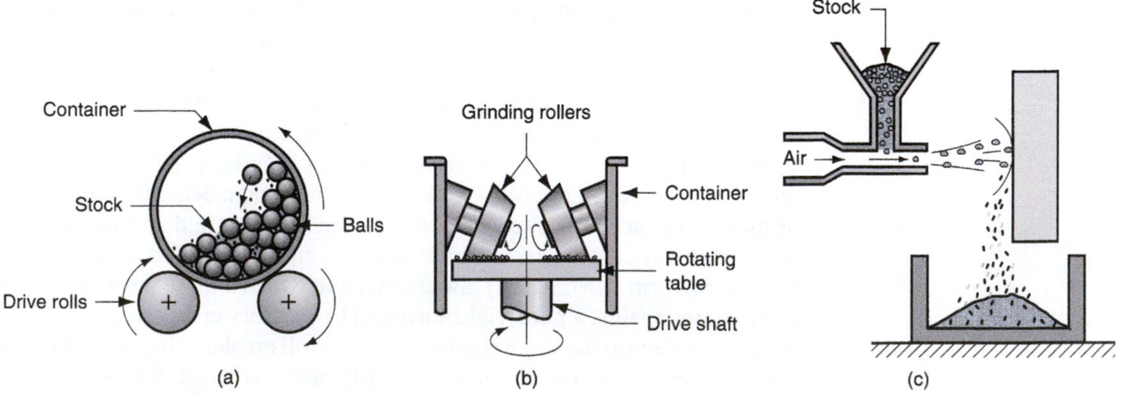

FIGURE 11.3 Mechanical methods of producing ceramic powders: (a) ball mill, (b) roller mill, and (c) impact grinding. (Credit: *Fundamentals of Modern Manufacturing,* 4th Edition by Mikell P. Groover, 2010. Reprinted with permission of John Wiley & Sons, Inc.)

balls, pebbles, or rods [1]. Examples of grinding include (a) ball mill, (b) roller mill, and (c) impact grinding, illustrated in Figure 11.3.

In a ***ball mill***, hard spheres mixed with the stock to be comminuted are tumbled inside a rotating cylindrical container. The rotation causes the balls and stock to be carried up the container wall, and then pulled back down by gravity to accomplish a grinding action by a combination of impact and attrition. These operations are often carried out with water added to the mixture, so that the ceramic is in the form of a slurry. In a ***roller mill***, stock is compressed against a flat horizontal grinding table by rollers riding over the table surface. Although not clearly shown in our sketch, the pressure of the grinding rollers against the table is regulated by mechanical springs or hydraulic-pneumatic means. In ***impact grinding***, which seems to be less frequently used, particles of stock are thrown against a hard flat surface, either in a high velocity air stream or in a high-speed slurry. The impact fractures the pieces into smaller particles.

The plastic paste required for shaping consists of ceramic powders and water. Clay is usually the main ingredient in the paste because it has ideal forming characteristics. The more water there is in the mixture, the more plastic and easily formed is the clay paste. However, when the formed part is later dried and fired, shrinkage occurs that can lead to cracking in the product. To address this problem, other ceramic raw materials that do not shrink on drying and firing are usually added to the paste, often in significant amounts. Also, other components can be included to serve special functions. Thus, the ingredients of the ceramic paste can be divided into the following three categories [3]: (1) clay, which provides the consistency and plasticity required for shaping; (2) nonplastic raw materials, such as alumina and silica, which do not shrink in drying and firing but unfortunately reduce plasticity in the mixture during forming; and (3) other ingredients, such as fluxes that melt (vitrify) during firing and promote sintering of the ceramic material, and wetting agents that improve mixing of ingredients.

These ingredients must be thoroughly mixed, either wet or dry. The ball mill often serves this purpose in addition to its grinding function. Also, the proper amounts of powder and water in the paste must be attained, so water must be added or removed, depending on the prior condition of the paste and its desired final consistency.

11.1.2 SHAPING PROCESSES

The optimum proportions of powder and water depend on the shaping process used. Some shaping processes require high fluidity; others act on a composition that contains very low water content. At about 50% water by volume, the mixture is a slurry that flows like a liquid. As the water content is reduced, increased pressure is required on the paste to produce a similar flow. Thus, the shaping processes can be divided according to the consistency of the mixture: (1) slip casting, in which the mixture is a slurry with 25% to 40% water; (2) plastic-forming methods that shape the clay in a plastic condition at 15% to 25% water; (3) semi-dry pressing, in which the clay is moist (10% to 15% water) but has low plasticity; and (4) dry pressing, where the clay is basically dry, containing less than 5% water. Dry clay has no plasticity. Each category includes several different shaping processes.

Slip Casting In slip casting, a suspension of ceramic powders in water, called a ***slip***, is poured into a porous plaster of paris ($CaSO_4$–$2H_2O$) mold so that water from the mix is gradually absorbed into the plaster to form a firm layer of clay at the mold surface. The composition of the slip is typically 25% to 40% water, the remainder being clay often mixed with other ingredients. It must be sufficiently fluid to flow into the crevices of the mold cavity, yet lower water content is desirable for faster production rates. Slip casting has two principal variations: drain casting and solid casting. In ***drain casting***, which is the traditional process, the mold is inverted to drain excess slip after the semisolid layer has

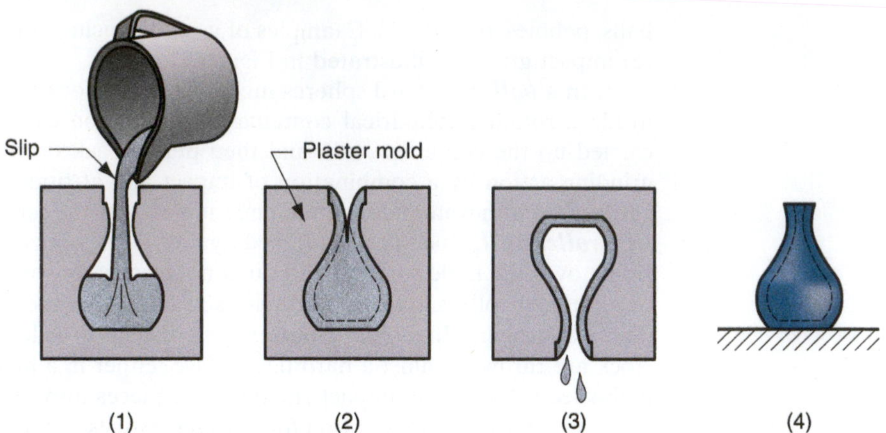

FIGURE 11.4 Sequence of steps in drain casting, a form of slip casting: (1) slip is poured into mold cavity; (2) water is absorbed into plaster mold to form a firm layer; (3) excess slip is poured out; and (4) part is removed from mold and trimmed. (Credit: *Fundamentals of Modern Manufacturing*, 4th Edition by Mikell P. Groover, 2010. Reprinted with permission of John Wiley & Sons, Inc.)

been formed, thus leaving a hollow part in the mold; the mold is then opened and the part removed. The sequence, which is very similar to slush casting of metals, is illustrated in Figure 11.4. It is used to make tea pots, vases, art objects, and other hollowware products. In ***solid casting***, used to produce solid products, adequate time is allowed for the entire body to become firm. The mold must be periodically resupplied with additional slip to account for shrinkage due to absorbed water.

Plastic Forming This category includes a variety of methods, both manual and mechanized. They all require the starting mixture to have a plastic consistency, which is generally achieved with 15% to 25% water. Manual methods generally make use of clay at the upper end of the range because it provides a material that is more easily formed; however, this is accompanied by greater shrinkage in drying. Mechanized methods generally employ a mixture with lower water content so that the starting clay is stiffer.

Although manual forming methods date back thousands of years, they are still used today by skilled artisans, either in production or for artworks. ***Hand modeling*** involves the creation of the ceramic product by manipulating the mass of plastic clay into the desired geometry. In addition to art pieces, patterns for plaster molds in slip casting are often made this way. ***Hand molding*** is a similar method, only a mold or form is used to define portions of the geometry. ***Hand throwing*** on a potter's wheel is another refinement of the handicraft methods. The ***potter's wheel*** is a round table that rotates on a vertical spindle, powered either by motor or foot-operated treadle. Ceramic products of circular cross section can be formed on the rotating table by throwing and shaping the clay, sometimes using a mold to provide the internal shape.

Strictly speaking, use of a motor-driven potter's wheel is a mechanized method. However, most mechanized clay-forming methods are characterized by much less manual participation than the hand-throwing method described above. These more mechanized methods include jiggering, plastic pressing, and extrusion. ***Jiggering*** is an extension of the potter's wheel methods, in which hand throwing is replaced by mechanized techniques. It is used to produce large numbers of identical items such as houseware plates and bowls. Although there are variations in the tools and methods used, reflecting different levels of automation and refinements to the basic process, a typical sequence is as follows, depicted in Figure 11.5: (1) a wet clay slug is placed on a convex mold; (2) a forming tool is pressed into the slug to provide the initial rough shape—the

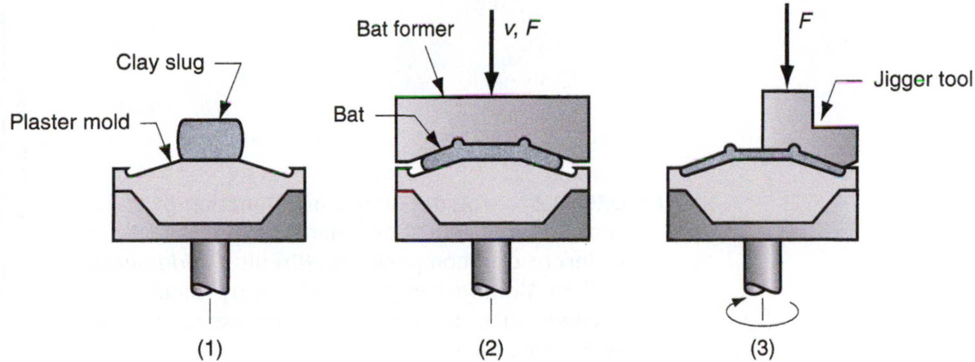

FIGURE 11.5 Sequence in jiggering: (1) wet clay slug is placed on a convex mold; (2) batting; and (3) a jigger tool imparts the final product shape. Symbols v and F indicate motion (v = velocity) and applied force, respectively. (Credit: *Fundamentals of Modern Manufacturing*, 4th Edition by Mikell P. Groover, 2010. Reprinted with permission of John Wiley & Sons, Inc.)

operation is called ***batting*** and the workpiece thus created is called a ***bat;*** and (3) a heated jigger tool is used to impart the final contoured shape to the product by pressing the profile into the surface during rotation of the workpart. The reason for heating the tool is to produce steam from the wet clay that prevents sticking. Closely related to jiggering is ***jolleying***, in which the basic mold shape is concave, rather than convex [8]. In both of these processes, a rolling tool is sometimes used in place of the nonrotating jigger (or jolley) tool; this rolls the clay into shape, avoiding the need to first bat the slug.

Plastic pressing is a forming process in which a plastic clay slug is pressed between upper and lower molds, contained in metal rings. The molds are made of a porous material such as gypsum, so that when a vacuum is drawn on the backs of the mold halves, moisture is removed from the clay. The mold sections are then opened, using positive air pressure to prevent sticking of the part in the mold. Plastic pressing achieves a higher production rate than jiggering and is not limited to radially symmetric parts.

Extrusion is used in ceramics processing to produce long sections of uniform cross section, which are then cut to required piece length. The extrusion equipment utilizes a screw-type action to assist in mixing the clay and pushing the plastic material through the die opening. This production sequence is widely used to make hollow bricks, shaped tiles, drain pipes, tubes, and insulators. It is also used to make the starting clay slugs for other ceramics processing methods such as jiggering and plastic pressing.

Semi-Dry Pressing In semi-dry pressing, the proportion of water in the starting clay is typically 10% to 15%. This results in low plasticity, precluding the use of plastic forming methods that require a very plastic clay. Semi-dry pressing uses high pressure to overcome the material's low plasticity and force it to flow into a die cavity. Flash is often formed due to excess clay being squeezed between the die sections.

Dry Pressing The main distinction between semi-dry and dry pressing is the moisture content of the starting mix. The moisture content of the clay in dry pressing is typically below 5%. Binders are usually added to the dry powder mix to provide sufficient strength in the pressed part for subsequent handling. Lubricants are also added to prevent die sticking during pressing and ejection. Because dry clay has no plasticity and is very abrasive, there are differences in die design and operating procedures, compared to semi-dry pressing. The dies must be made of hardened tool steel or cemented tungsten carbide to reduce wear. Since dry clay will not flow during pressing, the geometry of the part must be relatively simple, and the amount and distribution of starting powder in the die cavity

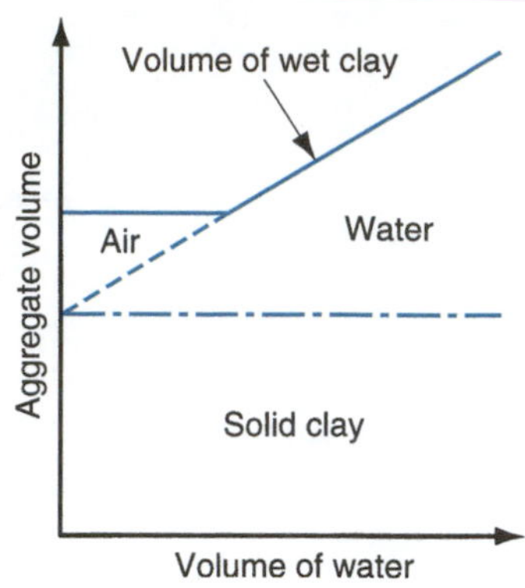

FIGURE 11.6 Volume of clay as a function of water content. Relationship shown here is typical; it varies for different clay compositions. (Credit: *Fundamentals of Modern Manufacturing*, 4th Edition by Mikell P. Groover, 2010. Reprinted with permission of John Wiley & Sons, Inc.)

must be right. No flash is formed in dry pressing, and no drying shrinkage occurs, so drying time is eliminated and good accuracy can be achieved in the dimensions of the final product. The process sequence in dry pressing is similar to semi-dry pressing. Typical products include bathroom tile, electrical insulators, and refractory brick.

11.1.3 DRYING

Water plays an important role in most of the traditional ceramics shaping processes. Thereafter, it serves no purpose and must be removed from the body of the clay piece before firing. Shrinkage is a problem during this step in the processing sequence because water contributes volume to the piece, and when it is removed, the volume is reduced. The effect can be seen in Figure 11.6. As water is initially added to dry clay, it simply replaces the air in the pores between ceramic grains, and there is no volumetric change. Increasing the water content above a certain point causes the grains to become separated and the volume to grow, resulting in a wet clay that has plasticity and formability. As more water is added, the mixture eventually becomes a liquid suspension of clay particles in water.

The reverse of this process occurs in drying. As water is removed from the wet clay, the volume of the piece shrinks. The drying process occurs in two stages, as depicted in Figure 11.7. In the first stage, the rate of drying is rapid and constant, as water is evaporated from the surface of the clay into the surrounding air and water from the interior migrates by capillary action toward the surface to replace it. It is during this stage that shrinkage occurs, with the associated risk of warping and cracking due to variations in drying in different sections of the piece. In the second stage of drying, the moisture content has been reduced to where the ceramic grains are in contact, and little or no further shrinkage occurs. The drying process slows, and this is seen in the decreasing rate in the plot.

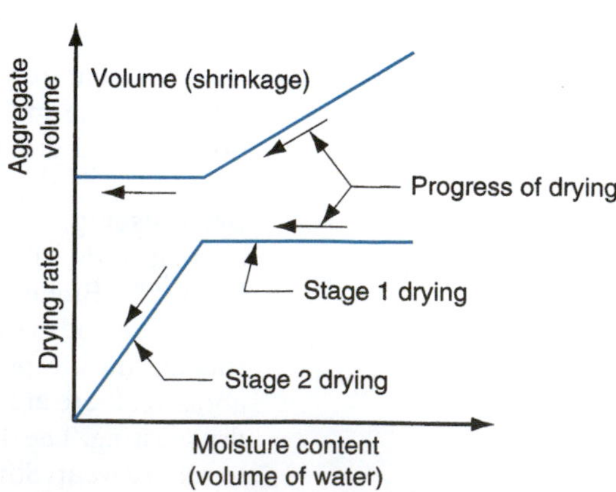

FIGURE 11.7 Typical drying rate curve and associated volume reduction (drying shrinkage) for a ceramic body in drying. Drying rate in the second stage of drying is depicted here as a straight line (constant rate decrease as a function of water content); the function is variously shown as concave or convex in the literature [3], [8]. (Credit: *Fundamentals of Modern Manufacturing*, 4th Edition by Mikell P. Groover, 2010. Reprinted with permission of John Wiley & Sons, Inc.)

In production, drying is usually accomplished in drying chambers in which temperature and humidity are controlled to achieve the proper drying schedule. Care must be taken so that water is not removed too rapidly, lest large moisture gradients be set up in the piece, making it more prone to crack. Heating is usually by a combination of convection and radiation, using infrared sources. Typical drying times range between a quarter hour for thin sections to several days for very thick sections.

11.1.4 FIRING (SINTERING)

After shaping but prior to firing, the ceramic piece is said to be ***green*** (same term as in powder metallurgy), meaning not fully processed or treated. The green piece lacks hardness and strength; it must be fired to fix the part shape and achieve hardness and strength in the finished ware. ***Firing*** is the heat treatment process that sinters the ceramic material; it is performed in a furnace called a ***kiln***. In ***sintering***, bonds are developed between the ceramic grains, and this is accompanied by densification and reduction of porosity. Therefore, shrinkage occurs in the polycrystalline material in addition to the shrinkage that has already occurred in drying. Sintering in ceramics is basically the same mechanism as in powder metallurgy. In the firing of traditional ceramics, certain chemical reactions between the components in the mixture may also take place, and a glassy phase also forms among the crystals that acts as a binder. Both of these phenomena depend on the chemical composition of the ceramic material and the firing temperatures used.

Unglazed ceramic ware is fired only once; glazed products are fired twice. ***Glazing*** refers to the application of a ceramic surface coating to make the piece more impervious to water and enhance its appearance. The usual processing sequence with glazed ware is (1) fire the ware once before glazing to harden the body of the piece, (2) apply the glaze, and (3) fire the piece a second time to harden the glaze.

11.2 PROCESSING OF NEW CERAMICS

Most of the traditional ceramics are based on clay, which possesses a unique capacity to be plastic when mixed with water but hard when dried and fired. Clay consists of various formulations of hydrous aluminum silicate, usually mixed with other ceramic materials, to form a rather complex chemistry. New ceramics (Section 2.2.2) are based on simpler chemical compounds, such as oxides, carbides, and nitrides. These materials do not possess the plasticity and formability of traditional clay when mixed with water. Accordingly, other ingredients must be combined with the ceramic powders to achieve plasticity and other desirable properties during forming, so that conventional shaping methods can be used. The new ceramics are generally designed for applications that require higher strength, hardness, and other properties not found in the traditional ceramic materials. These requirements have motivated the introduction of several new processing techniques not previously used for traditional ceramics.

The manufacturing sequence for the new ceramics can be summarized in the following steps: (1) preparation of starting materials, (2) shaping, (3) sintering, and (4) finishing. Although the sequence is nearly the same as for the traditional ceramics, the details are often quite different, as we shall see in the following.

11.2.1 PREPARATION OF STARTING MATERIALS

Since the strength specified for these materials is usually much greater than for traditional ceramics, the starting powders must be more homogeneous in size and composition, and particle size must be smaller (strength of the resulting ceramic product is inversely related to grain size). All of this means that greater control of the starting powders is required. Powder preparation includes mechanical and chemical methods. The mechanical methods consist of the same ball mill grinding operations used for traditional ceramics. The trouble with these methods is that the ceramic particles become contaminated from the materials used in the balls and walls of the mill. This compromises the purity of the ceramic powders and results in microscopic flaws that reduce the strength of the final product.

Two chemical methods are used to achieve greater homogeneity in the powders of new ceramics: freeze drying and precipitation from solution. In ***freeze drying***, salts of the appropriate starting chemistry are dissolved in water and the solution is sprayed to form small droplets, which are rapidly frozen. The water is then removed from the droplets in a vacuum chamber, and the resulting freeze-dried salt is decomposed by heating to form the ceramic powders. Freeze drying is not applicable to all ceramics, because in some cases a suitable water-soluble salt cannot be identified as the starting material.

Precipitation from solution is another preparation method used for new ceramics. In the typical process, the desired ceramic compound is dissolved from the starting mineral, thus permitting impurities to be filtered out. An intermediate compound is then precipitated from solution, which is converted into the desired compound by heating. An example of the precipitation method is the ***Bayer process*** for producing high purity alumina (also used in the production of aluminum). In this process, aluminum oxide is dissolved from the mineral bauxite so that iron compounds and other impurities can be removed. Then, aluminum hydroxide ($Al(OH)_3$) is precipitated from solution and reduced to Al_2O_3 by heating.

Further preparation of the powders includes classification by size and mixing before shaping. Very fine powders are required for new ceramics applications, and so the grains must be separated and classified according to size. Thorough mixing of the particles, especially when different ceramic powders are combined, is required to avoid segregation.

Various additives are often combined with the starting powders, usually in small amounts. The additives include (1) ***plasticizers*** to improve plasticity and workability; (2) ***binders*** to bond the ceramic particles into a solid mass in the final product, (3) ***wetting agents*** for better mixing; (4) ***deflocculants***, which help to prevent clumping and premature bonding of the powders; and (5) ***lubricants***, to reduce friction between ceramic grains during forming and to reduce sticking during mold release.

11.2.2 SHAPING

Many of the shaping processes for new ceramics are borrowed from powder metallurgy (PM) and traditional ceramics. The press and sinter methods discussed in Section 10.2 have been adapted to the new ceramic materials. And some of the traditional ceramics forming techniques (Section 11.1.2) are used to shape the new ceramics, including slip casting, extrusion, and dry pressing. The following processes are not normally associated with the forming of traditional ceramics, although several are associated with PM.

Hot Pressing Hot pressing is similar to dry pressing, except that the process is carried out at elevated temperatures, so that sintering of the product is accomplished simultaneously with pressing. This eliminates the need for a separate firing step in the sequence. Higher densities and finer grain size are obtained, but die life is reduced by the hot abrasive particles against the die surfaces.

Isostatic Pressing Isostatic pressing of ceramics is the same process used in powder metallurgy (Section 10.3.1). It uses hydrostatic pressure to compact the ceramic powders

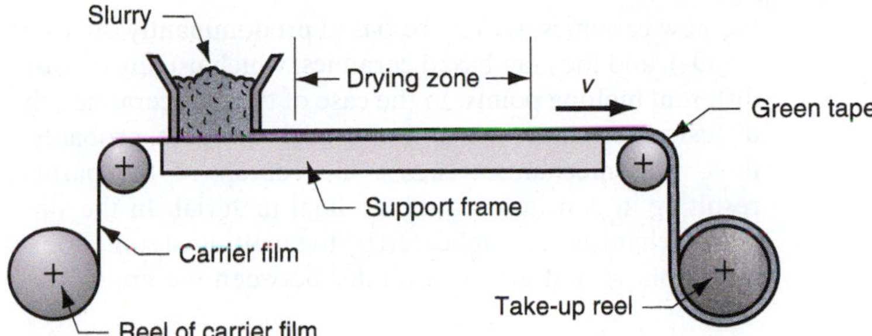

FIGURE 11.8 The doctor-blade process, used to fabricate thin ceramic sheets. Symbol *v* indicates motion (*v* = velocity). (Credit: *Fundamentals of Modern Manufacturing*, 4th Edition by Mikell P. Groover, 2010. Reprinted with permission of John Wiley & Sons, Inc.)

from all directions, thus avoiding the problem of nonuniform density in the final product that is often observed in the traditional uniaxial pressing method.

Doctor-Blade Process This process is used for making thin sheets of ceramic. One common application of the sheets is in the electronics industry as a substrate material for integrated circuits. The process is diagrammed in Figure 11.8. A ceramic slurry is introduced onto a moving carrier film such as cellophane. Thickness of the ceramic on the carrier is determined by a wiper, called a ***doctor-blade***. As the slurry moves down the line, it is dried into a flexible green ceramic tape. At the end of the line, a take-up spool reels in the tape for later processing. In its green condition, the tape can be cut or otherwise shaped before firing.

Powder Injection Molding (PIM) This is the same as the PM process (Section 10.3.2), except that the powders are ceramic rather than metallic. Ceramic particles are mixed with a thermoplastic polymer that acts as a carrier and provides the proper flow characteristics at molding temperatures. The mix is then heated and injected into a mold cavity. Upon cooling, which hardens the polymer, the mold is opened and the part is removed. Because the temperatures needed to plasticize the carrier are much lower than those required for sintering the ceramic, the piece is green after molding. Before sintering, the plastic binder must be removed. This is called ***debinding***, which is usually accomplished by a combination of thermal and solvent treatments.

Applications of ceramic PIM are currently inhibited by difficulties in debinding and sintering. Burning off the polymer is relatively slow, and its removal significantly weakens the green strength of the molded part. Warping and cracking often occur during sintering. Further, ceramic products made by powder injection molding are especially vulnerable to microstructural flaws that limit their strength.

11.2.3 SINTERING

Since the plasticity needed to shape the new ceramics is not normally based on a water mixture, the drying step so commonly required to remove water from the traditional green ceramics can be omitted in the processing of most new ceramic products. The sintering step, however, is still very much required to obtain maximum possible strength and hardness. The functions of sintering are the same as before: (1) to bond individual grains into a solid mass, (2) to increase density, and (3) to reduce or eliminate porosity.

Temperatures around 80% to 90% of the melting temperature of the material are commonly used in sintering ceramics. Sintering mechanisms differ somewhat between the new ceramics, which are based predominantly on a single chemical compound (e.g., Al_2O_3), and the clay-based ceramics, which usually consist of several compounds having different melting points. In the case of the new ceramics, the sintering mechanism is mass diffusion across the contacting particle surfaces, probably accompanied by some plastic flow. This mechanism causes the centers of the particles to move closer together, resulting in densification of the final material. In the sintering of traditional ceramics, this mechanism is complicated by the melting of some constituents and the formation of a glassy phase that acts as a binder between the grains.

11.2.4 FINISHING

Parts made of new ceramics sometimes require finishing. In general, these operations have one or more of the following purposes: (1) to increase dimensional accuracy, (2) to improve surface finish, and (3) to make minor changes in part geometry. Finishing operations usually involve grinding and other abrasive processes (Chapter 18). Diamond abrasives must be used to cut the hardened ceramic materials.

11.3 PROCESSING OF CERMETS

Many metal-matrix composites (MMCs) and ceramic-matrix composites (CMCs) are processed by particulate processing methods. The most prominent examples are cemented carbides and other cermets.

11.3.1 CEMENTED CARBIDES

The cemented carbides are a family of composite materials consisting of carbide ceramic particles imbedded in a metallic binder. They are classified as metal-matrix composites because the metallic binder is the matrix that holds the bulk material together; however, the carbide particles constitute the largest proportion of the composite material, normally ranging between 80% and 96% by volume. Cemented carbides are technically classified as cermets, although they are often distinguished from the other materials in this class.

The most important cemented carbide is tungsten carbide in a cobalt binder (WC–Co). Generally included within this category are certain mixtures of WC, TiC, and TaC in a Co matrix, in which tungsten carbide is the major component. Other cemented carbides include titanium carbide in nickel (TiC–Ni) and chromium carbide in nickel (Cr_3C_2–Ni). These composites were discussed in Section 2.4.2, and the carbide ingredients were described in Section 2.2.2. In our present discussion, we are concerned with the particulate processing of cemented carbide.

To provide a strong and pore-free part, the carbide powders must be sintered with a metal binder. Cobalt works best with WC, while nickel is better with TiC and Cr_3C_2. The usual proportion of binder metal is from around 4% up to 20%. Powders of carbide and binder metal are thoroughly mixed wet in a ball mill (or other suitable mixing machine) to form a homogeneous sludge. Milling also serves to refine particle size. The sludge is then dried in a vacuum or controlled atmosphere to prevent oxidation in preparation for compaction.

Compaction Various methods are used to shape the powder mix into a green compact of the desired geometry. The most common process is cold pressing, described earlier and

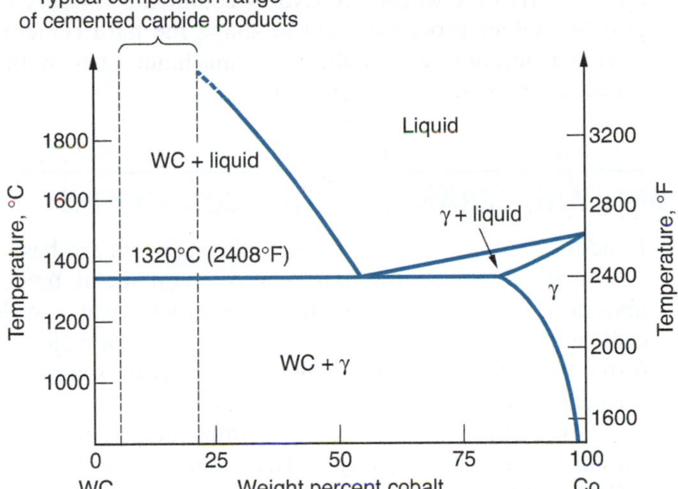

FIGURE 11.9 WC–Co phase diagram. Source: [7]. (Credit: *Fundamentals of Modern Manufacturing*, 4th Edition by Mikell P. Groover, 2010. Reprinted with permission of John Wiley & Sons, Inc.)

used for high production of cemented carbide parts such as cutting tool inserts. The dies used in cold pressing must be made oversized to account for shrinkage during sintering. Linear shrinkage can be 20% or more. For high production, the dies themselves are made with WC–Co liners to reduce wear, due to the abrasive nature of carbide particles. For smaller quantities, large flat sections are sometimes pressed and then cut into smaller pieces of the specified size.

Other compaction methods used for cemented carbide products include ***isostatic pressing*** and ***hot pressing*** for large pieces, such as draw dies and ball mill balls; and ***extrusion***, for long sections of circular, rectangular, or other cross section. Each of these processes has been described previously, either in this or the preceding chapter.

Sintering Although it is possible to sinter WC and TiC without a binder metal, the resulting material is somewhat less than 100% of true density. Use of a binder yields a structure that is virtually free of porosity.

Sintering of WC–Co involves liquid phase sintering (Section 10.3.5). The process can be explained with reference to the binary phase diagram for these constituents in Figure 11.9. The typical composition range for commercial cemented carbide products is identified in the diagram. The usual sintering temperatures for WC–Co are in the range 1370°C to 1425°C (2500°F to 2600°F), which is below cobalt's melting point of 1495°C (2716°F). Thus, the pure binder metal does not melt at the sintering temperature. However, as the phase diagram shows, WC dissolves in Co in the solid state. During the heat treatment, WC is gradually dissolved into the gamma phase, and its melting point is reduced so that melting finally occurs. As the liquid phase forms, it flows and wets the WC particles, further dissolving the solid. The presence of the molten metal also serves to remove gases from the internal regions of the compact. These mechanisms combine to effect a rearrangement of the remaining WC particles into a closer packing, which results in significant densification and shrinkage of the WC–Co mass. Later, during cooling in the sintering cycle, the dissolved carbide is precipitated and deposited onto the existing crystals to form a coherent WC skeleton, throughout which is imbedded the Co binder.

Secondary Operations Subsequent processing is usually required after sintering to achieve adequate dimensional control of cemented carbide parts. Grinding with a diamond abrasive wheel is the most common secondary operation performed for this purpose. Other processes used to shape the hard cemented carbides include electric discharge machining and ultrasonic machining, two nontraditional material removal processes discussed in Chapter 19.

11.3.2 OTHER CERMETS AND CERAMIC-MATRIX COMPOSITES

In addition to cemented carbides, other cermets are based on oxide ceramics such as Al_2O_3 and MgO. Chromium is a common metal binder used in these composite materials. The ceramic-to-metal proportions cover a wider range than those of the cemented carbides; in some cases, the metal is the major ingredient. These cermets are formed into useful products by the same basic shaping methods used for cemented carbides.

The current technology of ceramic-matrix composites (Section 2.4.2) includes ceramic materials (e.g., Al_2O_3, BN, Si_3N_4, and glass) reinforced by fibers of carbon, SiC, or Al_2O_3. If the fibers are whiskers (fibers consisting of single crystals), these CMCs can be processed by particulate methods used for new ceramics (Section 11.2).

11.4 PRODUCT DESIGN CONSIDERATIONS

Ceramic materials have special properties that make them attractive to designers if the application is right. The following design recommendations, compiled from Bralla [2] and other sources, apply to both new and traditional ceramic materials, although designers are more likely to find opportunities for new ceramics in engineered products. In general, the same guidelines apply to cemented carbides.

- Ceramic materials are several times stronger in compression than in tension; accordingly, ceramic components should be designed to be subjected to compressive stresses, not tensile stresses.
- Ceramics are brittle and possess almost no ductility. Ceramic parts should not be used in applications that involve impact loading or high stresses that might cause fracture.
- Although many of the ceramic shaping processes allow complex geometries to be formed, it is desirable to keep shapes simple for both economic and technical reasons. Deep holes, channels, and undercuts should be avoided, as should large cantilevered projections.
- Outside edges and corners should have radii or chamfers; likewise, inside corners should have radii. This guideline is, of course, violated in cutting tool applications, in which the cutting edge must be sharp in order to function. The cutting edge is often fabricated with a very small radius or chamfer to protect it from microscopic chipping, which could lead to failure.
- Part shrinkage in drying and firing (for traditional ceramics) and sintering (for new ceramics) may be significant and must be taken into account by the designer in dimensioning and tolerancing. This is mostly a problem for manufacturing engineers, who must determine appropriate size allowances so that the final dimensions will be within the tolerances specified.
- Screw threads in ceramic parts should be avoided. They are difficult to fabricate and do not have adequate strength in service after fabrication.

REFERENCES

[1] Bhowmick, A. K. Bradley Pulverizer Company, Allentown, Pennsylvania, personal communication, February 1992.

[2] Bralla, J. G. (Editor-in-Chief). ***Design for Manufacturability Handbook***, 2nd ed. McGraw-Hill Book Company, New York, 1999.

[3] Hlavac, J. ***The Technology of Glass and Ceramics***. Elsevier Scientific Publishing Company, New York, 1983.

[4] Kingery, W. D., Bowen, H. K., and Uhlmann, D. R. ***Introduction to Ceramics***, 2nd ed. John Wiley & Sons, Inc., New York, 1995.

[5] Rahaman, M. N. ***Ceramic Processing***. CRC Taylor & Francis, Boca Raton, Florida, 2007.

[6] Richerson, D. W. ***Modern Ceramic Engineering: Properties, Processing, and Use in Design***, 3rd ed. CRC Taylor & Francis, Boca Raton, Florida, 2006.

[7] Schwarzkopf, P., and Kieffer, R. ***Cemented Carbides***. The Macmillan Company, New York, 1960.

[8] Singer, F., and Singer, S. S. ***Industrial Ceramics***. Chemical Publishing Company, New York, 1963.

[9] Somiya, S. (ed.). ***Advanced Technical Ceramics***. Academic Press, Inc., San Diego, California, 1989.

REVIEW QUESTIONS

11.1. What is the difference between the traditional ceramics and the new ceramics, as far as raw materials are concerned?

11.2. List the basic steps in the traditional ceramics processing sequence.

11.3. What is the technical difference between crushing and grinding in the preparation of traditional ceramic raw materials?

11.4. Describe the slip casting process in traditional ceramics processing.

11.5. List and briefly describe some of the plastic forming methods used to shape traditional ceramic products.

11.6. What is the process of jiggering?

11.7. What is the difference between dry pressing and semi-dry pressing of traditional ceramic parts?

11.8. What happens to a ceramic material when it is sintered?

11.9. What is the name given to the furnace used to fire ceramic ware?

11.10. Why is the drying step, so important in the processing of traditional ceramics, usually not required in processing of new ceramics?

11.11. What is the freeze drying process used to make certain new ceramic powders?

11.12. Describe the doctor-blade process.

11.13. What are some design recommendations for ceramic parts?

Part IV Metal Forming and Sheet Metalworking

12 FUNDAMENTALS OF METAL FORMING

Chapter Contents

12.1 Overview of Metal Forming

12.2 Material Behavior in Metal Forming

12.3 Temperature in Metal Forming

12.4 Friction and Lubrication in Metal Forming

Metal forming includes a large group of manufacturing processes in which plastic deformation is used to change the shape of metal workpieces. Deformation results from the use of a tool, usually called a ***die*** in metal forming, which applies stresses that exceed the yield strength of the metal. The metal therefore deforms to take a shape determined by the geometry of the die. Metal forming dominates the class of shaping operations identified in Chapter 1 as the ***deformation processes*** (Figure 1.3).

Stresses applied to plastically deform the metal are usually compressive. However, some forming processes stretch the metal, while others bend the metal, and still others apply shear stresses to the metal. To be successfully formed, a metal must possess certain properties. Desirable properties include low yield strength and high ductility. These properties are affected by temperature. Ductility is increased and yield strength is reduced when work temperature is raised. The effect of temperature gives rise to distinctions between cold working, warm working, and hot working. Friction is an additional factor that affects performance in metal forming. We examine all of these issues in this chapter, but first let us provide an overview of the metal forming processes.

12.1 OVERVIEW OF METAL FORMING

Metal forming processes can be classified into two basic categories: bulk deformation processes and sheet

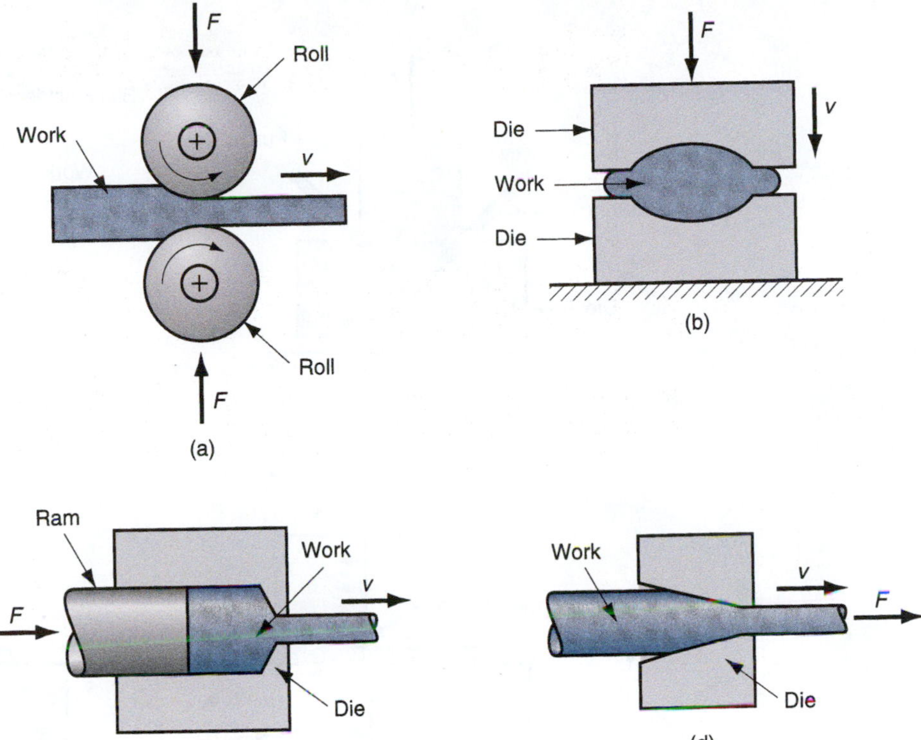

FIGURE 12.1 Basic bulk deformation processes: (a) rolling, (b) forging, (c) extrusion, and (d) drawing. Relative motion in the operations is indicated by *v*; forces are indicated by *F*. (Credit: *Fundamentals of Modern Manufacturing*, 4th Edition by Mikell P. Groover, 2010. Reprinted with permission of John Wiley & Sons, Inc.)

metalworking processes. These two categories are covered in detail in Chapters 13 and 14, respectively. Each category includes several major classes of shaping operations, as described briefly in this section.

Bulk Deformation Processes Bulk deformation processes are generally characterized by significant deformations and massive shape changes, and the surface area–to-volume of the work is relatively small. The term ***bulk*** describes the workparts that have this low area-to-volume ratio. Starting work shapes for these processes include cylindrical billets and rectangular bars. Figure 12.1 illustrates the following basic operations in bulk deformation:

- ***Rolling***. This is a compressive deformation process in which the thickness of a slab or plate is reduced by two opposing cylindrical tools called rolls. The rolls rotate so as to draw the work into the gap between them and squeeze it.
- ***Forging***. In forging, a workpiece is compressed between two opposing dies, so that the die shapes are imparted to the work. Forging is traditionally a hot working process, but many types of forging are performed cold.
- ***Extrusion***. This is a compression process in which the work metal is forced to flow through a die opening, thereby taking the shape of the opening as its own cross section.
- ***Drawing***. In this forming process, the diameter of a round wire or bar is reduced by pulling it through a die opening.

Sheet Metalworking Sheet metalworking processes are forming and cutting operations performed on metal sheets, strips, and coils. The surface area–to-volume ratio of the starting metal is high; thus, this ratio is a useful means to distinguish bulk deformation from sheet metal processes. ***Pressworking*** is the term often applied to sheet metal

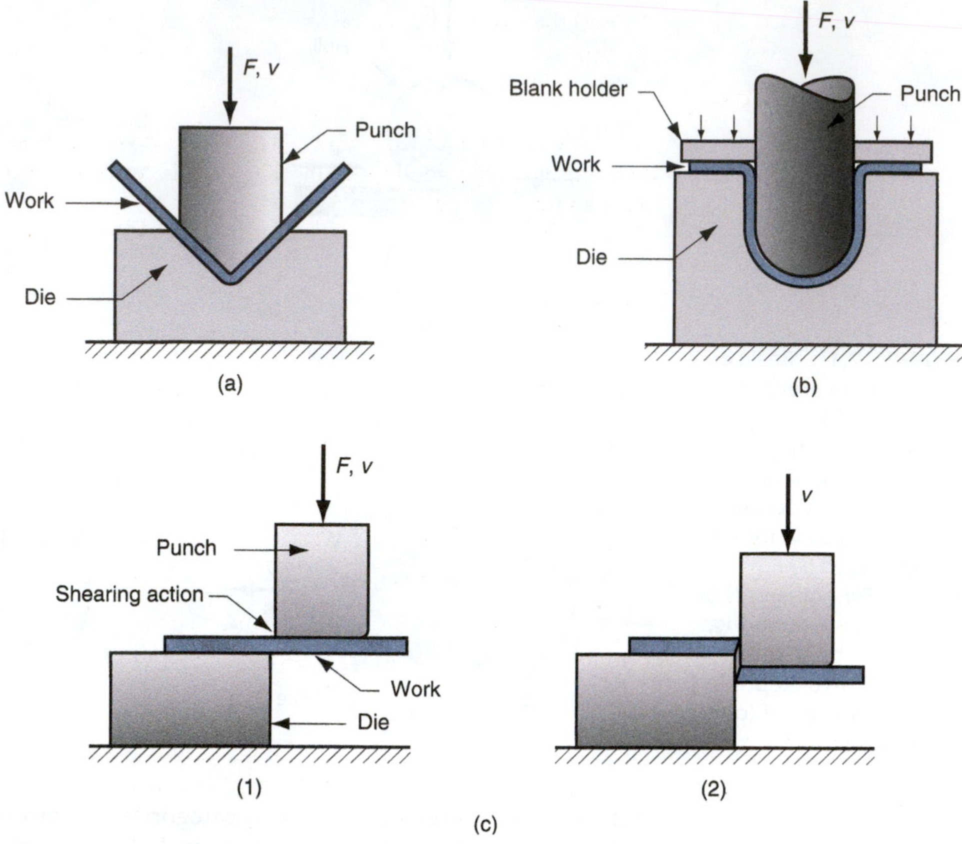

FIGURE 12.2 Basic sheet metalworking operations: (a) bending, (b) drawing, and (c) shearing: (1) as punch first contacts sheet, and (2) after cutting. Force and relative motion in these operations are indicated by F and v. (Credit: *Fundamentals of Modern Manufacturing*, 4th Edition by Mikell P. Groover, 2010. Reprinted with permission of John Wiley & Sons, Inc.)

operations because the machines used to perform these operations are presses (presses of various types are also used in other manufacturing processes). A part produced in a sheet metal operation is often called a ***stamping***.

Sheet metal operations are always performed as cold working processes and are usually accomplished using a set of tools called a ***punch*** and ***die***. The punch is the positive portion and the die is the negative portion of the tool set. The basic sheet metal operations are sketched in Figure 12.2 and are defined as follows:

- ***Bending***. Bending involves straining of a metal sheet or plate to take an angle along a (usually) straight axis.
- ***Drawing***. In sheet metalworking, drawing refers to the forming of a flat metal sheet into a hollow or concave shape, such as a cup, by stretching the metal. A blankholder is used to hold down the blank while the punch pushes into the sheet metal, as shown in Figure 12.2(b). To distinguish this operation from bar and wire drawing, the terms ***cup drawing*** or ***deep drawing*** are often used.
- ***Shearing***. This process seems somewhat out of place in a list of deformation processes, because it involves cutting rather than forming. A shearing operation

cuts the work using a punch and die, as in Figure 12.2(c). Although it is not a forming process, it is included here because it is a necessary and very common operation in sheet metalworking.

The sheet metalworking classification also includes several shaping processes that do not use punch and die tooling. These include stretch forming, roll bending, spinning, and bending of tube stock.

12.2 MATERIAL BEHAVIOR IN METAL FORMING

Considerable insight about the behavior of metals during forming can be obtained from the stress–strain curve. The typical stress–strain curve for most metals is divided into an elastic region and a plastic region (Section 3.1.1). In metal forming, the plastic region is of primary interest because the material is plastically and permanently deformed in these processes.

The typical stress–strain relationship for a metal exhibits elasticity below the yield point and strain hardening above it. Figures 3.4 and 3.5 indicate this behavior in linear and logarithmic axes. In the plastic region, the metal's behavior is expressed by the flow curve:

$$\sigma = K\epsilon^n$$

where K = the strength coefficient, MPa (lb/in^2); and n is the strain hardening exponent. The stress σ and strain ϵ in the flow curve are true stress and true strain. The flow curve is generally valid as a relationship that defines a metal's plastic behavior in cold working. Typical values of K and n for different metals at room temperature are listed in Table 3.4.

Flow Stress The flow curve describes the stress–strain relationship in the region in which metal forming takes place. It indicates the flow stress of the metal—the strength property that determines forces and power required to accomplish a particular forming operation. For most metals at room temperature, the stress–strain plot of Figure 3.5 indicates that as the metal is deformed, its strength increases due to strain hardening. The stress required to continue deformation must be increased to match this increase in strength. ***Flow stress*** is defined as the instantaneous value of stress required to continue deforming the material—to keep the metal "flowing." It is the yield strength of the metal as a function of strain, which can be expressed:

$$Y_f = K\epsilon^n \tag{12.1}$$

where Y_f = flow stress, MPa (lb/in^2).

In the individual forming operations discussed in the following two chapters, the instantaneous flow stress can be used to analyze the process as it is occurring. For example, in certain forging operations, the instantaneous force during compression can be determined from the flow stress value. Maximum force can be calculated based on the flow stress that results from the final strain at the end of the forging stroke.

In other cases, the analysis is based on the average stresses and strains that occur during deformation rather than instantaneous values. Extrusion represents this case, Figure 12.1(c). As the billet is reduced in cross section to pass through the extrusion die opening, the metal gradually strain hardens to reach a maximum value. Rather than determine a sequence of instantaneous stress–strain values during the reduction, which

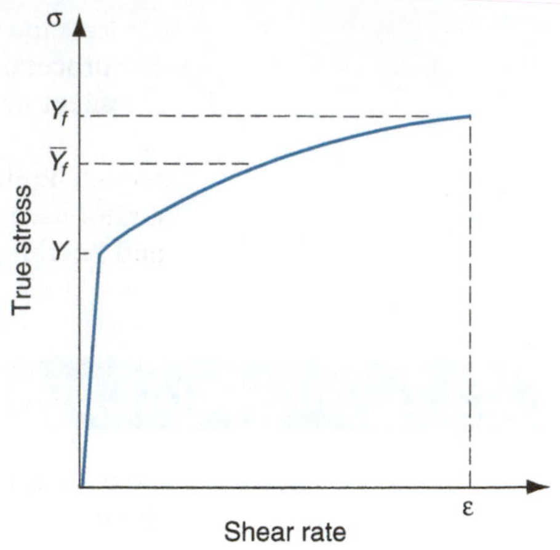

FIGURE 12.3 Stress–strain curve indicating location of average flow stress $\overline{Y}_f$ in relation to yield strength Y and final flow stress Y_f. (Credit: *Fundamentals of Modern Manufacturing*, 4th Edition by Mikell P. Groover, 2010. Reprinted with permission of John Wiley & Sons, Inc.)

would be not only difficult but also of limited interest, it is more useful to analyze the process based on the average flow stress during deformation.

Average Flow Stress The average flow stress (also called the ***mean flow stress***) is the average value of stress over the stress–strain curve from the beginning of strain to the final (maximum) value that occurs during deformation. The value is illustrated in the stress–strain plot of Figure 12.3. The average flow stress is determined by integrating the flow curve equation, Eq. (12.1), between zero and the final strain value defining the range of interest. This yields the equation:

$$\overline{Y}_f = \frac{K\epsilon^n}{1+n} \tag{12.2}$$

where $\overline{Y}_f$ = average flow stress, MPa (lb/in^2); and ϵ = maximum strain value during the deformation process.

We make extensive use of the average flow stress in our study of the bulk deformation processes in the following chapter. Given values of K and n for the work material, a method of computing final strain will be developed for each process. Based on this strain, Eq. (12.2) can be used to determine the average flow stress to which the metal is subjected during the operation.

12.3 TEMPERATURE IN METAL FORMING

The flow curve is a valid representation of stress–strain behavior of a metal during plastic deformation, particularly for cold working operations. For any metal, the values of K and n depend on temperature. Strength and strain hardening are both reduced at higher temperatures. These property changes are important because they result in lower forces and power during forming. In addition, ductility is increased at higher temperatures, which allows greater plastic deformation of the work metal. Three temperature ranges used in metal forming can be distinguished: cold, warm, and hot working.

Cold Working Cold working (also known as ***cold forming***) is metal forming performed at room temperature or slightly above. Significant advantages of cold forming compared to hot working are (1) greater accuracy, meaning closer tolerances can be achieved; (2) better surface finish; (3) higher strength and hardness of the part due to strain hardening; (4) grain flow during deformation provides the opportunity for desirable directional properties to be obtained in the resulting product; and (5) no heating of the work is required, which saves on furnace and fuel costs and permits higher production rates. Owing to this combination of advantages, many cold forming processes have become important mass-production operations. They provide close tolerances and good surfaces, minimizing the amount of machining required so that these operations can be classified as net shape or near net shape processes (Section 1.2.1).

There are certain disadvantages or limitations associated with cold forming operations: (1) higher forces and power are required to perform the operation; (2) care must be taken to ensure that the surfaces of the starting workpiece are free of scale and dirt; and (3) ductility and strain hardening of the work metal limit the amount of forming that can be done to the part. In some operations, the metal must be annealed (Section 20.1) to allow further deformation to be accomplished. In other cases, the metal is simply not ductile enough to be cold worked.

To overcome the strain hardening problem and reduce force and power requirements, many forming operations are performed at elevated temperatures. There are two elevated temperature ranges involved, giving rise to the terms warm working and hot working.

Warm Working Because plastic deformation properties are normally enhanced by increasing workpiece temperature, forming operations are sometimes performed at temperatures somewhat above room temperature but below the recrystallization temperature. The term ***warm working*** is applied to this second temperature range. The dividing line between cold working and warm working is often expressed in terms of the melting point for the metal. The dividing line is usually taken to be $0.3T_m$, where T_m is the melting point (absolute temperature) for the particular metal.

The lower strength and strain hardening at the intermediate temperatures, as well as higher ductility, provide warm working with the following advantages over cold working: (1) lower forces and power, (2) more intricate work geometries possible, and (3) need for annealing may be reduced or eliminated.

Hot Working Hot working (also called ***hot forming***) involves deformation at temperatures above the recrystallization temperature (Section 3.3). The recrystallization temperature for a given metal is about one-half of its melting point on the absolute scale. In practice, hot working is usually carried out at temperatures somewhat above $0.5T_m$. The work metal continues to soften as temperature is increased beyond $0.5T_m$, thus enhancing the advantage of hot working above this level. However, the deformation process itself generates heat, which increases work temperatures in localized regions of the part. This can cause melting in these regions, which is highly undesirable. Also, scale formation on the work surface is accelerated at higher temperatures. Accordingly, hot working temperatures are usually maintained within the range $0.5T_m$ to $0.75T_m$.

The most significant advantage of hot working is the capability to produce substantial plastic deformation of the metal—far more than is possible with cold working or warm working. The principal reason for this is that the flow curve of the hot-worked metal has a strength coefficient that is substantially less than at room temperature, the strain hardening exponent is zero (at least theoretically), and the ductility of the metal is significantly increased. All of this results in the following advantages relative to cold working: (1) the shape of the workpart can be significantly altered, (2) lower forces and power are required to deform the metal, (3) metals that usually fracture in cold working can be hot formed, (4) strength properties are generally isotropic because of the absence of the oriented grain structure typically created in cold working, and (5) no strengthening of the part occurs from work hardening. This last advantage may seem inconsistent, because strengthening of the metal is often considered an advantage for cold working. However, there are applications in which it is undesirable for the metal to be work hardened because it reduces ductility, for example, if the part is to be subsequently processed by cold forming. Disadvantages of hot working include (1) lower dimensional accuracy, (2) higher total energy required (due to the thermal energy to heat the workpiece), (3) work surface oxidation (scale), (4) poorer surface finish, and (5) shorter tool life.

Recrystallization of the metal in hot working involves atomic diffusion, which is a time-dependent process. Metal forming operations are often performed at high speeds that do not allow sufficient time for complete recrystallization of the grain structure during the deformation cycle itself. However, because of the high temperatures, recrystallization eventually does occur. It may occur immediately following the forming process or later, as the workpiece cools. Even though recrystallization may occur after the actual deformation, its eventual occurrence, and the substantial softening of the metal at high temperatures, are the features that distinguish hot working from warm working or cold working.

Isothermal Forming Certain metals, such as highly alloyed steels, many titanium alloys, and high-temperature nickel alloys, possess good hot hardness, a property that makes them useful for high-temperature service. However, this very property that makes them attractive in these applications also makes them difficult to form with conventional methods. The problem is that when these metals are heated to their hot working temperatures and then come in contact with the relatively cold forming tools, heat is quickly transferred away from the part surfaces, thus raising the strength in these regions. The variations in temperature and strength in different regions of the workpiece cause irregular flow patterns in the metal during deformation, leading to high residual stresses and possible surface cracking.

Isothermal forming refers to forming operations that are carried out in such a way as to eliminate surface cooling and the resulting thermal gradients in the workpart. It is accomplished by preheating the tools that come in contact with the part to the same temperature as the work metal. This weakens the tools and reduces tool life, but it avoids the problems described above when these difficult metals are formed by conventional methods. In some cases, isothermal forming represents the only way in which these work materials can be formed. The procedure is most closely associated with forging, and we discuss isothermal forging in the following chapter.

12.4 FRICTION AND LUBRICATION IN METAL FORMING

Friction in metal forming arises because of the close contact between the tool and work surfaces and the high pressures that drive the surfaces together in these operations. In most metal forming processes, friction is undesirable for the following reasons: (1) metal flow in the work is retarded, causing residual stresses and sometimes defects in the product; (2) forces and power to perform the operation are increased; and (3) tool wear can lead to loss of dimensional accuracy, resulting in defective parts and requiring replacement of the tooling. Because tools in metal forming are generally expensive, tool wear is a major concern. Friction and tool wear are more severe in hot working because of the much harsher environment.

Friction in metal forming is different from that encountered in most mechanical systems, such as gear trains, shafts and bearings, and other components involving relative motion between surfaces. These other cases are generally characterized by low contact pressures, low to moderate temperatures, and ample lubrication to minimize

TABLE 12.1 Typical values of temperature (relative to melting point T_m) and coefficient of friction in cold, warm, and hot working

Category	Temperature Range	Coefficient of Friction
Cold working	$\leq 0.3T_m$	0.1
Warm working	$0.3T_m$–$0.5T_m$	0.2
Hot working	$0.5T_m$–$0.75T_m$	0.4–0.5

Compiled from various sources.

metal-to-metal contact. By contrast, the metal forming environment features high pressures between a hardened tool and a soft workpart, plastic deformation of the softer material, and high temperatures (at least in hot working). These conditions can result in relatively high coefficients of friction in metalworking, even in the presence of lubricants. Typical values of coefficient of friction for the three categories of metal forming are listed in Table 12.1.

If the coefficient of friction becomes large enough, a condition known as sticking occurs. ***Sticking*** in metalworking (also called ***sticking friction***) is the tendency for the two surfaces in relative motion to adhere to each other rather than slide. It means that the friction stress between the surfaces exceeds the shear flow stress of the work metal, thus causing the metal to deform by a shear process beneath the surface rather than slip at the surface. Sticking occurs in metal forming operations and is a prominent problem in rolling; we discuss it in that context in the following chapter.

Metalworking lubricants are applied to the tool—work interface in many forming operations to reduce the harmful effects of friction. Benefits include reduced sticking, forces, power, and tool wear; and better surface finish on the product. Lubricants also serve other functions, such as removing heat from the tooling. Considerations in choosing an appropriate metalworking lubricant include (1) type of forming process (rolling, forging, sheet metal drawing, and so on), (2) whether used in hot working or cold working, (3) work material, (4) chemical reactivity with the tool and work metals (it is generally desirable for the lubricant to adhere to the surfaces to be most effective in reducing friction), (5) ease of application, (6) toxicity, (7) flammability, and (8) cost.

Lubricants used for cold working operations include [4], [7] mineral oils, fats and fatty oils; water-based emulsions; soaps; and other coatings. Hot working is sometimes performed dry for certain operations and materials (e.g., hot rolling of steel and extrusion of aluminum). When lubricants are used in hot working, they include mineral oils, graphite, and glass. Molten glass becomes an effective lubricant for hot extrusion of steel alloys. Graphite contained in water or mineral oil is a common lubricant for hot forging of various work materials. More detailed treatments of lubricants in metalworking are found in references [7] and [9].

REFERENCES

[1] Altan, T., Oh, S.-I., and Gegel, H. L. ***Metal Forming: Fundamentals and Applications***. ASM International, Materials Park, Ohio, 1983.

[2] Cook, N. H. ***Manufacturing Analysis***. Addison-Wesley Publishing Company, Inc., Reading, Massachusetts, 1966.

[3] Hosford, W. F., and Caddell, R. M. ***Metal Forming: Mechanics and Metallurgy***, 3rd ed. Cambridge University Press, Cambridge, United Kingdom, 2007.

[4] Lange, K. ***Handbook of Metal Forming***. Society of Manufacturing Engineers, Dearborn, Michigan, 2006.

[5] Lenard, J. G. ***Metal Forming Science and Practice***. Elsevier Science, Amsterdam, The Netherlands, 2002.

[6] Mielnik, E. M. ***Metalworking Science and Engineering***. McGraw-Hill, Inc., New York, 1991.

[7] Nachtman, E. S., and Kalpakjian, S. ***Lubricants and Lubrication in Metalworking Operations***. Marcel Dekker, Inc., New York, 1985.

[8] Wagoner, R. H., and Chenot, J.-L. ***Fundamentals of Metal Forming***. John Wiley & Sons, Inc., New York, 1997.

[9] Wick, C.,et al. (eds.). ***Tool and Manufacturing Engineers Handbook***, 4th ed., Vol. **II**, ***Forming***. Society of Manufacturing Engineers, Dearborn, Michigan, 1984.

REVIEW QUESTIONS

12.1. What are the differences between bulk deformation processes and sheet metal processes?

12.2. Extrusion is a fundamental shaping process. Describe it.

12.3. Why is the term *pressworking* often used for sheet metal processes?

12.4. What is the difference between deep drawing and bar drawing?

12.5. Indicate the mathematical equation for the flow curve.

12.6. How does increasing temperature affect the parameters in the flow curve equation?

12.7. Indicate some of the advantages of cold working relative to warm and hot working.

12.8. What is isothermal forming?

12.9. Why is friction generally undesirable in metal forming operations?

12.10. What is sticking friction in metalworking?

PROBLEMS

12.1. The strength coefficient = 550 MPa and strain-hardening exponent = 0.22 for a certain metal. During a forming operation, the final true strain that the metal experiences = 0.85. Determine the flow stress at this strain and the average flow stress that the metal experienced during the operation.

12.2. A metal has a flow curve with strength coefficient = 850 MPa and strain-hardening exponent = 0.30. A tensile specimen of the metal with gage length = 100 mm is stretched to a length = 157 mm. Determine the flow stress at the new length and the average flow stress that the metal has been subjected to during the deformation.

12.3. A particular metal has a flow curve with strength coefficient = 35,000 lb/in^2 and strain-hardening exponent = 0.26. A tensile specimen of the metal with gage length = 2.0 in is stretched to a length = 3.3 in. Determine the flow stress at this new length and the average flow stress that the metal has been subjected to during deformation.

12.4. The strength coefficient and strain-hardening exponent of a certain test metal are 40,000 lb/in^2 and 0.19, respectively. A cylindrical specimen of the metal with starting diameter = 2.5 in and length = 3.0 in is compressed to a length of 1.5 in. Determine the flow stress at this compressed length and the average flow stress that the metal has experienced during deformation.

12.5. For a certain metal, the strength coefficient = 700 MPa and strain-hardening exponent = 0.27. Determine the average flow stress that the metal experiences if it is subjected to a stress that is equal to its strength coefficient K.

12.6. Determine the value of the strain-hardening exponent for a metal that will cause the average flow stress to be 3/4 of the final flow stress after deformation.

12.7. The strength coefficient = 35,000 lb/in^2 and strain-hardening exponent = 0.40 for a metal used in a forming operation in which the workpart is reduced in cross-sectional area by stretching. If the average flow stress on the part is 20,000 lb/in^2, determine the amount of reduction in cross-sectional area experienced by the part.

12.8. In a tensile test, two pairs of values of stress and strain were measured for the specimen metal after it had yielded: (1) true stress = 217 MPa and true strain = 0.35, and (2) true stress = 259 MPa and true strain = 0.68. Based on these data points, determine the strength coefficient and strain-hardening exponent.

12.9. The following stress and strain values were measured in the plastic region during a tensile test carried out on a new experimental metal: (1) true stress = 43,608 lb/in^2 and true strain = 0.27 in/in, and (2) true stress = 52,048 lb/in^2 and true strain = 0.85 in/in. Based on these data points, determine the strength coefficient and strain-hardening exponent.

13 BULK DEFORMATION PROCESSES IN METALWORKING

Chapter Contents

13.1 Rolling
- 13.1.1 Flat Rolling and Its Analysis
- 13.1.2 Shape Rolling
- 13.1.3 Rolling Mills
- 13.1.4 Other Processes Related to Rolling

13.2 Forging
- 13.2.1 Open-Die Forging
- 13.2.2 Impression-Die Forging
- 13.2.3 Flashless Forging
- 13.2.4 Forging Hammers, Presses, and Dies
- 13.2.5 Other Processes Related to Forging

13.3 Extrusion
- 13.3.1 Types of Extrusion
- 13.3.2 Analysis of Extrusion
- 13.3.3 Extrusion Dies and Presses
- 13.3.4 Other Extrusion Processes
- 13.3.5 Defects in Extruded Products

13.4 Wire and Bar Drawing
- 13.4.1 Analysis of Drawing
- 13.4.2 Drawing Practice

The deformation processes described in this chapter accomplish significant shape change in metal parts whose initial form is bulk rather than sheet. The starting forms include cylindrical bars and billets, rectangular billets and slabs, and similar elementary geometries. The bulk deformation processes refine the starting shapes, sometimes improving mechanical properties, and always adding value. Deformation processes work by stressing the metal sufficiently to cause it to plastically flow into the desired shape.

Bulk deformation processes are performed as cold, warm, and hot working operations. Cold and warm working are appropriate when the shape change is less severe, and there is a need to improve mechanical properties and achieve good finish on the part. Hot working is generally required when massive deformation of large workparts is involved.

The commercial and technological importance of bulk deformation processes derives from the following:

- When performed as hot working operations, they can achieve significant change in the shape of the workpart.
- When performed as cold working operations, they can be used not only to shape the product, but also to increase its strength through strain hardening.
- These processes produce little or no waste as a byproduct of the operation. Some bulk deformation operations are ***near net shape*** or ***net shape*** processes; they achieve final product geometry with little or no subsequent machining.

The four basic bulk deformation processes are (1) rolling, (2) forging, (3) extrusion, and (4) wire and bar drawing. The chapter also documents some of the operations that are related to these four basic processes.

13.1 ROLLING

Rolling is a deformation process in which the thickness of the work is reduced by compressive forces exerted by two opposing rolls. The rolls rotate as illustrated in Figure 13.1 to pull and simultaneously squeeze the work between them. The basic process shown in our figure is flat rolling, used to reduce the thickness of a rectangular cross section. A closely related process is shape rolling, in which a square cross section is formed into a shape such as an I-beam.

Most rolling processes are very capital intensive, requiring massive pieces of equipment, called rolling mills, to perform them. The high investment cost requires the mills to be used for production in large quantities of standard items such as sheets and plates. Most rolling is carried out by hot working, called ***hot rolling***, owing to the large amount of deformation required. Hot-rolled metal is generally free of residual stresses, and its properties are isotropic. Disadvantages of hot rolling are that the product cannot be held to close tolerances, and the surface has a characteristic oxide scale.

Steelmaking provides the most common application of rolling mill operations. Let us follow the sequence of steps in a steel rolling mill to illustrate the variety of products made. Similar steps occur in other basic metal industries. The work starts out as a cast steel ingot that has just solidified. While it is still hot, the ingot is placed in a furnace where it remains for many hours until it has reached a uniform temperature throughout, so that the metal will flow consistently during rolling. For steel, the desired temperature for rolling is around 1200°C (2200°F). The heating operation is called ***soaking***, and the furnaces in which it is carried out are called ***soaking pits***.

From soaking, the ingot is moved to the rolling mill, where it is rolled into one of three intermediate shapes called blooms, billets, or slabs. A ***bloom*** has a square cross section 150 mm × 150 mm (6 in × 6 in) or larger. A ***slab*** is rolled from an ingot or a bloom and has a rectangular cross section of width 250 mm (10 in) or more and thickness 40 mm (1.5 in) or more. A ***billet*** is rolled from a bloom and is square with dimensions 40 mm (1.5 in) on a side or larger. These intermediate shapes are subsequently rolled into final product shapes.

Blooms are rolled into structural shapes and rails for railroad tracks. Billets are rolled into bars and rods. These shapes are the raw materials for machining, wire drawing, forging, and other metalworking processes. Slabs are rolled into plates, sheets, and strips. Hot-rolled plates are used in shipbuilding, bridges, boilers, welded structures for various heavy machines, tubes and pipes, and many other products. Figure 13.2 shows some of these rolled steel products. Further flattening of hot-rolled plates and sheets is often accomplished by ***cold rolling***, in order to prepare them for sheet-metal operations (Chapter 14). Cold rolling strengthens the metal and permits a tighter tolerance on thickness. In addition, the surface of the cold-rolled sheet is absent of scale and generally superior to the corresponding hot-rolled product. These characteristics make cold-rolled

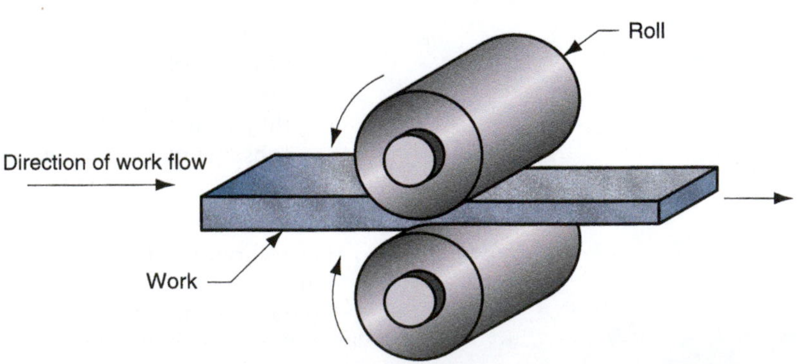

FIGURE 13.1 The rolling process (specifically, flat rolling). (Credit: *Fundamentals of Modern Manufacturing*, 4th Edition by Mikell P. Groover, 2010. Reprinted with permission of John Wiley & Sons, Inc.)

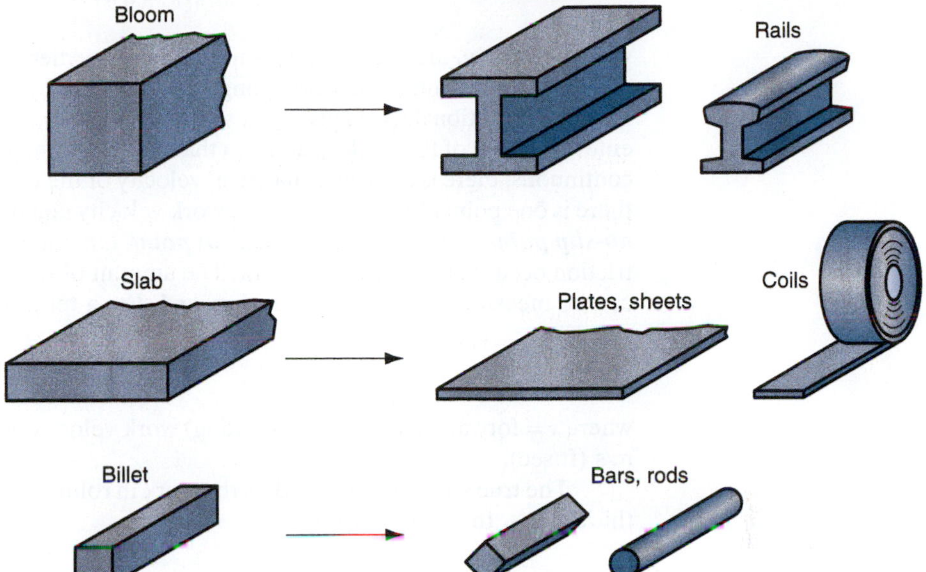

FIGURE 13.2 Some of the steel products made in a rolling mill. (Credit: *Fundamentals of Modern Manufacturing*, 4th Edition by Mikell P. Groover, 2010. Reprinted with permission of John Wiley & Sons, Inc.)

sheets, strips, and coils ideal for stampings, exterior panels, and other parts of products ranging from automobiles to appliances and office furniture.

13.1.1 FLAT ROLLING AND ITS ANALYSIS

Flat rolling is illustrated in Figures 13.1 and 13.3. It involves the rolling of slabs, strips, sheets, and plates—workparts of rectangular cross section in which the width is greater than the thickness. In flat rolling, the work is squeezed between two rolls so that its thickness is reduced by an amount called the ***draft***:

$$d = t_o - t_f \tag{13.1}$$

where d = draft, mm (in); t_o = starting thickness, mm (in); and t_f = final thickness, mm (in). Draft is sometimes expressed as a fraction of the starting stock thickness, called the ***reduction***:

$$r = \frac{d}{t_o} \tag{13.2}$$

where r = reduction. When a series of rolling operations are used, reduction is taken as the sum of the drafts divided by the original thickness.

In addition to thickness reduction, rolling usually increases work width. This is called ***spreading***, and it tends to be most pronounced with low width-to-thickness ratios and low coefficients of friction. Conservation of matter is preserved, so the volume of metal exiting the rolls equals the volume entering:

$$t_o w_o L_o = t_f w_f L_f \tag{13.3}$$

where w_o and w_f are the before and after work widths, mm (in); and L_o and L_f are the before and after work lengths, mm (in). Similarly, before and after volume rates of material flow must be the same, so the before and after velocities can be related:

$$t_o w_o v_o = t_f w_f v_f \tag{13.4}$$

where v_o and v_f are the entering and exiting velocities of the work.

The rolls contact the work along an arc defined by the angle θ. Each roll has radius R, and its rotational speed gives it a surface velocity v_r. This velocity is greater than the entering speed of the work v_o and less than its exiting speed v_f. Because the metal flow is continuous, there is a gradual change in velocity of the work between the rolls. However, there is one point along the arc where work velocity equals roll velocity. This is called the ***no-slip point***, also known as the ***neutral point***. On either side of this point, slipping and friction occur between roll and work. The amount of slip between the rolls and the work can be measured by means of the ***forward slip***, a term used in rolling that is defined:

$$s = \frac{v_f - v_r}{v_r} \tag{13.5}$$

where s = forward slip; v_f = final (exiting) work velocity, m/s (ft/sec); and v_r = roll speed, m/s (ft/sec).

The true strain experienced by the work in rolling is based on before and after stock thicknesses. In equation form,

$$\epsilon = \ln \frac{t_o}{t_f} \tag{13.6}$$

The true strain can be used to determine the average flow stress $\overline{Y}_f$ applied to the work material in flat rolling. Recall from the previous chapter, Eq. 12.2, that

$$\overline{Y}_f = \frac{K\epsilon^n}{1+n} \tag{13.7}$$

The average flow stress is used to compute estimates of force and power in rolling.

Friction in rolling occurs with a certain coefficient of friction, and the compression force of the rolls, multiplied by this coefficient of friction, results in a friction force between the rolls and the work. On the entrance side of the no-slip point, friction force is in one direction, and on the other side it is in the opposite direction. However, the two forces are not equal. The friction force on the entrance side is greater, so that the net force pulls the work through the rolls. If this were not the case, rolling would not be possible. There is a limit to the maximum possible draft that can be accomplished in flat rolling with a given coefficient of friction, defined by:

$$d_{max} = \mu^2 R \tag{13.8}$$

where d_{max} = maximum draft, mm (in); μ = coefficient of friction; and R = roll radius mm (in). The equation indicates that if friction were zero, draft would be zero, and it would be impossible to accomplish the rolling operation.

Coefficient of friction in rolling depends on lubrication, work material, and working temperature. In cold rolling, the value is around 0.1; in warm working, a typical value is around 0.2; and in hot rolling, μ is around 0.4 [16]. Hot rolling is often characterized by a condition called ***sticking***, in which the hot work surface adheres to the rolls over the contact arc. This condition often occurs in the rolling of steels and high-temperature alloys. When sticking occurs, the coefficient of friction can be as high as 0.7. The consequence of sticking is that the surface layers of the work are restricted to move at the same speed as the roll speed v_r; and below the surface, deformation is more severe to allow passage of the piece through the roll gap.

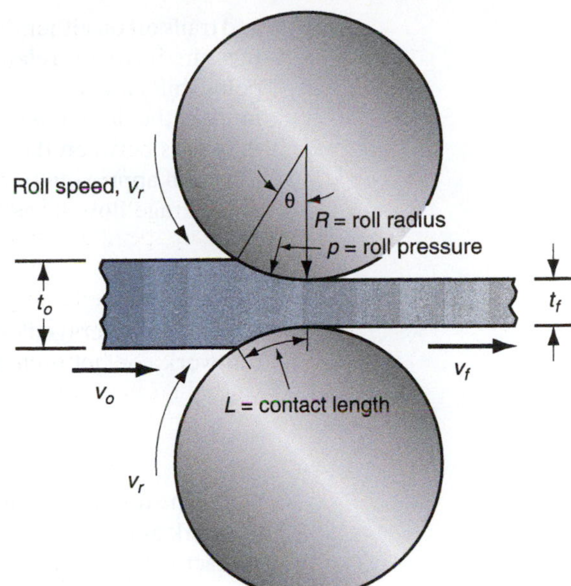

FIGURE 13.3 Side view of flat rolling, indicating before and after thicknesses, work velocities, angle of contact with rolls, and other features. (Credit: *Fundamentals of Modern Manufacturing*, 4th Edition by Mikell P. Groover, 2010. Reprinted with permission of John Wiley & Sons, Inc.)

Given a coefficient of friction sufficient to perform rolling, roll force F required to maintain separation between the two rolls can be computed by integrating the unit roll pressure (shown as p in Figure 13.3) over the roll-work contact area. This can be expressed:

$$F = w \int_0^L p dL \tag{13.9}$$

where F = rolling force, N (lb); w = the width of the work being rolled, mm (in); p = roll pressure, MPa (lb/in^2); and L = length of contact between rolls and work, mm (in). The integration requires two separate terms, one for either side of the neutral point. Variation in roll pressure along the contact length is significant. A sense of this variation can be obtained from the plot in Figure 13.4. Pressure reaches a maximum at the neutral point,

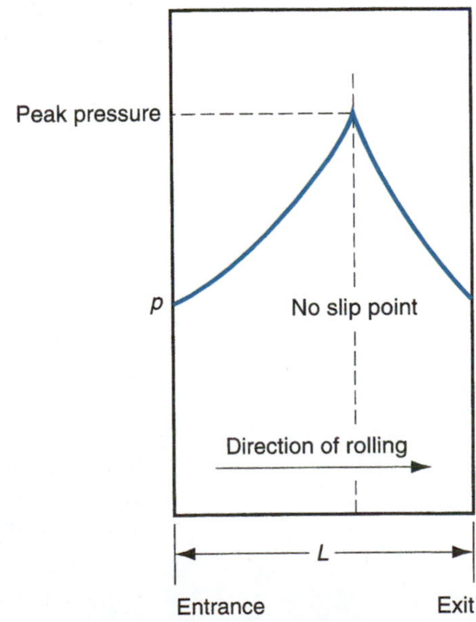

FIGURE 13.4 Typical variation in pressure along the contact length in flat rolling. The peak pressure is located at the neutral point. The area beneath the curve, representing the integration in Eq. (13.9), is the roll force F. (Credit: *Fundamentals of Modern Manufacturing*, 4th Edition by Mikell P. Groover, 2010. Reprinted with permission of John Wiley & Sons, Inc.)

and trails off on either side to the entrance and exit points. As friction increases, maximum pressure increases relative to entrance and exit values. As friction decreases, the neutral point shifts away from the entrance and toward the exit in order to maintain a net pull force in the direction of rolling. Otherwise, with low friction, the work would slip rather than pass between the rolls.

An approximation of the results obtained by Eq. (13.9) can be calculated based on the average flow stress experienced by the work material in the roll gap. That is,

$$F = \overline{Y}_f wL \tag{13.10}$$

where $\overline{Y}_f$ = average flow stress from Eq. (13.7), MPa (lb/in^2); and the product wL is the roll-work contact area, mm^2 (in^2). Contact length can be approximated by

$$L = \sqrt{R(t_o - t_f)} \tag{13.11}$$

The torque in rolling can be estimated by assuming that the roll force is centered on the work as it passes between the rolls, and that it acts with a moment arm of one-half the contact length L. Thus, torque for each roll is

$$T = 0.5\,FL \tag{13.12}$$

The power required to drive each roll is the product of torque and angular velocity. Angular velocity is $2\pi N$, where N = rotational speed of the roll. Thus, the power for each roll is $2\pi NT$. Substituting Eq. (13.12) for torque in this expression for power, and doubling the value to account for the fact that a rolling mill consists of two powered rolls, we get the following expression:

$$P = 2\pi NFL \tag{13.13}$$

where P = power, J/s or W (in-lb/min); N = rotational speed, 1/s (rev/min); F = rolling force, N (lb); and L = contact length, m (in).

Example 13.1 Flat Rolling

A 300-mm-wide strip 25 mm thick is fed through a rolling mill with two powered rolls each of radius = 250 mm. The work thickness is to be reduced to 22 mm in one pass at a roll speed of 50 rev/min. The work material has a flow curve defined by K = 275 MPa and n = 0.15, and the coefficient of friction between the rolls and the work is assumed to be 0.12. Determine if the friction is sufficient to permit the rolling operation to be accomplished. If so, calculate the roll force, torque, and horsepower.

Solution: The draft attempted in this rolling operation is

$$d = 25 - 22 = 3\text{ mm}$$

From Eq. (13.8), the maximum possible draft for the given coefficient of friction is

$$d_{max} = (0.12)^2(250) = 3.6\text{ mm}$$

Because the maximum allowable draft exceeds the attempted reduction, the rolling operation is feasible. To compute rolling force, we need the contact length L and the average flow stress $\overline{Y}_f$. The contact length is given by Eq. (13.11):

$$L = \sqrt{250\,(25 - 22)} = 27.4\text{ mm}$$

$\overline{Y}_f$ is determined from the true strain:

$$\epsilon = \ln\frac{25}{22} = 0.128$$

$$\overline{Y}_f = \frac{275(0.128)^{0.15}}{1.15} = 175.7 \text{ MPa}$$

Rolling force is determined from Eq. (13.10):

$$F = 175.7(300)(27.4) = 1{,}444{,}786 \text{ N}$$

Torque required to drive each roll is given by Eq. (13.12):

$$T = 0.5(1{,}444{,}786)(27{,}4)(10^{-3}) = 19{,}786 \text{ N-m}$$

and the power is obtained from Eq. (13.13):

$$P = 2\pi(50)(1{,}444{,}786)(27.4)(10^{-3}) = 12{,}432{,}086\text{N-m/min} = 207{,}201\text{N-m/s(W)}$$

For comparison, let us convert this to horsepower (we note that one horsepower = 745.7 W):

$$HP = \frac{207{,}201}{745.7} = 278 \text{ hp}$$

It can be seen from this example that large forces and power are required in rolling. Inspection of Eqs. (13.10) and (13.13) indicates that force and/or power to roll a strip of a given width and work material can be reduced by any of the following: (1) using hot rolling rather than cold rolling to reduce strength and strain hardening (K and n) of the work material; (2) reducing the draft in each pass; (3) using a smaller roll radius R to reduce force; and (4) using a lower rolling speed N to reduce power. ■

13.1.2 SHAPE ROLLING

In shape rolling, the work is deformed into a contoured cross section. Products made by shape rolling include construction shapes such as I-beams, L-beams, and U-channels; rails for railroad tracks; and round and square bars and rods (see Figure 13.2). The process is accomplished by passing the work through rolls that have the reverse of the desired shape.

Most of the principles that apply in flat rolling are also applicable to shape rolling. Shaping rolls are more complicated; the work, usually starting as a square shape, requires a gradual transformation through several rolls to achieve the final cross section. Designing the sequence of intermediate shapes and corresponding rolls is called ***roll-pass design***. Its goal is to achieve uniform deformation throughout the cross section in each reduction. Otherwise, certain portions of the work are reduced more than others, causing greater elongation in these sections. The consequence of nonuniform reduction can be warping and cracking of the rolled product. Both horizontal and vertical rolls are utilized to achieve consistent reduction of the work material.

13.1.3 ROLLING MILLS

Various rolling mill configurations are available to deal with the variety of applications and technical problems in the rolling process. The basic rolling mill consists of two

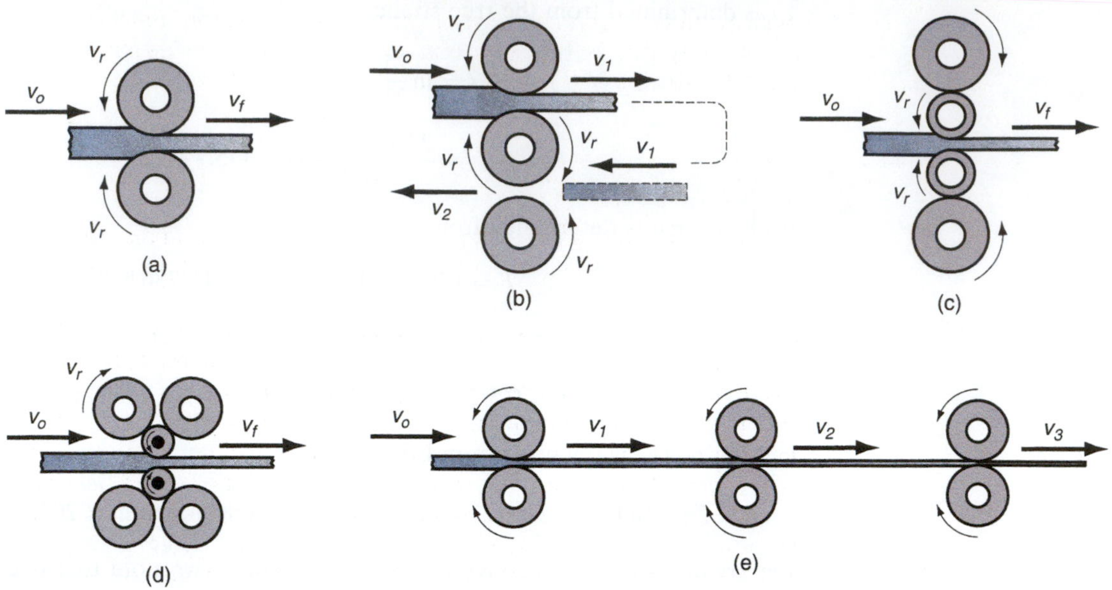

FIGURE 13.5 Various configurations of rolling mills: (a) 2-high, (b) 3-high, (c) 4-high, (d) cluster mill, and (e) tandem rolling mill. (Credit: *Fundamentals of Modern Manufacturing*, 4th Edition by Mikell P. Groover, 2010. Reprinted with permission of John Wiley & Sons, Inc.)

opposing rolls and is referred to as a ***two-high*** rolling mill, shown in Figure 13.5(a). The rolls in these mills have diameters in the range 0.6 to 1.4 m (2.0 to 4.5 ft). The two-high configuration can be either reversing or nonreversing. In the ***nonreversing mill***, the rolls always rotate in the same direction, and the work always passes through from the same side. The ***reversing mill*** allows the direction of roll rotation to be reversed, so that the work can be passed through in either direction. This permits a series of reductions to be made through the same set of rolls, simply by passing through the work from opposite directions multiple times. The disadvantage of the reversing configuration is the significant angular momentum possessed by large rotating rolls and the associated technical problems involved in reversing the direction.

Several alternative arrangements are illustrated in Figure 13.5. In the ***three-high*** configuration, Figure 13.5(b), there are three rolls in a vertical column, and the direction of rotation of each roll remains unchanged. To achieve a series of reductions, the work can be passed through from either side by raising or lowering the strip after each pass. The equipment in a three-high rolling mill becomes more complicated, because an elevator mechanism is needed to raise and lower the work.

As several of the previous equations indicate, advantages are gained in reducing roll diameter. Roll-work contact length is reduced with a lower roll radius, and this leads to lower forces, torque, and power. The ***four-high*** rolling mill uses two smaller-diameter rolls to contact the work and two backing rolls behind them, as in Figure 13.5(c). Owing to the high roll forces, these smaller rolls would deflect elastically between their end bearings as the work passes through unless the larger backing rolls were used to support them. Another roll configuration that allows smaller working rolls against the work is the ***cluster rolling mill***, Figure 13.5(d).

To achieve higher throughput rates in standard products, a ***tandem rolling mill*** is often used. This configuration consists of a series of rolling stands, as represented in Figure 13.5(e). Although only three stands are shown in our sketch, a typical tandem rolling mill may have eight or ten stands, each making a reduction in thickness or a refinement in shape of the work passing through. With each rolling step, work velocity increases, and the problem of synchronizing the roll speeds at each stand is a significant one.

13.1.4 OTHER PROCESSES RELATED TO ROLLING

Several other bulk deformation processes use rolls to form the workpart. The operations include thread rolling, ring rolling, gear rolling, and roll piercing.

Thread Rolling Thread rolling is used to form threads on cylindrical parts by rolling them between two dies. It is the most important commercial process for mass-producing external threaded components (e.g., bolts and screws). The competing process is threading (cutting the threads, Section 16.2.2). Most thread rolling operations are performed by cold working in thread rolling machines. These machines are equipped with special dies that determine the size and form of the thread. The dies are of two types: (1) flat dies, which reciprocate relative to each other, as illustrated in Figure 13.6; and (2) round dies, which rotate relative to each other to accomplish the rolling action.

Production rates in thread rolling can be high, ranging up to eight parts per second for small bolts and screws. Not only are these rates significantly higher than threading, but there are other advantages over machining as well: (1) better material utilization, (2) stronger threads due to work hardening, (3) smoother surface, and (4) better fatigue resistance due to compressive stresses introduced by rolling.

Ring Rolling Ring rolling is a deformation process in which a thick-walled ring of smaller diameter is rolled into a thin-walled ring of larger diameter. The before and after views of the process are illustrated in Figure 13.7. As the thick-walled ring is compressed, the deformed material elongates, causing the diameter of the ring to be enlarged. Ring rolling is usually performed as a hot-working process for large rings and as a cold-working process for smaller rings.

Applications of ring rolling include ball and roller bearing races, steel tires for railroad wheels, and rings for pipes, pressure vessels, and rotating machinery. The ring walls are not limited to rectangular cross sections; the process permits rolling of more complex shapes. There are several advantages of ring rolling over alternative methods of making the same parts: raw material savings, ideal grain orientation for the application, and strengthening through cold working.

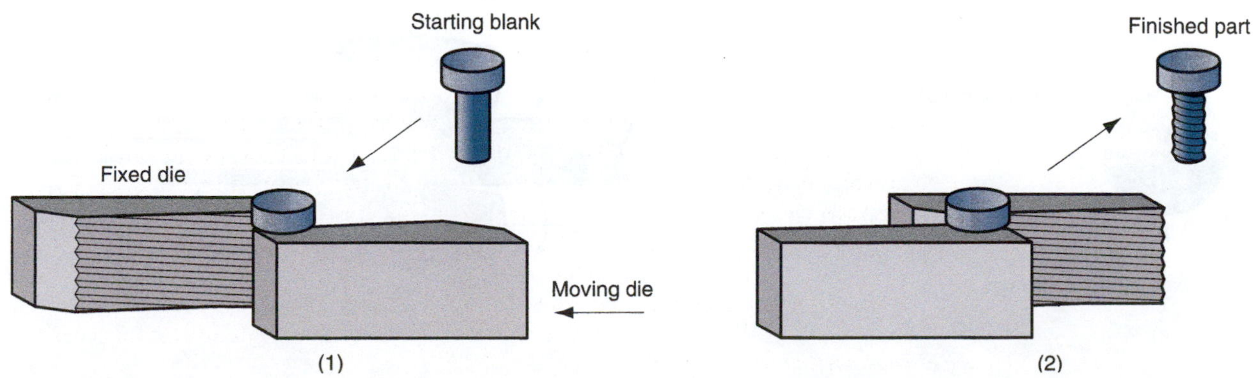

FIGURE 13.6 Thread rolling with flat dies: (1) start of cycle and (2) end of cycle. (Credit: *Fundamentals of Modern Manufacturing*, 4th Edition by Mikell P. Groover, 2010. Reprinted with permission of John Wiley & Sons, Inc.)

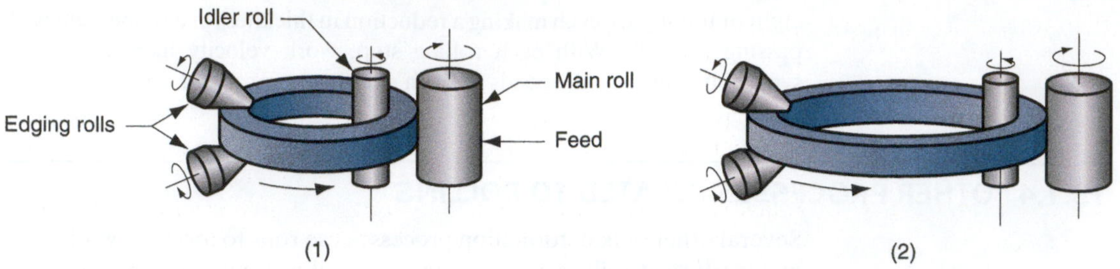

FIGURE 13.7 Ring rolling used to reduce the wall thickness and increase the diameter of a ring: (1) start and (2) completion of process. (Credit: *Fundamentals of Modern Manufacturing*, 4th Edition by Mikell P. Groover, 2010. Reprinted with permission of John Wiley & Sons, Inc.)

Gear Rolling Gear rolling is a cold working process to produce certain gears. The automotive industry is an important user of these products. The setup in gear rolling is similar to thread rolling, except that the deformed features of the cylindrical blank or disk are oriented parallel to its axis (or at an angle in the case of helical gears) rather than spiraled as in thread rolling. Alternative production methods for gears include machining. Advantages of gear rolling over machining are similar to those of thread rolling: higher production rates, better strength and fatigue resistance, and less material waste.

Roll Piercing Roll piercing is a specialized hot working process for making seamless thick-walled tubes. It utilizes two opposing rolls, and hence, it is grouped with the rolling processes. The process is based on the principle that when a solid cylindrical part is compressed on its circumference, as in Figure 13.8(a), high tensile stresses are developed at its center. If compression is high enough, an internal crack is formed. In roll piercing, this principle is exploited by the setup shown in Figure 13.8(b). Compressive stresses on a solid cylindrical billet are applied by two rolls, whose axes are oriented at slight angles ($\sim 6^\circ$) from the axis of the billet, so that their rotation tends to pull the billet through the rolls. A mandrel is used to control the size and finish of the hole created by the action. The terms ***rotary tube piercing*** and ***Mannesmann process*** are also used for this tube-making operation.

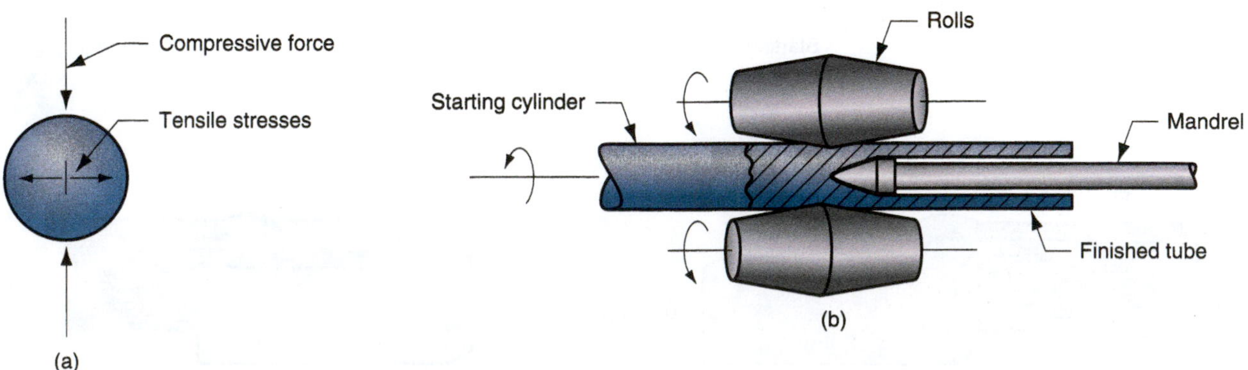

FIGURE 13.8 Roll piercing: (a) formation of internal stresses and cavity by compression of cylindrical part; and (b) setup of Mannesmann roll mill for producing seamless tubing. (Credit: *Fundamentals of Modern Manufacturing*, 4th Edition by Mikell P. Groover, 2010. Reprinted with permission of John Wiley & Sons, Inc.)

13.2 FORGING

Forging is a deformation process in which the work is compressed between two dies, using either impact or gradual pressure to form the part. It is the oldest of the metal forming operations, dating back to perhaps 5000 BCE. Today, forging is an important industrial process used to make a variety of high-strength components for automotive, aerospace, and other applications. These components include engine crankshafts and connecting rods, gears, aircraft structural components, and jet engine turbine parts. In addition, steel and other basic metals industries use forging to establish the basic form of large components that are subsequently machined to final shape and dimensions.

Forging is carried out in many different ways. One way to classify the operations is by working temperature. Most forging operations are performed hot or warm, owing to the significant deformation demanded by the process and the need to reduce strength and increase ductility of the work metal. However, cold forging is also very common for certain products. The advantage of cold forging is the increased strength that results from strain hardening of the component.

Either impact or gradual pressure is used in forging. The distinction derives more from the type of equipment used than differences in process technology. A forging machine that applies an impact load is called a ***forging hammer***, while one that applies gradual pressure is called a ***forging press***.

Another difference among forging operations is the degree to which the flow of the work metal is constrained by the dies. By this classification, there are three types of forging operations, shown in Figure 13.9: (a) open-die forging, (b) impression-die forging, and (c) flashless forging. In ***open-die forging***, the work is compressed between

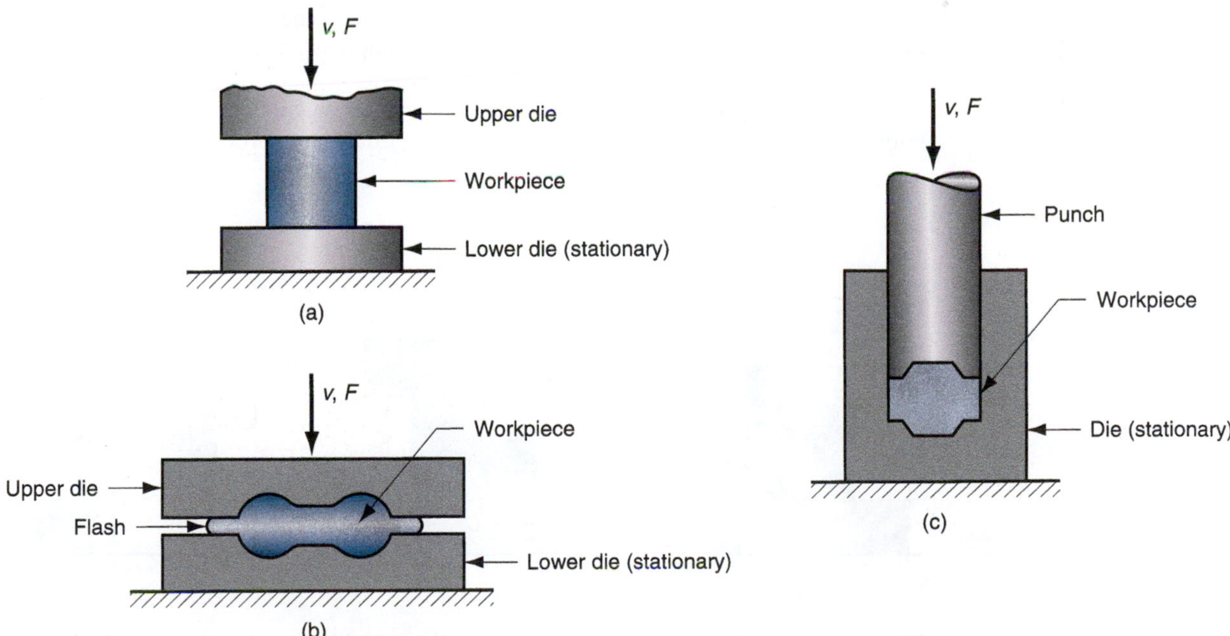

FIGURE 13.9 Three types of forging operation illustrated by cross-sectional sketches: (a) open-die forging, (b) impression-die forging, and (c) flashless forging. (Credit: *Fundamentals of Modern Manufacturing*, 4th Edition by Mikell P. Groover, 2010. Reprinted with permission of John Wiley & Sons, Inc.)

two flat (or almost flat) dies, thus allowing the metal to flow without constraint in a lateral direction relative to the die surfaces. In ***impression-die forging***, the die surfaces contain a shape or impression that is imparted to the work during compression, thus constraining metal flow to a significant degree. In this type of operation, a portion of the work metal flows beyond the die impression to form ***flash***, as shown in the figure. Flash is excess metal that must be trimmed off later. In ***flashless forging***, the work is completely constrained within the die and no excess flash is produced. The volume of the starting workpiece must be controlled very closely so that it matches the volume of the die cavity.

13.2.1 OPEN-DIE FORGING

The simplest case of open-die forging involves compression of a workpart of cylindrical cross section between two flat dies, much in the manner of a compression test (Section 3.1.2). This forging operation, known as ***upsetting*** or ***upset forging***, reduces the height of the work and increases its diameter.

Analysis of Open-Die Forging If open-die forging is carried out under ideal conditions of no friction between work and die surfaces, then homogeneous deformation occurs, and the radial flow of the material is uniform throughout its height, as pictured in Figure 13.10. Under these ideal conditions, the true strain experienced by the work during the process can be determined by

$$\epsilon = \ln \frac{h_o}{h} \tag{13.14}$$

where h_o = starting height of the work, mm (in); and h = the height at some intermediate point in the process, mm (in). At the end of the compression stroke, h = its final value h_f, and the true strain reaches its maximum value.

Estimates of force to perform upsetting can be calculated. The force required to continue the compression at any given height h during the process can be obtained by

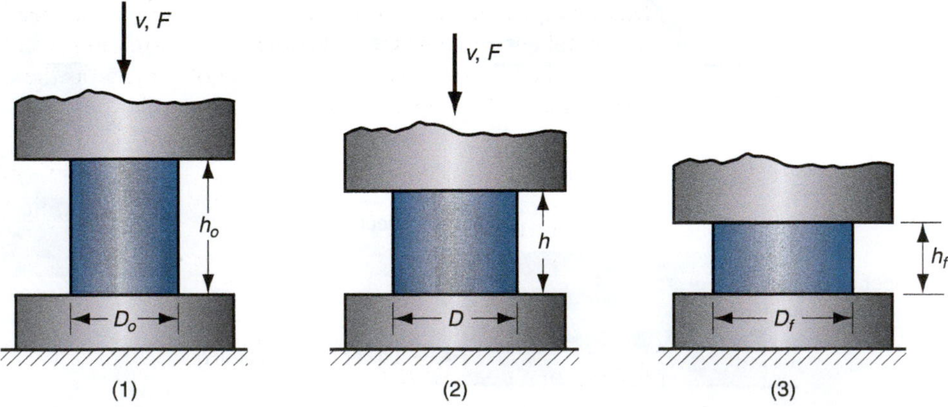

FIGURE 13.10 Homogeneous deformation of a cylindrical workpart under ideal conditions in an open-die forging operation: (1) start of process with workpiece at its original length and diameter, (2) partial compression, and (3) final size. (Credit: *Fundamentals of Modern Manufacturing*, 4th Edition by Mikell P. Groover, 2010. Reprinted with permission of John Wiley & Sons, Inc.)

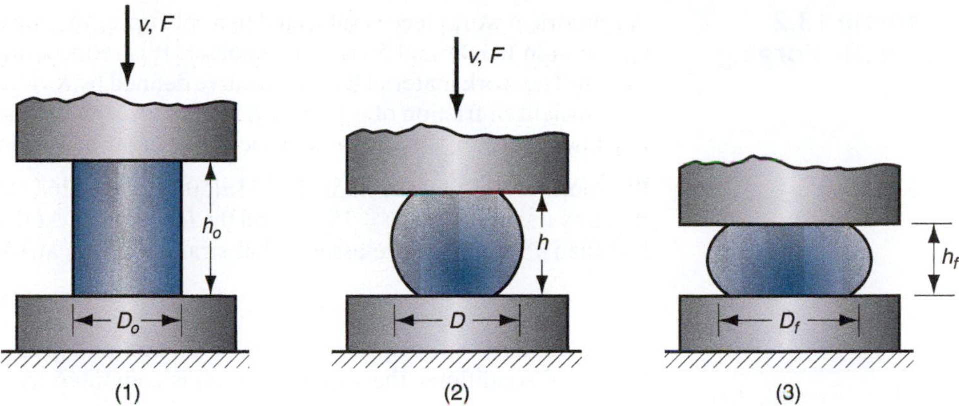

FIGURE 13.11 Actual deformation of a cylindrical workpart in open-die forging, showing pronounced barreling: (1) start of process, (2) partial deformation, and (3) final shape. (Credit: *Fundamentals of Modern Manufacturing*, 4th Edition by Mikell P. Groover, 2010. Reprinted with permission of John Wiley & Sons, Inc.)

multiplying the corresponding cross-sectional area by the flow stress:

$$F = Y_f A \tag{13.15}$$

where F = force, lb (N); A = cross-sectional area of the part, mm^2 (in^2); and Y_f = flow stress corresponding to the strain given by Eq. (13.14), MPa (lb/in^2). Area A continuously increases during the operation as height is reduced. Flow stress Y_f also increases as a result of work hardening, except when the metal is perfectly plastic (e.g., in hot working). In this case, the strain-hardening exponent $n = 0$, and flow stress Y_f equals the metal's yield strength Y. Force reaches a maximum value at the end of the forging stroke, when both area and flow stress are at their highest values.

An actual upsetting operation does not occur quite as shown in Figure 13.10 because friction opposes the flow of work metal at the die surfaces. This creates the barreling effect shown in Figure 13.11. When performed on a hot workpart with cold dies, the barreling effect is even more pronounced. This results from a higher coefficient of friction typical in hot working and heat transfer at and near the die surfaces, which cools the metal and increases its resistance to deformation. The hotter metal in the middle of the part flows more readily than the cooler metal at the ends. These effects are more significant as the diameter-to-height ratio of the workpart increases, due to the greater contact area at the work–die interface.

All of these factors cause the actual upsetting force to be greater than what is predicted by Eq. (13.15). As an approximation, we can apply a shape factor to Eq. (13.15) to account for effects of the D/h ratio and friction:

$$F = K_f Y_f A \tag{13.16}$$

where F, Y_f, and A have the same definitions as in the previous equation; and K_f is the forging shape factor, defined as

$$K_f = 1 + \frac{0.4\mu D}{h} \tag{13.17}$$

where μ = coefficient of friction; D = workpart diameter or other dimension representing contact length with die surface, mm (in); and h = workpart height, mm (in).

Example 13.2 Open-Die Forging

A cylindrical workpiece is subjected to a cold upset forging operation. The starting piece is 75 mm in height and 50 mm in diameter. It is reduced in the operation to a height of 36 mm. The work material has a flow curve defined by $K = 350$ MPa and $n = 0.17$. Assume a coefficient of friction of 0.1. Determine the force as the process begins, at intermediate heights of 62 mm, 49 mm, and at the final height of 36 mm.

Solution: Workpiece volume $V = 75\pi(50^2/4) = 147{,}262$ mm^3. At the moment contact is made by the upper die, $h = 75$ mm and the force $F = 0$. At the start of yielding, h is slightly less than 75 mm, and we assume that strain = 0.002, at which the flow stress is

$$Y_f = K\epsilon^n = 350(0.002)^{0.17} = 121.7\text{ MPa}$$

The diameter is still approximately $D = 50$ mm and area $A = \pi(50^2/4) = 1963.5$ mm^2. For these conditions, the shape factor K_f is computed as

$$K_f = 1 + \frac{0.4(0.1)(50)}{75} = 1.027$$

The forging force is

$$F = 1.027(121.7)(1963.5) = 245{,}410\text{ N}$$

At $h = 62$ mm,

$$\epsilon = \ln\frac{75}{62} = \ln(1.21) = 0.1904$$

$$Y_f = 350(0.1904)^{17} = 264.0\text{ MPa}$$

Assuming constant volume, and neglecting barreling,

$$A = 147{,}262/62 = 2375.2\text{ mm}^2 \text{ and } D = \sqrt{\frac{4(2375.2)}{\pi}} = 55.0\text{ mm}$$

$$K_f = 1 + \frac{0.4(0.1)(55)}{62} = 1.035$$

$$F = 1.035(264)(2375.2) = 649{,}303\text{ N}$$

Similarly, at $h = 49$ mm, $F = 955{,}642$ N; and at $h = 36$ mm, $F = 1{,}467{,}422$ N. The load-stroke curve in Figure 13.12 was developed from the values in this example. ■

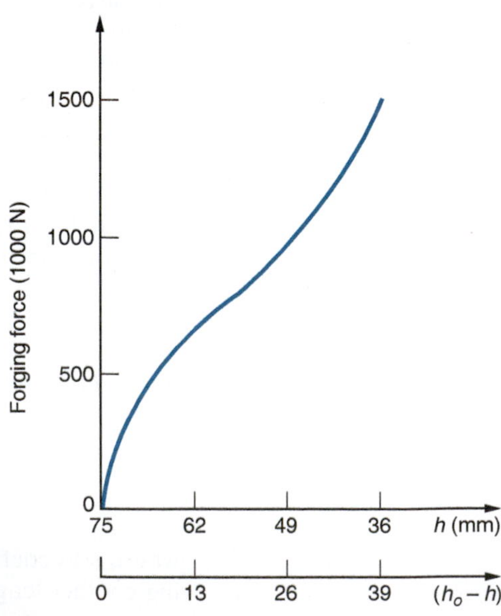

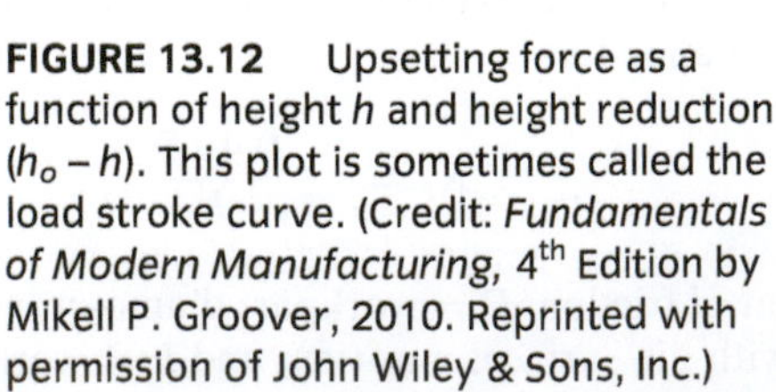
FIGURE 13.12 Upsetting force as a function of height h and height reduction $(h_o - h)$. This plot is sometimes called the load stroke curve. (Credit: *Fundamentals of Modern Manufacturing*, 4th Edition by Mikell P. Groover, 2010. Reprinted with permission of John Wiley & Sons, Inc.)

Open-Die Forging Practice Open-die hot forging is an important industrial process. Shapes generated by open-die operations are simple; examples include shafts, disks, and rings. In some applications, the dies have slightly contoured surfaces that help to shape the work. In addition, the work must often be manipulated (e.g., rotated in steps) to effect the desired shape change. Skill of the human operator is a factor in the success of these operations. An example of open-die forging in the steel industry is the shaping of a large square cast ingot into a round cross section. Open-die forging operations produce rough forms, and subsequent operations are required to refine the parts to final geometry and dimensions. An important contribution of open-die hot-forging is that it creates a favorable grain flow and metallurgical structure in the metal.

13.2.2 IMPRESSION-DIE FORGING

Impression-die forging is performed with dies that contain the inverse of the desired shape of the part. The process is illustrated in a three-step sequence in Figure 13.13. The raw workpiece is shown as a cylindrical part similar to that used in the previous open-die operation. As the die closes to its final position, flash is formed by metal that flows beyond the die cavity and into the small gap between the die plates. Although this flash must be cut away from the part in a subsequent trimming operation, it actually serves an important function during impression-die forging. As the flash begins to form in the die gap, friction resists continued flow of metal into the gap, thus constraining the bulk of the work material to remain in the die cavity. In hot forging, metal flow is further restricted because the thin flash cools quickly against the die plates, thereby increasing its resistance to deformation. Restricting metal flow in the gap causes the compression pressures on the part to increase significantly, thus forcing the material to fill the sometimes intricate details of the die cavity to ensure a high-quality product.

Several forming steps are often required in impression-die forging to transform the starting blank into the desired final geometry. Separate cavities in the die are needed for each step. The beginning steps are designed to redistribute the metal in the workpart to achieve a uniform deformation and desired metallurgical structure in the subsequent steps. The final steps bring the part to its final geometry. In addition, when drop forging is used, several blows of the hammer may be required for each step. When impression-die

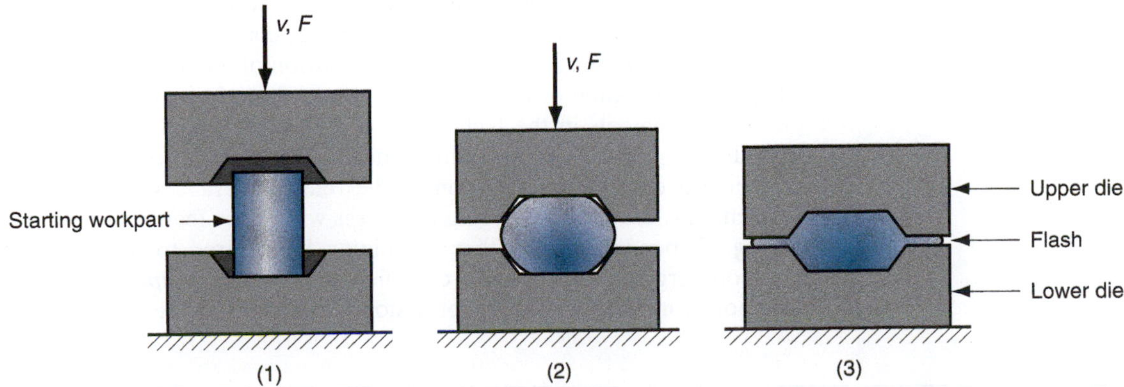

FIGURE 13.13 Sequence in impression-die forging: (1) just prior to initial contact with raw workpiece, (2) partial compression, and (3) final die closure, causing flash to form in gap between die plates. (Credit: *Fundamentals of Modern Manufacturing,* 4th Edition by Mikell P. Groover, 2010. Reprinted with permission of John Wiley & Sons, Inc.)

TABLE 13.1 Typical K_f values for various part shapes in impression-die and flashless forging

Impression-die forging	K_f	Flashless forging	K_f
Simple shapes with flash	6.0	Coining (top and bottom surfaces)	6.0
Complex shapes with flash	8.0	Complex shapes	8.0
Very complex shapes with flash	10.0		

Compiled from [13], [19], and other sources.

drop forging is done manually, as it often is, considerable operator skill is required under adverse conditions to achieve consistent results.

Because of flash formation in impression-die forging and the more complex part shapes made with these dies, forces in this process are significantly greater and more difficult to analyze than in open-die forging. Relatively simple formulas and design factors are often used to estimate forces in impression-die forging. The force formula is the same as previous Eq. (13.16) for open-die forging, but its interpretation is slightly different:

$$F = K_f Y_f A \tag{13.18}$$

where F = maximum force in the operation, N (lb); A = projected area of the part including flash, mm^2 (in^2); Y_f = flow stress of the material, MPa (lb/in^2); and K_f = forging shape factor. In hot forging, the appropriate value of Y_f is the yield strength of the metal at the elevated temperature. In other cases, selecting the proper value of flow stress is difficult because the strain varies throughout the workpiece for complex shapes. K_f in Eq. (13.18) is a factor intended to account for increases in force required to forge part shapes of various complexities. Table 13.1 indicates the range of values of K_f for different part geometries. Obviously, the problem of specifying the proper K_f value for a given workpart limits the accuracy of the force estimate.

Eq. (13.18) applies to the maximum force during the operation, since this is the load that will determine the required capacity of the press or hammer used in the operation. The maximum force is reached at the end of the forging stroke, when the projected area is greatest and friction is maximum.

Impression-die forging is not capable of close tolerance work, and machining is often required to achieve the accuracies needed. The basic geometry of the part is obtained from the forging process, with machining performed on those portions of the part that require precision finishing (e.g., holes, threads, and surfaces that mate with other components). The advantages of forging, compared to machining the part completely, are higher production rates, conservation of metal, greater strength, and favorable grain orientation of the metal that results from forging.

Improvements in the technology of impression-die forging have resulted in the capability to produce forgings with thinner sections, more complex geometries, drastic reductions in draft requirements on the dies, closer tolerances, and the virtual elimination of machining allowances. Forging processes with these features are known as ***precision forging***. Depending on whether machining is required to finish the part geometry, precision forgings are properly classified as ***near net shape*** or ***net shape*** processes. Common work metals used for precision forging include aluminum and titanium.

13.2.3 FLASHLESS FORGING

In flashless forging, illustrated in Figure 13.14, the raw workpiece is completely contained within the die cavity during compression, and no flash is formed. Process control requirements are more demanding in flashless forging than in impression-die forging.

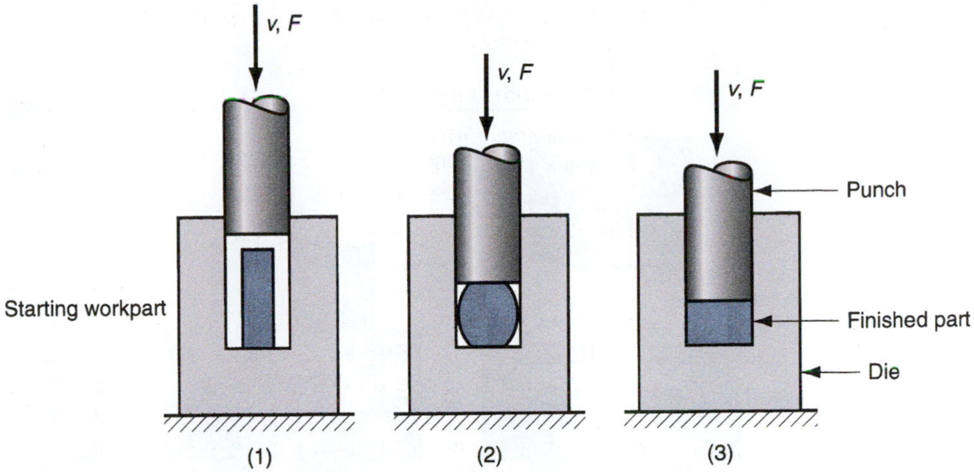

FIGURE 13.14 Flashless forging: (1) just before initial contact with workpiece, (2) partial compression, and (3) final punch and die closure. Symbols *v* and *F* indicate motion (*v* = velocity) and applied force, respectively. (Credit: *Fundamentals of Modern Manufacturing*, 4th Edition by Mikell P. Groover, 2010. Reprinted with permission of John Wiley & Sons, Inc.)

Most important is that the volume of the starting workpiece must equal the space in the die cavity within a very close tolerance. If the starting blank is too large, excessive pressures may cause damage to the die or press. If the blank is too small, the cavity will not be filled. Because of the special demands made by flashless forging, the process lends itself best to part geometries that are usually simple and symmetrical, and to work materials such as aluminum and magnesium and their alloys. Flashless forging is often classified as a ***precision forging*** process [5].

Forces in flashless forging reach values comparable to those in impression-die forging. Estimates of these forces can be computed using the same methods as for impression-die forging: Eq. (13.18) and Table 13.1.

Coining is a special application of closed-die forging in which fine details in the die are impressed into the top and bottom surfaces of the workpart. There is little flow of metal in coining, yet the pressures required to reproduce the surface details in the die cavity are high, as indicated by the value of K_f in Table 13.1. A common application of coining is, of course, in the minting of coins, shown in Figure 13.15. The process is also used to provide good surface finish and dimensional accuracy on workparts made by other operations.

13.2.4 FORGING HAMMERS, PRESSES, AND DIES

Equipment used in forging consists of forging machines, classified as hammers or presses, and forging dies, which are the special tooling used in these machines. In addition, auxiliary equipment is needed, such as furnaces to heat the work, mechanical devices to load and unload the work, and trimming stations to cut away the flash in impression-die forging.

Forging Hammers Forging hammers operate by applying an impact loading against the work. Shown in Figures 13.16 and 13.17, the term ***drop hammer*** is often used for these machines, owing to the means of delivering impact energy. Drop hammers are most frequently used for impression-die forging. The upper portion of the forging die is attached to the ram, and the lower portion is attached to the anvil. In the operation, the

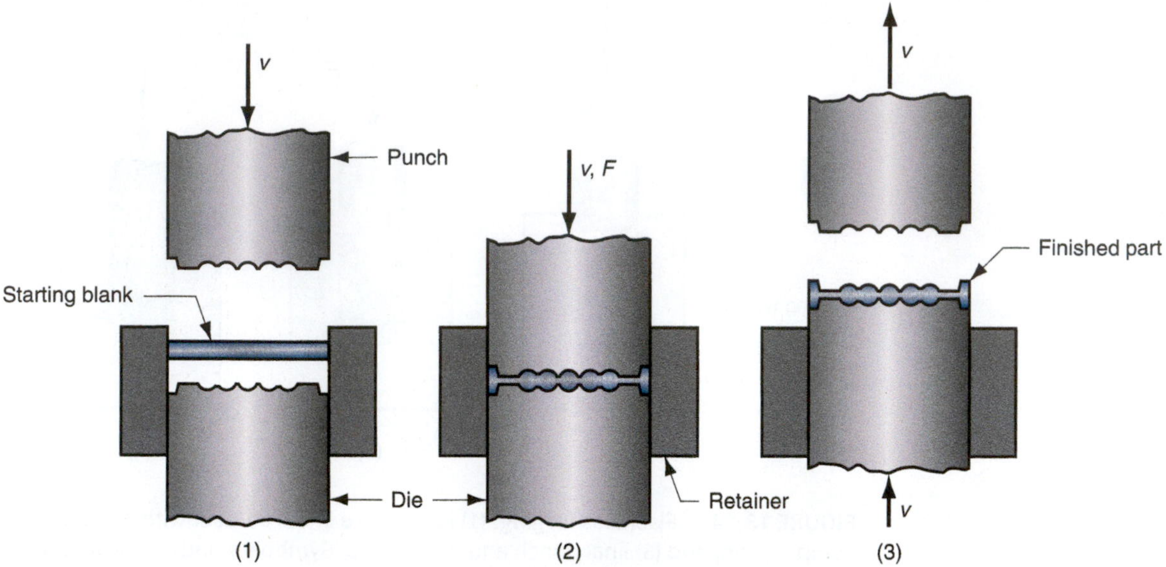

FIGURE 13.15 Coining operation: (1) start of cycle, (2) compression stroke, and (3) ejection of finished part. (Credit: *Fundamentals of Modern Manufacturing,* 4th Edition by Mikell P. Groover, 2010. Reprinted with permission of John Wiley & Sons, Inc.)

work is placed on the lower die, and the ram is lifted and then dropped. When the upper die strikes the work, the impact energy causes the part to assume the shape of the die cavity. Several blows of the hammer are often required to achieve the desired change in geometry. One of the disadvantages of drop hammers is that a large amount of the impact energy is transmitted through the anvil and into the floor of the building.

FIGURE 13.16 Drop forging hammer, fed by conveyor and heating units at the right of the scene. Photo courtesy of Ajax-Ceco. (Credit: *Fundamentals of Modern Manufacturing,* 4th Edition by Mikell P. Groover, 2010. Reprinted with permission of John Wiley & Sons, Inc.)

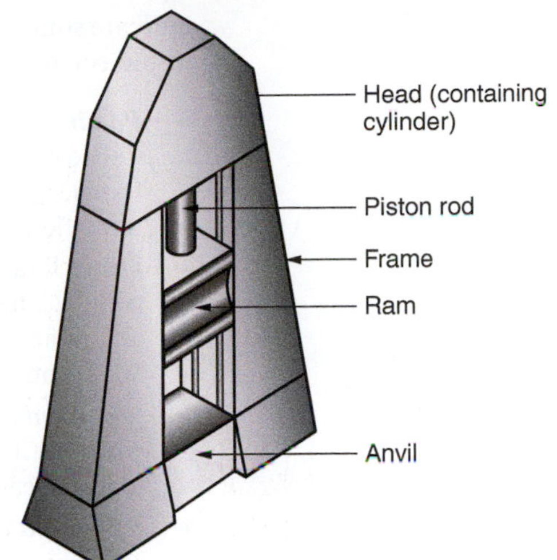

FIGURE 13.17 Diagram showing details of a drop hammer for impression-die forging. (Credit: *Fundamentals of Modern Manufacturing*, 4th Edition by Mikell P. Groover, 2010. Reprinted with permission of John Wiley & Sons, Inc.)

Forging Presses Presses apply gradual pressure, rather than sudden impact, to accomplish the forging operation. Forging presses include mechanical presses, hydraulic presses, and screw presses. ***Mechanical presses*** operate by means of eccentrics, cranks, or knuckle joints, which convert the rotating motion of a drive motor into the translation motion of the ram. These mechanisms are very similar to those used in stamping presses (Section 14.5.2). Mechanical presses typically achieve very high forces at the bottom of the forging stroke. ***Hydraulic presses*** use a hydraulically driven piston to actuate the ram. ***Screw presses*** apply force by a screw mechanism that drives the vertical ram. Both screw drive and hydraulic drive operate at relatively low ram speeds and can provide a constant force throughout the stroke. These machines are therefore suitable for forging (and other forming) operations that require a long stroke.

Forging Dies Proper die design is important in the success of a forging operation. Parts to be forged must be designed based on knowledge of the principles and limitations of this process. Our purpose here is to describe some of the terminology and guidelines used in the design of forgings and forging dies. Design of open dies is generally straightforward because the dies are relatively simple in shape. Our comments apply to impression dies and closed dies. Figure 13.18 defines some of the terminology in an impression die. In the following, we discuss the terminology and

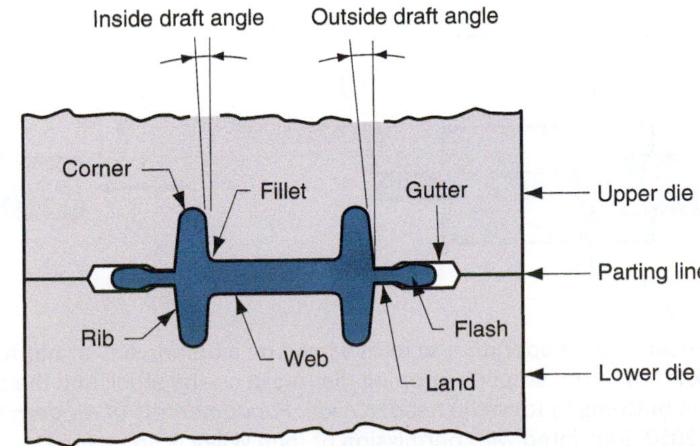

FIGURE 13.18 Terminology for a conventional impression die in forging. (Credit: *Fundamentals of Modern Manufacturing*, 4th Edition by Mikell P. Groover, 2010. Reprinted with permission of John Wiley & Sons, Inc.)

indicate some of the principles and limitations that must be considered in part design or in the selection of forging as the process to manufacture the part [5]:

- ➢ ***Parting line***. The parting line is the plane that divides the upper die from the lower die. Called the flash line in impression-die forging, it is the plane where the two die halves meet. Its selection by the die designer affects grain flow in the part, required load, and flash formation.
- ➢ ***Draft***. Draft is the amount of taper on the sides of the part required to remove it from the die. The term also applies to the taper on the sides of the die cavity. Typical draft angles are 3° on aluminum and magnesium parts and 5° to 7° on steel parts. Draft angles on precision forgings are near zero.
- ➢ ***Webs and ribs***. A web is a thin portion of the forging that is parallel to the parting line, while a rib is a thin portion that is perpendicular to the parting line. These part features cause difficulty in metal flow as they become thinner.
- ➢ ***Fillet and corner radii***. Fillet and corner radii are illustrated in Figure 13.18. Small radii tend to limit metal flow and increase stresses on die surfaces during forging.
- ➢ ***Flash***. Flash formation plays a critical role in impression-die forging by causing pressure buildup inside the die to promote filling of the cavity. This pressure buildup is controlled by designing a flash land and gutter into the die, as pictured in Figure 13.18. The land determines the surface area along which lateral flow of metal occurs, thereby controlling the pressure increase inside the die. The gutter permits excess metal to escape without causing the forging load to reach extreme values.

13.2.5 OTHER PROCESSES RELATED TO FORGING

In addition to the conventional forging operations discussed in the preceding sections, other metal forming operations are closely associated with forging.

Upsetting and Heading Upsetting (also called ***upset forging***) is a deformation operation in which a cylindrical workpart is increased in diameter and reduced in length. This operation was analyzed in our discussion of open-die forging (Section 13.2.1). However, as an industrial operation, it can also be performed in closed dies, as shown in Figure 13.19.

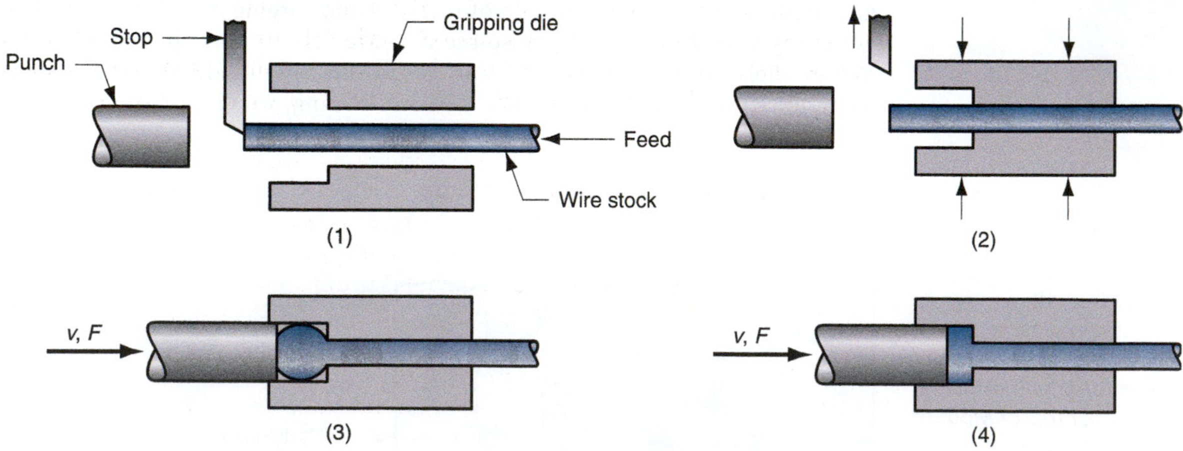

FIGURE 13.19 An upset forging operation to form a head on a bolt or similar hardware item. The cycle is as follows: (1) Wire stock is fed to the stop; (2) gripping dies close on the stock and the stop is retracted; (3) punch moves forward; and (4) bottoms to form the head. (Credit: *Fundamentals of Modern Manufacturing*, 4th Edition by Mikell P. Groover, 2010. Reprinted with permission of John Wiley & Sons, Inc.)

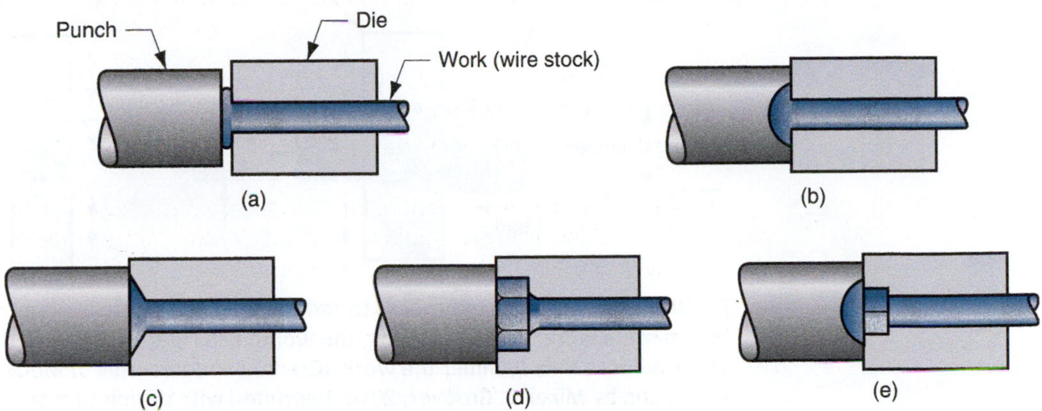

FIGURE 13.20 Examples of heading (upset forging) operations: (a) heading a nail using open dies, (b) round head formed by punch, (c) and (d) heads formed by die, and (e) carriage bolt head formed by punch and die. (Credit: *Fundamentals of Modern Manufacturing*, 4th Edition by Mikell P. Groover, 2010. Reprinted with permission of John Wiley & Sons, Inc.)

Upsetting is widely used in the fastener industry to form heads on nails, bolts, and similar hardware products. In these applications, the term ***heading*** is often used to denote the operation. Figure 13.20 illustrates a variety of heading applications, indicating various possible die configurations. Owing to widespread use of the products made in these applications, more parts are produced by upsetting than by any other forging operation. It is performed as a mass-production operation—cold, warm, or hot—on special upset forging machines, called headers or formers. These machines are usually equipped with horizontal slides, rather than vertical slides as in conventional forging hammers and presses. Long wire or bar stock is fed into the machines, the end of the stock is upset forged, and then the piece is cut to length to make the desired hardware item. For bolts and screws, thread rolling (Section 13.1.4) is used to form the threads.

There are limits on the amount of deformation that can be achieved in upsetting, usually defined as the maximum length of stock to be forged. The maximum length that can be upset in one blow is three times the diameter of the starting stock. Otherwise, the metal bends or buckles instead of compressing properly to fill the cavity.

Swaging and Radial Forging Swaging and radial forging are forging processes used to reduce the diameter of a tube or solid rod. Swaging is often performed on the end of a workpiece to create a tapered section. The ***swaging*** process, shown in Figure 13.21, is accomplished by means of rotating dies that hammer a workpiece radially inward to taper it as the piece is fed into the dies. A mandrel is sometimes required to control the shape and size of the internal diameter of tubular parts that are swaged. ***Radial forging*** is similar to swaging in its action against the work and is used to create similar part shapes. The difference is that in radial forging the dies do not rotate around the workpiece; instead the work is rotated as it feeds into the hammering dies.

Hubbing Hubbing is a deformation process in which a hardened steel form is pressed into a soft steel (or other soft metal) block. The process is often used to make mold cavities for plastic molding and die casting, as suggested by Figure 13.22. The hardened steel form, called the ***hub***, is machined to the geometry of the part to be molded. Substantial pressures are required to force the hub into the soft block, and this is usually

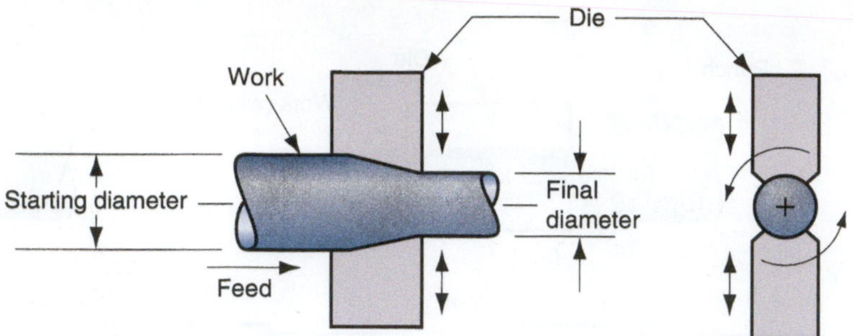

FIGURE 13.21 Swaging process to reduce solid rod stock; the dies rotate as they hammer the work. In radial forging, the workpiece rotates while the dies remain in a fixed orientation as they hammer the work. (Credit: *Fundamentals of Modern Manufacturing,* 4th Edition by Mikell P. Groover, 2010. Reprinted with permission of John Wiley & Sons, Inc.)

accomplished by a hydraulic press. Complete formation of the die cavity in the block often requires several steps—hubbing followed by annealing to recover the work metal from strain hardening. When significant amounts of material are deformed in the block, as shown in our figure, the excess must be machined away. The advantage of hubbing in this application is that it is generally easier to machine the positive form than the mating negative cavity. This advantage is multiplied in cases where more than one cavity are made in the die block.

Isothermal Forging Isothermal forging is a term applied to a hot-forging operation in which the workpart is maintained at or near its starting elevated temperature during deformation, usually by heating the forging dies to the same elevated temperature. By avoiding chill of the workpiece on contact with the cold die surfaces as in conventional forging, the metal flows more readily and the force required to perform the process is reduced. Isothermal forging is more expensive than conventional forging and is usually reserved for difficult-to-forge metals, such as titanium and superalloys, and for complex part shapes. The process is sometimes carried out in a vacuum to avoid rapid oxidation of

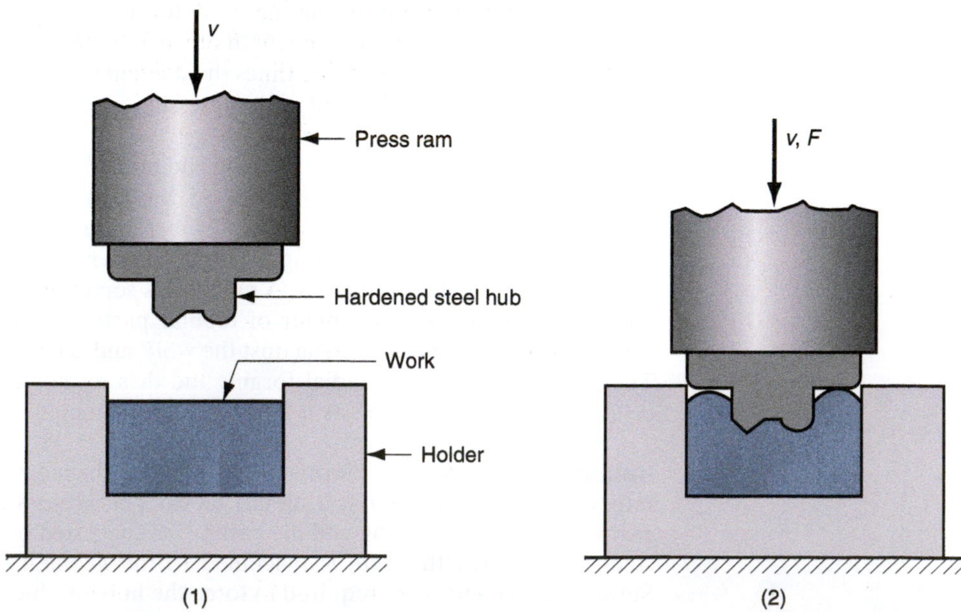

FIGURE 13.22 Hubbing: (1) before deformation, and (2) as the process is completed. Note that the excess material formed by the penetration of the hub must be machined away. (Credit: *Fundamentals of Modern Manufacturing,* 4th Edition by Mikell P. Groover, 2010. Reprinted with permission of John Wiley & Sons, Inc.)

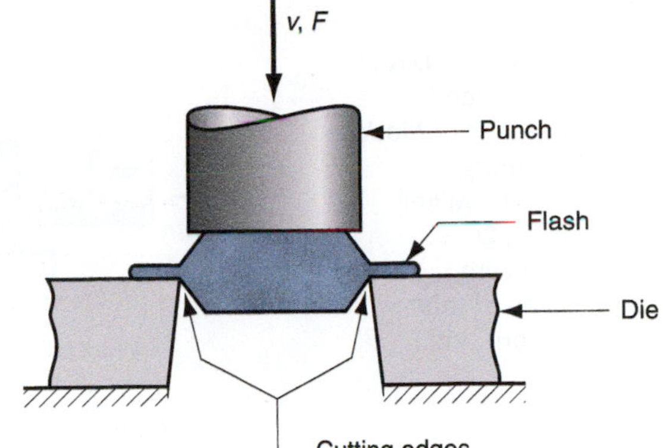

FIGURE 13.23 Trimming operation (shearing process) to remove the flash after impression-die forging. (Credit: *Fundamentals of Modern Manufacturing*, 4th Edition by Mikell P. Groover, 2010. Reprinted with permission of John Wiley & Sons, Inc.)

the die material. Similar to isothermal forging is ***hot-die forging***, in which the dies are heated to a temperature that is somewhat below that of the work metal.

Trimming Trimming is an operation used to remove flash on the workpart in impression-die forging. In most cases, trimming is accomplished by shearing, as in Figure 13.23, in which a punch forces the work through a cutting die, the cutting edges for which have the profile of the desired part. Trimming is usually done while the work is still hot, which means that a separate trimming press is included at each forging hammer or press. In cases where the work might be damaged by the cutting process, trimming may be done by alternative methods, such as grinding or sawing.

13.3 EXTRUSION

Extrusion is a compression process in which the work metal is forced to flow through a die opening to produce a desired cross-sectional shape. The process can be likened to squeezing toothpaste out of a toothpaste tube. Extrusion dates from around 1800. There are several advantages of the modern process: (1) a variety of shapes are possible, especially with hot extrusion; (2) grain structure and strength properties are enhanced in cold and warm extrusion; (3) fairly close tolerances are possible, especially in cold extrusion; and (4) in some extrusion operations, little or no wasted material is created. However, a limitation is that the cross section of the extruded part must be uniform throughout its length.

13.3.1 TYPES OF EXTRUSION

Extrusion is carried out in various ways. One important distinction is between direct extrusion and indirect extrusion. Another classification is by working temperature: cold, warm, or hot extrusion. Finally, extrusion is performed as either a continuous process or a discrete process.

Direct versus Indirect Extrusion Direct extrusion (also called ***forward extrusion***) is illustrated in Figure 13.24. A metal billet is loaded into a container, and a ram compresses the material, forcing it to flow through one or more openings in a die at the opposite end of the container. As the ram approaches the die, a small portion of the billet remains that cannot be forced through the die opening. This extra portion, called the ***butt***, is separated from the product by cutting it just beyond the exit of the die.

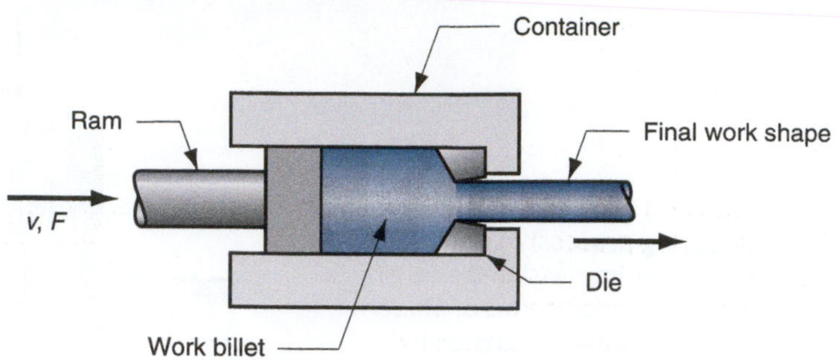

FIGURE 13.24 Direct extrusion. (Credit: *Fundamentals of Modern Manufacturing,* 4th Edition by Mikell P. Groover, 2010. Reprinted with permission of John Wiley & Sons, Inc.)

One of the problems in direct extrusion is the significant friction that exists between the work surface and the walls of the container as the billet is forced to slide toward the die opening. This friction causes a substantial increase in the ram force required in direct extrusion. In hot extrusion, the friction problem is aggravated by the presence of an oxide layer on the surface of the billet. This oxide layer can cause defects in the extruded product. To address these problems, a dummy block is often used between the ram and the work billet. The diameter of the dummy block is slightly smaller than the billet diameter, so that a narrow ring of work metal (mostly the oxide layer) is left in the container, leaving the final product free of oxides.

Hollow sections (e.g., tubes) are possible in direct extrusion by the process setup in Figure 13.25. The starting billet is prepared with a hole parallel to its axis. This allows passage of a mandrel that is attached to the dummy block. As the billet is compressed, the material is forced to flow through the clearance between the mandrel and the die opening. The resulting cross section is tubular. Semi-hollow cross-sectional shapes are usually extruded in the same way.

The starting billet in direct extrusion is usually round in cross section, but the final shape is determined by the shape of the die opening. Obviously, the largest dimension of the die opening must be smaller than the diameter of the billet.

In ***indirect extrusion***, also called ***backward extrusion*** and ***reverse extrusion***, Figure 13.26(a), the die is mounted to the ram rather than at the opposite end of

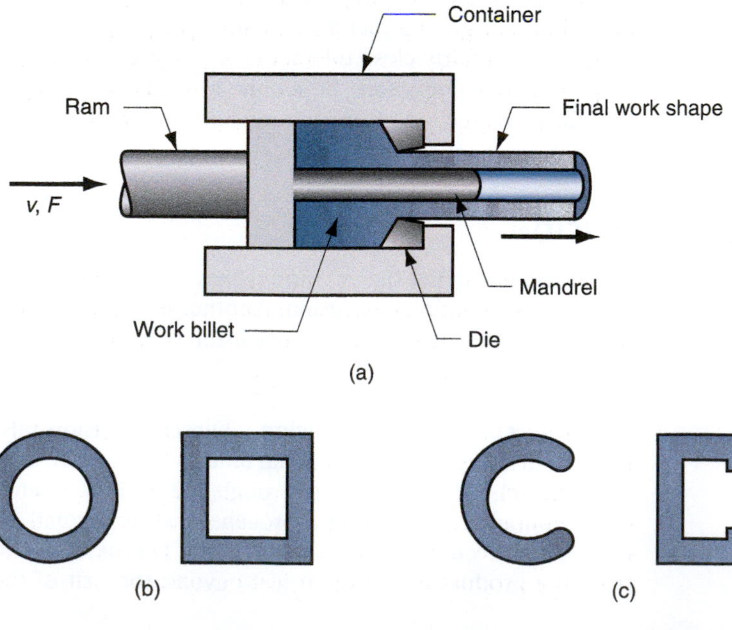

FIGURE 13.25 (a) Direct extrusion to produce a hollow or semi-hollow cross section; (b) hollow and (c) semi-hollow cross sections. (Credit: *Fundamentals of Modern Manufacturing,* 4th Edition by Mikell P. Groover, 2010. Reprinted with permission of John Wiley & Sons, Inc.)

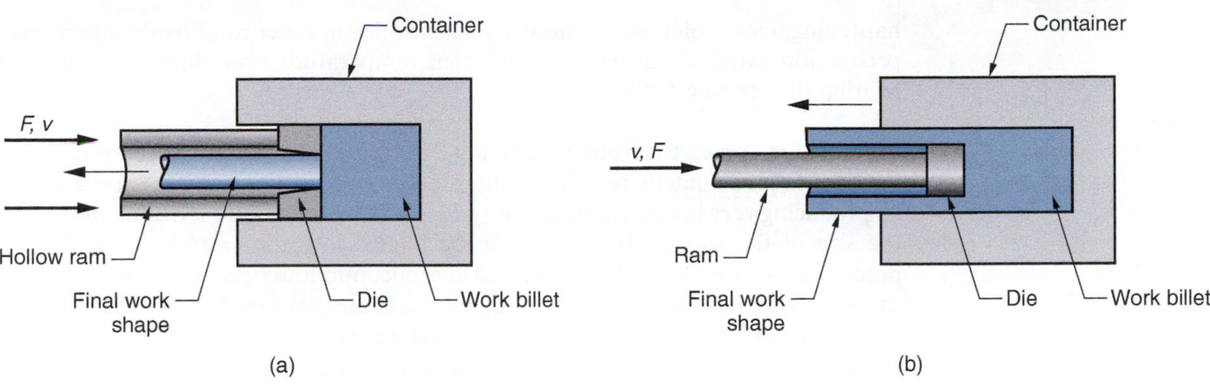

FIGURE 13.26 Indirect extrusion to produce (a) a solid cross section and (b) a hollow cross section. (Credit: *Fundamentals of Modern Manufacturing*, 4th Edition by Mikell P. Groover, 2010. Reprinted with permission of John Wiley & Sons, Inc.)

the container. As the ram penetrates into the work, the metal is forced to flow through the clearance in a direction opposite to the motion of the ram. Because the billet is not forced to move relative to the container, there is no friction at the container walls, and the ram force is therefore lower than in direct extrusion. Limitations of indirect extrusion are imposed by the lower rigidity of the hollow ram and the difficulty in supporting the extruded product as it exits the die.

Indirect extrusion can produce hollow (tubular) cross sections, as in Figure 13.26(b). In this method, the ram is pressed into the billet, forcing the material to flow around the ram and take a cup shape. There are practical limitations on the length of the extruded part that can be made by this method. Support of the ram becomes a problem as work length increases.

Hot versus Cold Extrusion Extrusion can be performed either hot or cold, depending on work metal and amount of strain to which it is subjected during deformation. Metals that are typically extruded hot include aluminum, copper, magnesium, zinc, tin, and their alloys. These same metals are sometimes extruded cold. Steel alloys are usually extruded hot, although the softer, more ductile grades are sometimes cold extruded (e.g., low carbon steels and stainless steel). Aluminum is probably the most ideal metal for extrusion (hot and cold), and many commercial aluminum products are made by this process (structural shapes, door and window frames, etc.).

Hot extrusion involves prior heating of the billet to a temperature above its recrystallization temperature. This reduces strength and increases ductility of the metal, permitting more extreme size reductions and more complex shapes to be achieved in the process. Additional advantages include reduction of ram force, increased ram speed, and reduction of grain flow characteristics in the final product. Cooling of the billet as it contacts the container walls is a problem, and ***isothermal extrusion*** is sometimes used to overcome this problem. Lubrication is critical in hot extrusion for certain metals (e.g., steels), and special lubricants have been developed that are effective under the harsh conditions in hot extrusion. Glass is sometimes used as a lubricant in hot extrusion; in addition to reducing friction, it also provides effective thermal insulation between the billet and the extrusion container.

Cold extrusion and warm extrusion are generally used to produce discrete parts, often in finished (or near finished) form. The term ***impact extrusion*** is used to indicate high-speed cold extrusion, and this method is described in more detail in Section 13.3.4. Some important advantages of cold extrusion include increased strength due to strain hardening, close tolerances, improved surface finish, absence of oxide layers, and high production rates. Cold extrusion at room temperature also eliminates the need for heating the starting billet.

Continuous versus Discrete Processing A true continuous process operates in steady-state mode for an indefinite period of time. Some extrusion operations approach this ideal by producing very long sections in one cycle, but these operations are ultimately limited by the size of the starting billet that can be loaded into the extrusion container. These processes are more accurately described as semicontinuous operations. In nearly all cases, the long section is cut into smaller lengths in a subsequent sawing or shearing operation.

In a discrete extrusion operation, a single part is produced in each extrusion cycle. Impact extrusion is an example of the discrete processing case.

13.3.2 ANALYSIS OF EXTRUSION

Let us use Figure 13.27 as a reference in discussing some of the parameters in extrusion. The diagram assumes that both billet and extrudate are round in cross section. One important parameter is the ***extrusion ratio***, also called the ***reduction ratio***. The ratio is defined:

$$r_x = \frac{A_o}{A_f} \tag{13.19}$$

where r_x = extrusion ratio; A_o = cross-sectional area of the starting billet, mm^2 (in^2); and A_f = final cross-sectional area of the extruded section, mm^2 (in^2). The ratio applies for both direct and indirect extrusion. The value of r_x can be used to determine true strain in extrusion, given that ideal deformation occurs with no friction and no redundant work:

$$\epsilon = \ln r_x = \ln\frac{A_o}{A_f} \tag{13.20}$$

Under the assumption of ideal deformation (no friction and no redundant work), the pressure applied by the ram to compress the billet through the die opening depicted in our figure can be computed as follows:

$$p = \overline{Y}_f \ln r_x \tag{13.21}$$

where $\overline{Y}_f$ = average flow stress during deformation, MPa (lb/in^2). For convenience, we restate Eq. (12.2) from the previous chapter:

$$\overline{Y}_f = \frac{K\epsilon^n}{1+n}$$

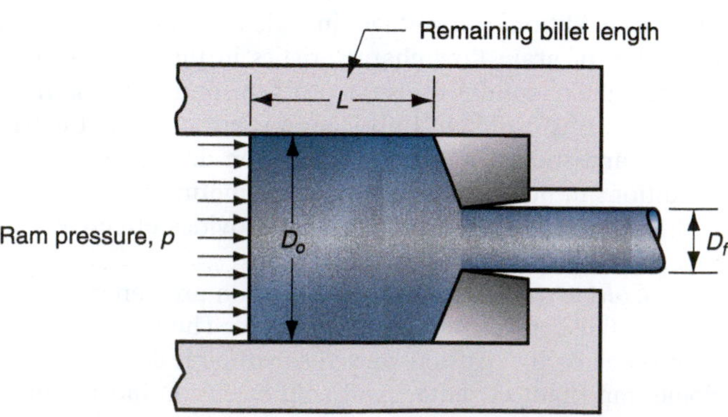

FIGURE 13.27 Pressure and other variables in direct extrusion. (Credit: *Fundamentals of Modern Manufacturing,* 4th Edition by Mikell P. Groover, 2010. Reprinted with permission of John Wiley & Sons, Inc.)

In fact, extrusion is not a frictionless process, and the previous equations grossly underestimate the strain and pressure in an extrusion operation. Friction exists between the die and the work as the billet squeezes down and passes through the die opening. In direct extrusion, friction also exists between the container wall and the billet surface. The effect of friction is to increase the strain experienced by the metal. Thus, the actual pressure is greater than that given by Eq. (13.21), which assumes no friction.

Various methods have been suggested to calculate the actual true strain and associated ram pressure in extrusion [1], [3], [6], [11], [12], and [19]. The following empirical equation proposed by Johnson [11] for estimating extrusion strain has gained considerable recognition:

$$\epsilon_x = a + b \ln r_x \tag{13.22}$$

where ϵ_x = extrusion strain; and a and b are empirical constants for a given die angle. Typical values of these constants are: $a = 0.8$ and $b = 1.2$ to 1.5. Values of a and b tend to increase with increasing die angle.

The ram pressure to perform ***indirect extrusion*** can be estimated based on Johnson's extrusion strain formula as follows:

$$p = \overline{Y}_f \epsilon_x \tag{13.23a}$$

where $\overline{Y}_f$ is calculated based on ideal strain from Eq. (13.20), rather than extrusion strain in Eq. (13.22).

In ***direct extrusion***, the effect of friction between the container walls and the billet causes the ram pressure to be greater than for indirect extrusion. We can write the following expression which isolates the friction force in the direct extrusion container:

$$\frac{p_f \pi D_o^2}{4} = \mu p_c \pi D_o L$$

where p_f = additional pressure required to overcome friction, MPa (lb/in^2); $\pi D_o{}^2/4$ = billet cross-sectional area, mm^2 (in^2); μ = coefficient of friction at the container wall; p_c = pressure of the billet against the container wall, MPa (lb/in^2); and $\pi D_o L$ = area of the interface between billet and container wall, mm^2 (in^2). The right-hand side of this equation indicates the billet-container friction force, and the left-hand side gives the additional ram force to overcome that friction. In the worst case, sticking occurs at the container wall so that friction stress equals shear yield strength of the work metal:

$$\mu p_c \pi D_o L = Y_s \pi D_o L$$

where Y_s = shear yield strength, MPa (lb/in^2). If we assume that $Y_s = \overline{Y}_f/2$, then p_f reduces to the following:

$$p_f = \overline{Y}_f \frac{2L}{D_o}$$

Based on this reasoning, the following formula can be used to compute ram pressure in direct extrusion:

$$p = \overline{Y}_f \left(\epsilon_x + \frac{2L}{D_o} \right) \tag{13.23b}$$

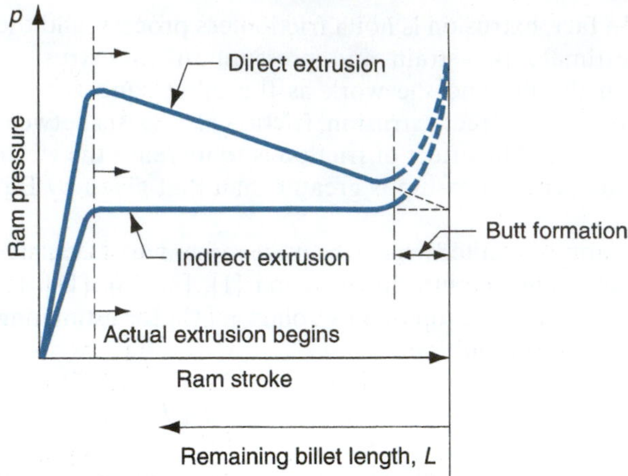

FIGURE 13.28 Typical plots of ram pressure versus ram stroke (and remaining billett length) for direct and indirect extrusion. The higher values in direct extrusion result from friction at the container wall. The shape of the initial pressure buildup at the beginning of the plot depends on die angle (higher die angles cause steeper pressure buildups). The pressure increase at the end of the stroke is related to formation of the butt. (Credit: *Fundamentals of Modern Manufacturing,* 4th Edition by Mikell P. Groover, 2010. Reprinted with permission of John Wiley & Sons, Inc.)

where the term $2L/D_o$ accounts for the additional pressure due to friction at the container-billet interface. L is the portion of the billet length remaining to be extruded, and D_o is the original diameter of the billet. Note that p is reduced as the remaining billet length decreases during the process. Typical plots of ram pressure as a function of ram stroke for direct and indirect extrusion are presented in Figure 13.28. Eq. (13.23b) probably overestimates ram pressure. With good lubrication, ram pressures would be lower than values calculated by this equation.

Ram force in indirect or direct extrusion is simply pressure p from Eqs. (13.23a) or (13.23b), respectively, multiplied by billet area A_o:

$$F = pA_o \tag{13.24}$$

where F = ram force in extrusion, N (lb). Power required to carry out the extrusion operation is simply

$$P = Fv \tag{13.25}$$

where P = power, J/s (in-lb/min); F = ram force, N (lb); and v = ram velocity, m/s (in/min).

Example 13.3 Extrusion Pressures

A billet 75 mm long and 25 mm in diameter is to be extruded in a direct extrusion operation with extrusion ratio $r_x = 4.0$. The extrudate has a round cross section. The die angle (half-angle) = 90°. The work metal has a strength coefficient = 415 MPa, and strain-hardening exponent = 0.18. Use the Johnson formula with $a = 0.8$ and $b = 1.5$ to estimate extrusion strain. Determine the pressure applied to the end of the billet as the ram moves forward.

Solution: Let us examine the ram pressure at billet lengths of L = 75 mm (starting value), L = 50 mm, L = 25 mm, and L = 0. We compute the ideal true strain, extrusion strain using Johnson's formula, and average flow stress:

$$\epsilon = \ln r_x = \ln 4.0 = 1.3863$$

$$\epsilon_x = 0.8 + 1.5(1.3863) = 2.8795$$

$$\overline{Y}_f = \frac{415(1.3863)^{0.18}}{1.18} = 373 \text{ MPa}$$

$L = 75$ mm: With a die angle of 90°, the billet metal is assumed to be forced through the die opening almost immediately; thus, our calculation assumes that maximum pressure is reached at the billet length of 75 mm. For die angles less than 90°, the pressure would build to a maximum as in Figure 13.28 as the starting billet is squeezed into the cone-shaped portion of the extrusion die. Using Eq. (13.23b),

$$p = 373\left(2.8795 + 2\frac{75}{25}\right) = 3312 \text{ MPa}$$

$$L = 50 \text{ mm} : p = 373\left(2.8795 + 2\frac{50}{25}\right) = 2566 \text{ MPa}$$

$$L = 25 \text{ mm} : p = 373\left(2.8795 + 2\frac{25}{25}\right) = 1820 \text{ MPa}$$

$L = 0$: Zero length is a hypothetical value in direct extrusion. In reality, it is impossible to squeeze all of the metal through the die opening. Instead, a portion of the billet (the "butt") remains unextruded and the pressure begins to increase rapidly as L approaches zero. This increase in pressure at the end of the stroke is seen in the plot of ram pressure versus ram stroke in Figure 13.28. Calculated below is the hypothetical minimum value of ram pressure that would result at $L = 0$.

$$p = 373\left(2.8795 + 2\frac{0}{25}\right) = 1074 \text{ MPa}$$

This is also the value of ram pressure that would be associated with indirect extrusion throughout the length of the billet. ■

13.3.3 EXTRUSION DIES AND PRESSES

Important factors in an extrusion die are die angle and orifice shape. Die angle, more precisely die half-angle, is shown as α in Figure 13.29(a). For low angles, surface area of the die is large, leading to increased friction at the die-billet interface. Higher friction results in larger ram force. On the other hand, a large die angle causes more turbulence in the metal flow during reduction, increasing the ram force required. Thus, the effect of die angle on ram force is a U-shaped function, as in Figure 13.29(b). An optimum die angle exists, as suggested by our hypothetical plot. The optimum angle depends on various factors (e.g., work material, billet temperature, and lubrication) and is therefore difficult to determine for a given extrusion job. Die designers rely on rules of thumb and judgment to decide the appropriate angle.

Our previous equations for ram pressure, Eqs. (13.23a, b), apply to a circular die orifice. The shape of the die orifice affects the ram pressure required to perform an extrusion operation. A complex cross section, such as the one shown in Figure 13.30, requires a higher pressure and greater force than a circular shape. The effect of the die orifice shape can be assessed by the die ***shape factor***, defined as the ratio of the pressure required to extrude a cross section of a given shape relative to the extrusion pressure for a

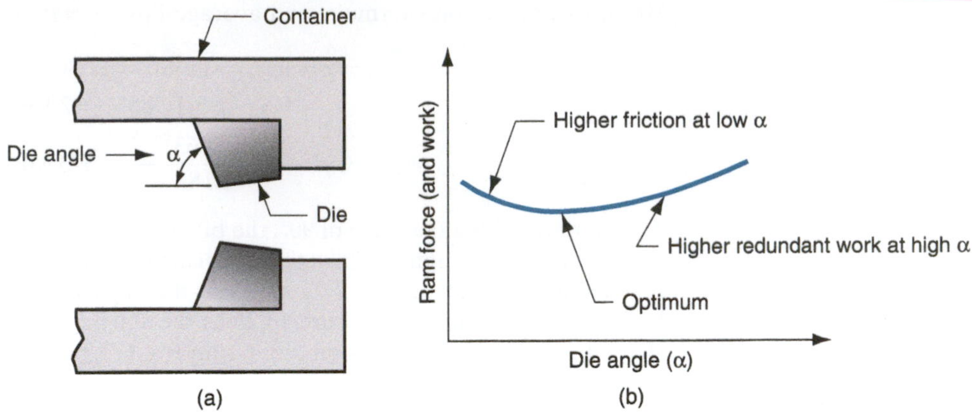

FIGURE 13.29 (a) Definition of die angle in direct extrusion; (b) effect of die angle on ram force. (Credit: *Fundamentals of Modern Manufacturing*, 4th Edition by Mikell P. Groover, 2010. Reprinted with permission of John Wiley & Sons, Inc.)

round cross section of the same area. We can express the shape factor as follows:

$$K_x = 0.98 + 0.02\left(\frac{C_x}{C_c}\right)^{2.25} \tag{13.26}$$

where K_x = die shape factor in extrusion; C_x = perimeter of the extruded cross section, mm (in); and C_c = perimeter of a circle of the same area as the extruded shape, mm (in).

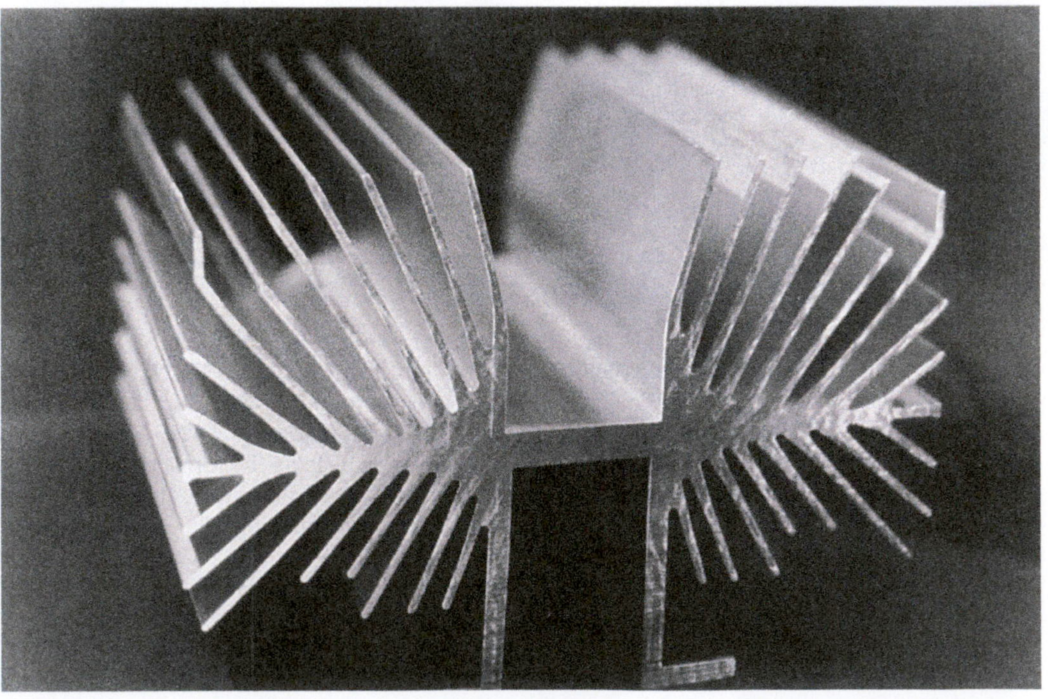

FIGURE 13.30 A complex extruded cross section for a heat sink. Photo courtesy of Aluminum Company of America. (Credit: *Fundamentals of Modern Manufacturing*, 4th Edition by Mikell P. Groover, 2010. Reprinted with permission of John Wiley & Sons, Inc.)

Eq. (13.26) is based on empirical data in Altan et al. [1] over a range of C_x/C_c values from 1.0 to about 6.0. The equation may be invalid much beyond the upper limit of this range.

As indicated by Eq. (13.26), the shape factor is a function of the perimeter of the extruded cross section divided by the perimeter of a circular cross section of equal area. A circular shape is the simplest shape, with a value of $K_x = 1.0$. Hollow, thin-walled sections have higher shape factors and are more difficult to extrude. The increase in pressure is not included in our previous pressure equations, Eqs. (13.23a, b), which apply only to round cross sections. For shapes other than round, the corresponding expression for indirect extrusion is

$$p = K_x \overline{Y}_f \epsilon_x \tag{13.27a}$$

and for direct extrusion,

$$p = K_x \overline{Y}_f \left(\epsilon_x + \frac{2L}{D_o} \right) \tag{13.27b}$$

where p = extrusion pressure, MPa (lb/in^2); K_x = shape factor; and the other terms have the same interpretation as before. Values of pressure given by these equations can be used in Eq. (13.24) to determine ram force.

Die materials used for hot extrusion include tool and alloy steels. Important properties of these die materials include high wear resistance, high hot hardness, and high thermal conductivity to remove heat from the process. Die materials for cold extrusion include tool steels and cemented carbides. Wear resistance and ability to retain shape under high stress are desirable properties. Carbides are used when high production rates, long die life, and good dimensional control are required.

Extrusion presses are either horizontal or vertical, depending on orientation of the work axis. Horizontal types are more common. Extrusion presses are usually hydraulically driven. This drive is especially suited to semicontinuous production of long sections, as in direct extrusion. Mechanical drives are often used for cold extrusion of individual parts, such as in impact extrusion.

13.3.4 OTHER EXTRUSION PROCESSES

Direct and indirect extrusion are the principal methods of extrusion. Other extrusion operations are unique. We examine two of them in this section.

Impact Extrusion Impact extrusion is performed at higher speeds and shorter strokes than conventional extrusion. It is used to make individual components. As the name suggests, the punch impacts the workpart rather than simply applying pressure to it. Impacting can be carried out as forward extrusion, backward extrusion, or combinations of these. Some representative examples are shown in Figure 13.31.

Impact extrusion is usually done cold on a variety of metals. Backward impact extrusion is most common. Products made by this process include toothpaste tubes and battery cases. As indicated by these examples, very thin walls are possible on impact extruded parts. The high-speed characteristics of impacting permit large reductions and high production rates, making this an important commercial process.

Hydrostatic Extrusion One of the problems in direct extrusion is friction along the billet-container interface. This problem can be addressed by surrounding the billet with fluid inside the container and pressurizing the fluid by the forward motion of the ram, as

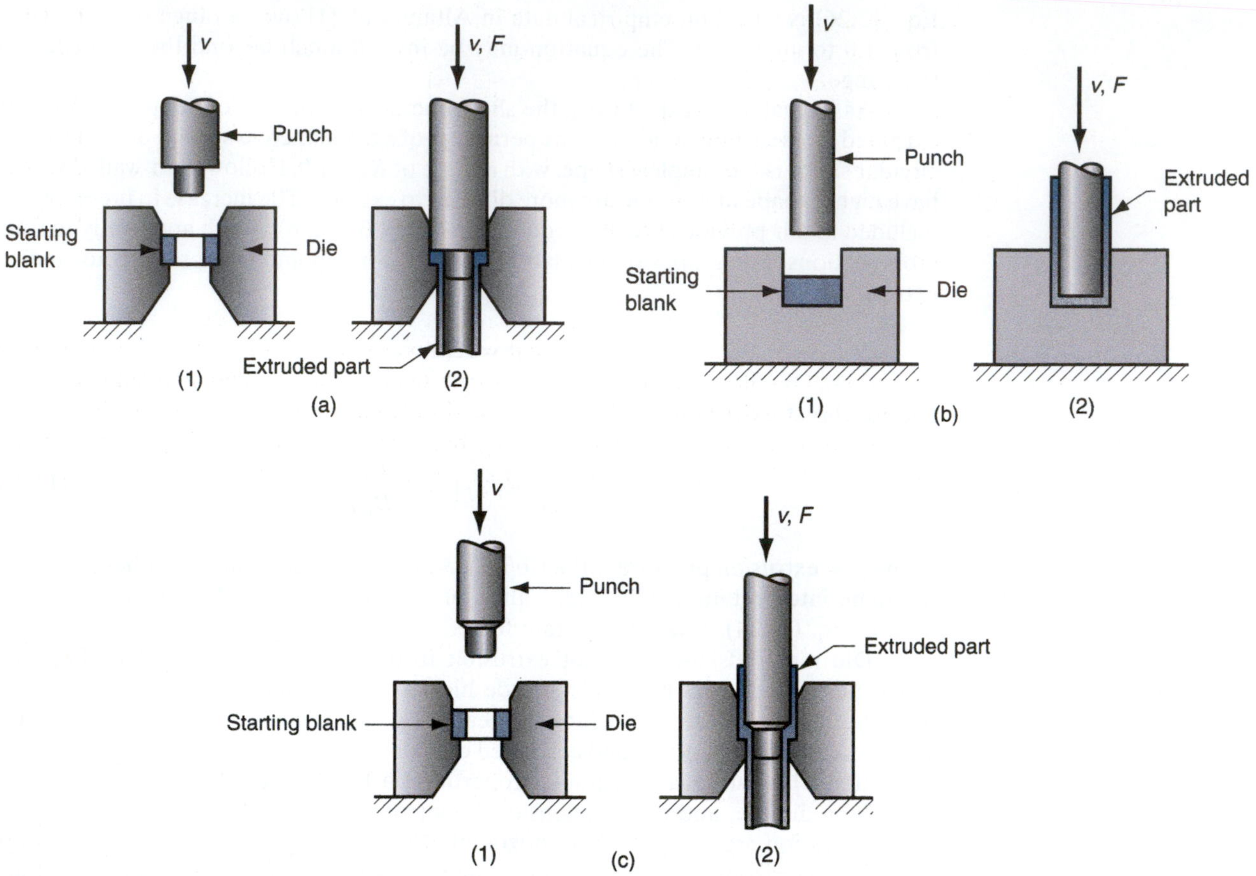

FIGURE 13.31 Several examples of impact extrusion: (a) forward, (b) backward, and (c) combination of forward and backward. (Credit: *Fundamentals of Modern Manufacturing,* 4th Edition by Mikell P. Groover, 2010. Reprinted with permission of John Wiley & Sons, Inc.)

in Figure 13.32. This way, there is no friction inside the container, and friction at the die opening is reduced. Consequently, ram force is significantly lower than in direct extrusion. The fluid pressure acting on all surfaces of the billet gives the process its name. It can be carried out at room temperature or at elevated temperatures. Special fluids and procedures must be used at elevated temperatures. Hydrostatic extrusion is an adaptation of direct extrusion.

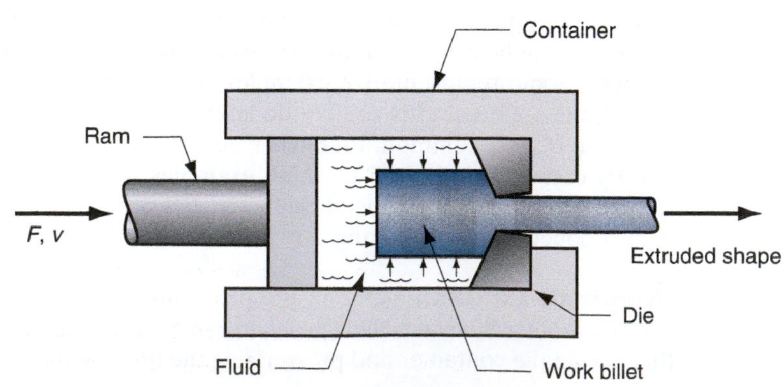

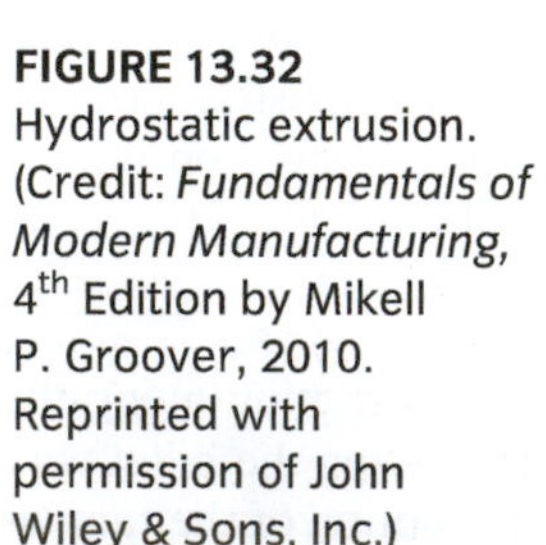

FIGURE 13.32
Hydrostatic extrusion. (Credit: *Fundamentals of Modern Manufacturing,* 4th Edition by Mikell P. Groover, 2010. Reprinted with permission of John Wiley & Sons, Inc.)

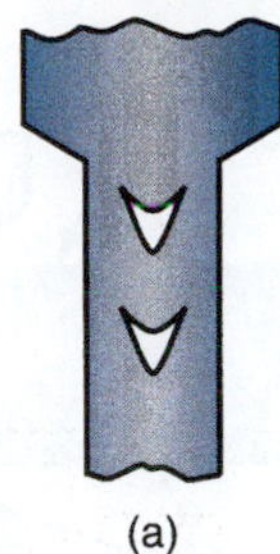

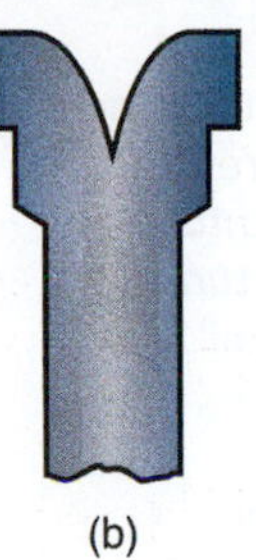

FIGURE 13.33 Some common defects in extrusion: (a) centerburst, (b) piping, and (c) surface cracking. (Credit: *Fundamentals of Modern Manufacturing*, 4th Edition by Mikell P. Groover, 2010. Reprinted with permission of John Wiley & Sons, Inc.)

Hydrostatic pressure on the work increases the material's ductility. Accordingly, this process can be used on metals that would be too brittle for conventional extrusion operations. Ductile metals can also be hydrostatically extruded, and high reduction ratios are possible on these materials. One of the disadvantages of the process is the required preparation of the starting work billet. The billet must be formed with a taper at one end to fit snugly into the die entry angle. This establishes a seal to prevent fluid from squirting out the die hole when the container is initially pressurized.

13.3.5 DEFECTS IN EXTRUDED PRODUCTS

Owing to the considerable deformation associated with extrusion operations, a number of defects can occur in extruded products. The defects can be classified into the following categories, illustrated in Figure 13.33:

(a) ***Centerburst***. This defect is an internal crack that develops as a result of tensile stresses along the centerline of the workpart during extrusion. Although tensile stresses may seem unlikely in a compression process such as extrusion, they tend to occur under conditions that cause large deformation in the regions of the work away from the central axis. The significant material movement in these outer regions stretches the material along the center of the work. If stresses are great enough, bursting occurs. Conditions that promote centerburst are high die angles, low extrusion ratios, and impurities in the work metal that serve as starting points for crack defects. The difficult aspect of centerburst is its detection. It is an internal defect that is usually not noticeable by visual observation. Other names sometimes used for this defect include ***arrowhead fracture***, ***center cracking***, and ***chevron cracking***.

(b) ***Piping***. Piping is a defect associated with direct extrusion. As in Figure 13.33(b), it is the formation of a sink hole in the end of the billet. The use of a dummy block whose diameter is slightly less than that of the billet helps to avoid piping. Other names given to this defect include ***tailpipe*** and ***fishtailing***.

(c) ***Surface cracking***. This defect results from high workpart temperatures that cause cracks to develop at the surface. They often occur when extrusion speed is too high, leading to high strain rates and associated heat generation. Other factors contributing to surface cracking are high friction and surface chilling of high temperature billets in hot extrusion.

13.4 WIRE AND BAR DRAWING

In the context of bulk deformation, drawing is an operation in which the cross section of a bar, rod, or wire is reduced by pulling it through a die opening, as in Figure 13.34. The general features of the process are similar to those of extrusion. The difference is that the work is pulled through the die in drawing, whereas it is pushed through the die in

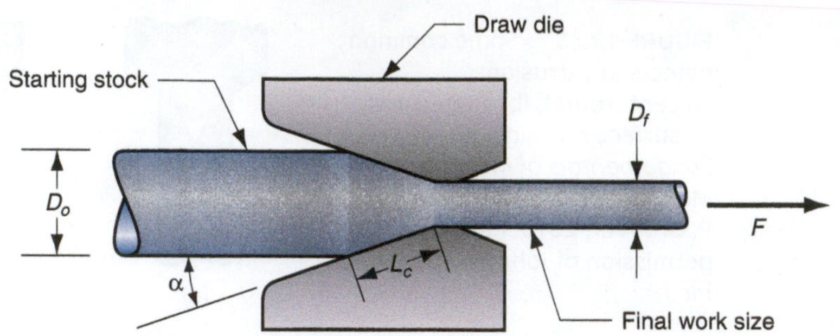

FIGURE 13.34 Drawing of bar, rod, or wire. (Credit: *Fundamentals of Modern Manufacturing*, 4th Edition by Mikell P. Groover, 2010. Reprinted with permission of John Wiley & Sons, Inc.)

extrusion. Although the presence of tensile stresses is obvious in drawing, compression also plays a significant role because the metal is squeezed down as it passes through the die opening. For this reason, the deformation that occurs in drawing is sometimes referred to as indirect compression. Drawing is also a term used in sheet metalworking (Section 14.3). The term ***wire and bar drawing*** is used to distinguish the drawing process discussed here from the sheet metal process of the same name.

The basic difference between bar drawing and wire drawing is the stock size that is processed. ***Bar drawing*** is the term used for large diameter bar and rod stock, while ***wire drawing*** applies to small diameter stock. Wire sizes down to 0.03 mm (0.001 in) are possible in wire drawing. Although the mechanics of the process are the same for the two cases, the methods, equipment, and even the terminology are somewhat different.

Bar drawing is generally accomplished as a ***single-draft*** operation—the stock is pulled through one die opening. Because the beginning stock has a large diameter, it is in the form of a straight cylindrical piece. This limits the length of the work that can be drawn, necessitating a batch type operation. By contrast, wire is drawn from coils consisting of several hundred (or even several thousand) feet of wire and is passed through a series of draw dies. The number of dies varies typically between 4 and 12. The term ***continuous drawing*** is used to describe this type of operation because of the long production runs that are achieved with the wire coils, which can be butt-welded each to the next to make the operation truly continuous.

In a drawing operation, the change in size of the work is usually given by the area reduction, defined as follows:

$$r = \frac{A_o - A_f}{A_o} \tag{13.28}$$

where r = area reduction in drawing; A_o = original area of work, mm^2 (in^2); and A_f = final area, mm^2 (in^2). Area reduction is often expressed as a percentage.

In bar drawing, rod drawing, and in drawing of large diameter wire for upsetting and heading operations, the term *draft* is used to denote the before and after difference in size of the processed work. The ***draft*** is simply the difference between original and final stock diameters:

$$d = D_o - D_f \tag{13.29}$$

where d = draft, mm (in); D_o = original diameter of work, mm (in); and D_f = final work diameter, mm (in).

13.4.1 ANALYSIS OF DRAWING

In this section, we consider the mechanics of wire and bar drawing. How are stresses and forces computed in the process? We also consider how large a reduction is possible in a drawing operation.

Mechanics of Drawing If no friction or redundant work occurred in drawing, true strain could be determined as follows:

$$\epsilon = \ln\frac{A_o}{A_f} = \ln\frac{1}{1-r} \tag{13.30}$$

where A_o and A_f are the original and final cross-sectional areas of the work, as previously defined; and r = drawing reduction as given by Eq. (13.28). The stress that results from this ideal deformation is given by

$$\sigma = \overline{Y}_f \epsilon = \overline{Y}_f \ln\frac{A_o}{A_f} \tag{13.31}$$

where $\overline{Y}_f = \frac{K\epsilon^n}{1+n}$ = average flow stress based on the value of strain given by Eq. (13.30).

Because friction is present in drawing and the work metal experiences inhomogeneous deformation, the actual stress is larger than provided by Eq. (13.31). In addition to the ratio A_o/A_f, other variables that influence draw stress are die angle and coefficient of friction at the work–die interface. A number of methods have been proposed for predicting draw stress based on values of these parameters [1], [3], and [19]. We present the equation suggested by Schey [19]:

$$\sigma_d = \overline{Y}_f\left(1 + \frac{\mu}{\tan\alpha}\right)\phi \ln\frac{A_o}{A_f} \tag{13.32}$$

where σ_d = draw stress, MPa (lb/in^2); μ = die–work coefficient of friction; α = die angle (half-angle) as defined in Figure 13.34; and ϕ is a factor that accounts for inhomogeneous deformation, which is determined as follows for a round cross section:

$$\phi = 0.88 \pm 0.12\frac{D}{L_c} \tag{13.33}$$

where D = average diameter of work during drawing, mm (in); and L_c = contact length of the work with the draw die in Figure 13.34, mm (in). Values of D and L_c can be determined from the following:

$$D = \frac{D_o + D_f}{2} \tag{13.34a}$$

$$L_c = \frac{D_o - D_f}{2\sin\alpha} \tag{13.34b}$$

The corresponding draw force is then the area of the drawn cross section multiplied by the draw stress:

$$F = A_f\sigma_d = A_f\overline{Y}_f\left(1 + \frac{\mu}{\tan\alpha}\right)\phi \ln\frac{A_o}{A_f} \tag{13.35}$$

where F = draw force, N (lb); and the other terms are defined above. The power required in a drawing operation is the draw force multiplied by exit velocity of the work.

Example 13.4 Stress and Force in Wire Drawing

Wire is drawn through a draw die with entrance angle = 15°. Starting diameter is 2.5 mm and final diameter = 2.0 mm. The coefficient of friction at the work–die interface = 0.07. The metal has a strength coefficient K = 205 MPa and a strain-hardening exponent n = 0.20. Determine the draw stress and draw force in this operation.

Solution: The values of D and L_c for Eq. (13.33) can be determined using Eqs. (13.34a, b). $D = 2.25$ mm and $L_c = 0.966$ mm. Thus,

$$\phi = 0.88 + 0.12\frac{2.25}{0.966} = 1.16$$

The areas before and after drawing are computed as $A_o = 4.91$ mm^2 and $A_f = 3.14$ mm^2. The resulting true strain $\epsilon = \ln(4.91/3.14) = 0.446$, and the average flow stress in the operation is computed:

$$\overline{Y}_f = \frac{205(0.446)^{0.20}}{1.20} = 145.4 \text{ MPa}$$

Draw stress is given by Eq. (13.32):

$$\sigma_d = (145.4)\left(1 + \frac{0.07}{\tan 15}\right)(1.16)(0.446) = 94.1 \text{ MPa}$$

Finally, the draw force is this stress multiplied by the cross-sectional area of the exiting wire:

$$F = 94.1(3.14) = 295.5 \text{ N}$$

■

Maximum Reduction per Pass A question that may occur to the reader is: Why is more than one step required to achieve the desired reduction in wire drawing? Why not take the entire reduction in a single pass through one die, as in extrusion? The answer can be explained as follows. From the preceding equations, it is clear that as the reduction increases, draw stress increases. If the reduction is large enough, draw stress will exceed the yield strength of the exiting metal. When that happens, the drawn wire will simply elongate instead of new material being squeezed through the die opening. For wire drawing to be successful, maximum draw stress must be less than the yield strength of the exiting metal.

It is a straightforward matter to determine this maximum draw stress and the resulting maximum possible reduction that can be made in one pass, under certain assumptions. Let us assume a perfectly plastic metal ($n = 0$), no friction, and no redundant work. In this ideal case, the maximum possible draw stress is equal to the yield strength of the work material. Expressing this using the equation for draw stress under conditions of ideal deformation, Eq. (13.31), and setting $\overline{Y}_f = Y$ (because $n = 0$),

$$\sigma_d = \overline{Y}_f \ln\frac{A_o}{A_f} = Y \ln\frac{A_o}{A_f} = Y \ln\frac{1}{1-r} = Y$$

This means that $\ln(A_o/A_f) = \ln(1/(1-r)) = 1$. That is, $\epsilon_{max} = 1.0$. In order for ϵ_{max} to be zero, then $A_o/A_f = 1/(1-r)$ must equal the natural logarithm base e. Accordingly, the maximum possible area ratio is

$$\frac{A_o}{A_f} = e = 2.7183 \qquad (13.36)$$

and the maximum possible reduction is

$$r_{max} = \frac{e-1}{e} = 0.632 \qquad (13.37)$$

The value given by Eq. (13.37) is often used as the theoretical maximum reduction possible in a single draw, even though it ignores (1) the effects of friction and redundant work, which would reduce the maximum possible value, and (2) strain hardening, which would increase the maximum possible reduction because the exiting wire would be stronger than the starting metal. In practice, draw reductions per pass are quite below the theoretical limit. Reductions of 0.50 for single-draft bar drawing and 0.30 for multiple-draft wire drawing seem to be the upper limits in industrial operations.

13.4.2 DRAWING PRACTICE

Drawing is usually performed as a cold-working operation. It is most frequently used to produce round cross sections, but squares and other shapes are also drawn. Wire drawing is an important industrial process, providing commercial products such as electrical wire and cable; wire stock for fences, coat hangers, and shopping carts; and rod stock to produce nails, screws, rivets, springs, and other hardware items. Bar drawing is used to produce metal bars for machining, forging, and other processes.

Advantages of drawing in these applications include (1) close dimensional control, (2) good surface finish, (3) improved mechanical properties such as strength and hardness, and (4) adaptability to economical batch or mass production. Drawing speeds are as high as 50 m/s (10,000 ft/min) for very fine wire. In the case of bar drawing to provide stock for machining, the operation improves the machinability of the bar (Section 17.5).

Drawing Equipment Bar drawing is accomplished on a machine called a ***draw bench***, consisting of an entry table, die stand (which contains the draw die), carriage, and exit rack. The arrangement is shown in Figure 13.35. The carriage is used to pull the stock through the draw die. It is powered by hydraulic cylinders or motor-driven chains. The die stand is often designed to hold more than one die, so that several bars can be pulled simultaneously through their respective dies.

Wire drawing is done on continuous drawing machines that consist of multiple draw dies, separated by accumulating drums between the dies, as in Figure 13.36. Each drum, called a ***capstan***, is motor-driven to provide the proper pull force to draw the wire stock through the upstream die. It also maintains a modest tension on the wire as it proceeds to the next draw die in the series. Each die provides a certain amount of

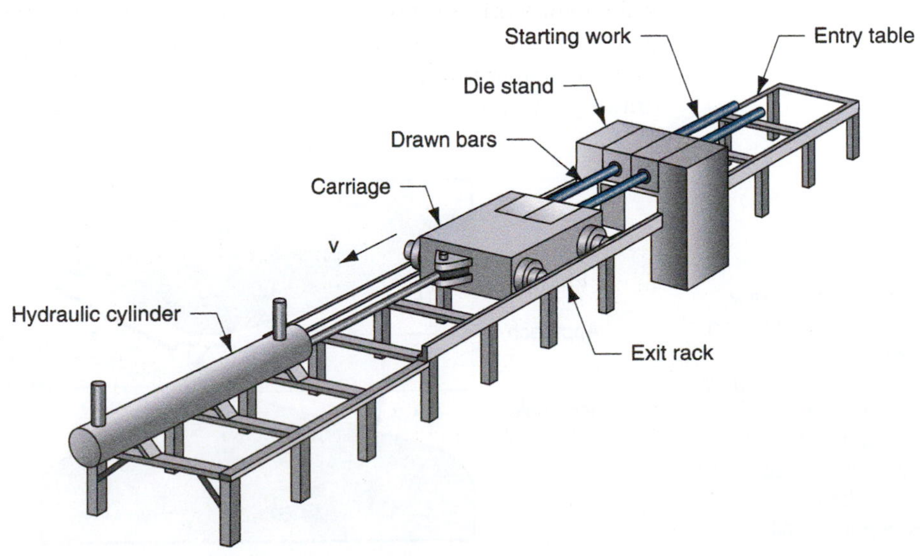

FIGURE 13.35 Hydraulically operated draw bench for drawing metal bars. (Credit: *Fundamentals of Modern Manufacturing*, 4th Edition by Mikell P. Groover, 2010. Reprinted with permission of John Wiley & Sons, Inc.)

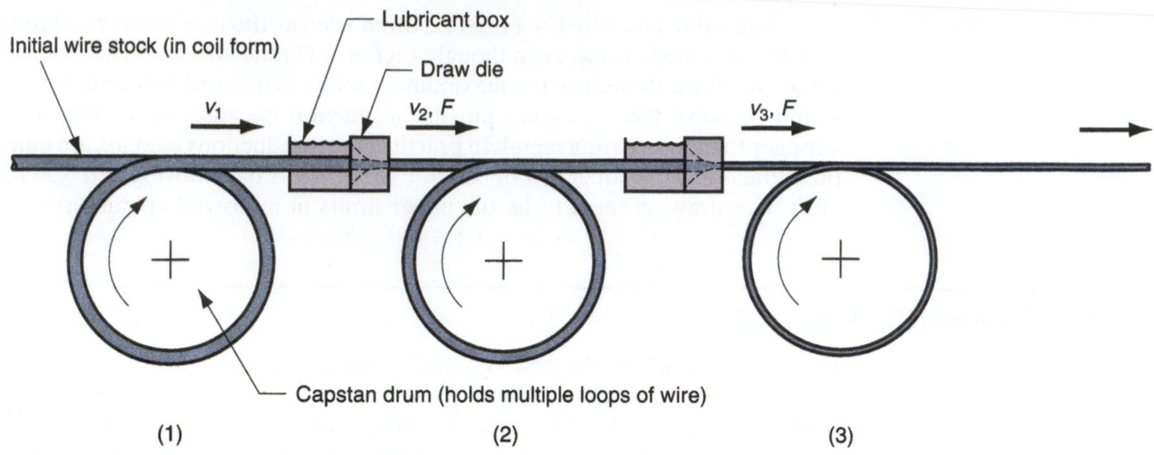

FIGURE 13.36 Continuous drawing of wire. (Credit: *Fundamentals of Modern Manufacturing,* 4th Edition by Mikell P. Groover, 2010. Reprinted with permission of John Wiley & Sons, Inc.)

reduction in the wire, so that the desired total reduction is achieved by the series. Depending on the metal to be processed and the total reduction, annealing of the wire is sometimes required between groups of dies in the series.

Draw Dies Figure 13.37 identifies the features of a typical draw die. Four regions of the die can be distinguished: (1) entry, (2) approach angle, (3) bearing surface (land), and (4) back relief. The ***entry*** region is usually a bell-shaped mouth that does not contact the work. Its purpose is to funnel the lubricant into the die and prevent scoring of work and die surfaces. The ***approach*** is where the drawing process occurs. It is cone-shaped with an angle (half-angle) normally ranging from about 6° to 20°. The proper angle varies according to work material. The ***bearing surface***, or ***land***, determines the size of the final drawn stock. Finally, the ***back relief*** is the exit zone. It is provided with a back relief angle (half-angle) of about 30°. Draw dies are made of tool steels or cemented carbides. Dies for high-speed wire drawing operations frequently use inserts made of diamond (both synthetic and natural) for the wear surfaces.

Preparation of the Work Prior to drawing, the beginning stock must be properly prepared. This involves three steps: (1) annealing, (2) cleaning, and (3) pointing. The purpose of annealing is to increase the ductility of the stock to accept deformation during drawing. As previously mentioned, annealing is sometimes needed between steps in

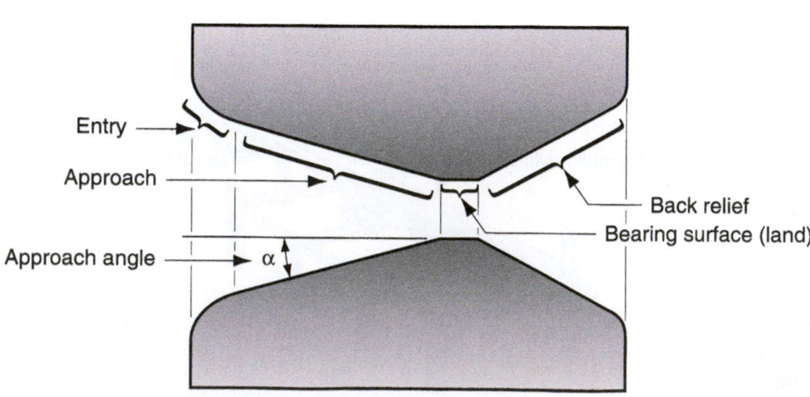

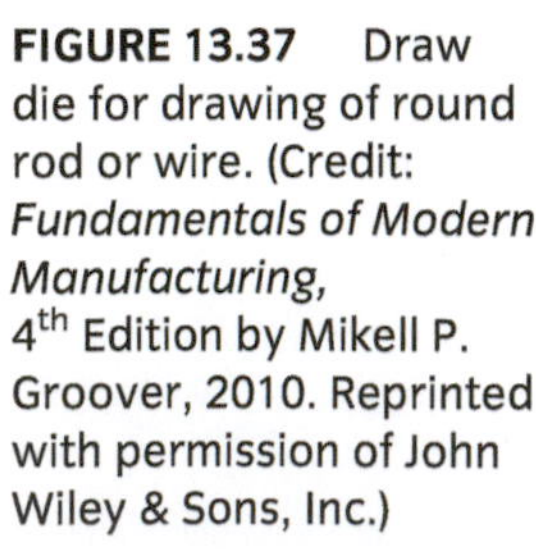
FIGURE 13.37 Draw die for drawing of round rod or wire. (Credit: *Fundamentals of Modern Manufacturing,* 4th Edition by Mikell P. Groover, 2010. Reprinted with permission of John Wiley & Sons, Inc.)

continuous drawing. Cleaning of the stock is required to prevent damage of the work surface and draw die. It involves removal of surface contaminants (e.g., scale and rust) by means of chemical pickling or shot blasting. In some cases, prelubrication of the work surface is accomplished subsequent to cleaning.

Pointing involves the reduction in diameter of the starting end of the stock so that it can be inserted through the draw die to start the process. This is usually accomplished by swaging, rolling, or turning. The pointed end of the stock is then gripped by the carriage jaws or other device to initiate the drawing process.

REFERENCES

[1] Altan, T., Oh, S-I., and Gegel, H. L. ***Metal Forming: Fundamentals and Applications***. ASM International, Materials Park, Ohio, 1983.

[2] ***ASM Handbook***, Vol. **14A**, ***Metalworking: Bulk Forming***. ASM International, Materials Park, Ohio, 2005.

[3] Avitzur, B. ***Metal Forming: Processes and Analysis***. Robert E. Krieger Publishing Company, Huntington, New York, 1979.

[4] Black, J. T., and Kohser, R. A. ***DeGarmo's Materials and Processes in Manufacturing***, 10th ed. John Wiley & Sons, Inc., Hoboken, New Jersey, 2008.

[5] Byrer, T. G., et al. (eds.). ***Forging Handbook***. Forging Industry Association, Cleveland, Ohio; and American Society for Metals, Materials Park, Ohio, 1985.

[6] Cook, N. H. ***Manufacturing Analysis***. Addison-Wesley Publishing Company, Inc., Reading, Massachusetts, 1966.

[7] Groover, M. P. "An Experimental Study of the Work Components and Extrusion Strain in the Cold Forward Extrusion of Steel." ***Research Report***. Bethlehem Steel Corporation, 1966.

[8] Harris, J. N. ***Mechanical Working of Metals***. Pergamon Press, Oxford, England, 1983.

[9] Hosford, W. F., and Caddell, R. M. ***Metal Forming: Mechanics and Metallurgy***, 3rd ed. Cambridge University Press, Cambridge, United Kingdom, 2007.

[10] Jensen, J. E. (ed.). ***Forging Industry Handbook***. Forging Industry Association, Cleveland, Ohio, 1970.

[11] Johnson, W. "The Pressure for the Cold Extrusion of Lubricated Rod Through Square Dies of Moderate Reduction at Slow Speeds." ***Journal of the Institute of Metals***, Vol. 85, 1956.

[12] Kalpakjian, S. ***Mechanical Processing of Materials***. D. Van Nostrand Company, Inc., Princeton, New Jersey, 1967, Chapter 5.

[13] Kalpakjian, S., and Schmid, S. R. ***Manufacturing Processes for Engineering Materials***, 6th ed. Pearson Prentice Hall, Upper Saddle River, New Jersey, 2010.

[14] Lange, K. ***Handbook of Metal Forming***. Society of Manufacturing Engineers, Dearborn, Michigan, 2006.

[15] Laue, K., and Stenger, H. ***Extrusion: Processes, Machinery, and Tooling***. American Society for Metals, Materials Park, Ohio, 1981.

[16] Mielnik, E. M. ***Metalworking Science and Engineering***. McGraw-Hill, Inc., New York, 1991.

[17] Roberts, W. L. ***Hot Rolling of Steel***. Marcel Dekker, Inc., New York, 1983.

[18] Roberts, W. L. ***Cold Rolling of Steel***. Marcel Dekker, Inc., New York, 1978.

[19] Schey, J. A. ***Introduction to Manufacturing Processes***, 3rd ed. McGraw-Hill Book Company, New York, 2000.

[20] Wick, C., et al. (eds.). ***Tool and Manufacturing Engineers Handbook***, 4th ed., Vol. II, ***Forming***. Society of Manufacturing Engineers, Dearborn, Michigan, 1984.

REVIEW QUESTIONS

13.1. What are the reasons why the bulk deformation processes are important commercially and technologically?

13.2. Name the four basic bulk deformation processes.

13.3. What is rolling in the context of the bulk deformation processes?

13.4. List some of the products produced on a rolling mill.

13.5. What is draft in a rolling operation?

13.6. What is sticking in a hot rolling operation?

13.7. Identify some of the ways in which force in flat rolling can be reduced.

13.8. What is a two-high rolling mill?

13.9. What is a reversing mill in rolling?

13.10. What is forging?

13.11. One way to classify forging operations is by the degree to which the work is constrained in the die. By this classification, name the three basic types.

13.12. Why is flash desirable in impression-die forging?

13.13. What is a trimming operation in the context of impression-die forging?

13.14. What are the two basic types of forging equipment?

13.15. What is isothermal forging?

13.16. What is extrusion?

13.17. Distinguish between direct and indirect extrusion.

13.18. Name some products that are produced by extrusion.

13.19. Why is friction a factor in determining the ram force in direct extrusion but not a factor in indirect extrusion?

13.20. What are wire drawing and bar drawing?

13.21. Although the workpiece in a wire drawing operation is obviously subjected to tensile stresses, how do compressive stresses also play a role in the process?

13.22. In a wire drawing operation, why must the drawing stress never exceed the yield strength of the work metal?

PROBLEMS

13.1. A 42.0-mm-thick plate made of low carbon steel is to be reduced to 34.0 mm in one pass in a rolling operation. As the thickness is reduced, the plate widens by 4%. The yield strength of the steel plate is 174 MPa and the tensile strength is 290 MPa. The entrance speed of the plate is 15.0 m/min. The roll radius is 325 mm and the rotational speed is 49.0 rev/min. Determine (a) the minimum required coefficient of friction that would make this rolling operation possible, (b) exit velocity of the plate, and (c) forward slip.

13.2. A 2.0-in-thick slab is 10.0 in wide and 12.0 ft long. Thickness is to be reduced in three steps in a hot rolling operation. Each step will reduce the slab to 75% of its previous thickness. It is expected that for this metal and reduction, the slab will widen by 3% in each step. If the entry speed of the slab in the first step is 40 ft/min, and roll speed is the same for the three steps, determine: (a) length and (b) exit velocity of the slab after the final reduction.

13.3. A series of cold rolling operations are used to reduce the thickness of a plate from 50 mm down to 25 mm in a reversing two-high mill. Roll diameter = 700 mm and coefficient of friction between rolls and work = 0.15. The specification is that the draft is to be equal on each pass. Determine (a) minimum number of passes required, and (b) draft for each pass?

13.4. In the previous problem, suppose that the percent reduction were specified to be equal for each pass, rather than the draft. (a) What is the minimum number of passes required? (b) What is the draft for each pass?

13.5. A plate that is 250 mm wide and 25 mm thick is to be reduced in a single pass in a two-high rolling mill to a thickness of 20 mm. The roll has a radius = 500 mm, and its speed = 30 m/min. The work material has a strength coefficient = 240 MPa and a strain-hardening exponent = 0.2. Determine (a) roll force, (b) roll torque, and (c) power required to accomplish this operation.

13.6. Solve Problem 13.5 using a roll radius = 250 mm.

13.7. A 4.50-in-thick slab that is 9 in wide and 24 in long is to be reduced in a single pass in a two-high rolling mill to a thickness of 3.87 in. The roll rotates at a speed of 5.50 rev/min and has a radius of 17.0 in. The work material has a strength coefficient = 30,000 lb/in^2 and a strain-hardening exponent = 0.15. Determine (a) roll force, (b) roll torque, and (c) power required to accomplish this operation.

13.8. A single-pass rolling operation reduces a 20 mm thick plate to 18 mm. The starting plate is 200 mm wide. Roll radius = 250 mm and rotational speed = 12 rev/min. The work material has a strength coefficient = 600 MPa and a strength coefficient = 0.22. Determine (a) roll force, (b) roll torque, and (c) power required for this operation.

13.9. A hot rolling mill has rolls of diameter = 24 in. It can exert a maximum force = 400,000 lb. The mill has a maximum horsepower = 100 hp. It is desired to reduce a 1.5 in thick plate by the maximum possible draft in one pass. The starting plate is 10 in wide. In the heated condition, the work material has a strength coefficient = 20,000 lb/in^2 and a strain-hardening exponent = zero. Determine (a) maximum possible draft, (b) associated true strain, and (c) maximum speed of the rolls for the operation.

13.10. A cylindrical part is warm upset forged in an open die. The initial diameter is 45 mm and the initial height is 40 mm. The height after forging is 25 mm. The coefficient of friction at the die–work interface is 0.20. The yield strength of the work material is 285 MPa, and its flow curve is defined by a strength coefficient of 600 MPa and a strain-hardening exponent of 0.12. Determine the force in the operation (a) just as the yield point is reached (yield at strain = 0.002), (b) at a height of 35 mm, (c) at a height of 30 mm, and (d) at a height of 25 mm. Use of a spreadsheet calculator is recommended.

13.11. A cylindrical workpart with diameter = 2.5 in and height = 2.5 in is upset forged in an open die to a height = 1.5 in. Coefficient of friction at the die–work interface = 0.10. The work material has a flow curve defined by: $K = 40{,}000$ lb/in^2 and $n = 0.15$. Yield strength = 15,750 lb/in^2. Determine the instantaneous force in the operation (a) just as the yield point is reached (yield at strain = 0.002), (b) at height $h = 2.3$ in, (c) $h = 2.1$ in, (d) $h = 1.9$ in, (e) $h = 1.7$ in, and (f) $h = 1.5$ in. Use of a spreadsheet calculator is recommended.

13.12. A cold heading operation is performed to produce the head on a steel nail. The strength coefficient for this steel is 600 MPa, and the strain-hardening exponent is 0.22. Coefficient of friction at the die–work interface is 0.14. The wire stock out of which the nail is made is 5.00 mm in diameter. The head is to have a diameter of 9.5 mm and a thickness of 1.6 mm. The final length of the nail is 120 mm. (a) What length of stock must project out of the die in order to provide sufficient volume of material for this upsetting operation? (b) Compute the maximum force that the punch must apply to form the head in this open-die operation.

13.13. Obtain a large common nail (flat head). Measure the head diameter and thickness, as well as the diameter of the nail shank. (a) What stock length must project out of the die in order to provide sufficient material to produce the nail? (b) Using appropriate values for strength coefficient and strain-hardening exponent for the metal out of which the nail is made (Table 3.4), compute the maximum force in the heading operation to form the head.

13.14. A hot upset forging operation is performed in an open die. The starting workpart has a diameter = 25 mm and height = 50 mm. The part is upset to an average diameter = 50 mm. The work metal at this elevated temperature yields at 85 MPa ($n = 0$). Coefficient of friction at the die–work interface = 0.40. Determine (a) final height of the part, and (b) maximum force in the operation.

13.15. A hydraulic forging press is capable of exerting a maximum force = 1,000,000 N. A cylindrical workpart is to be cold upset forged. The starting part has diameter = 30 mm and height = 30 mm. The flow curve of the metal is defined by $K =$ 400 MPa and $n = 0.2$. Determine the maximum reduction in height to which the part can be compressed with this forging press, if the coefficient of friction = 0.1. Use of a spreadsheet calculator is recommended.

13.16. A part is designed to be hot forged in an impression die. The projected area of the part, including flash, is 16 in^2. After trimming, the part has a projected area of 10 in^2. Part geometry is complex. As heated the work material yields at 10,000 lb/in^2, and has no tendency to strain harden. At room temperature, the material yields at 25,000 lb/in^2. Determine the maximum force required to perform the forging operation.

13.17. A connecting rod is designed to be hot forged in an impression die. The projected area of the part is 6,500 mm^2. The design of the die will cause flash to form during forging, so that the area, including flash, will be 9,000 mm^2. The part geometry is considered to be complex. As heated the work material yields at 75 MPa, and has no tendency to strain harden. Determine the maximum force required to perform the operation.

13.18. A cylindrical billet that is 100 mm long and 50 mm in diameter is reduced by indirect (backward) extrusion to a 20 mm diameter. The die angle is 90°. In the Johnson equation, $a = 0.8$ and $b = 1.4$. In the flow curve for the work metal, the strength coefficient = 800 MPa and strain-hardening exponent = 0.13. Determine (a) extrusion ratio, (b) true strain (homogeneous deformation), (c) extrusion strain, (d) ram pressure, and (e) ram force.

13.19. A 3.0-in-long cylindrical billet whose diameter = 1.5 in is reduced by indirect extrusion to a diameter = 0.375 in. Die angle = 90°. In the Johnson equation, $a = 0.8$ and $b = 1.5$. In the flow curve for the work metal, $K = 75{,}000$ lb/in^2 and $n = 0.25$. Determine (a) extrusion ratio, (b) true strain (homogeneous deformation), (c) extrusion strain, (d) ram pressure, (e) ram force, and (f) power if the ram speed = 20 in/min.

13.20. A billet that is 75 mm long with diameter = 35 mm is direct extruded to a diameter of 20 mm. The extrusion die has a die angle = 75°. For the work metal, $K = 600$ MPa and $n = 0.25$. In the Johnson extrusion strain equation, $a = 0.8$ and $b = 1.4$. Determine (a) extrusion ratio, (b) true strain (homogeneous deformation), (c) extrusion strain, and (d) ram pressure and force at $L = 70$, 60, 50, 40, 30, 20, and 10 mm. Use of a spreadsheet calculator is recommended for part (d).

13.21. A 2.0-in-long billet with diameter = 1.25 in is direct extruded to a diameter of 0.50 in. The extrusion die angle = 90°. For the work metal, $K = 45{,}000$ lb/in^2, and $n = 0.20$. In the Johnson extrusion strain equation, $a = 0.8$ and $b = 1.5$. Determine (a) extrusion ratio, (b) true strain (homogeneous deformation), (c) extrusion strain, and (d) ram pressure at $L = 2.0, 1.5, 1.0, 0.5$ and zero in. Use of a spreadsheet calculator is recommended for part (d).

13.22. An indirect extrusion process starts with an aluminum billet with diameter = 2.0 in and length = 3.0 in. Final cross section after extrusion is a square with 1.0 in on a side. The die angle = 90°. The operation is performed cold and the strength coefficient of the metal $K = 26{,}000$ lb/in^2 and strain-hardening exponent $n = 0.20$. In the Johnson extrusion strain equation, $a = 0.8$ and $b = 1.2$. (a) Compute the extrusion ratio, true strain, and extrusion strain. (b) What is the shape factor of the product? (c) If the butt left in the container at the end of the stroke is 0.5 in thick, what is the length of the extruded section? (d) Determine the ram pressure in the process.

13.23. A cup-shaped part is backward extruded from an aluminum slug that is 50 mm in diameter. The final dimensions of the cup are: OD = 50 mm, ID = 40 mm, height = 100 mm, and thickness of base = 5 mm. Determine (a) extrusion ratio, (b) shape factor, and (c) height of starting slug required to achieve the final dimensions. (d) If the metal has flow curve parameters $K = 400$ MPa and $n = 0.25$, and the constants in the Johnson extrusion strain equation are: $a = 0.8$ and $b = 1.5$, determine the extrusion force.

13.24. Determine the shape factor for each of the extrusion die orifice shapes in Figure P13.24.

13.25. A direct extrusion operation produces the cross section shown in Figure P13.24(a) from a brass billet whose diameter = 125 mm and length = 350 mm. The flow curve parameters of the brass are $K = 700$ MPa and $n = 0.35$. In the Johnson strain equation, $a = 0.7$ and $b = 1.4$. Determine (a) the extrusion ratio, (b) the shape factor, (c) the force required to drive the ram forward during extrusion at the point in the process when the billet length remaining in the container = 300 mm, and (d) the length of the extruded section at the end of the operation if the volume of the butt left in the container is 600,000 mm^3.

13.26. In a direct extrusion operation the cross section shown in Figure P13.24(b) is produced from a copper billet whose diameter = 100 mm and length = 500 mm. In the flow curve for copper, the strength coefficient = 300 MPa and strain-hardening exponent = 0.50. In the Johnson strain equation, $a = 0.8$ and $b = 1.5$. Determine (a) the extrusion ratio, (b) the shape factor, (c) the force required to drive the ram forward during extrusion at the point in the process when the billet length remaining in the container = 450 mm, and (d) the length of the extruded section at the end of the operation if the volume of the butt left in the container is 350,000 mm^3.

13.27. A direct extrusion operation produces the cross section shown in Figure P13.24(c) from an aluminum billet whose diameter = 150 mm and length = 500 mm. The flow curve parameters for the aluminum are $K = 240$ MPa and $n = 0.16$. In the Johnson strain equation, $a = 0.8$ and $b = 1.2$. Determine (a) the extrusion ratio, (b) the shape factor, (c) the force required to drive the ram forward during extrusion at the point in the process when the billet length remaining in the container = 400 mm, and (d) the length of the extruded section at the end of the operation if the volume of the butt left in the container is 600,000 mm^3.

13.28. A direct extrusion operation produces the cross section shown in Figure P13.24(d) from an aluminum billet whose diameter = 150 mm and length = 900 mm. The flow curve parameters for the

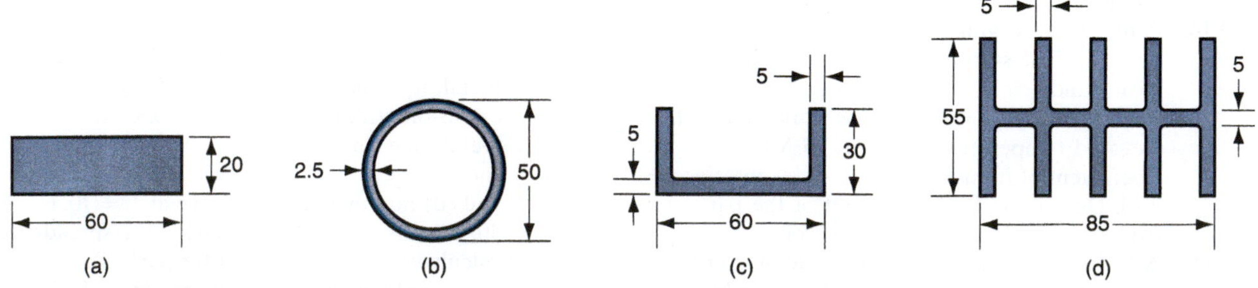

FIGURE P13.24 Cross-sectional shapes for Problem 13.24 (dimensions are mm): (a) rectangular bar, (b) tube, (c) channel, and (d) cooling fins. (Credit: *Fundamentals of Modern Manufacturing*, 4th Edition by Mikell P. Groover, 2010. Reprinted with permission of John Wiley & Sons, Inc.)

aluminum are $K = 240$ MPa and $n = 0.16$. In the Johnson strain equation, $a = 0.8$ and $b = 1.5$. Determine (a) the extrusion ratio, (b) the shape factor, (c) the force required to drive the ram forward during extrusion at the point in the process when the billet length remaining in the container = 850 mm, and (d) the length of the extruded section at the end of the operation if the volume of the butt left in the container is 600,000 mm^3.

13.29. A spool of wire has a starting diameter of 2.5 mm. It is drawn through a die with an opening that is 2.1 mm. The entrance angel of the die is 18° degrees. Coefficient of friction at the work–die interface is 0.08. The work metal has a strength coefficient of 450 MPa and a strain-hardening coefficient of 0.26. The drawing is performed at room temperature. Determine (a) area reduction, (b) draw stress, and (c) draw force required for the operation.

13.30. Rod stock that has an initial diameter of 0.50 in is drawn through a draw die with an entrance angle of 13°. The final diameter of the rod is = 0.375 in. The metal has a strength coefficient of 40,000 lb/in^2 and a strain-hardening exponent of 0.20. Coefficient of friction at the work–die interface = 0.1. Determine (a) area reduction, (b) draw force for the operation, and (c) horsepower to perform the operation if the exit velocity of the stock = 2 ft/sec.

13.31. Bar stock of initial diameter = 90 mm is drawn with a draft = 15 mm. The draw die has an entrance angle = 18°, and the coefficient of friction at the work–die interface = 0.08. The metal behaves as a perfectly plastic material with yield stress = 105 MPa. Determine (a) area reduction, (b) draw stress, (c) draw force required for the operation, and (d) power to perform the operation if exit velocity = 1.0 m/min.

13.32. Wire stock of initial diameter = 0.125 in is drawn through two dies each providing a 0.20 area reduction. The starting metal has a strength coefficient = 40,000 lb/in^2 and a strain-hardening exponent = 0.15. Each die has an entrance angle of 12°, and the coefficient of friction at the work–die interface is estimated to be 0.10. The motors driving the capstans at the die exits can each deliver 1.50 hp at 90% efficiency. Determine the maximum possible speed of the wire as it exits the second die.

14 SHEET METALWORKING

Chapter Contents

14.1 Cutting Operations
- 14.1.1 Shearing, Blanking, and Punching
- 14.1.2 Engineering Analysis of Sheet-Metal Cutting
- 14.1.3 Other Sheet-Metal-Cutting Operations

14.2 Bending Operations
- 14.2.1 V-Bending and Edge Bending
- 14.2.2 Engineering Analysis of Bending
- 14.2.3 Other Bending and Forming Operations

14.3 Drawing
- 14.3.1 Mechanics of Drawing
- 14.3.2 Engineering Analysis of Drawing
- 14.3.3 Other Drawing Operations
- 14.3.4 Defects in Drawing

14.4 Other Sheet-Metal-Forming Operations
- 14.4.1 Operations Performed with Metal Tooling
- 14.4.2 Rubber Forming Processes

14.5 Dies and Presses for Sheet-Metal Processes
- 14.5.1 Dies
- 14.5.2 Presses

14.6 Sheet-Metal Operations Not Performed on Presses
- 14.6.1 Stretch Forming
- 14.6.2 Roll Bending and Roll Forming
- 14.6.3 Spinning
- 14.6.4 High-Energy-Rate Forming

Sheet metalworking includes cutting and forming operations performed on relatively thin sheets of metal. Typical sheet-metal thicknesses are between 0.4 mm (1/64 in) and 6 mm (1/4 in). When thickness exceeds about 6 mm, the stock is usually referred to as plate rather than sheet. The sheet or plate stock used in sheet metalworking is produced by flat rolling (Section 13.1.1). The most commonly used sheet metal is low carbon steel (0.06% to 0.15% C typical). Its low cost and good formability, combined with sufficient strength for most product applications, make it ideal as a starting material.

The commercial importance of sheet metalworking is significant. Consider the number of consumer and industrial products that include sheet or plate metal parts: automobile and truck bodies, airplanes, railway cars, locomotives, farm and construction equipment, appliances, office furniture, and more. Although these examples are conspicuous because they have sheet-metal exteriors, many of their internal components are also made of sheet or plate stock. Sheet-metal parts are generally characterized by high strength, good dimensional accuracy, good surface finish, and relatively low cost. For components that must be made in large quantities, economical mass-production operations can be designed to process the parts. Aluminum beverage cans are a prime example.

Sheet-metal processing is usually performed at room temperature (cold working). The exceptions are when the stock is thick, the metal is brittle, or the deformation is significant. These are usually cases of warm working rather than hot working.

Most sheet-metal operations are performed on machine tools called ***presses***. The term ***stamping press*** is used to distinguish these presses from forging and extrusion presses. The tooling that performs sheet metalwork is called a ***punch and die***; the term ***stamping die*** is also used. The sheet-metal products are called ***stampings***. To facilitate mass production, the sheet metal is often presented to the press as long strips or coils. Various types of punch-and-die tooling and stamping presses are described in Section 14.5. Final sections of the chapter cover various operations that do not utilize conventional punch-and–die tooling, and most of them are not performed on stamping presses.

The three major categories of sheet-metal processes are (1) cutting, (2) bending, and (3) drawing. Cutting is used to separate large sheets into smaller pieces, to cut out part perimeters, and to punch holes in parts. Bending and drawing are used to form sheet-metal parts into their required shapes.

14.1 CUTTING OPERATIONS

Cutting of sheet metal is accomplished by a shearing action between two sharp cutting edges. The shearing action is depicted in the four stop-action sketches of Figure 14.1, in which the upper cutting edge (the punch) sweeps down past a stationary lower cutting edge (the die). As the punch begins to push into the work, ***plastic deformation*** occurs in the surfaces of the sheet. As the punch moves downward, ***penetration*** occurs in which the punch compresses the sheet and cuts into the metal. This penetration zone is generally about one-third the thickness of the sheet. As the punch continues to travel into the work, ***fracture*** is initiated in the work at the two cutting edges. If the clearance between the punch and die is correct, the two fracture lines meet, resulting in a clean separation of the work into two pieces.

The sheared edges of the sheet have characteristic features as in Figure 14.2. At the top of the cut surface is a region called the ***rollover***. This corresponds to the depression made by the punch in the work prior to cutting. It is where initial plastic deformation occurred in the work. Just below the rollover is a relatively smooth region called the ***burnish***. This results from penetration of the punch into the work before fracture began. Beneath the burnish is the ***fractured zone***, a relatively rough surface of the cut edge where continued downward movement of the punch caused fracture of the metal. Finally, at the bottom of the edge is a ***burr***, a sharp corner on the edge caused by elongation of the metal during final separation of the two pieces.

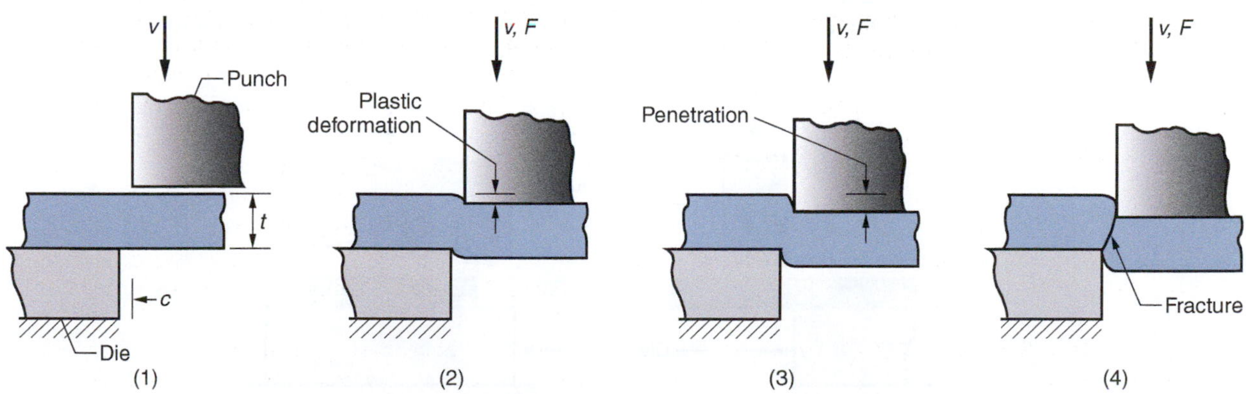

FIGURE 14.1 Shearing of sheet metal between two cutting edges: (1) just before the punch contacts work; (2) punch begins to push into work, causing plastic deformation; (3) punch compresses and penetrates into work causing a smooth cut surface; and (4) fracture is initiated at the opposing cutting edges that separate the sheet. Symbols v and F indicate motion and applied force, respectively, t = stock thickness, c = clearance. (Credit: *Fundamentals of Modern Manufacturing*, 4th Edition by Mikell P. Groover, 2010. Reprinted with permission of John Wiley & Sons, Inc.)

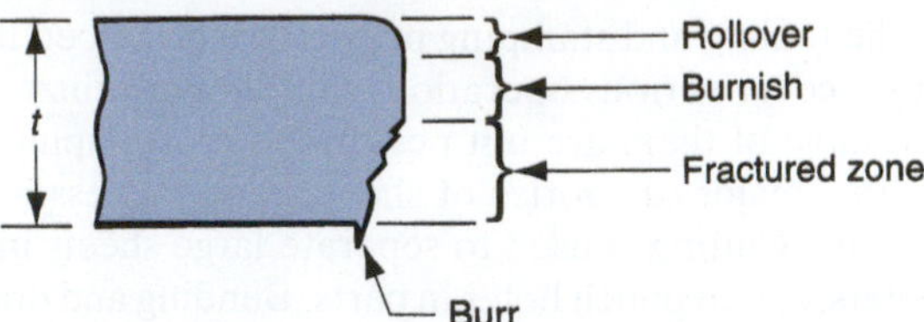

FIGURE 14.2 Characteristic sheared edges of the work. (Credit: *Fundamentals of Modern Manufacturing*, 4th Edition by Mikell P. Groover, 2010. Reprinted with permission of John Wiley & Sons, Inc.)

14.1.1 SHEARING, BLANKING, AND PUNCHING

The three most important operations in pressworking that cut metal by the shearing mechanism just described are shearing, blanking, and punching.

Shearing is a sheet-metal-cutting operation along a straight line between two cutting edges, as shown in Figure 14.3(a). Shearing is typically used to cut large sheets into smaller sections for subsequent pressworking operations. It is performed on a machine called a ***power shears***, or ***squaring shears***. The upper blade of the power shears is often inclined, as shown in Figure 14.3(b), to reduce the required cutting force.

Blanking involves cutting of the sheet metal along a closed outline in a single step to separate the piece from the surrounding stock, as in Figure 14.4(a). The part that is cut out is the desired product in the operation and is called the ***blank***. ***Punching*** is similar to blanking except that it produces a hole, and the separated piece is scrap, called the ***slug***. The remaining stock is the desired part. The distinction is illustrated in Figure 14.4(b).

14.1.2 ENGINEERING ANALYSIS OF SHEET-METAL CUTTING

Process parameters in sheet-metal cutting are clearance between punch and die, stock thickness, type of metal and its strength, and length of the cut. Let us define these parameters and some of the relationships among them.

Clearance The clearance c in a shearing operation is the distance between the punch and die, as shown in Figure 14.1(a). Typical clearances in conventional pressworking

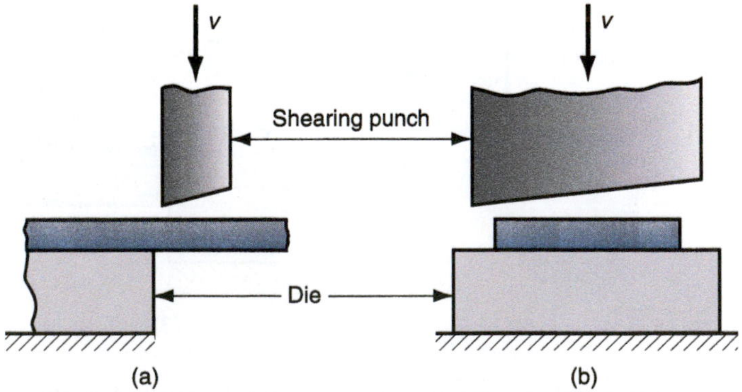

FIGURE 14.3 Shearing operation: (a) side view of the shearing operation; (b) front view of power shears equipped with inclined upper cutting blade. Symbol v indicates motion. (Credit: *Fundamentals of Modern Manufacturing*, 4th Edition by Mikell P. Groover, 2010. Reprinted with permission of John Wiley & Sons, Inc.)

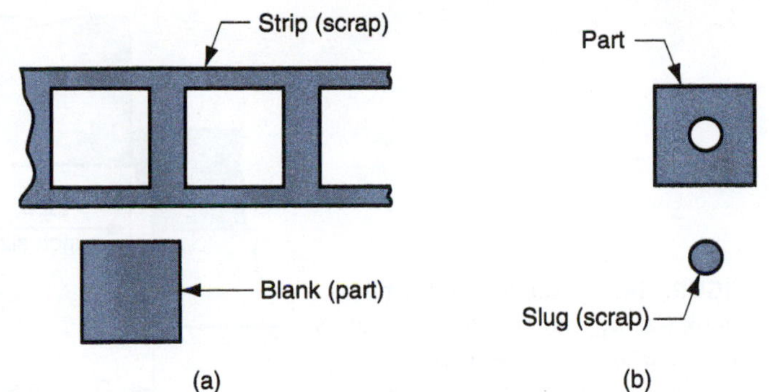

FIGURE 14.4 (a) Blanking and (b) punching. (Credit: *Fundamentals of Modern Manufacturing*, 4th Edition by Mikell P. Groover, 2010. Reprinted with permission of John Wiley & Sons, Inc.)

range between 4% and 8% of the sheet-metal thickness t. The correct clearance depends on sheet-metal type and thickness. The recommended clearance can be calculated by the following formula:

$$c = A_c t \tag{14.1}$$

where c = clearance, mm (in); A_c = clearance allowance; and t = stock thickness, mm (in). The clearance allowance is determined according to type of metal. For convenience, metals are classified into three groups given in Table 14.1, with an associated allowance value for each group.

These calculated clearance values can be applied to conventional blanking and hole-punching operations to determine the proper punch and die sizes. The die opening must always be larger than the punch size (obviously). Whether to add the clearance value to the die size or subtract it from the punch size depends on whether the part being cut out is a blank or a slug, as illustrated in Figure 14.5 for a circular part. Because of the geometry of the sheared edge, the outer dimension of the part cut out of the sheet will be larger than the hole size. Thus, punch-and-die sizes for a round blank of diameter D_b are determined as:

$$\text{Blanking punch diameter} = D_b - 2_c \tag{14.2a}$$

$$\text{Blanking die diameter} = D_b \tag{14.2b}$$

Punch-and-die sizes for a round hole of diameter D_h are determined as:

$$\text{Hole punch diameter} = D_h \tag{14.3a}$$

$$\text{Hole die diameter} = D_h + 2c \tag{14.3b}$$

In order for the slug or blank to drop through the die, the die opening must have an ***angular clearance*** (see Figure 14.6) of 0.25° to 1.5° on each side.

TABLE 14.1 Clearance allowance value for three sheet-metal groups

Metal Group	A_c
1100S and 5052S aluminum alloys, all tempers.	0.045
2024ST and 6061ST aluminum alloys; brass, all tempers; soft cold-rolled steel, soft stainless steel.	0.060
Cold-rolled steel, half-hard; stainless steel, half-hard and full-hard.	0.075

Compiled from [3].

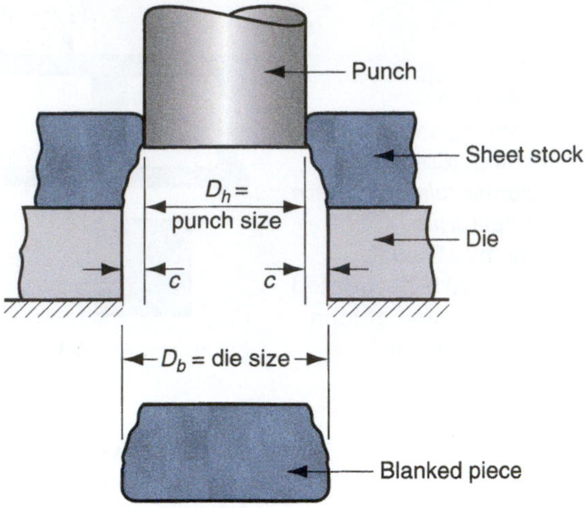

FIGURE 14.5 Die size determines blank size D_b; punch size determines hole size D_h.; c = clearance. (Credit: *Fundamentals of Modern Manufacturing*, 4th Edition by Mikell P. Groover, 2010. Reprinted with permission of John Wiley & Sons, Inc.)

Cutting Forces Estimates of cutting force are important because this force determines the size (tonnage) of the press needed. Cutting force F in sheet metalworking can be determined by

$$F = StL \tag{14.4}$$

where S = shear strength of the sheet metal, MPa (lb/in^2); t = stock thickness, mm (in), and L = length of the cut edge, mm (in). In blanking, punching, slotting, and similar operations, L is the perimeter length of the blank or hole being cut. The minor effect of clearance in determining the value of L can be neglected. If shear strength is unknown, an alternative way of estimating the cutting force is to use the tensile strength:

$$F = 0.7(TS)tL \tag{14.5}$$

where TS = ultimate tensile strength MPa (lb/in^2).

These equations for estimating cutting force assume that the entire cut along the sheared edge length L is made at the same time. In this case the cutting force will be a maximum. It is possible to reduce the maximum force by using an angled cutting edge on the punch or die, as in Figure 14.3(b). The angle (called the ***shear angle***), spreads the cut over time and reduces the force experienced at any one moment. However, the total energy required in the operation is the same, whether it is concentrated into a brief moment or distributed over a longer time period.

Example 14.1 Blanking Clearance and Force

A round disk of 150-mm diameter is to be blanked from a strip of 3.2-mm, half-hard cold-rolled steel whose shear strength = 310 MPa. Determine (a) the appropriate punch-and-die diameters and (b) blanking force.

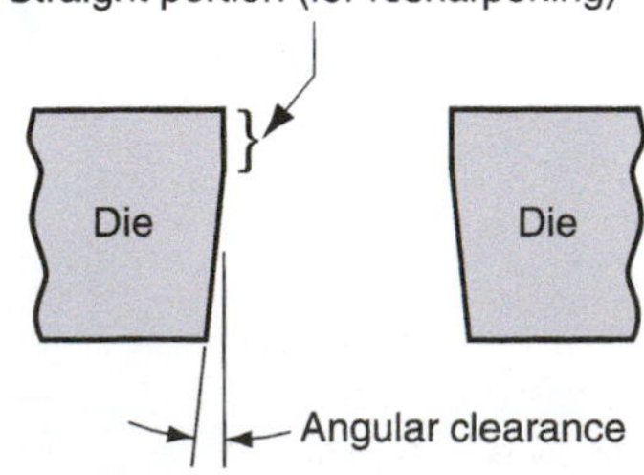

FIGURE 14.6 Angular clearance. (Credit: *Fundamentals of Modern Manufacturing*, 4th Edition by Mikell P. Groover, 2010. Reprinted with permission of John Wiley & Sons, Inc.)

Solution: (a) From Table 14.1, the clearance allowance for half-hard cold-rolled steel is $A_c = 0.075$. Accordingly,

$$c = 0.075(3.2\text{ mm}) = \mathbf{0.24\text{ mm}}$$

The blank is to have a diameter = 150 mm, and die size determines blank size. Therefore,

$$\text{Die opening diameter} = \mathbf{150.00\text{ mm}}$$
$$\text{Punch diameter} = 150 - 2(0.24) = \mathbf{149.52\text{ mm}}$$

(b) To determine the blanking force, we assume that the entire perimeter of the part is blanked at one time. The length of the cut edge is

$$L = \pi D_b = 150\pi = 471.2\text{ mm}$$

and the force is

$$F = 310(471.2)(3.2) = \mathbf{467,469\text{ N}}\ (\sim 53\text{ tons})$$

14.1.3 OTHER SHEET-METAL-CUTTING OPERATIONS

In addition to shearing, blanking, and punching, there are several other cutting operations in pressworking. The cutting mechanism in each case involves the same shearing action discussed earlier.

Cutoff and Parting Cutoff is a shearing operation in which blanks are separated from a sheet-metal strip by cutting the opposite sides of the part in sequence, as shown in Figure 14.7(a). With each cut, a new part is produced. The features of a cutoff operation that distinguish it from a conventional shearing operation are (1) the cut edges are not necessarily straight, and (2) the blanks can be nested on the strip in such a way that scrap is avoided.

Parting involves cutting a sheet-metal strip by a punch with two cutting edges that match the opposite sides of the blank, as shown in Figure 14.7(b). This might be required because the part outline has an irregular shape that precludes perfect nesting of the blanks on the strip. Parting is less efficient than cutoff in the sense that it results in some wasted material.

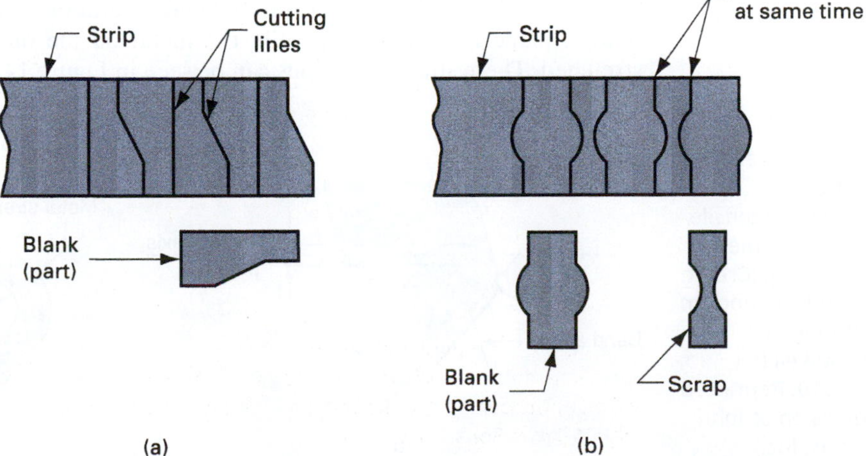

FIGURE 14.7 (a) Cutoff and (b) parting. (Credit: *Fundamentals of Modern Manufacturing*, 4th Edition by Mikell P. Groover, 2010. Reprinted with permission of John Wiley & Sons, Inc.)

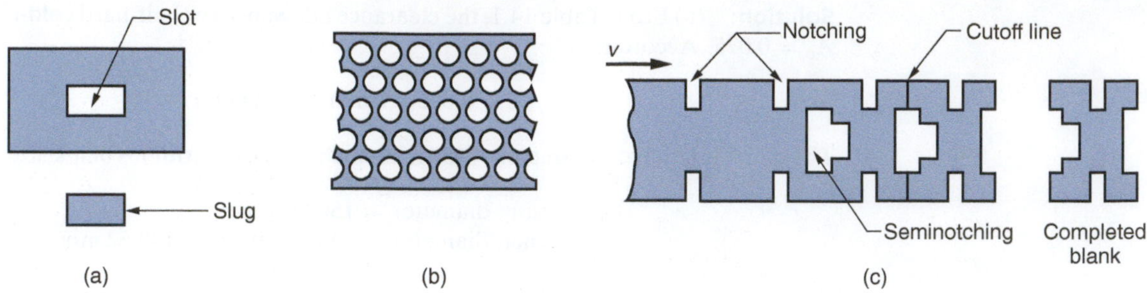

FIGURE 14.8 (a) Slotting, (b) perforating, (c) notching and seminotching. Symbol *v* indicates motion of strip. (Credit: *Fundamentals of Modern Manufacturing*, 4th Edition by Mikell P. Groover, 2010. Reprinted with permission of John Wiley & Sons, Inc.)

Slotting, Perforating, and Notching *Slotting* is the term sometimes used for a punching operation that cuts out an elongated or rectangular hole, as pictured in Figure 14.8(a). ***Perforating*** involves the simultaneous punching of a pattern of holes in sheet metal, as in Figure 14.8(b). The hole pattern is usually for decorative purposes, or to allow passage of light, gas, or fluid.

To obtain the desired outline of a blank, portions of the sheet metal are often removed by notching and seminotching. ***Notching*** involves cutting out a portion of metal from the side of the sheet or strip. ***Seminotching*** removes a portion of metal from the interior of the sheet. These operations are depicted in Figure 14.8(c). Seminotching might seem to the reader to be the same as a punching or slotting operation. The difference is that the metal removed by seminotching creates part of the blank outline, while punching and slotting create holes in the blank.

Trimming Trimming is a cutting operation performed on a formed part to remove excess metal and establish size. The term has the same basic meaning here as in forging (Section 13.2.5). A typical example in sheet metalwork is trimming the upper portion of a deep drawn cup to leave the desired dimensions on the cup.

14.2 BENDING OPERATIONS

Bending in sheet-metal work is defined as the straining of the metal around a straight axis, as in Figure 14.9. During the bending operation, the metal on the inside of the neutral plane is compressed, while the metal on the outside of the neutral plane is stretched. These strain conditions can be seen in Figure 14.9(b). The metal is plastically

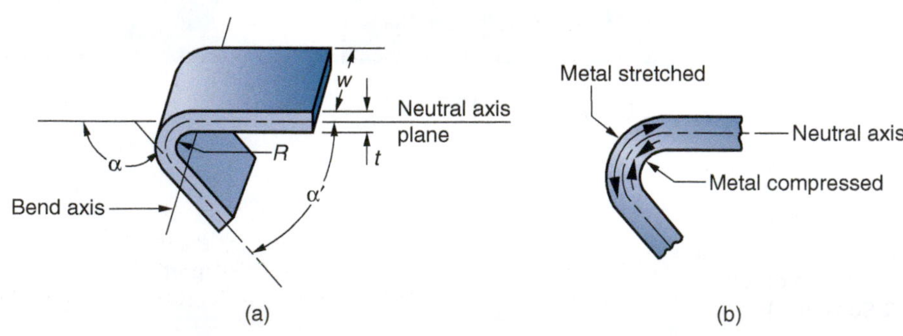

FIGURE 14.9 (a) Bending of sheet metal; (b) both compression and tensile elongation of the metal occur in bending. (Credit: *Fundamentals of Modern Manufacturing*, 4th Edition by Mikell P. Groover, 2010. Reprinted with permission of John Wiley & Sons, Inc.)

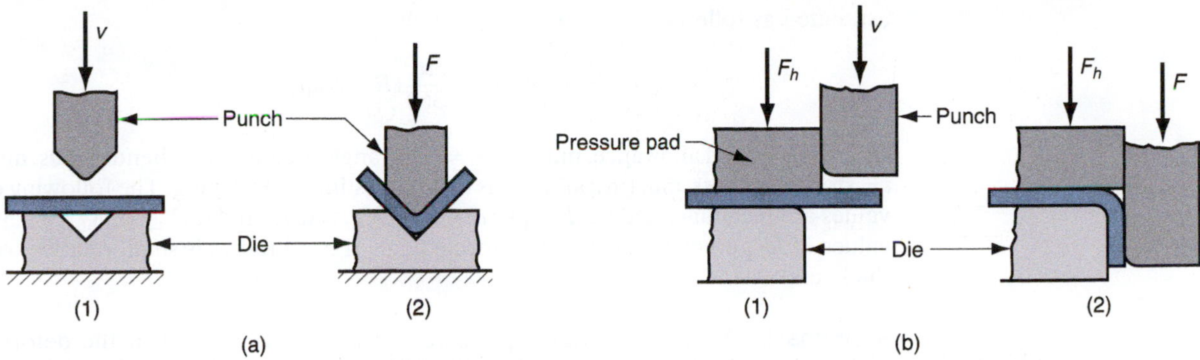

FIGURE 14.10 Two common bending methods: (a) V-bending and (b) edge bending; (1) before and (2) after bending. Symbols: v = motion, F = applied bending force, F_h = blank. (Credit: *Fundamentals of Modern Manufacturing*, 4th Edition by Mikell P. Groover, 2010. Reprinted with permission of John Wiley & Sons, Inc.)

deformed so that the bend takes a permanent set upon removal of the stresses that caused it. Bending produces little or no change in the thickness of the sheet metal.

14.2.1 V-BENDING AND EDGE BENDING

Bending operations are performed using punch-and-die tooling. The two common bending methods and associated tooling are V-bending, performed with a V-die; and edge bending, performed with a wiping die. These methods are illustrated in Figure 14.10.

In ***V-bending***, the sheet metal is bent between a V-shaped punch and–die. Included angles ranging from very obtuse to very acute can be made with V-dies. V-bending is generally used for low-production operations. It is often performed on a press brake (Section 14.5.2), and the associated V-dies are relatively simple and inexpensive.

Edge bending involves cantilever loading of the sheet metal. A pressure pad is used to apply a force F_h to hold the base of the part against the die, while the punch forces the part to yield and bend over the edge of the die. In the setup shown in Figure 14.10(b), edge bending is limited to bends of 90° or less. More complicated wiping dies can be designed for bend angles greater than 90°. Because of the pressure pad, wiping dies are more complicated and costly than V-dies and are generally used for high-production work.

14.2.2 ENGINEERING ANALYSIS OF BENDING

Some of the important terms in sheet-metal bending are identified in Figure 14.9. The metal of thickness t is bent through an angle called the bend angle α. This results in a sheet-metal part with an included angle α', where $\alpha + \alpha' = 180°$. The bend radius R is normally specified on the inside of the part, rather than at the neutral axis, and is determined by the radius on the tooling used to perform the operation. The bend is made over the width of the workpiece w.

Bend Allowance If the bend radius is small relative to stock thickness, the metal tends to stretch during bending. It is important to be able to estimate the amount of stretching that occurs, if any, so that the final part length will match the specified dimension. The problem is to determine the length of the neutral axis before bending to account for stretching of the final bent section. This length is called the ***bend allowance***, and it can be estimated as follows:

$$A_b = 2\pi \frac{\alpha}{360}(R + K_{ba}t) \tag{14.6}$$

where A_b = bend allowance, mm (in); α = bend angle, degrees; R = bend radius, mm (in); t = stock thickness, mm (in); and K_{ba} is factor to estimate stretching. The following design values are recommended for K_{ba} [3]: if $R < 2t$, $K_{ba} = 0.33$; and if $R \geq 2t$, $K_{ba} = 0.50$. The values of K_{ba} predict that stretching occurs only if bend radius is small relative to sheet thickness.

Springback When the bending pressure is removed at the end of the deformation operation, elastic energy remains in the bent part, causing it to recover partially toward its original shape. This elastic recovery is called ***springback***, defined as the increase in included angle of the bent part relative to the included angle of the forming tool after the tool is removed. This is illustrated in Figure 14.11 and is expressed:

$$SB = \frac{\alpha' - \alpha'_b}{\alpha'_b} \tag{14.7}$$

where SB = springback; α' = included angle of the sheet-metal part, degrees; and α_b' = included angle of the bending tool, degrees. Although not as obvious, an increase in the bend radius also occurs due to elastic recovery. The amount of springback increases with modulus of elasticity E and yield strength Y of the work metal.

Compensation for springback can be accomplished by several methods. Two common methods are overbending and bottoming. In ***overbending***, the punch angle and radius are fabricated slightly smaller than the specified angle on the final part so that the sheet metal springs back to the desired value. ***Bottoming*** involves squeezing the part at the end of the stroke, thus plastically deforming it in the bend region.

Bending Force The force required to perform bending depends on the geometry of the punch and die and the strength, thickness, and length of the sheet metal. The maximum

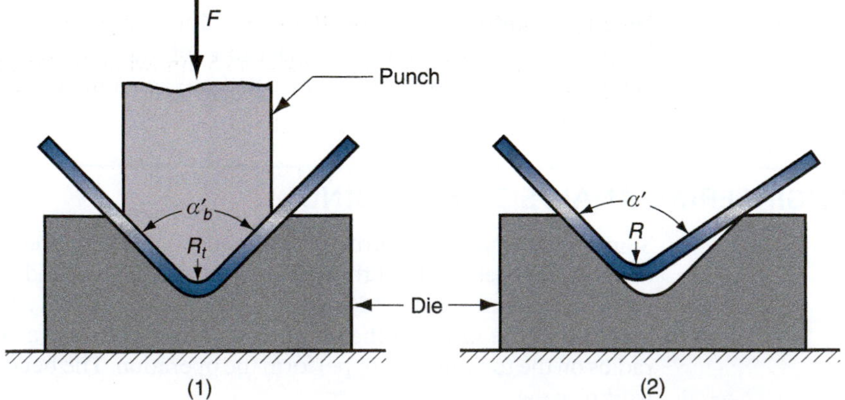

FIGURE 14.11 Springback in bending shows itself as a decrease in bend angle and an increase in bend radius: (1) during the operation, the work is forced to take the radius R_t and included angle α_b' = determined by the bending tool (punch in V-bending); (2) after the punch is removed, the work springs back to radius R and included angle α'. Symbol: F = applied bending force. (Credit: *Fundamentals of Modern Manufacturing,* 4th Edition by Mikell P. Groover, 2010. Reprinted with permission of John Wiley & Sons, Inc.)

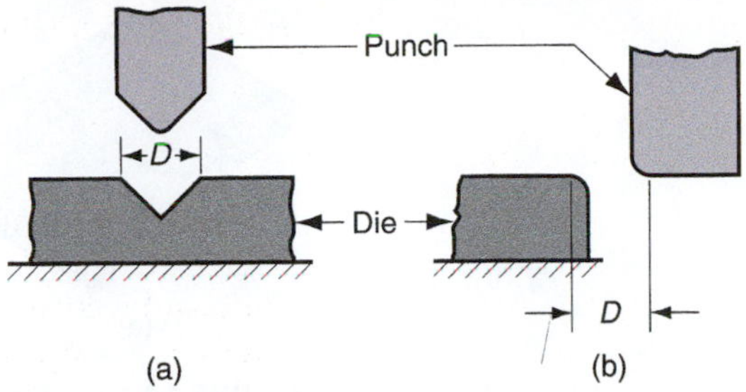

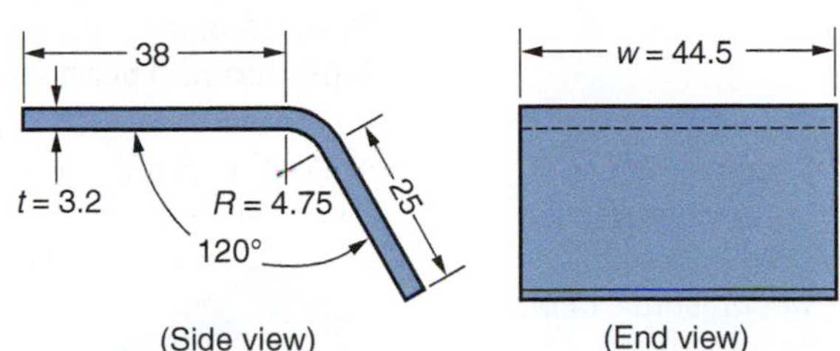

FIGURE 14.12 Die opening dimension *D*: (a) V-die, (b) wiping die. (Credit: *Fundamentals of Modern Manufacturing*, 4th Edition by Mikell P. Groover, 2010. Reprinted with permission of John Wiley & Sons, Inc.)

FIGURE 14.13 Sheet-metal part of Example 14.2 (dimensions in mm). (Credit: *Fundamentals of Modern Manufacturing*, 4th Edition by Mikell P. Groover, 2010. Reprinted with permission of John Wiley & Sons, Inc.)

bending force can be estimated by means of the following equation:

$$F = \frac{K_{bf}(TS)wt^2}{D} \tag{14.8}$$

where F = bending force, N (lb); TS = tensile strength of the sheet metal, MPa (lb/in^2); w = width of part in the direction of the bend axis, mm (in); t = stock thickness, mm (in); and D = die opening dimension as defined in Figure 14.12, mm (in). Eq. (14.8) is based on bending of a simple beam in mechanics, and K_{bf} is a constant that accounts for differences encountered in an actual bending process. Its value depends on type of bending: for V-bending, $K_{bf} = 1.33$; and for edge bending, $K_{bf} = 0.33$.

Example 14.2 Sheet-Metal Bending

A sheet-metal blank is to be bent as shown in Figure 14.13. The metal has a modulus of elasticity = 205(10^3) MPa, yield strength = 275 MPa, and tensile strength = 450 MPa. Determine (a) the starting blank size and (b) the bending force if a V-die is used with a die opening dimension = 25 mm.

Solution: (a) The starting blank = 44.5 mm wide. Its length = $38 + A_b + 25$ (mm). For the included angle $\alpha' = 120°$, the bend angle $\alpha = 60°$. The value of K_{ba} in Eq. (14.6) = 0.33 since R/t = 4.75/3.2 = 1.48 (less than 2.0).

$$A_b = 2\pi \frac{60}{360}(4.75 + 0.33 \times 3.2) = 6.08\,\text{mm}$$

Length of the blank is therefore 38 + 6.08 + 25 = 69.08 mm.
(b) Force is obtained from Eq. (14.8) using $K_{bf} = 1.33$.

$$F = \frac{1.33(450)(44.5)(3.2)^2}{2.5} = 10,909\,\text{N}$$

■

14.2.3 OTHER BENDING AND FORMING OPERATIONS

Some sheet-metal operations involve bending over a curved axis rather than a straight axis, or they have other features that differentiate them from the bending operations described above.

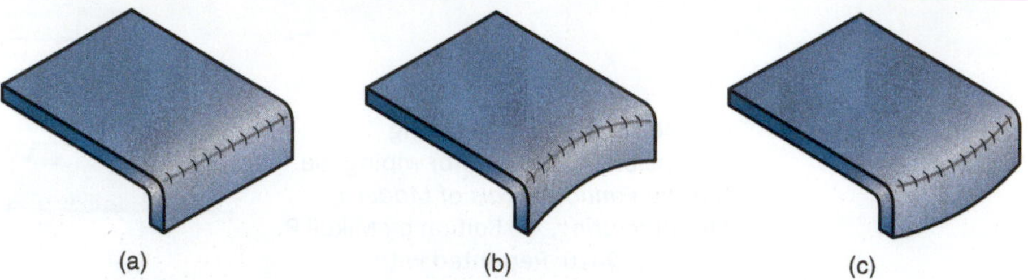

FIGURE 14.14 Flanging: (a) straight flanging, (b) stretch flanging, and (c) shrink flanging. (Credit: *Fundamentals of Modern Manufacturing*, 4th Edition by Mikell P. Groover, 2010. Reprinted with permission of John Wiley & Sons, Inc.)

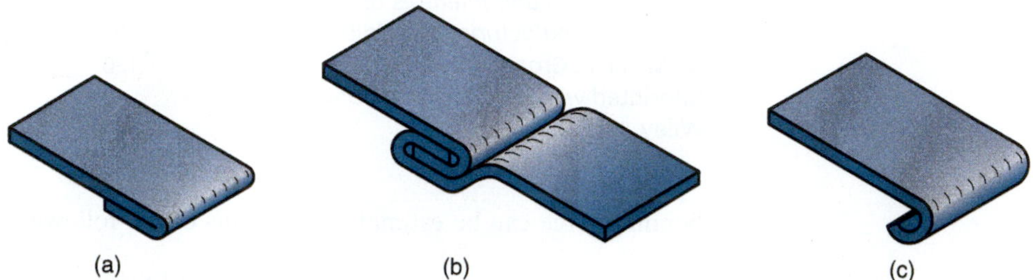

FIGURE 14.15 (a) Hemming, (b) seaming, and (c) curling. (Credit: *Fundamentals of Modern Manufacturing*, 4th Edition by Mikell P. Groover, 2010. Reprinted with permission of John Wiley & Sons, Inc.)

Flanging, Hemming, Seaming, and Curling Flanging is a bending operation in which the edge of a sheet-metal part is bent at a 90° angle (usually) to form a rim or flange. It is often used to strengthen or stiffen sheet metal. The flange can be formed over a straight bend axis, as in Figure 14.14(a), or it can involve some stretching or shrinking of the metal, as in (b) and (c).

Hemming involves bending the edge of the sheet over on itself, in more than one bending step. This is often done to eliminate the sharp edge on the piece, to increase stiffness, and to improve appearance. ***Seaming*** is a related operation in which two sheet-metal edges are assembled. Hemming and seaming are illustrated in Figure 14.15(a) and (b).

Curling, also called ***beading***, forms the edges of the part into a roll or curl, as in Figure 14.15(c). As in hemming, it is done for purposes of safety, strength, and aesthetics. Examples of products in which curling is used include hinges, pots and pans, and pocket-watch cases. These examples show that curling can be performed over straight or curved bend axes.

14.3 DRAWING

Drawing is a sheet-metal-forming operation used to make cup-shaped, box-shaped, or other complex-curved and concave parts. It is performed by placing a piece of sheet metal over a die cavity and then pushing the metal into the opening with a punch, as in Figure 14.16. The blank must usually be held down flat against the die by a blankholder. Common parts made by drawing include beverage cans, ammunition shells, sinks, cooking pots, and automobile body panels.

14.3.1 MECHANICS OF DRAWING

Drawing of a cup-shaped part is the basic drawing operation, with dimensions and parameters as pictured in Figure 14.16. A blank of diameter D_b is drawn into a die cavity

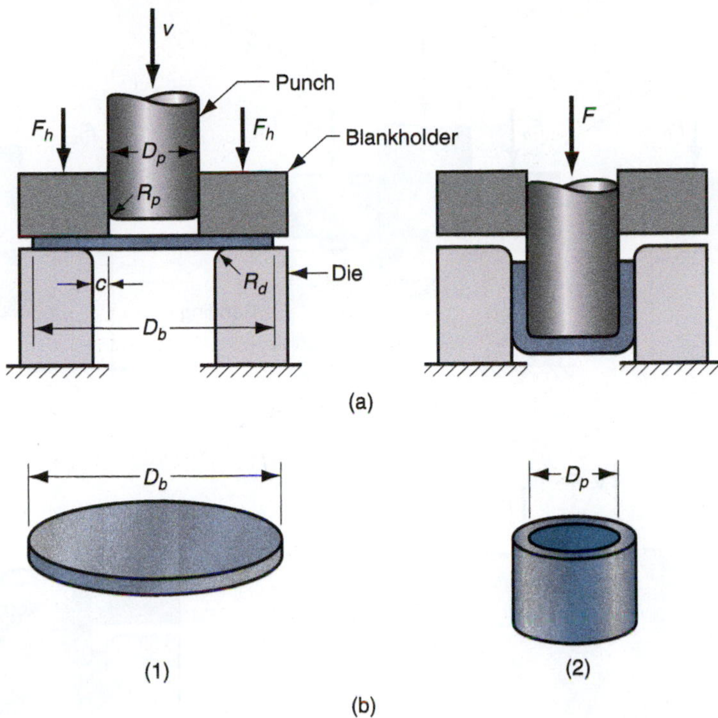

FIGURE 14.16 (a) Drawing of a cup-shaped part: (1) start of operation before punch contacts work, and (2) near end of stroke; and (b) corresponding workpart: (1) starting blank, and (2) drawn part. Symbols: c = clearance, D_b = blank diameter, D_p = punch diameter, R_d = die corner radius, R_p = punch corner radius, F = drawing force, F_h = holding force. (Credit: *Fundamentals of Modern Manufacturing*, 4th Edition by Mikell P. Groover, 2010. Reprinted with permission of John Wiley & Sons, Inc.)

by means of a punch with diameter D_p. The punch and die must have corner radii, given by R_p and R_d. If the punch and die were to have sharp corners (R_p and $R_d = 0$), a hole-punching operation (and not a very good one) would be accomplished rather than a drawing operation. The sides of the punch and die are separated by a clearance c. This clearance in drawing is about 10% greater than the stock thickness:

$$c = 1.1\,t \tag{14.9}$$

The punch applies a downward force F to accomplish the deformation of the metal, and a downward holding force F_h is applied by the blankholder, as shown in the sketch.

As the punch proceeds downward toward its final bottom position, the work experiences a complex sequence of stresses and strains as it is gradually formed into the shape defined by the punch-and-die cavity. The stages in the deformation process are illustrated in Figure 14.17. As the punch first begins to push into the work, the metal is subjected to a ***bending*** operation. The sheet is simply bent over the corner of the punch and the corner of the die, as in Figure 14.17(2). The outside perimeter of the blank moves in toward the center in this first stage, but only slightly.

As the punch moves further down, a ***straightening*** action occurs in the metal that was previously bent over the die radius, as in Figure 14.17(3). The metal at the bottom of the cup, as well as along the punch radius, has been moved downward with the punch, but the metal that was bent over the die radius must now be straightened in order to be pulled into the clearance to form the wall of the cylinder. At the same time, more metal must be added to replace that being used in the cylinder wall. This new metal comes from the

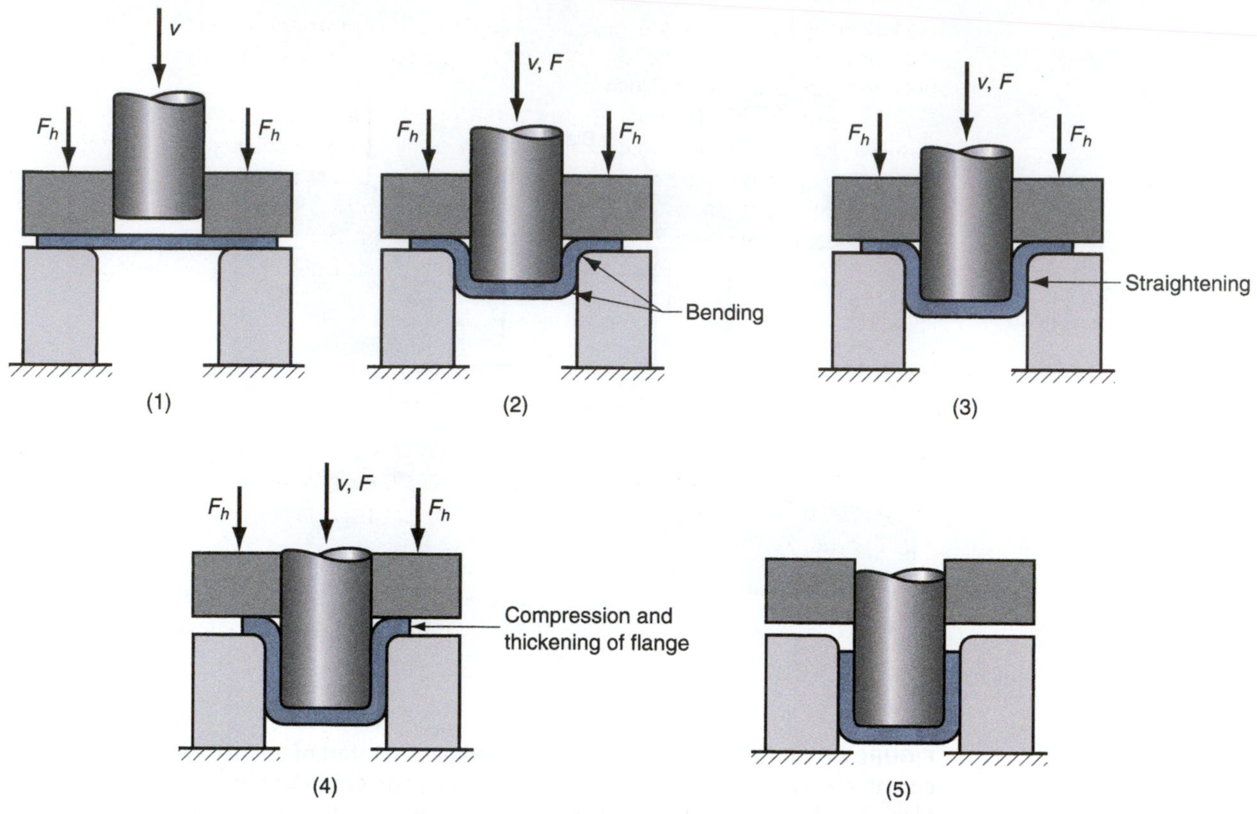

FIGURE 14.17 Stages in deformation of the work in deep drawing: (1) punch just before initial contact with work, (2) bending, (3) straightening, (4) friction and compression, and (5) final cup shape showing effects of thinning in the cup walls. Symbols: v = motion of punch, F = punch force, F_h = blankholder force. (Credit: *Fundamentals of Modern Manufacturing,* 4th Edition by Mikell P. Groover, 2010. Reprinted with permission of John Wiley & Sons, Inc.)

outside edge of the blank. The metal in the outer portions of the blank is pulled or ***drawn*** toward the die opening to resupply the previously bent and straightened metal now forming the cylinder wall. This type of metal flow through a constricted space gives the drawing process its name.

During this stage of the process, friction and compression play important roles in the flange of the blank. In order for the material in the flange to move toward the die opening, ***friction*** between the sheet metal and the surfaces of the blankholder and the die must be overcome. Initially, static friction is involved until the metal starts to slide; then, after metal flow begins, dynamic friction governs the process. The magnitude of the holding force applied by the blankholder, as well as the friction conditions at the two interfaces, are determining factors in the success of this aspect of the drawing operation. Lubricants or drawing compounds are generally used to reduce friction forces. In addition to friction, ***compression*** is also occurring in the outer edge of the blank. As the metal in this portion of the blank is drawn toward the center, the outer perimeter becomes smaller. Because the volume of metal remains constant, the metal is squeezed and becomes thicker as the perimeter is reduced. This often results in wrinkling of the remaining flange of the blank, especially when thin sheet metal is drawn, or when the blankholder force is too low. It is a condition that cannot be corrected once it has occurred. The friction and compression effects are illustrated in Figure 14.17(4).

The holding force applied by the blankholder is now seen to be a critical factor in deep drawing. If it is too small, wrinkling occurs. If it is too large, it prevents the metal from flowing properly toward the die cavity, resulting in stretching and possible tearing of the sheet metal. Determining the proper holding force involves a delicate balance between these opposing factors.

Progressive downward motion of the punch results in a continuation of the metal flow caused by drawing and compression. In addition, some ***thinning*** of the cylinder wall occurs, as in Figure 14.17(5). The force being applied by the punch is opposed by the metal in the form of deformation and friction in the operation. A portion of the deformation involves stretching and thinning of the metal as it is pulled over the edge of the die opening. Up to 25% thinning of the side wall may occur in a successful drawing operation, mostly near the base of the cup.

14.3.2 ENGINEERING ANALYSIS OF DRAWING

It is important to assess the limitations on the amount of drawing that can be accomplished. This is often guided by simple measures that can be readily calculated for a given operation. In addition, drawing force and holding force are important process variables. Finally, the starting blank size must be determined.

Measures of Drawing One of the measures of the severity of a deep drawing operation is the ***drawing ratio*** DR. This is most easily defined for a cylindrical shape as the ratio of blank diameter D_b to punch diameter D_p. In equation form,

$$DR = \frac{D_b}{D_p} \tag{14.10}$$

The drawing ratio provides an indication, albeit a crude one, of the severity of a given drawing operation. The greater the ratio, the more severe the operation. An approximate upper limit on the drawing ratio is a value of 2.0. The actual limiting value for a given operation depends on punch-and-die corner radii (R_p and R_d), friction conditions, depth of draw, and characteristics of the sheet metal (e.g., ductility, degree of directionality of strength properties in the metal).

Another way to characterize a given drawing operation is by the ***reduction*** r, where

$$r = \frac{D_b - D_p}{D_b} \tag{14.11}$$

It is very closely related to drawing ratio. Consistent with the previous limit on DR ($DR \leq 2.0$), the value of reduction r should be less than 0.50.

A third measure in deep drawing is the ***thickness-to-diameter ratio*** t/D_b (thickness of the starting blank t divided by the blank diameter D_b). Often expressed as a percentage, it is desirable for the t/D_b ratio to be greater than 1%. As t/D_b decreases, tendency for wrinkling (Section 14.3.4) increases.

In cases where these limits on drawing ratio, reduction, and t/D_b ratio are exceeded by the design of the drawn part, the blank must be drawn in two or more steps, sometimes with annealing between the steps.

Example 14.3 Cup Drawing

A drawing operation is used to form a cylindrical cup with inside diameter = 75 mm and height = 50 mm. The starting blank size = 138 mm and the stock thickness = 2.4 mm. Based on these data, is the operation feasible?

Solution: To assess feasibility, we determine the drawing ratio, reduction, and thickness-to-diameter ratio.

$$DR = 138/75 = 1.84$$
$$r = (138 - 75)/138 = 0.457 = 45.7\%$$
$$t/D_b = 2.4/138 = 0.017 = 1.7\%$$

According to these measures, the drawing operation is feasible. The drawing ratio is less than 2.0, the reduction is less than 50%, and the t/D_b ratio is greater than 1%. These are general guidelines frequently used to indicate technical feasibility. ■

Forces The ***drawing force*** required to perform a given operation can be estimated roughly by the formula:

$$F = \pi D_p t (TS)\left(\frac{D_b}{D_p} - 0.7\right) \tag{14.12}$$

where F = drawing force, N (lb); t = original blank thickness, mm (in); TS = tensile strength, MPa (lb/in^2); and D_b and D_p are the starting blank diameter and punch diameter, respectively, mm (in). The constant 0.7 is a correction factor to account for friction. Eq. (14.12) estimates the maximum force in the operation. The drawing force varies throughout the downward movement of the punch, usually reaching its maximum value at about one-third the length of the punch stroke.

The ***holding force*** is an important factor in a drawing operation. As a rough approximation, the holding pressure can be set at a value = 0.015 of the yield strength of the sheet metal [8]. This value is then multiplied by that portion of the starting area of the blank that is to be held by the blankholder. In equation form,

$$F_h = 0.015 Y \pi D_b^2 - (D_p + 2.2t + 2R_d)^2 \tag{14.13}$$

where F_h = holding force in drawing, N (lb); Y = yield strength of the sheet metal, MPa (lb/in^2); t = starting stock thickness, mm (in); R_d = die corner radius, mm (in); and the other terms have been previously defined. The holding force is usually about one-third the drawing force [10].

Example 14.4 Forces in Drawing

For the drawing operation of Example 14.3, determine (a) drawing force and (b) holding force, given that the tensile strength of the sheet metal (low-carbon steel) = 300 MPa and yield strength = 175 MPa. The die corner radius = 6 mm.

Solution: (a) Maximum drawing force is given by Eq. (14.12):

$$F = \pi(75)(2.4)(300)\left(\frac{138}{75} - 0.7\right) = 193,396\ \text{N}$$

(b) Holding force is estimated by Eq. (14.13):

$$F_h = 0.015(175)\pi(138^2 - (75 + 2.2 \times 2.4 + 2 \times 6)^2) = 86,824\ \text{N}$$

■

Blank Size Determination For the final dimensions to be achieved on the cylindrical drawn shape, the correct starting blank diameter is needed. It must be large enough to supply sufficient metal to complete the cup. Yet if there is too much material, unnecessary waste will result. For drawn shapes other than cylindrical cups, the same problem of estimating the starting blank size exists, only the shape of the blank may be other than round.

The following is a reasonable method for estimating the starting blank diameter in a deep drawing operation that produces a round part (e.g., cylindrical cup and more complex shapes so long as they are axisymmetric). Because the volume of the final product is the same as that of the starting sheet-metal blank, then the blank diameter can be calculated by setting the initial blank volume equal to the final volume of the product and solving for diameter D_b. To facilitate the calculation, it is often assumed that negligible thinning of the part wall occurs.

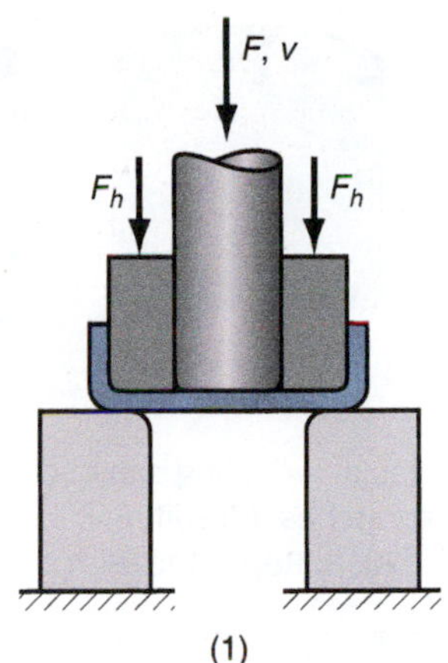

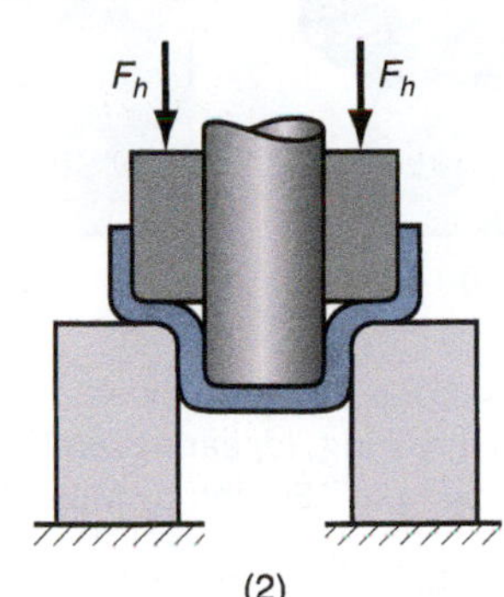

FIGURE 14.18 Redrawing of a cup: (1) start of redraw, and (2) end of stroke. Symbols: v = punch velocity, F = applied punch force, F_h = blankholder force. (Credit: *Fundamentals of Modern Manufacturing*, 4th Edition by Mikell P. Groover, 2010. Reprinted with permission of John Wiley & Sons, Inc.)

14.3.3 OTHER DRAWING OPERATIONS

Our discussion has focused on a conventional cup-drawing operation that produces a simple cylindrical shape in a single step and uses a blankholder to facilitate the process. Let us consider some of the variations of this basic operation.

Redrawing If the shape change required by the part design is too severe (drawing ratio is too high), complete forming of the part may require more than one drawing step. The second drawing step, and any further drawing steps if needed, are referred to as ***redrawing***. A redrawing operation is illustrated in Figure 14.18.

A related operation is ***reverse drawing***, in which a drawn part is positioned face down on the die so that the second drawing operation produces a configuration such as that shown in Figure 14.19. Although it may seem that reverse drawing would produce a more severe deformation than redrawing, it is actually easier on the metal. The reason is that the sheet metal is bent in the same direction at the outside and inside corners of the die in reverse drawing; while in redrawing the metal is bent in the opposite directions at the two corners. Because of this difference, the metal experiences less strain hardening in reverse drawing and the drawing force is lower.

Drawing of Shapes Other than Cylindrical Cups Many products require drawing of shapes other than cylindrical cups. The variety of drawn shapes include square or

FIGURE 14.19 Reverse drawing: (1) start and (2) completion. Symbols: v = punch velocity, F = applied punch force, F_h = blankholder force. (Credit: *Fundamentals of Modern Manufacturing*, 4th Edition by Mikell P. Groover, 2010. Reprinted with permission of John Wiley & Sons, Inc.)

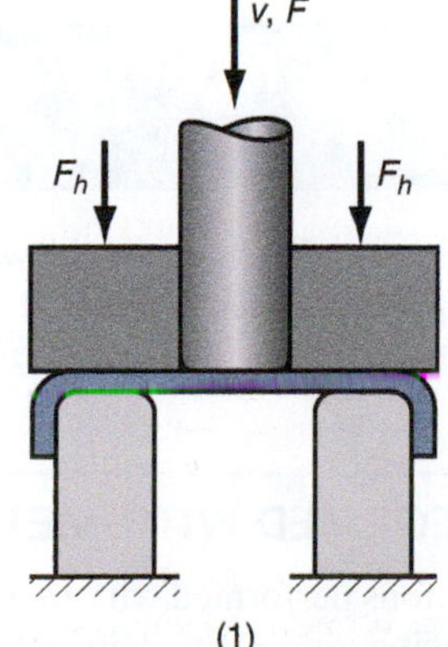

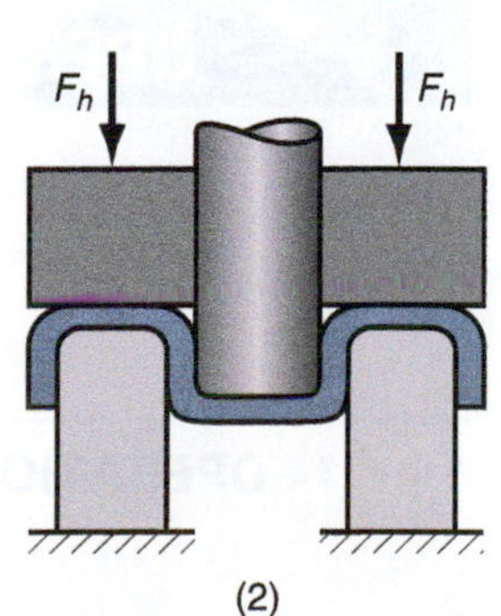

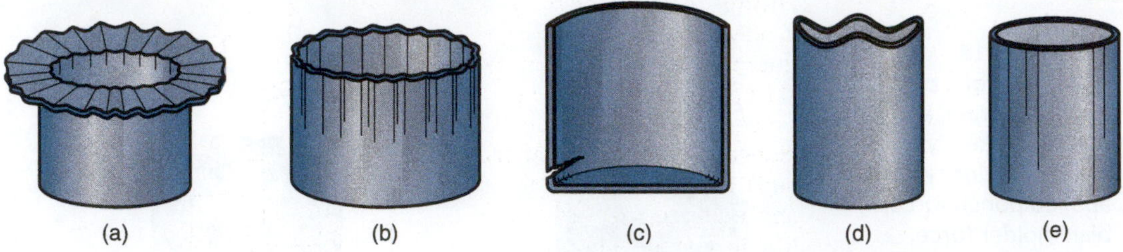

FIGURE 14.20 Common defects in drawn parts: (a) wrinkling can occur either in the flange or (b) in the wall, (c) tearing, (d) earing, and (e) surface scratches. (Credit: *Fundamentals of Modern Manufacturing*, 4th Edition by Mikell P. Groover, 2010. Reprinted with permission of John Wiley & Sons, Inc.)

rectangular boxes (as in sinks), stepped cups, cones, cups with spherical rather than flat bases, and irregular curved forms (as in automobile body panels). Each of these shapes presents unique technical problems in drawing. Eary and Reed [2] provide a detailed discussion of the drawing of these kinds of shapes.

14.3.4 DEFECTS IN DRAWING

Sheet-metal drawing is a more complex operation than cutting or bending, and more things can go wrong. A number of defects can occur in a drawn product, some of which we have already alluded to. Following is a list of common defects, with sketches in Figure 14.20:

(a) ***Wrinkling in the flange.*** Wrinkling in a drawn part consists of a series of ridges that form radially in the undrawn flange of the workpart due to compressive buckling.

(b) ***Wrinkling in the wall.*** If and when the wrinkled flange is drawn into the cup, these ridges appear in the vertical wall.

(c) ***Tearing.*** Tearing is an open crack in the vertical wall, usually near the base of the drawn cup, due to high tensile stresses that cause thinning and failure of the metal at this location. This type of failure can also occur as the metal is pulled over a sharp die corner.

(d) ***Earing.*** This is the formation of irregularities (called ***ears***) in the upper edge of a deep drawn cup, caused by anisotropy in the sheet metal. If the material is perfectly isotropic, ears do not form.

(e) ***Surface scratches.*** Surface scratches can occur on the drawn part if the punch and die are not smooth or if lubrication is insufficient.

14.4 OTHER SHEET-METAL-FORMING OPERATIONS

In addition to bending and drawing, several other sheet-metal forming operations can be accomplished on conventional presses. We classify these as (1) operations performed with metal tooling and (2) operations performed with flexible rubber tooling.

14.4.1 OPERATIONS PERFORMED WITH METAL TOOLING

Operations performed with metal tooling include (1) ironing, (2) coining and embossing, (3) lancing, and (4) twisting.

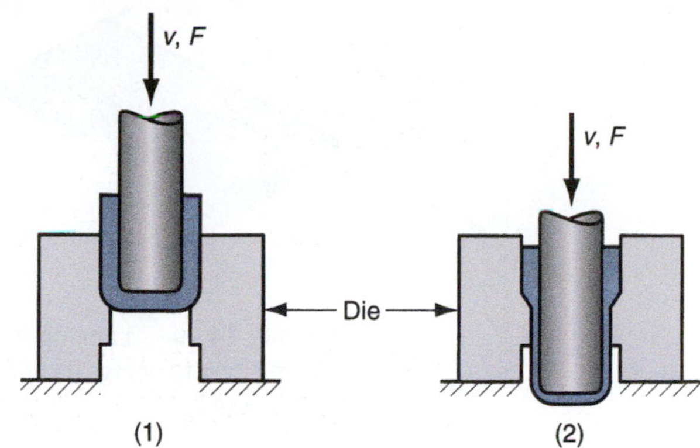

FIGURE 14.21 Ironing to achieve a more uniform wall thickness in a drawn cup: (1) start of process; (2) during process. Note thinning and elongation of walls. Symbols *v* and *F* indicate motion and applied force, respectively. (Credit: *Fundamentals of Modern Manufacturing,* 4th Edition by Mikell P. Groover, 2010. Reprinted with permission of John Wiley & Sons, Inc.)

Ironing In deep drawing the flange is compressed by the squeezing action of the blank perimeter seeking a smaller circumference as it is drawn toward the die opening. Because of this compression, the sheet metal near the outer edge of the blank becomes thicker as it moves inward. If the thickness of this stock is greater than the clearance between the punch and die, it will be squeezed to the size of the clearance, a process known as ***ironing***.

Sometimes ironing is performed as a separate step that follows drawing. This case is illustrated in Figure 14.21. Ironing makes the cylindrical cup more uniform in wall thickness. The drawn part is therefore longer and more efficient in terms of material usage. Beverage cans and artillery shells, two very high-production items, include ironing among their processing steps to achieve economy in material usage.

Coining and Embossing Coining is a bulk deformation operation discussed in the previous chapter. It is frequently used in sheet-metal work to form indentations and raised sections in the part. The indentations result in thinning of the sheet metal, and the raised sections result in thickening of the metal.

Embossing is a forming operation used to create indentations in the sheet, such as raised (or indented) lettering or strengthening ribs, as depicted in Figure 14.22. Some stretching and thinning of the metal are involved. This operation may seem similar to coining. However, embossing dies possess matching cavity contours, the punch containing the positive contour and the die containing the negative; whereas coining dies may have quite different cavities in the two die halves, thus causing more significant metal deformation than embossing.

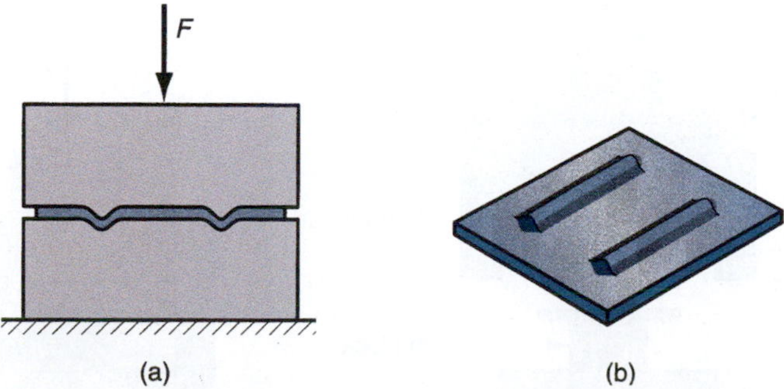

FIGURE 14.22 Embossing: (a) cross section of punch-and-die configuration during pressing; (b) finished part with embossed ribs. (Credit: *Fundamentals of Modern Manufacturing,* 4th Edition by Mikell P. Groover, 2010. Reprinted with permission of John Wiley & Sons, Inc.)

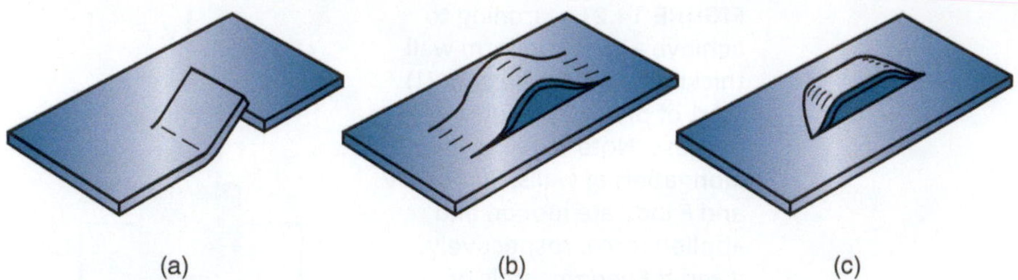

FIGURE 14.23 Lancing in several forms: (a) cutting and bending; (b) and (c) two types of cutting and forming. (Credit: *Fundamentals of Modern Manufacturing*, 4th Edition by Mikell P. Groover, 2010. Reprinted with permission of John Wiley & Sons, Inc.)

Lancing Lancing is a combined cutting and bending or cutting and forming operation performed in one step to partially separate the metal from the sheet. Several examples are shown in Figure 14.23. Among other applications, lancing is used to make louvers in sheet-metal air vents for heating and air conditioning systems in buildings.

Twisting Twisting subjects the sheet metal to a torsion loading rather than a bending load, thus causing a twist in the sheet over its length. This type of operation has limited applications. It is used to make such products as fan and propeller blades. It can be performed in a conventional punch and die which has been designed to deform the part in the required twist shape.

14.4.2 RUBBER FORMING PROCESSES

The two operations discussed here are performed on conventional presses, but the tooling is unusual in that it uses a flexible element (made of rubber or similar material) to effect the forming operation. The operations are (1) the Guerin process, and (2) hydroforming.

Guerin Process The Guerin process uses a thick rubber pad (or other flexible material) to form sheet metal over a positive form block, as in Figure 14.24. The rubber pad is confined in a steel container. As the ram descends, the rubber gradually surrounds the

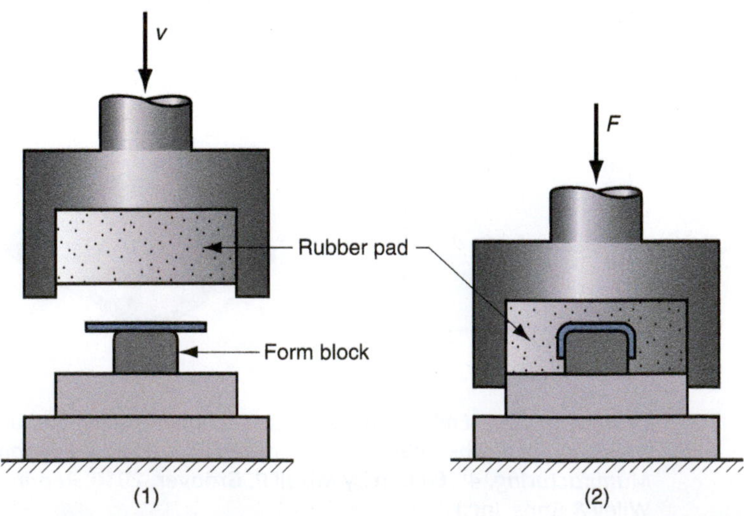

FIGURE 14.24 Guerin process: (1) before and (2) after. Symbols v and F indicate motion and applied force, respectively. (Credit: *Fundamentals of Modern Manufacturing*, 4th Edition by Mikell P. Groover, 2010. Reprinted with permission of John Wiley & Sons, Inc.)

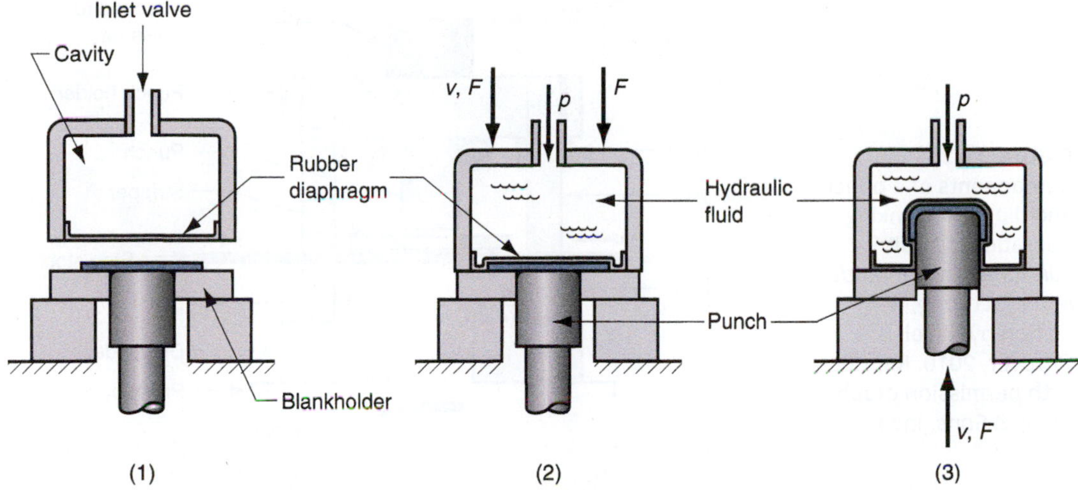

FIGURE 14.25 Hydroform process: (1) start-up, no fluid in cavity; (2) press closed, cavity pressurized with hydraulic fluid; (3) punch pressed into work to form part. Symbols: v = velocity, F = applied force, p = hydraulic pressure. (Credit: *Fundamentals of Modern Manufacturing*, 4th Edition by Mikell P. Groover, 2010. Reprinted with permission of John Wiley & Sons, Inc.)

sheet, applying pressure to deform it to the shape of the form block. It is limited to relatively shallow forms, because the pressures developed by the rubber—up to about 10 MPa (1500 lb/in^2)—are not sufficient to prevent wrinkling in deeper formed parts.

The advantage of the Guerin process is the relatively low cost of the tooling. The form block can be made of wood, plastic, or other materials that are easy to shape, and the rubber pad can be used with different form blocks. These factors make rubber forming attractive in small-quantity production, such as the aircraft industry, where the process was developed.

Hydroforming Hydroforming is similar to the Guerin process; the difference is that it substitutes a rubber diaphragm filled with hydraulic fluid in place of the thick rubber pad, as illustrated in Figure 14.25. This allows the pressure that forms the workpart to be increased—to around 100 MPa (15,000 lb/in^2)—thus, preventing wrinkling in deep-formed parts. In fact, deeper draws can be achieved with the hydroform process than with conventional deep drawing. This is because the uniform pressure in hydroforming forces the work to contact the punch throughout its length, thus increasing friction and reducing the tensile stresses that cause tearing at the base of the drawn cup.

14.5 DIES AND PRESSES FOR SHEET-METAL PROCESSES

In this section we examine the punch-and-die tooling and production equipment used in conventional sheet-metal processing.

14.5.1 DIES

Nearly all of the preceding pressworking operations are performed with conventional punch-and-die tooling. The tooling is referred to as a ***die***. It is custom-designed for the

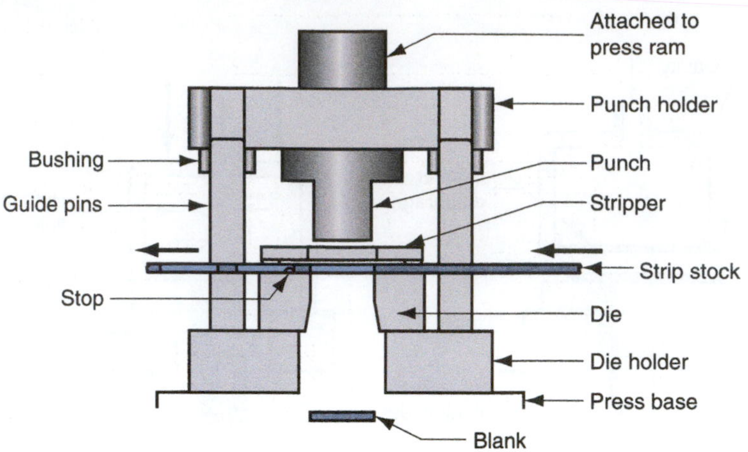

FIGURE 14.26 Components of a punch and die for a blanking operation. (Credit: *Fundamentals of Modern Manufacturing*, 4th Edition by Mikell P. Groover, 2010. Reprinted with permission of John Wiley & Sons, Inc.)

particular part to be produced. The term ***stamping die*** is sometimes used for high-production dies. Typical materials for stamping dies are tool steels (Section 2.1.1).

Components of a Stamping Die The components of a stamping die to perform a simple blanking operation are illustrated in Figure 14.26. The working components are the ***punch*** and ***die***, which perform the cutting operation. They are attached to the upper and lower portions of the ***die set***, respectively called the ***punch holder*** (or ***upper shoe***) and ***die holder*** (***lower shoe***). The die set also includes guide pins and bushings to ensure proper alignment between the punch and die during the stamping operation. The die holder is attached to the base of the press, and the punch holder is attached to the ram. Actuation of the ram accomplishes the pressworking operation.

In addition to these components, a die used for blanking or hole-punching must include a means of preventing the sheet metal from sticking to the punch when it is retracted upward after the operation. The newly created hole in the stock is the same size as the punch, and it tends to cling to the punch on its withdrawal. The device in the die that strips the sheet metal from the punch is called a ***stripper***. It is often a simple plate attached to the die as in Figure 14.26, with a hole slightly larger than the punch diameter.

For dies that process strips or coils of sheet metal, a device is required to stop the sheet metal as it advances through the die between press cycles. That device is called (try to guess) a ***stop***. Stops range from simple solid pins located in the path of the strip to block its forward motion, to more complex mechanisms synchronized to rise and retract with the actuation of the press. The simpler stop is shown in Figure 14.26.

There are other components in pressworking dies, but the preceding description provides an introduction to the terminology.

Types of Stamping Dies Aside from differences in stamping dies related to the operations they perform (e.g., cutting, bending, drawing), other differences deal with the number of separate operations to be performed in each press actuation and how they are accomplished.

The stamping die considered above performs a single blanking operation with each stroke of the press and is called a ***simple die***. Other dies that perform a single operation include V-dies (Section 14.2.1). More complicated pressworking dies include compound dies, combination dies, and progressive dies. A ***compound die*** performs two operations at a single station, such as blanking and punching, or blanking and drawing [2]. A good example is a compound die that blanks and punches a washer. A ***combination die*** is less common; it performs two operations at two different stations in the die. Examples of

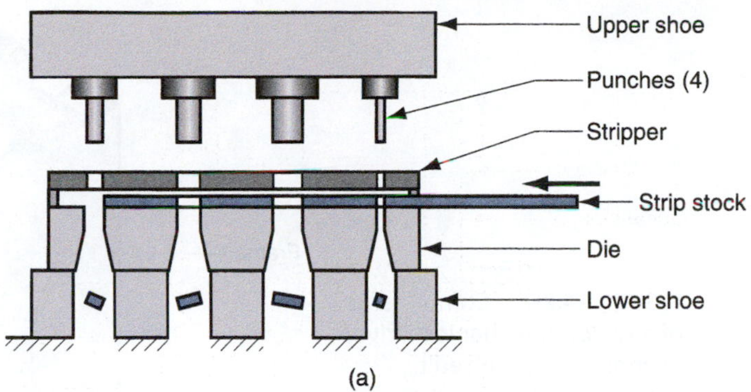

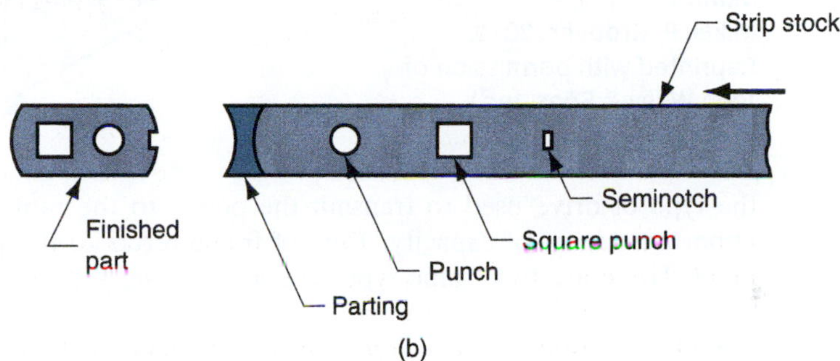

FIGURE 14.27 (a) Progressive die and (b) associated strip development. (Credit: *Fundamentals of Modern Manufacturing*, 4th Edition by Mikell P. Groover, 2010. Reprinted with permission of John Wiley & Sons, Inc.)

applications include blanking two different parts (e.g., right-hand and left-hand parts), or blanking and then bending the same part [2].

A ***progressive die*** performs two or more operations on a sheet-metal coil at two or more stations with each press stroke. The part is fabricated progressively. The coil is fed from one station to the next and different operations (e.g., punching, notching, bending, and blanking) are performed at each station. When the part exits the final station, it has been completed and separated (cut) from the remaining coil. Design of a progressive die begins with the layout of the part on the strip or coil and the determination of which operations are to be performed at each station. The result of this procedure is called the ***strip development***. A progressive die and associated strip development are illustrated in Figure 14.27. Progressive dies can have a dozen or more stations. They are the most complicated and most costly stamping dies, economically justified only for complex parts requiring multiple operations at high-production rates.

14.5.2 PRESSES

A press used for sheet metalworking is a machine tool with a stationary ***bed*** and a powered ***ram*** (or ***slide***) that can be driven toward and away from the bed to perform various cutting and forming operations. A typical press, with principal components labeled, is diagrammed in Figure 14.28. The relative positions of the bed and ram are established by the ***frame***, and the ram is driven by mechanical or hydraulic power. When a die is mounted in the press, the punch holder is attached to the ram, and the die holder is attached to a ***bolster plate*** of the press bed.

Presses are available in a variety of capacities, power systems, and frame types. The capacity of a press is its ability to deliver the required force and energy to accomplish the stamping operation. This is determined by the physical size of the press and by its power system. The power system refers to whether mechanical or hydraulic power is used and

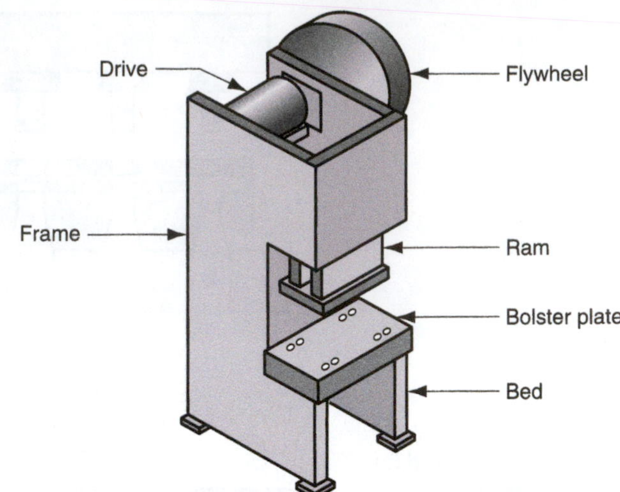

FIGURE 14.28 Components of a typical (mechanical drive) stamping press. (Credit: *Fundamentals of Modern Manufacturing*, 4th Edition by Mikell P. Groover, 2010. Reprinted with permission of John Wiley & Sons, Inc.)

the type of drive used to transmit the power to the ram. Production rate is another important aspect of capacity. Type of frame refers to the physical construction of the press. There are two frame types in common use: gap frame and straight-sided frame.

Gap Frame Presses The ***gap frame*** has the general configuration of the letter C and is often referred to as a ***C-frame***. Gap frame presses provide good access to the die, and they are usually open in the back to permit convenient ejection of stampings or scrap. The types of gap frame press include (a) solid gap frame, (b) open back inclinable, (c) press brake, and (d) turret press.

The ***solid gap frame*** (sometimes called simply a ***gap press***) has one-piece construction, as shown in Figure 14.28. Presses with this frame are rigid, yet the C-shape allows convenient access from the sides for feeding strip or coil stock. They are available in a range of sizes, with capacities up to around 9,000 kN (1,000 tons). The ***open back inclinable*** (OBI) press has a C-frame assembled to a base in such a way that the frame can be tilted back to various angles so that the stampings fall through the rear opening by gravity. Capacities of OBI presses range between one ton and around 2,250 kN (250 tons). They can be operated at high speeds—up to around 1,000 strokes per minute.

The ***press brake*** is a gap frame press with a very wide bed. The model in Figure 14.29 has a bed width of 9.15 m (30 ft). This allows a number of separate dies (simple V-bending dies are typical) to be set up in the bed, so that small quantities of stampings can be made economically. These low quantities of parts, sometimes requiring multiple bends at different angles, necessitate a manual operation. For a part requiring a series of bends, the operator moves the starting piece of sheet metal through the desired sequence of bending dies, actuating the press at each die, to complete the work needed.

Whereas press brakes are well adapted to bending operations, ***turret presses*** are suited to situations in which a sequence of punching, notching, and related cutting operations must be accomplished on sheet-metal parts, as in Figure 14.30. Turret presses have a C-frame, although this construction is not obvious in Figure 14.31. The conventional ram and punch is replaced by a turret containing many punches of different sizes and shapes. The turret works by indexing (rotating) to the position holding the punch to perform the required operation. Beneath the punch turret is a corresponding die turret that positions the die opening for each punch. Between the punch and die is the sheet-metal blank, held by an $x - y$ positioning system that operates by computer numerical control (Section 29.1). The blank is moved to the required coordinate position for each cutting operation.

FIGURE 14.29 Press brake. Photo courtesy of Strippit, Inc.

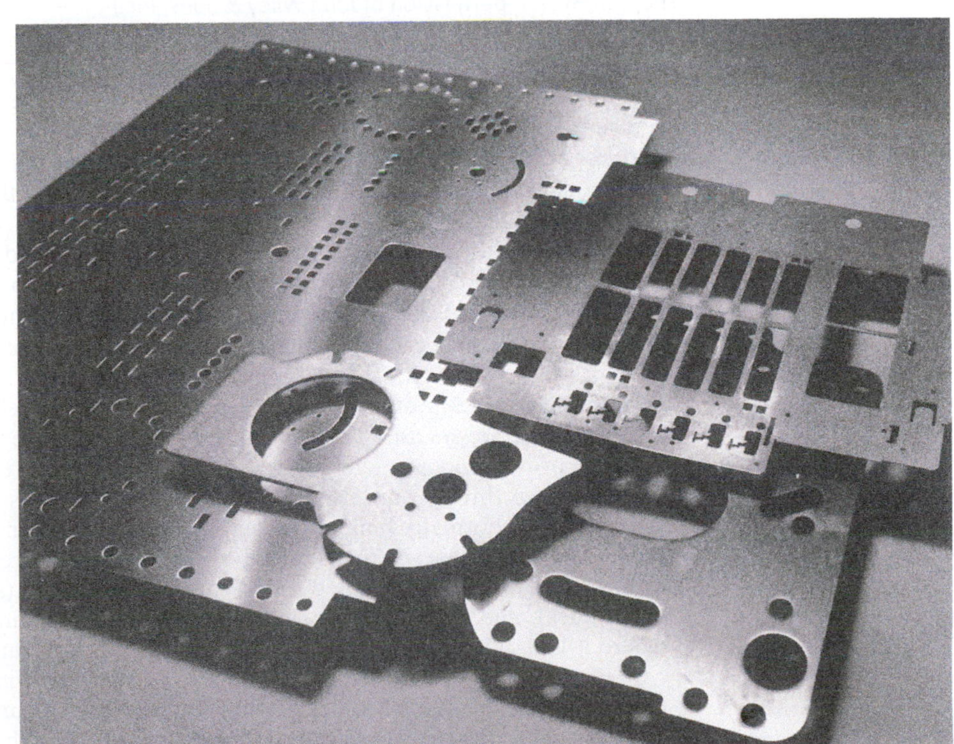

FIGURE 14.30 Several sheet-metal parts produced on a turret press, showing variety of possible hole shapes. Photo courtesy of Strippit, Inc. (Credit: *Fundamentals of Modern Manufacturing*, 4th Edition by Mikell P. Groover, 2010. Reprinted with permission of John Wiley & Sons, Inc.)

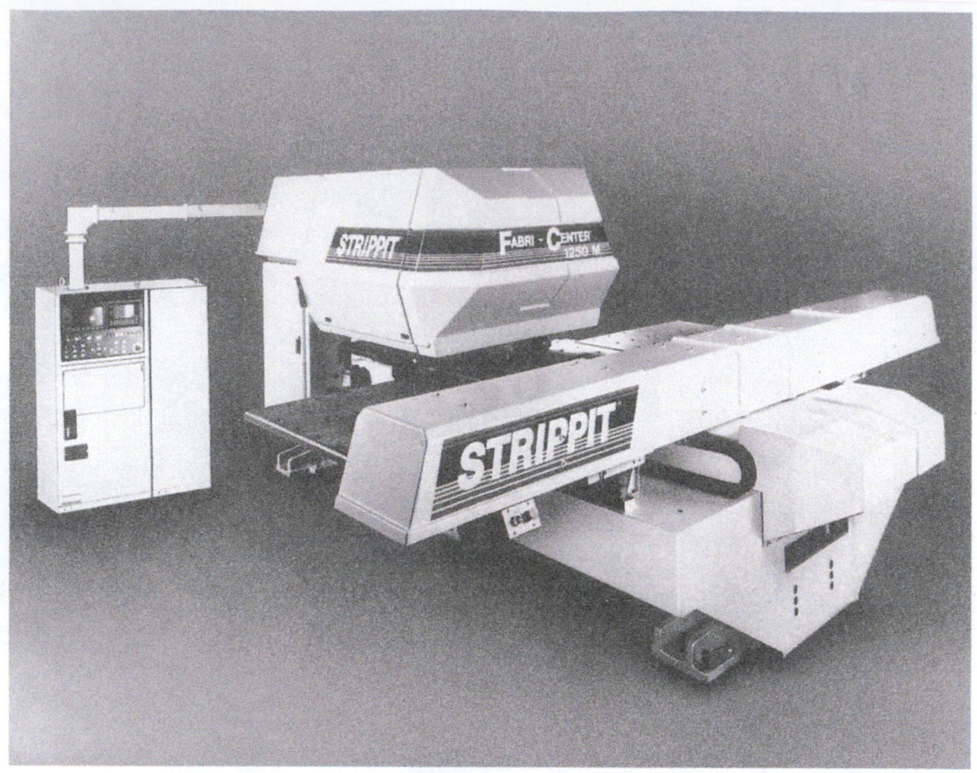

FIGURE 14.31 Computer numerical control turret press. Photo courtesy of Strippit, Inc. (Credit: *Fundamentals of Modern Manufacturing,* 4th Edition by Mikell P. Groover, 2010. Reprinted with permission of John Wiley & Sons, Inc.)

Straight-Sided Frame Presses For jobs requiring high tonnage, press frames with greater structural rigidity are needed. Straight-sided presses have full sides, giving it a box-like appearance as in Figure 14.32. This construction increases the strength and stiffness of the frame. As a result, capacities up to 35,000 kN (4000 tons) are available in straight-sided presses for sheet metalwork. Large presses of this frame type are also used for forging (Section 13.2).

In all of these presses, gap frame and straight-sided frame, the size is closely correlated to tonnage capacity. Larger presses are built to withstand higher forces in pressworking. Press size is also related to the speed at which it can operate. Smaller presses are generally capable of higher production rates than larger presses.

Power and Drive Systems Power systems on presses are either hydraulic or mechanical. ***Hydraulic presses*** use a large piston and cylinder to drive the ram. This power system typically provides longer ram strokes than mechanical drives and can develop the full tonnage force throughout the entire stroke. However, it is slower. Its application for sheet metal is normally limited to deep drawing and other forming operations where these load-stroke characteristics are advantageous. These presses are available with one or more independently operated slides, called single action (single slide), double action (two slides), and so on. Double-action presses are useful in deep drawing operations where it is required to separately control the punch force and the blankholder force.

Several types of drive mechanisms are used on ***mechanical presses***. These include eccentric, crankshaft, and knuckle joint, illustrated in Figure 14.33. They convert the rotational motion of a drive motor into the linear motion of the ram. A ***flywheel*** is used to

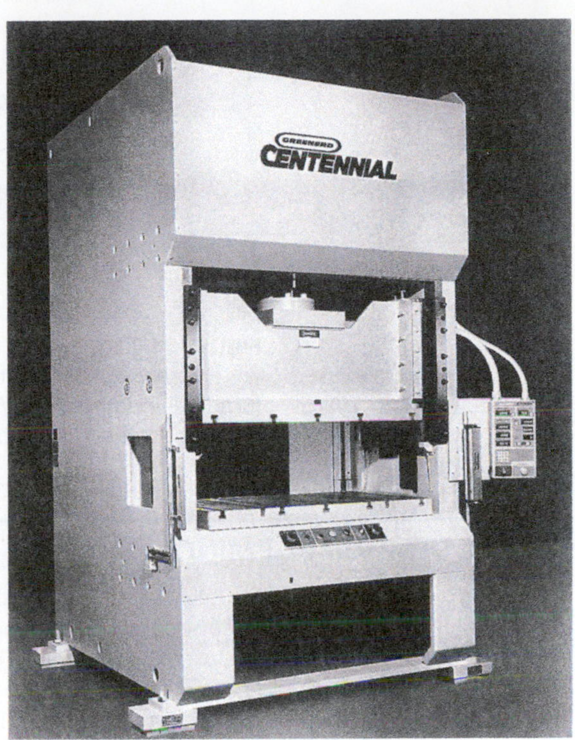

FIGURE 14.32 Straight-sided frame press. Photo courtesy Greenerd Press & Machine Company, Inc. (Credit: *Fundamentals of Modern Manufacturing*, 4th Edition by Mikell P. Groover, 2010. Reprinted with permission of John Wiley & Sons, Inc.)

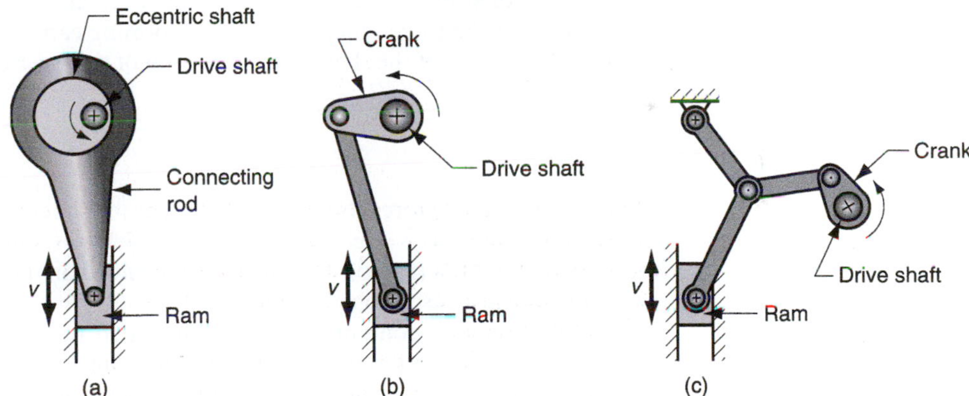

FIGURE 14.33 Types of drives for sheet-metal presses: (a) eccentric, (b) crankshaft, and (c) knuckle joint. (Credit: *Fundamentals of Modern Manufacturing*, 4th Edition by Mikell P. Groover, 2010. Reprinted with permission of John Wiley & Sons, Inc.)

store the energy of the drive motor for use in the stamping operation. Mechanical presses using these drives achieve very high forces at the bottom of their strokes, and are therefore quite suited to blanking and punching operations. The knuckle joint delivers very high force when it bottoms, and is therefore often used in coining operations.

14.6 SHEET-METAL OPERATIONS NOT PERFORMED ON PRESSES

A number of sheet-metal operations are not performed on conventional stamping presses. In this section we examine several of these processes: (1) stretch forming, (2) roll bending and roll forming, (3) spinning, and (4) high-energy-rate forming processes.

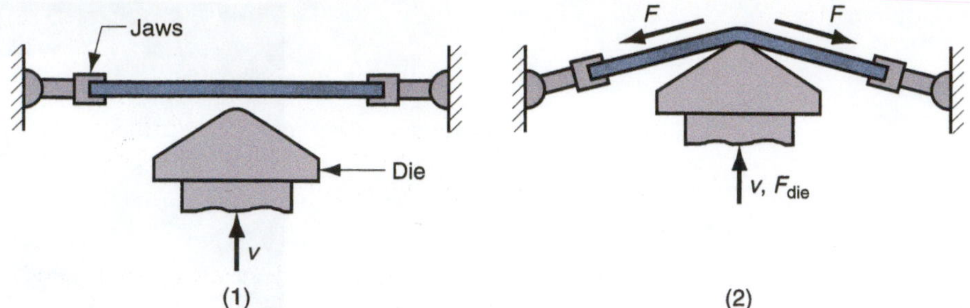

FIGURE 14.34 Stretch forming: (1) start of process; (2) form die is pressed into the work with force F_{die}, causing it to be stretched and bent over the form. F = stretching force. (Credit: *Fundamentals of Modern Manufacturing*, 4th Edition by Mikell P. Groover, 2010. Reprinted with permission of John Wiley & Sons, Inc.)

14.6.1 STRETCH FORMING

Stretch forming is a sheet-metal deformation process in which the sheet metal is intentionally stretched and simultaneously bent to achieve shape change. The process is illustrated in Figure 14.34 for a relatively simple and gradual bend. The workpart is gripped by one or more jaws on each end and then stretched and bent over a positive die containing the desired form. The metal is stressed in tension to a level above its yield point. When the tension loading is released, the metal has been plastically deformed. The combination of stretching and bending results in relatively little springback in the part. An estimate of the force required in stretch forming can be obtained by multiplying the cross-sectional area of the sheet in the direction of pulling by the flow stress of the metal. In equation form,

$$F = LtY_f \tag{14.14}$$

where F = stretching force, N (lb); L = length of the sheet in the direction perpendicular to stretching, mm (in); t = instantaneous stock thickness, mm (in); and Y_f = flow stress of the work metal, MPa (lb/in^2). The die force F_{die} shown in the figure can be determined by balancing vertical force components.

More complex contours than that shown in our figure are possible by stretch forming, but there are limitations on how sharp the curves in the sheet can be. Stretch forming is widely used in the aircraft and aerospace industries to economically produce large sheet-metal parts in the low quantities characteristic of those industries.

14.6.2 ROLL BENDING AND ROLL FORMING

The operations described in this section use rolls to form sheet metal. ***Roll bending*** is an operation in which (usually) large sheet-metal or plate-metal parts are formed into curved sections by means of rolls. One possible arrangement of the rolls is pictured in Figure 14.35. As the sheet passes between the rolls, the rolls are brought toward each other to a configuration that achieves the desired radius of curvature on the work. Components for large storage tanks and pressure vessels are fabricated by roll bending. The operation can also be used to bend structural shapes, railroad rails, and tubes.

A related operation is ***roll straightening*** in which nonflat sheets (or other cross-sectional forms) are straightened by passing them between a series of rolls. The rolls

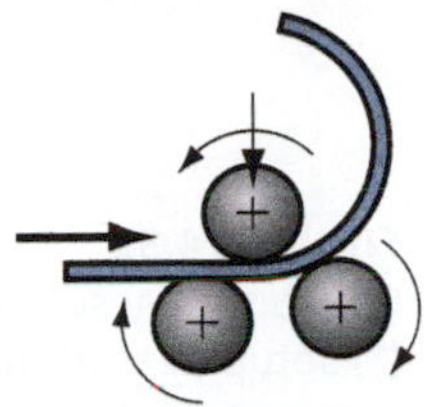

FIGURE 14.35 Roll bending. (Credit: *Fundamentals of Modern Manufacturing*, 4th Edition by Mikell P. Groover, 2010. Reprinted with permission of John Wiley & Sons, Inc.)

subject the work to a sequence of decreasing small bends in opposite directions, thus causing it to be straight at the exit.

Roll forming (also called ***contour roll forming***) is a continuous bending process in which opposing rolls are used to produce long sections of formed shapes from coil or strip stock. Several pairs of rolls are usually required to progressively accomplish the bending of the stock into the desired shape. The process is illustrated in Figure 14.36 for a U-shaped section. Products made by roll forming include channels, gutters, metal siding sections (for homes), pipes and tubing with seams, and various structural sections. Although roll forming has the general appearance of a rolling operation (and the tooling certainly looks similar), the difference is that roll forming involves bending rather than compressing the work.

14.6.3 SPINNING

Spinning is a metal-forming process in which an axially symmetric part is gradually shaped over a mandrel or form by means of a rounded tool or roller. The tool or roller applies a very localized pressure (almost a point contact) to deform the work by axial and radial motions over the surface of the part. Geometries produced by spinning include cups, cones, hemispheres, and tubes. In this section, we describe the conventional spinning process. As illustrated in Figure 14.37, a sheet-metal disk is held against the end of a rotating mandrel of the desired inside shape of the final part, while the tool or roller deforms the metal against the mandrel. In some cases, the starting workpart is other than a flat disk. The process requires a series of steps, as indicated in the figure, to complete the shaping of the part. The tool position is controlled either by a human operator, using a fixed fulcrum to achieve the required leverage, or by an automatic method such as numerical control. These alternatives are ***manual spinning*** and ***power spinning***. Power spinning has the capability to apply higher forces to the operation, resulting in faster cycle times and greater work size capacity. It also achieves better process control than manual spinning.

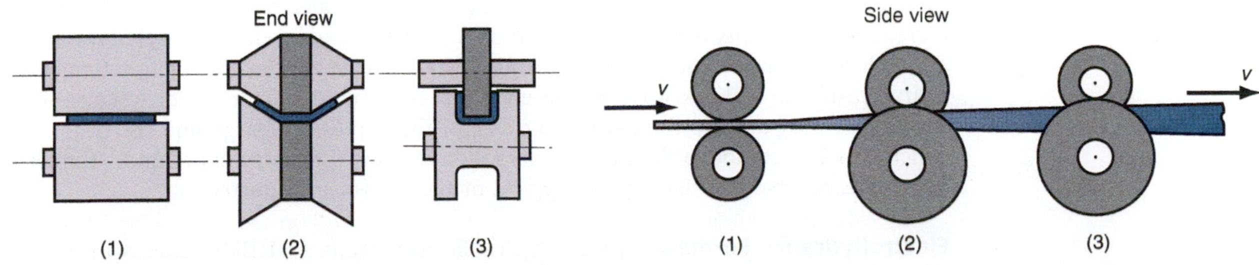

FIGURE 14.36 Roll forming of a continuous channel section: (1) straight rolls, (2) partial form, and (3) final form. (Credit: *Fundamentals of Modern Manufacturing*, 4th Edition by Mikell P. Groover, 2010. Reprinted with permission of John Wiley & Sons, Inc.)

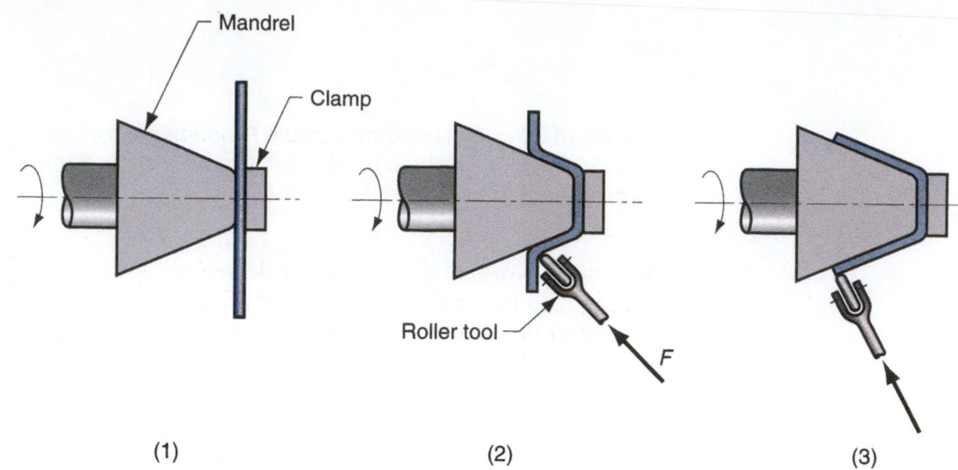

FIGURE 14.37 Conventional spinning: (1) setup at start of process; (2) during spinning; and (3) completion of process. (Credit: *Fundamentals of Modern Manufacturing*, 4th Edition by Mikell P. Groover, 2010. Reprinted with permission of John Wiley & Sons, Inc.)

Conventional spinning bends the metal around a moving circular axis to conform to the outside surface of the axisymmetric mandrel. The thickness of the metal therefore remains unchanged (more or less) relative to the starting disk thickness. The diameter of the disk must therefore be somewhat larger than the diameter of the resulting part. The required starting diameter can be figured by assuming constant volume, before and after spinning.

Applications of conventional spinning include production of conical and curved shapes in low quantities. Very large diameter parts—up to 5 m (15 ft) or more—can be made by spinning. Alternative sheet-metal processes would require excessively high die costs. The form mandrel in spinning can be made of wood or other soft materials that are easy to shape. It is therefore a low-cost tool compared to the punch and die required for deep drawing, which might be a substitute process for some parts.

14.6.4 HIGH-ENERGY-RATE FORMING

Several processes have been developed to form metals using large amounts of energy applied in a very short time. Owing to this feature, these operations are called ***high-energy-rate forming*** (HERF) processes. They include explosive forming, electrohydraulic forming, and electromagnetic forming.

Explosive Forming Explosive forming involves the use of an explosive charge to form sheet (or plate) metal into a die cavity. One method of implementing the process is illustrated in Figure 14.38. The workpart is clamped and sealed over the die, and a vacuum is created in the cavity beneath. The apparatus is then placed in a large vessel of water, and an explosive charge is positioned a certain distance above the work. Detonation of the charge results in a shock wave whose energy is transmitted by the water to cause rapid forming of the part into the cavity. The size of the explosive charge and the distance at which it is placed above the part are largely a matter of art and experience. Explosive forming is reserved for large parts, typical of the aerospace industry.

Electrohydraulic Forming Electrohydraulic forming is a HERF process in which a shock wave to deform the work into a die cavity is generated by the discharge of electrical energy between two electrodes submerged in a transmission fluid (water). Owing to its principle of operation, this process is also called ***electric discharge forming***. The setup for the process is illustrated in Figure 14.39. Electrical energy is accumulated in large capacitors and then released to the electrodes. Electrohydraulic forming is similar to

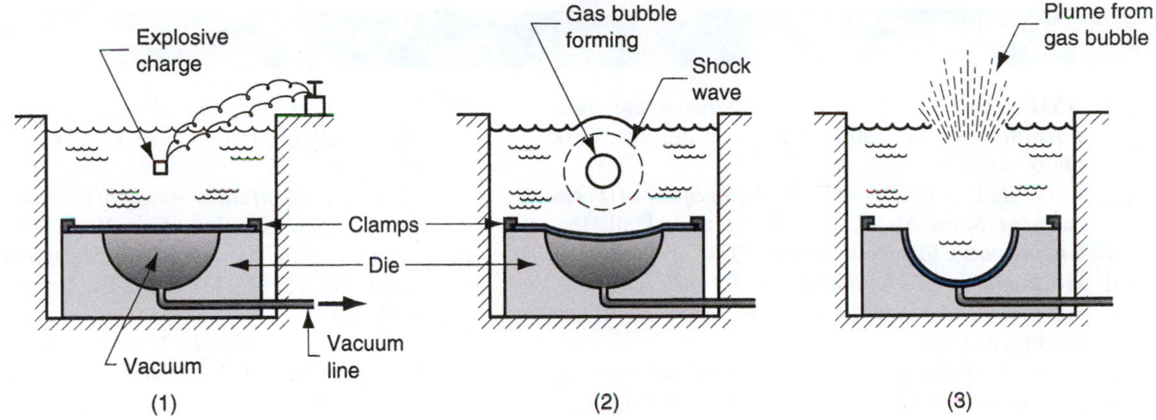

FIGURE 14.38 Explosive forming: (1) setup, (2) explosive is detonated, and (3) shock wave forms part and plume escapes water surface. (Credit: *Fundamentals of Modern Manufacturing*, 4th Edition by Mikell P. Groover, 2010. Reprinted with permission of John Wiley & Sons, Inc.)

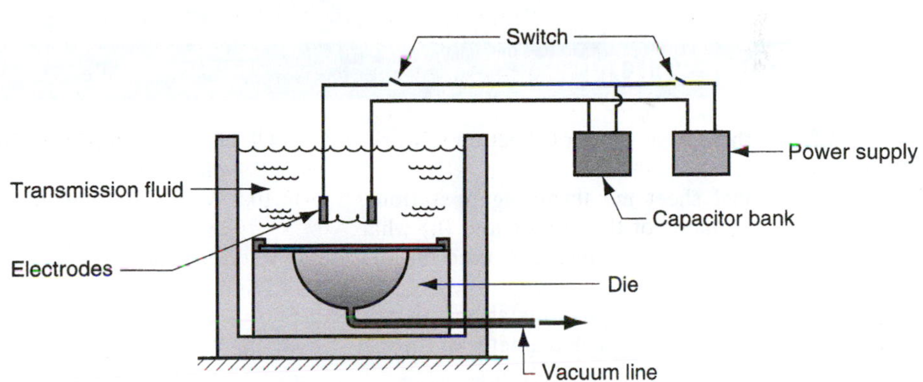

FIGURE 14.39 Electrohydraulic forming setup. (Credit: *Fundamentals of Modern Manufacturing*, 4th Edition by Mikell P. Groover, 2010. Reprinted with permission of John Wiley & Sons, Inc.)

explosive forming. The difference is in the method of generating the energy and the smaller amounts of energy that are released. This limits electrohydraulic forming to much smaller part sizes.

Electromagnetic Forming Electromagnetic forming, also called ***magnetic pulse forming***, is a process in which sheet metal is deformed by the mechanical force of an electromagnetic field induced in the workpart by an energized coil. The coil, energized by a capacitor, produces a magnetic field. This generates eddy currents in the work that produce their own magnetic field. The induced field opposes the primary field, producing a mechanical force that deforms the part into the surrounding cavity. Developed in the 1960s, electromagnetic forming is the most widely used HERF process [10]. It is typically used to form tubular parts, as illustrated in Figure 14.40.

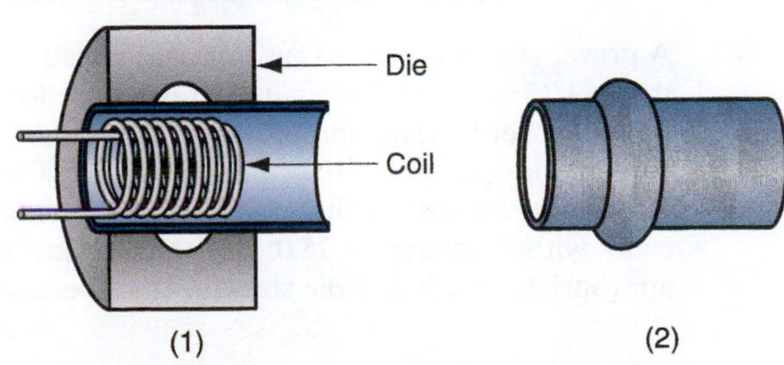

FIGURE 14.40 Electromagnetic forming: (1) setup in which coil is inserted into tubular workpart surrounded by die; (2) formed part. (Credit: *Fundamentals of Modern Manufacturing*, 4th Edition by Mikell P. Groover, 2010. Reprinted with permission of John Wiley & Sons, Inc.)

REFERENCES

[1] ***ASM Handbook***, Vol. 14B, ***Metalworking: Sheet Forming***. ASM International, Materials Park, Ohio, 2006.

[2] Eary, D. F., and Reed, E. A. ***Techniques of Pressworking Sheet Metal***, 2nd ed. Prentice-Hall, Inc., Englewood Cliffs, New Jersey, 1974.

[3] Hoffman, E. G. ***Fundamentals of Tool Design***, 2nd ed. Society of Manufacturing Engineers, Dearborn, Michigan, 1984.

[4] Hosford, W. F., and Caddell, R. M. ***Metal Forming: Mechanics and Metallurgy***, 3rd ed. Cambridge University Press, Cambridge, United Kingdom, 2007.

[5] Kalpakjian, S. ***Manufacturing Processes for Engineering Materials***, 4th ed. Prentice Hall/Pearson, Upper Saddle River, New Jersey, 2003.

[6] Lange, K., et al. (eds.). ***Handbook of Metal Forming***. Society of Manufacturing Engineers, Dearborn, Michigan, 1995.

[7] Mielnik, E. M. ***Metalworking Science and Engineering***. McGraw-Hill, Inc., New York, 1991.

[8] Schey, J. A. ***Introduction to Manufacturing Processes***, 3rd ed. McGraw-Hill Book Company, New York, 2000.

[9] Spitler, D., Lantrip, J., Nee, J., and Smith, D. A. ***Fundamentals of Tool Design***, 5th ed. Society of Manufacturing Engineers, Dearborn, Michigan, 2003.

[10] Wick, C., et al. (eds.). ***Tool and Manufacturing Engineers Handbook***, 4th ed., Vol. II, ***Forming***. Society of Manufacturing Engineers, Dearborn, Michigan, 1984.

REVIEW QUESTIONS

14.1. Identify the three basic types of sheet metalworking operations.

14.2. In conventional sheet metalworking operations, (a) what is the name of the tooling and (b) what is the name of the machine tool used in the operations?

14.3. In blanking of a circular sheet-metal part, is the clearance applied to the punch diameter or the die diameter?

14.4. What is the difference between a cutoff operation and a parting operation?

14.5. What is the difference between a notching operation and a seminotching operation?

14.6. Describe each of the two types of sheet-metal-bending operations: V-bending and edge bending.

14.7. For what is the bend allowance intended to compensate?

14.8. What is springback in sheet-metal bending?

14.9. Define drawing in the context of sheet metalworking.

14.10. What are some of the simple measures used to assess the feasibility of a proposed cup-drawing operation?

14.11. Distinguish between redrawing and reverse drawing.

14.12. What are some of the possible defects in drawn sheet-metal parts?

14.13. What is an embossing operation?

14.14. What is stretch forming?

14.15. What are the two basic categories of structural frames used in stamping presses?

14.16. What are the relative advantages and disadvantages of mechanical presses versus hydraulic presses in sheet metalworking?

14.17. What is the Guerin process?

14.18. Distinguish between roll bending and roll forming.

PROBLEMS

14.1. A power shears is used to cut soft cold-rolled steel that is 4.75 mm thick. At what clearance should the shears be set to yield an optimum cut?

14.2. A blanking operation is to be performed on 2.0 mm thick cold-rolled steel (half-hard). The part is circular with diameter = 75.0 mm. Determine the appropriate punch-and-die sizes for this operation.

14.3. A compound die will be used to blank and punch a large washer out of 6061ST aluminum alloy sheet stock 3.50 mm thick. The outside diameter of the washer is 50.0 mm and the inside diameter is 15.0 mm. Determine (a) the punch-and-die sizes for the blanking operation, and (b) the punch-and-die sizes for the punching operation.

14.4. A blanking die is to be designed to blank a rectangular part that is 105 mm by 37.5 mm. The sheet metal is 4 mm thick stainless steel (half-hard). Determine the dimensions of the blanking punch and the die opening.

14.5. Determine the blanking force required in Problem 14.2, if the shear strength of the steel = 325 MPa and the tensile strength is 450 MPa.

14.6. Determine the minimum tonnage press to perform the blanking and punching operation in Problem 14.3. The aluminum sheet metal has a tensile strength = 310 MPa, a strength coefficient of 350 MPa, and a strain-hardening exponent of 0.12. (a) Assume that blanking and punching occur simultaneously. (b) Assume the punches are staggered so that punching occurs first, then blanking.

14.7. Determine the tonnage requirement for the blanking operation in Problem 14.4, given that the stainless steel has a yield strength = 500 MPa, a shear strength = 600 MPa, and a tensile strength = 700 MPa.

14.8. A bending operation is to be performed on the part shown in Figure 14.13, except that the dimensions are changed as follows: stock thickness = 5.0 mm instead of 3.2 mm, inside bend radius = 8.0 mm instead of 4.75 mm, and included angle = 65° instead of 120°. Other dimensions are the same. Determine the blank size required.

14.9. Solve Problem 14.8, except that the inside bend radius R = 11.0 mm.

14.10. A bending operation is to be performed on 4.0-mm-thick cold-rolled steel sheet that is 25 mm wide and 100 mm long. The sheet is bent along the 25 mm direction, so that the bend is 25 mm long. The resulting sheet metal part has an acute angle of 30° and a bend radius of 6 mm. Determine (a) the bend allowance and (b) the length of the neutral axis of the part after the bend. (Hint: the length of the neutral axis before the bend = 100.0 mm.)

14.11. Determine the bending force required in Problem 14.8 if the bend is to be performed in a V-die with a die opening dimension of 30 mm. The material has a tensile strength of 600 MPa and a shear strength of 430 MPa.

14.12. Solve Problem 14.11, except that the operation is performed using a wiping die with die opening dimension = 22 mm.

14.13. A sheet-metal part 3.0 mm thick and 20.0 mm long is bent to an included angle = 60° and a bend radius = 7.5 mm in a V-die. The metal has a yield strength = 220 MPa and a tensile strength = 340 MPa. Compute the required force to bend the part, given that the die opening dimension = 15 mm.

14.14. Derive an expression for the reduction r in drawing as a function of drawing ratio DR.

14.15. A cup is to be drawn in a deep drawing operation. The height of the cup is 75 mm and its inside diameter = 100 mm. Sheet-metal thickness = 2 mm. If the blank diameter = 225 mm, determine (a) drawing ratio, (b) reduction, and (c) thickness-to-diameter ratio. (d) Does the operation seem feasible?

14.16. Solve Problem 14.15, except that the starting blank size diameter = 175 mm.

14.17. A deep drawing operation is performed in which the inside of the cylindrical cup has a diameter of 4.25 in and a height = 2.65 in. Stock thickness = 3/16 in, and starting blank diameter = 7.7 in. Punch-and-die radii = 5/32 in. The metal has a tensile strength = 65,000 lb/in^2, a yield strength = 32,000 lb/in^2, and a shear strength = 40,000 lb/in^2. Determine (a) drawing ratio, (b) reduction, (c) drawing force, and (d) blankholder force.

14.18. Solve Problem 14.17, except that the stock thickness t = 1/8 in.

14.19. A cup-drawing operation is performed in which the inside diameter = 80 mm and the height = 50 mm. Stock thickness = 3.0 mm, and starting blank diameter = 150 mm. Punch-and-die radii = 4 mm. Tensile strength = 400 MPa and yield strength = 180 MPa for this sheet metal. Determine (a) drawing ratio, (b) reduction, (c) drawing force, and (d) blankholder force.

14.20. A deep drawing operation is performed on a sheet-metal blank that is 1/8 in thick. The height (inside dimension) of the cup = 3.8 in and the diameter (inside dimension) = 5.0 in. Assuming the punch radius = 0, compute the starting diameter of the blank to complete the operation with no material left in the flange. Is the operation feasible (ignoring the fact that the punch radius is too small)?

14.21. Solve Problem 14.20, except use a punch radius = 0.375 in.

14.22. A drawing operation is performed on 3.0 mm stock. The part is a cylindrical cup with height = 50 mm and inside diameter = 70 mm. Assume the corner radius on the punch is zero. (a) Find the required starting blank size D_b. (b) Is the drawing operation feasible?

14.23. Solve Problem 14.22, except that the height = 60 mm.

14.24. Solve Problem 14.23, except that the corner radius on the punch = 10 mm.

14.25. The foreman in the drawing section of the shop brings to you several samples of parts that have been drawn in the shop. The samples have various defects. One has ears, another has wrinkles, and still a third has torn sections at its base. What are the causes of each of these defects and what remedies would you propose?

Part V Material Removal Processes

15 THEORY OF METAL MACHINING

Chapter Contents

15.1 Overview of Machining Technology

15.2 Theory of Chip Formation in Metal Machining

15.2.1 The Orthogonal Cutting Model

15.2.2 Actual Chip Formation

15.3 Force Relationships and the Merchant Equation

15.3.1 Forces in Metal Cutting

15.3.2 The Merchant Equation

15.4 Power and Energy Relationships in Machining

15.5 Cutting Temperature

15.5.1 Analytical Methods to Compute Cutting Temperatures

15.5.2 Measurement of Cutting Temperature

The ***material removal processes*** are a family of shaping operations (Figure 1.3) in which excess material is removed from a starting workpart so that what remains is the desired final geometry. The most important branch of the family is ***conventional machining***, in which a sharp cutting tool is used to mechanically cut the material to achieve the desired geometry. The three principal machining processes are turning, drilling, and milling. Other machining operations include shaping, planing, broaching, and sawing. This chapter begins our coverage of machining, which runs through Chapter 17.

Another group of material removal processes is the ***abrasive processes***, which mechanically remove material by the action of hard, abrasive particles. This process group, which includes grinding, is covered in Chapter 18. Finally, there are the ***nontraditional processes***, which use various energy forms other than a sharp cutting tool or abrasive particles to remove material. The energy forms include mechanical, electrochemical, thermal, and chemical.[1] The nontraditional processes are discussed in Chapter 19.

Machining is a manufacturing process in which a sharp cutting tool is used to cut away material to leave the desired part shape. The predominant cutting action in machining involves shear deformation of the work material to form a chip; as the chip is removed, a new surface is exposed.

[1] Some of the mechanical energy forms in the nontraditional processes involve the use of abrasive particles, and so they overlap with the abrasive processes in Chapter 18.

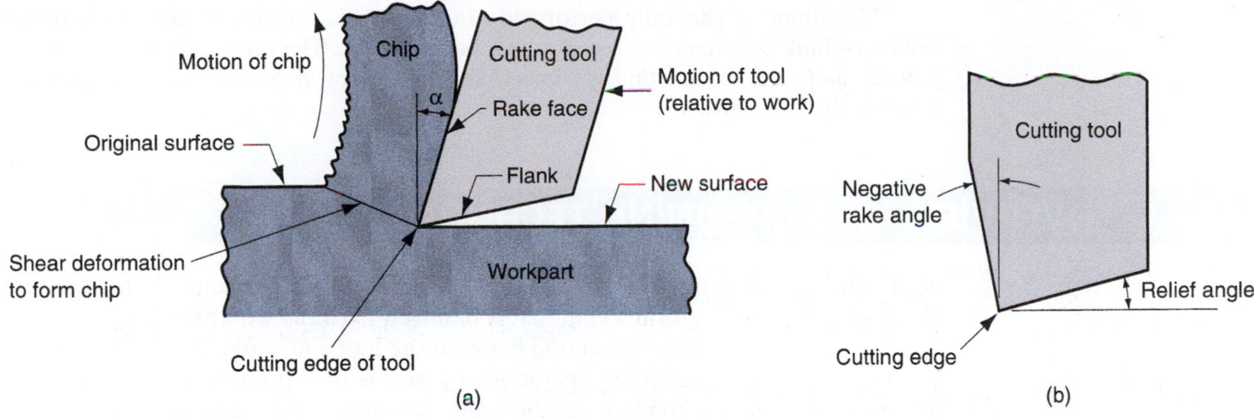

FIGURE 15.1 (a) A cross-sectional view of the machining process. (b) Tool with negative rake angle; compare with positive rake angle in (a). (Credit: *Fundamentals of Modern Manufacturing,* 4th Edition by Mikell P. Groover, 2010. Reprinted with permission of John Wiley & Sons, Inc.)

Machining is most frequently applied to shape metals. The process is illustrated in the diagram of Figure 15.1.

Machining is one of the most important manufacturing processes. The Industrial Revolution and the growth of the manufacturing-based economies of the world can be traced largely to the development of the various machining operations. Machining is important commercially and technologically for several reasons:

- ***Variety of work materials***. Machining can be applied to a wide variety of work materials. Virtually all solid metals can be machined. Plastics and plastic composites can also be cut by machining. Ceramics pose difficulties because of their high hardness and brittleness; however, most ceramics can be successfully cut by the abrasive machining processes (Chapter 18).
- ***Variety of part shapes and geometric features***. Machining can be used to create any regular geometries, such as flat planes, round holes, and cylinders. By introducing variations in tool shapes and tool paths, irregular geometries can be created, such as screw threads and T-slots. By combining several machining operations in sequence, shapes of almost unlimited complexity and variety can be produced.
- ***Dimensional accuracy***. Machining can produce dimensions to very close tolerances. Some machining processes can achieve tolerances of ± 0.025 mm (± 0.001 in), much more accurate than most other processes.
- ***Good surface finishes***. Machining is capable of creating very smooth surface finishes. Roughness values less than 0.4 microns (16 μin) can be achieved in conventional machining operations. Some abrasive processes can achieve even better finishes.

On the other hand, certain disadvantages are associated with machining and other material removal processes:

- ***Wasteful of material***. Machining is inherently wasteful of material. The chips generated in a machining operation are wasted material. Although these chips can usually be recycled, they represent waste in terms of the unit operation.
- ***Time consuming***. A machining operation generally takes more time to shape a given part than alternative shaping processes such as casting or forging.

Machining is generally performed after other manufacturing processes such as casting or bulk deformation (e.g., forging, bar drawing). The other processes create the general shape of the starting workpart, and machining provides the final geometry, dimensions, and finish.

15.1 OVERVIEW OF MACHINING TECHNOLOGY

Machining is not just one process; it is a group of processes. The common feature is the use of a cutting tool to form a chip that is removed from the workpart. To perform the operation, relative motion is required between the tool and work. This relative motion is achieved in most machining operations by means of a primary motion, called the ***cutting speed***, and a secondary motion, called the ***feed***. The shape of the tool and its penetration into the work surface, combined with these motions, produces the desired geometry of the resulting work surface.

Types of Machining Operations There are many kinds of machining operations, each of which is capable of generating a certain part geometry and surface texture. We discuss these operations in considerable detail in Chapter 16, but for now it is appropriate to identify and define the three most common types: turning, drilling, and milling, illustrated in Figure 15.2.

In ***turning***, a cutting tool with a single cutting edge is used to remove material from a rotating workpiece to generate a cylindrical shape, as in Figure 15.2(a). The speed motion in turning is provided by the rotating workpart, and the feed motion is achieved by the cutting tool moving slowly in a direction parallel to the axis of rotation of the workpiece. ***Drilling*** is used to create a round hole. It is accomplished by a rotating tool

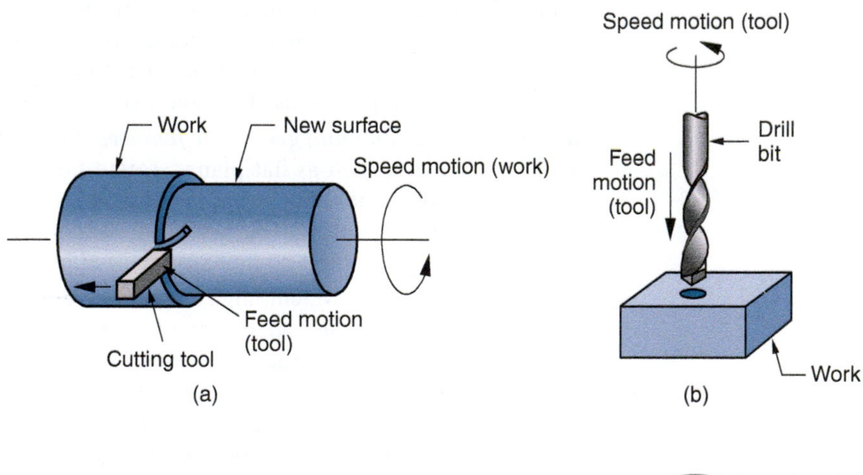

FIGURE 15.2 The three most common types of machining processes: (a) turning, (b) drilling, and two forms of milling: (c) peripheral milling, and (d) face milling. (Credit: *Fundamentals of Modern Manufacturing*, 4th Edition by Mikell P. Groover, 2010. Reprinted with permission of John Wiley & Sons, Inc.)

that typically has two cutting edges. The tool is fed in a direction parallel to its axis of rotation into the workpart to form the round hole, as in Figure 15.2(b). In ***milling***, a rotating tool with multiple cutting edges is fed slowly across the work material to generate a plane or straight surface. The direction of the feed motion is perpendicular to the tool's axis of rotation. The speed motion is provided by the rotating milling cutter. The two basic forms of milling are peripheral milling and face milling, as in Figure 15.2(c) and (d).

Other conventional machining operations include shaping, planing, broaching, and sawing (Section 16.6). Also, grinding and similar abrasive operations are often included within the category of machining. These processes commonly follow the conventional machining operations and are used to achieve a superior surface finish on the workpart.

The Cutting Tool A cutting tool has one or more sharp cutting edges and is made of a material that is harder than the work material. The cutting edge serves to separate a chip from the parent work material, as in Figure 15.1. Connected to the cutting edge are two surfaces of the tool: the rake face and the flank. The rake face, which directs the flow of the newly formed chip, is oriented at a certain angle called the ***rake angle*** α. It is measured relative to a plane perpendicular to the work surface. The rake angle can be positive, as in Figure 15.1(a), or negative as in (b). The flank of the tool provides a clearance between the tool and the newly generated work surface, thus protecting the surface from abrasion, which would degrade the finish. This flank surface is oriented at an angle called the ***relief angle***.

Most cutting tools in practice have more complex geometries than those in Figure 15.1. There are two basic types, examples of which are illustrated in Figure 15.3: (a) single-point tools and (b) multiple-cutting-edge tools. A ***single-point tool*** has one cutting edge and is used for operations such as turning. In addition to the tool features shown in Figure 15.1, there is one tool point from which the name of this cutting tool is derived. During machining, the point of the tool penetrates below the original work surface of the part. The point is usually rounded to a certain radius, called the nose radius. ***Multiple-cutting-edge tools*** have more than one cutting edge and usually achieve their motion relative to the workpart by rotating. Drilling and milling use rotating multiple-cutting-edge tools. Figure 15.3(b) shows a helical milling cutter used in peripheral milling. Although the shape is quite different from a single-point tool, many elements of tool geometry are similar. Single-point and multiple-cutting-edge tools and the materials used in them are discussed in more detail in Chapter 17.

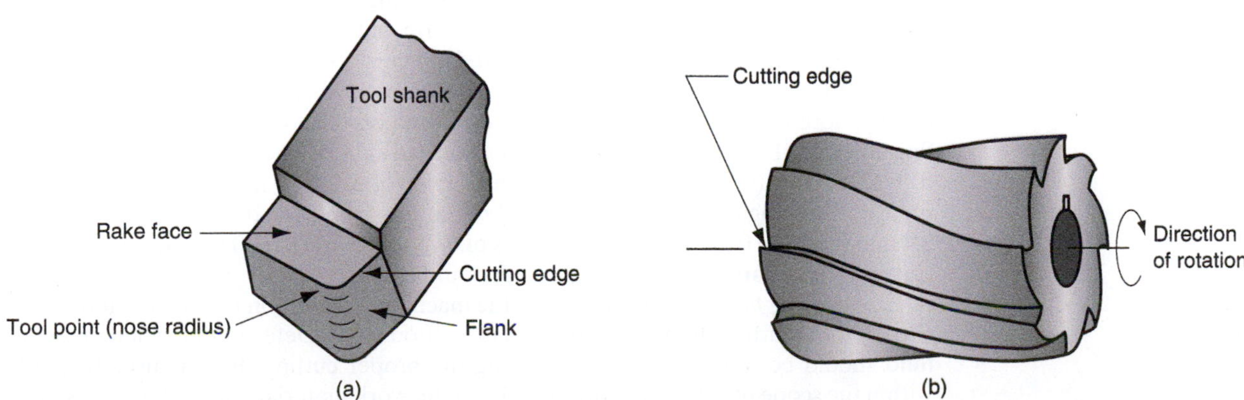

FIGURE 15.3 (a) A single-point tool showing rake face, flank, and tool point; and (b) a helical milling cutter, representative of tools with multiple cutting edges. (Credit: *Fundamentals of Modern Manufacturing,* 4th Edition by Mikell P. Groover, 2010. Reprinted with permission of John Wiley & Sons, Inc.)

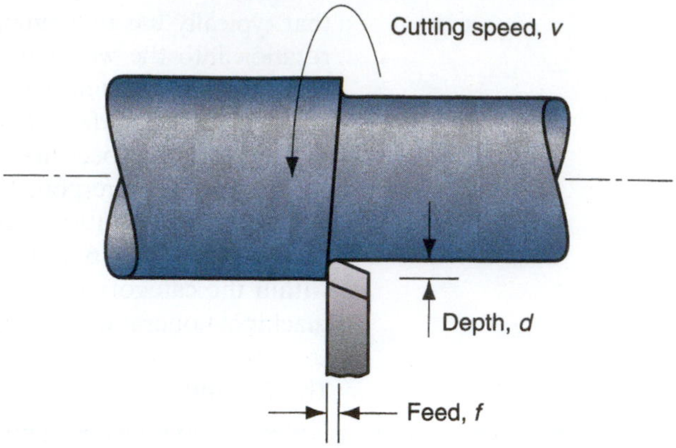

FIGURE 15.4 Cutting speed, feed, and depth of cut for a turning operation. (Credit: *Fundamentals of Modern Manufacturing*, 4th Edition by Mikell P. Groover, 2010. Reprinted with permission of John Wiley & Sons, Inc.)

Cutting Conditions Relative motion is required between the tool and work to perform a machining operation. The primary motion is accomplished at a certain ***cutting speed*** v. In addition, the tool must be moved laterally across the work. This is a much slower motion, called the ***feed*** f. The remaining dimension of the cut is the penetration of the cutting tool below the original work surface, called the ***depth of cut*** d. Collectively, speed, feed, and depth of cut are called the ***cutting conditions***. They form the three dimensions of the machining process, and for certain operations (e.g., most single-point tool operations), they can be used to calculate the material removal rate for the process:

$$R_{MR} = vfd \tag{15.1}$$

where R_{MR} = material removal rate, mm^3/s (in^3/min); v = cutting speed, m/s (ft/min), which must be converted to mm/s (in/min); f = feed, mm (in); and d = depth of cut, mm (in).

The cutting conditions for a turning operation are depicted in Figure 15.4. Typical units used for cutting speed are m/s (ft/min). Feed in turning is expressed in mm/rev (in/rev), and depth of cut is expressed in mm (in). In other machining operations, interpretations of the cutting conditions may differ. For example, in a drilling operation, depth is interpreted as the depth of the drilled hole.

Machining operations usually divide into two categories, distinguished by purpose and cutting conditions: roughing cuts and finishing cuts. ***Roughing*** cuts are used to remove large amounts of material from the starting workpart as rapidly as possible, in order to produce a shape close to the desired form, but leaving some material on the piece for a subsequent finishing operation. ***Finishing*** cuts are used to complete the part and achieve the final dimensions, tolerances, and surface finish. In production machining jobs, one or more roughing cuts are usually performed on the work, followed by one or two finishing cuts. Roughing operations are performed at high feeds and depths—feeds of 0.4 to 1.25 mm/rev (0.015 to 0.050 in/rev) and depths of 2.5 to 20 mm (0.100 to 0.750 in) are typical. Finishing operations are carried out at low feeds and depths—feeds of 0.125 to 0.4 mm (0.005 to 0.015 in/rev) and depths of 0.75 to 2.0 mm (0.030 to 0.075 in) are typical. Cutting speeds are lower in roughing than in finishing.

A ***cutting fluid*** is often applied to the machining operation to cool and lubricate the cutting tool (cutting fluids are discussed in Section 17.4). Determining whether a cutting fluid should be used, and, if so, choosing the proper cutting fluid, is usually included within the scope of cutting conditions. Given the work material and tooling, the selection of these conditions is very influential in determining the success of a machining operation.

Machine Tools A machine tool is used to hold the workpart, position the tool relative to the work, and provide power for the machining process at the speed, feed, and depth that have been set. By controlling the tool, work, and cutting conditions, machine tools permit parts to be made with great accuracy and repeatability, to tolerances of 0.025 mm (0.001 in) and better. The term ***machine tool*** applies to any power-driven machine that performs a machining operation, including grinding. The term is also applied to machines that perform metal forming and pressworking operations (Chapters 13 and 14).

The traditional machine tools used to perform turning, drilling, and milling are lathes, drill presses, and milling machines, respectively. Conventional machine tools are usually tended by a human operator, who loads and unloads the workparts, changes cutting tools, and sets the cutting conditions. Many modern machine tools are designed to accomplish their operations with a form of automation called computer numerical control (Section 29.1).

15.2 THEORY OF CHIP FORMATION IN METAL MACHINING

The geometry of most practical machining operations is somewhat complex. A simplified model of machining is available that neglects many of the geometric complexities, yet describes the mechanics of the process quite well. It is called the ***orthogonal*** cutting model, Figure 15.5. Although an actual machining process is three-dimensional, the orthogonal model has only two dimensions that play active roles in the analysis.

15.2.1 THE ORTHOGONAL CUTTING MODEL

By definition, orthogonal cutting uses a wedge-shaped tool in which the cutting edge is perpendicular to the direction of cutting speed. As the tool is forced into the material, the chip is formed by shear deformation along a plane called the ***shear plane***, which is oriented at an angle ϕ with the surface of the work. Only at the sharp cutting edge of the tool does failure of the material occur, resulting in separation of the chip from the parent material. Along the shear plane, where the bulk of the mechanical energy is consumed in machining, the material is plastically deformed.

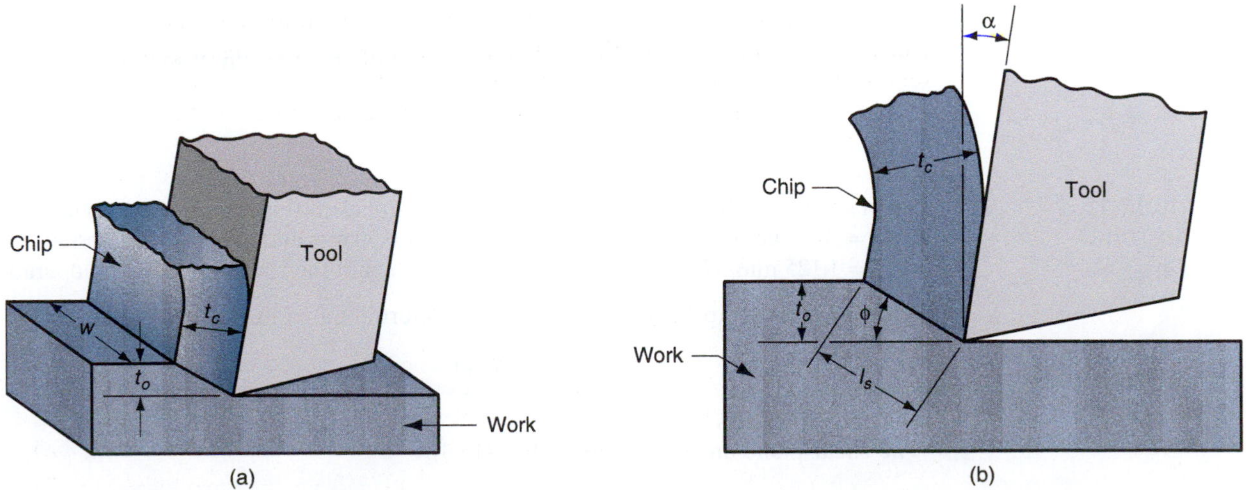

FIGURE 15.5 Orthogonal cutting: (a) as a three-dimensional process, and (b) how it reduces to two dimensions in the side view. (Credit: *Fundamentals of Modern Manufacturing,* 4th Edition by Mikell P. Groover, 2010. Reprinted with permission of John Wiley & Sons, Inc.)

The tool in orthogonal cutting has only two elements of geometry: (1) rake angle and (2) clearance angle. As indicated previously, the rake angle α determines the direction that the chip flows as it is formed from the workpart; and the clearance angle provides a small clearance between the tool flank and the newly generated work surface.

During cutting, the cutting edge of the tool is positioned a certain distance below the original work surface. This corresponds to the thickness of the chip prior to chip formation, t_o. As the chip is formed along the shear plane, its thickness increases to t_c. The ratio of t_o to t_c is called the ***chip thickness ratio*** (or simply the ***chip ratio***) r:

$$r = \frac{t_o}{t_c} \tag{15.2}$$

Since the chip thickness after cutting is always greater than the corresponding thickness before cutting, the chip ratio will always be less than 1.0.

In addition to t_o, the orthogonal cut has a width dimension w, as shown in Figure 15.5(a), even though this dimension does not contribute much to the analysis in orthogonal cutting.

The geometry of the orthogonal cutting model allows us to establish an important relationship between the chip thickness ratio, the rake angle, and the shear plane angle. Let l_s be the length of the shear plane. We can make the substitutions: $t_o = l_s \sin \phi$, and $t_c = l_s \cos(\phi - \alpha)$. Thus,

$$r = \frac{l_s \sin \phi}{l_s \cos(\phi - \alpha)} = \frac{\sin \phi}{\cos(\phi - \alpha)}$$

This can be rearranged to determine ϕ as follows:

$$\tan \phi = \frac{r \cos \alpha}{1 - r \sin \alpha} \tag{15.3}$$

The shear strain that occurs along the shear plane can be estimated by examining Figure 15.6. Part (a) shows shear deformation approximated by a series of parallel plates sliding against one another to form the chip. Consistent with our definition of shear strain (Section 3.1.4), each plate experiences the shear strain shown in Figure 15.6(b). Referring to part (c), this can be expressed as

$$\gamma = \frac{AC}{BD} = \frac{AD + DC}{BD}$$

which can be reduced to the following definition of shear strain in metal cutting:

$$\gamma = \tan(\phi - \alpha) + \cot \phi \tag{15.4}$$

Example 15.1 Orthogonal Cutting

In a machining operation that approximates orthogonal cutting, the cutting tool has a rake angle = 10°. The chip thickness before the cut $t_o = 0.50$ mm and the chip thickness after the cut $t_c = 1.125$ mm. Calculate the shear plane angle and the shear strain in the operation.

Solution: The chip thickness ratio can be determined from Eq. (15.2):

$$r = \frac{0.50}{1.125} = 0.444$$

The shear plane angle is given by Eq. (15.3):

$$\tan \phi = \frac{0.444 \cos 10}{1 - 0.444 \sin 10} = 0.4738$$

$$\phi = 25.4^\circ$$

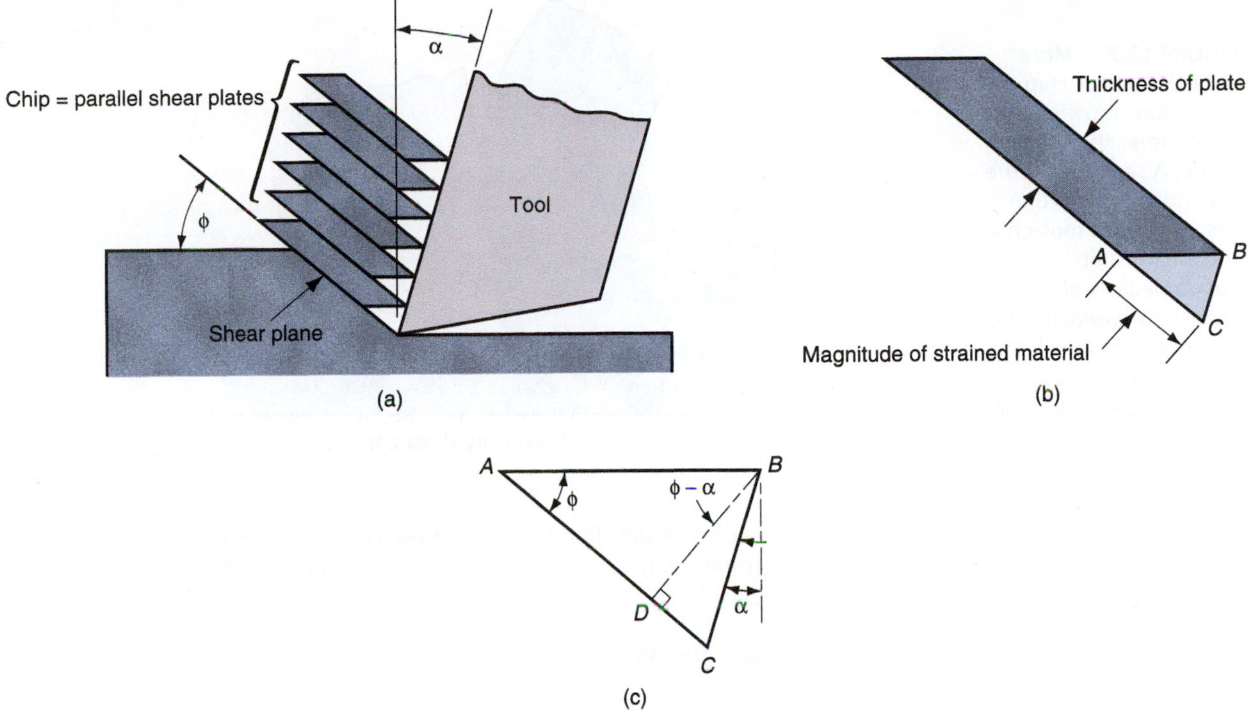

FIGURE 15.6 Shear strain during chip formation: (a) chip formation depicted as a series of parallel plates sliding relative to each other; (b) one of the plates isolated to illustrate the definition of shear strain based on this parallel plate model; and (c) shear strain triangle used to derive Eq. (15.4). (Credit: *Fundamentals of Modern Manufacturing*, 4th Edition by Mikell P. Groover, 2010. Reprinted with permission of John Wiley & Sons, Inc.)

Finally, the shear strain is calculated from Eq. (15.4):

$$\gamma = \tan(25.4 - 10) + \cot 25.4$$

$$\gamma = 0.275 + 2.111 = 2.386$$

15.2.2 ACTUAL CHIP FORMATION

We should note that there are differences between the orthogonal model and an actual machining process. First, the shear deformation process does not occur along a plane, but within a zone. If shearing were to take place across a plane of zero thickness, it would imply that the shearing action must occur instantaneously as it passes through the plane, rather than over some finite (although brief) time period. For the material to behave in a realistic way, the shear deformation must occur within a thin shear zone. This more realistic model of the shear deformation process in machining is illustrated in Figure 15.7. Metal-cutting experiments have indicated that the thickness of the shear zone is only a few thousandths of an inch. Since the shear zone is so thin, there is not a great loss of accuracy in most cases by referring to it as a plane.

Second, in addition to shear deformation that occurs in the shear zone, another shearing action occurs in the chip after it has been formed. This additional shear is referred to as secondary shear to distinguish it from primary shear. Secondary shear results from friction between the chip and the tool as the chip slides along the rake face of the tool. Its effect increases with increased friction between the tool and chip. The primary and secondary shear zones can be seen in Figure 15.7.

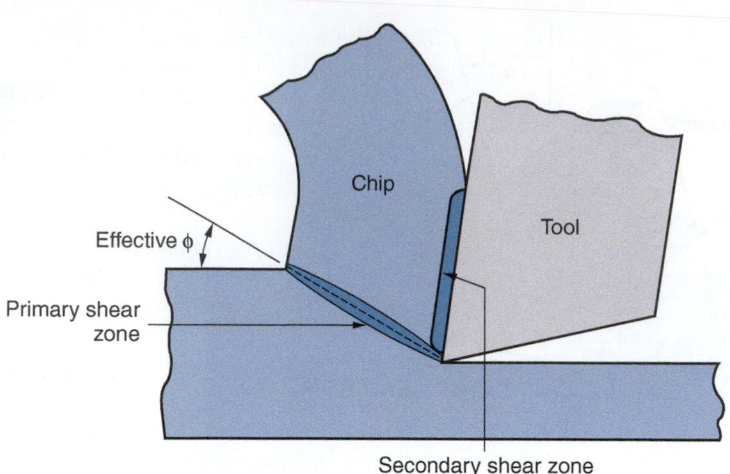

FIGURE 15.7 More realistic view of chip formation, showing shear zone rather than shear plane. Also shown is the secondary shear zone resulting from tool–chip friction. (Credit: *Fundamentals of Modern Manufacturing*, 4th Edition by Mikell P. Groover, 2010. Reprinted with permission of John Wiley & Sons, Inc.)

Third, formation of the chip depends on the type of material being machined and the cutting conditions of the operation. Four basic types of chip can be distinguished, illustrated in Figure 15.8:

- ***Discontinuous chip***. When relatively brittle materials (e.g., cast irons) are machined at low cutting speeds, the chips often form into separate segments (sometimes the segments are loosely attached). This tends to impart an irregular texture to the machined surface. High tool–chip friction and large feed and depth of cut promote the formation of this chip type.
- ***Continuous chip***. When ductile work materials are cut at high speeds and relatively small feeds and depths, long continuous chips are formed. A good surface finish typically results when this chip type is formed. A sharp cutting edge on the tool and low tool–chip friction encourage the formation of continuous chips. Long, continuous chips (as in turning) can cause problems with regard to chip disposal and/or tangling about the tool. To solve these problems, turning tools are often equipped with chip breakers (Section 17.3.1).
- ***Continuous chip with built-up edge***. When machining ductile materials at low-to-medium cutting speeds, friction between tool and chip tends to cause portions of the work material to adhere to the rake face of the tool near the cutting edge. This

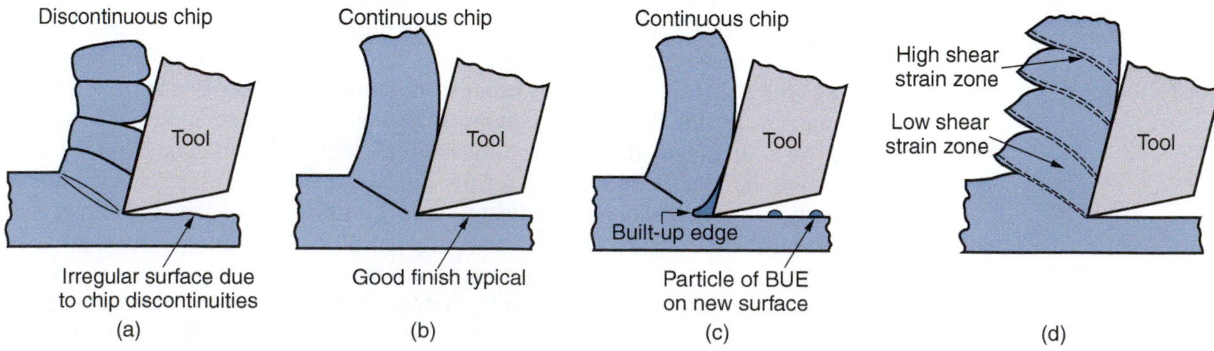

FIGURE 15.8 Four types of chip formation in metal cutting: (a) discontinuous, (b) continuous, (c) continuous with built-up edge, (d) serrated. (Credit: *Fundamentals of Modern Manufacturing*, 4th Edition by Mikell P. Groover, 2010. Reprinted with permission of John Wiley & Sons, Inc.)

formation is called a built-up edge (BUE). The formation of a BUE is cyclical; it forms and grows, then becomes unstable and breaks off. Much of the detached BUE is carried away with the chip, sometimes taking portions of the tool rake face with it, which reduces the life of the cutting tool. Portions of the detached BUE that are not carried off with the chip become imbedded in the newly created work surface, causing the surface to become rough.

The preceding chip types were first classified by Ernst in the late 1930s [13]. Since then, the available metals used in machining, cutting tool materials, and cutting speeds have all increased, and a fourth chip type has been identified:

- ***Serrated chips*** (the term ***shear-localized*** is also used for this fourth chip type). These chips are semicontinuous in the sense that they possess a sawtooth appearance that is produced by a cyclical chip formation of alternating high shear strain followed by low shear strain. This fourth type of chip is most closely associated with certain difficult-to-machine metals such as titanium alloys, nickel-base superalloys, and austenitic stainless steels when they are machined at higher cutting speeds. However, the phenomenon is also found with more common work metals (e.g., steels) when they are cut at high speeds [13].[2]

15.3 FORCE RELATIONSHIPS AND THE MERCHANT EQUATION

Several forces can be defined relative to the orthogonal cutting model. Based on these forces, shear stress, coefficient of friction, and certain other relationships can be defined.

15.3.1 FORCES IN METAL CUTTING

Consider the forces acting on the chip during orthogonal cutting in Figure 15.9(a). The forces applied against the chip by the tool can be separated into two mutually perpendicular components: friction force and normal force to friction. The ***friction force*** F is the frictional force resisting the flow of the chip along the rake face of the tool. The ***normal force to friction*** N is perpendicular to the friction force. These two components can be used to define the coefficient of friction between the tool and the chip:

$$\mu = \frac{F}{N} \tag{15.5}$$

The friction force and its normal force can be added vectorially to form a resultant force R, which is oriented at an angle β, called the friction angle. The friction angle is related to the coefficient of friction as

$$\mu = \tan\beta \tag{15.6}$$

In addition to the tool forces acting on the chip, there are two force components applied by the workpiece on the chip: shear force and normal force to shear. The ***shear force*** F_s is the force that causes shear deformation to occur in the shear plane, and the ***normal force to shear*** F_n is perpendicular to the shear force. Based on the shear force, we can define the shear stress that acts along the shear plane between the work and the chip:

$$\tau = \frac{F_s}{A_s} \tag{15.7}$$

[2] A more complete description of the serrated chip type can be found in Trent & Wright [13], pp. 348–367.

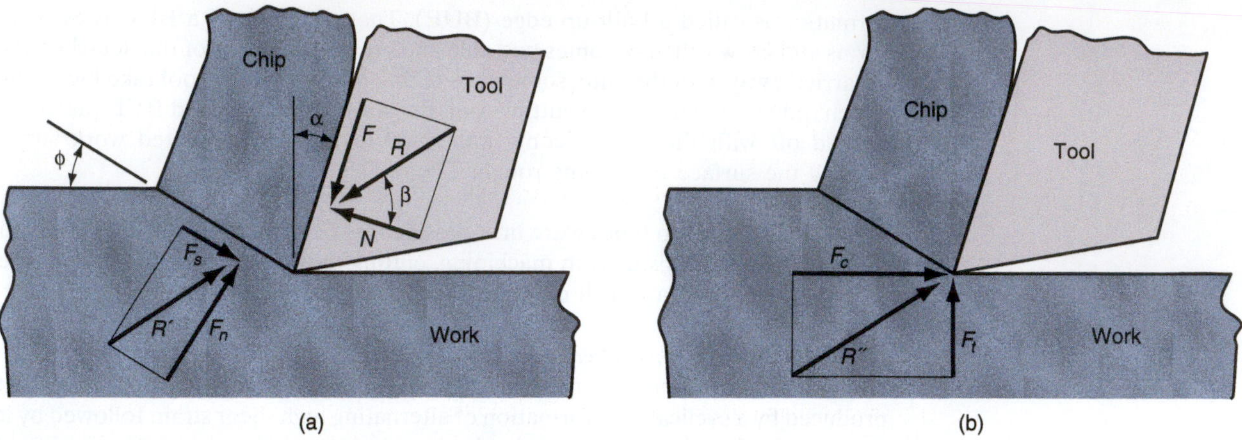

FIGURE 15.9 Forces in metal cutting: (a) forces acting on the chip in orthogonal cutting, and (b) forces acting on the tool that can be measured. (Credit: *Fundamentals of Modern Manufacturing*, 4th Edition by Mikell P. Groover, 2010. Reprinted with permission of John Wiley & Sons, Inc.)

where A_s = area of the shear plane. This shear plane area can be calculated as

$$A_s = \frac{t_o w}{\sin\phi} \tag{15.8}$$

The shear stress in Eq. (15.7) represents the level of stress required to perform the machining operation. Therefore, this stress is equal to the shear strength of the work material ($\tau = S$) under the conditions at which cutting occurs.

Vector addition of the two force components F_s and F_n yields the resultant force R'. In order for the forces acting on the chip to balance, this resultant R' must be equal in magnitude, opposite in direction, and collinear with the resultant R.

None of the four force components F, N, F_s, and F_n can be directly measured in a machining operation, because the directions in which they are applied vary with different tool geometries and cutting conditions. However, it is possible for the cutting tool to be instrumented using a force measuring device called a dynamometer, so that two additional force components acting against the tool can be directly measured: cutting force and thrust force. The ***cutting force*** F_c is in the direction of cutting, the same direction as the cutting speed v, and the ***thrust force*** F_t is perpendicular to the cutting force and is associated with the chip thickness before the cut t_o. The cutting force and thrust force are shown in Figure 15.9(b) together with their resultant force R''. The respective directions of these forces are known, so the force transducers in the dynamometer can be aligned accordingly.

The following equations can be derived to relate the four force components that cannot be measured to the two forces that can be measured:

$$F = F_c \sin\alpha + F_t \cos\alpha \tag{15.9}$$

$$N = F_c \cos\alpha - F_t \sin\alpha \tag{15.10}$$

$$F_s = F_c \cos\phi - F_t \sin\phi \tag{15.11}$$

$$F_n = F_c \sin\phi + F_t \cos\phi \tag{15.12}$$

If cutting force and thrust force are known, these four equations can be used to calculate estimates of shear force, friction force, and normal force to friction. Based on these force estimates, shear stress and coefficient of friction can be determined.

Note that in the special case of orthogonal cutting when the rake angle $\alpha = 0$, Eqs. (15.9) and (15.10) reduce to $F = F_t$ and $N = F_c$, respectively. Thus, in this special case, friction force and its normal force could be directly measured by the dynamometer.

Example 15.2 Shear Stress in Machining

Suppose in Example 15.1 that cutting force and thrust force are measured during an orthogonal cutting operation: $F_c = 1559$ N and $F_t = 1271$ N. The width of the orthogonal cutting operation $w = 3.0$ mm. Based on these data, determine the shear strength of the work material.

Solution: From Example 15.1, rake angle $\alpha = 10°$, and shear plane angle $\phi = 25.4°$. Shear force can be computed from Eq. (15.11):

$$F_s = 1559 \cos 25.4 - 1271 \sin 25.4 = 863\,\text{N}$$

The shear plane area is given by Eq. (15.8):

$$A_s = \frac{(0.5)(3.0)}{\sin 25.4} = 3.497\ \text{mm}^2$$

Thus the shear stress, which equals the shear strength of the work material, is

$$\tau = S = \frac{863}{3.497} = 247\,\text{N/mm}^2 = 247\,\text{MPa}$$

■

This example demonstrates that cutting force and thrust force are related to the shear strength of the work material. The relationships can be established in a more direct way. Recalling from Eq. (15.7) that the shear force $F_s = S\,A_s$, the following equations can be derived:

$$F_c = \frac{St_o w \cos(\beta - \alpha)}{\sin\phi \cos(\phi + \beta - \alpha)} = \frac{F_s \cos(\beta - \alpha)}{\cos(\phi + \beta - \alpha)} \tag{15.13}$$

and

$$F_t = \frac{S_t w \sin(\beta - \alpha)}{\sin\phi \cos(\phi + \beta - \alpha)} = \frac{F_s \sin(\beta - \alpha)}{\cos(\phi + \beta - \alpha)} \tag{15.14}$$

These equations allow one to estimate cutting force and thrust force in an orthogonal cutting operation if the shear strength of the work material is known.

15.3.2 THE MERCHANT EQUATION

One of the important relationships in metal cutting was derived by Eugene Merchant [10]. Its derivation was based on the assumption of orthogonal cutting, but its general validity extends to three-dimensional machining operations. Merchant started with the definition of shear stress expressed in the form of the following relationship derived by combining Eqs. (15.7), (15.8), and (15.11):

$$\tau = \frac{F_c \cos\phi - F_t \sin\phi}{(t_o w / \sin\phi)} \tag{15.15}$$

Merchant reasoned that, out of all the possible angles emanating from the cutting edge of the tool at which shear deformation could occur, there is one angle ϕ that predominates. This is the angle at which shear stress is just equal to the shear strength of the work material, and so shear deformation occurs at this angle. For all other possible shear angles, the shear stress is less than the shear strength, so chip formation cannot occur at these other angles. In effect, the work material will select a shear plane angle that minimizes energy. This angle can be determined by taking the derivative of the shear stress S in Eq. (15.15) with respect to ϕ and setting the derivative to zero. Solving for ϕ, we get the relationship named after Merchant:

$$\phi = 45 + \frac{\alpha}{2} - \frac{\beta}{2} \tag{15.16}$$

Among the assumptions in the Merchant equation is that shear strength of the work material is a constant, unaffected by strain rate, temperature, and other factors. Because this assumption is violated in practical machining operations, Eq. (15.16) must be considered an approximate relationship rather than an accurate mathematical equation. Let us nevertheless consider its application in the following example.

Example 15.3 Estimating Friction Angle

Using the data and results from our previous examples, determine (a) the friction angle and (b) the coefficient of friction.

Solution: (a) From Example 15.1, $\alpha = 10°$, and $\phi = 25.4°$. Rearranging Eq. (15.16), the friction angle can be estimated:

$$\beta = 2(45) + 10 - 2(25.4) = 49.2°$$

(b) The coefficient of friction is given by Eq. (15.6):

$$\mu = \tan 49.2 = 1.16$$

■

Lessons Based on the Merchant Equation The real value of the Merchant equation is that it defines the general relationship between rake angle, tool–chip friction, and shear plane angle. The shear plane angle can be increased by (1) increasing the rake angle and (2) decreasing the friction angle (and coefficient of friction) between the tool and the chip. Rake angle can be increased by proper tool design, and friction angle can be reduced by using a lubricant cutting fluid.

The importance of increasing the shear plane angle can be seen in Figure 15.10. If all other factors remain the same, a higher shear plane angle results in a smaller shear plane area. Since the shear strength is applied across this area, the shear force required to form the chip will decrease when the shear plane area is reduced. A greater shear plane angle results in lower cutting energy, lower power requirements, and lower cutting temperature. These are good reasons to try to make the shear plane angle as large as possible during machining.

Approximation of Turning by Orthogonal Cutting The orthogonal model can be used to approximate turning and certain other single-point machining operations so long as the feed in these operations is small relative to depth of cut. Thus, most of the cutting will take place in the direction of the feed, and cutting on the point of the tool will be negligible. Figure 15.11 indicates the conversion from one cutting situation to the other.

The interpretation of cutting conditions is different in the two cases. The chip thickness before the cut t_o in orthogonal cutting corresponds to the feed f in turning, and the width of cut w in orthogonal cutting corresponds to the depth of cut d in turning. In addition, the thrust force F_t in the orthogonal model corresponds to the feed force F_f in turning. Cutting speed and cutting force have the same meanings in the two cases. Table 15.1 summarizes the conversions.

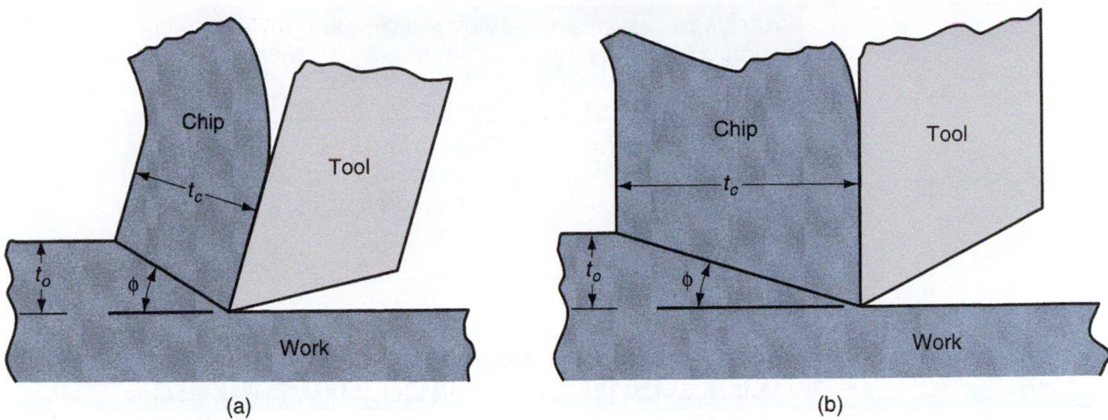

FIGURE 15.10 Effect of shear plane angle ϕ: (a) higher ϕ with a resulting lower shear plane area; (b) smaller ϕ with a corresponding larger shear plane area. Note that the rake angle is larger in (a), which tends to increase shear angle according to the Merchant equation. (Credit: *Fundamentals of Modern Manufacturing,* 4th Edition by Mikell P. Groover, 2010. Reprinted with permission of John Wiley & Sons, Inc.)

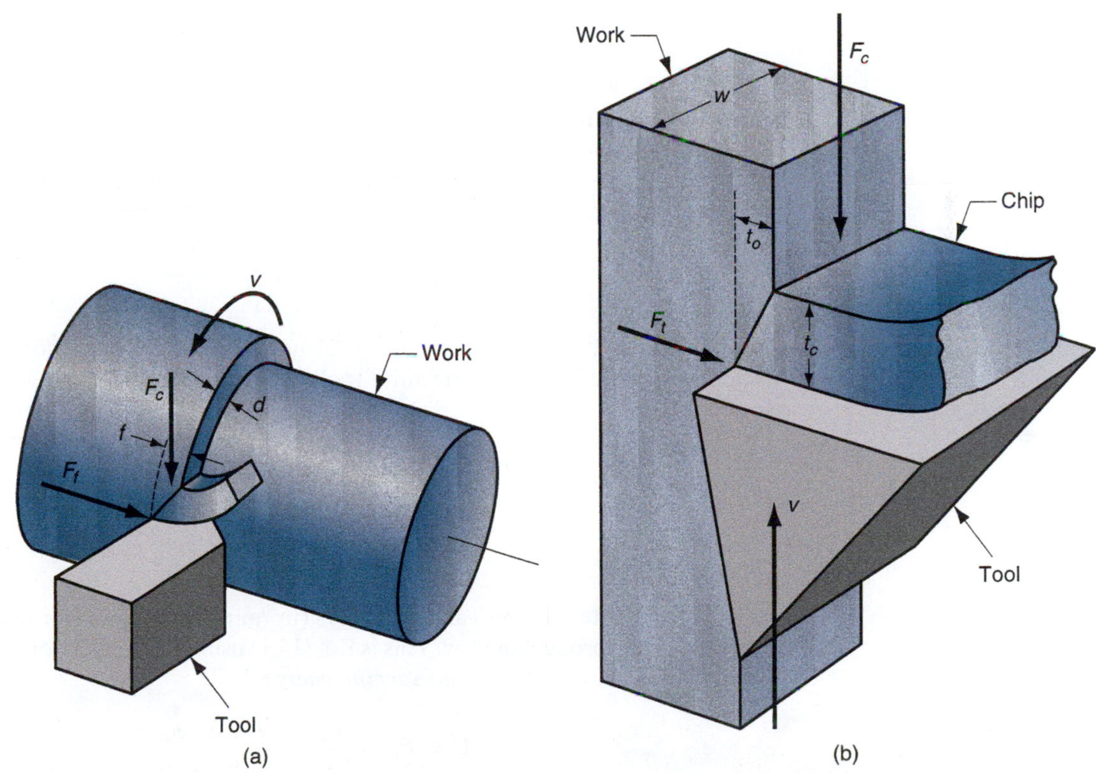

FIGURE 15.11 Approximation of turning by the orthogonal model: (a) turning; and (b) the corresponding orthogonal cutting. (Credit: *Fundamentals of Modern Manufacturing,* 4th Edition by Mikell P. Groover, 2010. Reprinted with permission of John Wiley & Sons, Inc.)

TABLE 15.1 Conversion key: turning operation vs. orthogonal cutting.

Turning Operation	Orthogonal Cutting Model
Feed $f =$	Chip thickness before cut t_o
Depth $d =$	Width of cut w
Cutting speed $v =$	Cutting speed v
Cutting force $F_c =$	Cutting force F_c
Feed force $F_f =$	Thrust force F_t

15.4 POWER AND ENERGY RELATIONSHIPS IN MACHINING

A machining operation requires power. The cutting force in a production machining operation might exceed 1000 N (several hundred pounds), as suggested by Example 15.2. Typical cutting speeds are several hundred m/min. The product of cutting force and speed gives the power (energy per unit time) required to perform a machining operation:

$$P_c = F_c v \tag{15.17}$$

where P_c = cutting power, N-m/s or W (ft-lb/min); F_c = cutting force, N (lb); and v = cutting speed, m/s (ft/min). In U.S. customary units, power is traditionally expressed as horsepower by dividing ft-lb/min by 33,000. Hence,

$$HP_c = \frac{F_c v}{33,000} \tag{15.18}$$

where HP_c = cutting horsepower, hp. The gross power required to operate the machine tool is greater than the power delivered to the cutting process because of mechanical losses in the motor and drive train in the machine. These losses can be accounted for by the mechanical efficiency of the machine tool:

$$P_g = \frac{P_c}{E} \quad \text{or} \quad HP_g = \frac{HP_c}{E} \tag{15.19}$$

where P_g = gross power of the machine tool motor, W; HP_g = gross horsepower; and E = mechanical efficiency of the machine tool. Typical values of E for machine tools are around 90%.

It is often useful to convert power into power per unit volume rate of metal cut. This is called the ***unit power***, P_u (or ***unit horsepower***, HP_u), defined:

$$P_u = \frac{P_c}{R_{MR}} \quad \text{or} \quad HP_u = \frac{HP_c}{R_{MR}} \tag{15.20}$$

where R_{MR} = material removal rate, mm^3/s (in^3/min). The material removal rate can be calculated as the product of vt_ow. This is Eq. (15.1) using the conversions from Table 15.1. Unit power is also known as the ***specific energy*** U.

$$U = P_u = \frac{P_c}{R_{MR}} = \frac{F_c v}{v t_o w} = \frac{F_c}{t_o w} \tag{15.21}$$

The units for specific energy are typically N-m/mm^3(in-lb/in^3). However, the last expression in Eq. (15.21) suggests that the units might be reduced to N/mm^2 (lb/in^2). It is more meaningful to retain the units as N-m/mm^3 or J/mm^3 (in-lb/in^3).

Example 15.4 Power Relationships in Machining

Continuing with our previous examples, let us determine cutting power and specific energy in the machining operation if the cutting speed = 100 m/min. Summarizing the data and results from previous examples, $t_o = 0.50$ mm, $w = 3.0$ mm, $F_c = 1557$ N.

Solution: From Eq. (15.18), power in the operation is

$$P_c = (1557\,\text{N})(100\,\text{m/min}) = 155{,}700\,\text{N-m/min} = 155{,}700\,\text{J/min} = 2595\,\text{J/s} = 2595\,\text{W}$$

Specific energy is calculated from Eq. (15.21):

$$U = \frac{155{,}700}{100(10^3)(3.0)(0.5)} = \frac{155{,}700}{150{,}000} = 1.038\,\text{N-m/min}^3$$

Unit power and specific energy provide a useful measure of how much power (or energy) is required to remove a unit volume of metal during machining. Using this measure, different work materials can be compared in terms of their power and energy requirements. Table 15.2 presents a listing of unit horsepower and specific energy values for selected work materials.

The values in Table 15.2 are based on two assumptions: (1) the cutting tool is sharp, and (2) the chip thickness before the cut $t_o = 0.25$ mm (0.010 in). If these assumptions are not met, some adjustments must be made. For worn tools, the power required to perform the cut is greater, and this is reflected in higher specific energy and unit horsepower values. As an approximate guide, the values in the table should be multiplied by a factor between 1.00 and 1.25, depending on the degree of dullness of the tool. For sharp tools, the factor is 1.00. For tools in a finishing operation that are nearly worn out, the factor is around 1.10, and for tools in a roughing operation that are nearly worn out, the factor is 1.25.

Chip thickness before the cut t_o also affects the specific energy and unit horsepower values. As t_o is reduced, unit power requirements increase. This relationship is referred to as the ***size effect***. For example, grinding, in which the chips are extremely small by

TABLE 15.2 Values of unit horsepower and specific energy for selected work materials using sharp cutting tools and chip thickness before the cut $t_o = 0.25$ mm (0.010 in).

Material	Brinell Hardness	Specific Energy U or Unit Power P_u		Unit Horsepower
		N-m/mm^3	in-lb/in^3	HP_u hp/(in^3/min)
Carbon steel	150–200	1.6	240,000	0.6
	201–250	2.2	320,000	0.8
	251–300	2.8	400,000	1.0
Alloy steels	200–250	2.2	320,000	0.8
	251–300	2.8	400,000	1.0
	301–350	3.6	520,000	1.3
	351–400	4.4	640,000	1.6
Cast irons	125–175	1.1	160,000	0.4
	175–250	1.6	240,000	0.6
Stainless steel	150–250	2.8	400,000	1.0
Aluminum	50–100	0.7	100,000	0.25
Aluminum alloys	100–150	0.8	120,000	0.3
Brass	100–150	2.2	320,000	0.8
Bronze	100–150	2.2	320,000	0.8
Magnesium alloys	50–100	0.4	60,000	0.15

Data compiled from [6], [8], [11], and other sources.

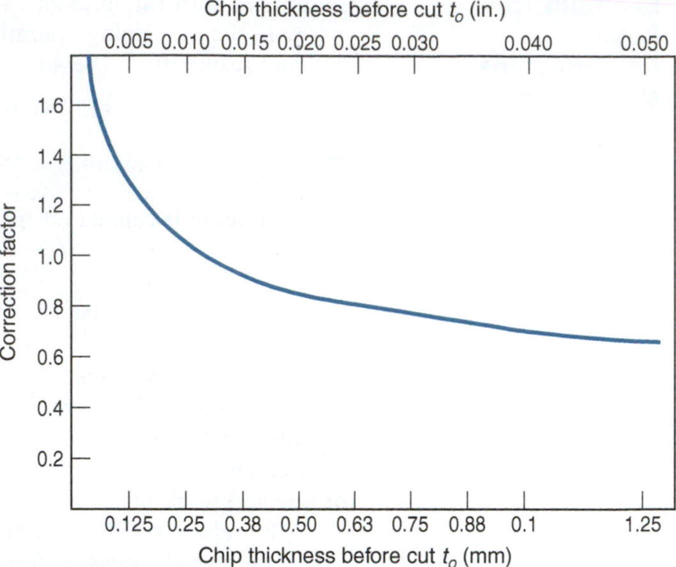

FIGURE 15.12 Correction factor for unit horsepower and specific energy when values of chip thickness before the cut t_o are different from 0.25 mm (0.010 in). (Credit: *Fundamentals of Modern Manufacturing,* 4th Edition by Mikell P. Groover, 2010. Reprinted with permission of John Wiley & Sons, Inc.)

comparison to most other machining operations, requires very high specific energy values. The U and HP_u values in Table 15.2 can still be used to estimate horsepower and energy for situations in which t_o is not equal to 0.25 mm (0.010 in) by applying a correction factor to account for any difference in chip thickness before the cut. Figure 15.12 provides values of this correction factor as a function of t_o. The unit horsepower and specific energy values in Table 15.2 should be multiplied by the appropriate correction factor when t_o differs from 0.25 mm (0.010 in).

In addition to tool sharpness and size effect, other factors also influence the values of specific energy and unit horsepower for a given operation. These other factors include rake angle, cutting speed, and cutting fluid. As rake angle or cutting speed are increased, or when cutting fluid is added, the U and HP_u values are reduced slightly. For our purposes in the end-of-chapter exercises, the effects of these additional factors can be ignored.

15.5 CUTTING TEMPERATURE

Of the total energy consumed in machining, nearly all of it (~98%) is converted into heat. This heat can cause temperatures to be very high at the tool–chip interface—over 600°C (1100°F) is not unusual. The remaining energy (~2%) is retained as elastic energy in the chip.

Cutting temperatures are important because high temperatures (1) reduce tool life, (2) produce hot chips that pose safety hazards to the machine operator, and (3) can cause inaccuracies in workpart dimensions due to thermal expansion of the work material. In this section, we discuss methods to calculate and measure temperatures in machining.

15.5.1 ANALYTICAL METHODS TO COMPUTE CUTTING TEMPERATURES

There are several analytical methods to calculate estimates of cutting temperature. References [3], [5], [9], and [15] present some of these approaches. We describe the method by Cook [5], which was derived using experimental data for a variety of work materials to establish parameter values for the resulting equation. The equation can be used to predict the increase in temperature at the tool–chip interface during machining:

$$\Delta T = \frac{0.4U}{\rho C}\left(\frac{vt_o}{K}\right)^{0.333} \quad (15.22)$$

where ΔT = mean temperature rise at the tool–chip interface, C° (F°); U = specific energy in the operation, N-m/mm^3 or J/mm^3 (in-lb/in^3); v = cutting speed, m/s (in/sec); t_o = chip thickness before the cut, m (in); ρC = volumetric specific heat of the work material, J/mm^3-C (in-lb/in^3-F); K = thermal diffusivity of the work material, m^2/s (in^2/sec).

Example 15.5 Cutting Temperature

For the specific energy obtained in Example 15.4, calculate the increase in temperature above ambient temperature of 20°C. Use the given data from the previous examples in this chapter: v = 100 m/min, t_o = 0.50 mm. In addition, the volumetric specific heat for the work material = 3.0 (10^{-3}) J/mm^3-C, and thermal diffusivity = 50 (10^{-6}) m^2/s (or 50 mm^2/s).

Solution: Cutting speed must be converted to mm/s: v = (100 m/min)(10^3 mm/m)/(60 s/min) = 1667 mm/s. Eq. (15.22) can now be used to compute the mean temperature rise:

$$\Delta T = \frac{0.4(1.038)}{3.0(10^3)}{}^\circ C\left(\frac{1667(0.5)}{50}\right)^{0.333} = (138.4)(2.552) = 353^\circ C$$

15.5.2 MEASUREMENT OF CUTTING TEMPERATURE

Experimental methods have been developed to measure temperatures in machining. The most frequently used measuring technique is the ***tool–chip thermocouple***, which consists of the tool and the chip as the two dissimilar metals forming the thermocouple junction. By properly connecting electrical leads to the tool and workpart (which is connected to

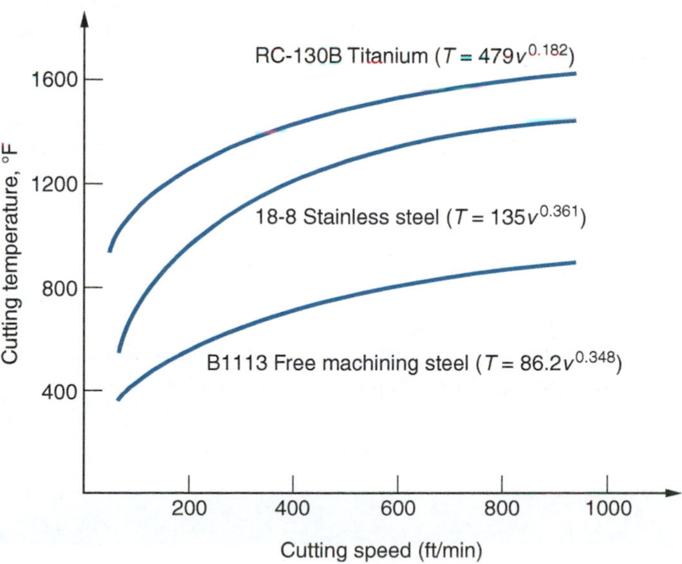

FIGURE 15.13 Experimentally measured cutting temperatures plotted against speed for three work materials, indicating general agreement with Eq. (15.23). Based on data in [9].[3] (Credit: *Fundamentals of Modern Manufacturing*, 4th Edition by Mikell P. Groover, 2010. Reprinted with permission of John Wiley & Sons, Inc.)

[3] The units reported in the Loewen and Shaw ASME paper [9] were °F for cutting temperature and ft/min for cutting speed. We have retained those units in the plots and equations of our figure.

the chip), the voltage generated at the tool–chip interface during cutting can be monitored using a recording potentiometer or other appropriate data-collection device. The voltage output of the tool–chip thermocouple (measured in mV) can be converted into the corresponding temperature value by means of calibration equations for the particular tool–work combination.

The tool–chip thermocouple has been utilized by researchers to investigate the relationship between temperature and cutting conditions such as speed and feed. Trigger [14] determined the speed–temperature relationship to be of the following general form:

$$T = Kv^m \tag{15.23}$$

where T = measured tool–chip interface temperature and v = cutting speed. The parameters K and m depend on cutting conditions (other than v) and work material. Figure 15.13 plots temperature versus cutting speed for several work materials, with equations of the form of Eq. (15.23) determined for each material. A similar relationship exists between cutting temperature and feed; however, the effect of feed on temperature is not as strong as cutting speed. These empirical results tend to support the general validity of the Cook equation: Eq. (15.22).

REFERENCES

[1] ***ASM Handbook***, Vol. **16**, ***Machining***. ASM International, Materials Park, Ohio, 1989.

[2] Black, J, and Kohser, R. ***DeGarmo's Materials and Processes in Manufacturing***, 10th ed. John Wiley & Sons, Inc. Hoboken, New Jersey, 2008.

[3] Boothroyd, G., and Knight, W. A. ***Fundamentals of Metal Machining and Machine Tools***, 3rd ed. CRC Taylor and Francis, Boca Raton, Florida, 2006.

[4] Chao, B. T., and Trigger, K. J. "Temperature Distribution at the Tool-Chip Interface in Metal Cutting," ***ASME Transactions***, Vol. 77, October 1955, pp. 1107–1121.

[5] Cook, N. "Tool Wear and Tool Life," ***ASME Transactions, J. Engrg. for Industry***, Vol. 95, November 1973, pp. 931–938.

[6] Drozda, T. J., and Wick, C. (eds.). ***Tool and Manufacturing Engineers Handbook***, 4th ed., Vol. I, ***Machining***. Society of Manufacturing Engineers, Dearborn, Michigan, 1983.

[7] Kalpakjian, S., and Schmid, R. ***Manufacturing Processes for Engineering Materials***, 4th ed. Prentice Hall/Pearson, Upper Saddle River, New Jersey, 2003.

[8] Lindberg, R. A. ***Processes and Materials of Manufacture***, 4th ed. Allyn and Bacon, Inc., Boston, Massachusetts, 1990.

[9] Loewen, E. G., and Shaw, M. C. "On the Analysis of Cutting Tool Temperatures," ***ASME Transactions***, Vol. **76**, No. 2, February 1954, pp. 217–225.

[10] Merchant, M. E. "Mechanics of the Metal Cutting Process: II. Plasticity Conditions in Orthogonal Cutting," ***Journal of Applied Physics***, Vol. **16**, June 1945 pp. 318–324.

[11] Schey, J. A. ***Introduction to Manufacturing Processes***, 3rd ed. McGraw-Hill Book Company, New York, 1999.

[12] Shaw, M. C. ***Metal Cutting Principles***, 2nd ed. Oxford University Press, Inc., Oxford, England, 2005.

[13] Trent, E. M., and Wright, P. K. ***Metal Cutting***, 4th ed. Butterworth Heinemann, Boston, Massachusetts, 2000.

[14] Trigger, K. J. "Progress Report No. 2 on Tool-Chip Interface Temperatures," ***ASME Transactions***, Vol. **71**, No. 2, February 1949, pp. 163–174.

[15] Trigger, K. J., and Chao, B. T. "An Analytical Evaluation of Metal Cutting Temperatures," ***ASME Transactions***, Vol. **73**, No. 1, January 1951, pp. 57–68.

REVIEW QUESTIONS

15.1. What are the three basic categories of material removal processes?

15.2. What distinguishes machining from other manufacturing processes?

15.3. Identify some of the reasons why machining is commercially and technologically important.

15.4. Name the three most common machining processes.

15.5. What are the two basic categories of cutting tools in machining? Give two examples of machining operations that use each of the tooling types.

15.6. What are the parameters of a machining operation that are included within the scope of cutting conditions?

15.7. Explain the difference between roughing and finishing operations in machining.

15.8. What is a machine tool?

15.9. What is an orthogonal cutting operation?

15.10. Why is the orthogonal cutting model useful in the analysis of metal machining?

15.11. Name and briefly describe the four types of chips that occur in metal cutting.

15.12. Identify the four forces that act upon the chip in the orthogonal metal-cutting model but cannot be measured directly in an operation.

15.13. Identify the two forces that can be measured in the orthogonal metal-cutting model.

15.14. What is the relationship between the coefficient of friction and the friction angle in the orthogonal cutting model?

15.15. Describe in words what the Merchant equation tells us.

15.16. How is the power required in a cutting operation related to the cutting force?

15.17. What is the specific energy in metal machining?

15.18. What does the term *size effect* mean in metal cutting?

15.19. What is a tool–chip thermocouple?

PROBLEMS

15.1. In an orthogonal cutting operation, the tool has a rake angle = 15°. The chip thickness before the cut = 0.30 mm and the cut yields a deformed chip thickness = 0.65 mm. Calculate (a) the shear plane angle and (b) the shear strain for the operation.

15.2. In an orthogonal cutting operation, the 0.250 in wide tool has a rake angle of 5°. The lathe is set so the chip thickness before the cut is 0.010 in. After the cut, the deformed chip thickness is measured to be 0.027 in. Calculate (a) the shear plane angle and (b) the shear strain for the operation.

15.3. In a turning operation, spindle speed is set to provide a cutting speed of 1.8 m/s. The feed and depth of cut are 0.30 mm and 2.6 mm, respectively. The tool rake angle is 8°. After the cut, the deformed chip thickness is measured to be 0.49 mm. Determine (a) shear plane angle, (b) shear strain, and (c) material removal rate. Use the orthogonal cutting model as an approximation of the turning process.

15.4. The cutting force and thrust force in an orthogonal cutting operation are 1470 N and 1589 N, respectively. The rake angle = 5°, the width of the cut = 5.0 mm, the chip thickness before the cut = 0.6, and the chip thickness ratio = 0.38. Determine (a) the shear strength of the work material and (b) the coefficient of friction in the operation.

15.5. The cutting force and thrust force have been measured in an orthogonal cutting operation to be 300 lb and 291 lb, respectively. The rake angle = 10°, width of cut = 0.200 in, chip thickness before the cut = 0.015, and chip thickness ratio = 0.4. Determine (a) the shear strength of the work material and (b) the coefficient of friction in the operation.

15.6. In an orthogonal cutting operation, the rake angle = −5°, chip thickness before the cut = 0.2 mm and width of cut = 4.0 mm. The chip ratio = 0.4. Determine (a) the chip thickness after the cut, (b) shear angle, (c) friction angle, (d) coefficient of friction, and (e) shear strain.

15.7. The shear strength of a certain work material = 50,000 lb/in^2. An orthogonal cutting operation is performed using a tool with a rake angle = 20° at the following cutting conditions: cutting speed = 100 ft/min, chip thickness before the cut = 0.015 in, and width of cut = 0.150 in. The resulting chip thickness ratio = 0.50. Determine (a) the shear plane angle, (b) shear force, (c) cutting force and thrust force, and (d) friction force.

15.8. A carbon steel bar with 7.64 in diameter has a tensile strength of 65,000 lb/in^2 and a shear strength of 45,000 lb/in^2. The diameter is reduced using a turning operation at a cutting speed of 400 ft/min. The feed is 0.011 in/rev and the depth of cut is 0.120 in. The rake angle on the tool in the direction of chip flow is 13°. The cutting conditions result in a chip ratio of 0.52. Using the orthogonal model as an approximation of turning, determine (a) the shear plane angle, (b) shear force, (c) cutting force and feed force, and (d) coefficient of friction between the tool and chip.

15.9. Low-carbon steel having a tensile strength of 300 MPa and a shear strength of 220 MPa is cut in a turning operation with a cutting speed of 3.0 m/s. The feed is 0.20 mm/rev and the depth of cut is 3.0 mm. The rake angle of the tool is 5° in the direction of chip flow. The resulting chip ratio is 0.45. Using the orthogonal model as an approximation of turning, determine (a) the shear plane angle, (b) shear force, (c) cutting force and feed force.

15.10. A turning operation is made with a rake angle of 10°, a feed of 0.010 in/rev, and a depth of cut

= 0.100 in. The shear strength of the work material is known to be 50,000 lb/in^2, and the chip thickness ratio is measured after the cut to be 0.40. Determine the cutting force and the feed force. Use the orthogonal cutting model as an approximation of the turning process.

15.11. In a turning operation on stainless steel with hardness = 200 HB, cutting speed = 200 m/min, feed = 0.25 mm/rev, and depth of cut = 7.5 mm. How much power will the lathe draw in performing this operation if its mechanical efficiency = 90%. Use Table 15.2 to obtain the appropriate specific energy value.

15.12. In Problem 15.11, compute the lathe power requirements if feed is changed to 0.50 mm/rev.

15.13. In a turning operation on aluminum, cutting speed = 900 ft/min, feed = 0.020 in/rev, and depth of cut = 0.250 in. What horsepower is required of the drive motor, if the lathe has a mechanical efficiency = 87%? Use Table 15.2 to obtain the appropriate unit horsepower value.

15.14. In a turning operation on plain carbon steel whose Brinell hardness = 275 HB, the cutting speed is set at 200 m/min and depth of cut = 6.0 mm. The lathe motor is rated at 25 kW, and its mechanical efficiency = 90%. Using the appropriate specific energy value from Table 15.2, determine the maximum feed that can be set for this operation. Use of a spreadsheet calculator is recommended for the iterative calculations required in this problem.

15.15. A turning operation is to be performed on a 20 hp lathe that has an 87% efficiency rating. The roughing cut is made on alloy steel whose hardness is in the range 325 to 335 HB. The cutting speed is 375 ft/min, feed is 0.030 in/rev, and depth of cut is 0.150 in. Based on these values, can the job be performed on the 20 hp lathe? Use Table 15.2 to obtain the appropriate unit horsepower value.

15.16. In a turning operation on low-carbon steel (175 BHN), cutting speed = 400 ft/min, feed = 0.010 in/rev, and depth of cut = 0.075 in. The lathe has a mechanical efficiency = 0.85. Based on the unit horsepower values in Table 15.2, determine (a) the horsepower consumed by the turning operation and (b) the horsepower that must be generated by the lathe.

15.17. Solve Problem 15.16 except that the feed = 0.0075 in/rev and the work material is stainless steel (Brinell Hardness = 240 HB).

15.18. A turning operation is carried out on aluminum (100 BHN). Cutting speed = 5.6 m/s, feed = 0.25 mm/rev, and depth of cut = 2.0 mm. The lathe has a mechanical efficiency = 0.85. Based on the specific energy values in Table 15.2, determine (a) the cutting power and (b) gross power in the turning operation, in Watts.

15.19. A turning operation is performed on an engine lathe using a tool with zero rake angle in the direction of chip flow. The work material is an alloy steel with hardness = 325 Brinell Hardness. The feed is 0.015 in/rev, depth of cut is 0.125 in, and cutting speed is 300 ft/min. After the cut, the chip thickness ratio is measured to be 0.45. (a) Using the appropriate value of specific energy from Table 15.2, compute the horsepower at the drive motor, if the lathe has an efficiency = 85%. (b) Based on horsepower, compute your best estimate of the cutting force for this turning operation. Use the orthogonal cutting model as an approximation of the turning process.

15.20. In a turning operation on an aluminum alloy workpiece, feed = 0.020 in/rev and depth of cut = 0.250 in. The motor horsepower of the lathe is 20 hp and it has a mechanical efficiency = 92%. The unit horsepower value = 0.25 hp/(in^3/min) for this aluminum grade. What is the maximum cutting speed that can be used on this job?

15.21. Orthogonal cutting is performed on a metal whose mass specific heat = 1.0 J/g-C, density = 2.9 g/cm^3, and thermal diffusivity = 0.8 cm^2/s. The cutting speed is 4.5 m/s, uncut chip thickness is 0.25 mm, and width of cut is 2.2 mm. The cutting force is measured at 1170 N. Using Cook's equation, determine the cutting temperature if the ambient temperature = 22°C.

15.22. Consider a turning operation performed on steel whose hardness = 225 HB at a speed = 3.0 m/s, feed = 0.25 mm, and depth = 4.0 mm. Using values of thermal properties found in the tables and definitions of Sections 3.6 and 3.7, and the appropriate specific energy value from Table 15.2, compute an estimate of cutting temperature using Cook's equation. Assume ambient temperature = 20°C.

15.23. An orthogonal cutting operation is performed on a certain metal whose volumetric specific heat = 110 in-lb/in^3-F, and thermal diffusivity = 0.140 in^2/sec. The cutting speed = 350 ft/min, chip thickness before the cut = 0.008 in, and width of cut = 0.100 in. The cutting force is measured at 200 lb. Using Cook's equation, determine the cutting temperature if the ambient temperature = 70°F.

15.24. During a turning operation, a tool–chip thermocouple was used to measure cutting temperature. The following temperature data were collected during the cuts at three different cutting speeds (feed and depth were held constant): (1) v = 100 m/min, T = 505°C, (2) v = 130 m/min, T = 552°C, (3) v = 160 m/min, T = 592°C. Determine an equation for temperature as a function of cutting speed that is in the form of the Trigger equation, Eq. (15.23).

16 MACHINING OPERATIONS AND MACHINE TOOLS

Chapter Contents

16.1 Machining and Part Geometry

16.2 Turning and Related Operations
- 16.2.1 Cutting Conditions in Turning
- 16.2.2 Operations Related to Turning
- 16.2.3 The Engine Lathe
- 16.2.4 Other Lathes and Turning Machines
- 16.2.5 Boring Machines

16.3 Drilling and Related Operations
- 16.3.1 Cutting Conditions in Drilling
- 16.3.2 Operations Related to Drilling
- 16.3.3 Drill Presses

16.4 Milling
- 16.4.1 Types of Milling Operations
- 16.4.2 Cutting Conditions in Milling
- 16.4.3 Milling Machines

16.5 Machining Centers and Turning Centers

16.6 Other Machining Operations
- 16.6.1 Shaping and Planing
- 16.6.2 Broaching
- 16.6.3 Sawing

16.7 High-Speed Machining

16.8 Tolerances and Surface Finish
- 16.8.1 Tolerances in Machining
- 16.8.2 Surface Finish in Machining

16.9 Product Design Considerations in Machining

Machining is the most versatile and accurate of all manufacturing processes in its capability to produce a diversity of part geometries and geometric features. Casting can also produce a variety of shapes, but it lacks the precision and accuracy of machining. In this chapter, we describe the important machining operations and the machine tools used to perform them.

16.1 MACHINING AND PART GEOMETRY

Machined parts can be classified as rotational or nonrotational, Figure 16.1. A ***rotational*** workpart has a cylindrical or disklike shape. The characteristic operation that produces this geometry is one in which a cutting tool removes material from a rotating workpart. Examples include turning and boring. Drilling is closely related except that an internal cylindrical shape is created and the tool rotates (rather than the work) in most drilling operations. A ***nonrotational*** (also called ***prismatic***) workpart is blocklike or platelike, as in Figure 16.1(b). This geometry is achieved by linear motions of the workpart, combined with either rotating or linear tool motions. Operations in this category include milling, shaping, planing, and sawing.

Each machining operation produces a characteristic geometry due to two factors: (1) the relative motions between the tool and the workpart and (2) the shape of the cutting tool. We classify these operations that create part shape as generating and forming. In ***generating***, the geometry of the workpart is determined by the feed trajectory of the cutting tool. The path followed by the tool during its feed motion is imparted to the work surface in order to create shape. Examples of generating include

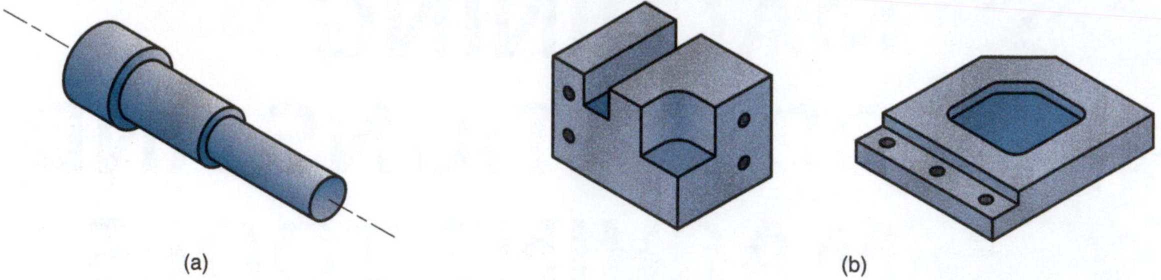

FIGURE 16.1 Machined parts are classified as (a) rotational, or (b) nonrotational, shown here by block and flat parts. (Credit: *Fundamentals of Modern Manufacturing,* 4th Edition by Mikell P. Groover, 2010. Reprinted with permission of John Wiley & Sons, Inc.)

straight turning, taper turning, contour turning, peripheral milling, and profile milling, all illustrated in Figure 16.2. In each of these operations, material removal is accomplished by the speed motion in the operation, but part shape is determined by the feed motion. The feed trajectory may involve variations in depth or width of cut during the operation. For example, in the contour turning and profile milling operations shown in

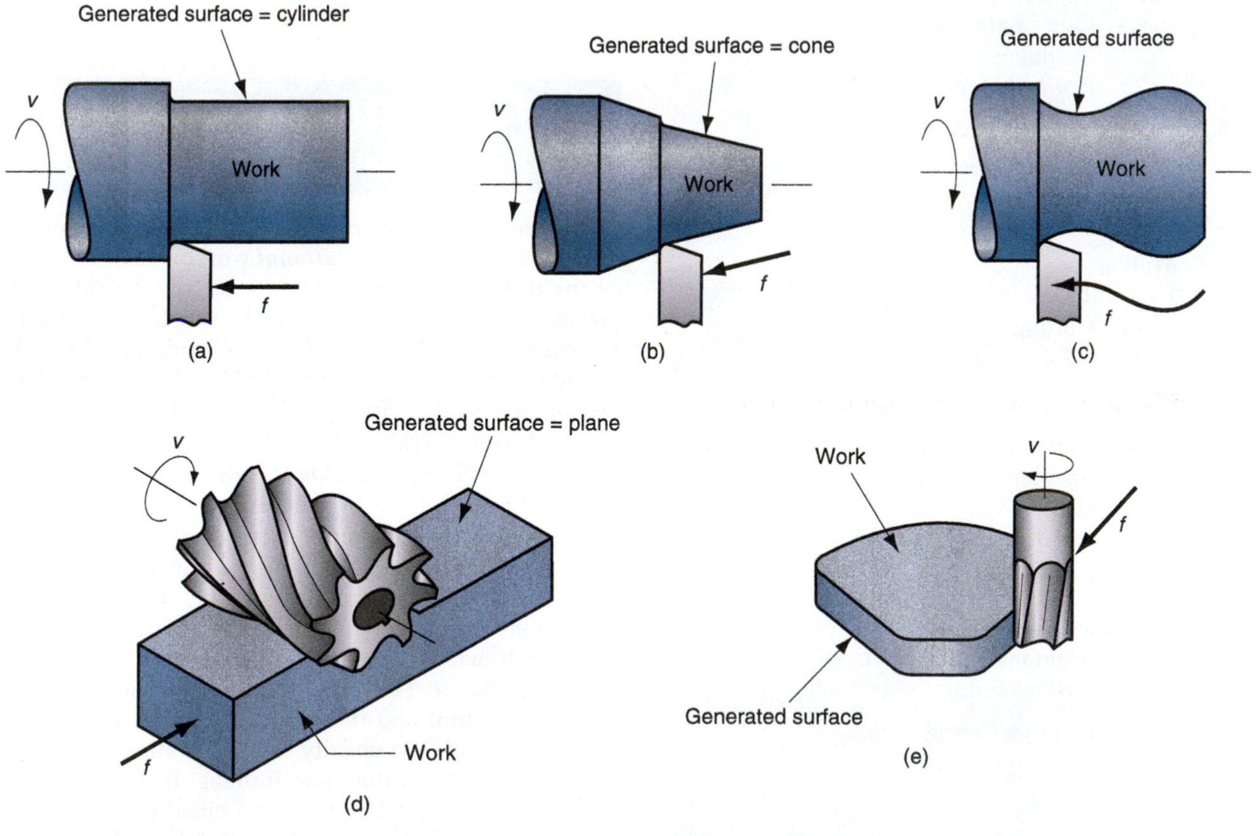

FIGURE 16.2 Generating shape in machining: (a) straight turning, (b) taper turning, (c) contour turning, (d) plain milling, and (e) profile milling. (Credit: *Fundamentals of Modern Manufacturing,* 4th Edition by Mikell P. Groover, 2010. Reprinted with permission of John Wiley & Sons, Inc.)

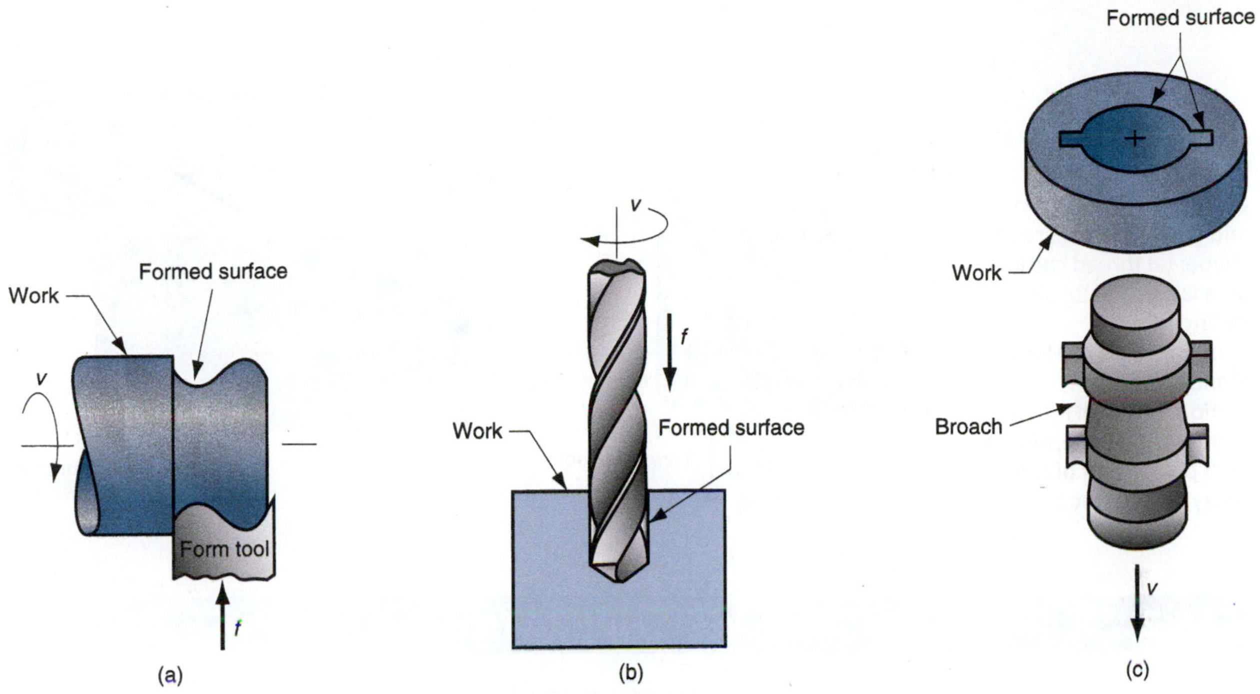

FIGURE 16.3 Forming to create shape in machining: (a) form turning, (b) drilling, and (c) broaching. (Credit: *Fundamentals of Modern Manufacturing*, 4th Edition by Mikell P. Groover, 2010. Reprinted with permission of John Wiley & Sons, Inc.)

our figure, the feed motion results in changes in depth and width, respectively, as cutting proceeds.

In ***forming***, the shape of the part is created by the geometry of the cutting tool. In effect, the cutting edge of the tool has the reverse of the shape to be produced on the part surface. Form turning, drilling, and broaching are examples of this case. In these operations, illustrated in Figure 16.3, the shape of the cutting tool is imparted to the work in order to create part geometry. The cutting conditions in forming usually include the primary speed motion combined with a feeding motion that is directed into the work. Depth of cut in this category of machining usually refers to the final penetration into the work after the feed motion has been completed.

Forming and generating are sometimes combined in one operation, as illustrated in Figure 16.4 for thread cutting on a lathe and slotting on a milling machine. In thread cutting, the pointed shape of the cutting tool determines the form of the threads, but the large feed rate generates the threads. In slotting (also called slot milling), the width of the cutter determines the width of the slot, but the feed motion creates the slot.

Machining is classified as a secondary process. In general, secondary processes follow basic processes, whose purpose is to establish the initial shape of a workpiece. Examples of basic processes include casting, forging, and bar rolling (to produce rod and bar stock). The shapes produced by these processes usually require refinement by secondary processes. Machining operations serve to transform the starting shapes into the final geometries specified by the part designer. For example, bar stock is the initial shape, but the final geometry after a series of machining operations is a shaft.

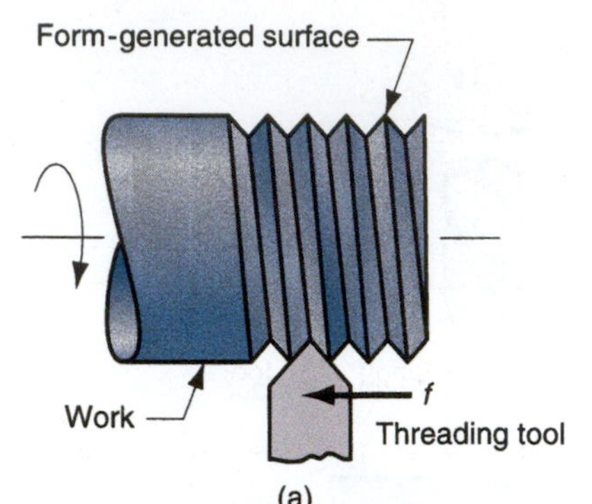

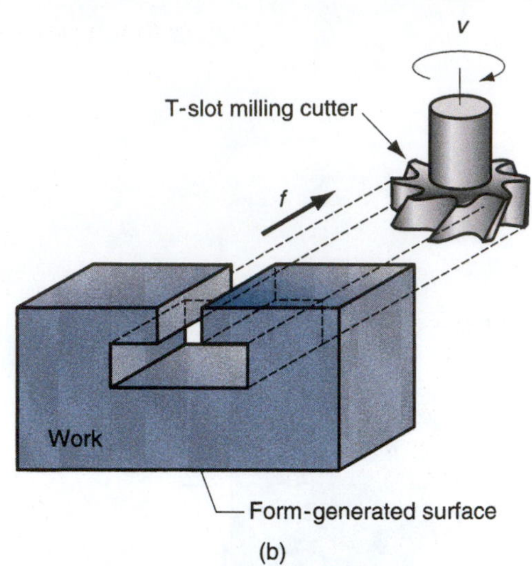

FIGURE 16.4 Combination of forming and generating to create shape: (a) thread cutting on a lathe, and (b) slot milling. (Credit: *Fundamentals of Modern Manufacturing*, 4th Edition by Mikell P. Groover, 2010. Reprinted with permission of John Wiley & Sons, Inc.)

16.2 TURNING AND RELATED OPERATIONS

Turning is a machining process in which a single-point tool removes material from the surface of a rotating workpiece. The tool is fed linearly in a direction parallel to the axis of rotation to generate a cylindrical geometry, as illustrated in Figures 16.2(a) and 16.5. Single-point tools used in turning and other machining operations are discussed in Section 17.3.1. Turning is traditionally carried out on a machine tool called a ***lathe***, which provides power to turn the part at a given rotational speed and to feed the tool at a specified rate and depth of cut.

16.2.1 CUTTING CONDITIONS IN TURNING

The rotational speed in turning is related to the desired cutting speed at the surface of the cylindrical workpiece by the equation:

$$N = \frac{v}{\pi D_o} \tag{16.1}$$

where N = rotational speed, rev/min; v = cutting speed, m/min (ft/min); and D_o = original diameter of the part, m (ft).

The turning operation reduces the diameter of the work from its original diameter D_o to a final diameter D_f, as determined by the depth of cut d:

$$D_f = D_o - 2d \tag{16.2}$$

The feed in turning is generally expressed in mm/rev (in/rev). This feed can be converted to a linear travel rate in mm/min (in/min) by the formula:

$$f_r = Nf \tag{16.3}$$

where f_r = feed rate, mm/min (in/min); and f = feed, mm/rev (in/rev).

The time to machine from one end of a cylindrical workpart to the other is given by:

$$T_m = \frac{L}{f_r} \tag{16.4}$$

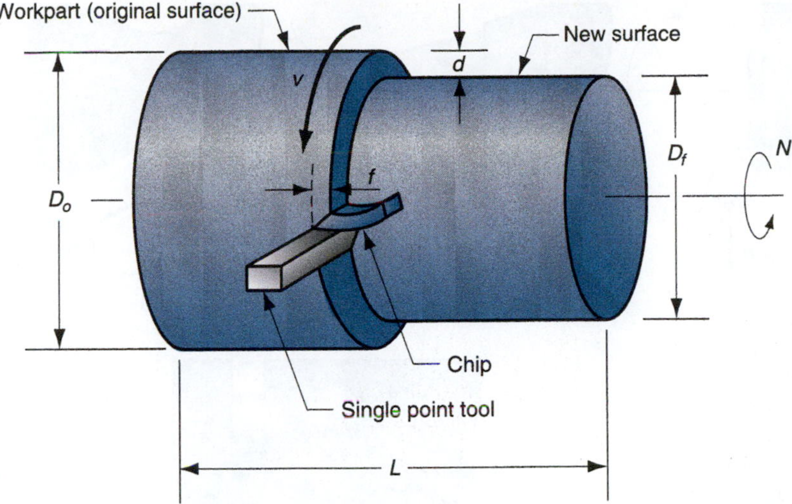

FIGURE 16.5 Turning operation. (Credit: *Fundamentals of Modern Manufacturing*, 4th Edition by Mikell P. Groover, 2010. Reprinted with permission of John Wiley & Sons, Inc.)

where T_m = machining time, min; and L = length of the cylindrical workpart, mm (in). A more direct computation of the machining time is provided by the following equation:

$$T_m = \frac{\pi D_o L}{fv} \tag{16.5}$$

where D_o = work diameter, mm (in); L = workpart length, mm (in); f = feed, mm/rev (in/rev); and v = cutting speed, mm/min (in/min). As a practical matter, a small distance is usually added to the workpart length at the beginning and end of the piece to allow for approach and overtravel of the tool. Thus, the duration of the feed motion past the work will be longer than T_m.

The volumetric rate of material removal can be most conveniently determined by the following equation:

$$R_{MR} = vfd \tag{16.6}$$

where R_{MR} = material removal rate, mm³/min (in³/min). In using this equation, the units for f are expressed simply as mm (in), in effect neglecting the rotational character of turning. Also, care must be exercised to ensure that the units for speed are consistent with those for f and d.

16.2.2 OPERATIONS RELATED TO TURNING

A variety of other machining operations can be performed on a lathe in addition to turning; these include the following, illustrated in Figure 16.6:

(a) ***Facing***. The tool is fed radially into the rotating work on one end to create a flat surface on the end.
(b) ***Taper turning***. Instead of feeding the tool parallel to the axis of rotation of the work, the tool is fed at an angle, thus creating a tapered cylinder or conical shape.
(c) ***Contour turning***. Instead of feeding the tool along a straight line parallel to the axis of rotation as in turning, the tool follows a contour that is other than straight, thus creating a contoured form in the turned part.

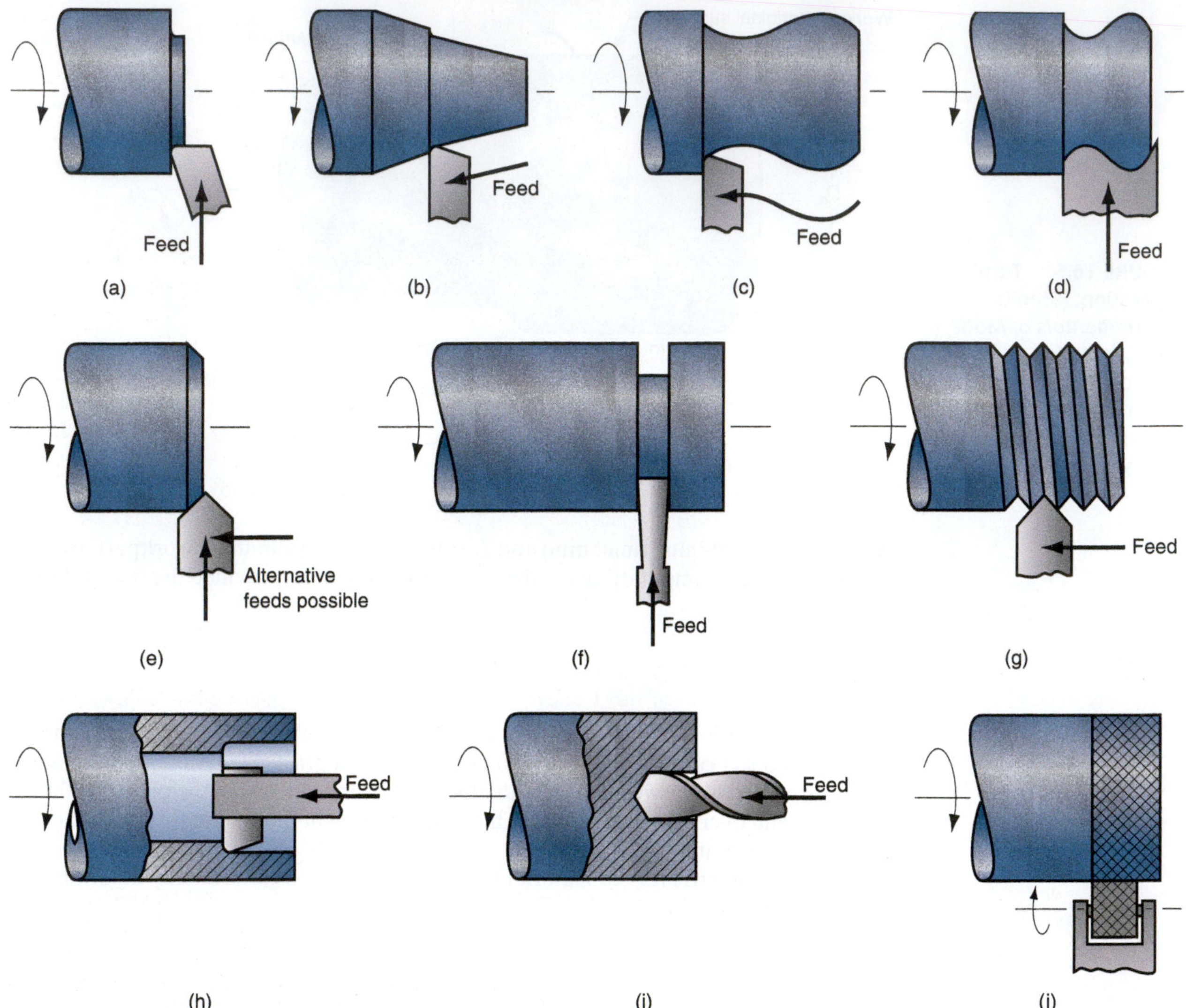

FIGURE 16.6 Machining operations other than turning that are performed on a lathe: (a) facing, (b) taper turning, (c) contour turning, (d) form turning, (e) chamfering, (f) cutoff, (g) threading, (h) boring, (i) drilling, and (j) knurling. (Credit: *Fundamentals of Modern Manufacturing*, 4th Edition by Mikell P. Groover, 2010. Reprinted with permission of John Wiley & Sons, Inc.)

(d) ***Form turning***. In this operation, sometimes called ***forming***, the tool has a shape that is imparted to the work by plunging the tool radially into the work.

(e) ***Chamfering***. The cutting edge of the tool is used to cut an angle on the corner of the cylinder, creating what is called a ''chamfer.''

(f) ***Cutoff***. The tool is fed radially into the rotating work at some location along its length to cut off the end of the part. This operation is sometimes referred to as ***parting***.

(g) ***Threading***. A pointed tool is fed linearly across the outside surface of the rotating workpart in a direction parallel to the axis of rotation at a large effective feed rate, thus creating threads in the cylinder.

(h) ***Boring***. A single-point tool is fed linearly, parallel to the axis of rotation, on the inside diameter of an existing hole in the part.

(i) ***Drilling***. Drilling can be performed on a lathe by feeding the drill into the rotating work along its axis. ***Reaming*** can be performed in a similar way.

(j) ***Knurling***. This is not a machining operation because it does not involve cutting of material. Instead, it is a metal forming operation used to produce a regular cross-hatched pattern in the work surface.

Most lathe operations use single-point tools (Section 17.3.1). Turning, facing, taper turning, contour turning, chamfering, and boring are all performed with single-point tools. A threading operation is accomplished using a single-point tool designed with a geometry that shapes the thread. Certain operations require tools other than single-point. Form turning is performed with a specially designed tool called a form tool. The profile shape ground into the tool establishes the shape of the workpart. A cutoff tool is basically a form tool. Drilling is accomplished by a drill bit (Section 17.3.2). Knurling is performed by a knurling tool, consisting of two hardened forming rolls, each mounted between centers. The forming rolls have the desired knurling pattern on their surfaces. To perform knurling, the tool is pressed against the rotating workpart with sufficient pressure to impress the pattern onto the work surface.

16.2.3 THE ENGINE LATHE

The basic lathe used for turning and related operations is an ***engine lathe***. It is a versatile machine tool, manually operated, and widely used in low and medium production.

Engine Lathe Technology Figure 16.7 is a sketch of an engine lathe showing its principal components. The ***headstock*** contains the drive unit to rotate the spindle, which rotates the work. Opposite the headstock is the ***tailstock***, in which a center is mounted to support the other end of the workpiece.

The cutting tool is held in a ***tool post*** fastened to the ***cross-slide***, which is assembled to the ***carriage***. The carriage is designed to slide along the ***ways*** of the lathe in order to feed the tool parallel to the axis of rotation. The ways are like tracks along which the carriage rides, and they are made with great precision to achieve a high degree of

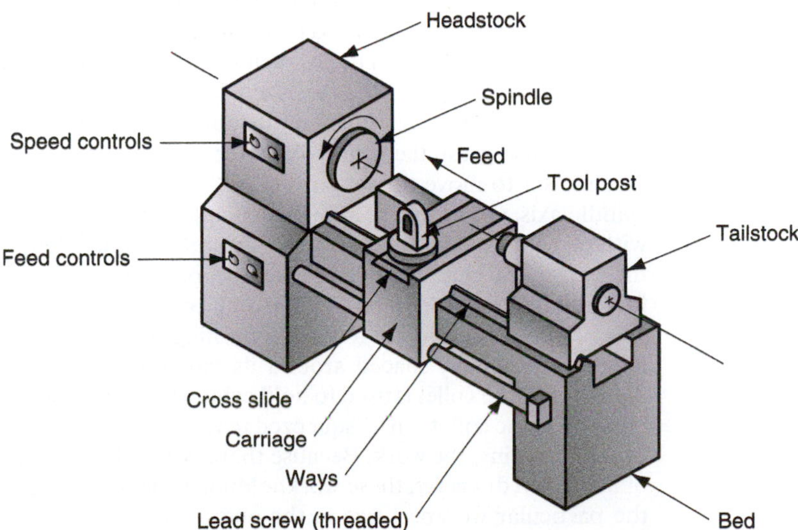

FIGURE 16.7 Diagram of an engine lathe, indicating its principal components. (Credit: *Fundamentals of Modern Manufacturing*, 4th Edition by Mikell P. Groover, 2010. Reprinted with permission of John Wiley & Sons, Inc.)

parallelism relative to the spindle axis. The ways are built into the ***bed*** of the lathe, providing a rigid frame for the machine tool.

The carriage is driven by a leadscrew that rotates at the proper speed to obtain the desired feed rate. The cross-slide is designed to feed in a direction perpendicular to the carriage movement. Thus, by moving the carriage, the tool can be fed parallel to the work axis to perform straight turning; or by moving the cross-slide, the tool can be fed radially into the work to perform facing, form turning, or cutoff operations.

The conventional engine lathe and most other machines described in this section are ***horizontal turning machines***; that is, the spindle axis is horizontal. This is appropriate for the majority of turning jobs, in which the length is greater than the diameter. For jobs in which the diameter is large relative to length and the work is heavy, it is more convenient to orient the work so that it rotates about a vertical axis; these are ***vertical turning machines***.

The size of a lathe is designated by swing and maximum distance between centers. The ***swing*** is the maximum workpart diameter that can be rotated in the spindle, determined as twice the distance between the centerline of the spindle and the ways of the machine. The actual maximum size of a cylindrical workpiece that can be accommodated on the lathe is smaller than the swing because the carriage and cross-slide assembly are in the way. The ***maximum distance between centers*** indicates the maximum length of a workpiece that can be mounted between headstock and tailstock centers. For example, a 350 mm × 1.2 m (14 in × 48 in) lathe designates that the swing is 350 mm (14 in) and the maximum distance between centers is 1.2 m (48 in).

Methods of Holding the Work in a Lathe Four common methods are used to hold workparts in turning. These workholding methods consist of various mechanisms to grasp the work, center and support it in position along the spindle axis, and rotate it. The methods, illustrated in Figure 16.8, are (a) mounting the work between centers, (b) chuck, (c) collet, and (d) face plate.

Holding the work ***between centers*** refers to the use of two centers, one in the headstock and the other in the tailstock, as in Figure 16.8(a). This method is appropriate for parts with large length-to-diameter ratios. At the headstock center, a device called a ***dog*** is attached to the outside of the work and is used to drive the rotation from the spindle. The tailstock center has a cone-shaped point, which is inserted into a tapered hole in the end of the work. The tailstock center is either a "live" center or a "dead" center. A ***live center*** rotates in a bearing in the tailstock, so that there is no relative rotation between the work and the live center, hence, no friction between the center and the workpiece. In contrast, a ***dead center*** is fixed to the tailstock, so that it does not rotate; instead, the workpiece rotates about it. Because of friction and the heat buildup that results, this setup is normally used at lower rotational speeds. The live center can be used at higher speeds.

The ***chuck***, Figure 16.8(b), is available in several designs, with three or four jaws to grasp the cylindrical workpart on its outside diameter. The jaws are often designed so they can also grasp the inside diameter of a tubular part. A ***self-centering*** chuck has a mechanism to move the jaws in or out simultaneously, thus centering the work at the spindle axis. Other chucks allow independent operation of each jaw. Chucks can be used with or without a tailstock center. For parts with low length-to-diameter ratios, holding the part in the chuck in a cantilever fashion is usually sufficient to withstand the cutting forces. For long workbars, the tailstock center is needed for support.

A ***collet*** consists of a tubular bushing with longitudinal slits running over half its length and equally spaced around its circumference, as in Figure 16.8(c). The inside diameter of the collet is used to hold cylindrical work such as barstock. Owing to the slits, one end of the collet can be squeezed to reduce its diameter and provide a secure grasping pressure against the work. Because there is a limit to the reduction obtainable in a collet of any given diameter, these workholding devices must be made in various sizes to match the particular workpart size in the operation.

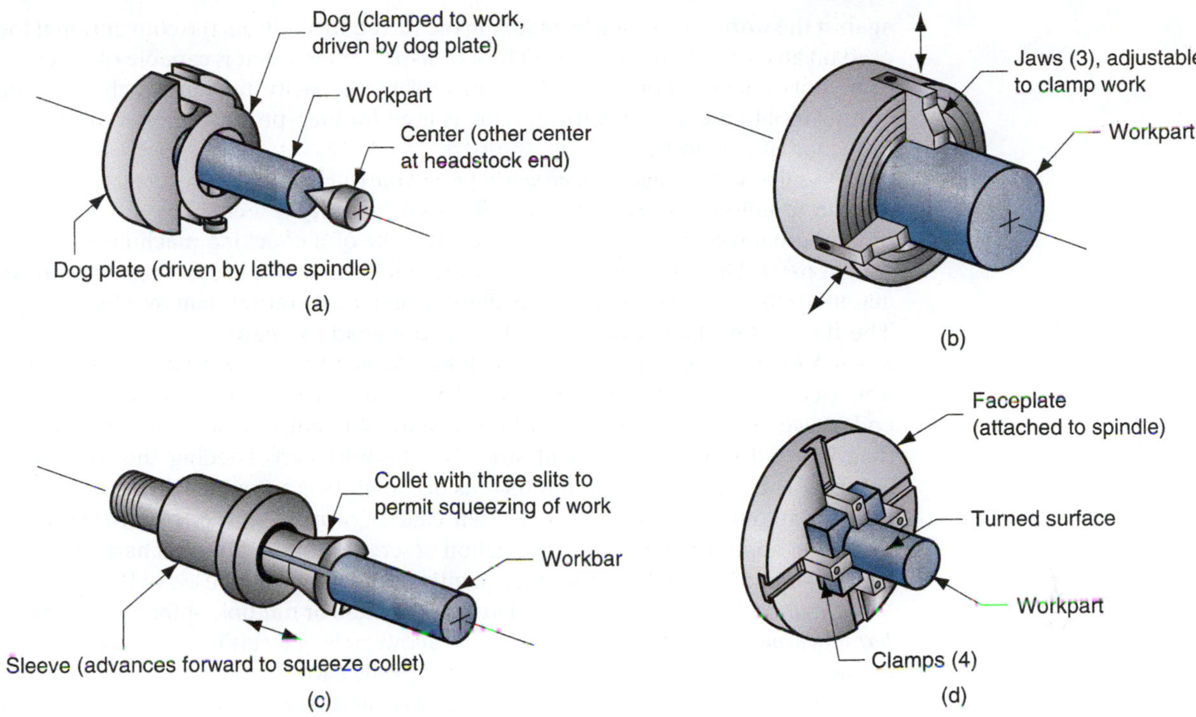

FIGURE 16.8 Four workholding methods used in lathes: (a) mounting the work between centers using a dog, (b) three-jaw chuck, (c) collet, and (d) faceplate for noncylindrical workparts. (Credit: *Fundamentals of Modern Manufacturing*, 4th Edition by Mikell P. Groover, 2010. Reprinted with permission of John Wiley & Sons, Inc.)

A ***face plate***, Figure 16.8(d), is a workholding device that fastens to the lathe spindle and is used to grasp parts with irregular shapes. Because of their irregular shape, these parts cannot be held by other workholding methods. The faceplate is therefore equipped with the custom-designed clamps for the particular geometry of the part.

16.2.4 OTHER LATHES AND TURNING MACHINES

In addition to the engine lathe, other turning machines have been developed to satisfy particular functions or to automate the turning process. Among these machines are (1) toolroom lathe, (2) speed lathe, (3) turret lathe, (4) chucking machine, (5) automatic screw machine, and (6) numerically controlled lathe.

The toolroom lathe and speed lathe are closely related to the engine lathe. The ***toolroom lathe*** is smaller and has a wider available range of speeds and feeds. It is also built for higher accuracy, consistent with its purpose of fabricating components for tools, fixtures, and other high-precision devices.

The ***speed lathe*** is simpler in construction than the engine lathe. It has no carriage and cross-slide assembly, and therefore no leadscrew to drive the carriage. The cutting tool is held by the operator using a rest attached to the lathe for support. The speeds are higher on a speed lathe, but the number of speed settings is limited. Applications of this machine type include wood turning, metal spinning (Section 14.6.3), and polishing operations (Section 18.2.4).

A ***turret lathe*** is a manually operated lathe in which the tailstock is replaced by a turret that holds up to six cutting tools. These tools can be rapidly brought into action against the work one by one by indexing the turret. In addition, the conventional tool post used on an engine lathe is replaced by a four-sided turret that is capable of indexing up to four tools into position. Hence, because of the capacity to quickly change from one cutting tool to the next, the turret lathe is used for high-production work that requires a sequence of cuts to be made on the part.

As the name suggests, a ***chucking machine*** (nicknamed ***chucker***) uses a chuck in its spindle to hold the workpart. The tailstock is absent on a chucker, so parts cannot be mounted between centers. This restricts the use of a chucking machine to short, lightweight parts. The setup and operation are similar to a turret lathe except that the feeding actions of the cutting tools are controlled automatically rather than by a human operator. The function of the operator is to load and unload the parts.

A ***bar machine*** is similar to a chucking machine except that a collet is used (instead of a chuck), which permits long bar stock to be fed through the headstock into position. At the end of each machining cycle, a cutoff operation separates the new part. The bar stock is then indexed forward to present stock for the next part. Feeding the stock as well as indexing and feeding the cutting tools is accomplished automatically. Owing to its high level of automatic operation, it is often called an ***automatic bar machine***. One of its important applications is in the production of screws and similar small hardware items; the name ***automatic screw machine*** is frequently used for machines used in these applications.

Bar machines can be classified as single spindle or multiple spindle. A ***single spindle bar machine*** has one spindle that normally allows only one cutting tool to be used at a time on the single workpart being machined. Thus, while each tool is cutting the work, the other tools are idle. (Turret lathes and chucking machines are also limited by this sequential, rather than simultaneous, tool operation). To increase cutting tool utilization and production rate, ***multiple spindle bar machines*** are available. These machines have more than one spindle, so multiple parts are machined simultaneously by multiple tools. For example, a six-spindle automatic bar machine works on six parts at a time, as in Figure 16.9. At the end of each machining cycle, the spindles (including collets and workbars) are indexed (rotated) to the next position. In our illustration, each part is cut sequentially by five sets of cutting tools, which takes six cycles (position 1 is for advancing the bar stock to a "stop"). With this arrangement, a part is completed at the end of each cycle. As a result, a six-spindle automatic screw machine has a very high production rate.

The sequencing and actuation of the motions on screw machines and chucking machines have traditionally been controlled by cams and other mechanical devices. The modern form of control is ***computer numerical control*** (CNC), in which the machine tool operations are controlled by a "program of instructions" consisting of alphanumeric code (Section 29.1). CNC provides a more sophisticated and versatile means of control than mechanical devices. CNC has led to the development of machine tools capable of more complex machining cycles and part geometries, and a higher level of automated operation than conventional screw machines and chucking machines. The CNC lathe is an example of these machines in turning. It is especially useful for contour turning operations and close tolerance work. Today, automatic chuckers and bar machines are implemented by CNC.

16.2.5 BORING MACHINES

Boring is similar to turning. It uses a single-point tool against a rotating workpart. The difference is that boring is performed on the inside diameter of an existing hole rather than the outside diameter of an existing cylinder. In effect, boring is an internal turning operation. Machine tools used to perform boring operations are called ***boring machines*** (also ***boring mills***). One might expect that boring machines would have features in

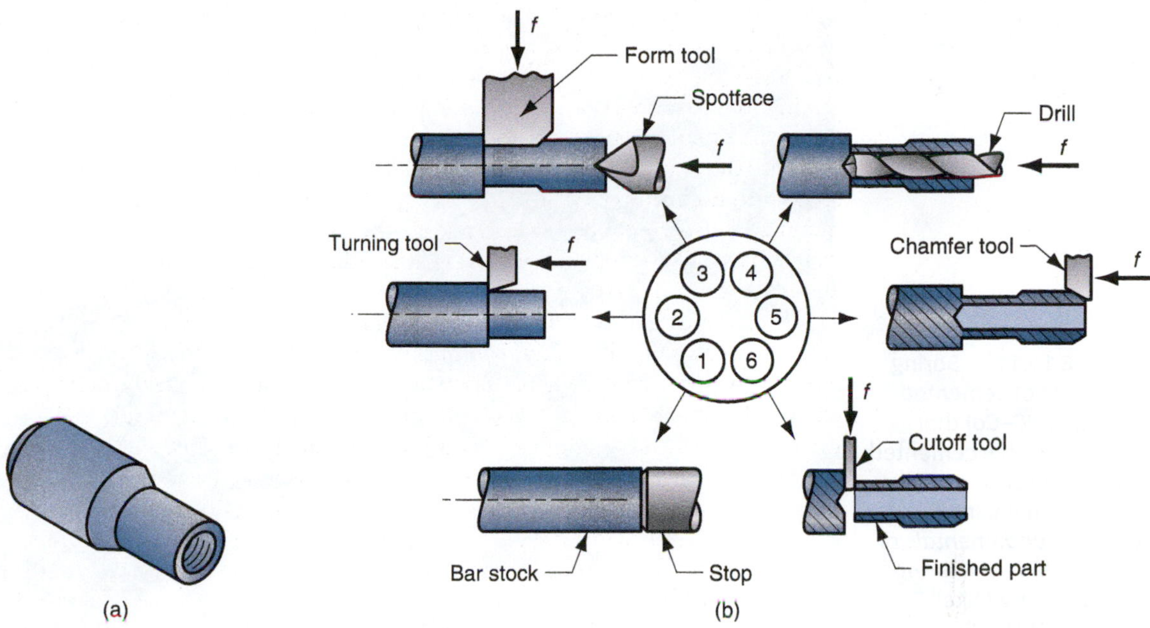

FIGURE 16.9 (a) Type of part produced on a six-spindle automatic bar machine; and (b) sequence of operations to produce the part: (1) feed stock to stop, (2) turn main diameter, (3) form second diameter and spotface, (4) drill, (5) chamfer, and (6) cutoff. (Credit: *Fundamentals of Modern Manufacturing,* 4th Edition by Mikell P. Groover, 2010. Reprinted with permission of John Wiley & Sons, Inc.)

common with turning machines; indeed, as previously indicated, lathes are sometimes used to accomplish boring.

Boring mills can be horizontal or vertical. The designation refers to the orientation of the axis of rotation of the machine spindle or workpart. In a ***horizontal boring*** operation, the setup can be arranged in either of two ways. The first setup is one in which the work is fixtured to a rotating spindle, and the tool is attached to a cantilevered boring bar that feeds into the work, as illustrated in Figure 16.10(a). The boring bar in this setup must be very stiff to avoid deflection and vibration during cutting. To achieve high stiffness, boring bars are often made of cemented carbide, whose modulus of elasticity approaches 620×10^3 MPa (90×10^6 lb/in^2). Figure 16.11 shows a carbide boring bar.

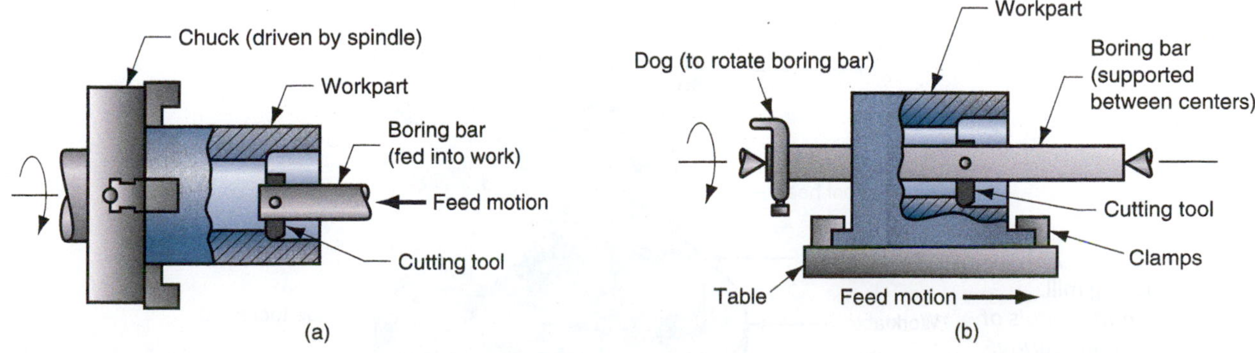

FIGURE 16.10 Two forms of horizontal boring: (a) boring bar is fed into a rotating workpart, and (b) work is fed past a rotating boring bar. (Credit: *Fundamentals of Modern Manufacturing,* 4th Edition by Mikell P. Groover, 2010. Reprinted with permission of John Wiley & Sons, Inc.)

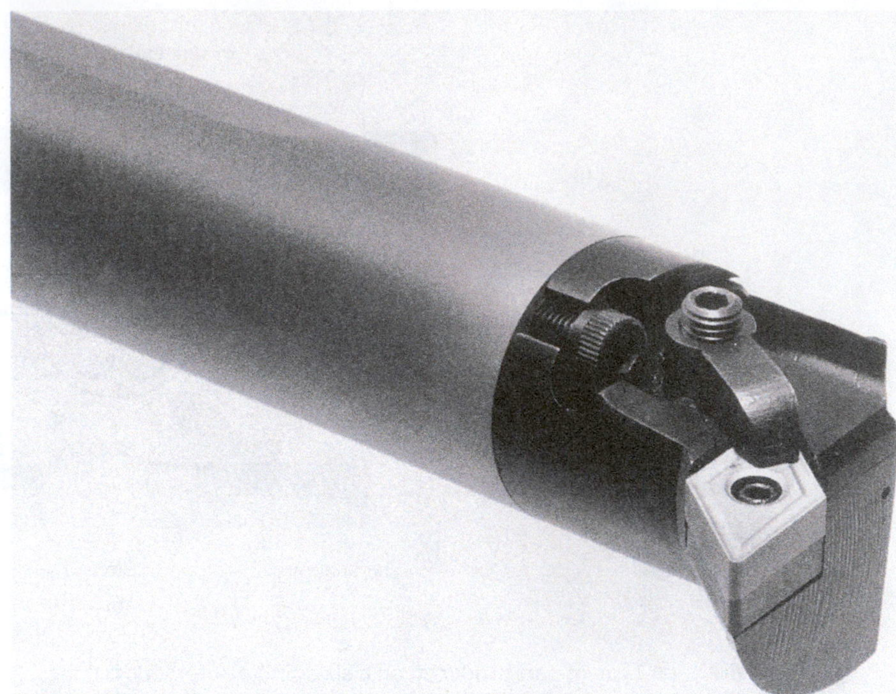

FIGURE 16.11 Boring bar made of cemented carbide (WC–Co) that uses indexable cemented carbide inserts. Courtesy of Kennametal Inc. (Credit: *Fundamentals of Modern Manufacturing*, 4th Edition by Mikell P. Groover, 2010. Reprinted with permission of John Wiley & Sons, Inc.)

The second possible setup is one in which the tool is mounted to a boring bar, and the boring bar is supported and rotated between centers. The work is fastened to a feeding mechanism that feeds it past the tool. This setup, Figure 16.10(b), can be used to perform a boring operation on a conventional engine lathe.

A ***vertical boring machine*** (VBM) is used for large, heavy workparts with large diameters; usually the workpart diameter is greater than its length. As in Figure 16.12, the part is clamped to a worktable that rotates relative to the machine base. Worktables up to 40 ft in diameter are available. The typical boring machine can position and feed several

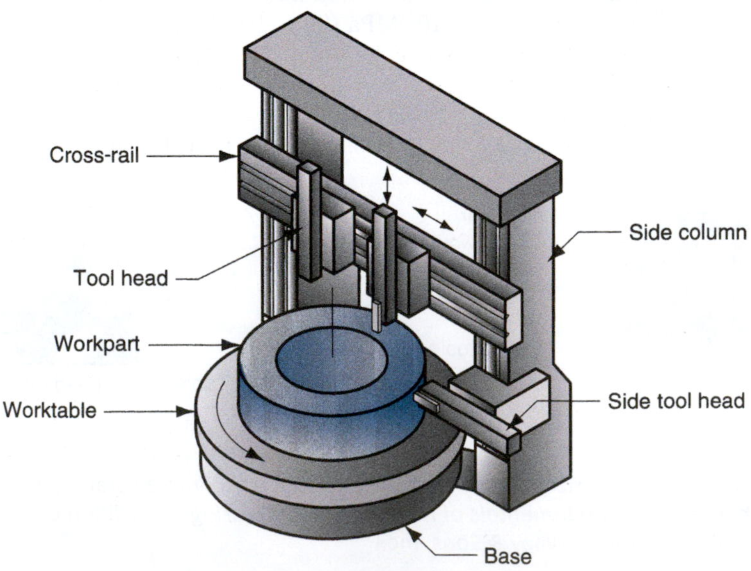

FIGURE 16.12 A vertical boring mill. (Credit: *Fundamentals of Modern Manufacturing*, 4th Edition by Mikell P. Groover, 2010. Reprinted with permission of John Wiley & Sons, Inc.)

cutting tools simultaneously. The tools are mounted on tool heads that can be fed horizontally and vertically relative to the worktable. One or two heads are mounted on a horizontal cross-rail assembled to the machine tool housing above the worktable. The cutting tools mounted above the work can be used for facing and boring. In addition to the tools on the cross-rail, one or two additional tool heads can be mounted on the side columns of the housing to enable turning on the outside diameter of the work.

The tool heads used on a vertical boring machine often include turrets to accommodate several cutting tools. This results in a loss of distinction between this machine and a ***vertical turret lathe*** (VTL). Some machine tool builders make the distinction that the VTL is used for work diameters up to 2.5 m (100 in), while the VBM is used for larger diameters [9]. Also, vertical boring mills are often applied to one-of-a-kind jobs, while vertical turret lathes are used for batch production.

16.3 DRILLING AND RELATED OPERATIONS

Drilling, Figure 16.3(b), is a machining operation used to create a round hole in a workpart. This contrasts with boring, which can only be used to enlarge an existing hole. Drilling is usually performed with a rotating cylindrical tool that has two cutting edges on its working end. The tool is called a ***drill*** or ***drill bit***. The most common drill bit is the twist drill, described in Section 17.3.2. The rotating drill feeds into the stationary workpart to form a hole whose diameter is equal to the drill diameter. Drilling is customarily performed on a ***drill press***, although other machine tools also perform this operation.

16.3.1 CUTTING CONDITIONS IN DRILLING

The cutting speed in a drilling operation is the surface speed at the outside diameter of the drill. It is specified in this way for convenience, even though nearly all of the cutting is actually performed at lower speeds closer to the axis of rotation. To set the desired cutting speed in drilling, it is necessary to determine the rotational speed of the drill. Letting N represent the spindle rev/min,

$$N = \frac{v}{\pi D} \tag{16.7}$$

where v = cutting speed, mm/min (in/min); and D = the drill diameter, mm (in). In some drilling operations, the workpiece is rotated about a stationary tool, but the same formula applies.

Feed f in drilling is specified in mm/rev (in/rev). Recommended feeds are roughly proportional to drill diameter; higher feeds are used with larger diameter drills. Since there are (usually) two cutting edges at the drill point, the uncut chip thickness (chip load) taken by each cutting edge is half the feed. Feed can be converted to feed rate using the same equation as for turning:

$$f_r = Nf \tag{16.8}$$

where f_r = feed rate, mm/min (in/min).

Drilled holes are either through holes or blind holes, Figure 16.13. In ***through holes***, the drill exits the opposite side of the work; in ***blind holes***, it does not. The machining time required to drill a through hole can be determined by the following formula:

$$T_m = \frac{t + A}{f_r} \tag{16.9}$$

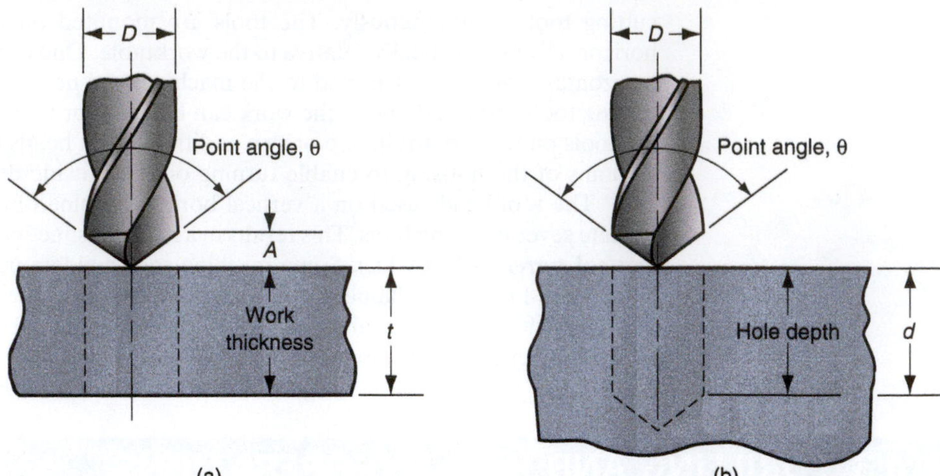

FIGURE 16.13 Two hole types: (a) through hole and (b) blind hole. (Credit: *Fundamentals of Modern Manufacturing*, 4th Edition by Mikell P. Groover, 2010. Reprinted with permission of John Wiley & Sons, Inc.)

where T_m = machining (drilling) time, min; t = work thickness, mm (in); f_r = feed rate, mm/min (in/min); and A = an approach allowance that accounts for the drill point angle, representing the distance the drill must feed into the work before reaching full diameter, Figure 16.13(a). This allowance is given by:

$$A = 0.5\,D \tan\left(90 - \frac{\theta}{2}\right) \tag{16.10}$$

where A = approach allowance, mm (in); and θ = drill point angle. In drilling a through hole, the feed motion usually proceeds slightly beyond the opposite side of the work, thus making the actual duration of the cut greater than T_m in Eq. (16.9) by a small amount.

In a blind hole, hole depth d is defined as the distance from the work surface to the depth of the full diameter, Figure 16.13(b). Thus, for a blind hole, machining time is given by:

$$T_m = \frac{d + A}{f_r} \tag{16.11}$$

where A = the approach allowance by Eq. (16.10).

The rate of metal removal in drilling is determined as the product of the drill cross-sectional area and the feed rate:

$$R_{MR} = \frac{\pi D^2 f_r}{4} \tag{16.12}$$

This equation is valid only after the drill reaches full diameter and excludes the initial approach of the drill into the work.

16.3.2 OPERATIONS RELATED TO DRILLING

Several operations related to drilling are illustrated in Figure 16.14 and described in this section. Most of the operations follow drilling; a hole must be made first by drilling, and then the hole is modified by one of the other operations. Centering and spot facing are exceptions to this rule. All of the operations use rotating tools.

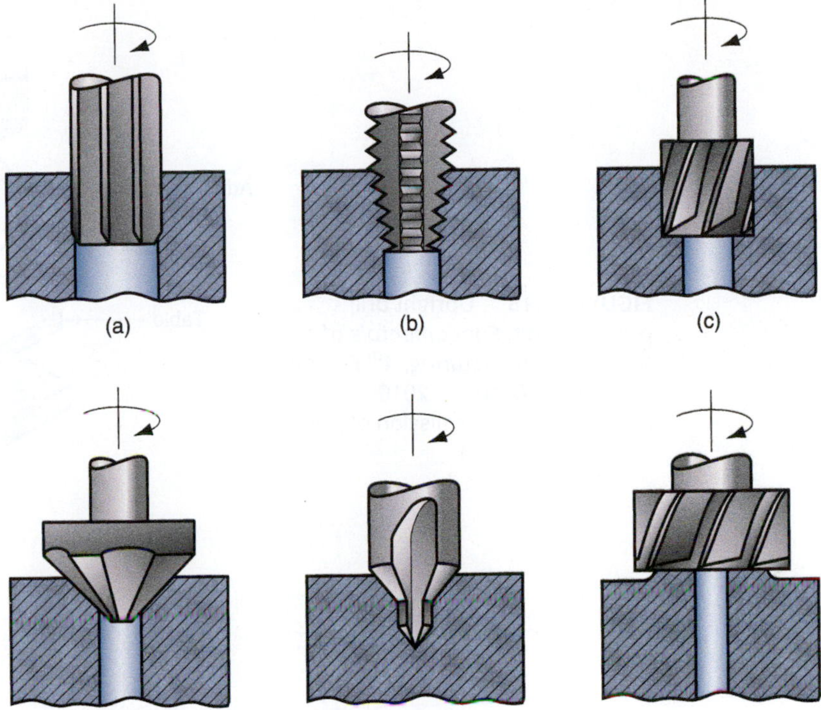

FIGURE 16.14 Machining operations related to drilling: (a) reaming, (b) tapping, (c) counterboring, (d) countersinking, (e) center drilling, and (f) spot facing. (Credit: *Fundamentals of Modern Manufacturing*, 4th Edition by Mikell P. Groover, 2010. Reprinted with permission of John Wiley & Sons, Inc.)

(a) ***Reaming***. Reaming is used to slightly enlarge a hole, to provide a better tolerance on its diameter, and to improve its surface finish. The tool is called a ***reamer***, and it usually has straight flutes.

(b) ***Tapping***. This operation is performed by a ***tap*** and is used to provide internal screw threads on an existing hole.

(c) ***Counterboring***. Counterboring provides a stepped hole, in which a larger diameter follows a smaller diameter partially into the hole. A counterbored hole is used to seat a bolt head into a hole so the head does not protrude above the surface.

(d) ***Countersinking***. This is similar to counterboring, except that the step in the hole is cone-shaped for flat head screws and bolts.

(e) ***Centering***. Also called center drilling, this operation drills a starting hole to accurately establish its location for subsequent drilling. The tool is called a ***center drill***.

(f) ***Spot facing***. Spot facing is similar to milling. It is used to provide a flat machined surface on the workpart in a localized area.

16.3.3 DRILL PRESSES

The standard machine tool for drilling is the drill press. There are various types of drill press, the most basic of which is the upright drill, Figure 16.15. The ***upright drill*** stands on the floor and consists of a table for holding the workpart, a drilling head with powered spindle for the drill bit, and a base and column for support. A similar drill press, but smaller, is the ***bench drill***, which is mounted on a table or bench rather than the floor.

The ***radial drill***, Figure 16.16, is a large drill press designed to cut holes in large parts. It has a radial arm along which the drilling head can be moved and clamped. The head therefore can be positioned along the arm at locations that are a significant distance

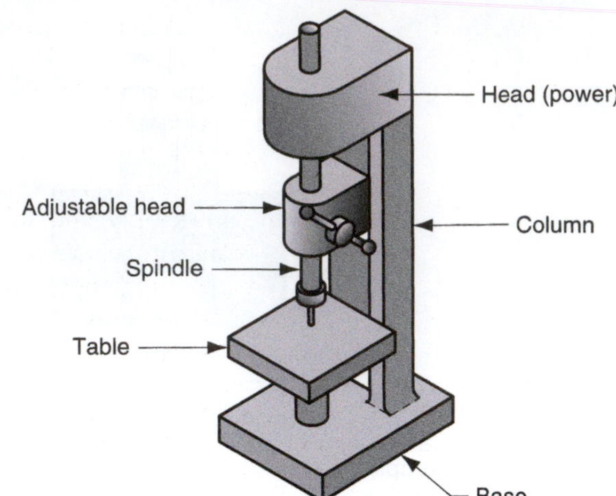

FIGURE 16.15 Upright drill press. (Credit: *Fundamentals of Modern Manufacturing*, 4th Edition by Mikell P. Groover, 2010. Reprinted with permission of John Wiley & Sons, Inc.)

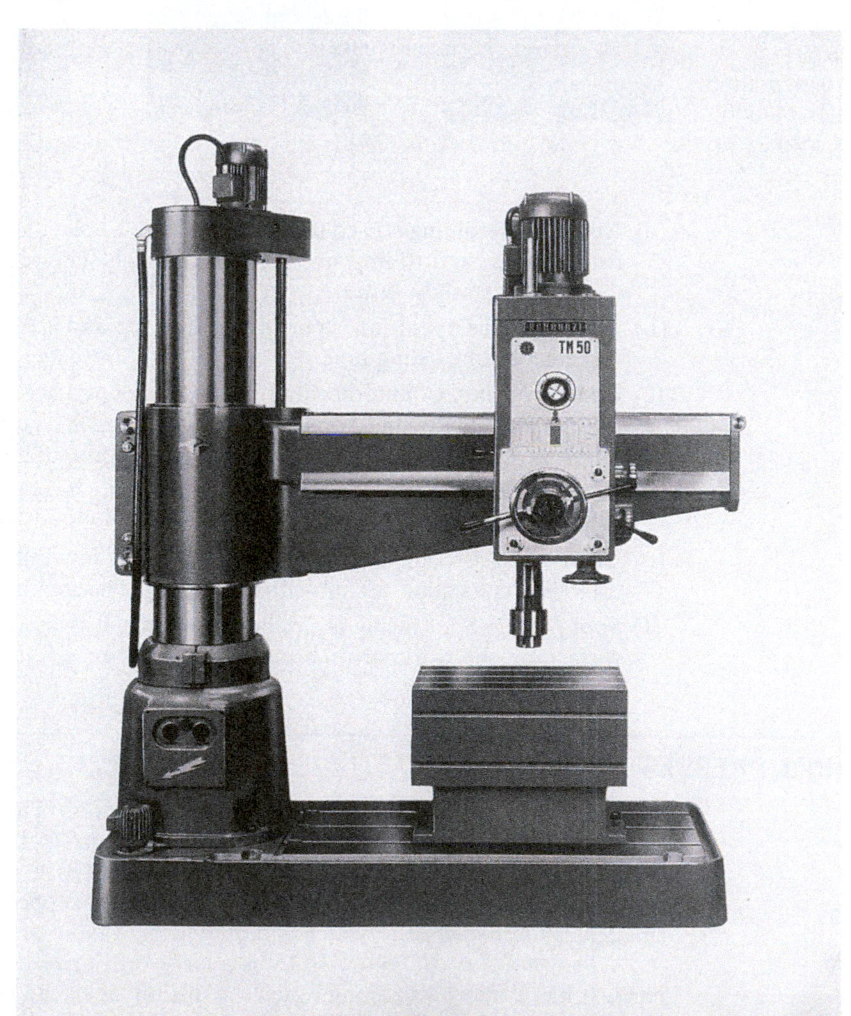

FIGURE 16.16 Radial drill press. Courtesy of Willis Machinery and Tools. (Credit: *Fundamentals of Modern Manufacturing*, 4th Edition by Mikell P. Groover, 2010. Reprinted with permission of John Wiley & Sons, Inc.)

from the column to accommodate large work. The radial arm can also be swiveled about the column to drill parts on either side of the worktable.

The ***gang drill*** is a drill press consisting basically of two to six upright drills connected together in an in-line arrangement. Each spindle is powered and operated independently, and they share a common worktable, so that a series of drilling and related operations can be accomplished in sequence (e.g., centering, drilling, reaming, tapping) simply by sliding the workpart along the worktable from one spindle to the next. A related machine is the ***multiple-spindle drill***, in which several drill spindles are connected together to drill multiple holes simultaneously into the workpart.

In addition, ***computer numerical control drill presses*** are available to control the positioning of the holes in the workparts. These drill presses are often equipped with turrets to hold multiple tools that can be indexed under control of the CNC program. The term ***CNC turret drill*** is used for these machine tools.

Workholding on a drill press is accomplished by clamping the part in a vise, fixture, or jig. A ***vise*** is a general-purpose workholding device possessing two jaws that grasp the work in position. A ***fixture*** is a workholding device that is usually custom-designed for the particular workpart. The fixture can be designed to achieve higher accuracy in positioning the part relative to the machining operation, faster production rates, and greater operator convenience in use. A ***jig*** is a workholding device that is also specially designed for the workpart. The distinguishing feature between a jig and a fixture is that the jig provides a means of guiding the tool during the drilling operation. A fixture does not have this tool guidance feature. A jig used for drilling is called a ***drill jig***.

16.4 MILLING

Milling is a machining operation in which a workpart is fed past a rotating cylindrical tool with multiple cutting edges, as illustrated in Figure 16.2(d) and (e). (In rare cases, a tool with one cutting edge, called a ***fly-cutter***, is used). The axis of rotation of the cutting tool is perpendicular to the direction of feed. This orientation between the tool axis and the feed direction is one of the features that distinguishes milling from drilling. In drilling, the cutting tool is fed in a direction parallel to its axis of rotation. The cutting tool in milling is called a ***milling cutter*** and the cutting edges are called teeth. Aspects of milling cutter geometry are discussed in Section 17.3.2. The conventional machine tool that performs this operation is a ***milling machine***.

The geometric form created by milling is a plane surface. Other work geometries can be created either by means of the cutter path or the cutter shape. Owing to the variety of shapes possible and its high production rates, milling is one of the most versatile and widely used machining operations.

Milling is an ***interrupted cutting*** operation; the teeth of the milling cutter enter and exit the work during each revolution. This interrupted cutting action subjects the teeth to a cycle of impact force and thermal shock on every rotation. The tool material and cutter geometry must be designed to withstand these conditions.

16.4.1 TYPES OF MILLING OPERATIONS

There are two basic types of milling operations, shown in Figure 16.17: (a) peripheral milling and (b) face milling. Most milling operations create geometry by generating the shape (Section 16.1).

Peripheral Milling In peripheral milling, also called ***plain milling***, the axis of the tool is parallel to the surface being machined, and the operation is performed by cutting edges

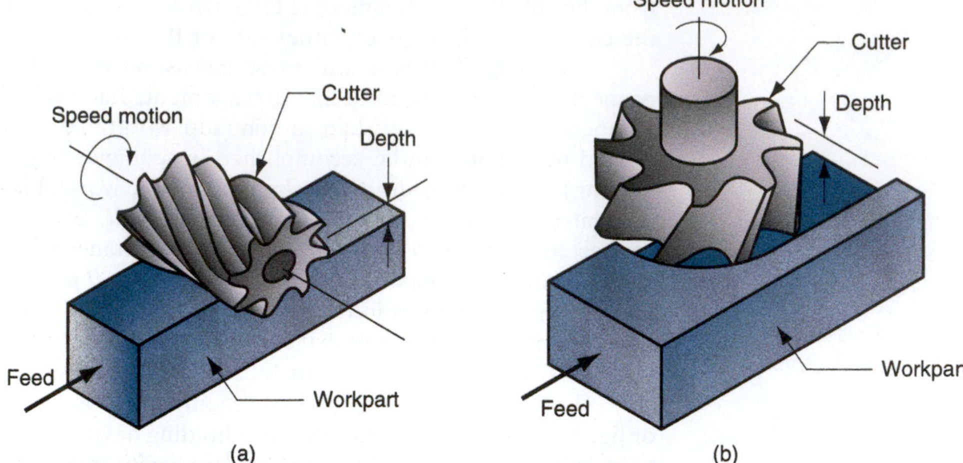

FIGURE 16.17 Two basic types of milling operations: (a) peripheral or plain milling and (b) face milling. (Credit: *Fundamentals of Modern Manufacturing*, 4th Edition by Mikell P. Groover, 2010. Reprinted with permission of John Wiley & Sons, Inc.)

on the outside periphery of the cutter. Several types of peripheral milling are shown in Figure 16.18: (a) ***slab milling***, the basic form of peripheral milling in which the cutter width extends beyond the workpiece on both sides; (b) ***slotting***, also called ***slot milling***, in which the width of the cutter is less than the workpiece width, creating a slot in the work—when the cutter is very thin, this operation can be used to mill narrow slots or cut a workpart in two, called ***saw milling***; (c) ***side milling***, in which the cutter machines the side of the workpiece; (d) ***straddle milling***, the same as side milling, only cutting takes place on both sides of the work; and (e) ***form milling***, in which the milling teeth have a special profile that determines the shape of the slot that is cut in the work. Form milling is therefore classified as a forming operation (Section 16.1).

In peripheral milling, the direction of cutter rotation distinguishes two forms of milling: up milling and down milling, illustrated in Figure 16.19. In ***up milling***, also called ***conventional milling***, the direction of motion of the cutter teeth is opposite the feed direction when the teeth cut into the work. It is milling "against the feed." In ***down milling***, also called ***climb milling***, the direction of cutter motion is the same as the feed direction when the teeth cut the work. It is milling "with the feed."

The relative geometries of these two forms of milling result in differences in their cutting actions. In up milling, the chip formed by each cutter tooth starts out very thin and increases in thickness during the sweep of the cutter. In down milling, each chip starts out

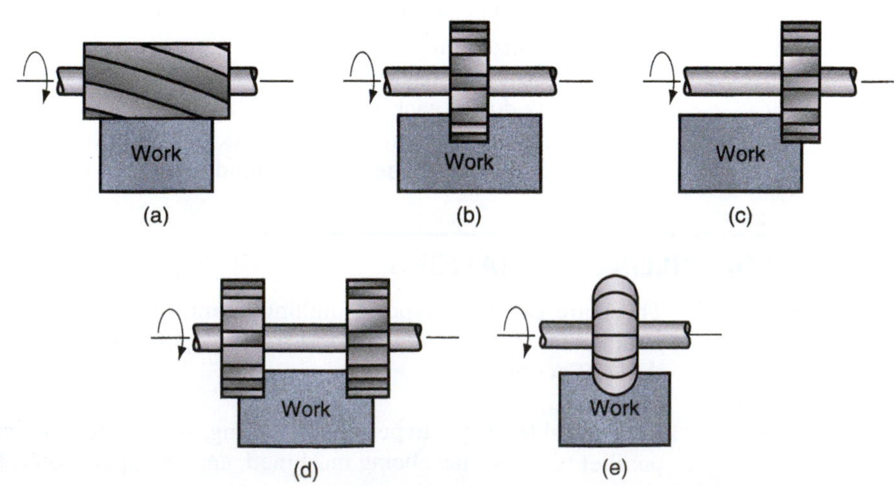

FIGURE 16.18 Peripheral milling: (a) slab milling, (b) slotting, (c) side milling, (d) straddle milling, and (e) form milling. (Credit: *Fundamentals of Modern Manufacturing*, 4th Edition by Mikell P. Groover, 2010. Reprinted with permission of John Wiley & Sons, Inc.)

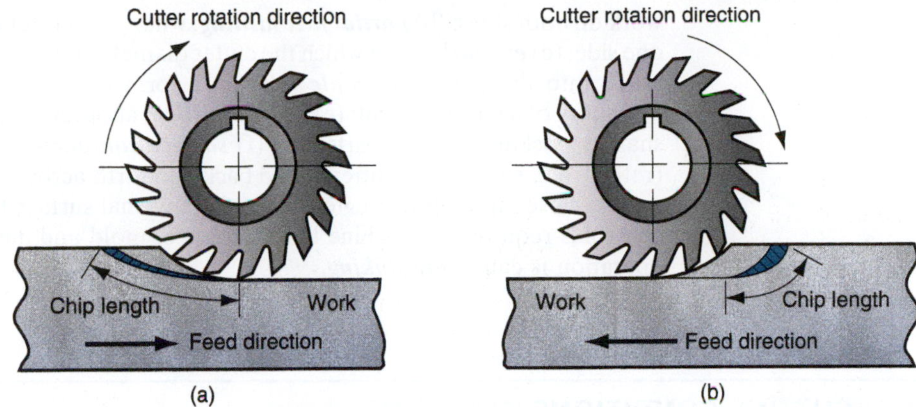

FIGURE 16.19 Two forms of peripheral milling operation with 20-tooth cutter: (a) up milling, and (b) down milling. (Credit: *Fundamentals of Modern Manufacturing*, 4th Edition by Mikell P. Groover, 2010. Reprinted with permission of John Wiley & Sons, Inc.)

thick and reduces in thickness throughout the cut. The length of a chip in down milling is less than in up milling (the difference is exaggerated in our figure). This means that the cutter is engaged in the work for less time per volume of material cut, and this tends to increase tool life in down milling.

The cutting force direction is tangential to the periphery of the cutter for the teeth that are engaged in the work. In up milling, this has a tendency to lift the workpart as the cutter teeth exit the material. In down milling, this cutter force direction is downward, tending to hold the work against the milling machine table.

Face Milling In face milling, the axis of the cutter is perpendicular to the surface being milled, and machining is performed by cutting edges on both the end and outside periphery of the cutter. As in peripheral milling, various forms of face milling exist, several of which are shown in Figure 16.20: (a) ***conventional face milling***, in which the diameter of the cutter is greater than the workpart width, so the cutter overhangs the

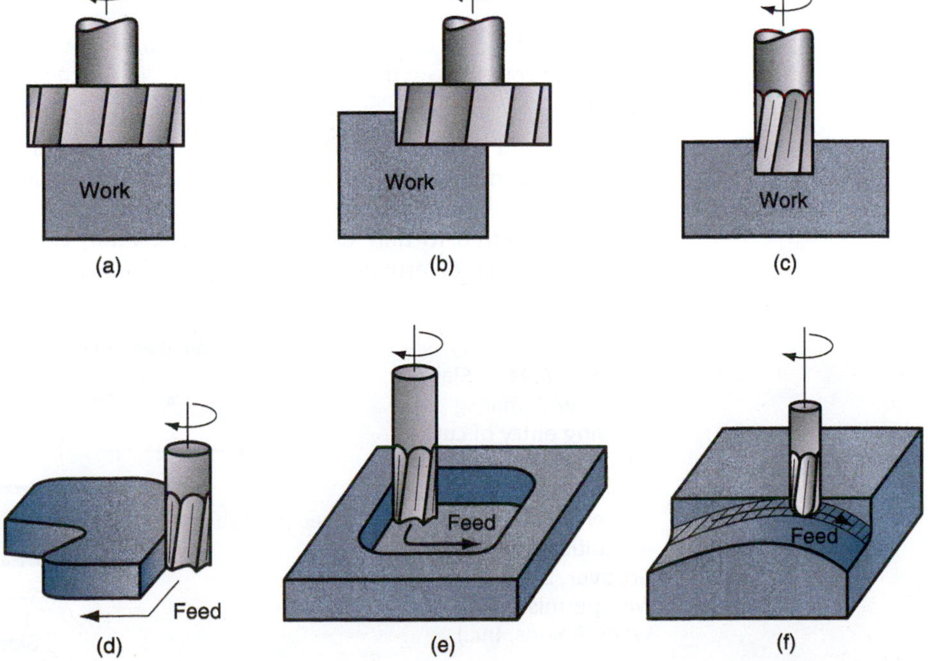

FIGURE 16.20 Face milling: (a) conventional face milling, (b) partial face milling, (c) end milling, (d) profile milling, (e) pocket milling, and (f) surface contouring. (Credit: *Fundamentals of Modern Manufacturing*, 4th Edition by Mikell P. Groover, 2010. Reprinted with permission of John Wiley & Sons, Inc.)

work on both sides; (b) ***partial face milling***, where the cutter overhangs the work on only one side; (c) ***end milling***, in which the cutter diameter is less than the work width, so a slot is cut into the part; (d) ***profile milling***, a form of end milling in which the outside periphery of a flat part is cut; (e) ***pocket milling***, another form of end milling used to mill shallow pockets into flat parts; and (f) ***surface contouring***, in which a ball-nose cutter (rather than square-end cutter) is fed back and forth across the work along a curvilinear path at close intervals to create a three-dimensional surface form. The same basic cutter control is required to machine the contours of mold and die cavities, in which case the operation is called ***die sinking***.

16.4.2 CUTTING CONDITIONS IN MILLING

The cutting speed is determined at the outside diameter of a milling cutter. This can be converted to spindle rotation speed using a formula that should now be familiar:

$$N = \frac{v}{\pi D} \tag{16.13}$$

The feed f in milling is usually given as a feed per cutter tooth; called the ***chip load***, it represents the size of the chip formed by each cutting edge. This can be converted to feed rate by taking into account the spindle speed and the number of teeth on the cutter as follows:

$$f_r = N n_t f \tag{16.14}$$

where f_r = feed rate, mm/min (in/min); N = spindle speed, rev/min; n_t = number of teeth on the cutter; and f = chip load in mm/tooth (in/tooth).

Material removal rate in milling is determined using the product of the cross-sectional area of the cut and the feed rate. Accordingly, if a slab-milling operation is cutting a workpiece with width w at a depth d, the material removal rate is:

$$R_{MR} = w d f_r \tag{16.15}$$

This neglects the initial entry of the cutter before full engagement. Eq. 16.15 can be applied to end milling, side milling, face milling, and other milling operations, making the proper adjustments in the computation of cross-sectional area of cut.

The time required to mill a workpiece of length L must account for the approach distance required to fully engage the cutter. First, consider the case of slab milling, Figure 16.21. To determine the time to perform a slab milling operation, the approach

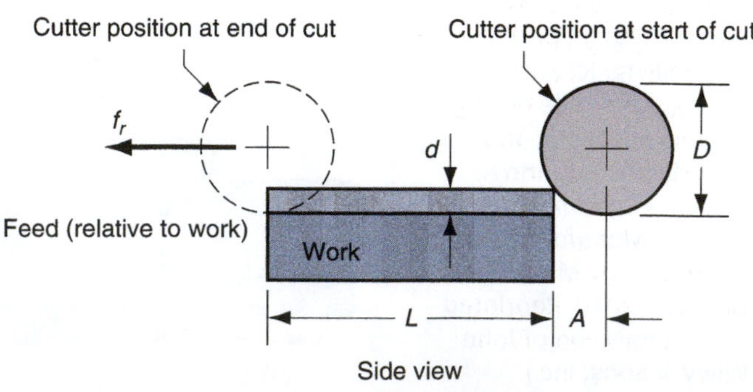

FIGURE 16.21 Slab (peripheral) milling showing entry of cutter into the workpiece. (Credit: *Fundamentals of Modern Manufacturing*, 4th Edition by Mikell P. Groover, 2010. Reprinted with permission of John Wiley & Sons, Inc.)

distance A to reach full cutter depth is given by:

$$A = \sqrt{d(D - d)} \tag{16.16}$$

where d = depth of cut, mm (in); and D = diameter of the milling cutter, mm (in). The time T_m in which the cutter is engaged milling the workpiece is therefore:

$$T_m = \frac{L + A}{f_r} \tag{16.17}$$

For face milling, let us consider the two possible cases pictured in Figure 16.22. The first case is when the cutter is centered over a rectangular workpiece as in Figure 16.22(a). The cutter feeds from right to left across the workpiece. In order for the cutter to reach the full width of the work, it must travel an approach distance given by:

$$A = 0.5\left(D - \sqrt{D^2 - w^2}\right) \tag{16.18}$$

where D = cutter diameter, mm (in) and w = width of the workpiece, mm (in). If $D = w$, then Eq. (16.18) reduces to $A = 0.5D$. And if $D < w$, then a slot is cut into the work and $A = 0.5D$.

The second case is when the cutter is offset to one side of the work, as in Figure 16.22(b). In this case, the approach distance is given by:

$$A = \sqrt{w(D - w)} \tag{16.19}$$

where w = width of the cut, mm (in). In either case, the machining time is given by:

$$T_m = \frac{L + A}{f_r} \tag{16.20}$$

It should be emphasized in all of these milling scenarios that T_m represents the time the cutter teeth are engaged in the work, making chips. Approach and overtravel distances are usually added at the beginning and end of each cut to allow access to the work for loading and unloading. Thus the actual duration of the cutter feed motion is likely to be greater than T_m.

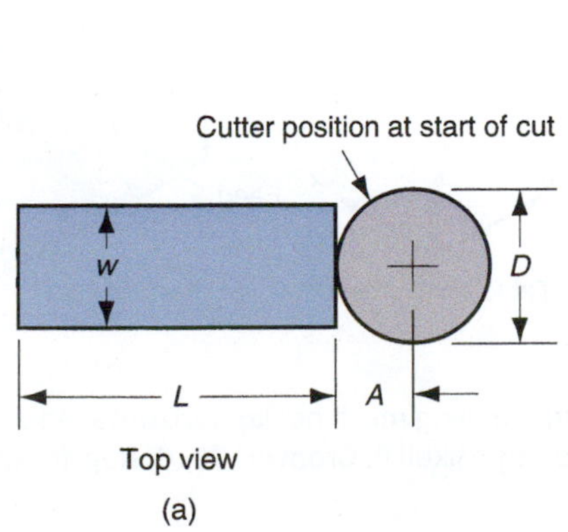

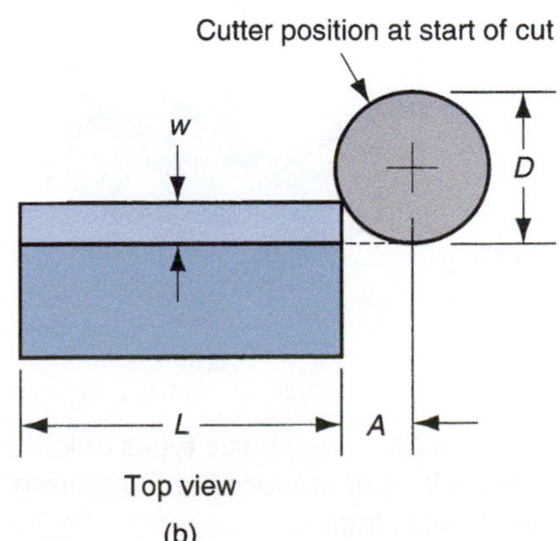

FIGURE 16.22 Face milling showing approach and overtravel distances for two cases: (a) when cutter is centered over the workpiece, and (b) when cutter is offset to one side over the work. (Credit: *Fundamentals of Modern Manufacturing*, 4th Edition by Mikell P. Groover, 2010. Reprinted with permission of John Wiley & Sons, Inc.)

16.4.3 MILLING MACHINES

Milling machines must provide a rotating spindle for the cutter and a table for fastening, positioning, and feeding the workpart. Various machine tool designs satisfy these requirements. To begin with, milling machines can be classified as horizontal or vertical. A ***horizontal milling machine*** has a horizontal spindle, and this design is well suited for performing peripheral milling (e.g., slab milling, slotting, side and straddle milling) on workparts that are roughly cube shaped. A ***vertical milling machine*** has a vertical spindle, and this orientation is appropriate for face milling, end milling, surface contouring, and die sinking on relatively flat workparts.

Other than spindle orientation, milling machines can be classified into the following types: (1) knee-and-column, (2) bed type, (3) planer type, (4) tracer mills, and (5) CNC milling machines.

The ***knee-and-column milling machine*** is the basic machine tool for milling. It derives its name from the fact that its two main components are a ***column*** that supports the spindle, and a ***knee*** (roughly resembling a human knee) that supports the worktable. It is available as either a horizontal or a vertical machine, as illustrated in Figure 16.23. In the horizontal version, an arbor usually supports the cutter. The ***arbor*** is basically a shaft that holds the milling cutter and is driven by the spindle. An overarm is provided on horizontal machines to support the arbor. On vertical knee-and-column machines, milling cutters can be mounted directly in the spindle without an arbor.

One of the features of the knee-and-column milling machine that makes it so versatile is its capability for worktable feed movement in any of the x–y–z axes. The worktable can be moved in the x-direction, the saddle can be moved in the y-direction, and the knee can be moved vertically to achieve the z-movement.

Two special knee-and-column machines should be identified. One is the ***universal*** milling machine, Figure 16.24(a), which has a table that can be swiveled in a horizontal plane (about a vertical axis) to any specified angle. This facilitates the cutting of angular shapes and helixes on workparts. Another special machine is the ***ram mill***, Figure 16.24(b), in which the toolhead containing the spindle is located on the end of a horizontal ram; the ram can be adjusted in and out over the worktable to locate the cutter relative to the work. The toolhead can also be swiveled to achieve an angular

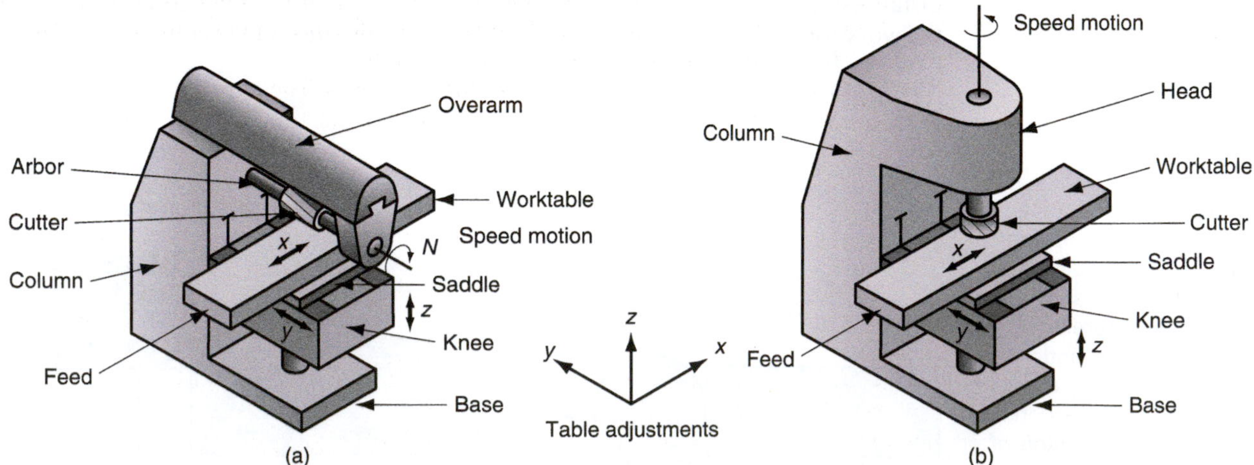

FIGURE 16.23 Two basic types of knee-and-column milling machine: (a) horizontal and (b) vertical. (Credit: *Fundamentals of Modern Manufacturing*, 4th Edition by Mikell P. Groover, 2010. Reprinted with permission of John Wiley & Sons, Inc.)

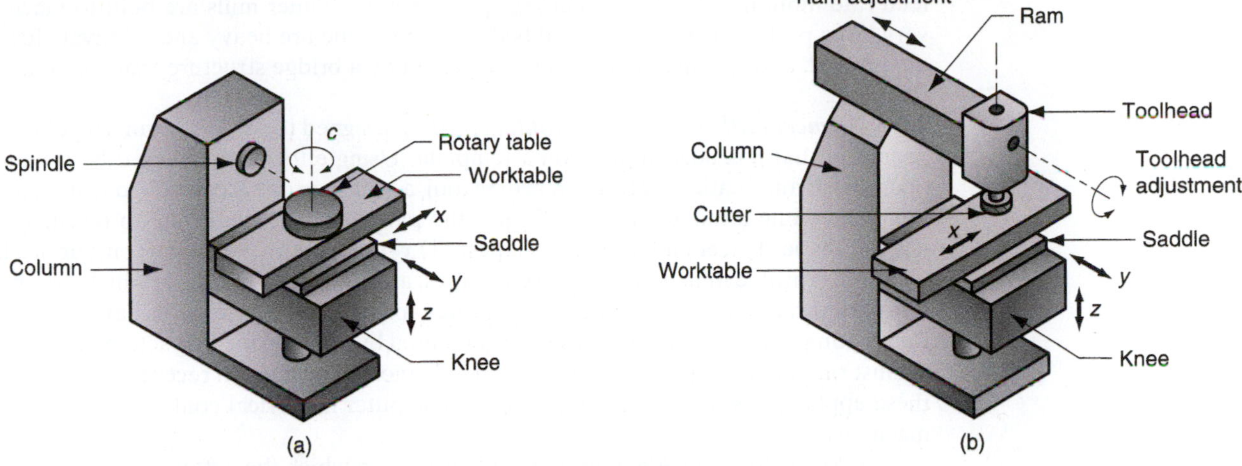

FIGURE 16.24 Special types of knee-and-column milling machine: (a) universal—overarm, arbor, and cutter omitted for clarity: and (b) ram type. (Credit: *Fundamentals of Modern Manufacturing*, 4th Edition by Mikell P. Groover, 2010. Reprinted with permission of John Wiley & Sons, Inc.)

orientation of the cutter with respect to the work. These features provide considerable versatility in machining a variety of work shapes.

Bed-type milling machines are designed for high production. They are constructed with greater rigidity than knee-and-column machines, thus permitting them to achieve heavier feed rates and depths of cut needed for high material removal rates. The characteristic construction of the bed-type milling machine is shown in Figure 16.25. The worktable is mounted directly to the bed of the machine tool, rather than using the less rigid knee-type design. This construction limits the possible motion of the table to longitudinal feeding of the work past the milling cutter. The cutter is mounted in a spindle head that can be adjusted vertically along the machine column. Single spindle bed machines are called ***simplex*** mills, as in Figure 16.25, and are available in either horizontal or vertical models. ***Duplex*** mills use two spindle heads. The heads are usually positioned horizontally on opposite sides of the bed to perform simultaneous operations during one feeding pass of the work. ***Triplex*** mills add a third spindle mounted vertically over the bed to further increase machining capability.

Planer type mills are the largest milling machines. Their general appearance and construction are those of a large planer (see Figure 16.31); the difference is that milling is performed instead of planing. Accordingly, one or more milling heads are substituted for the single-point cutting tools used on planers, and the motion of the work past the tool is a

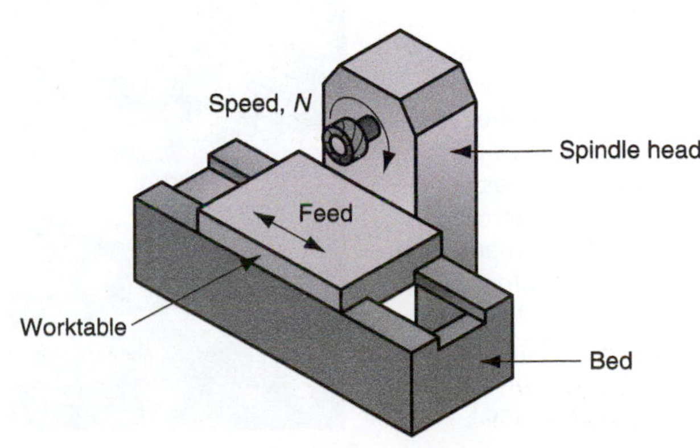

FIGURE 16.25 Simplex bed-type milling machine horizontal spindle. (Credit: *Fundamentals of Modern Manufacturing*, 4th Edition by Mikell P. Groover, 2010. Reprinted with permission of John Wiley & Sons, Inc.)

feed rate motion rather than a cutting speed motion. Planer mills are built to machine very large parts. The worktable and bed of the machine are heavy and relatively low to the ground, and the milling heads are supported by a bridge structure that spans across the table.

A ***tracer mill***, also called a ***profiling mill***, is designed to reproduce an irregular part geometry that has been created on a template. Using either manual feed by a human operator or automatic feed by the machine tool, a tracing probe is controlled to follow the template, while a milling head duplicates the path taken by the probe to machine the desired shape. Tracer mills are of two types: (1) ***x–y tracing***, in which the contour of a flat template is profile milled using two-axis control; and (2) ***x–y–z tracing***, in which the probe follows a three-dimensional pattern using three-axis control. Tracer mills have been used for creating shapes that cannot easily be generated by a simple feeding action of the work against the milling cutter. Applications include molds and dies. In recent years, many of these applications have been taken over by computer numerical control (CNC) milling machines.

CNC milling machines are milling machines in which the cutter path is controlled by alphanumerical data rather than a physical template. They are especially suited to profile milling, pocket milling, surface contouring, and die sinking operations, in which two or three axes of the worktable must be simultaneously controlled to achieve the required cutter path. An operator is normally required to change cutters as well as load and unload workparts.

16.5 MACHINING CENTERS AND TURNING CENTERS

A ***machining center***, illustrated in Figure 16.26, is a highly automated machine tool capable of performing multiple machining operations under computer numerical control in one setup with minimal human attention. Workers are needed to load and unload parts, which usually takes considerable less time than the machine cycle time, so one

FIGURE 16.26 A universal machining center. Capability to orient the workhead makes this a five-axis machine. Courtesy of Cincinnati Milacron. (Credit: *Fundamentals of Modern Manufacturing*, 4th Edition by Mikell P. Groover, 2010. Reprinted with permission of John Wiley & Sons, Inc.)

worker may be able to tend more than one machine. Typical operations performed on a machining center are milling and drilling, which use rotating cutting tools.

The typical features that distinguish a machining center from conventional machine tools and make it so productive include:

- ***Multiple operations in one setup***. Most workparts require more than one operation to completely machine the specified geometry. Complex parts may require dozens of separate machining operations, each requiring its own machine tool, setup, and cutting tool. Machining centers are capable of performing most or all of the operations at one location, thus minimizing setup time and production lead time.
- ***Automatic tool changing***. To change from one machining operation to the next, the cutting tools must be changed. This is done on a machining center under CNC program control by an automatic tool-changer designed to exchange cutters between the machine tool spindle and a ***tool storage carousels***. Capacities of these carousels commonly range from 16 to 80 cutting tools. The machine in Figure 16.26 has two storage carousels on the left side of the column.
- ***Pallet shuttles***. Some machining centers are equipped with pallet shuttles, which are automatically transferred between the spindle position and the loading station, as shown in Figure 16.26. Parts are fixtured on pallets that are attached to the shuttles. In this arrangement, the operator can be unloading the previous part and loading the next part while the machine tool is engaged in machining the current part. Nonproductive time on the machine is thereby reduced.
- ***Automatic workpart positioning***. Many machining centers have more than three axes. One of the additional axes is often designed as a rotary table to position the part at some specified angle relative to the spindle. The rotary table permits the cutter to perform machining on four sides of the part in a single setup.

Machining centers are classified as horizontal, vertical, or universal. The designation refers to spindle orientation. Horizontal machining centers (HMCs) normally machine cube-shaped parts, in which the four vertical sides of the cube can be accessed by the cutter. Vertical machining centers (VMCs) are suited to flat parts on which the tool can machine the top surface. Universal machining centers have workheads that swivel their spindle axes to any angle between horizontal and vertical, as in Figure 16.26.

Success of CNC machining centers led to the development of CNC turning centers. A modern ***CNC turning center***, Figure 16.27, is capable of performing various turning and related operations, contour turning, and automatic tool indexing, all under computer control. In addition, the most sophisticated turning centers can accomplish (1) workpart gaging (checking key dimensions after machining), (2) tool monitoring (sensors to indicate when the tools are worn), (3) automatic tool changing when tools become worn, and even (4) automatic workpart changing at the completion of the work cycle [16].

Another type of machine tool related to machining centers and turning centers is the ***CNC mill-turn center***. This machine has the general configuration of a turning center; in addition, it can position a cylindrical workpart at a specified angle so that a rotating cutting tool (e.g., milling cutter) can machine features into the outside surface of the part, as illustrated in Figure 16.28. An ordinary turning center does not have the capability to stop the workpart at a defined angular position, and it does not possess rotating tool spindles.

Further progress in machine tool technology has taken the mill-turn center one step further by integrating additional capabilities into a single machine. The additional capabilities include (1) combining milling, drilling, and turning with grinding, welding, and inspection operations, all in one machine tool; (2) using multiple spindles simultaneously, either on a single workpiece or two different workpieces; and (3) automating the part handling function by adding industrial robots to the machine [2], [16]. The terms

FIGURE 16.27 CNC four-axis turning center. Courtesy of Cincinnati Milacron. (Credit: *Fundamentals of Modern Manufacturing,* 4th Edition by Mikell P. Groover, 2010. Reprinted with permission of John Wiley & Sons, Inc.)

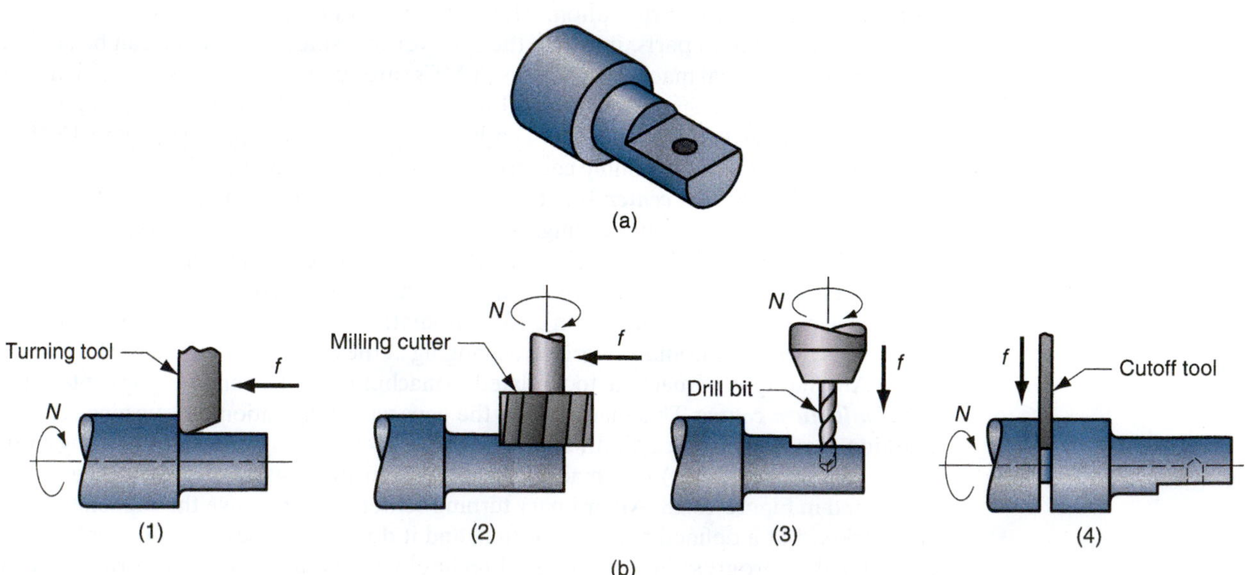

FIGURE 16.28 Operation of a mill-turn center: (a) example part with turned, milled, and drilled surfaces; and (b) sequence of operations on a mill-turn center: (1) turn second diameter, (2) mill flat with part in programmed angular position, (3) drill hole with part in same programmed position, and (4) cutoff. (Credit: *Fundamentals of Modern Manufacturing,* 4th Edition by Mikell P. Groover, 2010. Reprinted with permission of John Wiley & Sons, Inc.)

multitasking machine and ***multifunction machine*** are sometimes used for these machine tools.

16.6 OTHER MACHINING OPERATIONS

In addition to turning, drilling, and milling, several other machining operations should be included in our survey: (1) shaping and planing, (2) broaching, and (3) sawing.

16.6.1 SHAPING AND PLANING

Shaping and planing are similar operations, both involving the use of a single-point cutting tool moved linearly relative to the workpart. In conventional shaping and planing, a straight, flat surface is created by this action. The difference between the two operations is illustrated in Figure 16.29. In shaping, the speed motion is accomplished by moving the cutting tool; while in planing, the speed motion is accomplished by moving the workpart.

Cutting tools used in shaping and planing are single-point tools (Section 17.3.1). Unlike turning, interrupted cutting occurs in shaping and planing, subjecting the tool to an impact loading upon entry into the work. In addition, these machine tools are limited to low speeds due to their start-and-stop motion. The conditions normally dictate use of high-speed steel cutting tools.

Shaping Shaping is performed on a machine tool called a ***shaper***, Figure 16.30. The components of the shaper include a ***ram***, which moves relative to a ***column*** to provide the cutting motion, and a worktable that holds the part and accomplishes the feed motion. The motion of the ram consists of a forward stroke to achieve the cut, and a return stroke during which the tool is lifted slightly to clear the work and then reset for the next pass. On completion of each return stroke, the worktable is advanced laterally relative to the ram trajectory to feed the part. Feed is specified in mm/stroke (in/stroke). The drive mechanism for the ram can be either hydraulic or mechanical. Hydraulic drive has greater flexibility in adjusting the stroke length and a more uniform speed during the forward stroke, but it is more expensive than a mechanical drive unit. Both mechanical and hydraulic drives are

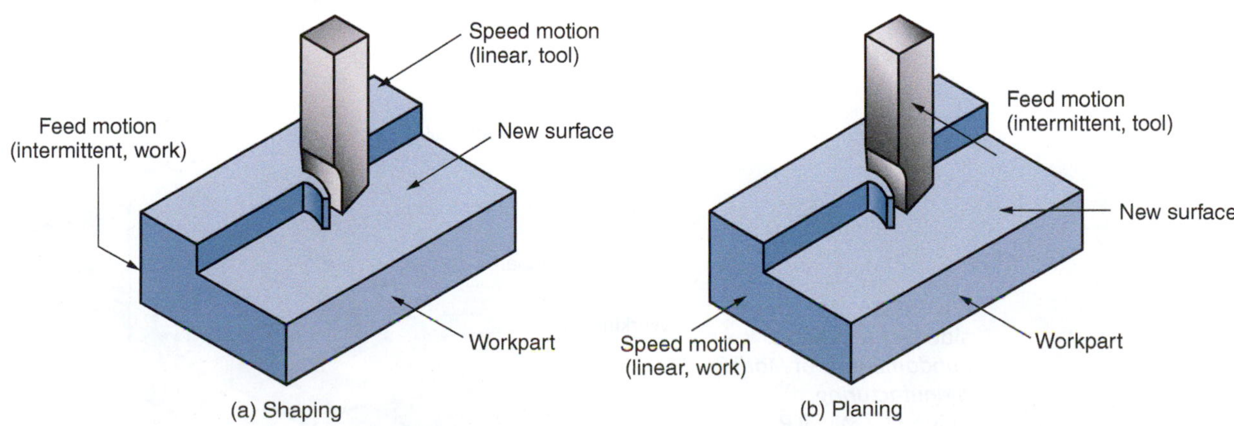

FIGURE 16.29 (a) Shaping, and (b) planing. (Credit: *Fundamentals of Modern Manufacturing*, 4th Edition by Mikell P. Groover, 2010. Reprinted with permission of John Wiley & Sons, Inc.)

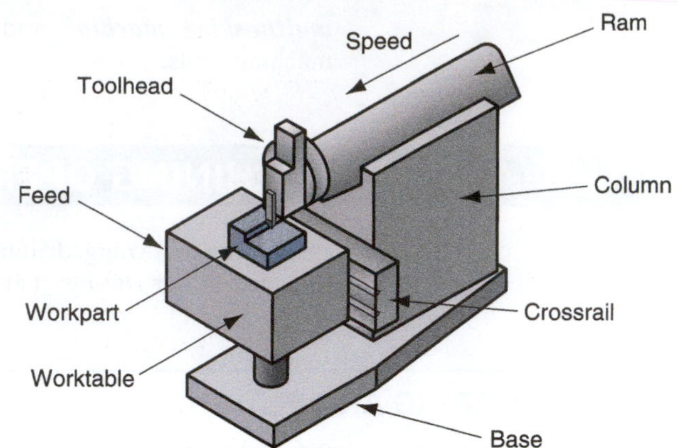

FIGURE 16.30 Components of a shaper. (Credit: *Fundamentals of Modern Manufacturing*, 4th Edition by Mikell P. Groover, 2010. Reprinted with permission of John Wiley & Sons, Inc.)

designed to achieve higher speeds on the return (noncutting) stroke than on the forward (cutting) stroke, thereby increasing the proportion of time spent cutting.

Planing The machine tool for planing is a ***planer***. Cutting speed is achieved by a reciprocating worktable that moves the part past the single-point cutting tool. The construction and motion capability of a planer permit much larger parts to be machined than on a shaper. Planers can be classified as open side planers or double-column planers. The ***open-side planer***, also known as a ***single-column planer***, Figure 16.31, has a single column supporting the cross-rail on which a toolhead is mounted. Another toolhead can also be mounted and fed along the vertical column. Multiple toolheads permit more than one cut to be taken on each pass. At the completion of each stroke, each toolhead is moved relative to the cross-rail (or column) to achieve the intermittent feed motion. The configuration of the open-side planer permits very wide workparts to be machined.

A ***double-column planer*** has two columns, one on either side of the base and worktable. The columns support the cross-rail, on which one or more toolheads are mounted. The two columns provide a more rigid structure for the operation; however, the two columns limit the width of the work that can be handled on this machine.

Shaping and planing can be used to machine shapes other than flat surfaces. The restriction is that the cut surface must be straight. This allows the cutting of grooves, slots, gear teeth, and other shapes as illustrated in Figure 16.32. Special machines and tool geometries must be specified to cut some of these shapes. An important example is the ***gear shaper***, a vertical shaper with a specially designed rotary feed table and synchronized tool head used to generate teeth on spur gears.

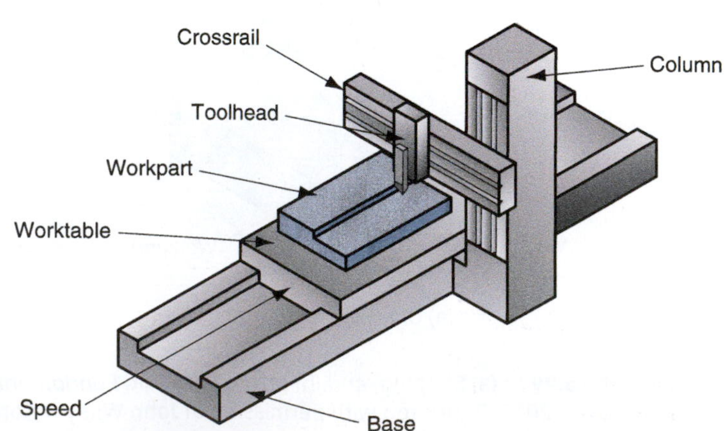

FIGURE 16.31 Open side planer. (Credit: *Fundamentals of Modern Manufacturing*, 4th Edition by Mikell P. Groover, 2010. Reprinted with permission of John Wiley & Sons, Inc.)

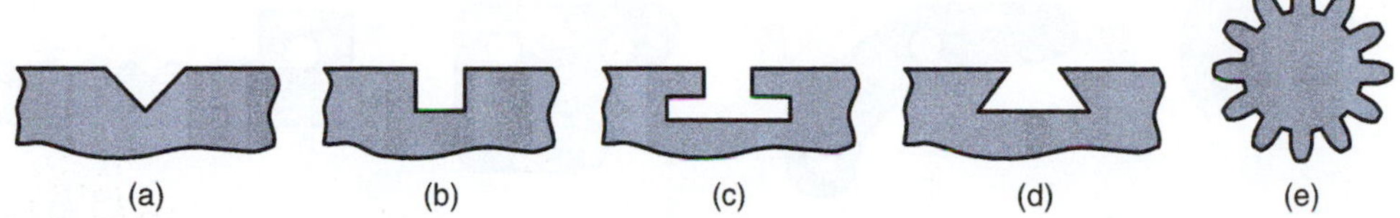

FIGURE 16.32 Types of shapes that can cut by shaping and planing: (a) V-groove, (b) square groove, (c) T-slot, (d) dovetail slot, and (e) gear teeth. (Credit: *Fundamentals of Modern Manufacturing,* 4th Edition by Mikell P. Groover, 2010. Reprinted with permission of John Wiley & Sons, Inc.)

16.6.2 BROACHING

Broaching is performed using a multiple-teeth cutting tool by moving the tool linearly relative to the work in the direction of the tool axis, as in Figure 16.33. The machine tool is called a ***broaching machine***, and the cutting tool is called a ***broach***. In certain jobs for which broaching can be used, it is a highly productive method of machining. Advantages include good surface finish, close tolerances, and a variety of work shapes. Owing to the complicated and often custom-shaped geometry of the broach, tooling is expensive.

There are two principal types of broaching: external (also called surface broaching) and internal. ***External broaching*** is performed on the outside surface of the work to create a certain cross-sectional shape on the surface. Figure 16.34(a) shows some possible cross sections that can be formed by external broaching. ***Internal broaching*** is accomplished on the internal surface of a hole in the part. Accordingly, a starting hole must be present in the part so as to insert the broach at the beginning of the broaching stroke. Figure 16.34(b) indicates some of the shapes that can be produced by internal broaching.

The basic function of a broaching machine is to provide a precise linear motion of the tool past a stationary work position, but there are various ways in which this can be done. Most broaching machines can be classified as either vertical or horizontal machines. The ***vertical broaching machine*** is designed to move the broach along a vertical path, while the ***horizontal broaching machine*** has a horizontal tool path. Most broaching machines pull the broach past the work. However, there are exceptions to this pull action. One exception is a relatively simple type called a ***broaching press***, used only for internal broaching, that pushes the tool through the workpart. Another exception is the ***continuous broaching machine***, in which the workparts are fixtured to an endless belt loop and moved past a stationary broach. Because of its continuous operation, this machine can be used only for surface broaching.

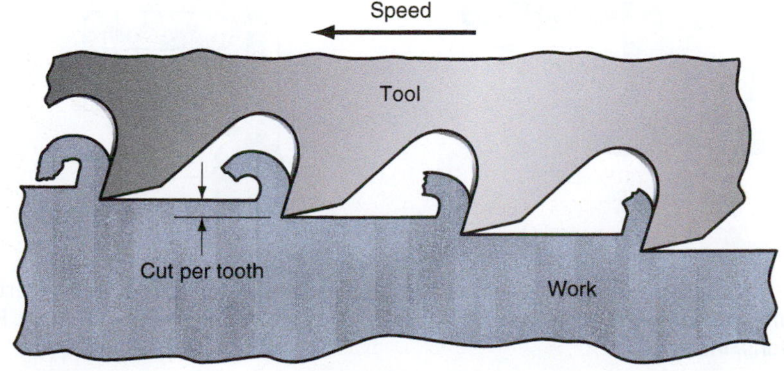

FIGURE 16.33 The broaching operation. (Credit: *Fundamentals of Modern Manufacturing,* 4th Edition by Mikell P. Groover, 2010. Reprinted with permission of John Wiley & Sons, Inc.)

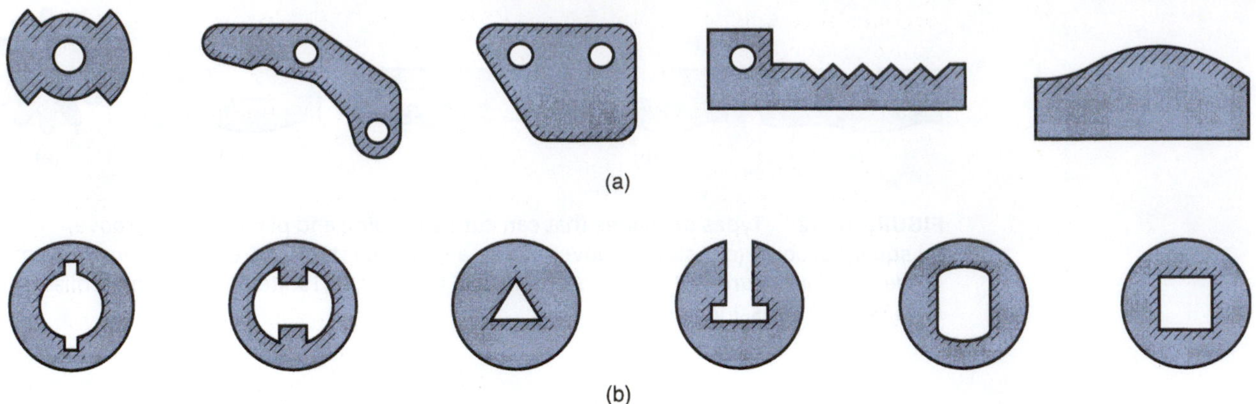

FIGURE 16.34 Work shapes that can be cut by: (a) external broaching, and (b) internal broaching. Cross-hatching indicates the surfaces broached. (Credit: *Fundamentals of Modern Manufacturing,* 4th Edition by Mikell P. Groover, 2010. Reprinted with permission of John Wiley & Sons, Inc.)

16.6.3 SAWING

Sawing is a process in which a narrow slit is cut into the work by a tool consisting of a series of narrowly spaced teeth. Sawing is normally used to separate a workpart into two pieces, or to cut off an unwanted portion of a part. These operations are often referred to as ***cutoff*** operations. Since many factories require cutoff operations at some point in the production sequence, sawing is an important manufacturing process.

In most sawing operations, the work is held stationary and the ***saw blade*** is moved relative to it. There are three basic types of sawing, as in Figure 16.35, according to the type of blade motion involved: (a) hacksawing, (b) bandsawing, and (c) circular sawing.

Hacksawing, Figure 16.35(a), involves a linear reciprocating motion of the saw against the work. This method of sawing is often used in cutoff operations. Cutting is

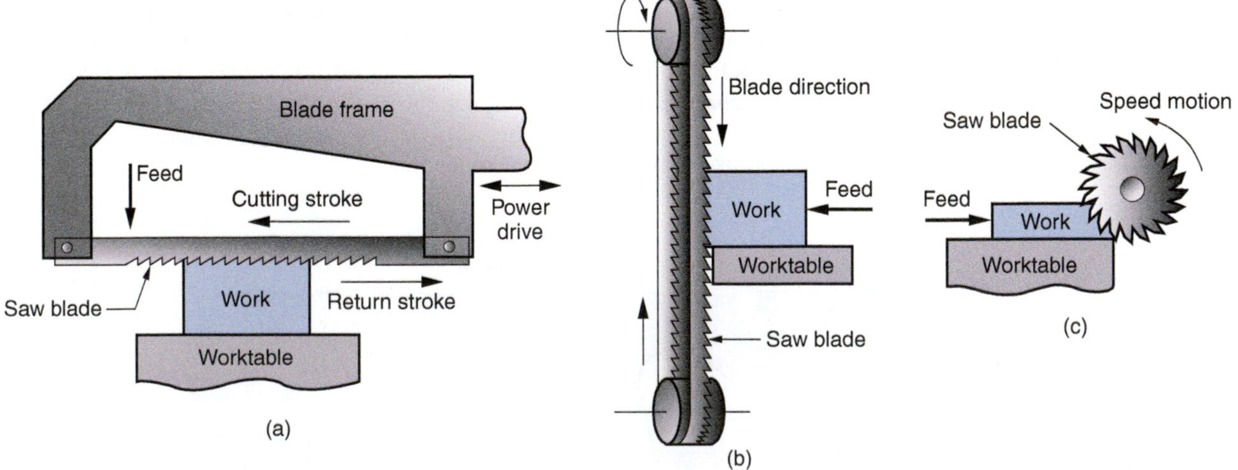

FIGURE 16.35 Three types of sawing operations: (a) power hacksaw, (b) bandsaw (vertical), and (c) circular saw. (Credit: *Fundamentals of Modern Manufacturing,* 4th Edition by Mikell P. Groover, 2010. Reprinted with permission of John Wiley & Sons, Inc.)

accomplished only on the forward stroke of the saw blade. Because of this intermittent cutting action, hacksawing is inherently less efficient than the other sawing methods, both of which are continuous. The ***hacksaw*** blade is a thin straight tool with cutting teeth on one edge. Hacksawing can be done either manually or with a power hacksaw. A ***power hacksaw*** provides a drive mechanism to operate the saw blade at a desired speed; it also applies a given feed rate or sawing pressure.

Bandsawing involves a linear continuous motion, using a ***bandsaw blade*** made in the form of an endless flexible loop with teeth on one edge. The sawing machine is a ***bandsaw***, which provides a pulley-like drive mechanism to continuously move and guide the bandsaw blade past the work. Bandsaws are classified as vertical or horizontal. The designation refers to the direction of saw blade motion during cutting. Vertical bandsaws are used for cutoff as well as other operations such as contouring and slotting. ***Contouring*** on a bandsaw involves cutting a part profile from flat stock. ***Slotting*** is the cutting of a thin slot into a part, an operation for which bandsawing is well suited. Contour sawing and slotting are operations in which the work is fed into the saw blade.

Vertical bandsaw machines can be operated either manually, in which the operator guides and feeds the work past the bandsaw blade, or automatically, in which the work is power fed past the blade. Recent innovations in bandsaw design have permitted the use of CNC to perform contouring of complex outlines. Some of the details of the vertical bandsawing operation are illustrated in Figure 16.35(b). Horizontal bandsaws are normally used for cutoff operations as alternatives to power hacksaws.

Circular sawing, Figure 16.35(c), uses a rotating saw blade to provide a continuous motion of the tool past the work. Circular sawing is often used to cut long bars, tubes, and similar shapes to specified length. The cutting action is similar to a slot milling operation, except that the saw blade is thinner and contains many more cutting teeth than a slot milling cutter. Circular sawing machines have powered spindles to rotate the saw blade and a feeding mechanism to drive the rotating blade into the work.

Two operations related to circular sawing are abrasive cutoff and friction sawing. In ***abrasive cutoff***, an abrasive disk is used to perform cutoff operations on hard materials that would be difficult to saw with a conventional saw blade. In ***friction sawing***, a steel disk is rotated against the work at very high speeds, resulting in friction heat that causes the material to soften sufficiently to permit penetration of the disk through the work. The cutting speeds in both of these operations are much faster than in circular sawing.

16.7 HIGH-SPEED MACHINING

One persistent trend throughout the history of metal machining has been the use of higher and higher cutting speeds. In recent years, there has been renewed interest in this area due to its potential for faster production rates, shorter lead times, reduced costs, and improved part quality. In its simplest definition, ***high-speed machining*** (HSM) means using cutting speeds that are significantly higher than those used in conventional machining operations. Some examples of cutting speed values for conventional and HSM are presented in Table 16.1, according to data compiled by Kennametal Inc.[1]

Other definitions of HSM have been developed to deal with the wide variety of work materials and tool materials used in machining. One popular HSM definition is the ***DN ratio***—the bearing bore diameter (mm) multiplied by the maximum spindle speed (rev/min). For high-speed machining, the typical DN ratio is between 500,000 and 1,000,000. This definition allows larger diameter bearings to fall within the HSM range, even though they operate at lower rotational speeds than smaller bearings. Typical HSM

[1] Kennametal Inc. is a leading cutting tool producer.

TABLE 16.1 Comparison of cutting speeds used in conventional versus high-speed machining for selected work materials.

	Solid Tools (end mills, drills)[a]				Indexable Tools (face mills)[a]			
	Conventional Speed		High Cutting Speed		Conventional Speed		High Cutting Speed	
Work Material	**m/min**	**ft/min**	**m/min**	**ft/min**	**m/min**	**ft/min**	**m/min**	**ft/min**
Aluminum	300+	1000+	3000+	10,000+	600+	2000+	3600+	12,000+
Cast iron, soft	150	500	360	1200	360	1200	1200	4000
Cast iron, ductile	105	350	250	800	250	800	900	3000
Steel, free machining	105	350	360	1200	360	1200	600	2000
Steel, alloy	75	250	250	800	210	700	360	1200
Titanium	40	125	60	200	45	150	90	300

[a]Solid tools are made of one solid piece, indexable tools use indexable inserts. Appropriate tool materials include cemented carbide and coated carbide of various grades for all materials, ceramics for all materials, polycrystalline diamond tools for aluminum, and cubic boron nitride for steels (see Section 17.2 for discussion of these tool materials).
Source: Kennametal Inc. [3].

spindle velocities range between 8000 and 35,000 rpm, although some spindles today are designed to rotate at 100,000 rpm.

Another HSM definition is based on the ratio of horsepower to maximum spindle speed, or ***hp/rpm ratio***. Conventional machine tools usually have a higher hp/rpm ratio than machines equipped for high-speed machining. By this metric, the dividing line between conventional machining and HSM is around 0.005 hp/rpm. Thus, high-speed machining includes 50 hp spindles capable of 10,000 rpm (0.005 hp/rpm) and 15 hp spindles that can rotate at 30,000 rpm (0.0005 hp/rpm).

Requirements for high-speed machining include the following: (1) high-speed spindles using special bearings designed for high rpm operation; (2) high feed rate capability, typically around 50 m/min (2000 in/min); (3) CNC motion controls with "look-ahead" features that allow the controller to see upcoming directional changes and to make adjustments to avoid undershooting or overshooting the desired tool path; (4) balanced cutting tools, toolholders, and spindles to minimize vibration effects; (5) coolant delivery systems that provide pressures much greater than in conventional machining; and (6) chip control and removal systems to cope with the much larger metal removal rates in HSM. Also important are the cutting tool materials. As listed in Table 16.1, various tool materials are used for high-speed machining, and these materials are discussed in the following chapter.

16.8 TOLERANCES AND SURFACE FINISH

Machining operations are used to produce parts with defined geometries to tolerances and surface finishes specified by the product designer. In this section we examine these issues of tolerance and surface finish in machining.

16.8.1 TOLERANCES IN MACHINING

There is variability in any manufacturing process, and tolerances are used to set permissible limits on this variability (Section 4.1.1). Machining is often selected when tolerances are close, because it is more accurate than most other shape-making processes.

TABLE 16.2 Typical tolerances and surface roughness values (arithmetic average) achievable in machining operations

Machining Operation	Tolerance Capability—Typical mm	in	Surface Roughness AA—Typical μm	μ-in	Machining Operation	Tolerance Capability—Typical mm	in	Surface Roughness AA—Typical μm	μ-in
Turning, boring			0.8	32	Reaming			0.4	16
Diameter $D < 25$ mm	±0.025	±0.001			Diameter $D < 12$ mm	±0.025	±0.001		
25 mm $< D <$ 50 mm	±0.05	±0.002			12 mm $< D <$ 25 mm	±0.05	±0.002		
Diameter $D > 50$ mm	±0.075	±0.003			Diameter $D > 25$ mm	±0.075	±0.003		
Drilling*			0.8	32	Milling			0.4	16
Diameter $D < 2.5$ mm	±0.05	±0.002			Peripheral	±0.025	±0.001		
2.5 mm $< D <$ 6 mm	±0.075	±0.003			Face	±0.025	±0.001		
6 mm $< D <$ 12 mm	±0.10	±0.004			End	±0.05	±0.002		
12 mm $< D <$ 25 mm	±0.125	±0.005			Shaping, slotting	±0.025	±0.001	1.6	63
Diameter $D > 25$ mm	±0.20	±0.008			Planing	±0.075	±0.003	1.6	63
Broaching	±0.025	±0.001	0.2	8	Sawing	±0.50	±0.02	6.0	250

*Drilling tolerances are typically expressed as biased bilateral tolerances (e.g., + 0.010/– 0.002). Values in this table are expressed as closest bilateral tolerance (e.g., ±0.006).
Compiled from various sources, including [8], [9], [10], [21], and other sources.

Table 16.2 indicates typical tolerances that can be achieved for most machining operations. It should be mentioned that the values in this tabulation represent ideal conditions, yet conditions that are readily achievable in a modern factory. If the machine tool is old and worn, process variability will likely be greater than the ideal, and these tolerances would be difficult to maintain. On the other hand, newer machine tools can achieve closer tolerances than those listed.

Tighter tolerances usually mean higher costs. For example, if the product designer specifies a tolerance of ±0.10 mm on a hole diameter of 6.0 mm, this tolerance could be achieved by a drilling operation, according to Table 16.2. However, if the designer specifies a tolerance of ±0.025 mm, then an additional reaming operation is needed to satisfy this tighter requirement. This is not to suggest that looser tolerances are always good. It often happens that closer tolerances and lower variability in the machining of the individual components will lead to fewer problems in assembly, final product testing, field service, and customer acceptance. Although these costs are not always as easy to quantify as direct manufacturing costs, they can nevertheless be significant. Tighter tolerances that push a factory to achieve better control over its manufacturing processes may lead to lower total operating costs for the company over the long run.

16.8.2 SURFACE FINISH IN MACHINING

Since machining is often the manufacturing process that determines the final geometry and dimensions of the part, it is also the process that determines the part's surface texture (Section 4.2.2). Table 16.2 lists typical surface roughness values that can be achieved in various machining operations. These finishes should be readily achievable by modern, well-maintained machine tools.

Let us examine how surface finish is determined in a machining operation. The roughness of a machined surface depends on many factors that can be grouped as follows: (1) geometric factors, (2) work material factors, and (3) vibration and machine tool factors. Our discussion of surface finish in this section examines these factors and their effects.

Geometric Factors These are the machining parameters that determine the surface geometry of a machined part. They include (1) type of machining operation; (2) cutting tool geometry, most importantly nose radius; and (3) feed. The surface geometry that would result from these factors is referred to as the "ideal" or "theoretical" surface roughness, which is the finish that would be obtained in the absence of work material, vibration, and machine tool factors.

Type of operation refers to the machining process used to generate the surface. For example, peripheral milling, facing milling, and shaping all produce a flat surface; however, the surface geometry is different for each operation because of differences in tool shape and the way the tool interacts with the surface. A sense of the differences can be seen in Figure 4.4 showing various possible lays of a surface.

Tool geometry and feed combine to form the surface geometry. The shape of the tool point is the important tool geometry factor. The effects can be seen for a single-point tool in Figure 16.36. With the same feed, a larger nose radius causes the feed marks to be less pronounced, thus leading to a better finish. If two feeds are compared with the same nose radius, the larger feed increases the separation between feed marks, leading to an increase in the value of ideal surface roughness. If feed rate is large enough and the nose radius is small enough so that the end cutting edge participates in creating the new surface, then the end cutting-edge angle will affect surface geometry. In this case, a higher ECEA will result in a higher surface roughness value. In theory, a zero ECEA would yield a perfectly smooth surface; however, imperfections in the tool, work material, and machining process preclude achieving such an ideal finish.

The effects of nose radius and feed can be combined in an equation to predict the ideal average roughness for a surface produced by a single-point tool. The equation

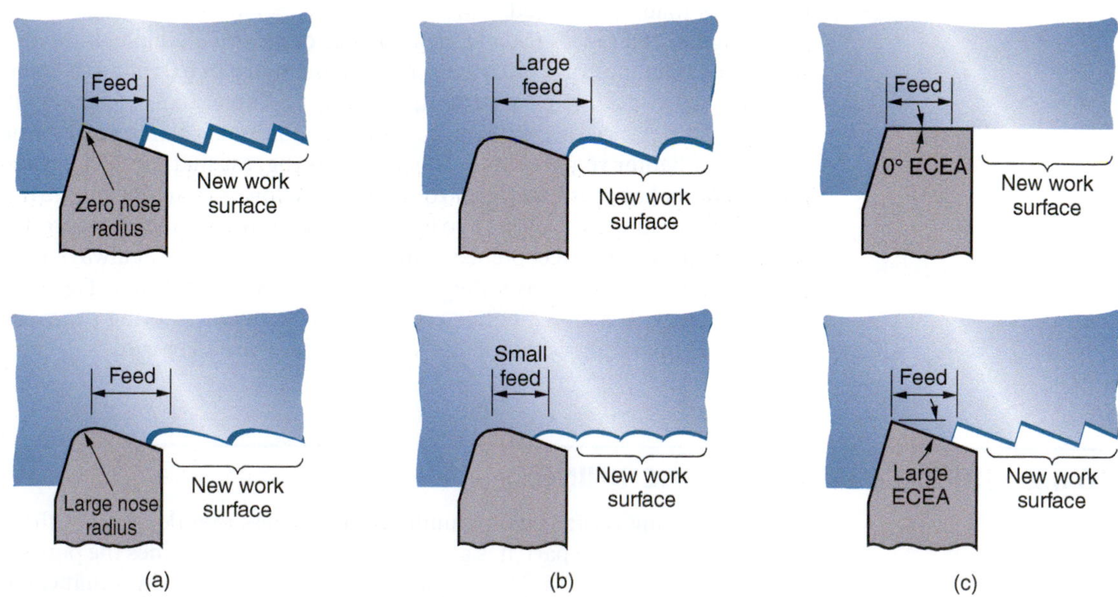

FIGURE 16.36 Effect of geometric factors in determining the theoretical finish on a work surface for single-point tools: (a) effect of nose radius, (b) effect of feed, and (c) effect of end cutting-edge angle. (Credit: *Fundamentals of Modern Manufacturing*, 4th Edition by Mikell P. Groover, 2010. Reprinted with permission of John Wiley & Sons, Inc.)

applies to operations such as turning, shaping, and planing:

$$R_i = \frac{f^2}{32NR} \tag{16.21}$$

where R_i = theoretical arithmetic average surface roughness, mm (in); f = feed, mm (in); and NR = nose radius on the tool point, mm (in). The equation assumes that the nose radius is not zero and that feed and nose radius will be the principal factors that determine the geometry of the surface. The values for R_i will be in units of mm (in), which can be converted to μm (μ-in). Eq. (16.21) can also be used to estimate the ideal surface roughness in face milling with insert tooling, using f to represent the chip load (feed per tooth).

Eq. (16.21) assumes a sharp cutting tool. As the tool wears, the shape of the cutting point changes, which is reflected in the geometry of the work surface. For slight amounts of tool wear, the effect is not noticeable. However, when tool wear becomes significant, especially nose radius wear, surface roughness deteriorates compared with the ideal values given by the preceding equations.

Work Material Factors Achieving the ideal surface finish is not possible in most machining operations because of factors related to the work material and its interaction with the tool. Work material factors that affect finish include (1) built-up edge effects—as the BUE cyclically forms and breaks away, particles are deposited on the newly created work surface, causing it to have a rough "sandpaper" texture; (2) damage to the surface caused by the chip curling back into the work; (3) tearing of the work surface during chip formation when machining ductile materials; (4) cracks in the surface caused by discontinuous chip formation when machining brittle materials; and (5) friction between the tool flank and the newly generated work surface. These work material factors are influenced by cutting speed and rake angle, such that an increase in cutting speed or rake angle generally improves surface finish.

The work material factors usually cause the actual surface finish to be worse than the ideal. An empirical ratio can be developed to convert the ideal roughness value into an estimate of the actual surface roughness value. This ratio takes into account BUE formation, tearing, and other factors. The value of the ratio depends on cutting speed as well as work material. Figure 16.37 shows the ratio of actual to ideal surface roughness as a function of speed for several classes of work material.

The procedure for predicting the actual surface roughness in a machining operation is to compute the ideal surface roughness value and then multiply this value by the ratio of actual to ideal roughness for the appropriate class of work material. This can be summarized as:

$$R_a = r_{ai} R_i \tag{16.22}$$

where R_a = the estimated value of actual roughness; r_{ai} = ratio of actual to ideal surface finish from Figure 16.37, and R_i = ideal roughness value from Eq. (16.21).

Example 16.1 Surface Roughness

A turning operation is performed on C1008 steel (a relatively ductile material) using a tool with a nose radius = 1.2 mm. Cutting speed = 100 m/min, and feed = 0.25 mm/rev. Compute an estimate of the surface roughness in this operation.

Solution: The ideal surface roughness can be calculated from Eq. (16.21):

$$R_i = (0.25)^2/(32 \times 1.2) = 0.0016\,\text{mm} = 1.6\,\mu\text{m}$$

From the chart in Figure 16.37, the ratio of actual to ideal roughness for ductile metals at 100 m/min is approximately 1.25. Accordingly, the actual surface roughness for the operation would be (approximately)

$$R_a = 1.25 \times 1.6 = 2.0\,\mu\text{m}$$

■

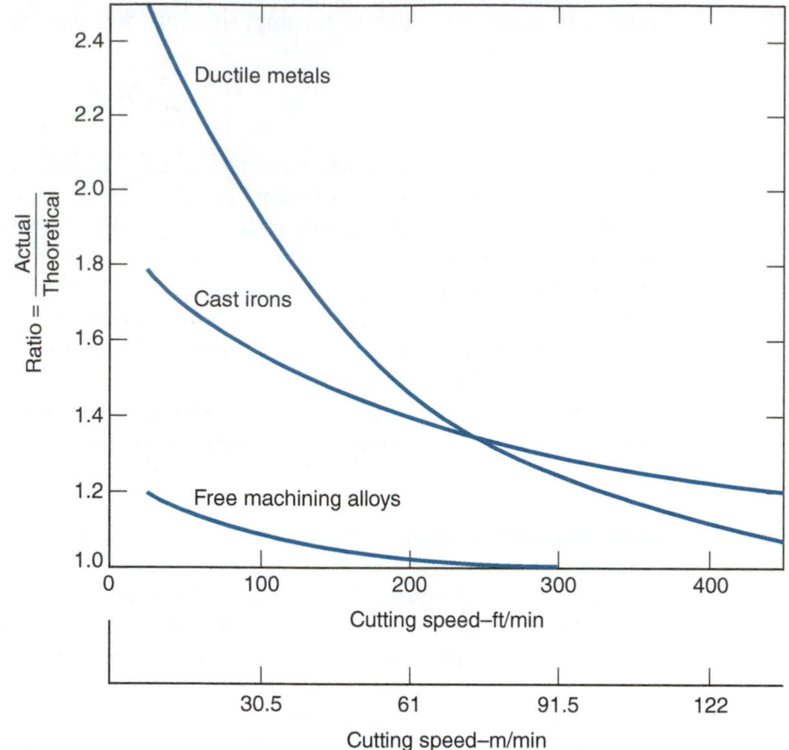

FIGURE 16.37 Ratio of actual surface roughness to ideal surface roughness for several classes of materials. *Source*: General Electric Co. data [19]. (Credit: *Fundamentals of Modern Manufacturing*, 4th Edition by Mikell P. Groover, 2010. Reprinted with permission of John Wiley & Sons, Inc.)

Vibration and Machine Tool Factors These factors are related to the machine tool, tooling, and setup in the operation. They include chatter or vibration in the machine tool or cutting tool; deflections in the fixturing, often resulting in vibration; and backlash in the feed mechanism, particularly on older machine tools. If these machine tool factors can be minimized or eliminated, the surface roughness in machining will be determined primarily by geometric and work material factors described above.

Chatter or vibration in a machining operation can result in pronounced waviness in the work surface. When chatter occurs, a distinctive noise occurs that can be recognized by any experienced machinist. Possible steps to reduce or eliminate vibration include (1) adding stiffness and/or damping to the setup, (2) operating at speeds that do not cause cyclical forces whose frequency approaches the natural frequency of the machine tool system, (3) reducing feeds and depths to reduce forces in cutting, and (4) changing the cutter design to reduce forces. Workpiece geometry can sometimes play a role in chatter. Thin cross sections tend to increase the likelihood of chatter, requiring additional supports to alleviate the condition.

16.9 PRODUCT DESIGN CONSIDERATIONS IN MACHINING

Several aspects of product design have already been considered in our preceding discussion of tolerance and surface finish. In this section, we present some design guidelines for machining, compiled from sources [5], [8], and [21]:

- If possible, parts should be designed that do not need machining. If this is not possible, then minimize the amount of machining required on the parts. In general, a

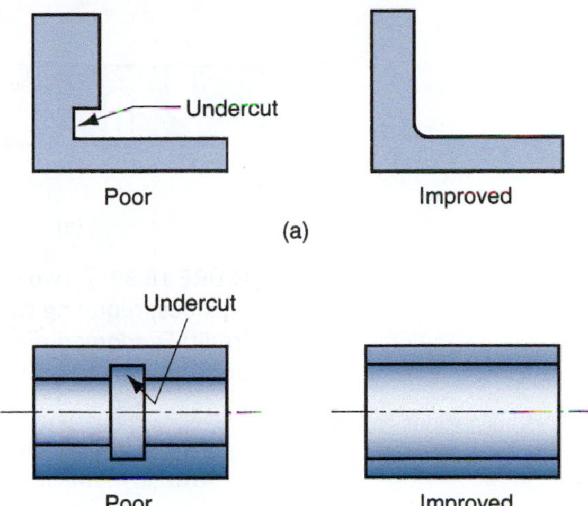

FIGURE 16.38 Two machined parts with undercuts: cross sections of (a) bracket and (b) rotational part. Also shown is how the part design might be improved. (Credit: *Fundamentals of Modern Manufacturing*, 4th Edition by Mikell P. Groover, 2010. Reprinted with permission of John Wiley & Sons, Inc.)

lower-cost product is achieved through the use of net shape processes such as precision casting, closed-die forging, or (plastic) molding; or near net shape processes such as impression-die forging. Reasons why machining may be required include close tolerances; good surface finish; and special geometric features such as threads, precision holes, cylindrical sections with a high degree of roundness, and similar shapes that cannot be achieved except by machining.

- Tolerances should be specified to satisfy functional requirements, but process capabilities should also be considered. See Table 16.2 for tolerance capabilities in machining. Excessively close tolerances add cost but may not add value to the part. As tolerances become tighter (smaller), product costs generally increase due to additional processing, fixturing, inspection, sortation, rework, and scrap.
- Surface finish should be specified to meet functional and/or aesthetic requirements, but better finishes generally increase processing costs by requiring additional operations such as grinding or lapping.
- Machined features such as sharp corners, edges, and points should be avoided; they are often difficult to accomplish by machining. Sharp internal corners require pointed cutting tools that tend to break during machining. Sharp external corners and edges tend to create burrs and are dangerous to handle.
- Deep holes that must be bored should be avoided. Deep hole boring requires a long boring bar. Boring bars must be stiff, and this often requires use of high modulus materials such as cemented carbide, which is expensive.
- Machined parts should be designed so they can be produced from standard available stock. Choose exterior dimensions equal to or close to the standard stock size to minimize machining; for example, rotational parts with outside diameters that are equal to standard bar stock diameters.
- Parts should be designed to be rigid enough to withstand forces of cutting and workholder clamping. Machining of long narrow parts, large flat parts, parts with thin walls, and similar shapes should be avoided if possible.
- Undercuts as in Figure 16.38 should be avoided since they often require additional setups and operations and/or special tooling; they can also lead to stress concentrations in service.
- Materials with good machinability should be selected by the designer (Section 17.5). As a rough guide, the machinability rating of a material correlates with the allowable

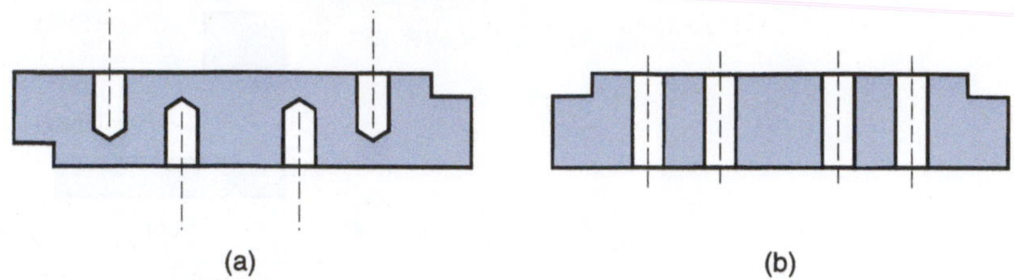

FIGURE 16.39 Two parts with similar hole features: (a) holes that must be machined from two sides, requiring two setups, and (b) holes that can all be machined from one side. (Credit: *Fundamentals of Modern Manufacturing*, 4th Edition by Mikell P. Groover, 2010. Reprinted with permission of John Wiley & Sons, Inc.)

cutting speed and production rate that can be used. Thus, parts made of materials with low machinability cost more to produce. Parts that are hardened by heat treatment must usually be finish ground or machined with higher cost tools after hardening to achieve final size and tolerance.

- Machined parts should be designed with features that can be produced in a minimum number of setups—one setup if possible. This usually means geometric features that can be accessed from one side of the part (see Figure 16.39).
- Machined parts should be designed with features that can be achieved with standard cutting tools. This means avoiding unusual hole sizes, threads, and features with unusual shapes requiring special form tools. In addition, it is helpful to design parts such that the number of individual cutting tools needed in machining is minimized; this often allows the part to be completed in one setup on a machine such as a machining center (Section 16.5).

REFERENCES

[1] Aronson, R. B. "Spindles are the Key to HSM," ***Manufacturing Engineering***, October 2004, pp. 67–80.

[2] Aronson, R. B. "Multitalented Machine Tools," ***Manufacturing Engineering***, January 2005, pp. 65–75.

[3] Ashley, S. "High-speed Machining Goes Mainstream," ***Mechanical Engineering***, May 1995 pp. 56–61.

[4] ***ASM Handbook***, Vol. **16**, ***Machining***. ASM International, Materials Park, Ohio, 1989.

[5] Bakerjian, R. (ed.). ***Tool and Manufacturing Engineers Handbook***, 4th ed., Vol. **VI**, ***Design for Manufacturability***. Society of Manufacturing Engineers, Dearborn, Michigan, 1992.

[6] Black, J., and Kohser, R. ***DeGarmo's Materials and Processes in Manufacturing***, 10th ed. John Wiley & Sons, Inc., Hoboken, New Jersey, 2008.

[7] Boston, O. W. ***Metal Processing***, 2nd ed. John Wiley & Sons, Inc., New York, 1951.

[8] Bralla, J. G. (ed.). ***Design for Manufacturability Handbook***, 2nd ed. McGraw-Hill Book Company, New York, 1998.

[9] Drozda, T. J., and Wick, C. (eds.). ***Tool and Manufacturing Engineers Handbook***, 4th ed. Vol. **I**, ***Machining***. Society of Manufacturing Engineers, Dearborn, Michigan, 1983.

[10] Early, D. F., and Johnson, G. E. ***Process Engineering for Manufacturing***. Prentice-Hall, Inc., Englewood Cliffs, New Jersey, 1962.

[11] Kalpakjian, S., and Schmid, S. R. ***Manufacturing Engineering and Technology***, 4th ed. Prentice Hall, Upper Saddle River, New Jersey, 2003.

[12] Kalpakjian, S., and Schmid, S. R. ***Manufacturing Processes for Engineering Materials***, 6th ed. Pearson Prentice Hall, Upper Saddle River, New Jersey, 2010.

[13] Krar, S. F., and Ratterman, E. ***Superabrasives: Grinding and Machining with CBN and Diamond***. McGraw-Hill, Inc., New York, 1990.

[14] Lindberg, R. A. ***Processes and Materials of Manufacture***, 4th ed. Allyn and Bacon, Inc., Boston, 1990.

[15] Marinac, D. "Smart Tool Paths for HSM," ***Manufacturing Engineering***, November 2000, pp. 44–50.

[16] Mason, F., and Freeman, N. B. "Turning Centers Come of Age," Special Report 773, ***American Machinist***, February 1985, pp. 97–116.

[17] ***Modern Metal Cutting***. AB Sandvik Coromant, Sandvik, Sweden, 1994.

[18] Ostwald, P. F., and Munoz, J. ***Manufacturing Processes and Systems***, 9th ed. John Wiley & Sons, Inc. New York, 1997.

[19] ***Surface Finish***. Machining Development Service, Publication A-5, General Electric Company, Schenectady, New York (no date).

[20] Trent, E. M., and Wright, P. K. ***Metal Cutting***, 4th ed. Butterworth Heinemann, Boston, 2000.

[21] Trucks, H. E., and Lewis, G. ***Designing for Economical Production***, 2nd ed. Society of Manufacturing Engineers, Dearborn, Michigan, 1987.

[22] Witkorski, M., and Bingeman, A. "The Case for Multiple Spindle HMCs," ***Manufacturing Engineering***, March 2004, pp. 139–148.

REVIEW QUESTIONS

16.1. What are the differences between rotational parts and prismatic parts in machining?

16.2. Distinguish between generating and forming when machining workpart geometries.

16.3. Give two examples of machining operations in which generating and forming are combined to create workpart geometry.

16.4. Describe the turning process.

16.5. What is the difference between threading and tapping?

16.6. How does a boring operation differ from a turning operation?

16.7. What is meant by the designation 12 × 36 inch lathe?

16.8. Name the various ways in which a workpart can be held in a lathe.

16.9. What is the difference between a live center and a dead center, when these terms are used in the context of workholding in a lathe?

16.10. How does a turret lathe differ from an engine lathe?

16.11. What is a blind hole?

16.12. What is the distinguishing feature of a radial drill press?

16.13. What is the difference between peripheral milling and face milling?

16.14. Describe profile milling.

16.15. What is pocket milling?

16.16. Describe the difference between up milling and down milling.

16.17. How does a universal milling machine differ from a conventional knee-and-column machine?

16.18. What is a machining center?

16.19. What is the difference between a machining center and a turning center?

16.20. What can a mill-turn center do that a conventional turning center cannot do?

16.21. How do shaping and planing differ?

16.22. What is the difference between internal broaching and external broaching?

16.23. Identify the three basic forms of sawing operation.

16.24. Why do costs tend to increase when better surface finish is required on a machined part?

16.25. What are the basic factors that affect surface finish in machining?

16.26. What are the parameters that have the greatest influence in determining the ideal surface roughness R_i in a turning operation?

16.27. Name some of the steps that can be taken to reduce or eliminate vibrations in machining.

PROBLEMS

16.1. A cylindrical workpart 200 mm in diameter and 700 mm long is to be turned in an engine lathe. Cutting speed = 2.30 m/s, feed = 0.32 mm/rev, and depth of cut = 1.80 mm. Determine (a) cutting time, and (b) metal removal rate.

16.2. In a production turning operation, the foreman has decreed that a single pass must be completed on the cylindrical workpiece in 5.0 min. The piece is 400 mm long and 150 mm in diameter. Using a feed = 0.30 mm/rev and a depth of cut = 4.0 mm, what cutting speed must be used to meet this machining time requirement?

16.3. A tapered surface is to be turned on an automatic lathe. The workpiece is 750 mm long with minimum and maximum diameters of 100 mm and 200 mm at opposite ends. The automatic controls on the lathe permit the surface speed to be maintained at a constant value of 200 m/min by adjusting the rotational speed as a function of workpiece diameter. Feed = 0.25 mm/rev and depth of cut =

3.0 mm. The rough geometry of the piece has already been formed, and this operation will be the final cut. Determine (a) the time required to turn the taper and (b) the rotational speeds at the beginning and end of the cut.

16.4. A cylindrical work bar with 4.5 in diameter and 52 in length is chucked in an engine lathe and supported at the opposite end using a live center. A 46.0 in portion of the length is to be turned to a diameter of 4.25 in one pass at a speed of 450 ft/min. The metal removal rate should be 6.75 in^3/min. Determine (a) the required depth of cut, (b) the required feed, and (c) the cutting time.

16.5. The end of a large tubular workpart is to be faced on a NC vertical boring mill. The part has an outside diameter of 38.0 in and an inside diameter of 24.0 in. If the facing operation is performed at a rotational speed of 40.0 rev/min, feed of 0.015 in/rev, and depth of cut of 0.180 in, determine (a) the cutting time to complete the facing operation and (b) the cutting speeds and metal removal rates at the beginning and end of the cut.

16.6. A drilling operation is to be performed with a 12.7 mm diameter twist drill in a steel workpart. The hole is a blind hole at a depth of 60 mm and the point angle is 118°. The cutting speed is 25 m/min and the feed is 0.30 mm/rev. Determine (a) the cutting time to complete the drilling operation, and (b) metal removal rate during the operation, after the drill bit reaches full diameter.

16.7. A gun-drilling operation is used to drill a 9/64-in diameter hole to a certain depth. It takes 4.5 minutes to perform the drilling operation using high-pressure fluid delivery of coolant to the drill point. The current spindle speed = 4000 rev/min, and feed = 0.0017 in/rev. In order to improve the surface finish in the hole, it has been decided to increase the speed by 20% and decrease the feed by 25%. How long will it take to perform the operation at the new cutting conditions?

16.8. A peripheral milling operation is performed on the top surface of a rectangular workpart, which is 400 mm long by 60 mm wide. The milling cutter, which is 80 mm in diameter and has five teeth, overhangs the width of the part on both sides. Cutting speed = 70 m/min, chip load = 0.25 mm/tooth, and depth of cut = 5.0 mm. Determine (a) the actual machining time to make one pass across the surface and (b) the maximum material removal rate during the cut.

16.9. A face milling operation is used to machine 6.0 mm from the top surface of a rectangular piece of aluminum 300 mm long by 125 mm wide in a single pass. The cutter follows a path that is centered over the workpiece. It has four teeth and is 150 mm in diameter. Cutting speed = 2.8 m/s, and chip load = 0.27 mm/tooth. Determine (a) the actual machining time to make the pass across the surface and (b) the maximum metal removal rate during cutting.

16.10. A slab milling operation is performed on the top surface of a steel rectangular workpiece 12.0 in long by 2.5 in wide. The helical milling cutter, which has a 3.0 in diameter and ten teeth, is set up to overhang the width of the part on both sides. Cutting speed is 125 ft/min, feed is 0.006 in/tooth, and depth of cut = 0.300 in. Determine (a) the actual machining time to make one pass across the surface and (b) the maximum metal removal rate during the cut. (c) If an additional approach distance of 0.5 in is provided at the beginning of the pass (before cutting begins), and an overtravel distance is provided at the end of the pass equal to the cutter radius plus 0.5 in, what is the duration of the feed motion?

16.11. A face milling operation is performed on the top surface of a steel rectangular workpiece 12.0 in long by 2.5 in wide. The milling cutter follows a path that is centered over the workpiece. It has five teeth and a 3.0 in diameter. Cutting speed = 250 ft/min, feed = 0.006 in/tooth, and depth of cut = 0.150 in. Determine (a) the actual cutting time to make one pass across the surface and (b) the maximum metal removal rate during the cut. (c) If an additional approach distance of 0.5 in is provided at the beginning of the pass (before cutting begins), and an overtravel distance is provided at the end of the pass equal to the cutter radius plus 0.5 in, what is the duration of the feed motion?

16.12. Solve Problem 16.11, except that the workpiece is 5.0 in wide and the cutter is offset to one side so that the swath cut by the cutter = 1.0 in wide. This is called partial face milling, Figure 16.20(b).

16.13. In a turning operation on cast iron, the nose radius on the single-point tool = 1.5 mm, feed = 0.22 mm/rev, and speed = 1.8 m/s. Compute an estimate of the surface roughness for this cut.

16.14. A turning operation uses a 2/64 in nose radius cutting tool on a free machining steel with a feed rate = 0.010 in/rev and a cutting speed = 300 ft/min. Determine the surface roughness for this cut.

16.15. A single-point HSS tool with a 3/64 in nose radius is used in a shaping operation on a ductile steel workpart. The cutting speed is 120 ft/min. The feed is 0.014 in/stroke and depth of cut is 0.135 in. Determine the surface roughness for this operation.

16.16. A part to be turned in an engine lathe must have a surface finish of 1.6 μm. The part is made of a free-machining aluminum alloy. Cutting speed = 150 m/min, and depth of cut = 4.0 mm. The nose radius on the tool = 0.75 mm. Determine the feed that will achieve the specified surface finish.

16.17. Solve previous Problem 16.16, except that the part is made of cast iron instead of aluminum and the cutting speed is reduced to 100 m/min.

16.18. A face milling operation is to be performed on a cast iron part to finish the surface to 36 μ-in. The cutter uses four inserts and its diameter is 3.0 in. The cutter rotates at 475 rev/min. To obtain the best possible finish, a type of carbide insert with 4/64 in nose radius is to be used. Determine the required feed rate (in/min) that will achieve the 36 μ-in finish.

16.19. A face milling operation is not yielding the required surface finish on the work. The cutter is a four-tooth insert type face milling cutter. The machine shop foreman thinks the problem is that the work material is too ductile for the job, but this property tests well within the ductility range for the material specified by the designer. Without knowing any more about the job, what changes in (a) cutting conditions and (b) tooling would you suggest to improve the surface finish?

17 CUTTING-TOOL TECHNOLOGY AND RELATED TOPICS

Chapter Contents

17.1 Tool Life
- 17.1.1 Tool Wear
- 17.1.2 Tool Life and the Taylor Tool Life Equation

17.2 Tool Materials
- 17.2.1 High-Speed Steel and Its Predecessors
- 17.2.2 Cast Cobalt Alloys
- 17.2.3 Cemented Carbides, Cermets, and Coated Carbides
- 17.2.4 Ceramics
- 17.2.5 Synthetic Diamonds and Cubic Boron Nitride

17.3 Tool Geometry
- 17.3.1 Single-Point Tool Geometry
- 17.3.2 Multiple-Cutting-Edge Tools

17.4 Cutting Fluids
- 17.4.1 Types of Cutting Fluids
- 17.4.2 Application of Cutting Fluids

17.5 Machinability

17.6 Machining Economics
- 17.6.1 Selecting Feed and Depth of Cut
- 17.6.2 Cutting Speed

Machining operations are accomplished using cutting tools. The high forces and temperatures during machining create a very harsh environment for the tool. If cutting force becomes too high, the tool fractures. If cutting temperature becomes too high, the tool material softens and fails. If neither of these conditions causes the tool to fail, continual wear of the cutting edge ultimately leads to failure.

Cutting-tool technology has two principal aspects: tool material and tool geometry. The first is concerned with developing materials that can withstand the forces, temperatures, and wearing action in the machining process. The second deals with optimizing the geometry of the cutting tool for the tool material and for a given operation. It is appropriate to begin by considering tool life, which is a prerequisite for much of our subsequent discussion on tool materials. The chapter also includes several additional topics that are related to cutting-tool technology: cutting fluids, machinability, and machining economics.

17.1 TOOL LIFE

As suggested by our opening paragraph, there are three possible modes by which a cutting tool can fail in machining:

1. ***Fracture failure***. This mode of failure occurs when the cutting force at the tool point becomes excessive, causing it to fail suddenly by brittle fracture.
2. ***Temperature failure***. This failure occurs when the cutting temperature is too high for the tool material, causing the material at the tool point to soften, which leads to plastic deformation and loss of the sharp edge.
3. ***Gradual wear***. Gradual wearing of the cutting edge causes loss of tool shape, reduction in cutting efficiency, acceleration of wearing as the tool becomes heavily worn, and finally tool failure in a manner similar to a temperature failure.

Fracture and temperature failures result in premature loss of the cutting tool. These two modes of failure are therefore undesirable. Of the three possible tool failures, gradual wear is preferred because it leads to the longest possible use of the tool, with the associated economic advantage of that longer use.

Product quality must also be considered when attempting to control the mode of tool failure. When the tool point fails suddenly during a cut, it often causes damage to the work surface. This damage requires either rework of the surface or possible scrapping of the part. The damage can be avoided by selecting cutting conditions that favor gradual wearing of the tool rather than fracture or temperature failure, and by changing the tool before the final catastrophic loss of the cutting edge occurs.

17.1.1 TOOL WEAR

Gradual wear occurs at two principal locations on a cutting tool: the top rake face and the flank. Accordingly, two main types of tool wear can be distinguished: crater wear and flank wear, illustrated in Figure 17.1 for a single-point tool. ***Crater wear*** consists of a cavity in the rake face of the tool that forms and grows from the action of the chip sliding against the surface. High stresses and temperatures characterize the tool–chip contact interface, contributing to the wearing action. The crater can be measured either by its depth or its area. ***Flank wear*** occurs on the flank, or relief face, of the tool. It results from rubbing between the newly generated work surface and the flank face adjacent to the cutting edge. Flank wear is measured by the width of the wear band, FW. This wear band is sometimes called the flank wear ***land***.

Certain features of flank wear can be identified. First, an extreme condition of flank wear often appears on the cutting edge at the location corresponding to the original surface of the workpart. This is called ***notch wear***. It occurs because the original work surface is harder and/or more abrasive than the internal material, which could be caused by work hardening from cold drawing or previous machining, sand particles in the surface from casting, or other reasons. As a consequence of the harder surface, wear is accelerated at this location. A second region of flank wear that can be identified is ***nose radius wear***; this occurs on the nose radius leading into the end cutting edge.

The mechanisms that cause wear at the tool–chip and tool–work interfaces in machining can be summarized as follows:

➢ ***Abrasion***. This is a mechanical wearing action caused by hard particles in the work material gouging and removing small portions of the tool. This abrasive action occurs in both flank wear and crater wear; it is a significant cause of flank wear.

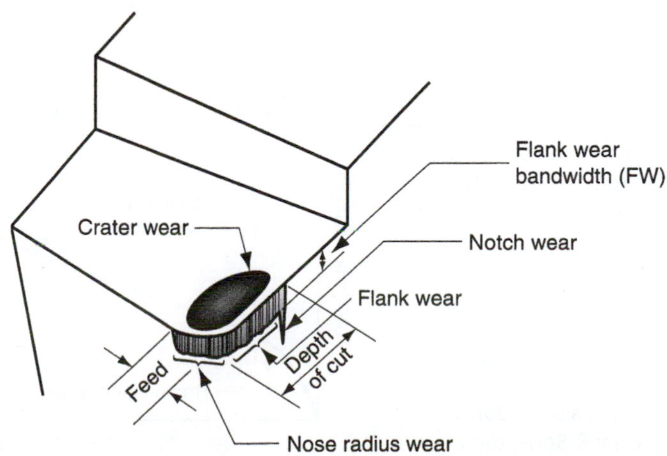

FIGURE 17.1 Diagram of worn cutting tool, showing the principal locations and types of wear that occur. (Credit: *Fundamentals of Modern Manufacturing*, 4th Edition by Mikell P. Groover, 2010. Reprinted with permission of John Wiley & Sons, Inc.)

- ***Adhesion***. When two metals are forced into contact under high pressure and temperature, adhesion (welding) occurs between them. These conditions are present between the chip and the rake face of the tool. As the chip flows across the tool, small particles of the tool adhere to the chip and are broken away from the surface, resulting in attrition of the surface.
- ***Diffusion***. This is a process in which an exchange of atoms takes place across a close contact boundary between two materials. In the case of tool wear, diffusion occurs at the tool–chip boundary, causing the tool surface to become depleted of the atoms responsible for its hardness. As this process continues, the tool surface becomes more susceptible to abrasion and adhesion. Diffusion is believed to be a principal mechanism of crater wear.
- ***Chemical reactions***. The high temperatures and clean surfaces at the tool–chip interface in machining at high speeds can result in chemical reactions, in particular, oxidation, on the rake face of the tool. The oxidized layer, being softer than the parent tool material, is sheared away, exposing new material to sustain the reaction process.
- ***Plastic deformation***. Another mechanism that contributes to tool wear is plastic deformation of the cutting edge. The cutting forces acting on the cutting edge at high temperature cause the edge to deform plastically, making it more vulnerable to abrasion of the tool surface. Plastic deformation contributes mainly to flank wear.

Most of these tool-wear mechanisms are accelerated at higher cutting speeds and temperatures. Diffusion and chemical reaction are especially sensitive to elevated temperature.

17.1.2 TOOL LIFE AND THE TAYLOR TOOL LIFE EQUATION

As cutting proceeds, the various wear mechanisms result in increasing levels of wear on the cutting tool. The general relationship of tool wear versus cutting time is shown in Figure 17.2. Although the relationship shown is for flank wear, a similar relationship occurs for crater wear. Three regions can usually be identified in the typical wear growth curve. The first is the ***break-in period***, in which the sharp cutting edge wears rapidly at the beginning of its use. This first region occurs within the first few minutes of cutting. The break-in period is followed by wear that occurs at a fairly uniform rate. This is the ***steady-state wear*** region. In our figure, this region is pictured as a linear function of time,

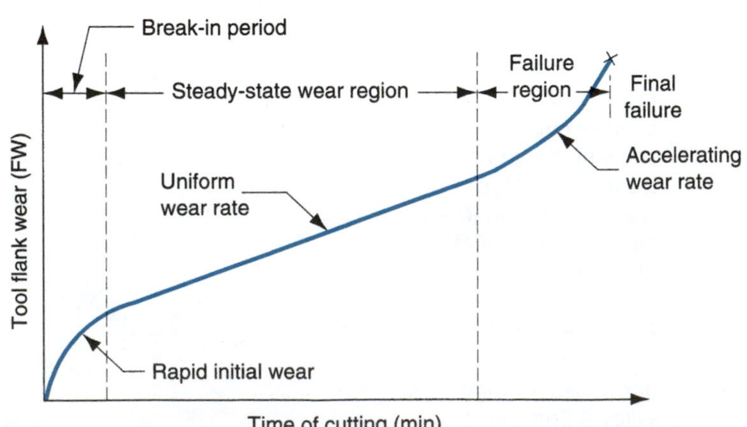

FIGURE 17.2 Tool wear as a function of cutting time. Flank wear (FW) is used here as the measure of tool wear. Crater wear follows a similar growth curve. (Credit: *Fundamentals of Modern Manufacturing*, 4th Edition by Mikell P. Groover, 2010. Reprinted with permission of John Wiley & Sons, Inc.)

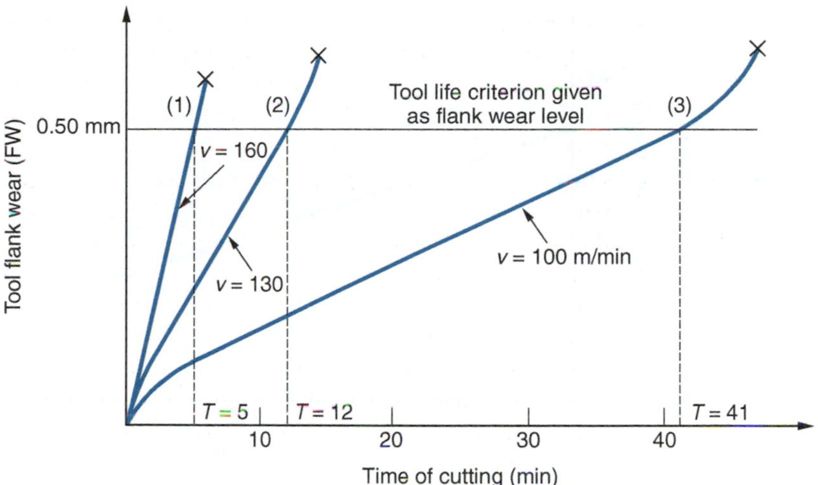

FIGURE 17.3 Effect of cutting speed on tool flank wear (FW) for three cutting speeds. Hypothetical values of speed and tool life are shown for a tool life criterion of 0.50-mm flank wear. (Credit: *Fundamentals of Modern Manufacturing*, 4th Edition by Mikell P. Groover, 2010. Reprinted with permission of John Wiley & Sons, Inc.)

although there are deviations from the straight line in actual machining. Finally, wear reaches a level at which the wear rate begins to accelerate. This marks the beginning of the ***failure region***, in which cutting temperatures are higher, and the general efficiency of the machining process is reduced. If allowed to continue, the tool finally fails by temperature failure.

The slope of the tool-wear curve in the steady-state region is affected by work material and cutting conditions. Harder work materials cause the wear rate (slope of the tool-wear curve) to increase. Increased speed, feed, and depth of cut have a similar effect, with speed being the most important of the three. If the tool-wear curves are plotted for several different cutting speeds, the results appear as in Figure 17.3. As cutting speed is increased, wear rate increases so the same level of wear is reached in less time.

Tool life is defined as the length of cutting time that the tool can be used. Operating the tool until final catastrophic failure is one way of defining tool life. This is indicated in Figure 17.3 by the end of each tool-wear curve. However, in production, it is often a disadvantage to use the tool until this failure occurs because of difficulties in resharpening the tool and problems with work surface quality. As an alternative, a level of tool wear can be selected as a criterion of tool life, and the tool is replaced when wear reaches that level. A convenient tool life criterion is a certain flank-wear value, such as 0.5 mm (0.020 in), illustrated as the horizontal line on the graph. When each of the three wear curves intersects that line, the life of the corresponding tool is defined as ended. If the intersection points are projected down to the time axis, the values of tool life can be identified, as we have done.

If the tool life values for the three wear curves in Figure 17.3 are plotted on a natural log–log graph of cutting speed versus tool life, the resulting relationship is a straight line as shown in Figure 17.4.[1]

The discovery of this relationship around 1900 is credited to F. W. Taylor. It can be expressed in equation form and is called the Taylor tool life equation:

$$vT^n = C \tag{17.1}$$

[1] The reader may have noted in Figure 17.4 that the dependent variable (tool life) has been plotted on the horizontal axis and the independent variable (cutting speed) on the vertical axis. Although this is a reversal of the normal plotting convention, it nevertheless is the way the Taylor tool life relationship is usually presented.

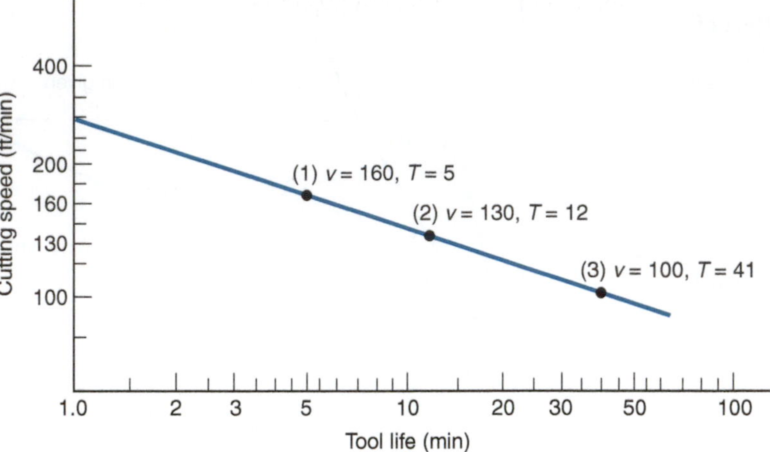

FIGURE 17.4 Natural log–log plot of cutting speed vs. tool life. (Credit: *Fundamentals of Modern Manufacturing*, 4th Edition by Mikell P. Groover, 2010. Reprinted with permission of John Wiley & Sons, Inc.)

where v = cutting speed, m/min (ft/min); T = tool life, min; and n and C are parameters whose values depend on feed, depth of cut, work material, tooling (material in particular), and the tool life criterion used. The value of n is relative constant for a given tool material, whereas the value of C depends on tool material, work material, and cutting conditions.

Basically, Eq. (17.1) states that higher cutting speeds result in shorter tool lives. Relating the parameters n and C to Figure 17.4, n is the slope of the plot (expressed in linear terms rather than in the scale of the axes), and C is the intercept on the speed axis. C represents the cutting speed that results in a 1-min tool life.

The problem with Eq. (17.1) is that the units on the right-hand side of the equation are not consistent with the units on the left-hand side. To make the units consistent, the equation should be expressed in the form:

$$vT^n = C(T_{\text{ref}}^n) \tag{17.2}$$

where T_{ref} = a reference value for C. T_{ref} is simply 1 min when m/min (ft/min) and minutes are used for v and T, respectively. The advantage of Eq. (17.2) is seen when it is desired to use the Taylor equation with units other than m/min (ft/min) and minutes—for example, if cutting speed were expressed as m/sec and tool life as sec. In this case, T_{ref} would be 60 sec and C would therefore be the same speed value as in Eq. (17.1), although converted to units of m/sec. The slope n would have the same numerical value as in Eq. (17.1).

Example 17.1 Taylor Tool Life Equation

Determine the values of C and n in the plot of Figure 17.4, using two of the three points on the curve and solving simultaneous equations of the form of Eq. (17.1).

Solution: Choosing the two extreme points: v = 160 m/min, T = 5 min; and v = 100 m/min, T = 41 min; we have:

$$160(5)^n = C$$
$$100(41)^n = C$$

Setting the left-hand sides of each equation equal,

$$160(5)^n = 100(41)^n$$

Taking the natural logarithms of each term,

$$\begin{aligned}\ln(160) + n\ln(5) &= \ln(100) + n\ln(41)\\ 5.0752 + 1.6094\,n &= 4.6052 + 3.7136\,n\\ 0.4700 &= 2.1042\,n\\ n = \frac{0.4700}{2.1042} &= 0.223\end{aligned}$$

Substituting this value of n into either starting equation, we obtain the value of C:

$$C = 160(5)^{0.223} = 229$$

or

$$C = 100(41)^{0.223} = 229$$

The Taylor tool life equation for the data of Figure 17.4 is therefore:

$$vT^{0.223} = 229$$

■

An expanded version of Eq. (17.2) can be formulated to include the effects of feed, depth of cut, and even work material hardness:

$$vT^n f^m d^p H^p = KT^n_{\text{ref}} f^m_{\text{ref}} d^p_{\text{ref}} H^q_{\text{ref}} \tag{17.3}$$

where f = feed, mm (in); d = depth of cut, mm (in); H = hardness, expressed in an appropriate hardness scale; m, p, and q are exponents whose values are experimentally determined for the conditions of the operation; K = a constant analogous to C in Eq. (17.2); and f_{ref}, d_{ref}, and H_{ref} are reference values for feed, depth of cut, and hardness. The values of m and p, the exponents for feed and depth, are less than 1.0. This indicates the greater effect of cutting speed on tool life, because the exponent of v is 1.0. After speed, feed is next in importance, so m has a value greater than p. The exponent for work hardness, q, is also less than 1.0.

Perhaps the greatest difficulty in applying Eq. (17.3) in a practical machining operation is the tremendous amount of machining data that would be required to determine the parameters of the equation. Variations in work materials and testing conditions also cause difficulties by introducing statistical variations in the data. Eq. (17.3) is valid in indicating general trends among its variables, but not in its ability to accurately predict tool life performance. To reduce these problems and make the scope of the equation more manageable, some of the terms are usually eliminated. For example, omitting depth and hardness reduces Eq. (17.3) to the following:

$$vT^n f^m = KT^n_{\text{ref}} f^m_{\text{ref}} \tag{17.4}$$

where the terms have the same meaning as before, except that the constant K will have a slightly different interpretation.

Although flank wear is the tool life criterion in our preceding discussion of the Taylor equation, this criterion is not very practical in a factory environment because of the difficulties and time required to measure flank wear. Instead, some of the alternative tool life criteria that are used in production machining operations include: (1) visual inspection of the cutting edge by the machine operator to determine when to change the tool, (2) degradation of the surface finish on the work, (3) changing the tool after a certain number of workpieces have been completed, and (4) changing the tool when a certain cumulative cutting time for that tool has been reached.

17.2 TOOL MATERIALS

The three modes of tool failure allow us to identify three important properties required in a tool material:

- ***Toughness***. To avoid fracture failure, the tool material must possess high toughness. Toughness is the capacity of a material to absorb energy without failing. It is usually characterized by a combination of strength and ductility in the material.
- ***Hot hardness***. Hot hardness is the ability of a material to retain its hardness at high temperatures. This is required because of the high-temperature environment in which the tool operates.
- ***Wear resistance***. Hardness is the single most important property needed to resist abrasive wear. All cutting-tool materials must be hard. However, wear resistance in metal cutting depends on more than just tool hardness, because of the other tool-wear mechanisms. Other characteristics affecting wear resistance include surface finish on the tool (a smoother surface means a lower coefficient of friction), chemistry of tool and work materials, and whether a cutting fluid is used.

Cutting-tool materials achieve this combination of properties in varying degrees. In this section, the following cutting-tool materials are discussed: (1) high-speed steel and its predecessors, plain carbon and low alloy steels, (2) cast cobalt alloys, (3) cemented carbides, cermets, and coated carbides, (4) ceramics, (5) synthetic diamond and cubic boron nitride. Before examining these individual materials, a brief overview and technical comparison will be helpful. Commercially, the most important tool materials are high-speed steel and cemented carbides, cermets, and coated carbides. These two categories account for more than 90% of the cutting tools used in machining operations.

Table 17.1 and Figure 17.5 present data on properties of various tool materials. The properties are those related to the requirements of a cutting tool: hardness, toughness, and hot hardness. Table 17.1 lists room temperature hardness and transverse rupture strength for selected materials. Transverse rupture strength (Section 3.1.3) is a property used to indicate toughness for hard materials. Figure 17.5 shows hardness as a function of temperature for several of the tool materials discussed in this section.

TABLE 17.1 Typical hardness values (at room temperature) and transverse rupture strengths for various tool materials.[a]

Material	Hardness	Transverse Rupture Strength MPa	Transverse Rupture Strength lb/in^2
Plain carbon steel	60 HRC	5200	750,000
High-speed steel	65 HRC	4100	600,000
Cast cobalt alloy	65 HRC	2250	325,000
Cemented carbide (WC)			
Low Co content	93 HRA, 1800 HK	1400	200,000
High Co content	90 HRA, 1700 HK	2400	350,000
Cermet (TiC)	2400 HK	1700	250,000
Alumina (Al_2O_3)	2100 HK	400	60,000
Cubic boron nitride	5000 HK	700	100,000
Polycrystalline diamond	6000 HK	1000	150,000
Natural diamond	8000 HK	1500	215,000

Compiled from [7], [12], [20], and other sources.

[a]*Note*: The values of hardness and TRS are intended to be comparative and typical. Variations in properties result from differences in composition and processing.

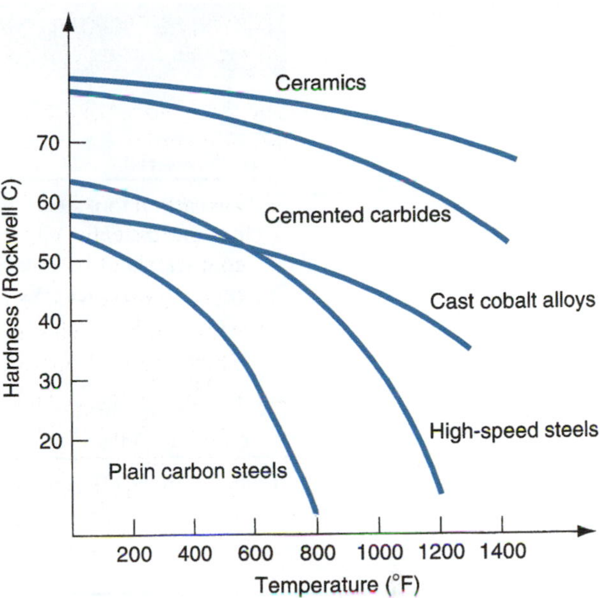

FIGURE 17.5 Typical hot hardness relationships for selected tool materials. Plain carbon steel shows a rapid loss of hardness as temperature increases. High-speed steel is substantially better, while cemented carbides and ceramics are significantly harder at elevated temperatures. (Credit: *Fundamentals of Modern Manufacturing*, 4th Edition by Mikell P. Groover, 2010. Reprinted with permission of John Wiley & Sons, Inc.)

In addition to these property comparisons, it is useful to compare the materials in terms of the parameters n and C in the Taylor tool life equation. In general, the development of new cutting-tool materials has resulted in increases in the values of these two parameters. Table 17.2 lists representative values of n and C in the Taylor tool life equation for selected cutting-tool materials.

The chronological development of tool materials has generally followed a path in which new materials have permitted higher and higher cutting speeds to be achieved. Table 17.3 identifies the cutting-tool materials, together with their approximate year of introduction and typical maximum allowable cutting speeds at which they can be used. Dramatic increases in machining productivity have been made possible due to advances in tool material technology, as indicated in our table. Machine tool practice has not always kept pace with cutting-tool technology. Limitations on horsepower, machine tool rigidity, spindle bearings, and the widespread use of older equipment in industry have acted to underutilize the possible upper speeds permitted by available cutting tools.

TABLE 17.2 Representative values of *n* and *C* in the Taylor tool life equation, Eq. (17.1), for selected tool materials

		C			
		Nonsteel Cutting		Steel Cutting	
Tool Material	*n*	m/min	(ft/min)	m/min	ft/min
Plain carbon tool steel	0.1	70	(200)	20	60
High-speed steel	0.125	120	(350)	70	200
Cemented carbide	0.25	900	(2700)	500	1500
Cermet	0.25			600	2000
Coated carbide	0.25			700	2200
Ceramic	0.6			3000	10,000

Compiled from [7], [12], and other sources.

The parameter values are approximated for turning at feed = 0.25 mm/rev (0.010 in/rev) and depth = 2.5 mm (0.100 in). Nonsteel cutting refers to easy-to-machine metals such as aluminum, brass, and cast iron. Steel cutting refers to the machining of mild (unhardened) steel. It should be noted that significant variations in these values can be expected in practice.

TABLE 17.3 Cutting-tool materials with their approximate dates of initial use and allowable cutting speeds

Tool Material	Year of Initial Use	Allowable Cutting Speed[a] Nonsteel Cutting m/min	ft/min	Steel Cutting m/min	ft/min
Plain carbon tool steel	1800s	Below 10	Below 30	Below 5	Below 15
High-speed steel	1900	25–65	75–200	17–33	50–100
Cast cobalt alloys	1915	50–200	150–600	33–100	100–300
Cemented carbides (WC)	1930	330–650	1000–2000	100–300	300–900
Cermets (TiC)	1950s			165–400	500–1200
Ceramics (Al_2O_3)	1955			330–650	1000–2000
Synthetic diamonds	1954, 1973	390–1300	1200–4000		
Cubic boron nitride	1969			500–800	1500–2500
Coated carbides	1970			165–400	500–1200

[a]Compiled from [12], [16], [19], [21], and other sources.

17.2.1 HIGH-SPEED STEEL AND ITS PREDECESSORS

Before the development of high-speed steel, plain carbon steel and Mushet's steel were the principal tool materials for metal cutting. Today, these steels are rarely used in industrial machining applications. The plain carbon steels used as cutting tools could be heat-treated to achieve relatively high hardness (Rockwell C 60), because of their fairly high-carbon content. However, because of low alloying levels, they possess poor hot hardness (Figure 17.5), which renders them unusable in metal cutting except at speeds too low to be practical by today's standards. Mushet's steel contained alloying elements tungsten (4% to 12%) and manganese (2% to 4%) in addition to carbon. It can be considered a predecessor to high-speed steel.

High-speed steel (HSS) is a highly alloyed tool steel capable of maintaining hardness at elevated temperatures better than high carbon and low alloy steels. Its good hot hardness permits tools made of HSS to be used at higher cutting speeds. Compared with the other tool materials at the time of its development, it was truly deserving of its name "high speed." A wide variety of high-speed steels are available, but they can be divided into two basic types: (1) tungsten-type, designated T-grades by the American Iron and Steel Institute (AISI); and (2) molybdenum-type, designated M-grades by AISI.

Tungsten-type HSS contains tungsten (W) as its principal alloying ingredient. Additional alloying elements are chromium (Cr), and vanadium (V). One of the original and best-known HSS grades is T1, or 18-4-1 high-speed steel, containing 18% W, 4% Cr, and 1% V. ***Molybdenum HSS*** grades contain combinations of tungsten and molybdenum (Mo), plus the same additional alloying elements as in the T-grades. Cobalt (Co) is sometimes added to HSS to enhance hot hardness. Of course, high-speed steel contains carbon, the element common to all steels. Typical alloying contents and functions of each alloying element in HSS are listed in Table 17.4.

Commercially, high-speed steel is one of the most important cutting-tool materials in use today, despite the fact that it was introduced more than a century ago. HSS is especially suited to applications involving complicated tool geometries, such as drills, taps, milling cutters, and broaches. These complex shapes are generally easier and less expensive to produce from unhardened HSS than other tool materials. They can then be heat-treated so that cutting-edge hardness is very good (Rockwell C 65), whereas toughness of the internal portions of the tool is also good. HSS cutters possess better toughness than any of the harder nonsteel tool materials used for machining, such as cemented carbides and ceramics. Even for single-point tools, HSS is popular among

TABLE 17.4 Typical contents and functions of alloying elements in high-speed steel

Alloying Element	Typical Content in HSS, % by Weight	Functions in High-Speed Steel
Tungsten	T-type HSS: 12–20 M-type HSS: 1.5–6	Increases hot hardness Improves abrasion resistance through formation of hard carbides in HSS
Molybdenum	T-type HSS: none M-type HSS: 5–10	Increases hot hardness Improves abrasion resistance through formation of hard carbides in HSS
Chromium	3.75–4.5	Depth hardenability during heat treatment Improves abrasion resistance through formation of hard carbides in HSS corrosion resistance (minor effect)
Vanadium	1–5	Combines with carbon for wear resistance Retards grain growth for better toughness
Cobalt	0–12	Increases hot hardness
Carbon	0.75–1.5	Principal hardening element in steel Provides available carbon to form carbides with other alloying elements for wear resistance

Compiled from [1], [10], [12], and other sources.

machinists because of the ease with which a desired tool geometry can be ground into the tool point. Over the years, improvements have been made in the metallurgical formulation and processing of HSS so that this class of tool material remains competitive in many applications. Also, HSS tools, drills in particular, are often coated with a thin film of titanium nitride (TiN) to provide significant increases in cutting performance. Physical vapor deposition (Section 21.5.1) is commonly used to coat these HSS tools.

17.2.2 CAST COBALT ALLOYS

Cast cobalt alloy cutting tools consist of cobalt, around 40% to 50%; chromium, about 25% to 35%; and tungsten, usually 15% to 20%; with trace amounts of other elements. These tools are made into the desired shape by casting in graphite molds and then grinding to final size and cutting-edge sharpness. High hardness is achieved as cast, an advantage over HSS, which requires heat treatment to achieve its hardness. Wear resistance of the cast cobalts is better than high-speed steel, but not as good as cemented carbide. Toughness of cast cobalt tools is better than carbides but not as good as HSS. Hot hardness also lies between the other two materials.

As might be expected from their properties, applications of cast cobalt tools are generally between those of high-speed steel and cemented carbides. They are capable of heavy roughing cuts at speeds greater than HSS and feeds greater than carbides. Work materials include both steels and nonsteels, as well as nonmetallic materials such as plastics and graphite. Today, cast cobalt alloy tools are not nearly as important commercially as either high-speed steel or cemented carbides. They were introduced around 1915 as a tool material that would allow higher cutting speeds than HSS. The carbides were subsequently developed and proved to be superior to the cast Co alloys in most cutting situations.

17.2.3 CEMENTED CARBIDES, CERMETS, AND COATED CARBIDES

Cermets are defined as composites of ***cer***amic and ***met***allic materials (Section 2.4.2). Technically speaking, cemented carbides are included within this definition; however, cermets based on WC–Co, including WC–TiC–TaC–Co, are known as carbides (cemented carbides) in common usage. In cutting-tool terminology, the term *cermet* is applied to ceramic-metal composites containing TiC, TiN, and certain other ceramics not including WC. One of the advances in cutting-tool materials involves the application of a very thin coating to a WC–Co substrate. These tools are called coated carbides. Thus, we have three important and closely related tool materials to discuss: (1) cemented carbides, (2) cermets, and (3) coated carbides.

Cemented Carbides Cemented carbides (also called ***sintered carbides***) are a class of hard tool material formulated from tungsten carbide (WC) using powder metallurgy techniques with cobalt (Co) as the binder (Section 11.3.1). There may be other carbide compounds in the mixture, such as titanium carbide (TiC) and/or tantalum carbide (TaC), in addition to WC.

The first cemented carbide cutting tools were made of WC–Co and could be used to machine cast irons and nonsteel materials at cutting speeds faster than those possible with high-speed steel and cast cobalt alloys. However, when the straight WC–Co tools were used to cut steel, crater wear occurred rapidly, leading to early tool failure. A strong chemical affinity exists between steel and the carbon in WC, resulting in accelerated wear by diffusion and chemical reaction at the tool–chip interface for this work–tool combination. Consequently, straight WC–Co tools cannot be used effectively to machine steel. It was subsequently discovered that additions of titanium carbide and tantalum carbide to the WC–Co mix significantly retarded the rate of crater wear when cutting steel. These new WC–TiC–TaC–Co tools could be used for steel machining. The result is that cemented carbides are divided into two basic types: (1) nonsteel-cutting grades, consisting of only WC–Co; and (2) steel-cutting grades, with combinations of TiC and TaC added to the WC–Co.

The general properties of the two types of cemented carbides are similar: (1) high compressive strength but low-to-moderate tensile strength; (2) high hardness (90 to 95 HRA); (3) good hot hardness; (4) good wear resistance; (5) high thermal conductivity; (6) high modulus of elasticity—E values up to around 600×10^3 MPa (90×10^6 lb/in^2); and (7) toughness lower than high-speed steel.

Nonsteel-cutting grades refer to those cemented carbides that are suitable for machining aluminum, brass, copper, magnesium, titanium, and other nonferrous metals; anomalously, gray cast iron is included in this group of work materials. In the nonsteel-cutting grades, grain size and cobalt content are the factors that influence properties of the cemented carbide material. The typical grain size found in conventional cemented carbides ranges between 0.5 and 5 μm (20 and 200 μ-in). As grain size is increased, hardness and hot hardness decrease, but transverse rupture strength increases.[2] The typical cobalt content in cemented carbides used for cutting tools is 3% to 12%. As cobalt content increases, transverse rupture strength (TRS, Section 3.1.3) improves at the expense of hardness and wear resistance. Cemented carbides with low percentages of cobalt content (3% to 6%) have high hardness and low TRS, whereas carbides with high Co (6% to 12%) have high TRS but lower hardness (Table 17.1). Accordingly, cemented carbides with higher cobalt are used for roughing operations and interrupted cuts (such as milling), while carbides with lower cobalt (therefore, higher hardness and wear resistance) are used in finishing cuts.

Steel-cutting grades are used for low carbon, stainless, and other alloy steels. For these carbide grades, titanium carbide and/or tantalum carbide is substituted for some of

[2] The effect of grain size (GS) on transverse rupture strength (TRS) is more complicated than we are reporting. Published data indicate that the effect of GS on TRS is influenced by cobalt content. At lower Co contents (less than 10%), TRS does indeed increase as GS increases, but at higher Co contents (greater than 10%) TRS decreases as GS increases [10], [18].

TABLE 17.5 The ANSI C-grade classification system for cemented carbides

Machining Application	Nonsteel-cutting Grades	Steel-cutting Grades	Cobalt and Properties
Roughing	C1	C5	High Co for max. toughness
General purpose	C2	C6	Medium to high Co
Finishing	C3	C7	Medium to low Co
Precision finishing	C4	C8	Low Co for max. hardness
Work materials	Al, brass, Ti, cast iron	Carbon & alloy steels	
Typical ingredients	WC–Co	WC–TiC–TaC–Co	

Source: [12].

the tungsten carbide. TiC is the more popular additive in most applications. Typically, from 10% to 25% of the WC might be replaced by combinations of TiC and TaC. This composition increases the crater-wear resistance for steel cutting, but tends to adversely affect flank-wear resistance for nonsteel-cutting applications. That is why two basic categories of cemented carbide are needed.

One of the important developments in cemented carbide technology in recent years is the use of very fine grain sizes (submicron sizes) of the various carbide ingredients (WC, TiC, and TaC). Although small grain size is usually associated with higher hardness but lower transverse rupture strength, the decrease in TRS is reduced or reversed at the submicron particle sizes. Therefore, these ultrafine grain carbides possess high hardness combined with good toughness.

Since the two basic types of cemented carbide were introduced in the 1920s and 1930s, the increasing number and variety of engineering materials have complicated the selection of the most appropriate cemented carbide for a given machining application. To address the problem of grade selection, two classification systems have been developed: (1) the ANSI[3] C-grade system, developed in the United States starting around 1942; and (2) the ISO R513-1975(E) system, introduced by the International Organization for Standardization (ISO) around 1964. In the C-grade system, summarized in Table 17.5, machining grades of cemented carbide are divided into two basic groups, corresponding to nonsteel-cutting and steel-cutting categories. Within each group there are four levels, corresponding to roughing, general purpose, finishing, and precision finishing.

The ISO R513-1975(E) system, titled "Application of Carbides for Machining by Chip Removal," classifies all machining grades of cemented carbides into three basic groups, each with its own letter and color code, as summarized in Table 17.6. Within each group, the grades are numbered on a scale that ranges from maximum hardness to maximum toughness. Harder grades are used for finishing operations (high speeds, low feeds and depths), whereas tougher grades are used for roughing operations. The ISO classification system can also be used to recommend applications for cermets and coated carbides.

Cermets Although cemented carbides are technically classified as cermet composites, the term ***cermet*** in cutting-tool technology is generally reserved for combinations of TiC, TiN, and titanium carbonitride (TiCN), with nickel and/or molybdenum as binders. Some of the cermet chemistries are more complex (e.g., ceramics such as Ta_xNb_yC and binders such as Mo_2C). However, cermets exclude metallic composites that are primarily based on WC–Co. Applications of cermets include high-speed finishing and semifinishing of steels, stainless steels, and cast irons. Higher speeds are generally allowed with these tools compared with steel-cutting carbide grades. Lower feeds are typically used so that better surface finish is achieved, often eliminating the need for grinding.

[3] ANSI = American National Standards Institute.

TABLE 17.6 ISO R513-1975(E) "Application of Carbides for Machining by Chip Removal."

Group	Carbide Type	Work Materials	Number Scheme (Cobalt and Properties)
P (blue)	Highly alloyed WC–TiC–TaC–Co	Steel, steel castings, ductile cast iron (ferrous metals with long chips)	P01 (low Co for maximum hardness) to P50 (high Co for maximum toughness)
M (yellow)	Alloyed WC–TiC–TaC–Co	Free-cutting steel, gray cast iron, austenitic stainless steel, superalloys	M10 (low Co for maximum hardness) to M40 (high Co for maximum toughness)
K (red)	Straight WC–Co	Nonferrous metals and alloys, gray cast iron (ferrous metals with short chips), nonmetallics	K01 (low Co for maximum hardness) to K40 (high Co for maximum toughness)

Source: [12].

Coated Carbides The development of coated carbides around 1970 represented a significant advance in cutting-tool technology. ***Coated carbides*** are a cemented carbide insert coated with one or more thin layers of wear-resistant material, such as titanium carbide, titanium nitride, and/or aluminum oxide (Al_2O_3). The coating is applied to the substrate by chemical vapor deposition or physical vapor deposition (Section 21.5). The coating thickness is only 2.5 to 13 μm (0.0001 to 0.0005 in). It has been found that thicker coatings tend to be brittle, resulting in cracking, chipping, and separation from the substrate.

The first generation of coated carbides had only a single layer coating (TiC, TiN, or Al_2O_3). More recently, coated inserts have been developed that consist of multiple layers. The first layer applied to the WC–Co base is usually TiN or TiCN because of good adhesion and similar coefficient of thermal expansion. Additional layers of various combinations of TiN, TiCN, Al_2O_3, and TiAlN are subsequently applied (see Figure 21.5).

Coated carbides are used to machine cast irons and steels in turning and milling operations. They are best applied at high cutting speeds in situations in which dynamic force and thermal shock are minimal. If these conditions become too severe, as in some interrupted cut operations, chipping of the coating can occur, resulting in premature tool failure. In this situation, uncoated carbides formulated for toughness are preferred. When properly applied, coated carbide tools usually permit increases in allowable cutting speeds compared with uncoated cemented carbides.

Use of coated carbide tools has expanded to nonferrous metal and nonmetal applications for improved tool life and higher cutting speeds. Different coating materials are required, such as chromium carbide (CrC), zirconium nitride (ZrN), and diamond [15].

17.2.4 CERAMICS

Cutting tools made from ceramics were first used commercially in the United States in the mid-1950s, although their development and use in Europe dates back to the early 1900s. Today's ceramic cutting tools are composed primarily of fine-grained ***aluminum oxide*** (Al_2O_3), pressed and sintered into inserts at high pressures and temperatures with no binder. The aluminum oxide is usually very pure (99% is typical), although some manufacturers add other oxides (such as zirconium oxide) in small amounts. In producing ceramic tools, it is important to use a very fine grain size in the alumina powder, and to maximize density of the mix through high-pressure compaction to improve the material's low toughness.

Aluminum oxide cutting tools are most successful in high-speed turning of cast iron and steel. Applications also include finish turning of hardened steels using high cutting speeds, low feeds and depths, and a rigid work setup. Many premature fracture failures of ceramic tools result from nonrigid machine tool setups, which subject the tools to mechanical shock. When properly applied, ceramic cutting tools can be used to obtain very good surface finish. Ceramics are not recommended for heavy interrupted cut operations (e.g., rough milling) because of their low toughness. In addition to its use as inserts in conventional machining operations, Al_2O_3 is widely used as an abrasive in grinding and other abrasive processes (Chapter 18).

Other commercially available ceramic cutting-tool materials include silicon nitride (SiN), ***sialon*** (silicon nitride and aluminum oxide, SiN–Al_2O_3), aluminum oxide and titanium carbide (Al_2O_3–TiC), and aluminum oxide reinforced with single crystal-whiskers of silicon carbide. These tools are usually intended for special applications, a discussion of which is beyond our scope.

17.2.5 SYNTHETIC DIAMONDS AND CUBIC BORON NITRIDE

Diamond is the hardest material known. By some measures of hardness, diamond is three to four times as hard as tungsten carbide or aluminum oxide. Since high hardness is one of the desirable properties of a cutting tool, it is natural to think of diamonds for machining and grinding applications. Synthetic diamond cutting tools are made of sintered polycrystalline diamond (SPD), which dates from the early 1970s. ***Sintered polycrystalline diamond*** is fabricated by sintering fine-grained diamond crystals under high temperatures and pressures into the desired shape. Little or no binder is used. The crystals have a random orientation and this adds considerable toughness to the SPD tools compared with single crystal diamonds. Tool inserts are typically made by depositing a layer of SPD about 0.5 mm (0.020 in) thick on the surface of a cemented carbide base. Very small inserts have also been made of 100% SPD.

Applications of diamond cutting tools include high-speed machining of nonferrous metals and abrasive nonmetals such as fiberglass, graphite, and wood. Machining of steel, other ferrous metals, and nickel-based alloys with SPD tools is not practical because of the chemical affinity that exists between these metals and carbon (a diamond, after all, is carbon).

Next to diamond, ***cubic boron nitride*** is the hardest material known, and its fabrication into cutting-tool inserts is basically the same as SPD, that is, coatings on WC–Co inserts. Cubic boron nitride (symbolized cBN) does not react chemically with iron and nickel as SPD does; therefore, the applications of cBN-coated tools are for machining steel and nickel-based alloys. Both SPD and cBN tools are expensive, as one might expect, and the applications must justify the additional tooling cost.

17.3 TOOL GEOMETRY

A cutting tool must possess a shape that is suited to the machining operation. One important way to classify cutting tools is according to the machining process. Thus, we have turning tools, cutoff tools, milling cutters, drill bits, reamers, taps, and many other cutting tools that are named for the operation in which they are used, each with its own tool geometry—in some cases quite unique.

As indicated in Section 15.1, cutting tools can be divided into single-point tools and multiple-cutting-edge tools. Single-point tools are used in turning, boring, shaping, and planing. Multiple-cutting-edge tools are used in drilling, reaming, tapping, milling, broaching, and sawing. Many of the principles that apply to single-point tools also apply to the other cutting-tool types, simply because the mechanism of chip formation is basically the same for all machining operations.

17.3.1 SINGLE-POINT TOOL GEOMETRY

The general shape of a single-point cutting tool is illustrated in Figure 15.3(a). Figure 17.6 shows a more detailed diagram. In the orthogonal model of metal cutting (Section 15.2.1), the rake angle of a cutting tool is treated as one parameter. In a single-point tool, the orientation of the rake face is defined by two angles, ***back rake angle*** (α_b) and ***side rake angle*** (α_s). Together, these angles are influential in determining the direction of chip flow across the rake face. The flank surface of the tool is defined by the ***end relief angle*** (ERA) and ***side relief angle*** (SRA). These angles determine the amount of clearance between the tool and the freshly cut work surface. The cutting edge of a single-point tool is divided into two sections, side cutting edge and end cutting edge. These two sections are separated by the tool point, which has a certain radius, called the nose radius. The ***side cutting-edge angle*** (SCEA) determines the entry of the tool into the work and can be used to reduce the sudden force the tool experiences as it enters a workpart. ***Nose radius*** (NR) determines to a large degree the texture of the surface generated in the operation. A very pointed tool (small nose radius) results in very pronounced feed marks on the surface (Section 16.8.2). ***End cutting-edge angle*** (ECEA) provides a clearance between the trailing edge of the tool and the newly generated work surface, thus reducing rubbing and friction against the surface.

In all, there are seven elements of tool geometry for a single-point tool. When specified in the following order, they are collectively called the ***tool geometry signature:*** back rake angle, side rake angle, end relief angle, side relief angle, end cutting-edge angle, side cutting-edge angle, and nose radius. For example, a single-point tool used in turning might have the following signature: 5, 5, 7, 7, 20, 15, 2/64 in.

Chip Breakers Chip disposal is a problem that is often encountered in turning and other continuous operations. Long, stringy chips are frequently generated, especially when turning ductile materials at high speeds. These chips cause a hazard to the machine operator and the workpart finish, and they interfere with automatic operation of the turning process. ***Chip breakers*** are frequently used with single-point tools to force the chips to curl more tightly than they would naturally be inclined to do, thus causing them to fracture. There are two principal forms of chip breaker design commonly used on

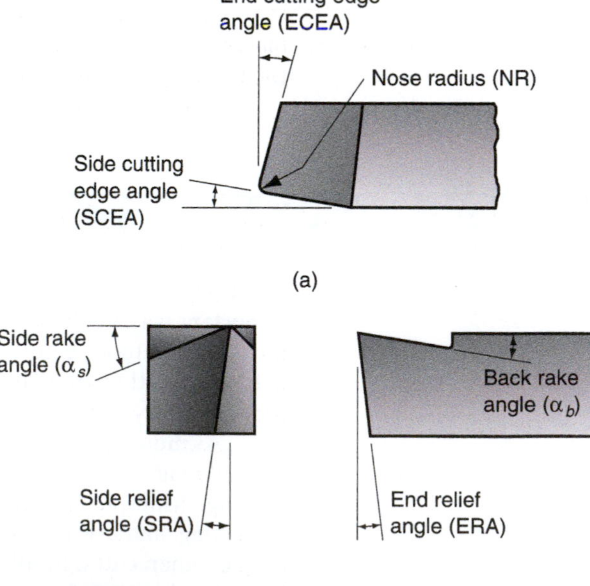

FIGURE 17.6 (a) Seven elements of single-point tool geometry, and (b) the tool signature convention that defines the seven elements. (Credit: *Fundamentals of Modern Manufacturing*, 4th Edition by Mikell P. Groover, 2010. Reprinted with permission of John Wiley & Sons, Inc.)

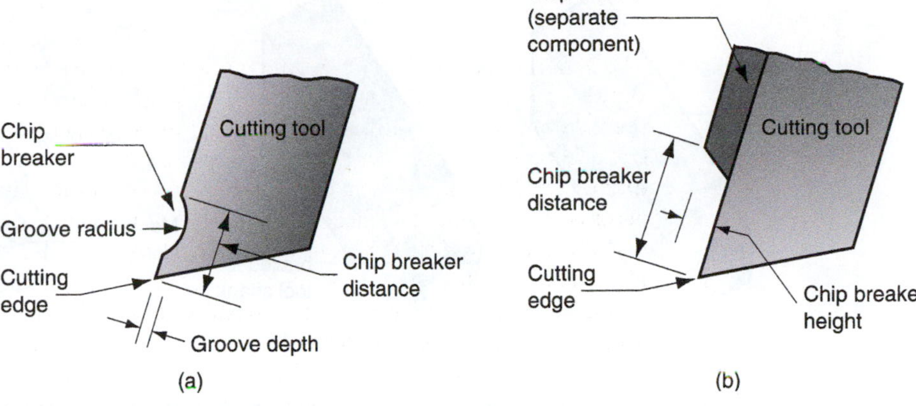

FIGURE 17.7 Two methods of chip breaking in single-point tools: (a) groove-type and (b) obstruction-type chip breakers. (Credit: *Fundamentals of Modern Manufacturing*, 4th Edition by Mikell P. Groover, 2010. Reprinted with permission of John Wiley & Sons, Inc.)

single-point turning tools, illustrated in Figure 17.7: (a) groove-type chip breaker designed into the cutting tool itself, and (b) obstruction-type chip breaker designed as an additional device on the rake face of the tool. The chip breaker distance can be adjusted in the obstruction-type device for different cutting conditions.

Effect of Tool Material on Tool Geometry It was noted in our discussion of the Merchant equation (Section 15.3.2) that a positive rake angle is generally desirable because it reduces cutting forces, temperature, and power consumption. High-speed steel cutting tools are almost always ground with positive rake angles, typically ranging from +5° to +20°. HSS has good strength and toughness, so that the thinner cross section of the tool created by high positive rake angles does not usually cause a problem with tool breakage. HSS tools are predominantly made of one piece. The heat treatment of high-speed steel can be controlled to provide a hard cutting edge while maintaining a tough inner core.

With the development of the very hard tool materials (e.g., cemented carbides and ceramics), changes in tool geometry were required. As a group, these materials have higher hardness and lower toughness than HSS. Also, their shear and tensile strengths are low relative to their compressive strengths, and their properties cannot be manipulated through heat treatment like those of HSS. Finally, cost per unit weight for these very hard materials is higher than the cost of HSS. These factors have affected cutting-tool design for the very hard tool materials in several ways.

First, the very hard materials must be designed with either negative rake or small positive angles. This change tends to load the tool more in compression and less in shear, thus favoring the high compressive strength of these harder materials. Cemented carbides, for example, are used with rake angles typically in the range from –5° to +10°. Ceramics have rake angles between –5° and –15°. Relief angles are made as small as possible (5° is typical) to provide as much support for the cutting edge as possible.

Another difference is the way in which the cutting edge of the tool is held in position. The alternative ways of holding and presenting the cutting edge for a single-point tool are illustrated in Figure 17.8. The geometry of a HSS tool is ground from a solid shank, as shown in part (a) of the figure. The higher cost and differences in properties and processing of the harder tool materials have given rise to the use of inserts that are either brazed or mechanically clamped to a toolholder. Part (b) shows a brazed insert, in which a cemented carbide insert is brazed to a tool shank. The shank is made of tool steel for strength and toughness. Part (c) illustrates one possible design for mechanically clamping an insert in a toolholder. Mechanical clamping is used for cemented carbides, ceramics, and the other hard materials. The significant advantage of the mechanically clamped insert is that each insert contains multiple cutting edges. When an edge wears out, the insert is unclamped,

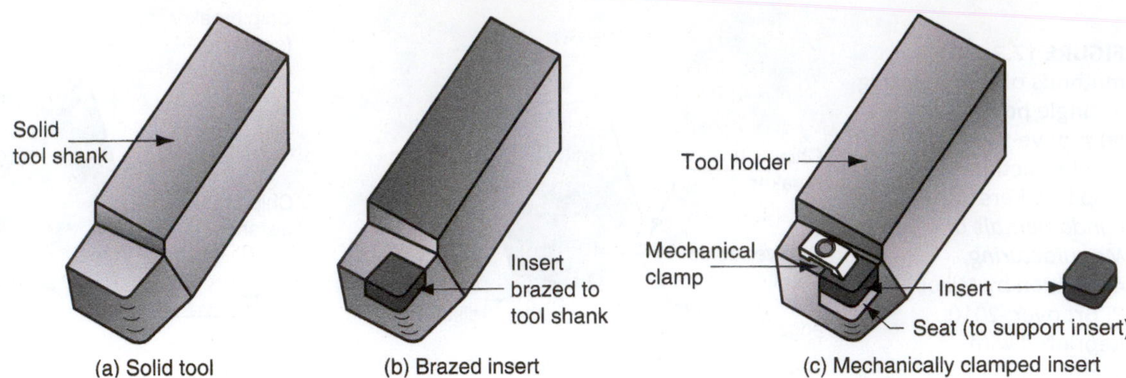

FIGURE 17.8 Three ways of holding and presenting the cutting edge for a single-point tool: (a) solid tool, typical of HSS; (b) brazed insert, one way of holding a cemented carbide insert; and (c) mechanically clamped insert, used for cemented carbides, ceramics, and other very hard tool materials. (Credit: *Fundamentals of Modern Manufacturing*, 4th Edition by Mikell P. Groover, 2010. Reprinted with permission of John Wiley & Sons, Inc.)

indexed (rotated in the toolholder) to the next edge, and reclamped in the toolholder. When all of the cutting edges are worn, the insert is discarded and replaced.

Inserts Cutting-tool inserts are widely used in machining because they are economical and adaptable to many different types of machining operations: turning, boring, threading, milling, and even drilling. They are available in a variety of shapes and sizes for the variety of cutting situations encountered in practice. A square insert is shown in Figure 17.8(c). Other common shapes used in turning operations are displayed in Figure 17.9. In general, the largest point angle should be selected for strength and economy. Round inserts possess large point angles (and large nose radii) just because of their shape. Inserts with large point

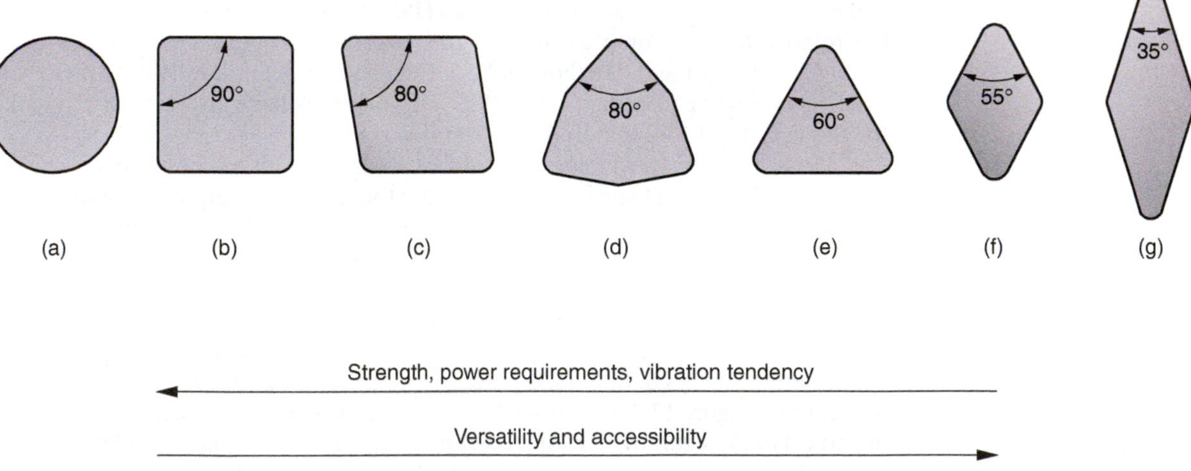

FIGURE 17.9 Common insert shapes: (a) round, (b) square, (c) rhombus with two 80° point angles, (d) hexagon with three 80° point angles, (e) triangle (equilateral), (f) rhombus with two 55° point angles, (g) rhombus with two 35° point angles. Also shown are typical features of the geometry. Strength, power requirements, and tendency for vibration increase as we move to the left; while versatility and accessibility tend to be better with the geometries at the right. (Credit: *Fundamentals of Modern Manufacturing*, 4th Edition by Mikell P. Groover, 2010. Reprinted with permission of John Wiley & Sons, Inc.)

angles are inherently stronger and less likely to chip or break during cutting, but they require more power, and there is a greater likelihood of vibration. The economic advantage of round inserts is that they can be indexed multiple times for more cuts per insert. Square inserts present four cutting edges, triangular shapes have three edges, whereas rhombus shapes have only two. Fewer edges are a cost disadvantage. If both sides of the insert can be used (e.g., in most negative rake angle applications), then the number of cutting edges is doubled. Rhombus shapes are used (especially with acute point angles) because of their versatility and accessibility when a variety of operations are to be performed. These shapes can be more readily positioned in tight spaces and can be used not only for turning but also for facing, Figure 16.6(a), and contour turning, Figure 16.6(c).

17.3.2 MULTIPLE-CUTTING-EDGE TOOLS

Most multiple-cutting-edge tools are used in machining operations in which the tool is rotated. In this section, standard cutting tools for drilling and milling are described.

Drills Various cutting tools are available for hole making, but the ***twist drill*** is by far the most common. It comes in diameters ranging from about 0.15 mm (0.006 in) to as large as 75 mm (3.0 in). Twist drills are widely used in industry to produce holes rapidly and economically.

The standard twist drill geometry is illustrated in Figure 17.10. The body of the drill has two spiral ***flutes*** (the spiral gives the twist drill its name). The angle of the spiral flutes is called the ***helix angle***, a typical value of which is around 30°. While drilling, the flutes act as passageways for extraction of chips from the hole. Although it is desirable for the flute openings to be large to provide maximum clearance for the chips, the body of the drill must be supported over its length. This support is provided by the ***web***, which is the thickness of the drill between the flutes.

The point of the twist drill has a conical shape. A typical value for the ***point angle*** is 118°. The point can be designed in various ways, but the most common design is a ***chisel edge***, as in Figure 17.10. Connected to the chisel edge are two cutting edges (sometimes called lips) that lead into the flutes. The portion of each flute adjacent to the cutting edge acts as the rake face of the tool.

The cutting action of the twist drill is complex. The rotation and feeding of the drill bit result in relative motion between the cutting edges and the workpiece to form the chips. The cutting speed along each cutting edge varies as a function of the distance from the axis of rotation. Accordingly, the efficiency of the cutting action varies, being most efficient at the outer diameter of the drill and least efficient at the center. In fact, the

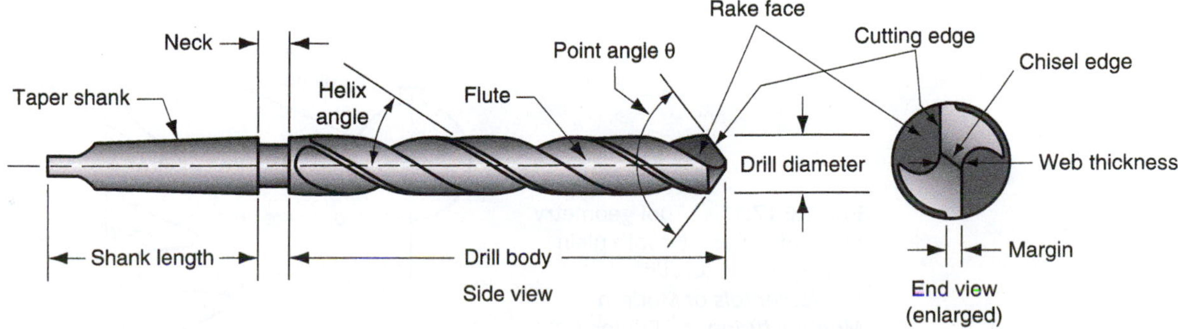

FIGURE 17.10 Standard geometry of a twist drill. (Credit: *Fundamentals of Modern Manufacturing*, 4th Edition by Mikell P. Groover, 2010. Reprinted with permission of John Wiley & Sons, Inc.)

relative velocity at the drill point is zero, so no cutting takes place. Instead, the chisel edge of the drill point pushes aside the material at the center as it penetrates into the hole, requiring a large thrust force to drive the twist drill forward into the hole. Also, at the beginning of the operation, the rotating chisel edge tends to wander on the surface of the workpart before starting the hole, causing loss of positional accuracy. Various alternative drill point designs have been developed to address this problem.

Chip removal can be a problem in drilling. The cutting action takes place inside the hole, and the flutes must provide sufficient clearance throughout the length of the drill to allow the chips to be extracted from the hole. As the chip is formed it is forced through the flutes to the work surface. Friction makes matters worse in two ways. In addition to the usual friction in metal cutting between the chip and the rake face of the cutting edge, friction also results from rubbing between the outside diameter of the drill bit and the newly formed hole. This increases the temperature of the drill and work. Delivery of cutting fluid to the drill point to reduce the friction and heat is difficult because the chips are flowing in the opposite direction. Because of chip removal and heat, a twist drill is normally limited to a hole depth of about four times its diameter. Some twist drills are designed with internal holes running their lengths, through which cutting fluid can be pumped to the hole near the drill point, thus delivering the fluid directly to the cutting operation. An alternative approach with twist drills that do not have fluid holes is to use a "pecking" procedure during the drilling operation. In this procedure, the drill is periodically withdrawn from the hole to clear the chips before proceeding deeper.

Twist drills are normally made of high-speed steel. The geometry of the drill is fabricated before heat treatment, and then the outer shell of the drill (cutting edges and friction surfaces) is hardened while retaining an inner core that is relatively tough. Grinding is used to sharpen the cutting edges and shape the drill point.

More information on hole-making tools (twist drills as well as other types) can be found in several of our references [3] and [12].

Milling Cutters Classification of milling cutters is closely associated with the milling operations described in Section 16.4.1. The major types of milling cutters are the following:

- ***Plain milling cutters***. These are used for peripheral or slab milling. As Figures 16.17(a) and 16.18(a) indicate, they are cylinder shaped with several rows of teeth. The cutting edges are usually oriented at a helix angle (as in the figures) to reduce impact on entry into the work, and these cutters are called ***helical milling cutters***. Tool geometry elements of a plain milling cutter are shown in Figure 17.11.

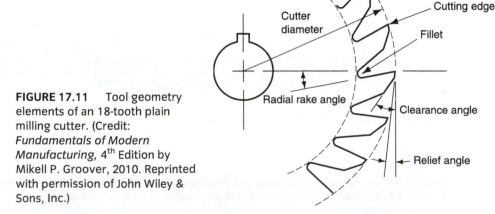

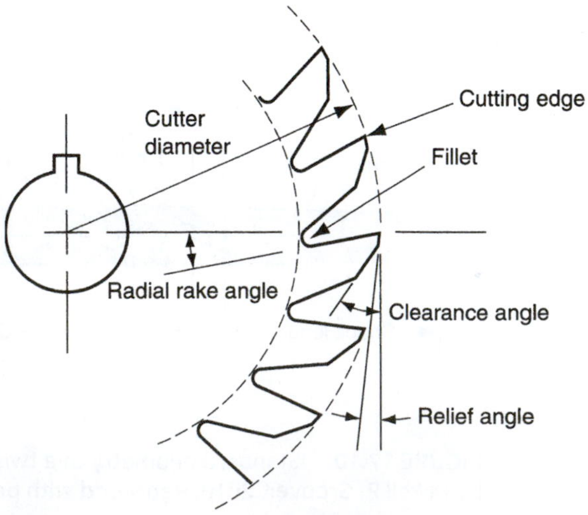

FIGURE 17.11 Tool geometry elements of an 18-tooth plain milling cutter. (Credit: *Fundamentals of Modern Manufacturing*, 4th Edition by Mikell P. Groover, 2010. Reprinted with permission of John Wiley & Sons, Inc.)

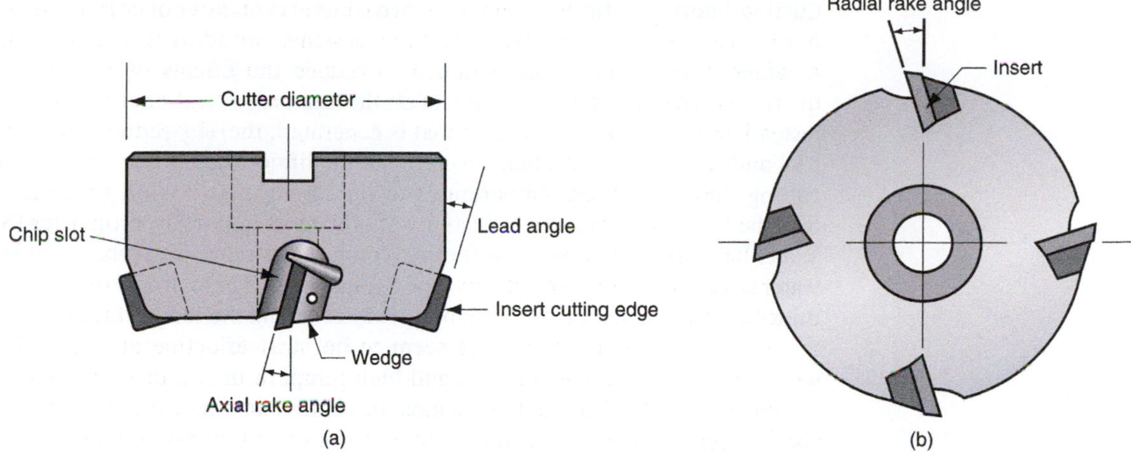

FIGURE 17.12 Tool geometry elements of a four-tooth face milling cutter: (a) side view and (b) bottom view. (Credit: *Fundamentals of Modern Manufacturing*, 4th Edition by Mikell P. Groover, 2010. Reprinted with permission of John Wiley & Sons, Inc.)

➢ ***Form milling cutters***. These are peripheral milling cutters in which the cutting edges have a special profile that is to be imparted to the work. An important application is in gear making, in which the form milling cutter is shaped to cut the slots between adjacent gear teeth, thereby leaving the geometry of the gear teeth.

➢ ***Face milling cutters***. These are designed with teeth that cut on both the periphery as well as the end of the cutter. Face milling cutters can be made of HSS, as in Figure 16.17(b), or they can be designed to use cemented carbide inserts. Figure 17.12 shows a four-tooth face milling cutter that uses inserts.

➢ ***End milling cutters***. As shown in Figure 16.20(c), an end milling cutter looks like a drill bit, but close inspection indicates that it is designed for primary cutting with its peripheral teeth rather than its end. (A drill bit cuts only on its end as it penetrates into the work.) End mills are designed with square ends, ends with radii, and ball ends. End mills can be used for face milling, profile milling and pocketing, cutting slots, engraving, surface contouring, and die sinking.

17.4 CUTTING FLUIDS

A ***cutting fluid*** is any liquid or gas that is applied directly to the machining operation to improve cutting performance. Cutting fluids address two main problems: (1) heat generation at the shear zone and friction zone, and (2) friction at the tool–chip and tool–work interfaces. In addition to removing heat and reducing friction, cutting fluids provide additional benefits, such as washing away chips (especially in grinding and milling), reducing the temperature of the workpart for easier handling, reducing cutting forces and power requirements, improving dimensional stability of the workpart, and improving surface finish.

17.4.1 TYPES OF CUTTING FLUIDS

A variety of cutting fluids are commercially available. It is appropriate to discuss them first according to function and then to classify them according to chemical formulation.

Cutting-Fluid Functions There are two general categories of cutting fluids, corresponding to the two main problems they are designed to address: coolants and lubricants. ***Coolants*** are cutting fluids designed to reduce the effects of heat in the machining operation. They have a limited effect on the amount of heat energy generated in cutting; instead, they carry away the heat that is generated, thereby reducing the temperature of tool and workpiece. This helps to prolong the life of the cutting tool. The capacity of a cutting fluid to reduce temperatures in machining depends on its thermal properties. Specific heat and thermal conductivity are the most important properties (Section 3.7.1). Water has high specific heat and thermal conductivity relative to other liquids, which is why water is used as the base in coolant-type cutting fluids. These properties allow the coolant to draw heat away from the operation, thereby reducing the temperature of the cutting tool.

Coolant-type cutting fluids seem to be most effective at relatively high cutting speeds in which heat generation and high temperatures are problems. They are most effective on tool materials that are most susceptible to temperature failures, such as high-speed steels, and are used frequently in turning and milling operations in which large amounts of heat are generated.

Lubricants are usually oil-based fluids (because oils possess good lubricating qualities) formulated to reduce friction at the tool–chip and tool–work interfaces. Lubricant cutting fluids operate by ***extreme pressure lubrication***, a special form of lubrication that involves formation of thin solid salt layers on the hot, clean metal surfaces through chemical reaction with the lubricant. Compounds of sulfur, chlorine, and phosphorus in the lubricant cause the formation of these surface layers, which act to separate the two metal surfaces (i.e., chip and tool). These extreme pressure films are significantly more effective in reducing friction in metal cutting than conventional lubrication, which is based on the presence of liquid films between the two surfaces.

Lubricant-type cutting fluids are most effective at lower cutting speeds. They tend to lose their effectiveness at speeds above about 120 m/min (400 ft/min) because the motion of the chip at these high speeds prevents the cutting fluid from reaching the tool–chip interface. In addition, high cutting temperatures at these speeds cause the oils to vaporize before they can lubricate. Machining operations such as drilling and tapping usually benefit from lubricants. In these operations, built-up edge formation is retarded, and torque on the tool is reduced.

Although the principal purpose of a lubricant is to reduce friction, it also reduces the temperature in the operation through several mechanisms. First, the specific heat and thermal conductivity of the lubricant help to remove heat from the operation, thereby reducing temperatures. Second, because friction is reduced, the heat generated from friction is also reduced. Third, a lower coefficient of friction means a lower friction angle. According to Merchant's equation, Eq. 15.16, a lower friction angle causes the shear plane angle to increase, hence reducing the amount of heat energy generated in the shear zone.

There is typically an overlapping effect between the two types of cutting fluids. Coolants are formulated with ingredients that help reduce friction. And lubricants have thermal properties that, although not as good as those of water, act to remove heat from the cutting operation. Cutting fluids (both coolants and lubricants) manifest their effect on the Taylor tool life equation through higher C values. Increases of 10% to 40% are typical. The slope n is not significantly affected.

Chemical Formulation of Cutting Fluids There are four categories of cutting fluids according to chemical formulation: (1) cutting oils, (2) emulsified oils, (3) semichemical fluids, and (4) chemical fluids. All of these cutting fluids provide both coolant and lubricating functions. The cutting oils are most effective as lubricants, whereas the other three categories are more effective as coolants because they are primarily water.

Cutting oils are based on oil derived from petroleum, animal, marine, or vegetable origin. Mineral oils (petroleum based) are the principal type because of their abundance and generally desirable lubricating characteristics. To achieve maximum lubricity, several types of oils are often combined in the same fluid. Chemical additives are also mixed with the oils to increase lubricating qualities. These additives contain compounds of sulfur, chlorine, and phosphorus, and are designed to react chemically with the chip and tool surfaces to form solid films (extreme pressure lubrication) that help to avoid metal-to-metal contact between the two.

Emulsified oils consist of oil droplets suspended in water. The fluid is made by blending oil (usually mineral oil) in water using an emulsifying agent to promote blending and stability of the emulsion. A typical ratio of water to oil is 30:1. Chemical additives based on sulfur, chlorine, and phosphorus are often used to promote extreme pressure lubrication. Because they contain both oil and water, the emulsified oils combine cooling and lubricating qualities in one cutting fluid.

Chemical fluids are chemicals in a water solution rather than oils in emulsion. The dissolved chemicals include compounds of sulfur, chlorine, and phosphorus, plus wetting agents. The chemicals are intended to provide some degree of lubrication to the solution. Chemical fluids provide good coolant qualities but their lubricating qualities are less than the other cutting-fluid types. ***Semichemical fluids*** have small amounts of emulsified oil added to increase the lubricating characteristics of the cutting fluid. In effect, they are a hybrid class between chemical fluids and emulsified oils.

17.4.2 APPLICATION OF CUTTING FLUIDS

Cutting fluids are applied to machining operations in various ways. In this section we consider these application techniques. We also consider the problem of cutting-fluid contamination and what steps can be taken to address this problem.

Application Methods The most common method is ***flooding***, sometimes called flood-cooling because it is generally used with coolant-type cutting fluids. In flooding, a steady stream of fluid is directed at the tool–work or tool–chip interface of the machining operation. A second method of delivery is ***mist application***, primarily used for water-based cutting fluids. In this method the fluid is directed at the operation in the form of a high-speed mist carried by a pressurized air stream. Mist application is generally not as effective as flooding in cooling the tool. However, because of the high-velocity air stream, mist application may be more effective in delivering the cutting fluid to areas that are difficult to access by conventional flooding.

Manual application by means of a squirt can or paint brush is sometimes used for applying lubricants in tapping and other operations in which cutting speeds are low and friction is a problem. It is generally not preferred by most production machine shops because of its variability in application.

Cutting-Fluid Filtration and Dry Machining Over time, cutting fluids can become contaminated with a variety of foreign substances, such as tramp oil (machine oil, hydraulic fluid, etc.), garbage (cigarette butts, food, etc.), small chips, molds, fungi, and bacteria. In addition to causing odors and health hazards, contaminated cutting fluids do not perform their lubricating function as well. Alternative ways of dealing with this problem are to (1) replace the cutting fluid at regular and frequent intervals (perhaps twice per month); (2) use a filtration system to continuously or periodically clean the fluid; or (3) dry machining, that is, machine without cutting fluids. Because of growing concern about environmental pollution and associated legislation, disposing of old fluids has become both costly and contrary to the general public welfare.

Filtration systems are being installed in numerous machine shops today to solve the contamination problem. Advantages of these systems include (1) prolonged cutting-fluid life between changes—instead of replacing the fluid once or twice per month, coolant lives of 1 year have been reported; (2) reduced fluid disposal cost, since disposal is much less frequent when a filter is used; (3) cleaner cutting fluid for better working environment and reduced health hazards; (4) lower machine tool maintenance; and (5) longer tool life. There are various types of filtration systems for filtering cutting fluids. For the interested reader, filtration systems and the benefits of using them are discussed in reference [12].

The third alternative is called ***dry machining***, meaning that no cutting fluid is used. Dry machining avoids the problems of cutting-fluid contamination, disposal, and filtration, but can lead to problems of its own: (1) overheating the tool, (2) operating at lower cutting speeds and production rates to prolong tool life, and (3) absence of chip removal benefits in grinding and milling. Cutting-tool producers have developed certain grades of carbides and coated carbides for use in dry machining.

17.5 MACHINABILITY

Properties of the work material have a significant influence on the success of the machining operation. These properties and other characteristics of the work are often summarized in the term "machinability." ***Machinability*** denotes the relative ease with which a material (usually a metal) can be machined using appropriate tooling and cutting conditions [14].

Various criteria are used to assess machinability, the most important of which is (1) tool life because of its economic significance in a machining operation. Other criteria include (2) cutting forces, (3) power, (4) surface finish, and (5) ease of chip disposal. Although machinability generally refers to the work material, it should be recognized that machining performance depends on more than just material. The type of machining operation, tooling, and cutting conditions are also important factors. In addition, the machinability criterion is a source of variation. One work material may yield a longer tool life, whereas another material provides a better surface finish. All of these factors make evaluation of machinability difficult.

Machinability testing usually involves a comparison of work materials. The machining performance of a test material is measured relative to that of a base (standard) material. The relative performance is expressed as an index number, called the machinability rating (MR). The base material used as the standard is given a machinability rating of 1.00. B1112 steel is often used as the base material in machinability comparisons. Materials that are easier to machine than the base have ratings greater than 1.00, and materials that are more difficult to machine have ratings less than 1.00. Machinability ratings are often expressed as percentages rather than index numbers. Let us illustrate how a machinability rating might be determined using a tool life test as the basis of comparison.

Example 17.2 Machinability

A series of tool life tests are conducted on two work materials under identical cutting conditions, varying only speed in the test procedure. The first material, defined as the base material, yields a Taylor tool life equation $vT^{0.28} = 350$, and the other material (test material) yields a Taylor equation $vT^{0.27} = 440$, where speed is in m/min and tool life is in min. Determine the machinability rating of the test material using the cutting speed that provides a 60-min tool life as the basis of comparison. This speed is denoted by v_{60}.

Solution: The base material has a machinability rating = 1.0. Its v_{60} value can be determined from the Taylor tool life equation as follows:

$$v_{60} = (350/60^{0.28}) = 111 \text{ m/min}$$

The cutting speed at a 60-min tool life for the test material is determined similarly:

$$v_{60} = (440/60^{0.27}) = 146\ \mathrm{m}/min$$

Accordingly, the machinability rating can be calculated as:

$$\mathrm{MR\ (for\ the\ test\ material)} = \frac{146}{111} = 1.31\ (131\%)$$

Many work material factors affect machining performance. Important mechanical properties include hardness and strength. As hardness increases, abrasive wear of the tool increases so that tool life is reduced. Strength is usually indicated as tensile strength, even though machining involves shear stresses. Of course, shear strength and tensile strength are correlated. As work material strength increases, cutting forces, specific energy, and cutting temperature increase, making the material more difficult to machine. On the other hand, very low hardness can be detrimental to machining performance. For

TABLE 17.7 Approximate values of Brinell Hardness and typical machinability ratings for selected work materials

Work Material	Brinell Hardness	Machinability Rating[a]	Work Material	Brinell Hardness	Machinability Rating[a]
Base steel: B1112	180 – 220	1.00	Tool steel (unhardened)	200 – 250	0.30
Low carbon steel:	130 – 170	0.50	Cast iron		
C1008, C1010, C1015			Soft	60	0.70
Medium carbon steel:	140 – 210	0.65	Medium hardness	200	0.55
C1020, C1025, C1030			Hard	230	0.40
High carbon steel:	180 – 230	0.55	Super alloys		
C1040, C1045, C1050			Inconel	240 – 260	0.30
Alloy steels24[b]			Inconel X	350 – 370	0.15
1320, 1330, 3130, 3140	170 – 230	0.55	Waspalloy	250 – 280	0.12
4130	180 – 200	0.65	Titanium		
4140	190 – 210	0.55	Plain	160	0.30
4340	200 – 230	0.45	Alloys	220 – 280	0.20
4340 (casting)	250 – 300	0.25	Aluminum		
6120, 6130, 6140	180 – 230	0.50	2-S, 11-S, 17-S	soft	5.00[c]
8620, 8630	190 – 200	0.60	Aluminum alloys (soft)	soft	2.00[d]
B1113	170 – 220	1.35	Aluminum alloys (hard)	hard	1.25[d]
Free machining steels	160 – 220	1.50	Copper	soft	0.60
Stainless steel			Brass	soft	2.00[d]
301, 302	170 – 190	0.50	Bronze	soft	0.65[d]
304	160 – 170	0.40			
316, 317	190 – 200	0.35			
403	190 – 210	0.55			
416	190 – 210	0.90			

Values are estimated average values based on [2], [5], [6], [12], and other sources. Ratings represent relative cutting speeds for a given tool life (see Example 17.2).

[a]Machinability ratings are often expressed as percents (index number × 100%).

[b]Our list of alloy steels is by no means complete. We have attempted to include some of the more common alloys and to indicate the range of machinability ratings among these steels.

[c]The machinability of aluminum varies widely. It is expressed here as MR = 5.00, but the range is probably from 3.00 to 10.00 or more.

[d]Aluminum alloys, brasses, and bronzes also vary significantly in machining performance. Different grades have different machinability ratings. For each case, we have attempted to reduce the variation to a single average value to indicate relative performance with other work materials.

example, low carbon steel, which has relatively low hardness, is often too ductile to machine well. High ductility causes tearing of the metal as the chip is formed, resulting in poor finish, and problems with chip disposal. Cold drawing is often used on low carbon bars to increase surface hardness and promote chip-breaking during cutting.

A metal's chemistry has an important effect on properties; and in some cases, chemistry affects the wear mechanisms that act on the tool material. Through these relationships, chemistry affects machinability. Carbon content has a significant effect on the properties of steel. As carbon is increased, strength and hardness of the steel increases; this reduces machining performance. Many alloying elements added to steel to enhance properties are detrimental to machinability. Chromium, molybdenum, and tungsten form carbides in steel, which increase tool wear and reduce machinability. Manganese and nickel add strength and toughness to steel, which reduce machinability. Certain elements can be added to steel to improve machining performance, such as lead, sulfur, and phosphorus. The additives have the effect of reducing the coefficient of friction between the tool and chip, thereby reducing forces, temperature, and built-up edge formation. Better tool life and surface finish result from these effects. Steel alloys formulated to improve machinability are referred to as ***free machining steels***.

Similar relationships exist for other work materials. Table 17.7 lists selected metals and their approximate machinability ratings. These ratings are intended to summarize the machining performance of the materials, with emphasis on the tool life criterion.

17.6 MACHINING ECONOMICS

One of the practical problems in machining is selecting the proper cutting conditions for a given operation. This is one of the tasks in process planning (Section 28.2). For each operation, decisions must be made about machine tool, cutting tool(s), and cutting conditions. These decisions must give due consideration to workpart machinability, part geometry, surface finish, and so forth.

17.6.1 SELECTING FEED AND DEPTH OF CUT

Cutting conditions in a machining operation consist of speed, feed, depth of cut, and cutting fluid (whether a cutting fluid is to be used and, if so, what type of cutting fluid). Tooling considerations are usually the dominant factor in decisions about cutting fluids (Section 17.4). Depth of cut is often predetermined by workpiece geometry and operation sequence. Many jobs require a series of roughing operations followed by a final finishing operation. In the roughing operations, depth is made as large as possible within the limitations of available horsepower, machine tool and setup rigidity, strength of the cutting tool, and so on. In the finishing cut, depth is set to achieve the final dimensions for the part.

The problem then reduces to selection of feed and speed. In general, values of these parameters should be decided in the order: ***feed first, speed second***. Determining the appropriate feed rate for a given machining operation depends on the following factors:

- ***Tooling***. What type of tooling will be used? Harder tool materials (e.g., cemented carbides, ceramics, etc.) tend to fracture more readily than high-speed steel. These tools are normally used at lower feed rates. HSS can tolerate higher feeds because of its greater toughness.
- ***Roughing or finishing***. Roughing operations involve high feeds, typically 0.5 to 1.25 mm/rev (0.020 to 0.050 in/rev) for turning; finishing operations involve low feeds, typically 0.125 to 0.4 mm/rev (0.005 to 0.015 in/rev) for turning.

- ***Constraints on feed in roughing***. If the operation is roughing, how high can the feed rate be set? To maximize metal removal rate, feed should be set as high as possible. Upper limits on feed are imposed by cutting forces, setup rigidity, and sometimes horsepower.
- ***Surface finish requirements in finishing***. If the operation is finishing, what is the desired surface finish? Feed is an important factor in surface finish, and computations like those in Example 16.1 can be used to estimate the feed that will produce a desired surface roughness.

17.6.2 CUTTING SPEED

Selection of cutting speed is based on making the best use of the cutting tool, which normally means choosing a speed that provides a high metal removal rate yet suitably long tool life. Mathematical formulas have been derived to determine optimal cutting speed for a machining operation, given that the various time and cost components of the operation are known. The original derivation of these ***machining economics*** equations is credited to W. Gilbert [13]. The formulas allow the optimal cutting speed to be calculated for either of two objectives: (1) maximum production rate or (2) minimum unit cost. Both objectives seek to achieve a balance between material removal rate and tool life. The formulas are based on the Taylor tool life equation that applies to the tool used in the operation, as well as feed, depth of cut, and work material. The derivations are illustrated here for a turning operation. Similar derivations can be developed for other types of machining operations [4].

Maximizing Production Rate For maximum production rate, the speed that minimizes production cycle time per workpiece is determined. This is equivalent to maximizing production rate. In turning, there are three time elements that contribute to the total production cycle time for one part:

1. ***Part handling time*** T_h. This is the time the operator spends loading the part into the machine tool at the beginning of the production cycle and unloading the part after machining is completed. Any additional time required to reposition the tool for the start of the next cycle should also be included here.
2. ***Machining time*** T_m. This is the time the tool is actually engaged in cutting metal during the cycle.
3. ***Tool change time*** T_t. At the end of the tool life, the tool must be changed, which takes time. This time must be apportioned over the number of parts cut during the tool life. Let n_p = the number of pieces cut in one tool life (the number of pieces cut with one cutting edge until the tool is changed); thus, the tool change time per part = T_t/n_p.

The sum of these three time elements gives the total cycle time per unit for the production cycle:

$$T_c = T_h + T_m + \frac{T_t}{n_p} \tag{17.5}$$

where T_c = production cycle time per piece, min; and the other terms are defined above. This cycle time is a function of cutting speed. As cutting speed is increased, T_m decreases and T_t/n_p increases; T_h is unaffected by speed. These relationships are shown in Figure 17.13.

The cycle time per part is minimized at a certain value of cutting speed, which can be determined by recasting Eq. (17.5) as a function of speed. Machining time in a straight

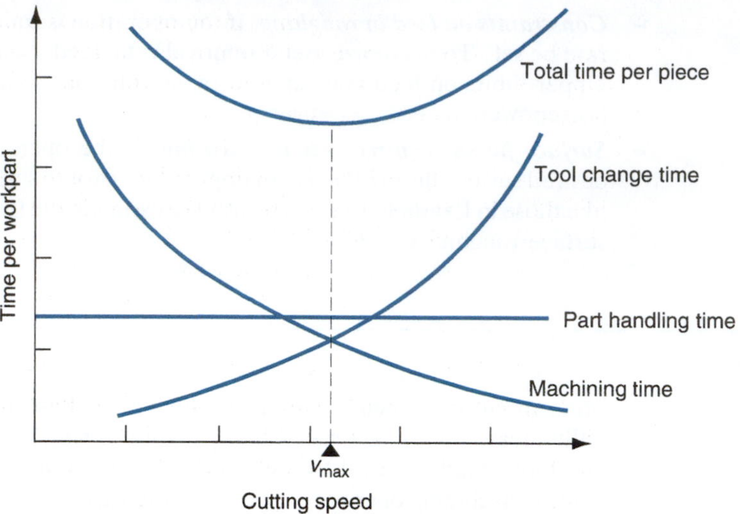

FIGURE 17.13 Time elements in a machining cycle plotted as a function of cutting speed. Total cycle time per piece is minimized at a certain value of cutting speed. This is the speed for maximum production rate. (Credit: *Fundamentals of Modern Manufacturing*, 4th Edition by Mikell P. Groover, 2010. Reprinted with permission of John Wiley & Sons, Inc.)

turning operation is given by previous Eq. (16.5):

$$T_m = \frac{\pi D L}{v f}$$

where T_m = machining time, min; D = workpart diameter, mm (in); L = workpart length, mm (in); f = feed, mm/rev (in/rev); and v = cutting speed, mm/min for consistency of units (in/min for consistency of units).

The number of pieces per tool n_p is also a function of speed. It can be shown that:

$$n_p = \frac{T}{T_m} \tag{17.6}$$

where T = tool life, min/tool; and T_m = machining time per part, min/pc. Both T and T_m are functions of speed; hence, the ratio is a function of speed:

$$n_p = \frac{f C^{1/n}}{\pi D L v^{1/n-1}} \tag{17.7}$$

The effect of this relation is to cause the T_t/n_p term in Eq. (17.5) to increase as cutting speed increases. Substituting Eqs. (16.5) and (17.7) into Eq. (17.5) for T_c, we have:

$$T_c = T_h + \frac{\pi D L}{f v} + \frac{T_t(\pi D L v^{1/n-1})}{f C^{1/n}} \tag{17.8}$$

The cycle time per piece is a minimum at the cutting speed at which the derivative of Eq. (17.8) is zero: $dT_c/dv = 0$. Solving this equation yields the cutting speed for maximum production rate in the operation:

$$v_{\max} = \frac{C}{\left[\left(\frac{1}{n} - 1\right) T_t\right]^n} \tag{17.9}$$

where $v_{\max}$ is expressed in m/min (ft/min). The corresponding tool life for maximum production rate is:

$$T_{\max} = \left(\frac{1}{n} - 1\right) T_t \tag{17.10}$$

Minimizing Cost per Unit For minimum cost per unit, the speed that minimizes production cost per piece for the operation is determined. To derive the equations for this case, we begin with the four cost components that determine total cost of producing one part during a turning operation:

1. ***Cost of part handling time***. This is the cost of the time the operator spends loading and unloading the part. Let $C_o =$ the cost rate (e.g., \$/min) for the operator and machine. Thus, the cost of part handling time $= C_oT_h$.
2. ***Cost of machining time***. This is the cost of the time the tool is engaged in machining. Using C_o again to represent the cost per minute of the operator and machine tool, the cutting time cost $= C_oT_m$.
3. ***Cost of tool change time***. The cost of tool change time $= C_oT_t/n_p$.
4. ***Tooling cost***. In addition to the tool change time, the tool itself has a cost that must be added to the total operation cost. This cost is the cost per cutting edge C_t, divided by the number of pieces machined with that cutting edge n_p. Thus, tool cost per workpiece is given by C_t/n_p.

Tooling cost requires an explanation, because it is affected by different tooling situations. For disposable inserts (e.g., cemented carbide inserts), tool cost is determined as:

$$C_t = \frac{P_t}{n_e} \tag{17.11}$$

where $C_t =$ cost per cutting edge, \$/tool life; $P_t =$ price of the insert, \$/insert; and $n_e =$ number of cutting edges per insert. This depends on the insert type; for example, triangular inserts that can be used only one side (positive rake tooling) have three edges/insert; if both sides of the insert can be used (negative rake tooling), there are six edges/insert; and so forth.

For regrindable tooling (e.g., high-speed steel solid shank tools, brazed carbide tools), the tool cost includes purchase price plus cost to regrind:

$$C_t = \frac{P_t}{n_g} + T_gC_g \tag{17.12}$$

where $C_t =$ cost per tool life, \$/tool life; $P_t =$ purchase price of the solid shank tool or brazed insert, \$/tool; $n_g =$ number of tool lives per tool, which is the number of times the tool can be sharpened before it can no longer be used (5 to 10 times for roughing tools and 10 to 20 times for finishing tools); $T_g =$ time to grind or regrind the tool, min/tool life; and $C_g =$ grinder's rate, \$/min.

The sum of the four cost components gives the total cost per unit product C_c for the machining cycle:

$$C_c = C_oT_h + C_oT_m + \frac{C_oT_t}{n_p} + \frac{C_t}{n_p} \tag{17.13}$$

C_c is a function of cutting speed, just as T_c is a function of v. The relationships for the individual terms and total cost as a function of cutting speed are shown in Figure 17.14. Eq. (17.13) can be rewritten in terms of v to yield:

$$C_c = C_oT_h + \frac{C_o\pi DL}{fv} + \frac{(C_oT_t + C_t)(\pi DLv^{1/n-1})}{fC^{1/n}} \tag{17.14}$$

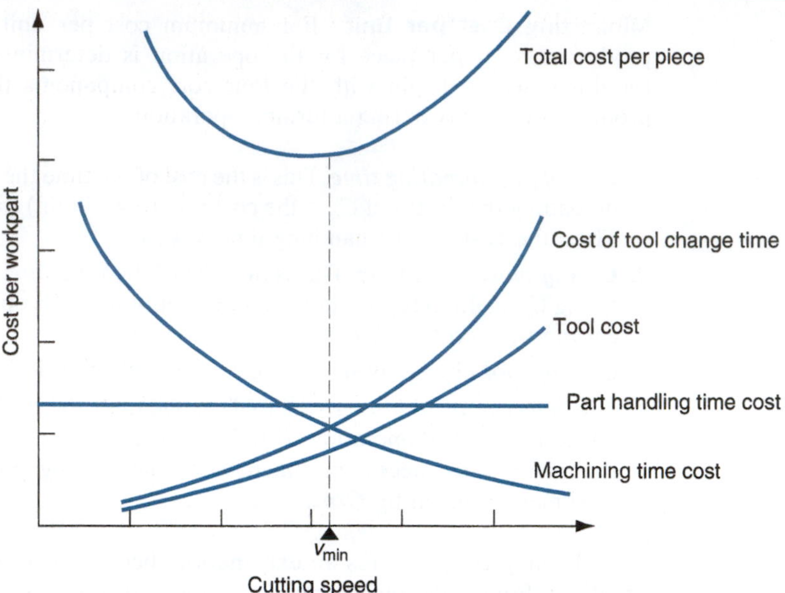

FIGURE 17.14 Cost components in a machining operation plotted as a function of cutting speed. Total cost per piece is minimized at a certain value of cutting speed. This is the speed for minimum cost per piece. (Credit: *Fundamentals of Modern Manufacturing*, 4th Edition by Mikell P. Groover, 2010. Reprinted with permission of John Wiley & Sons, Inc.)

The cutting speed that obtains minimum cost per piece for the operation can be determined by taking the derivative of Eq. (17.14) with respect to v, setting it to zero, and solving for $v_{\min}$:

$$v_{\min} = C\left(\frac{n}{1-n} \cdot \frac{C_o}{C_o T_t + C_t}\right)^n \tag{17.15}$$

The corresponding tool life is given by:

$$T_{\min} = \left(\frac{1}{n} - 1\right)\left(\frac{C_o T_t + C_t}{C_o}\right) \tag{17.16}$$

Example 17.3 Determining Cutting Speeds in Machining Economics

A turning operation is to be performed with HSS tooling on mild steel, with Taylor tool life parameters $n = 0.125$, $C = 70$ m/min (Table 17.2). Workpart length = 500 mm and diameter = 100 mm. Feed = 0.25 mm/rev. Handling time per piece = 5.0 min, and tool change time = 2.0 min. Cost of machine and operator = \$30/hr, and tooling cost = \$3 per cutting edge. Find: (a) cutting speed for maximum production rate, and (b) cutting speed for minimum cost.

Solution: (a) Cutting speed for maximum production rate is given by Eq. (17.9):

$$v_{\max} = 70\left(\frac{0.125}{0.875} \cdot \frac{1}{2}\right)^{0.125} = 50\,\text{m/min}$$

(b) Converting C_o = \$30/hr to \$0.5/min, minimum cost cutting speed is given by Eq. (17.15):

$$v_{\min} = 70\left(\frac{0.125}{0.875} \cdot \frac{0.5}{0.5(2) + 3.00}\right)^{0.125} = 42\,\text{m/min}$$

■

Example 17.4 Production Rate and Cost in Machining Economics

Determine the hourly production rate and cost per piece for the two cutting speeds computed in Example 17.3.

Solution: (a) For the cutting speed for maximum production, v_{max} = 50 m/min, machining time per piece and tool life are calculated as follows:

$$\text{Machining time } T_m = \frac{\pi(0.5)(0.1)}{(0.25)(10^{-3})(50)} = 12.57\ \text{min/pc}$$

$$\text{Tool life } T = \left(\frac{70}{50}\right)^8 = 14.76\ \text{min/cutting edge}$$

From these values the number of pieces per tool can be determined: $n_p = 14.76/12.57 = 1.17$. Use $n_p = 1$ to avoid tool failure during the second piece. From Eq. (17.5), average production cycle time for the operation is:

$$T_c = 5.0 + 12.57 + 2.0/1 = 19.57\ \text{min/pc}$$

Corresponding hourly production rate R_p = 60/19.57 = 3.1 pc/hr. From Eq. (17.13), average cost per piece for the operation is:

$$C_c = 0.5(5.0) + 0.5(12.57) + 0.5(2.0)/1 + 3.00/1 = \$12.79/\text{pc}$$

(b) Cutting speed for minimum production cost per piece is v_{min} = 42 m/min, and the machining time per piece and tool life are:

$$\text{Machining time } T_m = \frac{\pi(0.5)(0.1)}{(0.25)(10^{-3})(42)} = 14.96\ \text{min/pc}$$

$$\text{Tool life } T = \left(\frac{70}{42}\right)^8 = 59.54\ \text{min/cutting edge}$$

The number of pieces per tool $n_p = 59.54/14.96 = 3.98 \rightarrow$ Use $n_p = 3$ to avoid failure during the fourth workpiece. Average production cycle time for the operation is:

$$T_c = 5.0 + 14.96 + 2.0/3 = 20.63\ \text{min/pc}$$

Corresponding hourly production rate R_p = 60/20.63 = 2.9 pc/hr. Average cost per piece for the operation is:

$$C_c = 0.5(5.0) + 0.5(14.96) + 0.5(2.0)/3 + 3.00/3 = \$11.32/\text{pc}$$

Note that production rate is greater for v_{max} and cost per piece is minimum for v_{min}. ■

Some Comments on Machining Economics Some practical observations can be made relative to these optimum cutting speed equations. First, as the values of C and n increase in the Taylor tool life equation, the optimum cutting speed increases by either Eq. (17.9) or Eq. (17.15). Cemented carbides and ceramic cutting tools should be used at speeds that are significantly higher than for high-speed steel tools.

Second, as the tool change time and/or tooling cost (T_{tc} and C_t) increase, the cutting speed equations yield lower values. Lower speeds allow the tools to last longer, and it is wasteful to change tools too frequently if either the cost of tools or the time to change them is high. An important effect of this tool cost factor is that disposable inserts usually possess a substantial economic advantage over regrindable tooling. Even though the cost per insert is significant, the number of edges per insert is large enough and the time required to change the cutting edge is low enough that disposable tooling generally achieves higher production rates and lower costs per unit product.

Third, v_{max} is always greater than v_{min}. The C_t/n_p term in Eq. (17.15) has the effect of pushing the optimum speed value to the left in Figure 17.14, resulting in a lower value than in Figure 17.13. Rather than taking the risk of cutting at a speed above v_{max} or below v_{min}, some machine shops strive to operate in the interval between v_{min} and v_{max}—an interval sometimes referred to as the "high-efficiency range."

The procedures outlined for selecting feeds and speeds in machining are often difficult to apply in practice. The best feed rate is difficult to determine because the relationships between feed and surface finish, force, horsepower, and other constraints are not readily available for each machine tool. Experience, judgment, and experimentation are required to select the proper feed. The optimum cutting speed is difficult to calculate because the Taylor equation parameters C and n are not usually known without prior testing. Testing of this kind in a production environment is expensive.

REFERENCES

[1] ***ASM Handbook***, Vol. **16**, ***Machining***. ASM International, Materials Park, Ohio, 1989.

[2] Bakerjian, R. (ed.). ***Tool and Manufacturing Engineers Handbook***, 4th ed., Vol. **VI**, ***Design for Manufacturability***. Society of Manufacturing Engineers, Dearborn, Michigan, 1992.

[3] Black, J., and Kohser, R. ***DeGarmo's Materials and Processes in Manufacturing***, 10th ed. John Wiley & Sons, Inc., Hoboken, New Jersey, 2008.

[4] Boothroyd, G., and Knight, W. A. ***Fundamentals of Metal Machining and Machine Tools***, 3rd ed. CRC Taylor and Francis, Boca Raton, Florida, 2006.

[5] Boston, O. W. ***Metal Processing***, 2nd ed. John Wiley & Sons, Inc.New York, 1951.

[6] Bralla, J. G. (ed.). ***Design for Manufacturability Handbook***, 2nd ed. McGraw-Hill Book Company, New York, 1998.

[7] Brierley, R. G., and Siekman, H. J. ***Machining Principles and Cost Control***. McGraw-Hill Book Company, New York, 1964.

[8] Carnes, R., and Maddock, G. "Tool Steel Selection," ***Advanced Materials & Processes***, June 2004, pp. 37–40.

[9] Cook, N. H. "Tool Wear and Tool Life," ***ASME Transactions, J. Engrg. for Industry***, Vol. **95**, November 1973, pp. 931–938.

[10] Davis, J. R. (ed.). ***ASM Specialty Handbook: Tool Materials***. ASM International, Materials Park, Ohio, 1995.

[11] Destephani, J. "The Science of pCBN," ***Manufacturing Engineering***, January 2005, pp. 53–62.

[12] Drozda, T. J., and Wick, C. (eds.). ***Tool and Manufacturing Engineers Handbook***, 4th ed. Vol. **I**, Machining. Society of Manufacturing Engineers, Dearborn, Michigan, 1983.

[13] Gilbert, W. W. "Economics of Machining." ***Machining—Theory and Practice***. American Society for Metals, Materials Park, Ohio, 1950, pp. 465–485.

[14] Groover, M. P. "A Survey on the Machinability of Metals." ***Technical Paper MR76-269***. Society of Manufacturing Engineers, Dearborn, Michigan, 1976.

[15] Koelsch, J. R. "Beyond TiN," ***Manufacturing Engineering***, October 1992, pp. 27–32.

[16] Krar, S. F., and Ratterman, E. ***Superabrasives: Grinding and Machining with CBN and Diamond***. McGraw-Hill, Inc., New York, 1990.

[17] ***Machining Data Handbook***, 3rd ed., Vols. **I** and **II**. Metcut Research Associates, Inc., Cincinnati, Ohio, 1980.

[18] ***Modern Metal Cutting***. AB Sandvik Coromant, Sandvik, Sweden, 1994.

[19] Owen, J. V. "Are Cermets for Real?" *Manufacturing Engineering*, October 1991, pp. 28–31.

[20] Pfouts, W. R. "Cutting Edge Coatings," *Manufacturing Engineering*, July 2000, pp. 98–107.

[21] Shaw, M. C. ***Metal Cutting Principles***, 2nd ed. Oxford University Press, Inc., Oxford, England, 2005.

[22] Spitler, D., Lantrip, J., Nee, J.,and Smith, D. A. ***Fundamentals of Tool Design***, 5th ed. Society of Manufacturing Engineers, Dearborn, Michigan, 2003.

[23] Van Voast, J. ***United States Air Force Machinability Report***, Vol. **3**. Curtis-Wright Corporation, 1954.

REVIEW QUESTIONS

17.1. What are the two principal aspects of cutting-tool technology?

17.2. Name the three modes of tool failure in machining.

17.3. What are the two principal locations on a cutting tool where tool wear occurs?

17.4. Identify the mechanisms by which cutting tools wear during machining.

17.5. What is the physical interpretation of the parameter C in the Taylor tool life equation?

17.6. What are some of the tool life criteria used in production machining operations?

17.7. Identify three desirable properties of a cutting-tool material.

17.8. What are the principal alloying ingredients in high-speed steel?

17.9. What is the difference in ingredients between steel cutting grades and nonsteel cutting grades of cemented carbides?

17.10. Identify some of the common compounds that form the thin coatings on the surface of coated carbide inserts.

17.11. Name the seven elements of tool geometry for a single-point cutting tool.

17.12. Identify the alternative ways by which a cutting tool is held in place during machining.

17.13. Name the two main categories of cutting fluid according to function.

17.14. Name the four categories of cutting fluid according to chemistry.

17.15. What are the methods by which cutting fluids are applied in a machining operation?

17.16. Dry machining is being considered by machine shops because of certain problems inherent in the use of cutting fluids. What are those problems associated with the use of cutting fluids?

17.17. What are some of the new problems introduced by dry machining?

17.18. Define machinability.

17.19. What are the criteria by which machinability is commonly assessed in a production machining operation?

17.20. What are the factors on which the selection of feed in a machining operation should be based?

17.21. The unit cost in a machining operation is the sum of four cost terms. The first three terms are: (1) part load/unload cost, (2) cost of time the tool is actually cutting the work, and (3) cost of the time to change the tool. What is the fourth term?

17.22. Which cutting speed is always lower for a given machining operation, cutting speed for minimum cost or cutting speed for maximum production rate? Why?

PROBLEMS

17.1. Flank-wear data were collected in a series of turning tests using a coated carbide tool on hardened alloy steel at a feed of 0.30 mm/rev and a depth of 4.0 mm. At a speed of 125 m/min, flank wear = 0.12 mm at 1 min, 0.27 mm at 5 min, 0.45 mm at 11 min, 0.58 mm at 15 min, 0.73 at 20 min, and 0.97 mm at 25 min. At a speed of 165 m/min, flank wear = 0.22 mm at 1 min, 0.47 mm at 5 min, 0.70 mm at 9 min, 0.80 mm at 11 min, and 0.99 mm at 13 min. The last value in each case is when final tool failure occurred. (a) On a single piece of linear graph paper, plot flank wear as a function of time for both speeds. Using 0.75 mm of flank wear as the criterion of tool failure, determine the tool lives for the two cutting speeds. (b) On a piece of natural log–log paper, plot your results determined in the previous part. From the plot, determine the values of n and C in the Taylor tool life equation. (c) As a comparison, calculate the values of n and C in the Taylor equation solving simultaneous equations. Are the resulting n and C values the same?

17.2. Solve Problem 17.1, except that the tool life criterion is 0.50 mm of flank land wear rather than 0.75 mm.

17.3. Tool life tests on a lathe have resulted in the following data: (1) at a cutting speed of 375 ft/min, the tool life was 5.5 min; (2) at a cutting speed of 275 ft/min, the tool life was 53 min. (a) Determine the parameters n and C in the Taylor tool life equation. (b) Based on the n and C values, what is the likely tool material used in this operation? (c) Using your equation, compute the tool life that corresponds to a cutting speed of 300 ft/min. (d) Compute the cutting speed that corresponds to a tool life $T = 10$ min.

17.4. Tool life tests in turning yield the following data: (1) when cutting speed is 100 m/min, tool life is 10 min; (2) when cutting speed is 75 m/min, tool life is 30 min. (a) Determine the n and C values in the Taylor tool life equation. Based on your equation, compute (b) the tool life for a speed of 110 m/min, and (c) the speed corresponding to a tool life of 15 min.

17.5. Turning tests have resulted in 1-min tool life at a cutting speed = 4.0 m/s and a 20-min tool life at a speed = 2.0 m/s. (a) Find the n and C values in the Taylor tool life equation. (b) Project how long the tool would last at a speed of 1.0 m/s.

17.6. In a production turning operation, the workpart is 125 mm in diameter and 300 mm long. A feed of 0.225 mm/rev is used in the operation. If cutting speed = 3.0 m/s, the tool must be changed every 5 workparts; but if cutting speed = 2.0 m/s, the tool can be used to produce 25 pieces between tool changes. Determine the Taylor tool life equation for this job.

17.7. For the tool life plot of Figure 17.4, show that the middle data point (v = 130 m/min, T = 12 min) is consistent with the Taylor equation determined in Example 17.1 in the text.

17.8. In the tool-wear plots of Figure 17.3, complete failure of the cutting tool is indicated by the end of each wear curve. Using complete failure as the criterion of tool life instead of 0.50 mm flank wear, the resulting data are: (1) v = 160 m/min, T = 5.75 min; (2) v = 130 m/min, T = 14.25 min; and (3) v = 100 m/min, T = 47 min. Determine the parameters n and C in the Taylor tool life equation for this data.

17.9. A series of turning tests are performed to determine the parameters n, m, and K in the expanded version of the Taylor equation, Eq. (17.4). The following data were obtained during the tests: (1) cutting speed = 1.9 m/s, feed = 0.22 mm/rev, tool life = 10 min; (2) cutting speed = 1.3 m/s, feed = 0.22 mm/rev, tool life = 47 min; and (3) cutting speed = 1.9 m/s, feed = 0.32 mm/rev, tool life = 8 min. (a) Determine n, m, and K. (b) Using your equation, compute the tool life when the cutting speed is 1.5 m/s and the feed is 0.28 mm/rev.

17.10. A drilling operation is performed in which 0.5 in diameter holes are drilled through cast iron plates that are 1.0 in thick. Sample holes have been drilled to determine the tool life at two cutting speeds. At 80 surface ft/min, the tool lasted for exactly 50 holes. At 120 surface ft/min, the tool lasted for exactly 5 holes. The feed of the drill was 0.003 in/rev. (Ignore effects of drill entrance and exit from the hole. Consider the depth of cut to be exactly 1.00 in, corresponding to the plate thickness.) Determine the values of n and C in the Taylor tool life equation for the above sample data, where cutting speed v is expressed in ft/min, and tool life T is expressed in min.

17.11. The outside diameter of a cylinder made of titanium alloy is to be turned. The starting diameter is 400 mm and the length is 1100 mm. The feed is 0.35 mm/rev and the depth of cut is 2.5 mm. The cut will be made with a cemented carbide cutting tool whose Taylor tool life parameters are: n = 0.24 and C = 450. Units for the Taylor equation are min for tool life and m/min for cutting speed. Compute the cutting speed that will allow the tool life to be just equal to the cutting time for this part.

17.12. The workpart in a turning operation is 88 mm in diameter and 400 mm long. A feed of 0.25 mm/rev is used in the operation. If cutting speed = 3.5 m/s, the tool must be changed every 3 workparts; but if cutting speed = 2.5 m/s, the tool can be used to produce 20 pieces between tool changes. Determine the cutting speed that will allow the tool to be used for 50 parts between tool changes.

17.13. The outside diameter of a cylinder made of a steel alloy is to be turned. The starting diameter is 300 mm and the length is 625 mm. The feed is 0.35 mm/rev and the depth of cut is 2.5 mm. The cut will be made with a cemented carbide cutting tool whose Taylor tool life parameters are: n = 0.24 and C = 450. Units for the Taylor equation are min for tool life and m/min for cutting speed. Compute the cutting speed that will allow the tool life to be just equal to the cutting time for three of these parts.

17.14. In a turning operation using high-speed steel tooling, cutting speed = 110 m/min. The Taylor tool life equation has parameters n = 0.140 and C = 150 (m/min) when the operation is conducted dry. When a coolant is used in the operation, the value of C is increased by 15%. Determine the percent increase in tool life that results if the cutting speed is maintained at 110 m/min.

17.15. A production turning operation on a steel workpiece normally operates at a cutting speed of 125 ft/min using high-speed steel tooling with no cutting fluid. The appropriate n and C values in the Taylor equation are given in Table 17.2 in the text. It has been found that the use of a coolant-type cutting fluid will allow an increase of 25 ft/min in the speed without any effect on tool life. If it can be assumed that the effect of the cutting fluid is simply to increase the constant C by 25, what would be the increase in tool life if the original cutting speed of 125 ft/min were used in the operation?

17.16. A machinability rating is to be determined for a new work material using the cutting speed for a 60-min tool life as the basis of comparison. For the base material (B1112 steel), test data resulted in Taylor equation parameter values of n = 0.29 and C = 500, where speed is in m/min and tool life is min. For the new material, the parameter values were n = 0.21 and C = 400. These results were obtained using cemented carbide tooling. (a) Compute a machinability rating for the new material. (b) Suppose the machinability criterion were the cutting speed for a 10-min tool life rather than the present criterion. Compute the machinability rating for this case. (c) What do the results of the two calculations show about the difficulties in machinability measurement?

17.17. A machinability rating is to be determined for a new work material. For the base material (B1112), test data resulted in a Taylor equation with parameters $n = 0.29$ and $C = 490$. For the new material, the Taylor parameters were $n = 0.23$ and $C = 430$. Units in both cases are: speed in m/min and tool life in min. These results were obtained using cemented carbide tooling. (a) Compute a machinability rating for the new material using cutting speed for a 30-min tool life as the basis of comparison. (b) If the machinability criterion were tool life for a cutting speed of 150 m/min, what is the machinability rating for the new material?

17.18. A high-speed steel tool is used to turn a steel workpart that is 300 mm long and 80 mm in diameter. The parameters in the Taylor equation are: $n = 0.13$ and $C = 75$ (m/min) for a feed of 0.4 mm/rev. The operator and machine tool rate = \$30.00/hr, and the tooling cost per cutting edge = \$4.00. It takes 2.0 min to load and unload the workpart and 3.50 min to change tools. Determine (a) cutting speed for maximum production rate, (b) tool life in min of cutting, and (c) cycle time and cost per unit of product.

17.19. Solve Problem 17.18, except that in part (a), determine cutting speed for minimum cost.

17.20. A cemented carbide tool is used to turn a part with a length of 14.0 in and diameter = 4.0 in. The parameters in the Taylor equation are: $n = 0.25$ and $C = 1000$ (ft/min). The rate for the operator and machine tool = \$45.00/hr, and the tooling cost per cutting edge = \$2.50. It takes 2.5 min to load and unload the workpart and 1.50 min to change tools. The feed = 0.015 in/rev. Determine (a) cutting speed for maximum production rate, (b) tool life in min of cutting, and (c) cycle time and cost per unit of product.

17.21. Solve Problem 17.20, except that in part (a), determine cutting speed for minimum cost.

17.22. Disposable and regrindable tooling are to be compared. The same grade of cemented carbide tooling is available in two forms for turning operations in a certain machine shop: disposable inserts and brazed inserts. The parameters in the Taylor equation for this grade are: $n = 0.25$ and $C = 300$ (m/min) under the cutting conditions considered here. For the disposable inserts, price of each insert = \$6.00, there are four cutting edges per insert, and the tool change time = 1.0 min (this is an average of the time to index the insert and the time to replace it when all edges have been used). For the brazed insert, the price of the tool = \$30.00 and it is estimated that it can be used a total of 15 times before it must be scrapped. The tool change time for the regrindable tooling = 3.0 min. The standard time to grind or regrind the cutting edge is 5.0 min, and the grinder is paid at a rate = \$20.00/hr. Machine time on the lathe costs \$24.00/hr. The workpart to be used in the comparison is 375 mm long and 62.5 mm in diameter, and it takes 2.0 min to load and unload the work. The feed = 0.30 mm/rev. For the two tooling cases, compare (a) cutting speeds for minimum cost, (b) tool lives, (c) cycle time and cost per unit of production. Which tool would you recommend?

17.23. Solve Problem 17.22, except that in part (a), determine the cutting speeds for maximum production rate.

17.24. Three tool materials are to be compared for the same finish turning operation on a batch of 150 steel parts: high-speed steel, cemented carbide, and ceramic. For the high-speed steel tool, the Taylor equation parameters are: $n = 0.130$ and $C = 80$ (m/min). The price of the HSS tool is \$20.00 and it is estimated that it can be ground and reground 15 times at a cost of \$2.00 per grind. Tool change time is 3 min. Both carbide and ceramic tools are in insert form and can be held in the same mechanical toolholder. The Taylor equation parameters for the cemented carbide are: $n = 0.30$ and $C = 650$ (m/min); and for the ceramic: $n = 0.6$ and $C = 3{,}500$ (m/min). The cost per insert for the carbide is \$8.00 and for the ceramic is \$10.00. There are 6 cutting edges per insert in both cases. Tool change time is 1.0 min for both tools. The time to change a part is 2.5 min. The feed is 0.30 mm/rev, and depth of cut is 3.5 mm. The cost of machine time is \$40/hr. The part is 73.0 mm in diameter and 250 mm in length. Setup time for the batch is 2.0 hr. For the three tooling cases, compare: (a) cutting speeds for minimum cost, (b) tool lives, (c) cycle time, (d) cost per production unit, (e) total time to complete the batch and production rate. (f) What is the proportion of time spent actually cutting metal for each tooling? Use of a spreadsheet calculator is recommended.

17.25. Solve Problem 17.24, except that in parts (a) and (b), determine the cutting speeds and tool lives for maximum production rate. Use of a spreadsheet calculator is recommended.

17.26. A vertical boring mill is used to bore the inside diameter of a large batch of tube-shaped parts. The diameter = 28.0 in and the length of the bore = 14.0 in. Current cutting conditions are: speed = 200 ft/min, feed = 0.015 in/rev, and depth = 0.125 in. The parameters of the Taylor equation for the cutting tool in the operation are: $n = 0.23$ and $C = 850$ (ft/min). Tool change time = 3.0 min, and tooling cost = $3.50 per cutting edge. The time required to load and unload the parts = 12.0 min, and the cost of machine time on this boring mill = $42.00/hr. Management has decreed that the production rate must be increased by 25%. Is that possible? Assume that feed must remain unchanged in order to achieve the required surface finish. What is the current production rate and the maximum possible production rate for this job?

17.27. As indicated in Section 17.4, the effect of a cutting fluid is to increase the value of C in the Taylor tool life equation. In a certain machining situation using HSS tooling, the C value is increased from $C = 200$ to $C = 225$ due to the use of the cutting fluid. The n value is the same with or without fluid at $n = 0.125$. Cutting speed used in the operation is $v = 125$ ft/min. Feed = 0.010 in/rev and depth = 0.100 in. The effect of the cutting fluid can be to either increase cutting speed (at the same tool life) or increase tool life (at the same cutting speed). (a) What is the cutting speed that would result from using the cutting fluid if tool life remains the same as with no fluid? (b) What is the tool life that would result if the cutting speed remained at 125 ft/min? (c) Economically, which effect is better, given that tooling cost = $2.00 per cutting edge, tool change time = 2.5 min, and operator and machine rate = $30/hr? Justify your answer with calculations, using cost per cubic in of metal machined as the criterion of comparison. Ignore effects of workpart handling time.

18 GRINDING AND OTHER ABRASIVE PROCESSES

Chapter Contents

18.1 Grinding
- 18.1.1 The Grinding Wheel
- 18.1.2 Analysis of the Grinding Process
- 18.1.3 Application Considerations in Grinding
- 18.1.4 Grinding Operations and Grinding Machines

18.2 Related Abrasive Processes
- 18.2.1 Honing
- 18.2.2 Lapping
- 18.2.3 Superfinishing
- 18.2.4 Polishing and Buffing

Abrasive machining involves material removal by the action of hard, abrasive particles that are usually in the form of a bonded wheel. Grinding is the most important abrasive process. Other traditional abrasive processes include honing, lapping, superfinishing, polishing, and buffing. The abrasive machining processes are generally used as finishing operations, although some abrasive processes are capable of high material removal rates rivaling those of conventional machining operations.

The use of abrasives to shape parts is probably the oldest material removal process. Abrasive processes are important commercially and technologically for the following reasons:

- They can be used on all types of materials ranging from soft metals to hardened steels and hard nonmetallic materials such as ceramics and silicon.
- Some of these processes can produce extremely fine surface finishes, to 0.025 μm (1 μ-in).
- For certain abrasive processes, dimensions can be held to extremely close tolerances.

Abrasive water jet cutting and ultrasonic machining are also abrasive processes, because material removal is accomplished by means of abrasives. However, they are commonly classified as nontraditional processes and are covered in the following chapter.

18.1 GRINDING

Grinding is a material removal process carried out by abrasive particles that are contained in a bonded grinding wheel rotating at very high surface speeds. The grinding wheel is usually disk-shaped, and is precisely balanced for high rotational speeds.

Grinding can be likened to the milling process. Cutting occurs on either the periphery or the face of the grinding wheel, similar to peripheral and face milling. Peripheral grinding is much more common than face grinding. The rotating grinding wheel consists of many cutting teeth (the abrasive particles), and the work is fed relative to the wheel to accomplish material removal. Despite these similarities, there are significant differences between grinding and milling: (1) the abrasive grains in the wheel are much smaller and more numerous than the teeth on a milling cutter; (2) cutting speeds in grinding are much higher than in milling; (3) the abrasive grits in a grinding wheel are randomly oriented and possess on average a very high negative rake angle; and (4) a grinding wheel is self-sharpening—as the wheel wears, the abrasive particles become dull and either fracture to create fresh cutting edges or are pulled out of the surface of the wheel to expose new grains.

18.1.1 THE GRINDING WHEEL

A grinding wheel consists of abrasive particles and bonding material. The bonding material holds the particles in place and establishes the shape and structure of the wheel. These two ingredients and the way they are fabricated determine the five basic parameters of a grinding wheel: (1) abrasive material, (2) grain size, (3) bonding material, (4) wheel grade, and (5) wheel structure. To achieve the desired performance in a given application, each of the parameters must be carefully selected.

Abrasive Material Different abrasive materials are appropriate for grinding different work materials. General properties of an abrasive material used in grinding wheels include high hardness, wear resistance, toughness, and friability. Hardness, wear resistance, and toughness are desirable properties of any cutting-tool material. ***Friability*** refers to the capacity of the abrasive material to fracture when the cutting edge of the grain becomes dull, thereby exposing a new sharp edge.

The development of grinding abrasives is described in our historical note. Today, the abrasive materials of greatest commercial importance are aluminum oxide, silicon carbide, cubic boron nitride, and diamond. They are briefly described in Table 18.1, which also provides their relative hardness values.

TABLE 18.1 Abrasives of greatest importance in grinding

Abrasive	Description	Knoop Hardness
Aluminum oxide (Al_2O_3)	Most common abrasive material, used to grind steel and other ferrous, high-strength alloys.	2100
Silicon carbide (SiC)	Harder than Al_2O_3, but not as tough. Applications include ductile metals such as aluminum, brass, and stainless steel, as well as brittle materials such as some cast irons and certain ceramics. Cannot be used effectively for grinding steel because of the strong chemical affinity between the carbon in SiC and the iron in steel.	1800
Cubic boron nitride (cBN)	When used as an abrasive, cBN is produced under the trade name Borazon by the General Electric Company. cBN grinding wheels are used for hard materials such as hardened tool steels and aerospace alloys.	5000
Diamond	Diamond abrasives occur naturally and are also made synthetically. Diamond wheels are generally used in grinding applications on hard, abrasive materials such as ceramics, cemented carbides, and glass.	7000

Compiled from [11] and other sources.

Grain Size The grain size of the abrasive particle is important in determining surface finish and material removal rate. Small grit sizes produce better finishes, whereas larger grain sizes permit larger material removal rates. Thus, a choice must be made between these two objectives when selecting abrasive grain size. The selection of grit size also depends to some extent on the hardness of the work material. Harder work materials require smaller grain sizes to cut effectively, whereas softer materials require larger grit sizes.

The grit size is measured using a screen mesh procedure, as explained in the appendix of Chapter 10. In this procedure, smaller grit sizes have larger numbers and vice versa. Grain sizes used in grinding wheels typically range between 8 and 250. Grit size 8 is very coarse and size 250 is very fine. Even finer grit sizes are used for lapping and superfinishing (Section 18.2).

Bonding Materials The bonding material holds the abrasive grains and establishes the shape and structural integrity of the grinding wheel. Desirable properties of the bond material include strength, toughness, hardness, and temperature resistance. The bonding material must be able to withstand the centrifugal forces and high temperatures experienced by the grinding wheel, resist shattering in shock loading of the wheel, and hold the abrasive grains rigidly in place to accomplish the cutting action while allowing those grains that are worn to be dislodged so that new grains can be exposed. Bonding materials commonly used in grinding wheels are identified and briefly described in Table 18.2.

Wheel Structure and Wheel Grade Wheel structure refers to the relative spacing of the abrasive grains in the wheel. In addition to the abrasive grains and bond material, grinding wheels contain air gaps or pores, as illustrated in Figure 18.1. The volumetric proportions of grains, bond material, and pores can be expressed as:

$$P_g + P_b + P_p = 1.0 \tag{18.1}$$

where P_g = proportion of abrasive grains in the total wheel volume, P_b = proportion of bond material, and P_p = proportion of pores (air gaps).

Wheel structure is measured on a scale that ranges between "open" and "dense." An open structure is one in which P_p is relatively large, and P_g is relatively small. That is, there are more pores and fewer grains per unit volume in a wheel of open structure. By contrast, a dense structure is one in which P_p is relatively small, and P_g is larger.

TABLE 18.2 Bonding materials used in grinding wheels

Bonding Material	Description
Vitrified bond	Consists chiefly of baked clay and ceramic materials. Most grinding wheels in common use are vitrified bonded wheels. They are strong and rigid, resistant to elevated temperatures, and relatively unaffected by water and oil that might be used in grinding fluids.
Silicate bond	Consists of sodium silicate (Na_2SO_3). Applications are generally limited to situations in which heat generation must be minimized, such as grinding cutting tools.
Rubber bond	Most flexible of the bonding materials and used as a bonding material in cutoff wheels.
Resinoid bond	Consists of various thermosetting resin materials, such as phenol-formaldehyde. It has very high strength and is used for rough grinding and cutoff operations.
Shellac bond	Relatively strong but not rigid; often used in applications requiring a good finish.
Metallic bond	Metal, usually bronze, is the common bond material for diamond and cBN grinding wheels. Particulate processing (Chapters 10 and 11) is used to bond the metal matrix and abrasive grains to the outside periphery of the wheel, thus conserving the costly abrasive materials.

Compiled from [11] and other sources.

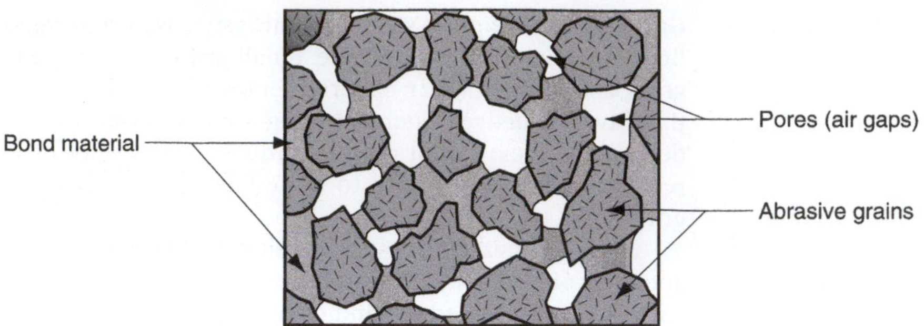

FIGURE 18.1 Typical structure of a grinding wheel. (Credit: *Fundamentals of Modern Manufacturing*, 4th Edition by Mikell P. Groover, 2010. Reprinted with permission of John Wiley & Sons, Inc.)

Generally, open structures are recommended for situations in which clearance for chips must be provided. Dense structures are used to obtain better surface finish and dimensional control.

Wheel grade indicates the grinding wheel's bond strength in retaining the abrasive grits during cutting. This depends largely on the amount of bonding material present in the wheel structure—P_b in Eq. (18.1). Grade is measured on a scale that ranges between soft and hard. "Soft" wheels lose grains readily, whereas "hard" wheels retain their abrasive grains. Soft wheels are generally used for applications requiring low material removal rates and grinding of hard work materials. Hard wheels are typically used to achieve high stock removal rates and for grinding of relative soft work materials.

Grinding Wheel Specification The preceding parameters can be concisely designated in a standard grinding wheel marking system defined by the American National Standards Institute (ANSI) [3]. This marking system uses numbers and letters to specify abrasive type, grit size, grade, structure, and bond material. Table 18.3 presents an abbreviated version of the ANSI Standard, indicating how the numbers and letters are interpreted. The standard also provides for additional identifications that might be used by the grinding wheel manufacturers. The ANSI Standard for diamond and cubic boron nitride grinding wheels is slightly different than for conventional wheels. The marking system for these newer grinding wheels is presented in Table 18.4.

TABLE 18.3 Marking system for conventional grinding wheels as defined by ANSI Standard B74.13-1977

30	A	46	H	6	V	XX

Manufacturer's private marking for wheel (optional)

Bond type: B = Resinoid, BF = Resinoid reinforced, E = Shellac, R = Rubber, RF = Rubber reinforced, S = Silicate, V = Vitrified.

Structure: Scale ranges from 1 to 15: 1 = very dense structure, 15 = very open structure

Grade: Scale ranges from A to Z: A = soft, M = medium, Z = hard

Grain size: Coarse = grit sizes 8 to 24, Medium = grit sizes 30 to 60, Fine = grit sizes 70 to 180, Very fine = grit sizes 220 to 600.

Abrasive type: A = aluminum oxide, C = silicon carbide.

Prefix: Manufacturer's symbol for abrasive (optional)

TABLE 18.4 Marking system for diamond and cubic boron nitride grinding wheels as defined by ANSI Standard B74.13-1977

XX	D	150	P	YY	M	ZZ	3

- ***Depth of abrasive*** = working depth of abrasive section in mm (shown) or inches, as in Figure 18.2(c).
- ***Bond modification*** = manufacturer's notation of special bond type or modification.
- ***Bond type***: B = Resin, M = metal, V = Vitrified
- ***Concentration***: Manufacturer's designation. May be number or symbol.
- ***Grade***: Scale ranges from A to Z: A = soft, M = medium, Z = hard
- ***Grain size***: Coarse = grit sizes 8 to 24, Medium = grit sizes 30 to 60, Fine = grit sizes 70 to 180, Very fine = grit sizes 220 to 600.
- ***Abrasive type***: D = diamond, B = cubic boron nitride.
- ***Prefix***: Manufacturer's symbol for abrasive (optional)

Grinding wheels come in a variety of shapes and sizes, as shown in Figure 18.2. Configurations (a), (b), and (c) are peripheral grinding wheels, in which material removal is accomplished by the outside circumference of the wheel. A typical abrasive cutoff wheel is shown in (d), which also involves peripheral cutting. Wheels (e), (f), and (g) are face grinding wheels, in which the flat face of the wheel removes material from the work surface.

18.1.2 ANALYSIS OF THE GRINDING PROCESS

The cutting conditions in grinding are characterized by very high speeds and very small cut size, compared to milling and other traditional machining operations. Using surface grinding to illustrate, Figure 18.3(a) shows the principal features of the process. The peripheral speed of the grinding wheel is determined by the rotational speed of the wheel:

$$v = \pi D N \tag{18.2}$$

where v = surface speed of wheel, m/min (ft/min); N = spindle speed, rev/min; and D = wheel diameter, m (ft).

Depth of cut d, called the ***infeed***, is the penetration of the wheel below the original work surface. As the operation proceeds, the grinding wheel is fed laterally across the surface on each pass by the work. This is called the ***crossfeed***, and it determines the width of the grinding path w in Figure 18.3(a). This width, multiplied by depth d determines the cross-sectional area of the cut. In most grinding operations, the work moves past the wheel at a certain speed v_w, so that the material removal rate is:

$$R_{MR} = v_w w d \tag{18.3}$$

Each grain in the grinding wheel cuts an individual chip whose longitudinal shape before cutting is shown in Figure 18.3(b) and whose assumed cross-sectional shape is triangular, as in Figure 18.3(c). At the exit point of the grit from the work, where the chip cross section is largest, this triangle has height t and width w'.

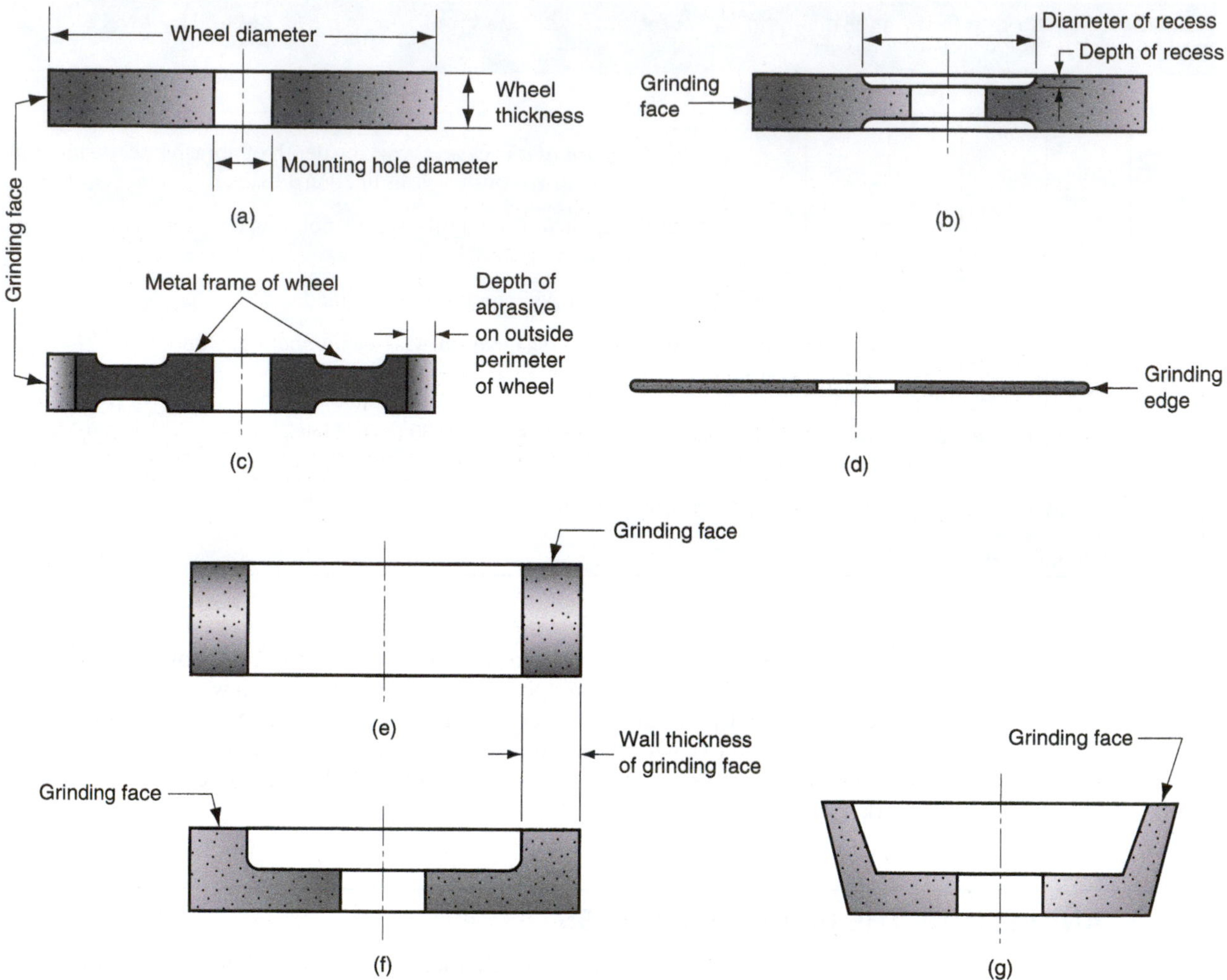

FIGURE 18.2 Some of the standard grinding wheel shapes: (a) straight, (b) recessed two sides, (c) metal wheel frame with abrasive bonded to outside circumference, (d) abrasive cutoff wheel, (e) cylinder wheel, (f) straight cup wheel, and (g) flaring cup wheel. (Credit: *Fundamentals of Modern Manufacturing*, 4th Edition by Mikell P. Groover, 2010. Reprinted with permission of John Wiley & Sons, Inc.)

In a grinding operation, we are interested in how the cutting conditions combine with the grinding wheel parameters to affect (1) surface finish, (2) forces and energy, (3) temperature of the work surface, and (4) wheel wear.

Surface Finish Most commercial grinding is performed to achieve a surface finish that is superior to that which can be accomplished with conventional machining. The surface finish of the workpart is affected by the size of the individual chips formed during grinding. One obvious factor in determining chip size is grit size—smaller grit sizes yield better finishes.

Let us examine the dimensions of an individual chip. From the geometry of the grinding process in Figure 18.3, it can be shown that the average length of a chip is given by:

$$l_c = \sqrt{Dd} \tag{18.4}$$

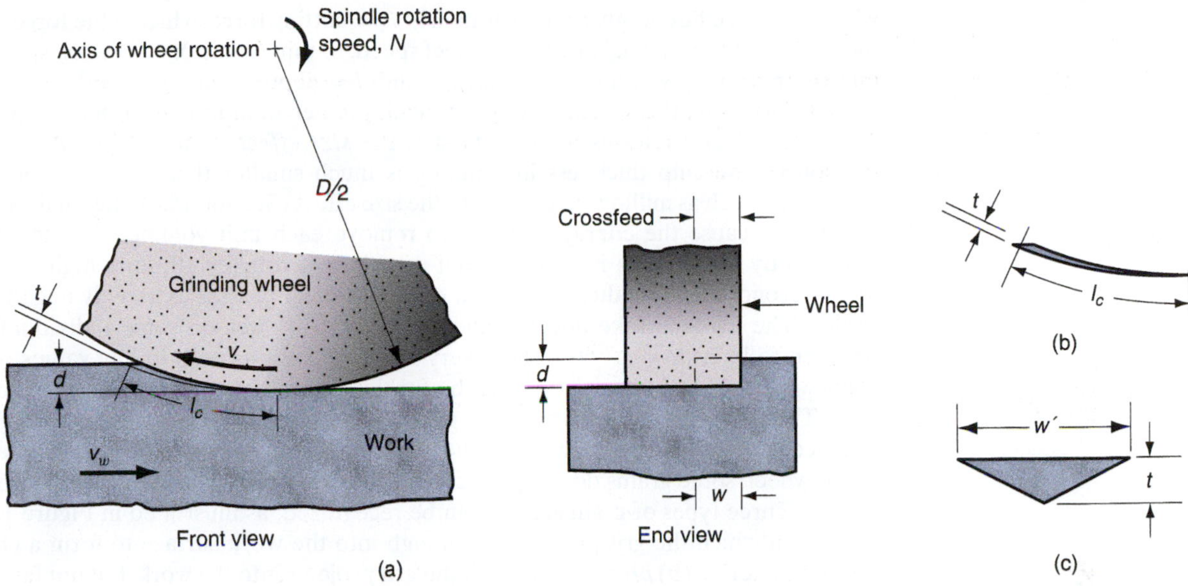

FIGURE 18.3 (a) The geometry of surface grinding, showing the cutting conditions; (b) assumed longitudinal shape; and (c) cross section of a single chip. (Credit: *Fundamentals of Modern Manufacturing*, 4th Edition by Mikell P. Groover, 2010. Reprinted with permission of John Wiley & Sons, Inc.)

where l_c is the length of the chip, mm (in); D = wheel diameter, mm (in); and d = depth of cut, or infeed, mm (in). This assumes the chip is formed by a grit that acts throughout the entire sweep arc shown in the diagram.

Figure 18.3(c) shows the assumed cross section of a chip in grinding. The cross-sectional shape is triangular with width w' being greater than the thickness t by a factor called the grain aspect ratio r_g, defined by:

$$r_g = \frac{w'}{t} \tag{18.5}$$

Typical values of grain aspect ratio are between 10 and 20.

The number of active grits (cutting teeth) per square inch on the outside periphery of the grinding wheel is denoted by C. In general, smaller grain sizes give larger C values. C is also related to the wheel structure. A denser structure means more grits per area. Based on the value of C, the number of chips formed per time n_c is given by:

$$n_c = vwC \tag{18.6}$$

where v = wheel speed, mm/min (in/min); w = crossfeed, mm (in); and C = grits per area on the grinding wheel surface, grits/mm^2 (grits/in^2). It stands to reason that surface finish will be improved by increasing the number of chips formed per unit time on the work surface for a given width w. Therefore, according to Eq. (18.6), increasing v and/or C will improve finish.

Forces and Energy If the force required to drive the work past the grinding wheel were known, the specific energy in grinding could be determined as:

$$U = \frac{F_c v}{v_w w d} \tag{18.7}$$

where U = specific energy, J/mm^3 (in-lb/in^3); F_c = cutting force, which is the force to drive the work past the wheel, N (lb); v = wheel speed, m/min (ft/min); v_w = work speed, mm/min (in/min); w = width of cut, mm (in); and d = depth of cut, mm (in).

In grinding, the specific energy is much greater than in conventional machining. There are several reasons for this. First is the ***size effect*** in machining. As discussed previously, the chip thickness in grinding is much smaller than for other machining operations, such as milling. According to the size effect (Section 15.4), the small chip sizes in grinding cause the energy required to remove each unit volume of material to be significantly higher than in conventional machining—roughly 10 times higher.

Second, the individual grains in a grinding wheel possess extremely negative rake angles. The average rake angle is about –30°, with values on some individual grains believed to be as low as –60°. These very low rake angles result in low values of shear plane angle and high shear strains, both of which mean higher energy levels in grinding.

Third, specific energy is higher in grinding because not all of the individual grits are engaged in actual cutting. Because of the random positions and orientations of the grains in the wheel, some grains do not project far enough into the work surface to accomplish cutting. Three types of grain actions can be recognized, as illustrated in Figure 18.4: (a) ***cutting***, in which the grit projects far enough into the work surface to form a chip and remove material; (b) ***plowing***, in which the grit projects into the work, but not far enough to cause cutting; instead, the work surface is deformed and energy is consumed without any material removal; and (c) ***rubbing***, in which the grit contacts the surface during its sweep, but only rubbing friction occurs, thus consuming energy without removing any material.

The size effect, negative rake angles, and ineffective grain actions combine to make the grinding process inefficient in terms of energy consumption per volume of material removed.

Using the specific energy relationship in Eq. (18.7), and assuming that the cutting force acting on a single grain in the grinding wheel is proportional to $r_g t$, it can be shown [10] that:

$$F'_c = K_1 \left(\frac{r_g v_w}{vC}\right)^{0.5} \left(\frac{d}{D}\right)^{0.25} \tag{18.8}$$

where F'_c is the cutting force acting on an individual grain, K_1 is a constant of proportionality that depends on the strength of the material being cut and the sharpness of the individual grain, and the other terms have been previously defined. The practical significance of this relationship is that F'_c affects whether an individual grain will be pulled out of the grinding wheel, an important factor in the wheel's capacity to "resharpen" itself.

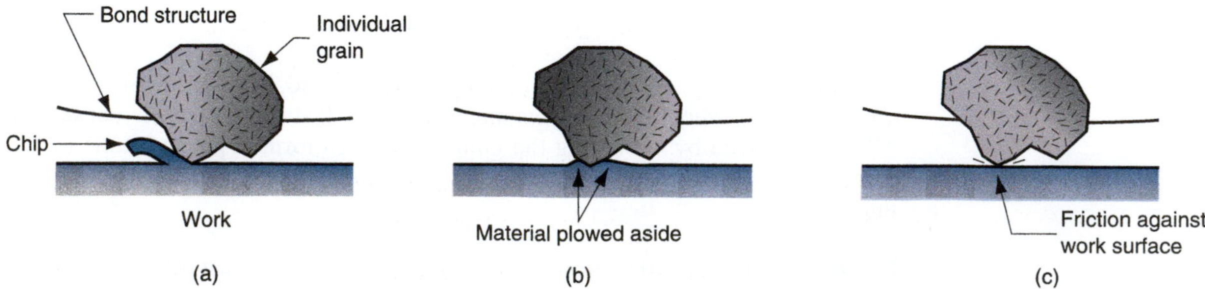

FIGURE 18.4 Three types of grain action in grinding: (a) cutting, (b) plowing, and (c) rubbing. (Credit: *Fundamentals of Modern Manufacturing*, 4th Edition by Mikell P. Groover, 2010. Reprinted with permission of John Wiley & Sons, Inc.)

Referring back to our discussion on wheel grade, a hard wheel can be made to appear softer by increasing the cutting force acting on an individual grain through appropriate adjustments in v_w, v, and d, according to Eq. (18.8).

Temperatures at the Work Surface Because of the size effect, high negative rake angles, and plowing and rubbing of the abrasive grits against the work surface, the grinding process is characterized by high temperatures. Unlike conventional machining operations in which most of the heat generated in the process is carried off in the chip, much of the heat in grinding remains in the ground surface, resulting in high work surface temperatures. The high surface temperatures have several possible damaging effects, primarily surface burns and cracks. The burn marks show themselves as discolorations on the surface caused by oxidation. Grinding burns are often a sign of metallurgical damage immediately beneath the surface. The surface cracks are perpendicular to the wheel speed direction. They indicate an extreme case of thermal damage to the work surface.

A second harmful thermal effect is softening of the work surface. Many grinding operations are carried out on parts that have been heat-treated to obtain high hardness. High grinding temperatures can cause the surface to lose some of its hardness. Third, thermal effects in grinding can cause residual stresses in the work surface, possibly decreasing the fatigue strength of the part.

It is important to understand what factors influence work surface temperatures in grinding. Experimentally, it has been observed that surface temperature is dependent on energy per surface area ground (closely related to specific energy U). Because this varies inversely with chip thickness, it can be shown that surface temperature T_s is related to grinding parameters as follows [10]:

$$T_s = K_2 d^{0.75} \left(\frac{r_g C v}{v_w} \right)^{0.5} D^{0.25} \tag{18.9}$$

where K_2 = a constant of proportionality. The practical implication of this relationship is that surface damage owing to high work temperatures can be mitigated by decreasing depth of cut d, wheel speed v, and number of active grits per square inch on the grinding wheel C, or by increasing work speed v_w. In addition, dull grinding wheels and wheels that have a hard grade and dense structure tend to cause thermal problems. Of course, using a cutting fluid can also reduce grinding temperatures.

Wheel Wear Grinding wheels wear, just as conventional cutting tools wear. Three mechanisms are recognized as the principal causes of wear in grinding wheels: (1) grain fracture, (2) attritious wear, and (3) bond fracture. ***Grain fracture*** occurs when a portion of the grain breaks off, but the rest of the grain remains bonded in the wheel. The edges of the fractured area become new cutting edges on the grinding wheel. The tendency of the grain to fracture is called ***friability***. High friability means that the grains fracture more readily due to the cutting forces on the grains F'_c.

Attritious wear involves dulling of the individual grains, resulting in flat spots and rounded edges. Attritious wear is analogous to tool wear in a conventional cutting tool. It is caused by similar physical mechanisms including friction and diffusion, as well as chemical reactions between the abrasive material and the work material in the presence of very high temperatures.

Bond fracture occurs when the individual grains are pulled out of the bonding material. The tendency toward this mechanism depends on wheel grade, among other factors. Bond fracture usually occurs because the grain has become dull due to attritious wear, and the resulting cutting force is excessive. Sharp grains cut more efficiently with lower cutting forces; hence, they remain attached in the bond structure.

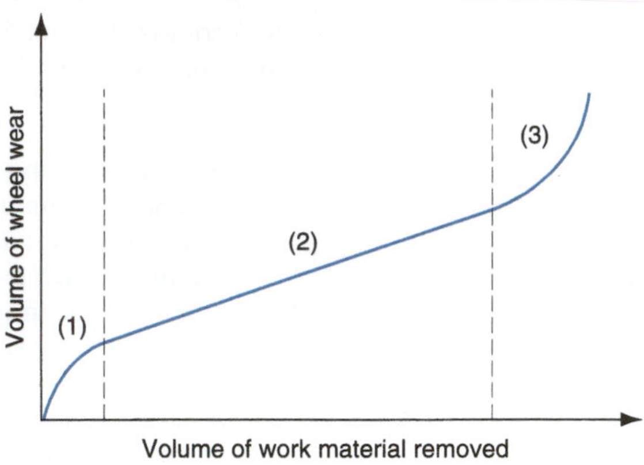

FIGURE 18.5 Typical wear curve of a grinding wheel. Wear is conveniently plotted as a function of volume of material removed, rather than as a function of time. Based on [16]. (Credit: *Fundamentals of Modern Manufacturing*, 4th Edition by Mikell P. Groover, 2010. Reprinted with permission of John Wiley & Sons, Inc.)

The three mechanisms combine to cause the grinding wheel to wear as depicted in Figure 18.5. Three wear regions can be identified. In the first region, the grains are initially sharp, and wear is accelerated because of grain fracture. This corresponds to the "break-in" period in conventional tool wear. In the second region, the wear rate is fairly constant, resulting in a linear relationship between wheel wear and volume of metal removed. This region is characterized by attritious wear, with some grain and bond fracture. In the third region of the wheel wear curve, the grains become dull, and the amount of plowing and rubbing increases relative to cutting. In addition, some of the chips become clogged in the pores of the wheel. This is called ***wheel loading***, and it impairs the cutting action and leads to higher heat and work surface temperatures. As a consequence, grinding efficiency decreases, and the volume of wheel removed increases relative to the volume of metal removed.

The ***grinding ratio*** is a term used to indicate the slope of the wheel wear curve. Specifically,

$$GR = \frac{V_w}{V_g} \tag{18.10}$$

where GR = the grinding ratio, V_w = the volume of work material removed, and V_g = the corresponding volume of the grinding wheel that is worn in the process. The grinding ratio has the most significance in the linear wear region of Figure 18.5. Typical values of GR range between 95 and 125 [5], which is about five orders of magnitude less than the analogous ratio in conventional machining. Grinding ratio is generally increased by increasing wheel speed v. The reason for this is that the size of the chip formed by each grit is smaller with higher speeds, so the amount of grain fracture is reduced. Because higher wheel speeds also improve surface finish, there is a general advantage in operating at high grinding speeds. However, when speeds become too high, attritious wear and surface temperatures increase. As a result, the grinding ratio is reduced and the surface finish is impaired. This effect was originally reported by Krabacher [14], as in Figure 18.6.

When the wheel is in the third region of the wear curve, it must be resharpened by a procedure called ***dressing***, which consists of (1) breaking off the dulled grits on the outside periphery of the grinding wheel in order to expose fresh sharp grains and (2) removing chips that have become clogged in the wheel. It is accomplished by a rotating disk, an abrasive stick, or another grinding wheel operating at high speed, held against the wheel being dressed as it rotates. Although dressing sharpens the wheel, it does not guarantee the shape of the wheel. ***Truing*** is an alternative procedure that not only sharpens the wheel, but also restores its cylindrical shape and ensures that it is straight across its outside perimeter. The procedure uses a diamond-pointed tool (other

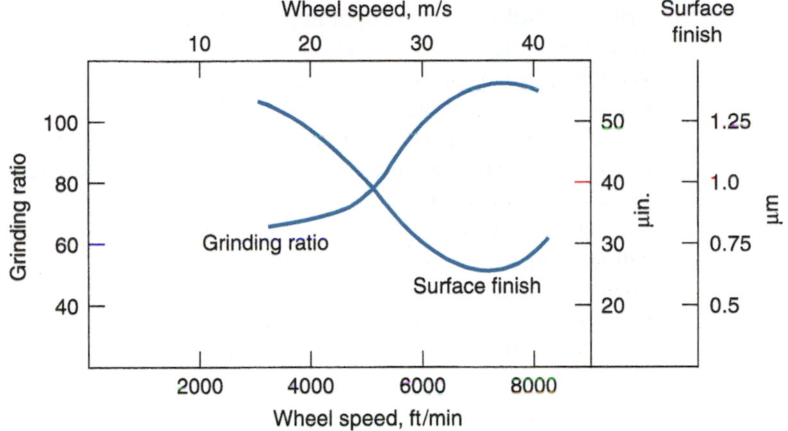

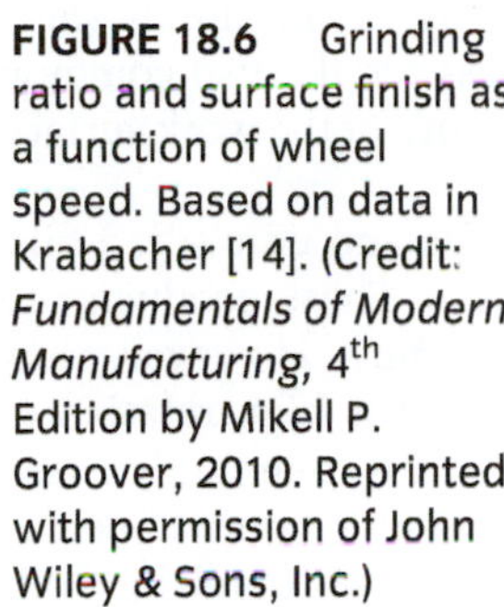
FIGURE 18.6 Grinding ratio and surface finish as a function of wheel speed. Based on data in Krabacher [14]. (Credit: *Fundamentals of Modern Manufacturing*, 4th Edition by Mikell P. Groover, 2010. Reprinted with permission of John Wiley & Sons, Inc.)

types of truing tools are also used) that is fed slowly and precisely across the wheel as it rotates. A very light depth is taken (0.025 mm or less) against the wheel.

18.1.3 APPLICATION CONSIDERATIONS IN GRINDING

In this section, we attempt to bring together the previous discussion of wheel parameters and theoretical analysis of grinding and consider their practical application. We also consider grinding fluids, which are commonly used in grinding operations.

Application Guidelines Many variables in grinding affect the performance and success of the operation. The guidelines listed in Table 18.5 are helpful in sorting out the many complexities and selecting the proper wheel parameters and grinding conditions.

TABLE 18.5 Application guidelines for grinding

Application Problem or Objective	Recommendation or Guideline
Grinding steel and most cast irons	Select aluminum oxide as the abrasive.
Grinding most nonferrous metals	Select silicon carbide as the abrasive.
Grinding hardened tool steels and certain aerospace alloys	Select cubic boron nitride as the abrasive.
Grinding hard abrasive materials such as ceramics, cemented carbides, and glass	Select diamond as the abrasive.
Grinding soft metals	Select a large grit size and harder grade wheel.
Grinding hard metals	Select a small grit size and softer grade wheel.
Optimize surface finish	Select a small grit size and dense wheel structure. Use high wheel speeds (v), lower work speeds (v_w).
Maximize material removal rate	Select a large grit size, more open wheel structure, and vitrified bond.
To minimize heat damage, cracking, and warping of the work surface	Maintain sharpness of the wheel. Dress the wheel frequently. Use lighter depths of cut (d), lower wheel speeds (v), and faster work speeds (v_w).
If the grinding wheel glazes and burns	Select wheel with a soft grade and open structure.
If the grinding wheel breaks down too rapidly	Select wheel with a hard grade and dense structure.

Compiled from [8], [11], and [16].

Grinding Fluids The proper application of cutting fluids has been found to be effective in reducing the thermal effects and high work surface temperatures described previously. When used in grinding operations, cutting fluids are called grinding fluids. The functions performed by grinding fluids are similar to those performed by cutting fluids (Section 17.4). Reducing friction and removing heat from the process are the two common functions. In addition, washing away chips and reducing temperature of the work surface are very important in grinding.

Types of grinding fluids by chemistry include grinding oils and emulsified oils. The grinding oils are derived from petroleum and other sources. These products are attractive because friction is such an important factor in grinding. However, they pose hazards in terms of fire and operator health, and their cost is high relative to emulsified oils. In addition, their capacity to carry away heat is less than fluids based on water. Accordingly, mixtures of oil in water are most commonly recommended as grinding fluids. These are usually mixed with higher concentrations than emulsified oils used as conventional cutting fluids. In this way, the friction reduction mechanism is emphasized.

18.1.4 GRINDING OPERATIONS AND GRINDING MACHINES

Grinding is traditionally used to finish parts whose geometries have already been created by other operations. Accordingly, grinding machines have been developed to grind plain flat surfaces, external and internal cylinders, and contour shapes such as threads. The contour shapes are often created by special formed wheels that have the opposite of the desired contour to be imparted to the work. Grinding is also used in tool rooms to form the geometries on cutting tools. In addition to these traditional uses, applications of grinding are expanding to include more high-speed, high material removal operations. Our discussion of operations and machines in this section includes the following types: (1) surface grinding, (2) cylindrical grinding, (3) centerless grinding, (4) creep feed grinding, and (5) other grinding operations.

Surface Grinding Surface grinding is normally used to grind plain flat surfaces. It is performed using either the periphery of the grinding wheel or the flat face of the wheel. Because the work is normally held in a horizontal orientation, peripheral grinding is performed by rotating the wheel about a horizontal axis, and face grinding is performed by rotating the wheel about a vertical axis. In either case, the relative motion of the workpart is achieved by reciprocating the work past the wheel or by rotating it. These possible combinations of wheel orientations and workpart motions provide the four types of surface grinding machines illustrated in Figure 18.7.

Of the four types, the horizontal spindle machine with reciprocating worktable is the most common, shown in Figure 18.8. Grinding is accomplished by reciprocating the work longitudinally under the wheel at a very small depth (infeed) and by feeding the wheel transversely into the work a certain distance between strokes. In these operations, the width of the wheel is usually less than that of the workpiece.

In addition to its conventional application, a grinding machine with horizontal spindle and reciprocating table can be used to form special contoured surfaces by employing a formed grinding wheel. Instead of feeding the wheel transversely across the work as it reciprocates, the wheel is ***plunge-fed*** vertically into the work. The shape of the formed wheel is therefore imparted to the work surface.

Grinding machines with vertical spindles and reciprocating tables are set up so that the wheel diameter is greater than the work width. Accordingly, these operations can be performed without using a transverse feed motion. Instead, grinding is accomplished by reciprocating the work past the wheel, and feeding the wheel vertically into the work to

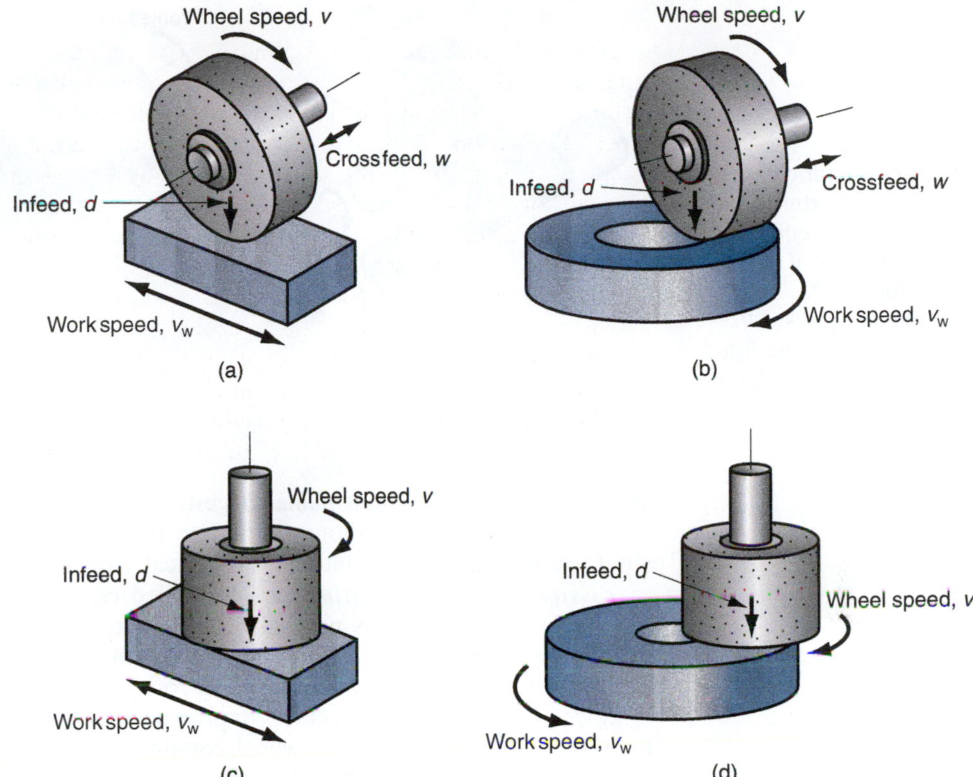

FIGURE 18.7 Four types of surface grinding: (a) horizontal spindle with reciprocating worktable, (b) horizontal spindle with rotating worktable, (c) vertical spindle with reciprocating worktable, and (d) vertical spindle with rotating worktable. (Credit: *Fundamentals of Modern Manufacturing*, 4th Edition by Mikell P. Groover, 2010. Reprinted with permission of John Wiley & Sons, Inc.)

the desired dimension. This configuration is capable of achieving a very flat surface on the work.

Of the two types of rotary table grinding in Figure 18.7(b) and (d), the vertical spindle machines are more common. Owing to the relatively large surface contact area between wheel and workpart, vertical spindle-rotary table grinding machines are capable of high metal removal rates when equipped with appropriate grinding wheels.

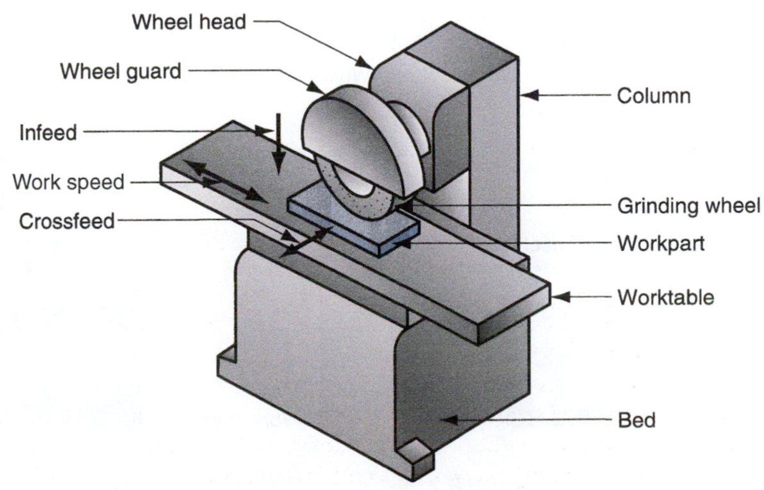

FIGURE 18.8 Surface grinder with horizontal spindle and reciprocating worktable. (Credit: *Fundamentals of Modern Manufacturing*, 4th Edition by Mikell P. Groover, 2010. Reprinted with permission of John Wiley & Sons, Inc.)

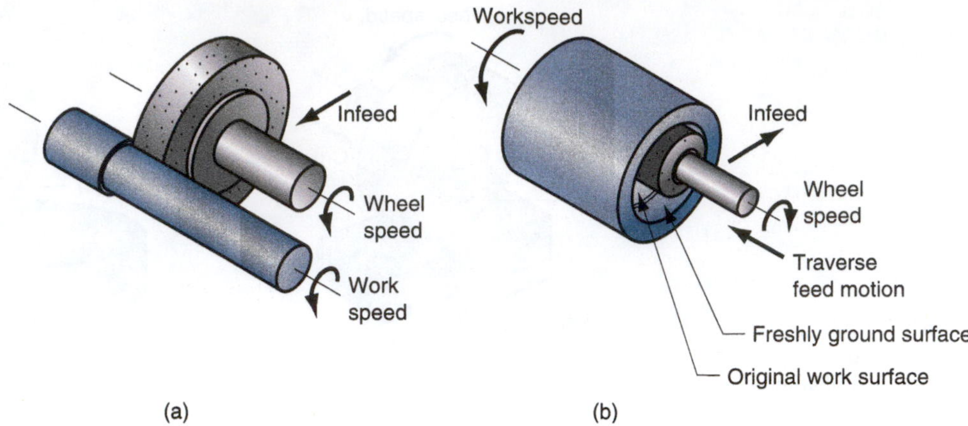

FIGURE 18.9 Two types of cylindrical grinding: (a) external, and (b) internal. (Credit: *Fundamentals of Modern Manufacturing,* 4th Edition by Mikell P. Groover, 2010. Reprinted with permission of John Wiley & Sons, Inc.)

Cylindrical Grinding As its name suggests, cylindrical grinding is used for rotational parts. These grinding operations divide into two basic types, Figure 18.9: (a) external cylindrical grinding and (b) internal cylindrical grinding.

External cylindrical grinding (also called ***center-type grinding*** to distinguish it from centerless grinding) is performed much like a turning operation. The grinding machines used for these operations closely resemble a lathe in which the tool post has been replaced by a high-speed motor to rotate the grinding wheel. The cylindrical workpiece is rotated between centers to provide a surface speed of 18 to 30 m/min (60 to 100 ft/min) [16], and the grinding wheel, rotating at 1200 to 2000 m/min (4000 to 6500 ft/min), is engaged to perform the cut. There are two types of feed motion possible, traverse feed and plunge-cut, shown in Figure 18.10. In traverse feed, the grinding wheel is fed in a direction parallel to the axis of rotation of the workpart. The infeed is set within a range typically from 0.0075 to 0.075 mm (0.0003 to 0.003 in). A longitudinal reciprocating motion is sometimes given to either the work or the wheel to improve surface finish. In plunge-cut, the grinding wheel is fed radially into the work. Formed grinding wheels use this type of feed motion.

External cylindrical grinding is used to finish parts that have been machined to approximate size and heat treated to desired hardness. The parts include axles, crankshafts, spindles, bearings and bushings, and rolls for rolling mills. The grinding operation produces the final size and required surface finish on these hardened parts.

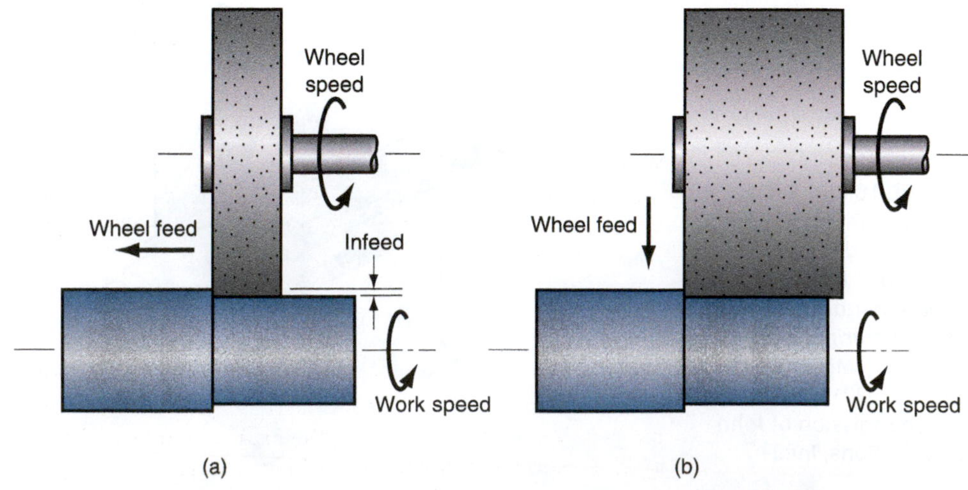

FIGURE 18.10 Two types of feed motion in external cylindrical grinding: (a) traverse feed, and (b) plunge-cut. (Credit: *Fundamentals of Modern Manufacturing,* 4th Edition by Mikell P. Groover, 2010. Reprinted with permission of John Wiley & Sons, Inc.)

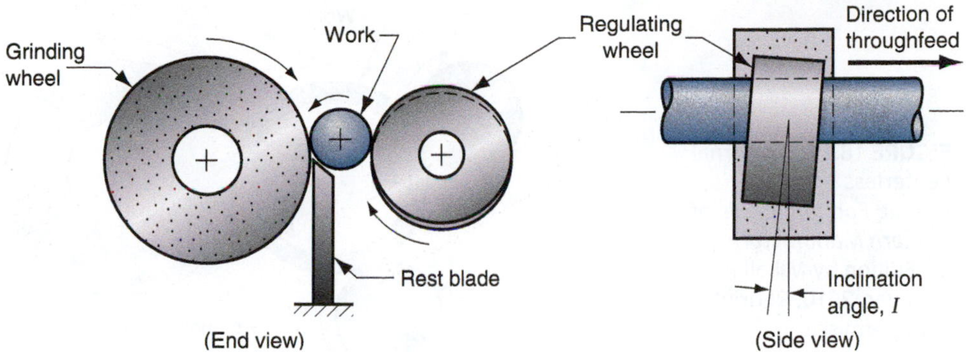

FIGURE 18.11 External centerless grinding. (Credit: *Fundamentals of Modern Manufacturing*, 4th Edition by Mikell P. Groover, 2010. Reprinted with permission of John Wiley & Sons, Inc.)

Internal cylindrical grinding operates somewhat like a boring operation. The workpiece is usually held in a chuck and rotated to provide surface speeds of 20 to 60 m/min (75 to 200 ft/min) [16]. Wheel surface speeds similar to external cylindrical grinding are used. The wheel is fed in either of two ways: traverse feed, Figure 18.9(b), or plunge feed. Obviously, the wheel diameter in internal cylindrical grinding must be smaller than the original bore hole. This often means that the wheel diameter is quite small, necessitating very high rotational speeds in order to achieve the desired surface speed. Internal cylindrical grinding is used to finish the hardened inside surfaces of bearing races and bushing surfaces.

Centerless Grinding Centerless grinding is an alternative process for grinding external and internal cylindrical surfaces. As its name suggests, the workpiece is not held between centers. This results in a reduction in work handling time; hence, centerless grinding is often used for high-production work. The setup for ***external centerless grinding*** (Figure 18.11), consists of two wheels: the grinding wheel and a regulating wheel. The workparts, which may be many individual short pieces or long rods (e.g., 3 to 4 m long), are supported by a rest blade and fed through between the two wheels. The grinding wheel does the cutting, rotating at surface speeds of 1200 to 1800 m/min (4000 to 6000 ft/min). The regulating wheel rotates at much lower speeds and is inclined at a slight angle I to control throughfeed of the work. The following equation can be used to predict throughfeed rate, based on inclination angle and other parameters of the process [16]:

$$f_r = \pi D_r N_r \sin I \tag{18.11}$$

where f_r = throughfeed rate, mm/min (in/min); D_r = diameter of the regulating wheel, mm (in); N_r = rotational speed of the regulating wheel, rev/min; and I = inclination angle of the regulating wheel.

The typical setup in ***internal centerless grinding*** is shown in Figure 18.12. In place of the rest blade, two support rolls are used to maintain the position of the work. The regulating wheel is tilted at a small inclination angle to control the feed of the work past the grinding wheel. Because of the need to support the grinding wheel, throughfeed of the work as in external centerless grinding is not possible. Therefore, this grinding operation cannot achieve the same high production rates as in the external centerless process. Its advantage is that it is capable of providing very close concentricity between internal and external diameters on a tubular part such as a roller bearing race.

Creep Feed Grinding A relatively new form of grinding is creep feed grinding, developed around 1958. Creep feed grinding is performed at very high depths of cut and very low feed rates; hence, the name creep feed. The comparison with conventional surface grinding is illustrated in Figure 18.13.

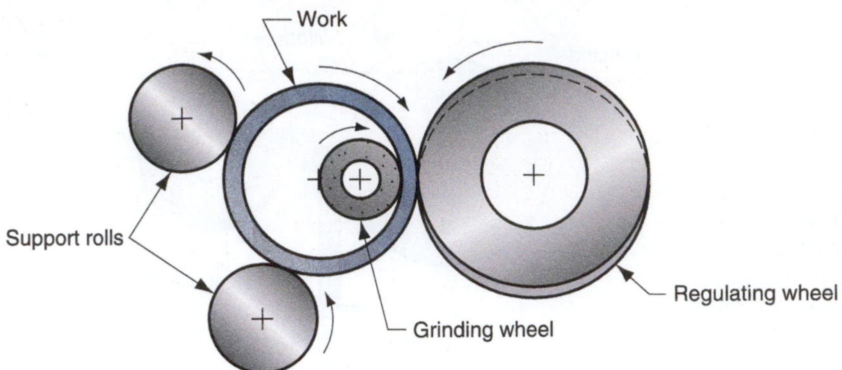

FIGURE 18.12 Internal centerless grinding. (Credit: *Fundamentals of Modern Manufacturing*, 4th Edition by Mikell P. Groover, 2010. Reprinted with permission of John Wiley & Sons, Inc.)

Depths of cut in creep feed grinding are 1000 to 10,000 times greater than in conventional surface grinding, and the feed rates are reduced by about the same proportion. However, material removal rate and productivity are increased in creep feed grinding because the wheel is continuously cutting. This contrasts with conventional surface grinding in which the reciprocating motion of the work results in significant lost time during each stroke.

Creep feed grinding can be applied in both surface grinding and external cylindrical grinding. Surface grinding applications include grinding of slots and profiles. The process seems especially suited to those cases in which depth-to-width ratios are relatively large. The cylindrical applications include threads, formed gear shapes, and other cylindrical components. The term ***deep grinding*** is used in Europe to describe these external cylindrical creep feed grinding applications.

The introduction of grinding machines designed with special features for creep feed grinding has spurred interest in the process. The features include [11] high static and dynamic stability, highly accurate slides, two to three times the spindle power of conventional grinding machines, consistent table speeds for low feeds, high-pressure grinding fluid delivery systems, and dressing systems capable of dressing the grinding wheels during the process. Typical advantages of creep feed grinding include (1) high material removal rates, (2) improved accuracy for formed surfaces, and (3) reduced temperatures at the work surface.

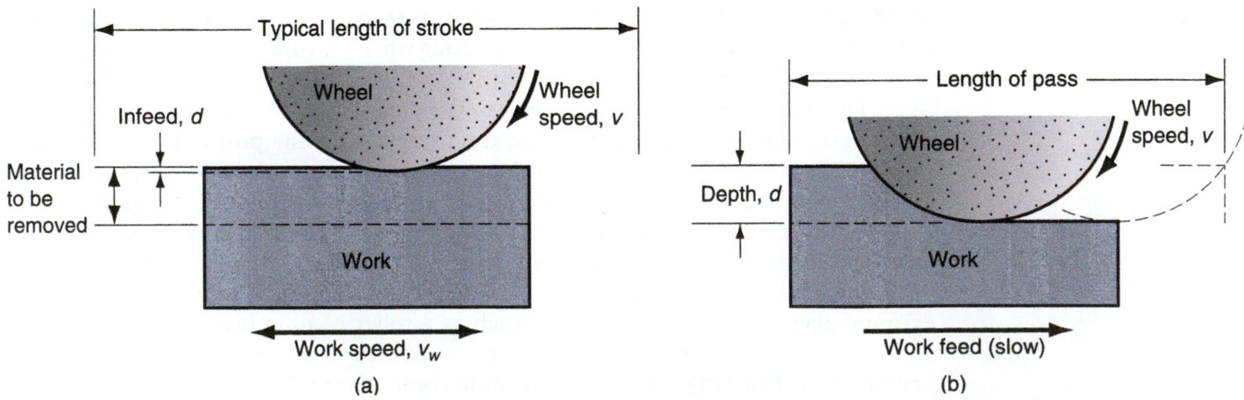

FIGURE 18.13 Comparison of (a) conventional surface grinding and (b) creep feed grinding. (Credit: *Fundamentals of Modern Manufacturing*, 4th Edition by Mikell P. Groover, 2010. Reprinted with permission of John Wiley & Sons, Inc.)

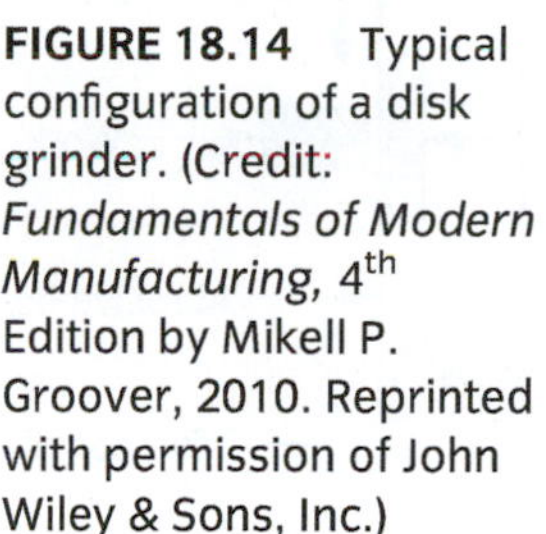

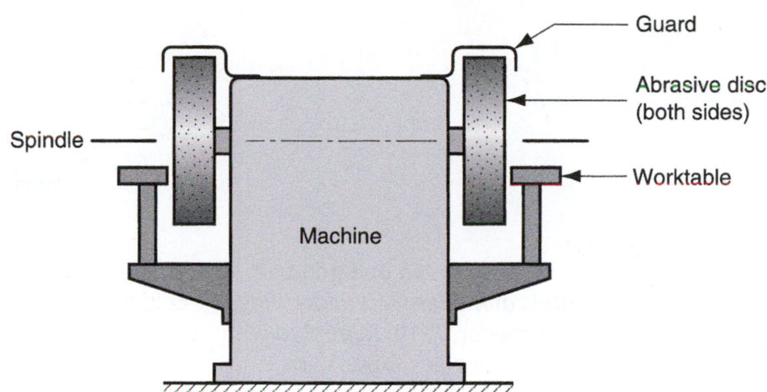

FIGURE 18.14 Typical configuration of a disk grinder. (Credit: *Fundamentals of Modern Manufacturing*, 4th Edition by Mikell P. Groover, 2010. Reprinted with permission of John Wiley & Sons, Inc.)

Other Grinding Operations Several other grinding operations should be briefly mentioned to complete our coverage. These include tool grinding, jig grinding, disk grinding, snag grinding, and abrasive belt grinding.

Cutting tools are made of hardened tool steel and other hard materials. ***Tool grinders*** are special grinding machines of various designs to sharpen and recondition cutting tools. They have devices for positioning and orienting the tools to grind the desired surfaces at specified angles and radii. Some tool grinders are general purpose, while others cut the unique geometries of specific tool types. General-purpose tool and cutter grinders use special attachments and adjustments to accommodate a variety of tool geometries. Single-purpose tool grinders include gear cutter sharpeners, milling cutter grinders of various types, broach sharpeners, and drill point grinders.

Jig grinders are grinding machines traditionally used to grind holes in hardened steel parts to high accuracies. The original applications included pressworking dies and tools. Although these applications are still important, jig grinders are used today in a broader range of applications in which high accuracy and good finish are required on hardened components. Numerical control is available on modern jig grinding machines to achieve automated operation.

Disk grinders are grinding machines with large abrasive disks mounted on either end of a horizontal spindle as in Figure 18.14. The work is held (usually manually) against the flat surface of the wheel to accomplish the grinding operation. Some disk grinding machines have double opposing spindles. By setting the disks at the desired separation, the workpart can be fed automatically between the two disks and ground simultaneously on opposite sides. Advantages of the disk grinder are good flatness and parallelism at high production rates.

The ***snag grinder*** is similar in configuration to a disk grinder. The difference is that the grinding is done on the outside periphery of the wheel rather than on the side flat surface. The grinding wheels are therefore different in design than those in disk grinding. Snag grinding is generally a manual operation, used for rough grinding operations such as removing the flash from castings and forgings, and smoothing weld joints.

Abrasive belt grinding uses abrasive particles bonded to a flexible (cloth) belt. A typical setup is illustrated in Figure 18.15. Support of the belt is required when the work is pressed against it, and this support is provided by a roll or platen located behind the belt. A flat platen is used for work that will have a flat surface. A soft platen can be used if it is desirable for the abrasive belt to conform to the general contour of the part during grinding. Belt speed depends on the material being ground; a range of 750 to 1700 m/min (2500 to 5500 ft/min) is typical [16]. Owing to improvements in abrasives and bonding materials, abrasive belt grinding is being used increasingly for heavy stock removal rates, rather than light grinding, which was its traditional application. The term ***belt sanding***

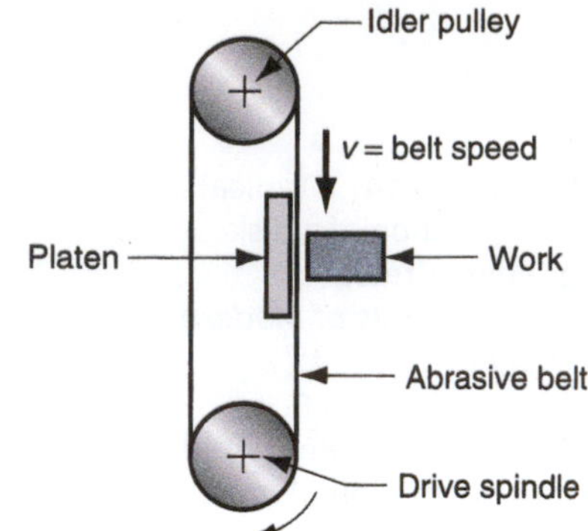

FIGURE 18.15 Abrasive belt grinding. (Credit: *Fundamentals of Modern Manufacturing,* 4th Edition by Mikell P. Groover, 2010. Reprinted with permission of John Wiley & Sons, Inc.)

refers to the light grinding applications in which the workpart is pressed against the belt to remove burrs and high spots, and to produce an improved finish quickly by hand.

18.2 RELATED ABRASIVE PROCESSES

Other abrasive processes include honing, lapping, superfinishing, polishing, and buffing. They are used exclusively as finishing operations. The initial part shape is created by some other process; then the part is finished by one of these operations to achieve superior surface finish. The usual part geometries and typical surface roughness values for these processes are indicated in Table 18.6. For comparison, we also present corresponding data for grinding.

Another class of finishing operations, called mass finishing (Section 21.1.2), is used to finish parts in bulk rather than individually. These mass finishing methods are also used for cleaning and deburring.

18.2.1 HONING

Honing is an abrasive process performed by a set of bonded abrasive sticks. A common application is to finish the bores of internal combustion engines. Other applications include bearings, hydraulic cylinders, and gun barrels. Surface finishes of around 0.12 μm (5 μ-in) or slightly better are typically achieved in these applications. In addition, honing

TABLE 18.6 Usual part geometries for honing, lapping, superfinishing, polishing, and buffing

Process	Usual Part Geometry	Surface Roughness	
		μm	**μ-in**
Grinding, medium grit size	Flat, external cylinders, round holes	0.4–1.6	16–63
Grinding, fine grit size	Flat, external cylinders, round holes	0.2–0.4	8–16
Honing	Round hole (e.g., engine bore)	0.1–0.8	4–32
Lapping	Flat or slightly spherical (e.g., lens)	0.025–0.4	1–16
Superfinishing	Flat surface, external cylinder	0.013–0.2	0.5–8
Polishing	Miscellaneous shapes	0.025–0.8	1–32
Buffing	Miscellaneous shapes	0.013–0.4	0.5–16

Compiled from [4], [7], and [16], and other sources.

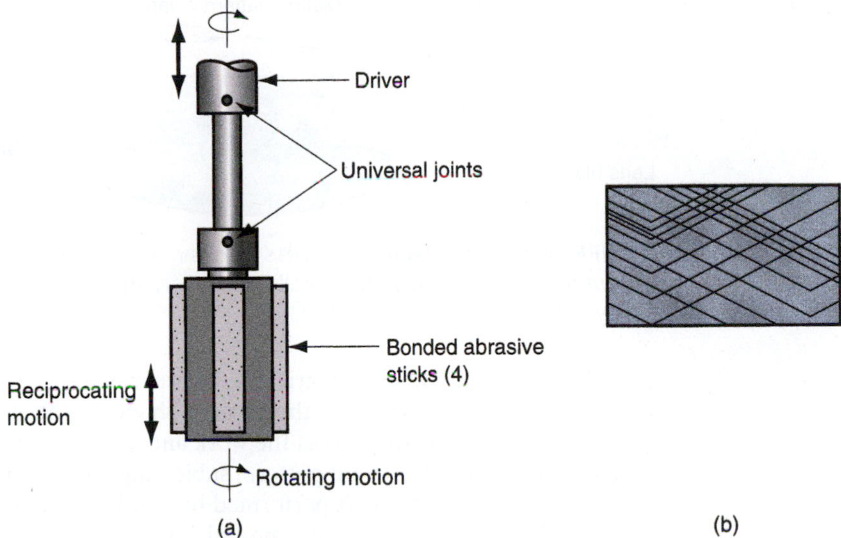

FIGURE 18.16 The honing process: (a) the honing tool used for internal bore surface, and (b) cross-hatched surface pattern created by the action of the honing tool. (Credit: *Fundamentals of Modern Manufacturing*, 4th Edition by Mikell P. Groover, 2010. Reprinted with permission of John Wiley & Sons, Inc.)

produces a characteristic cross-hatched surface that tends to retain lubrication during operation of the component, thus contributing to its function and service life.

The honing process for an internal cylindrical surface is illustrated in Figure 18.16. The honing tool consists of a set of bonded abrasive sticks. Four sticks are used on the tool shown in the figure, but the number depends on hole size. Two to four sticks would be used for small holes (e.g., gun barrels), and a dozen or more would be used for larger diameter holes. The motion of the honing tool is a combination of rotation and linear reciprocation, regulated in such a way that a given point on the abrasive stick does not trace the same path repeatedly. This rather complex motion accounts for the cross-hatched pattern on the bore surface. Honing speeds are 15 to 150 m/min (50 to 500 ft/min) [4]. During the process, the sticks are pressed outward against the hole surface to produce the desired abrasive cutting action. Hone pressures of 1 to 3 MPa (150 to 450 lb/in^2) are typical. The honing tool is supported in the hole by two universal joints, thus causing the tool to follow the previously defined hole axis. Honing enlarges and finishes the hole but cannot change its location.

Grit sizes in honing range between 30 and 600. The same trade-off between better finish and faster material removal rates exists in honing as in grinding. The amount of material removed from the work surface during a honing operation may be as much as 0.5 mm (0.020 in), but is usually much less than this. A cutting fluid must be used in honing to cool and lubricate the tool and to help remove the chips.

18.2.2 LAPPING

Lapping is an abrasive process used to produce surface finishes of extreme accuracy and smoothness. It is used in the production of optical lenses, metallic bearing surfaces, gages, and other parts requiring very good finishes. Metal parts that are subject to fatigue loading or surfaces that must be used to establish a seal with a mating part are often lapped.

Instead of a bonded abrasive tool, lapping uses a fluid suspension of very small abrasive particles between the workpiece and the lapping tool. The process is illustrated in Figure 18.17 as applied in lens-making. The fluid with abrasives is referred to as the ***lapping compound*** and has the general appearance of a chalky paste. The fluids used to make the compound include oils and kerosene. Common abrasives are aluminum oxide

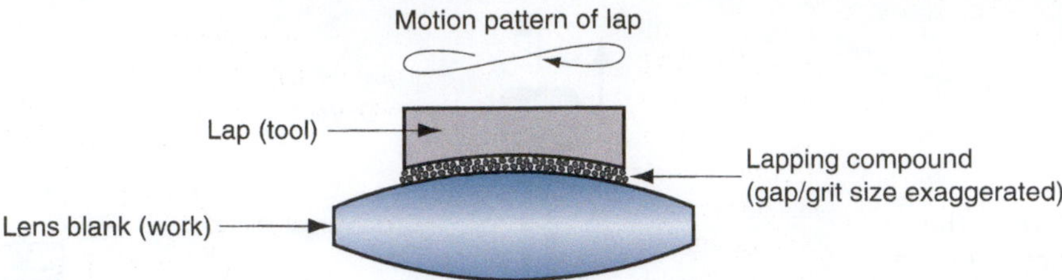

FIGURE 18.17 The lapping process in lens-making. (Credit: *Fundamentals of Modern Manufacturing*, 4th Edition by Mikell P. Groover, 2010. Reprinted with permission of John Wiley & Sons, Inc.)

and silicon carbide with typical grit sizes between 300 and 600. The lapping tool is called a ***lap***, and it has the reverse of the desired shape of the workpart. To accomplish the process, the lap is pressed against the work and moved back and forth over the surface in a figure-eight or other motion pattern, subjecting all portions of the surface to the same action. Lapping is sometimes performed by hand, but lapping machines accomplish the process with greater consistency and efficiency.

Materials used to make the lap range from steel and cast iron to copper and lead. Wood laps have also been made. Because a lapping compound is used rather than a bonded abrasive tool, the mechanism by which this process works is somewhat different than grinding and honing. It is hypothesized that two alternative cutting mechanisms are at work in lapping [4]. The first mechanism is that the abrasive particles roll and slide between the lap and the work, with very small cuts occurring in both surfaces. The second mechanism is that the abrasives become imbedded in the lap surface and the cutting action is very similar to grinding. It is likely that lapping is a combination of these two mechanisms, depending on the relative hardnesses of the work and the lap. For laps made of soft materials, the embedded grit mechanism is emphasized; and for hard laps, the rolling and sliding mechanism dominates.

18.2.3 SUPERFINISHING

Superfinishing is an abrasive process similar to honing. Both processes use a bonded abrasive stick moved with a reciprocating motion and pressed against the surface to be finished. Superfinishing differs from honing in the following respects [4]: (1) the strokes are shorter, 5 mm (3/16 in); (2) higher frequencies are used, up to 1500 strokes per minute; (3) lower pressures are applied between the tool and the surface, below 0.28 MPa (40 lb/in^2); (4) workpiece speeds are lower, 15 m/min (50 ft/min) or less; and (5) grit sizes are generally smaller. The relative motion between the abrasive stick and the work surface is varied so that individual grains do not retrace the same path. A cutting fluid is used to cool the work surface and wash away chips. In addition, the fluid tends to separate the abrasive stick from the work surface after a certain level of smoothness is achieved, thus preventing further cutting action. The result of these operating conditions is mirror-like finishes with surface roughness values around 0.025 μm (1 μ-in). Superfinishing can be used to finish flat and external cylindrical surfaces. The process is illustrated in Figure 18.18 for the latter geometry.

18.2.4 POLISHING AND BUFFING

Polishing is used to remove scratches and burrs and to smooth rough surfaces by means of abrasive grains attached to a polishing wheel rotating at high speed—around 2300 m/min

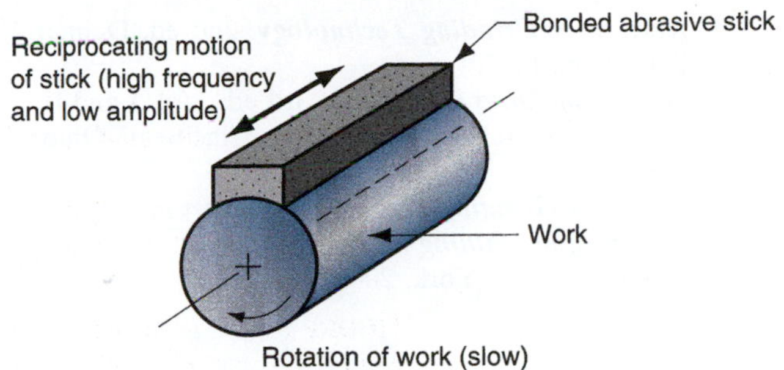

FIGURE 18.18 Superfinishing on an external cylindrical surface. (Credit: *Fundamentals of Modern Manufacturing*, 4th Edition by Mikell P. Groover, 2010. Reprinted with permission of John Wiley & Sons, Inc.)

(7500 ft/min). The wheels are made of canvas, leather, felt, and even paper; thus, the wheels are somewhat flexible. The abrasive grains are glued to the outside periphery of the wheel. After the abrasives have been worn down and used up, the wheel is replenished with new grits. Grit sizes of 20 to 80 are used for rough polishing, 90 to 120 for finish polishing, and above 120 for fine finishing. Polishing operations are often accomplished manually.

Buffing is similar to polishing in appearance, but its function is different. Buffing is used to provide attractive surfaces with high luster. Buffing wheels are made of materials similar to those used for polishing wheels—leather, felt, cotton, etc.—but buffing wheels are generally softer. The abrasives are very fine and are contained in a buffing compound that is pressed into the outside surface of the wheel while it rotates. This contrasts with polishing in which the abrasive grits are glued to the wheel surface. As in polishing, the abrasive particles must be periodically replenished. Buffing is usually done manually, although machines have been designed to perform the process automatically. Speeds are generally 2400 to 5200 m/min (8000 to 17,000 ft/min).

REFERENCES

[1] Aronson, R. B. "More Than a Pretty Finish," ***Manufacturing Engineering,*** *February* 2005, pp. 57–69.

[2] Andrew, C., Howes, T. D., and Pearce, T. R. A. ***Creep Feed Grinding***. Holt, Rinehart and Winston, Ltd., London, 1985.

[3] ***ANSI Standard B74. 13-1977***, "Markings for Identifying Grinding Wheels and Other Bonded Abrasives." American National Standards Institute, New York, 1977.

[4] Armarego, E. J. A., and Brown, R. H. ***The Machining of Metals***. Prentice-Hall, Inc., Englewood Cliffs, New Jersey, 1969.

[5] Bacher, W. R., and Merchant, M. E. "On the Basic Mechanics of the Grinding Process," ***Transactions ASME***, Series B, Vol. **80**, No. 1, 1958, pp. 141.

[6] Black, J., and Kohser, R. ***DeGarmo's Materials and Processes in Manufacturing***, 10th ed. John Wiley & Sons, Inc.,Hoboken, New Jersey, 2008.

[7] Black, P. H. ***Theory of Metal Cutting***. McGraw-Hill Book Company, Inc., New York, 1961.

[8] Boothroyd, G., and Knight, W. A. ***Fundamentals of Metal Machining and Machine Tools***, 3rd ed. CRC Taylor and Francis, Boca Raton, Florida, 2006.

[9] Boston, O. W. ***Metal Processing***, 2nd ed. John Wiley & Sons, Inc., New York, 1951.

[10] Cook, N. H. ***Manufacturing Analysis***. Addison-Wesley Publishing Company, Inc., Reading, Massachusetts, 1966.

[11] Drozda, T. J., and Wick, C. (eds.). ***Tool and Manufacturing Engineers Handbook***, 4th ed., Vol. **I**, ***Machining.*** Society of Manufacturing Engineers, Dearborn, Michigan, 1983.

[12] Early, D. F., and Johnson, G. E. ***Process Engineering for Manufacturing***. Prentice-Hall, Inc., Englewood Cliffs, New Jersey, 1962.

[13] Kaiser, R. "The Facts about Grinding," ***Manufacturing Engineering***, Vol. **125**, No. 3, September 2000, pp. 78–85.

[14] Krabacher, E. J. "Factors Influencing the Performance of Grinding Wheels," ***Transactions ASME***, *Series B*, Vol. **81**, No. 3, 1959, pp. 187–199.

[15] Krar, S. F. ***Grinding Technology***, 2nd ed. Delmar Publishers, 1995.

[16] ***Machining Data Handbook***, 3rd ed., Vol. **I** and **II**. Metcut Research Associates, Inc., Cincinnati, Ohio, 1980.

[17] Malkin, S. ***Grinding Technology: Theory and Applications of Machining with Abrasives***, 2nd ed. Industrial Press, New York, 2008.

[18] Phillips, D. "Creeping Up," ***Cutting Tool Engineering***, Vol. **52**, No. 3, March 2000, pp. 32–43.

[19] Rowe, W. ***Principles of Modern Grinding Technology***. William Andrew, Elsevier Applied Science Publishers, New York, 2009.

[20] Salmon, S. "Creep-Feed Grinding Is Surprisingly Versatile," ***Manufacturing Engineering***, November 2004, pp. 59–64.

REVIEW QUESTIONS

18.1. Why are abrasive processes technologically and commercially important?

18.2. What are the five basic parameters of a grinding wheel?

18.3. What are some of the principal abrasive materials used in grinding wheels?

18.4. What is wheel structure?

18.5. What is wheel grade?

18.6. Why are specific energy values so much higher in grinding than in traditional machining processes such as milling?

18.7. Grinding creates high temperatures. How is temperature harmful in grinding?

18.8. What are the three mechanisms of grinding wheel wear?

18.9. What is dressing, in reference to grinding wheels?

18.10. What is truing, in reference to grinding wheels?

18.11. What abrasive material would one select for grinding a cemented carbide cutting tool?

18.12. What are the functions of a grinding fluid?

18.13. What is centerless grinding?

18.14. How does creep feed grinding differ from conventional grinding?

18.15. How does abrasive belt grinding differ from a conventional surface grinding operation?

18.16. Name some of the abrasive operations available to achieve very good surface finishes.

PROBLEMS

18.1. In a surface grinding operation wheel diameter = 150 mm and infeed = 0.07 mm. Wheel speed = 1450 m/min, workspeed = 0.25 m/s, and crossfeed = 5 mm. The number of active grits per area of wheel surface = 0.75 grits/mm^2. Determine (a) average length per chip, (b) metal removal rate, and (c) number of chips formed per unit time for the portion of the operation when the wheel is engaged in the work.

18.2. The following conditions and settings are used in a certain surface grinding operation: wheel diameter = 6.0 in, infeed = 0.003 in, wheel speed = 4750 ft/min, workspeed = 50 ft/min, and crossfeed = 0.20 in. The number of active grits per square inch of wheel surface = 500. Determine (a) average length per chip, (b) metal removal rate, and (c) number of chips formed per unit time for the portion of the operation when the wheel is engaged in the work.

18.3. An internal cylindrical grinding operation is used to finish an internal bore from an initial diameter of 250.00 mm to a final diameter of 252.5 mm. The bore is 125 mm long. A grinding wheel with an initial diameter of 150.00 mm and a width of 20.00 mm is used. After the operation, the diameter of the grinding wheel has been reduced to 149.75 mm. Determine the grinding ratio in this operation.

18.4. In a surface grinding operation performed on hardened plain carbon steel, the grinding wheel has a diameter = 200 mm and width = 25 mm. The wheel rotates at 2400 rev/min, with a depth of cut (infeed) = 0.05 mm/pass and a crossfeed = 3.50 mm. The reciprocating speed of the work is 6 m/min, and the operation is performed dry. Determine (a) length of contact between the wheel and the work and (b) volume rate of metal removed. (c) If there are 64 active grits/cm^2 of wheel surface, estimate the number of chips formed per unit time. (d) What is the average volume per chip? (e) If the tangential cutting force on the work = 25 N, compute the specific energy in this operation.

18.5. An 8-in diameter grinding wheel, 1.0 in wide, is used in a surface grinding job performed on a flat piece of heat-treated 4340 steel. The wheel rotates to achieve a surface speed of 5000 ft/min, with a depth of cut (infeed) = 0.002 in per pass and a crossfeed = 0.15 in. The reciprocating speed of the work is 20 ft/min, and the operation is performed dry. (a) What is the length of contact between the wheel and the work? (b) What is the volume rate of metal removed? (c) If there are 300 active grits/in^2 of wheel surface, estimate the number of chips formed per unit time. (d) What is the average volume per chip? (e) If the tangential cutting force on the workpiece = 7.3 lbs, what is the specific energy calculated for this job?

18.6. A surface grinding operation is being performed on a 6150 steel workpart (annealed, approximately 200 BHN). The designation on the grinding wheel is C-24-D-5-V. The wheel diameter = 7.0 in and its width = 1.00 in. Rotational speed = 3000 rev/min. The depth (infeed) = 0.002 in per pass, and the crossfeed = 0.5 in. Workspeed = 20 ft/min. This operation has been a source of trouble right from the beginning. The surface finish is not as good as the 16 μ-in specified on the part print, and there are signs of metallurgical damage on the surface. In addition, the wheel seems to become clogged almost as soon as the operation begins. In short, nearly everything that can go wrong with the job has gone wrong. (a) Determine the rate of metal removal when the wheel is engaged in the work. (b) If the number of active grits per square inch = 200, determine the average chip length and the number of chips formed per time. (c) What changes would you recommend in the grinding wheel to help solve the problems encountered? Explain why you made each recommendation.

18.7. The grinding wheel in a centerless grinding operation has a diameter = 200 mm, and the regulating wheel diameter = 125 mm. The grinding wheel rotates at 3000 rev/min and the regulating wheel rotates at 200 rev/min. The inclination angle of the regulating wheel = 2.5°. Determine the throughfeed rate of cylindrical workparts that are 25.0 mm in diameter and 175 mm long.

18.8. A centerless grinding operation uses a regulating wheel that is 150 mm in diameter and rotates at 500 rev/min. At what inclination angle should the regulating wheel be set, if it is desired to feed a workpiece with length = 3.5 m and diameter = 18 mm through the operation in exactly 30 sec?

18.9. In a certain centerless grinding operation, the grinding wheel diameter = 8.5 in, and the regulating wheel diameter = 5.0 in. The grinding wheel rotates at 3500 rev/min and the regulating wheel rotates at 150 rev/min. The inclination angle of the regulating wheel = 3°. Determine the throughfeed rate of cylindrical workparts that have the following dimensions: diameter = 1.25 in and length = 8.0 in.

18.10. It is desired to compare the cycle times required to grind a particular workpiece using traditional surface grinding and using creep feed grinding. The workpiece is 200 mm long, 30 mm wide, and 75 mm thick. To make a fair comparison, the grinding wheel in both cases is 250 mm in diameter, 35 mm in width, and rotates at 1500 rev/min. It is desired to remove 25 mm of material from the surface. When traditional grinding is used, the infeed is set at 0.025 mm, and the wheel traverses twice (forward and back) across the work surface during each pass before resetting the infeed. There is no crossfeed since the wheel width is greater than the work width. Each pass is made at a workspeed of 12 m/min, but the wheel overshoots the part on both sides. With acceleration and deceleration, the wheel is engaged in the work for 50% of the time on each pass. When creep feed grinding is used, the depth is increased by 1000 and the forward feed is decreased by 1000. How long will it take to complete the grinding operation (a) with traditional grinding and (b) with creep feed grinding?

18.11. An aluminum alloy is to be ground in an external cylindrical grinding operation to obtain a good surface finish. Specify the appropriate grinding wheel parameters and the grinding conditions for this job.

18.12. A high-speed steel broach (hardened) is to be resharpened to achieve a good finish. Specify the appropriate parameters of the grinding wheel for this job.

19 NONTRADITIONAL MACHINING PROCESSES

Chapter Contents

19.1 Mechanical Energy Processes
- 19.1.1 Ultrasonic Machining
- 19.1.2 Processes Using Water Jets
- 19.1.3 Other Nontraditional Abrasive Processes

19.2 Electrochemical Machining Processes
- 19.2.1 Electrochemical Machining
- 19.2.2 Electrochemical Deburring and Grinding

19.3 Thermal Energy Processes
- 19.3.1 Electric Discharge Processes
- 19.3.2 Electron Beam Machining
- 19.3.3 Laser Beam Machining

19.4 Chemical Machining
- 19.4.1 Mechanics and Chemistry of Chemical Machining
- 19.4.2 CHM Processes

19.5 Application Considerations

Conventional machining processes (i.e., turning, drilling, milling) use a sharp cutting tool to form a chip from the work by shear deformation. In addition to these conventional methods, there is a group of processes that uses other mechanisms to remove material. The term ***nontraditional machining*** refers to this group that removes excess material by various techniques involving mechanical, thermal, electrical, or chemical energy (or combinations of these energies). They do not use a sharp cutting tool in the conventional sense.

The nontraditional processes have been developed since World War II largely in response to new and unusual machining requirements that could not be satisfied by conventional methods. These requirements, and the resulting commercial and technological importance of the nontraditional processes, include:

- The need to machine newly developed metals and nonmetals. These new materials often have special properties (e.g., high strength, high hardness, high toughness) that make them difficult or impossible to machine by conventional methods.
- The need for unusual and/or complex part geometries that cannot easily be accomplished and in some cases are impossible to achieve by conventional machining.
- The need to avoid surface damage that often accompanies the stresses created by conventional machining.

Many of these requirements are associated with the aerospace and electronics industries, which have become increasingly important in recent decades.

There are literally dozens of nontraditional machining processes, most of which are unique in their range of applications. In the present chapter, we discuss those that are most important commercially. More detailed discussions of these nontraditional methods are presented in several of the references.

The nontraditional processes are often classified according to principal form of energy used to effect material removal. By this classification, there are four types:

1. ***Mechanical.*** Mechanical energy in some form other than the action of a conventional cutting tool is used in these nontraditional processes. Erosion of the work material by a high velocity stream of abrasives or fluid (or both) is a typical form of mechanical action in these processes.
2. ***Electrical.*** These nontraditional processes use electrochemical energy to remove material; the mechanism is the reverse of electroplating.
3. ***Thermal.*** These processes use thermal energy to cut or shape the workpart. The thermal energy is generally applied to a very small portion of the work surface, causing that portion to be removed by fusion and/or vaporization. The thermal energy is generated by the conversion of electrical energy.
4. ***Chemical.*** Most materials (metals particularly) are susceptible to chemical attack by certain acids or other etchants. In chemical machining, chemicals selectively remove material from portions of the workpart, while other portions of the surface are protected by a mask.

19.1 MECHANICAL ENERGY PROCESSES

In this section we examine several of the nontraditional processes that utilize mechanical energy other than a sharp cutting tool: (1) ultrasonic machining, (2) water jet processes, and (3) other abrasive processes.

19.1.1 ULTRASONIC MACHINING

Ultrasonic machining (USM) is a nontraditional machining process in which abrasives contained in a slurry are driven at high velocity against the work by a tool vibrating at low amplitude and high frequency. The amplitudes are around 0.075 mm (0.003 in), and the frequencies are approximately 20,000 Hz. The tool oscillates in a direction perpendicular to the work surface, and is fed slowly into the work, so that the shape of the tool is formed in the part. However, it is the action of the abrasives, impinging against the work surface, that performs the cutting. The general arrangement of the USM process is depicted in Figure 19.1.

Common tool materials used in USM include soft steel and stainless steel. Abrasive materials in USM include boron nitride, boron carbide, aluminum oxide, silicon carbide,

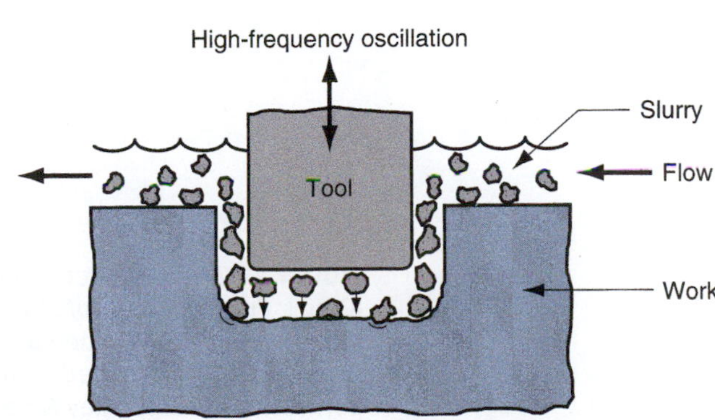

FIGURE 19.1 Ultrasonic machining. (Credit: *Fundamentals of Modern Manufacturing*, 4th Edition by Mikell P. Groover, 2010. Reprinted with permission of John Wiley & Sons, Inc.)

and diamond. Grit size (Chapter 10, Appendix A10.1) ranges between 100 and 2000. The vibration amplitude should be set approximately equal to the grit size, and the gap size should be maintained at about two times grit size. To a significant degree, grit size determines the surface finish on the new work surface. In addition to surface finish, material removal rate is an important performance variable in ultrasonic machining. For a given work material, the removal rate in USM increases with increasing frequency and amplitude of vibration.

The cutting action in USM operates on the tool as well as the work. As the abrasive particles erode the work surface, they also erode the tool, thus affecting its shape. It is therefore important to know the relative volumes of work material and tool material removed during the process—similar to the grinding ratio (Section 18.1.2). This ratio of stock removed to tool wear varies for different work materials, ranging from around 100:1 for cutting glass down to about 1:1 for cutting tool steel.

The slurry in USM consists of a mixture of water and abrasive particles. Concentration of abrasives in water ranges from 20% to 60% [5]. The slurry must be continuously circulated in order to bring fresh grains into action at the tool–work gap. It also washes away chips and worn grits created by the cutting process.

The development of ultrasonic machining was motivated by the need to machine hard, brittle work materials, such as ceramics, glass, and carbides. It is also successfully used on certain metals such as stainless steel and titanium. Shapes obtained by USM include nonround holes, holes along a curved axis, and coining operations, in which an image pattern on the tool is imparted to a flat work surface.

19.1.2 PROCESSES USING WATER JETS

The two processes described in this section remove material by means of high-velocity streams of water or a combination of water and abrasives.

Water Jet Cutting Water jet cutting (WJC) uses a fine, high-pressure, high-velocity stream of water directed at the work surface to cause cutting of the work, as illustrated in Figure 19.2. To obtain the fine stream of water, a small nozzle opening of diameter 0.1 to 0.4 mm (0.004 to 0.016 in) is used. To provide the stream with sufficient energy for cutting, pressures up to 400 MPa (60,000 lb/in^2) are used, and the jet reaches velocities up to 900 m/s (3,000 ft/sec). The fluid is pressurized to the desired level by a hydraulic pump.

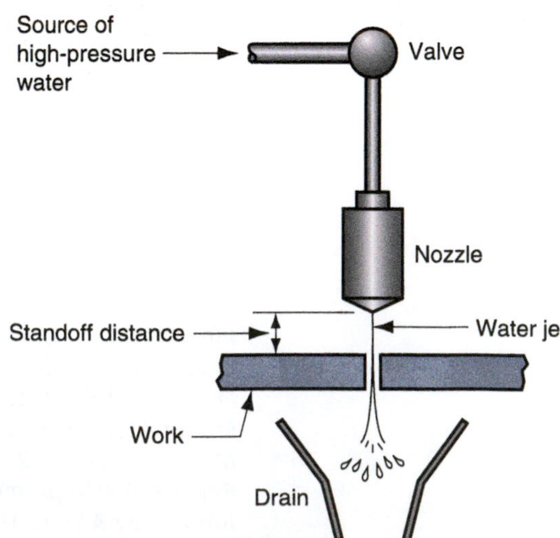

FIGURE 19.2 Water jet cutting. (Credit: *Fundamentals of Modern Manufacturing*, 4th Edition by Mikell P. Groover, 2010. Reprinted with permission of John Wiley & Sons, Inc.)

The nozzle unit consists of a holder made of stainless steel, and a jewel nozzle made of sapphire, ruby, or diamond. Diamond lasts the longest but costs the most. Filtration systems must be used in WJC to separate the swarf produced during cutting.

Cutting fluids in WJC are polymer solutions, preferred because of their tendency to produce a coherent stream. Cutting fluids were discussed before in the context of conventional machining (Section 17.4), but never has the term been more appropriately applied than in WJC.

Important process parameters include standoff distance, nozzle opening diameter, water pressure, and cutting feed rate. As shown in Figure 19.2, the ***standoff distance*** is the separation between the nozzle opening and the work surface. It is generally desirable for this distance to be small to minimize dispersion of the fluid stream before it strikes the surface. A typical standoff distance is 3.2 mm (0.125 in). Size of the nozzle orifice affects the precision of the cut; smaller openings are used for finer cuts on thinner materials. To cut thicker stock, thicker jet streams and higher pressures are required. The cutting feed rate refers to the velocity at which the WJC nozzle is traversed along the cutting path. Typical feed rates range from 5 mm/s (12 in/min) to more than 500 mm/s (1200 in/min), depending on work material and its thickness [5]. The WJC process is usually automated using computer numerical control or industrial robots to manipulate the nozzle unit along the desired trajectory.

Water jet cutting can be used effectively to cut narrow slits in flat stock such as plastic, textiles, composites, floor tile, carpet, leather, and cardboard. Robotic cells have been installed with WJC nozzles mounted as the robot's tool to follow cutting patterns that are irregular in three dimensions, such as cutting and trimming of automobile dashboards prior to assembly [9]. In these applications, advantages of WJC include: (1) no crushing or burning of the work surface typical in other mechanical or thermal processes, (2) minimum material loss because of the narrow cut slit, (3) no environmental pollution, and (4) ease of automating the process. A limitation of WJC is that the process is not suitable for cutting brittle materials (e.g., glass) because of their tendency to crack during cutting.

Abrasive Water Jet Cutting When WJC is used on metallic workparts, abrasive particles must usually be added to the jet stream to facilitate cutting. This process is therefore called ***abrasive water jet cutting*** (AWJC). Introduction of abrasive particles into the stream complicates the process by adding to the number of parameters that must be controlled. Among the additional parameters are abrasive type, grit size, and flow rate. Aluminum oxide, silicon dioxide, and garnet (a silicate mineral) are typical abrasive materials, at grit sizes ranging between 60 and 120. The abrasive particles are added to the water stream at approximately 0.25 kg/min (0.5 lb/min) after it has exited the WJC nozzle.

The remaining process parameters include those that are common to WJC: nozzle opening diameter, water pressure, and standoff distance. Nozzle orifice diameters are 0.25 to 0.63 mm (0.010 to 0.025 in)—somewhat larger than in water jet cutting to permit higher flow rates and more energy to be contained in the stream prior to injection of abrasives. Water pressures are about the same as in WJC. Standoff distances are somewhat less to minimize the effect of dispersion of the cutting fluid that now contains abrasive particles. Typical standoff distances are between 1/4 and 1/2 of those in WJC.

19.1.3 OTHER NONTRADITIONAL ABRASIVE PROCESSES

Two additional mechanical energy processes utilize abrasives to accomplish deburring, polishing, or other operations in which very little material is removed.

Abrasive Jet Machining Not to be confused with AWJC is the process called abrasive jet machining (AJM), a material removal process caused by the action of a high-velocity

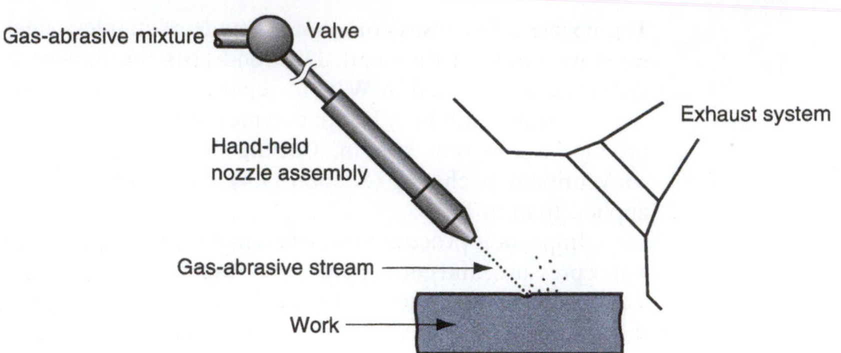

FIGURE 19.3 Abrasive jet machining (AJM). (Credit: *Fundamentals of Modern Manufacturing*, 4th Edition by Mikell P. Groover, 2010. Reprinted with permission of John Wiley & Sons, Inc.)

stream of gas containing small abrasive particles, as in Figure 19.3. The gas is dry, and pressures of 0.2 to 1.4 MPa (25 to 200 lb/in^2) are used to propel it through nozzle orifices of diameter 0.075 to 1.0 mm (0.003 to 0.040 in) at velocities of 2.5 to 5.0 m/s (500 to 1000 ft/min). Gases include dry air, nitrogen, carbon dioxide, and helium.

The process is usually performed manually by an operator who directs the nozzle at the work. Typical distances between nozzle tip and work surface range between 3 mm and 75 mm (0.125 in and 3 in). The workstation must be set up to provide proper ventilation for the operator.

AJM is normally used as a finishing process rather than a production cutting process. Applications include deburring, trimming and deflashing, cleaning, and polishing. Cutting is accomplished successfully on hard, brittle materials (e.g., glass, silicon, mica, and ceramics) that are in the form of thin flat stock. Typical abrasives used in AJM include aluminum oxide (for aluminum and brass), silicon carbide (for stainless steel and ceramics), and glass beads (for polishing). Grit sizes are small, 15 to 40 μm (0.0006 to 0.0016 in) in diameter, and must be uniform in size for a given application. It is important not to recycle the abrasives because used grains become fractured (and therefore smaller in size), worn, and contaminated.

Abrasive Flow Machining This process was developed in the 1960s to deburr and polish difficult-to-reach areas using abrasive particles mixed in a viscoelastic polymer that is forced to flow through or around the part surfaces and edges. The polymer has the consistency of putty. Silicon carbide is a typical abrasive. Abrasive flow machining (AFM) is particularly well suited for internal passageways that are often inaccessible by conventional methods. The abrasive-polymer mixture, called the media, flows past the target regions of the part under pressures ranging between 0.7 and 20 MPa (100 and 3000 lb/in^2). In addition to deburring and polishing, other AFM applications include forming radii on sharp edges, removing rough surfaces on castings, and other finishing operations. These applications are found in industries such as aerospace, automotive, and die-making. The process can be automated to economically finish hundreds of parts per hour.

A common setup is to position the workpart between two opposing cylinders, one containing media and the other empty. The media is forced to flow through the part from the first cylinder to the other, and then back again, as many times as necessary to achieve the desired material removal and finish.

19.2 ELECTROCHEMICAL MACHINING PROCESSES

An important group of nontraditional processes uses electrical energy to remove material. This group is identified by the term ***electrochemical processes***, because electrical energy is used in combination with chemical reactions to accomplish material removal. In effect, these processes are the reverse of electroplating (Section 21.3.1). The work material must be a conductor in the electrochemical machining processes.

19.2.1 ELECTROCHEMICAL MACHINING

The basic process in this group is electrochemical machining (ECM). Electrochemical machining removes metal from an electrically conductive workpiece by anodic dissolution, in which the shape of the workpiece is obtained by a formed electrode tool in close proximity to, but separated from, the work by a rapidly flowing electrolyte. ECM is basically a deplating operation. As illustrated in Figure 19.4, the workpiece is the anode, and the tool is the cathode. The principle underlying the process is that material is deplated from the anode (the positive pole) and deposited onto the cathode (the negative pole) in the presence of an electrolyte bath. The difference in ECM is that the electrolyte bath flows rapidly between the two poles to carry off the deplated material, so that it does not become plated onto the tool.

The electrode tool, usually made of copper, brass, or stainless steel, is designed to possess approximately the inverse of the desired final shape of the part. An allowance in the tool size must be provided for the gap that exists between the tool and the work. To accomplish metal removal, the electrode is fed into the work at a rate equal to the rate of metal removal from the work. Metal removal rate is determined by Faraday's First Law, which states that the amount of chemical change produced by an electric current (i.e., the amount of metal dissolved) is proportional to the quantity of electricity passed (current × time):

$$V = Cit \tag{19.1}$$

where V = volume of metal removed, mm^3 (in^3); C = a constant called the specific removal rate that depends on atomic weight, valence, and density of the work material, mm^3/amp- s (in^3/amp-min); I = current, amps; and t = time, s (min).

Based on Ohm's law, current $I = E/R$, where E = voltage and R = resistance. Under the conditions of the ECM operation, resistance is given by:

$$R = \frac{gr}{A} \tag{19.2}$$

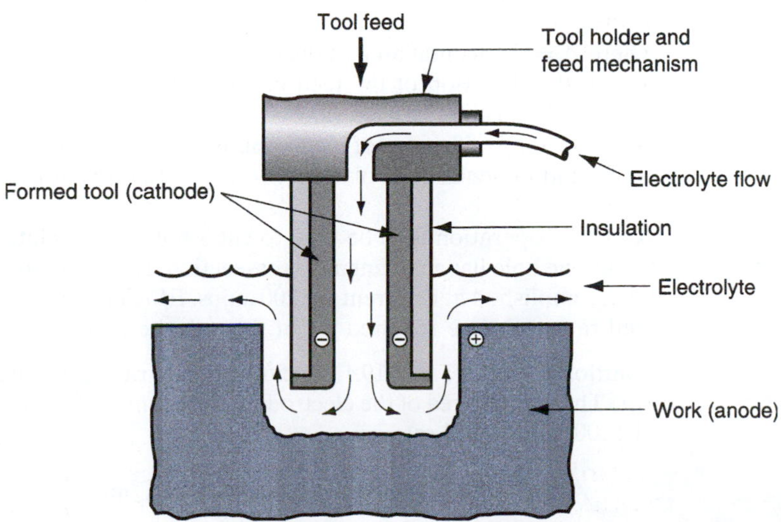

FIGURE 19.4 Electrochemical machining (ECM). (Credit: *Fundamentals of Modern Manufacturing*, 4th Edition by Mikell P. Groover, 2010. Reprinted with permission of John Wiley & Sons, Inc.)

TABLE 19.1 Typical values of specific removal rate C for selected work materials in electrochemical machining

	Specific Removal Rate C			Specific Removal Rate C	
Work Material[a]	mm^3/amp-sec	in^3/amp-min	Work Material[a]	mm^3/amp-sec	in^3/amp-min
Aluminum (3)	3.44×10^{-2}	1.26×10^{-4}	Steels:		
Copper (1)	7.35×10^{-2}	2.69×10^{-4}	Low alloy	3.0×10^{-2}	1.1×10^{-4}
Iron (2)	3.69×10^{-2}	1.35×10^{-4}	High alloy	2.73×10^{-2}	1.0×10^{-4}
Nickel (2)	3.42×10^{-2}	1.25×10^{-4}	Stainless	2.46×10^{-2}	0.9×10^{-4}
			Titanium (4)	2.73×10^{-2}	1.0×10^{-4}

Compiled from data in [8].

[a]Most common valence given in parentheses () is assumed in determining specific removal rate C. For different valence, multiply C by most common valence and divide by actual valence.

where g = gap between electrode and work, mm (in); r = resistivity of electrolyte, ohm-mm (ohm-in); and A = surface area between work and tool in the working frontal gap, mm^2 (in^2). Substituting this expression for R into Ohm's law, we have:

$$I = \frac{EA}{gr} \tag{19.3}$$

And substituting this equation back into the equation defining Faraday's law,

$$V = \frac{CEAt}{gr} \tag{19.4}$$

It is convenient to convert this equation into an expression for feed rate, the rate at which the electrode (tool) can be advanced into the work. This conversion can be accomplished in two steps. First, divide Eq. (19.4) by At (area × time) to convert volume of metal removed into a linear travel rate:

$$\frac{V}{At} = f_r = \frac{CE}{gr} \tag{19.5}$$

where f_r = feed rate, mm/s (in/min). Second, substitute I/A in place of $E/(gr)$, as provided by Eq. (19.3). Thus, feed rate in ECM is:

$$f_r = \frac{CI}{A} \tag{19.6}$$

where A = the frontal area of the electrode, mm^2 (in^2). This is the projected area of the tool in the direction of the feed into the work. Values of specific removal rate C are presented in Table 19.1 for various work materials. We should note that this equation assumes 100% efficiency of metal removal. The actual efficiency is in the range 90% to 100% and depends on tool shape, voltage and current density, and other factors.

Example 19.1 Electrochemical Machining

An ECM operation is to be used to cut a hole into a plate of aluminum that is 12 mm thick. The hole has a rectangular cross section, 10 mm by 30 mm. The ECM operation will be accomplished at a current = 1200 amps. Efficiency is expected to be 95%. Determine feed rate and time required to cut through the plate.

Solution: From Table 19.1, specific removal rate C for aluminum = 3.44×10^{-2} mm^3/A-s. The frontal area of the electrode A = 10 mm × 30 mm = 300 mm^2. At a current level of 1200 amps, feed rate is:

$$f_r = 0.0344\,\text{mm}^3/\text{A-s}\left(\frac{1200}{300}\,\text{A/mm}^2\right) = 0.1376\,\text{mm/s}$$

At an efficiency of 95%, the actual feed rate is:

$$f_r = 0.1376\,\text{mm/s}(0.95) = 0.1307\,\text{mm/s}$$

Time to machine through the 12-mm plate is:

$$T_m = \frac{12.0}{0.1307} = 91.8\,\text{s} = 1.53\,\text{min}$$

■

The preceding equations indicate the important process parameters for determining metal removal rate and feed rate in electrochemical machining: gap distance g, electrolyte resistivity r, current I, and electrode frontal area A. Gap distance needs to be controlled closely. If g becomes too large, the electrochemical process slows down. However, if the electrode touches the work, a short circuit occurs, which stops the process altogether. As a practical matter, gap distance is usually maintained within a range 0.075 to 0.75 mm (0.003 to 0.030 in).

Water is used as the base for the electrolyte in ECM. To reduce electrolyte resistivity, salts such as NaCl or $NaNO_3$ are added in solution. In addition to carrying off the material that has been removed from the workpiece, the flowing electrolyte also serves the function of removing heat and hydrogen bubbles created in the chemical reactions of the process. The removed work material is in the form of microscopic particles that must be separated from the electrolyte through centrifuge, sedimentation, or other means. The separated particles form a thick sludge whose disposal is an environmental problem associated with ECM.

Large amounts of electrical power are required to perform ECM. As the equations indicate, rate of metal removal is determined by electrical power, specifically the current density that can be supplied to the operation. The voltage in ECM is kept relatively low to minimize arcing across the gap.

Electrochemical machining is generally used in applications where the work metal is very hard or difficult to machine, or where the workpart geometry is difficult (or impossible) to accomplish by conventional machining methods. Work hardness makes no difference in ECM, because the metal removal is not mechanical. Typical ECM applications include: (1) ***die sinking***, which involves the machining of irregular shapes and contours into forging dies, plastic molds, and other shaping tools; (2) multiple hole drilling, where many holes can be drilled simultaneously with ECM and conventional drilling would probably require the holes to be made sequentially; and (3) holes that are not round, since ECM does not use a rotating drill.

Advantages of ECM include (1) little surface damage to the workpart, (2) no burrs as in conventional machining, (3) low tool wear (the only tool wear results from the flowing electrolyte), and (4) relatively high metal removal rates for hard and difficult-to-machine metals. Disadvantages of ECM are (1) significant cost of electrical power to drive the operation and (2) problems of disposing of the electrolyte sludge.

19.2.2 ELECTROCHEMICAL DEBURRING AND GRINDING

Electrochemical deburring (ECD) is an adaptation of ECM designed to remove burrs or to round sharp corners on metal workparts by anodic dissolution. One possible setup for ECD is shown in Figure 19.5. The hole in the workpart has a sharp burr of the type that is produced in a conventional through-hole drilling operation. The electrode tool is designed to focus the metal removal action on the burr. Portions of the tool not being used for machining are insulated. The electrolyte flows through the hole to carry away the burr particles. The same ECM principles of operation also apply to ECD. However, since much less material is removed in electrochemical deburring, cycle times are much

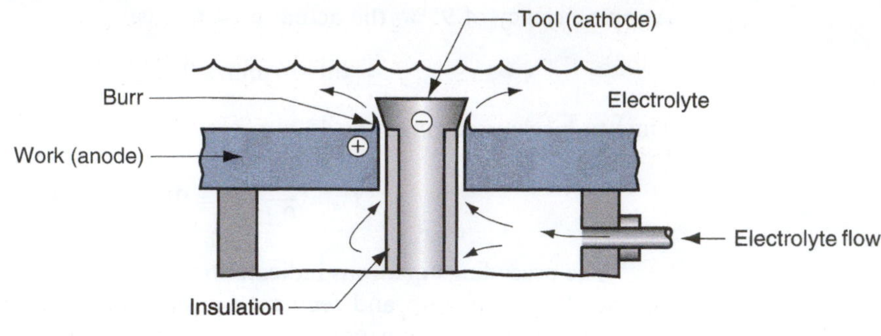

FIGURE 19.5 Electrochemical deburring (ECD). (Credit: *Fundamentals of Modern Manufacturing*, 4th Edition by Mikell P. Groover, 2010. Reprinted with permission of John Wiley & Sons, Inc.)

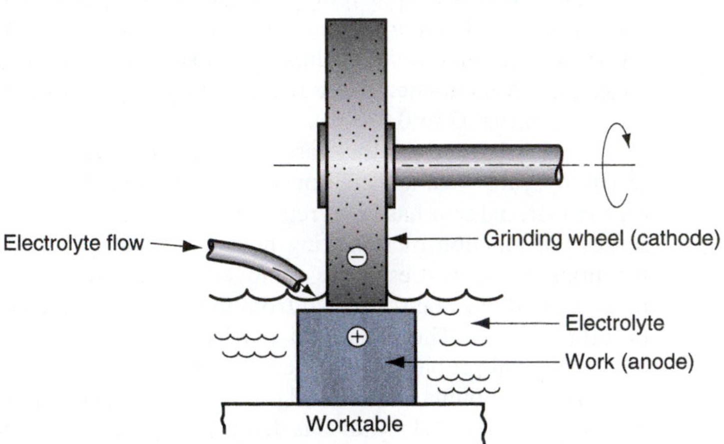

FIGURE 19.6 Electrochemical grinding (ECG). (Credit: *Fundamentals of Modern Manufacturing*, 4th Edition by Mikell P. Groover, 2010. Reprinted with permission of John Wiley & Sons, Inc.)

shorter. A typical cycle time in ECD is less than a minute. The time can be increased if it is desired to round the corner in addition to removing the burr.

Electrochemical grinding (ECG) is a special form of ECM in which a rotating grinding wheel with a conductive bond material is used to augment the anodic dissolution of the metal workpart surface, as illustrated in Figure 19.6. Abrasives used in ECG include aluminum oxide and diamond. The bond material is either metallic (for diamond abrasives) or resin bond impregnated with metal particles to make it electrically conductive (for aluminum oxide). The abrasive grits protruding from the grinding wheel at the contact with the workpart establish the gap distance in ECG. The electrolyte flows through the gap between the grains to play its role in electrolysis.

Deplating is responsible for 95% or more of the metal removal in ECG, and the abrasive action of the grinding wheel removes the remaining 5% or less, mostly in the form of salt films that have been formed during the electrochemical reactions at the work surface. Because most of the machining is accomplished by electrochemical action, the grinding wheel in ECG lasts much longer than a wheel in conventional grinding. The result is a much higher grinding ratio. In addition, dressing of the grinding wheel is required much less frequently. These are the significant advantages of the process. Applications of ECG include sharpening of cemented carbide tools and grinding of surgical needles, other thin wall tubes, and fragile parts.

19.3 THERMAL ENERGY PROCESSES

Material removal processes based on thermal energy are characterized by very high local temperatures—hot enough to remove material by fusion or vaporization. Because of the high temperatures, these processes cause physical and metallurgical damage to the new work surface. In some cases, the resulting finish is so poor that subsequent processing is required to smooth the surface. In this section we examine several nontraditional thermal energy processes that are commercially important: (1) electric discharge machining and electric discharge wire cutting, (2) electron beam machining, and (3) laser beam machining.

19.3.1 ELECTRIC DISCHARGE PROCESSES

Electric discharge processes remove metal by a series of discrete electrical discharges (sparks) that cause localized temperatures high enough to melt or vaporize the metal in the immediate vicinity of the discharge. The two main processes in this category are (1) electric discharge machining and (2) wire electric discharge machining. These processes can be used only on electrically conducting work materials.

Electric Discharge Machining Electric discharge machining (EDM) is one of the most widely used nontraditional processes. An EDM setup is illustrated in Figure 19.7. The shape of the finished work surface is produced by a formed electrode tool. The sparks occur across a small gap between tool and work surface. The EDM process must take place in the presence of a dielectric fluid, which creates a path for each discharge as the fluid becomes ionized in the gap. The discharges are generated by a pulsating direct current power supply connected to the work and the tool.

Figure 19.7(b) shows a close-up view of the gap between the tool and the work. The discharge occurs at the location where the two surfaces are closest. The dielectric fluid ionizes at this location to create a path for the discharge. The region in which discharge occurs is heated to extremely high temperatures, so that a small portion of the work surface is suddenly melted and removed. The flowing dielectric then flushes away the small particle (call it a "chip"). Since the surface of the work at the location of the previous discharge is now separated from the tool by a greater distance, this location is

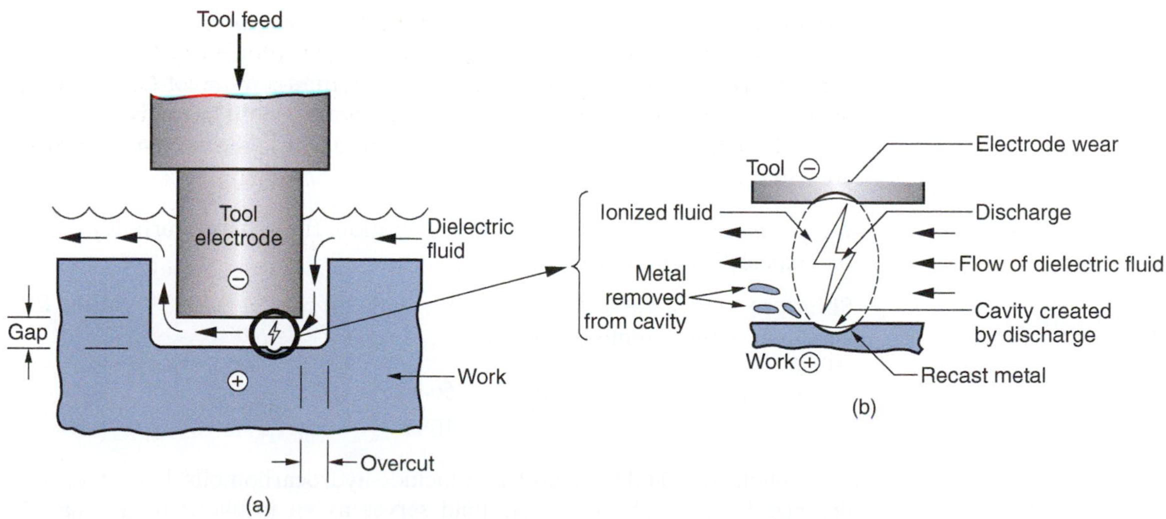

FIGURE 19.7 Electric discharge machining (EDM): (a) overall setup, and (b) close-up view of gap, showing discharge and metal removal. (Credit: *Fundamentals of Modern Manufacturing,* 4th Edition by Mikell P. Groover, 2010. Reprinted with permission of John Wiley & Sons, Inc.)

less likely to be the site of another spark until the surrounding regions have been reduced to the same level or below. Although the individual discharges remove metal at very localized points, they occur hundreds or thousands of times per second so that a gradual erosion of the entire surface occurs in the area of the gap.

Two important process parameters in EDM are discharge current and frequency of discharges. As either of these parameters is increased, metal removal rate increases. Surface roughness is also affected by current and frequency. The best surface finish is obtained in EDM by operating at high frequencies and low discharge currents. As the electrode tool penetrates into the work, overcutting occurs. ***Overcut*** in EDM is the distance by which the machined cavity in the workpart exceeds the size of the tool on each side, as illustrated in Figure 19.7(a). It is produced because the electrical discharges occur at the sides of the tool as well as its frontal area. The size of the overcut is several hundredths of a mm. Overcut increases with current and decreases at higher frequencies.

The high spark temperatures that melt the work also melt the tool, creating a small cavity in the surface opposite the cavity produced in the work. Tool wear is usually measured as the ratio of work material removed to tool material removed (similar to the grinding ratio). This wear ratio ranges between 1.0 and 100 or slightly above, depending on the combination of work and electrode materials. Electrodes are made of graphite, copper, brass, copper tungsten, silver tungsten, and other materials. The selection depends on the type of power supply circuit available on the EDM machine, the type of work material that is to be machined, and whether roughing or finishing is to be done. Graphite is preferred for many applications because of its melting characteristics. In fact, graphite does not melt. It vaporizes at very high temperatures, and the cavity created by the spark is generally smaller than for most other EDM electrode materials. Consequently, a high ratio of work material removed to tool wear is usually obtained with graphite tools.

The hardness and strength of the work material are not factors in EDM, since the process is not a contest of hardness between tool and work. The melting point of the work material is an important property, and metal removal rate can be related to melting point approximately by the following empirical formula, based on an equation described in Weller [16]:

$$R_{MR} = \frac{KI}{T_m{}^{1.23}} \tag{19.7}$$

where R_{MR} = metal removal rate, mm^3/s (in^3/min); K = constant of proportionality whose value = 664 in SI units (5.08 in U.S. customary units); I = discharge current, amps; and T_m = melting temperature of work metal, °C (°F). Melting points of selected metals are listed in Table 3.10.

Example 19.2 Electric Discharge Machining?

Copper is to be machined in an EDM operation. If discharge current = 25 amps, what is the expected metal removal rate?

Solution: From Table 3.10, the melting point of copper is 1083°C. Using Eq. (19.7), the anticipated metal removal rate is:

$$R_{MR} = \frac{664(25)}{1083^{1.23}} = 3.07 \text{ mm}^3/\text{s}$$

■

Dielectric fluids used in EDM include hydrocarbon oils, kerosene, and distilled or deionized water. The dielectric fluid serves as an insulator in the gap except when ionization occurs in the presence of a spark. Its other functions are to flush debris out of the gap and to remove heat from tool and workpart.

Applications of electric discharge machining include both tool fabrication and parts production. The tooling for many of the mechanical processes discussed in this book

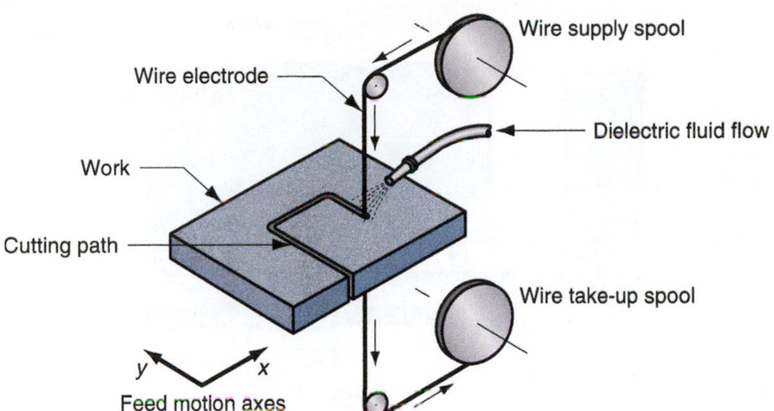

FIGURE 19.8 Electric discharge wire cutting (EDWC), also called wire EDM. (Credit: *Fundamentals of Modern Manufacturing*, 4th Edition by Mikell P. Groover, 2010. Reprinted with permission of John Wiley & Sons, Inc.)

are often made by EDM, including molds for plastic injection molding, extrusion dies, wire drawing dies, forging and heading dies, and sheet metal stamping dies. As in ECM, the term ***die sinking*** is used for operations in which a mold cavity is produced. For many of the applications, the materials used to fabricate the tooling are difficult (or impossible) to machine by conventional methods. Certain production parts also call for application of EDM. Examples include delicate parts that are not rigid enough to withstand conventional cutting forces, hole drilling where the axis of the hole is at an acute angle to the surface so that a conventional drill would be unable to start the hole, and production machining of hard and exotic metals.

Electric Discharge Wire Cutting Electric discharge wire cutting (EDWC), commonly called ***wire EDM***, is a special form of electric discharge machining that uses a small diameter wire as the electrode to cut a narrow kerf in the work. The cutting action in wire EDM is achieved by thermal energy from electric discharges between the electrode wire and the workpiece. Wire EDM is illustrated in Figure 19.8. The workpiece is fed past the wire in order to achieve the desired cutting path, somewhat in the manner of a bandsaw operation. Numerical control is used to control the workpart motions during cutting. As it cuts, the wire is slowly and continuously advanced between a supply spool and a take-up spool to present a fresh electrode of constant diameter to the work. This helps to maintain a constant kerf width during cutting. As in EDM, wire EDM must be carried out in the presence of a dielectric. This is applied by nozzles directed at the tool–work interface as in our figure, or the workpart is submerged in a dielectric bath.

Wire diameters range from 0.076 to 0.30 mm (0.003 to 0.012 in), depending on required kerf width. Materials used for the wire include brass, copper, tungsten, and molybdenum. Dielectric fluids include deionized water or oil. As in EDM, an overcut exists in wire EDM that makes the kerf larger than the wire diameter, as shown in Figure 19.9. This overcut is in the range 0.020 to 0.050 mm (0.0008 to 0.002 in). Once cutting conditions have been established for a given cut, the overcut remains fairly constant and predictable.

Although EDWC seems similar to a bandsaw operation, its precision far exceeds that of a bandsaw. The kerf is much narrower, corners can be made much sharper, and the cutting forces against the work are nil. In addition, hardness and toughness of the work material do not affect cutting performance. The only requirement is that the work material must be electrically conductive.

The special features of wire EDM make it ideal for making components for stamping dies. Since the kerf is so narrow, it is often possible to fabricate punch and die in a single cut out of a single block of tool steel. Other tools and parts with intricate outline shapes, such as lathe form tools, extrusion dies, and flat templates, are also made with electric discharge wire cutting.

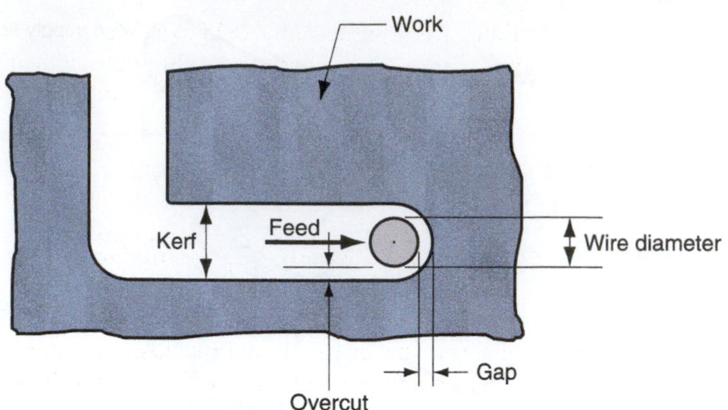

FIGURE 19.9 Definition of kerf and overcut in electric discharge wire cutting. (Credit: *Fundamentals of Modern Manufacturing,* 4th Edition by Mikell P. Groover, 2010. Reprinted with permission of John Wiley & Sons, Inc.)

19.3.2 ELECTRON BEAM MACHINING

Electron beam machining (EBM) is one of several industrial processes that use electron beams. In addition to machining, other applications of the technology include heat treating and welding. ***Electron beam machining*** uses a high-velocity stream of electrons focused on the workpiece surface to remove material by melting and vaporization. A schematic of the EBM process is illustrated in Figure 19.10. An electron beam gun generates a continuous stream of electrons that is accelerated to about 75% of the speed of light and focused through an electromagnetic lens on the work surface. The lens is capable of reducing the area of the beam to a diameter as small as 0.025 mm (0.001 in). On impinging the surface, the kinetic energy of the electrons is converted into thermal energy of extremely high density that melts or vaporizes the material in a very localized area.

Electron beam machining is used for a variety of high-precision cutting applications on any known material. Applications include drilling of extremely small diameter holes—down to 0.05 mm (0.002 in) diameter, drilling of holes with very high depth-to-diameter ratios—more than 100:1, and cutting of slots that are only about 0.025 mm (0.001 in) wide. These cuts can be made to very close tolerances with no cutting forces or tool wear. The process is ideal for micromachining and is generally limited to cutting

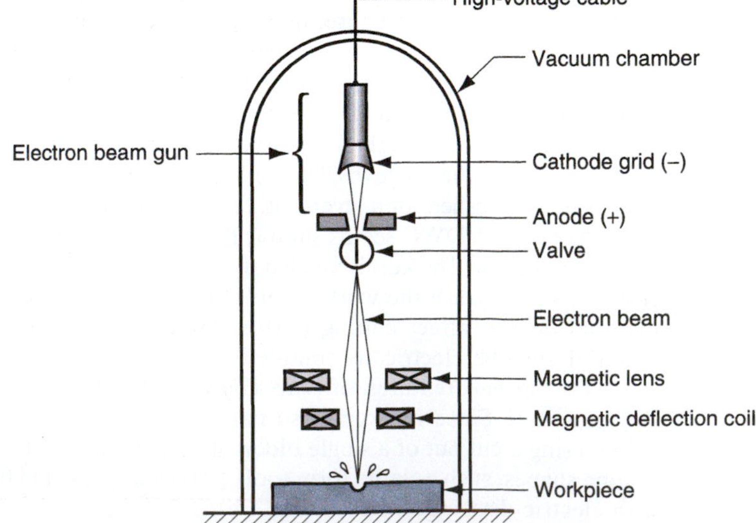

FIGURE 19.10 Electron beam machining (EBM). (Credit: *Fundamentals of Modern Manufacturing,* 4th Edition by Mikell P. Groover, 2010. Reprinted with permission of John Wiley & Sons, Inc.)

operations in thin parts—in the range 0.25 to 6.3 mm (0.010 to 0.250 in) thick. EBM must be carried out in a vacuum chamber to eliminate collision of the electrons with gas molecules. Other limitations include the high energy levels and expensive equipment.

19.3.3 LASER BEAM MACHINING

Lasers are being used for a variety of industrial applications, including heat treatment, welding, and measurement, as well as scribing, cutting, and drilling, which are described here. The term ***laser*** stands for ***l***ight ***a***mplification by ***s***timulated ***e***mission of ***r***adiation. A laser is an optical transducer that converts electrical energy into a highly coherent light beam. A laser light beam has several properties that distinguish it from other forms of light. It is monochromatic (theoretically, the light has a single wave length) and highly collimated (the light rays in the beam are almost perfectly parallel). These properties allow the light generated by a laser to be focused, using conventional optical lenses, onto a very small spot with resulting high power densities. Depending on the amount of energy contained in the light beam, and its degree of concentration at the spot, the various laser processes identified above can be accomplished.

Laser beam machining (LBM) uses the light energy from a laser to remove material by vaporization and ablation. The setup for LBM is illustrated in Figure 19.11. The types of lasers used in LBM are carbon dioxide gas lasers and solid-state lasers (of which there are several types). In laser beam machining, the energy of the coherent light beam is concentrated not only optically but also in terms of time. The light beam is pulsed so that the released energy results in an impulse against the work surface that produces a combination of evaporation and melting, with the melted material evacuating the surface at high velocity.

LBM is used to perform various types of drilling, slitting, slotting, scribing, and marking operations. Drilling small diameter holes is possible—down to 0.025 mm (0.001 in). For larger holes, above 0.50-mm (0.020-in) diameter, the laser beam is controlled to cut the outline of the hole. LBM is not considered a mass production process, and it is

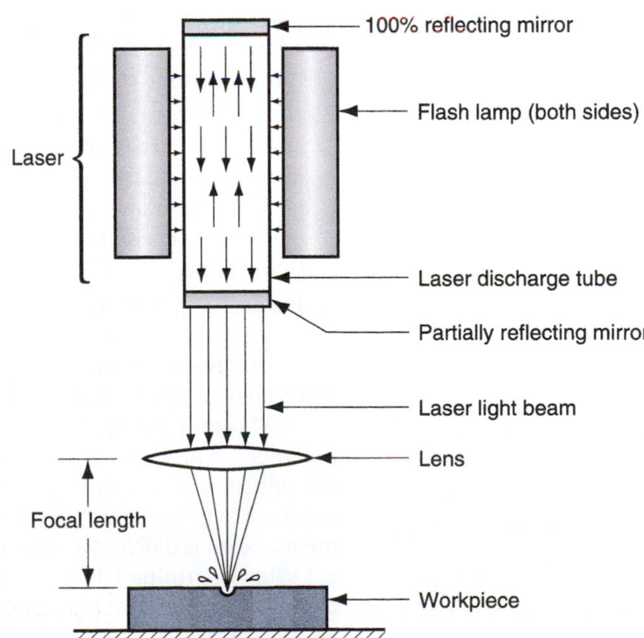

FIGURE 19.11 Laser beam machining (LBM). (Credit: *Fundamentals of Modern Manufacturing*, 4th Edition by Mikell P. Groover, 2010. Reprinted with permission of John Wiley & Sons, Inc.)

generally used on thin stock. The range of work materials that can be machined by LBM is virtually unlimited. Ideal properties of a material for LBM include high light energy absorption, poor reflectivity, good thermal conductivity, low specific heat, low heat of fusion, and low heat of vaporization. Of course, no material has this ideal combination of properties. The actual list of work materials processed by LBM includes metals with high hardness and strength, soft metals, ceramics, glass and glass epoxy, plastics, rubber, cloth, and wood.

19.4 CHEMICAL MACHINING

Chemical machining (CHM) is a nontraditional process in which material is removed by means of a strong chemical etchant. Applications as an industrial process began shortly after World War II in the aircraft industry. The use of chemicals to remove unwanted material from a workpart can be accomplished in several ways, and different terms have been developed to distinguish the applications. These terms include chemical milling, chemical blanking, chemical engraving, and photochemical machining (PCM). They all utilize the same mechanism of material removal, and it is appropriate to discuss the general characteristics of chemical machining before defining the individual processes.

19.4.1 MECHANICS AND CHEMISTRY OF CHEMICAL MACHINING

The chemical machining process consists of several steps. Differences in applications and the ways in which the steps are implemented account for the different forms of CHM. The steps are:

1. ***Cleaning***. The first step is a cleaning operation to ensure that material will be removed uniformly from the surfaces to be etched.
2. ***Masking***. A protective coating called a maskant is applied to certain portions of the part surface. This maskant is made of a material that is chemically resistant to the etchant (the term ***resist*** is used for this masking material). It is therefore applied to those portions of the work surface that are not to be etched.
3. ***Etching***. This is the material removal step. The part is immersed in an etchant that chemically attacks those portions of the part surface that are not masked. The usual method of attack is to convert the work material (e.g., a metal) into a salt that dissolves in the etchant and is thereby removed from the surface. When the desired amount of material has been removed, the part is withdrawn from the etchant and washed to stop the process.
4. ***Demasking***. The maskant is removed from the part.

The two steps in chemical machining that involve significant variations in methods, materials, and process parameters are masking and etching—steps 2 and 3.

Maskant materials include neoprene, polyvinylchloride, polyethylene, and other polymers. Masking can be accomplished by any of three methods: (1) cut and peel, (2) photographic resist, and (3) screen resist. The ***cut-and-peel*** method applies the maskant over the entire part by dipping, painting, or spraying. The resulting thickness of the maskant is 0.025 to 0.125 mm (0.001 to 0.005 in). After the maskant has hardened, it is cut using a scribing knife and peeled away in the areas of the work surface that are to be etched. The maskant cutting operation is performed by hand, usually guiding the knife with a template. The cut-and-peel method is generally used for large workparts, low production quantities, and where accuracy is not a critical factor. This method cannot hold tolerances tighter than ±0.125 mm (±0.005 in) except with extreme care.

As the name suggests, the ***photographic resist*** method (called the ***photoresist*** method for short) uses photographic techniques to perform the masking step. The masking materials contain photosensitive chemicals. They are applied to the work surface and exposed to light through a negative image of the desired areas to be etched. These areas of the maskant can then be removed from the surface using photographic developing techniques. This procedure leaves the desired surfaces of the part protected by the maskant and the remaining areas unprotected, vulnerable to chemical etching. Photoresist masking techniques are normally applied where small parts are produced in high quantities, and close tolerances are required. Tolerances closer than ±0.0125 mm (±0.0005 in) can be held [16].

The ***screen resist*** method applies the maskant by means of silk screening methods. In these methods, the maskant is painted onto the workpart surface through a silk or stainless steel mesh. Embedded in the mesh is a stencil that protects those areas to be etched from being painted. The maskant is thus painted onto the work areas that are not to be etched. The screen resist method is generally used in applications that are between the other two masking methods in terms of accuracy, part size, and production quantities. Tolerances of ±0.075 mm (±0.003 in) can be achieved with this masking method.

Selection of the ***etchant*** depends on work material to be etched, desired depth and rate of material removal, and surface finish requirements. The etchant must also be matched with the type of maskant that is used to ensure that the maskant material is not chemically attacked by the etchant. Table 19.2 lists some of the work materials machined by CHM together with the etchants that are generally used on these materials. Also included in the table are penetration rates and etch factors. These parameters are explained next.

Material removal rates in CHM are generally indicated as penetration rates, mm/min (in/min), since rate of chemical attack of the work material by the etchant is directed into the surface. The penetration rate is unaffected by surface area. Penetration rates listed in Table 19.2 are typical values for the given material and etchant.

Depths of cut in chemical machining are as much as 12.5 mm (0.5 in) for aircraft panels made out of metal plates. However, many applications require depths that are only several hundredths of a mm. Along with the penetration into the work, etching also occurs sideways under the maskant, as illustrated in Figure 19.12. The effect is referred to as the ***undercut,*** and it must be accounted for in the design of the mask in order for the resulting cut to have the specified dimensions. For a given work material, the undercut is

TABLE 19.2 Common work materials and etchants in CHM, with typical penetration rates and etch factors

Work Material	Etchant	Penetration Rates		Etch Factor
		mm/min	in/min	
Aluminum	$FeCl_3$	0.020	0.0008	1.75
and alloys	NaOH	0.025	0.001	1.75
Copper and alloys	$FeCl_3$	0.050	0.002	2.75
Magnesium and alloys	H_2SO_4	0.038	0.0015	1.0
Silicon	HNO_3: HF: H_2O	very slow		NA
Mild steel	HCl:HNO_3	0.025	0.001	2.0
	$FeCl_3$	0.025	0.001	2.0
Titanium	HF	0.025	0.001	1.0
and alloys	HF: HNO_3	0.025	0.001	1.0

Compiled from [5], [8], and [16].
NA = data not available.

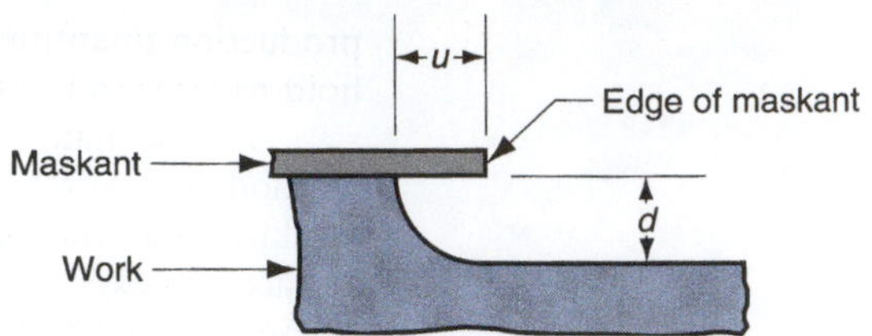

FIGURE 19.12 Undercut in chemical machining. (Credit: *Fundamentals of Modern Manufacturing*, 4th Edition by Mikell P. Groover, 2010. Reprinted with permission of John Wiley & Sons, Inc.)

directly related to the depth of cut. The constant of proportionality for the material is called the etch factor, defined as:

$$F_e = \frac{d}{u} \tag{19.8}$$

where F_e = etch factor; d = depth of cut, mm (in); and u = undercut, mm (in). The dimensions u and d are defined in Figure 19.12. Different work materials have different etch factors in chemical machining. Some typical values are presented in Table 19.2. The etch factor can be used to determine the dimensions of the cutaway areas in the maskant, so that the specified dimensions of the etched areas on the part can be achieved.

19.4.2 CHM PROCESSES

In this section, we describe the principal chemical machining processes: (1) chemical milling, (2) chemical blanking, (3) chemical engraving, and (4) photochemical machining.

Chemical Milling Chemical milling was the first CHM process to be commercialized. During World War II, an aircraft company in the United States began to use chemical milling to remove metal from aircraft components. They referred to their process as the "chem-mill" process. Today, chemical milling is still used largely in the aircraft industry, to remove material from aircraft wing and fuselage panels for weight reduction. It is applicable to large parts where substantial amounts of metal are removed during the process. The cut-and-peel maskant method is employed. A template is generally used that takes into account the undercut that will result during etching. The sequence of processing steps is illustrated in Figure 19.13.

Chemical Blanking Chemical blanking uses chemical erosion to cut very thin sheet metal parts—down to 0.025 mm (0.001 in) thick and/or for intricate cutting patterns.

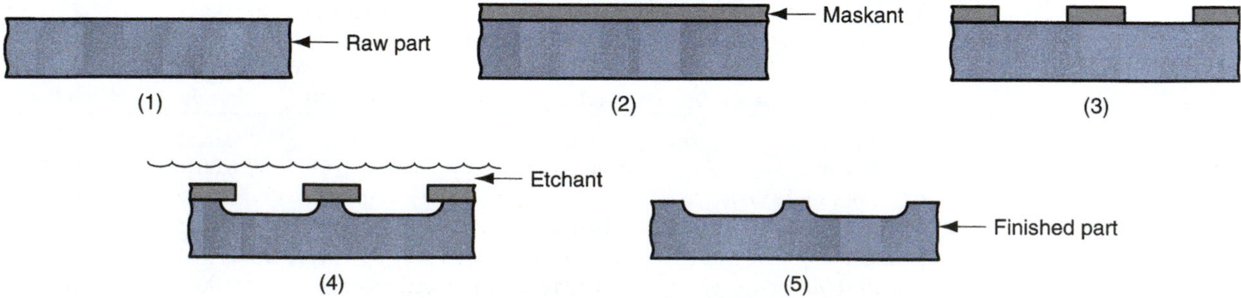

FIGURE 19.13 Sequence of processing steps in chemical milling: (1) clean raw part, (2) apply maskant, (3) scribe, cut, and peel the maskant from areas to be etched, (4) etch, and (5) remove maskant and clean to yield finished part. (Credit: *Fundamentals of Modern Manufacturing*, 4th Edition by Mikell P. Groover, 2010. Reprinted with permission of John Wiley & Sons, Inc.)

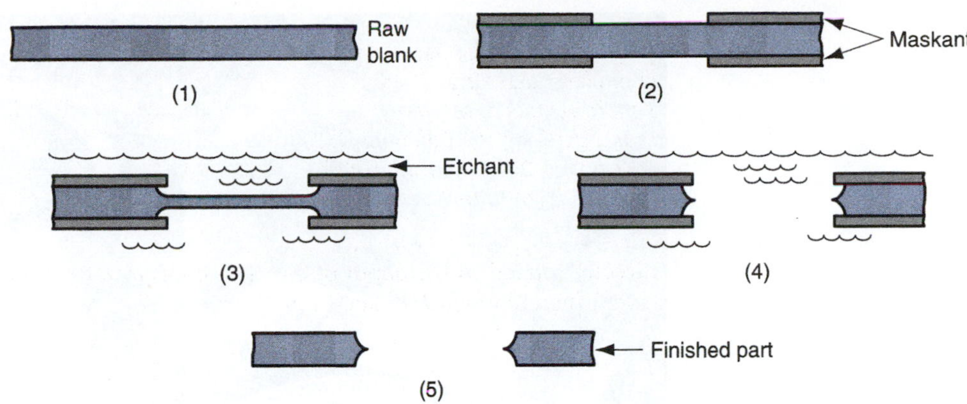

FIGURE 19.14 Sequence of processing steps in chemical blanking: (1) clean raw part, (2) apply resist (maskant) by painting through screen, (3) etch (partially completed), (4) etch (completed), and (5) remove resist and clean to yield finished part. (Credit: *Fundamentals of Modern Manufacturing*, 4th Edition by Mikell P. Groover, 2010. Reprinted with permission of John Wiley & Sons, Inc.)

In both instances, conventional punch-and-die methods do not work because the stamping forces damage the sheet metal, or the tooling cost would be prohibitive, or both. Chemical blanking produces parts that are burr-free, an advantage over conventional shearing operations.

Methods used for applying the maskant in chemical blanking are either the photoresist method or the screen resist method. For small and/or intricate cutting patterns and close tolerances, the photoresist method is used. Tolerances as close as ±0.0025 mm (±0.0001 in) can be held on 0.025 mm (0.001 in) thick stock using the photoresist method of masking. As stock thickness increases, more generous tolerances must be allowed. Screen resist masking methods are not nearly so accurate as photoresist. The small work size in chemical blanking excludes the cut-and-peel maskant method.

Using the screen resist method to illustrate, the steps in chemical blanking are shown in Figure 19.14. Since chemical etching takes place on both sides of the part in chemical blanking, it is important that the masking procedure provides accurate registration between the two sides. Otherwise, the erosion into the part from opposite directions will not line up. This is especially critical with small part sizes and intricate patterns.

Application of chemical blanking is generally limited to thin materials and/or intricate patterns for reasons given above. Maximum stock thickness is around 0.75 mm (0.030 in). Also, hardened and brittle materials can be processed by chemical blanking where mechanical methods would surely fracture the work. Figure 19.15 presents a sampling of parts produced by the chemical blanking process.

Chemical Engraving Chemical engraving is a chemical machining process for making name plates and other flat panels with lettering and/or artwork on one side. These plates and panels would otherwise be made using a conventional engraving machine or similar process. Chemical engraving can be used to make panels with either recessed lettering or raised lettering, simply by reversing the portions of the panel to be etched. Masking is done by either the photoresist or screen resist methods. The sequence in chemical engraving is similar to the other CHM processes, except that a filling operation follows etching. The purpose of filling is to apply paint or other coating into the recessed areas that have been created by etching. Then, the panel is immersed in a solution that dissolves the resist but does not attack the coating material. Thus, when the resist is

FIGURE 19.15 Parts made by chemical blanking. Courtesy of Buckbee-Mears St. Paul. (Credit: *Fundamentals of Modern Manufacturing,* 4th Edition by Mikell P. Groover, 2010. Reprinted with permission of John Wiley & Sons, Inc.)

removed, the coating remains in the etched areas but not in the areas that were masked. The effect is to highlight the pattern.

Photochemical Machining Photochemical machining (PCM) is chemical machining in which the photoresist method of masking is used. The term can therefore be applied correctly to chemical blanking and chemical engraving when these methods use the photographic resist method. PCM is employed in metalworking when close tolerances and/or intricate patterns are required on flat parts. Photochemical processes are also used extensively in the electronics industry to produce intricate circuit designs on semiconductor wafers.

Figure 19.16 shows the sequence of steps in photochemical machining as it is applied to chemical blanking. There are various ways to photographically expose the desired image onto the resist. The figure shows the negative in contact with the surface of the resist during exposure. This is contact printing, but other photographic printing methods are available that expose the negative through a lens system to enlarge or reduce the size of the pattern printed on the resist surface. Photoresist materials in current use are sensitive to ultraviolet light but not to light of other wavelengths. Therefore, with proper lighting in the factory, there is no need to carry out the processing steps in a dark room environment. Once the masking operation is accomplished, the remaining steps in the procedure are similar to the other chemical machining methods.

In photochemical machining, the term corresponding to etch factor is ***anisotropy***, which is defined as the depth of cut d divided by the undercut u (see Figure 19.12). This is the same definition as in Eq. (19.8).

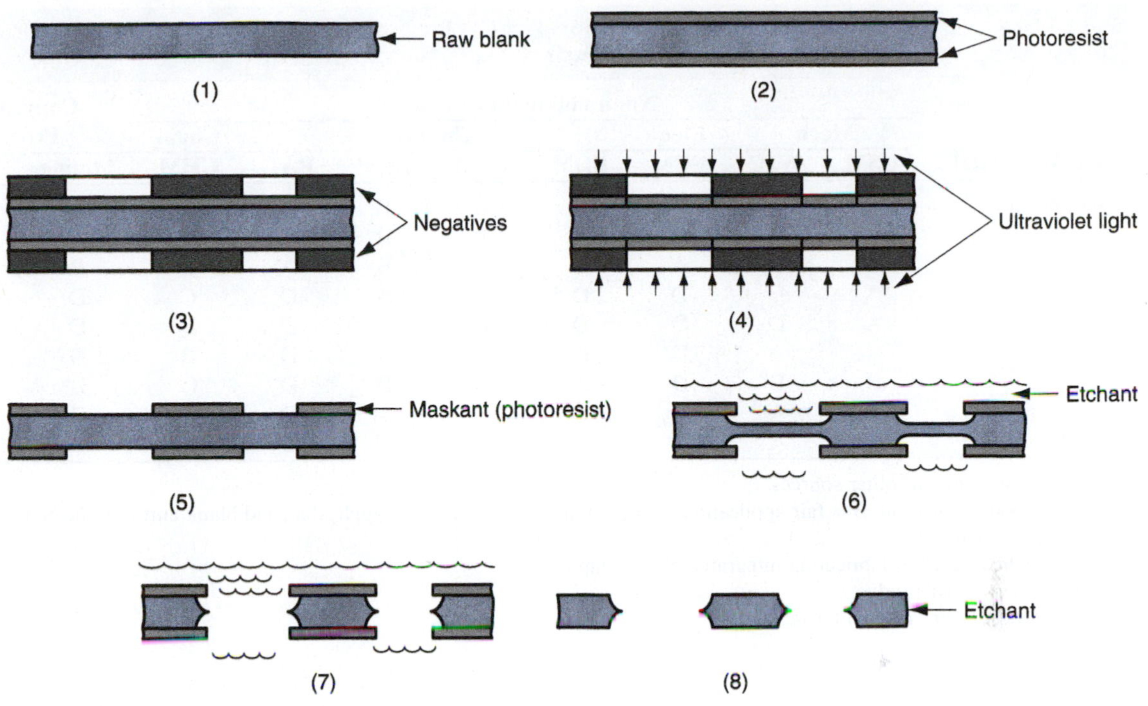

FIGURE 19.16 Sequence of processing steps in photochemical machining: (1) clean raw part, (2) apply resist (maskant) by dipping, spraying, or painting, (3) place negative on resist, (4) expose to ultraviolet light, (5) develop to remove resist from areas to be etched, (6) etch (shown partially etched), (7) etch (completed), (8) remove resist and clean to yield finished part. (Credit: *Fundamentals of Modern Manufacturing*, 4th Edition by Mikell P. Groover, 2010. Reprinted with permission of John Wiley & Sons, Inc.)

TABLE 19.3 Workpart geometric features and appropriate nontraditional processes

Geometric Feature	Likely Process
Very small holes. Diameters less than 0.125 mm (0.005 in), in some cases down to 0.025 mm (0.001 in), generally smaller than the diameter range of conventional drill bits.	EBM, LBM
Holes with large depth-to-diameter ratios, e.g., $d/D > 20$. Except for gun drilling, these holes cannot be machined in conventional drilling operations.	ECM, EDM
Holes that are not round. Nonround holes cannot be drilled with a rotating drill bit.	EDM, ECM
Narrow slots in slabs and plates of various materials. The slots are not necessarily straight. In some cases, the slots have extremely intricate shapes.	EBM, LBM, WJC, wire EDM, AWJC
Micromachining. In addition to cutting small holes and narrow slits, there are other material removal applications where the workpart and/or areas to be cut are very small.	PCM, LBM, EBM
Shallow pockets and surface details in flat parts. There is a significant range in the sizes of the parts in this category, from microscopic integrated circuit chips to large aircraft panels.	CHM
Special contoured shapes for mold and die applications. These applications are sometimes referred to as die-sinking.	EDM, ECM

TABLE 19.4 Applicability of selected nontraditional machining processes to various work materials. For comparison, conventional milling and grinding are included in the compilation

	Nontraditional Processes								Conventional Processes	
	Mech		Elec	Thermal				Chem		
Work Material	USM	WJC	ECM	EDM	EBM	LBM	PAC	CHM	Milling	Grinding
Aluminum	C	C	B	B	B	B	A	A	A	A
Steel	B	D	A	A	B	B	A	A	A	A
Super alloys	C	D	A	A	B	B	A	B	B	B
Ceramic	A	D	D	D	A	A	D	C	D	C
Glass	A	D	D	D	B	B	D	B	D	C
Silicon [a]			D	D	B	B	D	B	D	B
Plastics	B	B	D	D	B	B	D	C	B	C
Cardboard[b]	D	A	D	D			D	D	D	D
Textiles[c]	D	A	D	D			D	D	D	D

Compiled from [16] and other sources.

Key: A = good application, B = fair application, C = poor application, D = not applicable, and blank entries indicate no data available during compilation.

[a]Refers to silicon used in fabricating integrated circuit chips.

[b]Includes other paper products.

[c]Includes felt, leather, and similar materials.

19.5 APPLICATION CONSIDERATIONS

Typical applications of nontraditional processes include special geometric features and work materials that cannot be readily processed by conventional techniques. Some of the special workpart shapes for which nontraditional processes are well suited are listed in Table 19.3 along with the nontraditional processes that are likely to be appropriate.

As a group, the nontraditional processes can be applied to nearly all work materials, metals and nonmetals. However, certain processes are not suited to certain work materials. Table 19.4 relates applicability of the nontraditional processes to various types of materials. Several of the processes can be used on metals but not nonmetals. For example, ECM and EDM require work materials that are electrical conductors. This generally limits their applicability to metal parts. Chemical machining depends on the availability of an appropriate etchant for the given work material. Since metals are more susceptible to chemical attack by various etchants, CHM is commonly used to process metals. With some exceptions, USM, AJM, EBM, and LBM can be used on both metals and nonmetals. WJC is generally limited to the cutting of plastics, cardboards, textiles, and other materials that do not possess the strength of metals.

REFERENCES

[1] Aronson, R. B. "Waterjets Move into the Mainstream," ***Manufacturing Engineering***, *April* 2005, pp. 69–74.

[2] Bellows, G., and Kohls, J. B. "Drilling without Drills," Special Report 743, *American Machinist*, March 1982, pp. 173–188.

[3] Benedict, G. F. ***Nontraditional Manufacturing Processes***. Marcel Dekker, Inc., New York, 1987.

[4] Dini, J. W. "Fundamentals of Chemical Milling," Special Report 768, ***American Machinist***, July 1984 pp. 99–114.

[5] Drozda, T. J., and Wick, C. (eds.). ***Tool and Manufacturing Engineers Handbook***, 4th ed., Vol. **I**, ***Machining***. Society of Manufacturing Engineers, Dearborn, Michigan, 1983.

[6] El-Hofy, H. ***Advanced Machining Processes: Nontraditional and Hybrid Machining Processes***. McGraw-Hill Professional, New York, 2005.

[7] Guitrau, E. "Sparking Innovations," ***Cutting Tool Engineering***, Vol. **52**, No. 10, October 2000, pp. 36–43.

[8] ***Machining Data Handbook***, 3rd ed., Vol. **2**. Machinability Data Center, Metcut Research Associates Inc., Cincinnati, Ohio, 1980.

[9] Mason, F. "Water Jet Cuts Instrument Panels," ***American Machinist & Automated Manufacturing***, July 1988, pp. 126–127.

[10] McGeough, J. A. ***Advanced Methods of Machining***. Chapman and Hall, London, England, 1988.

[11] Pande, P. C., and Shan, H. S. ***Modern Machining Processes***. Tata McGraw-Hill Publishing Company, New Delhi, India, 1980.

[12] Vaccari, J. A. "The Laser's Edge in Metalworking," Special Report 768, ***American Machinist***, August 1984, pp. 99–114.

[13] Vaccari, J. A. "Thermal Cutting," Special Report 778, ***American Machinist***, July 1988, pp. 111–126.

[14] Vaccari, J. A. "Advances in Laser Cutting," ***American Machinist & Automated Manufacturing***, March 1988, pp. 59–61.

[15] Waurzyniak, P. "EDM's Cutting Edge," ***Manufacturing Engineering***, Vol. **123**, No. 5, November 1999, pp. 38–44.

[16] Weller, E. J. (ed.). ***Nontraditional Machining Processes***, 2nd ed. Society of Manufacturing Engineers, Dearborn, Michigan, 1984.

[17] www.engineershandbook.com/MfgMethods

REVIEW QUESTIONS

19.1. Why are the nontraditional material removal processes important?

19.2. There are four categories of nontraditional machining processes, based on principal energy form. Name the four categories.

19.3. How does the ultrasonic machining process work?

19.4. Describe the water jet cutting process.

19.5. What is the difference between water jet cutting, abrasive water jet cutting, and abrasive jet cutting?

19.6. Name the three main types of electrochemical machining.

19.7. Identify the two significant disadvantages of electrochemical machining.

19.8. How does increasing discharge current affect metal removal rate and surface finish in electric discharge machining?

19.9. What is meant by the term *overcut* in electric discharge machining?

19.10. Name the four principal steps in chemical machining.

19.11. What are the three methods of performing the masking step in chemical machining?

19.12. What is a photoresist in chemical machining?

PROBLEMS

19.1. For the following application, identify one or more nontraditional machining processes that might be used, and present arguments to support your selection. Assume that either the part geometry or the work material (or both) preclude the use of conventional machining. The application is a matrix of 0.1 mm (0.004 in) diameter holes in a plate of 3.2 mm (0.125 in) thick hardened tool steel. The matrix is rectangular, 75 by 125 mm (3.0 by 5.0 in) with the separation between holes in each direction = 1.6 mm (0.0625 in).

19.2. For the following application, identify one or more nontraditional machining processes that might be used, and present arguments to support your selection. Assume that either the part geometry or the work material (or both) preclude the use of conventional machining. The application is an engraved aluminum printing plate to be used in an offset printing press to make 275 by 350 mm (11 by 14 in) posters of Lincoln's Gettysburg address.

19.3. For the following application, identify one or more nontraditional machining processes that might be used, and present arguments to support your selection. Assume that either the part geometry or the work material (or both) preclude the use of conventional machining. The application is a through hole in the shape of the letter L in a 12.5 mm (0.5 in) thick plate of glass. The size of the "L" is 25 by 15 mm (1.0 by 0.6 in) and the width of the hole is 3 mm (1/8 in).

19.4. For the following application, identify one or more nontraditional machining processes that might be used, and present arguments to support your selection. Assume that either the part geometry or the work material (or both) preclude the use of conventional machining. The application is a blind hole in the shape of the letter G in a 50 mm (2.0 in) cube of steel. The overall size of the "G" is 25 by 19 mm (1.0 by 0.75 in), the depth of the hole is 3.8 mm (0.15 in), and its width is 3 mm (1/8 in).

19.5. A furniture company that makes upholstered chairs and sofas must cut large quantities of fabrics. Many of these fabrics are strong and wear-resistant, which properties make them difficult to cut. What nontraditional process(es) would you recommend to the company for this application? Justify your answer by indicating the characteristics of the process that make it attractive.

19.6. The frontal working area of the electrode in an ECM operation is 2000 mm^2. The applied current = 1800 amps and the voltage = 12 volts. The material being cut is nickel (valence = 2). (a) If the process is 90% efficient, determine the rate of metal removal in mm^3/min. (b) If the resistivity of the electrolyte = 140 ohm-mm, determine the working gap.

19.7. In an electrochemical machining operation, the frontal working area of the electrode is 2.5 in^2. The applied current = 1500 amps, and the voltage = 12 volts. The work material is pure aluminum. (a) If the ECM process is 90 percent efficient, determine the rate of metal removal in in^3/hr. (b) If the resistivity of the electrolyte = 6.2 ohm-in, determine the working gap.

19.8. A square hole is to be cut using ECM through a plate of pure copper (valence = 1) that is 20 mm thick. The hole is 25 mm on each side, but the electrode used to cut the hole is slightly less that 25 mm on its sides to allow for overcut, and its shape includes a hole in its center to permit the flow of electrolyte and to reduce the area of the cut. This tool design results in a frontal area of 200 mm^2. The applied current = 1000 amps. Using an efficiency of 95%, determine how long it will take to cut the hole.

19.9. A 3.5 in diameter through hole is to be cut in a block of pure iron (valence = 2) by electrochemical machining. The block is 2.0 in thick. To speed the cutting process, the electrode tool will have a center hole of 3.0 in which will produce a center core that can be removed after the tool breaks through. The outside diameter of the electrode is undersized to allow for overcut. The overcut is expected to be 0.005 in on a side. If the efficiency of the ECM operation is 90%, what current will be required to complete the cutting operation in 20 minutes?

19.10. An electric discharge machining operation is performed on two work materials: tungsten and tin. Determine the amount of metal removed in the operation after one hour at a discharge current of 20 amps for each of these metals. Use metric units and express the answers in mm^3/hr. Use Table 3.10 to determine melting temperatures for tungsten and tin.

19.11. An electric discharge machining operation is being performed on two work materials: iron and zinc. Determine the amount of metal removed in the operation after one hour at a discharge amperage = 15 amps for each of these metals. Use U.S. Customary units and express the answer in in^3/hr. Use Table 3.10 to determine melting temperatures for iron and zinc.

19.12. Suppose the hole in Problem 19.9 were to be cut using EDM rather than ECM. Using a discharge current = 20 amps (which would be typical for EDM), how long would it take to cut the hole? Use Table 3.10 to determine the melting temperature of pure iron.

19.13. A metal removal rate of 0.01 in^3/min is achieved in a certain EDM operation on a pure copper workpart. What metal removal rate would be achieved on nickel in this EDM operation, if the same discharge current were used? Use Table 3.10 to determine the melting temperatures of copper and nickel.

19.14. In a wire EDM operation performed on 7-mm-thick C1080 steel using a tungsten wire electrode whose diameter = 0.125 mm, past experience suggests that the overcut will be 0.02 mm, so that the kerf width will be 0.165 mm. Using a discharge current = 10 amps, what is the allowable feed rate that can be used in the operation? Use a melting temperature of 1500°C for 1080 steel.

19.15. A wire EDM operation is to be performed on a slab of $^3/_4$-in-thick aluminum using a brass wire electrode whose diameter = 0.005 in. It is anticipated that the overcut will be 0.001 in, so that the kerf width will be 0.007 in. Using a discharge current = 7 amps, what is the expected allowable feed rate that can be used in the operation? The melting temperature of aluminum is 1220°F.

19.16. Chemical milling is used in an aircraft plant to create pockets in wing sections made of an aluminum alloy. The starting thickness of one workpart of interest is 20 mm. A series of rectangular-shaped pockets 12 mm deep are to be etched with dimensions 200 mm by 400 mm. The corners of each rectangle are radiused to 15 mm. The part is an aluminum alloy and the etchant is NaOH. Use Table 19.2 to determine the penetration rate and etch factor for this combination. Determine (a) metal removal rate in mm^3/min, (b) time required to etch to the specified depth, and (c) required dimensions of the opening in the cut-and-peel maskant to achieve the desired pocket size on the part.

19.17. In a chemical milling operation on a flat mild steel plate, it is desired to cut an ellipse-shaped pocket to a depth of 0.4 in. The semiaxes of the ellipse are $a = 9.0$ in and $b = 6.0$ in. A solution of hydrochloric and nitric acids will be used as the etchant. Use Table 19.2 to determine the penetration rate and etch factor for this combination. Determine (a) metal removal rate in in^3/hr, (b) time required to etch to depth, and (c) required dimensions of the opening in the cut-and-peel maskant required to achieve the desired pocket size on the part.

19.18. In a certain chemical blanking operation, a sulfuric acid etchant is used to remove material from a sheet of magnesium alloy. The sheet is 0.25 mm thick. The screen resist method of masking was used to permit high production rates to be achieved. As it turns out, the process is producing a large proportion of scrap. Specified tolerances of ± 0.025 mm are not being achieved. The foreman in the CHM department complains that there must be something wrong with the sulfuric acid. "Perhaps the concentration is incorrect," he suggests. Analyze the problem and recommend a solution.

19.19. In a chemical blanking operation, stock thickness of the aluminum sheet is 0.015 in. The pattern to be cut out of the sheet is a hole pattern, consisting of a matrix of 0.100 in diameter holes. If photochemical machining is used to cut these holes, and contact printing is used to make the resist (maskant) pattern, determine the diameter of the holes that should be used in the pattern.

Part VI Property Enhancing and Surface Processing Operations

20 HEAT TREATMENT OF METALS

Chapter Contents

20.1 Annealing

20.2 Martensite Formation in Steel
20.2.1 The Time-Temperature-Transformation Curve
20.2.2 The Heat Treatment Process
20.2.3 Hardenability

20.3 Precipitation Hardening

20.4 Surface Hardening

The manufacturing processes covered in the preceding chapters involve the creation of part geometry. We now consider processes that either enhance the properties of the workpart (Chapter 20) or apply some surface treatment to it, such as cleaning or coating (Chapter 21). Property-enhancing operations are performed to improve mechanical or physical properties of the work material. They do not alter part geometry, at least not intentionally. The most important property-enhancing operations are heat treatments. ***Heat treatment*** involves various heating and cooling procedures performed to effect microstructural changes in a material, which in turn affect its mechanical properties. Its most common applications are on metals, discussed in this chapter. Similar treatments are performed on glass-ceramics (Section 2.2.3), tempered glass (Section 7.3.1), and powder metals and ceramics (Sections 10.2.3 and 11.2.3).

Heat treatment operations can be performed on a metallic workpart at various times during its manufacturing sequence. In some cases, the treatment is applied before shaping (e.g., to soften the metal so that it can be more easily formed while hot). In other cases, heat treatment is used to relieve the effects of strain hardening that occur during forming, so that the material can be subjected to further deformation. Heat treatment can also be accomplished at or near the end of the sequence to achieve the final strength and hardness required in the finished product. The principal heat treatments are annealing, martensite formation in steel, precipitation hardening, and surface hardening.

20.1 ANNEALING

Annealing consists of heating the metal to a suitable temperature, holding at that temperature for a certain time (called ***soaking***), and slowly cooling. It is performed on a metal for any of the following reasons: (1) to reduce hardness and brittleness, (2) to alter microstructure so that desirable mechanical properties can be obtained, (3) to soften metals for improved machinability or formability, (4) to recrystallize cold-worked (strain-hardened) metals, and (5) to relieve residual stresses induced by prior processes. Different terms are used in annealing, depending on the details of the process and the temperature used relative to the recrystallization temperature of the metal being treated.

Full annealing is associated with ferrous metals (usually low and medium carbon steels); it involves heating the alloy into the austenite region, followed by slow cooling in the furnace to produce coarse pearlite. ***Normalizing*** involves similar heating and soaking cycles, but the cooling rates are faster. The steel is allowed to cool in air to room temperature. This results in fine pearlite, higher strength and hardness, but lower ductility than the full anneal treatment.

Cold-worked parts are often annealed to reduce effects of strain hardening and increase ductility. The treatment allows the strain-hardened metal to recrystallize partially or completely, depending on temperatures, soaking periods, and cooling rates. When annealing is performed to allow for further cold working of the part, it is called a ***process anneal***. When performed on the completed (cold-worked) part to remove the effects of strain hardening and where no subsequent deformation will be accomplished, it is simply called an ***anneal***. The process itself is pretty much the same, but different terms are used to indicate the purpose of the treatment.

If annealing conditions permit full recovery of the cold-worked metal to its original grain structure, then ***recrystallization*** has occurred. After this type of anneal, the metal has the new geometry created by the forming operation, but its grain structure and associated properties are essentially the same as before cold working. The conditions that tend to favor recrystallization are higher temperature, longer holding time, and slower cooling rate. If the annealing process only permits partial return of the grain structure toward its original state, it is termed a ***recovery anneal***. Recovery allows the metal to retain most of the strain hardening obtained in cold working, but the toughness of the part is improved.

The preceding annealing operations are performed primarily to accomplish functions other than stress relief. However, annealing is sometimes performed solely to relieve residual stresses in the workpiece. Called ***stress-relief annealing***, it helps to reduce distortion and dimensional variations that might otherwise occur in the stressed parts.

20.2 MARTENSITE FORMATION IN STEEL

The iron–carbon phase diagram in Figure 2.1 indicates the phases of iron and iron carbide (cementite) present under equilibrium conditions. It assumes that cooling from high temperature is slow enough to permit austenite to decompose into a mixture of

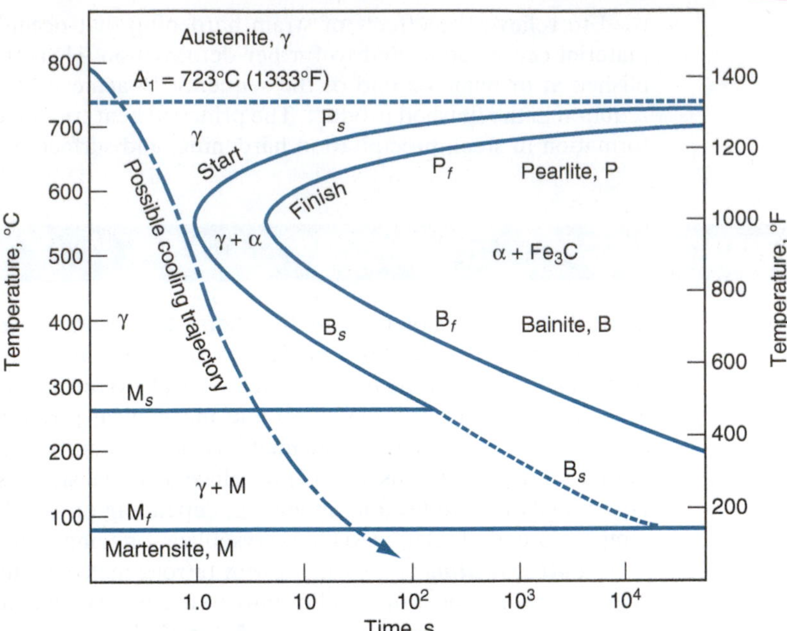

FIGURE 20.1 The TTT curve, showing the transformation of austenite into other phases as a function of time and temperature for a composition of about 0.80% C steel. The cooling trajectory shown here yields martensite. (Credit: *Fundamentals of Modern Manufacturing*, 4th Edition by Mikell P. Groover, 2010. Reprinted with permission of John Wiley & Sons, Inc.)

ferrite and cementite (Fe_3C) at room temperature. This decomposition reaction requires diffusion and other processes that depend on time and temperature to transform the metal into its preferred final form. However, under conditions of rapid cooling, so that the equilibrium reaction is inhibited, austenite transforms into a nonequilibrium phase called martensite. ***Martensite*** is a hard, brittle phase that gives steel its unique ability to be strengthened to very high levels.

20.2.1 THE TIME-TEMPERATURE-TRANSFORMATION CURVE

The nature of the martensite transformation can best be understood using the time-temperature-transformation curve (TTT curve) for eutectoid steel, illustrated in Figure 20.1. The TTT curve shows how cooling rate affects the transformation of austenite into various possible phases. The phases can be divided between (1) alternative forms of ferrite and cementite and (2) martensite. Time is displayed (logarithmically for convenience) along the horizontal axis, and temperature is scaled on the vertical axis. The curve is interpreted by starting at time zero in the austenite region (somewhere above the A_1 temperature line for the given composition) and proceeding downward and to the right along a trajectory representing how the metal is cooled as a function of time. The TTT curve shown in the figure is for a specific composition of steel (0.80% carbon). The shape of the curve is different for other compositions.

At slow cooling rates, the trajectory proceeds through the region indicating transformation into pearlite or bainite, which are alternative forms of ferrite–carbide mixtures. Because these transformations take time, the TTT diagram shows two lines—the start and finish of the transformation as time passes, indicated for the different phase regions by the subscripts *s* and *f*, respectively. ***Pearlite*** is a mixture of ferrite and carbide phases in the form of thin parallel plates. It is obtained by slow cooling from austenite, so that the cooling trajectory passes through P_s above the "nose" of the TTT curve. ***Bainite*** is an alternative mixture of the same phases that can be produced by initial rapid cooling to a temperature somewhat above M_s, so that the nose of the TTT curve is avoided; this is followed by much slower cooling to pass through B_s and into the ferrite–carbide region. Bainite has a needle-like or feather-like structure consisting of fine carbide regions.

If cooling occurs at a sufficiently rapid rate (indicated by the dashed line in Figure 20.1), austenite is transformed into martensite. ***Martensite*** is a unique phase consisting of an iron–carbon solution whose composition is the same as the austenite from which it was derived. The face-centered cubic structure of austenite is transformed into the body-centered tetragonal (BCT) structure of martensite almost instantly—without the time-dependent diffusion process needed to separate ferrite and iron carbide in the preceding transformations.

During cooling, the martensite transformation begins at a certain temperature M_s, and finishes at a lower temperature M_f, as shown in our TTT diagram. At points between these two levels, the steel is a mixture of austenite and martensite. If cooling is stopped at a temperature between the M_s and M_f lines, the austenite will transform to bainite as the time-temperature trajectory crosses the B_s threshold. The level of the M_s line is influenced by alloying elements, including carbon. In some cases, the M_s line is depressed below room temperature, making it impossible for these steels to form martensite by traditional heat-treating methods.

The extreme hardness of martensite results from the lattice strain created by carbon atoms trapped in the BCT structure, thus providing a barrier to slip. Figure 20.2 shows the significant effect that the martensite transformation has on the hardness of steel for increasing carbon contents.

20.2.2 THE HEAT TREATMENT PROCESS

The heat treatment to form martensite consists of two steps: austenitizing and quenching. These steps are often followed by tempering to produce tempered martensite. ***Austenitizing*** involves heating the steel to a sufficiently high temperature that it is converted entirely or partially to austenite. This temperature can be determined from the phase diagram for the particular alloy composition. The transformation to austenite involves a phase change, which requires time as well as heat. Accordingly, the steel must be held at

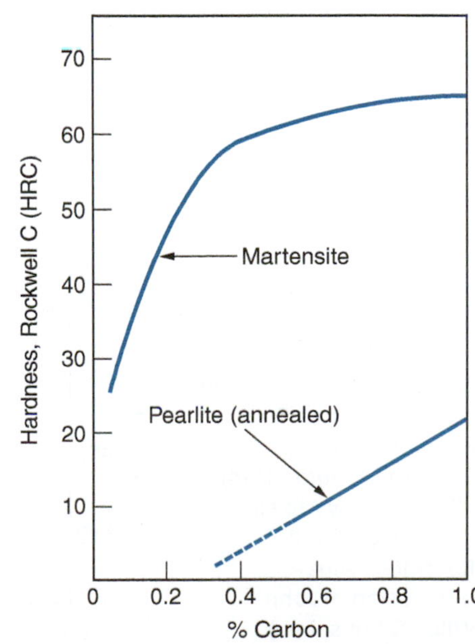

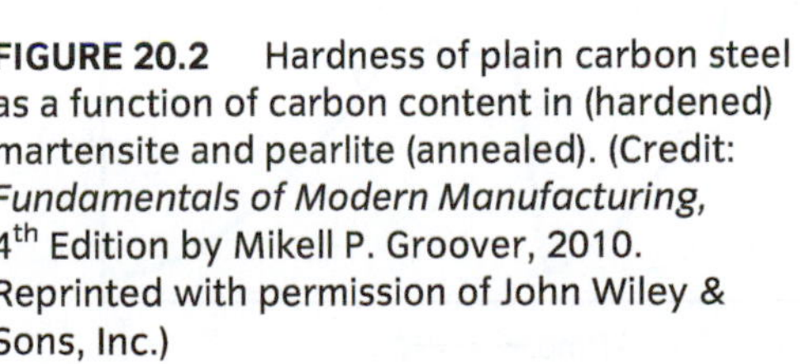

FIGURE 20.2 Hardness of plain carbon steel as a function of carbon content in (hardened) martensite and pearlite (annealed). (Credit: *Fundamentals of Modern Manufacturing*, 4th Edition by Mikell P. Groover, 2010. Reprinted with permission of John Wiley & Sons, Inc.)

the elevated temperature for a sufficient period of time to allow the new phase to form and the required homogeneity of composition to be achieved.

The ***quenching*** step involves cooling the austenite rapidly enough to avoid passing through the nose of the TTT curve, as indicated in the cooling trajectory shown in Figure 20.1. The cooling rate depends on the quenching medium and the rate of heat transfer within the steel workpiece. Various quenching media are used in commercial heat treatment practice: (1) brine—saltwater, usually agitated; (2) freshwater—still, not agitated; (3) still oil; and (4) air. Quenching in agitated brine provides the fastest cooling of the heated part surface, whereas air quench is the slowest. Trouble is, the more effective the quenching media is at cooling, the more likely it is to cause internal stresses, distortion, and cracks in the product.

The rate of heat transfer within the part depends largely on its mass and geometry. A large cubic shape will cool much more slowly than a small, thin sheet. The coefficient of thermal conductivity k of the particular composition is also a factor in the flow of heat in the metal. There is considerable variation in k for different grades of steel; for example, plain low carbon steel has a typical k value equal to 0.046 J/sec-mm-C (2.2 Btu/hr-in-F), whereas a highly alloyed steel might have one-third that value.

Martensite is hard and brittle. ***Tempering*** is a heat treatment applied to hardened steels to reduce brittleness, increase ductility and toughness, and relieve stresses in the martensite structure. It involves heating and soaking at a temperature below the austenitizing level for about one hour, followed by slow cooling. This results in precipitation of very fine carbide particles from the martensitic iron–carbon solution, and gradually transforms the crystal structure from BCT to BCC. This new structure is called ***tempered martensite***. A slight reduction in strength and hardness accompanies the improvement in ductility and toughness. The temperature and time of the tempering treatment control the degree of softening in the hardened steel, because the change from untempered to tempered martensite involves diffusion.

Taken together, the three steps in the heat treatment of steel to form tempered martensite can be pictured as in Figure 20.3. There are two heating and cooling cycles, the first to produce martensite and the second to temper the martensite.

20.2.3 HARDENABILITY

Hardenability refers to the relative capacity of a steel to be hardened by transformation to martensite. It is a property that determines the depth below the quenched surface to which the steel is hardened, or the severity of the quench required to achieve a certain hardness penetration. Steels with good hardenability can be hardened more deeply below

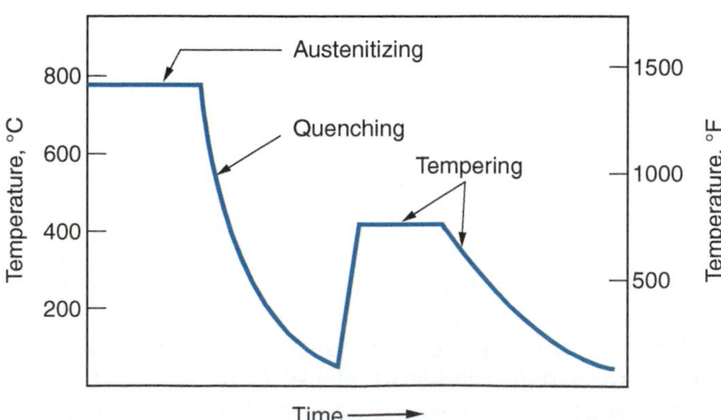

FIGURE 20.3 Typical heat treatment of steel: austenitizing, quenching, and tempering. (Credit: *Fundamentals of Modern Manufacturing*, 4th Edition by Mikell P. Groover, 2010. Reprinted with permission of John Wiley & Sons, Inc.)

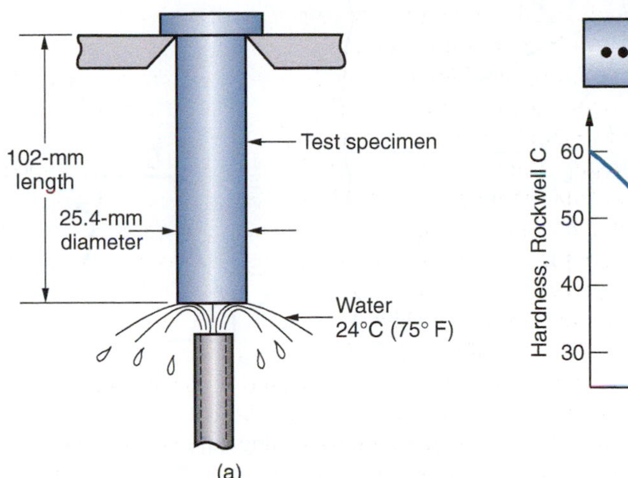

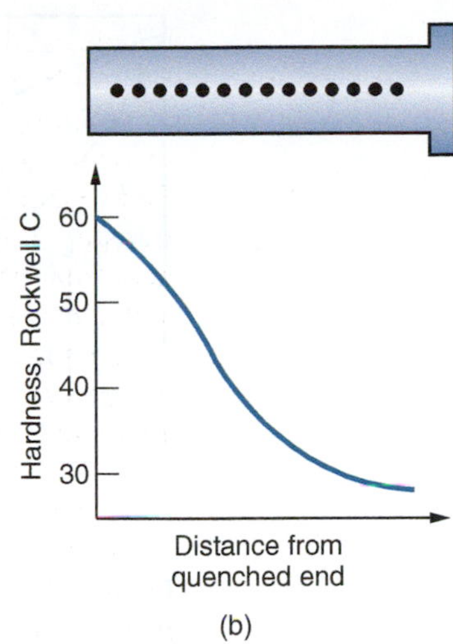

FIGURE 20.4 The Jominy end-quench test: (a) setup of the test, showing end quench of the test specimen; and (b) typical pattern of hardness readings as a function of distance from quenched end. (Credit: *Fundamentals of Modern Manufacturing*, 4th Edition by Mikell P. Groover, 2010. Reprinted with permission of John Wiley & Sons, Inc.)

the surface and do not require high cooling rates. Hardenability does not refer to the maximum hardness that can be attained in the steel; that depends on the carbon content.

The hardenability of a steel is increased through alloying. Alloying elements having the greatest effect are chromium, manganese, molybdenum, and nickel, to a lesser extent. The mechanism by which these alloying ingredients operate is to extend the time before the start of the austenite-to-pearlite transformation in the TTT diagram. In effect, the TTT curve is moved to the right, thus permitting slower quenching rates during quenching. Therefore, the cooling trajectory is able to follow a less hastened path to the M_s line, more easily avoiding the nose of the TTT curve.

The most common method for measuring hardenability is the ***Jominy end-quench test***. The test involves heating a standard specimen of diameter = 25.4 mm (1.0 in) and length = 102 mm (4.0 in) into the austenite range, and then quenching one end with a stream of cold water while the specimen is supported vertically as shown in Figure 20.4(a). The cooling rate in the test specimen decreases with increased distance from the quenched end. Hardenability is indicated by the hardness of the specimen as a function of distance from quenched end, as in Figure 20.4(b).

20.3 PRECIPITATION HARDENING

Precipitation hardening involves the formation of fine particles (precipitates) that act to block the movement of dislocations and thus strengthen and harden the metal. It is the principal heat treatment for strengthening alloys of aluminum, copper, magnesium, nickel, and other nonferrous metals. Precipitation hardening can also be used to strengthen certain steel alloys. When applied to steels, the process is called ***maraging*** (an abbreviation of martensite and aging), and the steels are called maraging steels.

The necessary condition that determines whether an alloy system can be strengthened by precipitation hardening is the presence of a sloping solvus line, as shown in the phase diagram of Figure 20.5(a). A composition that can be precipitation hardened is one that contains two phases at room temperature, but which can be heated to a temperature that dissolves the second phase. Composition C satisfies this requirement. The heat treatment process consists of three steps, illustrated in Figure 20.5(b): (1) ***solution treatment***, in which the alloy is heated to a temperature T_s above the solvus line into the alpha phase region and held for a period sufficient to dissolve the beta phase; (2) ***quenching*** to room

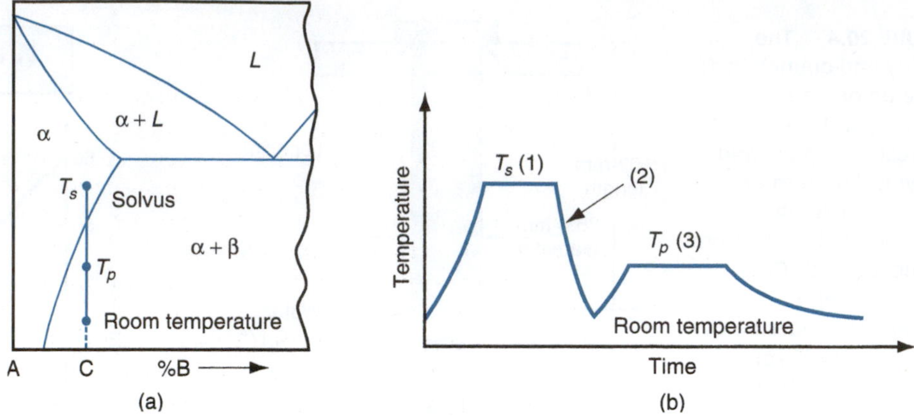

FIGURE 20.5 Precipitation hardening: (a) phase diagram of an alloy system consisting of metals A and B that can be precipitation hardened; and (b) heat treatment: (1) solution treatment, (2) quenching, and (3) precipitation treatment. (Credit: *Fundamentals of Modern Manufacturing*, 4th Edition by Mikell P. Groover, 2010. Reprinted with permission of John Wiley & Sons, Inc.)

temperature to create a supersaturated solid solution; and (3) ***precipitation treatment***, in which the alloy is heated to a temperature T_p, below T_s, to cause precipitation of fine particles of the beta phase. This third step is called ***aging***, and for this reason the whole heat treatment is sometimes called ***age hardening***. However, aging can occur in some alloys at room temperature, and so the term ***precipitation hardening*** seems more precise for the three-step heat treatment process under discussion here. When the aging step is performed at room temperature, it is called ***natural aging***. When it is accomplished at an elevated temperature, as in our figure, the term ***artificial aging*** is often used.

It is during the aging step that high strength and hardness are achieved in the alloy. The combination of temperature and time during the precipitation treatment (aging) is critical in bringing out the desired properties in the alloy. At higher precipitation treatment temperatures, as in Figure 20.6(a), the hardness peaks in a relatively short time; whereas at lower temperatures, as in Figure 20.6(b), more time is required to harden the alloy but its maximum hardness is likely to be greater than in the first case. As seen in the plot, continuation of the aging process results in a reduction in hardness and strength properties, called ***overaging***. Its overall effect is similar to annealing.

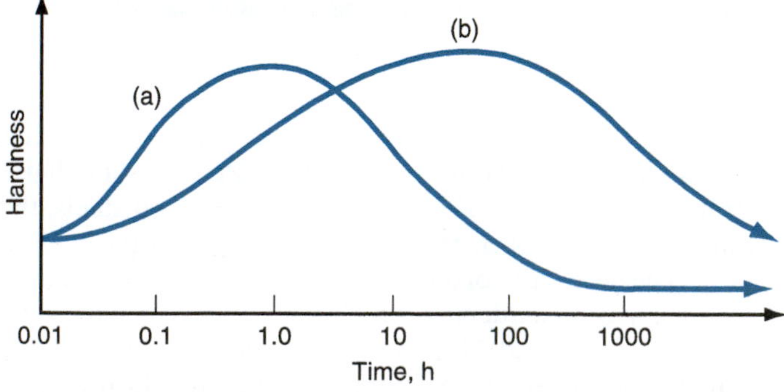

FIGURE 20.6 Effect of temperature and time during precipitation treatment (aging): (a) high precipitation temperature; and (b) lower precipitation temperature. (Credit: *Fundamentals of Modern Manufacturing*, 4th Edition by Mikell P. Groover, 2010. Reprinted with permission of John Wiley & Sons, Inc.)

20.4 SURFACE HARDENING

Surface hardening refers to any of several thermochemical treatments applied to steels in which the composition of the part surface is altered by addition of carbon, nitrogen, or other elements. The most common treatments are carburizing, nitriding, and carbonitriding. These processes are commonly applied to low carbon steel parts to achieve a hard, wear-resistant outer shell while retaining a tough inner core. The term ***case hardening*** is often used for these treatments.

Carburizing is the most common surface-hardening treatment. It involves heating a part of low carbon steel in the presence of a carbon-rich environment so that C is diffused into the surface. In effect the surface is converted to high carbon steel, capable of higher hardness than the low-C core. The carbon-rich environment can be created in several ways. One method involves the use of carbonaceous materials such as charcoal or coke packed in a closed container with the parts. This process, called ***pack carburizing***, produces a relatively thick layer on the part surface, ranging from around 0.6 to 4 mm (0.025 to 0.150 in). Another method, called ***gas carburizing***, uses hydrocarbon fuels such as propane (C_3H_8) inside a sealed furnace to diffuse carbon into the parts. The case thickness in this treatment is thin, 0.13 to 0.75 mm (0.005 to 0.030 in). Another process is ***liquid carburizing***, which employs a molten salt bath containing sodium cyanide (NaCN), barium chloride ($BaCl_2$), and other compounds to diffuse carbon into the steel. This process produces surface layer thicknesses generally between those of the other two treatments. Typical carburizing temperatures are 875° to 925°C (1600° to 1700°F), well into the austenite range.

Carburizing followed by quenching produces a case hardness of around 60 HRC. However, because the internal regions of the part consist of low carbon steel, and its hardenability is low, it is unaffected by the quench and remains relatively tough and ductile to withstand impact and fatigue stresses.

Nitriding is a treatment in which nitrogen is diffused into the surfaces of special alloy steels to produce a thin hard casing without quenching. To be most effective, the steel must contain certain alloying ingredients such as aluminum (0.855 to 1.5%) or chromium (5% or more). These elements form nitride compounds that precipitate as very fine particles in the casing to harden the steel. Nitriding methods include: ***gas nitriding***, in which the steel parts are heated in an atmosphere of ammonia (NH_3) or other nitrogen-rich gas mixture; and ***liquid nitriding***, in which the parts are dipped in molten cyanide salt baths. Both processes are carried out at around 500°C (950°F). Case thicknesses range as low as 0.025 mm (0.001 in) and up to around 0.5 mm (0.020 in), with hardnesses up to HRC 70.

As its name suggests, ***carbonitriding*** is a treatment in which both carbon and nitrogen are absorbed into the steel surface, usually by heating in a furnace containing carbon and ammonia. Case thicknesses are usually 0.07 to 0.5 mm (0.003 to 0.020 in), with hardnesses comparable with those of the other two treatments.

REFERENCES

[1] ***ASM Handbook***, Vol. **4**, ***Heat Treating***. ASM International, Materials Park, Ohio, 1991.

[2] Babu, S. S., and Totten, G. E. ***Steel Heat Treatment Handbook***, 2nd ed. CRC Taylor & Francis, Boca Raton, FL, 2006.

[3] Brick, R. M., Pense, A. W., and Gordon, R. B. ***Structure and Properties of Engineering Materials***, 4th ed. McGraw-Hill Book Company, New York, 1977.

[4] Chandler, H. (ed.). ***Heat Treater's Guide: Practices and Procedures for Irons and Steels***. ASM International, Materials Park, Ohio, 1995.

[5] Chandler, H. (ed.). ***Heat Treater's Guide: Practices and Procedures for Nonferrous Alloys***. ASM International, Materials Park, Ohio, 1996.

[6] Dossett, J. L., and Boyer, H. E. ***Practical Heat Treating***, 2nd ed. ASM International, Materials Park, Ohio, 2006.

[7] Flinn, R. A., and Trojan, P. K. ***Engineering Materials and Their Applications***, 5th ed. John Wiley & Sons, Inc., New York, 1995.

[8] Guy, A. G., and Hren, J. J. ***Elements of Physical Metallurgy***, 3rd ed. Addison-Wesley Publishing Co., Reading, Massachusetts, 1974.

[9] Ostwald, P. F., and Munoz, J. ***Manufacturing Processes and Systems***, 9th ed. John Wiley & Sons, New York, 1997.

[10] Vaccari, J. A. "Fundamentals of Heat Treating," Special Report 737, ***American Machinist***, September 1981, pp. 185–200.

[11] Wick, C. and Veilleux, R. F. (eds.). ***Tool and Manufacturing Engineers Handbook***, 4th ed., Vol. **3**, ***Materials, Finishing, and Coating***. Section 2: Heat Treatment. Society of Manufacturing Engineers, Dearborn, Michigan, 1985.

REVIEW QUESTIONS

20.1. Why are metals heat-treated?

20.2. Identify the important reasons why metals are annealed.

20.3. What is the most important heat treatment for hardening steels?

20.4. What is the mechanism by which carbon strengthens steel during heat treatment?

20.5. What information is conveyed by the TTT curve?

20.6. What function is served by tempering of martensite?

20.7. Define hardenability.

20.8. Name some of the elements that have the greatest effect on the hardenability of steel.

20.9. Indicate how the hardenability alloying elements in steel affect the TTT curve.

20.10. Define precipitation hardening.

20.11. How does carburizing work?

21 SURFACE PROCESSING OPERATIONS

Chapter Contents

21.1 Industrial Cleaning Processes
- 21.1.1 Chemical Cleaning
- 21.1.2 Mechanical Cleaning and Surface Treatments

21.2 Diffusion and Ion Implantation
- 21.2.1 Diffusion
- 21.2.2 Ion Implantation

21.3 Plating and Related Processes
- 21.3.1 Electroplating
- 21.3.2 Electroforming
- 21.3.3 Electroless Plating
- 21.3.4 Hot Dipping

21.4 Conversion Coating
- 21.4.1 Chemical Conversion Coatings
- 21.4.2 Anodizing

21.5 Vapor Deposition Processes
- 21.5.1 Physical Vapor Deposition
- 21.5.2 Chemical Vapor Deposition

21.6 Organic Coatings
- 21.6.1 Application Methods
- 21.6.2 Powder Coating

The processes discussed in this chapter operate on the surfaces of parts and/or products. The major categories of surface processing operations are (1) cleaning, (2) surface treatments, and (3) coating and thin film deposition. Cleaning refers to industrial processes that remove soils and contaminants that result from previous processing or the factory environment. They include both chemical and mechanical cleaning methods. Surface treatments are mechanical and physical operations that alter the part surface in some way, such as improving its finish or impregnating it with atoms of a foreign material to change its chemistry and physical properties.

Coating and thin film deposition include various processes that apply a layer of material to a surface. Products made of metal are almost always coated by electroplating (e.g., chrome plating), painting, or other process. Principal reasons for coating a metal are to (1) provide corrosion protection, (2) enhance product appearance (e.g., providing a specified color or texture), (3) increase wear resistance and/or reduce friction of the surface, (4) increase electrical conductivity, (5) increase electrical resistance, and (6) prepare a metallic surface for subsequent processing. Nonmetallic materials are also sometimes coated. Examples include (1) plastic parts coated to give them a metallic appearance; (2) antireflection coatings on optical glass lenses; and (3) certain coating and deposition processes used in the fabrication of semiconductor chips and printed circuit boards. In all cases, good adhesion must be achieved between coating and substrate, and for this to occur the substrate surface must be very clean.

21.1 INDUSTRIAL CLEANING PROCESSES

Most workparts must be cleaned one or more times during their manufacturing sequence. Chemical and/or mechanical processes are used to accomplish this cleaning.

Chemical cleaning methods use chemicals to remove unwanted oils and soils from the workpiece surface. Mechanical cleaning involves removal of substances from a surface by mechanical operations of various kinds. These operations often serve other functions such as removing burrs, improving smoothness, adding luster, and enhancing surface properties.

21.1.1 CHEMICAL CLEANING

A typical surface is covered with various films, oils, dirt, and other contaminants (Section 4.2.1). Although some of these substances may operate in a beneficial way (such as the oxide film on aluminum), it is usually desirable to remove contaminants from the surface. In this section, we survey the principal chemical cleaning processes used in industry.

Some of the important reasons why manufactured parts (and products) must be cleaned are (1) to prepare the surface for subsequent industrial processing, such as a coating application or adhesive bonding; (2) to improve hygiene conditions for workers and customers; (3) to remove contaminants that might chemically react with the surface; and (4) to enhance appearance and performance of the product.

There is no single cleaning method that can be used for all cleaning tasks. Just as various soaps and detergents are required for different household jobs (laundry, dishwashing, pot scrubbing, bathtub cleaning, and so forth), various cleaning methods are also needed to solve different cleaning problems in industry. To select the best cleaning method, one must first identify what must be cleaned. Surface contaminants found in the factory usually divide into one of the following categories: (1) oil and grease, which includes lubricants used in metalworking; (2) solid particles such as metal chips, abrasive grits, shop dirt, dust, and similar materials; (3) buffing and polishing compounds; and (4) oxide films, rust, and scale.

The major chemical cleaning methods are (1) alkaline cleaning, (2) emulsion cleaning, (3) solvent cleaning, (4) acid cleaning, and (5) ultrasonic cleaning. In some cases, chemical action is augmented by other energy forms; for example, ultrasonic cleaning uses high-frequency mechanical vibrations combined with chemical cleaning.

Alkaline cleaning is the most widely used industrial cleaning method. As its name indicates, it employs an alkali to remove oils, grease, wax, and various types of particles (metal chips, silica, carbon, and light scale) from a metallic surface. Alkaline cleaning solutions consist of water-soluble salts such as sodium and potassium hydroxide (NaOH, KOH), sodium carbonate (Na_2CO_3), borax ($Na_2B_4O_7$), phosphates, and silicates of sodium and potassium, combined with dispersants and surfactants in water. The cleaning method is commonly by immersion or spraying, usually at temperatures of 50°C to 95°C (120°F to 200°F). Following application of the alkaline solution, a water rinse is used to remove the alkali residue. Metal surfaces cleaned by alkaline solutions are typically electroplated or conversion coated.

Electrolytic cleaning, also called ***electrocleaning***, is a related process in which a 3- to 12-V direct current is applied to an alkaline cleaning solution. The electrolytic action results in the generation of gas bubbles at the part surface, causing a scrubbing action that aids in removal of tenacious dirt films.

Emulsion cleaning uses organic solvents (oils) dispersed in an aqueous solution. The use of suitable emulsifiers (soaps) results in a two-phase cleaning fluid (oil-in-water), which functions by dissolving or emulsifying the soils on the part surface. The process can be used on either metal or nonmetallic parts. Emulsion cleaning must be followed by alkaline cleaning to eliminate all residues of the organic solvent prior to plating.

In ***solvent cleaning***, organic soils such as oil and grease are removed from a metallic surface by means of chemicals that dissolve the soils. Common application techniques include hand-wiping, immersion, spraying, and vapor degreasing. ***Vapor degreasing*** uses hot vapors of solvents to dissolve and remove oil and grease on part surfaces. The common solvents include trichlorethylene (C_2HCl_3), methylene chloride (CH_2Cl_2), and perchlorethylene (C_2Cl_4), all of which have relatively low boiling points.[1] The vapor degreasing process consists of heating the liquid solvent to its boiling point in a container to produce hot vapors. Parts to be cleaned are immersed in the vapor, which condenses on the relatively cold part surfaces, dissolving the contaminants and dripping to the bottom of the container. Condensing coils near the top of the container prevent any vapors from escaping the container into the surrounding atmosphere. This is important because these solvents are classified as hazardous air pollutants under the 1992 Clean Air Act [10].

Acid cleaning removes oils and light oxides from metal surfaces by soaking, spraying, or manual brushing or wiping. The process is carried out at ambient or elevated temperatures. Common cleaning fluids are acid solutions combined with water-miscible solvents, wetting and emulsifying agents. Cleaning acids include hydrochloric (HCl), nitric (HNO_3), phosphoric (H_3PO_4), and sulfuric (H_2SO_4), the selection depending on the base metal and purpose of the cleaning. For example, phosphoric acid produces a light phosphate film on the metallic surface, which can be a useful preparation for painting. A closely related cleaning process is ***acid pickling***, which involves a more severe treatment to remove thicker oxides, rusts, and scales; it generally results in some etching of the metallic surface, which serves to improve organic paint adhesion.

Ultrasonic cleaning combines chemical cleaning and mechanical agitation of the fluid to provide a highly effective method for removing surface contaminants. The cleaning fluid is generally an aqueous solution containing alkaline detergents. The mechanical agitation is produced by high-frequency vibrations of sufficient amplitude to cause cavitation—formation of low-pressure vapor bubbles or cavities. As the vibration wave passes a given point in the liquid, the low-pressure region is followed by a high-pressure front that implodes the cavity, thereby producing a shock wave capable of penetrating contaminant particles adhering to the work surface. This rapid cycle of cavitation and implosion occurs throughout the liquid medium, thus making ultrasonic cleaning effective even on complex and intricate internal shapes. The cleaning process is performed at frequencies between 20 and 45 kHz, and the cleaning solution is usually at an elevated temperature, typically 65°C to 85°C (150°F to 190°F).

21.1.2 MECHANICAL CLEANING AND SURFACE TREATMENTS

Mechanical cleaning involves the physical removal of soils, scales, or films from the surface of the workpart by means of abrasives or similar mechanical action. The processes used for mechanical cleaning often serve other functions in addition to cleaning, such as deburring and improving surface finish.

Blast Finishing and Shot Peening Blast finishing uses the high-velocity impact of particulate media to clean and finish a surface. The most well known of these methods is ***sand blasting***, which uses grits of sand (SiO_2) as the blasting media. Various other media are also used in blast finishing, including hard abrasives such as aluminum oxide (Al_2O_3) and silicon carbide (SiC), and soft media such as nylon beads and crushed nut shells. The media is propelled at the target surface by pressurized air or centrifugal force. In some applications, the process is performed wet, in which fine particles in a water slurry are directed under hydraulic pressure at the surface.

[1] The highest boiling point of the three solvents is 121°C (250°F) for C_2Cl_4.

In ***shot peening***, a high-velocity stream of small cast steel pellets (called ***shot***) is directed at a metallic surface with the effect of cold working and inducing compressive stresses into the surface layers. Shot peening is used primarily to improve fatigue strength of metal parts. Its purpose is therefore different from blast finishing, although surface cleaning is accomplished as a by-product of the operation.

Tumbling and Other Mass Finishing Tumbling, vibratory finishing, and similar operations comprise a group of finishing processes known as mass finishing methods. ***Mass finishing*** involves the finishing of parts in bulk by a mixing action inside a container, usually in the presence of an abrasive media. The mixing causes the parts to rub against the media and each other to achieve the desired finishing action. Mass finishing methods are used for deburring, descaling, deflashing, polishing, radiusing, burnishing, and cleaning. The parts include stampings, castings, forgings, extrusions, and machined parts. Even plastic and ceramic parts are sometimes subjected to these mass finishing operations to achieve desired finishing results. The parts processed by these methods are usually small and are therefore uneconomical to finish individually.

Mass finishing methods include tumbling, vibratory finishing, and several techniques that utilize centrifugal force. ***Tumbling*** (also called ***barrel finishing*** and ***tumbling barrel finishing***) involves the use of a horizontally oriented barrel of hexagonal or octagonal cross-section in which parts are mixed by rotating the barrel at speeds of 10 to 50 rev/min. Finishing is performed by a "landslide" action of the media and parts as the barrel revolves. As pictured in Figure 21.1, the contents rise in the barrel due to rotation, followed by a tumbling down of the top layer due to gravity. This cycle of rising and tumbling occurs continuously and, over time, subjects all of the parts to the same desired finishing action. However, because only the top layer of parts is being finished at any moment, barrel finishing is a relatively slow process compared to other mass finishing methods. It often takes several hours of tumbling to complete the processing. Other drawbacks of barrel finishing include high noise levels and large floor space requirements.

Vibratory finishing was introduced in the late 1950s as an alternative to tumbling. The vibrating vessel subjects all parts to agitation with the abrasive media, as opposed to only the top layer as in barrel finishing. Consequently, processing times for vibratory finishing are significantly reduced. The open tubs used in this method permit inspection of the parts during processing, and noise is reduced.

Most of the ***media*** in these operations are abrasive; however, some media perform nonabrasive finishing operations such as burnishing and surface hardening. The media may be natural or synthetic materials. Natural media include corundum, granite, limestone, and even hardwood. The problem with these materials is that they are generally softer (and therefore wear more rapidly) and nonuniform in size (and sometimes clog in the

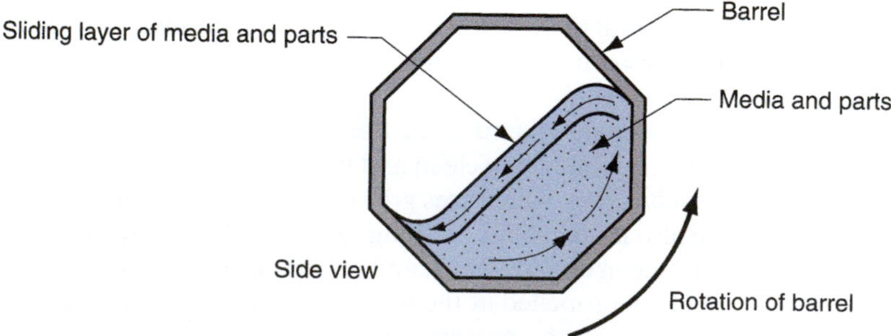

FIGURE 21.1 Diagram of tumbling (barrel finishing) operation showing "landslide" action of parts and abrasive media to finish the parts. (Credit: *Fundamentals of Modern Manufacturing*, 4th Edition by Mikell P. Groover, 2010. Reprinted with permission of John Wiley & Sons, Inc.)

workparts). Synthetic media can be made with greater consistency, both in size and hardness. These materials include Al_2O_3 and SiC, compacted into a desired shape and size using a bonding material such as a polyester resin. The shapes for these media include spheres, cones, angle-cut cylinders, and other regular geometric forms. Steel media are also used in mass finishing for burnishing, surface hardening, and light deburring operations. Selection of media is based on part size and shape, as well as finishing requirements.

In most mass finishing processes, a compound is used with the media. The mass finishing ***compound*** is a combination of chemicals for specific functions such as cleaning, cooling, rust inhibiting (of steel parts and steel media), and enhancing brightness and color of the parts (especially in burnishing).

21.2 DIFFUSION AND ION IMPLANTATION

In this section we discuss two processes in which the surface of a substrate is impregnated with foreign atoms that alter its chemistry and properties.

21.2.1 DIFFUSION

Diffusion involves the alteration of surface layers of a material by diffusing atoms of a different material (usually an element) into the surface. The diffusion process impregnates the surface layers of the substrate with the foreign element, but the surface still contains a high proportion of substrate material. The characteristic of a diffusion impregnated surface is that the diffused element has a maximum percentage at the surface and rapidly declines with distance below the surface. The diffusion process has important applications in metallurgy and semiconductor manufacture.

In metallurgical applications, diffusion is used to alter the surface chemistry of metals in a number of processes and treatments. One important example is surface hardening, typified by ***carburizing***, ***nitriding***, and ***carbonitriding*** (Section 20.4) In these treatments, one or more elements (C and/or Ni) are diffused into the surface of iron or steel. In other diffusion processes, corrosion resistance and/or high-temperature oxidation resistance are the main objectives. Aluminizing and siliconizing are important examples. ***Aluminizing***, also known as ***calorizing***, involves diffusion of aluminum into carbon steel, alloy steels, and alloys of nickel and cobalt. ***Siliconizing*** is a treatment in which silicon is diffused into a steel part surface to create a layer with good corrosion and wear resistance and moderate heat resistance.

In semiconductor processing, diffusion of an impurity element into the surface of a silicon chip is used to change the electrical properties at the surface to create devices such as transistors and diodes.

21.2.2 ION IMPLANTATION

Ion implantation is an alternative to diffusion when the latter method is not feasible because of the high temperatures required. The ion implantation process involves embedding atoms of one (or more) foreign element(s) into a substrate surface using a high-energy beam of ionized particles. The result is an alteration of the chemical and physical properties of the layers near the substrate surface. Penetration of atoms produces a much thinner altered layer than diffusion.

Advantages of ion implantation include (1) low-temperature processing, (2) good control and reproducibility of penetration depth of impurities, and (3) solubility limits can be exceeded without precipitation of excess atoms. Ion implantation finds some of its applications as a substitute for certain coating processes, where its advantages include (4) no problems with waste disposal as in electroplating and many coating processes, and (5) no discontinuity between coating and substrate. Principal applications of ion implantation are in modifying metal surfaces to improve properties and fabrication of semiconductor devices.

21.3 PLATING AND RELATED PROCESSES

Plating involves the coating of a thin metallic layer onto the surface of a substrate material. The substrate is usually metallic, although methods are available to plate plastic and ceramic parts. The most familiar and widely used plating technology is electroplating.

21.3.1 ELECTROPLATING

Electroplating, also known as ***electrochemical plating***, is an electrolytic process in which metal ions in an electrolyte solution are deposited onto a cathode workpart. The setup is shown in Figure 21.2. The anode is generally made of the metal being plated and thus serves as the source of the plate metal. Direct current from an external power supply is passed between the anode and the cathode. The electrolyte is an aqueous solution of acids, bases, or salts; it conducts electric current by the movement of plate metal ions in solution. For optimum results, parts must be chemically cleaned just prior to electroplating.

Principles of Electroplating Electrochemical plating is based on two physical laws first stated by British scientist Michael Faraday. Briefly for our purposes, the laws are (1) the mass of a substance liberated in electrolysis is proportional to the quantity of electricity passed through the cell; and (2) the mass of the material liberated is proportional to its electrochemical equivalent (ratio of atomic weight to valence). The effects can be summarized in the equation:

$$V = CIt \tag{21.1}$$

where V = volume of metal plated, mm^3 (in^3); C = plating constant, which depends on electrochemical equivalent and density, mm^3/amp-s (in^3/amp-min); I = current, amps; and t = time during which current is applied, s (min). The product It (current × time) is the electrical charge passed in the cell, and the value of C indicates the amount of plating material deposited onto the cathodic workpart per electrical charge.

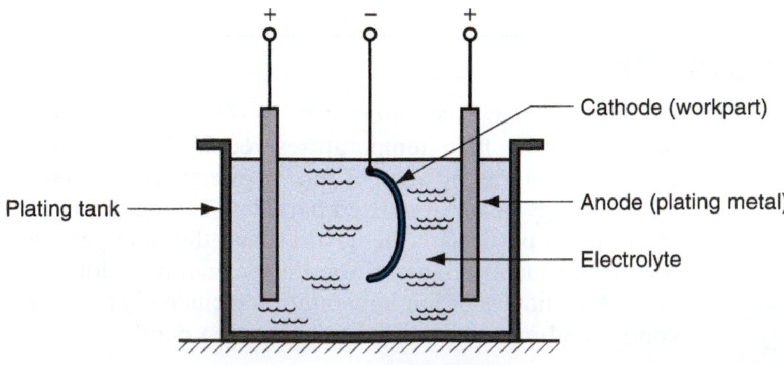

FIGURE 21.2 Setup for electroplating. (Credit: *Fundamentals of Modern Manufacturing*, 4th Edition by Mikell P. Groover, 2010. Reprinted with permission of John Wiley & Sons, Inc.)

TABLE 21.1 Typical cathode efficiencies in electroplating and values of plating constant *C*.

Plate Metal[a]	Electrolyte	Cathode Efficiency %	Plating Constant C^a mm³/amp-s	Plating Constant C^a in³/amp-min
Cadmium (2)	Cyanide	90	6.73×10^{-2}	2.47×10^{-4}
Chromium (3)	Chromium-acid-sulfate	15	2.50×10^{-2}	0.92×10^{-4}
Copper (1)	Cyanide	98	7.35×10^{-2}	2.69×10^{-4}
Gold (1)	Cyanide	80	10.6×10^{-2}	3.87×10^{-4}
Nickel (2)	Acid sulfate	95	3.42×10^{-2}	1.25×10^{-4}
Silver (1)	Cyanide	100	10.7×10^{-2}	3.90×10^{-4}
Tin (4)	Acid sulfate	90	4.21×10^{-2}	1.54×10^{-4}
Zinc (2)	Chloride	95	4.75×10^{-2}	1.74×10^{-4}

Compiled from [17].

[a]Most common valence given in parenthesis ( ); this is the value assumed in determining the plating constant *C*. For a different valence, compute the new *C* by multiplying *C* value in the table by the most common valence and then dividing by the new valence.

For most plating metals, not all of the electrical energy in the process is used for deposition; some energy may be consumed in other reactions, such as the liberation of hydrogen at the cathode. This reduces the amount of metal plated. The actual amount of metal deposited on the cathode (workpart) divided by the theoretical amount given by Eq. (21.1) is called the ***cathode efficiency***. Taking the cathode efficiency into account, a more realistic equation for determining the volume of metal plated is:

$$V = ECIt \tag{21.2}$$

where E = cathode efficiency, and the other terms are defined as before. Typical values of cathode efficiency E and plating constant C for different metals are presented in Table 21.1. The average plating thickness can be determined from the following:

$$d = \frac{V}{A} \tag{21.3}$$

where d = plating depth or thickness, mm (in); V = volume of plate metal from Eq. (21.2); and A = surface area of plated part, mm² (in²).

Example 21.1 Electroplating

A steel part with surface area $A = 125$ cm² is to be nickel plated. What average plating thickness will result if 12 amps are applied for 15 min in an acid sulfate electrolyte bath?

Solution: From Table 21.1, the cathode efficiency for nickel is $E = 0.95$ and the plating constant $C = 3.42(10^{-2})$ mm³/amp-s. Using Eq. (21.2), the total amount of plating metal deposited onto the part surface in 15 min is given by:

$$V = 0.95(3.42 \times 10^{-2})(12)(15)(60) = 350.9 \text{ mm}^3$$

This is spread across an area $A = 125 \text{ cm}^2 = 12{,}500 \text{ mm}^2$, so the average plate thickness is:

$$d = \frac{350.9}{12500} = \mathbf{0.028\ mm}$$

Methods and Applications A variety of equipment is available for electroplating, the choice depending on part size and geometry, throughput requirements, and plating metal. The principal methods are (1) barrel plating, (2) rack plating, and (3) strip plating. ***Barrel plating*** is performed in rotating barrels that are oriented either horizontally or at an oblique angle (35°). The method is suited to the plating of many small parts in a batch. Electrical contact is maintained through the tumbling action of the parts themselves and by means of an externally connected conductor that projects into the barrel. There are limitations to barrel plating; the tumbling action inherent in the process may damage soft metal parts, threaded components, parts requiring good finishes, and heavy parts with sharp edges.

Rack plating is used for parts that are too large, heavy, or complex for barrel plating. The racks are made of heavy-gauge copper wire, formed into suitable shapes for holding the parts and conducting current to them. The racks are fabricated so that workparts can be hung on hooks, or held by clips, or loaded into baskets. To avoid plating of the copper itself, the racks are covered with insulation except in locations where part contact occurs. The racks containing the parts are moved through a sequence of tanks that perform the electroplating operation. ***Strip plating*** is a high-production method in which the work consists of a continuous strip that is pulled through the plating solution by means of a take-up reel. Plated wire is an example of a suitable application. Small sheet-metal stampings held in a long strip can also be plated by this method. The process can be set up so that only specific regions of the parts are plated, for example, contact points plated with gold on electrical connectors.

Common coating metals in electroplating include zinc, nickel, tin, and chromium. Steel is the most common substrate metal. Precious metals (gold, silver, platinum) are plated on jewelry. Gold is also used for electrical contacts.

Zinc-plated steel products include fasteners, wire goods, electric switch boxes, and various sheet-metal parts. The zinc coating serves as a sacrificial barrier to the corrosion of the steel beneath. An alternative process for coating zinc onto steel is galvanizing (Section 21.3.4). ***Nickel plating*** is used for corrosion resistance and decorative purposes over steel, brass, zinc die castings, and other metals. Applications include automotive trim and other consumer goods. Nickel is also used as a base coat under a much thinner chrome plate. ***Tin plate*** is still widely used for corrosion protection in "tin cans" and other food containers. Tin plate is also used to improve solderability of electrical components. ***Chromium plate*** (popularly known as ***chrome plate***) is valued for its decorative appearance and is widely used in automotive products, office furniture, and kitchen appliances. It also produces one of the hardest of all electroplated coatings, and so it is widely used for parts requiring wear resistance (e.g., hydraulic pistons and cylinders, piston rings, aircraft engine components, and thread guides in textile machinery).

21.3.2 ELECTROFORMING

This process is virtually the same as electroplating but its purpose is quite different. Electroforming involves electrolytic deposition of metal onto a pattern until the required thickness is achieved; the pattern is then removed to leave the formed part. Whereas typical plating thickness is only about 0.05 mm (0.002 in) or less, electroformed parts are often substantially thicker, so the production cycle is proportionally longer.

Patterns used in electroforming are either solid or expendable. Solid patterns have a taper or other geometry that permits removal of the electroplated part. Expendable patterns are destroyed during part removal; they are used when part shape precludes a solid pattern. Expendable patterns are either fusible or soluble. The fusible type is made of low-melting alloys, plastic, wax, or other material that can be removed by melting.

When nonconductive materials are used, the pattern must be metallized to accept the electrodeposited coating. Soluble patterns are made of a material that can be readily dissolved by chemicals; for example, aluminum can be dissolved in sodium hydroxide (NaOH).

Electroformed parts are commonly fabricated of copper, nickel, and nickel cobalt alloys. Applications include fine molds for lenses, compact discs (CDs), and videodiscs (DVDs); copper foil used to produce blank printed circuit boards; and plates for embossing and printing. Molds for compact discs and videodiscs represent a demanding application because the surface details that must be imprinted on the disc are measured in μm (1 μm = 10^{-6} m). These details are readily obtained in the mold by electroforming.

21.3.3 ELECTROLESS PLATING

Electroless plating is a plating process driven entirely by chemical reactions—no external source of electric current is required. Deposition of metal onto a part surface occurs in an aqueous solution containing ions of the desired plating metal. The process uses a reducing agent, and the workpart surface acts as a catalyst for the reaction.

The metals that can be electroless plated are limited; and for those that can be processed by this technique, the cost is generally greater than electrochemical plating. The most common electroless plating metal is nickel and certain of its alloys (Ni–Co, Ni–P, and Ni–B). Copper and, to a lesser degree, gold are also used as plating metals. Nickel plating by this process is used for applications requiring high resistance to corrosion and wear. Electroless copper plating is used to plate through holes of printed circuit boards. Cu can also be plated onto plastic parts for decorative purposes. Advantages sometimes cited for electroless plating include: (1) uniform plate thickness on complex part geometries (a problem with electroplating); (2) the process can be used on both metallic and nonmetallic substrates; and (3) no need for a DC power supply to drive the process.

21.3.4 HOT DIPPING

Hot dipping is a process in which a metal substrate is immersed in a molten bath of a second metal; upon removal, the second metal is coated onto the first. Of course, the first metal must possess a higher melting temperature than the second. The most common substrate metals are steel and iron. Zinc, aluminum, tin, and lead are the common coating metals. Hot dipping works by forming transition layers of varying alloy compositions. Next to the substrate are normally intermetallic compounds of the two metals; at the exterior are solid solution alloys consisting predominantly of the coating metal. The transition layers provide excellent adhesion of the coating.

The primary purpose of hot dipping is corrosion protection. Two mechanisms normally operate to provide this protection: (1) barrier protection—the coating simply serves as a shield for the metal beneath; and (2) sacrificial protection—the coating corrodes by a slow electrochemical process to preserve the substrate.

Hot dipping goes by different names, depending on the coating metal: ***galvanizing*** is when zinc (Zn) is coated onto steel or iron; ***aluminizing*** refers to coating of aluminum (Al) onto a substrate; and ***tinning*** is coating of tin (Sn). Galvanizing is by far the most important hot dipping process, dating back about 200 years. It is applied to finished steel and iron parts in a batch process; and to sheet, strip, piping, tubing, and wire in an automated continuous process. Coating thickness is typically 0.04 to 0.09 mm (0.0016 to

0.0035 in). Thickness is controlled largely by immersion time. Bath temperature is maintained at around 450°C (850°F).

Commercial use of aluminizing is on the rise, gradually increasing in market share relative to galvanizing. Hot-dipped aluminum coatings provide excellent corrosion protection, in some cases five times more effective than galvanizing [17]. Tin plating by hot dipping provides a nontoxic corrosion protection for steel in applications for food containers, dairy equipment, and soldering applications. Hot dipping has gradually been overtaken by electroplating as the preferred commercial method for plating of tin onto steel.

21.4 CONVERSION COATING

Conversion coating refers to a family of processes in which a thin film of oxide, phosphate, or chromate is formed on a metallic surface by chemical or electrochemical reaction. Immersion and spraying are the two common methods of exposing the metal surface to the reacting chemicals. The common metals treated by conversion coating are steel (including galvanized steel), zinc, and aluminum. However, nearly any metal product can benefit from the treatment. Important reasons for using a conversion coating process are to (1) provide corrosion protection, (2) prepare the surface for painting, (3) increase wear resistance, (4) permit the surface to better hold lubricants for metal forming processes, (5) increase electrical resistance of surface, and (6) provide a decorative finish [17].

Conversion coating processes divide into two categories: (1) chemical treatments, which involve a chemical reaction only, and (2) anodizing, which consists of an electrochemical reaction to produce an oxide coating (anodize is a contraction of ***anodic oxidize***).

21.4.1 CHEMICAL CONVERSION COATINGS

These processes expose the base metal to certain chemicals that form thin, nonmetallic surface films. Similar reactions occur in nature; the oxidation of iron and aluminum are examples. Whereas rusting is progressively destructive of iron, formation of a thin Al_2O_3 coating on aluminum protects the base metal. It is the purpose of these chemical conversion treatments to accomplish the latter effect. The two main processes are phosphate and chromate coating.

Phosphate coating transforms the base metal surface into a protective phosphate film by exposure to solutions of certain phosphate salts (e.g., Zn, Mg, and Ca) together with dilute phosphoric acid (H_3PO_4). The coatings range in thickness from 0.0025 mm to 0.05 mm (0.0001 in to 0.002 in). The most common base metals are zinc and steel, including galvanized steel. The phosphate coating serves as a useful preparation for painting in the automotive and large appliance industries.

Chromate coating converts the base metal into various forms of chromate films using aqueous solutions of chromic acid, chromate salts, and other chemicals. Metals treated by this method include aluminum, cadmium, copper, magnesium, and zinc (and their alloys). Immersion of the base part is the common method of application. Chromate conversion coatings are somewhat thinner than phosphate, typically less than 0.0025 mm (0.0001 in). Usual reasons for chromate coating are (1) corrosion protection, (2) base for painting, and (3) decorative purposes. Chromate coatings can be clear or colorful; available colors include olive drab, bronze, yellow, or bright blue.

21.4.2 ANODIZING

Although the previous processes are normally performed without electrolysis, anodizing is an electrolytic treatment that produces a stable oxide layer on a metallic surface. Its most common applications are with aluminum and magnesium, but it is also applied to zinc, titanium, and other less common metals. Anodized coatings are used primarily for decorative purposes; they also provide corrosion protection.

It is instructive to compare anodizing to electroplating, since they are both electrolytic processes. Two differences stand out. (1) In electrochemical plating, the workpart to be coated is the cathode in the reaction. By contrast, in anodizing, the work is the anode, whereas the processing tank is cathodic. (2) In electroplating, the coating is grown by adhesion of ions of a second metal to the base metal surface. In anodizing, the surface coating is formed through chemical reaction of the substrate metal into an oxide layer.

Anodized coatings usually range in thickness between 0.0025 mm and 0.075 mm (0.0001 in and 0.003 in). Dyes can be incorporated into the anodizing process to create a wide variety of colors; this is especially common in aluminum anodizing. Very thick coatings up to 0.25 mm (0.010 in) can also be formed on aluminum by a special process called ***hard anodizing***; these coatings are noted for high resistance to wear and corrosion.

21.5 VAPOR DEPOSITION PROCESSES

The vapor deposition processes form a thin coating on a substrate by either condensation or chemical reaction of a gas onto the surface of the substrate. The two categories of processes that fall under this heading are physical vapor deposition and chemical vapor deposition.

21.5.1 PHYSICAL VAPOR DEPOSITION

Physical vapor deposition (PVD) is a group of thin film processes in which a material is converted into its vapor phase in a vacuum chamber and condensed onto a substrate surface as a very thin layer. PVD can be used to apply a wide variety of coating materials: metals, alloys, ceramics and other inorganic compounds, and even certain polymers. Possible substrates include metals, glass, and plastics. Thus, PVD represents a versatile coating technology, applicable to an almost unlimited combination of coating substances and substrate materials.

Applications of PVD include thin decorative coatings on plastic and metal parts such as trophies, toys, pens and pencils, watchcases, and interior trim in automobiles. The coatings are thin films of aluminum (around 150 nm) coated with clear lacquer to give a high gloss silver or chrome appearance. Another use of PVD is to apply antireflection coatings of magnesium fluoride (MgF_2) onto optical lenses. PVD is applied in the fabrication of electronic devices, principally for depositing metal to form electrical connections in integrated circuits. Finally, PVD is widely used to coat titanium nitride (TiN) onto cutting tools and plastic injection molds for wear resistance.

All physical vapor deposition processes consist of the following steps: (1) synthesis of the coating vapor, (2) vapor transport to the substrate, and (3) condensation of vapors onto the substrate surface. These steps are generally carried out inside a vacuum chamber, so evacuation of the chamber must precede the actual PVD process.

Synthesis of the coating vapor can be accomplished by any of several methods, such as electric resistance heating or ion bombardment to vaporize an existing solid (or liquid).

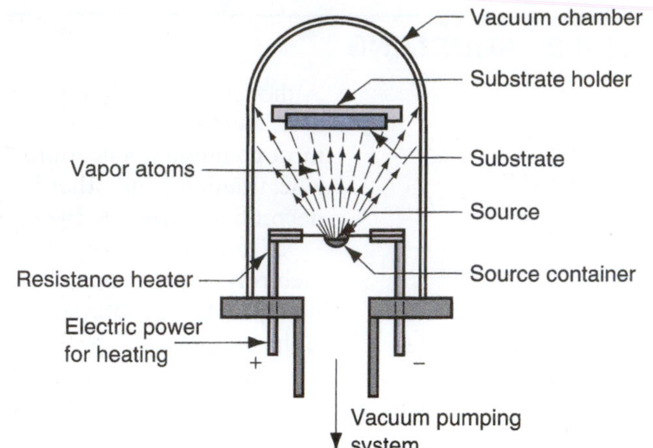

FIGURE 21.3 Setup for vacuum evaporation PVD. (Credit: *Fundamentals of Modern Manufacturing*, 4th Edition by Mikell P. Groover, 2010. Reprinted with permission of John Wiley & Sons, Inc.)

These and other variations result in several PVD processes. They are grouped into three principal types: (1) vacuum evaporation, (2) sputtering, and (3) ion plating.

Vacuum Evaporation Certain materials (mostly pure metals) can be deposited onto a substrate by first transforming them from solid to vapor state in a vacuum and then letting them condense on the substrate surface. The setup for the vacuum evaporation process is shown in Figure 21.3. The material to be deposited, called the source, is heated to a sufficiently high temperature that it evaporates. Since heating is accomplished in a vacuum, the temperature required for vaporization is significantly below the corresponding temperature required at atmospheric pressure. Also, the absence of air in the chamber prevents oxidation of the source material at the heating temperatures.

Various methods can be used to heat and vaporize the material. A container must be provided to hold the source material before vaporization. Among the important vaporization methods are resistance heating and electron beam bombardment. ***Resistance heating*** is the simplest technology. A refractory metal (e.g., W, Mo) is formed into a suitable container to hold the source material. Current is applied to heat the container, which then heats the material in contact with it. In ***electron beam evaporation***, a stream of electrons at high velocity is directed to bombard the surface of the source material to cause vaporization.

Whatever the vaporization technique, evaporated atoms leave the source and follow straight-line paths until they collide with other gas molecules or strike a solid surface. The vacuum inside the chamber virtually eliminates other gas molecules, thus reducing the probability of collisions with source vapor atoms. The substrate surface to be coated is usually positioned relative to the source so that it is the likely solid surface on which the vapor atoms will be deposited. A mechanical manipulator is sometimes used to rotate the substrate so that all surfaces are coated. Upon contact with the relative cool substrate surface, the energy level of the impinging atoms is suddenly reduced to the point where they cannot remain in a vapor state; they condense and become attached to the solid surface, forming a deposited thin film.

Sputtering If the surface of a solid (or liquid) is bombarded by atomic particles of sufficiently high energy, individual atoms of the surface may acquire enough energy due to the collision that they are ejected from the surface by transfer of momentum. This is the process known as sputtering. The most convenient form of high-energy particle is an ionized gas, such as argon, energized by means of an electric field to form a plasma. As a PVD process, ***sputtering*** involves bombardment of the cathodic coating material with argon ions (Ar^+), causing surface atoms to escape and then be deposited onto a substrate, forming a thin film on the substrate surface. The substrate must be placed close to the

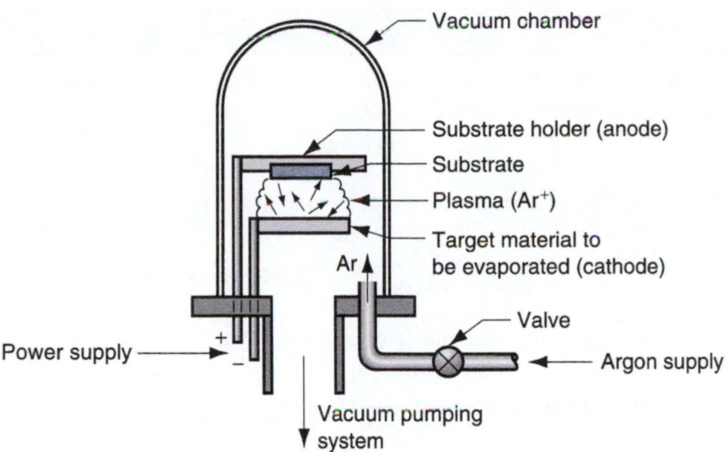

FIGURE 21.4 One possible setup for sputtering, a form of physical vapor deposition. (Credit: *Fundamentals of Modern Manufacturing*, 4th Edition by Mikell P. Groover, 2010. Reprinted with permission of John Wiley & Sons, Inc.)

cathode and is usually heated to improve bonding of the coating atoms. A typical arrangement is shown in Figure 21.4.

Whereas vacuum evaporation is generally limited to metals, sputtering can be applied to nearly any material—metallic and nonmetallic elements; alloys, ceramics, and polymers. Films of alloys and compounds can be sputtered without changing their chemical compositions. Films of chemical compounds can also be deposited by employing reactive gases that form oxides, carbides, or nitrides with the sputtered metal.

Drawbacks of sputtering PVD include (1) slow deposition rates and (2) since the ions bombarding the surface are a gas, traces of the gas can usually be found in the coated films, and the entrapped gases sometimes affect mechanical properties adversely.

Ion Plating Ion plating uses a combination of sputtering and vacuum evaporation to deposit a thin film onto a substrate. The process works as follows. The substrate is set up to be the cathode in the upper part of the chamber, and the source material is placed below it. A vacuum is then established in the chamber. Argon gas is admitted and an electric field is applied to ionize the gas (Ar^+) and establish a plasma. This results in ion bombardment (sputtering) of the substrate so that its surface is scrubbed to a condition of atomic cleanliness (interpret this as "very clean"). Next, the source material is heated sufficiently to generate coating vapors. The heating methods used here are similar to those used in vacuum evaporation: resistance heating and electron beam bombardment. The vapor molecules pass through the plasma and coat the substrate. Sputtering is continued during deposition, so that the ion bombardment consists not only of the original argon ions but also source material ions that have been energized while being subjected to the same energy field as the argon. The effect of these processing conditions is to produce films of uniform thickness and excellent adherence to the substrate.

Ion plating is applicable to parts having irregular geometries, due to the scattering effects that exist in the plasma field. An example of interest here is TiN coating of high-speed steel cutting tools (e.g., drill bits). In addition to coating uniformity and good adherence, other advantages include high deposition rates, high film densities, and the capability to coat the inside walls of holes and other hollow shapes.

21.5.2 CHEMICAL VAPOR DEPOSITION

PVD involves deposition of a coating by condensation onto a substrate from the vapor phase; it is strictly a physical process. By comparison, ***chemical vapor deposition*** (CVD) involves the interaction between a mixture of gases and the surface of a heated substrate, causing chemical decomposition of some of the gas constituents and formation of a solid

TABLE 21.2 Some examples of reactions in chemical vapor deposition.

1. The ***Mond process*** includes a CVD process for decomposition of nickel from nickel carbonyl ($Ni(CO)_4$), which is an intermediate compound formed in reducing nickel ore:

$$Ni(CO)_4 \xrightarrow{200^\circ C\ (400^\circ F)} Ni + 4CO \tag{21.4}$$

2. Coating of titanium carbide (TiC) onto a substrate of cemented tungsten carbide (WC–Co) to produce a high-performance cutting tool:

$$TiCl_4 + CH_4 \xrightarrow[\text{excess } H_2]{1000^\circ C\ (1800^\circ F)} TiC + 4HCl \tag{21.5}$$

3. Coating of titanium nitride (TiN) onto a substrate of cemented tungsten carbide (WC–Co) to produce a high-performance cutting tool:

$$TiCl_4 + 0.5N_2 + 2H_2 \xrightarrow{900^\circ C\ (1650^\circ F)} TiN + 4HCl \tag{21.6}$$

4. Coating of aluminum oxide (Al_2O_3) onto a substrate of cemented tungsten carbide (WC–Co) to produce a high-performance cutting tool:

$$2AlCl_3 + 3CO_2 + 3H_2 \xrightarrow{500^\circ C\ (900^\circ F)} Al_2O_3 + 3CO + 6HCl \tag{21.7}$$

5. Coating of silicon nitride (Si_3N_4) onto silicon (Si), a process in semiconductor manufacturing:

$$3SiF_4 + 4NH_3 \xrightarrow{1000^\circ C\ (1800^\circ F)} Si_3N_4 + 12\ HF \tag{21.8}$$

6. Coating of silicon dioxide (SiO_2) onto silicon (Si), a process in semiconductor manufacturing:

$$2SiCl_3 + 3H_2O + 0.5O_2 \xrightarrow{900^\circ C\ (1600^\circ F)} 2SiO_2 + 6\ HCl \tag{21.9}$$

7. Coating of the refractory metal tungsten (W) onto a substrate, such as a jet engine turbine blade:

$$WF_6 + 3H_2 \xrightarrow{600^\circ C\ (1100^\circ F)} W + 6HF \tag{21.10}$$

Compiled from [6], [13], and [17].

film on the substrate. The reactions take place in an enclosed reaction chamber. The reaction product (either a metal or a compound) nucleates and grows on the substrate surface to form the coating. Most CVD reactions require heat. However, depending on the chemicals involved, the reactions can be driven by other possible energy sources, such as ultraviolet light or plasma. CVD includes a wide range of pressures and temperatures; and it can be applied to a great variety of coating and substrate materials.

Industrial metallurgical processes based on chemical vapor deposition date back to the 1800s (e.g., the Mond process in Table 21.2). Modern interest in CVD is focused on its coating applications such as coated cemented carbide tools, solar cells, depositing refractory metals on jet engine turbine blades, and other applications where resistance to wear, corrosion, erosion, and thermal shock are important. In addition, CVD is an important technology in integrated circuit fabrication.

Advantages typically cited for CVD include: (1) capability to deposit refractory materials at temperatures below their melting or sintering temperatures; (2) control of grain size is possible; (3) the process is carried out at atmospheric pressure—it does not require vacuum equipment; and (4) good bonding of coating to substrate surface [1]. Disadvantages include: (1) corrosive and/or toxic nature of chemicals generally necessitates a closed chamber as well as special pumping and disposal equipment; (2) certain reaction ingredients are relatively expensive; and (3) material utilization is low.

CVD Materials and Reactions In general, metals that are readily electroplated are not good candidates for CVD, owing to the hazardous chemicals that must be used and the

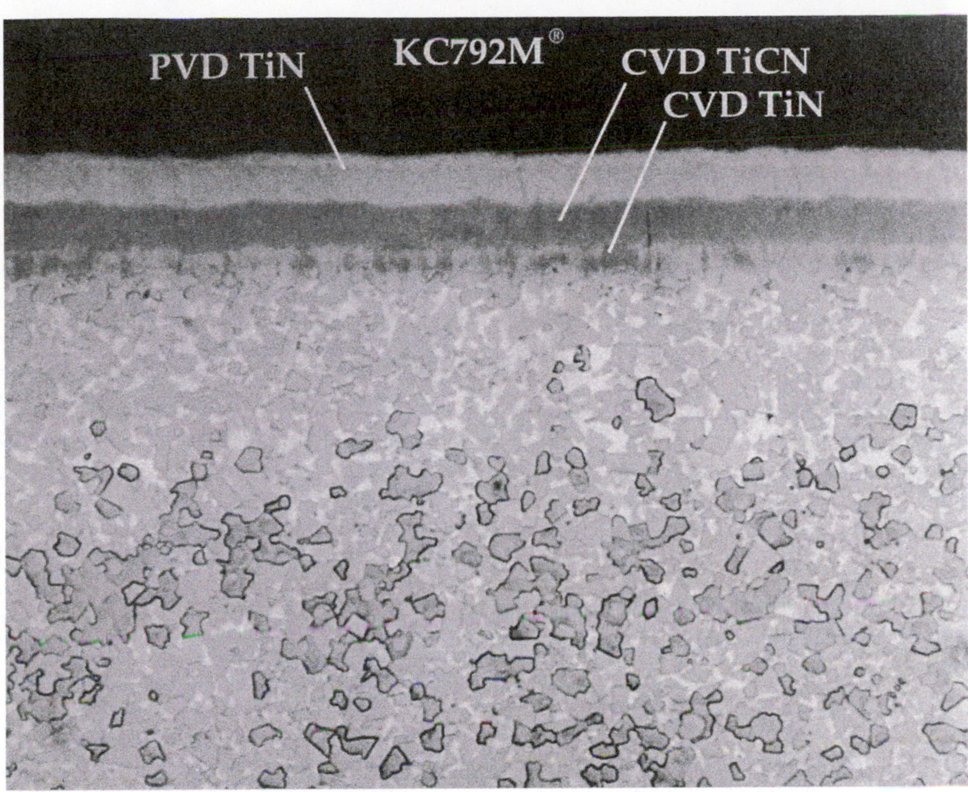

FIGURE 21.5 Photomicrograph of the cross section of a coated carbide cutting tool (Kennametal Grade KC792M); CVD was used to coat TiN and TiCN onto the surface of a WC–Co substrate, followed by a TiN coating applied by PVD. Photo courtesy of Kennametal Inc. (Credit: *Fundamentals of Modern Manufacturing,* 4th Edition by Mikell P. Groover, 2010. Reprinted with permission of John Wiley & Sons, Inc.)

costs of safeguarding against them. Metals suitable for coating by CVD include tungsten, molybdenum, titanium, vanadium, and tantalum. Chemical vapor deposition is especially suited to the deposition of compounds, such as aluminum oxide (Al_2O_3), silicon dioxide (SiO_2), silicon nitride (Si_3N_4), titanium carbide (TiC), and titanium nitride (TiN). Figure 21.5 illustrates the application of both CVD and PVD to provide multiple wear-resistant coatings on a cemented carbide cutting tool.

The commonly used reacting gases or vapors are metallic hydrides (MH_x), chlorides (MCl_x), fluorides (MF_x), and carbonyls ($M(CO)_x$), where M = the metal to be deposited and x is used to balance the valences in the compound. Other gases, such as hydrogen (H_2), nitrogen (N_2), methane (CH_4), carbon dioxide (CO_2), and ammonia (NH_3) are used in some of the reactions. Table 21.2 presents some examples of CVD reactions that result in deposition of a metal or ceramic coating onto a suitable substrate. Typical temperatures at which these reactions are carried out are also given.

Processing Equipment Chemical vapor deposition processes are carried out in a reactor, which consists of (1) reactant supply system, (2) deposition chamber, and (3) recycle/disposal system. Although reactor configurations differ depending on the application, one possible CVD reactor is illustrated in Figure 21.6. The purpose of the reactant supply system is to deliver reactants to the deposition chamber in the proper proportions. Different types of supply system are required, depending on whether the reactants are delivered as gas, liquid, or solid (e.g., pellets, powders).

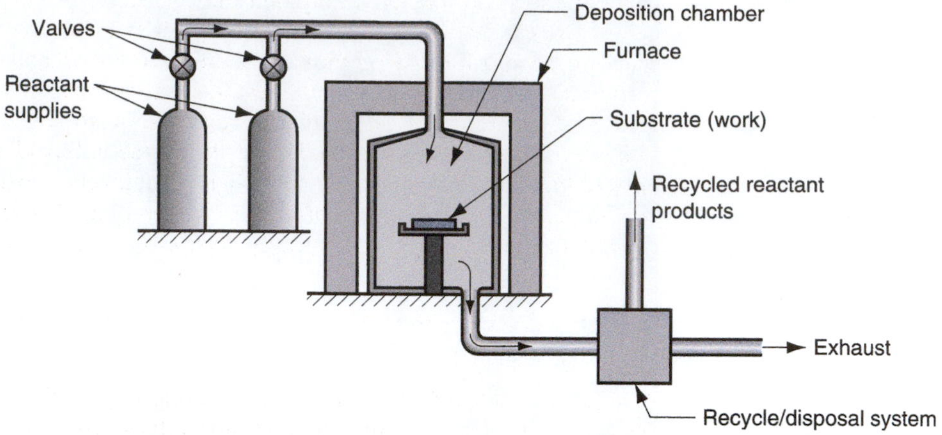

FIGURE 21.6 A typical reactor used in chemical vapor deposition. (Credit: *Fundamentals of Modern Manufacturing*, 4th Edition by Mikell P. Groover, 2010. Reprinted with permission of John Wiley & Sons, Inc.)

The deposition chamber contains the substrates and chemical reactions that lead to deposition of reaction products onto the substrate surfaces. Deposition occurs at elevated temperatures, and the substrate must be heated by induction heating, radiant heat, or other means. Deposition temperatures for different CVD reactions range from 250°C to 1950°C (500°F to 3500°F), so the chamber must be designed to meet these temperature demands.

The third component of the reactor is the recycle/disposal system, whose function is to render harmless the by-products of the CVD reaction. This includes collection of materials that are toxic, corrosive, and/or flammable, followed by proper processing and disposition.

21.6 ORGANIC COATINGS

Organic coatings are polymers and resins, produced either naturally or synthetically, usually formulated to be applied as liquids that dry or harden as thin surface films on substrate materials. These coatings are valued for the variety of colors and textures possible, their capacity to protect the substrate surface, low cost, and ease with which they can be applied. In this section, we consider the compositions of organic coatings and the methods to apply them. Although most organic coatings are applied in liquid form, some are applied as powders; we consider this alternative in Section 21.6.2.

Organic coatings are formulated to contain the following: (1) binders, which give the coating its properties; (2) dyes or pigments, which lend color to the coating; (3) solvents, to dissolve the polymers and resins and add proper fluidity to the liquid; and (4) additives.

Binders in organic coatings are polymers and resins that determine the solid-state properties of the coating, such as strength, physical properties, and adhesion to the substrate surface. The binder holds the pigments and other ingredients in the coating during and after application to the surface. The most common binders in organic coatings are natural oils (used to produce oil-based paints), and resins of polyesters, polyurethanes, epoxies, acrylics, and cellulosics.

Dyes and pigments provide color to the coating. ***Dyes*** are soluble chemicals that color the coating liquid but do not conceal the surface beneath. Thus, dye-colored coatings are generally transparent or translucent. ***Pigments*** are solid particles of uniform, microscopic size that are dispersed in the coating liquid but insoluble in it. They not only color the coating, but they also hide the surface below. Since pigments are particulate matter, they also tend to strengthen the coating.

Solvents are used to dissolve the binder and certain other ingredients in the liquid coating composition. Common solvents used in organic coatings are aliphatic and aromatic hydrocarbons, alcohols, esters, ketones, and chlorinated solvents. Different solvents are required for different binders. ***Additives*** in organic coatings include surfactants (to facilitate spreading on the surface), biocides and fungicides, thickeners, freeze/thaw stabilizers, heat and light stabilizers, coalescing agents, plasticizers, defoamers, and catalysts to promote cross-linking. These ingredients are formulated to obtain a wide variety of coatings, such as paints, lacquers, and varnishes.

21.6.1 APPLICATION METHODS

The method of applying an organic coating to a surface depends on factors such as composition of the coating liquid, required thickness of the coating, production rate and cost considerations, part size, and environmental requirements. For any of the application methods, it is of utmost importance that the surface be properly prepared. This includes cleaning and possible treatment of the surface such as phosphate coating. In some cases, metallic surfaces are plated prior to organic coating for maximum corrosion protection.

With any coating method, transfer efficiency is a critical measure. ***Transfer efficiency*** is the proportion of paint supplied to the process that is actually deposited onto the work surface. Some methods yield as low as a 30% transfer efficiency (meaning that 70% of the paint is wasted and cannot be recovered).

Available methods of applying liquid organic coatings include brushing and rolling, spray coating, immersion, and flow coating. In some cases, several successive coatings are applied to the substrate surface to achieve the desired result. An automobile car body is an important example; the following is a typical sequence applied to the sheet-metal car body in a mass-production automobile: (1) phosphate coat applied by dipping, (2) primer coat applied by dipping, (3) color paint coat applied by spray coating, and (4) clear coat (for high gloss and added protection) applied by spraying.

Brushing and ***rolling*** are the two most familiar application methods to most people. They have a high transfer efficiency—approaching 100%. Manual brushing and rolling methods are suited to low production but not mass production. While brushing is quite versatile, rolling is limited to flat surfaces.

Spray coating is a widely used production method for applying organic coatings. The process forces the coating liquid to atomize into a fine mist immediately prior to deposition onto the part surface. When the droplets hit the surface, they spread and flow together to form a uniform coating within the localized region of the spray. If done properly, spray coating provides a uniform coating over the entire work surface.

Spray coating can be performed manually in spray painting booths, or it can be set up as an automated process. Transfer efficiency is relatively low (as low as 30%) with these methods. Efficiency can be improved by ***electrostatic spraying***, in which the workpart is grounded electrically and the atomized droplets are electrostatically charged. This causes the droplets to be drawn to the part surfaces, increasing transfer efficiencies to values up to 90% [17]. Spraying is utilized extensively in the automotive industry for applying external paint coats to car bodies. It is also used for coating appliances and other consumer products.

Immersion applies large amounts of liquid coating to the workpart and allows the excess to drain off and be recycled. The simplest method is ***dip coating***, in which a part is immersed in an open tank of liquid coating material; when the part is withdrawn, the excess liquid drains back into the tank. A variation of dip coating is ***electrocoating***, in which the part is electrically charged and then dipped into a paint bath that has been given an opposite charge. This improves adhesion and permits use of water-based paints (which reduce fire and pollution hazards).

In ***flow coating***, workparts are moved through an enclosed paint booth, where a series of nozzles shower the coating liquid onto the part surfaces. Excess liquid drains back into a sump, which allows it to be reused.

Once applied, the organic coating must convert from liquid to solid. The term ***drying*** is often used to describe this conversion process. Many organic coatings dry by evaporation of their solvents. However, in order to form a durable film on the substrate surface, a further conversion is necessary, called curing. ***Curing*** involves a chemical change in the organic resin in which polymerization or cross-linking occurs to harden the coating.

The type of resin determines the type of chemical reaction that takes place in curing. The principal methods by which curing is effected in organic coatings are (1) ***ambient temperature curing***, which involves evaporation of the solvent and oxidation of the resin (most lacquers cure by this method); (2) ***elevated temperature curing***, in which elevated temperatures are used to accelerate solvent evaporation, as well as polymerization and cross-linking of the resin; (3) ***catalytic curing***, in which the starting resins require reactive agents mixed immediately prior to application to bring about polymerization and cross-linking (epoxy and polyurethane paints are examples); and (4) ***radiation curing***, in which various forms of radiation, such as microwaves, ultraviolet light, and electron beams, are required to cure the resin [17].

21.6.2 POWDER COATING

The organic coatings discussed above are liquid systems consisting of resins that are soluble (or at least miscible) in a suitable solvent. Powder coatings are different. They are applied as dry, finely pulverized, solid particles that are melted on the surface to form a uniform liquid film, after which they resolidify into a dry coating. Powder coating systems have grown significantly in commercial importance among organic coatings since the mid-1970s.

Powder coatings are classified as thermoplastic or thermosetting. Common thermoplastic powders include polyvinylchloride, nylon, polyester, polyethylene, and polypropylene. They are generally applied as relatively thick coatings, 0.08 mm to 0.30 mm (0.003 in to 0.012 in). Common thermosetting coating powders are epoxy, polyester, and acrylic. They are applied as uncured resins that polymerize and cross-link on heating or reaction with other ingredients. Coating thicknesses are typically 0.025 mm to 0.075 mm (0.001 in to 0.003 in).

There are two principal application methods for powder coatings: spraying and fluidized bed. In the ***spraying*** method, an electrostatic charge is given to each particle in order to attract it to an electrically grounded part surface. Several spray gun designs are available to impart the charge to the powders. The spray guns can be operated manually or by industrial robots. Compressed air is used to propel the powders to the nozzle. The powders are dry when sprayed, and any excess particles that do not attach to the surface can be recycled (unless multiple paint colors are mixed in the same spray booth). Powders can be sprayed onto a part at room temperature, followed by heating of the part to melt the powders; or they can be sprayed onto a part that has been heated to above the melting point of the powder, which usually provides a thicker coating.

The ***fluidized bed*** is a less commonly used alternative to electrostatic spraying. In this method, the workpart to be coated is preheated and passed through a fluidized bed, in which powders are suspended (fluidized) by an airstream. The powders attach themselves to the part surface to form the coating. In some implementations of this coating method, the powders are electrostatically charged to increase attraction to the grounded part surface.

REFERENCES

[1] ***ASM Handbook***, Vol. **5**, ***Surface Engineering***. ASM International, Materials Park, Ohio, 1993.

[2] Budinski, K. G. ***Surface Engineering for Wear Resistance***. Prentice Hall, Inc., Englewood Cliffs, New Jersey, 1988.

[3] Durney, L. J. (ed.). ***The Graham's Electroplating Engineering Handbook***, 4th ed. Chapman & Hall, London, 1996.

[4] Freeman, N. B. "A New Look at Mass Finishing," Special Report 757, *American Machinist*, August 1983, pp. 93–104.

[5] George, J. ***Preparation of Thin Films***. Marcel Dekker, Inc., New York, 1992.

[6] Hocking, M. G., Vasantasree, V., and Sidky, P. S. ***Metallic and Ceramic Coatings***. Addison-Wesley Longman, Ltd., Reading, Massachusetts, 1989.

[7] ***Metal Finishing***, Guidebook and Directory Issue. Metals and Plastics Publications, Inc., Hackensack, New Jersey, 2000.

[8] Morosanu, C. E. ***Thin Films by Chemical Vapour Deposition***. Elsevier, Amsterdam, Holland, 1990.

[9] Murphy, J. A. (ed.). ***Surface Preparation and Finishes for Metals***. McGraw-Hill Book Company, New York, 1971.

[10] Sabatka, W. "Vapor Degreasing." www.pfonline.com.

[11] Satas, D. (ed.). ***Coatings Technology Handbook***, 2nd ed. Marcel Dekker, Inc., New York, 2000.

[12] Stuart, R. V. ***Vacuum Technology, Thin Films, and Sputtering***. Academic Press, New York, 1983.

[13] Sze, S. M. ***VLSI Technology***, 2nd ed. McGraw-Hill Book Company, New York, 1988.

[14] Tracton, A. A. (ed.). ***Coatings Technology Handbook***, 3rd ed. CRC Taylor & Francis, Boca Raton, Florida, 2006.

[15] Tucker, Jr., R. C. "Surface Engineering Technologies," ***Advanced Materials & Processes***, April 2002, pp. 36–38.

[16] Tucker, Jr., R. C. "Considerations in the Selection of Coatings," ***Advanced Materials & Processes***, March 2004, pp. 25–28.

[17] Wick, C., and Veilleux, R. (eds.). ***Tool and Manufacturing Engineers Handbook***, 4th ed., Vol III, ***Materials, Finishes, and Coating***. Society of Manufacturing Engineers, Dearborn, Michigan, 1985.

REVIEW QUESTIONS

21.1. What are some of the important reasons why manufactured parts must be cleaned?

21.2. Mechanical surface treatments are often performed for reasons other than or in addition to cleaning. What are the reasons?

21.3. What are the basic types of contaminants that must be cleaned from metallic surfaces in manufacturing?

21.4. Name some of the important chemical cleaning methods.

21.5. In addition to surface cleaning, what is the main function performed by shot peening?

21.6. What is meant by the term *mass finishing*?

21.7. What is the difference between diffusion and ion implantation?

21.8. What is calorizing?

21.9. Why are metals coated?

21.10. Identify the most common types of coating processes.

21.11. What is the most commonly plated substrate metal?

21.12. One of the mandrel types in electroforming is a solid mandrel. How is the part removed from a solid mandrel?

21.13. How does electroless plating differ from electrochemical plating?

21.14. What is a conversion coating?

21.15. How does anodizing differ from other conversion coatings?

21.16. What is physical vapor deposition?

21.17. What is the difference between physical vapor deposition (PVD) and chemical vapor deposition (CVD)?

21.18. Name the commonly used coating materials deposited by PVD onto cutting tools.

21.19. What are some of the advantages of chemical vapor deposition?

21.20. Identify the four major types of ingredients in organic coatings.

21.21. What is meant by the term *transfer efficiency* in organic coating technology?

21.22. Describe the principal methods by which organic coatings are applied to a surface.

21.23. The terms *drying* and *curing* have different meanings; indicate the distinction.

PROBLEMS

21.1. What volume (cm^3) and weight (g) of zinc will be deposited onto a cathodic workpart if 10 amps of current are applied for one hour?

21.2. A sheet-metal steel part with surface area = 100 cm^2 is to be zinc plated. What average plating thickness will result if 15 amps are applied for 12 minutes in a chloride electrolyte solution?

21.3. A sheet-metal steel part with surface area = 15.0 in^2 is to be chrome plated. What average plating thickness will result if 15 amps are applied for 10 min in a chromic acid–sulfate bath?

21.4. Twenty-five jewelry pieces, each with a surface area = 0.5 in^2 are to be gold plated in a batch plating operation. (a) What average plating thickness will result if 8 amps are applied for 10 min in a cyanide bath? (b) What is the value of the gold that will be plated onto each piece if one ounce of gold is valued at $1000? The density of gold = 0.698 lb/in^3.

21.5. A part made of sheet steel is to be nickel plated. The part is a rectangular flat plate that is 0.075 cm thick and whose face dimensions are 14 cm by 19 cm. The plating operation is carried out in an acid sulfate electrolyte, using a current = 20 amps for a duration = 30 min. Determine the average thickness of the plated metal resulting from this operation.

21.6. A steel sheet-metal part has total surface area = 36 in^2. How long will it take to deposit a copper plating (assume valence = +1) of thickness = 0.001 in onto the surface if 15 amps of current are applied?

Part VII Joining and Assembly Processes

22 FUNDAMENTALS OF WELDING

Chapter Contents

22.1 Overview of Welding Technology
- 22.1.1 Types of Welding Processes
- 22.1.2 Welding as a Commercial Operation

22.2 The Weld Joint
- 22.2.1 Types of Joints
- 22.2.2 Types of Welds

22.3 Physics of Welding
- 22.3.1 Power Density
- 22.3.2 Heat Balance in Fusion Welding

22.4 Features of a Fusion-Welded Joint

In this part of the book, we consider processes that are used to join two or more parts into an assembled entity. These processes are labeled in the lower stem of Figure 1.3. The term ***joining*** is generally used for welding, brazing, soldering, and adhesive bonding, which form a permanent joint between the parts—a joint that cannot easily be separated. The term ***assembly*** usually refers to mechanical methods of fastening parts together. Some of these methods allow for easy disassembly, while others do not. We begin our coverage of the joining and assembly processes with two chapters on welding. Brazing, soldering, and adhesive bonding are discussed in Chapter 24, and mechanical assembly is presented in Chapter 25.

Welding is a materials joining process in which two or more parts are coalesced at their contacting surfaces by a suitable application of heat and/or pressure. Many welding processes are accomplished by heat alone, with no pressure applied; others by a combination of heat and pressure; and still others by pressure alone, with no external heat supplied. In some welding processes a ***filler*** material is added to facilitate coalescence. The assemblage of parts that are joined by welding is called a ***weldment***. Welding is most commonly associated with metal parts, but the process is also used for joining plastics. Our discussion of welding will focus on metals.

The commercial and technological importance of welding derives from the following:

- Welding provides a permanent joint. The welded components become a single entity.

- The welded joint can be stronger than the parent materials if a filler metal is used that has strength properties superior to those of the parents, and if proper welding techniques are used.
- Welding is usually the most economical way to join components in terms of material usage and fabrication costs. Alternative mechanical methods of assembly require more complex shape alterations (e.g., drilling of holes) and addition of fasteners (e.g., rivets or bolts). The resulting mechanical assembly is usually heavier than a corresponding weldment.
- Welding is not restricted to the factory environment. It can be accomplished "in the field."

Although welding has the advantages indicated above, it also has certain limitations and drawbacks (or potential drawbacks):

- Most welding operations are performed manually and are expensive in terms of labor cost. Many welding operations are considered "skilled trades," and the labor to perform these operations may be scarce.
- Most welding processes are inherently dangerous because they involve the use of high energy.
- Because welding accomplishes a permanent bond between the components, it does not allow for convenient disassembly. If the product must occasionally be disassembled (e.g., for repair or maintenance), then welding should not be used as the assembly method.
- The welded joint can suffer from certain quality defects that are difficult to detect. The defects can reduce the strength of the joint.

22.1 OVERVIEW OF WELDING TECHNOLOGY

Welding involves localized coalescence or joining together of two metallic parts at their faying surfaces. The ***faying surfaces*** are the part surfaces in contact or close proximity that are to be joined. Welding is usually performed on parts made of the same metal, but some welding operations can be used to join dissimilar metals.

22.1.1 TYPES OF WELDING PROCESSES

Some 50 different welding processes have been cataloged by the American Welding Society. They use various types or combinations of energy to provide the required power. We can divide the welding processes into two major groups: (1) fusion welding and (2) solid-state welding.

Fusion Welding Fusion-welding processes use heat to melt the base metals. In many fusion-welding operations, a filler metal is added to the molten pool to facilitate the process and provide bulk and strength to the welded joint. A fusion-welding operation in which no filler metal is added is referred to as an ***autogenous*** weld. The fusion category includes the most widely used welding processes, which can be organized into the following general groups (initials in parentheses are designations of the American Welding Society):

- ***Arc welding*** (AW). Arc welding refers to a group of welding processes in which heating of the metals is accomplished by an electric arc, as shown in Figure 22.1.

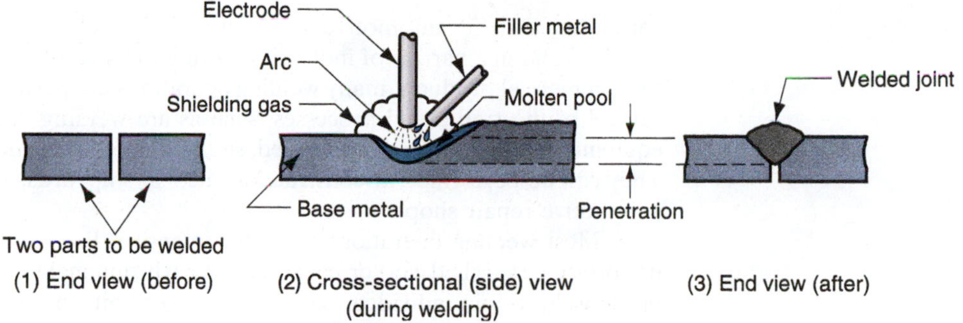

FIGURE 22.1 Basics of arc welding: (1) before the weld; (2) during the weld (the base metal is melted and filler metal is added to the molten pool); and (3) the completed weldment. There are many variations of the arc-welding process. (Credit: *Fundamentals of Modern Manufacturing*, 4th Edition by Mikell P. Groover, 2010. Reprinted with permission of John Wiley & Sons, Inc.)

Some arc-welding operations also apply pressure during the process and most utilize a filler metal.

- ***Resistance welding*** (RW). Resistance welding achieves coalescence using heat from electrical resistance to the flow of a current passing between the faying surfaces of two parts held together under pressure.
- ***Oxyfuel gas welding*** (OFW). These joining processes use an oxyfuel gas, such as a mixture of oxygen and acetylene, to produce a hot flame for melting the base metal and filler metal, if one is used.
- Other fusion-welding processes. Other welding processes that produce fusion of the metals joined include ***electron beam welding*** and ***laser beam welding***.

Solid-State Welding Solid-state welding refers to joining processes in which coalescence results from application of pressure alone or a combination of heat and pressure. If heat is used, the temperature in the process is below the melting point of the metals being welded. No filler metal is utilized. Representative welding processes in this group include:

- ***Diffusion welding*** (DFW). Two surfaces are held together under pressure at an elevated temperature and the parts coalesce by solid-state diffusion.
- ***Friction welding*** (FRW). Coalescence is achieved by the heat of friction between two surfaces.
- ***Ultrasonic welding*** (USW). Moderate pressure is applied between the two parts and an oscillating motion at ultrasonic frequencies is used in a direction parallel to the contacting surfaces. The combination of normal and vibratory forces results in shear stresses that remove surface films and achieve atomic bonding of the surfaces.

In Chapter 23, we describe the various welding processes in greater detail. The preceding survey should provide a sufficient framework for our discussion of welding terminology and principles in the present chapter.

22.1.2 WELDING AS A COMMERCIAL OPERATION

The principal applications of welding are (1) construction, such as buildings and bridges; (2) piping, pressure vessels, boilers, and storage tanks; (3) shipbuilding; (4) aircraft and aerospace; and (5) automotive and railroad [1]. Welding is performed in a variety of locations and in a variety of industries. Owing to its versatility as an assembly technique for commercial products, many welding operations are performed in factories. However, several of the traditional processes, such as arc welding and oxyfuel gas welding, use equipment that can be readily moved, so these operations are not limited to the factory. They can be performed at construction sites, in shipyards, at customers' plants, and in automotive repair shops.

Most welding operations are labor intensive. For example, arc welding is usually performed by a skilled worker, called a ***welder***, who manually controls the path or placement of the weld to join individual parts into a larger unit. In factory operations in which arc welding is manually performed, the welder often works with a second worker, called a ***fitter***. It is the fitter's job to arrange the individual components for the welder prior to making the weld. Welding fixtures and positioners are used for this purpose. A ***welding fixture*** is a device for clamping and holding the components in fixed position for welding. It is custom-fabricated for the particular geometry of the weldment and therefore must be economically justified on the basis of the quantities of assemblies to be produced. A ***welding positioner*** is a device that holds the parts and also moves the assemblage to the desired position for welding. This differs from a welding fixture that only holds the parts in a single fixed position. The desired position is usually one in which the weld path is flat and horizontal.

The Safety Issue Welding is inherently dangerous to human workers. Strict safety precautions must be practiced by those who perform these operations. The high temperatures of the molten metals in welding are an obvious danger. In gas welding, the fuels (e.g., acetylene) are a fire hazard. Most of the processes use high energy to cause melting of the part surfaces to be joined. In many welding processes, electrical power is the source of thermal energy, so there is the hazard of electrical shock to the worker. Certain welding processes have their own particular perils. For example, arc welding emits ultraviolet radiation that is injurious to human vision. The welder must wear a special helmet that includes a dark viewing window. This window filters out the dangerous radiation but is so dark that it renders the welder virtually blind, except when the arc is struck. Sparks, spatters of molten metal, smoke, and fumes add to the risks associated with welding operations. Ventilation facilities must be used to exhaust the dangerous fumes generated by some of the fluxes and molten metals used in welding. If the operation is performed in an enclosed area, special ventilation suits or hoods are required.

Automation in Welding Because of the hazards of manual welding, and in efforts to increase productivity and improve product quality, various forms of mechanization and automation have been developed. The categories include machine welding, automatic welding, and robotic welding.

Machine welding can be defined as mechanized welding with equipment that performs the operation under the continuous supervision of an operator. It is normally accomplished by a welding head that is moved by mechanical means relative to a stationary work, or by moving the work relative to a stationary welding head. The human worker must continually observe and interact with the equipment to control the operation.

If the equipment is capable of performing the operation without control by a human operator, it is referred to as ***automatic welding***. A human worker is usually present to oversee the process and detect variations from normal conditions. What distinguishes automatic welding from machine welding is a weld cycle controller to regulate the arc movement and workpiece positioning without continuous human attention. Automatic welding requires a welding fixture and/or positioner to position the work relative to the welding head. It also requires a higher degree of consistency and accuracy in the component parts used in the weldment. For these reasons, automatic welding can be justified only for large-quantity production.

In ***robotic welding***, an industrial robot or programmable manipulator is used to automatically control the movement of the welding head relative to the work. The versatile reach of the robot arm permits the use of relatively simple fixtures, and the robot's capacity to be reprogrammed for new part configurations allows this form of automation to be justified for relatively low production quantities. A typical robotic arc-welding cell consists of two welding fixtures and a human fitter to load and unload parts while the robot welds. In addition to arc welding, industrial robots are also used in automobile final assembly plants to perform resistance welding on car bodies.

22.2 THE WELD JOINT

Welding produces a solid connection between two pieces, called a weld joint. A ***weld joint*** is the junction of the edges or surfaces of parts that have been joined by welding. This section covers two classifications related to weld joints: (1) types of joints and (2) the types of welds used to join the pieces that form the joints.

22.2.1 TYPES OF JOINTS

There are five basic types of joints for bringing two parts together for joining. The five joint types are not limited to welding; they apply to other joining and fastening techniques as well. With reference to Figure 22.2, the five joint types can be defined as follows:

(a) ***Butt joint***. In this joint type, the parts lie in the same plane and are joined at their edges.
(b) ***Corner joint***. The parts in a corner joint form a right angle and are joined at the corner of the angle.
(c) ***Lap joint***. This joint consists of two overlapping parts.
(d) ***Tee joint***. In a tee joint, one part is perpendicular to the other in the approximate shape of the letter "T."
(e) ***Edge joint***. The parts in an edge joint are parallel with at least one of their edges in common, and the joint is made at the common edge(s).

22.2.2 TYPES OF WELDS

Each of the preceding joints can be made by welding. It is appropriate to distinguish between the joint type and the way in which it is welded—the weld type. Differences among weld types are in geometry (joint type) and welding process.

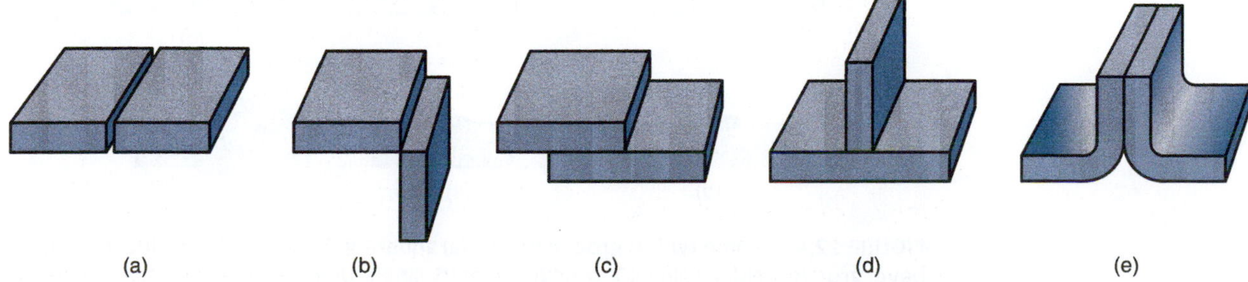

FIGURE 22.2 Five basic types of joints: (a) butt, (b) corner, (c) lap, (d) tee, and (e) edge. (Credit: *Fundamentals of Modern Manufacturing*, 4th Edition by Mikell P. Groover, 2010. Reprinted with permission of John Wiley & Sons, Inc.)

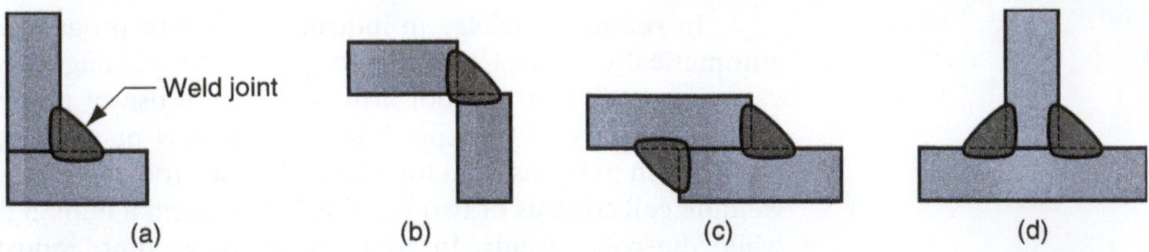

FIGURE 22.3 Various forms of fillet welds: (a) inside single fillet corner joint; (b) outside single fillet corner joint; (c) double fillet lap joint; and (d) double fillet tee joint. Dashed lines show the original part edges. (Credit: *Fundamentals of Modern Manufacturing*, 4th Edition by Mikell P. Groover, 2010. Reprinted with permission of John Wiley & Sons, Inc.)

A ***fillet weld*** is used to fill in the edges of plates created by corner, lap, and tee joints, as in Figure 22.3. Filler metal is used to provide a cross section approximately the shape of a right triangle. It is the most common weld type in arc and oxyfuel welding because it requires minimum edge preparation—the basic square edges of the parts are used. Fillet welds can be single or double (i.e., welded on one side or both) and can be continuous or intermittent (i.e., welded along the entire length of the joint or with unwelded spaces along the length).

Groove welds usually require that the edges of the parts be shaped into a groove to facilitate weld penetration. The grooved shapes include square, bevel, V, U, and J, in single or double sides, as shown in Figure 22.4. Filler metal is used to fill in the joint, usually by arc or oxyfuel welding. Preparation of the part edges beyond the basic square edge, although requiring additional processing, is often done to increase the strength of the welded joint or where thicker parts are to be welded. Although most closely associated with a butt joint, groove welds are used on all joint types except lap.

Plug welds and ***slot welds*** are used for attaching flat plates, as shown in Figure 22.5, using one or more holes or slots in the top part and then filling with filler metal to fuse the two parts together.

Spot welds and seam welds, used for lap joints, are diagrammed in Figure 22.6. A ***spot weld*** is a small fused section between the surfaces of two sheets or plates. Multiple spot welds are typically required to join the parts. It is most closely associated with resistance welding. A ***seam weld*** is similar to a spot weld, except it consists of a more or less continuously fused section between the two sheets or plates.

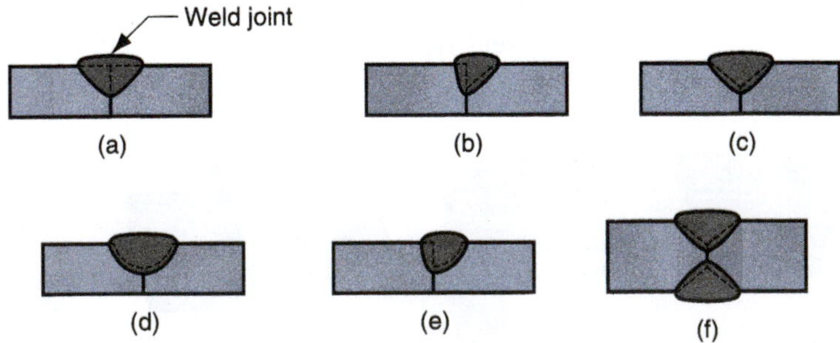

FIGURE 22.4 Some typical groove welds: (a) square groove weld, one side; (b) single bevel groove weld; (c) single V-groove weld; (d) single U-groove weld; (e) single J-groove weld; (f) double V-groove weld for thicker sections. Dashed lines show the original part edges. (Credit: *Fundamentals of Modern Manufacturing*, 4th Edition by Mikell P. Groover, 2010. Reprinted with permission of John Wiley & Sons, Inc.)

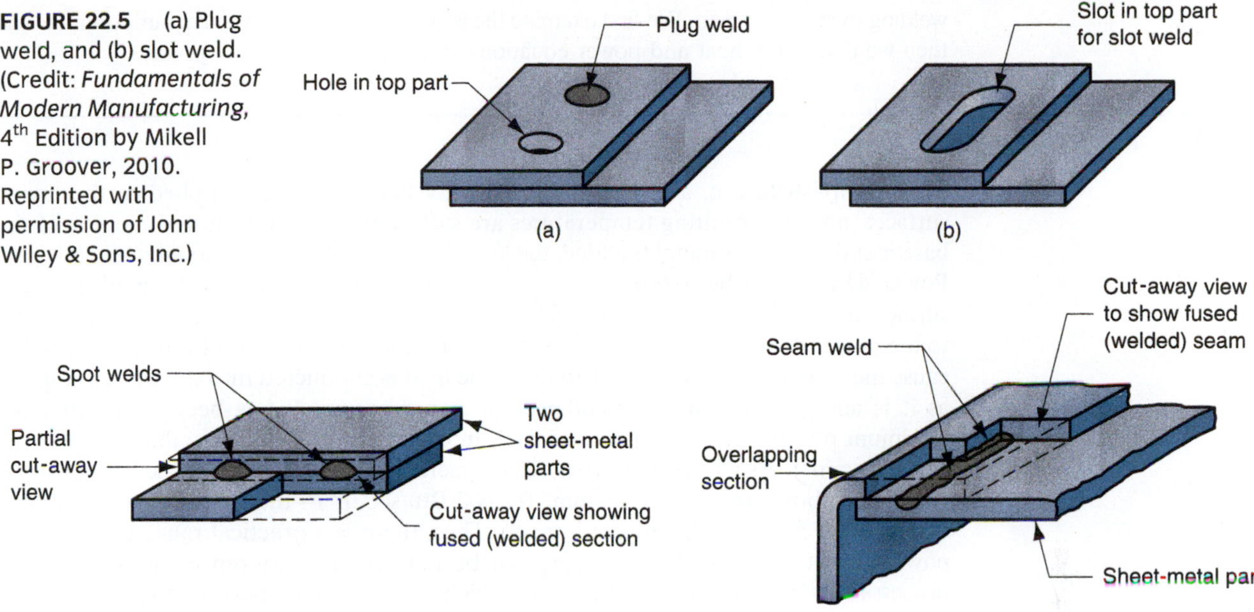

FIGURE 22.5 (a) Plug weld, and (b) slot weld. (Credit: *Fundamentals of Modern Manufacturing,* 4th Edition by Mikell P. Groover, 2010. Reprinted with permission of John Wiley & Sons, Inc.)

FIGURE 22.6 (a) Spot weld, and (b) seam weld. (Credit: *Fundamentals of Modern Manufacturing,* 4th Edition by Mikell P. Groover, 2010. Reprinted with permission of John Wiley & Sons, Inc.)

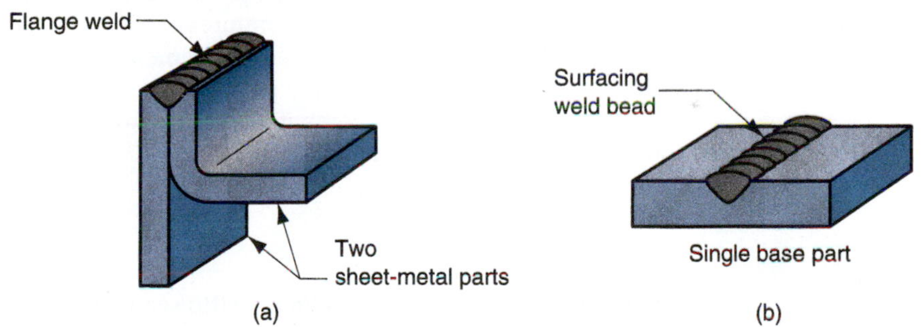

FIGURE 22.7 (a) Flange weld, and (b) surfacing weld. (Credit: *Fundamentals of Modern Manufacturing,* 4th Edition by Mikell P. Groover, 2010. Reprinted with permission of John Wiley & Sons, Inc.)

Flange welds and surfacing welds are shown in Figure 22.7. A ***flange weld*** is made on the edges of two (or more) parts, usually sheet metal or thin plate, at least one of the parts being flanged as in Figure 22.7(a). A ***surfacing weld*** is not used to join parts, but rather to deposit filler metal onto the surface of a base part in one or more weld beads. The weld beads can be made in a series of overlapping parallel passes, thereby covering large areas of the base part. The purpose is to increase the thickness of the plate or to provide a protective coating on the surface.

22.3 PHYSICS OF WELDING

Although several coalescing mechanisms are available for welding, fusion is by far the most common means. In this section, we consider the physical relationships that allow fusion welding to be performed. We first examine the issue of power density and its importance, and then we define the heat and power equations that describe a welding process.

22.3.1 POWER DENSITY

To accomplish fusion, a source of high-density heat energy is supplied to the faying surfaces, and the resulting temperatures are sufficient to cause localized melting of the base metals. If a filler metal is added, the heat density must be high enough to melt it also. Power density can be defined as the heating power transferred to the work per unit surface area, W/mm^2 (Btu/sec-in^2). The time to melt the metal is inversely proportional to the power density. At low power densities, a significant amount of time is required to cause melting. If power density is too low, the heat is conducted into the work as rapidly as it is added at the surface, and melting never occurs. It has been found that the minimum power density required to melt most metals in welding is about 10 W/mm^2 (6 Btu/sec-in^2). As power density increases, melting time is reduced. If power density is too high—above around 10^5 W/mm^2 (60,000 Btu/sec-in^2)—the localized temperatures vaporize the metal in the affected region. Thus, there is a practical range of values for power density within which welding can be performed. Differences among welding processes in this range are (1) the rate at which welding can be performed and/or (2) the size of the region that can be welded. Table 22.1 provides a comparison of power densities for the major fusion-welding processes. Oxyfuel gas welding is capable of developing large amounts of heat, but the power density is relatively low because it is spread over a large area. Oxyacetylene gas, the hottest of the OFW fuels, burns at a top temperature of around 3500°C (6300°F). By comparison, arc welding produces high energy over a smaller area, resulting in local temperatures of 5500°C to 6600°C (10,000°F to 12,000°F). For metallurgical reasons, it is desirable to melt the metal with minimum energy, and high power densities are generally preferable.

Power density can be computed as the power entering the surface divided by the corresponding surface area:

$$PD = \frac{P}{A} \tag{22.1}$$

where PD = power density, W/mm^2 (Btu/sec-in^2); P = power entering the surface, W (Btu/sec); and A = surface area over which the energy is entering, mm^2 (in^2). The issue is more complicated than indicated by Eq. (22.1). One complication is that the power source (e.g., the arc) is moving in many welding processes, which results in preheating ahead of the operation and postheating behind it. Another complication is that power density is not uniform throughout the affected surface; it is distributed as a function of area, as demonstrated by the following example.

TABLE 22.1 Comparison of several fusion-welding processes on the basis of their power densities.

	Approximate Power Density	
Welding Process	**W/mm^2**	**Btu/sec-in^2**
Oxyfuel welding	10	6
Arc welding	50	30
Resistance welding	1000	600
Laser beam welding	9000	5000
Electron beam welding	10,000	6000

Compiled from [1] and other sources.

Example 22.1 Power Density in Welding

A heat source transfers 3000 W to the surface of a metal part. The heat impinges the surface in a circular area, with intensities varying inside the circle. The distribution is as follows: 70% of the power is transferred within a circle of diameter = 5 mm, and 90% is transferred within a concentric circle of diameter = 12 mm. What are the power densities in (a) the 5-mm-diameter inner circle and (b) the 12-mm-diameter ring that lies around the inner circle?

Solution: (a) The inner circle has an area $A = \frac{\pi(5)^2}{4} = 19.63\ \text{mm}^2$.

The power inside this area $P = 0.70 \times 3000 = 2100$ W.

Thus the power density $PD = \frac{2100}{19.63} = 107\ \text{W/mm}^2$.

(b) The area of the ring outside the inner circle is $A = \frac{\pi(12^2 - 5^2)}{4} = 93.4\ \text{mm}^2$.

The power in this region $P = 0.9(3000) - 2100 = 600$ W.

The power density is therefore $PD = \frac{600}{93.4} = 6.4\ \text{W/mm}^2$.

Observation The power density seems high enough for melting in the inner circle, but probably not sufficient in the ring that lies outside this inner circle. ■

22.3.2 HEAT BALANCE IN FUSION WELDING

The quantity of heat required to melt a given volume of metal depends on (1) the heat to raise the temperature of the solid metal to its melting point, which depends on the metal's volumetric specific heat, (2) the melting point of the metal, and (3) the heat to transform the metal from solid to liquid phase at the melting point, which depends on the metal's heat of fusion. To a reasonable approximation, this quantity of heat can be estimated by the following equation [5]:

$$U_m = KT_m^2 \tag{22.2}$$

where U_m = the unit energy for melting (i.e., the quantity of heat required to melt a unit volume of metal starting from room temperature), J/mm^3 (Btu/in^3); T_m = melting point of the metal on an absolute temperature scale, °K (°R); and K = constant whose value is 3.33×10^{-6} when the Kelvin scale is used (and $K = 1.467 \times 10^{-5}$ for the Rankine temperature scale). Absolute melting temperatures for selected metals are presented in Table 22.2.

Not all of the energy generated at the heat source is used to melt the weld metal. There are two heat transfer mechanisms at work, both of which reduce the amount of generated heat that is used by the welding process. The first mechanism involves the transfer of heat between the heat source and the surface of the work. This process has a certain ***heat transfer factor*** f_1, defined as the ratio of the actual heat received by the workpiece divided by the total heat generated at the source. The second mechanism involves the conduction of heat away from the weld area to be dissipated throughout the work metal, so that only a portion of the heat transferred to the surface is available for melting. This ***melting factor*** f_2 is the proportion of heat received at the work surface that can be used for melting. The combined effect of these two factors is to reduce the heat energy available for welding as follows:

$$H_w = f_1 f_2 H \tag{22.3}$$

where H_w = net heat available for welding, J (Btu), f_1 = heat transfer factor, f_2 = the melting factor, and H = the total heat generated by the welding process, J (Btu).

TABLE 22.2 Melting temperatures on the absolute temperature scale for selected metals.

Metal	Melting Temperature °K[a]	Melting Temperature °R[a]	Metal	Melting Temperature °K[a]	Melting Temperature °R[b]
Aluminum alloys	930	1680	Steels		
Cast iron	1530	2760	Low carbon	1760	3160
Copper and alloys			Medium carbon	1700	3060
Pure	1350	2440	High carbon	1650	2960
Brass, navy	1160	2090	Low alloy	1700	3060
Bronze (90 Cu–10 Sn)	1120	2010	Stainless steels		
Inconel	1660	3000	Austenitic	1670	3010
Magnesium	940	1700	Martensitic	1700	3060
Nickel	1720	3110	Titanium	2070	3730

Based on values in [2].
[a]Kelvin scale = Centigrade (Celsius) temperature + 273.
[b]Rankine scale = Fahrenheit temperature + 460.

The factors f_1 and f_2 range in value between zero and one. It is appropriate to separate f_1 and f_2 in concept, even though they act in concert during the welding process. The heat transfer factor f_1 is determined largely by the welding process and the capacity to convert the power source (e.g., electrical energy) into usable heat at the work surface. Arc-welding processes are relatively efficient in this regard, while oxyfuel gas-welding processes are relatively inefficient.

The melting factor f_2 depends on the welding process, but it is also influenced by the thermal properties of the metal, joint configuration, and work thickness. Metals with high thermal conductivity, such as aluminum and copper, present a problem in welding because of the rapid dissipation of heat away from the heat contact area. The problem is exacerbated by welding heat sources with low power densities (e.g., oxyfuel welding) because the heat input is spread over a larger area, thus facilitating conduction into the work. In general, a high power density combined with a low conductivity work material results in a high melting factor.

We can now write a balance equation between the energy input and the energy needed for welding:

$$H_w = U_m V \tag{22.4}$$

where H_w = net heat energy used by the welding operation, J (Btu); U_m = unit energy required to melt the metal, J/mm^3 (Btu/in^3); and V = the volume of metal melted, mm^3 (in^3).

Most welding operations are rate processes; that is, the net heat energy H_w is delivered at a given rate, and the weld bead is made at a certain travel velocity. This is characteristic for example of most arc-welding operations, many oxyfuel gas-welding operations, and even some resistance-welding operations. It is therefore appropriate to express Eq. (22.4) as a rate balance equation:

$$R_{Hw} = U_m R_{WV} \tag{22.5}$$

where R_{Hw} = rate of heat energy delivered to the operation for welding, J/s = W (Btu/min); and R_{WV} = volume rate of metal welded, mm^3/s (in^3/min). In the welding of a continuous bead, the volume rate of metal welded is the product of weld area A_w and travel velocity v. Substituting these terms into the above equation, the rate balance equation can now be expressed as:

$$R_{Hw} = f_1 f_2 R_H = U_m A_w v \tag{22.6}$$

where f_1 and f_2 are the heat transfer and melting factors; R_H = rate of input energy generated by the welding power source, W (Btu/min); A_w = weld cross-sectional area, mm^2 (in^2); and v = the travel velocity of the welding operation, mm/s (in/min). In Chapter 23, we examine how the power density in Eq. (22.1) and the input energy rate for Eq. (22.6) are generated for some of the individual welding processes.

Example 22.2 Welding Travel Speed

The power source in a particular welding setup generates 3500 W that can be transferred to the work surface with a heat transfer factor = 0.7. The metal to be welded is low carbon steel, whose melting temperature, from Table 22.2, is 1760°K. The melting factor in the operation is 0.5. A continuous fillet weld is to be made with a cross-sectional area = 20 mm^2. Determine the travel speed at which the welding operation can be accomplished.

Solution: Let us first find the unit energy required to melt the metal U_m from Eq. (22.2).

$$U_m = 3.33(10^{-6}) \times 1760^2 = 10.3\ \text{J/mm}^3$$

Rearranging Eq. (22.6) to solve for travel velocity, we have $v = \dfrac{f_1 f_2 R_H}{U_m A_w}$, and solving for the conditions of the problem, $v = \dfrac{0.7\,(0.5)\,(3500)}{10.3\,(20)} = 5.95\ \text{mm/s}$. ■

22.4 FEATURES OF A FUSION-WELDED JOINT

Most weld joints are fusion welded. As illustrated in the cross-sectional view of Figure 22.8(a), a typical fusion-weld joint in which filler metal has been added consists of several zones: (1) fusion zone, (2) weld interface, (3) heat-affected zone, and (4) unaffected base metal zone.

The ***fusion zone*** consists of a mixture of filler metal and base metal that has completely melted. This zone is characterized by a high degree of homogeneity among the component metals that have been melted during welding. The mixing of these components is motivated largely by convection in the molten weld pool. Solidification in the fusion zone has similarities to a casting process. In welding the mold is formed by the unmelted edges or surfaces of the components being welded. The significant difference between solidification in casting and in welding is that epitaxial grain growth occurs in welding. The reader may recall that in casting, the metallic grains are formed from the melt by nucleation of solid particles at the mold wall, followed by grain growth. In welding, by contrast, the nucleation stage of solidification is avoided by the mechanism of ***epitaxial grain growth,*** in which atoms from the molten pool solidify on preexisting lattice sites of the adjacent solid base metal. Consequently, the grain structure in the fusion zone near the heat affected zone tends to mimic the crystallographic orientation of the surrounding heat-affected zone. Further into the fusion zone, a preferred orientation develops in which the grains are roughly perpendicular to the boundaries of the weld interface. The resulting structure in the solidified fusion zone tends to feature coarse columnar grains, as depicted in Figure 22.8(b). The grain structure depends on various factors, including welding process, metals being welded (e.g., identical metals vs. dissimilar metals welded), whether a filler metal is used, and the feed rate at which

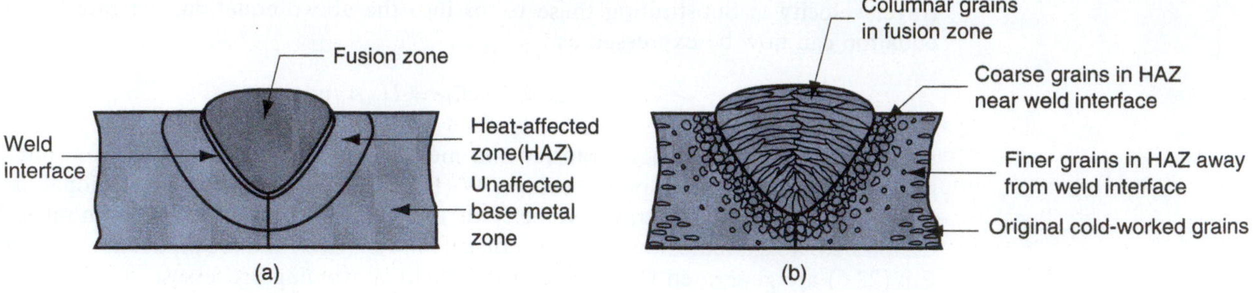

FIGURE 22.8 Cross section of a typical fusion-welded joint: (a) principal zones in the joint and (b) typical grain structure. (Credit: *Fundamentals of Modern Manufacturing,* 4th Edition by Mikell P. Groover, 2010. Reprinted with permission of John Wiley & Sons, Inc.)

welding is accomplished. A detailed discussion of welding metallurgy is beyond the scope of this text, and interested readers can consult any of several references [1], [4], [5].

The second zone in the weld joint is the ***weld interface***, a narrow boundary that separates the fusion zone from the heat-affected zone. The interface consists of a thin band of base metal that was melted or partially melted (localized melting within the grains) during the welding process but then immediately solidified before any mixing with the metal in the fusion zone. Its chemical composition is therefore identical to that of the base metal.

The third zone in the typical fusion weld is the ***heat-affected zone*** (HAZ). The metal in this zone has experienced temperatures that are below its melting point, yet high enough to cause microstructural changes in the solid metal. The chemical composition in the heat-affected zone is the same as the base metal, but this region has been heat-treated due to the welding temperatures so that its properties and structure have been altered. The amount of metallurgical damage in the HAZ depends on factors such as the amount of heat input and peak temperatures reached, distance from the fusion zone, length of time the metal has been subjected to the high temperatures, cooling rate, and the metal's thermal properties. The effect on mechanical properties in the heat-affected zone is usually negative, and it is in this region of the weld joint that welding failures often occur.

As the distance from the fusion zone increases, the ***unaffected base metal zone*** is finally reached, in which no metallurgical change has occurred. Nevertheless, the base metal surrounding the HAZ is likely to be in a state of high residual stress, the result of shrinkage in the fusion zone.

REFERENCES

[1] ***ASM Handbook,*** Vol. **6**, ***Welding, Brazing, and Soldering.*** ASM International, Materials Park, Ohio, 1993.

[2] Cary, H. B., and Helzer, S. C. ***Modern Welding Technology,*** 6th ed. Pearson/Prentice-Hall, Upper Saddle River, New Jersey, 2005.

[3] Datsko, J. ***Material Properties and Manufacturing Processes.*** John Wiley & Sons, Inc., New York, 1966.

[4] Messler, R. W., Jr. ***Principles of Welding: Processes, Physics, Chemistry, and Metallurgy.*** John Wiley & Sons, Inc., New York, 1999.

[5] ***Welding Handbook,*** 9th ed. Vol. **1**, ***Welding Science and Technology.*** American Welding Society, Miami, Florida, 2007.

[6] Wick, C., and Veilleux, R. F. ***Tool and Manufacturing Engineers Handbook,*** 4th ed., Vol. IV, ***Quality Control and Assembly.*** Society of Manufacturing Engineers, Dearborn, Michigan, 1987.

REVIEW QUESTIONS

22.1. What are the advantages and disadvantages of welding compared to other types of assembly operations?

22.2. What is meant by the term *faying surface*?

22.3. Define the term *fusion weld*.

22.4. What is the fundamental difference between a fusion weld and a solid-state weld?

22.5. What is an autogenous weld?

22.6. Discuss the reasons why most welding operations are inherently dangerous.

22.7. What is the difference between machine welding and automatic welding?

22.8. Name and sketch the five joint types.

22.9. Define and sketch a fillet weld.

22.10. Define and sketch a groove weld.

22.11. Why is a surfacing weld different from the other weld types?

22.12. Why is it desirable to use energy sources for welding that have high heat densities?

22.13. What is the unit melting energy in welding, and what are the factors on which it depends?

22.14. Define and distinguish the two terms *heat transfer factor* and *melting factor* in welding.

22.15. What is the heat-affected zone (HAZ) in a fusion weld?

PROBLEMS

22.1. A heat source can transfer 3500 J/sec to a metal part surface. The heated area is circular, and the heat intensity decreases as the radius increases, as follows: 70% of the heat is concentrated in a circular area that is 3.75 mm in diameter. Is the resulting power density enough to melt metal?

22.2. A welding heat source is capable of transferring 150 Btu/min to the surface of a metal part. The heated area is approximately circular, and the heat intensity decreases with increasing radius as follows: 50% of the power is transferred within a circle of diameter = 0.1 in and 75% is transferred within a concentric circle of diameter = 0.25 in. What are the power densities in (a) the 0.1-in diameter inner circle and (b) the 0.25-in diameter ring that lies around the inner circle? (c) Are these power densities sufficient for melting metal?

22.3. Compute the unit energy for melting for the following metals: (a) aluminum and (b) plain low carbon steel.

22.4. Compute the unit energy for melting for the following metals: (a) copper and (b) titanium.

22.5. A fillet weld has a cross-sectional area of 25.0 mm^2 and is 300 mm long. (a) What quantity of heat (in joules) is required to accomplish the weld, if the metal to be welded is low carbon steel? (b) How much heat must be generated at the welding source, if the heat transfer factor is 0.75 and the melting factor = 0.63?

22.6. A groove weld has a cross-sectional area = 0.045 in^2 and is 10 in long. (a) What quantity of heat (in Btu) is required to accomplish the weld, if the metal to be welded is medium carbon steel? (b) How much heat must be generated at the welding source, if the heat transfer factor = 0.9 and the melting factor = 0.7?

22.7. Solve the previous problem, except that the metal to be welded is aluminum, and the corresponding melting factor is half the value for steel.

22.8. The welding power generated in a particular arc-welding operation = 3000 W. This is transferred to the work surface with a heat transfer factor = 0.9. The metal to be welded is copper whose melting point is given in Table 22.2. Assume that the melting factor = 0.25. A continuous fillet weld is to be made with a cross-sectional area = 15.0 mm^2. Determine the travel speed at which the welding operation can be accomplished.

22.9. Solve the previous problem except that the metal to be welded is high carbon steel, the cross-sectional area of the weld = 25.0 mm^2, and the melting factor = 0.6.

22.10. A welding operation on an aluminum alloy makes a groove weld. The cross-sectional area of the weld is 30.0 mm^2. The welding velocity is 4.0 mm/sec. The heat transfer factor is 0.92 and the melting factor is 0.48. The melting temperature of the aluminum alloy is 650°C. Determine the rate of heat generation required at the welding source to accomplish this weld.

22.11. The power source in a particular welding operation generates 125 Btu/min, which is transferred to the work surface with heat transfer factor = 0.8. The melting point for the metal to be welded = 1800°F and its melting factor = 0.5. A continuous fillet weld is to be made with a cross-sectional area = 0.04 in^2. Determine the travel speed at which the welding operation can be accomplished.

22.12. In a certain welding operation to make a fillet weld, the cross-sectional area = 0.025 in^2 and the travel speed = 15 in/min. If the heat transfer factor = 0.95 and melting factor = 0.5, and the melting point = 2000°F for the metal to be welded, determine the rate of heat generation required at the heat source to accomplish this weld.

23 WELDING PROCESSES

Chapter Contents

23.1 Arc Welding
- 23.1.1 General Technology of Arc Welding
- 23.1.2 AW Processes—Consumable Electrodes
- 23.1.3 AW Processes—Nonconsumable Electrodes

23.2 Resistance Welding
- 23.2.1 Power Source in Resistance Welding
- 23.2.2 Resistance-Welding Processes

23.3 Oxyfuel Gas Welding
- 23.3.1 Oxyacetylene Welding
- 23.3.2 Alternative Gases for Oxyfuel Welding

23.4 Other Fusion-Welding Processes

23.5 Solid-State Welding
- 23.5.1 General Considerations in Solid-State Welding
- 23.5.2 Solid-State Welding Processes

23.6 Weld Quality

23.7 Design Considerations in Welding

Welding processes divide into two major categories: (1) ***fusion welding***, in which coalescence is accomplished by melting the two parts to be joined, in some cases adding filler metal to the joint; and (2) ***solid-state welding***, in which heat and/or pressure are used to achieve coalescence, but no melting of the base metals occurs and no filler metal is added.

Fusion welding is by far the more important category. It includes (1) arc welding, (2) resistance welding, (3) oxyfuel gas welding, and (4) other fusion-welding processes—ones that cannot be classified as any of the first three types. Fusion-welding processes are discussed in the first four sections of this chapter. Section 23.5 covers solid-state welding. And in the final two sections of the chapter, we examine weld quality and design for welding.

23.1 ARC WELDING

Arc welding (AW) is a fusion-welding process in which coalescence of the metals is achieved by the heat of an electric arc between an electrode and the work. The same basic process is also used in arc cutting. A generic AW process is shown in Figure 23.1. An electric arc is a discharge of electric current across a gap in a circuit. It is sustained by the presence of a thermally ionized column of gas (called a ***plasma***) through which current flows. To initiate the arc in an AW process, the electrode is brought into contact with the work and then quickly separated from it by a short distance. The electric energy from the arc thus formed produces temperatures of 5500°C (10,000°F) or higher, sufficiently hot to melt any metal. A pool of molten metal, consisting of base metal(s) and filler metal (if one is used) is formed near the tip of the electrode. In most arc-welding processes, filler metal is added during the operation to increase the volume and strength of the weld joint. As the electrode is moved along the joint, the molten weld pool solidifies in its wake.

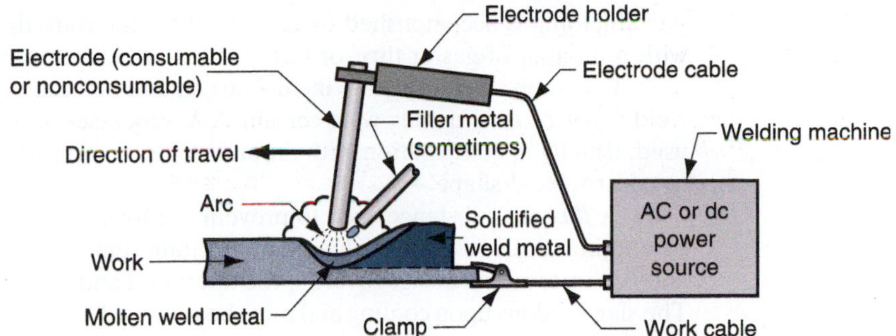

FIGURE 23.1 The basic configuration and electrical circuit of an arc-welding process. (Credit: *Fundamentals of Modern Manufacturing*, 4th Edition by Mikell P. Groover, 2010. Reprinted with permission of John Wiley & Sons, Inc.)

Movement of the electrode relative to the work is accomplished by either a human welder (manual welding) or by mechanical means (i.e., machine welding, automatic welding, or robotic welding). One of the troublesome aspects of manual arc welding is that the quality of the weld joint depends on the skill and work ethic of the human welder. Productivity is also an issue. It is often measured as ***arc time*** (also called ***arc-on time***)—the proportion of hours worked that arc welding is being accomplished:

$$\text{Arc time} = (\text{time arc is on})/(\text{hours worked}) \tag{23.1}$$

This definition can be applied to an individual welder or to a mechanized workstation. For manual welding, arc time is usually around 20%. Frequent rest periods are needed by the welder to overcome fatigue in manual arc welding, which requires hand-eye coordination under stressful conditions. Arc time increases to about 50% (more or less, depending on the operation) for machine, automatic, and robotic welding.

23.1.1 GENERAL TECHNOLOGY OF ARC WELDING

Before describing the individual AW processes, it is instructional to examine some of the general technical issues that apply to these processes.

Electrodes Electrodes used in AW processes are classified as consumable or nonconsumable. Consumable electrodes provide the source of filler metal in arc welding. These electrodes are available in two principal forms: rods (also called sticks) and wire. Welding rods are typically 450 mm (18 in) long or less. The problem with consumable welding rods, at least in production welding operations, is that they must be changed periodically, reducing arc time of the welder. Consumable weld wire has the advantage that it can be continuously fed into the weld pool from spools containing long lengths of wire, thus avoiding the frequent interruptions that occur when using welding sticks. In both rod and wire forms, the electrode is consumed by the arc during the welding process and added to the weld joint as filler metal.

Nonconsumable electrodes are made of tungsten (or carbon, rarely), which resists melting by the arc. Despite its name, a nonconsumable electrode is gradually depleted during the welding process (vaporization is the principal mechanism), analogous to the gradual wearing of a cutting tool in a machining operation. For AW processes that utilize nonconsumable electrodes, any filler metal used in the operation must be supplied by means of a separate wire that is fed into the weld pool.

Arc Shielding At the high temperatures in arc welding, the metals being joined are chemically reactive to oxygen, nitrogen, and hydrogen in the air. The mechanical properties of the weld joint can be seriously degraded by these reactions. Thus, some means to shield the arc from the surrounding air is provided in nearly all AW processes.

Arc shielding is accomplished by covering the electrode tip, arc, and molten weld pool with a blanket of gas or flux, or both, which inhibit exposure of the weld metal to air.

Common shielding gases include argon and helium, both of which are inert. In the welding of ferrous metals with certain AW processes, oxygen and carbon dioxide are used, usually in combination with Ar and/or He, to produce an oxidizing atmosphere or to control weld shape.

A ***flux*** is a substance used to prevent the formation of oxides and other unwanted contaminants, or to dissolve them and facilitate removal. During welding, the flux melts and becomes a liquid slag, covering the operation and protecting the molten weld metal. The slag hardens upon cooling and must be removed later by chipping or brushing. Flux is usually formulated to serve several additional functions: (1) provide a protective atmosphere for welding, (2) stabilize the arc, and (3) reduce spattering.

The method of flux application differs for each process. The delivery techniques include (1) pouring granular flux onto the welding operation, (2) using a stick electrode coated with flux material in which the coating melts during welding to cover the operation, and (3) using tubular electrodes in which flux is contained in the core and released as the electrode is consumed. These techniques are discussed further in our descriptions of the individual AW processes.

Power Source in Arc Welding Both direct current (DC) and alternating current (AC) are used in arc welding. AC machines are less expensive to purchase and operate, but are generally restricted to welding of ferrous metals. DC equipment can be used on all metals with good results and is generally noted for better arc control.

In all arc-welding processes, power to drive the operation is the product of the current I passing through the arc and the voltage E across it. This power is converted into heat, but not all of the heat is transferred to the surface of the work. Convection, conduction, radiation, and spatter account for losses that reduce the amount of usable heat. The effect of the losses is expressed by the heat-transfer factor f_1 (Section 22.3). Some representative values of f_1 for several AW processes are given in Table 23.1. Heat-transfer factors are greater for AW processes that use consumable electrodes because most of the heat consumed in melting the electrode is subsequently transferred to the work as molten metal. The process with the lowest f_1 value in Table 23.1 is gas tungsten arc welding, which uses a nonconsumable electrode. Melting factor f_2 (Section 22.3) further reduces the available heat for welding. The resulting power balance in arc welding is defined by:

$$R_{Hw} = f_1 f_2 I E = U_m A_w v \tag{23.2}$$

where E = voltage, V; I = current, A; and the other terms were defined in Section 22.3. The units of R_{Hw} are Watts (current multiplied by voltage), which equal J/s. This can be converted to Btu/sec by recalling that 1 Btu = 1055 J, and thus 1 Btu/sec = 1055 W.

TABLE 23.1 Heat-transfer factors for several arc-welding processes.[a]

Arc-Welding Process	Typical Heat-Transfer Factor f_1
Shielded metal arc welding	0.9
Gas metal arc welding	0.9
Flux-cored arc welding	0.9
Submerged arc welding	0.95
Gas tungsten arc welding	0.7

Compiled from [1].

[a]The arc-welding processes are described in Sections 23.1.2 and 23.1.3.

Example 23.1 Power in Arc Welding

A gas tungsten arc-welding operation is performed at a current of 300 A and voltage of 20 V. The melting factor $f_2 = 0.5$, and the unit melting energy for the metal $U_m = 10$ J/mm^3. Determine (a) power in the operation, (b) rate of heat generation at the weld, and (c) volume rate of metal welded.

Solution: (a) The power in this arc-welding operation is:

$$P = IE = (300\ \text{A})(20\ \text{V}) = 6000\text{W}$$

(b) From Table 23.1, the heat-transfer factor $f_1 = 0.7$. The rate of heat used for welding is given by:

$$R_{Hw} = f_1 f_2 I\,E = (0.7)(0.5)(6000) = 2100\text{W} = 2100\ \text{J/s}$$

(c) The volume rate of metal welded is:

$$R_{VW} = (2100\ \text{J/s})/(10\ \text{J/mm}^3) = 210\ \text{mm}^3/\text{s}$$

■

23.1.2 AW PROCESSES—CONSUMABLE ELECTRODES

A number of important arc-welding processes use consumable electrodes. These are discussed in this section. Symbols for the welding processes are those used by the American Welding Society.

Shielded Metal Arc Welding Shielded metal arc welding (SMAW) is an AW process that uses a consumable electrode consisting of a filler metal rod coated with chemicals that provide flux and shielding. The process is illustrated in Figures 23.2 and 23.3. The welding stick (SMAW is sometimes called ***stick welding***) is typically 225 to 450 mm (9 to

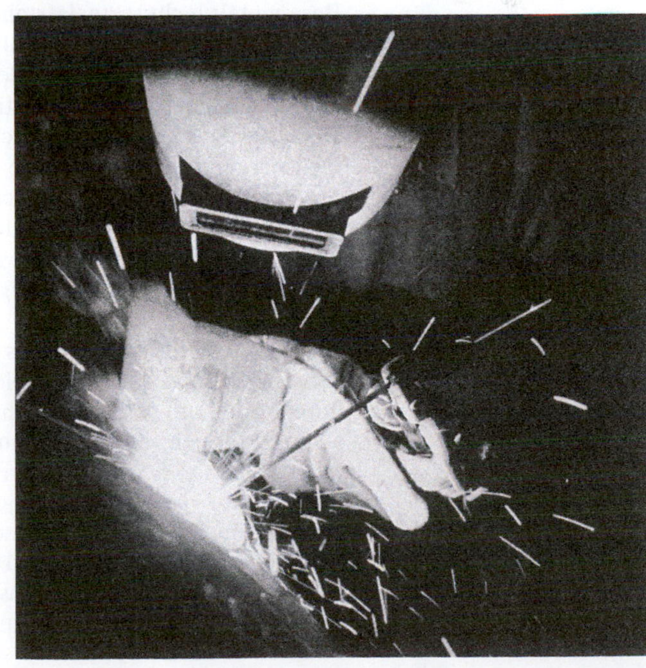

FIGURE 23.2 Shielded metal arc welding (stick welding) performed by a (human) welder. Photo courtesy of Hobart Brothers Company. (Credit: *Fundamentals of Modern Manufacturing*, 4th Edition by Mikell P. Groover, 2010. Reprinted with permission of John Wiley & Sons, Inc.)

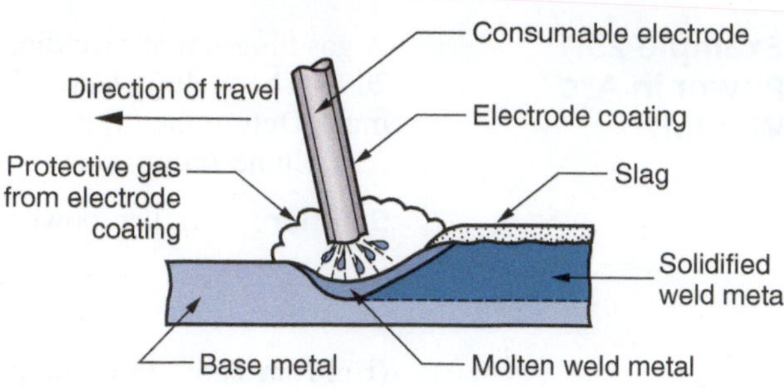

FIGURE 23.3 Shielded metal arc welding (SMAW). (Credit: *Fundamentals of Modern Manufacturing*, 4th Edition by Mikell P. Groover, 2010. Reprinted with permission of John Wiley & Sons, Inc.)

18 in) long and 2.5 to 9.5 mm (3/32 to 3/8 in) in diameter. The filler metal used in the rod must be compatible with the metal to be welded, the composition usually being very close to that of the base metal. The coating consists of powdered cellulose (i.e., cotton and wood powders) mixed with oxides, carbonates, and other ingredients, held together by a silicate binder. Metal powders are also sometimes included in the coating to increase the amount of filler metal and to add alloying elements. The heat of the welding process melts the coating to provide a protective atmosphere and slag for the welding operation. It also helps to stabilize the arc and regulate the rate at which the electrode melts.

During operation the bare metal end of the welding stick (opposite the welding tip) is clamped in an electrode holder that is connected to the power source. The holder has an insulated handle so that it can be held and manipulated by a human welder. Currents typically used in SMAW range between 30 and 300 A at voltages from 15 to 45 V. Selection of the proper power parameters depends on the metals being welded, electrode type and length, and depth of weld penetration required. Power supply, connecting cables, and electrode holder can be bought for a few thousand dollars.

Shielded metal arc welding is usually performed manually. Common applications include construction, pipelines, machinery structures, shipbuilding, job shop fabrication, and repair work. It is preferred over oxyfuel welding for thicker sections—above 5 mm (3/16 in)—because of its higher power density. The equipment is portable and low cost, making SMAW highly versatile and probably the most widely used of the AW processes. Base metals include steels, stainless steels, cast irons, and certain nonferrous alloys. It is not used or seldom used for aluminum and its alloys, copper alloys, and titanium.

A disadvantage of shielded metal arc welding as a production operation is the use of the consumable electrode stick. As the sticks are used up, they must periodically be changed. This reduces the arc time with this welding process. Another limitation is the current level that can be used. Because the electrode length varies during the operation and this length affects the resistance heating of the electrode, current levels must be maintained within a safe range or the coating will overheat and melt prematurely when starting a new welding stick. Some of the other AW processes overcome the limitations of welding stick length in SMAW by using a continuously fed wire electrode.

Gas Metal Arc Welding Gas metal arc welding (GMAW) is an AW process in which the electrode is a consumable bare metal wire, and shielding is accomplished by flooding the arc with a gas. The bare wire is fed continuously and automatically from a spool through the welding gun, as illustrated in Figure 23.4. Wire diameters ranging from 0.8 mm to 6.5 mm (1/32 in to 1/4 in) are used in GMAW, the size depending on the thickness of the parts being joined and the desired deposition rate. Gases used for shielding include inert gases such as argon and helium, and active gases such as carbon dioxide. Selection of gases (and mixtures of gases) depends on the metal being welded, as well as other factors. Inert gases are used for welding aluminum alloys and stainless steels, while CO_2 is commonly used for welding low and medium carbon steels. The combination of bare electrode wire and shielding gases eliminates the slag covering on the

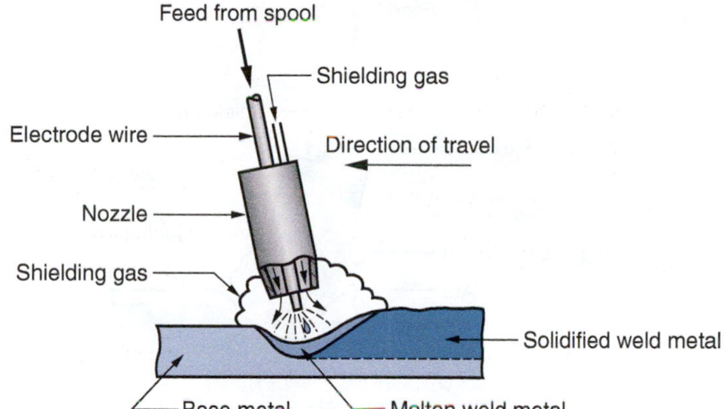

FIGURE 23.4 Gas metal arc welding (GMAW). (Credit: *Fundamentals of Modern Manufacturing*, 4th Edition by Mikell P. Groover, 2010. Reprinted with permission of John Wiley & Sons, Inc.)

weld bead and thus precludes the need for manual grinding and cleaning of the slag. The GMAW process is therefore ideal for making multiple welding passes on the same joint.

The various metals on which GMAW is used and the variations of the process itself have given rise to a variety of names for gas metal arc welding. When the process was first introduced in the late 1940s, it was applied to the welding of aluminum using inert gas (argon) for arc shielding. The name applied to this process was ***MIG welding*** (for ***m***etal ***i***nert ***g***as welding). When the same welding process was applied to steel, it was found that inert gases were expensive and CO_2 was used as a substitute. Hence, the term ***CO_2 welding*** was applied. Refinements in GMAW for steel welding have led to the use of gas mixtures, including CO_2 and argon, and even oxygen and argon.

GMAW is widely used in fabrication operations in factories for welding a variety of ferrous and nonferrous metals. Because it uses continuous weld wire rather than welding sticks, it has a significant advantage over SMAW in terms of arc time when performed manually. For the same reason, it also lends itself to automation of arc welding. The electrode stubs remaining after stick welding also waste filler metal, so the utilization of electrode material is higher with GMAW. Other features of GMAW include higher deposition rates than SMAW and good versatility.

Flux-Cored Arc Welding This arc-welding process was developed in the early 1950s as an adaptation of shielded metal arc welding to overcome the limitations imposed by the use of stick electrodes. Flux-cored arc welding (FCAW) is an arc-welding process in which the electrode is a continuous consumable tubing that contains flux and other ingredients in its core. Other ingredients may include deoxidizers and alloying elements. The tubular flux-cored "wire" is flexible and can therefore be supplied in the form of coils to be continuously fed through the arc-welding gun. There are two versions of FCAW: (1) self-shielded and (2) gas shielded. In the first version of FCAW to be developed, arc shielding was provided by a flux core, thus leading to the name ***self-shielded flux-cored arc welding***. The core in this form of FCAW includes not only fluxes but also ingredients that generate shielding gases for protecting the arc. The second version of FCAW, developed primarily for welding steels, obtains arc shielding from externally supplied gases, similar to gas metal arc welding. This version is called ***gas-shielded flux-cored arc welding***. Because it utilizes an electrode containing its own flux together with separate shielding gases, it might be considered a hybrid of SMAW and GMAW. Shielding gases typically employed are carbon dioxide for mild steels or mixtures of argon and carbon dioxide for stainless steels. Figure 23.5 illustrates the FCAW process, with the gas (optional) distinguishing between the two types.

FCAW has advantages similar to GMAW, due to continuous feeding of the electrode. It is used primarily for welding steels and stainless steels over a wide stock

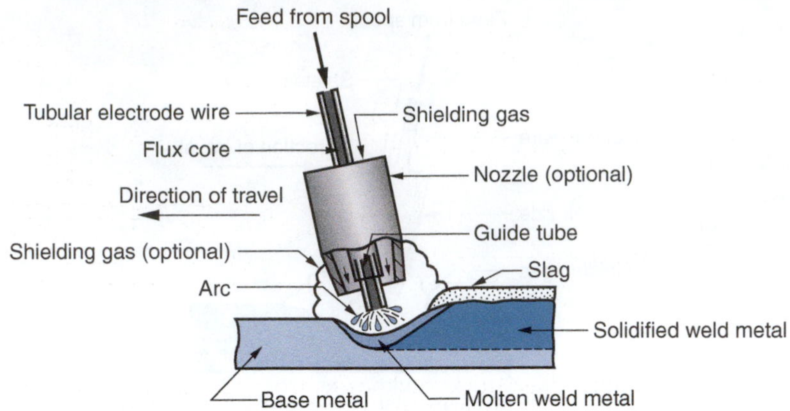

FIGURE 23.5 Flux-cored arc welding. The presence or absence of externally supplied shielding gas distinguishes the two types: (1) self-shielded, in which the core provides the ingredients for shielding; and (2) gas shielded, in which external shielding gases are supplied. (Credit: *Fundamentals of Modern Manufacturing*, 4th Edition by Mikell P. Groover, 2010. Reprinted with permission of John Wiley & Sons, Inc.)

thickness range. It is noted for its capability to produce very-high-quality weld joints that are smooth and uniform.

Electrogas Welding Electrogas welding (EGW) is an AW process that uses a continuous consumable electrode (either flux-cored wire or bare wire with externally supplied shielding gases) and molding shoes to contain the molten metal. The process is primarily applied to vertical butt welding, as pictured in Figure 23.6. When the flux-cored electrode wire is employed, no external gases are supplied, and the process can be considered a special application of self-shielded FCAW. When a bare electrode wire is used with shielding gases from an external source, it is considered a special case of GMAW. The molding shoes are water cooled to prevent their being added to the weld pool. Together with the edges of the parts being welded, the shoes form a container, almost like a mold cavity, into which the molten metal from the electrode and base parts is gradually added. The process is performed automatically, with a moving weld head travelling vertically upward to fill the cavity in a single pass.

Principal applications of electrogas welding are steels (low- and medium-carbon, low-alloy, and certain stainless steels) in the construction of large storage tanks and in

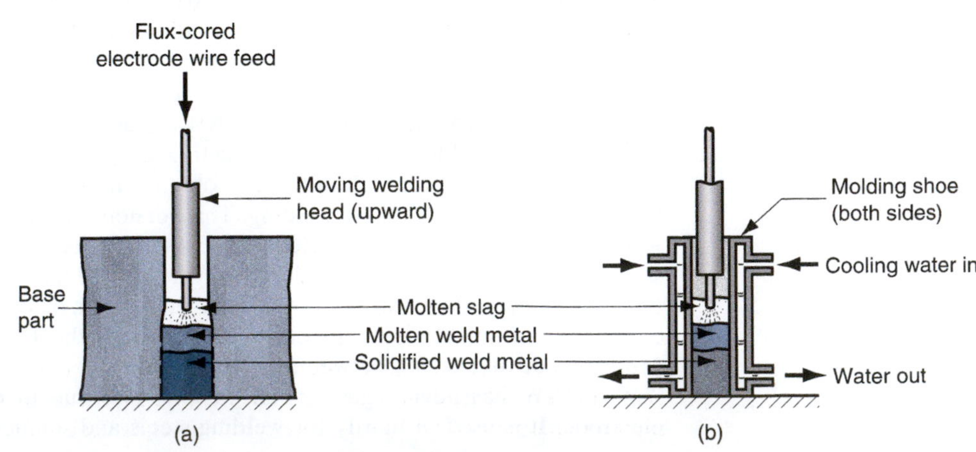

FIGURE 23.6 Electrogas welding using flux-cored electrode wire: (a) front view with molding shoe removed for clarity, and (b) side view showing molding shoes on both sides. (Credit: *Fundamentals of Modern Manufacturing*, 4th Edition by Mikell P. Groover, 2010. Reprinted with permission of John Wiley & Sons, Inc.)

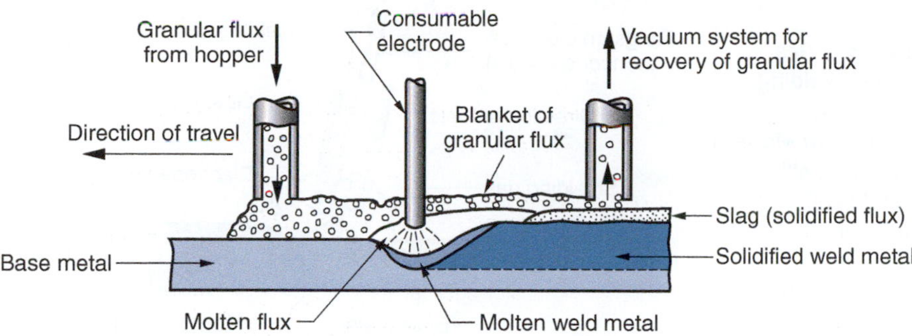

FIGURE 23.7 Submerged arc welding (SAW). (Credit: *Fundamentals of Modern Manufacturing*, 4th Edition by Mikell P. Groover, 2010. Reprinted with permission of John Wiley & Sons, Inc.)

shipbuilding. Stock thicknesses from 12 mm to 75 mm (0.5 in to 3.0 in) are within the capacity of EGW. In addition to butt welding, it can also be used for fillet and groove welds, always in a vertical orientation. Specially designed molding shoes must sometimes be fabricated for the joint shapes involved.

Submerged Arc Welding This process, developed during the 1930s, was one of the first AW processes to be automated. Submerged arc welding (SAW) is an arc-welding process that uses a continuous, consumable bare wire electrode, and arc shielding is provided by a cover of granular flux. The electrode wire is fed automatically from a coil into the arc. The flux is introduced into the joint slightly ahead of the weld arc by gravity from a hopper, as shown in Figure 23.7. The blanket of granular flux completely submerges the welding operation, preventing sparks, spatter, and radiation that are so hazardous in other AW processes. Thus, the welding operator in SAW need not wear the somewhat cumbersome face shield required in the other operations (safety glasses and protective gloves, of course, are required). The portion of the flux closest to the arc is melted, mixing with the molten weld metal to remove impurities and then solidifying on top of the weld joint to form a glass-like slag. The slag and unfused flux granules on top provide good protection from the atmosphere and good thermal insulation for the weld area, resulting in relatively slow cooling and a high-quality weld joint, noted for toughness and ductility. As depicted in our sketch, the unfused flux remaining after welding can be recovered and reused. The solid slag covering the weld must be chipped away, usually by manual means.

Submerged arc welding is widely used in steel fabrication for structural shapes (e.g., welded I-beams); longitudinal and circumferential seams for large diameter pipes, tanks, and pressure vessels; and welded components for heavy machinery. In these kinds of applications, steel plates of 25-mm (1.0-in) thickness and heavier are routinely welded by this process. Low-carbon, low-alloy, and stainless steels can be readily welded by SAW; but not high-carbon steels, tool steels, and most nonferrous metals. Because of the gravity feed of the granular flux, the parts must always be in a horizontal orientation, and a backup plate is often required beneath the joint during the welding operation.

23.1.3 AW PROCESSES—NONCONSUMABLE ELECTRODES

The AW processes discussed above use consumable electrodes. Gas tungsten arc welding and plasma arc welding use nonconsumable electrodes.

Gas Tungsten Arc Welding Gas tungsten arc welding (GTAW) is an AW process that uses a nonconsumable tungsten electrode and an inert gas for arc shielding. The term ***TIG welding*** (*t*ungsten *i*nert *g*as welding) is often applied to this process (in Europe, ***WIG welding*** is the term—the chemical symbol for tungsten is W, for Wolfram). GTAW can be implemented with or without a filler metal. Figure 23.8 illustrates the latter case.

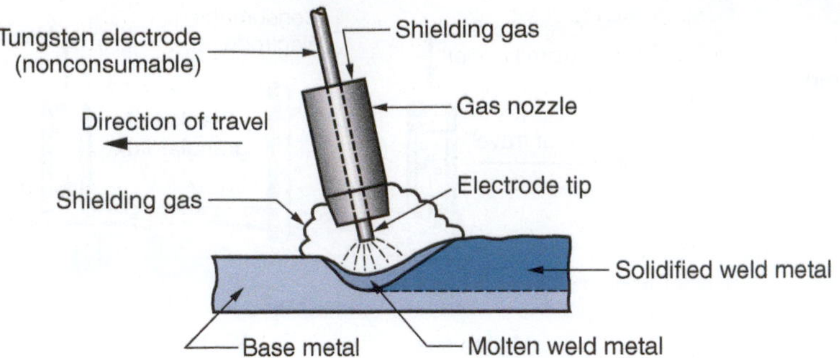

FIGURE 23.8 Gas tungsten arc welding (GTAW). (Credit: *Fundamentals of Modern Manufacturing,* 4th Edition by Mikell P. Groover, 2010. Reprinted with permission of John Wiley & Sons, Inc.)

When a filler metal is used, it is added to the weld pool from a separate rod or wire, being melted by the heat of the arc rather than transferred across the arc as in the consumable electrode AW processes. Tungsten is a good electrode material due to its high melting point of 3410°C (6170°F). Typical shielding gases include argon, helium, or a mixture of these gas elements.

GTAW is applicable to nearly all metals in a wide range of stock thicknesses. It can also be used for joining various combinations of dissimilar metals. Its most common applications are for aluminum and stainless steel. Cast irons, wrought irons, and of course tungsten are difficult to weld by GTAW. In steel welding applications, GTAW is generally slower and more costly than the consumable electrode AW processes, except when thin sections are involved and very-high-quality welds are required. When thin sheets are TIG welded to close tolerances, filler metal is usually not added. The process can be performed manually or by machine and automated methods for all joint types. Advantages of GTAW in the applications to which it is suited include high-quality welds, no weld spatter because no filler metal is transferred across the arc, and little or no postweld cleaning because no flux is used.

Plasma Arc Welding Plasma arc welding (PAW) is a special form of gas tungsten arc welding in which a constricted plasma arc is directed at the weld area. In PAW, a tungsten electrode is contained in a specially designed nozzle that focuses a high-velocity stream of inert gas (e.g., argon or argon–hydrogen mixtures) into the region of the arc to form a high-velocity, intensely hot plasma arc stream, as in Figure 23.9. Argon, argon–hydrogen, and helium are also used as the arc-shielding gases.

Temperatures in plasma arc welding reach 17,000°C (30,000°F) or greater, hot enough to melt any known metal. The reason why temperatures are so high in PAW (significantly higher than those in GTAW) derives from the constriction of the arc. Although the typical power levels used in PAW are below those used in GTAW, the power is highly concentrated to produce a plasma jet of small diameter and very high power density.

Plasma arc welding was introduced around 1960 but was slow to catch on. In recent years its use is increasing as a substitute for GTAW in applications such as automobile

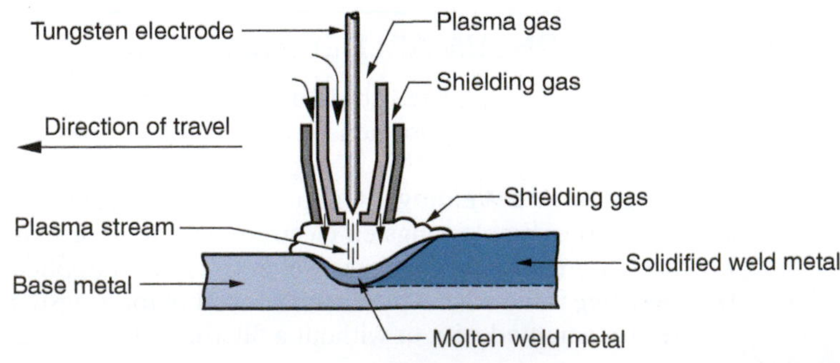

FIGURE 23.9 Plasma arc welding (PAW). (Credit: *Fundamentals of Modern Manufacturing,* 4th Edition by Mikell P. Groover, 2010. Reprinted with permission of John Wiley & Sons, Inc.)

subassemblies, metal cabinets, door and window frames, and home appliances. Owing to the special features of PAW, its advantages in these applications include good arc stability, better penetration control than most other AW processes, high travel speeds, and excellent weld quality. The process can be used to weld almost any metal, including tungsten. Difficult-to-weld metals with PAW include bronze, cast irons, lead, and magnesium. Other limitations include high equipment cost and larger torch size than other AW operations, which tends to restrict access in some joint configurations.

23.2 RESISTANCE WELDING

Resistance welding (RW) is a group of fusion-welding processes that uses a combination of heat and pressure to accomplish coalescence, the heat being generated by electrical resistance to current flow at the junction to be welded. The principal components in resistance welding are shown in Figure 23.10 for a resistance spot-welding operation, the most widely used process in the group. The components include workparts to be welded (usually sheet metal parts), two opposing electrodes, a means of applying pressure to squeeze the parts between the electrodes, and an AC power supply from which a controlled current can be applied. The operation results in a fused zone between the two parts, called a ***weld nugget*** in spot welding.

By comparison to arc welding, resistance welding uses no shielding gases, flux, or filler metal; and the electrodes that conduct electrical power to the process are non-consumable. RW is classified as fusion welding because the applied heat causes melting of the faying surfaces.

23.2.1 POWER SOURCE IN RESISTANCE WELDING

The heat energy supplied to the welding operation depends on current flow, resistance of the circuit, and length of time the current is applied. This can be expressed by the equation

$$H = I^2Rt \tag{23.3}$$

where H = heat generated, J (to convert to Btu divide by 1055); I = current, A; R = electrical resistance, Ω; and t = time, s.

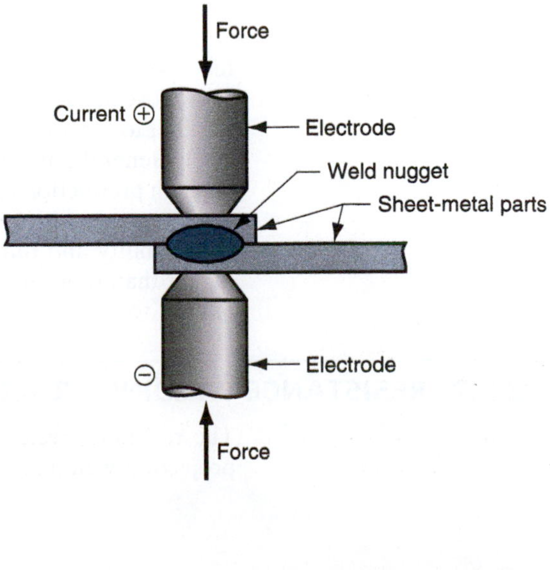

FIGURE 23.10 Resistance welding (RW), showing the components in spot welding, the predominant process in the RW group. (Credit: *Fundamentals of Modern Manufacturing*, 4th Edition by Mikell P. Groover, 2010. Reprinted with permission of John Wiley & Sons, Inc.)

The current used in resistance welding operations is very high (5000 to 20,000 A, typically), although voltage is relatively low (usually below 10 V). The duration t of the current is short in most processes, perhaps lasting 0.1 to 0.4 seconds in a typical spot-welding operation.

The reason why such a high current is used in RW is because (1) the squared term in Eq. (23.3) amplifies the effect of current, and (2) the resistance is very low (around 0.0001 Ω). Resistance in the welding circuit is the sum of (1) resistance of the electrodes, (2) resistances of the workparts, (3) contact resistances between electrodes and workparts, and (4) contact resistance of the faying surfaces. Thus, heat is generated in all of these regions of electrical resistance. The ideal situation is for the faying surfaces to be the largest resistance in the sum, because this is the desired location of the weld. The resistance of the electrodes is minimized by using metals with very low resistivities, such as copper. Also, the electrodes are often water cooled to dissipate the heat that is generated there. The workpart resistances are a function of the resistivities of the base metals and the part thicknesses. The contact resistances between the electrodes and the parts are determined by the contact areas (i.e., size and shape of the electrode) and the condition of the surfaces (e.g., cleanliness of the work surfaces and scale on the electrode). Finally, the resistance at the faying surfaces depends on surface finish, cleanliness, contact area, and pressure. No paint, oil, dirt, or other contaminants should be present to separate the contacting surfaces.

Example 23.2 Resistance Welding

A resistance spot-welding operation is performed on two pieces of 1.5-mm-thick sheet steel using 12,000 A for a 0.20 s duration. The electrodes are 6 mm in diameter at the contacting surfaces. Resistance is assumed to be 0.0001 Ω, and the resulting weld nugget is 6 mm in diameter and 2.5 mm thick. The unit melting energy for the metal $U_m = 12.0$ J/mm^3. What portion of the heat generated was used to form the weld nugget, and what portion was dissipated into the work metal, electrodes, and surrounding air?

Solution: The heat generated in the operation is given by Eq. (23.3) as:

$$H = (12,000)^2(0.0001)(0.2) = 2880 \text{ J}$$

The volume of the weld nugget (assumed disc-shaped) is

$$V = 2.5\frac{\pi(6)^2}{4} = 70.7 \text{ mm}^3.$$

The heat required to melt this volume of metal is $H_w = 70.7(12.0) = 848$ J. The remaining heat, $2880 - 848 = 2032$ J (70.6% of the total), is lost into the work metal, electrodes, and surrounding air. In effect, this loss represents the combined effect of the heat-transfer factor f_1 and the melting factor f_2 (Section 22.3). ■

Success in resistance welding depends on pressure as well as heat. The principal functions of pressure in RW are to (1) force contact between the electrodes and the workparts and between the two work surfaces current is applied and (2) press the faying surfaces together to accomplish coalescence when the proper welding temperature has been reached.

General advantages of resistance welding include: (1) no filler metal is required, (2) high production rates are possible, (3) lends itself to mechanization and automation, (4) operator skill level is lower than that required for arc welding, and (5) good repeatability and reliability. Drawbacks are (1) equipment cost is high—usually much higher than most arc-welding operations, and (2) types of joints that can be welded are limited to lap joints for most RW processes.

23.2.2 RESISTANCE-WELDING PROCESSES

The resistance-welding processes of most commercial importance are spot, seam, and projection welding.

Resistance Spot Welding Resistance spot welding (RSW) is by far the predominant process in this group. It is widely used in mass production of automobiles, appliances, metal furniture, and other products made of sheet metal. If one considers that a typical car body has approximately 10,000 individual spot welds, and that the annual production of automobiles throughout the world is measured in tens of millions of units, the economic importance of resistance spot welding can be appreciated.

Resistance spot welding is an RW process in which fusion of the faying surfaces of a lap joint is achieved at one location by opposing electrodes. The process is used to join sheet-metal parts of thickness 3 mm (0.125 in) or less, using a series of spot welds, in situations where an airtight assembly is not required. The size and shape of the weld spot is determined by the electrode tip, the most common electrode shape being round, but hexagonal, square, and other shapes are also used. The resulting weld nugget is typically 5 mm to 10 mm (0.2 in to 0.4 in) in diameter, with a heat-affected zone extending slightly beyond the nugget into the base metals. If the weld is made properly, its strength will be comparable to that of the surrounding metal. The steps in a spot welding cycle are depicted in Figure 23.11.

Materials used for RSW electrodes consist of two main groups: (1) copper-based alloys and (2) refractory metal compositions such as copper and tungsten combinations. The second group is noted for superior wear resistance. As in most manufacturing processes, the tooling in spot welding gradually wears out as it is used. Whenever practical, the electrodes are designed with internal passageways for water cooling.

Because of its widespread industrial use, various machines and methods are available to perform spot-welding operations. The equipment includes rocker-arm and press-type spot-welding machines, and portable spot-welding guns. ***Rocker-arm spot welders***, shown in Figure 23.12, have a stationary lower electrode and a movable

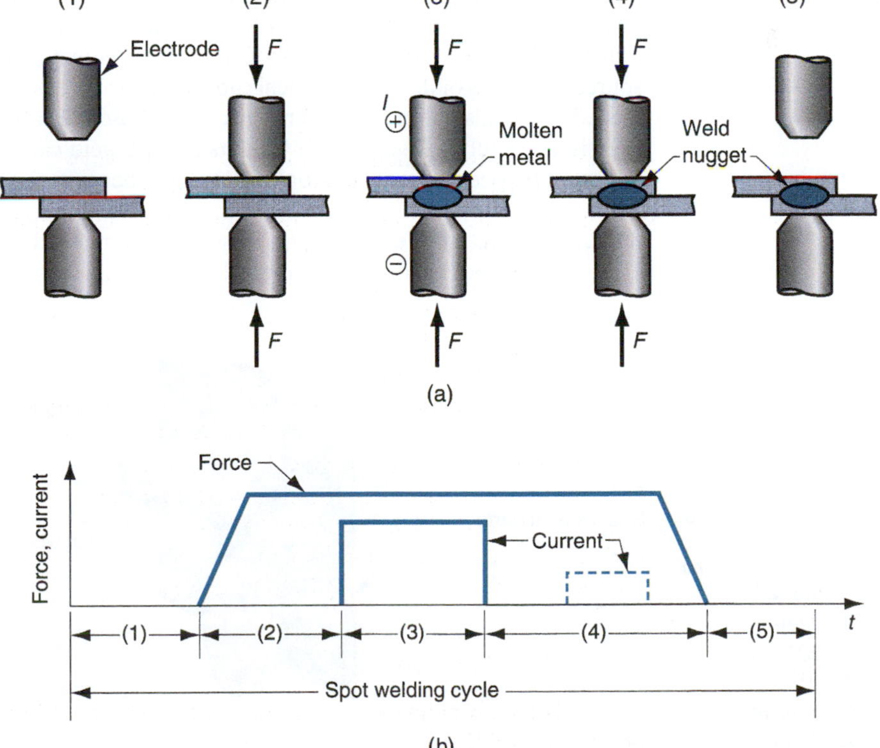

FIGURE 23.11 (a) Steps in a spot-welding cycle, and (b) plot of squeezing force and current during cycle. The sequence is: (1) parts inserted between open electrodes, (2) electrodes close and force is applied, (3) weld time—current is switched on, (4) current is turned off but force is maintained or increased (a reduced current is sometimes applied near the end of this step for stress relief in the weld region), and (5) electrodes are opened, and the welded assembly is removed. (Credit: *Fundamentals of Modern Manufacturing*, 4th Edition by Mikell P. Groover, 2010. Reprinted with permission of John Wiley & Sons, Inc.)

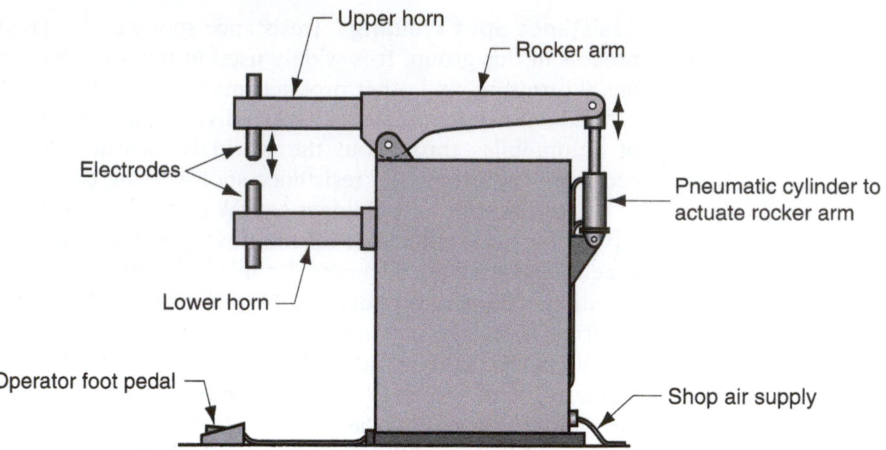

FIGURE 23.12 Rocker-arm spot-welding machine. (Credit: *Fundamentals of Modern Manufacturing*, 4th Edition by Mikell P. Groover, 2010. Reprinted with permission of John Wiley & Sons, Inc.)

upper electrode that can be raised and lowered for loading and unloading the work. The upper electrode is mounted on a rocker arm (hence, the name) whose movement is controlled by a foot pedal operated by the worker. Modern machines can be programmed to control force and current during the weld cycle.

Press-type spot welders are intended for larger work. The upper electrode has a straight-line motion provided by a vertical press that is pneumatically or hydraulically powered. The press action permits larger forces to be applied, and the controls usually permit programming of complex weld cycles.

The previous two machine types are both stationary spot welders, in which the work is brought to the machine. For large, heavy work it is difficult to move and position the part into stationary machines. For these cases, ***portable spot-welding guns*** are available in various sizes and configurations. These devices consist of two opposing electrodes contained in a pincer mechanism. Each unit is lightweight so that it can be held and manipulated by a human worker or an industrial robot. The gun is connected to its own power and control source by means of flexible electrical cables and air hoses. Water cooling for the electrodes, if needed, can also be provided through a water hose. Portable spot-welding guns are widely used in automobile final assembly plants to spot-weld car bodies. Some of these guns are operated by people, but industrial robots have become the preferred technology.

Resistance Seam Welding In resistance seam welding (RSEW), the stick-shaped electrodes in spot welding are replaced by rotating wheels, as shown in Figure 23.13, and

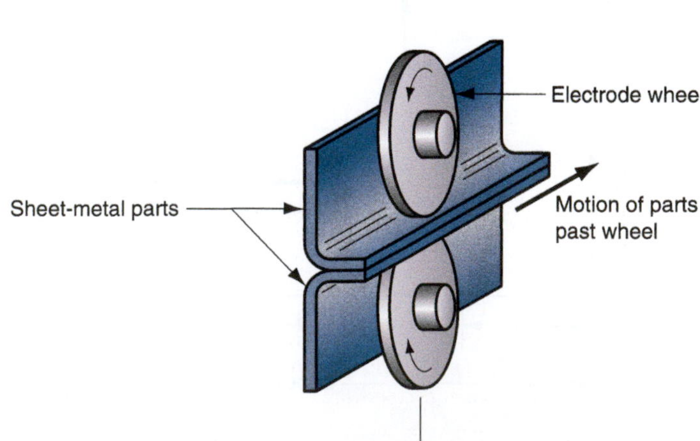

FIGURE 23.13 Resistance seam welding (RSEW). (Credit: *Fundamentals of Modern Manufacturing*, 4th Edition by Mikell P. Groover, 2010. Reprinted with permission of John Wiley & Sons, Inc.)

a series of overlapping spot welds are made along the lap joint. The process is capable of producing air-tight joints, and its industrial applications include the production of gasoline tanks, automobile mufflers, and various other fabricated sheet-metal containers. Technically, RSEW is the same as spot welding, except that the wheel electrodes introduce certain complexities. Because the operation is usually carried out continuously, rather than discretely, the seams should be along a straight or uniformly curved line. Sharp corners and similar discontinuities are difficult to deal with. Also, warping of the parts becomes more of a factor in resistance seam welding, and fixtures are required to hold the work in position and minimize distortion.

The spacing between the weld nuggets in resistance seam welding depends on the motion of the electrode wheels relative to the application of the weld current. In the usual method of operation, called ***continuous motion welding***, the wheel is rotated continuously at a constant velocity, and current is turned on at timing intervals consistent with the desired spacing between spot welds along the seam. Frequency of the current discharges is normally set so that overlapping weld spots are produced. But if the frequency is reduced sufficiently, then there will be spaces between the weld spots, and this method is termed ***roll spot welding***. In another variation, the welding current remains on at a constant level (rather than being pulsed) so that a truly continuous welding seam is produced. These variations are depicted in Figure 23.14.

An alternative to continuous motion welding is ***intermittent motion welding***, in which the electrode wheel is periodically stopped to make the spot weld. The amount of wheel rotation between stops determines the distance between weld spots along the seam, yielding patterns similar to (a) and (b) in Figure 23.14.

Seam-welding machines are similar to press-type spot welders, except that electrode wheels are used rather than rod-shaped electrodes. Cooling of the work and wheels is often necessary in RSEW, and this is accomplished by directing water at the top and underside of the workpart surfaces near the electrode wheels.

Resistance Projection Welding Resistance projection welding (RPW) is an RW process in which coalescence occurs at one or more relatively small contact points on the parts. These contact points are determined by the design of the parts to be joined, and may consist of projections, embossments, or localized intersections of the parts. A typical case in which two sheet-metal parts are welded together is described in Figure 23.15. The part on top has been fabricated with two embossed points to contact the other part at the start of the process. It might be argued that the embossing

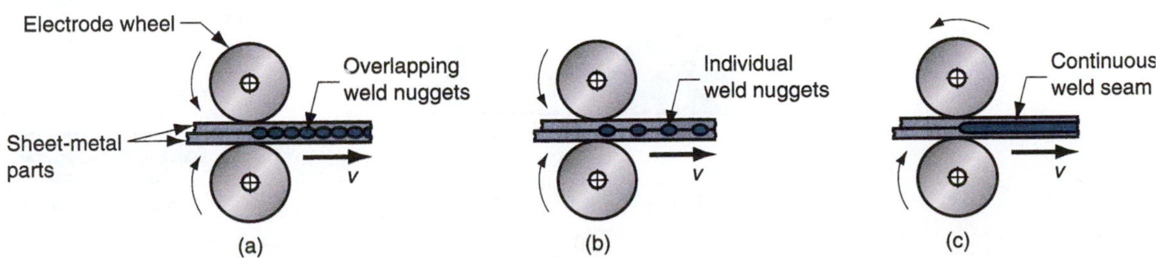

FIGURE 23.14 Different types of seams produced by electrode wheels: (a) conventional resistance seam welding, in which overlapping spots are produced; (b) roll spot welding; and (c) continuous resistance seam. (Credit: *Fundamentals of Modern Manufacturing*, 4th Edition by Mikell P. Groover, 2010. Reprinted with permission of John Wiley & Sons, Inc.)

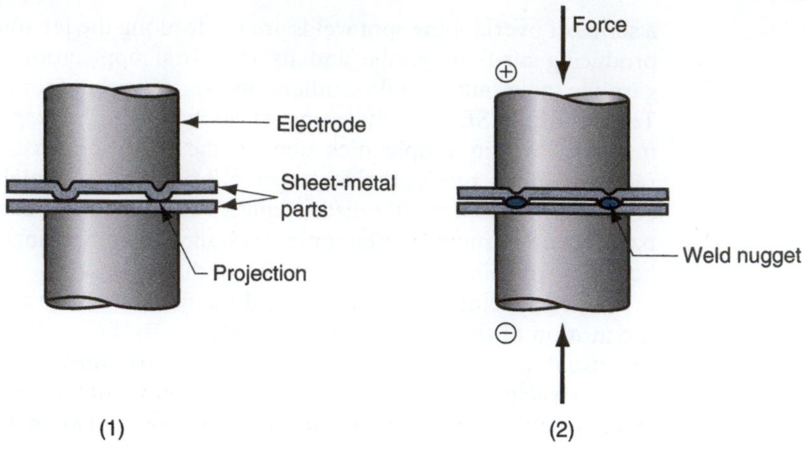

FIGURE 23.15 Resistance projection welding (RPW): (1) at start of operation, contact between parts is at projections; and (2) when current is applied, weld nuggets similar to those in spot welding are formed at the projections. (Credit: *Fundamentals of Modern Manufacturing,* 4th Edition by Mikell P. Groover, 2010. Reprinted with permission of John Wiley & Sons, Inc.)

operation increases the cost of the part, but this increase may be more than offset by savings in welding cost.

There are variations of resistance projection welding, two of which are shown in Figure 23.16. In one variation, fasteners with machined or formed projections can be permanently joined to sheet or plate by RPW, facilitating subsequent assembly operations. Another variation, called ***cross-wire welding***, is used to fabricate welded wire products such as wire fence, shopping carts, and stove grills. In this process, the contacting surfaces of the round wires serve as the projections to localize the resistance heat for welding.

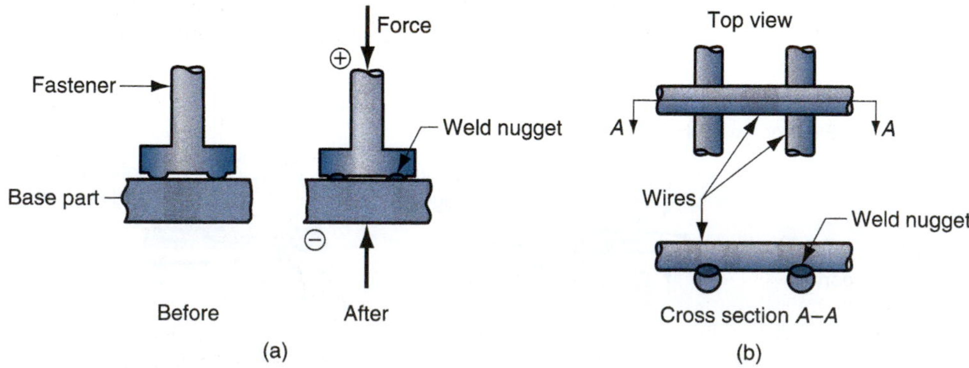

FIGURE 23.16 Two variations of resistance projection welding: (a) welding of a machined or formed fastener onto a sheet-metal part; and (b) cross-wire welding. (Credit: *Fundamentals of Modern Manufacturing,* 4th Edition by Mikell P. Groover, 2010. Reprinted with permission of John Wiley & Sons, Inc.)

23.3 OXYFUEL GAS WELDING

Oxyfuel gas welding (OFW) is the term used to describe the group of FW operations that burn various fuels mixed with oxygen to perform welding. The OFW processes employ several types of gases, which is the primary distinction among the members of this group. Oxyfuel gas is also commonly used in cutting torches to cut and separate metal plates and other parts. The most important OFW process is oxyacetylene welding.

23.3.1 OXYACETYLENE WELDING

Oxyacetylene welding (OAW) is a fusion-welding process performed by a high-temperature flame from combustion of acetylene and oxygen. The flame is directed by a welding torch. A filler metal is sometimes added, and pressure is occasionally applied in OAW between the contacting part surfaces. A typical OAW operation is sketched in Figure 23.17. When filler metal is used, it is typically in the form of rods with diameters ranging from 1.6 mm to 9.5 mm (1/16 in to 3/8 in). Composition of the filler must be similar to that of the base metals. The filler is often coated with a ***flux*** that helps to clean the surfaces and prevent oxidation, thus creating a better weld joint.

Acetylene (C_2H_2) is the most popular fuel among the OFW group because it is capable of higher temperatures than any of the others—up to 3480°C (6300°F). The flame in OAW is produced by the chemical reaction of acetylene and oxygen in two stages. The first stage is defined by the reaction:

$$C_2H_2 + O_2 \rightarrow 2CO + H_2 + \text{heat} \tag{23.4a}$$

the products of which are both combustible, which leads to the second-stage reaction:

$$2CO + H_2 + 1.5O_2 \rightarrow 2CO_2 + H_2O + \text{heat} \tag{23.4b}$$

The two stages of combustion are visible in the oxyacetylene flame emitted from the torch. When the mixture of acetylene and oxygen is in the ratio 1:1, as described in Eq. (23.4), the resulting ***neutral flame*** is shown in Figure 23.18. The first-stage reaction is seen as the inner cone of the flame (which is bright white), while the second-stage reaction is exhibited by the outer envelope (which is nearly colorless but with tinges ranging from blue to orange). The maximum temperature of the flame is reached at the tip of the inner cone; the second-stage temperatures are somewhat below those of the inner cone. During welding, the outer envelope spreads out and covers the work surfaces being joined, thus shielding them from the surrounding atmosphere.

Total heat liberated during the two stages of combustion is 55×10^6 J/m^3 (1470 Btu/ft^3) of acetylene. However, because of the temperature distribution in the flame, the way in

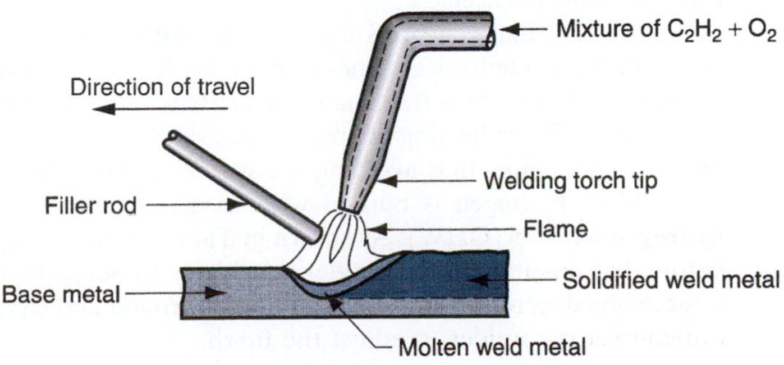

FIGURE 23.17 A typical oxyacetylene welding operation (OAW). (Credit: *Fundamentals of Modern Manufacturing*, 4th Edition by Mikell P. Groover, 2010. Reprinted with permission of John Wiley & Sons, Inc.)

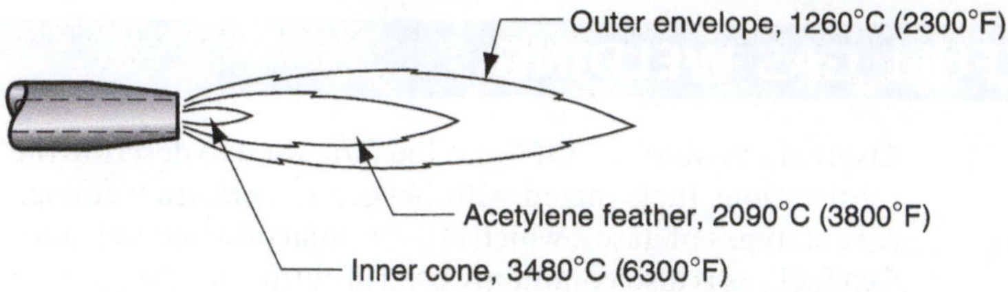

FIGURE 23.18 The neutral flame from an oxyacetylene torch, indicating temperatures achieved. (Credit: *Fundamentals of Modern Manufacturing*, 4th Edition by Mikell P. Groover, 2010. Reprinted with permission of John Wiley & Sons, Inc.)

which the flame spreads over the work surface, and losses to the air, power densities and heat-transfer factors in oxyacetylene welding are relatively low; $f_1 = 0.10$ to 0.30.

The combination of acetylene and oxygen is highly flammable, and the environment in which OAW is performed is therefore hazardous. Some of the dangers relate specifically to the acetylene. Pure C_2H_2 is a colorless, odorless gas. For safety reasons, commercial acetylene is processed to have a characteristic garlic odor. One of the physical limitations of the gas is that it is unstable at pressures much above 1 atm (0.1 MPa or 15 lb/in^2). Accordingly, acetylene storage cylinders are packed with a porous filler material (such as asbestos, balsa wood, and other materials) saturated with acetone (CH_3COCH_3). Acetylene dissolves in liquid acetone; in fact, acetone dissolves about 25 times its own volume of acetylene, thus providing a relatively safe means of storing this welding gas. The welder wears eye and skin protection (goggles, gloves, and protective clothing) as an additional safety precaution, and different screw threads are standard on the acetylene and oxygen cylinders and hoses to avoid accidental connection of the wrong gases. Proper maintenance of the equipment is imperative. OAW equipment is relatively inexpensive and portable. It is therefore an economical, versatile process that is well suited to low-quantity production and repair jobs. It is rarely used to weld sheet and plate stock thicker than 6.4 mm (1/4 in) because of the advantages of arc welding in such applications. Although OAW can be mechanized, it is usually performed manually and is hence dependent on the skill of the welder to produce a high-quality weld joint.

23.3.2 ALTERNATIVE GASES FOR OXYFUEL WELDING

Several members of the OFW group are based on gases other than acetylene. Most of the alternative fuels are listed in Table 23.2, together with their burning temperatures and combustion heats. For comparison, acetylene is included in the list. Although oxyacetylene is the most common OFW fuel, each of the other gases can be used in certain applications—typically limited to welding of sheet metal and metals with low melting temperatures, and brazing (Section 24.1). In addition, some users prefer these alternative gases for safety reasons.

The fuel that competes most closely with acetylene in burning temperature and heating value is methylacetylene-propadiene. It is a fuel developed by the Dow Chemical Company sold under the trade name ***MAPP*** (we are grateful to Dow for the abbreviation). MAPP (C_3H_4) has heating characteristics similar to acetylene and can be stored under pressure as a liquid, thus avoiding the special storage problems associated with C_2H_2.

When hydrogen is burned with oxygen as the fuel, the process is called ***oxy-hydrogen welding*** (OHW). As shown in Table 23.2, the welding temperature in OHW is below that possible in oxyacetylene welding. In addition, the color of the flame is not affected by differences in the mixture of hydrogen and oxygen, and therefore it is more difficult for the welder to adjust the torch.

TABLE 23.2 Gases used in oxyfuel welding and/or cutting, with flame temperatures and heats of combustion.

	Temperature[a]		Heat of Combustion	
Fuel	°C	°F	MJ/m^3	Btu/ft^3
Acetylene (C_2H_2)	3087	5589	54.8	1470
MAPP[b] (C_3H_4)	2927	5301	91.7	2460
Hydrogen (H_2)	2660	4820	12.1	325
Propylene[c] (C_3H_6)	2900	5250	89.4	2400
Propane (C_3H_8)	2526	4579	93.1	2498
Natural gas[d]	2538	4600	37.3	1000

Compiled from [10].

[a]Neutral flame temperatures are compared because this is the flame that would most commonly be used for welding.

[b]MAPP is the commercial abbreviation for methylacetylene-propadiene.

[c]Propylene is used primarily in flame cutting.

[d]Data are based on methane gas (CH_4); natural gas consists of ethane (C_2H_6) as well as methane; flame temperature and heat of combustion vary with composition.

Other fuels used in OFW include propane and natural gas. Propane (C_3H_8) is more closely associated with brazing, soldering, and cutting operations than with welding. Natural gas consists mostly of ethane (C_2H_6) and methane (CH_4). When mixed with oxygen, it achieves a high temperature flame and is becoming more common in small welding shops.

23.4 OTHER FUSION-WELDING PROCESSES

Some fusion-welding processes cannot be classified as arc, resistance, or oxyfuel welding. Each of these other processes uses a unique technology to develop heat for melting; and typically, the applications are unique.

Electron-Beam Welding Electron-beam welding (EBW) is a fusion-welding process in which the heat for welding is produced by a highly focused, high-intensity stream of electrons impinging against the work surface. The equipment is similar to that used for electron-beam machining (Section 19.3.2). The electron-beam gun operates at high voltage to accelerate the electrons (e.g., 10 to 150 kV typical), and beam currents are low (measured in milliamps). The power in EBW is not exceptional, but power density is. High power density is achieved by focusing the electron beam on a very small area of the work surface, so that the power density PD is based on:

$$PD = \frac{f_1 EI}{A} \tag{23.5}$$

where PD = power density, W/mm^2 (W/in^2, which can be converted to $Btu/sec\text{-}in^2$ by dividing by 1055.); f_1 = heat-transfer factor (typical values for EBW range from 0.8 to 0.95 [9]); E = accelerating voltage, V; I = beam current, A; and A = the work surface area on which the electron beam is focused, mm^2 (in^2). Typical weld areas for EBW range from 13×10^{-3} to 2000×10^{-3} mm^2 (20×10^{-6} to 3000×10^{-6} in^2).

The process had its beginnings in the 1950s in the atomic power field. When first developed, welding had to be carried out in a vacuum chamber to minimize the disruption of the electron beam by air molecules. This requirement was, and still is, a serious inconvenience in production, due to the time required to evacuate the chamber prior to welding. The pump-down time, as it is called, can take as long as an hour, depending on the size of the chamber and the level of vacuum required. Today, EBW technology has progressed to where some operations are performed without a vacuum. Three categories can be distinguished: (1) ***high-vacuum welding*** (EBW-HV), in which welding is carried out in the same vacuum as beam generation; (2) ***medium-vacuum welding*** (EBW-MV), in which the operation is performed in a separate chamber where only a partial vacuum is achieved; and (3) ***nonvacuum welding*** (EBW-NV), in which welding is accomplished at or near atmospheric pressure. The pump-down time during workpart loading and unloading is reduced in medium-vacuum EBW and minimized in nonvacuum EBW, but there is a price paid for this advantage. In the latter two operations, the equipment must include one or more vacuum dividers (very small orifices that impede air flow but permit passage of the electron beam) to separate the beam generator (which requires a high vacuum) from the work chamber. Also, in nonvacuum EBW, the work must be located close to the orifice of the electron beam gun, approximately 13 mm (0.5 in) or less. Finally, the lower vacuum processes cannot achieve the high weld qualities and depth-to-width ratios accomplished by EBW-HV.

EBW can be used to weld any metals that can be arc welded, as well as certain refractory and difficult-to-weld metals that are not suited to AW. Work sizes range from thin foil to thick plate. EBW is applied mostly in the automotive, aerospace, and nuclear industries. In the automotive industry, electron-beam welded assemblies include aluminum manifolds, steel torque converters, catalytic converters, and transmission components. In these and other applications, electron-beam welding is noted for high-quality welds with deep and/or narrow profiles, limited heat-affected zone, and low thermal distortion. Welding speeds are high compared to other continuous welding operations. No filler metal is used, and no flux or shielding gases are needed. Disadvantages of EBW include high equipment cost, need for precise joint preparation and alignment, and the limitations associated with performing the process in a vacuum, as we have already discussed. In addition, there are safety concerns because EBW generates X-rays from which humans must be shielded.

Laser-Beam Welding Laser-beam welding (LBW) is a fusion-welding process in which coalescence is achieved by the energy of a highly concentrated, coherent light beam focused on the joint to be welded. The term ***laser*** is an acronym for *l*ight *a*mplification by *s*timulated *e*mission of *r*adiation. This same technology is used for laser-beam machining (Section 19.3.3). LBW is normally performed with shielding gases (e.g., helium, argon, nitrogen, and carbon dioxide) to prevent oxidation. Filler metal is not usually added.

LBW produces welds of high quality, deep penetration, and narrow heat-affected zone. These features are similar to those achieved in electron-beam welding, and the two processes are often compared. There are several advantages of LBW over EBW: No vacuum chamber is required, no X-rays are emitted, and laser beams can be focused and directed by optical lenses and mirrors. On the other hand, LBW does not possess the capability for the deep welds and high depth-to-width ratios of EBW. Maximum depth in laser welding is about 19 mm (0.75 in), whereas EBW can be used for weld depths of 50 mm (2 in) or more; and the depth-to-width ratios in LBW are typically limited to around 5:1. Because of the highly concentrated energy in the small area of the laser beam, the process is often used to join small parts.

Thermit Welding ***Thermit*** is a trademark name for ***thermite***, a mixture of aluminum powder and iron oxide that produces an exothermic reaction when ignited. It is used in incendiary bombs and for welding. As a welding process, the use of Thermit dates from

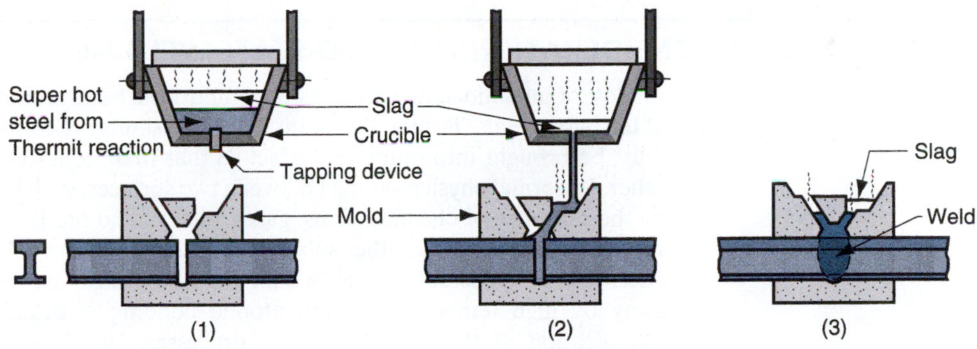

FIGURE 23.19 Thermit welding: (1) Thermit ignited; (2) crucible tapped, superheated metal flows into mold; (3) metal solidifies to produce weld joint. (Credit: *Fundamentals of Modern Manufacturing,* 4th Edition by Mikell P. Groover, 2010. Reprinted with permission of John Wiley & Sons, Inc.)

around 1900. Thermit welding (TW) is a fusion-welding process in which the heat for coalescence is produced by superheated molten metal from the chemical reaction of Thermit. Filler metal is obtained from the liquid metal; although the process is used for joining, it has more in common with casting than it does with welding.

Finely mixed powders of aluminum and iron oxide (in a 1:3 mixture), when ignited at a temperature of around 1300°C (2300°F), produce the following chemical reaction:

$$8\text{Al} + 3\text{Fe}_3\text{O}_4 \rightarrow 9\text{Fe} + 4\text{Al}_2\text{O}_3 + \text{heat} \tag{23.6}$$

The temperature from the reaction is around 2500°C (4500°F), resulting in superheated molten iron plus aluminum oxide that floats to the top as a slag and protects the iron from the atmosphere. In Thermit welding, the superheated iron (or steel if the mixture of powders is formulated accordingly) is contained in a crucible located above the joint to be welded, as indicated by our diagram of the TW process in Figure 23.19. After the reaction is complete (about 30 seconds, irrespective of the amount of Thermit involved), the crucible is tapped and the liquid metal flows into a mold built specially to surround the weld joint. Because the entering metal is so hot, it melts the edges of the base parts, causing coalescence upon solidification. After cooling, the mold is broken away, and the gates and risers are removed by oxyacetylene torch or other method.

Thermit welding has applications in joining of railroad rails (as in our figure), and repair of cracks in large steel castings and forgings such as ingot molds, large diameter shafts, frames for machinery, and ship rudders. The surface of the weld in these applications is often sufficiently smooth so that no subsequent finishing is required.

23.5 SOLID-STATE WELDING

In solid-state welding, coalescence of the part surfaces is achieved by (1) pressure alone, or (2) heat and pressure. For some solid-state processes, time is also a factor. If both heat and pressure are used, the amount of heat by itself is not sufficient to cause melting of the work surfaces. In some cases, the combination of heat and pressure, or the particular manner in which pressure alone is applied, generates sufficient energy to cause localized melting of the faying surfaces. Filler metal is not added in solid-state welding.

23.5.1 GENERAL CONSIDERATIONS IN SOLID-STATE WELDING

In most of the solid-state processes, a metallurgical bond is created with little or no melting of the base metals. To metallurgically bond two similar or dissimilar metals, the two metals must be brought into intimate contact so that their cohesive atomic forces attract each other. In normal physical contact between two surfaces, such intimate contact is prohibited by the presence of chemical films, gases, oils, and so on. In order for atomic bonding to succeed, these films and other substances must be removed. In fusion welding (as well as other joining processes such as brazing and soldering), the films are dissolved or burned away by high temperatures, and atomic bonding is established by the melting and solidification of the metals in these processes. But in solid-state welding, the films and other contaminants must be removed by other means to allow metallurgical bonding to take place. In some cases, a thorough cleaning of the surfaces is done just before the welding process; while in other cases, the cleaning action is accomplished as an integral part of bringing the part surfaces together. To summarize, the essential conditions for a successful solid-state weld are that the two surfaces must be very clean, and they must be brought into very close physical contact with each other to permit atomic bonding.

Welding processes that do not involve melting have several advantages over fusion-welding processes. If no melting occurs, then there is no heat-affected zone, and so the metal surrounding the joint retains its original properties. Many of these processes produce welded joints that comprise the entire contact interface between the two parts, rather than at distinct spots or seams, as in most fusion-welding operations. Also, some of these processes are quite applicable to bonding dissimilar metals, without concerns about differences in melting points, thermal expansions, conductivities, and other problems that usually arise when dissimilar metals are melted and then solidified during joining.

23.5.2 SOLID-STATE WELDING PROCESSES

The solid-state welding group includes the oldest joining process as well as some of the most modern. Each process in this group has its own unique way of creating the bond at the faying surfaces. We begin our coverage with forge welding, the first welding process.

Forge Welding Forge welding is of historic significance in the development of manufacturing technology. The process dates from about 1000 B.C.E, when blacksmiths of the ancient world learned to join two pieces of metal. Forge welding is a welding process in which the components to be joined are heated to hot working temperatures and then forged together by hammer or other means. Considerable skill was required by the craftsmen who practiced it to achieve a good weld by present-day standards. The process may be of historic interest; however, it is of minor commercial importance today, except for its variants that are discussed below.

Cold Welding Cold welding (CW) is a solid-state welding process accomplished by applying high pressure between clean contacting surfaces at room temperature. The faying surfaces must be exceptionally clean for CW to work, and cleaning is usually done by degreasing and wire brushing immediately before joining. Also, at least one of the metals to be welded, and preferably both, must be very ductile and free of work hardening. Metals such as soft aluminum and copper can be readily cold welded. The applied compression forces in the process result in cold working of the metal parts, reducing thickness by as much as 50%; but they also cause localized plastic deformation at the contacting surfaces, resulting in coalescence. For small parts, the forces may be applied by simple hand-operated tools. For heavier work, powered presses are required to exert the necessary force. No heat is applied from external sources in CW, but the

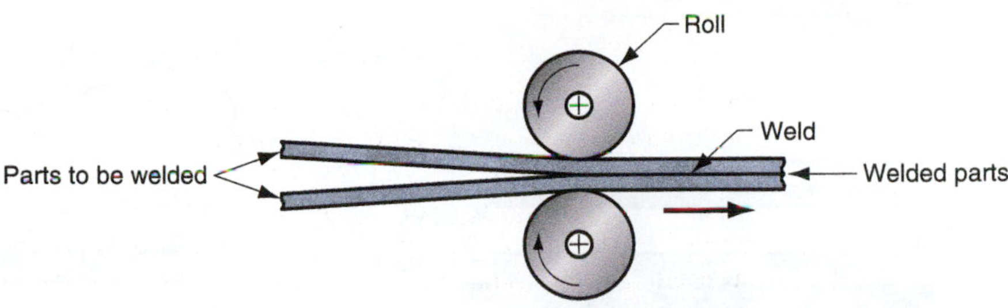

FIGURE 23.20 Roll welding (ROW). (Credit: *Fundamentals of Modern Manufacturing*, 4th Edition by Mikell P. Groover, 2010. Reprinted with permission of John Wiley & Sons, Inc.)

deformation process raises the temperature of the work somewhat. Applications of CW include making electrical connections.

Roll Welding Roll welding is a variation of either forge welding or cold welding, depending on whether external heating of the workparts is accomplished prior to the process. Roll welding (ROW) is a solid-state welding process in which pressure sufficient to cause coalescence is applied by means of rolls, either with or without external application of heat. The process is illustrated in Figure 23.20. If no external heat is supplied, the process is called ***cold-roll welding***; if heat is supplied, the term ***hot-roll welding*** is used. Applications of roll welding include cladding stainless steel to mild or low alloy steel for corrosion resistance, making bimetallic strips for measuring temperature, and producing "sandwich" coins for the U.S. mint.

Hot Pressure Welding Hot pressure welding (HPW) is another variation of forge welding in which coalescence occurs from the application of heat and pressure sufficient to cause considerable deformation of the base metals. The deformation disrupts the surface oxide film, thus leaving clean metal to establish a good bond between the two parts. Time must be allowed for diffusion to occur across the faying surfaces. The operation is usually carried out in a vacuum chamber or in the presence of a shielding medium. Principal applications of HPW are in the aerospace industry.

Diffusion Welding Diffusion welding (DFW) is a solid-state welding process that results from the application of heat and pressure, usually in a controlled atmosphere, with sufficient time allowed for diffusion and coalescence to occur. Temperatures are well below the melting points of the metals (about 0.5 T_m is the maximum), and plastic deformation at the surfaces is minimal. The primary mechanism of coalescence is solid-state diffusion, which involves migration of atoms across the interface between contacting surfaces. Applications of DFW include the joining of high-strength and refractory metals in the aerospace and nuclear industries. The process is used to join both similar and dissimilar metals, and in the latter case a filler layer of a different metal is often sandwiched between the two base metals to promote diffusion. The time for diffusion to occur between the faying surfaces can be significant, requiring more than an hour in some applications [10].

Explosion Welding Explosion welding (EXW) is a solid-state welding process in which rapid coalescence of two metallic surfaces is caused by the energy of a detonated explosive. It is commonly used to bond two dissimilar metals, in particular to clad one metal on top of a base metal over large areas. Applications include production of corrosion-resistant sheet and plate stock for making processing equipment in the chemical and petroleum industries. The term ***explosion cladding*** is used in this context. No filler metal is used in EXW, and no external heat is applied. Also, no diffusion occurs during the process (the time is too short). The nature of the bond is metallurgical, in many cases combined with a mechanical interlocking that results from a rippled or wavy interface between the metals.

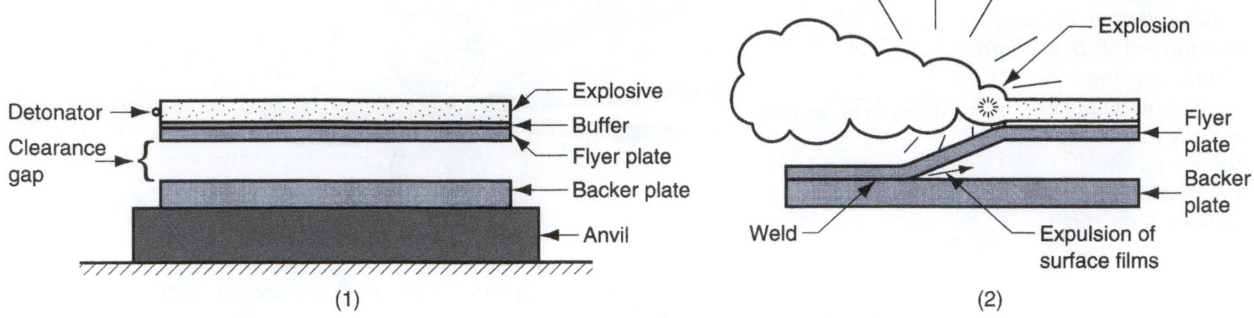

FIGURE 23.21 Explosive welding (EXW): (1) setup in the parallel configuration, and (2) during detonation of the explosive charge. (Credit: *Fundamentals of Modern Manufacturing,* 4th Edition by Mikell P. Groover, 2010. Reprinted with permission of John Wiley & Sons, Inc.)

The process for cladding one metal plate on another can be described with reference to Figure 23.21. In this setup, the two plates are in a parallel configuration, separated by a certain gap distance, with the explosive charge above the upper plate, called the ***flyer plate***. A buffer layer (e.g., rubber, plastic) is often used between the explosive and the flyer plate to protect its surface. The lower plate, called the ***backer*** metal, rests on an anvil for support. When detonation is initiated, the explosive charge propagates from one end of the flyer plate to the other, caught in the stop-action view shown in Figure 23.21(2). One of the difficulties in comprehending what happens in EXW is the common misconception that an explosion occurs instantaneously; it is actually a progressive reaction, although admittedly very rapid—propagating at rates as high as 8500 m/s (28,000 ft/sec). The resulting high-pressure zone propels the flyer plate to collide with the backer metal progressively at high velocity, so that it takes on an angular shape as the explosion advances, as illustrated in our sketch. The upper plate remains in position in the region where the explosive has not yet detonated. The high-speed collision, occurring in a progressive and angular fashion as it does, causes the surfaces at the point of contact to become fluid, and any surface films are expelled forward from the apex of the angle. The colliding surfaces are thus chemically clean, and the fluid behavior of the metal, which involves some interfacial melting, provides intimate contact between the surfaces, leading to metallurgical bonding. Variations in collision velocity and impact angle during the process can result in a wavy or rippled interface between the two metals. This kind of interface strengthens the bond because it increases the contact area and tends to mechanically interlock the two surfaces.

Friction Welding Friction welding is a widely used commercial process, amenable to automated production methods. The process was developed in the (former) Soviet Union and introduced into the United States around 1960. Friction welding (FRW) is a solid-state welding process in which coalescence is achieved by frictional heat combined with pressure. The friction is induced by mechanical rubbing between the two surfaces, usually by rotation of one part relative to the other, to raise the temperature at the joint interface to the hot working range for the metals involved. Then the parts are driven toward each other with sufficient force to form a metallurgical bond. The sequence is portrayed in Figure 23.22 for welding two cylindrical parts, the typical application. The axial compression force upsets the parts, and flash is produced by the displaced material. Any surface films that may have been on the contacting surfaces are expunged during the process. The flash must be subsequently trimmed (e.g., by turning) to provide a smooth surface in the weld region. When properly carried out, no melting occurs at the faying surfaces. No filler metal, flux, or shielding gases are normally used.

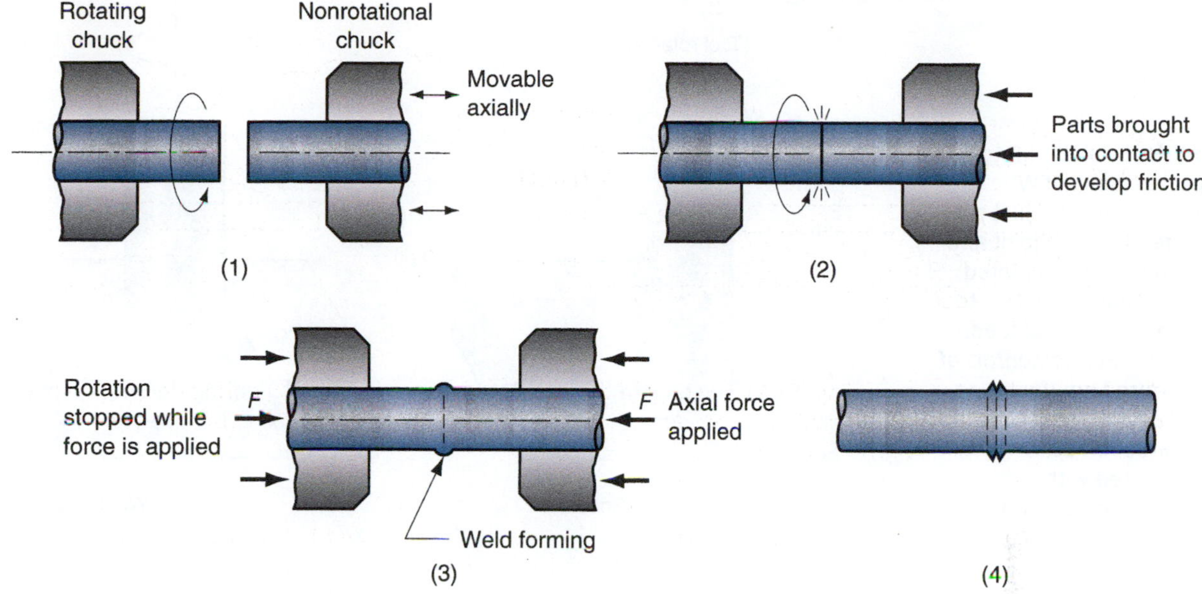

FIGURE 23.22 Friction welding (FRW): (1) rotating part, no contact; (2) parts brought into contact to generate friction heat; (3) rotation stopped and axial pressure applied; and (4) weld created. (Credit: *Fundamentals of Modern Manufacturing*, 4th Edition by Mikell P. Groover, 2010. Reprinted with permission of John Wiley & Sons, Inc.)

Nearly all FRW operations use rotation to develop the frictional heat for welding. There are two principal drive systems, distinguishing two types of FRW: (1) continuous-drive friction welding, and (2) inertia friction welding. In ***continuous-drive friction welding***, one part is driven at a constant rotational speed and forced into contact with the stationary part at a certain force level so that friction heat is generated at the interface. When the proper hot working temperature has been reached, braking is applied to stop the rotation abruptly, and simultaneously the pieces are forced together at forging pressures. In ***inertia friction welding***, the rotating part is connected to a flywheel, which is brought up to a predetermined speed. Then the flywheel is disengaged from the drive motor, and the parts are forced together. The kinetic energy stored in the flywheel is dissipated in the form of friction heat to cause coalescence at the abutting surfaces. The total cycle for these operations is about 20 seconds.

Machines used for friction welding have the appearance of an engine lathe. They require a powered spindle to turn one part at high speed, and a means of applying an axial force between the rotating part and the nonrotating part. With its short cycle times, the process lends itself to mass production. It is applied in the welding of various shafts and tubular parts in industries such as automotive, aircraft, farm equipment, petroleum, and natural gas. The process yields a narrow heat-affected zone and can be used to join dissimilar metals. However, at least one of the parts must be rotational, flash must usually be removed, and upsetting reduces the part lengths (which must be taken into consideration in product design).

The conventional friction-welding operations discussed above utilize a rotary motion to develop the required friction between faying surfaces. A more recent version of the process is ***linear friction welding***, in which a linear reciprocating motion is used to generate friction heat between the parts. This eliminates the requirement for at least one of the parts to be rotational (e.g., cylindrical, tubular).

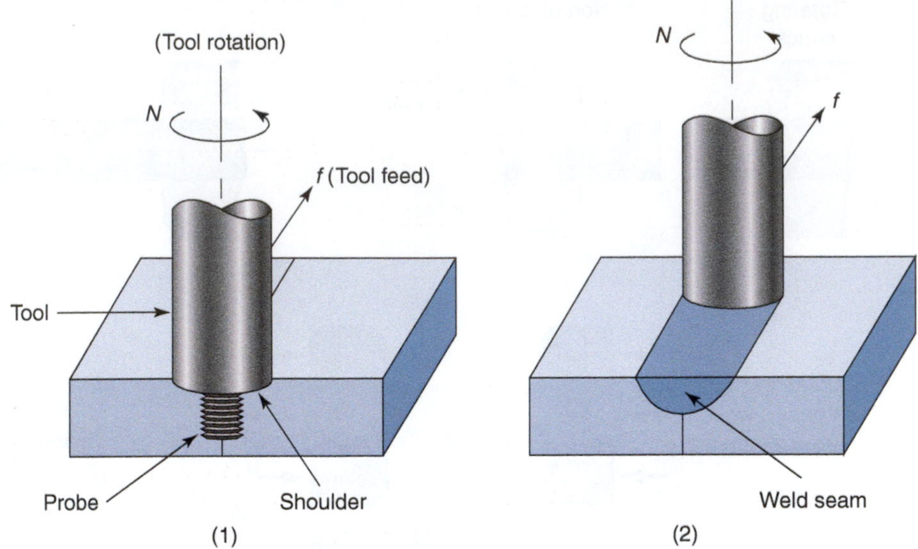

FIGURE 23.23 Friction stir welding (FSW): (1) rotating tool just prior to feeding into joint and (2) partially completed weld seam. Key: N = tool rotation, f = tool feed. (Credit: *Fundamentals of Modern Manufacturing*, 4th Edition by Mikell P. Groover, 2010. Reprinted with permission of John Wiley & Sons, Inc.)

Friction Stir Welding Friction stir welding (FSW), illustrated in Figure 23.23, is a solid-state welding process in which a rotating tool is fed along the joint line between two workpieces, generating friction heat and mechanically stirring the metal to form the weld seam. The process derives its name from this stirring or mixing action. FSW is distinguished from conventional FRW by the fact that friction heat is generated by a separate wear-resistant tool rather than by the parts themselves. FSW was developed in 1991 at The Welding Institute in Cambridge, United Kingdom.

The rotating tool is stepped, consisting of a cylindrical shoulder and a smaller probe projecting beneath it. During welding, the shoulder rubs against the top surfaces of the two parts, developing much of the friction heat, while the probe generates additional heat by mechanically mixing the metal along the butt surfaces. The probe has a geometry designed to facilitate the mixing action. The heat produced by the combination of friction and mixing does not melt the metal but softens it to a highly plastic condition. As the tool is fed forward along the joint, the leading surface of the rotating probe forces the metal around it and into its wake, developing forces that forge the metal into a weld seam. The shoulder serves to constrain the plasticized metal flowing around the probe.

The FSW process is used in the aerospace, automotive, railway, and shipbuilding industries. Typical applications are butt joints on large aluminum parts. Other metals, including steel, copper, and titanium, as well as polymers and composites have also been joined using FSW. Advantages in these applications include: (1) good mechanical properties of the weld joint; (2) avoidance of toxic fumes, warping, shielding issues, and other problems associated with arc welding; (3) little distortion or shrinkage; and (4) good weld appearance. Disadvantages include (1) an exit hole is produced when the tool is withdrawn from the work, and (2) heavy-duty clamping of the parts is required.

Ultrasonic Welding Ultrasonic welding (USW) is a solid-state welding process in which two components are held together under modest clamping force, and oscillatory shear stresses of ultrasonic frequency are applied to the interface to cause coalescence. The operation is illustrated in Figure 23.24 for lap welding, the typical application. The oscillatory motion between the two parts breaks down any surface films to allow intimate contact and strong metallurgical bonding between the surfaces. Although heating of the contacting surfaces occurs due to interfacial rubbing and plastic

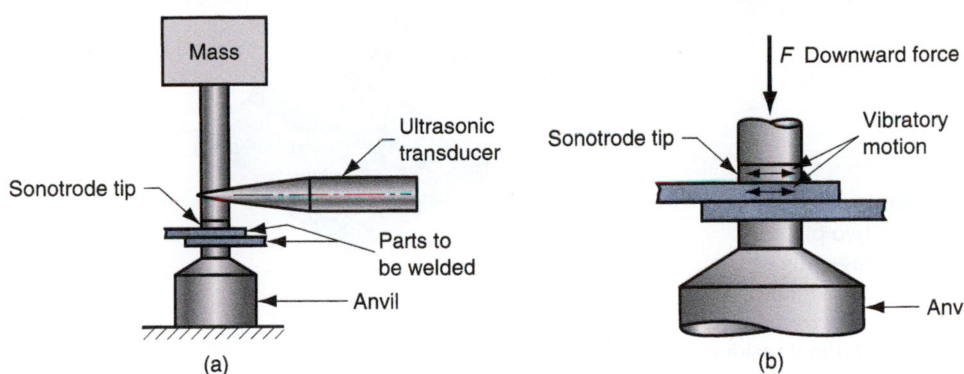

FIGURE 23.24 Ultrasonic welding (USW): (a) general setup for a lap joint; and (b) close-up of weld area. (Credit: *Fundamentals of Modern Manufacturing*, 4th Edition by Mikell P. Groover, 2010. Reprinted with permission of John Wiley & Sons, Inc.)

deformation, the resulting temperatures are well below the melting point. No filler metals, fluxes, or shielding gases are required in USW.

The oscillatory motion is transmitted to the upper workpart by means of a ***sonotrode***, which is coupled to an ultrasonic transducer. This device converts electrical power into high-frequency vibratory motion. Typical frequencies used in USW are 15 to 75 kHz, with amplitudes of 0.018 to 0.13 mm (0.0007 to 0.005 in). Clamping pressures are well below those used in cold welding and produce no significant plastic deformation between the surfaces. Welding times under these conditions are less than 1 second.

USW operations are generally limited to lap joints on soft materials such as aluminum and copper. Welding harder materials causes rapid wear of the sonotrode contacting the upper workpart. Workparts should be relatively small, and welding thicknesses less than 3 mm (1/8 in) is the typical case. Applications include wire terminations and splicing in electrical and electronics industries (eliminates the need for soldering), assembly of aluminum sheet-metal panels, welding of tubes to sheets in solar panels, and other tasks in small parts assembly.

23.6 WELD QUALITY

The purpose of any welding process is to join two or more components into a single structure. The physical integrity of the structure thus formed depends on the quality of the weld. Our discussion of weld quality deals primarily with arc welding, the most widely used welding process and the one for which the quality issue is the most critical and complex.

Residual Stresses and Distortion The rapid heating and cooling in localized regions of the work during fusion welding, especially arc welding, result in thermal expansion and contraction that cause residual stresses in the weldment. These stresses, in turn, can cause distortion and warping of the welded assembly.

The situation in welding is complicated because (1) heating is very localized, (2) melting of the base metals occurs in these local regions, and (3) the location of heating and melting is in motion (at least in arc welding). Consider, for example, butt welding of two plates by arc welding as shown in Figure 23.25(a). The operation begins at one end and travels to the opposite end. As it proceeds, a molten pool is formed from the base metal (and filler metal, if used) that quickly solidifies behind the moving arc. The portions of the work immediately adjacent to the weld bead become extremely hot and expand, while portions removed from the weld remain relatively cool. The weld pool quickly solidifies in the cavity between the two parts, and as it and the surrounding metal cool and contract, shrinkage occurs across the width of the weldment, as seen in Figure 23.25(b). The weld

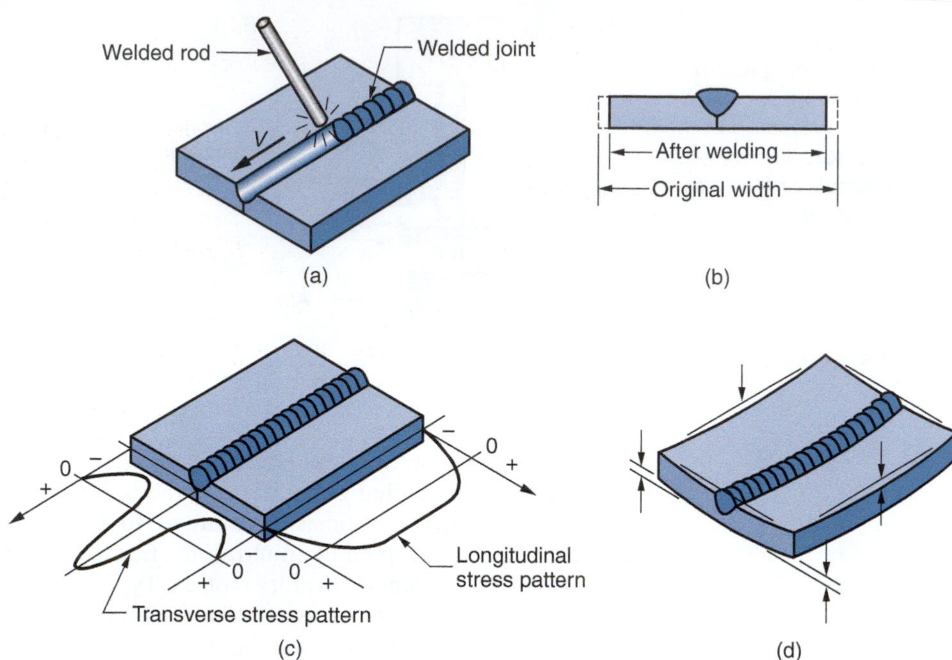

FIGURE 23.25 (a) Butt welding two plates; (b) shrinkage across the width of the welded assembly; (c) transverse and longitudinal residual stress pattern; and (d) likely warping in the welded assembly. (Credit: *Fundamentals of Modern Manufacturing,* 4th Edition by Mikell P. Groover, 2010. Reprinted with permission of John Wiley & Sons, Inc.)

seam is left in residual tension, and reactionary compressive stresses are set up in regions of the parts away from the weld. Residual stresses and shrinkage also occurs along the length of the weld bead. Because the outer regions of the base parts have remained relatively cool and dimensionally unchanged, while the weld bead has solidified from very high temperatures and then contracted, residual tensile stresses remain longitudinally in the weld bead. These transverse and longitudinal stress patterns are depicted in Figure 23.25(c). The net result of these residual stresses, transversely and longitudinally, is likely to cause warping in the welded assembly as shown in Figure 23.25(d).

The arc-welded butt joint in our example is only one of a variety of joint types and welding operations. Thermally induced residual stresses and the accompanying distortion are a potential problem in nearly all fusion-welding processes and in certain solid-state welding operations in which significant heating takes place. Following are some techniques to minimize warping in a weldment: (1) ***Welding fixtures*** can be used to physically restrain movement of the parts during welding. (2) ***Heat sinks*** can be used to rapidly remove heat from sections of the welded parts to reduce distortion. (3) ***Tack welding*** at multiple points along the joint can create a rigid structure prior to continuous seam welding. (4) ***Welding conditions*** (speed, amount of filler metal used, etc.) can be selected to reduce warping. (5) The base parts can be ***preheated*** to reduce the level of thermal stresses experienced by the parts. (6) ***Stress relief*** heat treatment can be performed on the welded assembly, either in a furnace for small weldments, or using methods that can be used in the field for large structures. (7) ***Proper design*** of the weldment itself can reduce the degree of warping.

Welding Defects In addition to residual stresses and distortion in the final assembly, other defects can occur in welding. Following is a brief description of each of the major categories, based on a classification in Cary [3]:

- ***Cracks.*** Cracks are fracture-type interruptions either in the weld itself or in the base metal adjacent to the weld. This is perhaps the most serious welding defect because it constitutes a discontinuity in the metal that significant reduces weld strength.

Welding cracks are caused by embrittlement or low ductility of the weld and/or base metal combined with high restraint during contraction. Generally, this defect must be repaired.

- ***Cavities***. These include various porosity and shrinkage voids. ***Porosity*** consists of small voids in the weld metal formed by gases entrapped during solidification. The shapes of the voids vary between spherical (blow holes) to elongated (worm holes). Porosity usually results from inclusion of atmospheric gases, sulfur in the weld metal, or contaminants on the surfaces. ***Shrinkage voids*** are cavities formed by shrinkage during solidification. Both of these cavity-type defects are similar to defects found in castings and emphasize the close kinship between casting and welding.
- ***Solid inclusions***. These are nonmetallic solid materials trapped inside the weld metal. The most common form is slag inclusions generated during arc-welding processes that use flux. Instead of floating to the top of the weld pool, globules of slag become encased during solidification of the metal. Another form of inclusion is metallic oxides that form during the welding of metals such as aluminum, which normally has a surface coating of Al_2O_3.
- ***Incomplete fusion***. Also known as ***lack of fusion***, this defect is simply a weld bead in which fusion has not occurred throughout the entire cross section of the joint. A related defect is ***lack of penetration***, which means that fusion has not penetrated deeply enough into the root of the joint.
- ***Imperfect shape or unacceptable contour***. The weld should have a certain desired profile for maximum strength, as indicated in Figure 23.26(a) for a single V-groove weld. This weld profile maximizes the strength of the welded joint and avoids incomplete fusion and lack of penetration. Some of the common defects in weld shape and contour are illustrated in Figure 23.26.
- ***Miscellaneous defects***. This category includes ***arc strikes***, in which the welder accidentally allows the electrode to touch the base metal next to the joint, leaving a scar on the surface; and ***excessive spatter***, in which drops of molten weld metal splash onto the surface of the base parts.

Inspection and Testing Methods A variety of inspection and testing methods are available to check the quality of the welded joint. Standardized procedures have been developed and specified over the years by engineering and trade societies such as the American Welding Society (AWS). For purposes of discussion, these inspection and testing procedures can be divided into three categories: (1) visual, (2) nondestructive, and (3) destructive.

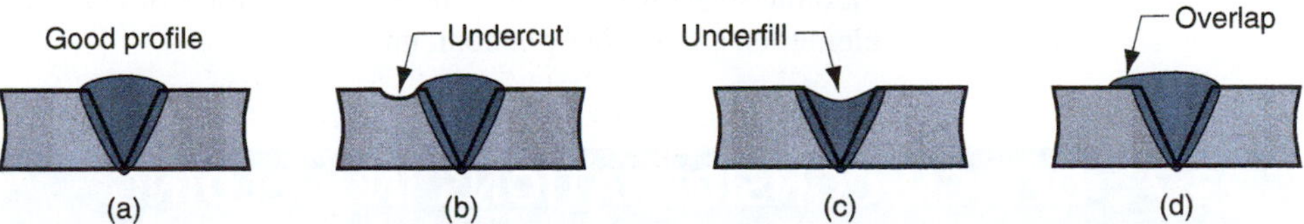

FIGURE 23.26 (a) Desired weld profile for single V-groove weld joint. Same joint but with several weld defects: (b) ***undercut***, in which a portion of the base metal part is melted away; (c) ***underfill***, a depression in the weld below the level of the adjacent base metal surface; and (d) ***overlap***, in which the weld metal spills beyond the joint onto the surface of the base part but no fusion occurs. (Credit: *Fundamentals of Modern Manufacturing*, 4th Edition by Mikell P. Groover, 2010. Reprinted with permission of John Wiley & Sons, Inc.)

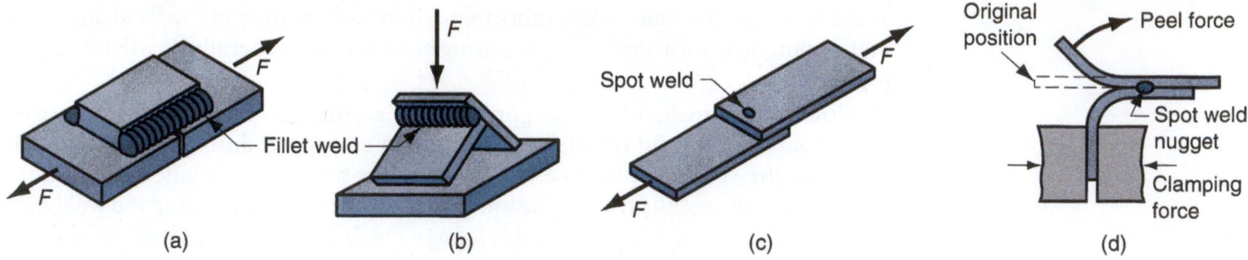

FIGURE 23.27 Mechanical tests used in welding: (a) tension–shear test of arc weldment, (b) fillet break test, (c) tension–shear test of spot weld, (d) peel test for spot weld. (Credit: *Fundamentals of Modern Manufacturing*, 4th Edition by Mikell P. Groover, 2010. Reprinted with permission of John Wiley & Sons, Inc.)

Visual inspection is no doubt the most widely used welding inspection method. An inspector visually examines the weldment for (1) conformance to dimensional specifications on the part drawing, (2) warping, and (3) cracks, cavities, incomplete fusion, and other visible defects. The welding inspector also determines if additional tests are warranted, usually in the nondestructive category. The limitation of visual inspection is that only surface defects are detectable; internal defects cannot be discovered by visual methods.

Nondestructive evaluation (NDE) includes various methods that do not damage the specimen being inspected. ***Dye-penetrant*** and ***fluorescent-penetrant tests*** are methods for detecting small defects such as cracks and cavities that are open to the surface. Fluorescent penetrants are highly visible when exposed to ultraviolet light, and their use is therefore more sensitive than dyes.

Several other NDE methods should be mentioned. ***Magnetic particle testing*** is limited to ferromagnetic materials. A magnetic field is established in the subject part, and magnetic particles (e.g., iron filings) are sprinkled on the surface. Subsurface defects such as cracks and inclusions reveal themselves by distorting the magnetic field, causing the particles to be concentrated in certain regions on the surface. ***Ultrasonic testing*** involves the use of high-frequency sound waves (>20 kHz) directed through the specimen. Discontinuities (e.g., cracks, inclusions, porosity) are detected by losses in sound transmission. ***Radiographic testing*** uses X-rays or gamma radiation to detect flaws internal to the weld metal. It provides a photographic film record of any defects.

Destructive testing methods in which the weld is destroyed either during the test or to prepare the test specimen. They include mechanical and metallurgical tests. ***Mechanical tests*** are similar in purpose to conventional testing methods such as tensile tests and shear tests (Chapter 3). The difference is that the test specimen is a weld joint. Figure 23.27 presents a sampling of the mechanical tests used in welding. ***Metallurgical tests*** involve the preparation of metallurgical specimens of the weldment to examine such features as metallic structure, defects, extent and condition of heat-affected zone, presence of other elements, and similar phenomena.

23.7 DESIGN CONSIDERATIONS IN WELDING

If an assembly is to be permanently welded, the designer should follow certain guidelines (compiled from [2], [3], and other sources):

➢ ***Design for welding***. The most basic guideline is that the product should be designed from the start as a welded assembly, and not as a casting or forging or other formed shape.

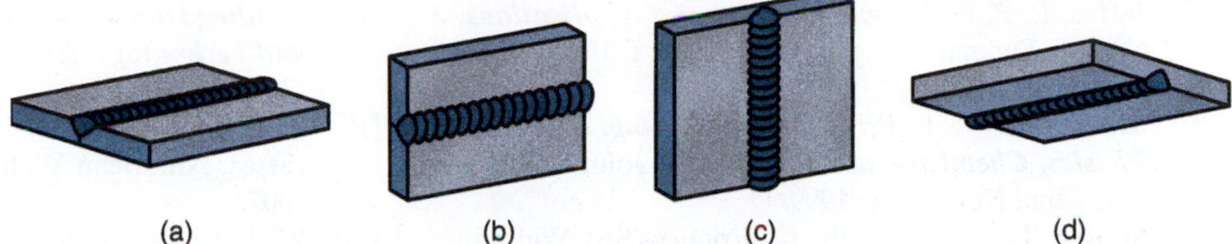

FIGURE 23.28 Welding positions (defined here for groove welds): (a) flat, (b) horizontal, (c) vertical, and (d) overhead. (Credit: *Fundamentals of Modern Manufacturing*, 4th Edition by Mikell P. Groover, 2010. Reprinted with permission of John Wiley & Sons, Inc.)

- ***Minimum parts***. Welded assemblies should consist of the fewest number of parts possible. For example, it is usually more cost efficient to perform simple bending operations on a part than to weld an assembly from flat plates and sheets.

The following guidelines apply to arc welding:

- ***Good fit-up of parts*** to be welded is important to maintain dimensional control and minimize distortion. Machining is sometimes required to achieve satisfactory fit-up.
- The assembly must provide access room to allow the welding gun to reach the welding area.
- Whenever possible, design of the assembly should allow ***flat welding*** to be performed, because this is the fastest and most convenient welding position. The possible welding positions are defined in Figure 23.28. The overhead position is the most difficult.

The following design guidelines apply to resistance spot welding:

- Low-carbon sheet steel up to 3.2 mm (0.125 in) is the ideal metal for resistance spot welding.
- Additional strength and stiffness can be obtained in large flat sheet-metal components by: (1) spot welding reinforcing parts into them, or (2) forming flanges and embossments into them.
- The spot-welded assembly must provide access for the electrodes to reach the welding area.
- Sufficient overlap of the sheet-metal parts is required for the electrode tip to make proper contact in spot welding. For example, for low-carbon sheet steel, the overlap distance should range from about six times stock thickness for thick sheets of 3.2 mm (0.125 in) to about 20 times thickness for thin sheets, such as 0.5 mm (0.020 in).

REFERENCES

[1] ***ASM Handbook***, Vol. **6**, ***Welding, Brazing, and Soldering***. ASM International, Materials Park, Ohio, 1993.

[2] Bralla, J. G. (Editor in Chief). ***Design for Manufacturability Handbook***, 2nd ed. McGraw-Hill Book Company, New York, 1998.

[3] Cary, H. B., and Helzer S. C. ***Modern Welding Technology***, 6th ed. Pearson/Prentice-Hall, Upper Saddle River, New Jersey, 2005.

[4] Galyen, J., Sear, G., and Tuttle, C. A. ***Welding, Fundamentals and Procedures***, 2nd ed. Prentice-Hall, Inc., Upper Saddle River, New Jersey, 1991.

[5] Jeffus, L. F. ***Welding: Principles and Applications***, 6th ed. Delmar Cengage Learning, Clifton Park, New York, 2007.

[6] Messler, R. W., Jr. ***Principles of Welding: Processes, Physics, Chemistry, and Metallurgy***. John Wiley & Sons, Inc. New York, 1999.

[7] Stotler, T., and Bernath, J. "Friction Stir Welding Advances," ***Advanced Materials and Processes***, March 2009, pp. 35–37.

[8] Stout, R. D., and Ott, C. D. ***Weldability of Steels***, 4th ed. Welding Research Council, New York, 1987.

[9] ***Welding Handbook***, 9th ed., Vol. **1**, ***Welding Science and Technology***. American Welding Society, Miami, Florida, 2007.

[10] ***Welding Handbook***, 9th ed., Vol. **2**, **Welding Processes**. American Welding Society, Miami, Florida, 2007.

[11] Wick, C., and Veilleux, R. F.(eds.). ***Tool and Manufacturing Engineers Handbook***, 4th ed., Vol. **IV**, ***Quality Control and Assembly***. Society of Manufacturing Engineers, Dearborn, Michigan, 1987.

REVIEW QUESTIONS

23.1. Name the principal groups of processes included in fusion welding.

23.2. What is the fundamental feature that distinguishes fusion welding from solid-state welding?

23.3. Define what an electrical arc is.

23.4. What do the terms *arcon time* and *arc time* mean?

23.5. Electrodes in arc welding are divided into two categories. Name and define the two types.

23.6. What are the two basic methods of arc shielding?

23.7. Why is the heat-transfer factor in arc-welding processes that utilize consumable electrodes greater than in those that use nonconsumable electrodes?

23.8. Describe the shielded metal arc-welding (SMAW) process.

23.9. Why is the shielded metal arc-welding (SMAW) process difficult to automate?

23.10. Describe submerged arc welding (SAW).

23.11. Why are the temperatures much higher in plasma arc welding than in other AW processes?

23.12. Define resistance welding.

23.13. Describe the sequence of steps in the cycle of a resistance spot-welding operation.

23.14. What is resistance-projection welding?

23.15. Describe cross-wire welding.

23.16. Why is the oxyacetylene welding process favored over the other oxyfuel welding processes?

23.17. Electron-beam welding has a significant disadvantage in high-production applications. What is that disadvantage?

23.18. Laser-beam welding and electron-beam welding are often compared because they both produce very high power densities. LBW has certain advantages over EBW. What are they?

23.19. There are several modern-day variations of forge welding, the original welding process. Name them.

23.20. What is friction stir welding (FSW), and how is it different from friction welding?

23.21. What is a sonotrode in ultrasonic welding?

23.22. Distortion (warping) is a serious problem in fusion welding, particularly arc welding. What are some of the techniques that can be taken to reduce the incidence and extent of distortion?

23.23. What are some of the important welding defects?

23.24. What are the three basic categories of inspection and testing techniques used for weldments? Name some typical inspections and/or tests in each category.

23.25. What are some of the design guidelines for weldments that are fabricated by arc welding?

PROBLEMS

23.1. A shielded metal arc-welding operation is performed on steel at a voltage = 30 volts and a current = 225 amps. The heat-transfer factor = 0.90 and melting factor = 0.75. The unit melting energy for steel = 10.2 J/mm^3. Determine (a) the rate of heat generation at the weld and (b) the volume rate of metal welded.

23.2. A GTAW operation is performed on low-carbon steel, whose unit melting energy is 10.3 J/mm^3. The welding voltage is 22 volts and the current is 135 amps. The heat-transfer factor is 0.7 and the melting factor is 0.65. If filler metal wire of 3.5 mm diameter is added to the operation, the final weld bead is composed of 60% volume of filler and 40% volume base metal. If the travel speed in the operation is 5 mm/sec, determine (a) cross-sectional area of the weld bead, and (b) the feed rate (mm/sec) at which the filler wire must be supplied.

23.3. A flux-cored arc-welding operation is performed to butt weld two austenitic stainless steel plates together. The welding voltage is 21 volts and the current is 185 amps. The cross-sectional area of the weld seam = 75 mm^2 and the melting factor of the stainless steel is assumed to be 0.60. Using tabular data and equations given in this and the preceding chapter, determine the likely value for travel speed v in the operation.

23.4. A gas metal arc-welding test is performed to determine the value of melting factor f_2 for a certain metal and operation. The welding voltage = 25 volts, current = 125 amps, and heat-transfer factor is assumed to be = 0.90, a typical value for GMAW. The rate at which the filler metal is added to the weld is 0.50 in^3 per minute, and measurements indicate that the final weld bead consists of 57% filler metal and 43% base metal. The unit melting energy for the metal is known to be 75 Btu/in^3. (a) Find the melting factor. (b) What is the travel speed if the cross-sectional area of the weld bead = 0.05 in^2?

23.5. A continuous weld is to be made around the circumference of a round steel tube of diameter = 6.0 ft, using a submerged arc-welding operation under automatic control at a voltage of 25 volts and current of 300 amps. The tube is slowly rotated under a stationary welding head. The heat-transfer factor for SAW is = 0.95 and the assumed melting factor = 0.7. The cross-sectional area of the weld bead is 0.12 in^2. If the unit melting energy for the steel = 150 Btu/in^3, determine (a) the rotational speed of the tube and (b) the time required to complete the weld.

23.6. An RSW operation is used to make a series of spot welds between two pieces of aluminum, each 2.0 mm thick. The unit melting energy for aluminum = 2.90 J/mm^3. Welding current = 6,000 amps, and time duration = 0.15 sec. Assume that the resistance = 75 micro-ohms. The resulting weld nugget measures 5.0 mm in diameter by 2.5 mm thick. How much of the total energy generated is used to form the weld nugget?

23.7. The unit melting energy for a certain sheet metal is 9.5 J/mm^3. Thickness of each of the two sheets to be spot welded is 3.5 mm. To achieve required strength, it is desired to form a weld nugget that is 5.5 mm in diameter and 5.0 mm thick. The weld duration will be set at 0.3 sec. If it is assumed that the electrical resistance between the surfaces is 140 micro-ohms, and that only one-third of the electrical energy generated will be used to form the weld nugget (the rest being dissipated), determine the minimum current level required in this operation.

23.8. A resistance spot-welding operation is performed on two pieces of 0.040 in thick sheet steel (low carbon). The unit melting energy for steel = 150 Btu/in^3. Current = 9500 A and time duration = 0.17 sec. This results in a weld nugget of diameter = 0.19 in and thickness = 0.060 in. Assume the resistance = 100 micro-ohms. Determine (a) the average power density in the interface area defined by the weld nugget, and (b) the proportion of energy generated that went into formation of the weld nugget.

23.9. A resistance seam-welding operation is performed on two pieces of 2.5-mm-thick austenitic stainless steel to fabricate a container. The weld current in the operation is 10,000 amps, the weld duration = 0.3 sec, and the resistance at the interface is 75 micro-ohms. Continuous-motion welding is used, with 200 mm diameter electrode wheels. The individual weld nuggets formed in this RSEW operation have diameter = 6 mm and thickness = 3 mm (assume the weld nuggets are disc-shaped). These weld nuggets must be contiguous to form a sealed seam. The power unit driving the process requires an off-time between spot welds of 1.0 s. Given these conditions, determine (a) the unit melting energy of stainless steel using the methods of the previous chapter, (b) the proportion of energy generated that goes into the formation of each weld nugget, and (c) the rotational speed of the electrode wheels.

23.10. The voltage in an EBW operation is 45 kV. The beam current is 60 milliamp. The electron beam is focused on a circular area that is 0.25 mm in diameter. The heat-transfer factor is 0.87. Calculate the average power density in the area in watt/mm^2.

23.11. An electron-beam welding operation is to be accomplished to butt weld two sheet-metal parts that are 3.0 mm thick. The unit melting energy = 5.0 J/mm^3. The weld joint is to be 0.35 mm wide, so that the cross section of the fused metal is 0.35 mm by 3.0 mm. If accelerating voltage = 25 kV, beam current = 30 milliamp, heat-transfer factor f_1 = 0.85, and melting factor f_2 = 0.75, determine the travel speed at which this weld can be made along the seam.

23.12. An electron-beam welding operation uses the following process parameters: accelerating voltage = 25 kV, beam current = 100 milliamp, and the circular area on which the beam is focused has a diameter = 0.020 in. If the heat-transfer factor = 90%, determine the average power density in the area in Btu/sec in^2.

24 BRAZING, SOLDERING, AND ADHESIVE BONDING

Chapter Contents

24.1 Brazing
- 24.1.1 Brazed Joints
- 24.1.2 Filler Metals and Fluxes
- 24.1.3 Brazing Methods

24.2 Soldering
- 24.2.1 Joint Designs in Soldering
- 24.2.2 Solders and Fluxes
- 24.2.3 Soldering Methods

24.3 Adhesive Bonding
- 24.3.1 Joint Design
- 24.3.2 Adhesive Types
- 24.3.3 Adhesive Application Technology

In this chapter, we consider three joining processes that are similar to welding in certain respects: brazing, soldering, and adhesive bonding. Brazing and soldering both use filler metals to join and bond two (or more) metal parts to provide a permanent joint. It is difficult, although not impossible, to disassemble the parts after a brazed or soldered joint has been made. In the spectrum of joining processes, brazing and soldering lie between fusion welding and solid-state welding. A filler metal is added in brazing and soldering as in most fusion-welding operations; however, no melting of the base metals occurs, which is similar to solid-state welding. Despite these anomalies, brazing and soldering are generally considered to be distinct from welding. Brazing and soldering are attractive compared to welding under circumstances where (1) the metals have poor weldability, (2) dissimilar metals are to be joined, (3) the intense heat of welding may damage the components being joined, (4) the geometry of the joint does not lend itself to any of the welding methods, and/or (5) high strength is not a requirement.

Adhesive bonding shares certain features in common with brazing and soldering. It utilizes the forces of attachment between a filler material and two closely spaced surfaces to bond the parts. The differences are that the filler material in adhesive bonding is not metallic, and the joining process is carried out at room temperature or only modestly above.

24.1 BRAZING

Brazing is a joining process in which a filler metal is melted and distributed by capillary action between the faying surfaces of the metal parts being joined. No melting of the base metals occurs in brazing; only the filler melts. In brazing, the filler metal (also called the ***brazing metal***) has a melting temperature (liquidus) that is above 450°C (840°F) but below the melting point (solidus) of the base metal(s) to be joined. If the joint is properly designed and the brazing operation has been properly performed, the brazed joint will be stronger than the filler metal out of which it has been formed upon solidification. This rather remarkable result is due to the small part clearances used in brazing, the metallurgical bonding that occurs between base and filler metals, and the geometric constrictions that are imposed on the joint by the base parts.

Brazing has several advantages compared to welding: (1) any metals can be joined, including dissimilar metals; (2) certain brazing methods can be performed quickly and consistently, thus permitting high cycle rates and automated production; (3) some methods allow multiple joints to be brazed simultaneously; (4) brazing can be applied to join thin-walled parts that cannot be welded; (5) in general, less heat and power are required than in fusion welding; (6) problems with the heat-affected zone (HAZ) in the base metal near the joint are reduced; and (7) joint areas that are inaccessible by many welding processes can be brazed, because capillary action draws the molten filler metal into the joint.

Disadvantages and limitations of brazing include: (1) joint strength is generally less than that of a welded joint; (2) although strength of a good brazed joint is greater than that of the filler metal, it is likely to be less than that of the base metals; (3) high service temperatures may weaken a brazed joint; and (4) the color of the metal in the brazed joint may not match the color of the base metal parts, a possible aesthetic disadvantage.

Brazing as a production process is widely used in a variety of industries, including automotive (e.g., joining tubes and pipes), electrical equipment (e.g., joining wires and cables), cutting tools (e.g., brazing cemented carbide inserts to shanks), and jewelry making. In addition, the chemical processing industry and plumbing and heating contractors join metal pipes and tubes by brazing. The process is used extensively for repair and maintenance work in nearly all industries.

24.1.1 BRAZED JOINTS

Brazed joints are commonly of two types: butt and lap (Section 22.2.1). However, the two types have been adapted for the brazing process in several ways. The conventional butt joint provides a limited area for brazing, thus jeopardizing the strength of the joint. To increase the faying areas in brazed joints, the mating parts are often scarfed or stepped or otherwise altered, as shown in Figure 24.1. Of course, additional processing is usually required in the making of the parts for these special joints. One of the particular difficulties associated with a scarfed joint is the problem of maintaining the alignment of the parts before and during brazing.

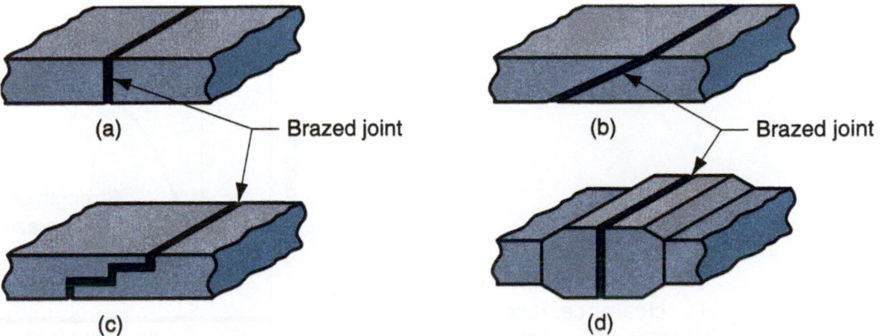

FIGURE 24.1 (a) Conventional butt joint, and adaptations of the butt joint for brazing: (b) scarf joint, (c) stepped butt joint, (d) increased cross section of the part at the joint. (Credit: *Fundamentals of Modern Manufacturing*, 4th Edition by Mikell P. Groover, 2010. Reprinted with permission of John Wiley & Sons, Inc.)

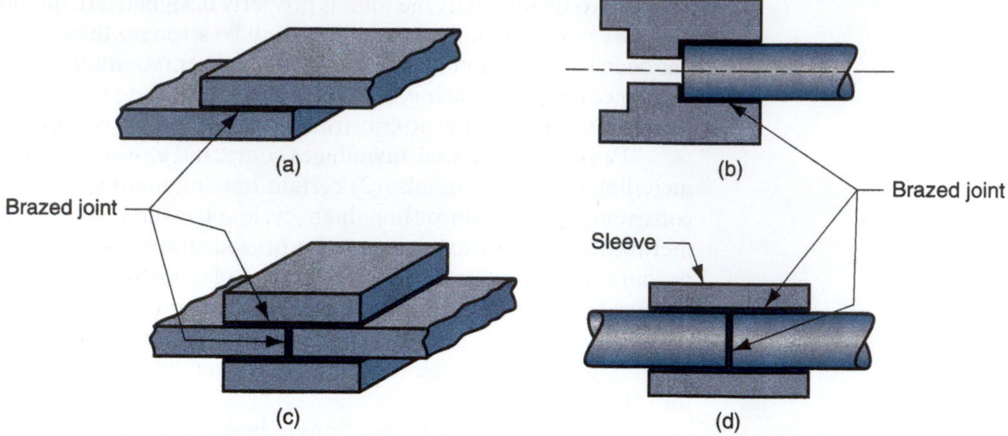

FIGURE 24.2 (a) Conventional lap joint, and adaptations of the lap joint for brazing: (b) cylindrical parts, (c) sandwiched parts, and (d) use of sleeve to convert butt joint into lap joint. (Credit: *Fundamentals of Modern Manufacturing*, 4th Edition by Mikell P. Groover, 2010. Reprinted with permission of John Wiley & Sons, Inc.)

Lap joints are more widely used in brazing, because they can provide a relatively large interface area between the parts. An overlap of at least three times the thickness of the thinner part is generally considered good design practice. Some adaptations of the lap joint for brazing are illustrated in Figure 24.2. An advantage of brazing over welding in lap joints is that the filler metal is bonded to the base parts throughout the entire interface area between the parts, rather than only at the edges (as in fillet welds made by arc welding) or at discrete spots (as in resistance spot welding).

Clearance between mating surfaces of the base parts is important in brazing. The clearance must be large enough so as not to restrict molten filler metal from flowing throughout the entire interface. Yet, if the joint clearance is too great, capillary action will be reduced and there will be areas between the parts where no filler metal is present. Joint strength is affected by clearance, as depicted in Figure 24.3. There is an optimum clearance value at which joint strength is maximized. The issue is complicated by the fact that the optimum depends on base and filler metals, joint configuration, and processing

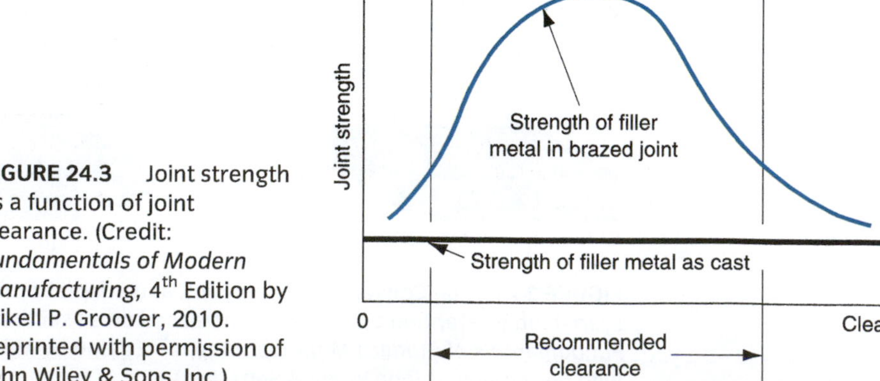

FIGURE 24.3 Joint strength as a function of joint clearance. (Credit: *Fundamentals of Modern Manufacturing*, 4th Edition by Mikell P. Groover, 2010. Reprinted with permission of John Wiley & Sons, Inc.)

conditions. Typical brazing clearances in practice are 0.025 mm to 0.25 mm (0.001 in to 0.010 in). These values represent the joint clearance at the brazing temperature, which may be different from room temperature clearance, depending on thermal expansion of the base metal(s).

Cleanliness of the joint surfaces prior to brazing is also important. Surfaces must be free of oxides, oils, and other contaminants to promote wetting and capillary attraction during the process, as well as bonding across the entire interface. Chemical treatments such as solvent cleaning (Section 21.1.1) and mechanical treatments such as wire brushing and sand blasting (Section 21.1.2) are used to clean the surfaces. After cleaning and during the brazing operation, fluxes are used to maintain surface cleanliness and promote wetting for capillary action in the clearance between faying surfaces.

24.1.2 FILLER METALS AND FLUXES

Common filler metals used in brazing are listed in Table 24.1 along with the principal base metals on which they are typically used. To qualify as a brazing metal, the following characteristics are needed: (1) melting temperature must be compatible with the base metal, (2) surface tension in the liquid phase must be low for good wettability, (3) fluidity of the molten metal must be high for penetration into the interface, (4) the metal must be capable of being brazed into a joint of adequate strength for the application, and (5) chemical and physical interactions with base metal (e.g., galvanic reaction) must be avoided. Filler metals are applied to the brazing operation in various ways, including wire, rod, sheets and strips, powders, pastes, preformed parts made of braze metal designed to fit a particular joint configuration, and cladding on one of the surfaces to be brazed. Several of these techniques are illustrated in Figure 24.4.

Brazing fluxes serve a similar purpose as in welding; they dissolve, combine with, and otherwise inhibit the formation of oxides and other unwanted by-products in the brazing process. Use of a flux does not substitute for the cleaning steps described above. Characteristics of a good flux include (1) low melting temperature, (2) low viscosity so that it can be displaced by the filler metal, (3) facilitates wetting, and (4) protects the joint until solidification of the filler metal. The flux should also be easy to remove after brazing. Common ingredients for brazing fluxes include borax, borates, fluorides, and chlorides. Wetting agents are also included in the mix to reduce surface tension of the molten filler metal and to improve wettability. Forms of flux include powders, pastes, and slurries. Alternatives to using a flux are to perform the operation in a vacuum or a reducing atmosphere that inhibits oxide formation.

TABLE 24.1 Common filler metals used in brazing and the base metals on which they are used.

Filler Metal	Typical Composition	Approximate Brazing Temperature °C	Approximate Brazing Temperature °F	Base Metals
Aluminum and silicon	90 Al, 10 Si	600	1100	Aluminum
Copper	99.9 Cu	1120	2050	Nickel copper
Copper and phosphorous	95 Cu, 5 P	850	1550	Copper
Copper and zinc	60 Cu, 40 Zn	925	1700	Steels, cast irons, nickel
Gold and silver	80 Au, 20 Ag	950	1750	Stainless steel, nickel alloys
Nickel alloys	Ni, Cr, others	1120	2050	Stainless steel, nickel alloys
Silver alloys	Ag, Cu, Zn, Cd	730	1350	Titanium, Monel, Inconel, tool steel, nickel

Compiled from [5] and [7].

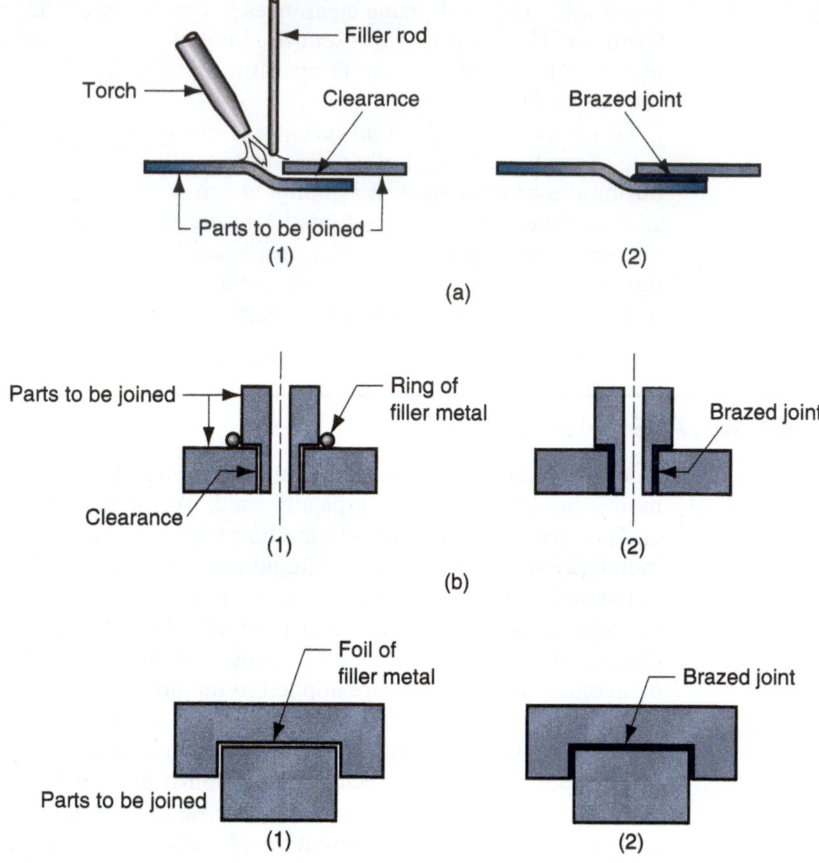

FIGURE 24.4 Several techniques for applying filler metal in brazing: (a) torch and filler rod; (b) ring of filler metal at entrance of gap; and (c) foil of filler metal between flat part surfaces. Sequence: (1) before, and (2) after. (Credit: *Fundamentals of Modern Manufacturing*, 4th Edition by Mikell P. Groover, 2010. Reprinted with permission of John Wiley & Sons, Inc.)

24.1.3 BRAZING METHODS

Various methods are used in brazing. Referred to as brazing processes, they are differentiated by their heating sources.

Torch Brazing In torch brazing, flux is applied to the part surfaces and a torch is used to direct a flame against the work in the vicinity of the joint. A reducing flame is typically used to inhibit oxidation. After the workpart joint areas have been heated to a suitable temperature, filler wire is added to the joint, usually in wire or rod form. Fuels used in torch brazing include acetylene, propane, and other gases, with air or oxygen. The selection of the mixture depends on heating requirements of the job. Torch brazing is often performed manually, and skilled workers must be employed to control the flame, manipulate the hand-held torches, and properly judge the temperatures; repair work is a common application. The method can also be used in mechanized production operations, in which parts and brazing metal are loaded onto a conveyor or indexing table and passed under one or more torches.

Furnace Brazing Furnace brazing uses a furnace to supply heat for brazing and is best suited for medium and high production. In medium production, usually in batches, the component parts and brazing metal are loaded into the furnace, heated to brazing temperature, and then cooled and removed. High-production operations use flow-through furnaces, in which parts are placed on a conveyor and are transported through the various heating and cooling sections. Temperature and atmosphere control are important in furnace brazing; the atmosphere must be neutral or reducing. Vacuum furnaces are sometimes used. Depending on the atmosphere and metals being brazed, the need for a flux may be eliminated.

Induction Brazing Induction brazing utilizes heat from electrical resistance to a high-frequency current induced in the work. The parts are preloaded with filler metal and placed in a high-frequency AC field—the parts do not directly contact the induction coil. Frequencies range from 5 kHz to 5 MHz. High-frequency power sources tend to provide surface heating, while lower frequencies cause deeper heat penetration into the work and are appropriate for heavier sections. The process can be used to meet low- to high-production requirements.

Resistance Brazing Heat to melt the filler metal in this process is obtained by resistance to flow of electrical current through the parts. As distinguished from induction brazing, the parts are directly connected to the electrical circuit in resistance brazing. The equipment is similar to that used in resistance welding, except that a lower power level is required for brazing. The parts with preplaced filler metal are held between electrodes while pressure and current are applied. Both induction and resistance brazing achieve rapid heating cycles and are used for relatively small parts. Induction brazing seems to be the more widely used of the two processes.

Dip Brazing In dip brazing, either a molten salt bath or a molten metal bath accomplishes heating. In both methods, assembled parts are immersed in the baths contained in a heating pot. Solidification occurs when the parts are removed from the bath. In the ***salt bath method***, the molten mixture contains fluxing ingredients and the filler metal is preloaded onto the assembly. In the ***metal bath method***, the molten filler metal is the heating medium; it is drawn by capillary action into the joint during submersion. A flux cover is maintained on the surface of the molten metal bath. Dip brazing achieves fast heating cycles and can be used to braze many joints on a single part or on multiple parts simultaneously.

Infrared Brazing This method uses heat from a high-intensity infrared (IR) lamp. Some IR lamps are capable of generating up to 5000 W of radiant heat energy, which can be directed at the workparts for brazing. The process is slower than most of the other processes reviewed above, and is generally limited to thin sections.

Braze Welding This process differs from the other brazing processes in the type of joint to which it is applied. As pictured in Figure 24.5, braze welding is used for filling a more conventional weld joint, such as the V-joint shown. A greater quantity of filler metal is deposited than in brazing, and no capillary action occurs. In braze welding, the joint consists entirely of filler metal; the base metal does not melt and is therefore not fused

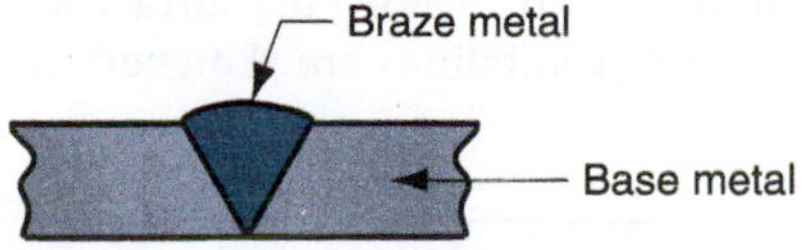

FIGURE 24.5 Braze welding. The joint consists of braze (filler) metal; no base metal is fused in the joint. (Credit: *Fundamentals of Modern Manufacturing*, 4th Edition by Mikell P. Groover, 2010. Reprinted with permission of John Wiley & Sons, Inc.)

into the joint as in a conventional fusion-welding process. The principal application of braze welding is repair work.

24.2 SOLDERING

Soldering is similar to brazing and can be defined as a joining process in which a filler metal with melting point (liquidus) not exceeding 450°C (840°F) is melted and distributed by capillary action between the faying surfaces of the metal parts being joined. As in brazing, no melting of the base metals occurs, but the filler metal wets and combines with the base metal to form a metallurgical bond. Details of soldering are similar to those of brazing, and many of the heating methods are the same. Surfaces to be soldered must be precleaned so they are free of oxides, oils, and so on. An appropriate flux must be applied to the faying surfaces, and the surfaces are heated. Filler metal, called ***solder***, is added to the joint, which distributes itself between the closely fitting parts.

In some applications, the solder is precoated onto one or both of the surfaces—a process called ***tinning***, irrespective of whether the solder contains any tin. Typical clearances in soldering range from 0.075 mm to 0.125 mm (0.003 in to 0.005 in), except when the surfaces are tinned, in which case a clearance of about 0.025 mm (0.001 in) is used. After solidification, the flux residue must be removed.

As an industrial process, soldering is most closely associated with electronics assembly. It is also used for mechanical joints, but not for joints subjected to elevated stresses or temperatures. Advantages attributed to soldering include (1) low energy input relative to brazing and fusion welding, (2) variety of heating methods available, (3) good electrical and thermal conductivity in the joint, (4) capability to make air-tight and liquid-tight seams for containers, and (5) easy to repair and rework.

The biggest disadvantages of soldering are (1) low joint strength unless reinforced by mechanical means and (2) possible weakening or melting of the joint in elevated temperature service.

24.2.1 JOINT DESIGNS IN SOLDERING

As in brazing, soldered joints are limited to lap and butt types, although butt joints should not be used in load-bearing applications. Some of the brazing adaptations of these joints also apply to soldering, and soldering technology has added a few more variations of its own to deal with the special part geometries that occur in electrical connections. In soldered mechanical joints of sheet-metal parts, the edges of the sheets are often bent over and interlocked before soldering, as shown in Figure 24.6, to increase joint strength.

For electronics applications, the principal function of the soldered joint is to provide an electrically conductive path between two parts being joined. Other design considerations in these types of soldered joints include problems with heat generation (from the electrical resistance of the joint) and vibration. Mechanical strength in a soldered electrical connection is often achieved by deforming one or both of the metal parts to accomplish a mechanical joint between them, or by making the surface area larger to provide maximum support by the solder. Several possibilities are sketched in Figure 24.7.

24.2.2 SOLDERS AND FLUXES

Solders and fluxes are the materials used in soldering. Both are critically important in the joining process.

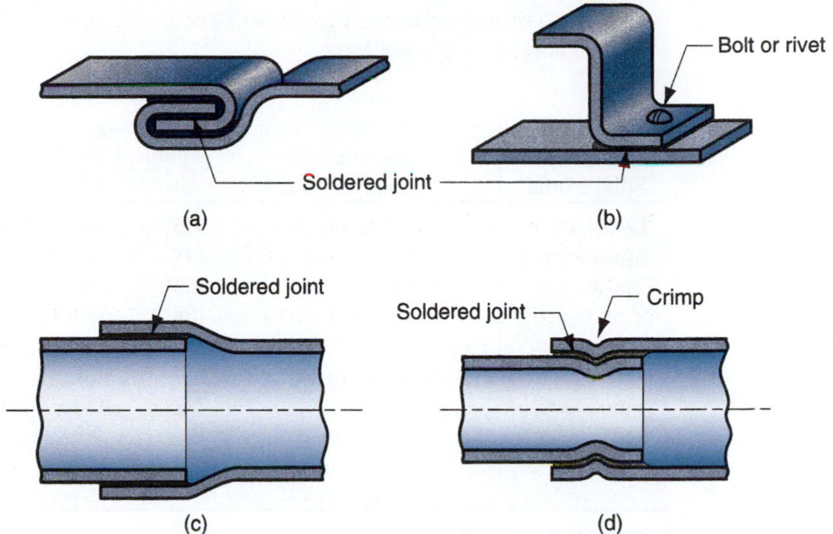

FIGURE 24.6 Mechanical interlocking in soldered joints for increased strength: (a) flat lock seam; (b) bolted or riveted joint; (c) copper pipe fittings—lap cylindrical joint; and (d) crimping (forming) of cylindrical lap joint. (Credit: *Fundamentals of Modern Manufacturing,* 4th Edition by Mikell P. Groover, 2010. Reprinted with permission of John Wiley & Sons, Inc.)

Solders Many solders are alloys of tin and lead, because both metals have low melting points. Their alloys possess a range of liquidus and solidus temperatures to achieve good control of the soldering process for a variety of applications. Lead is poisonous and its percentage is minimized in most solder compositions. Tin is chemically active at soldering temperatures and promotes the wetting action required for successful joining. In soldering copper, common in electrical connections, intermetallic compounds of copper and tin are formed that strengthen the bond. Silver and antimony are also sometimes used in soldering alloys. Table 24.2 lists various solder alloy compositions, indicating their approximate soldering temperatures and principal applications. Lead-free solders are becoming increasingly important as legislation to eliminate lead from solders is enacted.

Soldering Fluxes Soldering fluxes should do the following: (1) be molten at soldering temperatures, (2) remove oxide films and tarnish from the base part surfaces, (3) prevent

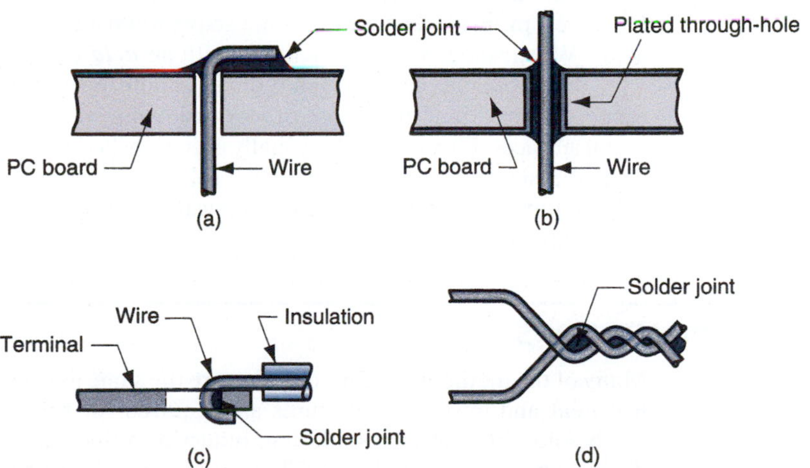

FIGURE 24.7 Techniques for securing the joint by mechanical means prior to soldering in electrical connections: (a) crimped lead wire on PC board; (b) plated through hole on PC board to maximize solder contact surface; (c) hooked wire on flat terminal; and (d) twisted wires. (Credit: *Fundamentals of Modern Manufacturing,* 4th Edition by Mikell P. Groover, 2010. Reprinted with permission of John Wiley & Sons, Inc.)

TABLE 24.2 Some common solder alloy compositions with their melting temperatures and applications.

Filler Metal	Approximate Composition	Approximate Melting Temperature °C	°F	Principal Applications
Lead-silver	96 Pb, 4 Ag	305	580	Elevated temperature joints
Tin-antimony	95 Sn, 5 Sb	238	460	Plumbing & heating
Tin-lead	63 Sn, 37 Pb	183[a]	361[a]	Electrical/electronics
	60 Sn, 40 Pb	188	370	Electrical/electronics
	50 Sn, 50 Pb	199	390	General purpose
	40 Sn, 60 Pb	207	405	Automobile radiators
Tin-silver	96 Sn, 4 Ag	221	430	Food containers
Tin-zinc	91 Sn, 9 Zn	199	390	Aluminum joining
Tin-silver-copper	95.5 Sn, 3.9 Ag, 0.6 Cu	217	423	Electronics: surface mount technology

Compiled from [2], [3], [4], and [13].
[a]Eutectic composition—lowest melting point of tin–lead compositions.

oxidation during heating, (4) promote wetting of the faying surfaces, (5) be readily displaced by the molten solder during the process, and (6) leave a residue that is noncorrosive and nonconductive. Unfortunately, there is no single flux that serves all of these functions perfectly for all combinations of solder and base metals. The flux formulation must be selected for a given application.

Soldering fluxes can be classified as organic or inorganic. ***Organic fluxes*** are made of either rosin (i.e., natural rosin such as gum wood, which is not water-soluble) or water-soluble ingredients (e.g., alcohols, organic acids, and halogenated salts). The water-soluble type facilitates cleanup after soldering. Organic fluxes are most commonly used for electrical and electronics connections. They tend to be chemically reactive at elevated soldering temperatures but relatively noncorrosive at room temperatures. ***Inorganic fluxes*** consist of inorganic acids (e.g., muriatic acid) and salts (e.g., combinations of zinc and ammonium chlorides) and are used to achieve rapid and active fluxing where oxide films are a problem. The salts become active when melted, but are less corrosive than the acids. When solder wire is purchased with an ***acid core***, it is in this category.

Both organic and inorganic fluxes should be removed after soldering, but it is especially important in the case of inorganic acids to prevent continued corrosion of the metal surfaces. Flux removal is usually accomplished using water solutions, except in the case of rosins, which require chemical solvents. Recent trends in industry favor water-soluble fluxes over rosins because chemical solvents used with rosins are harmful to the environment and to humans.

24.2.3 SOLDERING METHODS

Many of the methods used in soldering are the same as those used in brazing, except that less heat and lower temperatures are required for soldering. These methods include torch soldering, furnace soldering, induction soldering, resistance soldering, dip soldering, and infrared soldering. There are other soldering methods, not used in brazing, that should be described here. These methods are hand soldering, wave soldering, and reflow soldering.

Hand Soldering Hand soldering is performed manually using a hot soldering iron. A ***bit***, made of copper, is the working end of a soldering iron. Its functions are (1) to deliver

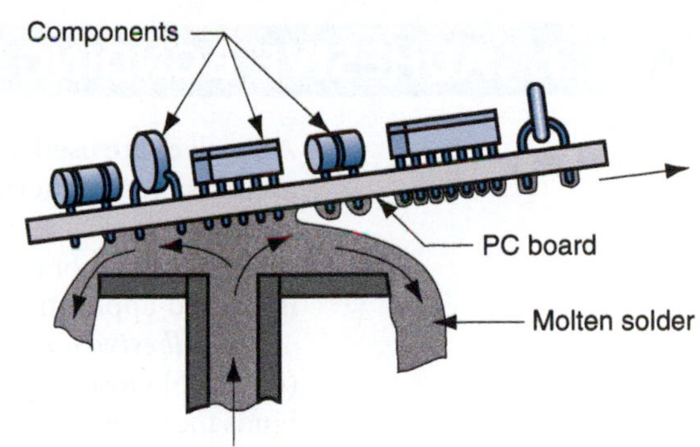

FIGURE 24.8 Wave soldering, in which molten solder is delivered up through a narrow slot onto the underside of a printed circuit board to connect the component lead wires. (Credit: *Fundamentals of Modern Manufacturing*, 4th Edition by Mikell P. Groover, 2010. Reprinted with permission of John Wiley & Sons, Inc.)

heat to the parts being soldered, (2) to melt the solder, (3) to convey molten solder to the joint, and (4) to withdraw excess solder. Most modern soldering irons are heated by electrical resistance. Some are designed as fast-heating ***soldering guns***, which are popular in electronics assembly for intermittent (on-off) operation actuated by a trigger. They are capable of making a solder joint in about a second.

Wave Soldering Wave soldering is a mechanized technique that allows multiple lead wires to be soldered to a printed circuit board (PCB) as it passes over a wave of molten solder. The typical setup is one in which a PCB, on which electronic components have been placed with their lead wires extending through the holes in the board, is loaded onto a conveyor for transport through the wave-soldering equipment. The conveyor supports the PCB on its sides, so that its underside is exposed to the processing steps, which consist of the following: (1) flux is applied using any of several methods, including foaming, spraying, or brushing; (2) preheating (using light bulbs, heating coils, and infrared devices) to evaporate solvents, activate the flux, and raise the temperature of the assembly; and (3) wave soldering, in which the liquid solder is pumped from a molten bath through a slit onto the bottom of the board to make the soldering connections between the lead wires and the metal circuit on the board. This third step is illustrated in Figure 24.8. The board is often inclined slightly, as depicted in the sketch, and a special tinning oil is mixed with the molten solder to lower its surface tension. Both of these measures help to inhibit buildup of excess solder and formation of "icicles" on the bottom of the board. Wave soldering is widely applied in electronics to produce printed circuit board assemblies.

Reflow Soldering This process is also widely used in electronics to assemble surface mount components to printed circuit boards. In the process, a solder paste consisting of solder powders in a flux binder is applied to spots on the board where electrical contacts are to be made between surface mount components and the copper circuit. The components are then placed on the paste spots, and the board is heated to melt the solder, forming mechanical and electrical bonds between the component leads and the copper on the circuit board.

Heating methods for reflow soldering include vapor phase reflow and infrared reflow. In ***vapor phase reflow soldering***, an inert fluorinated hydrocarbon liquid is vaporized by heating in an oven; it subsequently condenses on the board surface where it transfers its heat of vaporization to melt the solder paste and form solder joints on the printed circuit boards. In ***infrared reflow soldering***, heat from an infrared lamp is used to melt the solder paste and form joints between component leads and circuit areas on the board. Additional heating methods to reflow the solder paste include use of hot plates, hot air, and lasers.

24.3 ADHESIVE BONDING

Adhesives are used in a wide range of bonding and sealing applications for joining similar and dissimilar materials such as metals, plastics, ceramics, wood, paper, and cardboard. Although well established as a joining technique, adhesive bonding is considered a growth area among assembly technologies because of the tremendous opportunities for increased applications.

Adhesive bonding is a joining process in which a filler material is used to hold two (or more) closely spaced parts together by surface attachment. The filler material that binds the parts together is the ***adhesive***. It is a nonmetallic substance—usually a polymer. The parts being joined are called ***adherends***. Adhesives of greatest interest in engineering are ***structural adhesives***, which are capable of forming strong, permanent joints between strong, rigid adherends. A large number of commercially available adhesives are cured by various mechanisms and suited to the bonding of various materials. ***Curing*** refers to the process by which the adhesive's physical properties are changed from a liquid to a solid, usually by chemical reaction, to accomplish the surface attachment of the parts. The chemical reaction may involve polymerization, condensation, or vulcanization. Curing is often motivated by heat and/or a catalyst, and pressure is sometimes applied between the two parts to activate bonding. If heat is required, the curing temperatures are relatively low, and so the materials being joined are usually unaffected—an advantage for adhesive bonding. The curing or hardening of the adhesive takes time, called ***curing time*** or ***setting time***. In some cases this time is significant—generally a disadvantage in manufacturing.

Joint strength in adhesive bonding is determined by the strength of the adhesive itself and the strength of attachment between adhesive and each of the adherends. One of the criteria often used to define a satisfactory adhesive joint is that if a failure should occur due to excessive stresses, it occurs in one of the adherends rather than at an interface or within the adhesive itself. The strength of the attachment results from several mechanisms, all depending on the particular adhesive and adherends: (1) chemical bonding, in which the adhesive unites with the adherends and forms a primary chemical bond upon hardening; (2) physical interactions, in which secondary bonding forces result between the atoms of the opposing surfaces; and (3) mechanical interlocking, in which the surface roughness of the adherend causes the hardened adhesive to become entangled or trapped in its microscopic surface asperities.

For these adhesion mechanisms to operate with best results, the following conditions must prevail: (1) surfaces of the adherend must be clean—free of dirt, oil, and oxide films that would interfere with achieving intimate contact between adhesive and adherend; special preparation of the surfaces is often required; (2) the adhesive in its initial liquid form must achieve thorough wetting of the adherend surface; and (3) it is usually helpful for the surfaces to be other than perfectly smooth—a slightly roughened surface increases the effective contact area and promotes mechanical interlocking. In addition, the joint must be designed to exploit the particular strengths of adhesive bonding and avoid its limitations.

24.3.1 JOINT DESIGN

Adhesive joints are generally not as strong as those by welding, brazing, or soldering. Accordingly, consideration must be given to the design of joints that are adhesively bonded. The following design principles are applicable: (1) Joint contact area should be maximized. (2) Adhesive joints are strongest in shear and tension as in Figure 24.9(a) and (b), and joints should be designed so that the applied stresses are of these types.

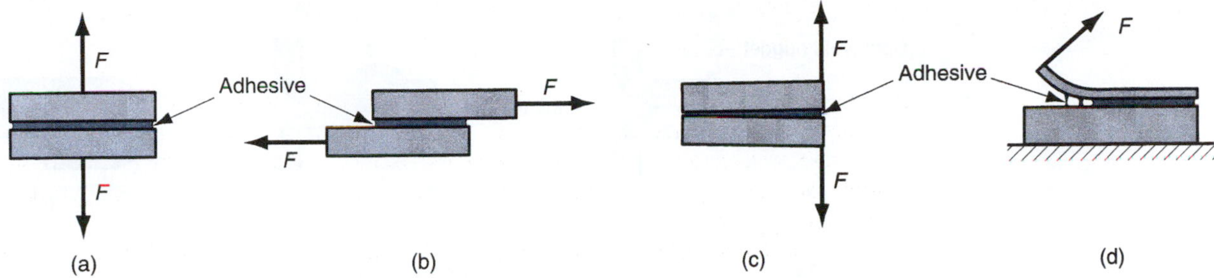

FIGURE 24.9 Types of stresses that must be considered in adhesive bonded joints: (a) tension, (b) shear, (c) cleavage, and (d) peeling. (Credit: *Fundamentals of Modern Manufacturing,* 4th Edition by Mikell P. Groover, 2010. Reprinted with permission of John Wiley & Sons, Inc.)

(3) Adhesive bonded joints are weakest in cleavage or peeling as in Figure 24.9(c) and (d), and adhesive bonded joints should be designed to avoid these types of stresses.

Typical joint designs for adhesive bonding that illustrate these design principles are presented in Figure 24.10. Some joint designs combine adhesive bonding with other joining methods to increase strength and/or provide sealing between the two components. Some of the possibilities are shown in Figure 24.11. For example, the combination of adhesive bonding and spot welding is called ***weldbonding***.

In addition to the mechanical configuration of the joint, the application must be selected so that the physical and chemical properties of adhesive and adherends are compatible under the service conditions to which the assembly will be subjected. Adherend materials include metals, ceramics, glass, plastics, wood, rubber, leather, cloth, paper, and cardboard. Note that the list includes materials that are rigid and flexible, porous and nonporous, metallic and nonmetallic, and that similar or dissimilar substances can be bonded together.

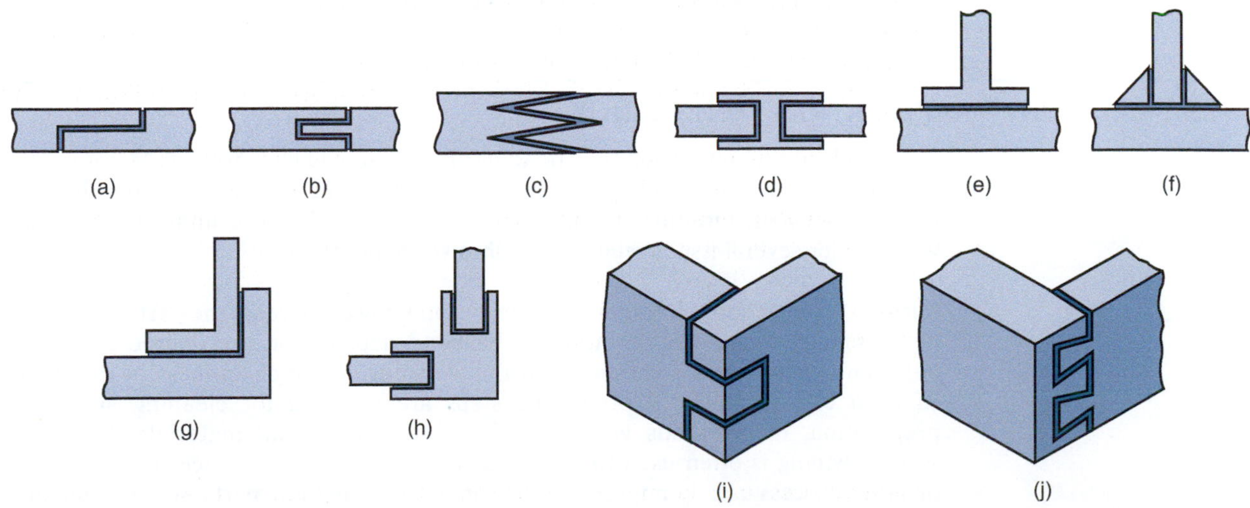

FIGURE 24.10 Some joint designs for adhesive bonding: (a) through (d) butt joints; (e) and (f) T-joints; and (g) through (j) corner joints. (Credit: *Fundamentals of Modern Manufacturing,* 4th Edition by Mikell P. Groover, 2010. Reprinted with permission of John Wiley & Sons, Inc.)

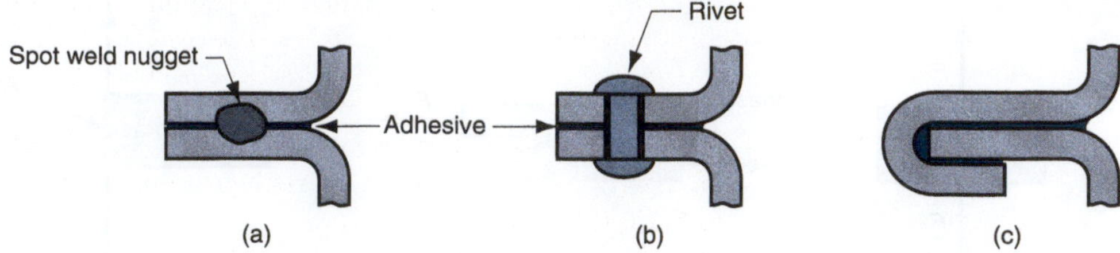

FIGURE 24.11 Adhesive bonding combined with other joining methods: (a) weldbonding—spot welded and adhesive bonded; (b) riveted (or bolted) and adhesive bonded; and (c) formed plus adhesive bonded. (Credit: *Fundamentals of Modern Manufacturing*, 4th Edition by Mikell P. Groover, 2010. Reprinted with permission of John Wiley & Sons, Inc.)

24.3.2 ADHESIVE TYPES

A large number of commercial adhesives are available. They can be classified into three categories: (1) natural, (2) inorganic, and (3) synthetic.

Natural adhesives are derived from natural sources (e.g., plants and animals), including gums, starch, dextrin, soy flour, and collagen. This category of adhesive is generally limited to low-stress applications, such as cardboard cartons, furniture, and bookbinding; or where large surface areas are involved (e.g., plywood). ***Inorganic adhesives*** are based principally on sodium silicate and magnesium oxychloride. Although relatively low in cost, they are also low in strength—a serious limitation in a structural adhesive.

Synthetic adhesives constitute the most important category in manufacturing. They include a variety of thermoplastic and thermosetting polymers. They are cured by various mechanisms, such as (1) mixing a catalyst or reactive ingredient with the polymer immediately prior to applying, (2) heating to initiate the chemical reaction, (3) radiation curing, such as ultraviolet light, and (4) curing by evaporation of water from the liquid or paste adhesive. In addition, some synthetic adhesives are applied as films or as pressure-sensitive coatings on the surface of one of the adherends.

24.3.3 ADHESIVE APPLICATION TECHNOLOGY

Industrial applications of adhesive bonding are widespread and growing. Major users are automotive, aircraft, building products, and packaging industries; other industries include footwear, furniture, bookbinding, electrical, and shipbuilding. In this section we consider several issues relating to adhesives application technology.

Surface Preparation In order for adhesive bonding to succeed, the part surfaces must be extremely clean. The strength of the bond depends on the degree of adhesion between adhesive and adherend, and this depends on the cleanliness of the surface. In most cases, additional processing steps are required for cleaning and surface preparation, the methods varying with different adherend materials. For metals, solvent wiping is often used for cleaning, and abrading the surface by sand blasting or other process usually improves adhesion. For nonmetallic parts, solvent cleaning is generally used, and the surfaces are sometimes mechanically abraded or chemically etched to increase roughness. It is desirable to accomplish the adhesive bonding process as soon as possible after these treatments, because surface oxidation and dirt accumulation increase with time.

Application Methods The actual application of the adhesive to one or both part surfaces is accomplished in a number of ways. The following list, though incomplete, provides a sampling of the techniques used in industry:

- ***Brushing***, performed manually, uses a stiff-bristled brush. Coatings are often uneven.
- ***Flowing***, using manually operated pressure-fed flow guns, has more consistent control than brushing.
- ***Manual rollers***, similar to paint rollers, are used to apply adhesive from a flat container.
- ***Silk screening*** involves brushing the adhesive through the open areas of the screen onto the part surface, so that only selected areas are coated.
- ***Spraying*** uses an air-driven (or airless) spray gun for fast application over large or difficult-to-reach areas.
- ***Automatic applicators*** include various automatic dispensers and nozzles for use on medium- and high-speed production applications.
- ***Roll coating*** is a mechanized technique in which a rotating roller is partially submersed in a pan of liquid adhesive and picks up a coating of the adhesive, which is then transferred to the work surface. Roll coating is used for applying adhesive onto thin, flexible materials (e.g., paper, cloth, leather) as well as wood, wood composite, cardboard, and similar materials with large surface areas.

Advantages and Limitations Advantages of adhesive bonding are: (1) the process is applicable to a wide variety of materials; (2) parts of different sizes and cross sections can be joined—fragile parts can be joined by adhesive bonding; (3) bonding occurs over the entire surface area of the joint, rather than in discrete spots or along seams as in fusion welding, thereby distributing stresses over the entire area; (4) some adhesives are flexible after bonding and are thus tolerant of cyclical loading and differences in thermal expansion of adherends; (5) low-temperature curing avoids damage to parts being joined; (6) sealing as well as bonding can be achieved; and (7) joint design is often simplified (e.g., two flat surfaces can be joined without providing special part features such as screw holes).

Principal limitations of this technology include: (1) joints are generally not as strong as other joining methods; (2) adhesive must be compatible with materials being joined; (3) service temperatures are limited; (4) cleanliness and surface preparation prior to application of adhesive are important; (5) curing times can impose a limit on production rates; and (6) inspection of the bonded joint is difficult.

REFERENCES

[1] Adams, R. S. (ed.). ***Adhesive Bonding: Science, Technology, and Applications***. CRC Taylor & Francis, Boca Raton, Florida, 2005.

[2] Bastow, E. "Five Solder Families and How They Work," ***Advanced Materials & Processes***, December 2003, pp. 26–29.

[3] Bilotta, A. J. ***Connections in Electronic Assemblies***. Marcel Dekker, Inc., New York, 1985.

[4] Bralla, J. G.(Editorin Chief). ***Design for Manufacturability Handbook***, 2nd ed. McGraw-Hill Book Company, New York, 1998.

[5] ***Brazing Manual***, 3rd ed. American Welding Society, Miami, Florida, 1976.

[6] Brockman, W., Geiss, P. L., Klingen, J., and Schroeder, K. B. ***Adhesive Bonding: Materials, Applications, and Technology***. John Wiley & Sons, Hoboken, New Jersey, 2009.

[7] Cary, H. B., and Helzer, S. C. ***Modern Welding Technology***, 6th ed. Pearson/Prentice-Hall, Upper Saddle River, New Jersey, 2005.

[8] Doyle, D. J. "The Sticky Six—Steps for Selecting Adhesives," ***Manufacturing Engineering***, June 1991 pp. 39–43.

[9] Driscoll, B., and Campagna, J. "Epoxy, Acrylic, and Urethane Adhesives," ***Advanced Materials & Processes***, August 2003, pp. 73–75.

[10] Hartshorn, S. R. (ed.). ***Structural Adhesives, Chemistry and Technology***. Plenum Press, New York, 1986.

[11] Humpston, G., and Jacobson, D. M. ***Principles of Brazing***. ASM International, Materials Park, Ohio, 2005.

[12] Humpston, G., and Jacobson, D. M. ***Principles of Soldering***. ASM International, Materials Park, Ohio, 2004.

[13] Lambert, L. P. ***Soldering for Electronic Assemblies***. Marcel Dekker, Inc., New York, 1988.

[14] Lincoln, B., Gomes, K. J., and Braden, J. F. ***Mechanical Fastening of Plastics***. Marcel Dekker, Inc., New York, 1984.

[15] Petrie, E. M. ***Handbook of Adhesives and Sealants***, 2nd ed. McGraw-Hill, New York, 2006.

[16] Schneberger, G. L. (ed.). ***Adhesives in Manufacturing***. CRC Taylor & Francis, Boca Raton, Florida, 1983.

[17] Shields, J. ***Adhesives Handbook***, 3rd ed. Butterworths Heinemann, Woburn, England, 1984.

[18] Skeist, I. (ed.). ***Handbook of Adhesives***, 3rd ed. Chapman & Hall, New York, 1990.

[19] ***Soldering Manual***, 2nd ed. American Welding Society, Miami, Florida, 1978.

[20] ***Welding Handbook***, 9th ed., Vol. **2**, ***Welding Processes***. American Welding Society, Miami, Florida, 2007.

[21] Wick, C.,and Veilleux, R. F.(eds.). ***Tool and Manufacturing Engineers Handbook***, 4th ed., Vol. **4**, ***Quality Control and Assembly***. Society of Manufacturing Engineers, Dearborn, Michigan, 1987.

REVIEW QUESTIONS

24.1. How do brazing and soldering differ from the fusion-welding processes?

24.2. How do brazing and soldering differ from the solid-state welding processes?

24.3. What is the technical difference between brazing and soldering?

24.4. Under what circumstances would brazing or soldering be preferred over welding?

24.5. What are the two joint types most commonly used in brazing?

24.6. Certain changes in joint configuration are usually made to improve the strength of brazed joints. What are some of these changes?

24.7. The molten filler metal in brazing is distributed throughout the joint by capillary action. What is capillary action?

24.8. What are the desirable characteristics of a brazing flux?

24.9. What is dip brazing?

24.10. Define braze welding.

24.11. What are some of the disadvantages and limitations of brazing?

24.12. What are the functions served by the bit of a soldering iron in hand soldering?

24.13. What is wave soldering?

24.14. List the advantages often attributed to soldering as an industrial joining process.

24.15. What are the disadvantages and drawbacks of soldering?

24.16. What is meant by the term *structural adhesive*?

24.17. An adhesive must cure in order to bond. What is meant by the term *curing*?

24.18. What are some of the methods used to cure adhesives?

24.19. Name the three basic categories of commercial adhesives.

24.20. What is an important precondition for the success of an adhesive bonding operation?

24.21. What are some of the methods used to apply adhesives in industrial production operations?

24.22. Identify some of the advantages of adhesive bonding compared to alternative joining methods.

24.23. What are some of the limitations of adhesive bonding?

25 MECHANICAL ASSEMBLY

Chapter Contents

25.1 Threaded Fasteners
- 25.1.1 Screws, Bolts, and Nuts
- 25.1.2 Other Threaded Fasteners and Related Hardware
- 25.1.3 Stresses and Strengths in Bolted Joints
- 25.1.4 Tools and Methods for Threaded Fasteners

25.2 Rivets

25.3 Assembly Methods Based on Interference Fits

25.4 Other Mechanical Fastening Methods

25.5 Molding Inserts and Integral Fasteners

25.6 Design for Assembly
- 25.6.1 General Principles of DFA
- 25.6.2 Design for Automated Assembly

Mechanical assembly consists of various methods to mechanically attach two (or more) parts together. In most cases, the method involves the use of discrete hardware components, called ***fasteners***, that are added to the parts during the assembly operation. In other cases, the method involves the shaping or reshaping of one of the components being assembled, and no separate fasteners are required. Many consumer products are produced using mechanical assembly: automobiles, large and small appliances, telephones, furniture, computers—even wearing apparel is ‘‘assembled’’ by mechanical means. In addition, industrial products such as airplanes, machine tools, and construction equipment almost always involve mechanical assembly.

Mechanical fastening methods can be divided into two major classes: (1) those that allow for disassembly, and (2) those that create a permanent joint. Threaded fasteners (e.g., screws, bolts, and nuts) are examples of the first class, and rivets illustrate the second. There are good reasons why mechanical assembly is often preferred over other joining processes discussed in previous chapters. The main reasons are (1) ease of assembly and (2) ease of disassembly (for the fastening methods that permit disassembly).

Mechanical assembly is usually accomplished by unskilled workers with a minimum of special tooling and in a relatively short time. The technology is simple, and the results are easily inspected. These factors are advantageous not only in the factory, but also during field installation. Large products that are too big and heavy to be transported completely assembled can be shipped in smaller subassemblies and then put together at the customer's site.

Ease of disassembly applies, of course, only to the mechanical fastening methods that permit disassembly. Periodic disassembly is required for many products so that maintenance and repair can be performed, for example, to replace worn-out components, make adjustments, and so forth. Permanent joining techniques such as welding do not allow for disassembly.

For purposes of organization, we divide mechanical assembly methods into the following categories: (1) threaded fasteners, (2) rivets, (3) interference fits, (4) other mechanical fastening methods, and (5) molded-in inserts and integral fasteners. These categories are described in Sections 25.1 through 25.5. In Section 25.6, we discuss an important topic in assembly: design for assembly.

25.1 THREADED FASTENERS

Threaded fasteners are discrete hardware components that have external or internal threads for assembly of parts. In nearly all cases, they permit disassembly. Threaded fasteners are the most important category of mechanical assembly; the common threaded fastener types are screws, bolts, and nuts.

25.1.1 SCREWS, BOLTS, AND NUTS

Screws and bolts are threaded fasteners that have external threads. There is a technical distinction between a screw and a bolt that is often blurred in popular usage. A ***screw*** is an externally threaded fastener that is generally assembled into a blind threaded hole. Some types, called ***self-tapping screws***, possess geometries that permit them to form or cut the matching threads in the hole. A ***bolt*** is an externally threaded fastener that is inserted through holes in the parts and "screwed" into a nut on the opposite side. A ***nut*** is an internally threaded fastener having standard threads that match those on bolts of the same diameter, pitch, and thread form. The typical assemblies that result from the use of screws and bolts are illustrated in Figure 25.1.

Screws and bolts come in a variety of standard sizes, threads, and shapes in both metric units (ISO standard) and U.S. customary units (ANSI standard).[1] The metric specification consists of the nominal major diameter, mm, followed by the pitch, mm. For example, a specification of 4-0.7 means a 4.0-mm major diameter and a pitch of 0.7 mm. The U.S. standard specifies either a number designating the major diameter (up to 0.2160 in) or the nominal major diameter in inches followed by the number of threads per inch. For example, the specification 1/4-20 indicates a major diameter of 0.25 in and 20 threads per inch.

Technical data on standard threaded fasteners can be found in design handbooks and fastener product catalogs. The United States has been gradually converting to metric

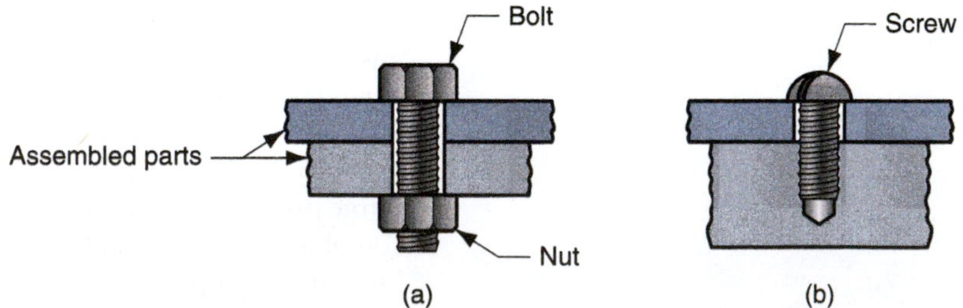

FIGURE 25.1 Typical assemblies using: (a) bolt and nut, and (b) screw. (Credit: *Fundamentals of Modern Manufacturing*, 4th Edition by Mikell P. Groover, 2010. Reprinted with permission of John Wiley & Sons, Inc.)

[1] ISO is the abbreviation for the International Standards Organization. ANSI is the abbreviation for the American National Standards Institute.

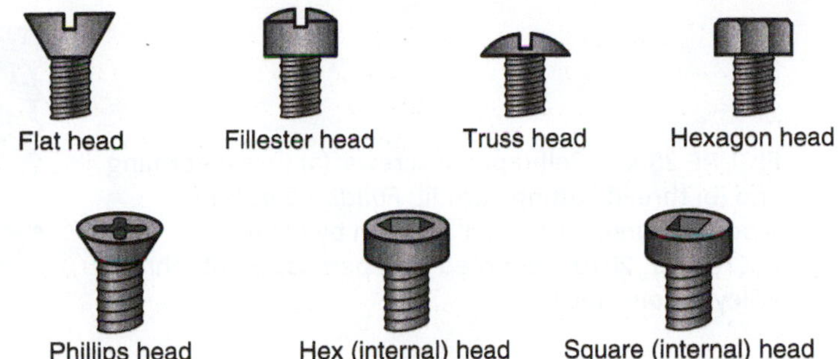

FIGURE 25.2 Various head styles available on screws and bolts. There are additional head styles not shown. (Credit: *Fundamentals of Modern Manufacturing,* 4th Edition by Mikell P. Groover, 2010. Reprinted with permission of John Wiley & Sons, Inc.)

thread sizes, which will reduce proliferation of specifications. It should be noted that differences among threaded fasteners have tooling implications in manufacturing. To use a particular type of screw or bolt, the assembly worker must have tools that are designed for that fastener type. For example, there are numerous head styles available on bolts and screws, the most common of which are shown in Figure 25.2. The geometries of these heads, as well as the variety of sizes available, require different hand tools (e.g., screwdrivers) for the worker. One cannot turn a hex-head bolt with a conventional flat-blade screwdriver.

Screws come in a greater variety of configurations than bolts, because their functions vary more. The types include machine screws, capscrews, setscrews, and self-tapping screws. ***Machine screws*** are the generic type, designed for assembly into tapped holes. They are sometimes assembled to nuts, and in this usage they overlap with bolts. ***Capscrews*** have the same geometry as machine screws but are made of higher strength metals and to closer tolerances. ***Setscrews*** are hardened and designed for assembly functions such as fastening collars, gears, and pulleys to shafts as shown in Figure 25.3(a). They come in various geometries, some of which are illustrated in Figure 25.3(b). A ***self-tapping screw*** (also called a ***tapping screw***) is designed to form or cut threads in a preexisting hole into which it is being turned. Figure 25.4 shows two of the typical thread geometries for self-tapping screws.

Most threaded fasteners are produced by cold forming (Section 13.1.4). Some are machined (Sections 16.2.2 and 16.3.2), but this is usually a more expensive thread-making process. A variety of materials are used to make threaded fasteners, steels being the most common because of their good strength and low cost. These include low and medium carbon as well as alloy steels. Fasteners made of steel are usually plated or coated for

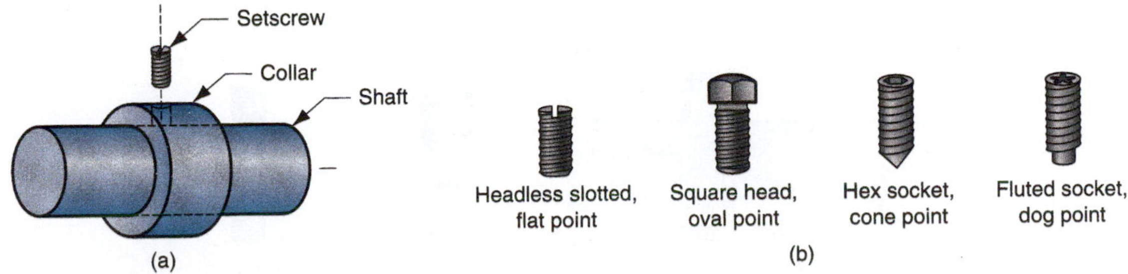

FIGURE 25.3 (a) Assembly of collar to shaft using a setscrew; (b) various setscrew geometries (head types and points). (Credit: *Fundamentals of Modern Manufacturing,* 4th Edition by Mikell P. Groover, 2010. Reprinted with permission of John Wiley & Sons, Inc.)

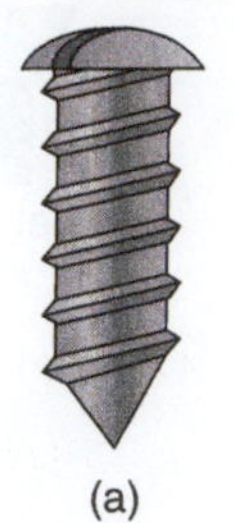

FIGURE 25.4 Self-tapping screws: (a) thread-forming and (b) thread-cutting. (Credit: *Fundamentals of Modern Manufacturing*, 4th Edition by Mikell P. Groover, 2010. Reprinted with permission of John Wiley & Sons, Inc.)

superficial resistance to corrosion. Nickel, chromium, zinc, black oxide, and similar coatings are used for this purpose. When corrosion or other factors deny the use of steel fasteners, other materials must be used, including stainless steels, aluminum alloys, nickel alloys, and plastics (however, plastics are suited to low-stress applications only).

25.1.2 OTHER THREADED FASTENERS AND RELATED HARDWARE

Additional threaded fasteners and related hardware include studs, screw thread inserts, captive threaded fasteners, and washers. A ***stud*** (in the context of fasteners) is an externally threaded fastener, but without the usual head possessed by a bolt. Studs can be used to assemble two parts using two nuts as shown in Figure 25.5(a). They are available with threads on one end or both as in Figure 25.5(b) and (c).

Screw thread inserts are internally threaded plugs or wire coils made to be inserted into an unthreaded hole and to accept an externally threaded fastener. They are assembled into weaker materials (e.g., plastic, wood, and lightweight metals such as magnesium) to provide strong threads. There are many designs of screw thread inserts, one example of which is illustrated in Figure 25.6. Upon subsequent assembly of the screw into the insert, the insert barrel expands into the sides of the hole, securing the assembly.

A ***washer*** is a hardware component often used with threaded fasteners to ensure tightness of the mechanical joint; in its simplest form, it is a flat thin ring of sheet metal. Washers serve various functions. They (1) distribute stresses that might otherwise be concentrated at the bolt or screw head and nut, (2) provide support for large clearance holes in the assembled parts, (3) increase spring tension, (4) protect part surfaces, (5) seal the joint, and (6) resist inadvertent unfastening [13]. Three washer types are illustrated in Figure 25.7.

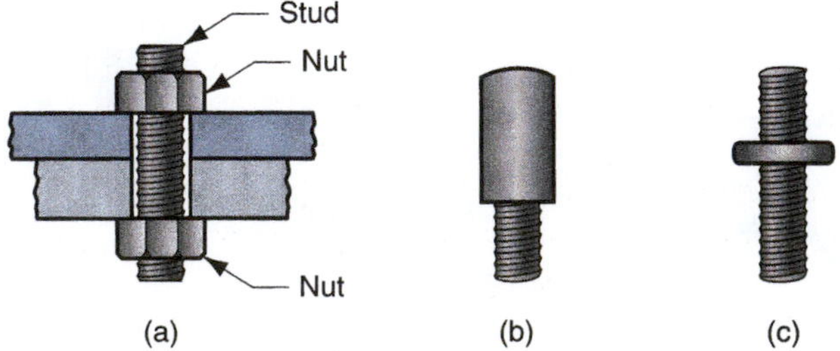

FIGURE 25.5 (a) Stud and nuts used for assembly. Other stud types: (b) threads on one end only and (c) double-end stud. (Credit: *Fundamentals of Modern Manufacturing*, 4th Edition by Mikell P. Groover, 2010. Reprinted with permission of John Wiley & Sons, Inc.)

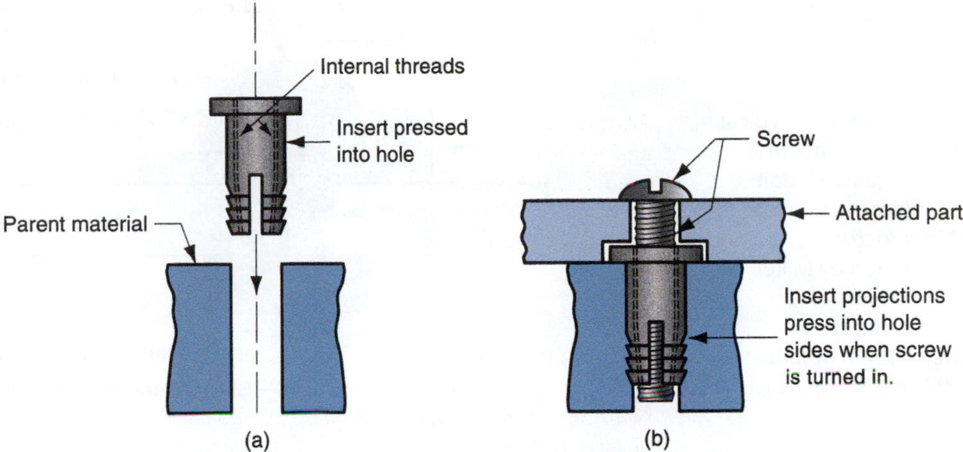

FIGURE 25.6 Screw thread inserts: (a) before insertion, and (b) after insertion into hole and screw is turned into the insert. (Credit: *Fundamentals of Modern Manufacturing,* 4th Edition by Mikell P. Groover, 2010. Reprinted with permission of John Wiley & Sons, Inc.)

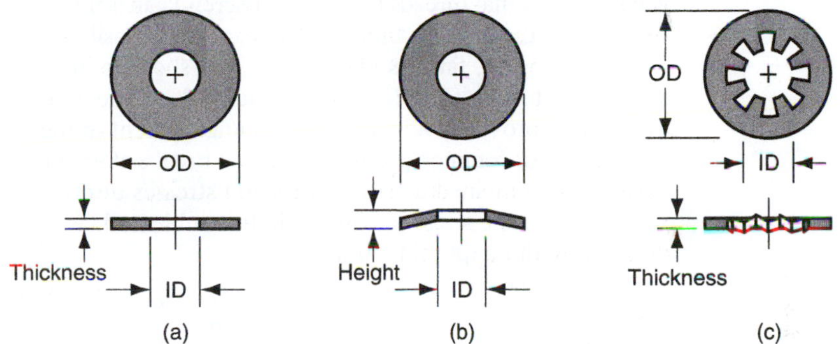

FIGURE 25.7 Types of washers: (a) plain (flat) washers; (b) spring washers, used to dampen vibration or compensate for wear; and (c) lockwasher designed to resist loosening of the bolt or screw. (Credit: *Fundamentals of Modern Manufacturing,* 4th Edition by Mikell P. Groover, 2010. Reprinted with permission of John Wiley & Sons, Inc.)

25.1.3 STRESSES AND STRENGTHS IN BOLTED JOINTS

Typical stresses acting on a bolted or screwed joint include both tensile and shear, as depicted in Figure 25.8. Shown in the figure is a bolt-and-nut assembly. Once tightened, the bolt is loaded in tension, and the parts are loaded in compression. In addition, forces may be acting in opposite directions on the parts, which results in a shear stress on the bolt cross section. Finally, there are stresses applied on the threads throughout their engagement length with the nut in a direction parallel to the axis of the bolt. These shear stresses can cause ***stripping*** of the threads. (This failure can also occur on the internal threads of a nut.)

The strength of a threaded fastener is generally specified by two measures: (1) tensile strength, which has the traditional definition (Section 3.1.1), and (2) proof strength. ***Proof strength*** is roughly equivalent to yield strength; specifically, it is the maximum tensile stress to which an externally threaded fastener can be subjected without

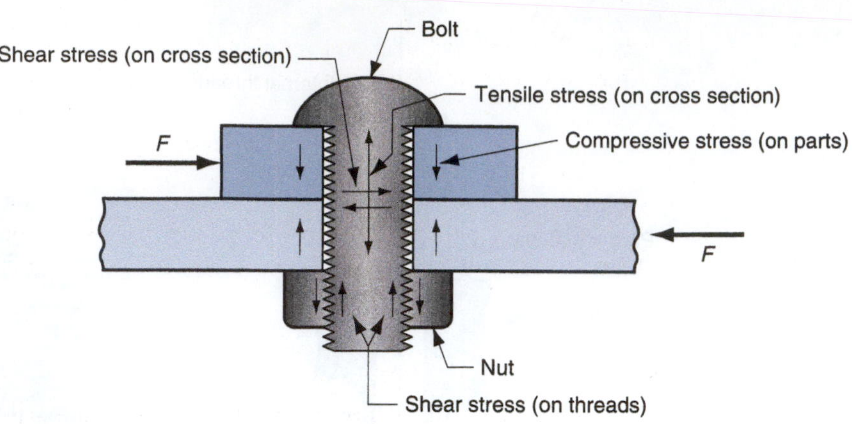

FIGURE 25.8 Typical stresses acting on a bolted joint. (Credit: *Fundamentals of Modern Manufacturing*, 4th Edition by Mikell P. Groover, 2010. Reprinted with permission of John Wiley & Sons, Inc.)

permanent deformation. Typical values of tensile and proof strength for steel bolts are given in Table 25.1.

The problem that can arise during assembly is that the threaded fasteners are overtightened, causing stresses that exceed the strength of the fastener material. Assuming a bolt-and-nut assembly as shown in Figure 25.8, failure can occur in one of the following ways: (1) external threads (e.g., bolt or screw) can strip, (2) internal threads (e.g., nut) can strip, or (3) the bolt can break due to excessive tensile stresses on its cross-sectional area. Thread stripping, failures (1) and (2), is a shear failure and occurs when the length of engagement is too short (less than about 60% of the nominal bolt diameter). This can be avoided by providing adequate thread engagement in the fastener design. Tensile failure (3) is the most common problem. The bolt breaks at about 85% of its rated tensile strength because of combined tensile and torsion stresses during tightening [2].

The tensile stress to which a bolt is subjected can be calculated as the tensile load divided by the applicable area:

$$\sigma = \frac{F}{A_s} \tag{25.1}$$

where σ = stress, MPa (lb/in^2); F = load, N (lb); and A_s = tensile stress area, mm^2 (in^2). This stress is compared to the bolt strength values listed in Table 25.1. The tensile stress area for a threaded fastener is the cross-sectional area of the minor diameter. This area can be calculated directly from one of the following equations [2], depending on whether the bolt is metric standard or American standard. For the metric standard (ISO), the formula is:

$$A_s = \frac{\pi}{4}(D - 0.9382p)^2 \tag{25.2}$$

TABLE 25.1 Typical values of proof and tensile strengths for steel bolts and screws, diameters range from 6.4 mm (0.25 in) to 38 mm (1.50 in).

	Proof Stress		Tensile Stress	
Material	**MPa**	**lb/in^2**	**MPa**	**lb/in^2**
Low/medium C steel	228	33,000	414	60,000
Alloy steel	830	120,000	1030	150,000

Source: [13].

where D = nominal size (basic major diameter) of the bolt or screw, mm; and p = thread pitch, mm. For the American standard (ANSI), the formula is:

$$A_s = \frac{\pi}{4}\left(D - \frac{0.9743}{n}\right)^2 \tag{25.3}$$

where D = nominal size (basic major diameter) of the bolt or screw, in; and n = the number of threads per inch.

25.1.4 TOOLS AND METHODS FOR THREADED FASTENERS

The basic function of the tools and methods for assembling threaded fasteners is to provide relative rotation between the external and internal threads, and to apply sufficient torque to secure the assembly. Available tools range from simple hand-held screwdrivers or wrenches to powered tools with sophisticated electronic sensors to ensure proper tightening. It is important that the tool match the screw or bolt and/or the nut in style and size, because there are so many sizes and styles available. Hand tools are usually made with a single point or blade, but powered tools are generally designed to use interchangeable bits. The powered tools operate by pneumatic, hydraulic, or electric power.

Whether a threaded fastener serves its intended purpose depends to a large degree on the amount of torque applied to tighten it. Once the bolt or screw (or nut) has been rotated until it is seated against the part surface, additional tightening will increase the tension in the fastener (and simultaneously the compression in the parts being held together); and the tightening will be resisted by an increasing torque. Thus, there is a correlation between the torque required to tighten the fastener and the tensile stress experienced by it. To achieve the desired function in the assembled joint (e.g., to improve fatigue resistance) and to lock the threaded fasteners, the product designer will often specify the tension force that should be applied. This force is called the ***preload***. The following relationship can be used to determine the required torque to obtain a specified preload [13]:

$$T = C_t DF \tag{25.4}$$

where T = torque, N-mm (lb-in); C_t = the torque coefficient whose value typically ranges between 0.15 and 0.25, depending on the thread surface conditions; D = nominal bolt or screw diameter, mm (in); and F = specified preload tension force, N (lb).

Various methods are employed to apply the required torque, including (1) operator feel—not very accurate, but adequate for most assemblies; (2) torque wrenches, which measure the torque as the fastener is being turned; (3) stall-motors, which are motorized wrenches designed to stall when the required torque is reached, and (4) torque-turn tightening, in which the fastener is initially tightened to a low torque level and then rotated a specified additional amount (e.g., a quarter turn).

25.2 RIVETS

Rivets are widely used for achieving a permanent mechanically fastened joint. Riveting is a fastening method that offers high production rates, simplicity, dependability, and low cost. Despite these apparent advantages, its applications have declined in recent decades in favor of threaded fasteners, welding, and adhesive bonding. Riveting is one of the primary fastening processes in the aircraft and aerospace industries for joining skins to channels and other structural members.

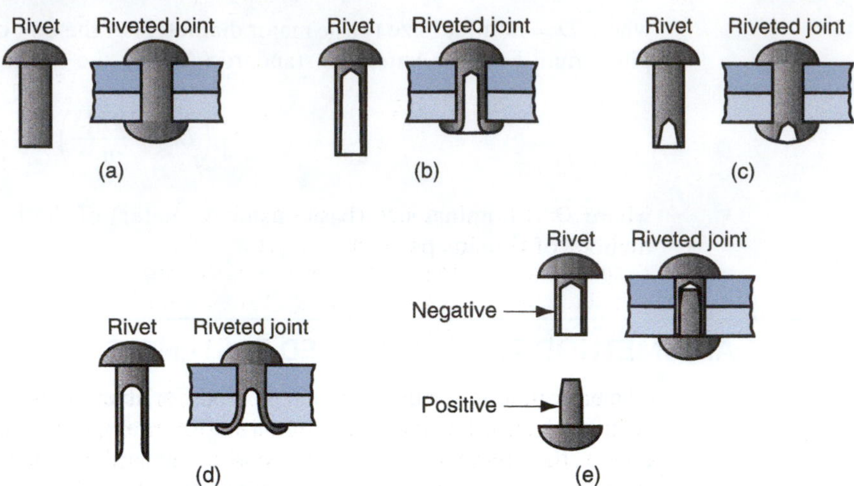

FIGURE 25.9 Five basic rivet types, also shown in assembled configuration: (a) solid, (b) tubular, (c) semitubular, (d) bifurcated, and (e) compression. (Credit: *Fundamentals of Modern Manufacturing,* 4th Edition by Mikell P. Groover, 2010. Reprinted with permission of John Wiley & Sons, Inc.)

A ***rivet*** is an unthreaded, headed pin used to join two (or more) parts by passing the pin through holes in the parts and then forming (upsetting) a second head in the pin on the opposite side. The deforming operation can be performed hot or cold (hot working or cold working), and by hammering or steady pressing. Once the rivet has been deformed, it cannot be removed except by breaking one of the heads. Rivets are specified by their length, diameter, head, and type. Rivet type refers to five basic geometries that affect how the rivet will be upset to form the second head. The five types are defined in Figure 25.9. In addition, there are special rivets for special applications.

Rivets are used primarily for lap joints. The clearance hole into which the rivet is inserted must be close to the diameter of the rivet. If the hole is too small, rivet insertion will be difficult, thus reducing production rate. If the hole is too large, the rivet will not fill the hole and may bend or compress during formation of the opposite head. Rivet design tables are available to specify the optimum hole sizes.

The tooling and methods used in riveting can be divided into the following categories: (1) impact, in which a pneumatic hammer delivers a succession of blows to upset the rivet; (2) steady compression, in which the riveting tool applies a continuous squeezing pressure to upset the rivet; and (3) a combination of impact and compression. Much of the equipment used in riveting is portable and manually operated. Automatic drilling-and-riveting machines are available for drilling the holes and then inserting and upsetting the rivets.

25.3 ASSEMBLY METHODS BASED ON INTERFERENCE FITS

Several assembly methods are based on mechanical interference between the two mating parts being joined. This interference, which occurs either during assembly or after the parts are joined, holds the parts together. The methods include press fitting, shrink and expansion fits, snap fits, and retaining rings.

Press Fitting A press fit assembly is one in which the two components have an interference fit between them. The typical case is where a pin (e.g., a straight cylindrical pin) of a certain diameter is pressed into a hole of a slightly smaller diameter. Standard pin sizes are commercially available to accomplish a variety of functions, such as (1) locating and locking the components—used to augment threaded fasteners by holding two (or more) parts in fixed alignment with each other; (2) pivot points, to permit rotation

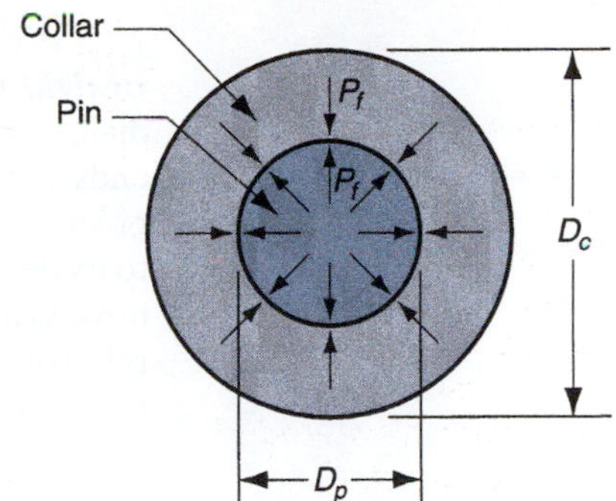

FIGURE 25.10 Cross section of a solid pin or shaft assembled to a collar by interference fit. (Credit: *Fundamentals of Modern Manufacturing*, 4th Edition by Mikell P. Groover, 2010. Reprinted with permission of John Wiley & Sons, Inc.)

of one component about the other; and (3) shear pins. Except for (3), the pins are normally hardened. Shear pins are made of softer metals so as to break under a sudden or severe shearing load to save the rest of the assembly. Other applications of press fitting include assembly of collars, gears, pulleys, and similar components onto shafts.

The pressures and stresses in an interference fit can be estimated using several applicable formulas. If the fit consists of a round solid pin or shaft inside a collar (or similar component), as depicted in Figure 25.10, and the components are made of the same material, the radial pressure between the pin and the collar can be determined by [13]:

$$p_f = \frac{Ei(D_c^2 - D_p^2)}{D_p D_c^2} \tag{25.5}$$

where p_f = radial or interference fit pressure, MPa (lb/in^2); E = modulus of elasticity for the material; i = interference between the pin (or shaft) and the collar; that is, the starting difference between the inside diameter of the collar hole and the outside diameter of the pin, mm (in); D_c = outside diameter of the collar, mm (in); and D_p = pin or shaft diameter, mm (in).

The maximum effective stress occurs in the collar at its inside diameter and can be calculated as:

$$\text{Max}\,\sigma_e = \frac{2p_f D_c^2}{D_c^2 - D_p^2} \tag{25.6}$$

where Max σ_e = the maximum effective stress, MPa (lb/in^2), and p_f is the interference fit pressure computed from Eq. (25.5).

In situations in which a straight pin or shaft is pressed into the hole of a large part with geometry other than that of a collar, we can alter the previous equations by taking the outside diameter D_c to be infinite, thus reducing the equation for interference pressure to:

$$p_f = \frac{Ei}{D_p} \tag{25.7}$$

and the corresponding maximum effective stress becomes:

$$\text{Max}\,\sigma_e = 2_{p_f} \tag{25.8}$$

In most cases, particularly for ductile metals, the maximum effective stress should be compared with the yield strength of the material, applying an appropriate safety factor, as in the following:

$$\text{Max}\,\sigma_e \leq \frac{Y}{SF} \tag{25.9}$$

where Y = yield strength of the material, and SF is the applicable safety factor.

Various pin geometries are available for interference fits. The basic type is a ***straight pin***, usually made from cold-drawn carbon steel wire or bar stock, ranging in diameter from 1.6 to 25 mm (1/16 to 1.0 in). They are unground, with chamfered or square ends (chamfered ends facilitate press fitting). ***Dowel pins*** are manufactured to more precise specifications than straight pins, and can be ground and hardened. They are used to fix the alignment of assembled components in dies, fixtures, and machinery. ***Taper pins*** possess a taper of 6.4 mm (0.25 in) per foot and are driven into the hole to establish a fixed relative position between the parts. Their advantage is that they can readily be driven back out of the hole.

Shrink and Expansion Fits These terms refer to the assembly of two parts that have an interference fit at room temperature. The typical case is a cylindrical pin or shaft assembled into a collar. To assemble by ***shrink fitting***, the external part is heated to enlarge it by thermal expansion, and the internal part either remains at room temperature or is cooled to contract its size. The parts are then assembled and brought back to room temperature, so that the external part shrinks, and if previously cooled, the internal part expands to form a strong interference fit. An ***expansion fit*** is when only the internal part is cooled to contract it for assembly; once inserted into the mating component, it warms to room temperature, expanding to create the interference assembly. These assembly methods are used to fit gears, pulleys, sleeves, and other components onto solid and hollow shafts.

Various methods are used to heat and/or cool the workparts. Heating equipment includes torches, furnaces, electric resistance heaters, and electric induction heaters. Cooling methods include conventional refrigeration, packing in dry ice, and immersion in cold liquids, including liquid nitrogen. The resulting change in diameter depends on the coefficient of thermal expansion and the temperature difference that is applied to the part. If we assume that the heating or cooling has produced a uniform temperature throughout the work, then the change in diameter is given by:

$$D_2 - D_1 = \alpha D_1(T_2 - T_1) \tag{25.10}$$

where α = the coefficient of linear thermal expansion, mm/mm-°C (in/in-°F) for the material (see Table 3.10); T_2 = the temperature to which the parts have been heated or cooled, °C (°F); T_1 = starting ambient temperature; D_2 = diameter of the part at T_2, mm (in); and D_1 = diameter of the part at T_1.

Equations (25.5) through (25.9) for computing interference pressures and effective stresses can be used to determine the corresponding values for shrink and expansion fits.

Snap Fits and Retaining Rings Snap fits are a variation of interference fits. A ***snap fit*** involves joining two parts in which the mating elements possess a temporary interference while being pressed together, but once assembled they interlock to maintain the assembly. A typical example is shown in Figure 25.11: As the parts are pressed together, the mating elements elastically deform to accommodate the interference, subsequently allowing the parts to snap together; once in position, the elements become connected mechanically so that they cannot easily be disassembled. The parts are usually designed so that a slight interference exists after assembly.

Advantages of snap-fit assembly include: (1) the parts can be designed with self-aligning features, (2) no special tooling is required, and (3) assembly can be accomplished very quickly. Snap fitting was originally conceived as a method that would be ideally suited to industrial robotics applications; however, it is no surprise that assembly techniques that are easier for robots are also easier for human assembly workers.

A ***retaining ring***, also known as a ***snap ring***, is a fastener that snaps into a circumferential groove on a shaft or tube to form a shoulder, as in Figure 25.12. The

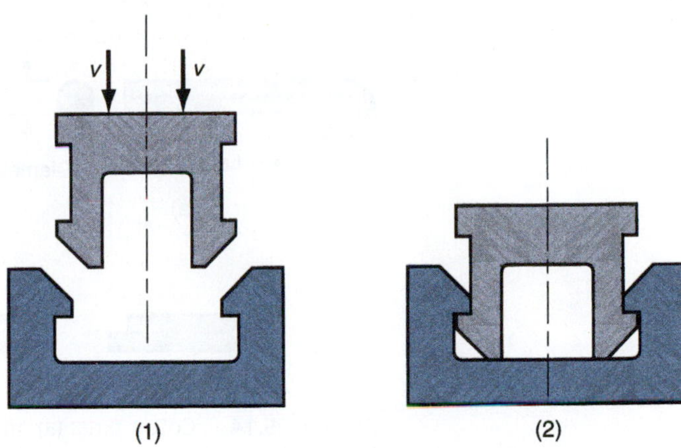

FIGURE 25.11 Snap-fit assembly, showing cross sections of two mating parts: (1) before assembly and (2) parts snapped together. (Credit: *Fundamentals of Modern Manufacturing,* 4th Edition by Mikell P. Groover, 2010. Reprinted with permission of John Wiley & Sons, Inc.)

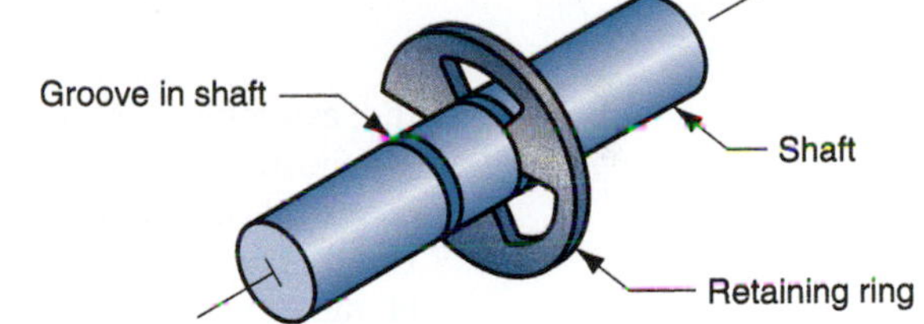

FIGURE 25.12 Retaining ring assembled into a groove on a shaft. (Credit: *Fundamentals of Modern Manufacturing,* 4th Edition by Mikell P. Groover, 2010. Reprinted with permission of John Wiley & Sons, Inc.)

assembly can be used to locate or restrict the movement of parts mounted on the shaft. Retaining rings are available for both external (shaft) and internal (bore) applications. They are made from either sheet metal or wire stock, heat-treated for hardness and stiffness. To assemble a retaining ring, a special pliers tool is used to elastically deform the ring so that it fits over the shaft (or into the bore) and then is released into the groove.

25.4 OTHER MECHANICAL FASTENING METHODS

In addition to the mechanical assembly techniques discussed above, there are several additional methods that involve the use of fasteners. These include stitching, stapling, sewing, and cotter pins.

Industrial stitching and stapling are similar operations involving the use of U-shaped metal fasteners. ***Stitching*** is a fastening operation in which a stitching machine is used to form the U-shaped stitches one at a time from steel wire and immediately drive them through the two parts to be joined. Figure 25.13 illustrates several types of wire stitches. The parts to be joined must be relatively thin, consistent with the stitch size, and

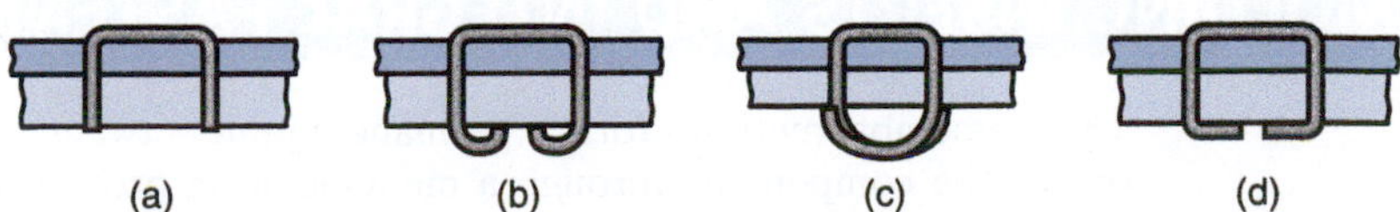

FIGURE 25.13 Common types of wire stitches: (a) unclinched, (b) standard loop, (c) bypass loop, and (d) flat clinch. (Credit: *Fundamentals of Modern Manufacturing,* 4th Edition by Mikell P. Groover, 2010. Reprinted with permission of John Wiley & Sons, Inc.)

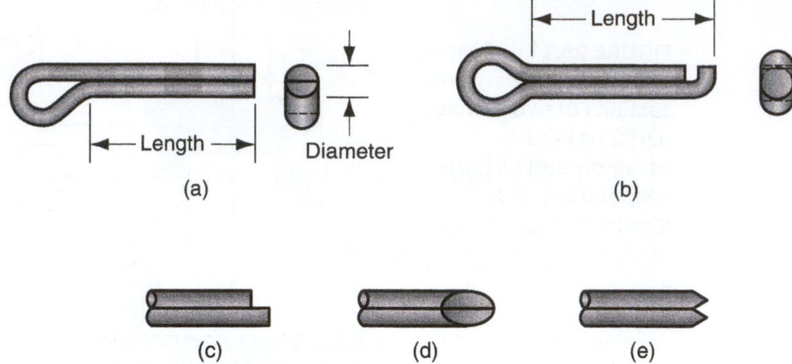

FIGURE 25.14 Cotter pins: (a) offset head, standard point; (b) symmetric head, hammerlock point; (c) square point; (d) mitered point; and (e) chisel point. (Credit: *Fundamentals of Modern Manufacturing*, 4th Edition by Mikell P. Groover, 2010. Reprinted with permission of John Wiley & Sons, Inc.)

the assembly can involve various combinations of metal and nonmetal materials. Applications of industrial stitching include light sheet-metal assembly, metal hinges, electrical connections, magazine binding, corrugated boxes, and final product packaging. Conditions that favor stitching in these applications are (1) high-speed operation, (2) elimination of the need for prefabricated holes in the parts, and (3) desirability of using fasteners that encircle the parts.

In ***stapling***, preformed U-shaped staples are punched through the two parts to be attached. The staples are supplied in convenient strips. The individual staples are lightly stuck together to form the strip, but they can be separated by the stapling tool for driving. The staples come with various point styles to facilitate their entry into the work. Staples are usually applied by means of portable pneumatic guns, into which strips containing several hundred staples can be loaded. Applications of industrial stapling include: furniture and upholstery, assembly of car seats, and various light-gage sheet-metal and plastic assembly jobs.

Sewing is a common joining method for soft, flexible parts such as cloth and leather. The method involves the use of a long thread or cord interwoven with the parts so as to produce a continuous seam between them. The process is widely used in the needle trades industry for assembling garments.

Cotter pins are fasteners formed from half-round wire into a single two-stem pin, as in Figure 25.14. They vary in diameter, ranging between 0.8 mm (0.031 in) and 19 mm (3/4 in), and in point style, several of which are shown in the figure. Cotter pins are inserted into holes in the mating parts and their legs are split to lock the assembly. They are used to secure parts onto shafts and similar applications.

25.5 MOLDING INSERTS AND INTEGRAL FASTENERS

These assembly methods form a permanent joint between parts by shaping or reshaping one of the components through a manufacturing process such as casting, molding, or sheet-metal forming.

Inserts in Moldings and Castings This method involves the placement of a component into a mold prior to plastic molding or metal casting, so that it becomes a permanent and integral part of the molding or casting. Inserting a separate component is preferable

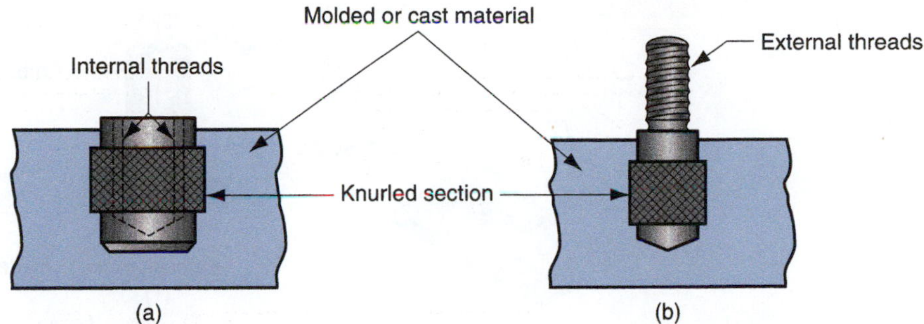

FIGURE 25.15 Examples of molded-in inserts: (a) threaded bushing and (b) threaded stud. (Credit: *Fundamentals of Modern Manufacturing*, 4th Edition by Mikell P. Groover, 2010. Reprinted with permission of John Wiley & Sons, Inc.)

to molding or casting its shape if the superior properties (e.g., strength) of the insert material are required, or the geometry achieved through the use of the insert is too complex or intricate to incorporate into the mold. Examples of inserts in molded or cast parts include internally threaded bushings and nuts, externally threaded studs, bearings, and electrical contacts. Two of these are illustrated in Figure 25.15. Internally threaded inserts must be placed into the mold with threaded pins to prevent the molding material from flowing into the threaded hole.

Placing inserts into a mold has certain disadvantages in production: (1) design of the mold becomes more complicated; (2) handling and placing the insert into the cavity takes time that reduces production rate; and (3) inserts introduce a foreign material into the casting or molding, and in the event of a defect, the cast metal or plastic cannot be easily reclaimed and recycled. Despite these disadvantages, use of inserts is often the most functional design and least-cost production method.

Integral Fasteners Integral fasteners involve deformation of component parts so they interlock and create a mechanically fastened joint. This assembly method is most common for sheet-metal parts. The possibilities, Figure 25.16, include: (a) ***lanced tabs*** to attach wires or shafts to sheet-metal parts; (b) ***embossed protrusions***, in which bosses are formed in one part and flattened over the mating assembled part; (c) ***seaming***, where the edges of two separate sheet-metal parts or the opposite edges of the same part are bent over to form the fastening seam—the metal must be ductile for the bending to be feasible; (d) ***beading***, in which a tube-shaped part is attached to a smaller shaft (or other round part) by deforming the outer diameter inward to cause an interference around the entire circumference; and (e) ***dimpling***—forming of simple round indentations in an outer part to retain an inner part.

Crimping, in which the edges of one part are deformed over a mating component, is another example of integral assembly. A common example involves squeezing the barrel of an electrical terminal around a wire.

25.6 DESIGN FOR ASSEMBLY

Design for assembly (DFA) has received much attention in recent years because assembly operations constitute a high labor cost for many manufacturing companies. The key to successful design for assembly can be simply stated [3]: (1) design the product with as few parts as possible, and (2) design the remaining parts so they are easy to assemble. The cost of assembly is determined largely during product design, because that

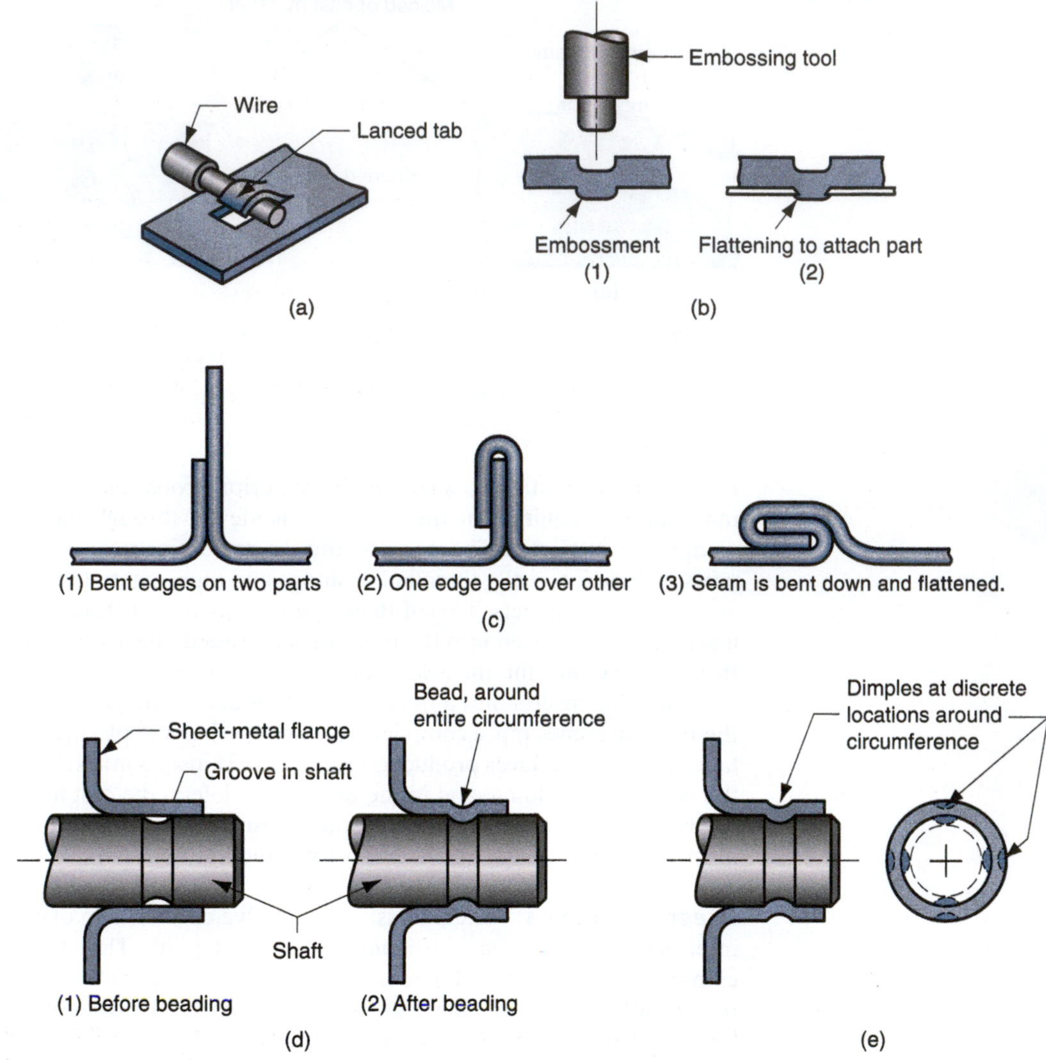

FIGURE 25.16 Integral fasteners: (a) lanced tabs to attach wires or shafts to sheet metal, (b) embossed protrusions, similar to riveting, (c) single-lock seaming, (d) beading, and (e) dimpling. Numbers in parentheses indicate sequence in (b), (c), and (d). (Credit: *Fundamentals of Modern Manufacturing*, 4th Edition by Mikell P. Groover, 2010. Reprinted with permission of John Wiley & Sons, Inc.)

is when the number of separate components in the product is determined, and decisions are made about how these components will be assembled. Once these decisions have been made, there is little that can be done in manufacturing to influence assembly costs (except, of course, to manage the operations well).

In this section we consider some of the principles that can be applied during product design to facilitate assembly. Most of the principles have been developed in the context of mechanical assembly, although some of them apply to the joining processes (i.e., welding, brazing, etc.). Much of the research in design for assembly has been motivated by the increasing use of automated assembly systems in industry. Accordingly, our discussion is divided into two sections, the first dealing with general principles of DFA, and the second concerned specifically with design for automated assembly.

25.6.1 GENERAL PRINCIPLES OF DFA

Most of the general principles apply to both manual and automated assembly. Their goal is to achieve the required design function by the simplest and lowest cost means. The following recommendations have been compiled from [1], [3], [4], and [6]:

- ***Use the fewest number of parts possible to reduce the amount of assembly required.*** This principle is implemented by combining functions within the same part that might otherwise be accomplished by separate components (e.g., using a plastic molded part instead of an assembly of sheet metal parts).
- ***Reduce the number of threaded fasteners required.*** Instead of using separate threaded fasteners, design the component to utilize snap fits, retaining rings, integral fasteners, and similar fastening mechanisms that can be accomplished more rapidly. Use threaded fasteners only where justified (e.g., where disassembly or adjustment is required).
- ***Standardize fasteners.*** This is intended to reduce the number of sizes and styles of fasteners required in the product. Ordering and inventory problems are reduced, the assembly worker does not have to distinguish between so many separate fasteners, the workstation is simplified, and the number of separate fastening tools is reduced.
- ***Reduce parts orientation difficulties.*** Orientation problems are generally reduced by designing a part to be symmetrical and minimizing the number of asymmetric features. This allows easier handling and insertion during assembly. This principle is illustrated in Figure 25.17.
- ***Avoid parts that tangle.*** Certain part configurations are more likely to become entangled in parts bins, frustrating assembly workers or jamming automatic feeders. Parts with hooks, holes, slots, and curls exhibit more of this tendency than parts without these features. See Figure 25.18.

25.6.2 DESIGN FOR AUTOMATED ASSEMBLY

Methods suitable for manual assembly are not necessarily the best methods for automated assembly. Some assembly operations readily performed by a human worker are quite difficult to automate (e.g., assembly using bolts and nuts). To automate the assembly process, parts fastening methods must be specified during product design that lend themselves to machine insertion and joining techniques and do not require the senses, dexterity, and intelligence of human assembly workers. Following are some

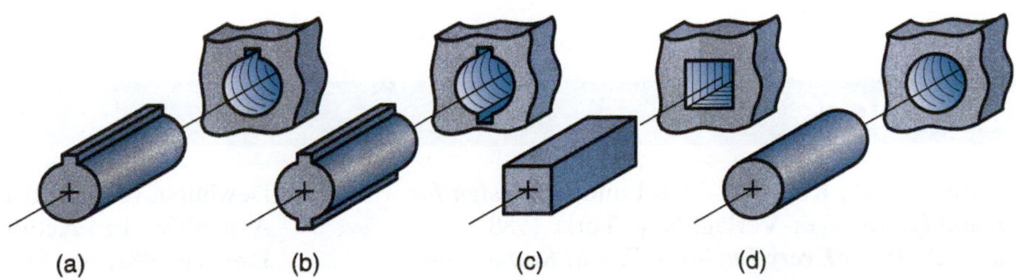

FIGURE 25.17 Symmetrical parts are generally easier to insert and assemble: (a) only one rotational orientation possible for insertion; (b) two possible orientations; (c) four possible orientations; and (d) infinite rotational orientations. (Credit: *Fundamentals of Modern Manufacturing*, 4th Edition by Mikell P. Groover, 2010. Reprinted with permission of John Wiley & Sons, Inc.)

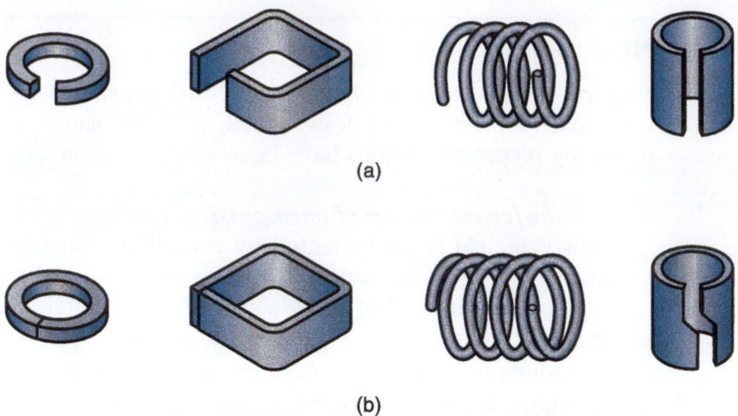

FIGURE 25.18 (a) Parts that tend to tangle and (b) parts designed to avoid tangling. (Credit: *Fundamentals of Modern Manufacturing*, 4th Edition by Mikell P. Groover, 2010. Reprinted with permission of John Wiley & Sons, Inc.)

recommendations and principles that can be applied in product design to facilitate automated assembly [6], [10]:

- ***Use modularity in product design***. Increasing the number of separate tasks that are accomplished by an automated assembly system will reduce the reliability of the system. To alleviate the reliability problem, Riley [10] suggests that the design of the product be modular in which each module or subassembly has a maximum of 12 or 13 parts to be produced on a single assembly system. Also, the subassembly should be designed around a base part to which other components are added.
- ***Reduce the need for multiple components to be handled at once***. The preferred practice for automated assembly is to separate the operations at different stations rather than to simultaneously handle and fasten multiple components at the same workstation.
- ***Limit the required directions of access***. This means that the number of directions in which new components are added to the existing subassembly should be minimized. Ideally, all components should be added vertically from above, if possible.
- ***High-quality components***. High performance of an automated assembly system requires that consistently good-quality components are added at each workstation. Poor-quality components cause jams in feeding and assembly mechanisms that result in downtime.
- ***Use of snap fit assembly***. This eliminates the need for threaded fasteners; assembly is by simple insertion, usually from above. It requires that the parts be designed with special positive and negative features to facilitate insertion and fastening.

REFERENCES

[1] Andreasen, M., Kahler, S., and Lund, T. ***Design for Assembly***. Springer-Verlag, New York, 1988.

[2] Blake, A. ***What Every Engineer Should Know About Threaded Fasteners***. Marcel Dekker, New York, 1986.

[3] Boothroyd, G., Dewhurst, P., and Knight, W. ***Product Design for Manufacture and Assembly***, 2nd ed. CRC Taylor & Francis, Boca Raton, Florida, 2001.

[4] Bralla, J. G.(Editor-in-Chief). ***Design for Manufacturability Handbook***, 2nd ed. McGraw-Hill Book Company, New York, 1998.

[5] Dewhurst, P., and Boothroyd, G. "Design for Assembly in Action," ***Assembly Engineering***, January 1987, pp. 64–68.

[6] Groover, M. P. ***Automation, Production Systems, and Computer Integrated Manufacturing***, 3rd ed. Pearson Prentice-Hall, Upper Saddle River, New Jersey, 2008.

[7] Groover, M. P., Weiss, M., Nagel, R. N., and Odrey, N. G. ***Industrial Robotics: Technology,***

Programming, and Applications. McGraw-Hill Book Company, New York, 1986.

[8] Nof, S. Y., Wilhelm, W. E., and Warnecke, H-J. ***Industrial Assembly***. Chapman & Hall, New York, 1997.

[9] Parmley, R. O. (ed.). ***Standard Handbook of Fastening and Joining***, 3rd ed. McGraw-Hill Company, New York, 1997.

[10] Riley, F. J. ***Assembly Automation, A Management Handbook***, 2nd ed. Industrial Press, New York, 1999.

[11] Speck, J. A. ***Mechanical Fastening, Joining, and Assembly***. Marcel Dekker, New York, 1997.

[12] Whitney, D. E. ***Mechanical Assemblies***. Oxford University Press, New York, 2004.

[13] Wick, C., and Veilleux, R. F. (eds.). ***Tool and Manufacturing Engineers Handbook***, 4th ed., Vol **IV**, ***Quality Control and Assembly.*** Society of Manufacturing Engineers, Dearborn, Michigan, 1987.

REVIEW QUESTIONS

25.1. How does mechanical assembly differ from the other methods of assembly discussed in previous chapters (e.g., welding, brazing, etc.)?

25.2. What are some of the reasons why assemblies must be sometimes disassembled?

25.3. What is the technical difference between a screw and a bolt?

25.4. What is a stud (in the context of threaded fasteners)?

25.5. What is torque-turn tightening?

25.6. Define *proof strength* as the term applies in threaded fasteners.

25.7. What are the three ways in which a threaded fastener can fail during tightening?

25.8. What is a rivet?

25.9. What is the difference between a shrink fit and expansion fit in assembly?

25.10. What are the advantages of snap fitting?

25.11. What is the difference between industrial stitching and stapling?

25.12. What are integral fasteners?

25.13. Identify some of the general principles and guidelines for design for assembly.

25.14. Identify some of the general principles and guidelines that apply specifically to automated assembly.

PROBLEMS

25.1. A 5 mm diameter bolt is to be tightened to produce a preload = 250 N. If the torque coefficient = 0.23, determine the torque that should be applied.

25.2. An alloy steel Metric 10 × 1.5 screw (10 mm diameter, pitch p = 1.5 mm) is to be turned into a threaded hole and tightened to one-half of its proof strength. According to Table 25.1, the proof strength = 830 MPa. Determine the maximum torque that should be used if the torque coefficient = 0.18.

25.3. A Metric 16 × 2 bolt (16 mm diameter, 2 mm pitch) is subjected to a torque of 15 N-m during tightening. If the torque coefficient is 0.24, determine the tensile stress on the bolt.

25.4. A 1/2-13 screw is to be preloaded to a tension force = 1000 lb. Torque coefficient = 0.22. Determine the torque that should be used to tighten the bolt.

25.5. A torque wrench is used on a 3/4-10 bolt in an automobile final assembly plant. A torque of 70 ft-lb is generated by the wrench. If the torque coefficient = 0.17, determine the tensile stress in the bolt.

25.6. The designer has specified that a 3/8-16 low-carbon bolt (3/8 in nominal diameter, 16 threads/in) in a certain application should be stressed to its proof stress of 33,000 lb/in^2 (from Table 25.1). Determine the maximum torque that should be used if the torque coefficient = 0.25.

25.7. A dowel pin made of steel (elastic modulus = 209,000 MPa) is to be press fitted into a steel collar. The pin has a nominal diameter of 16.0 mm, and the collar has an outside diameter of 27.0 mm. (a) Compute the radial pressure and the maximum effective stress if the interference between the shaft OD and the collar ID is 0.03 mm. (b) Determine the effect of increasing the outside diameter of the collar to 39.0 mm on the radial pressure and the maximum effective stress.

25.8. A gear made of aluminum (modulus of elasticity = 69,000 MPa) is press fitted onto an aluminum shaft. The gear has a diameter of 55 mm at the base of its teeth. The nominal internal diameter of the gear = 30 mm and the interference = 0.10 mm. Compute: (a) the radial pressure between the shaft and the gear, and (b) the maximum effective stress in the gear at its inside diameter.

25.9. A steel collar is press fitted onto a steel shaft. The modulus of elasticity of steel is 30×10^6 lb/in^2. The collar has an internal diameter of 2.498 in and the shaft has an outside diameter = 2.500 in. The outside diameter of the collar is 4.000 in. Determine the radial (interference) pressure on the assembly, and (b) the maximum effective stress in the collar at its inside diameter.

25.10. The yield strength of a certain metal = 50,000 lb/in^2 and its modulus of elasticity = 22×10^6 lb/in^2. It is to be used for the outer ring of a press-fit assembly with a mating shaft made of the same metal. The nominal inside diameter of the ring is 1.000 in and its outside diameter = 2.500 in. Using a safety factor = 2.0, determine the maximum interference that should be used with this assembly.

25.11. A shaft made of aluminum is 40.0 mm in diameter at room temperature (21°C). Its coefficient of thermal expansion = 24.8×10^{-6} mm/mm per °C. If it must be reduced in size by 0.20 mm in order to be expansion fitted into a hole, determine the temperature to which the shaft must be cooled.

25.12. A steel ring has an inside diameter = 30 mm and an outside diameter = 50 mm at room temperature (21°C). If the coefficient of thermal expansion of steel = 12.1×10^{-6} mm/mm per °C, determine the inside diameter of the ring when heated to 500°C.

25.13. A steel collar is to be heated from room temperature (70°F) to 700°F. Its inside diameter = 1.000 in, and its outside diameter = 1.625 in. If the coefficient of thermal expansion of the steel is = 6.7×10^{-6} in/in per °F, determine the increase in the inside diameter of the collar.

25.14. A steel collar whose outside diameter = 3.000 in at room temperature is to be shrink fitted onto a steel shaft by heating it to an elevated temperature while the shaft remains at room temperature. The shaft diameter = 1.500 in. For ease of assembly when the collar is heated to a temperature of 1000°F, the clearance between the shaft and the collar is to be 0.007 in. Determine (a) the initial inside diameter of the collar at room temperature so that this clearance is satisfied, (b) the radial pressure and (c) maximum effective stress on the resulting interference fit at room temperature (70°F). For steel, the elastic modulus = 30,000,000 lb/in^2 and coefficient of thermal expansion = 6.7×10^{-6} in/in per °F.

Part VIII Special Processing and Assembly Technologies

26 RAPID PROTOTYPING

Chapter Contents

26.1 Fundamentals of Rapid Prototyping

26.2 Rapid Prototyping Technologies

26.2.1 Liquid-Based Rapid Prototyping Systems

26.2.2 Solid-Based Rapid Prototyping Systems

26.2.3 Powder-Based Rapid Prototyping Systems

26.3 Application Issues in Rapid Prototyping

In this part of the book, we discuss several processing and assembly technologies that do not fit neatly into our classification scheme in Figure 1.3. They are technologies that have been adapted from conventional manufacturing and assembly operations or developed from scratch to serve the special functions or needs of designers and manufacturers. Rapid prototyping, covered in the present chapter, is a collection of processes used to fabricate a model, part, or tool in minimum possible time. Chapter 27 surveys some of the technologies used to produce very small parts and products. Microfabrication technologies are used to produce items measured in microns (10^{-6} m), while nanofabrication technologies produce items measured in nanometers (10^{-9} m).

Rapid prototyping (RP) is a family of fabrication methods to make engineering prototypes in minimum possible lead times based on a computer-aided design (CAD) model of the item. The traditional method of fabricating a prototype part is machining, which can require significant lead times—up to several weeks, sometimes longer, depending on part complexity, difficulty in ordering materials, and scheduling production equipment. A number of rapid prototyping techniques are now available that allow a part to be produced in hours or days rather than weeks, given that a computer model of the part has been generated on a CAD system.

26.1 FUNDAMENTALS OF RAPID PROTOTYPING

The special need that motivates the variety of rapid prototyping technologies arises because product designers would like to have a physical model of a new part or product design rather than a computer model or line drawing. The creation of a prototype is an integral step in the design procedure. A ***virtual prototype***, which is a computer model of the part design on a CAD system, may not be adequate for the designer to visualize the part. It certainly is not sufficient to conduct real physical tests on the part, although it is possible to perform simulated tests by finite element analysis or other methods. Using one of the available RP technologies, a solid physical part can be created in a relatively short time (hours if the company possesses the RP equipment or days if the part fabrication must be contracted to an outside firm specializing in RP). The designer can therefore visually examine and physically feel the part and begin to perform tests and experiments to assess its merits and shortcomings.

Available rapid prototyping technologies can be divided into two categories: (1) material removal processes and (2) material addition processes. The ***material removal RP*** alternative involves machining (Chapter 16), primarily milling and drilling, using a dedicated computer numerical control (CNC) machine that is available to the design department on short notice. To use CNC, a part program must be prepared from the CAD model (Section 29.1). The starting material is often a solid block of wax, which is very easy to machine, and the part and chips can be melted and resolidified for reuse when the current prototype is no longer needed. Other starting materials can also be used, such as wood, plastics, or metals (e.g., a machinable grade of aluminum or brass). The CNC machines used for rapid prototyping are often small, and the terms ***desktop milling*** or ***desktop machining*** are sometimes used for this technology.

The principal emphasis in this chapter is on ***material-addition RP*** technologies, all of which work by adding layers of material one at a time to build the solid part from bottom to top. Starting materials include (1) liquid monomers and polymers that are cured layer by layer into solid parts, (2) powders that are aggregated and bonded layer by layer, and (3) solid sheets that are laminated to create the solid part. In addition to starting material, the method of building and adding the layers also distinguishes the various material addition RP technologies. Some techniques use lasers to solidify the starting material, another deposits a soft plastic filament in the outline of each layer, while others bond solid layers together. There is a correlation between the starting material and the part-building techniques, as we shall see in our discussion of RP technologies.

The common approach to prepare the control instructions (part program) in all of the current material addition RP techniques involves the following steps [5]:

1. ***Geometric modeling***. This consists of modeling the component on a CAD system to define its enclosed volume. Solid modeling is the preferred technique because it provides a complete and unambiguous mathematical representation of the geometry. For rapid prototyping, the important issue is to distinguish the interior (mass) of the part from its exterior, and solid modeling provides for this distinction.
2. ***Tessellation of the geometric model***. [1] In this step, the CAD model is converted into a format that approximates its surfaces by triangles or polygons, with their vertices arranged to distinguish the object's interior from its exterior. The common tessellation

[1] More generally, the term **tessellation** refers to the laying out or creation of a mosaic, such as one consisting of small colored tiles affixed to a surface for decoration.

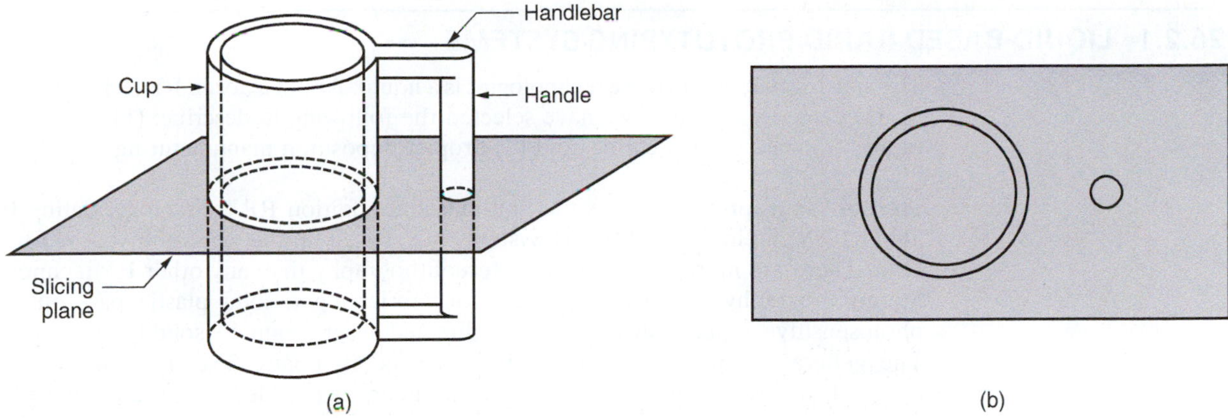

FIGURE 26.1 Conversion of a solid model of an object into layers (only one layer is shown). (Credit: *Fundamentals of Modern Manufacturing,* 4th Edition by Mikell P. Groover, 2010. Reprinted with permission of John Wiley & Sons, Inc.)

format used in rapid prototyping is STL,[2] which has become the de facto standard input format for nearly all RP systems.

3. ***Slicing of the model into layers.*** In this step, the model in STL file format is sliced into closely spaced parallel horizontal layers. Conversion of a solid model into layers is illustrated in Figure 26.1. These layers are subsequently used by the RP system to construct the physical model. By convention, the layers are formed in the x-y plane orientation, and the layering procedure occurs in the z-axis direction. For each layer, a curing path is generated, called the STI file, which is the path that will be followed by the RP system to cure (or otherwise solidify) the layer.

As our brief overview indicates, there are several different technologies used for material addition rapid prototyping. This heterogeneity has spawned several alternative names for rapid prototyping, including ***layer manufacturing, direct CAD manufacturing***, and ***solid freeform fabrication***. The term ***rapid prototyping and manufacturing*** (RPM) is also being used more frequently to indicate that the RP technologies can be applied to make production parts and production tooling, not just prototypes.

26.2 RAPID PROTOTYPING TECHNOLOGIES

The RP techniques currently available can be classified in various ways. Let us adopt a classification system recommended in [5] that is consistent with the way we classify other part-shaping processes in this book (after all, rapid prototyping is a part-shaping process). The classification method is based on the form of the starting material in the RP process: (1) liquid-based, (2) solid-based, and (3) powder-based. We discuss examples of each class in the following three sections.

[2] STL stands for STereoLithography, one of the primary technologies used for rapid prototyping, developed by 3D Systems Inc.

26.2.1 LIQUID-BASED RAPID PROTOTYPING SYSTEMS

The starting material in these technologies is a liquid. About a dozen RP technologies are in this category, of which we have selected the following to describe: (1) stereolithography, (2) solid ground curing, and (3) droplet deposition manufacturing.

Stereolithography This was the first material addition RP technology, dating from about 1988 and introduced by 3D Systems Inc., based on the work of inventor Charles Hull. There are more installations of stereolithography than any other RP technology. Stereolithography (STL) is a process for fabricating a solid plastic part out of a photosensitive liquid polymer using a directed laser beam to solidify the polymer. The general setup for the process is illustrated in Figure 26.2. Part fabrication is accomplished as a series of layers, in which one layer is added onto the previous layer to gradually build the desired three-dimensional geometry.

The stereolithography apparatus consists of (1) a platform that can be moved vertically inside a vessel containing the photosensitive polymer, and (2) a laser whose beam can be controlled in the *x-y* direction. At the start of the process, the platform is positioned vertically near the surface of the liquid photopolymer, and a laser beam is directed through a curing path that comprises an area corresponding to the base (bottom layer) of the part. This and subsequent curing paths are defined by the STI file (step 3 in preparing the control instructions described earlier). The action of the laser is to harden (cure) the photosensitive polymer where the beam strikes the liquid, forming a solid layer of plastic that adheres to the platform. When the initial layer is completed, the platform is lowered by a distance equal to the layer thickness, and a second layer is formed on top of the first by the laser, and so on. Before each new layer is cured, a wiper blade is passed over the viscous liquid resin to ensure that its level is the same throughout the surface. Each layer consists of its own area shape, so that the succession of layers, one on top of the previous, creates the solid part shape. Each layer is 0.076 to 0.50 mm (0.003 to 0.020 in) thick. Thinner layers provide better resolution and allow more intricate part shapes; but processing time is greater. Photopolymers are typically acrylic [13], although use of epoxy for STL has also been reported [10]. The starting materials are liquid

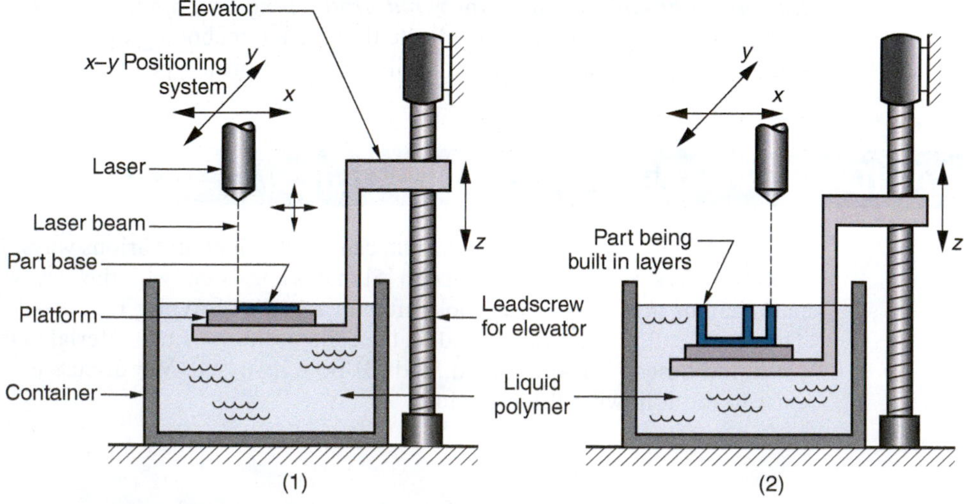

FIGURE 26.2 Stereolithography: (1) at the start of the process, in which the initial layer is added to the platform; and (2) after several layers have been added so that the part geometry gradually takes form. (Credit: *Fundamentals of Modern Manufacturing*, 4th Edition by Mikell P. Groover, 2010. Reprinted with permission of John Wiley & Sons, Inc.)

monomers. Polymerization occurs upon exposure to ultraviolet light produced by helium-cadmium or argon ion lasers. Scan speeds of STL lasers typically range between 500 and 2500 mm/s.

The time required to build the part by this layering process ranges from one hour for small parts of simple geometry up to several dozen hours for complex parts. Other factors that affect cycle time are scan speed and layer thickness. After all of the layers have been formed, the photopolymer is about 95% cured. The piece is therefore "baked" in a fluorescent oven to completely solidify the polymer. Excess polymer is removed with alcohol, and light sanding is sometimes used to improve smoothness and appearance.

Depending on its design and orientation, a part may contain overhanging features that have no means of support during the bottom-up approach used in stereolithography. For example, in the part of Figure 26.1, if the lower half of the handle and the lower handlebar were eliminated, the upper portion of the handle would be unsupported during fabrication. In these cases, extra pillars or webs may need to be added to the part simply for support purposes. Otherwise, the overhangs may float away or otherwise distort the desired part geometry. These extra features must be trimmed away after the process is completed.

Solid Ground Curing Like stereolithography, solid ground curing (SGC) works by curing a photosensitive polymer layer by layer to create a solid model based on CAD geometric data. Instead of using a scanning laser beam to accomplish the curing of a given layer, the entire layer is exposed to an ultraviolet light source through a mask that is positioned above the surface of the liquid polymer. The hardening process takes 2 to 3 seconds for each layer. SGC systems are sold under the name ***Solider system*** by Cubital Ltd.

The starting data in SGC is similar to that used in stereolithography: a CAD geometric model of the part that has been sliced into layers. For each layer, the step-by-step procedure in SGC is illustrated in Figure 26.3 and described here: (1) A mask is created on a glass plate by electrostatically charging a negative image of the layer onto the surface. The imaging technology is basically the same as that used in photocopiers. (2) A thin flat layer of liquid photopolymer is distributed over the surface of the work platform. (3) The mask is positioned above the liquid polymer surface and exposed by a high powered (e.g., 2000 W) ultraviolet lamp. The portions of the liquid polymer layer that are unprotected by the mask are solidified in about 2 seconds. The shaded areas of the layer remain in the liquid state. (4) The mask is removed, the glass plate is cleaned and made ready for a subsequent layer in step 1. Meanwhile, the liquid polymer remaining on the surface is removed in a wiping and vacuuming procedure. (5) The now-open areas of the layer are filled in with hot wax. When hardened, the wax acts to support overhanging sections of the part. (6) When the wax has cooled and solidified, the polymer-wax surface is milled to form a flat layer of specified thickness, ready to receive the next application of liquid photopolymer in step 2. Although we have described SGC as a sequential process, certain steps are accomplished in parallel. Specifically, the mask preparation step 1 for the next layer is performed simultaneously with the layer fabrication steps 2 through 6, using two glass plates during alternating layers.

The sequence for each layer takes about 90 seconds. Throughput time to produce a part by SGC is claimed to be about eight times faster than competing RP systems [5]. The solid cubic form created in SGC consists of solid polymer and wax. The wax provides support for fragile and overhanging features of the part during fabrication, but can be melted away later to leave the free-standing part. No post curing of the completed prototype model is required, as in stereolithography.

Droplet Deposition Manufacturing These systems operate by melting the starting material and shooting small droplets onto a previously formed layer. The liquid droplets cold weld to the surface to form a new layer. The deposition of droplets for each new

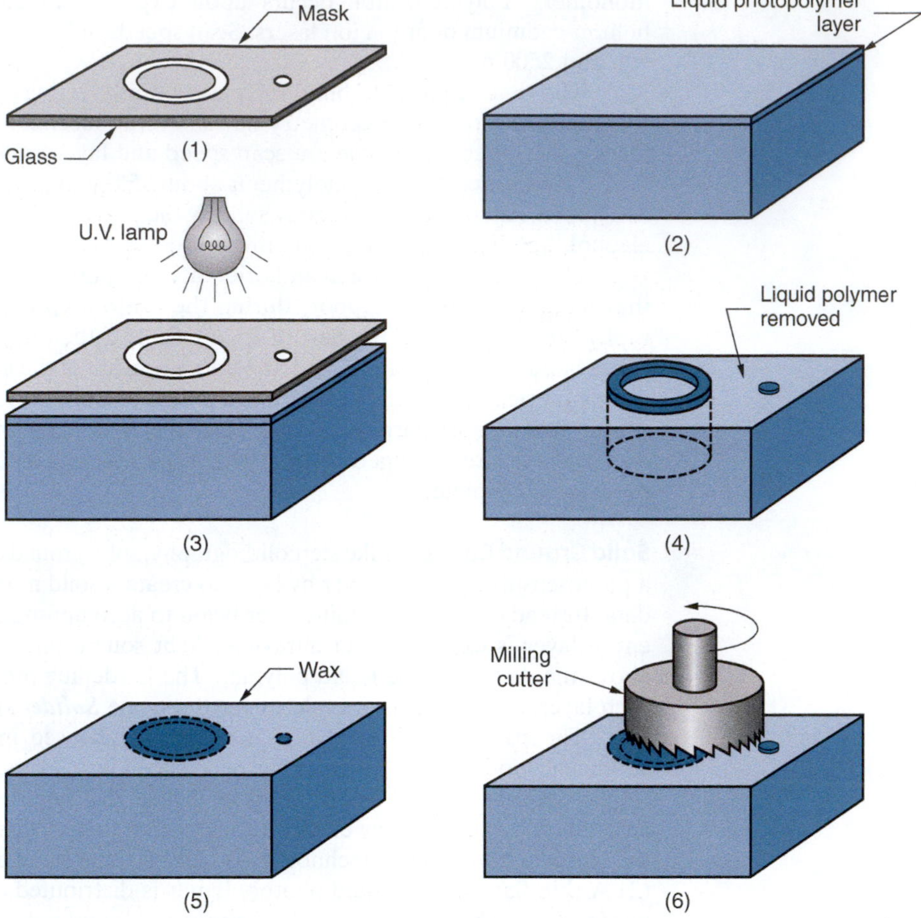

FIGURE 26.3 Solid ground curing process for each layer: (1) mask preparation, (2) applying liquid photopolymer layer, (3) mask positioning and exposure of layer, (4) uncured polymer removed from surface, (5) wax filling, (6) milling for flatness and thickness. (Credit: *Fundamentals of Modern Manufacturing,* 4th Edition by Mikell P. Groover, 2010. Reprinted with permission of John Wiley & Sons, Inc.)

layer is controlled by a moving x-y spray nozzle workhead whose path is based on a cross section of a CAD geometric model that has been sliced into layers (similar to the other RP systems described above). After each layer has been applied, the platform supporting the part is lowered a certain distance corresponding to the layer thickness, in preparation for the next layer. The term *droplet deposition manufacturing* (DDM) refers to the fact that small particles of work material are deposited as projectile droplets from the workhead nozzle.

Several commercial RP systems are based on this general operating principle, the differences being in the type of material that is deposited and the corresponding technique by which the workhead operates to melt and apply the material. An important criterion that must be satisfied by the starting material is that it be readily melted and solidified. Work materials used in DDM include wax and thermoplastics. Metals with a low melting point, such as tin, zinc, lead, and aluminum, have also been tested.

One of the more popular BPM systems is the Personal Modeler®, available from BPM Technology, Inc. Wax is commonly used as the work material. The ejector head operates using a piezoelectric oscillator that shoots droplets of wax at a rate of 10,000 to 15,000 per second. The droplets are of uniform size at about 0.076 mm (0.003 in) diameter, which flatten to about 0.05 mm (0.002 in) solidified thickness on impact against the existing part surface. After each layer has been deposited, the surface is milled or thermally smoothed to achieve accuracy in the z-direction. Layer thickness is about 0.09 mm (0.0035 in).

26.2.2 SOLID-BASED RAPID PROTOTYPING SYSTEMS

The common feature in these RP systems is that the starting material is solid. In this section we discuss two solid-based RP systems: (1) laminated-object manufacturing and (2) fused-deposition modeling.

Laminated-Object Manufacturing The principal company offering laminated-object manufacturing (LOM) systems is Helisys, Inc. Of interest is that much of the early research and development work on LOM was funded by National Science Foundation. The first commercial LOM unit was shipped in 1991.

Laminated-object manufacturing produces a solid physical model by stacking layers of sheet stock that are each cut to an outline corresponding to the cross-sectional shape of a CAD model that has been sliced into layers. The layers are bonded one on top of the previous one before cutting. After cutting, the excess material in the layer remains in place to support the part during building. Starting material in LOM can be virtually any material in sheet stock form, such as paper, plastic, cellulose, metals, or fiber-reinforced materials. Stock thickness is 0.05 to 0.50 mm (0.002 to 0.020 in). In LOM, the sheet material is usually supplied with adhesive backing as rolls that are spooled between two reels, as in Figure 26.4. Otherwise, the LOM process must include an adhesive coating step for each layer.

The data preparation phase in LOM consists of slicing the geometric model using the STL file for the given part. The slicing function is accomplished by LOMSlice™, the special software used in laminated-object manufacturing. Slicing the STL model in LOM is performed after each layer has been physically completed and the vertical height of the part has been measured. This provides a feedback correction to account for the actual thickness of the sheet stock being used, a feature unavailable on most other RP systems. With reference to Figure 26.4, the LOM process for each layer can be described as follows, picking up the action with a sheet of stock in place and bonded to the previous stack: (1) LOMSlice™ computes the cross-sectional perimeter of the STL model based on the measured height of the physical part at the current layer of completion. (2) A laser beam is used to cut along the perimeter, as well as to crosshatch the exterior portions of the sheet for subsequent removal. The cutting trajectory is controlled by means of an

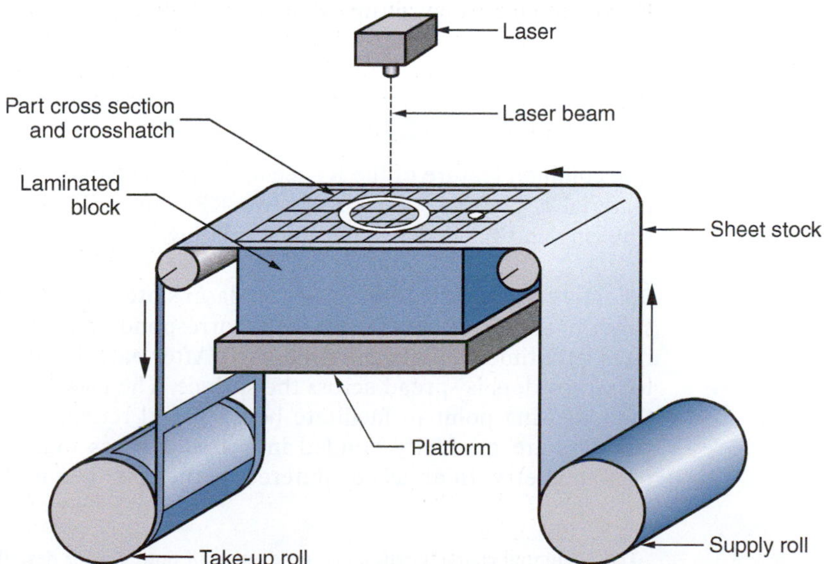

FIGURE 26.4 Laminated-object manufacturing. (Credit: *Fundamentals of Modern Manufacturing*, 4th Edition by Mikell P. Groover, 2010. Reprinted with permission of John Wiley & Sons, Inc.)

x-y positioning system. The cutting depth is controlled so that only the top layer is cut. (3) The platform holding the stack is lowered, and the sheet stock is advanced between supply roll and take-up spool for the next layer. The platform is then raised to a height consistent with the stock thickness and a heated roller moves across the new layer to bond it to the previous layer. The height of the physical stack is measured in preparation for the next slicing computation by LOMSlice™.

When all of the layers are completed, the new part is separated from the excess external material using a hammer, putty knife, and wood carving tools. The part can then be sanded to smooth and blend the layer edges. A sealing application is recommended, using a urethane, epoxy, or other polymer spray to prevent moisture absorption and damage. LOM part sizes can be relatively large among RP processes, with work volumes up to 800 mm × 500 mm by 550 mm (32 in × 20 in × 22 in). More common work volumes are 380 mm × 250 mm × 350 mm (15 in × 10 in × 14 in).

Fused-Deposition Modeling Fused-deposition modeling (FDM) is an RP process in which a filament of wax or polymer is extruded onto the existing part surface from a workhead to complete each new layer. The workhead is controlled in the x-y plane during each layer and then moved up by a distance equal to one layer in the z-direction. The starting material is a solid filament with typical diameter = 1.25 mm (0.050 in) fed from a spool into the workhead that heats the material to about 0.5°C (1°F) above its melting point before extruding it onto the part surface. The extrudate is solidified and cold welded to the cooler part surface in about 0.1 s. The part is fabricated from the base up, using a layer-by-layer procedure similar to other RP systems.

FDM was developed by Stratasys Inc., which sold its first machine in 1990. The starting data is a CAD geometric model that is processed by Stratasys's software modules QuickSlice® and SupportWork™. QuickSlice® is used to slice the model into layers, and SupportWork™ is used to generate any support structures that are required during the build process. If supports are needed, a dual extrusion head and a different material is used to create the supports. The second material is designed to readily be separated from the primary modeling material. The slice (layer) thickness can be set anywhere from 0.05 to 0.75 mm (0.002 to 0.030 in). About 400 mm of filament material can be deposited per second by the extrusion workhead in widths that can be set between 0.25 and 2.5 mm (0.010 to 0.100 in). Starting materials are wax and several polymers, including ABS, polyamide, polyethylene, and polypropylene. These materials are nontoxic, allowing the FDM machine to be set up in an office environment.

26.2.3 POWDER-BASED RAPID PROTOTYPING SYSTEMS

The common feature of the RP technologies described in this section is that the starting material is powder.[3] We discuss two RP systems in this category: (1) selective laser sintering and (2) three-dimensional printing.

Selective Laser Sintering Selective laser sintering (SLS) uses a moving laser beam to sinter heat-fusible powders in areas corresponding to the CAD geometric model one layer at a time to build the solid part. After each layer is completed, a new layer of loose powders is spread across the surface. The powders are preheated to just below their melting point to facilitate bonding and reduce distortion. Layer by layer, the powders are gradually bonded into a solid mass that forms the three-dimensional part geometry. In areas not sintered by the laser beam, the powders remain loose so

[3] The definition, characteristics, and production of powders are described in Chapters 10 and 11.

they can be poured out of the completed part. Meanwhile, they serve to support the solid regions of the part as fabrication proceeds. Layer thickness is 0.075 to 0.50 mm (0.003 to 0.020 in).

SLS was developed at the University of Texas (Austin) as an alternative to stereolithography, and SLS machines are currently marketed by DTM Corporation. It is a more versatile process than stereolithography in terms of possible work materials. Current materials used in selective laser sintering include polyvinylchloride, polycarbonate, polyester, polyurethane, ABS, nylon, and investment casting wax. These materials are less expensive than the photosensitive resins used in stereolithography. They are also nontoxic and can be sintered using low power (25 to 50 W) CO_2 lasers. Metal and ceramic powders are also being used in SLS.

Three-Dimensional Printing This RP technology was developed at Massachusetts Institute of Technology. Three-dimensional printing (3DP) builds the part in the usual layer-by-layer fashion using an ink-jet printer to eject an adhesive bonding material onto successive layers of powders. The binder is deposited in areas corresponding to the cross sections of the solid part, as determined by slicing the CAD geometric model into layers. The binder holds the powders together to form the solid part, while the unbonded powders remain loose to be removed later. While the loose powders are in place during the build process, they provide support for overhanging and fragile features of the part. When the build process is completed, the part is heat-treated to strengthen the bonding, followed by removal of the loose powders. To further strengthen the part, a sintering step can be applied to bond the individual powders.

The part is built on a platform whose level is controlled by a piston. Let us describe the process for one cross section with reference to Figure 26.5: (1) A layer of powder is spread on the existing part-in-process. (2) An ink-jet printing head moves across the surface, ejecting droplets of binder on those regions that are to become the solid part. (3) When the printing of the current layer is completed, the piston lowers the platform for the next layer.

Starting materials in 3DP are powders of ceramic, metal, or cermet, and binders that are polymeric or colloidal silica or silicon carbide [10], [13]. Typical layer thickness ranges from 0.10 to 0.18 mm (0.004 to 0.007 in). The ink-jet printing head moves across

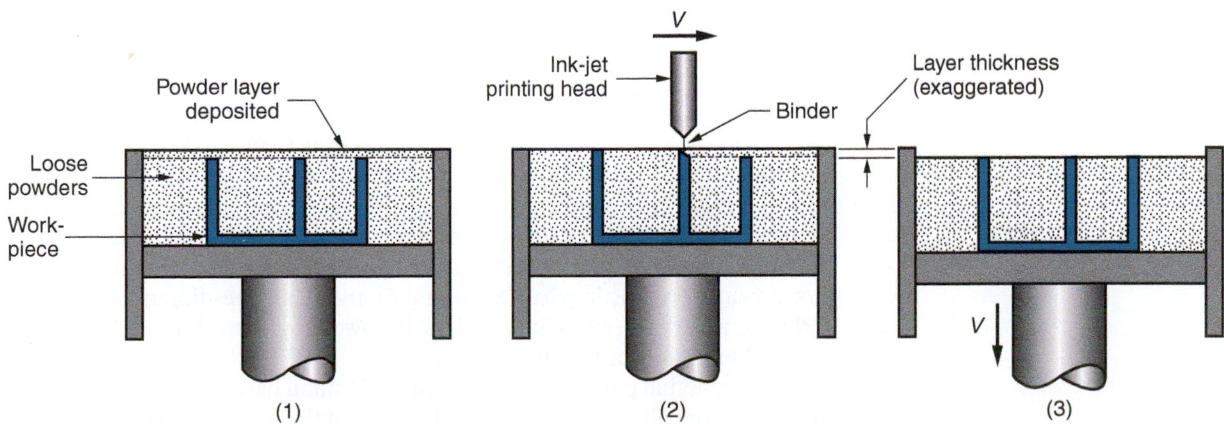

FIGURE 26.5 Three-dimensional printing: (1) powder layer is deposited, (2) ink-jet printing of areas that will become the part, and (3) piston is lowered for next layer (key: v = motion). (Credit: *Fundamentals of Modern Manufacturing*, 4th Edition by Mikell P. Groover, 2010. Reprinted with permission of John Wiley & Sons, Inc.)

the layer at a speed of about 1.5 m/s (59 in/sec), with ejection of liquid binder determined during the sweep by raster scanning. The sweep time, together with the spreading of the powders, permits a cycle time per layer of about 2 seconds [13].

26.3 APPLICATION ISSUES IN RAPID PROTOTYPING

Applications of rapid prototyping can be classified into three categories: (1) design, (2) engineering analysis and planning, and (3) tooling and manufacturing.

Design This was the initial application area for RP systems. Designers are able to confirm their design by building a real physical model in minimum time using rapid prototyping. The features and functions of the part can be communicated to others more easily using a physical model than by a paper drawing or displaying it on a CAD system monitor. Design benefits attributed to rapid prototyping include (1) reduced lead times to produce prototype components, (2) improved ability to visualize the part geometry because of its physical existence, (3) earlier detection and reduction of design errors, and (4) increased capability to compute mass properties of components and assemblies [2].

Engineering Analysis and Planning The existence of an RP-fabricated part allows for certain types of engineering analysis and planning activities to be accomplished that would be more difficult without the physical entity. Some of the possibilities are (1) comparison of different shapes and styles to optimize aesthetic appeal of the part, (2) analysis of fluid flow through different orifice designs in valves fabricated by RP, (3) wind tunnel testing of different streamline shapes using physical models created by RP, (4) stress analysis of a physical model, (5) fabrication of preproduction parts by RP as an aid in process planning and tool design, and (6) combining medical imaging technologies, such as magnetic resonance imaging (MRI), with RP to create models for doctors in planning surgical procedures or fabricating prostheses or implants.

Tooling and Manufacturing The trend in RP applications is toward its greater use in the fabrication of production tooling and for actual manufacture of parts. When RP is adopted to fabricate production tooling, the term ***rapid tool making*** (RTM) is often used. RTM applications divide into two approaches [4]: ***indirect*** RTM method, in which a pattern is created by RP and the pattern is used to fabricate the tool, and ***direct*** RTM method, in which RP is used to make the tool itself. Examples of indirect RTM include (1) use of an RP-fabricated part as the master in making a silicon rubber mold that is subsequently used as a production mold, (2) RP patterns to make the sand molds in sand casting (Section 6.1), (3) fabrication of patterns of low-melting point materials (e.g., wax) in limited quantities for investment casting (Section 6.2.3), and (4) making electrodes for EDM (Section 19.3.1) [5], [10]. Examples of direct RTM include (1) RP-fabricated mold cavity inserts that can be sprayed with metal to produce injection molds for a limited quantity of production plastic parts (Section 8.6) and (2) three-dimensional printing to create a die geometry in metallic powders followed by sintering and infiltration to complete the fabrication of the die [4], [5], [10].

Examples of actual part production include (1) small batch sizes of plastic parts that could not be economically injection molded because of the high cost of the mold, (2) parts with intricate internal geometries that could not be made using conventional technologies without assembly, and (3) one-of-a-kind parts such as bone replacements that must be made to correct size for each user [10].

Not all RP technologies can be used for all of these tooling and manufacturing examples. Interested readers should consult more complete treatments of the RP technologies for specific details on these and other examples.

Problems with Rapid Prototyping The principal problems with current RP technologies include (1) part accuracy, (2) limited variety of materials, and (3) mechanical performance of the fabricated parts.

Several sources of error limit part accuracy in RP systems: (1) mathematical, (2) process related, and (3) material related [13]. Mathematical errors include approximations of part surfaces used in RP data preparation and differences between the slicing thicknesses and actual layer thicknesses in the physical part. The latter differences result in z-axis dimensional errors. An inherent limitation in the physical part is the steps between layers, especially as layer thickness is increased, resulting in a staircase appearance for sloping part surfaces. Process-related errors are those that result from the particular part building technology used in the RP system. These errors degrade the shape of each layer as well as the registration between adjacent layers. Process errors can also affect the z-axis dimension. Finally, material-related errors include shrinkage and distortion. An allowance for shrinkage can be made by enlarging the CAD model of the part based on previous experience with the process and materials.

Current rapid prototyping systems are limited in the variety of materials they can process. For example, the most common RP technology, stereolithography, is limited to photosensitive polymers. In general, the materials used in RP systems are not as strong as the production part materials that will be used in the actual product. This limits the mechanical performance of the prototypes and the amount of realistic testing that can be done to verify the design during product development.

REFERENCES

[1] Ashley, S. "Rapid Prototyping Is Coming of Age," ***Mechanical Engineering***, July 1995 pp. 62–68.

[2] Bakerjian, R., and Mitchell, P. (eds.). ***Tool and Manufacturing Engineers Handbook***, 4th ed., Vol. **VI, *Design for Manufacturability.*** Society of Manufacturing Engineers, Dearborn, Michigan, 1992, Chapter 7.

[3] Destefani, J. "Plus or Minus," ***Manufacturing Engineering***, April 2005, pp. 93–97.

[4] Hilton, P. "Making the Leap to Rapid Tool Making," ***Mechanical Engineering***, July 1995 pp. 75–76.

[5] Kai, C. C., Fai, L. K., and Chu-Sing, L. ***Rapid Prototyping: Principles and Applications***, 2nd ed. World Scientific Publishing Co., Singapore, 2003.

[6] Kai, C. C., and Fai, L. K."Rapid Prototyping and Manufacturing: The Essential Link between Design and Manufacturing," Chapter 6 in ***Integrated Product and Process Development: Methods, Tools, and Technologies***, J. M. Usher, U. Roy, and H. R. Parsaei (eds.). John Wiley & Sons, Inc., New York, 1998, pp. 151–183.

[7] Kochan, D., Kai, C. C., and Zhaohui, D. "Rapid Prototyping Issues in the 21st Century," ***Computers in Industry***, Vol. **39**, 1999, pp. 3–10.

[8] Noorani, R. I. ***Rapid Prototyping: Principles and Applications***. John Wiley & Sons, Hoboken, New Jersey, 2006.

[9] Pacheco, J. M. ***Rapid Prototyping***, Report MTIAC SOAR- 93-01. Manufacturing Technology Information Analysis Center, IIT Research Institute, Chicago, Illinois, 1993.

[10] Pham, D. T., and Gault, R. S. "A Comparison of Rapid Prototyping Technologies," ***International Journal of Machine Tools and Manufacture***, Vol. **38**, 1998, pp. 1257–1287.

[11] Tseng, A. A., Lee, M. H., and Zhao, B. "Design and Operation of a Droplet Deposition System for Freeform Fabrication of Metal Parts," ***ASME Journal of Eng. Mat. Tech.***, Vol. **123**, No. 1, 2001.

[12] Wohlers, T. "Direct Digital Manufacturing," ***Manufacturing Engineering***, January 2009, pp. 73–81.

[13] Yan, X., and Gu, P. "A Review of Rapid Prototyping Technologies and Systems," ***Computer-Aided Design***, Vol. **28**, No. 4, 1996, pp. 307–318.

REVIEW QUESTIONS

26.1. What is rapid prototyping? Provide a definition of the term.

26.2. What are the three types of starting materials in rapid prototyping?

26.3. Besides the starting material, what other feature distinguishes the rapid prototyping technologies?

26.4. What is the common approach used in all of the material addition technologies to prepare the control instructions for the RP system?

26.5. Of all of the current rapid prototyping technologies, which one is the most widely used?

26.6. Describe the RP technology called solid ground curing.

26.7. Describe the RP technology called laminated-object manufacturing.

26.8. What is the starting material in fused-deposition modeling?

27 MICROFABRICATION AND NANOFABRICATION TECHNOLOGIES

Chapter Contents

27.1 Microsystem Products
27.1.1 Types of Microsystem Devices
27.1.2. Microsystem Applications

27.2 Microfabrication Processes
27.2.1 Silicon Layer Processes
27.2.2 LIGA Process
27.2.3 Other Microfabrication Processes

27.3 Nanotechnology Products

27.4 Scanning Probe Microscopes

27.5 Nanofabrication Processes
27.5.1 Top-Down Processing Approaches
27.5.2 Bottom-Up Processing Approaches

An important trend in engineering design and manufacturing is the growing number of products and/or components of products whose features sizes are measured in microns (1 μm = 10^{-3} mm = 10^{-6} m). Several terms have been applied to these miniaturized items. Figure 27.1 indicates the relative sizes and other factors associated with these terms. ***Microsystem technology*** (MST) is the general term that refers to the products as well as the fabrication technologies to produce them. ***Microelectromechanical systems*** (MEMS) emphasizes the miniaturization of systems consisting of both electronic and mechanical components. The word ***micromachines*** is sometimes used for these devices.

Nanotechnology refers to the fabrication and application of even smaller entities whose feature sizes range from less than 1 nm to 100 nm (1 nm = 10^{-3} μm = 10^{-6} mm = 10^{-9} m).[1] The entities include films, coatings, dots, lines, wires, tubes, structures, and systems. ***Nanoscience*** is the field of scientific study that is concerned with objects in the same size range. ***Nanoscale*** refers to dimensions within this range and slightly below, which overlaps on the lower end with the sizes of atoms and molecules. For example, the smallest atom is helium, with a diameter close to 0.1 nm. Uranium has a diameter of about 0.22 nm and is the largest of the naturally occurring atoms. Molecules tend to be larger because they consist of multiple atoms. Molecules made up of about 30 atoms are roughly 1 nm in size, depending on the elements involved. Thus, nanoscience is the study of matter at the atomic and molecular level, and nanotechnology involves the application of nanoscience to create useful products.

[1] The dividing line between nanotechnology and microsystem technology is considered to be 100 nm = 0.1 μm [8]. This is illustrated approximately in Figure 27.1

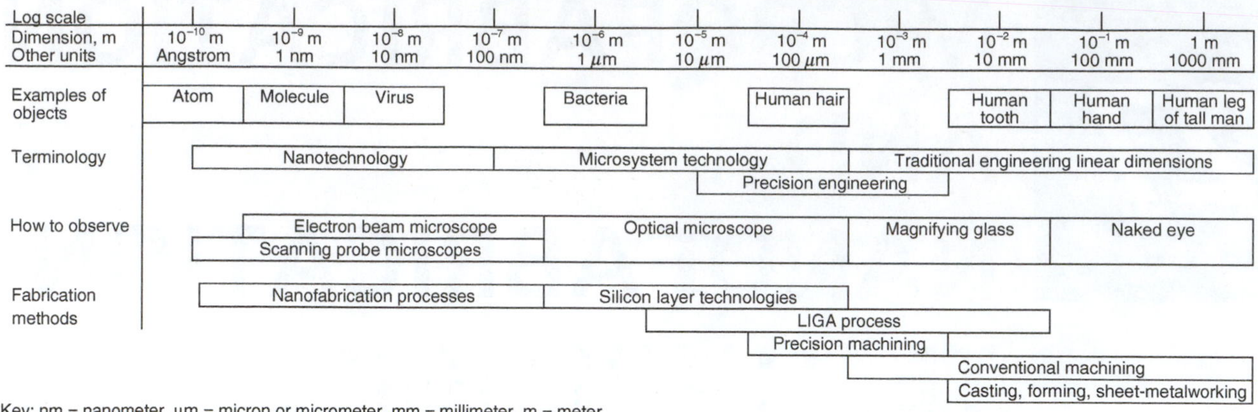

FIGURE 27.1 Terminology and relative sizes for microsystems and related technologies. (Credit: *Fundamentals of Modern Manufacturing*, 4th Edition by Mikell P. Groover, 2010. Reprinted with permission of John Wiley & Sons, Inc.)

The organization of our coverage of microfabrication and nanofabrication technologies begins with two sections on microsystem products and associated fabrication processes. We then explore nanotechnology products and their fabrication techniques.

27.1 MICROSYSTEM PRODUCTS

Designing small products comprised of even smaller parts and subassemblies means less material usage, lower power requirements, greater functionality per unit space, and accessibility to regions that are forbidden to larger products. In most cases, smaller products should mean lower prices because less material is used; however, the price of a given product is influenced by the costs of research, development, and production, and how these costs can be spread over the number of units sold. The economies of scale that result in lower-priced products have not yet fully been realized in microsystems technology, except for a limited number of cases that we shall examine in this section.

27.1.1 TYPES OF MICROSYSTEM DEVICES

Microsystem products can be classified by type of device (e.g., sensor, actuator) or by application area (e.g., medical, automotive). The device categories are as follows [6]:

- ***Microsensors***. A sensor is a device that detects or measures some physical phenomenon such as heat or pressure. It includes a transducer that converts one form of physical variable into another form (e.g., a piezoelectric device converts mechanical force into electrical current) plus the physical packaging and external connections. Most microsensors are fabricated on silicon substrates using the same processing technologies as those used for integrated circuits. Microscopic-sized sensors have been developed for measuring force, pressure, position, speed, acceleration, temperature, flow, and a variety of optical, chemical, environmental, and biological variables. Figure 27.2 shows a micrograph of a micro-accelerometer developed at Motorola Co.

FIGURE 27.2 Microscopic ratchet mechanism fabricated of silicon (Photo courtesy of Paul McWhorter).

- ***Microactuators***. Like a sensor, an actuator converts a physical variable of one type into another type, but the converted variable usually involves some mechanical action (e.g., a piezoelectric device oscillating in response to an alternating electrical field). An actuator causes a change in position or the application of force. Examples of microactuators include valves, positioners, switches, pumps, and rotational and linear motors [6].
- ***Microstructures and microcomponents***. These terms are used to denote a microsized part that is not a sensor or actuator. Examples of microstructures and microcomponents include microscopic gears, lenses, mirrors, nozzles, and beams. These items must be combined with other components (microscopic or otherwise) to provide a useful function. Figure 27.3 shows a microscopic gear alongside a human hair for comparison.
- ***Microsystems and micro-instruments***. These terms denote the integration of several of the preceding components together with the appropriate electronics package into a miniature system or instrument. Microsystems and micro-instruments tend to be very application specific; for example, microlasers, optical chemical analyzers, and microspectrometers. The economics of manufacturing these kinds of systems have tended to make commercialization difficult.

27.1.2 MICROSYSTEM APPLICATIONS

The preceding microdevices and systems have been applied in a wide variety of fields. Some important examples are the following:

- ***Ink-jet printing heads***. This is currently one of the largest applications of MST, because a typical ink-jet printer uses up several cartridges each year. Today's ink-jet printers possess resolutions of 1200 dots per inch (dpi), which converts to a nozzle separation of only about 21 μm, certainly in the microsystem range.

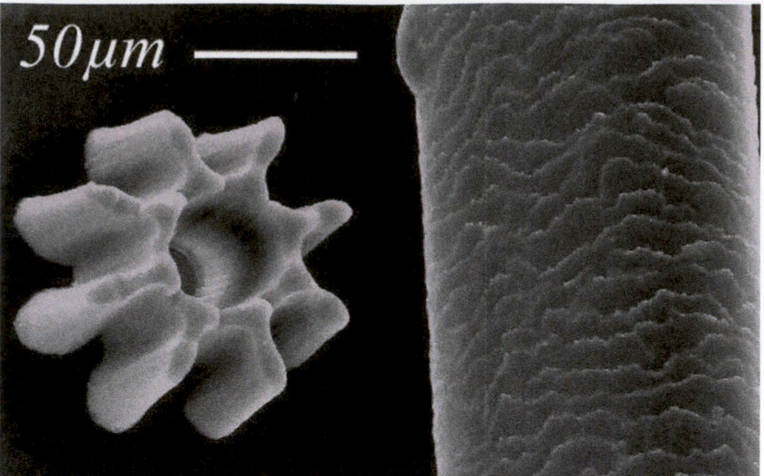

FIGURE 27.3 A microscopic gear and a human hair. The image was made using a scanning electron microscope. The gear is high-density polyethylene molded by a process similar to the LIGA process (Section 27.2.2), except that the mold cavity was fabricated using a focused ion beam instead of X-rays. Photo courtesy of M. Ali, International Islamic University Malaysia.

- ***Thin-film magnetic heads***. Read-write heads are key components in magnetic storage devices. Development of thin-film magnetic heads at IBM Corporation was an important breakthrough in digital storage technology as well as a significant success story for microfabrication technologies. The miniature size of the read-write head has enabled significant increases in bit densities of magnetic storage media. Thin-film read-write heads are produced annually in hundreds of millions of units, with a market of several billions of dollars per year.
- ***Compact discs and DVDs***. Compact discs (CDs) and digital versatile discs (DVDs)[2] are important commercial products today, as storage media for audio, video, games, and computer software and data applications. A CD disk, molded of polycarbonate, is 120 mm in diameter and 1.2 mm thick. The data consists of small pits (depressions) in a helical track that begins at a diameter of 46 mm and ends at about 117 mm. The tracks in the spiral are separated by about 1.6 μm. Each pit in the track is about 0.5 μm wide and about 0.8 μm to 3.5 μm long. The corresponding dimensions of DVDs are even smaller, permitting much higher data storage capacities.
- ***Automotive applications***. Microsensors and other microdevices are widely used in motor vehicles, consistent with the increased application of on-board electronics to accomplish control and safety functions for the vehicle. The functions include electronic engine control, cruise control, anti-lock braking systems, air-bag deployment, automatic transmission control, power steering, all-wheel drive, automatic stability control, on-board navigation systems, and remote locking and unlocking. These control systems and safety features require sensors and actuators, and a growing number of these are microscopic in size. There are currently 20 to 100 sensors installed in a modern automobile. In 1970 there were virtually no on-board sensors.

[2] The DVD was originally called a digital video disc because its primary applications were movie videos. However, DVDs of various formats are now used for data storage and other computer applications, video games, and high-quality audio.

- ***Medical applications***. Opportunities for using microsystems technology in this area are tremendous. Indeed, significant strides have already been made, and many of the traditional medical and surgical methods have already been transformed by MST. One of the driving forces behind the use of microscopic devices is the principle of minimal-invasive therapy, which involves the use of very small incisions or even available body orifices to access the medical problem of concern. Among the techniques based on miniaturization of medical instrumentation is the field of endoscopy,[3] now routinely used for diagnostic purposes and with growing applications in surgery. It is standard medical practice today to use endoscopic examination accompanied by laparoscopic surgery for hernia repair and removal of organs such as the gallbladder and the appendix. Growing use of similar procedures is expected in brain surgery, operating through one or more small holes drilled through the skull.
- ***Chemical and environmental applications***. A principal role of microsystem technology in chemical and environmental applications is the analysis of substances in order to measure trace amounts of chemicals or to detect harmful contaminants. A variety of chemical microsensors have been developed. They are capable of analyzing very small samples of the substance of interest.
- ***Scanning probe microscope***. This is a technology for measuring microscopic details of surfaces, allowing surface structures to be examined at the nanometer level. In order to operate in this dimensional range, the instruments require probes that are only a few μm in length and that scan the surface at a distance measured in nm. These probes are produced using microfabrication techniques.[4]

27.2 MICROFABRICATION PROCESSES

Many products in microsystem technology are based on silicon, and most of the processing techniques used to fabricate microsystems are borrowed from the microelectronics industry, which has been producing integrated circuits with features measured in microns for many years. There are several important reasons why silicon is a desirable material in MST: (1) The microdevices in MST often include electronic circuits, so both the circuit and the microdevice can be fabricated in combination on the same substrate. (2) In addition to its desirable electronic properties, silicon also possesses useful mechanical properties, such as high strength and elasticity, good hardness, and relatively low density. (3) The technologies for processing silicon are well established, owing to their widespread use in microelectronics. (4) Use of single-crystal silicon permits the production of physical features to very close tolerances.

Microsystem technology often requires silicon to be fabricated along with other materials in order to obtain a particular microdevice. For example, microactuators often consist of several components made of different materials. Accordingly, microfabrication techniques consist of more than just silicon processing. Our coverage of the microfabrication processes is organized into three sections: (1) silicon layering processes, (2) the LIGA process, and (3) other processes accomplished on a microscopic scale.

---

[3] The use of a small instrument (i.e., an endoscope) to visually examine the inside of a hollow body organ such as the rectum or colon.

[4] Scanning probe microscopes are discussed in Section 27.4.2.

27.2.1 SILICON LAYERING PROCESSES

The first application of silicon in microsystems technology was in the fabrication of Si piezoresistive sensors for the measurement of stress, strain, and pressure in the early 1960s [13]. Silicon is now widely used in MST to produce sensors, actuators, and other microdevices. The basic processing technologies are those used to produce integrated circuits. However, it should be noted that certain differences exist between the processing of ICs and the fabrication of microdevices:

1. The aspect ratios in microfabrication are generally much greater than in integrated circuit (IC) fabrication. ***Aspect ratio*** is defined as the height-to-width ratio of the features produced, as illustrated in Figure 27.4. Typical aspect ratios in semiconductor processing are about 1.0 or less, whereas in microfabrication the corresponding ratio might be as high as 400 [13].
2. The sizes of the devices made in microfabrication are often much larger than in IC processing, whereas the prevailing trend in microelectronics is inexorably toward miniaturization and greater circuit densities.
3. The structures produced in microfabrication often include cantilevers and bridges and other shapes requiring gaps between layers. These kinds of structures are uncommon in IC fabrication.

Notwithstanding these differences, let us nevertheless recognize that most of the silicon-processing steps used in microfabrication are the same or very similar to those used to produce ICs. After all, silicon is the same material whether it is used for integrated circuits or microdevices. The processing steps are listed in Table 27.1, together with brief descriptions. Many of these processing steps are discussed in previous chapters, as noted in the table. As in IC fabrication, the processes add, alter, or remove layers of material from a substrate according to geometric data contained in lithographic masks. Lithography is the fundamental technology that determines the shape of the microdevice being fabricated.

Regarding our preceding list of differences between IC fabrication and microdevice fabrication, the issue of aspect ratio should be addressed in more detail. The structures in IC processing are basically planar, whereas three-dimensional structures are more likely to be required in microsystems. The features of microdevices are likely to

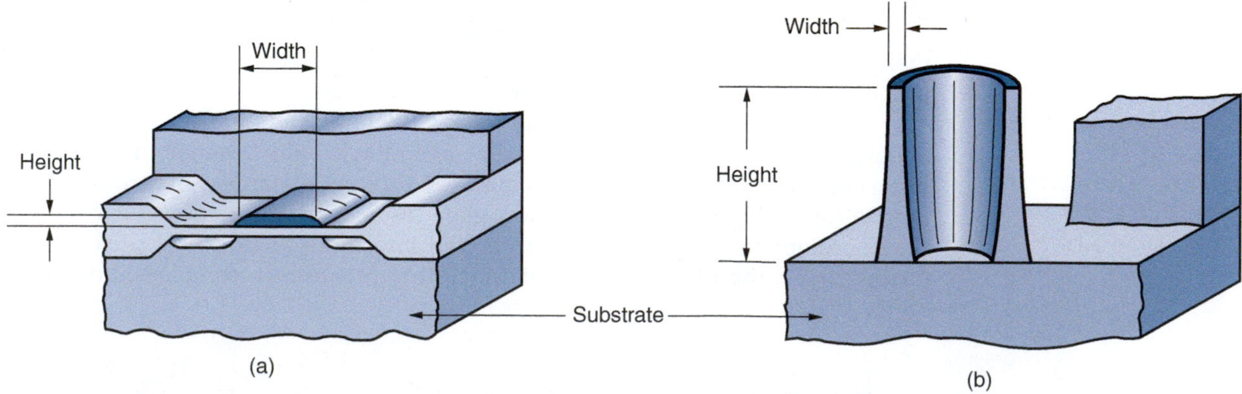

FIGURE 27.4 Aspect ratio (height-to-width ratio) typical in (a) fabrication of integrated circuits and (b) microfabricated components. (Credit: *Fundamentals of Modern Manufacturing*, 4th Edition by Mikell P. Groover, 2010. Reprinted with permission of John Wiley & Sons, Inc.)

TABLE 27.1 Silicon layering processes used in microfabrication.

Process	Brief Description
Lithography	Process used to expose a coating of resist on the surface of silicon or other substrate (e.g., silicon dioxide) to radiation. A mask containing the required pattern separates the radiation source from the resist, so that only the portions that are not blocked by the mask are exposed. Thus, the mask pattern is transferred to the resist, which is a polymer whose solubility to certain chemicals is altered by the radiation. The change in solubility permits regions of the resist corresponding to the mask pattern to be removed so that the substrate can be processed (e.g., etched, coated). The usual technique in microfabrication is photolithography, in which visible or ultraviolet (UV) light is the radiation source. See our description of photochemical machining in Section 19.4. Alternative lithography technologies include X-rays and electron beams.
Thermal oxidation	(Layer addition) Oxidation of silicon surface to form silicon dioxide layer.
Chemical vapor deposition	(Layer addition) Formation of a thin film on the surface of a substrate by chemical reactions or decomposition of gases (Section 21.5.2).
Physical vapor deposition	(Layer addition) Family of deposition processes in which a material is converted to vapor phase and condensed onto a substrate surface as a thin film (Section 21.5.1).
Electroplating and electroforming	(Layer addition) Electrolytic process in which metal ions in solution are deposited onto a cathode work material (Sections 21.3.1 and 21.3.2).
Electroless plating	(Layer addition) Deposition in an aqueous solution containing ions of the plating metal with no external electric current. Work surface acts as catalyst for the reaction (Section 21.3.3).
Thermal diffusion (doping)	(Layer alteration) Physical process in which atoms migrate from regions of high concentration into regions of low concentration (Section 21.2.1).
Ion implantation (doping)	(Layer alteration) Embedding atoms of one or more elements in a substrate using a high-energy beam of ionized particles (Section 21.2.2).
Wet etching	(Layer removal) Application of a chemical etchant in aqueous solution to etch away a target material, usually in conjunction with a mask pattern.
Dry etching	(Layer removal) Dry plasma etching using an ionized gas to etch a target material.

Credit: *Fundamentals of Modern Manufacturing*, 4th edition, by Mikell P. Groover. Reprinted by permission of John Wiley & Sons, Inc.

possess large height-to-width ratios. These three-dimensional features can be produced in single-crystal silicon by wet etching, provided the crystal structure is oriented to allow the etching process to proceed anisotropically. Chemical wet etching of polycrystalline silicon is isotropic. However, in single-crystal Si, the etching rate depends on the orientation of the lattice structure. In Figure 27.5, the three crystal faces of silicon's cubic lattice structure are illustrated. Certain etching solutions, such as potassium hydroxide (KOH) and sodium hydroxide (NaOH), have a very low etching rate in

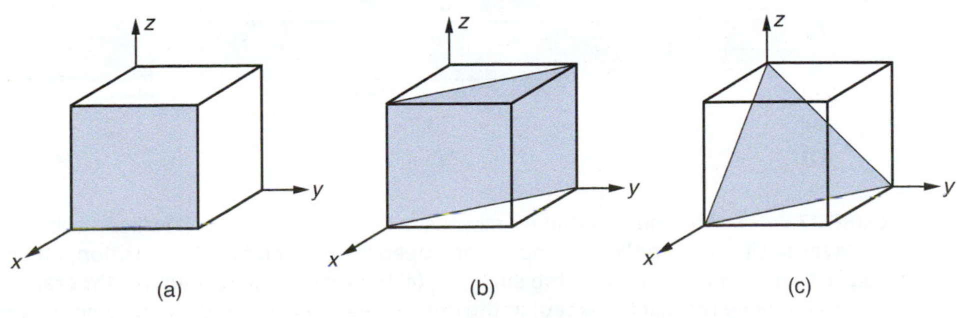

FIGURE 27.5 Three crystal faces in the silicon cubic lattice structure: (a) (100) crystal face, (b) (110) crystal face, and (c) (111) crystal face. (Credit: *Fundamentals of Modern Manufacturing*, 4th Edition by Mikell P. Groover, 2010. Reprinted with permission of John Wiley & Sons, Inc.)

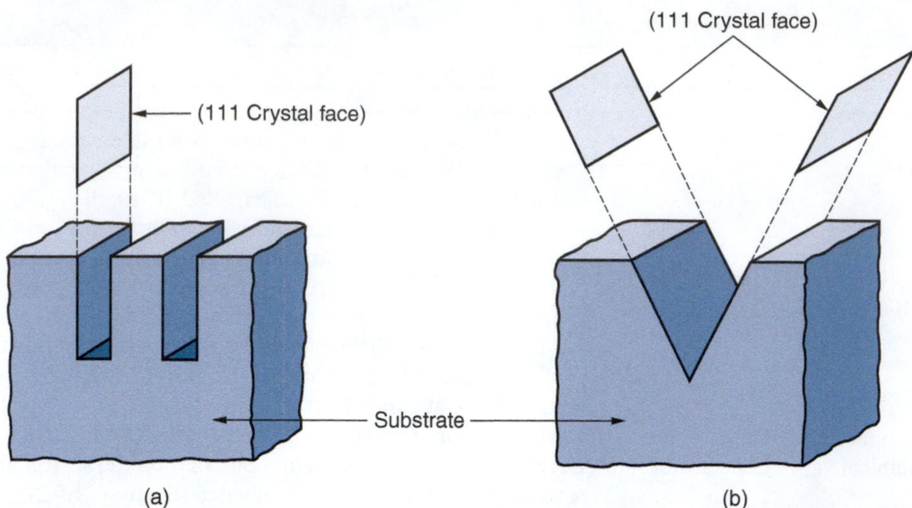

FIGURE 27.6 Several structures that can be formed in single-crystal silicon substrate by bulk micromachining: (a) (110) silicon and (b) (100) silicon. (Credit: *Fundamentals of Modern Manufacturing*, 4th Edition by Mikell P. Groover, 2010. Reprinted with permission of John Wiley & Sons, Inc.)

the direction of the (111) crystal face. This permits the formation of distinct geometric structures with sharp edges in a single-crystal Si substrate whose lattice is oriented to favor etch penetration vertically or at sharp angles into the substrate. Structures such as those in Figure 27.6 can be created using this procedure. The term ***bulk micromachining*** is used for the relatively deep wet-etching process into single-crystal silicon substrate (Si wafer); whereas the term ***surface micromachining*** refers to the planar structuring of the substrate surface, using much more shallow layering processes.

Bulk micromachining can be used to create thin membranes in a microstructure. However, a method is needed to control the etching penetration into the silicon, so as to leave the membrane layer. A common method used for this purpose is to dope (diffuse) boron atoms into the silicon substrate, which significantly reduces the etching rate of the silicon. The processing sequence is shown in Figure 27.7. In step (2), epitaxial deposition is used to apply the upper layer of silicon so that it will possess the same single-crystal structure and lattice orientation as the substrate. This is a requirement of bulk micromachining that will be used to provide the deeply etched region in subsequent processing.

Surface micromachining can be used to construct cantilevers, overhangs, and similar structures on a silicon substrate, as shown in part (5) of Figure 27.8. The cantilevered beams in the figure are parallel to but separated by a gap from the silicon surface.

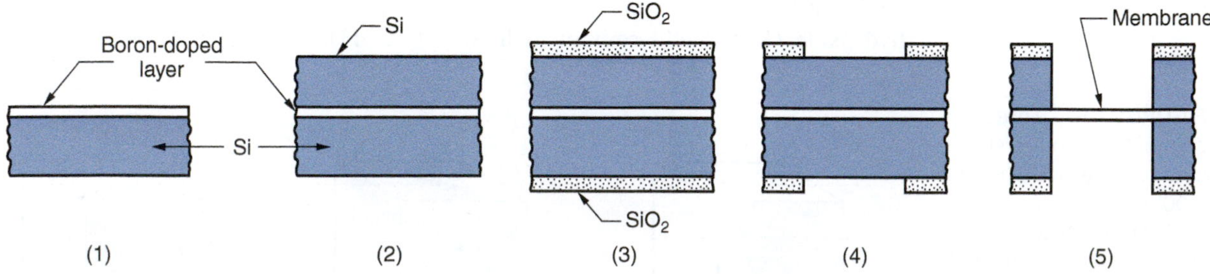

FIGURE 27.7 Formation of a thin membrane in a silicon substrate: (1) silicon substrate is doped with boron, (2) a thick layer of silicon is applied on top of the doped layer by epitaxial deposition, (3) both sides are thermally oxidized to form a SiO_2 resist on the surfaces, (4) the resist is patterned by lithography, and (5) anisotropic etching is used to remove the silicon except in the boron-doped layer. (Credit: *Fundamentals of Modern Manufacturing*, 4th Edition by Mikell P. Groover, 2010. Reprinted with permission of John Wiley & Sons, Inc.)

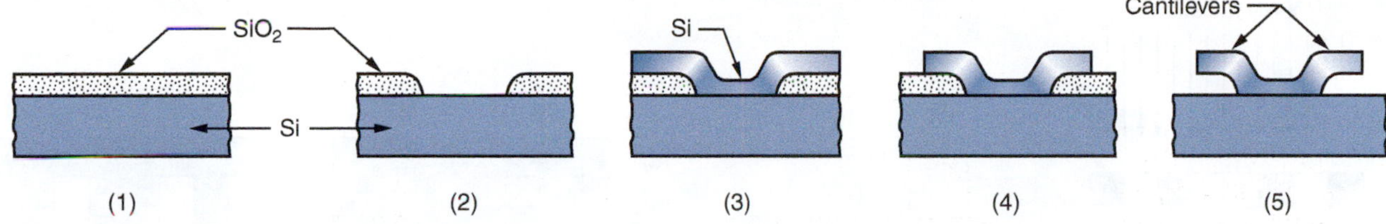

FIGURE 27.8 Surface micromachining to form cantilevers: (1) on the silicon substrate is formed a silicon dioxide layer, whose thickness will determine the gap size for the cantilevered member; (2) portions of the SiO_2 layer are etched using lithography; (3) a polysilicon layer is applied; (4) portions of the polysilicon layer are etched using lithography; and (5) the SiO_2 layer beneath the cantilevers is selectively etched. (Credit: *Fundamentals of Modern Manufacturing,* 4th Edition by Mikell P. Groover, 2010. Reprinted with permission of John Wiley & Sons, Inc.)

Gap size and beam thickness are in the micron range. The process sequence to fabricate this type of structure is depicted in the earlier parts of Figure 27.8.

A procedure called the ***lift-off technique*** is used in microfabrication to pattern metals such as platinum on a substrate. These structures are used in certain chemical sensors, but are difficult to produce by wet etching. The processing sequence in the lift-off technique is illustrated in Figure 27.9.

27.2.2 LIGA PROCESS

LIGA is an important process in MST. It was developed in Germany in the early 1980s. The letters ***LIGA*** stand for the German words ***LI***thographie (lithography, Table 27.1, in particular, X-ray lithography, although other lithographic exposure methods are also used, such as ion beams in Figure 27.3), ***G***alvanoformung (translated electrodeposition or electroforming, Section 21.3.2), and **A**bformtechnik (plastic molding, Section 8.6).

The LIGA processing steps are illustrated in Figure 27.10. Let us elaborate on the brief description provided in the figure's caption: (1) A thick layer of (X-ray) radiation-sensitive resist is applied to a substrate. Layer thickness can range between several microns to centimeters, depending on the size of the part(s) to be produced. The common resist material used in LIGA is polymethylmethacrylate (PMMA, an acrylic thermoplastic). The substrate must be a conductive material for the subsequent electrodeposition processes performed. The resist is exposed through a mask to high-energy X-ray radiation. (2) The irradiated areas of the resist are chemically removed from the substrate surface, leaving the unexposed portions standing as a three-dimensional plastic structure. (3) The regions where the resist has been removed are filled with metal using

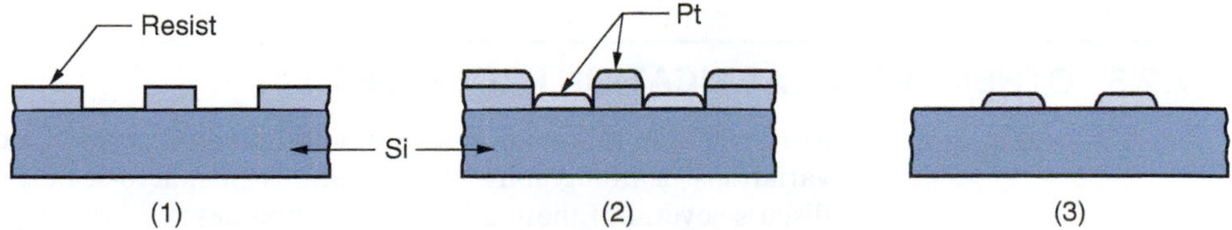

FIGURE 27.9 The lift-off technique: (1) resist is applied to substrate and structured by lithography; (2) platinum is deposited onto surfaces; and (3) resist is removed, taking with it the platinum on its surface but leaving the desired platinum microstructure. (Credit: *Fundamentals of Modern Manufacturing,* 4th Edition by Mikell P. Groover, 2010. Reprinted with permission of John Wiley & Sons, Inc.)

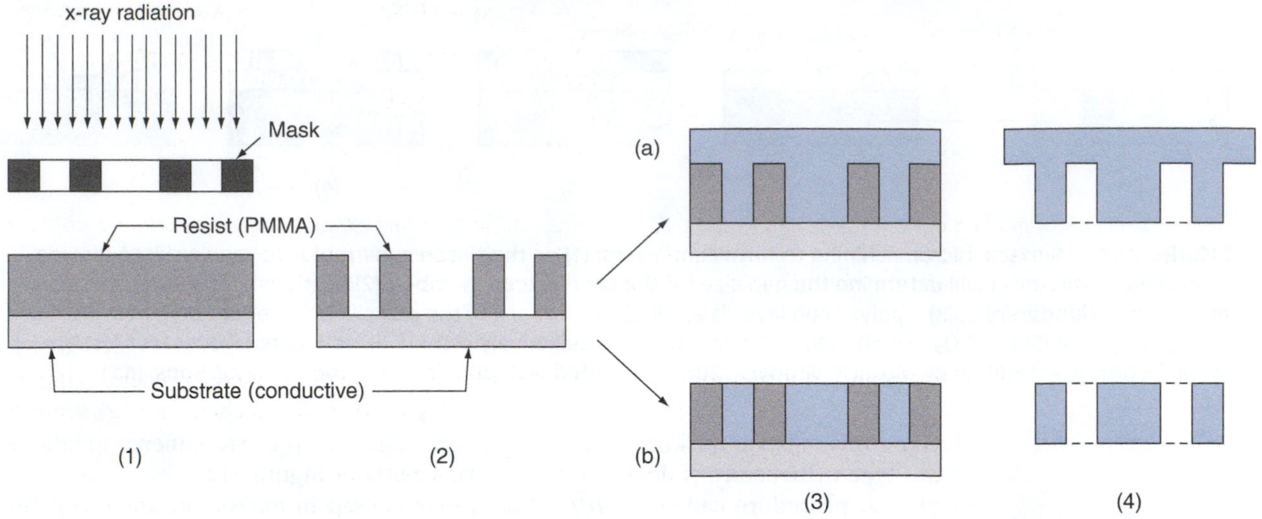

FIGURE 27.10 LIGA processing steps: (1) thick layer of resist is applied and exposed to X-rays through mask, (2) exposed portions of resist are removed, (3) electrodeposition is used to fill openings in resist, and (4) resist is stripped to provide (a) a mold or (b) a metal part. (Credit: *Fundamentals of Modern Manufacturing,* 4th Edition by Mikell P. Groover, 2010. Reprinted with permission of John Wiley & Sons, Inc.)

electrodeposition. Nickel is the common plating metal used in LIGA. (4) The remaining resist structure is stripped, yielding a three-dimensional metal structure. Depending on the geometry created, this metallic structure may be (a) the mold used for producing plastic parts by injection molding, reaction injection molding, or compression molding. In the case of injection molding, in which thermoplastic parts are produced, these parts may be used as "lost molds" in investment casting (Section 6.2.3). Alternatively, (b) the metal part may be a pattern for fabricating plastic molds that will be used to produce more metallic parts by electrodeposition.

As our description indicates, LIGA can produce parts by several different methods. This is one of the greatest advantages of this microfabrication process: (1) LIGA is a versatile process. Other advantages include: (2) high aspect ratios are possible—large height-to-width ratios in the fabricated part; (3) a wide range of part sizes is feasible, with heights ranging from micrometers to centimeters; and (4) close tolerances can be achieved. A significant disadvantage of LIGA is that it is a very expensive process, so large quantities of parts are usually required to justify its application. Also, the use of X-ray radiation is a disadvantage.

27.2.3 OTHER MICROFABRICATION PROCESSES

MST research is providing several additional fabrication techniques, most of which are variations of lithography or adaptations of macro-scale processes. In this section we discuss several of these additional techniques.

Soft Lithography This term is used for processes that utilize an elastomeric flat mold (similar to a rubber ink stamp) to create a pattern on a substrate surface. The sequence for creating the mold is illustrated in Figure 27.11. A master pattern is fabricated on a silicon surface using a lithography process such as ultraviolet photolithography. This master pattern is then used to produce the flat mold for the soft lithography process. The

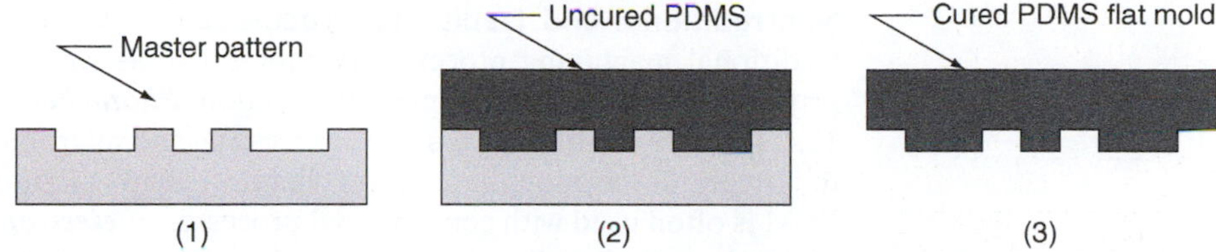

FIGURE 27.11 Steps in mold-making for soft lithography: (1) master pattern fabricated by traditional lithography, (2) polydimethylsiloxane flat mold is cast from the master pattern, and (3) cured flat mold is peeled off pattern for use. (Credit: *Fundamentals of Modern Manufacturing,* 4th Edition by Mikell P. Groover, 2010. Reprinted with permission of John Wiley & Sons, Inc.)

common mold material is polydimethylsiloxane (PDMS, a silicon rubber). After the PDMS has cured, it is peeled away from the pattern and attached to a substrate for support and handling.

Two of the soft lithography processes are micro-imprint lithography and micro-contact printing. In ***micro-imprint lithography***, the mold is pressed into the surface of a soft resist to displace the resist away from certain regions of the substrate for subsequent etching. The process sequence is illustrated in Figure 27.12. The flat mold consists of raised and depressed regions, and the raised regions correspond to areas on the resist surface that will be displaced to expose the substrate. The resist material is a thermoplastic polymer that has been softened by heating prior to pressing. The alteration of the resist layer is by mechanical deformation rather than electromagnetic radiation, as in the more traditional lithography methods. The compressed regions of the resist layer are subsequently removed by etching. The etching process also reduces the thickness of the remaining resist layer, but enough remains to protect the substrate for subsequent processing. Micro-imprint lithography can be set up for high production rates at modest cost. A mask is not required in the imprint procedure, although fabricating the mold requires an analogous preparation.

The same type of flat stamp can be used in a printing mode, in which case the process is called ***micro-contact printing***. In this form of soft lithography, the mold is used to transfer a pattern of a substance to a substrate surface, much like ink can be transferred to a paper surface. This process allows very thin layers to be fabricated onto the substrate.

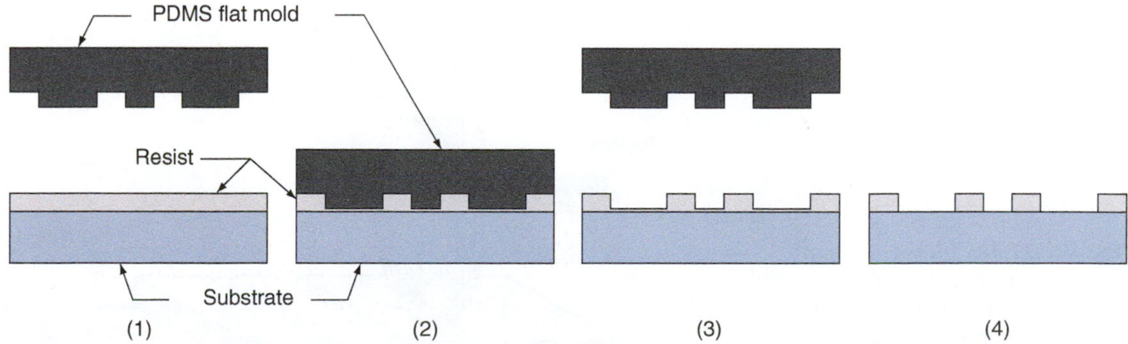

FIGURE 27.12 Steps in micro-imprint lithography: (1) mold is positioned above and (2) pressed into resist, (3) mold is lifted, and (4) remaining resist is removed from substrate surface in defined regions. (Credit: *Fundamentals of Modern Manufacturing,* 4th Edition by Mikell P. Groover, 2010. Reprinted with permission of John Wiley & Sons, Inc.)

Nontraditional and Traditional Processes in Microfabrication A number of nontraditional machining processes (Chapter 19), as well as conventional manufacturing processes, are important in microfabrication. ***Photochemical machining*** (PCM, Section 19.4.2) is an essential process in IC processing and microfabrication, but we have referred to it in our descriptions here as wet chemical etching (combined with photolithography). PCM is often used with conventional processes of ***electroplating***, ***electroforming***, and/or ***electroless plating*** (Section 21.3.3) to add layers of metallic materials according to microscopic pattern masks.

Other nontraditional processes capable of micro-level processing include [13]: (1) ***electric discharge machining***, used to cut holes as small as 0.3 mm in diameter with aspect ratios (depth-to-diameter) as high as 100; (2) ***electron-beam machining***, for cutting holes of diameter smaller than 100 μm in hard-to-machine materials; (3) ***laser-beam machining***, which can produce complex profiles and holes as small as 10 μm in diameter with aspect ratios (depth-to-width or depth-to-diameter) approaching 50; (4) ***ultrasonic machining***, capable of drilling holes in hard and brittle materials as small as 50 μm in diameter; and (5) ***wire electric discharge cutting***, or ***wire-EDM***, which can cut very narrow swaths with aspect ratios (depth-to-width) greater than 100.

Trends in conventional machining have included its capabilities for taking smaller and smaller cut sizes and associated tolerances. Referred to as ***ultra-high-precision machining***, the enabling technologies have included single-crystal diamond-cutting tools and position control systems with resolutions as fine as 0.01 μm [13]. Figure 27.13 depicts one reported application, the milling of grooves in aluminum foil using a single-point diamond fly-cutter. The aluminum foil is 100 μm thick, and the grooves are 85 μm wide and 70 μm deep. Similar ultra-high-precision machining is being applied today to produce products such as computer hard discs, photocopier drums, mold inserts for compact disk reader heads, and high-definition TV projection lenses.

Rapid Prototyping Technologies Several rapid prototyping (RP) methods (Chapter 26) have been adapted to produce micro-sized parts [20]. RP methods use a layer additive approach to build three-dimensional components, based on a CAD (computer-aided design) geometric model of the component. Each layer is very thin, typically as low as 0.05 mm thick, which approaches the scale of microfabrication technologies. By making the layers even thinner, microcomponents can be fabricated.

One RP approach is called ***electrochemical fabrication*** (EFAB), which involves the electrochemical deposition of metallic layers in specific areas that are determined by

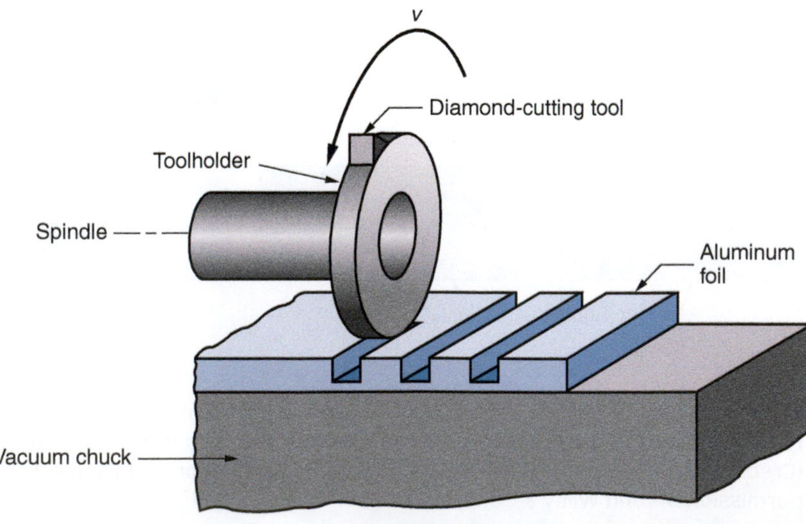

FIGURE 27.13 Ultra-high-precision milling of grooves in aluminum foil. (Credit: *Fundamentals of Modern Manufacturing*, 4th Edition by Mikell P. Groover, 2010. Reprinted with permission of John Wiley & Sons, Inc.)

pattern masks created by "slicing" a CAD model of the object to be made (Section 26.1). The deposited layers are generally 5 to 10 μm thick, with feature sizes as small as 20 μm in width. EFAB is carried out at temperatures below 60°C (140°F) and does not require a clean room environment. However, the process is slow, requiring about 40 minutes to apply each layer, or about 36 layers (a height between 180 μm and 360 μm) per 24-hour period. To overcome this disadvantage, the mask for each layer can contain multiple copies of the part slice pattern, permitting multiple parts to be produced simultaneously in a batch process.

Another RP approach, called ***microstereolithography***, is based on stereolithography (STL, Section 26.2.1), but the scale of the processing steps is reduced in size. Whereas the layer thickness in conventional stereolithography ranges between 75 μm and 500 μm, microstereolithography (MSTL) uses layer thicknesses between 10 μm and 20 μm typically, with even thinner layers possible. The laser spot size in STL is typically around 250 μm in diameter, whereas MSTL uses a spot size as small as 1 μm or 2 μm. Another difference in MSTL is that the work material is not limited to a photosensitive polymer. Researchers report success in fabricating three-dimensional microstructures from ceramic and metallic materials. The difference is that the starting material is a powder rather than a liquid.

27.3 NANOTECHNOLOGY PRODUCTS

Most products associated with nanotechnology are not just smaller versions of microsystem technology products; they also include new materials, coatings, and unique entities that are not included within the scope of MST. Nanoscale products and processes that have been around for a while include the following:

- The colorful stained-glass windows of churches built during the Middle Ages were based on gold particles of nanometer scale imbedded in the glass. Depending on their size, the particles can take on a variety of different colors.
- Traditional film photography has roots dating back more than 150 years and depends on the formation of silver nanoparticles to create the image in the photographic film.
- Nanoscale particles of carbon are used as reinforcing fillers in automobile tires.
- Catalytic converters required in the exhaust systems of modern automobiles make use of nanoscale coatings of platinum and palladium on a ceramic honeycomb structure. The metal coatings act as catalysts to convert harmful emission gases into harmless gases.

We should also mention that the fabrication technology for integrated circuits now includes feature sizes that are in the nanotech range. Of course, integrated circuits have been produced since the 1960s, but only in recent years have nanoscale features been achieved.

Other more recent products exploiting applications of nanotechnology include cosmetics, sun lotions, car polishes and waxes, coatings for eyeglass lenses, and scratch-resistant paints. All of these categories contain nanoscale particles (nanoparticles), which qualifies them as products of nanotechnology. For an extensive list of nanotech products, the interested reader can consult www.nanotechproject.org/inventories/consumer [33].

An important product category in microsystem technology is microelectromechanical systems (MEMS), which have found quite a few applications in the computer, medical, and automotive industries (Section 27.1.2). With the advent of nanotechnology, there has been growing interest in the notion of extending the development of these kinds

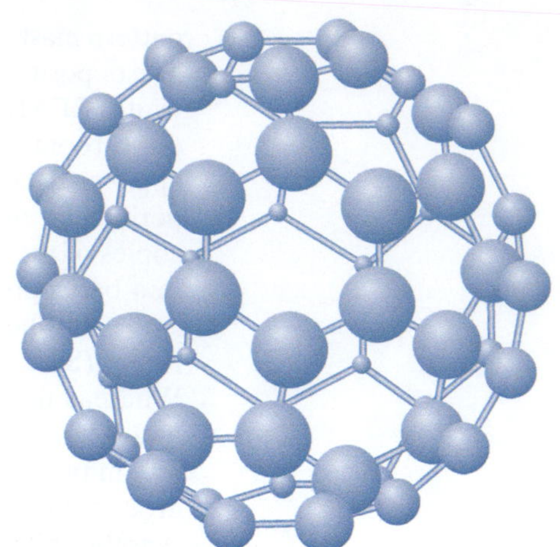

FIGURE 27.14 Fullerine structure of the C_{60} molecule. Reprinted by permission from [23]. (Credit: *Fundamentals of Modern Manufacturing*, 4th Edition by Mikell P. Groover, 2010. Reprinted with permission of John Wiley & Sons, Inc.)

of devices into the nanoscale range. ***Nanoelectromechanical systems*** (NEMS) are the sub-micron-sized counterparts of MEMS devices, only their smaller sizes would result in even greater potential advantages. An important NEMS structural product currently produced is the probe used in atomic force microscopes (Section 27.4). The sharp point on the probe is of nanoscale size. Nanosensors are another developing application. Nanosensors would be more accurate, faster responding, and operate with lower power requirements than larger sensors. Current NEMS sensor applications include accelerometers and chemical sensors. It has been suggested that multiple nanosensors could be distributed throughout the subject area to collect data, thus providing the benefit of multiple readings of the variable of interest, rather than using a single larger sensor at one location.

Two structures of significant scientific and commercial interest in nanotechnology are carbon buckyballs and nanotubes. They are basically graphite layers that have been formed into spheres and tubes, respectively.

The name ***buckyball*** refers to the molecule C_{60}, which contains exactly 60 carbon atoms and is shaped like a soccer ball, as in Figure 27.14. The original name of the molecule was ***buckministerfullerene***, after the architect/inventor R. Buckminister Fuller, who designed the geodesic dome that resembles the C_{60} structure. Today, C_{60} is simply called a ***fullerene***, which refers to any closed hollow carbon molecule that consists of 12 pentagonal and various numbers of hexagonal faces. In the case of C_{60}, the 60 atoms are arranged symmetrically into 12 pentagonal faces and 20 hexagonal faces to form a ball. These molecular balls can be bonded together to form crystals. The separation between any C_{60} molecule and its closest neighbor in the crystal structure is 1 nm.

Fullerenes are of interest for a number of reasons. One is their electrical properties and the capability to alter these properties. A C_{60} crystal has the properties of an insulator. However, when diffused with an alkaline metal such as potassium (forming K_3C_{60}), it is transformed into an electrical conductor. Moreover, it exhibits properties of a superconductor at temperatures of around 18°K. Another potential application area for the C_{60} fullerenes is in the medical field. The C_{60} molecule has many possible attachment points for focused drug treatments. Other possible medical applications for buckyballs include antioxidants, burn creams, and diagnostic imaging.

Carbon nanotubes (CNTs) are another molecular structure consisting of carbon atoms bonded together in the shape of a long tube. The atoms can be arranged into a number of alternative configurations, three of which are illustrated in Figure 27.15. The nanotubes shown in the figure are all single-walled nanotubes (SWNT), but multi-walled nanotubes (MWNT) can also be fabricated, which are tubes within a tube. A SWNT has a

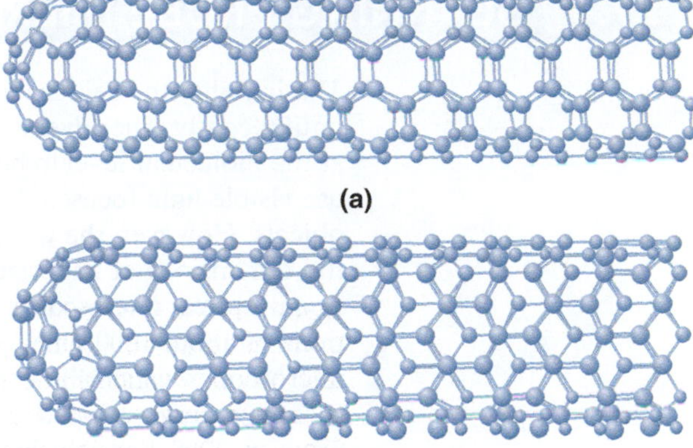

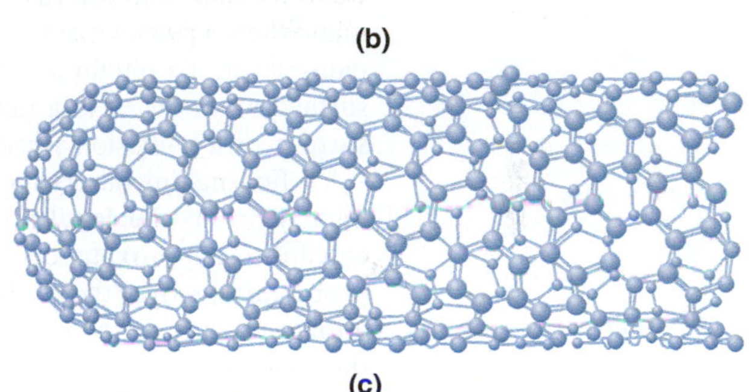

FIGURE 27.15 Several possible structures of carbon nanotubes: (a) armchair, (b) zigzag, and (c) chiral. Reprinted by permission from [23]. (Credit: *Fundamentals of Modern Manufacturing*, 4th Edition by Mikell P. Groover, 2010. Reprinted with permission of John Wiley & Sons, Inc.)

typical diameter of a few nanometers (down to 1 nm) and a length of around 100 nm, and it is closed at both ends.

The electrical properties of nanotubes are unusual. Depending on the structure and diameter, nanotubes can have metallic (conducting) or semiconducting properties. Conductivity of metallic nanotubes can be superior to that of copper by six orders of magnitude [8]. The explanation for this is that nanotubes contain few of the defects existing in metals that tend to scatter electrons, thus increasing electrical resistance. Because nanotubes have such low resistance, high currents do not increase their temperature the way metals heat up under the same electrical loads. Thermal conductivity of metallic nanotubes is also very high. These electrical and thermal properties are of significant interest to manufacturers of computers and integrated circuits because they may allow higher clock speeds of processors without the heat buildup problems currently encountered as the density of components on a silicon chip increases. Clock speeds 10^4 times faster than current-day processors may be possible [23], along with much higher densities.

Another electrical property of carbon nanotubes is field emission, in which electrons are emitted from the ends of the tubes at very high rates when an electrical field is applied parallel to the axis of a nanotube. The possible commercial applications of field emission properties of nanotubes include flat panel displays for televisions and computer monitors.

Mechanical properties are another reason for the interest in single-walled nanotubes. Compared with steel, density is only 1/6, modulus of elasticity is five times higher, and tensile strength is 100 times greater [8]. Yet, when SWNTs are bent, they exhibit great resilience to return to their previous shape without damage. These mechanical properties present opportunities for using them in applications ranging from reinforcing materials in polymer matrix composites to fiber cloths in bulletproof vests. Ironically, multi-walled nanotubes are not as strong.

27.4 SCANNING PROBE MICROSCOPES

The inability to "see" nanoscale objects has inhibited developments in nanotechnology until recently. The advent of scanning probe microscopes in the 1980s has allowed objects at the molecular level to be visualized and measured. Conventional optical microscopes use visible light focused through optical lenses to provide enlarged images of very small objects. However, the wavelength of visible light is 400 to 700 nm, which is greater than the dimensions of nanosized objects. Thus, these objects cannot be seen with conventional optical microscopes. The most powerful optical microscopes provide magnifications of about 1000 times, allowing resolutions of about 0.0002 mm (200 nm). Electron microscopes, which allow specimens to be visualized utilizing a beam of electrons instead of light, were developed in the 1930s. The electron beam can be considered as a form of wave motion, but one that has a much shorter effective wavelength. Today's electron microscopes permit magnifications of about 1,000,000 times and resolutions of about one nanometer. To obtain an image of a surface, the electron beam is scanned across the surface of an object in a raster pattern, similar to the way that a cathode-ray scans the surface of a television screen.

For making observations on the nanoscale level, an improvement over the electron microscope is the family of scanning probe instruments. They possess magnification capabilities approximately 10 times greater than an electron microscope. In a scanning probe microscope, the probe consists of a needle with a very sharp tip. The point size approaches the size of a single atom. In operation, the probe is moved along the surface of the specimen at a distance of only one nanometer or so, and any of several properties of the surface are measured, depending on the type of scanning probe device. The two scanning probe microscopes of greatest interest in nanotechnology are the scanning tunneling microscope and the atomic force microscope.

The ***scanning tunneling microscope*** (STM) was the first scanning probe instrument to be developed. It is called a tunneling microscope because its operation is based on the quantum mechanics phenomenon known as ***tunneling***, in which individual electrons in a solid material can jump beyond the surface of the solid into space. The probability of electrons being in this space beyond the surface decreases exponentially in proportion to the distance from the surface. This sensitivity to distance is exploited in the STM by positioning the probe tip very close to the surface (i.e., 1 nm) and applying a small voltage between the two. This causes electrons of surface atoms to be attracted to the small positive charge of the tip, and they tunnel across the gap to the probe. As the probe is moved along the surface, variations in the resulting current occur due to the positions of individual atoms on the surface. Alternatively, if the elevation of the tip above the surface is allowed to float by maintaining a constant current, then the vertical deflection of the tip can be measured as it traverses the surface. These variations in current or deflection can be used to create images or topographical maps of the surface on an atomic or molecular scale.

A limitation of the scanning tunneling microscope is that it can only be used on surfaces of conducting materials. By comparison, the ***atomic force microscope*** (AFM) can be used on any material; it uses a probe attached to a delicate cantilever that deflects due to the force exerted by the surface on the probe as it traverses the specimen surface. The AFM responds to various types of forces, depending on the application. The forces include mechanical due to physical contact of the probe with the specimen surface, and non-contact, such as van der Waals forces, capillary forces, magnetic forces,[5] and others. The vertical deflection of the probe is measured optically, based on the interference

[5] The term *magnetic force microscope* (MFM) is used when the forces are magnetic. The principle of operation is similar to that of the reading head on a hard disk drive.

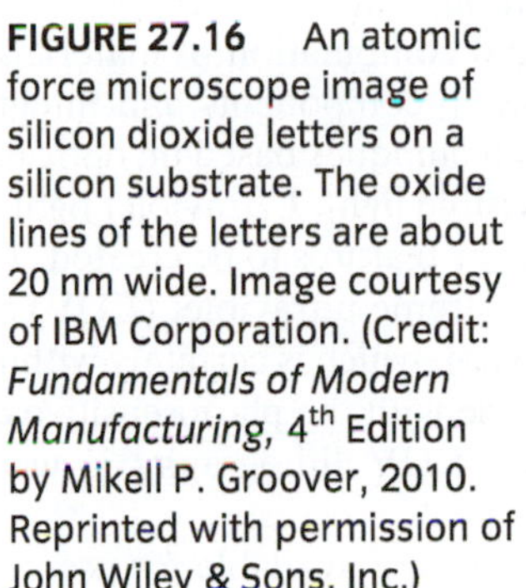
FIGURE 27.16 An atomic force microscope image of silicon dioxide letters on a silicon substrate. The oxide lines of the letters are about 20 nm wide. Image courtesy of IBM Corporation. (Credit: *Fundamentals of Modern Manufacturing*, 4th Edition by Mikell P. Groover, 2010. Reprinted with permission of John Wiley & Sons, Inc.)

pattern of a light beam or the reflection of a laser beam from the cantilever. Figure 27.16 shows an image generated by an AFM.

Our discussion here has focused on the use of scanning probe microscopes for observing surfaces. In Section 27.5.2, we describe applications of these instruments for manipulating individual atoms, molecules, and other nanoscale clusters of atoms or molecules.

27.5 NANOFABRICATION PROCESSES

Fabrication processes for creating products and structures whose feature sizes are in the nanometer range can be divided into two basic categories:

1. ***Top-down approaches***, which adapt some of the microfabrication techniques to nanoscale object sizes. They involve mostly subtractive processes (material removal) to achieve the desired geometry.
2. ***Bottom-up approaches***, in which atoms and molecules are manipulated and combined into larger structures. These might be described as additive processes because they construct the nanoscale entity from smaller components.

Our organization in this section is based on these two approaches. Since the processing methods associated with the top-down approaches have been discussed in Section 27.2, our coverage in Section 27.5.1 will emphasize how these processes must be modified for the nanoscale. Section 27.5.2 discusses the bottom-up approaches, which are perhaps of greater interest here because of their uniqueness and special relevance to nanotechnology.

27.5.1 TOP-DOWN PROCESSING APPROACHES

The top-down approaches for fabricating nanoscale objects involve the processing of bulk materials (e.g., silicon wafers) and thin films using lithographic techniques like those used in the fabrication of integrated circuits and microsystems. The top-down approaches also include other precision machining techniques (Section 27.2.3) that have been adapted for making nanostructures. The term ***nanomachining*** is used for these processes that involve material removal when applied in the sub-micron scale. Nanostructures have been machined out of materials such as silicon, silicon carbide, diamond, and silicon nitride [30]. Nanomachining must often be coupled with thin-film deposition processes such as physical vapor deposition and chemical vapor deposition (Section 21.5) to achieve the desired structure and combination of materials.

As the feature sizes of MST components and integrated circuits (ICs) become smaller and smaller, fabrication techniques based on optical lithography become limited because of the wavelengths of visible light. Ultraviolet light is currently used because its shorter wavelengths permit smaller features to be created. The current technology being used in IC fabrication is called extreme ultraviolet (EUV) lithography. It uses UV light with a wavelength as short as 13 nm, which is certainly within the nanotechnology range. However, technical problems arise with the photosensitive chemicals (photoresists) and laser focusing equipment when EUV lithography is used at these very short UV wavelengths.

Other lithography techniques are available for use in fabricating nanoscale structures. These include electron-beam lithography, X-ray lithography, and nano-imprint lithography. ***Electron-beam lithography*** (EBL) operates by directing a highly focused beam of electrons along the desired pattern across the surface of a material, thus exposing the surface areas using a sequential process without the need for a mask. Although EBL is capable of resolutions on the order of 10 nm, its sequential operation makes it relatively slow compared to masking techniques and thus unsuited to mass production. ***X-ray lithography*** can produce patterns with resolutions around 20 nm, and it uses masking techniques, which makes high production possible. However, X-rays are difficult to focus, the equipment is expensive for production applications, and X-rays are hazardous to humans.

Two processes known as soft lithography are described in Section 27.2.3: micro-imprint lithography and micro-contact printing. These same processes can be applied to nanofabrication, in which case they are called ***nano-imprint lithography*** and ***nano-contact printing***. Nano-imprint lithography can produce pattern resolutions of approximately 5 nm [30]. One of the original applications of nano-contact printing was to transfer a thin film of thiols (a family of organic compounds derived from hydrogen sulfide) onto a gold surface. The uniqueness of the application was that the film was only one molecule thick (called a monolayer, Section 27.5.2), which certainly qualifies as nanoscale.

27.5.2 BOTTOM-UP PROCESSING APPROACHES

In the bottom-up approaches, the starting materials are atoms, molecules, and ions. The processes bring these basic building blocks together, in some cases one at a time, to fabricate the desired nanoscale entity. Our coverage consists of three approaches that are of considerable interest in nanotechnology: (1) production of carbon nanotubes, (2) nanofabrication by scanning probe techniques, and (3) self-assembly.

Production of Carbon Nanotubes The remarkable properties and potential applications of carbon nanotubes are discussed in Section 27.3.1. Carbon nanotubes can be produced by several techniques. In the following paragraphs we discuss three: (1) laser evaporation, (2) carbon arc techniques, and (3) chemical vapor deposition.

In the ***laser evaporation method***, the starting raw material is a graphite workpiece containing small amounts of cobalt and nickel. These metal traces perform the role of catalyst, acting as nucleation sites for the subsequent formation of the nanotubes. The graphite is placed in a quartz tube filled with argon gas and heated to 1200°C (2200°F). A pulsed laser beam is focused on the workpiece, causing the carbon atoms to evaporate from the bulk graphite. The argon moves the carbon atoms out of the high-temperature region of the tube and into an area where a water-cooled copper apparatus is located. The carbon atoms condense on the cold copper, and as they do, they form nanotubes with diameters of 10 to 20 nm and lengths of about 100 μm.

The ***carbon arc technique*** uses two carbon electrodes that are 5 to 20 μm in diameter and separated by 1 mm. The electrodes are located in a partially evacuated container (about two-thirds of 1 atmospheric pressure) with helium flowing in it. To start the process, a voltage of about 25 V is applied across the two electrodes, causing carbon atoms to be ejected from the positive electrode and carried to the negative electrode where they form nanotubes. The structure of the nanotubes depends on whether a catalyst is used. If no catalyst is used, then multi-walled nanotubes are produced. If trace amounts of cobalt, iron, or nickel are placed in the interior of the positive electrode, then the process creates single-walled nanotubes that are 1 to 5 nm in diameter and about 1 μm long.

Chemical vapor deposition (Section 21.5.2) can be used to produce carbon nanotubes. In one variation of CVD, the starting work material is a hydrocarbon gas such as methane (CH_4). The gas is heated to 1100°C (2000°F), causing it to decompose and release carbon atoms. The atoms then condense on a cool substrate to form nanotubes with open ends rather than the closed ends characteristic of the other fabrication techniques. The substrate may contain iron or other metals that act as catalysts for the process. The metal catalyst acts as a nucleation site for creation of the nanotube, and it also controls the orientation of the structure. An alternative CVD process called HiPCO[6] starts with carbon monoxide (CO) and uses carbon pentacarbonyl ($Fe(CO)_5$) as the catalyst to produce high-purity single-walled nanotubes at 900°C to 1100°C (1700°F to 2000°F) and 30 atm to 50 atm [8].

Production of nanotubes by CVD has the advantage that it can be operated continuously, which makes it economically attractive for mass production.

Nanofabrication by Scanning Probe Techniques Scanning probe microscopy techniques are described in Section 27.4 in the context of measuring and observing nanometer-scale features and objects. In addition to viewing a surface, the scanning tunneling microscope (STM) and atomic force microscope (AFM) can also be used to manipulate individual atoms, molecules, or clusters of atoms or molecules that adhere to a substrate surface by the forces of adsorption (weak chemical bonds). Clusters of atoms or molecules are called ***nanoclusters***, and their size is just a few nanometers [30]. Figure 27.17(a) illustrates the variation in either current or deflection of the STM probe tip as it is moved across a surface upon which is located an adsorbed atom. As the tip moves over the surface immediately above the adsorbed atom, there is an increase in the signal. Although the bonding force that attracts the atom to the surface is weak, it is significantly greater than the force of attraction created by the tip, simply because the distance is greater. However, if the probe tip is moved close enough to the adsorbed atom so that its force of attraction is greater than the adsorption force, the atom will be dragged along the surface, as suggested in Figure 27.17(b). In this way, individual atoms or molecules can be manipulated to create various nanoscale structures. A notable STM example accomplished at the IBM Research Labs was the fabrication of the company logo out of xenon atoms adsorbed onto a nickel surface in an area 5 nm by 16 nm. This scale is considerably smaller than the lettering in Figure 27.16 (which is also nanoscale, as noted in the caption).

The manipulation of individual atoms or molecules by scanning tunneling microscopy techniques can be classified as lateral manipulation and vertical manipulation. In lateral manipulation, atoms or molecules are transferred horizontally along the surface

[6] HiPCO stands for high-pressure carbon monoxide decomposition process.

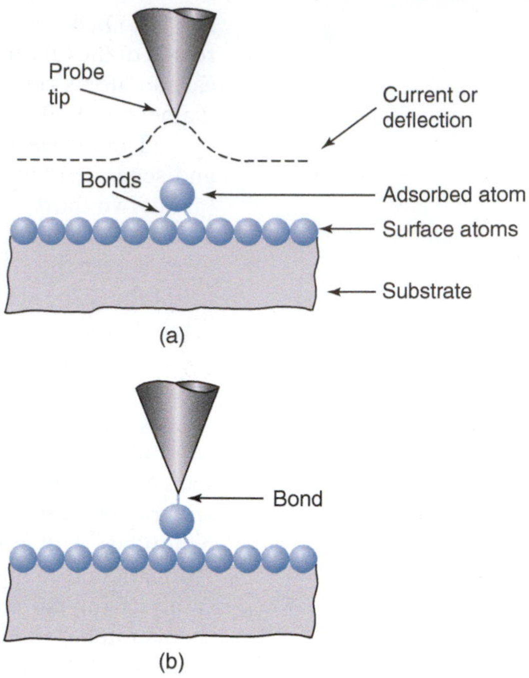

FIGURE 27.17 Manipulation of individual atoms by means of scanning tunneling microscopy techniques: (a) the probe tip is maintained a distance from the surface that is sufficient to avoid disturbance of adsorbed atom and (b) the probe tip is moved closer to the surface so that the adsorbed atom is attracted to the tip. (Credit: *Fundamentals of Modern Manufacturing*, 4th Edition by Mikell P. Groover, 2010. Reprinted with permission of John Wiley & Sons, Inc.)

by the attractive or repulsive forces exerted by the STM tip, as in Figure 27.17(b). In vertical manipulation, the atoms or molecules are lifted from the surface and deposited at a different location to form a structure. Although this kind of STM manipulation of atoms and molecules is of scientific interest, there are technological limitations that inhibit its commercial application, at least in high production of nanotech products. One of the limitations is that it must be carried out in a very high vacuum environment to prevent stray atoms or molecules from interfering with the process. Another limitation is that the surface of the substrate must be cooled to temperatures approaching absolute zero (−273°C or −460°F) in order to reduce thermal diffusion that would gradually distort the atomic structure being formed. These limitations make it a very slow and expensive process.

Another scanning probe technique, one that shows promise for practical applications, is called dip-pen nanolithography. In ***dip-pen nanolithography*** (DPN), the tip of an atomic force microscope is used to transfer molecules to a substrate surface by means of a solvent meniscus, as shown in Figure 27.18. The process is somewhat analogous to using an old-fashioned quill pen to transfer ink to a paper surface via capillary forces. In DPN, the AFM tip serves as the nib of the pen, and the substrate becomes the surface onto which the dissolved molecules (i.e., the ink) are deposited. The deposited molecules must have a chemical affinity for the substrate material, just as wet ink adheres to paper. DPN can be used to "write" patterns of molecules onto a surface, where the patterns are of submicron dimension. In addition, DPN can be used to deposit different types of molecules at different locations on the substrate surface.

Self-Assembly Self-assembly is a fundamental process in nature. The natural formation of a crystalline structure during the slow cooling of molten minerals is an example of nonliving self-assembly. The growth of living organisms is an example of biological self-assembly. In both instances, entities at the atomic and molecular level combine on their own into larger entities, proceeding in a constructive manner toward the creation of some deliberate thing. If the thing is a living organism, the intermediate entities are biological

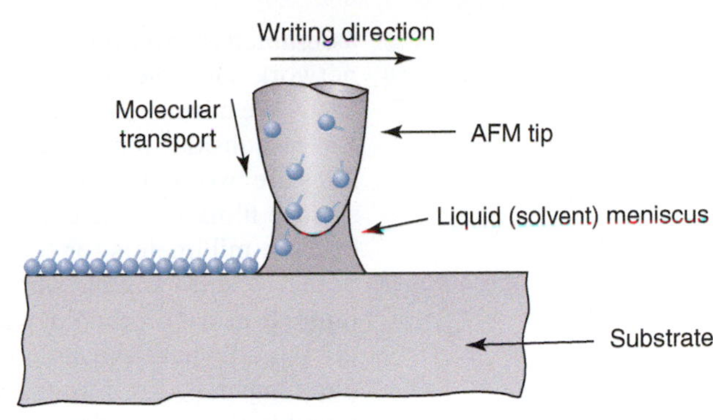

FIGURE 27.18 Dip-pen nanolithography, in which the tip of an atomic force microscope is used to deposit molecules through the water meniscus that forms naturally between the tip and the substrate. (Credit: *Fundamentals of Modern Manufacturing*, 4th Edition by Mikell P. Groover, 2010. Reprinted with permission of John Wiley & Sons, Inc.)

cells, and the organism is grown through an additive process that exhibits massive replication of individual cell formations, yet the final result is often remarkably intricate and complex (e.g., a human being).

One of the promising bottom-up approaches in nanotechnology involves the emulation of nature's self-assembly process to produce materials and systems that have nanometer-scale features or building blocks, but the final product may be larger than nanoscale. It may be of micro- or macro-scale size, at least in some of its dimensions. The term ***biomimetics*** describes this process of building artificial, nonbiological entities by imitating nature's methods. Desirable attributes of atomic or molecular self-assembly processes in nanotechnology include the following: (1) they can be carried out rapidly; (2) they occur automatically and do not require any central control; (3) they exhibit massive replication; and (4) they can be performed under mild environmental conditions (at or near atmospheric pressure and room temperature). Self-assembly is likely to be the most important of the nanofabrication processes because of its low cost, its capacity for producing structures over a range of sizes (from nanoscale to macroscale), and its general applicability to a wide variety of products [24].

An underlying principle behind self-assembly is the principle of minimum energy. Physical entities such as atoms and molecules seek out a state that minimizes the total energy of the system of which they are components. This principle has the following implications for self-assembly:

1. There must be some mechanism for the movement of the entities (e.g., atoms, molecules, ions) in the system, thus causing the entities to come into close proximity with one another. Possible mechanisms for this movement include diffusion, convection in a fluid, and electric fields.
2. There must be some form of molecular recognition among the entities. Molecular recognition refers to the tendency of one molecule (or atom or ion) to be attracted to and bind with another molecule (or atom or ion), for example, the way sodium and chlorine are attracted to each other to form table salt.
3. The molecular recognition among the entities causes them to join in such a way that the resulting physical arrangement of the entities achieves a state of minimum energy. The joining process involves chemical bonding, usually the weaker secondary types (e.g., van der Waals bonds).

A number of self-assembly nanofabrication techniques have been developed, most of which are still in the research stage. They include the following categories: (1) fabrication of nanoscale objects, including molecules, macromolecules, clusters of molecules, nanotubes, and crystals; and (2) formation of two-dimensional arrays such as self-assembled monolayers (surface films that are one molecule thick) and three-dimensional networks of molecules.

We have already discussed the production of nanotubes, which represent category 1. Let us consider the self-assembly of surface films as an example of category 2. Surface films are two-dimensional coatings formed on a solid (three-dimensional) substrate. Most surface films are inherently thin, yet the thickness is typically measured in micrometers or even millimeters (or fractions thereof), well above the nanometer scale. Of interest here are surface films whose thicknesses are measured in nanometers. Of particular interest in nanotechnology are surface films that self-assemble, are only one molecule thick, and where the molecules are organized in some orderly fashion. These types of films are called self-assembled monolayers (SAMs). Multilayered structures are also possible that possess order and are two or more molecules thick.

The substrate materials for self-assembled monolayers and multilayers include a variety of metallic and other inorganic materials. The list includes gold, silver, copper, silicon, and silicon dioxide. Noble metals have the advantage of not forming an oxide surface film that would interfere with the reactions that generate the desired layer of interest. Layering materials include thiols, sulfides, and disulfides. The layering material must be capable of being adsorbed onto the surface material. The typical process sequence in the formation of the monolayer of a thiol on gold is illustrated in Figure 27.19.[7] Layering molecules move freely above the substrate surface and are adsorbed onto the surface. Contact occurs between adsorbed molecules on the surface, and they form stable islands. The islands become larger and gradually join together through the addition of more molecules laterally on the surface, until the substrate is completely covered. Bonding to the gold surface is provided by the sulfur atom in the thiol, sulfide, or disulfide layer. In some applications, self-assembled monolayers can be formed into desired patterns or regions on the substrate surface using techniques such as nano-contact printing and dip-pen nanolithography.

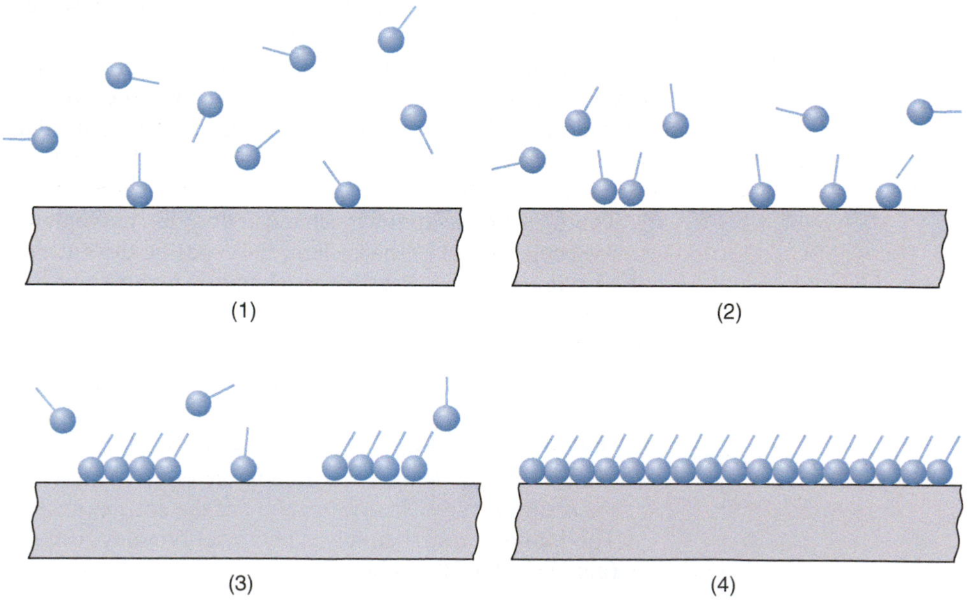

FIGURE 27.19 Typical sequence in the formation of a monolayer of a thiol onto a gold substrate: (1) some of the layering molecules in motion above the substrate become attracted to the surface, (2) they are adsorbed on the surface, (3) they form islands, (4) the islands grow until the surface is covered. Based on a figure in [10]. (Credit: *Fundamentals of Modern Manufacturing*, 4th Edition by Mikell P. Groover, 2010. Reprinted with permission of John Wiley & Sons, Inc.)

[7] We mentioned this combination of thiol on a gold surface in Section 27.5.1 in the context of nano-contact printing.

REFERENCES

[1] Baker, S., and Aston, A. "The Business of Nanotech," ***Business Week***, February 14, 2005, pp. 64–71.

[2] Balzani, V., Credi, A., and Venturi, M. ***Molecular Devices and Machines—A Journey into the Nano World***. Wiley-VCH Verlag GmbH & Co. KGaA, Weinheim, Germany, 2003.

[3] Bashir, R."Biologically Mediated Assembly of Artificial Nanostructures and Microstructures," Chapter 5 in ***Handbook of Nanoscience, Engineering, and Technology***, W. A. Goddard, III, D. W. Brenner, S. E. Lyshevskiand G. J. Iafrate (eds.). CRC Press, Boca Raton, Florida, 2003.

[4] Chaiko, D. J. "Nanocomposite Manufacturing," ***Advanced Materials & Processes***, June 2003, pp. 44–46.

[5] Drexler, K. E. ***Nanosystems: Molecular Machinery, Manufacturing, and Computation***. Wiley-Interscience, John Wiley & Sons, Inc., New York, 1992.

[6] Fatikow, S., and Rembold, U. ***Microsystem Technology and Microrobotics***. Springer-Verlag, Berlin, 1997.

[7] Fujita, H. (ed.). ***Micromachines as Tools for Nanotechnology***. Springer-Verlag, Berlin, 2003.

[8] Hornyak, G. L., Moore, J. J., Tibbals, H. F., and Dutta, J. ***Fundamentals of Nanotechnology***. CRC Taylor & Francis, Boca Raton, Florida, 2009.

[9] Jackson, M. J. ***Micro and Nanomanufacturing***. Springer, New York, 2007.

[10] Kohler, M., and Fritsche, W. ***Nanotechnology: An Introduction to Nanostructuring Techniques***. Wiley-VCH Verlag GmbH & Co. KGaA, Weinheim, Germany, 2004.

[11] Li, G., and Tseng, A. A. "Low Stress Packaging of a Micromachined Accelerometer," ***IEEE Transactions on Electronics Packaging Manufacturing***, Vol. **24**, No. 1, January 2001, pp. 18–25.

[12] Lyshevski, S. E."Nano- and Micromachines in NEMS and MEMS," Chapter 23 in ***Handbook of Nanoscience, Engineering, and Technology***, W. A. Goddard, III, D. W. Brenner, S. E. Lyshevski, and G. J. Iafrate (eds.). CRC Press, Boca Raton, Florida, 2003, pp. 23–27.

[13] Madou, M. ***Fundamentals of Microfabrication***. CRC Press, Boca Raton, Florida, 1997.

[14] Madou, M. ***Manufacturing Techniques for Microfabrication and Nanotechnology***. CRC Taylor & Francis, Boca Raton, Florida, 2009.

[15] Maynor, B. W., and Liu, J."Dip-Pen Lithography," ***Encyclopedia of Nanoscience and Nanotechnology***, American Scientific Publishers, 2004, pp. 429–441.

[16] Meyyappan, M., and Srivastava, D."Carbon Nanotubes," Chapter 18 in ***Handbook of Nanoscience, Engineering, and Technology***, W. A. Goddard, IIID. W. Brenner, S. E. Lyshevski, and G. J. Iafrate (eds.). CRC Press, Boca Raton, Florida, 2003, pp. 18–1 to 18–26.

[17] Morita, S., Wiesendanger, R., and Meyer, E. (eds.). ***Noncontact Atomic Force Microscopy***. Springer-Verlag, Berlin, 2002.

[18] National Research Council (NRC). ***Implications of Emerging Micro-and Nanotechnologies***. Committee on Implications of Emerging Micro- and Nanotechnologies, The National Academies Press, Washington, D.C., 2002.

[19] Nazarov, A. A., and Mulyukov, R. R."Nanostructured Materials," Chapter 22 in ***Handbook of Nanoscience, Engineering, and Technology***, W. A. Goddard, IIID. W. Brenner, S. E. Lyshevski, and G. J. Iafrate (eds.). CRC Press, Boca Raton, Florida, 2003, pp. 22–1 to 22–41.

[20] O'Connor, L., and Hutchinson, H. "Skyscrapers in a Microworld," ***Mechanical Engineering***, Vol. **122**, No. 3, March 2000, pp. 64-67.

[21] Paula, G. "An Explosion in Microsystems Technology," ***Mechanical Engineering***, Vol. **119**, No. 9, September 1997, pp. 71–74.

[22] Piner, R. D., Zhu, J., Xu, F., Hong, S., and Mirkin, C. A. "Dip-Pen Nanolithography," ***Science***, Vol. **283**, January 29, 1999, pp. 661–663.

[23] Poole, Jr., C. P., and Owens, F. J. ***Introduction to Nanotechnology***. Wiley-Interscience, John Wiley & Sons, Inc., Hoboken, New Jersey, 2003.

[24] Ratner, M., and Ratner, D. ***Nanotechnology: A Gentle Introduction to the Next Big Idea***. Prentice Hall PTR, Pearson Education, Inc., Upper Saddle River, New Jersey, 2003.

[25] Rietman, E. A. ***Molecular Engineering of Nanosystems***. Springer-Verlag, Berlin, 2000.

[26] Rubahn, H.-G. ***Basics of Nanotechnology***, 3rd ed. Wiley-VCH, Weinheim, Germany, 2008.

[27] Schmid, G. (ed.). ***Nanoparticles: From Theory to Application***. Wiley-VCH Verlag GmbH & Co. KGaA, Weinheim, Germany, 2004.

[28] Torres, C. M. S. (ed.). ***Alternative Lithography: Unleashing the Potentials of Nanotechnology***. Kluwer Academic/Plenum Publishers, New York, 2003.

[29] Tseng, A. A., and Mon, J-I."NSF 2001 Workshop on Manufacturing of Micro-Electro Mechanical Systems," in ***Proceedings of the 2001 NSF Design, Service, and Manufacturing Grantees and Research Conference***, National Science Foundation, 2001.

[30] Tseng, A. A. (ed.). ***Nanofabrication Fundamentals and Applications***. World Scientific, Singapore, 2008.

[31] Weber, A. "Nanotech: Small Products, Big Potential," ***Assembly***, February 2004, pp. 54–59.

[32] Website: en.wikipedia.org/wiki/nanotechnology.

[33] Website: www.nanotechproject.org/inventories/consumer.

[34] Website: www.research.ibm.com/nanscience.

[35] Website: www.zurich.ibm.com/st/atomic_manipulation.

REVIEW QUESTIONS

27.1. Define the term *microelectromechanical system*.

27.2. What is the approximate size scale in microsystem technology?

27.3. Why is it reasonable to believe that microsystem products would be available at lower costs than products of larger, more conventional size?

27.4. What are some of the basic types of microsystem devices?

27.5. Name some products that represent microsystem technology.

27.6. Why is silicon a desirable work material in microsystem technology?

27.7. What is meant by the term *aspect ratio* in microsystem technology?

27.8. What is the difference between bulk micromachining and surface micromachining?

27.9. What are the three steps in the LIGA process?

27.10. What is the range of feature sizes of entities associated with nanotechnology?

27.11. What is a buckyball?

27.12. What is a carbon nanotube?

27.13. What is a scanning probe instrument, and why is it so important in nanoscience and nanotechnology?

27.14. What is tunneling, as referred to in the scanning tunneling microscope?

27.15. What are the two basic categories of approaches used in nanofabrication?

27.16. Why is photolithography based on visible light not used in nanotechnology?

27.17. What are the lithography techniques used in nanofabrication?

27.18. How is nano-imprint lithography different from micro-imprint lithography?

27.19. What are the limitations of scanning tunneling microscope in nanofabrication that inhibit its commercial application?

27.20. What is self-assembly in nanofabrication?

27.21. What are the desirable features of atomic or molecular self-assembly processes in nanotechnology?

Part IX Systems Topics in Manufacturing

28 PRODUCTION SYSTEMS AND PROCESS PLANNING

Chapter Contents

28.1 Overview of Production Systems
- 28.1.1 Production Facilities
- 28.1.2 Manufacturing Support Systems

28.2 Process Planning
- 28.2.1 Traditional Process Planning
- 28.2.2 Make or Buy Decision
- 28.2.3 Computer-Aided Process Planning
- 28.2.4 Problem Solving and Continuous Improvement

28.3 Concurrent Engineering and Design for Manufacturability
- 28.3.1 Design for Manufacturing and Assembly
- 28.3.2 Concurrent Engineering

This final part of the book is concerned with topics related to the systems and procedures used in manufacturing. These systems include automated and computerized production equipment that perform unit operations, groupings of machines and workers usually associated with sequences of operations, practices that improve operational efficiency, and manufacturing support systems that plan and control the production operations and the quality of their output. In this chapter, we provide an overview of production systems, and how they are organized for different types of manufacturing operations. We then examine process planning, which is concerned with deciding how a given part or product is to be made. The chapter also includes two topics that are closely related to process planning: design for manufacturability and concurrent engineering. In Chapter 29, several topics related to factory automation and manufacturing systems are covered. The topics include numerical control, cellular manufacturing, lean production, and computer-integrated manufacturing. Finally, Chapter 30 discusses quality control and inspection systems. These topics are covered more thoroughly in reference [4].

28.1 OVERVIEW OF PRODUCTION SYSTEMS

To operate effectively, a manufacturing firm must have systems that allow it to efficiently accomplish its type of

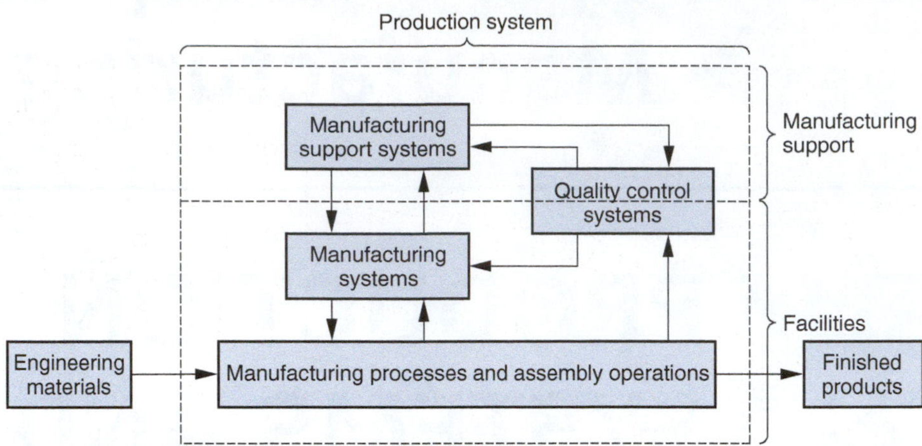

FIGURE 28.1 Overview of the production system and its components: manufacturing support systems, quality control systems, and manufacturing systems in the factory. (Credit: *Fundamentals of Modern Manufacturing,* 4th Edition by Mikell P. Groover, 2010. Reprinted with permission of John Wiley & Sons, Inc.)

production. Production systems consist of people, equipment, and procedures designed for the combination of materials and processes that constitute a firm's manufacturing operations. Production systems can be divided into two categories: (1) production facilities and (2) manufacturing support systems, as shown in Figure 28.1. ***Production facilities*** refer to the physical equipment and the arrangement of equipment in the factory. ***Manufacturing support systems*** are the procedures used by the company to manage production and solve the technical and logistics problems encountered in ordering materials, moving work through the factory, and ensuring that products meet quality standards. Both categories include people. People make these systems work. In general, direct labor workers are responsible for operating the manufacturing equipment; and professional staff workers are responsible for manufacturing support.

28.1.1 PRODUCTION FACILITIES

Production facilities consist of the factory and the equipment in the factory, which includes production machines, material handling equipment, and other hardware. The equipment comes in direct physical contact with the parts and/or assemblies as they are being made. The facilities "touch" the product. Facilities also include the way the equipment is arranged in the factory—the ***plant layout***. The equipment is usually organized into logical groupings, which are referred to as ***manufacturing systems*** in this text. They include automated machine tools, cells consisting of several production machines, and methods for reducing waste in manufacturing (e.g., lean production). We examine these kinds of manufacturing systems in Chapter 29.

A manufacturing company attempts to design its manufacturing systems and organize its factories to serve the particular mission of each plant in the most efficient way. Over the years, certain types of production facilities have come to be recognized as the most appropriate way to organize for a given combination of product variety and production quantity (Section 1.1.2). Different types of facilities are required for each of the three ranges of annual production quantities: low, medium, and high production.

Low-Quantity Production In the low-quantity range (1 to 100 units/year), the term ***job shop*** is often used to describe the type of production facility. A job shop makes low quantities of specialized and customized products. The products are typically complex, such as space capsules, prototype aircraft, and special machinery. The equipment in a job shop is general purpose, and the labor force is highly skilled.

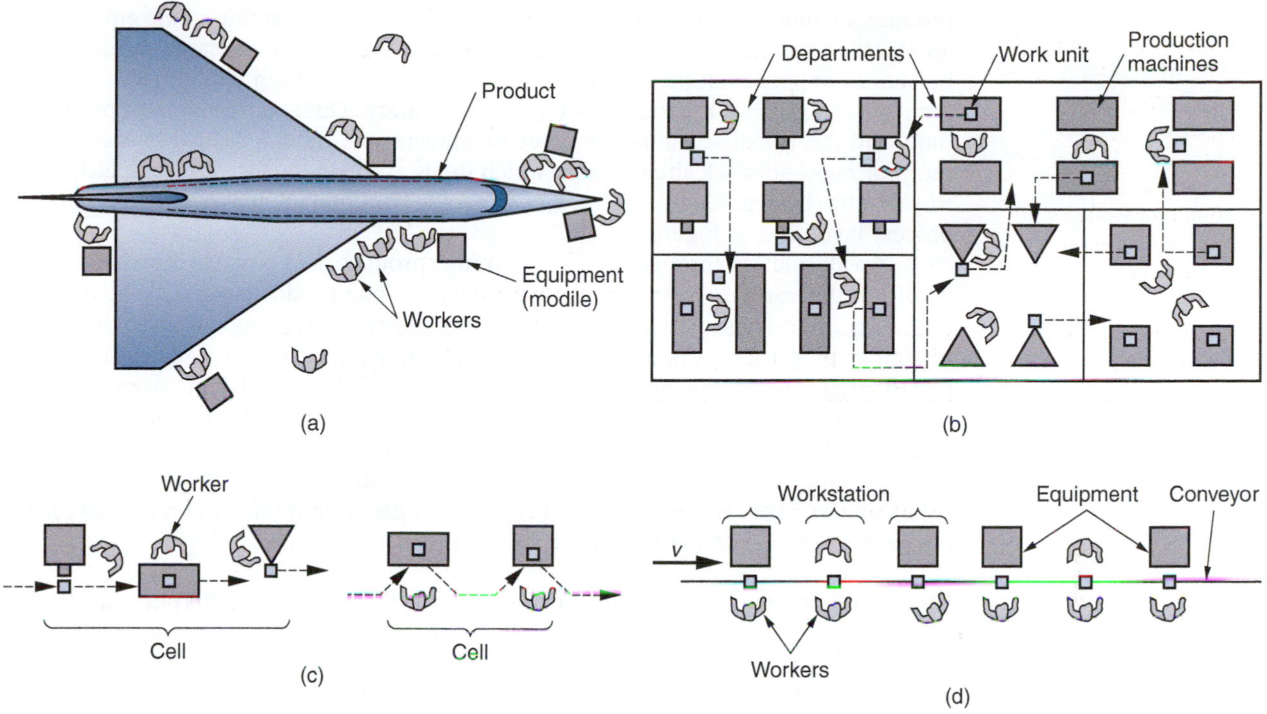

FIGURE 28.2 Various types of plant layout: (a) fixed-position layout, (b) process layout, (c) cellular layout, and (d) product layout. (Credit: *Fundamentals of Modern Manufacturing*, 4th Edition by Mikell P. Groover, 2010. Reprinted with permission of John Wiley & Sons, Inc.)

A job shop must be designed for maximum flexibility to deal with the wide product variations encountered (hard product variety, Section 1.1.2). If the product is large and heavy, and therefore difficult to move, it typically remains in a single location during its fabrication or assembly. Workers and processing equipment are brought to the product, rather than moving the product to the equipment. This type of layout is referred to as a ***fixed-position layout***, shown in Figure 28.2(a). In a pure situation, the product remains in a single location during its entire production. Examples of such products include ships, aircraft, locomotives, and heavy machinery. In actual practice, these items are usually built in large modules at single locations, and then the completed modules are brought together for final assembly using large-capacity cranes.

The individual components of these large products are often made in factories in which the equipment is arranged according to function or type. This arrangement is called a ***process layout***. The lathes are in one department, the milling machines are in another department, and so on, as in Figure 28.2(b). Different parts, each requiring a different operation sequence, are routed through the departments in the particular order needed for their processing, usually in batches. The process layout is noted for its flexibility; it can accommodate a great variety of operation sequences for different part configurations. Its disadvantage is that the machinery and methods to produce a part are not designed for high efficiency.

Medium-Quantity Production In the medium-quantity range (100 to 10,000 units annually), two different types of facility are distinguished, depending on product variety. When product variety is hard, the usual approach is ***batch production***, in which a batch of one product is made, after which the manufacturing equipment is changed over to produce a batch of the next product, and so on. The production rate of the equipment is greater than the demand rate for any single product type, and so the same equipment can be shared among multiple products. The changeover between production runs takes time—time to change tooling and set up the machinery. This setup time is lost production time, and this is a disadvantage of batch manufacturing. Batch production is commonly used for make-to-stock situations, in which items are manufactured to replenish inventory that has been gradually depleted by demand. The equipment is usually arranged in a process layout, as in Figure 28.2(b).

An alternative approach to medium-range production is possible if product variety is soft. In this case, extensive changeovers between one product style and the next may not be necessary. It is often possible to configure the manufacturing system so that groups of similar products can be made on the same equipment without significant lost time due to setup. The processing or assembly of different parts or products is accomplished in cells consisting of several workstations or machines. The term ***cellular manufacturing*** is often associated with this type of production. Each cell is designed to produce a limited variety of part configurations; that is, the cell specializes in the production of a given set of similar parts, according to the principles of ***group technology*** (Section 29.2). The layout is called a ***cellular layout***, depicted in Figure 28.2(c).

High Production The high-quantity range (10,000 to millions of units per year) is often referred to as ***mass production***, especially when annual quantities exceed 100,000. The situation is characterized by a high demand rate for the product, and the manufacturing system is dedicated to the production of that single item. Two categories of mass production can be distinguished: quantity production and flow line production. ***Quantity production*** involves the mass production of single parts on single pieces of equipment. It typically involves standard machines (e.g., stamping presses) equipped with special tooling (e.g., dies and material handling devices), in effect dedicating the equipment to the production of one part type. A process layout is typical in quantity production.

Flow line production involves multiple pieces of equipment or workstations arranged in sequence, and the work units are physically moved through the sequence to complete the product. The workstations and equipment are designed specifically for the product to maximize efficiency. The layout is called a ***product layout***, and the workstations are arranged into one long line, as in Figure 28.2(d), or into a series of connected line segments. The work is usually moved between stations by mechanized conveyor. At each station, a small amount of the total work is completed on each unit of product.

The most familiar example of flow line production is the assembly line, associated with products such as cars and household appliances. The pure case of flow line production occurs when there is no variation in the products made on the line. Every product is identical, and the line is referred to as a ***single model production line***. To successfully market a given product, it is often beneficial to introduce feature and model variations so that individual customers can choose the exact merchandise that appeals to them. From a production viewpoint, the feature differences represent a case of soft product variety. The term ***mixed-model production line*** applies to situations in which there is soft variety in the products made on the line. Modern automobile assembly is an example. Cars coming off the assembly line have variations in options and trim representing different models and in many cases different nameplates of the same basic car design.

28.1.2 MANUFACTURING SUPPORT SYSTEMS

To operate its facilities efficiently, a company must organize itself to design the processes and equipment, plan and control the production orders, and satisfy product quality requirements. These functions are accomplished by manufacturing support systems—

people and procedures by which a company manages its production operations. Specifically, the ***manufacturing support systems*** are the set of procedures and systems used by a company to solve the technical and logistics problems encountered in planning the processes, ordering materials, controlling production, and ensuring that the company's products meet required quality specifications. As with the manufacturing systems in the factory, the manufacturing support systems include people. People make the systems work. Unlike the manufacturing systems in the factory, most of the support systems do not directly contact the product during its processing and assembly. Instead, they plan and control the activities in the factory to ensure that the products are completed and delivered to the customer on time, in the right quantities, and to the highest quality standards.

Manufacturing support functions are often carried out in the firm by people organized into departments such as the following:

- ***Manufacturing engineering***. The manufacturing engineering department is responsible for process planning—deciding what production processes should be used to make the parts and assemble the products. Process planning and manufacturing engineering are covered in Section 28.2.
- ***Production planning and control***. This department is responsible for solving the logistics problem in manufacturing—ordering materials and purchased parts, scheduling production, and making sure that the operating departments have the necessary capacity to meet the production schedules.
- ***Quality control***. Producing high-quality products should be a top priority of any manufacturing firm in today's competitive environment. It means designing and building products that conform to specifications and satisfy or exceed customer expectations. Quality control and inspection systems are covered in Chapter 30.

28.2 PROCESS PLANNING

Process planning consists of determining the most appropriate manufacturing processes and the order in which they should be performed to produce a given part or product according to the specifications established by design engineering. If it is an assembled product, process planning includes deciding the appropriate sequence of assembly steps. The process plan must be developed within the limitations imposed by available processing equipment and productive capacity of the factory. Parts or subassemblies that cannot be made internally must be purchased from external suppliers. In some cases, items that can be produced internally may be purchased from outside vendors for economic or other reasons.

Process planning is usually accomplished in the manufacturing engineering department of an organization. ***Manufacturing engineering*** is a technical staff function that is concerned with planning the manufacturing processes for the economic production of high-quality products. Its principal role is to engineer the transition from design specification to physical product. It is a manufacturing support function whose overall goal is to optimize production in the organization. The manufacturing engineering department usually reports to the manager of manufacturing. In some companies the department is known by other names, such as process engineering or production engineering. Often included within manufacturing engineering are tool design, tool fabrication, and various technical support groups. The scope of manufacturing engineering includes many activities and responsibilities that depend on the type of production operations accomplished by the organization. These activities are discussed in this section. The types of production operations were discussed in Section 28.1.1.

TABLE 28.1 Decisions and details required in process planning.

Processes and sequence. The process plan should briefly describe all processing steps used on the work unit (e.g., part, assembly) in the order in which they are performed.

Equipment selection. In general, manufacturing engineers try to develop process plans that utilize existing equipment. When this is not possible, the component in question must be purchased (Section 28.2.2), or new equipment must be installed in the plant.

Tools, dies, molds, fixtures, *and* ***gages.*** The process planner must decide what tooling is needed for each process. Actual design is usually delegated to the tool design department, and fabrication is accomplished by the tool room.

Cutting tools *and* ***cutting conditions*** *for machining operations.* These are specified by the process planner, industrial engineer, shop foreman, or machine operator, often with reference to standard handbook recommendations.

Methods. Methods include hand and body motions, workplace layout, small tools, hoists for lifting heavy parts, and so forth. Methods must be specified for manual operations (e.g., assembly) and manual portions of machine cycles (e.g., loading and unloading a production machine). Methods planning is traditionally done by industrial engineers.

Work standards. Work measurement techniques are used to establish time standards for each operation.

Estimating production costs. This is often accomplished by cost estimators with help from the process planner.

Credit: *Fundamentals of Modern Manufacturing*, 4th edition, by Mikell P. Groover. Reprinted by permission of John Wiley & Sons, Inc

28.2.1 TRADITIONAL PROCESS PLANNING

Process planning is traditionally accomplished by manufacturing engineers who are knowledgeable in the particular processes used in the factory and are able to read engineering drawings. Based on their knowledge, skill, and experience, they develop the processing steps in the most logical sequence required to make each part. Table 28.1 lists the many details and decisions usually included within the scope of process planning. Some of these details are often delegated to specialists, such as tool designers; but manufacturing engineering is responsible for them.

Process Planning for Parts The processes needed to manufacture a given part are determined largely by the material out of which it is to be made. The material is selected by the product designer based on functional requirements. Once the material has been selected, the choice of possible processes is narrowed considerably.

A typical processing sequence to fabricate a discrete part consists of (1) a basic process, (2) one or more secondary processes, (3) operations to enhance physical properties, and (4) finishing operations, illustrated in Figure 28.3. Basic and secondary processes are shaping processes (Section 1.2.1), which create and/or alter the geometry of a starting workpart. A ***basic process*** establishes the initial geometry of the part. Examples include metal casting, forging, and sheet-metal rolling. In most cases, the starting geometry must be refined by a series of ***secondary processes***. These operations transform the basic shape into the final geometry. There is a correlation between the secondary processes that might be used and the basic process that provides the initial form. For example, when sand casting or forging are the basic processes, machining operations are generally the secondary processes. When a rolling mill produces strips or coils of sheet metal, the secondary processes are stamping operations such as blanking, punching, and bending. Selection of certain basic processes minimizes the need for secondary processes.

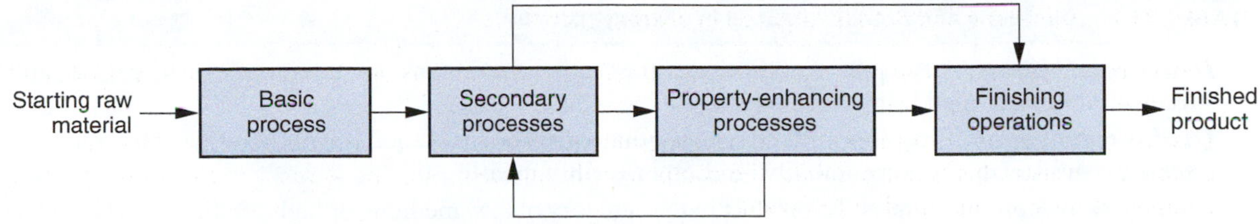

FIGURE 28.3 Typical sequence of processes required in part fabrication. (Credit: *Fundamentals of Modern Manufacturing*, 4th Edition by Mikell P. Groover, 2010. Reprinted with permission of John Wiley & Sons, Inc.)

For example, if plastic injection molding is the basic process, secondary operations are usually not required because molding is capable of providing the detailed geometric features with good dimensional accuracy.

Shaping operations are generally followed by operations to enhance physical properties and/or finish the product. ***Operations to enhance properties*** include heat-treating operations on metal components and glassware. In many cases, parts do not require these property-enhancing steps in their processing sequence. This is indicated by the alternate arrow path in our figure. ***Finishing operations*** are the final operations in the sequence; they usually provide a coating on the workpart (or assembly) surface. Examples of these processes are electroplating and painting.

In some cases, property-enhancing processes are followed by additional secondary operations before proceeding to finishing, as suggested by the return loop in Figure 28.3. An example is a machined part that is hardened by heat treatment. Prior to heat treatment, the part is left slightly oversized to allow for distortion. After hardening, it is reduced to final size and tolerance by finish grinding. Another example, again in metal parts fabrication, is when annealing is used to restore ductility to the metal after cold working to permit further deformation of the workpiece.

Table 28.2 presents some of the typical processing sequences for various materials and basic processes. The task of the process planner usually begins after the basic process has provided the initial shape of the part. Machined parts begin as bar stock or castings or forgings, and the basic processes for these starting shapes are often external to the fabricating plant. Stampings begin as sheet-metal coils or strips purchased from the mill. These are the raw materials supplied from external suppliers for the secondary processes and subsequent operations to be performed in the factory. Determining

TABLE 28.2 Some typical process sequences.

Basic Process	Secondary Process(es)	Property-Enhancing Processes	Finishing Operations
Sand casting	Machining	(none)	Painting
Die casting	(none, net shape)	(none)	Painting
Casting of glass	Pressing, blow molding	(none)	(none)
Injection molding	(none, net shape)	(none)	(none)
Rolling of bar stock	Machining	Heat treatment (optional)	Electroplating
Rolling of sheet metal	Blanking, bending, drawing	(none)	Electroplating
Forging	Machining (near net shape)	(none)	Painting
Extrusion of aluminum	Cut to length	(none)	Anodize
Atomize metal powders	Pressing of PM part	Sintering	Painting

Compiled from [4].

TABLE 28.3 Guidelines and considerations in deciding processes and their sequence in process planning.

Design requirements. The sequence of processes must satisfy the dimensions, tolerances, surface finish, and other specifications established by product design.

Quality requirements. Processes must be selected that satisfy quality requirements in terms of tolerances, surface integrity, consistency and repeatability, and other quality measures.

Production volume *and* ***rate.*** Is the product in the category of low, medium, or high production? The selection of processes and systems is strongly influenced by volume and production rate.

Available processes. If the product and its components are to be made in-house, the process planner must select processes and equipment already available in the factory.

Material utilization. It is desirable for the process sequence to make efficient use of materials and minimize waste. When possible, net shape or near net shape processes should be selected.

Precedence constraints. These are technological sequencing requirements that determine or restrict the order in which the processing steps can be performed. A hole must be drilled before it can be tapped; a powder-metal part must be pressed before sintering; a surface must be cleaned before painting; and so on.

Reference surfaces. Certain surfaces of the part must be formed (usually by machining) near the beginning of the sequence so they can serve as locating surfaces for other dimensions that are formed subsequently. For example, if a hole is to be drilled a certain distance from the edge of a given part, that edge must first be machined.

Minimize setups. The number of separate machine setups should be minimized. Wherever possible, operations should be combined at the same workstation. This saves time and reduces material handling.

Eliminate unnecessary steps. The process sequence should be planned with the minimum number of processing steps. Unnecessary operations should be avoided. Design changes should be requested to eliminate features not absolutely needed, thereby eliminating the processing steps associated with those features.

Flexibility. Where feasible, the process should be sufficiently flexible to accommodate engineering design changes. This is often a problem when special tooling must be designed to produce the part; if the part design is changed, the special tooling may be rendered obsolete.

Safety. Worker safety must be considered in process selection. This makes good economic sense, and it is the law (Occupational Safety and Health Act).

Minimum cost. The process sequence should be the production method that satisfies all of the above requirements and also achieves the lowest possible product cost.

the most appropriate processes and the order in which they must be accomplished relies on the skill, experience, and judgment of the process planner. Some of the basic guidelines and considerations used by process planners to make these decisions are outlined in Table 28.3.

The Route Sheet The process plan is prepared on a form called a ***route sheet***, an example of which is shown in Figure 28.4 (some companies use other names for this form). It is called a route sheet because it specifies the sequence of operations and equipment that will be visited by the part during its production. The route sheet is to the process planner what the engineering drawing is to the product designer. It is the official document that specifies the details of the process plan. The route sheet should include all manufacturing operations to be performed on the workpart, listed in the proper order in which they are to be accomplished. For each operation, the following should be listed: (1) a brief description of the operation indicating the work to be done, surfaces to be processed with references to the part drawing, and dimensions (and tolerances, if not specified on part drawing) to be achieved; (2) the equipment on which the work is to be

Part No: 031393	Part Name: Housing, valve			Rev. 2	Page 1 of 2		
Matl: 416 Stainless		Size: 2.0 dia × 5. long		Planner: MPG		Date: 3/13/XX	
No.	**Operation**	**Dept.**		**Machine**	**Tooling, gages**	**Setup time**	**Cycle time**
10	Face; rough & finish turn to 1.473 ± 0.003 dia. × 1.250 ± 0.003 length; face shoulder to 0.313 ± 0.002; finish turn to 1.875 ± 0.002 dia.; form 3 grooves at 0.125 width × 0.063 deep.	L		325	G857	1.0 h	8.22 m
20	Reverse; face to 4.750 ± 0.005 length; finish turn to 1.875 ± 0.002 dia.; drill 1.000 + 0.006, −0.002 dia. axial hole.	L		325		0.5 h	3.10 m
30	Drill & ream 3 radial holes at 0.375 ± 0.002 dia.	D		114	F511	0.3 h	2.50 m
40	Mill 0.500 ± 0.004 wide × 0.375 ± 0.003 deep slot.	M		240	F332	0.3 h	1.75 m
50	Mill 0.750 ± 0.004 wide × 0.375 ± 0.003 deep flat.	M		240	F333	0.3 h	1.60 m

FIGURE 28.4 Typical route sheet for specifying the process plan. (Credit: *Fundamentals of Modern Manufacturing*, 4th Edition by Mikell P. Groover, 2010. Reprinted with permission of John Wiley & Sons, Inc.)

performed; and (3) any special tooling required, such as dies, molds, cutting tools, jigs or fixtures, and gages. In addition, some companies include cycle time standards, setup times, and other data on the route sheet.

Sometimes a more detailed ***operation sheet*** is also prepared for each operation listed in the routing. This is retained in the particular department where the work is performed. It indicates the specific details of the operation, such as cutting speeds, feeds, and tools, and other instructions useful to the machine operator. Setup sketches are sometimes also included.

Process Planning for Assemblies For low production, assembly is generally done at individual workstations and a worker or team of workers performs the assembly elements to complete the product. In medium and high production, assembly is usually performed on production lines. In either case, there is a precedence order in which the work must be accomplished.

Process planning for assembly involves preparation of the assembly instructions that must be performed. For single stations, the documentation is similar to the processing route sheet in Figure 28.4. It contains a list of the assembly steps in the order in which they must be accomplished. For assembly-line production, process planning consists of assigning work elements to particular stations along the line, a procedure called ***line balancing***. In effect, the assembly line moves the work units through the individual stations, and the line balancing solution determines what assembly steps are performed at each station. As with process planning for parts, any tools and fixtures needed to accomplish a given assembly work element must be decided, and the workplace layout must be designed.

28.2.2 MAKE OR BUY DECISION

Inevitably, the question arises as to whether a given part should be purchased from an outside vendor or made internally. First of all, it should be recognized that virtually all manufacturers purchase their starting materials from suppliers. A machine shop buys bar stock from a metals distributor and castings from a foundry. A plastic molder obtains molding compound from a chemical company. A pressworking company purchases sheet metal from a rolling mill. Very few companies are vertically integrated all the way from raw materials to finished product.

Given that a company purchases at least some of its starting materials, it is reasonable to ask whether the company should purchase at least some of the parts that would otherwise be made in its own factory. The answer to the question is the ***make or buy decision***. The make versus buy question is probably appropriate to ask for every component used by the company.

Cost is the most important factor in deciding whether a part should be made in-house or purchased. If the vendor is significantly more proficient in the processes required to make the component, it is likely that the internal production cost will be greater than the purchase price even when a profit is included for the vendor. On the other hand, if purchasing the part results in idle equipment in the factory, then an apparent cost advantage for the vendor may be a disadvantage for the home factory. Consider the following example.

Example 28.1 Make or Buy Cost Comparison

Suppose that the quoted price for a certain component from a vendor is \$8.00 per unit for 1000 units. The same part made in the home factory would cost \$9.00. The cost breakdown on the make alternative is as follows:

$$\begin{aligned} \text{Unit material cost} &= \$2.25 \text{ per unit} \\ \text{Direct labor} &= \$2.00 \text{ per unit} \\ \text{Labor overhead at 150\%} &= \$3.00 \text{ per unit} \\ \text{Equipment fixed cost} &= \underline{\$1.75 \text{ per unit}} \\ \text{Total} &= \$9.00 \text{ per unit} \end{aligned}$$

Should the component by bought or made in-house?

Solution: Although the vendor's quote seems to favor the buy decision, let us consider the possible effect on the factory if we decide to accept the quote. The equipment fixed cost is an allocated cost based on an investment that has already been made. If it turns out that the equipment is rendered idle by the decision to buy the part, then one might argue that the fixed cost of \$1.75 continues even if the equipment is not in use. Similarly, the overhead cost of \$3.00 consists of factory floor space, indirect labor, and other costs that will also continue even if the part is bought. By this reasoning, the decision to purchase might cost the company as much as \$8.00 + \$1.75 + \$3.00 = \$12.75 per unit if it results in idle time in the factory on the machine that would have been used to make the part.

On the other hand, if the equipment can be used to produce other components for which the internal prices are less than the corresponding external quotes, then a buy decision makes good economic sense. ■

Make or buy decisions are rarely as clear as in Example 28.1. Some of the other factors that enter the decision are listed in Table 28.4. Although these factors appear to be subjective, they all have cost implications, either directly or indirectly. In recent years, major companies have placed strong emphasis on building close relationships with parts suppliers. This trend has been especially prevalent in the automobile industry, where

TABLE 28.4 Key factors in the make or buy decision.

Factor	Explanation and Effect on Make/Buy Decision
Process available in-house	If a given process is not available internally, then the obvious decision is to purchase. Vendors often develop proficiency in a limited set of processes that makes them cost competitive in external-internal comparisons. There are exceptions to this guideline, in which a company decides that, in its long-term strategy, it must develop a proficiency in a manufacturing process technology that it does not currently possess.
Production quantity	Number of units required. High volume tends to favor make decisions. Low quantities tend to favor buy decisions.
Product life	Long product life favors internal production.
Standard items	Standard catalog items, such as bolts, screws, nuts, and many other types of components are produced economically by suppliers specializing in those products. It is almost always better to purchase these standard items.
Supplier reliability	The reliable supplier gets the business.
Alternative source	In some cases, factories buy parts from vendors as an alternative source to their own production plants. This is an attempt to ensure uninterrupted supply of parts, or to smooth production in peak demand periods.

Credit: *Fundamentals of Modern Manufacturing*, 4th edition, by Mikell P. Groover. Reprinted by permission of John Wiley & Sons, Inc.

long-term agreements have been reached between each carmaker and a limited number of vendors who are able to deliver high-quality components reliably on schedule.

28.2.3 COMPUTER-AIDED PROCESS PLANNING

During the last several decades, there has been considerable interest in ***computer-aided process planning*** (CAPP)—automating the process planning function by means of computer systems. Shop people knowledgeable in manufacturing processes are gradually retiring. An alternative approach to process planning is needed, and CAPP systems provide this alternative. Computer-aided process planning systems are designed around either of two approaches: retrieval systems and generative systems.

Retrieval CAPP Systems Retrieval CAPP systems, also known as ***variant CAPP systems***, are based on group technology and parts classification and coding (Section 29.2.1). In these systems, a standard process plan is stored in computer files for each part code number. The code number identifies the unique characteristics of a part, such as material type, shape, major dimensions, tolerances, and production quantities. These kinds of data are the basis for determining what sequence of manufacturing processes should be used to fabricate the part. The standard plans are based on current part routings used in the factory, or on an ideal plan that is prepared for each part code number. Retrieval CAPP systems operate as indicated in Figure 28.5. The user begins by identifying the GT code of the part for which the process plan is to be determined. A search is made of the part family file to determine if a standard route sheet exists for the given part code. If the file contains a process plan for the part, it is retrieved and displayed for the user. The standard process plan is examined to determine whether modifications are necessary. Although the new part has the same code number, minor differences in the processes might be required to make the part. The standard plan is edited accordingly. The capacity to alter an existing process plan is why retrieval CAPP systems are also called variant systems.

If the file does not contain a standard process plan for the given code number, the user may search the file for a similar code number for which a standard routing exists. By

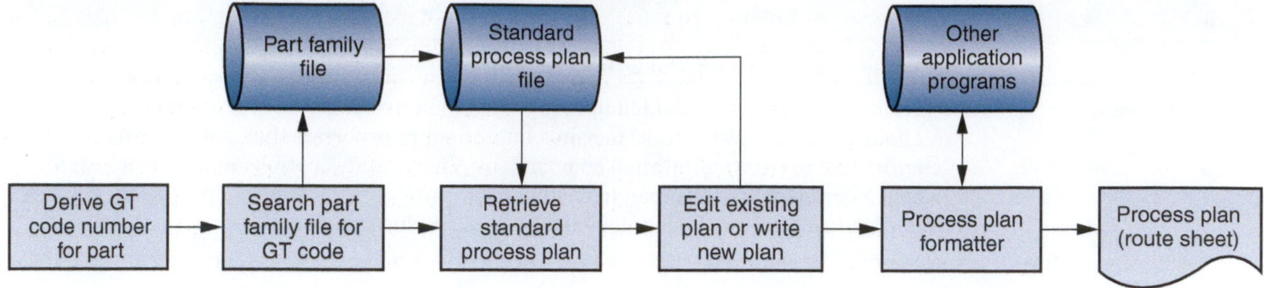

FIGURE 28.5 Operation of a retrieval computer-aided process planning system. Sources: [3], [4]. (Credit: *Fundamentals of Modern Manufacturing*, 4th Edition by Mikell P. Groover, 2010. Reprinted with permission of John Wiley & Sons, Inc.)

editing the existing process plan, or by starting from scratch, the user develops the process plan for the new part. This becomes the standard process plan for the new part code number.

The final step is the process plan formatter, which prints the route sheet in the proper format. The formatter may call other application programs: determining cutting conditions for machine tool operations, calculating standard times for machining cycles, or computing cost estimates.

Generative CAPP Systems Generative CAPP systems are an alternative to retrieval systems. Rather than retrieving and editing existing plans from a database, a generative system creates the process plan using systematic procedures that might be applied by a human planner. In a fully generative CAPP system, the process sequence is planned without human assistance and without predefined standard plans.

Designing a generative CAPP system is a problem in the field of expert systems, a branch of artificial intelligence. ***Expert systems*** are computer programs capable of solving complex problems that normally require a human who has years of education and experience. Process planning fits that definition. Several ingredients are required in a fully generative CAPP system:

1. ***Knowledge base***. The technical knowledge of manufacturing and the logic used by successful process planners must be captured and coded into a computer program. An expert system applied to process planning requires the knowledge and logic of human process planners to be incorporated into a knowledge base. Generative CAPP systems then use the knowledge base to solve process planning problems; that is, to create route sheets.
2. ***Computer-compatible part description***. Generative process planning requires a computer-compatible description of the part. The description contains all the pertinent data needed to plan the process sequence. Two possible descriptions are (1) the geometric model of the part developed on a CAD system during product design, or (2) a group technology code number of the part defining its features in significant detail.
3. ***Inference engine***. A generative CAPP system requires the capability to apply the planning logic and process knowledge contained in the knowledge base to a given part description. The CAPP system applies its knowledge base to solve a specific problem of planning the process for a new part. This problem-solving procedure is referred to as the inference engine in the terminology of expert systems. By using its knowledge base and inference engine, the CAPP system synthesizes a new process plan for each new part presented to it.

Benefits of CAPP Benefits of computer-automated process planning include the following: (1) process rationalization and standardization—automated process planning leads to more logical and consistent process plans than when traditional process planning is used; (2) increased productivity of process planners—the systematic approach and availability of standard process plans in the data files permit a greater number of process plans to be developed by the user; (3) reduced lead time to prepare process plans; (4) improved legibility compared to manually prepared route sheets; and (5) ability to interface CAPP programs with other application programs, such as cost estimating, work standards, and others.

28.2.4 PROBLEM SOLVING AND CONTINUOUS IMPROVEMENT

Problems arise in manufacturing that require technical staff support beyond what is normally available in the line organization of the production departments. Providing this technical support is one of the responsibilities of manufacturing engineering. The problems are usually specific to the particular technologies of the processes performed in the operating department. In machining, the problems may relate to selection of cutting tools, fixtures that do not work properly, parts with out-of-tolerance conditions, or nonoptimal cutting conditions. In plastic molding, the problems may be excessive flash, parts sticking in the mold, or any of several defects that can occur in a molded part. These problems are technical, and engineering expertise is often required to solve them.

In some cases, the solution may require a design change; for example, changing the tolerance on a part dimension to eliminate a finish grinding operation while still achieving the functionality of the part. The manufacturing engineer is responsible for developing the proper solution to the problem and proposing the engineering change to the design department.

In addition to solving current technical problems ("fire fighting," as it might be called), the manufacturing engineering department is also responsible for continuous improvement projects. Continuous improvement means constantly searching for and implementing ways to reduce cost, improve quality, and increase productivity in manufacturing. It is accomplished one project at a time. Depending on the type of problem area, it may involve a project team whose membership includes not only manufacturing engineers, but also other personnel such as product designers, quality engineers, and production workers.

One of the areas that is ripe for improvement is setup time. The procedures involved in changing over from one production setup to the next (i.e., in batch production) are time consuming and costly. Manufacturing engineers are responsible for analyzing changeover procedures and finding ways to reduce the time required to perform them.

28.3 CONCURRENT ENGINEERING AND DESIGN FOR MANUFACTURABILITY

Much of the process planning function described in Section 28.2 is preempted by decisions made in product design. Decisions on material, part geometry, tolerances, surface finish, grouping of parts into subassemblies, and assembly techniques limit the available manufacturing processes that can be used to make a given part. If the product engineer designs an aluminum sand casting with features that can be achieved only by machining (e.g., flat surfaces with good finishes, close tolerances, and threaded holes), then the process planner has no choice but to plan for sand casting followed by the required machining operations. If the product designer specifies a collection of sheet-metal stampings to be assembled by threaded fasteners, then the process planner must lay out the series of blanking, punching, and forming steps to fabricate the stampings and then assemble them. In both of these examples, a plastic molded part may be a superior design, both functionally and economically. It is important for the manufacturing engineer to act as an advisor to the design engineer in matters of manufacturability because manufacturability matters, not only to the production departments but to the design engineer. A product design that is functionally superior and at the same time can be produced at minimum cost holds the greatest promise of success in the marketplace. Successful careers in design engineering are built on successful products.

Terms often associated with this attempt to favorably influence the manufacturability of a product are ***design for manufacturing*** (DFM) and ***design for assembly*** (DFA). Of course, DFM and DFA are inextricably coupled, so let us refer to them as DFM/A. The scope of DFM/A is expanded in some companies to include not only manufacturability issues but also marketability, testability, serviceability, maintainability, and so forth. This broader view calls for inputs from many departments in addition to design and manufacturing engineering. The approach is called ***concurrent engineering***. Our discussion is organized around these two topics: DFM/A and concurrent engineering.

28.3.1 DESIGN FOR MANUFACTURING AND ASSEMBLY

Design for manufacturing and assembly is an approach to product design that systematically includes considerations of manufacturability and assemblability in the design. DFM/A includes organizational changes and design principles and guidelines.

To implement DFM/A, a company must change its organizational structure, either formally or informally, to provide closer interaction and better communication between design and manufacturing personnel. This is often accomplished by forming project teams consisting of product designers, manufacturing engineers, and other specialties (e.g., quality engineers, material scientists) to design the product. In some companies, design engineers are required to spend some career time in manufacturing to learn about the problems encountered in making things. Another possibility is to assign manufacturing engineers to the product design department as full-time consultants.

DFM/A also includes principles and guidelines that indicate how to design a given product for maximum manufacturability. Many of these are universal design guidelines, such as those presented in Table 28.5. They are rules of thumb that can be applied to nearly any product design situation. In addition, several of our chapters on manufacturing processes include DFM/A principles that are specific to those processes.

The guidelines are sometimes in conflict. For example, one guideline for part design is to make the geometry as simple as possible. Yet, in design for assembly, it is often desirable to combine features of several assembled parts into a single component to reduce part count and assembly time. However, combining multiple features has the effect of making the part geometry more complex. In these instances, design for manufacture conflicts with design for assembly, and a compromise must be found that achieves the best balance between opposing sides of the conflict.

Benefits typically cited for DFM/A include (1) shorter time to bring the product to market, (2) smoother transition into production, (3) fewer components in the final product, (4) easier assembly, (5) lower costs of production, (6) higher product quality, and (7) greater customer satisfaction [1].

TABLE 28.5 General principles and guidelines in design for manufacturing and assembly.

Minimize number of components. Assembly costs are reduced. The final product is more reliable because there are fewer connections. Disassembly for maintenance and field service is easier. Reduced part count usually means automation is easier to implement. Work-in-process is reduced, and there are fewer inventory control problems. Fewer parts need to be purchased, which reduces ordering costs.
Use standard commercially available components. Design time and effort are reduced. Design of custom-engineered components is avoided. There are fewer part numbers. Inventory control is facilitated. Quantity discounts may be possible.
Use common parts across product lines. There is an opportunity to apply group technology (Section 29.2). Implementation of manufacturing cells may be possible. Quantity discounts may be possible.
Design for ease of part fabrication. Net shape and near net shape processes may be feasible. Part geometry is simplified, and unnecessary features are avoided. Unnecessary surface finish requirements should be avoided; otherwise, additional processing may be needed.
Design parts with tolerances that are within process capability. Tolerances tighter than the process capability (Section 30.2) should be avoided; otherwise, additional processing or sortation will be required. Bilateral tolerances should be specified.
Design the product to be foolproof during assembly. Assembly should be unambiguous. Components should be designed so they can be assembled only one way. Special geometric features must sometimes be added to components to achieve foolproof assembly.
Minimize use of flexible components. Flexible components include parts made of rubber, belts, gaskets, cables, etc. Flexible components are generally more difficult to handle and assemble.
Design for ease of assembly. Part features such as chamfers and tapers should be designed on mating parts. Design the assembly using base parts to which other components are added. The assembly should be designed so that components are added from one direction, usually vertically. Threaded fasteners (screws, bolts, nuts) should be avoided where possible, especially when automated assembly is used; instead, fast assembly techniques such as snap fits and adhesive bonding should be employed. The number of distinct fasteners should be minimized.
Use modular design. Each subassembly should consist of five to fifteen parts. Maintenance and repair are facilitated. Automated and manual assembly are implemented more readily. Inventory requirements are reduced. Final assembly time is minimized.
Shape parts and products for ease of packaging. The product should be designed so that standard packaging cartons can be used, which are compatible with automated packaging equipment. Shipment to customer is facilitated.
Eliminate or reduce adjustment required. Adjustments are time-consuming in assembly. Designing adjustments into the product means more opportunities for out-of-adjustment conditions to arise.

Compiled from [1], [8].

28.3.2 CONCURRENT ENGINEERING

Concurrent engineering refers to an approach to product design in which companies attempt to reduce the elapsed time required to bring a new product to market by integrating design engineering, manufacturing engineering, and other functions in the company. The traditional approach to launch a new product tends to separate the two functions, as illustrated in Figure 28.6(a). The product design department develops the new design, sometimes with small regard for the manufacturing capabilities possessed by the company. There is little interaction between design engineers and manufacturing engineers who might provide advice on these capabilities and how the product design might be altered to accommodate them. It is as if a wall exists between the two functions;

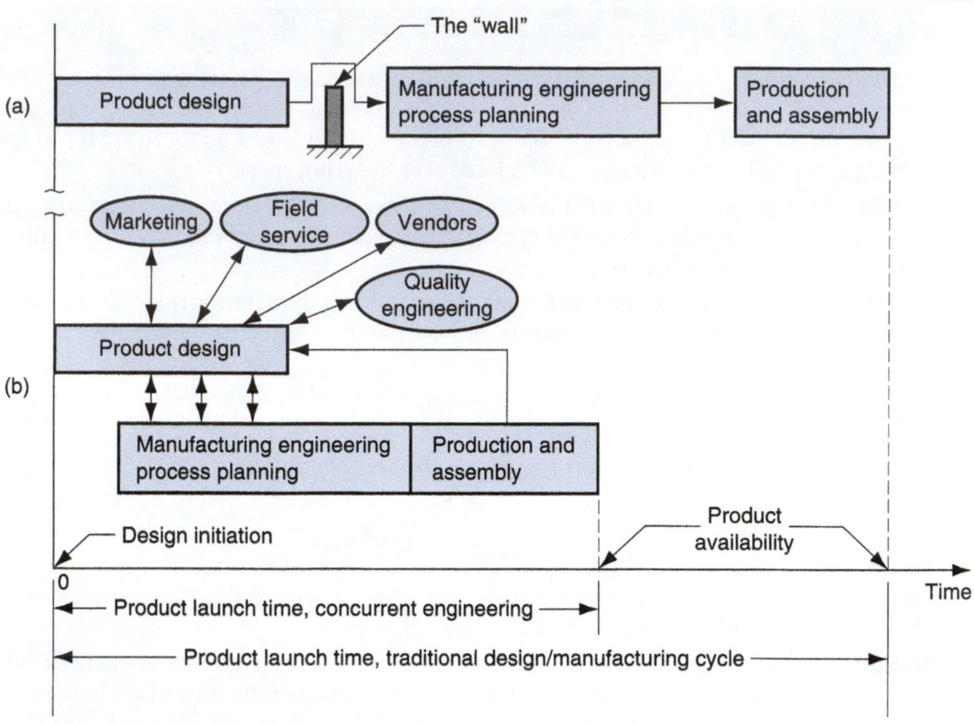

FIGURE 28.6 Comparison of: (a) traditional product development cycle, and (b) product development using concurrent engineering. (Credit: *Fundamentals of Modern Manufacturing*, 4th Edition by Mikell P. Groover, 2010. Reprinted with permission of John Wiley & Sons, Inc.)

when design engineering completes the design, it tosses the drawings and specifications over the wall so that process planning can commence.

In a company that practices concurrent engineering (also known as ***simultaneous engineering***), manufacturing planning begins while the product design is being developed, as depicted in Figure 28.6(b). Manufacturing engineering becomes involved early in the product development cycle. In addition, other functions are also involved, such as field service, quality engineering, the manufacturing departments, vendors supplying critical components, and in some cases customers who will use the product. All of these functions can contribute to a product design that not only performs well functionally, but is also manufacturable, assemblable, inspectable, testable, serviceable, maintainable, free of defects, and safe. All viewpoints have been combined to design a product of high quality that will deliver customer satisfaction. And through early involvement, rather than a procedure of reviewing the final design and suggesting changes after it is too late to conveniently make them, the total product development cycle is substantially reduced.

REFERENCES

[1] Bakerjian, R., and Mitchell, P. ***Tool and Manufacturing Engineers Handbook***, 4th ed. Vol. **VI**, ***Design for Manufacturability***. Society of Manufacturing Engineers, Dearborn, Michigan, 1992.

[2] Eary, D. F., and Johnson, G. E. ***Process Engineering for Manufacturing***. Prentice-Hall, Inc., Englewood Cliffs, New Jersey, 1962.

[3] Groover, M. P., and Zimmers, E. W., Jr. ***CAD/CAM: Computer-Aided Design and Manufacturing***. Prentice Hall, Englewood Cliffs, New Jersey, 1984.

[4] Groover, M. P. ***Automation, Production Systems, and Computer Integrated Manufacturing***, 3rd ed. Pearson Prentice Hall, Upper Saddle River, New Jersey, 2008.

[5] Kane, G. E."The Role of the Manufacturing Engineer." ***Technical Paper M M70-222.*** Society of Manufacturing Engineers, Dearborn, Michigan, 1970.

[6] Koenig, D. T. ***Manufacturing Engineering***. Hemisphere Publishing Corporation (Harper & Row, Publishers, Inc.), Washington, D.C., 1987.

[7] Kusiak, A. (ed.). ***Concurrent Engineering: Automation, Tools, and Techniques***. John Wiley & Sons, Inc. New York, 1993.

[8] Martin, J. M. "The Final Piece of the Puzzle," ***Manufacturing Engineering***, September 1988, pp. 46–51.

[9] Nevins, J. L., and Whitney, D. E. (eds.). ***Concurrent Design of Products and Processes***. McGraw-Hill, New York, 1989.

[10] Tanner, J. P. ***Manufacturing Engineering***, 2nd ed. CRC Taylor & Francis, Boca Raton, Florida, 1990.

[11] Usher, J. M., Roy, U., and Parsaei, H. R. (eds.). ***Integrated Product and Process Development***. John Wiley & Sons, Inc., New York, 1998.

[12] Veilleux, R. F., and Petro, L. W. ***Tool and Manufacturing Engineers Handbook***, 4th ed., Vol. **V**, ***Manufacturing Management***. Society of Manufacturing Engineers, Dearborn, Michigan, 1988.

REVIEW QUESTIONS

28.1. Provide a definition of the term *production systems*.

28.2. What are the two categories of production systems?

28.3. Name the four types of plant layouts typically found in factories. What types of production are each normally associated with?

28.4. Define the term *manufacturing support systems*.

28.5. What is process planning?

28.6. Define the term *manufacturing engineering*.

28.7. Identify some of the details and decisions that are included within the scope of process planning.

28.8. What is a route sheet?

28.9. What is the difference between a basic process and a secondary process?

28.10. What is a precedence constraint in process planning?

28.11. In the make or buy decision, why is it that purchasing a component from a vendor may cost more than producing the component internally, even though the quoted price from the vendor is lower than the internal price?

28.12. Identify some of the important factors that should enter into the make or buy decision.

28.13. Name three of the general principles and guidelines in design for manufacturability.

28.14. What is concurrent engineering?

29 SURVEY OF AUTOMATION AND MANUFACTURING SYSTEMS

Chapter Contents

29.1 Computer Numerical Control
- 29.1.1 The Technology of Numerical Control
- 29.1.2 Analysis of NC Positioning Systems
- 29.1.3 NC Part Programming
- 29.1.4 Applications of Numerical Control

29.2 Cellular Manufacturing
- 29.2.1 Part Families
- 29.2.2 Machine Cells

29.3 Flexible Manufacturing Systems and Cells
- 29.3.1 Integrating the FMS Components
- 29.3.2 Applications of Flexible Manufacturing Systems

29.4 Lean Production
- 29.4.1 Just-in-Time Production Systems
- 29.4.2 Other Approaches in Lean Production

29.5 Computer Integrated Manufacturing

In this chapter, we cover several systems topics that are related to manufacturing: computer numerical control, cellular manufacturing, flexible manufacturing systems, lean production, and computer-integrated manufacturing. ***Computer numerical control*** (CNC) is a widely used automation technology for controlling the operations of machine tools in batch production, in which the processing instructions are different for each batch. The instructions are entered into the controller by means of a program that is prepared specifically for the parts in each batch. CNC is used to control machines that operate independently or as components in larger manufacturing systems.

A ***manufacturing system*** can be defined as a collection of integrated equipment and human resources that performs processing and/or assembly operations on a starting work material, part, or set of parts. The integrated equipment consists of one or more production machines, material handling and positioning devices, and computer systems. Human resources are required either full-time or part-time to keep the systems working. The manufacturing systems are located in the factory, where they accomplish the value-added work on the part or product.

Manufacturing systems include both automated and manually operated systems. The distinction between the two categories is not always clear, because many manufacturing systems consist of both automated and manual work elements (e.g., a machine tool that operates on a processing cycle under CNC control but must be loaded and unloaded each cycle by a human worker). Some manufacturing systems consist of a single machine that is supported by auxiliary equipment (e.g., parts loaders, inspection equipment), whereas other systems consist of multiple workstations and/or machines whose operations are integrated by means of a material handling subsystem that moves parts or products between stations. In addition, many of these systems use computers to coordinate the actions of the stations and material handling equipment and to collect data on overall system performance. The integrated manufacturing systems discussed in this chapter are manufacturing cells (from which the term "cellular manufacturing" is derived), and flexible manufacturing systems.

In the final two sections, lean production and computer-integrated manufacturing (CIM) are discussed. Lean production is a plantwide approach aimed at eliminating waste and increasing productivity in manufacturing. CIM involves the extensive application of computer systems to plan and control the operations of a manufacturing enterprise.

A more detailed discussion of the topics in this chapter can be found in reference [7]. Our coverage begins with computer numerical control, the technology that enables the automation of machine tools.

29.1 COMPUTER NUMERICAL CONTROL

Numerical control (NC) is a form of programmable automation in which the mechanical actions of a piece of equipment are controlled by a program containing coded alphanumeric data. The data represent relative positions between a workhead and a workpart. The workhead is a tool or other processing element, and the workpart is the object being processed. The operating principle of NC is to control the motion of the workhead relative to the workpart and to control the sequence in which the motions are carried out. The first application of numerical control was in machining, and this is still an important application area. Two NC machine tools are shown in Chapter 16 (Figures 16.26 and 16.27).

29.1.1 THE TECHNOLOGY OF NUMERICAL CONTROL

In this section we define the components of a numerical control system, and then proceed to describe the coordinate axis system and motion controls.

Components of an NC System A numerical control system consists of three components: (1) part program, (2) machine control unit, and (3) processing equipment. The ***part program*** (the term commonly used in machine tool technology) is the detailed set of commands to be followed by the processing equipment for a particular workpart. Each command specifies a position or motion that is to be accomplished by the workhead relative to the workpart. A position is defined by its x-y-z coordinates. In machine tool applications, additional details in the NC program include spindle rotation speed, spindle direction, feed rate, tool change instructions, and other commands related to the operation. The part program is prepared by a ***part programmer***, a person who is familiar with the details of the programming language and also understands the technology of the processing equipment.

The ***machine control unit*** (MCU) in modern NC technology is a microcomputer that stores and executes the program by converting each command into actions by the processing equipment, one command at a time. The MCU consists of both hardware and software. The hardware includes the microcomputer, components to interface with the processing equipment, and certain feedback control elements. The MCU software includes control system software, calculation algorithms, and translation software to convert the NC part program into a usable format for the MCU. The MCU also permits the part program to be edited in case the program contains errors, or changes in cutting

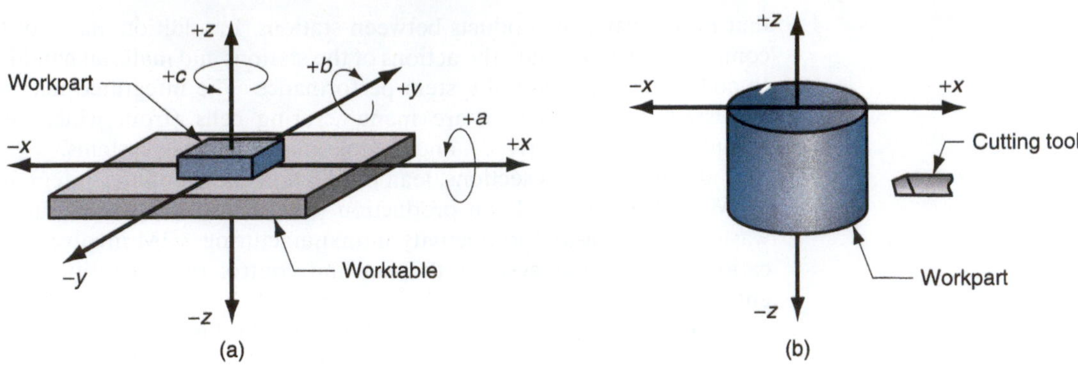

FIGURE 29.1 Coordinate systems used in numerical control: (a) for flat and prismatic work, and (b) for rotational work. (Credit: *Fundamentals of Modern Manufacturing*, 4th Edition by Mikell P. Groover, 2010. Reprinted with permission of John Wiley & Sons, Inc.)

conditions are required. Because the MCU is a computer, the term ***computer numerical control*** (CNC) is often used to distinguish this type of NC from its technological predecessors that were based entirely on hard-wired electronics.

The ***processing equipment*** accomplishes the sequence of processing steps to transform the starting workpart into a completed part. It operates under the control of the MCU according to the instructions in the part program. We survey the variety of applications and processing equipment in Section 29.1.4.

Coordinate System and Motion Control in NC A standard coordinate axis system is used to specify positions in numerical control. The system consists of the three linear axes (x, y, z) of the Cartesian coordinate system, plus three rotational axes (a, b, c), as shown in Figure 29.1(a). The rotational axes are used to rotate the workpart to present different surfaces for machining, or to orient the tool or workhead at some angle relative to the part. Most NC systems do not require all six axes. The simplest NC systems (e.g., plotters, pressworking machines for flat sheet-metal stock, and component insertion machines) are positioning systems whose locations can be defined in an x-y plane. Programming of these machines involves specifying a sequence of x-y coordinates. By contrast, some machine tools have five-axis control to shape complex workpart geometries. These systems typically include three linear axes plus two rotational axes.

The coordinates for a rotational NC system are illustrated in Figure 29.1(b). These systems are associated with turning operations on NC lathes. Although the work rotates, this is not one of the controlled axes in a conventional NC turning system. The cutting path of the tool relative to the rotating workpiece is defined in the x-z plane, as shown in our figure.

In many NC systems, the relative movements between the processing tool and the workpart are accomplished by fixing the part to a worktable and then controlling the positions and motions of the table relative to a stationary or semi-stationary workhead. Most machine tools and component insertion machines are based on this method of operation. In other systems, the workpart is held stationary and the workhead is moved along two or three axes. Flame cutters, x-y plotters, and coordinate measuring machines operate in this mode.

Motion control systems based on NC can be divided into two types: (1) point-to-point and (2) continuous path. ***Point-to-point systems***, also called ***positioning systems***, move the workhead (or workpiece) to a programmed location with no regard for the path taken to get to that location. Once the move is completed, some processing action is

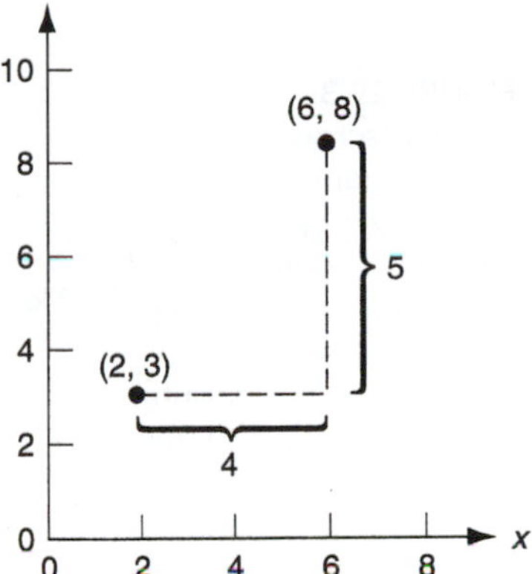

FIGURE 29.2 Absolute vs. incremental positioning. The workhead is at point (2,3) and is to be moved to point (6,8). In absolute positioning, the move is specified by $x = 6$, $y = 8$; while in incremental positioning, the move is specified by $x = 4$, $y = 5$. (Credit: *Fundamentals of Modern Manufacturing*, 4th Edition by Mikell P. Groover, 2010. Reprinted with permission of John Wiley & Sons, Inc.)

accomplished by the workhead at the location, such as drilling or punching a hole. Thus, the program consists of a series of point locations at which operations are performed.

Continuous path systems provide continuous simultaneous control of more than one axis, thus controlling the path followed by the tool relative to the part. This permits the tool to perform a process while the axes are moving, enabling the system to generate angular surfaces, two-dimensional curves, or three-dimensional contours in the workpart. This operating scheme is required in drafting machines, certain milling and turning operations, and flame cutting. In machining, continuous path control also goes by the name ***contouring***.

An important aspect of continuous path motion is ***interpolation***, which is concerned with calculating the path to be followed by the workhead relative to the part between two points. Two common forms of interpolation are linear and circular. ***Linear interpolation*** is used for straight-line paths, in which the part programmer specifies the coordinates of the beginning point and end point of the straight line as well as the feed rate to be used. The interpolator then computes the travel speeds of the two or three axes that will accomplish the specified trajectory. ***Circular interpolation*** allows the workhead to follow a circular arc by specifying the coordinates of its beginning and end points together with either the center or radius of the arc. The interpolator computes a series of small straight-line segments that approximate the arc within a defined tolerance.

Another aspect of motion control is concerned with whether the positions in the coordinate system are defined absolutely or incrementally. In ***absolute positioning***, the workhead locations are always defined with respect to the origin of the axis system. In ***incremental positioning***, the next workhead position is defined relative to the present location. The difference is illustrated in Figure 29.2.

29.1.2 ANALYSIS OF NC POSITIONING SYSTEMS

The function of the positioning system is to convert the coordinates specified in the NC part program into relative positions between the tool and workpart during processing. Let us consider how a simple positioning system, shown in Figure 29.3, might operate. The system consists of a worktable on which a workpart is fixtured. The purpose of the table is to move the part relative to a tool or workhead. To accomplish this purpose, the worktable is moved linearly by means of a rotating leadscrew that is driven by a motor. For simplicity, only one axis is shown in our sketch. To provide x-y capability, the system shown would be piggybacked on top of a second axis perpendicular to the first. The leadscrew has a certain pitch p, mm/thread (in/thread) or mm/rev (in/rev). Thus, the table is moved a distance equal to the leadscrew pitch for each revolution. The velocity at which the worktable moves is determined by the rotational speed of the leadscrew.

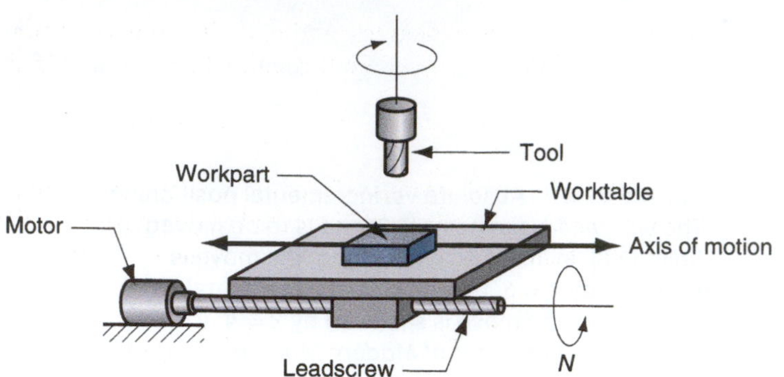

FIGURE 29.3 Motor and leadscrew arrangement in an NC positioning system. (Credit: *Fundamentals of Modern Manufacturing*, 4th Edition by Mikell P. Groover, 2010. Reprinted with permission of John Wiley & Sons, Inc.)

Two basic types of motion control are used in NC: (a) open loop and (b) closed loop, as shown in Figure 29.4. The difference is that an open-loop system operates without verifying that the desired position of the worktable has been achieved. A closed-loop control system uses feedback measurement to verify that the position of the worktable is indeed the location specified in the program. Open-loop systems are less expensive than closed-loop systems and are appropriate when the force resisting the actuating motion is minimal, as in point-to-point drilling, for example. Closed-loop systems are normally specified for machine tools that perform continuous path operations such as milling or turning, in which the resisting forces can be significant.

Open-Loop Positioning Systems To turn the leadscrew, an open-loop positioning system typically uses a stepping motor (a.k.a. stepper motor). In NC, the stepping motor is driven by a series of electrical pulses generated by the machine control unit. Each pulse causes the motor to rotate a fraction of one revolution, called the step angle. The allowable step angles must conform to the relationship:

$$\alpha = \frac{360}{n_s} \tag{29.1}$$

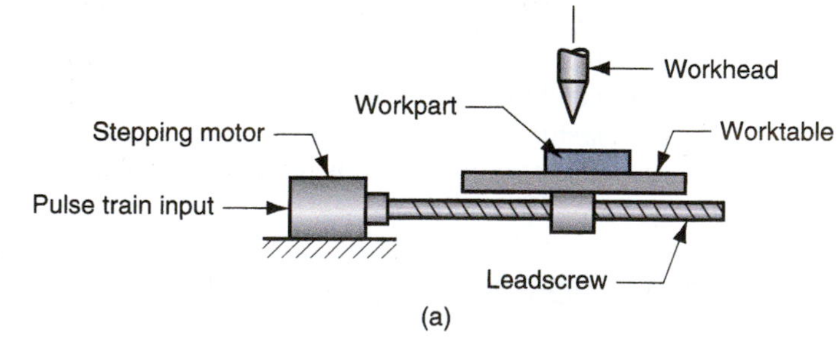

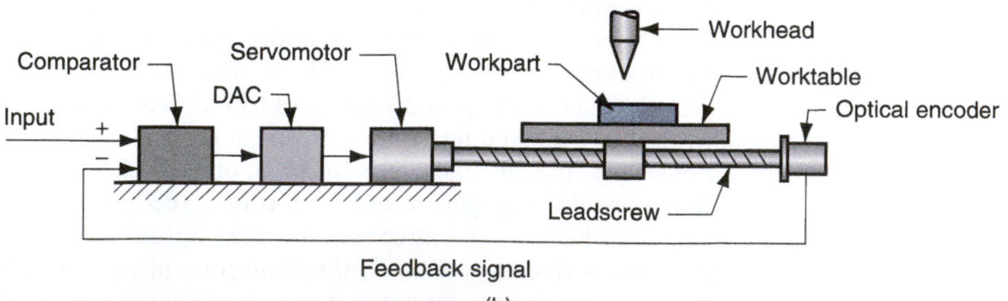

FIGURE 29.4 Two types of motion control in NC: (a) open loop and (b) closed loop. (Credit: *Fundamentals of Modern Manufacturing*, 4th Edition by Mikell P. Groover, 2010. Reprinted with permission of John Wiley & Sons, Inc.)

where α = step angle, degrees; and n_s = the number of step angles for the motor, which must be an integer. The angle through which the motor shaft rotates is given by:

$$A_m = \alpha n_p \tag{29.2}$$

where A_m = angle of motor shaft rotation, degrees; n_p = number of pulses received by the motor; and α = step angle, here defined as degrees/pulse. Finally, the rotational speed of the motor shaft is determined by the frequency of pulses sent to the motor:

$$N_m = \frac{60\alpha f_p}{360} \tag{29.3}$$

where N_m = speed of motor shaft rotation, rev/min; f_p = frequency of pulses driving the stepper motor, Hz (pulses/sec), the constant 60 converts pulses/sec to pulses/min; the constant 360 converts degrees of rotation to full revolutions; and α = step angle of the motor, as before.

The motor shaft drives the leadscrew that determines the position and velocity of the worktable. The connection is often designed using a gear reduction to increase the precision of table movement. However, the angle of rotation and rotational speed of the leadscrew are reduced by this gear ratio. The relationships are as follows:

$$A_m = r_g A_{ls} \tag{29.4a}$$

and

$$N_m = r_g N_{ls} \tag{29.4b}$$

where A_m and N_m are the angle of rotation, degrees, and rotational speed, rev/min, of the motor, respectively; A_{ls} and N_{ls} are the angle of rotation, degrees, and rotational speed, rev/min, of the leadscrew, respectively; and r_g = gear reduction between the motor shaft and the leadscrew; for example, a gear reduction of 2 means that the motor shaft rotates through two revolutions for each rotation of the leadscrew.

The linear position of the table in response to the rotation of the leadscrew depends on the leadscrew pitch p, and can be determined as follows:

$$x = \frac{pA_{ls}}{360}$$

where x = x-axis position relative to the starting position, mm (in); p = pitch of the leadscrew, mm/rev (in/rev); and $A_{ls}/360$ = the number of revolutions (and partial revolutions) of the leadscrew. By combining Eqs. (29.2), (29.4a), and (29.5) and rearranging, the number of pulses required to achieve a specified x-position increment in a point-to-point system can be found:

$$n_p = \frac{360 r_g x}{p\alpha} = \frac{r_g n_s A_{ls}}{360} \tag{29.6}$$

The velocity of the worktable in the direction of the leadscrew axis can be determined as follows:

$$v_t = f_r = N_{ls}\, p \tag{29.7}$$

where v_t = table travel speed, mm/min (in/min); f_r = table feed rate, mm/min (in/min); N_{ls} = rotational speed of the leadscrew, rev/min; and p = leadscrew pitch, mm/rev (in/rev).

The rotational speed of the leadscrew depends on the frequency of pulses driving the stepping motor:

$$N_{ls} = \frac{60 f_p}{n_s r_g} \tag{29.8}$$

where N_{ls} = leadscrew rotational speed, rev/min; f_p = pulse train frequency, Hz (pulses/sec); n_s = steps/rev, or pulses/rev, and r_g = gear reduction between the motor and the leadscrew. For a two-axis table with continuous path control, the relative velocities of the axes are coordinated to achieve the desired travel direction. Finally, the required pulse frequency to drive the table at a specified feed rate can be obtained by combining Eqs. (29.7) and (29.8) and rearranging to solve for f_p:

$$f_p = \frac{v_t n_s r_g}{60p} = \frac{f_r n_s r_g}{60p} = \frac{N_{ls} n_s r_g}{60} = \frac{N_m n_s}{60} \tag{29.9}$$

Example 29.1 Open-Loop Positioning

A stepping motor has 48 step angles. Its output shaft is coupled to a leadscrew with a 4:1 gear reduction (4 turns of the motor shaft for each turn of the leadscrew). The leadscrew pitch = 5.0 mm. The worktable of a positioning system is driven by the leadscrew. The table must move a distance of 75.0 mm from its current position at a travel speed of 400 mm/min. Determine (a) how many pulses are required to move the table the specified distance; and (b) the motor speed and (c) pulse frequency required to achieve the desired table speed.

Solution: (a) To move a distance x = 75 mm, the leadscrew must rotate through an angle calculated as follows:

$$A_{ls} = \frac{360x}{p} = \frac{360(75)}{5} = 5400^\circ$$

With 48 step angles and a gear reduction of 4, the number of pulses to move the table 75 mm is:

$$n_p = \frac{4(48)(5400)}{360} = 2880 \text{ pulses}$$

(b) Equation (29.7) can be used to find the leadscrew speed corresponding to the table speed of 400 mm/min,

$$N_{ls} = \frac{v_t}{p} = \frac{400}{5.0} = 80.0 \text{ rev/min}$$

The motor speed will be 4 times as fast:

$$N_m = r_g N_{ls} = 4(80) = 320 \text{ rev/min}$$

(c) Finally, the pulse rate is given by Eq. (29.9):

$$f_p = \frac{320(48)}{60} = 256 \text{ Hz}$$

■

Closed-Loop Positioning Systems Closed-loop NC systems, Figure 29.4(b), use servomotors and feedback measurements to ensure that the desired position is achieved.

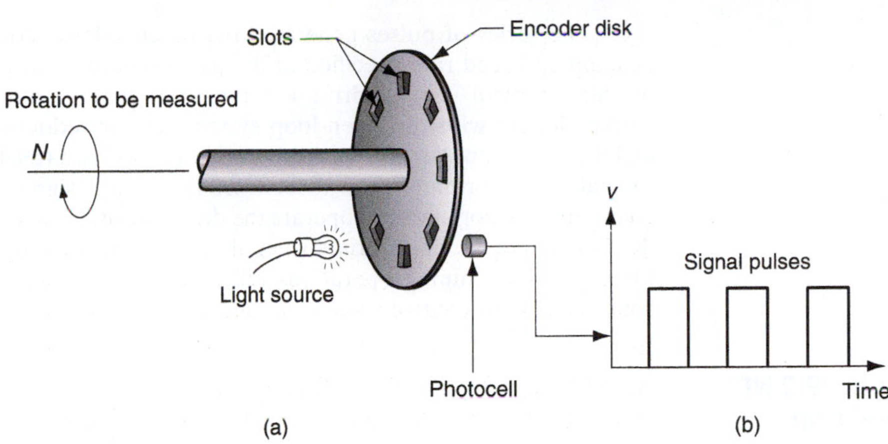

FIGURE 29.5 Optical encoder: (a) apparatus, and (b) series of pulses emitted to measure rotation of disk. (Credit: *Fundamentals of Modern Manufacturing*, 4th Edition by Mikell P. Groover, 2010. Reprinted with permission of John Wiley & Sons, Inc.)

A common feedback sensor used in NC is the optical rotary encoder, illustrated in Figure 29.5. It consists of a light source, a photocell, and a disk containing a series of slots through which the light source can shine to energize the photocell. The disk is connected to a rotating shaft, which in turn is connected directly to the leadscrew. As the leadscrew rotates, the slots cause the light source to be seen by the photocell as a series of flashes, which are converted into an equivalent series of electrical pulses. By counting the pulses and computing the frequency of the pulse train, the leadscrew angle and rotational speed can be determined, and thus worktable position and speed can be calculated using the pitch of the leadscrew.

The equations describing the operation of a closed-loop positioning system are analogous to those for an open-loop system. In the basic optical encoder, the angle between slots in the disk must satisfy the following requirement:

$$\alpha = \frac{360}{n_s} \tag{29.10}$$

where α = angle between slots, degrees/slot; and n_s = the number of slots in the disk, slots/rev; and 360 = degrees/rev. For a certain angular rotation of the leadscrew, the encoder generates a number of pulses given by:

$$n_p = \frac{A_{ls}}{\alpha} = \frac{A_{ls}n_s}{360} \tag{29.11}$$

where n_p = pulse count; A_{ls} = angle of rotation of the leadscrew, degrees; and α = angle between slots in the encoder, degrees/pulse. The pulse count can be used to determine the linear x-axis position of the worktable by factoring in the leadscrew pitch. Thus,

$$x = \frac{pn_p}{n_s} = \frac{pA_{ls}}{360} \tag{29.12}$$

Similarly, the feed rate at which the worktable moves is obtained from the frequency of the pulse train:

$$v_t = f_r = \frac{60pf_p}{n_s} \tag{29.13}$$

where v_t = table travel speed, mm/min (in/min); f_r = feed rate, mm/min (in/min); p = pitch, mm/rev (in/rev); f_p = frequency of the pulse train, Hz (pulses/sec); n_s = number of slots in the encoder disk, pulses/rev; and 60 converts seconds to minutes. The speed relationship given by Eq. (29.7) is also valid for a closed-loop positioning system.

The series of pulses generated by the encoder is compared with the coordinate position and feed rate specified in the part program, and the difference is used by the machine control unit to drive a servomotor that in turn drives the leadscrew and worktable. As with the open-loop system, a gear reduction between the servomotor and the leadscrew can also be used, so Eqs. (29.4a) and (29.4b) are applicable. A digital-to-analog converter is used to convert the digital signals used by the MCU into a continuous analog signal to operate the drive motor. Closed-loop NC systems of the type described here are appropriate when there is force resisting the movement of the table. Most metal-machining operations fall into this category, particularly those involving continuous path control such as milling and turning.

Example 29.2 NC Closed-Loop Positioning

An NC worktable is driven by a closed-loop positioning system consisting of a servomotor, leadscrew, and optical encoder. The leadscrew has a pitch = 5.0 mm and is coupled to the motor shaft with a gear ratio of 4:1 (four turns of the motor for each turn of the leadscrew). The optical encoder generates 100 pulses/rev of the leadscrew. The table has been programmed to move a distance of 75.0 mm at a feed rate = 400 mm/min. Determine (a) how many pulses are received by the control system to verify that the table has moved exactly 75.0 mm; and (b) the pulse rate and (c) motor speed that correspond to the specified feed rate.

Solution: (a) Rearranging Eq. (29.12) to find n_p,

$$n_p = \frac{x n_s}{p} = \frac{75(100)}{5} = 1500 \text{ pulses}$$

(b) The pulse rate corresponding to 400 mm/min can be obtained by rearranging Eq. (29.13):

$$f_p = \frac{f_r n_s}{60p} = \frac{400(100)}{60(5)} = 133.33 \text{ Hz}$$

(c) Leadscrew rotational speed is the table velocity divided by the pitch:

$$N_{ls} = \frac{f_r}{p} = 80 \text{ rev/min}$$

With a gear ratio $r_g = 4.0$, the motor speed $N = 4(80) = 320$ rev/min ■

29.1.3 NC PART PROGRAMMING

In machine tool applications, the task of programming the system is called NC part programming because the program is prepared for a given part. It is usually accomplished by someone familiar with the metalworking process who has learned the programming procedure for the particular equipment in the plant. For other processes, other terms may be used for programming, but the principles are similar and a trained individual is needed to prepare the program. Computer systems are used extensively to prepare NC programs.

Part programming requires the programmer to define the points, lines, and surfaces of the workpart in the axis system, and to control the movement of the cutting tool relative to these defined part features. Several part programming techniques are available, the most important of which are (1) manual part programming, (2) computer-assisted part programming, (3) CAD/CAM-assisted part programming, and (4) manual data input.

Manual Part Programming For simple point-to-point machining jobs, such as drilling operations, manual programming is often the easiest and most economical method. Manual part programming uses basic numerical data and special alphanumeric codes to define the steps in the process. For example, to perform a drilling operation, a command of the following type is entered:

n010 *x*70.0 *y*85.5 f175 s500

Each "word" in the statement specifies a detail in the drilling operation. The *n*-word (n010) is simply a sequence number for the statement. The *x*- and *y*-words indicate the *x* and *y* coordinate positions ($x = 70.0$ mm and $y = 85.5$ mm). The *f*-word and *s*-word specify the feed rate and spindle speed to be used in the drilling operation (feed rate = 175 mm/min and spindle speed = 500 rev/min). The complete NC part program consists of a sequence of statements similar to the above command.

Computer-Assisted Part Programming Computer-assisted part programming involves the use of a high-level programming language. It is suited to the programming of more complex jobs than manual programming. The first part programming language was APT (Automatically Programmed Tooling), developed as an extension of the original NC machine tool research and first used in production around 1960.

In APT, the part programming task is divided into two steps: (1) definition of part geometry and (2) specification of tool path and operation sequence. In step 1, the part programmer defines the geometry of the workpart by means of basic geometric elements such as points, lines, planes, circles, and cylinders. These elements are defined using APT geometry statements, such as:

```
P1 = POINT/25.0, 150.0
L1 = LINE/P1, P2
```

P1 is a point defined in the *x*-*y* plane located at $x = 25$ mm and $y = 150$ mm. L1 is a line that goes through points P1 and P2. Similar statements can be used to define circles, cylinders, and other geometry elements. Most workpart shapes can be described using statements like these to define their surfaces, corners, edges, and hole locations.

Specification of the tool path is accomplished with APT motion statements. A typical statement for point-to-point operation is:

```
GOTO/P1
```

This directs the tool to move from its current location to a position defined by P1, where P1 has been defined by a previous APT geometry statement. Continuous path commands use geometry elements such as lines, circles, and planes. For example, the command:

```
GORGT/L3, PAST, L4
```

directs the tool to go right (GORGT) along line L3 until it is positioned just past line L4 (of course, L4 must be a line that intersects L3).

Additional APT statements are used to define operating parameters such as feed rates, spindle speeds, tool sizes, and tolerances. When completed, the part programmer enters the APT program into the computer, where it is processed to generate low-level statements (similar to statements prepared in manual part programming) that can be used by a particular machine tool.

CAD/CAM-Assisted Part Programming The use of CAD/CAM takes computer-assisted part programming a step further by using a computer graphics system (CAD/CAM system) to interact with the programmer as the part program is being prepared. In the conventional use of APT, a complete program is written and then entered into the computer for processing. Many programming errors are not detected until computer processing. When a CAD/CAM system is used, the programmer receives immediate visual verification when each statement is entered to determine whether the statement is correct. When part geometry is entered by the programmer, the element is graphically displayed on the monitor. When the tool path is constructed, the programmer can see exactly how the motion commands will move the tool relative to the part. Errors can be corrected immediately rather than after the entire program has been written.

Interaction between programmer and programming system is a significant benefit of CAD/CAM-assisted programming. There are other important benefits of using CAD/CAM in NC part programming. First, the design of the product and its components may have been accomplished on a CAD/CAM system. The resulting design database, including the geometric definition of each part, can be retrieved by the NC programmer to use as the starting geometry for part programming. This retrieval saves valuable time compared to reconstructing the part from scratch using the APT geometry statements.

Second, special software routines are available in CAD/CAM-assisted part programming to automate portions of the tool path generation, such as profile milling around the outside periphery of a part, milling a pocket into the surface of a part, surface contouring, and certain point-to-point operations. These routines are called by the part programmer as special ***macro*** commands. Their use results in significant savings in programming time and effort.

Manual Data Input Manual data input (MDI) is a method in which a machine operator enters the part program in the factory. The method involves use of a CRT display with graphics capability at the machine tool controls. NC part programming statements are entered using a menu-driven procedure that requires minimum training of the machine tool operator. Because part programming is simplified and does not require a special staff of NC part programmers, MDI is a way for small machine shops to economically implement numerical control into their operations.

29.1.4 APPLICATIONS OF NUMERICAL CONTROL

Machining is an important application area for numerical control, but the operating principle of NC can be applied to other operations as well. There are many industrial processes in which the position of a workhead must be controlled relative to the part or product being worked on. We divide the applications into two categories: (1) machine tool applications, and (2) non-machine tool applications. It should be noted that the applications are not all identified by the name numerical control in their respective industries.

In the machine tool category, NC is widely used for ***machining operations*** such as turning, drilling, and milling (Sections 16.2, 16.3, and 16.4, respectively). The use of NC in these processes has motivated the development of highly automated machine tools called ***machining centers***, which change their own cutting tools to perform a variety of machining operations under NC program control (Section 16.5). In addition to machining, other numerically controlled machine tools include (1) grinding machines, (2) sheet-metal pressworking machines, (3) tube-bending machines, and (4) thermal cutting processes.

In the non-machine tool category, NC applications include (1) tape-laying machines and filament-winding machines for composites; (2) welding machines, both arc welding and resistance welding; (3) component-insertion machines in electronics assembly; (4) drafting machines; and (5) coordinate measuring machines for inspection.

Benefits of NC relative to manually operated equipment in these applications include: (1) reduced nonproductive time, which results in shorter cycle times; (2) lower manufacturing lead times; (3) simpler fixturing; (4) greater manufacturing flexibility; (5) improved accuracy; and (6) reduced human error.

29.2 CELLULAR MANUFACTURING

Cellular manufacturing refers to the use of work cells that specialize in the production of families of parts or products made in medium quantities. Parts and products in this quantity range are traditionally made in batches; however, batch production requires downtime for setup changeovers between batches, and the inventory carrying costs of the batches can be significant. Cellular manufacturing is based on an approach called ***group technology*** (GT), which minimizes the disadvantages of batch production by recognizing that, although the parts are different, they also possess similarities. When these similarities are exploited in production, operating efficiencies are improved. The improvement is typically achieved by organizing production around manufacturing cells. Each cell is designed to produce one part family (or a limited number of part families), thereby following the principle of specialization of operations. The cell includes special production equipment and custom-designed tools and fixtures, so that the production of the part families can be optimized. In effect, each cell becomes a factory within the factory.

29.2.1 PART FAMILIES

A central feature of cellular manufacturing and group technology is the part family. A part family is a group of parts that possess similarities in geometric shape and size, or in the processing steps used in their manufacture. It is not unusual for a factory that produces 10,000 different parts to be able to group most of those parts into 20 to 30 part families. In each part family the processing steps are similar. There are always differences among parts in a family, but the similarities are close enough that the parts can be grouped into the same family. Figures 29.6 and 29.7 show two different part families. The parts shown in Figure 29.6 have the same size and shape; however, their processing requirements are quite different because of differences in work material, production quantities, and design tolerances. Figure 29.7 shows several parts with geometries that differ, but their manufacturing requirements are quite similar.

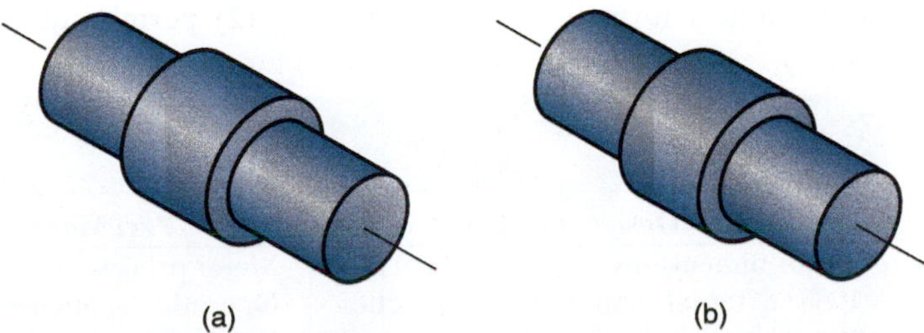

FIGURE 29.6 Two parts that are identical in shape and size but quite different in manufacturing: (a) 1,000,000 units/yr, tolerance = ±0.010 in., 1015 CR steel, nickel plate; and (b) 100/yr, tolerance = ±0.001 in, 18-8 stainless steel. (Credit: *Fundamentals of Modern Manufacturing,* 4th Edition by Mikell P. Groover, 2010. Reprinted with permission of John Wiley & Sons, Inc.)

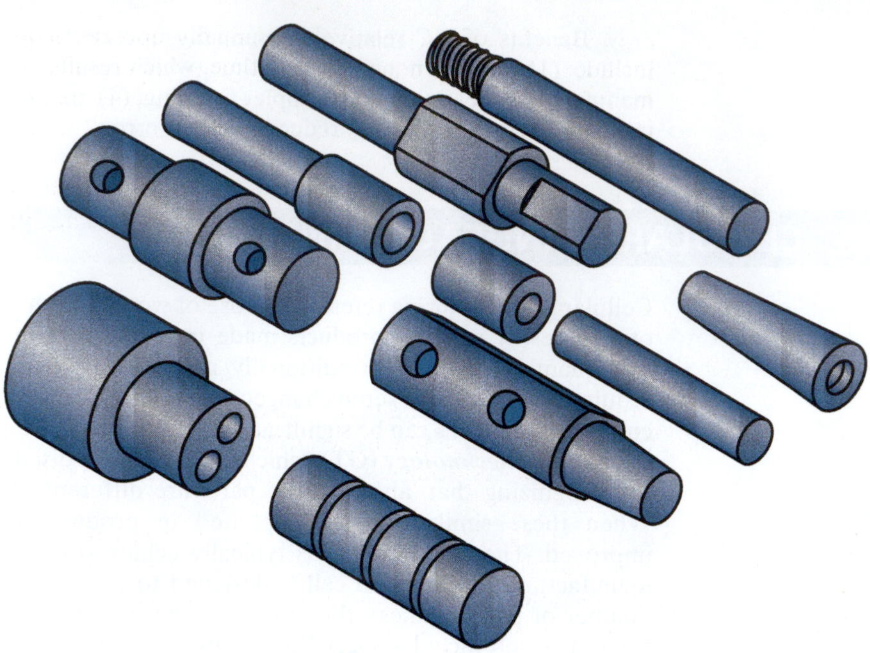

FIGURE 29.7 Ten parts that are different in size and shape, but quite similar in terms of manufacturing. All parts are machined from cylindrical stock by turning; some parts require drilling and/or milling. (Credit: *Fundamentals of Modern Manufacturing*, 4th Edition by Mikell P. Groover, 2010. Reprinted with permission of John Wiley & Sons, Inc.)

There are several ways by which part families are identified in industry. One method involves visual inspection of all the parts made in the factory (or photos of the parts) and using best judgment to group them into appropriate families. Another approach, called ***production flow analysis***, uses information contained on route sheets (Section 28.2.1) to classify parts. In effect, parts with similar manufacturing steps are grouped into the same family.

A third method, usually the most expensive but most useful, is parts classification and coding. ***Parts classification and coding*** involve the identification of similarities and differences among parts and relating these features by means of a numerical coding scheme. Most classification and coding systems are based on one of the following: (1) part design attributes, (2) part manufacturing attributes, and (3) both design and manufacturing attributes. Common part design and manufacturing attributes used in GT systems are presented in Table 29.1. Because each company produces a unique set of parts and products, a classification and coding system that may be satisfactory for one company is not necessarily appropriate for another company. Each company must design its own coding scheme. Parts classification and coding systems are described more thoroughly in several of our references: [6], [7], [8].

Benefits often cited for a well-designed classification and coding system include: (1) facilitates formation of part families, (2) permits quick retrieval of part design

TABLE 29.1 Design and manufacturing attributes typically included in a parts classification and coding system.

Part Design Attributes		Part Manufacturing Attributes	
Major dimensions	Material type	Major process	Major dimensions
Basic external shape	Part function	Operation sequence	Basic external shape
Basic internal shape	Tolerances	Batch size	Length/diameter ratio
Length/diameter ratio	Surface finish	Annual production	Material type
		Machine tools	Tolerances
		Cutting tools	Surface finish

Compiled from [6], [7], and [8].

drawings, (3) reduces design duplication because similar or identical part designs can be retrieved and reused rather than designed from scratch, (4) promotes design standardization, (5) improves cost estimating and cost accounting, (6) facilitates numerical control (NC) part programming by allowing new parts to use the same basic part program as existing parts in the same family, (7) allows sharing of tools and fixtures, and (8) aids computer-aided process planning (Section 28.2.3) because standard process plans can be correlated to part family code numbers, so that existing process plans can be reused or edited for new parts in the same family.

29.2.2 MACHINE CELLS

To fully exploit the similarities among parts in a family, production should be accomplished using machine cells designed to specialize in making that part family. One of the principles in designing a group technology machine cell is the composite part concept.

Composite Part Concept Members of a part family possess similar design and/or manufacturing features. There is usually a correlation between part design features and the manufacturing operations that produce those features. Round holes are made by drilling; cylindrical shapes are made by turning; and so on.

The ***composite part*** for a given family (not to be confused with a part made of composite material) is a hypothetical part that includes all of the design and manufacturing attributes of the family. In general, an individual part in the family will have some of the features that characterize the family, but not all of them. A production cell designed for the part family would include those machines required to make the composite part. Such a cell would be capable of producing any member of the family, simply by omitting those operations corresponding to features not possessed by the particular part. The cell would also be designed to allow for size variations within the family as well as feature variations.

To illustrate, consider the composite part in Figure 29.8(a). It represents a family of rotational parts with features defined in part (b) of the figure. Associated with each feature is a certain machining operation, as summarized in Table 29.2. A machine cell to produce this part family would be designed with the capability to accomplish all of the operations in the last column of the table.

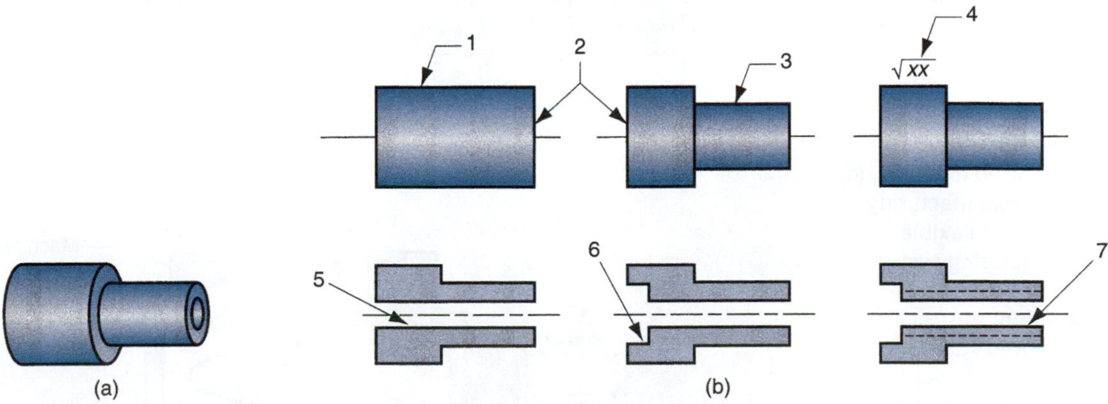

FIGURE 29.8 Composite part concept: (a) the composite part for a family of machined rotational parts, and (b) the individual features of the composite part. (Credit: *Fundamentals of Modern Manufacturing*, 4th Edition by Mikell P. Groover, 2010. Reprinted with permission of John Wiley & Sons, Inc.)

TABLE 29.2 Design features of the composite part in Figure 29.8 and the manufacturing operations required to create those features.

Label	Design Feature	Corresponding Manufacturing Operation
1	External cylinder	Turning
2	Face of cylinder	Facing
3	Cylindrical step	Turning
4	Smooth surface	External cylindrical grinding
5	Axial hole	Drilling
6	Counterbore	Bore, counterbore
7	Internal threads	Tapping

Machine Cell Designs Machine cells can be classified according to number of machines and level of automation. The possibilities are (a) single machine, (b) multiple machines with manual handling, (c) multiple machines with mechanized handling, (d) flexible manufacturing cell, or (e) flexible manufacturing system. These production cells are depicted in Figure 29.9.

The ***single machine cell*** has one machine that is manually operated. The cell would also include fixtures and tools to allow for feature and size variations within the part family produced by the cell. The machine cell required for the part family of Figure 29.8 would probably be of this type.

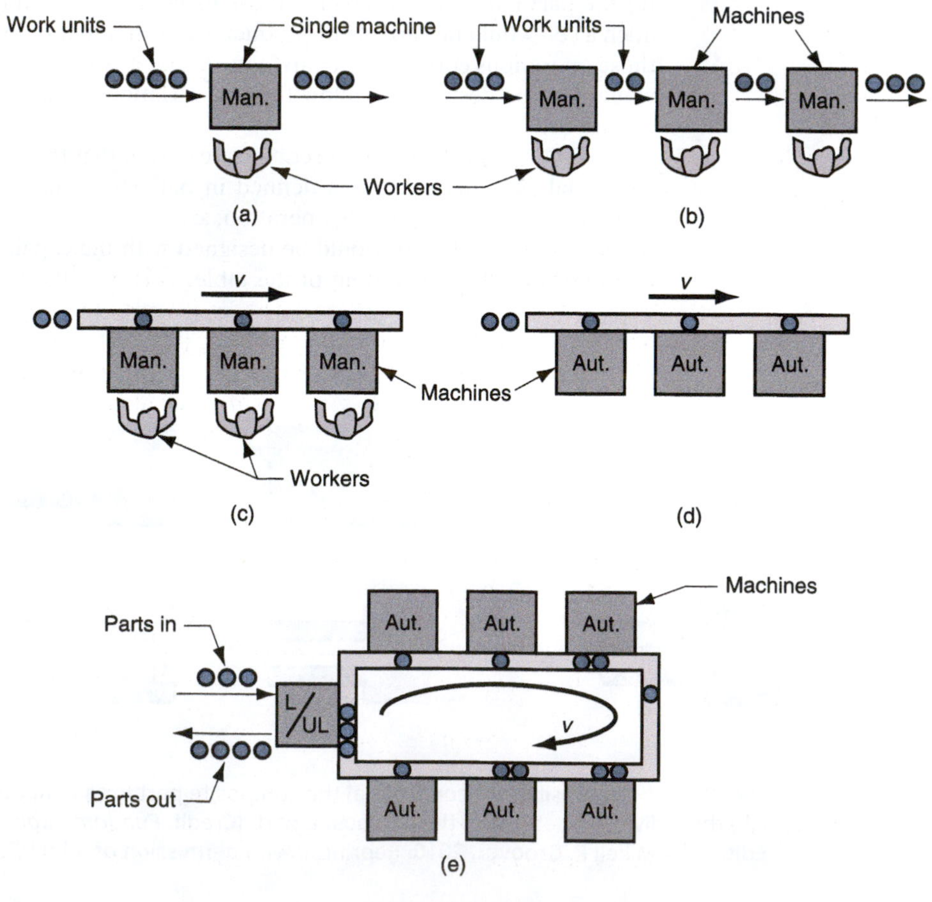

FIGURE 29.9 Types of group technology machine cells: (a) single machine, (b) multiple machines with manual handling, (c) multiple machines with mechanized handling, (d) flexible manufacturing cell, and (e) flexible manufacturing system. Key: Man = manual operation; Aut = automated station. (Credit: *Fundamentals of Modern Manufacturing*, 4th Edition by Mikell P. Groover, 2010. Reprinted with permission of John Wiley & Sons, Inc.)

Multiple machine cells have two or more manually operated machines. These cells are distinguished by the method of workpart handling in the cell, manual or mechanized. Manual handling means that parts are moved within the cell by workers, usually the machine operators. Mechanized handling refers to conveyorized transfer of parts from one machine to the next. This may be required by the size and weight of the parts made in the cell, or simply to increase production rate. Our sketch depicts the work flow as being a line; other layouts are also possible, such as U-shaped or loop.

Flexible manufacturing cells and ***flexible manufacturing systems*** consist of automated machines with automated handling. Given the special nature of these integrated manufacturing systems and their importance, we devote Section 29.3 to their discussion.

Benefits and Problems in Group Technology The use of machine cells and group technology provide substantial benefits to companies that have the discipline and perseverance to implement it. The potential benefits include the following: (1) GT promotes standardization of tooling, fixturing, and setups; (2) material handling is reduced because parts are moved within a machine cell rather than the entire factory; (3) production scheduling is simplified; (4) manufacturing lead time is reduced; (5) work-in-process is reduced; (6) process planning is simpler; (7) worker satisfaction usually improves working in a cell; and (8) higher quality work is accomplished.

Two main problems are encountered in implementing machine cells. One obvious problem is rearranging production machines in the plant into the appropriate machine cells. It takes time to plan and accomplish this rearrangement, and the machines are not producing during the changeover. The biggest problem in starting a GT program is identifying the part families. If the plant makes 10,000 different parts, reviewing all of the part drawings and grouping the parts into families are substantial tasks that consume a significant amount of time.

29.3 FLEXIBLE MANUFACTURING SYSTEMS AND CELLS

A flexible manufacturing system (FMS) is a highly automated GT machine cell, consisting of a group of processing stations (usually CNC machine tools), interconnected by an automated material handling and storage system, and controlled by an integrated computer system. An FMS is capable of processing a variety of different part styles simultaneously under NC program control at the different workstations.

An FMS relies on the principles of group technology. No manufacturing system can be completely flexible. It cannot produce an infinite range of parts or products. There are limits to how much flexibility can be incorporated into an FMS. Accordingly, a flexible manufacturing system is designed to produce parts (or products) within a range of styles, sizes, and processes. In other words, an FMS is capable of producing a single part family or a limited range of part families.

Flexible manufacturing systems vary in terms of number of machine tools and level of flexibility. When the system has only a few machines, the term ***flexible manufacturing cell*** (FMC) is sometimes used. Both cell and system are highly automated and computer controlled. The difference between an FMS and an FMC is not always clear, but it is sometimes based on the number of machines (workstations) included. The flexible manufacturing system consists of four or more machines, while a flexible manufacturing cell has three or fewer machines [7].

To qualify as being flexible, a manufacturing system should satisfy several criteria. The tests of flexibility in an automated production system are the capability to (1) process different part styles in a nonbatch mode, (2) accept changes in production schedule,

(3) respond gracefully to equipment malfunctions and breakdowns in the system, and (4) accommodate the introduction of new part designs. These capabilities are made possible by the use of a central computer that controls and coordinates the components of the system. The most important criteria are (1) and (2); criteria (3) and (4) are softer and can be implemented at various levels of sophistication.

29.3.1 INTEGRATING THE FMS COMPONENTS

An FMS consists of hardware and software that must be integrated into an efficient and reliable unit. It also includes human personnel. In this section we examine these components and how they are integrated.

Hardware Components FMS hardware includes workstations, a material handling system, and a central control computer. The workstations are CNC machines in a machining type system, plus inspection stations, parts cleaning and other stations, as required. A central chip conveyor system is often installed below floor level.

The material handling system is the means by which parts are moved between stations. The material handling system usually includes a limited capability to store parts. Handling systems suitable for automated manufacturing include roller conveyors, automated guided vehicles, and industrial robots. The most appropriate type depends on part size and geometry, as well as factors relating to economics and compatibility with other FMS components. Nonrotational parts are often moved in an FMS on pallet fixtures, so the pallets are designed for the particular handling system, and the fixtures are designed to accommodate the various part geometries in the family. Rotational parts are often handled by robots, if weight is not a limiting factor.

The handling system establishes the basic layout of the FMS. Five layout types can be distinguished: (1) in-line, (2) loop, (3) ladder, (4) open field, and (5) robot-centered cell. Types 1, 3, 4, and 5 are shown in Figure 29.10. Type 2 is shown in Figure 29.9(e). The ***in-line layout*** uses a linear transfer system to move parts between processing stations and load/unload station(s). The in-line transfer system is usually capable of two-directional movement; if not, then the FMS operates much like a transfer line, and the different part styles made on the system must follow the same basic processing sequence due to the one-direction flow. The ***loop layout*** consists of a conveyor loop with workstations located around its periphery. This configuration permits any processing sequence, because any station is accessible from any other station. This is also true for the ***ladder layout***, in which workstations are located on the rungs of the ladder. The ***open field layout*** is the most complex FMS configuration, and consists of several loops tied together. Finally, the robot-centered cell consists of a robot whose work volume includes the load/unload positions of the machines in the cell.

The FMS also includes a central computer that is interfaced to the other hardware components. In addition to the central computer, the individual machines and other components generally have microcomputers as their individual control units. The function of the central computer is to coordinate the activities of the components so as to achieve a smooth overall operation of the system. It accomplishes this function by means of software.

FMS Software and Control Functions FMS software consists of modules associated with the various functions performed by the manufacturing system. For example, one function involves downloading NC part programs to the individual machine tools; another function is concerned with controlling the material handling system; another is concerned with tool management; and so on. Table 29.3 lists the functions included in the operation of a typical FMS. Associated with each function is one or more

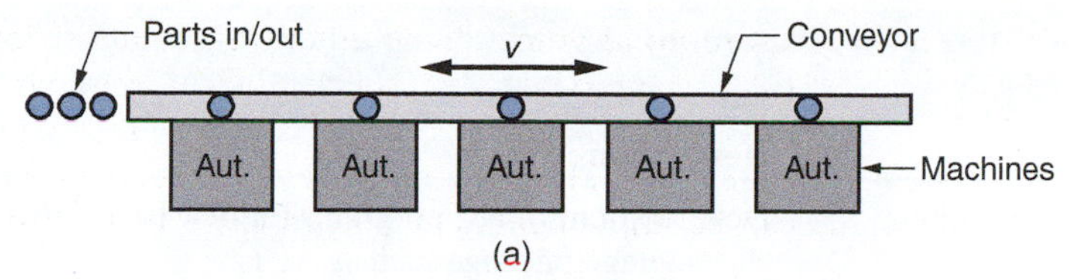

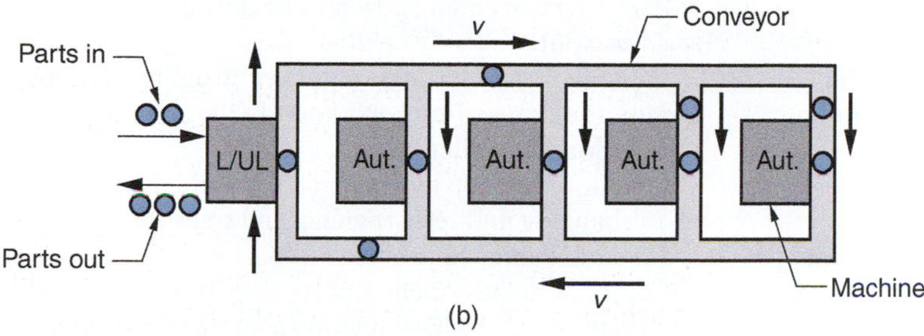

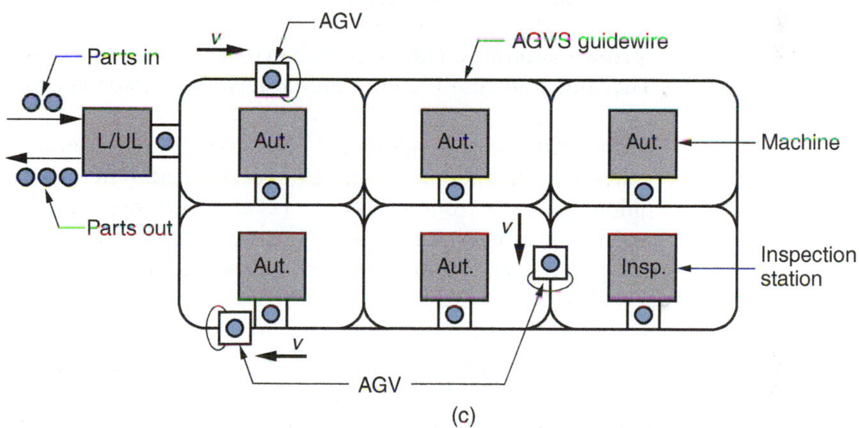

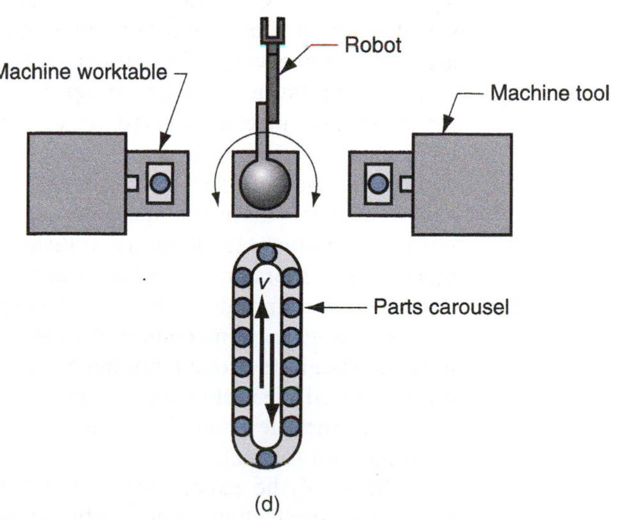

FIGURE 29.10 Four of the five FMS layout types: (a) in-line, (b) ladder, (c) open field, and (d) robot-centered cell. Key: Aut = automated station; L/UL = load/unload station; Insp = inspection station; AGV = automated guided vehicle; AGVS = automated guided vehicle system. (Credit: *Fundamentals of Modern Manufacturing*, 4th Edition by Mikell P. Groover, 2010. Reprinted with permission of John Wiley & Sons, Inc.)

TABLE 29.3 Typical computer functions implemented by application software modules in a flexible manufacturing system.

Function	Description
NC part programming	Development of NC programs for new parts introduced into the system. This includes a language package such as APT.
Production control	Product mix, machine scheduling, and other planning functions.
NC program download	Part program commands must be downloaded to individual stations from the central computer.
Machine control	Individual workstations require controls, usually computer numerical control.
Workpart control	Monitor status of each workpart in the system, status of pallet fixtures, orders on loading/unloading pallet fixtures.
Tool management	Functions include tool inventory control, tool status relative to expected tool life, tool changing and resharpening, and transport to and from tool grinding.
Transport control	Scheduling and control of workpart handling system.
System management	Compiles management reports on performance (utilization, piece counts, production rates, etc.). FMS simulation sometimes included.

software modules. Terms other than those in our table may be used in a given installation. The functions and modules are largely application specific.

Human Labor An additional component in the operation of a flexible manufacturing system or cell is human labor. Duties performed by human workers include: (1) loading and unloading parts from the system, (2) changing and setting cutting tools, (3) maintenance and repair of equipment, (4) NC part programming, (5) programming and operating the computer system, and (6) overall management of the system.

29.3.2 APPLICATIONS OF FLEXIBLE MANUFACTURING SYSTEMS

Flexible manufacturing systems are typically used for mid-volume, mid-variety production. If the part or product is made in high quantities with no style variations, then a transfer line or similar dedicated manufacturing system is most appropriate. If the parts are low volume with high variety, then a stand-alone NC machine or even manual methods would be more appropriate.

Flexible machining systems comprise the most common application of FMS technology. Owing to the inherent flexibilities and capabilities of computer numerical control, it is possible to connect several CNC machine tools to a small central computer, and to devise automated material handling methods for transferring parts between machines. Figure 29.11 shows a flexible machining system consisting of five CNC machining centers and an in-line transfer system to pick up parts from a central load/unload station and move them to the appropriate machining stations.

In addition to machining systems, other types of flexible manufacturing systems have also been developed, although the state of technology in these other processes has not permitted the rapid implementation that has occurred in machining. The other types of systems include assembly, inspection, and sheet-metal processing (punching, shearing, bending, and forming).

Most of the experience in flexible manufacturing systems has been gained in machining applications. For flexible machining systems, the benefits usually given are (1) higher machine utilization than a conventional machine shop—relative utilizations are 40% to 50% for conventional batch type operations and about 75% for a FMS due to better work handling, off-line setups, and improved scheduling; (2) reduced work-in-

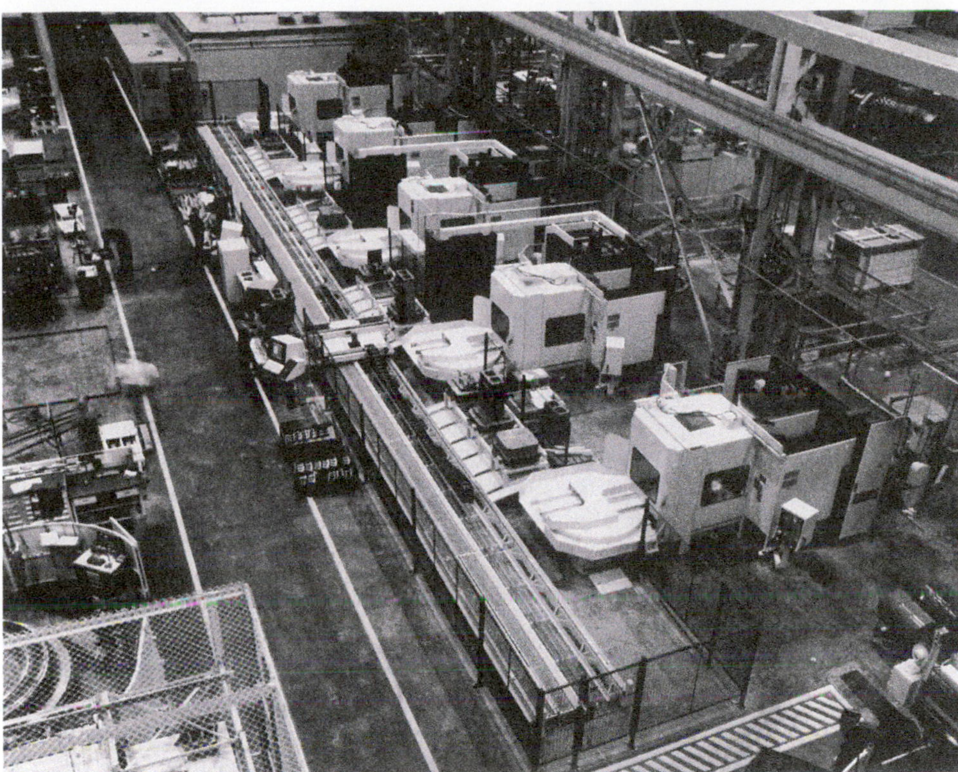

FIGURE 29.11 A five-station flexible manufacturing system. Photo courtesy of Cincinnati Milacron. (Credit: *Fundamentals of Modern Manufacturing*, 4th Edition by Mikell P. Groover, 2010. Reprinted with permission of John Wiley & Sons, Inc.)

process due to continuous production rather than batch production; (3) lower manufacturing lead times; and (4) greater flexibility in production scheduling.

29.4 LEAN PRODUCTION

In this section and the following, we cover systems that are broader in scope than the preceding topics in this chapter. Lean production is an approach that is applied to the entire factory, rather than to specific machines (numerical control) or groups of machines (cellular manufacturing and flexible manufacturing systems). ***Lean production*** can be defined as "an adaptation of mass production in which workers and work cells are made more flexible and efficient by adopting methods that reduce waste in all forms."[1] In a lean system, an attempt is made to identify and eliminate all activities that are wasteful of resources, which are the workers, tools and equipment, materials, time, and space in the factory. All activities in manufacturing can be classified into the following categories: (1) value-adding activities, which typically refer to the processing and assembly activities that add real value for the customer; (2) activities that support the value-adding activities, such as loading parts into a production machine; and (3) wasteful activities that do not add value nor do they support the value-adding activities. Examples of waste in manufacturing include production of defective parts, production of more parts than are needed (overproduction), workers waiting, excessive inventories, and unnecessary processing steps. Lean production aims to eliminate the wasteful activities in manufacturing so that only the value-adding and supporting activities are performed. The result is

[1] M. P. Groover, ***Automation, Production Systems, and Computer-Integrated Manufacturing*** [7], p. 834.

that fewer resources and less time are required, and the quality of the final product is improved. One of the foundations of lean production is just-in-time delivery of parts.

29.4.1 JUST-IN-TIME PRODUCTION SYSTEMS

A just-in-time (JIT) system is an approach to production scheduling in which materials or parts are delivered to the next workstation in the sequence of processing steps just prior to when those parts are needed at that station. When this approach is applied across the entire factory, it has the effect of minimizing work-in-process inventories and lead times. In addition, parts quality is improved because defective parts can force production to be stopped; thus, great emphasis is placed on avoiding defects in a JIT system.

Just-in-time procedures have proven most effective in high-volume repetitive manufacturing, such as the automobile industry [12]. The potential for in-process inventory accumulation in this type of manufacturing is significant because both the quantities of products and the number of components per product are large. A just-in-time system produces exactly the right number of each component required to satisfy the next operation in the manufacturing sequence just when that component is needed—"just in time." The ideal batch size is one part. As a practical matter, more than one part are produced at a time, but the batch size is kept small. Under JIT, producing too many units is to be avoided as much as producing too few units.

JIT requires a ***pull system*** of production control, in which the order to produce parts at a given workstation comes from the downstream station that uses those parts. As the supply of parts becomes exhausted at a given station, it "places an order" at the upstream workstation to replenish the supply. This order provides the authorization for the upstream station to produce the needed parts. This procedure, repeated at each workstation throughout the plant, has the effect of pulling parts through the production system. By contrast, a ***push system*** of production operates by supplying parts to each station in the plant, in effect driving the work from upstream stations to downstream stations. The risk in a push system is to overload the factory by scheduling more work than it can handle. This results in large queues of parts in front of machines that cannot keep up with arriving work.

One famous pull system is the ***kanban*** system used by Toyota Motors. Kanban (pronounced kahn-bahn) is a Japanese word meaning ***card***. The kanban system of production control is based on the use of cards to authorize production and work flow in the plant. There are two types of kanbans: (1) production kanbans, and (2) transport kanbans. A ***production kanban*** authorizes production of a batch of parts. The parts are placed in containers, so the batch must consist of just enough parts to fill the container. Production of additional parts is not permitted. The ***transport kanban*** authorizes movement of the container of parts to the next station in the sequence.

Refer to Figure 29.12 as we explain how two workstations, one that feeds the other, operate in a kanban system. The figure shows four stations, but B and C are the stations of interest here. Station B is the supplier in this pair, and station C is the consumer. Station C supplies downstream station D. B is supplied by upstream station A. When station C starts work on a full container of parts, a worker removes the transport kanban from that container and takes it back to B. The worker finds a full container of parts at B that have just been produced, removes the production kanban from that container, and places it on a rack at B. The worker then places the transport kanban in the full container, which authorizes its movement to station C. The production kanban on the rack at station B authorizes production of a new batch of parts. Station B produces more than one part style, perhaps for several other downstream stations in addition to C. The scheduling of work is determined by the order in which the production kanbans are placed on the rack.

FIGURE 29.12 Operation of a kanban system between workstations. (Credit: *Fundamentals of Modern Manufacturing*, 4th Edition by Mikell P. Groover, 2010. Reprinted with permission of John Wiley & Sons, Inc.)

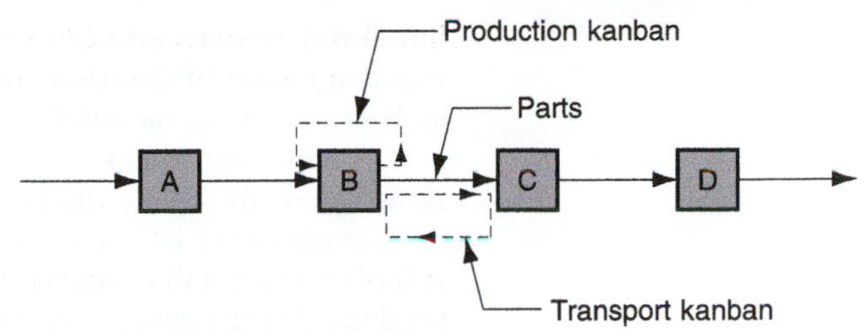

The kanban pull system between stations A and B and between stations C and D operates the same as it does between stations B and C. This system of production control eliminates unnecessary paperwork; the cards are used over and over rather than generating new production and transport orders every cycle. An apparent disadvantage is the considerable labor involvement in material handling (moving the cards and containers between stations); however, it is claimed that this promotes teamwork and cooperation among workers.

Success of just-in-time production requires near perfection in on-time delivery, parts quality, and equipment reliability. The small lot sizes and parts buffers used in JIT require parts to be delivered before stock-outs occur at downstream stations. Otherwise, production must be suspended at these stations for lack of parts. If the delivered parts are defective, they cannot be used in assembly. This tends to promote zero defects in parts fabrication. Workers inspect their own output to make sure it is right before it proceeds to the next operation. Low work-in-process also requires reliable production equipment. Machines that break down cannot be tolerated in a JIT production system. This emphasizes the need for reliable equipment designs and preventive maintenance.

Just-in-time extends to the material and component suppliers of the company. Suppliers must be held to the same standards of on-time delivery, zero defects, and other JIT requisites as the company itself. Some of the vendor policies used by companies to implement JIT include (1) reducing the total number of suppliers, (2) selecting suppliers with proven records for meeting quality and delivery standards, (3) establishing long-term partnerships with suppliers, and (4) selecting suppliers that are located near the company's manufacturing plant.

29.4.2 OTHER APPROACHES IN LEAN PRODUCTION

In addition to just-in-time deliveries, lean production systems include several additional approaches that are designed to reduce waste and increase efficiency. Several of these are described here.

Worker Involvement and Continuous Improvement Workers in a factory that practices lean production are given greater responsibilities and are trained to be flexible and versatile in the work they can do. This allows them to be assigned to a wide variety of tasks. The number of job classifications in the factory and work rules that forbid workers from performing work outside their classifications are minimized in a lean system. Workers are trained to inspect the quality of their own output and to deal with minor technical problems with their production equipment so that major stoppages are avoided. Also, workers participate in worker teams to solve continuous improvement problems faced by the company. As indicated in Section 28.2.4, continuous improvement refers to a constant search for ways to reduce cost, improve quality, and increase productivity. Workers can make important contributions to these improvement programs because they are intimately familiar with the operations.

Small Batch Sizes and Setup Reduction Two requirements for minimizing inventories in a just-in-time system are small batch sizes and short setup times. Smaller batch sizes cannot be economically justified unless setup times between batches are reduced. Some of the approaches used to reduce setup time include: (1) performing as much of the setup as possible while the previous job is still running; (2) using quick-acting clamping devices instead of bolts and nuts; (3) eliminating or minimizing adjustments in the setup; and (4) using group technology and cellular manufacturing so that similar part styles are produced on the same equipment.

Stop the Process Another approach in lean production is to design production machines that stop automatically when they operate abnormally. Abnormal operation could mean defective parts being made, broken or worn tooling, or producing more parts than required by the next production machine in the sequence. When an automatic stop device is activated, it draws attention to the machine, thus demanding that corrective action be taken to address the problem.

Error Prevention Mistakes and errors are common in manufacturing. Examples include processing steps omitted during the production sequence, components not added in final assembly, use of improper tools, and workparts located incorrectly in fixtures. In a lean system, design features are incorporated into the products and inexpensive devices are implemented at the workstations to prevent these kinds of mistakes.

Total Productive Maintenance It was mentioned in the previous section that reliable equipment is an important prerequisite of a just-in-time production system. Total productive maintenance (TPM) uses several approaches whose goal is to minimize lost production time due to equipment malfunctions. The approaches include: (1) workers performing routine maintenance and cleaning of their equipment; (2) preventive maintenance, whose purpose is to periodically replace worn components and tools to avoid equipment stoppages; and (3) predictive maintenance, which attempts to anticipate potential problems using computer monitoring and other predictive techniques. A successful TPM program maximizes equipment utilization by avoiding emergency breakdowns that disrupt production.

29.5 COMPUTER-INTEGRATED MANUFACTURING

Distributed computer networks are widely used in modern manufacturing plants to monitor and/or control the systems described in this chapter. Even though some of the operations are manually accomplished (e.g., manned cells), computer systems are utilized for production scheduling, data collection, recordkeeping, performance tracking, and other information-related functions. In the more automated systems (e.g., CNC machine tools, flexible manufacturing cells), computers directly control the operations. The term ***computer-integrated manufacturing*** (CIM) refers to the pervasive use of computer systems throughout the organization, not only to monitor and control the operations, but also to design the product, plan the manufacturing processes, and accomplish the business functions related to production. One might say that CIM is the ultimate integrated manufacturing system.

To begin, let us identify four general functions that have to be accomplished in most manufacturing enterprises: (1) product design, (2) manufacturing planning, (3) manufacturing control, and (4) business functions. Product design is usually an iterative process that includes recognition of a need for a product, problem definition, creative synthesis of a solution, analysis and optimization, evaluation, and documentation. The overall quality of the resulting design is likely to be the most important factor upon which the commercial success of a product depends. In addition, a very significant portion of the final cost of the product is determined by decisions made during product design. Manufacturing planning is concerned with converting the engineering drawings and specifications that define the product design into a plan for producing the product. Manufacturing planning includes decisions on which parts will be purchased (the "make-or-buy decision"), how each "make" part will be produced, the equipment that will be used, how the work will be scheduled, and so on. Most of these decisions are discussed in Chapter 28. Manufacturing control includes not only control of the individual processes and equipment in the plant, but also the supporting functions such as quality control and inspection, discussed in Chapter 30. Finally, the business functions include order entry, cost accounting, payroll, customer billing, and other business-oriented information activities related to manufacturing.

Computer systems play an important role in these four general functions, and their integration within the organization is a distinguishing feature of computer-integrated manufacturing, as depicted in Figure 29.13. Computer systems associated with product design are called CAD systems (for computer-aided design). Design systems and software include geometric modeling, engineering analysis such as finite element modeling, design review and evaluation, and automated drafting. Computer systems that support manufacturing planning are called CAM systems (for computer-aided manufacturing) and include computer-aided process planning, NC part programming, production scheduling, and manufacturing resource planning. Manufacturing control systems include those used in process control, shop floor control, inventory control, and computer-aided inspection for quality control. And computerized business systems are used for order entry, customer billing, and other business functions. Customer orders are entered by the company's sales force or by the customers themselves into the computerized order entry system. The orders include product specifications that provide the inputs to the design department. Based on these inputs, new products are designed on

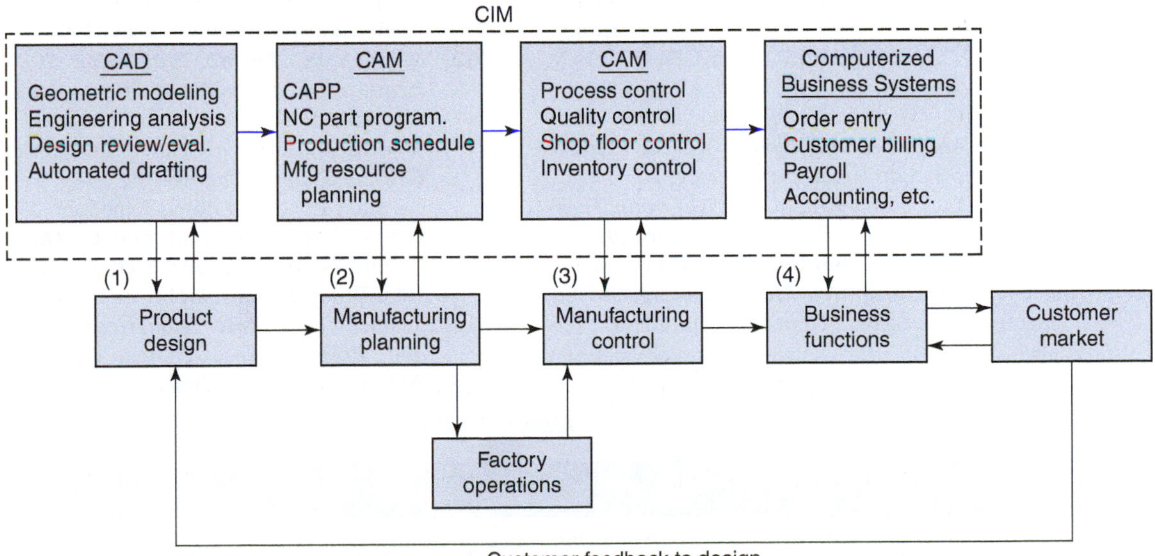

FIGURE 29.13 Four general functions in a manufacturing organization and how computer-integrated manufacturing systems support these functions. (Credit: *Fundamentals of Modern Manufacturing*, 4th Edition by Mikell P. Groover, 2010. Reprinted with permission of John Wiley & Sons, Inc.)

the company's CAD system. The design details serve as inputs to the manufacturing engineering group, where computer-aided process planning, computer-aided tool design, and related functions are performed in advance of actual production. The output from manufacturing engineering provides much of the input data required for manufacturing resource planning and production scheduling. Thus, computer-integrated manufacturing provides the information flows required to accomplish the actual production of the product.

Today, computer-integrated manufacturing is implemented in many companies using ***enterprise resource planning*** (ERP), an extension of manufacturing resource planning that organizes and integrates the information flows in a company through a single, central data base. The functions covered by ERP spread well beyond manufacturing operations; they include sales, marketing, purchasing, logistics, distribution, inventory control, finance, and human resources. ERP users within a company access and interact with the system using personal computers at their own workplaces, whether they are located in offices or in the factory.

REFERENCES

[1] Black, J. T. ***The Design of the Factory with a Future***. McGraw-Hill, New York, 1990.

[2] Black, J. T. "An Overview of Cellular Manufacturing Systems and Comparison to Conventional Systems," ***Industrial Engineering***, November 1983, pp. 36–84.

[3] Bollinger, J. G., and DuffieN. A. ***Computer Control of Machines and Processes***. Addison-Wesley Longman, Inc., New York, 1989.

[4] Chang, C-H, and Melkanoff, M. A. ***NC Machine Programming and Software Design***, 3rd ed. Prentice-Hall, Inc., Upper Saddle River, New Jersey, 2005.

[5] Chang, T-C, Wysk, R. A., and Wang, H-P. ***Computer-Aided Manufacturing***, 3rd ed. Prentice Hall, Upper Saddle River, New Jersey, 2005.

[6] Gallagher, C. C., and Knight, W. A. ***Group Technology***, Butterworth & Co., Ltd., London, 1973.

[7] Groover, M. P. ***Automation, Production Systems, and Computer Integrated Manufacturing***, 3rd ed. Pearson Prentice-Hall, Upper Saddle River, New Jersey, 2008.

[8] Ham, I., Hitomi, K., and Yoshida, T. ***Group Technology***. Kluwer Nijhoff Publishers, Hingham, Massachusetts, 1985.

[9] Houtzeel, A. "The Many Faces of Group Technology," ***American Machinist***, January 1979, pp. 115–120.

[10] Luggen, W. W. ***Flexible Manufacturing Cells and Systems***. Prentice Hall, Inc., Englewood Cliffs, New Jersey, 1991.

[11] Maleki, R. A. ***Flexible Manufacturing Systems: The Technology and Management***. Prentice Hall, Inc., Englewood Cliffs, New Jersey, 1991.

[12] Monden, Y. ***Toyota Production System***, 3rd ed. Engineering and Management Press, Norcross, Georgia, 1998.

[13] Moodie, C., Uzsoy, R., and Yih, Y. ***Manufacturing Cells: A Systems Engineering View***. Taylor & Francis, Ltd., London, United Kingdom, 1995.

[14] Parsai, H., Leep, H., and Jeon, G. ***The Principles of Group Technology and Cellular Manufacturing***. John Wiley & Sons, Hoboken, New Jersey, 2006.

[15] SeamesW. ***Computer Numerical Control, Concepts and Programming***. Delmar-Thomson Learning, Albany, New York, 2002.

REVIEW QUESTIONS

29.1. Define the term *manufacturing system*.

29.2. Identify and briefly describe the three basic components of a numerical control system.

29.3. What is the difference between point-to-point and continuous path in a motion control system?

29.4. What is the difference between absolute positioning and incremental positioning?

29.5. What is the difference between an open-loop positioning system and a closed-loop positioning system?

29.6. Under what circumstances is a closed-loop positioning system preferable to an open-loop system?
29.7. Explain the operation of an optical encoder.
29.8. What is manual data input in NC part programming?
29.9. Identify some of the non-machine tool applications of numerical control.
29.10. What are some of the benefits usually cited for NC compared to using manual alternative methods?
29.11. Define the term *group technology*.
29.12. What is a part family?
29.13. Define the term *cellular manufacturing*.
29.14. What is the composite part concept in group technology?
29.15. What is a flexible manufacturing system?
29.16. What are the criteria that should be satisfied to make an automated manufacturing system flexible?
29.17. Name some of the FMS software and control functions.
29.18. What are the advantages of FMS technology, compared to conventional batch operations?
29.19. Define the term *lean production*.
29.20. Identify the principal objective in a just-in-time production system.
29.21. How is a pull system distinguished from a push system in production and inventory control?
29.22. Define the term *computer-integrated manufacturing*.

PROBLEMS

29.1. A leadscrew with a 7.5 mm pitch drives a worktable in an NC positioning system. The leadscrew is powered by a stepping motor, which has 200 step angles. The worktable is programmed to move a distance of 120 mm from its present position at a travel speed of 300 mm/min. Determine (a) the number of pulses required to move the table the specified distance and (b) the required motor speed and pulse rate to achieve the desired table speed.

29.2. A stepping motor has 200 step angles. Its output shaft is directly coupled to leadscrew with pitch = 0.250 in. A worktable is driven by the leadscrew. The table must move a distance of 5.00 in from its present position at a travel speed of 20.0 in/min. Determine (a) the number of pulses required to move the table the specified distance and (b) the required motor speed and pulse rate to achieve the specified table speed.

29.3. A stepping motor with 100 step angles is coupled to a leadscrew through a gear reduction of 9:1 (9 rotations of the motor for each rotation of the leadscrew). The leadscrew has 5 threads/in. The worktable driven by the leadscrew must move a distance = 10.00 in at a feed rate of 30.0 in/min. Determine (a) number of pulses required to move the table, and (b) the required motor speed and pulse rate to achieve the desired table speed.

29.4. An NC machine tool table is powered by a servomotor, leadscrew, and optical encoder. The leadscrew has a pitch = 5.0 mm and is connected to the motor shaft with a gear ratio of 16:1 (16 turns of the motor for each turn of the leadscrew). The optical encoder is connected directly to the leadscrew and generates 200 pulses/rev of the leadscrew. The table must move a distance = 100 mm at a feed rate = 500 mm/min. Determine (a) the pulse count received by the control system to verify that the table has moved exactly 100 mm; and (b) the pulse rate and (c) motor speed that correspond to the feed rate of 500 mm/min.

29.5. The worktable of a numerical control machine tool is driven by a closed-loop positioning system that consists of a servomotor, leadscrew, and optical encoder. The leadscrew has 4 threads/in and is coupled directly to the motor shaft (gear ratio = 1:1). The optical encoder generates 200 pulses per motor revolution. The table has been programmed to move a distance of 7.5 in at a feed rate = 20.0 in/min. (a) How many pulses are received by the control system to verify that the table has moved the programmed distance? What are (b) the pulse rate and (c) motor speed that correspond to the specified feed rate?

29.6. A leadscrew coupled directly to a dc servomotor is used to drive one of the table axes of an NC milling machine. The leadscrew has 5 threads/in. The optical encoder attached to the leadscrew emits 100 pulses/rev of the leadscrew. The motor rotates at a maximum speed of 800 rev/min. Determine (a) the frequency of the pulse train emitted by the optical encoder when the servomotor operates at maximum speed and (b) the travel speed of the table at the maximum rpm of the motor.

29.7. Solve the previous problem only the servomotor is connected to the leadscrew through a gear box whose reduction ratio = 12:1 (12 revolutions of the motor for each revolution of the leadscrew).

29.8. A leadscrew connected directly to a DC servomotor is the drive system for a positioning table. The leadscrew pitch = 4 mm. The optical encoder attached to the leadscrew emits 250 pulses/rev of the leadscrew. Determine (a) the frequency of the pulse train emitted by the optical encoder when the servomotor operates at 14 rev/s and (b) the travel speed of the table at the operating speed of the motor.

29.9. A milling operation is performed on an NC machining center. Total travel distance = 300 mm in a direction parallel to one of the axes of the worktable. Cutting speed = 1.25 m/s and chip load = 0.05 mm. The end milling cutter has four teeth and its diameter = 20.0 mm. The axis uses a DC servomotor whose output shaft is coupled to a leadscrew with pitch = 6.0 mm. The feedback sensing device connected to the leadscrew is an optical encoder that emits 250 pulses per revolution. Determine (a) feed rate and time to complete the cut, and (b) rotational speed of the motor and the pulse rate of the encoder at the feed rate indicated.

29.10. An end milling operation is carried out along a straight-line path that is 325 mm long. The cut is in a direction parallel to the x-axis on an NC machining center. Cutting speed = 30 m/min and chip load = 0.06 mm. The end milling cutter has two teeth and its diameter = 16.0 mm. The x-axis uses a DC servomotor connected directly to a leadscrew whose pitch = 6.0 mm. The feedback sensing device is an optical encoder that emits 400 pulses per revolution. Determine (a) feed rate and time to complete the cut, and (b) rotational speed of the motor and the pulse rate of the encoder at the feed rate indicated.

29.11. A DC servomotor drives the x-axis of a NC milling machine table. The motor is coupled to the table leadscrew using a 4:1 gear reduction (4 turns of the motor for each turn of the leadscrew). The leadscrew pitch = 6.25 mm. An optical encoder is connected to the leadscrew. The optical encoder emits 500 pulses per revolution. To execute a certain programmed instruction, the table must move from point (x = 87.5 mm, y = 35.0) to point (x = 25.0 mm, y = 180.0 mm) in a straight-line trajectory at a feed rate = 200 mm/min. For the x-axis only, determine (a) the rotational speed of the motor and (b) frequency of the pulse train emitted by the optical encoder at the desired feed rate.

29.12. A DC servomotor drives the y-axis of a NC milling machine table. The motor is coupled to the table leadscrew with a gear reduction of 2:1 (2 turns of the motor shaft for each single rotation of the leadscrew). There are 2 threads per cm in the leadscrew. An optical encoder is directly connected to the leadscrew (1:1 gear ratio). The optical encoder emits 100 pulses per revolution. To execute a certain programmed instruction, the table must move from point (x = 25.0 mm, y = 28.0) to point (x = 155.0 mm, y = 275.0 mm) in a straight-line trajectory at a feed rate = 200 mm/min. For the y-axis only, determine (a) the rotational speed of the motor and (b) frequency of the pulse train emitted by the optical encoder at the desired feed rate.

30 QUALITY CONTROL AND INSPECTION

Chapter Contents

30.1 Product Quality

30.2 Process Capability and Tolerances

30.3 Statistical Process Control
30.3.1 Control Charts for Variables
30.3.2 Control Charts for Attributes
30.3.3 Interpreting the Charts

30.4 Quality Programs in Manufacturing
30.4.1 Total Quality Management
30.4.2 Six Sigma
30.4.3 ISO 9000

30.5 Inspection Principles

30.6 Modern Inspection Technologies
30.6.1 Coordinate Measuring Machines
30.6.2 Machine Vision
30.6.3 Other Noncontact Inspection Techniques

Traditionally, ***quality control*** (QC) has been concerned with detecting poor quality in manufactured products and taking corrective action to eliminate it. QC has often been limited to inspecting the product and its components, and deciding whether the dimensions and other features conformed to design specifications. If they did, the product was shipped. The modern view of quality control encompasses a broader scope of activities, including various quality programs such as statistical process control and Six Sigma as well as modern inspection technologies such as coordinate measuring machines and machine vision. In this chapter, we discuss these and other quality and inspection topics that are relevant today in modern manufacturing operations. Let us begin our coverage by defining product quality.

30.1 PRODUCT QUALITY

The American Society for Quality (ASQ) defines quality as "the totality of features and characteristics of a product or service that bear on its ability to satisfy given needs" [2]. In a manufactured product, quality has two aspects [4]: (1) product features and (2) freedom from deficiencies. ***Product features*** are the characteristics of the product that result from design. They are the functional and aesthetic features of the item intended to appeal to and provide satisfaction to the customer. In an automobile, these features include the size of the car, its styling, the finish of the body, gas mileage, reliability, reputation of the manufacturer, and similar aspects. They also include the available options for the customer to choose. The sum of a product's features usually defines its ***grade***, which relates to the level in the market at which the product is aimed. Cars (and most other products) come in different grades. Some cars provide basic transportation because that is what some customers want, while others are upscale for consumers willing to spend more to own a "better product." The features of a product are decided in design, and they generally determine the inherent cost of the product. Superior features and more of them mean higher cost.

Freedom from deficiencies means that the product does what it is supposed to do (within the limitations of its design features), that it is absent of defects and out-of-tolerance conditions, and that no parts are missing. This aspect of quality includes components and subassemblies of the product as well as the product itself. Freedom from deficiencies means conforming to design specifications, which is accomplished in manufacturing. Although the inherent cost to make a product is a function of its design, minimizing the product's cost to the lowest possible level within the limits set by its design is largely a matter of avoiding defects, tolerance deviations, and other errors during production. Costs of these deficiencies make a long list indeed: scrapped parts, larger lot sizes for scrap allowances, rework, reinspection, sortation, customer complaints and returns, warranty costs and customer allowances, lost sales, and lost goodwill in the marketplace.

Thus, product features are the aspect of quality for which the design department is responsible. Product features determine to a large degree the price that a company can charge for its products. Freedom from deficiencies is the quality aspect for which the manufacturing departments are responsible. The ability to minimize these deficiencies has an important influence on the cost of the product. These generalities oversimplify the way things work, because the responsibility for high quality extends well beyond the design and manufacturing functions in an organization.

30.2 PROCESS CAPABILITY AND TOLERANCES

In any manufacturing operation, variability exists in the process output. In a machining operation, which is one of the most accurate processes, the machined parts may appear to be identical, but close inspection reveals dimensional differences from one part to the next. Manufacturing variations can be divided into two types: random and assignable.

Random variations are caused by many factors: human variability within each operation cycle, variations in raw materials, machine vibration, and so on. Individually, these factors may not amount to much, but collectively the errors can be significant enough to cause trouble unless they are within the tolerances for the part. Random variations typically form a normal statistical distribution. The output of the process tends to cluster about the mean value, in terms of the product's quality characteristic of interest (e.g., length, diameter). A large proportion of the population of parts is centered around the mean, with fewer parts away from the mean. When the only variations in the process are of this type, the process is said to be in ***statistical control***. This kind of variability will continue so long as the process is operating normally. It is when the process deviates from this normal operating condition that variations of the second type appear.

Assignable variations indicate an exception from normal operating conditions. Something has occurred in the process that is not accounted for by random variations. Reasons for assignable variations include operator mistakes, defective raw materials, tool failures, machine malfunctions, and so on. Assignable variations in manufacturing usually betray themselves by causing the output to deviate from the normal distribution. The process is no longer in statistical control.

Process capability relates to the normal variations inherent in the output when the process is in statistical control. By definition, ***process capability*** equals ± 3 standard deviations about the mean output value (a total of 6 standard deviations),

$$PC = \mu \pm 3\sigma \tag{30.1}$$

where PC = process capability; μ = process mean, which is set at the nominal value of the product characteristic when bilateral tolerancing is used (Section 4.1.1); and σ = standard deviation of the process. Assumptions underlying this definition are (1) steady-state operation has been achieved and the process is in statistical control, and (2) the output is normally distributed. Under these assumptions, 99.73% of the parts produced will have output values that fall within $\pm 3.0\sigma$ of the mean. The process capability of a given manufacturing operation is not always known, and experiments must be conducted to assess it.

The issue of tolerances is critical to product quality. Design engineers tend to assign dimensional tolerances to components and assemblies based on their judgment of how size variations will affect function and performance. Conventional wisdom is that closer tolerances beget better performance. Small regard is given to the cost resulting from tolerances that are unduly tight relative to process capability. As tolerance is reduced, the cost of achieving the tolerance tends to increase because additional processing steps may be needed and/or more accurate and expensive production machines may be required. The design engineer should be aware of this relationship. Although primary consideration must be given to function in assigning tolerances, cost is also a factor, and any relief that can be given to the manufacturing departments in the form of wider tolerances without sacrificing product function is worthwhile.

Design tolerances must be compatible with process capability. It serves no useful purpose to specify a tolerance of ± 0.025 mm (± 0.001 in) on a dimension if the process capability is significantly wider than ± 0.025 mm (± 0.001 in). Either the tolerance should be opened further (if design functionality permits), or a different manufacturing process should be selected. Ideally, the specified tolerance should be greater than the process capability. If function and available processes prevent this, then a sortation operation must be included in the manufacturing sequence to inspect every unit and separate those that meet specification from those that do not.

Design tolerances can be specified as being equal to process capability as defined in Eq. (30.1). The upper and lower boundaries of this range are known as the ***natural tolerance limits***. When design tolerances are set equal to the natural tolerance limits, then 99.73% of the parts will be within tolerance and 0.27% will be outside the limits. Any increase in the tolerance range will reduce the percentage of defective parts.

The desire to achieve very-low-fraction defect rates has led to the popular notion of "six sigma" limits in quality control. Achieving six sigma limits virtually eliminates defects in a manufactured product, assuming the process is maintained within statistical control. As we shall see later in the chapter, Six Sigma quality programs do not quite live up to their names. Before addressing that issue, let us discuss a widely used quality control technique: statistical process control.

30.3 STATISTICAL PROCESS CONTROL

Statistical process control (SPC) involves the use of various statistical methods to assess and analyze variations in a process. SPC methods include simply keeping records of the production data, histograms, process capability analysis, and control charts. Control charts are the most widely used SPC method, and this section will focus on them.

The underlying principle in control charts is that the variations in any process divide into two types, as discussed in Section 30.2: (1) random variations, which are the only variations present if the process is in statistical control, and (2) assignable variations that indicate a departure from statistical control. It is the objective of a control chart to identify when the process has gone out of statistical control, thus signaling that some corrective action should be taken.

A ***control chart*** is a graphical technique in which statistics computed from measured values of a certain process characteristic are plotted over time to determine if the process remains in statistical control. The general form of the control chart is

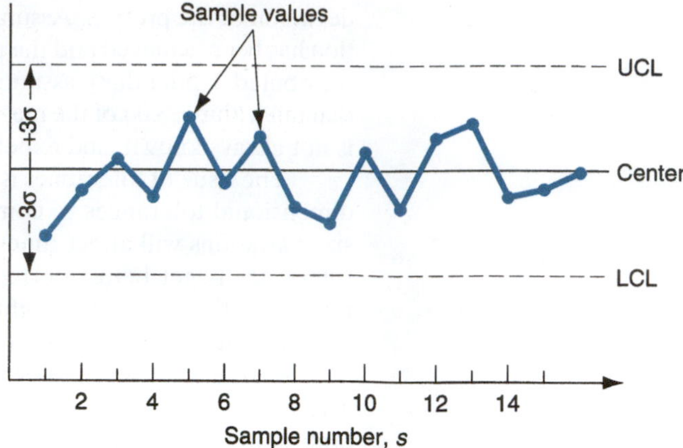

FIGURE 30.1 Control chart. (Credit: *Fundamentals of Modern Manufacturing*, 4th Edition by Mikell P. Groover, 2010. Reprinted with permission of John Wiley & Sons, Inc.)

illustrated in Figure 30.1. The chart consists of three horizontal lines that remain constant over time: a center, a lower control limit (LCL), and an upper control limit (UCL). The center is usually set at the nominal design value. The upper and lower control limits are generally set at ±3 standard deviations of the sample mean.

It is highly unlikely that a random sample drawn from the process will lie outside the upper or lower control limits while the process is in statistical control. Thus, if it happens that a sample value does fall outside these limits, it is interpreted to mean that the process is out of control. Therefore, an investigation is undertaken to determine the reason for the out-of-control condition, with appropriate corrective action to eliminate the condition. By similar reasoning, if the process is found to be in statistical control, and there is no evidence of undesirable trends in the data, then no adjustments should be made because they would introduce an assignable variation to the process. The philosophy, "If it ain't broke, don't fix it," is applicable to control charts.

There are two basic types of control charts: (1) control charts for variables and (2) control charts for attributes. Control charts for variables require a measurement of the quality characteristic of interest. Control charts for attributes simply require a determination of whether a part is defective or how many defects there are in the sample.

30.3.1 CONTROL CHARTS FOR VARIABLES

A process that is out of statistical control manifests this condition in the form of significant changes in process mean and/or process variability. Corresponding to these possibilities, there are two principal types of control charts for variables: $\bar{x}$ chart and R chart. The ***$\bar{x}$ chart*** (call it "x bar chart") is used to plot the average measured value of a certain quality characteristic for each of a series of samples taken from the production process. It indicates how the process mean changes over time. The ***R chart*** plots the range of each sample, thus monitoring the variability of the process and indicating whether it changes over time.

A suitable quality characteristic of the process must be selected as the variable to be monitored on the $\bar{x}$ and R charts. In a mechanical process, this might be a shaft diameter or other critical dimension. Measurements of the process itself must be used to construct the two control charts.

With the process operating smoothly and absent of assignable variations, a series of samples (e.g., $m = 20$ or more is generally recommended) of small size (e.g., $n = 4$, 5, or 6 parts per sample) are collected and the characteristic of interest is measured for

TABLE 30.1 Constants for the $\bar{x}$ and R charts.

Sample Size n	$\bar{x}$ Chart A_2	R Chart D_3	R Chart D_4
3	1.023	0	2.574
4	0.729	0	2.282
5	0.577	0	2.114
6	0.483	0	2.004
7	0.419	0.076	1.924
8	0.373	0.136	1.864
9	0.337	0.184	1.816
10	0.308	0.223	1.777

Source: [11].

each part. The following procedure is used to construct the center, LCL, and UCL for each chart:

1. Compute the mean $\bar{x}$ and range R for each of the m samples.
2. Compute the grand mean $\bar{\bar{x}}$, which is the mean of the $\bar{x}$ values for the m samples; this will be the center for the $\bar{x}$ chart.
3. Compute $\bar{R}$, which is the mean of the R values for the m samples; this will be the center for the R chart.
4. Determine the upper and lower control limits, UCL and LCL, for the $\bar{x}$ and R charts. The approach is based on statistical factors tabulated in Table 30.1 that have been derived specifically for these control charts. Values of the factors depend on sample size n. For the $\bar{x}$ chart:

$$\text{LCL} = \bar{\bar{x}} - A_2\bar{R} \quad \text{and} \quad \text{UCL} = \bar{\bar{x}} - A_2\bar{R} \tag{30.2}$$

and for the R chart:

$$\text{LCL} = D_3\bar{R} \quad \text{and} \quad \text{UCL} = D_4\bar{R} \tag{30.3}$$

Example 30.1 $\bar{x}$ and R Charts

Eight samples ($m = 8$) of size 4 ($n = 4$) have been collected from a manufacturing process that is in statistical control, and the dimension of interest has been measured for each part. It is desired to determine the values of the center, LCL, and UCL for $\bar{x}$ and R charts. The calculated $\bar{x}$ values (units are centimeters) for the eight samples are 2.008, 1.998, 1.993, 2.002, 2.001, 1.995, 2.004, and 1.999. The calculated R values (cm) are, respectively, 0.027, 0.011, 0.017, 0.009, 0.014, 0.020, 0.024, and 0.018.

Solution: The calculation of $\bar{x}$ and R values above comprise step 1 in our procedure. In step 2, we compute the grand mean of the sample averages.

$$\bar{\bar{x}} = (2.008 + 1.998 + \cdots + 1.999)/8 = 2.000$$

In step 3, the mean value of R is computed.

$$\bar{R} = (0.027 + 0.011 + \cdots + 0.018)/8 = 0.0175$$

In step 4, the values of LCL and UCL are determined based on factors in Table 30.1. First, using Eq. (30.2) for the $\bar{x}$ chart,

$$\text{LCL} = 2.000 - 0.729(0.0175) = 1.9872$$
$$\text{UCL} = 2.000 + 0.729(0.0175) = 2.0128$$

and for the R chart using Eq. (30.3),

$$\text{LCL} = 0(0.0175) = 0$$
$$\text{UCL} = 2.282(0.0175) = 0.0399$$

The two control charts are constructed in Figure 30.2 with the sample data plotted in the charts. ■

30.3.2 CONTROL CHARTS FOR ATTRIBUTES

Control charts for attributes do not use a measured quality variable; instead, they monitor the number of defects present in the sample or the fraction defect rate as the plotted statistic. Examples of these kinds of attributes include number of defects per automobile, fraction of bad parts in a sample, existence or absence of flash in plastic moldings, and number of flaws in a roll of sheet steel. The two principal types of control charts for attributes are the ***p chart***, which plots the fraction defect rate in successive samples; and the ***c chart***, which plots the number of defects, flaws, or other nonconformities per sample.

***p* Chart** In the p chart, the quality characteristic of interest is the proportion (p for proportion) of nonconforming or defective units. For each sample, this proportion p_i is the ratio of the number of nonconforming or defective items d_i over the number of

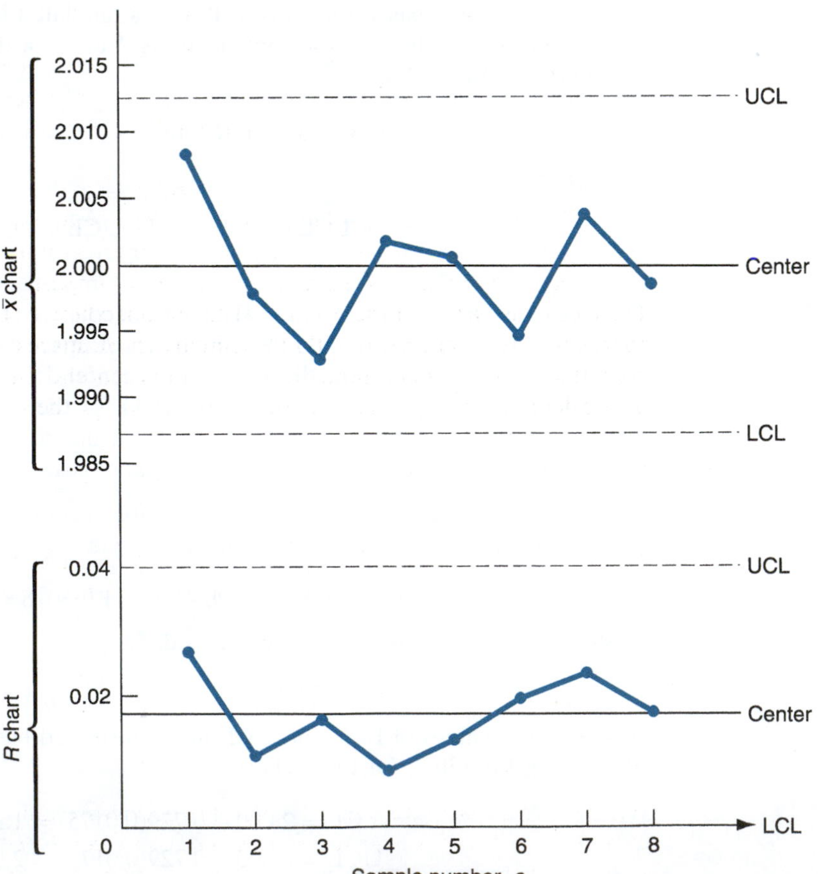

FIGURE 30.2 Control charts for Example 30.2. (Credit: *Fundamentals of Modern Manufacturing,* 4th Edition by Mikell P. Groover, 2010. Reprinted with permission of John Wiley & Sons, Inc.)

units in the sample n (we assume samples of equal size in constructing and using the control chart):

$$p_i = \frac{d_i}{n} \tag{30.4}$$

where i is used to identify the sample. If the p_i values for a sufficient number of samples are averaged, the mean value $\overline{p}$ is a reasonable estimate of the true value of p for the process. The p chart is based on the binomial distribution, where p is the probability of a nonconforming unit. The center in the p chart is the computed value of $\overline{p}$ for m samples of equal size n collected while the process is operating in statistical control.

$$\overline{p} = \frac{\sum_{i=1}^{m} p_i}{m} \tag{30.5}$$

The control limits are computed as three standard deviations on either side of the center. Thus,

$$\text{LCL} = \overline{p} - 3\sqrt{\frac{\overline{p}(1-\overline{p})}{n}} \quad \text{and} \quad \text{UCL} = \overline{p} + 3\sqrt{\frac{\overline{p}(1-\overline{p})}{n}} \tag{30.6}$$

where the standard deviation of $\overline{p}$ in the binomial distribution is given by:

$$\sigma_p = \sqrt{\frac{\overline{p}(1-\overline{p})}{n}}$$

If the value of $\overline{p}$ is relatively low and the sample size n is small, then the lower control limit computed by the first of these equations is likely to be a negative value. In this case, let LCL $= 0$ (the fraction defect rate cannot be less than zero).

c Chart In the c chart (c for count), the number of defects in the sample are plotted over time. The sample may be a single product such as an automobile, and $c =$ number of quality defects found during final inspection. Or the sample may be a length of carpeting at the factory prior to cutting, and $c =$ number of imperfections discovered in that strip. The c chart is based on the Poisson distribution, where $c =$ parameter representing the number of events occurring within a defined sample space (defects per car, imperfections per specified length of carpet). Our best estimate of the true value of c is the mean value over a large number of samples drawn while the process is in statistical control:

$$\overline{c} = \frac{\sum_{i=1}^{m} c_i}{m} \tag{30.7}$$

This value of $\overline{c}$ is used as the center for the control chart. In the Poisson distribution, the standard deviation is the square root of parameter c. Thus, the control limits are:

$$\text{LCL} = \overline{c} - 3\sqrt{\overline{c}} \quad \text{and} \quad \text{UCL} = \overline{c} + 3\sqrt{\overline{c}} \tag{30.8}$$

30.3.3 INTERPRETING THE CHARTS

When control charts are used to monitor production quality, random samples are drawn from the process of the same size n used to construct the charts. For $\overline{x}$ and R charts, the $\overline{x}$ and R values of the measured characteristic are plotted on the control chart.

By convention, the points are usually connected, as in our figures. To interpret the data, one looks for signs that indicate the process is not in statistical control. The most obvious sign is when $\bar{x}$ or R (or both) lie outside the LCL or UCL limits. This indicates an assignable cause such as bad starting materials, new operator, broken tooling, or similar factors. An out-of-limit $\bar{x}$ indicates a shift in the process mean. An out-of-limit R shows that the variability of the process has changed. The usual effect is that R increases, indicating variability has risen. Less obvious conditions may reveal process problems, even though the sample points lie within the $\pm 3\sigma$ limits. These conditions include: (1) trends or cyclical patterns in the data, which may mean wear or other factors that occur as a function of time; (2) sudden changes in average level of the data; and (3) points consistently near the upper or lower limits.

The same kinds of interpretations that apply to the $\bar{x}$ chart and R chart are also applicable to the p chart and c chart.

30.4 QUALITY PROGRAMS IN MANUFACTURING

Statistical process control is widely used for monitoring the quality of manufactured parts and products. Several additional quality programs are also used in industry, and in this section we briefly describe three: (1) total quality management, (2) Six Sigma, and (3) ISO 9000. These programs are not alternatives to statistical process control; in fact, the tools used in SPC are included within the methodologies of total quality management and Six Sigma.

30.4.1 TOTAL QUALITY MANAGEMENT

Total quality management (TQM) is a management approach to quality that pursues three main goals: (1) achieving customer satisfaction, (2) encouraging the involvement of the entire workforce, and (3) continuous improvement.

The customer and customer satisfaction are a central focus of TQM, and products are designed and manufactured with this focus in mind. The product must be designed with the features that customers want, and it must be manufactured free of deficiencies. Within the scope of customer satisfaction is the recognition that there are two categories of customers: (1) external customers and (2) internal customers. External customers are those who purchase the company's products and services. Internal customers are inside the company, such as the company's final assembly department, which is the customer of the parts production departments. For the total organization to be effective and efficient, satisfaction must be achieved in both categories of customers.

In TQM, worker involvement in the quality efforts of the organization extends from the top executives through all levels beneath. There is recognition of the important influence that product design has on product quality and how decisions made during design affect the quality that can be achieved in manufacturing. In addition, production workers are made responsible for the quality of their own output, rather than rely on inspectors to uncover defects after the parts are already produced. TQM training, including use of the tools of statistical process control, is provided to all workers. The pursuit of high quality is embraced by every member of the organization.

The third goal of TQM is continuous improvement; that is, adopting the attitude that it is always possible to make something better, whether it is a product or a process. Continuous improvement in an organization is generally implemented using worker teams that have been organized to solve specific problems that are identified in production. The problems are not limited to quality issues. They may include productivity, cost, safety, or any other area of interest to the organization. Team members are selected on the basis of their knowledge and expertise in the problem area. They are drawn from various departments and serve part-time on the team, meeting several times per month until they are able to make recommendations and/or solve the problem. Then the team is disbanded.

30.4.2 SIX SIGMA

The Six Sigma quality program originated and was first used at Motorola Corporation in the 1980s. It has been adopted by many other companies in the United States. Six Sigma is quite similar to total quality management in its emphasis on management involvement, worker teams to solve specific problems, and the use of SPC tools such as control charts. The major difference between Six Sigma and TQM is that Six Sigma establishes measurable targets for quality based on the number of standard deviations (sigma σ) away from the mean in the normal distribution. Six sigma implies near perfection in the process in the normal distribution. A process operating at the 6σ level in a Six Sigma program produces no more than 3.4 defects per million, where a defect is anything that might result in lack of customer satisfaction.

As in TQM, worker teams participate in problem-solving projects. A project requires the Six Sigma team to (1) define the problem, (2) measure the process and assess current performance, (3) analyze the process, (4) recommend improvements, and (5) develop a control plan to implement and sustain the improvements. The responsibility of management in Six Sigma is to identify important problems in their operations and sponsor the teams to address those problems.

Statistical Basis of Six Sigma An underlying assumption in Six Sigma is that the defects in any process can be measured and quantified. Once quantified, the causes of the defects can be identified, and improvements can be made to eliminate or reduce the defects. The effects of any improvements can be assessed using the same measurements in a before-and-after comparison. The comparison is often summarized as a sigma level; for example, the process is now operating at the 4.8-sigma level, whereas before it was only operating at the 2.6-sigma level. The relationship between sigma level and defects per million (DPM) is listed in Table 30.2 for a Six Sigma program. We see that the DPM was previously at 135,666 defects per 1,000,000 in our example, whereas it has now been reduced to 483 DPM.

A traditional measure for good process quality is $\pm 3\sigma$ (three sigma level). It implies that the process is stable and in statistical control, and the variable representing the output

TABLE 30.2 Sigma levels and corresponding defects per million in a Six Sigma program.

Sigma Level	Defects per Million[a]	Sigma Level	Defects per Million[a]
6.0σ	3.4	3.8σ	10,724
5.8σ	8.5	3.6σ	17,864
5.6σ	21	3.4σ	28,716
5.4σ	48	3.2σ	44,565
5.2σ	108	3.0σ	66,807
5.0σ	233	2.8σ	96,801
4.8σ	483	2.6σ	135,666
4.6σ	968	2.4σ	184,060
4.4σ	1,866	2.2σ	241,964
4.2σ	3,467	2.0σ	308,538
4.0σ	6,210	1.8σ	382,089

(Source: [3]).

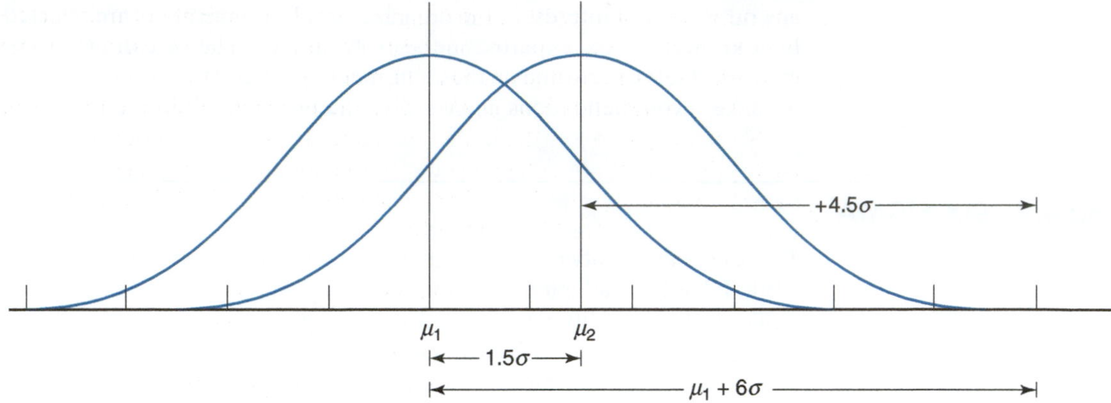

FIGURE 30.3 Normal distribution shift by 1.5σ from original mean and consideration of only one tail of the distribution (at right). Key: μ_1 = mean of original distribution, μ_2 = mean of shifted distribution, σ = standard distribution. (Credit: *Fundamentals of Modern Manufacturing*, 4th Edition by Mikell P. Groover, 2010. Reprinted with permission of John Wiley & Sons, Inc.)

of the process is normally distributed. Under these conditions, 99.73% of the output will be within the $\pm 3\sigma$ range, and 0.27% or 2700 parts per million will lie outside these limits (0.135% or 1350 parts per million beyond the upper limit and the same number beyond the lower limit). But wait a minute, if we look up 3.0 sigma in Table 30.2, we find that there are 66,807 defects per million. Why is there a difference between the standard normal distribution value (2700 DPM) and the value given in Table 30.2 (66,807 DPM)? There are two reasons for this discrepancy. First, the values in Table 30.2 refer to only one tail of the distribution, so that an appropriate comparison with the standard normal tables would only use one tail of the distribution (1350 DPM). Second, and much more significant, is that when Motorola devised the Six Sigma program, they considered the operation of processes over long periods of time, and processes over long periods tend to experience shifts from their original process means. To compensate for these shifts, Motorola decided to adjust the standard normal values by 1.5σ. To summarize, Table 30.2 includes only one tail of the normal distribution, and it shifts the distribution by 1.5 sigma relative to the standard normal distribution. These effects can be seen in Figure 30.3.

Measuring the Sigma Level In a Six Sigma project, the performance level of the process of interest is summarized by a sigma level. This is done at two points during the project: (1) after measurements have been taken of the process as it is currently operating and (2) after process improvements have been made to assess the effect of the improvements. This provides a before-and-after comparison. High sigma values represent good performance; low sigma values mean poor performance.

To find the sigma level, the number of defects per million must first be determined. There are three measures of defects per million used in Six Sigma. The first and most important is the defects per million opportunities (DPMO), which considers that there may be more than one type of defect that can occur in each unit (product or service). More complex products are likely to have more opportunities for defects, while simple products have fewer opportunities. Thus, DPMO accounts for the complexity of the product and allows entirely different kinds of products or services to be compared. Defects per million opportunities is calculated using the following equation:

$$\text{DPMO} = 1,000,000\frac{N_d}{N_u N_o} \tag{30.9}$$

where N_d = total number of defects found, N_u = number of units in the population of interest, and N_o = number of opportunities for a defect per unit. The constant 1,000,000 converts the ratio into defects per million.

Other measures besides DPMO are defects per million (DPM), which measures all of the defects found in the population, and defective units per million (DUPM), which counts the number of defective units in the population and recognizes that there may be more than one type of defect in any defective unit. The following two equations can be used to compute *DPM* and DUPM:

$$\text{DPM} = 1{,}000{,}000\frac{N_d}{N_u} \tag{30.10}$$

$$\text{DUPM} = 1{,}000{,}000\frac{N_{du}}{N_u} \tag{30.11}$$

where N_{du} = number of defective units in the population, and the other terms are the same as for Eq. (30.9). Once the values of DPMO, DPM, and DUPM have been determined, Table 30.2 can be used to convert these values to their corresponding sigma levels.

Example 30.2 Determining the Sigma Level of a Process

A final assembly plant that makes dishwashers inspects for 23 features that are considered important for overall quality. During the previous month, 9,056 dishwashers were produced. During inspection, 479 defects among the 23 features were found, and 226 dishwashers had one or more defect. Determine DPMO, DPM, and DUPM for these data and convert each to its corresponding sigma level.

Solution: Summarizing the data, $N_u = 9056$, $N_o = 23$, $N_d = 479$, and $N_{du} = 226$. Thus,

$$\text{DPMO} = 1{,}000{,}000\frac{479}{9056(23)} = 2300$$

The corresponding sigma level is about 4.3 from Table 30.2.

$$\text{DPM} = 1{,}000{,}000\frac{479}{9056} = 52{,}893$$

The corresponding sigma level is about 3.1.

$$\text{DUPM} = 1{,}000{,}000\frac{226}{9056} = 24{,}956$$

The corresponding sigma level is about 3.4. ■

30.4.3 ISO 9000

ISO 9000 is a set of international standards that relate to the quality of the products (or services, if applicable) delivered by a given facility. The standards were developed by the International Organization for Standardization (ISO), which is based in Geneva, Switzerland. ISO 9000 establishes standards for the systems and procedures used by the facility that determine the quality of its products. ISO 9000 is not a standard for the products themselves. Its focus is on systems and procedures, which include the facility's organizational structure, responsibilities, methods, and resources needed to manage quality. ISO 9000 is concerned with the activities used by the facility to ensure that its products achieve customer satisfaction.

ISO 9000 can be implemented in two ways, formally and informally. Formal implementation means that the facility becomes registered, which certifies that the facility meets the requirements of the standard. Registration is obtained through a third-party agency that conducts on-site inspections and reviews the facility's quality systems and procedures. A benefit of registration is that it qualifies the facility to do business with companies that require ISO 9000 registration, which is common in the European Economic Community where certain products are regulated and ISO 9000 registration is required for companies making these products.

Informal implementation of ISO 9000 means that the facility practices the standards or portions thereof simply to improve its quality systems. Even without formal certification, such improvements are worthwhile for companies desiring to deliver high-quality products.

30.5 INSPECTION PRINCIPLES

Inspection involves the use of measurement and gaging techniques to determine whether a product, its components, subassemblies, or starting materials conform to design specifications. The design specifications are established by the product designer, and for mechanical products they refer to dimensions, tolerances, surface finish, and similar features. Dimensions, tolerances, and surface finish were defined in Chapter 4, and many of the measuring instruments and gages for assessing these specifications were described in the appendix in Chapter 4.

Inspection is performed before, during, and after manufacturing. The incoming materials and starting parts are inspected upon receipt from suppliers; work units are inspected at various stages during their production; and the final product should be inspected prior to shipment to the customer.

We should clarify the distinction between inspection and testing, which is a closely related topic. Whereas inspection determines the quality of the product relative to design specifications, testing generally refers to the functional aspects of the product. Does the product operate the way it is supposed to operate? Will it continue to operate for a reasonable period of time? Will it operate in environments of extreme temperature and humidity? In quality control, ***testing*** is a procedure in which the product, subassembly, part, or material is observed under conditions that might be encountered during service. For example, a product might be tested by operating it for a certain period of time to determine whether it functions properly. If it passes the test, it is approved for shipment to the customer.

Testing of a component or material is sometimes damaging or destructive. In these cases, the items must be tested on a sampling basis. The expense of destructive testing is significant, and great efforts are made to develop methods that do not destroy the item. These methods are referred to as ***nondestructive testing*** or ***nondestructive evaluation***.

Inspections divide into two types: (1) ***inspection by variables***, in which the product or part dimensions of interest are measured by the appropriate measuring instruments; and (2) ***inspection by attributes***, in which the parts are gaged to determine whether they are within tolerance limits. The advantage of measuring a part dimension is that data are obtained about its actual value. The data might be recorded over time and used in control charts and to analyze trends in the manufacturing process. Adjustments can be made in the process based on the data so that future parts are produced closer to the nominal design value. When a part dimension is simply gaged, all that is known is whether it is within tolerance or too big or too small. On the positive side, gaging can be done quickly and at low cost.

Inspection procedures are often performed manually. The work is usually boring and monotonous, yet the need for precision and accuracy is high. Hours are sometimes required to measure the important dimensions of only one part. Because of the time and cost of manual inspection, statistical sampling procedures are generally used to reduce the need to inspect every part.

Sampling versus 100% Inspection When sampling inspection is used, the number of parts in the sample is generally small compared to the quantity of parts produced. The sample size may be only 1% of the production run. Because not all of the items in the population are measured, there is a risk in any sampling procedure that defective parts will slip through. One of the goals in statistical sampling is to define the expected risk; that is, to determine the average defect rate that will pass through the sampling procedure. The risk can be reduced by increasing the sample size and the frequency with which samples are collected. But the fact remains that 100% good quality cannot be guaranteed in a sampling inspection procedure.

Theoretically, the only way to achieve 100% good quality is by 100% inspection; thus, all defects are screened and only good parts pass through the inspection procedure. However, when 100% inspection is done manually, two problems are encountered. The first is the expense involved. Instead of dividing the cost of inspecting the sample over the number of parts in the production run, the unit inspection cost is applied to every part in the batch. Inspection cost sometimes exceeds the cost of making the part. Second, in 100% manual inspection, there are almost always errors associated with the procedure. The error rate depends on the complexity and difficulty of the inspection task and how much judgment is required by the human inspector. These factors are compounded by operator fatigue. Errors mean that a certain number of poor quality parts will be accepted and a certain number of good quality parts will be rejected. Therefore, 100% inspection using manual methods is no guarantee of 100% good quality product.

Automated 100% Inspection Automation of the inspection process offers a possible way to overcome the problems associated with 100% manual inspection. Automated 100% inspection can be integrated with the manufacturing process to accomplish some action relative to the process. The actions can be one or both of the following: (1) parts sortation and/or (2) feedback of data to the process. ***Parts sortation*** means separating parts into two or more quality levels. The basic sortation includes two levels: acceptable and unacceptable. Some situations require more than two levels, such as acceptable, reworkable, and scrap. Sortation and inspection may be combined in the same station. An alternative approach is to locate one or more inspections along the processing line, and instructions are sent to a sortation station at the end of the line indicating what action is required for each part.

Feedback of inspection data to the upstream manufacturing operation allows compensating adjustments to be made in the process to reduce variability and improve quality. If inspection measurements indicate that the output is drifting toward one of the tolerance limits (e.g., due to tool wear), corrections can be made to process parameters to move the output toward the nominal value. The output is thereby maintained within a smaller variability range than possible with sampling inspection methods.

Contact versus Noncontact Inspection There are a variety of measurement and gaging technologies available for inspection. The possibilities can be divided into contact and noncontact inspection. ***Contact inspection*** involves the use of a mechanical probe or other device that makes contact with the object being inspected. By its nature, contact inspection is usually concerned with measuring or gaging some physical dimension of the part. It is accomplished manually or automatically. Most of the traditional measuring and gaging devices described in the appendix for Chapter 4 relate to contact inspection. An example of an automated contact measuring system is the coordinate measuring machine (Section 30.6.1).

Noncontact inspection methods utilize a sensor located a certain distance from the object to measure or gage the desired feature(s). Typical advantages of noncontact inspection are (1) faster inspection cycles, and (2) avoidance of damage to the part that might result from contact. Noncontact methods can often be accomplished on the production line without any special handling. By contrast, contact inspection usually requires special positioning of the part, necessitating its removal from the production line. Also, noncontact inspection methods are inherently faster because they employ a stationary probe that does not require positioning for every part. By contrast, contact inspection requires positioning of the contact probe against the part, which takes time.

Noncontact inspection technologies can be classified as optical or nonoptical. Prominent among the optical methods is machine vision (Section 30.6.2). Nonoptical inspection sensors include electrical field techniques, radiation techniques, and ultrasonics (Section 30.6.3).

30.6 MODERN INSPECTION TECHNOLOGIES

Advanced technologies are substituting for manual measuring and gaging techniques in modern manufacturing plants. They include contact and noncontact sensing methods. We begin our coverage with an important contact inspection technology: coordinate measuring machines.

30.6.1 COORDINATE MEASURING MACHINES

A coordinate measuring machine (CMM) consists of a contact probe and a mechanism to position the probe in three dimensions relative to surfaces and features of a workpart. See Figure 30.4. The location coordinates of the probe can be accurately recorded as it contacts the part surface to obtain part geometry data.

In a CMM, the probe is fastened to a structure that allows movement of the probe relative to the part, which is fixtured on a worktable connected to the structure. The structure must be rigid to minimize deflections that contribute to measurement errors. The machine in Figure 30.4 has a bridge structure, one of the most common designs. An important component of a CMM is the contact probe and its operation. Modern "touch-trigger" probes have a sensitive electrical contact that emits a signal when the probe is deflected from its neutral position in the slightest amount. On contact, the coordinate positions are recorded by the CMM controller, adjusting for overtravel and probe size.

Positioning of the probe relative to the part can be accomplished either manually or under computer control. Methods of operating a CMM can be classified as (1) manual control, (2) manual computer-assisted, (3) motorized computer-assisted, and (4) direct computer control.

In ***manual control***, a human operator physically moves the probe along the axes to contact the part and record the measurements. The probe is free-floating for easy movement. Measurements are indicated by digital read-out, and the operator can record the measurement manually or automatically (paper print-out). Any trigonometric calculations must be made by the operator. The ***manual computer-assisted*** CMM is capable of computer data processing to perform these calculations. Types of computations include simple conversions from U.S. customary units to metric, determining the angle between two planes, and determining hole-center locations. The probe is still free-floating to permit the operator to bring it into contact with part surfaces.

Motorized computer-assisted CMMs power drive the probe along the machine axes under operator guidance. A joystick or similar device is used to control the motion. Low-power stepping motors and friction clutches are used to reduce the effects of collisions between probe and part. The ***direct computer-control*** CMM operates like a

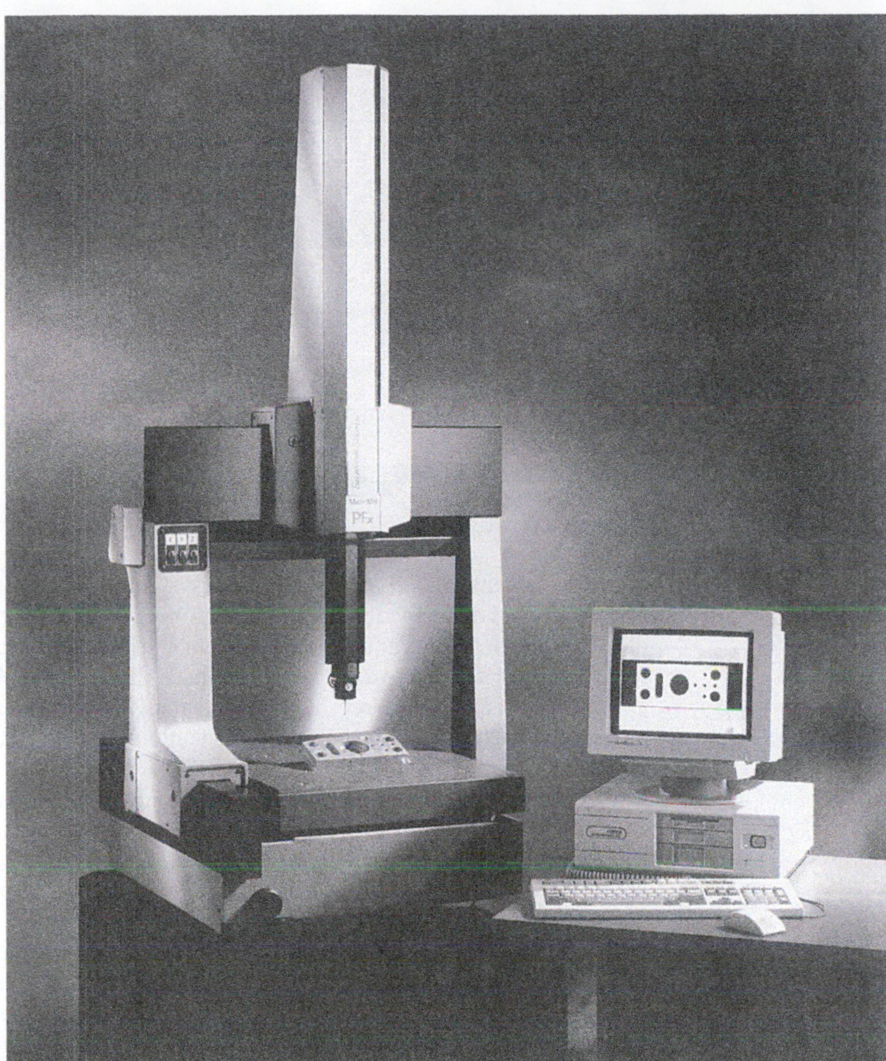

FIGURE 30.4 Coordinate measuring machine. Courtesy of Brown & Sharpe Manufacturing Company. (Credit: *Fundamentals of Modern Manufacturing*, 4th Edition by Mikell P. Groover, 2010. Reprinted with permission of John Wiley & Sons, Inc.)

CNC machine tool. It is a computerized inspection machine that operates under program control. The basic capability of a CMM is to determine coordinate values where its probe contacts the surface of a part. Computer control permits the CMM to accomplish more sophisticated measurements and inspections, such as (1) determining center location of a hole or cylinder, (2) defining a plane, (3) measuring flatness of a surface or parallelism between two surfaces, and (4) measuring an angle between two planes.

Advantages of using coordinate measuring machines over manual inspection methods include: (1) higher productivity—a CMM can perform complex inspection procedures in much less time than traditional manual methods; (2) greater inherent accuracy and precision than conventional methods; and (3) reduced human error through automation of the inspection procedure and associated computations [8].

30.6.2 MACHINE VISION

Machine vision involves the acquisition, processing, and interpretation of image data by computer for some useful application. Vision systems can be classified as two dimensional

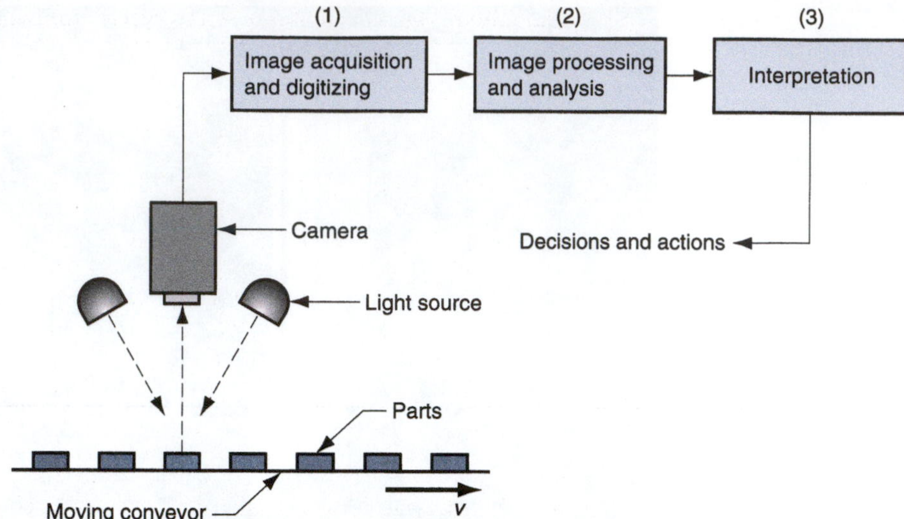

FIGURE 30.5 Operation of a machine vision system. (Credit: *Fundamentals of Modern Manufacturing*, 4th Edition by Mikell P. Groover, 2010. Reprinted with permission of John Wiley & Sons, Inc.)

or three dimensional. Two-dimensional systems view the scene as a two-dimensional image, which is quite adequate for applications involving a planar object. Examples include dimensional measuring and gaging, verifying the presence of components, and checking for features on a flat (or almost flat) surface. Three-dimensional vision systems are required for applications requiring a three-dimensional analysis of the scene, where contours or shapes are involved. The majority of current applications are two-dimensional, and our discussion will focus (excuse the pun) on this technology.

Operation of a machine vision system consists of three steps, depicted in Figure 30.5: (1) image acquisition and digitization, (2) image processing and analysis, and (3) interpretation.

Image acquisition and digitizing is accomplished by a video camera connected to a digitizing system to store the image data for subsequent processing. With the camera focused on the subject, an image is obtained by dividing the viewing area into a matrix of discrete picture elements (called ***pixels***), in which each element assumes a value proportional to the light intensity of that portion of the scene. The intensity value for each pixel is converted to its equivalent digital value by analog-to-digital conversion. Image acquisition and digitizing is depicted in Figure 30.6 for a ***binary vision*** system, in which the light intensity is reduced to either of two values (black or white = 0 or 1), as in Table 30.3. The pixel matrix in our illustration is only 12 × 12; a real vision system would have many more pixels for better resolution. Each set of pixel values is a ***frame***, which consists of the set of digitized pixel values. The frame is stored in computer memory.

(a)

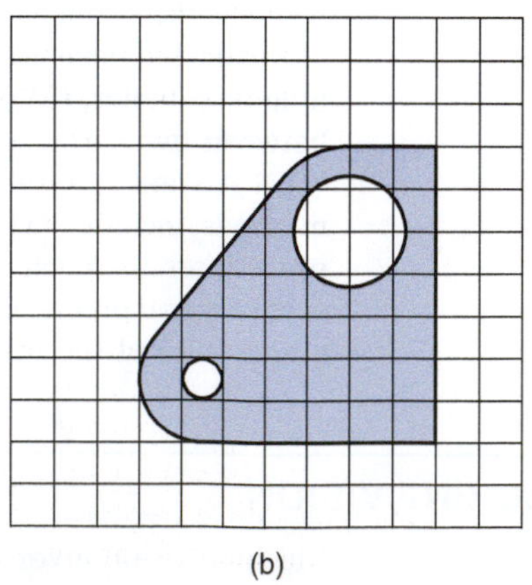

(b)

FIGURE 30.6 Image acquisition and digitizing: (a) the scene consists of a dark-colored part against a light background; (b) a 12 × 12 matrix of pixels imposed on the scene. (Credit: *Fundamentals of Modern Manufacturing*, 4th Edition by Mikell P. Groover, 2010. Reprinted with permission of John Wiley & Sons, Inc.)

TABLE 30.3 Pixel values in a binary vision system for the image in Figure 30.6.

1	1	1	1	1	1	1	1	1	1	1	1
1	1	1	1	1	1	1	1	1	1	1	1
1	1	1	1	1	1	1	1	1	1	1	1
1	1	1	1	1	1	1	0	0	0	1	1
1	1	1	1	1	1	0	1	1	0	1	1
1	1	1	1	1	0	0	1	1	0	1	1
1	1	1	1	0	0	0	0	0	0	1	1
1	1	1	0	0	0	0	0	0	0	1	1
1	1	1	0	1	0	0	0	0	0	1	1
1	1	1	1	0	0	0	0	0	0	1	1
1	1	1	1	1	1	1	1	1	1	1	1
1	1	1	1	1	1	1	1	1	1	1	1

The process of reading all the pixel values in a frame is performed 30 times per second in United States, 25 cycle/s in European systems.

The ***resolution*** of a vision system is its ability to sense fine details and features in the image. This depends on the number of pixels used. Common pixel arrays include 640 (horizontal) × 480 (vertical), 1024 × 768, or 1040 × 1392 picture elements. The more pixels in the vision system, the higher its resolution. However, system cost increases as pixel count increases. Also, time required to read the picture elements and process the data increases with number of pixels. In addition to binary vision systems, more sophisticated vision systems distinguish various gray levels in the image that permit them to determine surface characteristics such as texture. Called ***gray-scale vision***, these systems typically use 4, 6, or 8 bits of memory. Other vision systems can recognize colors.

The second function in machine vision is ***image processing and analysis***. The data for each frame must be analyzed within the time required to complete one scan (1/30 s or 1/25 s). Several techniques have been developed to analyze image data, including edge detection and feature extraction. ***Edge detection*** involves determining the locations of boundaries between an object and its surroundings. This is accomplished by identifying contrast in light intensity between adjacent pixels at the borders of the object. ***Feature extraction*** is concerned with determining feature values of an image. Many machine vision systems identify an object in the image by means of its features. Features of an object include area, length, width, or diameter of the object, perimeter, center of gravity, and aspect ratio. Feature extraction algorithms are designed to determine these features based on the object's area and boundaries. Area of an object can be determined by counting the number of pixels that make up the object. Length can be found by measuring the distance (in pixels) between two opposite edges of the part.

Interpretation of the image is the third function. It is accomplished by extracted features. Interpretation is usually concerned with recognizing the object—identifying the object in the image by comparing it to predefined models or standard values. One common interpretation technique is ***template matching***, which refers to methods that compare one or more features of an image with corresponding features of a model (template) stored in computer memory.

The interpretation function in machine vision is generally related to applications, which divide into four categories: (1) inspection, (2) part identification, (3) visual guidance and control, and (4) safety monitoring.

Inspection is the most important category, accounting for about 90% of all industrial applications. The applications are in mass production, where the time to program and install the system can be divided by many thousands of units. Typical inspection tasks include: (1) dimensional measurement or gaging, which involves measuring or gaging certain dimensions of parts or products moving along a conveyor; (2) verification functions, which include verifying presence of components in an assembled product, presence of a hole in a workpart, and similar tasks; and (3) identification of flaws and defects, such as identifying flaws in a printed label in the form of mislocation, poorly printed text, numbering, or graphics on the label.

Part identification applications include counting different parts flowing past on a conveyor, part sorting, and character recognition. ***Visual guidance and control*** involves a vision system interfaced with a robot or similar machine to control the movement of the machine. Examples include seam tracking in continuous arc welding, part positioning, part reorientation, and picking parts from a bin. In ***safety monitoring*** applications, the vision system monitors the production operation to detect irregularities that might indicate a hazardous condition to equipment or humans.

30.6.3 OTHER NONCONTACT INSPECTION TECHNIQUES

In addition to optical inspection methods, various nonoptical techniques are used in inspection. These include sensor techniques based on electrical fields, radiation, and ultrasonics.

Under certain conditions, ***electrical fields*** created by an electrical probe can be used for inspection. The fields include reluctance, capacitance, and inductance; they are affected by an object in the vicinity of the probe. In a typical application, the workpart is positioned in a fixed relationship to the probe. By measuring the effect of the object on the electrical field, an indirect measurement of certain part characteristics can be made, such as dimensional features, thickness of sheet material, and flaws (cracks and voids below the surface) in the material.

Radiation techniques use X-ray radiation to inspect metals and weldments. The amount of radiation absorbed by the metal object indicates thickness and presence of flaws in the part or welded section. For example, X-ray inspection is used to measure thickness of sheet metal in rolling (Section 13.1). Data from the inspection is used to adjust the gap between rolls in the rolling mill.

Ultrasonic techniques use high-frequency sound (>20,000 Hz) to perform various inspection tasks. One of the techniques analyses the ultrasonic waves emitted by a probe and reflected off the object. During the setup for the inspection procedure, an ideal test part is positioned in front of the probe to obtain a reflected sound pattern. This sound pattern is used as the standard against which production parts are subsequently compared. If the reflected pattern from a given part matches the standard, the part is accepted. If a match is not obtained, the part is rejected.

REFERENCES

[1] DeFeo, J. A., Gryna, F. M., and Chua, R. C. H. ***Juran's Quality Planning and Analysis for Enterprise Quality***, 5th ed. McGraw-Hill, New York, 2006.

[2] Evans, J. R., and Lindsay, W. M. ***The Management and Control of Quality***, 6th ed. Thomson/South-Western College Publishing Company, Mason, Ohio, 2005.

[3] Groover, M. P. ***Automation, Production Systems, and Computer Integrated Manufacturing***, 3rd ed. Prentice Hall, Upper Saddle River, New Jersey, 2008.

[4] Juran, J. M., and Gryna, F. M. ***Quality Planning and Analysis***, 3rd ed. McGraw-Hill, New York, 1993.

[5] Lochner, R. H., and Matar, J. E. ***Designing for Quality***. ASQC Quality Press, Milwaukee, Wisconsin, 1990.

[6] Montgomery, D. C. ***Introduction to Statistical Quality Control***, 6th ed. John Wiley & Sons, Inc., Hoboken, New Jersey, 2008.

[7] Pyzdek, T., and Keller, P. ***Quality Engineering Handbook***, 2nd ed. CRC Taylor & Francis, Boca Raton, Florida, 2003.

[8] Schaffer, G. H. "Taking the Measure of CMMs," Special Report 749, ***American Machinist***, October 1982, pp. 145–160.

[9] Schaffer, G. H. "Machine Vision: A Sense for CIM," Special Report 767, ***American Machinist***, June 1984 pp. 101–120.

[10] Taguchi, G., Elsayed, E. A., and Hsiang, T. C. ***Quality Engineering in Production Systems***. McGraw-Hill, New York, 1989.

[11] Wick, C., and Veilleux, R. F. *Tool and Manufacturing Engineers Handbook*, 4th ed., Vol. IV, ***Quality Control and Assembly.*** Society of Manufacturing Engineers, Dearborn, Michigan, 1987.

REVIEW QUESTIONS

30.1. What are the two principal aspects of product quality?

30.2. How is a process operating in statistical control distinguished from one that is not?

30.3. Define the term *process capability*.

30.4. What are the natural tolerance limits?

30.5. What is the difference between control charts for variables and control charts for attributes?

30.6. Identify the two types of control charts for variables.

30.7. What are the two basic types of control charts for attributes?

30.8. When interpreting a control chart, what does one look for to identify problems?

30.9. What are the three main goals in total quality management (TQM)?

30.10. What is the difference between external customers and internal customers in TQM? At what company was the Six Sigma quality program first used?

30.11. Why is the normal statistical table used in a Six Sigma program different from the standard normal tables found in textbooks on probability and statistics?

30.12. A Six Sigma program uses three measures of defects per million (DPM) to assess the performance of a given process. Name the three measures of DPM.

30.13. Automated inspection can be integrated with the manufacturing process to accomplish certain actions. What are these possible actions?

30.14. Give an example of a noncontact inspection technique.

30.15. What is a coordinate measuring machine?

30.16. What is a binary vision system?

30.17. Name some of the nonoptical noncontact sensor technologies available for inspection.

PROBLEMS

30.1. An automatic turning process is set up to produce parts with a mean diameter = 6.255 cm. The process is in statistical control and the output is normally distributed with a standard deviation = 0.004 cm. Determine the process capability.

30.2. A sheet-metal bending operation produces bent parts with an included angle = 92.1°. The process is in statistical control and the values of included angle are normally distributed with a standard deviation = 0.23°. The design specification on the angle = 90 ± 2°. Determine the process capability.

30.3. A plastic extrusion process produces round tubular extrudate with a mean outside diameter = 28.6 mm. The process is in statistical control and the output is normally distributed with standard deviation = 0.53 mm. Determine the process capability.

30.4. In 12 samples of size $n = 7$, the average value of the sample means is $\bar{\bar{x}} = 6.860$ cm for the dimension of interest, and the mean of the ranges of the samples is $\overline{R} = 0.027$ cm. Determine (a) lower and upper control limits for the $\bar{x}$ chart and (b) lower and upper control limits for the R chart.

30.5. In nine samples of size $n = 10$, the grand mean of the samples is $\bar{\bar{x}} = 100$ for the characteristic of interest, and the mean of the ranges of the samples is $\overline{R} = 8.5$. Determine (a) lower and upper control limits for the $\bar{x}$ chart and (b) lower and upper control limits for the R chart.

30.6. Ten samples of size $n = 8$ have been collected from a process in statistical control, and the dimension of interest has been measured for each part. The calculated values of $\bar{x}$ for each sample are (mm) 9.22, 9.15, 9.20, 9.28, 9.19, 9.12, 9.20, 9.24, 9.17, and 9.23. The values of R are (mm) 0.24, 0.17, 0.30, 0.26, 0.26, 0.19, 0.21, 0.32, 0.21, and 0.23, respectively. (a) Determine the values of the center, LCL, and UCL for the $\bar{x}$ and R charts. (b) Construct the control charts and plot the sample data on the charts.

30.7. Seven samples of 5 parts each have been collected from an extrusion process that is in statistical control, and the diameter of the extrudate has been measured for each part. The calculated values of $\bar{x}$ for each sample are (in) 1.002, 0.999, 0.995, 1.004, 0.996, 0.998, and 1.006. The values of R are (in) 0.010, 0.011, 0.014, 0.020, 0.008, 0.013, and 0.017, respectively. (a) Determine the values of the center, LCL, and UCL for $\bar{x}$ and R charts. (b) Construct the control charts and plot the sample data on the charts.

30.8. A p chart is to be constructed. Six samples of 25 parts each have been collected, and the average number of defects per sample was 2.75. Determine the center, LCL and UCL for the p chart.

30.9. Ten samples of equal size are taken to prepare a p chart. The total number of parts in these ten samples was 900 and the total number of defects counted was 117. Determine the center, LCL and UCL for the p chart.

30.10. The yield of good chips during a certain step in silicon processing of integrated circuits averages 91%. The number of chips per wafer is 200. Determine the center, LCL, and UCL for the p chart that might be used for this process.

30.11. Twelve cars were inspected after final assembly. The number of defects found ranged between 87 and 139 defect per car with an average of 116. Determine the center and upper and lower control limits for the c chart that might be used in this situation.

30.12. A foundry that casts turbine blades inspects for eight features that are considered critical-to-quality. During the previous month, 1,236 castings were produced. During inspection, 47 defects among the eight features were found, and 29 castings had one or more defects. Determine DPMO, DPM, and DUPM in a Six Sigma program for these data and convert each to its corresponding sigma level.

30.13. In the previous problem, if the foundry desired to improve its quality performance to the 5.0 sigma level in all three measures of DPM, how many defects and defective units would they produce in an annual production quantity of 15,000 castings? Assume the same eight features are used to assess quality.

30.14. The inspection department in an automobile final assembly plant inspects cars coming off the production line against 55 quality features considered important to customer satisfaction. The department counts the number of defects found per 100 cars, which is the same type of metric used by a national consumer advocate agency. During a one-month period, a total of 16,582 cars rolled off the assembly line. These cars included a total of 6045 defects of the 55 features, which translates to 36.5 defects per 100 cars. In addition, a total of 1955 cars had one or more of the defects during this month. Determine DPMO, DPM, and DUPM in a Six Sigma program for these data and convert each to its corresponding sigma level.

INDEX

Abrasive belt grinding, 449–450
Abrasive cutoff, 387
Abrasive flow machining, 460
Abrasive jet machining, 459–460
Abrasive processes, 336, 433–453, 457–460
Abrasive water jet cutting, 459
Abrasives, 29–31
Absolute positioning, 643
Accuracy, defined, 89
Acetylene, 537–539
Acid cleaning and pickling, 491
Acrylics, 38
Acrylonitrile-butadiene-styrene, 38
Adhesive bonding, 12, 564–567
Age hardening, 486
Alloys:
  aluminum, 24–25, 136
  cast cobalt, 407
  casting, 136
  copper, 26, 136
  defined, 6
  high speed steel, 406–407
  magnesium, 25, 136
  nickel, 26, 136
  steels, 19–22
  titanium, 26–27, 136
  zinc, 27, 136
Allowance:
  bending, 311–312
  drilling, 370
  machining (in casting), 137–138
  shearing, 307
  shrinkage (in casting), 108
Alumina, 7, 29–31
Aluminizing, 493, 497
Aluminum, 24–25
Aluminum oxide, 410, 434
Amino resins, 39
Angle of repose, 236
Angularity, 82
Annealing:
  glass, 146
  metals, 481
Anodizing, 12, 498–499
Antioch process, 122
APT, 649
Aramides, 38
Arbor, 378
Arc welding, 510–511, 522–531
Area reduction, 53
Assembly:
  defined, 12
  design for, 581–584
  mechanical, 569–584
  process planning, 631
Atomic force microscope, 614
Atomization, 217
Austenite, 17, 483
Austenitic stainless steel, 20
Austenitizing, 483
Autoclave, 196, 207
Autogenous weld, 510
Automated assembly, 583–584
Automated tape-laying machines, 206
Automated welding, 512–513
Automatic screw machine, 366
Average flow stress, 256, 266, 287, 295

Backward extrusion, 284
Bainite, 482
Ball mill, 241
Bambooing, 161
Banbury mixer, 194
Bandsaw, 387
Bar drawing, 293–299
Bar machine, 366
Barrel finishing, 492
Barreling, 58
Batch production, 625
Batting, 243
Bauxite, 24, 30
Bayer process, 246
Beading, 314, 581
Belt sanding, 449
Bend allowance, 311–312
Bending (sheet metal), 254, 310–315
Bending test, 59–60
Bernoulli's theorem, 102
Bilateral tolerance, 81
Billet, 262, 283
Biomimetics, 619
Blanking, 306
Blast finishing, 491
Blind hole, 369

Bloom, 262
Blow forming, 182
Blow molding, 177–180
Blowing, glass, 143
Blown-film extrusion, 163–164, 186
Bolt(s), 570
Borazon, 32
Boring, 363, 366–369
Boron nitride, 32
Brass, 26
Braze welding, 559–560
Brazing, 12, 554–560
Brinell hardness, 62–64
Broaching, 359, 385
Bronze, 26
Buckyball, 612
Buffing, 453
Built-up edge, 344
Bulk deformation, 253, 261–299
Bulk micromachining, 606
Burr, 305
Butadiene rubber, 41
Butyl rubber, 41

C-frame press, 326
CAD/CAM, 650
Calendering, 164, 194
Calipers, 90–91
Calorizing, 493
Capstan, 297–298
Carbide ceramics, 7, 31–32
Carbon:
  diamond, 411
  in high speed steel, 407
  in steel, 17–19
  iron-carbon alloy system, 17
Carbon arc technique, 617
Carbon black, 41, 193
Carbon buckyballs, 612
Carbon nanotubes, 612–613, 616–617
Carbonitriding, 487, 493
Carburizing, 487, 493
Case hardening, 487
Cast cobalt alloy, 407
Cast iron, 6–7, 17, 22–23, 136
Casting:
  advantages and disadvantages, 98
  defined, 97
  glass, 143
  heating and pouring, 101–103, 132–133
  molds, 98–100
  plastics, 185
  processes, 98–100, 113–129
  product design considerations, 137–138
  quality, 133–135
  rubber, 196
  solidification and cooling, 103–110
Cellular manufacturing, 626, 651–655
Cemented carbide:
  cutting tools, 408–409, 413
  defined, 31–32, 45
  processing of, 248–250
Cementite, 17
Centerburst (extrusion), 293
Centering, 371
Centerless grinding, 447
Centrifugal atomization, 217
Centrifugal casting, 128–129, 142, 212
Centrifugal spraying (glass), 145
Ceramic matrix composites, 45, 250
Ceramic-mold casting, 121–122
Ceramic(s):
  classification of, 7, 29, 238
  cutting tools, 410–411, 413
  defined, 7, 29
  hardness, 65
  glass, *see* Glass
  processing of, 238–248
  products, 30–31
  properties, 29
  raw materials, 29–30
Cermets:
  cutting tools, 408–409
  defined, 32, 45
  processing of, 248–250
Chamfering, 362
Chaplet, 115
Chemical blanking, 472–473
Chemical cleaning, 490–491
Chemical conversion coating, 498
Chemical engraving, 473–474
Chemical machining, 470–475
Chemical milling, 472
Chemical vapor deposition, 12, 501–504, 605, 617
Chill-roll extrusion, 163
Chip (metal cutting), 343–345
Chip breaker, 412–413
Chip thickness ratio, 342
Chloroprene rubber, 41
Chromate coating, 498
Chromium, 19, 20–21, 407
Chromium carbide, 31
Chuck, 364
Chucking machine, 366
Chvorinov's rule, 106–107, 109
Circularity, 82

Clay, 7, 241–243
Cleaning processes, 12, 489–493
Clearance:
  brazing, 556–557
  sheet-metal drawing, 315
  sheet-metal shearing, 306–307
CMM, *see* Coordinate measuring machine
CNC, *see* Numerical control
Coated carbides, 410
Coating:
  carbides, 410, 503
  plastic, 166–167
  processes, 12, 489, 498–499, 504–507
  rubber, 195
  wire, 160
Cobalt, 28
Coining, 224, 277, 321
Cold extrusion, 285–286
Cold forming, 256
Cold rolling, 262
Cold welding, 542–543
Cold working, 256–257, 261
Collet, 364
Comminution, 249
Compact discs, 602
Compaction, 220–222, 248–249
Composites:
  components, 43–44
  defined, 7–8, 42
  processing, 200–213
  properties, 42–44
  types, 43–46
Compression molding, 174–175, 195, 207–208
Compression properties, 57–59
Computer-aided process planning, 633–635
Computer integrated manufacturing, 662–664
Computer numerical control, *see* Numerical control
Concentricity, 82
Concurrent engineering, 636–638
Contact lamination, 203
Contact molding, 203
Continuity law, 103–104
Continuous improvement, 635, 661
Continuous laminating, 213
Continuous path, 643
Contour turning, 361
Contouring, 643
Control chart(s), 669–674
Conversion coating, 498–499
Coolants (machining), 418
Coordinate measuring machine, 680–681
Copper, 26
Core (casting), 114–115
Corundum, 30
Cotter pin, 580
Counterboring, 371
Countersinking, 371
Crater wear, 399
Creep feed grinding, 447–448
Crimping, 581
Cross-linking, 36–37, 39, 40
Cross-wire welding, 536
Cubic boron nitride, 32, 411, 434
Cup drawing, 254
Cupola, 130
Curing:
  adhesives, 564
  organic coatings, 506
  polymer composites, 206–207, 210
  polymers, 39, 41, 175
  rubbers, 196, 199–200
Curling, 314
Cutoff, 309, 362, 387
Cutoff length, 85
Cutting:
  glass, 147
  metal, *see* Machining
  polymer composites, 213
  sheet metal, 305–310
Cutting conditions (machining):
  defined, 340, 348–349
  drilling, 369–370
  grinding, 437–438
  milling, 376–377
  selection of, 422–428
  turning, 360–361
Cutting fluid(s), 340, 417–419
Cutting force, *see* Forces
Cutting speed, 338, 423–428
Cutting temperature, 352–354
Cutting tool(s):
  basic types, 339
  costs, 425, 427–428
  geometry, *see* Tool geometry
  grinding wheels, 434–437
  materials, 404–411
  technology, 398–417
  tool life, 398–403
Cylindrical grinding, 446–447
Cylindricity, 82

Danner process, 145
Dead center, 364
Deburring, 463–464
Deep drawing, 254
Deep grinding, 448

Defects:
casting, 133–135
drawing sheet metal, 320
extrusion (plastic), 160–162
extrusion (metal), 293
injection molding, 172
welding, 548–549
Deformation processes:
bulk, 253,
defined, 9, 10, 252
Density, 72, 237
Depth of cut, 340
Design considerations, *see* Product design considerations
Design for assembly, 581–584, 636–637
Design for assembly, 581–584
Design for manufacturing, 636–637
Desktop machining, 588
Devitrification, 33
Dial indicator, 93
Diamond, 411, 434
Die(s):
bar drawing, 298
extrusion (metal), 289–291
extrusion (plastic), 157–160
forging, 279–280
stamping, 323–325
Die casting, 125–127
Die sinking, 376, 463, 467
Die swell, 151
Diffusion, 493, 605
Diffusion welding, 511, 543
Dimensions, 80–81
Dimpling, 581
Dip casting, 196
Dip coating, 505
Dip-pen nanolithography, 618
Direct extrusion, 283–284, 287
Disc grinder, 449
Divider, 90
Doctor blade, 167, 247
Doping, 605
Draft:
bar drawing, 294
casting, 137
forging, 280
plastic molding, 188
rolling, 263
Drain casting, 241
Draw bench, 297
Drawing:
bar, 253
deep, 254
glass, 145–146
plastic filaments, 166
sheet metal, 254, 314–320
wire, 253
Drawing ratio, 317
Dressing (grinding), 442
Drill bit(s), 369, 415–516
Drill jig, 373
Drill press(es), 369, 371–373
Drilling, 11, 338–339, 363, 369–373
Droplet deposition manufacturing, 591–592
Dry machining, 420
Dry plasma etching, 605
Dry pressing, 243–244
Dry spinning, 166
Drying:
ceramics, 244–245
organic coatings, 506
Ductile iron, 23
Ductility, 53–54

EBM, *see* Electron beam machining
ECM, *see* Electrochemical machining
Edge bending, 311
EDM, *see* Electric discharge machining
Elastic limit, 52
Elastic modulus, *see* Modulus of elasticity
Elastic reservoir molding, 207–208
Elastomers:
defined, 7, 35, 40
processing technology, 192–200
products, 197–198, 200
properties, 40
types, 40–42
Electric discharge forming, 332
Electric discharge machining, 465–467, 610
Electric discharge wire cutting, 467
Electric furnaces, 131, 141
Electrochemical deburring, 463–464
Electrochemical fabrication, 610
Electrochemical grinding, 464
Electrochemical machining, 461–463
Elecrochemical plating, 494–496
Electroforming, 496–497, 605, 610
Electrogas welding, 528–529
Electrohydraulic forming, 332–333
Electroless plating, 497, 605, 610
Electromagnetic forming, 333
Electron-beam lithography, 616
Electron-beam machining, 468–469, 610
Electron-beam welding, 511, 539–540
Electroplating, 12, 494–496, 605, 610
Elongation, 53

Embossing, 321, 581
Encapsulation (plastic), 185
Engine lathe, 363–364
Engineering materials, 6–8, 15–46
Engineering stress-strain, 51–54
Enterprise resource planning, 664
Epoxies, 39
Etchant, 470, 471
Etch factor, 472, 474
Etching, 470, 605
Ethylene-propylene rubber, 41
Eutectic alloys, 106
Evaporative-foam process, 118
Expandable foam molding, 186
Expanded polystyrene process, 118–119
Expansion fit, 12, 578
Expendable mold, 99, 113
Expendable mold casting, 113–122
Expert systems, 634
Explosion welding, 543–544
Explosive forming, 332
Extrusion:
  cemented carbides, 249
  ceramics, 243
  metals, 10, 253, 283–293
  plastics, 152–162
  powdered metals, 226–227
  rubber, 194
Extrusion blow molding, 177–178

Face plate, 365
Facing, 361
Fastener(s), 569–576, 581
Faying surfaces, 510
Feed (cutting), 338, 340
Feldspar, 30
Ferrite, 17
Ferritic stainless steel, 20
Ferrous metals, 6, 16
Fiber-reinforced polymers:
  applications, 46
  defined, 45–46
  properties, 46
  shaping processes, 191,
Fibers:
  defined, 43, 165
  glass, 145–146
  in composites, 45–46
  materials, 46
  plastics, 165–166
  types, 44
Filament, 145–146, 165, 209
Filament winding, 209–210
Film, plastic, 162–164
Finishing:
  ceramics, 248
  glass, 147
  machining, 340
  powdered metals, 225
Firing (sintering), 29–30, 245
Fixture, 373, 512, 548
Flakes, 44, 202
Flanging, 314
Flank wear, 399
Flash:
  die casting, 126
  forging, 275, 280
  injection molding, 172
Flashless forging, 276–277
Flatness, 82
Flexible manufacturing system(s), 655–659
Flexure test, 60
Float process (glass), 144
Flow curve, 56, 59
Flow line production, 626
Flow stress, 255–256, 273
Fluid properties, 67–69
Fluidity, 67
Fluidized bed, 506
Flux, 524, 527, 529, 557, 561–562
Flux-cored arc welding, 527–528
Fly cutter, 373
Foam, polymers, 185–186
Foam injection molding, 173
Force(s):
  bar drawing, 295–296
  bending sheet metal, 312–313
  cutting sheet metal, 308–309
  drawing sheet metal, 318
  extrusion, 288
  forging, 273–274, 276–277
  grinding, 439–441
  machining, 345–347
  powder metals, 221
  rolling, 265–267
  stretch forming, 330
  wire drawing, 295–296
Forge welding, 542
Forging:
  metals, 10, 253, 271–280
  powered metals, 226–227
Forging hammer, 271, 277–278
Forging press, 271, 279
Form milling, 374
Form turning, 362
Forward extrusion, 283

Foundry, 98, 129–133
Fracture stress, 53
Free machining steel, 422
Freezing point, 73
Friction:
  forging, 275
  metal cutting, 344–345, 348
  metal extrusion, 284–285, 287–288
  metal forming, 258–259
  rolling, 264–266
  sheet metal drawing, 316
Friction sawing, 387
Friction stir welding, 546
Friction welding, 511, 544–545
Fullerene, 612
Furnaces:
  brazing, 559
  casting, 129–132
  electric, 131, 141
  glassworking, 141
Fused-deposition modeling, 594
Fusion welding, 510–511, 522–541

Gage blocks, 89–90
Gages and gaging, 89, 93–94
Galvanized steel, 27
Galvanizing, 498
Gang drill, 373
Gas atomization, 217
Gas metal arc welding, 526–527
Gas tungsten arc welding, 529–530
Gear shaper, 384
Geometry:
  machined parts, 357–360
  nontraditional processes, 456, 476
  tool, *see* Tool geometry
Glass:
  chemistry, 33–34, 140–141
  defined, 7, 32, 140
  fibers, 34, 145–146
  product design,
  products, 33–34
  properties, 33, 141
  shaping processes, 141–146
Glass-ceramics, 29, 34–35
Glassworking, 140–147
Glazing, 245
Gray cast iron, 22–23
Grinding, 11, 249–251, 433–450
Grinding fluids, 444
Grinding ratio, 442
Grinding wheels, 434–437
Gross domestic product, 1
Group technology, 626, 651, 655
Guerin process, 322–323

Hacksaw, 387
Hand lay-up, 204–205
Hand modeling, 242
Hand molding, 242
Hand throwing, 242
Hardenability, 484–485
Hardness:
  cutting tool materials, 404
  defined, 62
  materials, 64–65
  tests, 62–64
Heading, 280–281
Heat-affected zone, 520
Heat of fusion, 73
Heat treatment:
  castings, 133
  ceramics, 12, 245, 247–248
  glass, 12, 146–147
  metals, 12, 16, 480–487
  powdered metals, 12, 225
Hemming, 314
High-energy-rate forming, 332–333
High speed machining, 387–388
High speed steel, 21, 406–407, 413
High strength low-alloy steel, 19
Honing, 450–451
Hooke's Law, 52, 70
Hot dipping, 497–498
Hot extrusion, 285
Hot hardness, 66–67
Hot pressing, 227, 246, 249
Hot pressure welding, 543
Hot rolling, 262
Hot-runner mold, 171
Hot working, 67, 257–258, 261
Hubbing, 281–282
Hydroforming, 323
Hydrostatic extrusion, 291–292

Impact extrusion, 285, 291
Impact grinding, 241
Impregnation, 224–225
Impression-die forging, 275–276
Incremental positioning, 643
Indirect extrusion, 284–285, 287
Induction heating, 131–132
Industries, 3–4
Infiltration, 225
Injection blow molding, 179
Injection molding, 167–174, 195, 208

Ink-jet printing heads, 601
Inserts:
  cutting tool, 414–415
  molding, 580–581
  screw thread, 572
Inspection:
  casting, 135
  defined, 678
  instruments and gages, 89–96
  principles, 678–680
  technologies, 680–684
  welding, 549–550
Integral fasteners, 581
Interference fit(s), 576–579
Interpolation, 643
Investment casting, 119–121
Ion implantation, 493–494, 605
Ion plating, 501
Iron, 17, 28
Iron-carbon alloy system, 17
Ironing, 321
ISO 9000, 677–678
Isostatic pressing, 225–226, 246–247, 249
Isothermal extrusion, 285
Isothermal forging, 282–283
Isothermal forming, 258

Jig, 373
Jig grinder, 449
Jiggering, 242–243
Job shop, 624
Joining, 509
Joint(s):
  adhesive bonded, 564–565
  bolted, 573–575
  brazed, 555–557
  soldered, 560
  weld, 513–515, 519–520, 549
Jolleying, 243
Jominy end-quench test, 485
Just-in-time, 660–661

Kanban, 660
Kaolinite, 29
Kevlar, 38, 46, 201
Kiln, 245
Knoop hardness, 64
Knurling, 363

Ladles (casting), 132
Laminated-object manufacturing, 593–594
Lancing, 322, 581
Lapping, 451–452
Laser-beam machining, 469–470, 610
Laser-beam welding, 511, 540
Laser evaporation method, 616–617
Lathe, 363–366
Lay (in surface texture), 84
Layer processes, 604–607, 620
Lean production, 659–662
Lehr, 147
Lift-off technique, 607
LIGA process, 607–608
Limit dimensions, 81
Line balancing, 631
Liquid-metal forging, 127
Liquid phase sintering, 228
Liquidus, 73, 105
Lithography, 605
Live center, 364
Lost-foam process, 118
Lost pattern process, 118
Lost-wax process, 119
Low-pressure casting, 124
Lubricants and lubrication:
  ceramics, 246
  metal cutting, 348, 418–419
  metal forming, 259
  powdered metals, 220

Machinability, 393–394, 420–422
Machine cell(s), 653–655
Machine tools:
  defined, 12–13, 340–341
  drilling, 371–373
  machining centers, 380–381, 650
  milling, 378–380
  plastic extrusion, 152–154
  plastic molding, 167–168
  presses, *see* Presses
  turning and boring, 363–369
Machine vision, 681–684
Machine welding, 512
Machining:
  advantages and disadvantages, 337
  defined, 11, 336
  economics, 422–428
  high speed, 387–388
  machine tools, *see* Machine tools
  operations, 338–339, 357–387
  part geometry, 357–360
  powder metallurgy, 224
  product design considerations, 392–394
  theory, 336–354
Machining center, 380–381, 650
Machining economics, 422–428

Magnesium, 25
Magnetic pulse forming, 333
Make or buy decision, 632–633
Malleable iron, 23
Manganese, 19
Mannesmann process, 270
Manual data input, 650
Manufactured products, *see* Products, manufactured
Manufacturing (general), 1, 2, 15
Manufacturing capability, 5
Manufacturing engineering, 627
Manufacturing industries, 1, 3
Manufacturing processes, classification of, 8–12, 628–629
Manufacturing support systems, 626–627
Manufacturing systems, 624, 640
Maraging steel, 485
Martensite, 481–484
Martensitic stainless steel, 20–21
Masks and masking, 470–471
Mass finishing, 492–493
Mass production, 626
Material-addition RP, 588
Material removal processes, 9, 10–12, 336
Materials, engineering, *see* Engineering materials
Materials in manufacturing, 6–8
Measurement:
  conventional instruments, 89–94
  cutting temperatures, 353–354
  defined, 89
  surfaces, 94–96
Mechanical assembly, 12, 569–584
Mechanical cleaning, 491–493
Mechanical properties:
  hardness, 62–66
  stress-strain relationships, 50–62
  temperature effect, 66–67
Mechanical thermoforming, 184
Melamine formaldehyde, 39
Melt fracture, 160
Melt spinning, 165–166
Melting point, 73
MEMS, 599
Merchant equation, 347–350
Mesh count, 234–235
Metal cutting, *see* Machining
Metal forming, 252–259
Metal injection molding, 226
Metal matrix composites, 43, 45
Metals, 6–7, 16–28
Micro-contact printing, 609, 616
Microelectromechanical systems, 599
Microfabrication, 603–611
Micro-imprint lithography, 609, 616
Micromachining, 606
Micrometer, 91–93
Microsensors, 600
Microstereolithography, 611
Microsystems, 599, 601–603
MIG welding, 527
Milling, 11, 339, 373–380
Milling cutters, 416–417
Milling machine(s), 373, 378–380
Mill-turn center, 381
Modulus of elasticity, 52, 61, 70
Mold(s):
  casting, 98–100, 115–122
  plastic injection, 168–171
  polymer matrix composites, 203–208
  thermoforming, 182–184
Molding:
  compression molding, 174–175, 207–208
  injection, 167–174, 208
  polymer matrix composites, 203–208
  rubber, 195–196
  tires, 199–200
  transfer molding, 176–177, 208
Molding compounds, 202–203, 207
Molding inserts, 580–581
Molybdenum, 19, 407
Mond process, 502
Monolayer, 620
Multifunction machine, 383
Multitasking machine, 383
Mushy zone, 105

Nanofabrication processes, 615–620
Nano-imprint lithography, 616
Nanoscience, 599
Nanotechnology, 599, 611–613
Nanotubes, 612–613
Natural rubber, 40–41, 192–193
Natural tolerance limits, 669
NC, *see* Numerical control
Near net shape, 12, 98, 216, 261, 276
Necking, 53, 55, 59
Neoprene, 41
Net shape, 12, 35, 98, 149, 216, 261, 276
Newtonian fluid, 68, 150–151
Nickel, 19, 26, 28
Nitride ceramics, 7, 32
Nitriding, 487, 493
Noncrystalline structures,
Nonferrous metals, 7, 16, 23–28
Nontraditional processes, 11, 336, 456–476
Normalizing, 481
Notching, 310

Numerical control:
  applications, 650–651
  definition, 640–641
  drilling, 373
  filament winding, 210
  machining center, 380–381, 650
  milling, 380
  part programming, 641, 648–650
  punch press, 326
  tape-laying, 206
  technology, 641–650
  turning, 366, 381
Nut(s), 570
Nylon, 38

Open-back inclinable, 326
Open-die forging, 271–275
Open mold, 99, 201, 203–207
Operation sheet, 631
Optical encoder, 647
Organic coating, 504–507
Orthogonal cutting, 341–343, 347–348
Oxide ceramics, 31
Oxyacetylene welding, 537–538
Oxyfuel gas welding, 511, 537–539

Packing factor (powders), 237
Painting, 12, 504
Parallelism, 82
Part family, 651–653
Part geometry (machining), 357–360
Part programming, NC, 648–650
Particles, 44, 202
Particulate processing, 9–10, 215
Parting, 309
Parts classification and coding, 652
Pattern (casting), 100, 114–115
Pearlite, 482
Perforating, 310
Permanent mold, 100
Permanent mold casting, 122–129
Perpendicularity, 82
Phase(s):
  defined, 7
  in composites, 7, 43–44, 201–202
  in steel, 17
Phase diagrams:
  copper-nickel, 105
  iron-carbon, 17, 20
  WC-Co, 249
Phenol-formaldehyde, 40
Phenolics, 40
Phosphate coating, 498
Photochemical machining, 474, 610
Photoresist, 471
Physical properties, 72–76
Physical vapor deposition, 12, 499–501, 605
Piping (extrusion), 293
Planing, 384
Plant capacity, 6
Plant layout, 624–626
Plasma arc welding, 530–531
Plaster-mold casting, 121–122
Plastic deformation, 305
Plastic forming, 242
Plastic pressing, 243
Plastics, *see* Polymers
Plating processes, 494–498
Pointing (drawing), 299
Point-to-point, 642–643
Polishing, 452–453
Polyamides, 38
Polybutadiene, 41
Polycarbonate, 38
Polyesters, 38, 40
Polyethylene, 38
Polyethylene terephthalate, 38
Polymer matrix composites:
  defined, 7, 45–46
  processing, 191, 203–213
  starting materials, 201–203
Polymer melt, 71, 150–152, 154
Polymer(s):
  categories, 7, 35
  composites, *see* Polymer matrix composites
  defined, 7, 35
  product design considerations, 186–188
  products, 149–150
  properties, 35–36, 65
  rubbers, *see* Elastomers
  shape processing, 149–186
  structures, 36–38
  thermoplastics, *see* Thermoplastic polymers
  thermosets, *see* Thermosetting polymers
Polymethylmethacrylate, 38
Polypropylene, 38
Polystyrene, 39, 44, 186
Polyurethanes, 40, 42, 186
Polyvinylchloride, 39
Porosity (metal powders), 237
Positioning systems, 642–648
Potter's wheel, 242
Powder coating, 506–507
Powder injection molding, 226, 247
Powder metallurgy, 16, 215–231
Powders, 234–237

Power:
arc welding, 524–525
extrusion, 288
machining, 350–352
resistance welding, 531–532
rolling, 266–267
Power density (welding), 516–517, 539
Precipitation hardening, 485–486
Precision, defined, 89
Precision forging, 276–277
Precision gage blocks, 89–90
Preform molding, 207
Prepreg, 203–204, 209
Press(es):
drill, 371–373
extrusion (metal), 291
forging, 279
stamping, 304, 325–329
Press-and-blow, 143
Press brake, 326
Press fitting, 12, 576–578
Pressing:
ceramics, 243–244, 246–247
glass, 142
powder metallurgy, 215, 220–222
Pressure thermoforming, 182–184
Pressworking, 253
Process capability, 668–669
Process planning, 627–635
Processes, manufacturing, classification of, 8–12
Product design considerations:
assembly, 581–584
casting, 137–138
ceramics, 250
design for manufacturing, 637
glass, 147–148
machining, 392–394
plastics, 186–188
powder metallurgy, 229–231
welding, 550–551
Product variety, 5
Products, manufactured, 2, 4
Production capacity, 6
Production flow analysis, 652
Production line, 626
Production planning and control, 627
Production quantity, 4–5, 633
Production systems, 623–627
Properties:
fluid, *see* Fluid properties
general, 49–50
mechanical, *see* Mechanical properties
physical, *see* Physical properties
Property-enhancing processes, 12
Protractor, 94
Pseudoplastic, 69, 151
Pulforming, 211–212
Pultrusion, 210–211
Punch-and-die, 254, 304
Punching, 306

Quality:
casting, 133–135
defined, 667–668
plastic extrusion, 160–161
plastic injection molding, 172
programs, 674–678
weld, 547–550
Quality control, 627, 667
Quantity, production, *see* Production quantity
Quartz, 30, 140
Quenching, 484–485

Rack plating, 496
Radial drill, 371–373
Radial forging, 281
Rake angle, 339, 412
Rapid prototyping, 587–597
Rapid tool making, 596
Reaction injection molding, 173–174, 186, 208
Reaming, 363, 371
Recrystallization, 67, 258, 481
Recrystallization temperature, 67
Redrawing, 319
Reduction:
bar drawing, 294, 296–297
deep drawing, 317
extrusion, 286
rolling, 263
Reflow soldering, 563
Refractory ceramics, 30
Reinforcing agents, 46, 201–202
Relief angle, 339
Resin transfer molding, 208
Resistance projection welding, 535–536
Resistance welding, 511, 531–536
Retaining ring, 578–579
Reverse drawing, 319
Reverse extrusion, 284
Rheocasting, 128
Ring rolling, 269
Riser (casting), 99–100, 109–110
Rivet(s), 12, 575–576
Robotic welding, 513
Rockwell hardness, 63–64
Roll bending, 330–331

Roll coating, 567
Roll forming, 330–331
Roll piercing, 270
Roll welding, 543
Roller mill, 241
Rolling:
  gear, 270
  glass, 144
  metals, 253, 262–269
  powdered metals, 226–227
  ring, 269
  thread, 269
Rolling mills, 267–269
Rotary tube piercing, 270
Rotational molding, 180–181
Rotomolding, 180
Roughing, 340
Roughness, surface, *see* Surface roughness
Roundness, 82
Route sheet, 630–631
Rubber, *see* Elastomers
Rubber forming processes, 322–323
Rule, steel, 90

Sand blasting, 491
Sand casting, 100, 113–117
Sawing, 386–387
Scanning probe microscopes, 614–615, 617–618
Scanning tunneling microscope, 614
Screen resist, 471
Screw(s), 570–572
Screw thread inserts, 572
Seam welding, 514, 534–535
Seaming, 314, 581
Selective laser sintering, 594–595
Self-assembly, 618–620
Semi-dry pressing, 243
Seminotching, 310
Semipermanent-mold casting, 122
Semisolid metal casting, 127–128
Setup reduction, 662
Sewing, 580
Shape factor:
  extrusion (metal), 289–291
  extrusion (plastic), 157
  forging, 273, 276
Shape rolling, 267
Shaping, 383–384
Shaping processes, 8–12
Sharkskin, 161
Shear angle, 308, 341–342, 348
Shear modulus, 61
Shear plane, 341
Shear properties, 60–62
Shear strength, 62
Shearing, 254–255, 306
Sheet:
  metal, 262–263, 304
  metalworking, 253–255, 304–333
  plastic, 162–164
Shell molding, 117–118
Shielded metal arc welding, 525–526
Shot peening, 492
Shrink fit, 578
Shrinkage:
  casting, 107–108
  ceramics, 244, 250
  plastic molding, 171–172
Sialon, 411
Silica, 7, 29–30, 33, 140
Silicon carbide, 7, 30, 434
Silicon nitride, 32
Silicon processing, 604–607
Siliconizing, 493
Simultaneous engineering, 638
Single-point tool(s), 339
Sintered carbides, 408
Sintered polycrystalline diamond, 411
Sintering:
  cemented carbides, 249
  ceramics, 245, 247–248
  liquid phase, 228
  metal powders, 215–216, 222–224
Six sigma, 675–677
Size effect, 351–351, 440
Slab, 262
Slip casting, 241–242
Slit-die extrusion, 162–163
Slotting, 310, 374, 387
Slush casting, 124
Snag grinder, 449
Snap fit, 578
Snap ring, 578–579
Soaking, 262, 481
Soft lithography, 608–609, 616
Soldering, 12, 560–563
Solid ground curing, 591
Solid-state welding, 511, 522, 541–547
Solidification time (casting), 104–107, 109
Solidification processes:
  casting, 9, 98–100, 113–129
  defined, 9, 97
  glassworking, 140–146
  plastics, 149–186
  polymer-matrix composites, 200–213

Solidus, 73, 105
Spark sintering, 227–228
Specific energy (machining), 350–351
Specific gravity, 72
Specific heat, 75, 101
Speed lathe, 365
Spinning:
  glass, 142
  plastics, 165–166
  sheet metal, 331–332
Spot facing, 371
Spot welding, 514, 533–534
Spraying (coating), 167, 505
Spray-up, 205–206
Springback, 312
Sputtering, 500–501
Squareness, 82
Squeeze casting, 127
Stainless steel, 19–21, 26
Stamping, 254, 304
Stapling, 580
Statistical process control, 669–674
Steel(s):
  defined, 6, 17–18
  for casting, 136
  high speed, *see* High speed steel
  low alloy, 19
  plain carbon, 18–19
  stainless, 19–21
  tool, 21–22
Stepping motor, 644–646
Stereolithography, 590–591, 611
Stick welding, 525
Sticking (friction), 259, 264
Stitching, 579–580
Straightness, 82
Strain:
  defined, 52, 54–55, 57, 60
  metal extrusion, 286–287
  metal forging, 272
  metal machining, 342
  rolling, 264
  wire and bar drawing, 295–296
Strain hardening, 55
Strain hardening exponent, 56
Strength coefficient, 56
Strength-to-weight ratio, 42, 46, 72
Stress-strain relationship:
  compression, 57–59
  shear, 60–62
  tensile, 50–57
  types of, 56–57
Stretch blow molding, 179–180
Stretch forming, 330
Structural foam molding, 173, 186
Stud(s), 572
Styrene-butadiene rubber, 42
Submerged arc welding, 529
Super alloys, 28
Supercooled liquid, 73
Superfinishing, 452
Superheat (casting), 101
Surface finish, *see* Surface roughness
Surface grinding, 444–445
Surface hardening, 487
Surface integrity, 83–85, 96
Surface micromachining, 606
Surface plate, 90
Surface processing, 12, 489–507
Surface roughness:
  abrasive processes, 450
  casting, 137
  defined, 83–85
  grinding, 438–439
  machining, 389–393
  measurement of, 95–96
  manufacturing processes, 87
Surface technology, 82
Surface texture, 83–85
Surface treatments, 12, 489–507
Surfaces, 80–83
Surfacing weld, 515
Swaging, 281
Swell ratio (polymers), 152
Synthetic rubber, 41–42, 193
Systems, production, 623–627

Tantalum carbide, 31
Tape-laying machines, 206
Taper turning, 361
Tapping, 371
Taylor tool life equation, 400–403, 424–426
Technological processing capability, 5–6
Technology (defined), 1
Temperature:
  effect on properties, 66–67, 73
  grinding, 441
  machining, 352–354
  metal forming, 256–258
Tempering:
  glass, 147
  steel, 484
Tensile strength, 23, 53
Tensile test, 50–54
Testing:
  hardness, 62–64

inspection, 678
tensile, 50–54
torsion, 61–62
welds, 549–550
Thermal energy processes, 464–470
Thermal oxidation, 605
Thermal properties:
conductivity, 75
diffusivity, 75
expansion, 72–73
in manufacturing, 76
in metal cutting, 352–353
specific heat, 74–75
Thermit welding, 540–541
Thermoforming, 181–184
Thermoplastic elastomers, 42, 200
Thermoplastic polymers:
defined, 7, 37, 150
important thermoplastics, 37–39
properties, 37, 42
shaping processes, 152–172, 177–185
Thermosetting polymers:
defined, 7, 39, 150
important thermosets, 39–40
properties, 39
shaping processes, 173–177, 185
Thin-film magnetic heads, 602
Thixocasting, 128
Thixomolding, 128
Threaded fasteners, 12, 570–575
Threading, 362
Thread rolling, 269
Three-dimensional printing, 595–596
Three-plate mold, 170–171
Through hole, 369
TIG welding, 529
Time-temperature-transformation curve, 482–483
Time, machining:
drilling, 369–370
electrochemical machining, 462–463
milling, 377
minimizing, 423–424
turning, 360–361
Tinning, 497, 560
Tires, 197–200
Titanium, 26–27
Titanium carbide, 31, 45, 409–410
Titanium nitride, 32, 409–410
Tolerance(s):
casting, 137–138
defined, 80–81
machining, 388–389, 393
manufacturing processes, 86
natural tolerance limits, 669
plastic molding, 188
Tool-chip thermocouple, 353–354
Tool geometry:
effect on surface roughness, 390
milling cutters, 416–417
multiple-cutting edge, 339, 415–417
orthogonal cutting, 342
single-point, 339, 412–415
twist drill, 415–516
Tool grinders, 449
Tooling, general, 13
Tool life (machining), 400–403
Tool, machine, *see* Machine tool
Tool steels, 21–22
Tool wear (machining), 399–400
Tools, *see* Cutting tools or Dies
Torque-turn tightening, 575
Torque wrench, 575
Torsion test, 60
Total productive maintenance, 662
Total quality management, 674–675
Total solidification time, 104, 106–107, 109
Transfer molding, 176–177, 195, 208
Transverse rupture strength, 60, 405
Trimming, 133, 283, 310
True stress-strain, 54–56
Truing (grinding), 442
TTT curve, 482–483
Tube rolling, 212
Tumbling, 492–493
Tungsten, 407
Tungsten carbide:
cutting tools, 408–409
general, 31–32, 45
processing of, 248–250
Tunneling, 614
Turning, 11, 338, 360–369
Turning center, 381
Turret drill, 373
Turret lathe, 365
Turret press, 326
Twist drill, 415–516
Twisting, 322
Two-plate mold, 168–170
Two-roll mill, 194

Ultimate tensile strength, 53
Ultra-high precision machining, 610
Ultrasonic inspection, 684
Ultrasonic machining, 457–458, 610
Ultrasonic welding, 511, 546–547
Undercut, 472, 475

Unilateral tolerance, 81
Unit operation, 8
Unit power (machining), 350–351
Upset forging, 280
Upsetting, 280–281
Urea formaldehyde, 39

V-bending, 311
Vacuum evaporation, 500
Vacuum forming, 181
Vacuum permanent-mold casting, 124
Vacuum thermoforming, 181–182
Vanadium, 19, 407
Vapor degreasing, 491
Vapor deposition processes, 499–504
Vernier caliper, 91
Vibratory finishing, 492
Vickers hardness, 63–64
Viscoelasticity, 69–71, 151–152
Viscosity, 67–69, 150–151, 155
Vision, machine, 681–684
Vitreous, 29, 34
Volumetric specific heat, 75
Vulcanization, 40, 191, 196–197

Warm working, 257, 261
Washer, 572
Water atomization, 217
Water jet cutting, 458–459
Wave soldering, 563
Waviness (in surface texture), 83
Wear:
  cutting tool, 399–400
  grinding wheel, 441–443
Weldbonding, 565
Weld joints, 513–515, 519–520, 549
Welding:
  defects, 548–549
  definition and overview, 509–513
  design considerations, 550–551
  joints, 513–515, 519–520
  physics, 515–519
  processes, 510–511, 522–547
  quality, 547–550
Wet chemical etching, 605
Wet lay-up, 204
Wet spinning, 166
White cast iron, 23
Windshields (automobile), 147
Wire and cable coating, 160
Wire drawing, 293–299
Wire EDM, 467, 610
Work hardening, 55
Work holding:
  boring, 367–368
  drilling, 373
  turning, 364–365
Wrought metal, 16

X-ray inspection, 684
X-ray lithography, 616

Yield point, 52
Yield strength, 52

Zinc, 27–28

Standard Units Used in this Book

Units for both the System International (SI, metric) and United States Customary System (USCS) are listed in equations and tables throughout this textbook. Metric units are listed as the primary units and USCS units are given in parentheses.

Prefixes for SI units:

Prefix	Symbol	Multiplier	Example units (and symbols)
nano-	n	10^{-9}	nanometer (nm)
micro-	μ	10^{-6}	micrometer, micron (μm)
milli-	m	10^{-3}	millimeter (mm)
centi-	c	10^{-2}	centimeter (cm)
kilo-	k	10^{3}	kilometer (km)
mega-	M	10^{6}	megaPascal (MPa)
giga-	G	10^{9}	gigaPascal (GPa)

Table of Equivalencies between USCS and SI units:

Variable	SI units	USCS units	Equivalencies
Length	meter (m)	inch (in)	1.0 in = 25.4 mm = 0.0254 m
		foot (ft)	1.0 ft = 12.0 in = 0.3048 m = 304.8 mm
		yard	1.0 yard = 3.0 ft = 0.9144 m = 914.4 mm
		mile	1.0 mile = 5280 ft = 1609.34 m = 1.60934 km
		micro-inch (μ-in)	1.0 μ-in = 1.0×10^{-6} in = 25.4×10^{-3} μm
Area	m^2, mm^2	in^2, ft^2	1.0 in^2 = 645.16 mm^2
			1.0 ft^2 = 144 in^2 = 92.90×10^{-3} m^2
Volume	m^3, mm^3	in^3, ft^3	1.0 in^3 = 16,387 mm^3
			1.0 ft^2 = 1728 in^3 = 2.8317×10^{-2} m^3
Mass	kilogram (kg)	pound (lb)	1.0 lb = 0.4536 kg
		ton	1.0 ton (short) = 2,000 lb = 907.2 kg
Density	kg/m^3	lb/in^3	1.0 lb/in^3 = 27.68×10^3 kg/m^3
		lb/ft^3	1.0 lb/ft^3 = 16.0184 kg/m^3
Velocity	m/min	ft/min	1.0 ft/min = 0.3048 m/min = 5.08×10^{-3} m/s
	m/s	in/min	1.0 in/min = 25.4 mm/min = 0.42333 mm/s
Acceleration	m/s^2	ft/sec^2	1.0 ft/sec = 0.3048 m/s^2
Force	Newton (N)	pound (lb)	1.0 lb = 4.4482 N
Torque	N-m	ft-lb, in-lb	1.0 ft-lb = 12.0 in-lb = 1.356 N-m
			1.0 in-lb = 0.113 N-m
Pressure	Pascal (Pa)	lb/in^2	1.0 lb/in^2 = 6895 N/m^2 = 6895 Pa
Stress	Pascal (Pa)	lb/in^2	1.0 lb/in^2 = 6.895×10^{-3} N/mm^2 = 6.895×10^{-3} MPa
Energy, work	Joule (J)	ft-lb, in-lb	1.0 ft-lb = 1.356 N-m = 1.356 J
			1.0 in-lb = 0.113 N-m = 0.113 J
Heat energy	Joule (J)	British thermal unit (Btu)	1.0 Btu = 1055 J
Power	Watt (W)	Horsepower (hp)	1.0 hp = 33,000 ft-lb/min = 745.7 J/s = 745.7 W
			1.0 ft-lb/min = 2.2597×10^{-2} J/s = 2.2597×10^{-2} W
Specific heat	J/kg-°C	Btu/lb-°F	1.0 Btu/lb-°F = 1.0 Calorie/g-°C = 4,187 J/kg-°C
Thermal conductivity	J/s-mm-°C	Btu/hr-in -°F	1.0 Btu/hr-in -°F = 2.077×10^{-2} J/s-mm-°C
Thermal expansion	(mm/mm)/°C	(in/in)/°F	1.0 (in/in)/°F = 1.8 (mm/mm)/°C
Viscosity	Pa-s	lb-sec/in^2	1.0 lb-sec/in^2 = 6895 Pa-s = 6895 N-s/m^2

Conversion between USCS and SI

To convert from USCS to SI: To convert the value of a variable from USCS units to equivalent SI units, ***multiply*** the value to be converted by the right-hand side of the corresponding equivalency statement in the Table of Equivalencies.

Example: Convert a length $L = 3.25$ in to its equivalent value in millimeters.

Solution: The corresponding equivalency statement is: 1.0 in = 25.4 mm

$$L = 3.25 \text{ in x } (25.4 \text{ mm/in}) = \mathbf{82.55\ mm}$$

To convert from SI to USCS: To convert the value of a variable from SI units to equivalent USCS units, ***divide*** the value to be converted by the right-hand side of the corresponding equivalency statement in the Table of Equivalencies.

Example: Convert an area $A = 1000\ \text{mm}^2$ to its equivalent in square inches.

Solution: The corresponding equivalency statement is: $1.0\ \text{in}^2 = 645.16\ \text{mm}^2$

$$A = 1000\ \text{mm}^2/(645.16\ \text{mm}^2/\text{in}^2) = \mathbf{1.55\ in^2}$$